ANALYSIS AND DESIGN OF ANALOG INTEGRATED CIRCUITS

Fifth Edition

International Student Version

PAUL R. GRAY
University of California, Berkeley

PAUL J. HURST
University of California, Davis

STEPHEN H. LEWIS
University of California, Davis

ROBERT G. MEYER
University of California, Berkeley

WILEY

Analysis and Design of Analog Integrated Circuits
Fifth Edition

Authorised reprint by Wiley India Pvt. Ltd., 4435-36/7, Ansari Road, Daryaganj, New Delhi-110002.

Copyright © 2010 by John Wiley & Sons Inc. All rights reserved.

Cover Image: Getty Images/iStockphoto

No part of this book, including interior design, cover design, and icons, may be reproduced or transmitted in any form except with the permission of John Wiley & Sons, Inc., 111 River Street, Hoboken, NJ 07030, (201) 748-6011. All rights reserved.

Limit of Liability/ Disclaimer of Warranty: The publisher and the author make no representations or warranties with respect to the accuracy or completeness of the contents of this work and specifically disclaim all warranties, including without limitation warranties of fitness for a particular purpose. No warranty may be created or extended by sales or promotional materials. The advice and strategies contained herein may not be suitable for every situation. This work is sold with the understanding that the publisher is not engaged in rendering legal, accounting, or other professional services. If professional assistance is required, the services of a competent professional person should be sought. Neither the publisher nor the author shall be liable for damages arising herefrom. The fact that an organization or Website is referred to in this work as a citation and/or a potential source of further information does not mean that the author or the publisher endorses the information the organization or Website may provide or recommendations it may make. Further, readers should be aware that Internet Websites listed in this work may have changed or disappeared between when this work was written and when it is read.

TRADEMARKS: Wiley, the Wiley logo are trademarks or registered trademarks of John Wiley & Sons, Inc. and/or its affiliates, in the United States and other countries, and may not be used without written permission. All other trademarks are the property of their respective owners. John Wiley & Sons, Inc. is not associated with any product or vendor mentioned in this book.

Wiley also publishes its books in a variety of electronic formats. Some content that appears in print may not be available in electronic books.

This edition is authorized for sale in the Indian Sub-continent only.

Reprint: 2017

Printed at: Chaudhary Press, Delhi

ISBN: 978-81-265-2148-7

To Liz, Barbara, Robin, and Judy

Preface

Since the publication of the first edition of this book, the field of analog integrated circuits has developed and matured. The initial groundwork was laid in bipolar technology, followed by a rapid evolution of MOS analog integrated circuits. Thirty years ago, CMOS technologies were fast enough to support applications only at audio frequencies. However, the continuing reduction of the minimum feature size in integrated-circuit (IC) technologies has greatly increased the maximum operating frequencies, and CMOS technologies have become fast enough for many new applications as a result. For example, the bandwidth in some video applications is about 4 MHz, requiring bipolar technologies as recently as about 23 years ago. Now, however, CMOS easily can accommodate the required bandwidth for video and is being used for radio-frequency applications. Today, bipolar integrated circuits are used in some applications that require very low noise, very wide bandwidth, or driving low-impedance loads.

In this fifth edition, coverage of the bipolar 741 op amp has been replaced with a low-voltage bipolar op amp, the NE5234, with rail-to-rail common-mode input range and almost rail-to-rail output swing. Analysis of a fully differential CMOS folded-cascode operational amplifier (op amp) is now included in Chapter 12. The 560B phase-locked loop, which is no longer commercially available, has been deleted from Chapter 10.

The SPICE computer analysis program is now readily available to virtually all electrical engineering students and professionals, and we have included extensive use of SPICE in this edition, particularly as an integral part of many problems. We have used computer analysis as it is most commonly employed in the engineering design process—both as a more accurate check on hand calculations, and also as a tool to examine complex circuit behavior beyond the scope of hand analysis.

An in-depth look at SPICE as an indispensable tool for IC robust design can be found in The SPICE Book, 2nd ed., published by J. Wiley and Sons. This text contains many worked out circuit designs and verification examples linked to the multitude of analyses available in the most popular versions of SPICE. The SPICE Book conveys the role of simulation as an integral part of the design process, but not as a replacement for solid circuit-design knowledge.

This book is intended to be useful both as a text for students and as a reference book for practicing engineers. For class use, each chapter includes many worked problems; the problem sets at the end of each chapter illustrate the practical applications of the material in the text. All of the authors have extensive industrial experience in IC design and in the teaching of courses on this subject; this experience is reflected in the choice of text material and in the problem sets.

Although this book is concerned largely with the analysis and design of ICs, a considerable amount of material also is included on applications. In practice, these two subjects are closely linked, and a knowledge of both is essential for designers and users of ICs. The latter compose the larger group by far, and we believe that a working knowledge of IC design is a great advantage to an IC user. This is particularly apparent when the user must choose from among a number of competing designs to satisfy a particular need. An understanding of the IC structure is then useful in evaluating the relative desirability of the different designs under extremes of environment or in the presence of variations in supply voltage. In addition, the IC user is in a much better position to interpret a manufacturer's data if he or she has a working knowledge of the internal operation of the integrated circuit.

The contents of this book stem largely from courses on analog integrated circuits given at the University of California at the Berkeley and Davis campuses. The courses are senior-level electives and first-year graduate courses. The book is structured so that it can be used as the basic text for a sequence of such courses. The more advanced material is found at the end of each chapter or in an appendix so that a first course in analog integrated circuits can omit this material without loss of continuity. An outline of each chapter is given below with suggestions for material to be covered in such a first course. It is assumed that the course consists of three hours of lecture per week over a 15-week semester and that the students have a working knowledge of Laplace transforms and frequency-domain circuit analysis. It is also assumed that the students have had an introductory course in electronics so that they are familiar with the principles of transistor operation and with the functioning of simple analog circuits. Unless otherwise stated, each chapter requires three to four lecture hours to cover.

Chapter 1 contains a summary of bipolar transistor and MOS transistor device physics. We suggest spending one week on selected topics from this chapter, with the choice of topics depending on the background of the students. The material of Chapters 1 and 2 is quite important in IC design because there is significant interaction between circuit and device design, as will be seen in later chapters. A thorough understanding of the influence of device fabrication on device characteristics is essential.

Chapter 2 is concerned with the technology of IC fabrication and is largely descriptive. One lecture on this material should suffice if the students are assigned the chapter to read.

Chapter 3 deals with the characteristics of elementary transistor connections. The material on one-transistor amplifiers should be a review for students at the senior and graduate levels and can be assigned as reading. The section on two-transistor amplifiers can be covered in about three hours, with greatest emphasis on differential pairs. The material on device mismatch effects in differential amplifiers can be covered to the extent that time allows.

In Chapter 4, the important topics of current mirrors and active loads are considered. These configurations are basic building blocks in modern analog IC design, and this material should be covered in full, with the exception of the material on band-gap references and the material in the appendices.

Chapter 5 is concerned with output stages and methods of delivering output power to a load. Integrated-circuit realizations of Class A, Class B, and Class AB output stages are described, as well as methods of output-stage protection. A selection of topics from this chapter should be covered.

Chapter 6 deals with the design of operational amplifiers (op amps). Illustrative examples of dc and ac analysis in both MOS and bipolar op amps are performed in detail, and the limitations of the basic op amps are described. The design of op amps with improved characteristics in both MOS and bipolar technologies are considered. This key chapter on amplifier design requires at least six hours.

In Chapter 7, the frequency response of amplifiers is considered. The zero-value time-constant technique is introduced for the calculations of the -3-dB frequency of complex circuits. The material of this chapter should be considered in full.

Chapter 8 describes the analysis of feedback circuits. Two different types of analysis are presented: two-port and return-ratio analyses. Either approach should be covered in full with the section on voltage regulators assigned as reading.

Chapter 9 deals with the frequency response and stability of feedback circuits and should be covered up to the section on root locus. Time may not permit a detailed discussion of root locus, but some introduction to this topic can be given.

In a 15-week semester, coverage of the above material leaves about two weeks for Chapters 10, 11, and 12. A selection of topics from these chapters can be chosen as follows. Chapter 10 deals with nonlinear analog circuits and portions of this chapter up to Section 10.2 could be covered in a first course. Chapter 11 is a comprehensive treatment of noise in integrated circuits and material up to and including Section 11.4 is suitable. Chapter 12 describes fully differential operational amplifiers and common-mode feedback and may be best suited for a second course.

We are grateful to the following colleagues for their suggestions for and/or evaluation of this book: R. Jacob Baker, Bernhard E. Boser, A. Paul Brokaw, Iwen Chao, John N. Churchill, David W. Cline, Kenneth C. Dyer, Ozan E. Erdoğan, John W. Fattaruso, Weinan Gao, Edwin W. Greeneich, Alex Gros-Balthazard, Tünde Gyurics, Ward J. Helms, Kaveh Hosseini, Timothy H. Hu, Shafiq M. Jamal, John P. Keane, Haideh Khorramabadi, Pak Kim Lau, Thomas W. Matthews, Krishnaswamy Nagaraj, Khalil Najafi, Borivoje Nikolić, Keith O'Donoghue, Robert A. Pease, Lawrence T. Pileggi, Edgar Sánchez-Sinencio, Bang-Sup Song, Richard R. Spencer, Eric J. Swanson, Andrew Y. J. Szeto, Yannis P. Tsividis, Srikanth Vaidianathan, T. R. Viswanathan, Chorng-Kuang Wang, Dong Wang, and Mo Maggie Zhang. We are also grateful to Darrel Akers, Mu Jane Lee, Lakshmi Rao, Nattapol Sitthimahachaikul, Haoyue Wang, and Mo Maggie Zhang for help with proofreading, and to Chi Ho Law for allowing us to use on the cover of this book a die photograph of an integrated circuit he designed. Finally, we would like to thank the staffs at Wiley and Elm Street Publishing Services for their efforts in producing this edition.

The material in this book has been greatly influenced by our association with the late Donald O. Pederson, and we acknowledge his contributions.

Berkeley and Davis, CA, 2008

Paul R. Gray
Paul J. Hurst
Stephen H. Lewis
Robert G. Meyer

Contents

CHAPTER 1
Models for Integrated-Circuit Active Devices 1

1.1 Introduction 1
1.2 Depletion Region of a *pn* Junction 1
    1.2.1 Depletion-Region Capacitance 5
    1.2.2 Junction Breakdown 6
1.3 Large-Signal Behavior of Bipolar Transistors 8
    1.3.1 Large-Signal Models in the Forward-Active Region 8
    1.3.2 Effects of Collector Voltage on Large-Signal Characteristics in the Forward-Active Region 14
    1.3.3 Saturation and Inverse-Active Regions 16
    1.3.4 Transistor Breakdown Voltages 20
    1.3.5 Dependence of Transistor Current Gain β_F on Operating Conditions 23
1.4 Small-Signal Models of Bipolar Transistors 25
    1.4.1 Transconductance 26
    1.4.2 Base-Charging Capacitance 27
    1.4.3 Input Resistance 28
    1.4.4 Output Resistance 29
    1.4.5 Basic Small-Signal Model of the Bipolar Transistor 30
    1.4.6 Collector-Base Resistance 30
    1.4.7 Parasitic Elements in the Small-Signal Model 31
    1.4.8 Specification of Transistor Frequency Response 34
1.5 Large-Signal Behavior of Metal-Oxide-Semiconductor Field-Effect Transistors 38
    1.5.1 Transfer Characteristics of MOS Devices 38
    1.5.2 Comparison of Operating Regions of Bipolar and MOS Transistors 45
    1.5.3 Decomposition of Gate-Source Voltage 47
    1.5.4 Threshold Temperature Dependence 47
    1.5.5 MOS Device Voltage Limitations 48
1.6 Small-Signal Models of MOS Transistors 49
    1.6.1 Transconductance 50
    1.6.2 Intrinsic Gate-Source and Gate-Drain Capacitance 51
    1.6.3 Input Resistance 52
    1.6.4 Output Resistance 52
    1.6.5 Basic Small-Signal Model of the MOS Transistor 52
    1.6.6 Body Transconductance 53
    1.6.7 Parasitic Elements in the Small-Signal Model 54
    1.6.8 MOS Transistor Frequency Response 55
1.7 Short-Channel Effects in MOS Transistors 59
    1.7.1 Velocity Saturation from the Horizontal Field 59
    1.7.2 Transconductance and Transition Frequency 63
    1.7.3 Mobility Degradation from the Vertical Field 65
1.8 Weak Inversion in MOS Transistors 65
    1.8.1 Drain Current in Weak Inversion 66
    1.8.2 Transconductance and Transition Frequency in Weak Inversion 69
1.9 Substrate Current Flow in MOS Transistors 71
A.1.1 Summary of Active-Device Parameters 73

CHAPTER 2
Bipolar, MOS, and BiCMOS Integrated-Circuit Technology 78

2.1 Introduction 78
2.2 Basic Processes in Integrated-Circuit Fabrication 79
    2.2.1 Electrical Resistivity of Silicon 79
    2.2.2 Solid-State Diffusion 80
    2.2.3 Electrical Properties of Diffused Layers 82
    2.2.4 Photolithography 84
    2.2.5 Epitaxial Growth 86
    2.2.6 Ion Implantation 87
    2.2.7 Local Oxidation 87
    2.2.8 Polysilicon Deposition 87
2.3 High-Voltage Bipolar Integrated-Circuit Fabrication 88
2.4 Advanced Bipolar Integrated-Circuit Fabrication 92
2.5 Active Devices in Bipolar Analog Integrated Circuits 95
    2.5.1 Integrated-Circuit *npn* Transistors 96
    2.5.2 Integrated-Circuit *pnp* Transistors 107
2.6 Passive Components in Bipolar Integrated Circuits 115
    2.6.1 Diffused Resistors 115
    2.6.2 Epitaxial and Epitaxial Pinch Resistors 119
    2.6.3 Integrated-Circuit Capacitors 120
    2.6.4 Zener Diodes 121
    2.6.5 Junction Diodes 122
2.7 Modifications to the Basic Bipolar Process 123
    2.7.1 Dielectric Isolation 123
    2.7.2 Compatible Processing for High-Performance Active Devices 124
    2.7.3 High-Performance Passive Components 127
2.8 MOS Integrated-Circuit Fabrication 127
2.9 Active Devices in MOS Integrated Circuits 131
    2.9.1 n-Channel Transistors 131
    2.9.2 p-Channel Transistors 144
    2.9.3 Depletion Devices 144
    2.9.4 Bipolar Transistors 145
2.10 Passive Components in MOS Technology 146
    2.10.1 Resistors 146
    2.10.2 Capacitors in MOS Technology 148
    2.10.3 Latchup in CMOS Technology 151
2.11 BiCMOS Technology 152
2.12 Heterojunction Bipolar Transistors 153
2.13 Interconnect Delay 156
2.14 Economics of Integrated-Circuit Fabrication 156
    2.14.1 Yield Considerations in Integrated-Circuit Fabrication 157
    2.14.2 Cost Considerations in Integrated-Circuit Fabrication 159
A.2.1 SPICE Model-Parameter Files 162

CHAPTER 3
Single-Transistor and Multiple-Transistor Amplifiers 169

3.1 Device Model Selection for Approximate Analysis of Analog Circuits 170
3.2 Two-Port Modeling of Amplifiers 171
3.3 Basic Single-Transistor Amplifier Stages 173
    3.3.1 Common-Emitter Configuration 174
    3.3.2 Common-Source Configuration 178
    3.3.3 Common-Base Configuration 182
    3.3.4 Common-Gate Configuration 185
    3.3.5 Common-Base and Common-Gate Configurations with Finite r_0 187
        3.3.5.1 Common-Base and Common-Gate Input Resistance 187
        3.3.5.2 Common-Base and Common-Gate Output Resistance 189

- 3.3.6 Common-Collector Configuration (Emitter Follower) 191
- 3.3.7 Common-Drain Configuration (Source Follower) 194
- 3.3.8 Common-Emitter Amplifier with Emitter Degeneration 196
- 3.3.9 Common-Source Amplifier with Source Degeneration 199

3.4 Multiple-Transistor Amplifier Stages 201
- 3.4.1 The CC-CE, CC-CC, and Darlington Configurations 201
- 3.4.2 The Cascode Configuration 205
  - 3.4.2.1 The Bipolar Cascode 205
  - 3.4.2.2 The MOS Cascode 207
- 3.4.3 The Active Cascode 210
- 3.4.4 The Super Source Follower 212

3.5 Differential Pairs 214
- 3.5.1 The dc Transfer Characteristic of an Emitter-Coupled Pair 214
- 3.5.2 The dc Transfer Characteristic with Emitter Degeneration 216
- 3.5.3 The dc Transfer Characteristic of a Source-Coupled Pair 217
- 3.5.4 Introduction to the Small-Signal Analysis of Differential Amplifiers 220
- 3.5.5 Small-Signal Characteristics of Balanced Differential Amplifiers 223
- 3.5.6 Device Mismatch Effects in Differential Amplifiers 229
  - 3.5.6.1 Input Offset Voltage and Current 230
  - 3.5.6.2 Input Offset Voltage of the Emitter-Coupled Pair 230
  - 3.5.6.3 Offset Voltage of the Emitter-Coupled Pair: Approximate Analysis 231
  - 3.5.6.4 Offset Voltage Drift in the Emitter-Coupled Pair 233
  - 3.5.6.5 Input Offset Current of the Emitter-Coupled Pair 233
  - 3.5.6.6 Input Offset Voltage of the Source-Coupled Pair 234
  - 3.5.6.7 Offset Voltage of the Source-Coupled Pair: Approximate Analysis 235
  - 3.5.6.8 Offset Voltage Drift in the Source-Coupled Pair 236
  - 3.5.6.9 Small-Signal Characteristics of Unbalanced Differential Amplifiers 237

A.3.1 Elementary Statistics and the Gaussian Distribution 244

CHAPTER 4
Current Mirrors, Active Loads, and References 251

4.1 Introduction 251

4.2 Current Mirrors 251
- 4.2.1 General Properties 251
- 4.2.2 Simple Current Mirror 253
  - 4.2.2.1 Bipolar 253
  - 4.2.2.2 MOS 255
- 4.2.3 Simple Current Mirror with Beta Helper 258
  - 4.2.3.1 Bipolar 258
  - 4.2.3.2 MOS 260
- 4.2.4 Simple Current Mirror with Degeneration 260
  - 4.2.4.1 Bipolar 260
  - 4.2.4.2 MOS 261
- 4.2.5 Cascode Current Mirror 261
  - 4.2.5.1 Bipolar 261
  - 4.2.5.2 MOS 264
- 4.2.6 Wilson Current Mirror 272
  - 4.2.6.1 Bipolar 272
  - 4.2.6.2 MOS 275

4.3 Active Loads 276
- 4.3.1 Motivation 276
- 4.3.2 Common-Emitter–Common-Source Amplifier with Complementary Load 277
- 4.3.3 Common-Emitter–Common-Source Amplifier with Depletion Load 280
- 4.3.4 Common-Emitter–Common-Source Amplifier with Diode-Connected Load 282
- 4.3.5 Differential Pair with Current-Mirror Load 285
  - 4.3.5.1 Large-Signal Analysis 285
  - 4.3.5.2 Small-Signal Analysis 286
  - 4.3.5.3 Common-Mode Rejection Ratio 291

4.4 Voltage and Current References 297
  4.4.1 Low-Current Biasing 297
    4.4.1.1 Bipolar Widlar Current Source 297
    4.4.1.2 MOS Widlar Current Source 300
    4.4.1.3 Bipolar Peaking Current Source 301
    4.4.1.4 MOS Peaking Current Source 302
  4.4.2 Supply-Insensitive Biasing 303
    4.4.2.1 Widlar Current Sources 304
    4.4.2.2 Current Sources Using Other Voltage Standards 305
    4.4.2.3 Self-Biasing 307
  4.4.3 Temperature-Insensitive Biasing 315
    4.4.3.1 Band-Gap-Referenced Bias Circuits in Bipolar Technology 315
    4.4.3.2 Band-Gap-Referenced Bias Circuits in CMOS Technology 321

A.4.1 Matching Considerations in Current Mirrors 325
  A.4.1.1 Bipolar 325
  A.4.1.2 MOS 328
A.4.2 Input Offset Voltage of Differential Pair with Active Load 330
  A.4.2.1 Bipolar 330
  A.4.2.2 MOS 332

CHAPTER 5
Output Stages 341

5.1 Introduction 341
5.2 The Emitter Follower as an Output Stage 341
  5.2.1 Transfer Characteristics of the Emitter-Follower 341
  5.2.2 Power Output and Efficiency 344
  5.2.3 Emitter-Follower Drive Requirements 351
  5.2.4 Small-Signal Properties of the Emitter Follower 352
5.3 The Source Follower as an Output Stage 353
  5.3.1 Transfer Characteristics of the Source Follower 353
  5.3.2 Distortion in the Source Follower 355
5.4 Class B Push–Pull Output Stage 359
  5.4.1 Transfer Characteristic of the Class B Stage 360
  5.4.2 Power Output and Efficiency of the Class B Stage 362
  5.4.3 Practical Realizations of Class B Complementary Output Stages 366
  5.4.4 All-*npn* Class B Output Stage 373
  5.4.5 Quasi-Complementary Output Stages 376
  5.4.6 Overload Protection 377
5.5 CMOS Class AB Output Stages 379
  5.5.1 Common-Drain Configuration 380
  5.5.2 Common-Source Configuration with Error Amplifiers 381
  5.5.3 Alternative Configurations 388
    5.5.3.1 Combined Common-Drain Common-Source Configuration 388
    5.5.3.2 Combined Common-Drain Common-Source Configuration with High Swing 390
    5.5.3.3 Parallel Common-Source Configuration 390

CHAPTER 6
Operational Amplifiers with Single-Ended Outputs 400

6.1 Applications of Operational Amplifiers 401
  6.1.1 Basic Feedback Concepts 401
  6.1.2 Inverting Amplifier 402
  6.1.3 Noninverting Amplifier 404
  6.1.4 Differential Amplifier 404
  6.1.5 Nonlinear Analog Operations 405
  6.1.6 Integrator, Differentiator 406
  6.1.7 Internal Amplifiers 407
    6.1.7.1 Switched-Capacitor Amplifier 407
    6.1.7.2 Switched-Capacitor Integrator 412

- 6.2 Deviations from Ideality in Real Operational Amplifiers 415
  - 6.2.1 Input Bias Current 415
  - 6.2.2 Input Offset Current 416
  - 6.2.3 Input Offset Voltage 416
  - 6.2.4 Common-Mode Input Range 416
  - 6.2.5 Common-Mode Rejection Ratio (CMRR) 417
  - 6.2.6 Power-Supply Rejection Ratio (PSRR) 418
  - 6.2.7 Input Resistance 420
  - 6.2.8 Output Resistance 420
  - 6.2.9 Frequency Response 420
  - 6.2.10 Operational-Amplifier Equivalent Circuit 420
- 6.3 Basic Two-Stage MOS Operational Amplifiers 421
  - 6.3.1 Input Resistance, Output Resistance, and Open-Circuit Voltage Gain 422
  - 6.3.2 Output Swing 423
  - 6.3.3 Input Offset Voltage 424
  - 6.3.4 Common-Mode Rejection Ratio 427
  - 6.3.5 Common-Mode Input Range 427
  - 6.3.6 Power-Supply Rejection Ratio (PSRR) 430
  - 6.3.7 Effect of Overdrive Voltages 434
  - 6.3.8 Layout Considerations 435
- 6.4 Two-Stage MOS Operational Amplifiers with Cascodes 438
- 6.5 MOS Telescopic-Cascode Operational Amplifiers 439
- 6.6 MOS Folded-Cascode Operational Amplifiers 442
- 6.7 MOS Active-Cascode Operational Amplifiers 446
- 6.8 Bipolar Operational Amplifiers 448
  - 6.8.1 The dc Analysis of the NE5234 Operational Amplifier 452
  - 6.8.2 Transistors that Are Normally Off 467
  - 6.8.3 Small-Signal Analysis of the NE5234 Operational Amplifier 469
  - 6.8.4 Calculation of the Input Offset Voltage and Current of the NE5234 477

CHAPTER 7
Frequency Response of Integrated Circuits 490

- 7.1 Introduction 490
- 7.2 Single-Stage Amplifiers 490
  - 7.2.1 Single-Stage Voltage Amplifiers and the Miller Effect 490
    - 7.2.1.1 The Bipolar Differential Amplifier: Differential-Mode Gain 495
    - 7.2.1.2 The MOS Differential Amplifier: Differential-Mode Gain 499
  - 7.2.2 Frequency Response of the Common-Mode Gain for a Differential Amplifier 501
  - 7.2.3 Frequency Response of Voltage Buffers 503
    - 7.2.3.1 Frequency Response of the Emitter Follower 505
    - 7.2.3.2 Frequency Response of the Source Follower 511
  - 7.2.4 Frequency Response of Current Buffers 514
    - 7.2.4.1 Common-Base Amplifier Frequency Response 516
    - 7.2.4.2 Common-Gate Amplifier Frequency Response 517
- 7.3 Multistage Amplifier Frequency Response 518
  - 7.3.1 Dominant-Pole Approximation 518
  - 7.3.2 Zero-Value Time Constant Analysis 519
  - 7.3.3 Cascode Voltage-Amplifier Frequency Response 524
  - 7.3.4 Cascode Frequency Response 527
  - 7.3.5 Frequency Response of a Current Mirror Loading a Differential Pair 534
  - 7.3.6 Short-Circuit Time Constants 536
- 7.4 Analysis of the Frequency Response of the NE5234 Op Amp 539
  - 7.4.1 High-Frequency Equivalent Circuit of the NE5234 539
  - 7.4.2 Calculation of the −3-dB Frequency of the NE5234 540
  - 7.4.3 Nondominant Poles of the NE5234 542

xii Contents

7.5 Relation Between Frequency Response and Time Response 542

CHAPTER 8
Feedback 553

8.1 Ideal Feedback Equation 553
8.2 Gain Sensitivity 555
8.3 Effect of Negative Feedback on Distortion 555
8.4 Feedback Configurations 557
    8.4.1 Series-Shunt Feedback 557
    8.4.2 Shunt-Shunt Feedback 560
    8.4.3 Shunt-Series Feedback 561
    8.4.4 Series-Series Feedback 562
8.5 Practical Configurations and the Effect of Loading 563
    8.5.1 Shunt-Shunt Feedback 563
    8.5.2 Series-Series Feedback 569
    8.5.3 Series-Shunt Feedback 579
    8.5.4 Shunt-Series Feedback 583
    8.5.5 Summary 587
8.6 Single-Stage Feedback 587
    8.6.1 Local Series-Series Feedback 587
    8.6.2 Local Series-Shunt Feedback 591
8.7 The Voltage Regulator as a Feedback Circuit 593
8.8 Feedback Circuit Analysis Using Return Ratio 599
    8.8.1 Closed-Loop Gain Using Return Ratio 601
    8.8.2 Closed-Loop Impedance Formula Using Return Ratio 607
    8.8.3 Summary—Return-Ratio Analysis 612
8.9 Modeling Input and Output Ports in Feedback Circuits 613

CHAPTER 9
Frequency Response and Stability of Feedback Amplifiers 624

9.1 Introduction 624
9.2 Relation Between Gain and Bandwidth in Feedback Amplifiers 624
9.3 Instability and the Nyquist Criterion 626
9.4 Compensation 633
    9.4.1 Theory of Compensation 633
    9.4.2 Methods of Compensation 637
    9.4.3 Two-Stage MOS Amplifier Compensation 643
    9.4.4 Compensation of Single-Stage CMOS Op Amps 650
    9.4.5 Nested Miller Compensation 654
9.5 Root-Locus Techniques 664
    9.5.1 Root Locus for a Three-Pole Transfer Function 665
    9.5.2 Rules for Root-Locus Construction 667
    9.5.3 Root Locus for Dominant-Pole Compensation 676
    9.5.4 Root Locus for Feedback-Zero Compensation 677
9.6 Slew Rate 681
    9.6.1 Origin of Slew-Rate Limitations 681
    9.6.2 Methods of Improving Slew-Rate in Two-Stage Op Amps 685
    9.6.3 Improving Slew-Rate in Bipolar Op Amps 687
    9.6.4 Improving Slew-Rate in MOS Op Amps 688
    9.6.5 Effect of Slew-Rate Limitations on Large-Signal Sinusoidal Performance 692
A.9.1 Analysis in Terms of Return-Ratio Parameters 693
A.9.2 Roots of a Quadratic Equation 694

CHAPTER 10
Nonlinear Analog Circuits 704

10.1 Introduction 704
10.2 Analog Multipliers Employing the Bipolar Transistor 704
    10.2.1 The Emitter-Coupled Pair as a Simple Multiplier 704
    10.2.2 The dc Analysis of the Gilbert Multiplier Cell 706

10.2.3 The Gilbert Cell as an Analog Multiplier 708

10.2.4 A Complete Analog Multiplier 711

10.2.5 The Gilbert Multiplier Cell as a Balanced Modulator and Phase Detector 712

10.3 Phase-Locked Loops (PLL) 716

10.3.1 Phase-Locked Loop Concepts 716

10.3.2 The Phase-Locked Loop in the Locked Condition 718

10.3.3 Integrated-Circuit Phase-Locked Loops 727

10.4 Nonlinear Function Synthesis 731

CHAPTER 11
Noise in Integrated Circuits 736

11.1 Introduction 736

11.2 Sources of Noise 736

11.2.1 Shot Noise 736

11.2.2 Thermal Noise 740

11.2.3 Flicker Noise ($1/f$ Noise) 741

11.2.4 Burst Noise (*Popcorn Noise*) 742

11.2.5 Avalanche Noise 743

11.3 Noise Models of Integrated-Circuit Components 744

11.3.1 Junction Diode 744

11.3.2 Bipolar Transistor 745

11.3.3 MOS Transistor 746

11.3.4 Resistors 747

11.3.5 Capacitors and Inductors 747

11.4 Circuit Noise Calculations 748

11.4.1 Bipolar Transistor Noise Performance 750

11.4.2 Equivalent Input Noise and the Minimum Detectable Signal 754

11.5 Equivalent Input Noise Generators 756

11.5.1 Bipolar Transistor Noise Generators 757

11.5.2 MOS Transistor Noise Generators 762

11.6 Effect of Feedback on Noise Performance 764

11.6.1 Effect of Ideal Feedback on Noise Performance 764

11.6.2 Effect of Practical Feedback on Noise Performance 765

11.7 Noise Performance of Other Transistor Configurations 771

11.7.1 Common-Base Stage Noise Performance 771

11.7.2 Emitter-Follower Noise Performance 773

11.7.3 Differential-Pair Noise Performance 773

11.8 Noise in Operational Amplifiers 776

11.9 Noise Bandwidth 782

11.10 Noise Figure and Noise Temperature 786

11.10.1 Noise Figure 786

11.10.2 Noise Temperature 790

CHAPTER 12
Fully Differential Operational Amplifiers 796

12.1 Introduction 796

12.2 Properties of Fully Differential Amplifiers 796

12.3 Small-Signal Models for Balanced Differential Amplifiers 799

12.4 Common-Mode Feedback 804

12.4.1 Common-Mode Feedback at Low Frequencies 805

12.4.2 Stability and Compensation Considerations in a CMFB Loop 810

12.5 CMFB Circuits 811

12.5.1 CMFB Using Resistive Divider and Amplifier 812

12.5.2 CMFB Using Two Differential Pairs 816

12.5.3 CMFB Using Transistors in the Triode Region 819

12.5.4 Switched-Capacitor CMFB 821

12.6 Fully Differential Op Amps 823

12.6.1 A Fully Differential Two-Stage Op Amp 823

12.6.2 Fully Differential Telescopic Cascode Op Amp 833

- 12.6.3 Fully Differential Folded-Cascode Op Amp 834
- 12.6.4 A Differential Op Amp with Two Differential Input Stages 835
- 12.6.5 Neutralization 835
- 12.7 Unbalanced Fully Differential Circuits 838
- 12.8 Bandwidth of the CMFB Loop 844
- 12.9 Analysis of a CMOS Fully Differential Folded-Cascode Op Amp 845
  - 12.9.1 DC Biasing 848
  - 12.9.2 Low-Frequency Analysis 850
  - 12.9.3 Frequency and Time Responses in a Feedback Application 856

Index 871

Symbol Convention

Unless otherwise stated, the following symbol convention is used in this book. *Bias* or *dc* quantities, such as transistor collector current I_C and collector-emitter voltage V_{CE}, are represented by uppercase symbols with uppercase subscripts. Small-signal quantities, such as the incremental change in transistor collector current i_c, are represented by lowercase symbols with lowercase subscripts. Elements such as transconductance g_m in small-signal equivalent circuits are represented in the same way. Finally, quantities such as *total* collector current I_c, which represent the sum of the bias quantity *and* the signal quantity, are represented by an uppercase symbol with a lowercase subscript.

CHAPTER 1

Models for Integrated-Circuit Active Devices

1.1 Introduction

The analysis and design of integrated circuits depend heavily on the utilization of suitable models for integrated-circuit components. This is true in hand analysis, where fairly simple models are generally used, and in computer analysis, where more complex models are encountered. Since any analysis is only as accurate as the model used, it is essential that the circuit designer have a thorough understanding of the origin of the models commonly utilized and the degree of approximation involved in each.

This chapter deals with the derivation of large-signal and small-signal models for integrated-circuit devices. The treatment begins with a consideration of the properties of *pn* junctions, which are basic parts of most integrated-circuit elements. Since this book is primarily concerned with circuit analysis and design, no attempt has been made to produce a comprehensive treatment of semiconductor physics. The emphasis is on summarizing the basic aspects of semiconductor-device behavior and indicating how these can be modeled by equivalent circuits.

1.2 Depletion Region of a *pn* Junction

The properties of reverse-biased *pn* junctions have an important influence on the characteristics of many integrated-circuit components. For example, reverse-biased *pn* junctions exist between many integrated-circuit elements and the underlying substrate, and these junctions all contribute voltage-dependent parasitic capacitances. In addition, a number of important characteristics of active devices, such as breakdown voltage and output resistance, depend directly on the properties of the depletion region of a reverse-biased *pn* junction. Finally, the basic operation of the junction field-effect transistor is controlled by the width of the depletion region of a *pn* junction. Because of its importance and application to many different problems, an analysis of the depletion region of a reverse-biased *pn* junction is considered below. The properties of forward-biased *pn* junctions are treated in Section 1.3 when bipolar-transistor operation is described.

Consider a *pn* junction under reverse bias as shown in Fig. 1.1. Assume *constant doping densities* of N_D atoms/cm^3 in the *n*-type material and N_A atoms/cm^3 in the *p*-type material. (The characteristics of junctions with nonconstant doping densities will be described later.) Due to the difference in carrier concentrations in the *p*-type and *n*-type regions, there exists a region at the junction where the mobile holes and electrons have been removed, leaving the fixed acceptor and donor ions. Each acceptor atom carries a negative charge and each donor atom carries a positive charge, so that the region near the junction is one of significant space charge and resulting high electric field. This is called the *depletion* region or *space-charge*

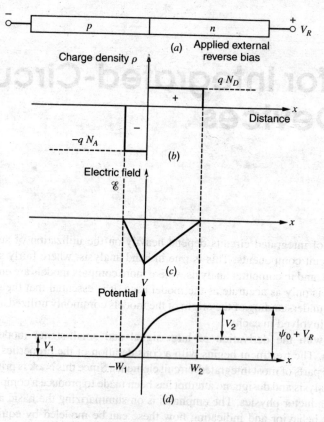

Figure 1.1 The abrupt junction under reverse bias V_R. (a) Schematic. (b) Charge density. (c) Electric field. (d) Electrostatic potential.

region. It is assumed that the edges of the depletion region are sharply defined as shown in Fig. 1.1, and this is a good approximation in most cases.

For zero applied bias, there exists a voltage ψ_0 across the junction called the *built-in potential*. This potential opposes the diffusion of mobile holes and electrons across the junction in equilibrium and has a value[1]

$$\psi_0 = V_T \ln \frac{N_A N_D}{n_i^2} \qquad (1.1)$$

where

$$V_T = \frac{kT}{q} \simeq 26 \text{ mV} \quad \text{at} \quad 300°\text{K}$$

the quantity n_i is the intrinsic carrier concentration in a pure sample of the semiconductor and $n_i \simeq 1.5 \times 10^{10} \text{cm}^{-3}$ at 300°K for silicon.

In Fig. 1.1 the built-in potential is augmented by the applied reverse bias, V_R, and the total voltage across the junction is $(\psi_0 + V_R)$. If the depletion region penetrates a distance W_1 into the p-type region and W_2 into the n-type region, then we require

$$W_1 N_A = W_2 N_D \qquad (1.2)$$

because the total charge per unit area on either side of the junction must be equal in magnitude but opposite in sign.

Poisson's equation in one dimension requires that

$$\frac{d^2V}{dx^2} = -\frac{\rho}{\epsilon} = \frac{qN_A}{\epsilon} \quad \text{for} \quad -W_1 < x < 0 \tag{1.3}$$

where ρ is the charge density, q is the electron charge (1.6×10^{-19} coulomb), and ϵ is the permittivity of the silicon (1.04×10^{-12} farad/cm). The permittivity is often expressed as

$$\epsilon = K_S \epsilon_0 \tag{1.4}$$

where K_S is the dielectric constant of silicon and ϵ_0 is the permittivity of free space (8.86×10^{-14} F/cm). Integration of (1.3) gives

$$\frac{dV}{dx} = \frac{qN_A}{\epsilon} x + C_1 \tag{1.5}$$

where C_1 is a constant. However, the electric field $\mathcal{E}$ is given by

$$\mathcal{E} = -\frac{dV}{dx} = -\left(\frac{qN_A}{\epsilon} x + C_1\right) \tag{1.6}$$

Since there is zero electric field outside the depletion region, a boundary condition is

$$\mathcal{E} = 0 \quad \text{for} \quad x = -W_1$$

and use of this condition in (1.6) gives

$$\mathcal{E} = -\frac{qN_A}{\epsilon}(x + W_1) = -\frac{dV}{dx} \quad \text{for} \quad -W_1 < x < 0 \tag{1.7}$$

Thus the dipole of charge existing at the junction gives rise to an electric field that varies linearly with distance.

Integration of (1.7) gives

$$V = \frac{qN_A}{\epsilon}\left(\frac{x^2}{2} + W_1 x\right) + C_2 \tag{1.8}$$

If the zero for potential is arbitrarily taken to be the potential of the neutral p-type region, then a second boundary condition is

$$V = 0 \quad \text{for} \quad x = -W_1$$

and use of this in (1.8) gives

$$V = \frac{qN_A}{\epsilon}\left(\frac{x^2}{2} + W_1 x + \frac{W_1^2}{2}\right) \quad \text{for} \quad -W_1 < x < 0 \tag{1.9}$$

At $x = 0$, we define $V = V_1$, and then (1.9) gives

$$V_1 = \frac{qN_A}{\epsilon}\frac{W_1^2}{2} \tag{1.10}$$

If the potential difference from $x = 0$ to $x = W_2$ is V_2, then it follows that

$$V_2 = \frac{qN_D}{\epsilon}\frac{W_2^2}{2} \tag{1.11}$$

and thus the total voltage across the junction is

$$\psi_0 + V_R = V_1 + V_2 = \frac{q}{2\epsilon}(N_A W_1^2 + N_D W_2^2) \tag{1.12}$$

Substitution of (1.2) in (1.12) gives

$$\psi_0 + V_R = \frac{qW_1^2 N_A}{2\epsilon}\left(1 + \frac{N_A}{N_D}\right) \tag{1.13}$$

From (1.13), the penetration of the depletion layer into the *p*-type region is

$$W_1 = \left[\frac{2\epsilon(\psi_0 + V_R)}{qN_A\left(1 + \dfrac{N_A}{N_D}\right)}\right]^{1/2} \tag{1.14}$$

Similarly,

$$W_2 = \left[\frac{2\epsilon(\psi_0 + V_R)}{qN_D\left(1 + \dfrac{N_D}{N_A}\right)}\right]^{1/2} \tag{1.15}$$

Equations 1.14 and 1.15 show that the depletion regions extend into the *p*-type and *n*-type regions in *inverse* relation to the impurity concentrations and in proportion to $\sqrt{\psi_0 + V_R}$. If either N_D or N_A is much larger than the other, the depletion region exists almost entirely in the *lightly doped* region.

■ **EXAMPLE**

An abrupt *pn* junction in silicon has doping densities $N_A = 10^{15}$ atoms/cm^3 and $N_D = 10^{16}$ atoms/cm^3. Calculate the junction built-in potential, the depletion-layer depths, and the maximum field with 10 V reverse bias.

From (1.1)

$$\psi_0 = 26 \ln \frac{10^{15} \times 10^{16}}{2.25 \times 10^{20}} \text{ mV} = 638 \text{ mV} \quad \text{at} \quad 300°\text{K}$$

From (1.14) the depletion-layer depth in the *p*-type region is

$$W_1 = \left(\frac{2 \times 1.04 \times 10^{-12} \times 10.64}{1.6 \times 10^{-19} \times 10^{15} \times 1.1}\right)^{1/2} = 3.5 \times 10^{-4} \text{cm}$$

$$= 3.5 \text{ µm} \quad (\text{where } 1 \text{ µm} = 1 \text{ micrometer} = 10^{-6} \text{ m})$$

The depletion-layer depth in the more heavily doped *n*-type region is

$$W_2 = \left(\frac{2 \times 1.04 \times 10^{-12} \times 10.64}{1.6 \times 10^{-19} \times 10^{16} \times 11}\right)^{1/2} = 0.35 \times 10^{-4} \text{ cm} = 0.35 \text{ µm}$$

Finally, from (1.7) the maximum field that occurs for $x = 0$ is

$$\mathscr{E}_{max} = -\frac{qN_A}{\epsilon}W_1 = -1.6 \times 10^{-19} \times \frac{10^{15} \times 3.5 \times 10^{-4}}{1.04 \times 10^{-12}}$$

$$= -5.4 \times 10^4 \text{ V/cm}$$

■ Note the large magnitude of this electric field.

1.2.1 Depletion-Region Capacitance

Since there is a *voltage-dependent charge* Q associated with the depletion region, we can calculate a small-signal capacitance C_j as follows:

$$C_j = \frac{dQ}{dV_R} = \frac{dQ}{dW_1}\frac{dW_1}{dV_R} \tag{1.16}$$

Now

$$dQ = AqN_A dW_1 \tag{1.17}$$

where A is the cross-sectional area of the junction. Differentiation of (1.14) gives

$$\frac{dW_1}{dV_R} = \left[\frac{\epsilon}{2qN_A\left(1+\frac{N_A}{N_D}\right)(\psi_0 + V_R)}\right]^{1/2} \tag{1.18}$$

Use of (1.17) and (1.18) in (1.16) gives

$$C_j = A\left[\frac{q\epsilon N_A N_D}{2(N_A + N_D)}\right]^{1/2}\frac{1}{\sqrt{\psi_0 + V_R}} \tag{1.19}$$

The above equation was derived for the case of reverse bias V_R applied to the diode. However, it is valid for positive bias voltages as long as the forward current flow is small. Thus, if V_D represents the bias on the junction (positive for forward bias, negative for reverse bias), then (1.19) can be written as

$$C_j = A\left[\frac{q\epsilon N_A N_D}{2(N_A + N_D)}\right]^{1/2}\frac{1}{\sqrt{\psi_0 - V_D}} \tag{1.20}$$

$$= \frac{C_{j0}}{\sqrt{1 - \frac{V_D}{\psi_0}}} \tag{1.21}$$

where C_{j0} is the value of C_j for $V_D = 0$.

Equations 1.20 and 1.21 were derived using the assumption of constant doping in the p-type and n-type regions. However, many practical diffused junctions more closely approach a *graded* doping profile as shown in Fig. 1.2. In this case, a similar calculation yields

$$C_j = \frac{C_{j0}}{\sqrt[3]{1 - \frac{V_D}{\psi_0}}} \tag{1.22}$$

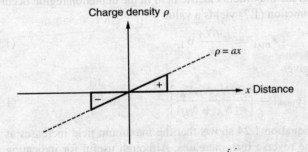

Figure 1.2 Charge density versus distance in a graded junction.

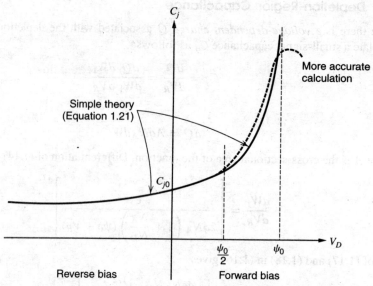

Figure 1.3 Behavior of *pn* junction depletion-layer capacitance C_j as a function of bias voltage V_D.

Note that both (1.21) and (1.22) predict values of C_j approaching infinity as V_D approaches ψ_0. However, the current flow in the diode is then appreciable and the equations are no longer valid. A more exact analysis[2,3] of the behavior of C_j as a function of V_D gives the result shown in Fig. 1.3. For forward bias voltages up to about $\psi_0/2$, the values of C_j predicted by (1.21) are very close to the more accurate value. As an approximation, some computer programs approximate C_j for $V_D > \psi_0/2$ by a linear extrapolation of (1.21) or (1.22).

■ **EXAMPLE**

If the zero-bias capacitance of a diffused junction is 3 pF and $\psi_0 = 0.5$ V, calculate the capacitance with 10 V reverse bias. Assume the doping profile can be approximated by an abrupt junction.

From (1.21)

$$C_j = \frac{3}{\sqrt{1 + \dfrac{10}{0.5}}} \text{pF} = 0.65 \text{ pF}$$

1.2.2 Junction Breakdown

From Fig. 1.1c it can be seen that the maximum electric field in the depletion region occurs at the junction, and for an abrupt junction (1.7) yields a value

$$\mathscr{E}_{max} = -\frac{qN_A}{\epsilon} W_1 \qquad (1.23)$$

Substitution of (1.14) in (1.23) gives

$$|\mathscr{E}_{max}| = \left[\frac{2qN_A N_D V_R}{\epsilon (N_A + N_D)}\right]^{1/2} \qquad (1.24)$$

where ψ_0 has been neglected. Equation 1.24 shows that the maximum field increases as the doping density increases and the reverse bias increases. Although useful for indicating the functional dependence of $\mathscr{E}_{max}$ on other variables, this equation is strictly valid for an ideal plane junction only. Practical junctions tend to have edge effects that cause somewhat higher values of $\mathscr{E}_{max}$ due to a concentration of the field at the curved edges of the junction.

Any reverse-biased *pn* junction has a small reverse current flow due to the presence of minority-carrier holes and electrons in the vicinity of the depletion region. These are swept across the depletion region by the field and contribute to the leakage current of the junction. As the reverse bias on the junction is increased, the maximum field increases and the carriers acquire increasing amounts of energy between lattice collisions in the depletion region. At a critical field $\mathscr{E}_{crit}$ the carriers traversing the depletion region acquire sufficient energy to create new hole-electron pairs in collisions with silicon atoms. This is called the *avalanche process* and leads to a sudden increase in the reverse-bias leakage current since the newly created carriers are also capable of producing avalanche. The value of $\mathscr{E}_{crit}$ is about 3×10^5 V/cm for junction doping densities in the range of 10^{15} to 10^{16} atoms/cm^3, but it increases slowly as the doping density increases and reaches about 10^6 V/cm for doping densities of 10^{18} atoms/cm^3.

A typical *I-V* characteristic for a junction diode is shown in Fig. 1.4, and the effect of avalanche breakdown is seen by the large increase in reverse current, which occurs as the reverse bias approaches the breakdown voltage *BV*. This corresponds to the maximum field $\mathscr{E}_{max}$ approaching $\mathscr{E}_{crit}$. It has been found empirically[4] that if the normal reverse bias current of the diode is I_R with no avalanche effect, then the actual reverse current near the breakdown voltage is

$$I_{RA} = MI_R \tag{1.25}$$

where M is the *multiplication factor* defined by

$$M = \frac{1}{1 - \left(\dfrac{V_R}{BV}\right)^n} \tag{1.26}$$

In this equation, V_R is the reverse bias on the diode and n has a value between 3 and 6.

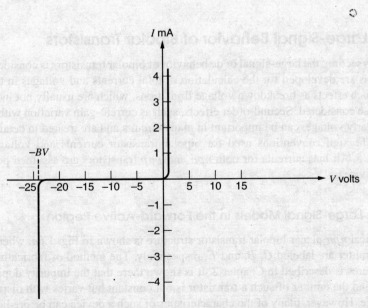

Figure 1.4 Typical *I-V* characteristic of a junction diode showing avalanche breakdown.

The operation of a *pn* junction in the breakdown region is not inherently destructive. However, the avalanche current flow must be limited by external resistors in order to prevent excessive power dissipation from occurring at the junction and causing damage to the device. Diodes operated in the avalanche region are widely used as voltage references and are called *Zener diodes*. There is another, related process called *Zener breakdown*,[5] which is different from the avalanche breakdown described above. Zener breakdown occurs only in very heavily doped junctions where the electric field becomes large enough (even with small reverse-bias voltages) to strip electrons away from the valence bonds. This process is called *tunneling*, and there is no multiplication effect as in avalanche breakdown. Although the Zener breakdown mechanism is important only for breakdown voltages below about 6 V, all breakdown diodes are commonly referred to as Zener diodes.

The calculations so far have been concerned with the breakdown characteristic of plane abrupt junctions. Practical diffused junctions differ in some respects from these results and the characteristics of these junctions have been calculated and tabulated for use by designers.[5] In particular, edge effects in practical diffused junctions can result in breakdown voltages as much as 50 percent below the value calculated for a plane junction.

■ EXAMPLE

An abrupt plane *pn* junction has doping densities $N_A = 5 \times 10^{15}$ atoms/cm^3 and $N_D = 10^{16}$ atoms/cm^3. Calculate the breakdown voltage if $\mathscr{E}_{crit} = 3 \times 10^5$ V/cm.

The breakdown voltage is calculated using $\mathscr{E}_{max} = \mathscr{E}_{crit}$ in (1.24) to give

$$BV = \frac{\epsilon(N_A + N_D)}{2qN_A N_D}\mathscr{E}_{crit}^2$$

$$= \frac{1.04 \times 10^{-12} \times 15 \times 10^{15}}{2 \times 1.6 \times 10^{-19} \times 5 \times 10^{15} \times 10^{16}} \times 9 \times 10^{10} \text{ V}$$

$$= 88 \text{ V}$$

■

1.3 Large-Signal Behavior of Bipolar Transistors

In this section, the large-signal or dc behavior of bipolar transistors is considered. Large-signal models are developed for the calculation of total currents and voltages in transistor circuits, and such effects as breakdown voltage limitations, which are usually not included in models, are also considered. Second-order effects, such as current-gain variation with collector current and Early voltage, can be important in many circuits and are treated in detail.

The sign conventions used for bipolar transistor currents and voltages are shown in Fig. 1.5. All bias currents for both *npn* and *pnp* transistors are assumed positive going into the device.

1.3.1 Large-Signal Models in the Forward-Active Region

A typical *npn* planar bipolar transistor structure is shown in Fig. 1.6a, where collector, base, and emitter are labeled C, B, and E, respectively. The method of fabricating such transistor structures is described in Chapter 2. It is shown there that the impurity doping density in the base and the emitter of such a transistor is not constant but varies with distance from the top surface. However, many of the characteristics of such a device can be predicted by analyzing the idealized transistor structure shown in Fig. 1.6b. In this structure, the base and emitter

1.3 Large-Signal Behavior of Bipolar Transistors

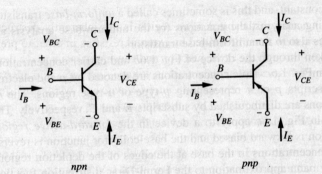

Figure 1.5 Bipolar transistor sign convention.

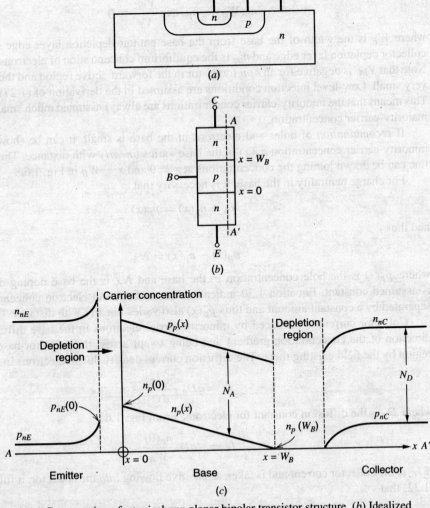

Figure 1.6 (a) Cross section of a typical *npn* planar bipolar transistor structure. (b) Idealized transistor structure. (c) Carrier concentrations along the cross section AA' of the transistor in (b). Uniform doping densities are assumed. (Not to scale.)

doping densities are assumed constant, and this is sometimes called a *uniform-base* transistor. Where possible in the following analyses, the equations for the uniform-base analysis are expressed in a form that applies also to nonuniform-base transistors.

A cross section AA' is taken through the device of Fig. 1.6b and carrier concentrations along this section are plotted in Fig. 1.6c. Hole concentrations are denoted by p and electron concentrations by n with subscripts p or n representing p-type or n-type regions. The n-type emitter and collector regions are distinguished by subscripts E and C, respectively. The carrier concentrations shown in Fig. 1.6c apply to a device in the *forward-active region*. That is, the base-emitter junction is forward biased and the base-collector junction is reverse biased. The minority-carrier concentrations in the base at the edges of the depletion regions can be calculated from a Boltzmann approximation to the Fermi-Dirac distribution function to give[6]

$$n_p(0) = n_{po} \exp \frac{V_{BE}}{V_T} \tag{1.27}$$

$$n_p(W_B) = n_{po} \exp \frac{V_{BC}}{V_T} \simeq 0 \tag{1.28}$$

where W_B is the width of the base from the base-emitter depletion layer edge to the base-collector depletion layer edge and n_{po} is the equilibrium concentration of electrons in the base. Note that V_{BC} is negative for an *npn* transistor in the forward-active region and thus $n_p(W_B)$ is very small. Low-level injection conditions are assumed in the derivation of (1.27) and (1.28). This means that the minority-carrier concentrations are always assumed much smaller than the majority-carrier concentration.

If *recombination* of holes and electrons in the base is small, it can be shown that[7] the minority-carrier concentration $n_p(x)$ in the base varies *linearly* with distance. Thus a straight line can be drawn joining the concentrations at $x = 0$ and $x = W_B$ in Fig. 1.6c.

For charge neutrality in the base, it is necessary that

$$N_A + n_p(x) = p_p(x) \tag{1.29}$$

and thus

$$p_p(x) - n_p(x) = N_A \tag{1.30}$$

where $p_p(x)$ is the hole concentration in the base and N_A is the base doping density that is assumed constant. Equation 1.30 indicates that the hole and electron concentrations are separated by a constant amount and thus $p_p(x)$ also varies linearly with distance.

Collector current is produced by minority-carrier electrons in the base diffusing in the direction of the concentration gradient and being swept across the collector-base depletion region by the field existing there. The diffusion current density due to electrons in the base is

$$J_n = qD_n \frac{dn_p(x)}{dx} \tag{1.31}$$

where D_n is the diffusion constant for electrons. From Fig. 1.6c

$$J_n = -qD_n \frac{n_p(0)}{W_B} \tag{1.32}$$

If I_C is the collector current and is taken as positive flowing *into* the collector, it follows from (1.32) that

$$I_C = qAD_n \frac{n_p(0)}{W_B} \tag{1.33}$$

where A is the cross-sectional area of the emitter. Substitution of (1.27) into (1.33) gives

$$I_C = \frac{qAD_n n_{po}}{W_B} \exp \frac{V_{BE}}{V_T} \quad (1.34)$$

$$= I_S \exp \frac{V_{BE}}{V_T} \quad (1.35)$$

where

$$I_S = \frac{qAD_n n_{po}}{W_B} \quad (1.36)$$

and I_S is a constant used to describe the transfer characteristic of the transistor in the forward-active region. Equation 1.36 can be expressed in terms of the base doping density by noting that[8] (see Chapter 2)

$$n_{po} = \frac{n_i^2}{N_A} \quad (1.37)$$

and substitution of (1.37) in (1.36) gives

$$I_S = \frac{qAD_n n_i^2}{W_B N_A} = \frac{qA\overline{D}_n n_i^2}{Q_B} \quad (1.38)$$

where $Q_B = W_B N_A$ is the number of doping atoms in the base per unit area of the emitter and n_i is the intrinsic carrier concentration in silicon. In this form (1.38) applies to both uniform- and nonuniform-base transistors and D_n has been replaced by $\overline{D}_n$, which is an average effective value for the electron diffusion constant in the base. This is necessary for nonuniform-base devices because the diffusion constant is a function of impurity concentration. Typical values of I_S as given by (1.38) are from 10^{-14} to 10^{-16} A.

Equation 1.35 gives the collector current as a function of base-emitter voltage. The base current I_B is also an important parameter and, at moderate current levels, consists of two major components. One of these (I_{B1}) represents recombination of holes and electrons in the base and is proportional to the minority-carrier charge Q_e in the base. From Fig. 1.6c, the minority-carrier charge in the base is

$$Q_e = \frac{1}{2} n_p(0) W_B q A \quad (1.39)$$

and we have

$$I_{B1} = \frac{Q_e}{\tau_b} = \frac{1}{2} \frac{n_p(0) W_B q A}{\tau_b} \quad (1.40)$$

where τ_b is the minority-carrier lifetime in the base. I_{B1} represents a flow of majority holes from the base lead into the base region. Substitution of (1.27) in (1.40) gives

$$I_{B1} = \frac{1}{2} \frac{n_{po} W_B q A}{\tau_b} \exp \frac{V_{BE}}{V_T} \quad (1.41)$$

The second major component of base current (usually the dominant one in integrated-circuit npn devices) is due to injection of holes from the base into the emitter. This current component depends on the gradient of minority-carrier holes in the emitter and is[9]

$$I_{B2} = \frac{qAD_p}{L_p} p_{nE}(0) \quad (1.42)$$

where D_p is the diffusion constant for holes and L_p is the diffusion length (assumed small) for holes in the emitter. $p_{nE}(0)$ is the concentration of holes in the emitter at the edge of the depletion region and is

$$p_{nE}(0) = p_{nEo} \exp \frac{V_{BE}}{V_T} \qquad (1.43)$$

If N_D is the donor atom concentration in the emitter (assumed constant), then

$$p_{nEo} \simeq \frac{n_i^2}{N_D} \qquad (1.44)$$

The emitter is deliberately doped much more heavily than the base, making N_D large and p_{nEo} small, so that the base-current component, I_{B2}, is minimized.

Substitution of (1.43) and (1.44) in (1.42) gives

$$I_{B2} = \frac{qAD_p}{L_p} \frac{n_i^2}{N_D} \exp \frac{V_{BE}}{V_T} \qquad (1.45)$$

The total base current, I_B, is the sum of I_{B1} and I_{B2}:

$$I_B = I_{B1} + I_{B2} = \left(\frac{1}{2} \frac{n_{po} W_B q A}{\tau_b} + \frac{qAD_p}{L_p} \frac{n_i^2}{N_D} \right) \exp \frac{V_{BE}}{V_T} \qquad (1.46)$$

Although this equation was derived assuming uniform base and emitter doping, it gives the correct functional dependence of I_B on device parameters for practical double-diffused nonuniform-base devices. Second-order components of I_B, which are important at low current levels, are considered later.

Since I_C in (1.35) and I_B in (1.46) are both proportional to $\exp(V_{BE}/V_T)$ in this analysis, the base current can be expressed in terms of collector current as

$$I_B = \frac{I_C}{\beta_F} \qquad (1.47)$$

where β_F is the forward current gain. An expression for β_F can be calculated by substituting (1.34) and (1.46) in (1.47) to give

$$\beta_F = \frac{\dfrac{qAD_n n_{po}}{W_B}}{\dfrac{1}{2}\dfrac{n_{po}W_B qA}{\tau_b} + \dfrac{qAD_p n_i^2}{L_p N_D}} = \frac{1}{\dfrac{W_B^2}{2\tau_b D_n} + \dfrac{D_p}{D_n}\dfrac{W_B}{L_p}\dfrac{N_A}{N_D}} \qquad (1.48)$$

where (1.37) has been substituted for n_{po}. Equation 1.48 shows that β_F is maximized by minimizing the base width W_B and maximizing the ratio of emitter to base doping densities N_D/N_A. Typical values of β_F for npn transistors in integrated circuits are 50 to 500, whereas lateral pnp transistors (to be described in Chapter 2) have values 10 to 100. Finally, the emitter current is

$$I_E = -(I_C + I_B) = -\left(I_C + \frac{I_C}{\beta_F}\right) = -\frac{I_C}{\alpha_F} \qquad (1.49)$$

where

$$\alpha_F = \frac{\beta_F}{1+\beta_F} \qquad (1.50)$$

1.3 Large-Signal Behavior of Bipolar Transistors

The value of α_F can be expressed in terms of device parameters by substituting (1.48) in (1.50) to obtain

$$\alpha_F = \cfrac{1}{1 + \cfrac{1}{\beta_F}} = \cfrac{1}{1 + \cfrac{W_B^2}{2\tau_b D_n} + \cfrac{D_p}{D_n}\cfrac{W_B}{L_p}\cfrac{N_A}{N_D}} \simeq \alpha_T \gamma \qquad (1.51)$$

where

$$\alpha_T = \cfrac{1}{1 + \cfrac{W_B^2}{2\tau_b D_n}} \qquad (1.51a)$$

$$\gamma = \cfrac{1}{1 + \cfrac{D_p}{D_n}\cfrac{W_B}{L_p}\cfrac{N_A}{N_D}} \qquad (1.51b)$$

The validity of (1.51) depends on $W_B^2/2\tau_b D_n \ll 1$ and $(D_p/D_n)(W_B/L_p)(N_A/N_D) \ll 1$, and this is always true if β_F is large [see (1.48)]. The term γ in (1.51) is called the *emitter injection efficiency* and is equal to the ratio of the electron current (*npn* transistor) injected into the base from the emitter to the total hole and electron current crossing the base-emitter junction. Ideally $\gamma \to 1$, and this is achieved by making N_D/N_A large and W_B small. In that case very little reverse injection occurs from base to emitter.

The term α_T in (1.51) is called the *base transport factor* and represents the fraction of carriers injected into the base (from the emitter) that reach the collector. Ideally $\alpha_T \to 1$ and this is achieved by making W_B small. It is evident from the above development that fabrication changes that cause α_T and γ to approach unity also maximize the value of β_F of the transistor.

The results derived above allow formulation of a large-signal model of the transistor suitable for bias-circuit calculations with devices in the forward-active region. One such circuit is shown in Fig. 1.7 and consists of a base-emitter diode to model (1.46) and a controlled collector-current generator to model (1.47). Note that the collector voltage ideally has no influence on the collector current and the collector node acts as a high-impedance current source. A simpler version of this equivalent circuit, which is often useful, is shown in Fig. 1.7b, where the input diode has been replaced by a battery with a value $V_{BE(on)}$, which is usually 0.6 to 0.7 V. This represents the fact that in the forward-active region the base-emitter voltage varies very little because of the steep slope of the exponential characteristic. In some circuits the temperature coefficient of $V_{BE(on)}$ is important, and a typical value for this is -2 mV/°C. The equivalent circuits of Fig. 1.7 apply for *npn* transistors. For *pnp* devices the corresponding equivalent circuits are shown in Fig. 1.8.

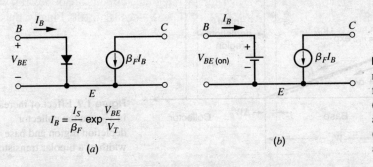

Figure 1.7 Large-signal models of *npn* transistors for use in bias calculations. (*a*) Circuit incorporating an input diode. (*b*) Simplified circuit with an input voltage source.

14 Chapter 1 ■ Models for Integrated-Circuit Active Devices

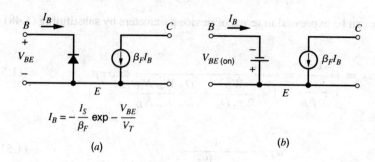

$$I_B = -\frac{I_S}{\beta_F} \exp -\frac{V_{BE}}{V_T}$$

(a)

(b)

Figure 1.8 Large-signal models of *pnp* transistors corresponding to the circuits of Fig. 1.7.

1.3.2 Effects of Collector Voltage on Large-Signal Characteristics in the Forward-Active Region

In the analysis of the previous section, the collector-base junction was assumed reverse biased and ideally had no effect on the collector currents. This is a useful approximation for first-order calculations, but is not strictly true in practice. There are occasions when the influence of collector voltage on collector current is important, and this will now be investigated.

The collector voltage has a dramatic effect on the collector current in two regions of device operation. These are the saturation (V_{CE} approaches zero) and breakdown (V_{CE} very large) regions that will be considered later. For values of collector-emitter voltage V_{CE} between these extremes, the collector current increases slowly as V_{CE} increases. The reason for this can be seen from Fig. 1.9, which is a sketch of the minority-carrier concentration in the base of the transistor. Consider the effect of changes in V_{CE} on the carrier concentration for constant V_{BE}. Since V_{BE} is constant, the change in V_{CB} equals the change in V_{CE} and this causes an increase in the collector-base depletion-layer width as shown. The change in the base width of the transistor, ΔW_B, equals the change in the depletion-layer width and causes an increase ΔI_C in the collector current.

From (1.35) and (1.38) we have

$$I_C = \frac{qA\overline{D}_n n_i^2}{Q_B} \exp \frac{V_{BE}}{V_T} \tag{1.52}$$

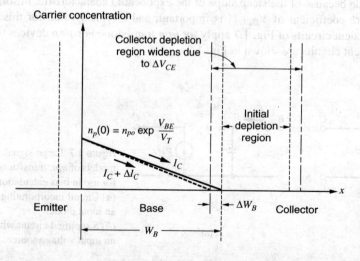

Figure 1.9 Effect of increases in V_{CE} on the collector depletion region and base width of a bipolar transistor.

Differentiation of (1.52) yields

$$\frac{\partial I_C}{\partial V_{CE}} = -\frac{qA\overline{D}_n n_i^2}{Q_B^2}\left(\exp\frac{V_{BE}}{V_T}\right)\frac{dQ_B}{dV_{CE}} \tag{1.53}$$

and substitution of (1.52) in (1.53) gives

$$\frac{\partial I_C}{\partial V_{CE}} = -\frac{I_C}{Q_B}\frac{dQ_B}{dV_{CE}} \tag{1.54}$$

For a uniform-base transistor $Q_B = W_B N_A$, and (1.54) becomes

$$\frac{\partial I_C}{\partial V_{CE}} = -\frac{I_C}{W_B}\frac{dW_B}{dV_{CE}} \tag{1.55}$$

Note that since the base width *decreases* as V_{CE} increases, dW_B/dV_{CE} in (1.55) is negative and thus $\partial I_C/\partial V_{CE}$ is positive. The magnitude of dW_B/dV_{CE} can be calculated from (1.18) for a uniform-base transistor. This equation predicts that dW_B/dV_{CE} is a function of the bias value of V_{CE}, but the variation is typically small for a reverse-biased junction and dW_B/dV_{CE} is often assumed constant. The resulting predictions agree adequately with experimental results.

Equation 1.55 shows that $\partial I_C/\partial V_{CE}$ is proportional to the collector-bias current and inversely proportional to the transistor base width. Thus narrow-base transistors show a greater dependence of I_C on V_{CE} in the forward-active region. The dependence of $\partial I_C/\partial V_{CE}$ on I_C results in typical transistor output characteristics as shown in Fig. 1.10. In accordance with the assumptions made in the foregoing analysis, these characteristics are shown for constant values of V_{BE}. However, in most integrated-circuit transistors the base current is dependent only on V_{BE} and not on V_{CE}, and thus constant-base-current characteristics can often be used in the following calculation. The reason for this is that the base current is usually dominated by the I_{B2} component of (1.45), which has no dependence on V_{CE}. Extrapolation of the characteristics of Fig. 1.10 back to the V_{CE} axis gives an intercept V_A called the Early voltage, where

$$V_A = \frac{I_C}{\dfrac{\partial I_C}{\partial V_{CE}}} \tag{1.56}$$

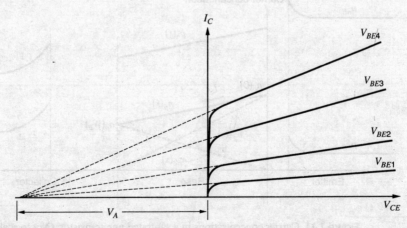

Figure 1.10 Bipolar transistor output characteristics showing the Earlyvoltage, V_A.

Substitution of (1.55) in (1.56) gives

$$V_A = -W_B \frac{dV_{CE}}{dW_B} \qquad (1.57)$$

which is a constant, independent of I_C. Thus all the characteristics extrapolate to the same point on the V_{CE} axis. The variation of I_C with V_{CE} is called the Early effect, and V_A is a common model parameter for circuit-analysis computer programs. Typical values of V_A for integrated-circuit transistors are 15 to 100 V. The inclusion of Early effect in dc bias calculations is usually limited to computer analysis because of the complexity introduced into the calculation. However, the influence of the Early effect is often dominant in small-signal calculations for high-gain circuits and this point will be considered later.

Finally, the influence of Early effect on the transistor large-signal characteristics in the forward-active region can be represented approximately by modifying (1.35) to

$$I_C = I_S \left(1 + \frac{V_{CE}}{V_A}\right) \exp \frac{V_{BE}}{V_T} \qquad (1.58)$$

This is a common means of representing the device output characteristics for computer simulation.

1.3.3 Saturation and Inverse-Active Regions

Saturation is a region of device operation that is usually avoided in analog circuits because the transistor gain is very low in this region. Saturation is much more commonly encountered in digital circuits, where it provides a well-specified output voltage that represents a logic state.

In saturation, both emitter-base and collector-base junctions are forward biased. Consequently, the collector-emitter voltage V_{CE} is quite small and is usually in the range 0.05 to 0.3 V. The carrier concentrations in a saturated npn transistor with uniform base doping are shown in Fig. 1.11. The minority-carrier concentration in the base at the edge of the depletion region is again given by (1.28) as

$$n_p(W_B) = n_{po} \exp \frac{V_{BC}}{V_T} \qquad (1.59)$$

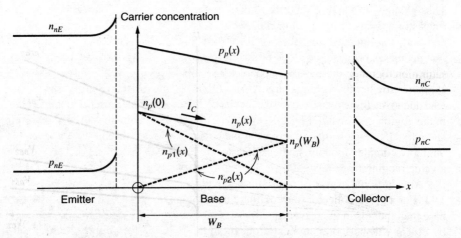

Figure 1.11 Carrier concentrations in a saturated npn transistor. (Not to scale.)

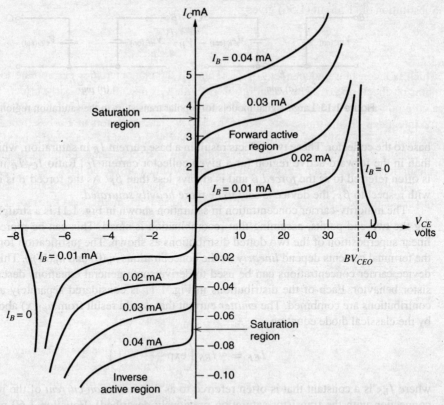

Figure 1.12 Typical I_C-V_{CE} characteristics for an *npn* bipolar transistor. Note the different scales for positive and negative currents and voltages.

but since V_{BC} is now positive, the value of $n_p(W_B)$ is no longer negligible. Consequently, changes in V_{CE} with V_{BE} held constant (which cause equal changes in V_{BC}) directly affect $n_p(W_B)$. Since the collector current is proportional to the slope of the minority-carrier concentration in the base [see (1.31)], it is also proportional to $[n_p(0) - n_p(W_B)]$ from Fig. 1.11. Thus changes in $n_p(W_B)$ directly affect the collector current, and the collector node of the transistor appears to have a *low impedance*. As V_{CE} is decreased in saturation with V_{BE} held constant, V_{BC} increases, as does $n_p(W_B)$ from (1.59). Thus, from Fig. 1.11, the collector current decreases because the slope of the carrier concentration decreases. This gives rise to the saturation region of the $I_C - V_{CE}$ characteristic shown in Fig. 1.12. The slope of the $I_C - V_{CE}$ characteristic in this region is largely determined by the resistance in series with the collector lead due to the finite resistivity of the *n*-type collector material. A useful model for the transistor in this region is shown in Fig. 1.13 and consists of a fixed voltage source to represent $V_{BE(\text{on})}$, and a fixed voltage source to represent the collector-emitter voltage $V_{CE(\text{sat})}$. A more accurate but more complex model includes a resistor in series with the collector. This resistor can have a value ranging from 20 to 500 Ω, depending on the device structure.

An additional aspect of transistor behavior in the saturation region is apparent from Fig. 1.11. For a given collector current, there is now a much larger amount of stored charge in the base than there is in the forward-active region. Thus the base-current contribution represented by (1.41) will be larger in saturation. In addition, since the collector-base junction is now forward biased, there is a new base-current component due to injection of carriers from the

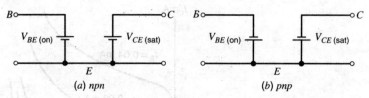

Figure 1.13 Large-signal models for bipolar transistors in the saturation region.

base to the collector. These two effects result in a base current I_B in saturation, which is larger than in the forward-active region for a given collector current I_C. Ratio I_C/I_B in saturation is often referred to as the *forced* β and is always less than β_F. As the forced β is made lower with respect to β_F, the device is said to be more *heavily saturated*.

The minority-carrier concentration in saturation shown in Fig. 1.11 is a straight line joining the two end points, assuming that recombination is small. This can be represented as a linear superposition of the two dotted distributions as shown. The justification for this is that the terminal currents depend *linearly* on the concentrations $n_p(0)$ and $n_p(W_B)$. This picture of device carrier concentrations can be used to derive some general equations describing transistor behavior. Each of the distributions in Fig. 1.11 is considered separately and the two contributions are combined. The *emitter* current that would result from $n_{p1}(x)$ above is given by the classical diode equation

$$I_{EF} = -I_{ES}\left(\exp\frac{V_{BE}}{V_T} - 1\right) \tag{1.60}$$

where I_{ES} is a constant that is often referred to as the *saturation current* of the junction (no connection with the transistor saturation previously described). Equation 1.60 predicts that the junction current is given by $I_{EF} \simeq I_{ES}$ with a reverse-bias voltage applied. However, in practice (1.60) is applicable only in the forward-bias region, since second-order effects dominate under reverse-bias conditions and typically result in a junction current several orders of magnitude larger than I_{ES}. The junction current that flows under reverse-bias conditions is often called the *leakage current* of the junction.

Returning to Fig. 1.11, we can describe the *collector* current resulting from $n_{p2}(x)$ alone as

$$I_{CR} = -I_{CS}\left(\exp\frac{V_{BC}}{V_T} - 1\right) \tag{1.61}$$

where I_{CS} is a constant. The total collector current I_C is given by I_{CR} plus the fraction of I_{EF} that reaches the collector (allowing for recombination and reverse emitter injection). Thus

$$I_C = \alpha_F I_{ES}\left(\exp\frac{V_{BE}}{V_T} - 1\right) - I_{CS}\left(\exp\frac{V_{BC}}{V_T} - 1\right) \tag{1.62}$$

where α_F has been defined previously by (1.51). Similarly, the total emitter current is composed of I_{EF} plus the fraction of I_{CR} that reaches the emitter with the transistor acting in an inverted mode. Thus

$$I_E = -I_{ES}\left(\exp\frac{V_{BE}}{V_T} - 1\right) + \alpha_R I_{CS}\left(\exp\frac{V_{BC}}{V_T} - 1\right) \tag{1.63}$$

where α_R is the ratio of emitter to collector current with the transistor operating *inverted* (i.e., with the collector-base junction forward biased and emitting carriers into the base and the emitter-base junction reverse biased and collecting carriers). Typical values of α_R are 0.5 to 0.8.

An inverse current gain β_R is also defined

$$\beta_R = \frac{\alpha_R}{1 - \alpha_R} \tag{1.64}$$

and has typical values 1 to 5. This is the current gain of the transistor when operated inverted and is much lower than β_F because the device geometry and doping densities are designed to maximize β_F. The inverse-active region of device operation occurs for V_{CE} negative in an *npn* transistor and is shown in Fig. 1.12. In order to display these characteristics adequately in the same figure as the forward-active region, the negative voltage and current scales have been expanded. The inverse-active mode of operation is rarely encountered in analog circuits.

Equations 1.62 and 1.63 describe *npn* transistor operation in the saturation region when V_{BE} and V_{BC} are both positive, and also in the forward-active and inverse-active regions. These equations are the *Ebers-Moll* equations. In the forward-active region, they degenerate into a form similar to that of (1.35), (1.47), and (1.49) derived earlier. This can be shown by putting V_{BE} positive and V_{BC} negative in (1.62) and (1.63) to obtain

$$I_C = \alpha_F I_{ES} \left(\exp \frac{V_{BE}}{V_T} - 1 \right) + I_{CS} \tag{1.65}$$

$$I_E = -I_{ES} \left(\exp \frac{V_{BE}}{V_T} - 1 \right) - \alpha_R I_{CS} \tag{1.66}$$

Equation 1.65 is similar in form to (1.35) except that leakage currents that were previously neglected have now been included. This minor difference is significant only at high temperatures or very low operating currents. Comparison of (1.65) with (1.35) allows us to identify $I_S = \alpha_F I_{ES}$, and it can be shown[10] in general that

$$\alpha_F I_{ES} = \alpha_R I_{CS} = I_S \tag{1.67}$$

where this expression represents a reciprocity condition. Use of (1.67) in (1.62) and (1.63) allows the Ebers-Moll equations to be expressed in the general form

$$I_C = I_S \left(\exp \frac{V_{BE}}{V_T} - 1 \right) - \frac{I_S}{\alpha_R} \left(\exp \frac{V_{BC}}{V_T} - 1 \right) \tag{1.62a}$$

$$I_E = -\frac{I_S}{\alpha_F} \left(\exp \frac{V_{BE}}{V_T} - 1 \right) + I_S \left(\exp \frac{V_{BC}}{V_T} - 1 \right) \tag{1.63a}$$

This form is often used for computer representation of transistor large-signal behavior.

The effect of leakage currents mentioned above can be further illustrated as follows. In the forward-active region, from (1.66)

$$I_{ES} \left(\exp \frac{V_{BE}}{V_T} - 1 \right) = -I_E - \alpha_R I_{CS} \tag{1.68}$$

Substitution of (1.68) in (1.65) gives

$$I_C = -\alpha_F I_E + I_{CO} \tag{1.69}$$

where

$$I_{CO} = I_{CS}(1 - \alpha_R \alpha_F) \tag{1.69a}$$

and I_{CO} is the collector-base leakage current with the emitter open. Although I_{CO} is given theoretically by (1.69a), in practice, surface leakage effects dominate when the collector-base junction is reverse biased and I_{CO} is typically several orders of magnitude larger than the value given by (1.69a). However, (1.69) is still valid if the appropriate measured value for I_{CO} is used. Typical values of I_{CO} are from 10^{-10} to 10^{-12} A at 25°C, and the magnitude doubles about every 8°C. As a consequence, these leakage terms can become very significant at high temperatures. For example, consider the base current I_B. From Fig. 1.5 this is

$$I_B = -(I_C + I_E) \cdot \quad (1.70)$$

If I_E is calculated from (1.69) and substituted in (1.70), the result is

$$I_B = \frac{1-\alpha_F}{\alpha_F} I_C - \frac{I_{CO}}{\alpha_F} \quad (1.71)$$

But from (1.50)

$$\beta_F = \frac{\alpha_F}{1-\alpha_F} \quad (1.72)$$

and use of (1.72) in (1.71) gives

$$I_B = \frac{I_C}{\beta_F} - \frac{I_{CO}}{\alpha_F} \quad (1.73)$$

Since the two terms in (1.73) have opposite signs, the effect of I_{CO} is to *decrease* the magnitude of the external base current at a given value of collector current.

■ **EXAMPLE**

If I_{CO} is 10^{-10} A at 24°C, estimate its value at 120°C.

Assuming that I_{CO} doubles every 8°C, we have

$$I_{CO}(120°C) = 10^{-10} \times 2^{12}$$
$$= 0.4 \ \mu A$$

■

1.3.4 Transistor Breakdown Voltages

In Section 1.2.2 the mechanism of avalanche breakdown in a *pn* junction was described. Similar effects occur at the base-emitter and base-collector junctions of a transistor and these effects limit the maximum voltages that can be applied to the device.

First consider a transistor in the common-base configuration shown in Fig. 1.14a and supplied with a constant emitter current. Typical $I_C - V_{CB}$ characteristics for an *npn* transistor in such a connection are shown in Fig. 1.14b. For $I_E = 0$ the collector-base junction breaks down at a voltage BV_{CBO}, which represents collector-base breakdown with the emitter open. For finite values of I_E, the effects of avalanche multiplication are apparent for values of V_{CB} below BV_{CBO}. In the example shown, the effective common-base current gain $\alpha_F = I_C/I_E$ becomes larger than unity for values of V_{CB} above about 60 V. Operation in this region (but below BV_{CBO}) can, however, be safely undertaken if the device power dissipation is not excessive. The considerations of Section 1.2.2 apply to this situation, and neglecting leakage currents, we can calculate the collector current in Fig. 1.14a as

$$I_C = -\alpha_F I_E M \quad (1.74)$$

where M is defined by (1.26) and thus

$$I_C = -\alpha_F I_E \frac{1}{1 - \left(\dfrac{V_{CB}}{BV_{CBO}}\right)^n} \quad (1.75)$$

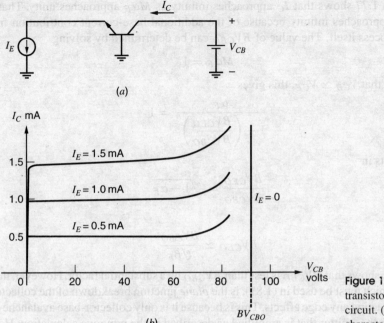

Figure 1.14 Common-base transistor connection. (a) Test circuit. (b) $I_C - V_{CB}$ characteristics.

One further point to note about the common-base characteristics of Fig. 1.14b is that for low values of V_{CB} where avalanche effects are negligible, the curves show very little of the Early effect seen in the common-emitter characteristics. Base widening still occurs in this configuration as V_{CB} is increased, but unlike the common-emitter connection, it produces little change in I_C. This is because I_E is now fixed instead of V_{BE} or I_B, and in Fig. 1.9, this means the slope of the minority-carrier concentration at the emitter edge of the base is fixed. Thus the collector current remains almost unchanged.

Now consider the effect of avalanche breakdown on the common-emitter characteristics of the device. Typical characteristics are shown in Fig. 1.12, and breakdown occurs at a value BV_{CEO}, which is sometimes called the sustaining voltage LV_{CEO}. As in previous cases, operation near the breakdown voltage is destructive to the device only if the current (and thus the power dissipation) becomes excessive.

The effects of avalanche breakdown on the common-emitter characteristics are more complex than in the common-base configuration. This is because hole-electron *pairs* are produced by the avalanche process and the holes are swept into the base, where they effectively contribute to the base current. In a sense, the avalanche current is then *amplified* by the transistor. The base current is still given by

$$I_B = -(I_C + I_E) \tag{1.76}$$

Equation 1.74 still holds, and substitution of this in (1.76) gives

$$I_C = \frac{M\alpha_F}{1 - M\alpha_F} I_B \tag{1.77}$$

where

$$M = \frac{1}{1 - \left(\dfrac{V_{CB}}{BV_{CBO}}\right)^n} \tag{1.78}$$

Equation 1.77 shows that I_C approaches infinity as $M\alpha_F$ approaches unity. That is, the effective β approaches infinity because of the additional base-current contribution from the avalanche process itself. The value of BV_{CEO} can be determined by solving

$$M\alpha_F = 1 \tag{1.79}$$

If we assume that $V_{CB} \simeq V_{CE}$, this gives

$$\frac{\alpha_F}{1 - \left(\dfrac{BV_{CEO}}{BV_{CBO}}\right)^n} = 1 \tag{1.80}$$

and this results in

$$\frac{BV_{CEO}}{BV_{CBO}} = \sqrt[n]{1 - \alpha_F}$$

and thus

$$BV_{CEO} \simeq \frac{BV_{CBO}}{\sqrt[n]{\beta_F}} \tag{1.81}$$

Equation 1.81 shows that BV_{CEO} is less than BV_{CBO} by a substantial factor. However, the value of BV_{CBO}, which must be used in (1.81), is the *plane* junction breakdown of the collector-base junction, neglecting any edge effects. This is because it is only collector-base avalanche current actually under the emitter that is amplified as described in the previous calculation. However, as explained in Section 1.2.2, the measured value of BV_{CBO} is usually determined by avalanche in the curved region of the collector, which is remote from the active base. Consequently, for typical values of $\beta_F = 100$ and $n = 4$, the value of BV_{CEO} is about one-half of the measured BV_{CBO} and not 30 percent as (1.81) would indicate.

Equation 1.81 explains the shape of the breakdown characteristics of Fig. 1.12 if the dependence of β_F on collector current is included. As V_{CE} is increased from zero with $I_B = 0$, the initial collector current is approximately $\beta_F I_{CO}$ from (1.73); since I_{CO} is typically several picoamperes, the collector current is very small. As explained in the next section, β_F is small at low currents, and thus from (1.81) the breakdown voltage is high. However, as avalanche breakdown begins in the device, the value of I_C increases and thus β_F increases. From (1.81) this causes a *decrease* in the breakdown voltage and the characteristic bends back as shown in Fig. 1.12 and exhibits a negative slope. At higher collector currents, β_F approaches a constant value and the breakdown curve with $I_B = 0$ becomes perpendicular to the V_{CE} axis. The value of V_{CE} in this region of the curve is usually defined to be BV_{CEO}, since this is the maximum voltage the device can sustain. The value of β_F to be used to calculate BV_{CEO} in (1.81) is thus the *peak* value of β_F. Note from (1.81) that high-β transistors will thus have low values of BV_{CEO}.

The base-emitter junction of a transistor is also subject to avalanche breakdown. However, the doping density in the emitter is made very large to ensure a high value of β_F [N_D is made large in (1.45) to reduce I_{B2}]. Thus the base is the more lightly doped side of the junction and determines the breakdown characteristic. This can be contrasted with the collector-base junction, where the collector is the more lightly doped side and results in typical values of BV_{CBO} of 20 to 80 V or more. The base is typically an order of magnitude more heavily doped than the collector, and thus the base-emitter breakdown voltage is much less than BV_{CBO} and is typically about 6 to 8 V. This is designated BV_{EBO}. The breakdown voltage for inverse-active operation shown in Fig. 1.12 is approximately equal to this value because the base-emitter junction is reverse biased in this mode of operation.

The base-emitter breakdown voltage of 6 to 8 V provides a convenient reference voltage in integrated-circuit design, and this is often utilized in the form of a *Zener* diode. However, care must be taken to ensure that all other transistors in a circuit are protected against reverse base-emitter voltages sufficient to cause breakdown. This is because, unlike collector-base breakdown, base-emitter breakdown *is* damaging to the device. It can cause a large degradation in β_F, depending on the duration of the breakdown-current flow and its magnitude.[11] If the device is used purely as a Zener diode, this is of no consequence, but if the device is an amplifying transistor, the β_F degradation may be serious.

■ **EXAMPLE**

If the collector doping density in a transistor is 2×10^{15} atoms/cm^3 and is much less than the base doping, calculate BV_{CEO} for $\beta = 100$ and $n = 4$. Assume $\mathscr{E}_{crit} = 3 \times 10^5$ V/cm.

The plane breakdown voltage in the collector can be calculated from (1.24) using $\mathscr{E}_{max} = \mathscr{E}_{crit}$:

$$BV_{CBO} = \frac{\epsilon (N_A + N_D)}{2qN_A N_D} \mathscr{E}_{crit}^2$$

Since $N_D \ll N_A$, we have

$$BV_{CBO}|_{plane} = \frac{\epsilon}{2qN_D} \mathscr{E}_{crit}^2 = \frac{1.04 \times 10^{-12}}{2 \times 1.6 \times 10^{-19} \times 2 \times 10^{15}} \times 9 \times 10^{10} \text{ V} = 146 \text{ V}$$

From (1.81)

$$BV_{CEO} = \frac{146}{\sqrt[4]{100}} \text{V} = 46 \text{ V}$$

■

1.3.5 Dependence of Transistor Current Gain β_F on Operating Conditions

Although most first-order analyses of integrated circuits make the assumption that β_F is constant, this parameter does in fact depend on the operating conditions of the transistor. It was shown in Section 1.3.2, for example, that increasing the value of V_{CE} increases I_C while producing little change in I_B, and thus the effective β_F of the transistor increases. In Section 1.3.4 it was shown that as V_{CE} approaches the breakdown voltage, BV_{CEO}, the collector current increases due to avalanche multiplication in the collector. Equation 1.77 shows that the effective current gain approaches infinity as V_{CE} approaches BV_{CEO}.

In addition to the effects just described, β_F also varies with both temperature and transistor collector current. This is illustrated in Fig. 1.15, which shows typical curves of β_F versus I_C at three different temperatures for an *npn* integrated circuit transistor. It is evident that β_F increases as temperature increases, and a typical temperature coefficient for β_F is +7000 ppm/°C (where ppm signifies *parts per million*). This temperature dependence of β_F is due to the effect of the extremely high doping density in the emitter,[12] which causes the emitter injection efficiency γ to increase with temperature.

The variation of β_F with collector current, which is apparent in Fig. 1.15, can be divided into three regions. Region I is the low-current region, where β_F decreases as I_C decreases. Region II is the midcurrent region, where β_F is approximately constant. Region III is the high-current region, where β_F decreases as I_C increases. The reasons for this behavior of β_F with I_C can be better appreciated by plotting base current I_B and collector current I_C on a log scale

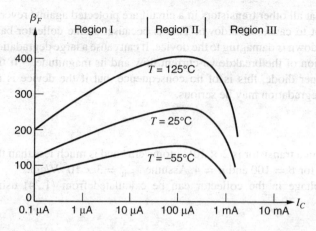

Figure 1.15 Typical curves of β_F versus I_C for an *npn* integrated-circuit transistor with 6 μm² emitter area.

as a function of V_{BE}. This is shown in Fig. 1.16, and because of the log scale on the vertical axis, the value of $\ln \beta_F$ can be obtained directly as the distance between the two curves.

At moderate current levels represented by region II in Figs. 1.15 and 1.16, both I_C and I_B follow the ideal behavior, and

$$I_C = I_S \exp \frac{V_{BE}}{V_T} \tag{1.82}$$

$$I_B \simeq \frac{I_S}{\beta_{FM}} \exp \frac{V_{BE}}{V_T} \tag{1.83}$$

where β_{FM} is the maximum value of β_F and is given by (1.48).

At low current levels, I_C still follows the ideal relationship of (1.82), and the decrease in β_F is due to an additional component in I_B, which is mainly due to recombination of carriers in the base-emitter depletion region and is present at any current level. However, at higher current levels the base current given by (1.83) dominates, and this additional component has

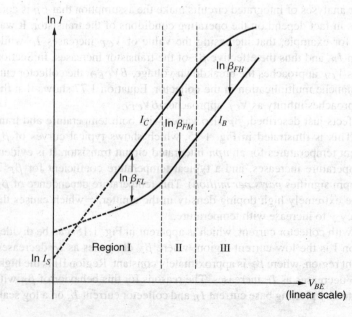

Figure 1.16 Base and collector currents of a bipolar transistor plotted on a log scale versus V_{BE} on a linear scale. The distance between the curves is a direct measure of $\ln \beta_F$.

little effect. The base current resulting from recombination in the depletion region is[5]

$$I_{BX} \simeq I_{SX} \exp \frac{V_{BE}}{mV_T} \tag{1.84}$$

where

$$m \simeq 2$$

At very low collector currents, where (1.84) dominates the base current, the current gain can be calculated from (1.82) and (1.84) as

$$\beta_{FL} \simeq \frac{I_C}{I_{BX}} = \frac{I_S}{I_{SX}} \exp \frac{V_{BE}}{V_T} \left(1 - \frac{1}{m}\right) \tag{1.85}$$

Substitution of (1.82) in (1.85) gives

$$\beta_{FL} \simeq \frac{I_S}{I_{SX}} \left(\frac{I_C}{I_S}\right)^{[1-(1/m)]} \tag{1.86}$$

If $m \simeq 2$, then (1.86) indicates that β_F is proportional to $\sqrt{I_C}$ at very low collector currents.

At high current levels, the base current I_B tends to follow the relationship of (1.83), and the decrease in β_F in region III is due mainly to a decrease in I_C below the value given by (1.82). (In practice the measured curve of I_B versus V_{BE} in Fig. 1.16 may also deviate from a straight line at high currents due to the influence of voltage drop across the base resistance.) The decrease in I_C is due partly to the effect of high-level injection, and at high current levels the collector current approaches[7]

$$I_C \simeq I_{SH} \exp \frac{V_{BE}}{2V_T} \tag{1.87}$$

The current gain in this region can be calculated from (1.87) and (1.83) as

$$\beta_{FH} \simeq \frac{I_{SH}}{I_S} \beta_{FM} \exp\left(-\frac{V_{BE}}{2V_T}\right) \tag{1.88}$$

Substitution of (1.87) in (1.88) gives

$$\beta_{FH} \simeq \frac{I_{SH}^2}{I_S} \beta_{FM} \frac{1}{I_C}$$

Thus β_F decreases rapidly at high collector currents.

In addition to the effect of high-level injection, the value of β_F at high currents is also decreased by the onset of the Kirk effect,[13] which occurs when the minority-carrier concentration in the collector becomes comparable to the donor-atom doping density. The base region of the transistor then stretches out into the collector and becomes greatly enlarged.

1.4 Small-Signal Models of Bipolar Transistors

Analog circuits often operate with signal levels that are small compared to the bias currents and voltages in the circuit. In these circumstances, *incremental* or *small-signal* models can be derived that allow calculation of circuit gain and terminal impedances without the necessity of including the bias quantities. A hierarchy of models with increasing complexity can be derived, and the more complex ones are generally reserved for computer analysis. Part of the designer's skill is knowing which elements of the model can be omitted when performing hand calculations on a particular circuit, and this point is taken up again later.

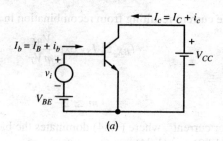

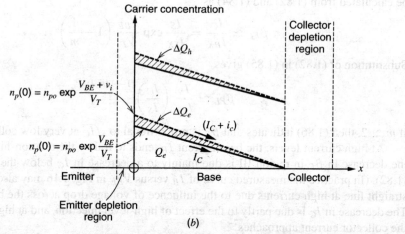

Figure 1.17 Effect of a small-signal input voltage applied to a bipolar transistor. (*a*) Circuit schematic. (*b*) Corresponding changes in carrier concentrations in the base when the device is in the forward-active region.

Consider the bipolar transistor in Fig. 1.17*a* with bias voltages V_{BE} and V_{CC} applied as shown. These produce a quiescent collector current, I_C, and a quiescent base current, I_B, and the device is in the *forward-active region*. A *small-signal* input voltage v_i is applied in series with V_{BE} and produces a small variation in base current i_b and a small variation in collector current i_c. Total values of base and collector currents are I_b and I_c, respectively, and thus $I_b = (I_B + i_b)$ and $I_c = (I_C + i_c)$. The carrier concentrations in the base of the transistor corresponding to the situation in Fig. 1.17*a* are shown in Fig. 1.17*b*. With only bias voltages applied, the carrier concentrations are given by the solid lines. Application of the small-signal voltage v_i causes $n_p(0)$ at the emitter edge of the base to increase, and produces the concentrations shown by the dotted lines. These pictures can now be used to derive the various elements in the small-signal equivalent circuit of the bipolar transistor.

1.4.1 Transconductance

The transconductance is defined as

$$g_m = \frac{dI_C}{dV_{BE}} \quad (1.89)$$

Since

$$\Delta I_C = \frac{dI_C}{dV_{BE}} \Delta V_{BE}$$

we can write

$$\Delta I_C = g_m \Delta V_{BE}$$

and thus

$$i_c = g_m v_i \tag{1.90}$$

The value of g_m can be found by substituting (1.35) in (1.89) to give

$$g_m = \frac{d}{dV_{BE}} I_S \exp \frac{V_{BE}}{V_T} = \frac{I_S}{V_T} \exp \frac{V_{BE}}{V_T} = \frac{I_C}{V_T} = \frac{qI_C}{kT} \tag{1.91}$$

The transconductance thus depends linearly on the bias current I_C and is 38 mA/V for $I_C = 1$ mA at 25°C for any bipolar transistor of either polarity (*npn* or *pnp*), of any size, and made of any material (Si, Ge, GaAs).

To illustrate the limitations on the use of small-signal analysis, the foregoing relation will be derived in an alternative way. The total collector current in Fig. 1.17a can be calculated using (1.35) as

$$I_c = I_S \exp \frac{V_{BE} + v_i}{V_T} = I_S \exp \frac{V_{BE}}{V_T} \exp \frac{v_i}{V_T} \tag{1.92}$$

But the collector bias current is

$$I_C = I_S \exp \frac{V_{BE}}{V_T} \tag{1.93}$$

and use of (1.93) in (1.92) gives

$$I_c = I_C \exp \frac{v_i}{V_T} \tag{1.94}$$

If $v_i < V_T$, the exponential in (1.94) can be expanded in a power series,

$$I_c = I_C \left[1 + \frac{v_i}{V_T} + \frac{1}{2} \left(\frac{v_i}{V_T} \right)^2 + \frac{1}{6} \left(\frac{v_i}{V_T} \right)^3 + \cdots \right] \tag{1.95}$$

Now the incremental collector current is

$$i_c = I_c - I_C \tag{1.96}$$

and substitution of (1.96) in (1.95) gives

$$i_c = \frac{I_C}{V_T} v_i + \frac{1}{2} \frac{I_C}{V_T^2} v_i^2 + \frac{1}{6} \frac{I_C}{V_T^3} v_i^3 + \cdots \tag{1.97}$$

If $v_i \ll V_T$, (1.97) reduces to (1.90), and the small-signal analysis is valid. The criterion for use of small-signal analysis is thus $v_i = \Delta V_{BE} \ll 26$ mV at 25°C. In practice, if ΔV_{BE} is less than 10 mV, the small-signal analysis is accurate within about 10 percent.

1.4.2 Base-Charging Capacitance

Figure 1.17b shows that the change in base-emitter voltage $\Delta V_{BE} = v_i$ has caused a change $\Delta Q_e = q_e$ in the minority-carrier charge in the base. By charge-neutrality requirements, there is an equal change $\Delta Q_h = q_h$ in the majority-carrier charge in the base. Since majority carriers are supplied by the base lead, the application of voltage v_i requires the supply of charge q_h to the base, and the device has an apparent input capacitance

$$C_b = \frac{q_h}{v_i} \tag{1.98}$$

The value of C_b can be related to fundamental device parameters as follows. If (1.39) is divided by (1.33), we obtain

$$\frac{Q_e}{I_C} = \frac{W_B^2}{2D_n} = \tau_F \tag{1.99}$$

The quantity τ_F has the dimension of time and is called the base transit time in the forward direction. Since it is the ratio of the charge in transit (Q_e) to the current flow (I_C), it can be identified as the average time per carrier spent in crossing the base. To a first order it is independent of operating conditions and has typical values 10 to 500 ps for integrated *npn* transistors and 1 to 40 ns for lateral *pnp* transistors. Practical values of τ_F tend to be somewhat lower than predicted by (1.99) for diffused transistors that have nonuniform base doping.[14] However, the functional dependence on base width W_B and diffusion constant D_n is as predicted by (1.99).

From (1.99)

$$\Delta Q_e = \tau_F \Delta I_C \tag{1.100}$$

But since $\Delta Q_e = \Delta Q_h$, we have

$$\Delta Q_h = \tau_F \Delta I_C \tag{1.101}$$

and this can be written

$$q_h = \tau_F i_c \tag{1.102}$$

Use of (1.102) in (1.98) gives

$$C_b = \tau_F \frac{i_c}{v_i} \tag{1.103}$$

and substitution of (1.90) in (1.103) gives

$$C_b = \tau_F g_m \tag{1.104}$$

$$= \tau_F \frac{qI_C}{kT} \tag{1.105}$$

Thus the small-signal, base-charging capacitance is proportional to the collector bias current.

In the inverse-active mode of operation, an equation similar to (1.99) relates stored charge and current via a time constant τ_R. This is typically orders of magnitude larger than τ_F because the device structure and doping are optimized for operation in the forward-active region. Since the saturation region is a combination of forward-active and inverse-active operation, inclusion of the parameter τ_R in a SPICE listing will model the large charge storage that occurs in saturation.

1.4.3 Input Resistance

In the forward-active region, the base current is related to the collector current by

$$I_B = \frac{I_C}{\beta_F} \tag{1.47}$$

Small changes in I_B and I_C can be related using (1.47):

$$\Delta I_B = \frac{d}{dI_C}\left(\frac{I_C}{\beta_F}\right) \Delta I_C \qquad (1.106)$$

and thus

$$\beta_0 = \frac{\Delta I_C}{\Delta I_B} = \frac{i_c}{i_b} = \left[\frac{d}{dI_C}\left(\frac{I_C}{\beta_F}\right)\right]^{-1} \qquad (1.107)$$

where β_0 is the *small-signal* current gain of the transistor. Note that if β_F is constant, then $\beta_F = \beta_0$. Typical values of β_0 are close to those of β_F, and in subsequent chapters little differentiation is made between these quantities. A single value of β is often assumed for a transistor and then used for both ac and dc calculations.

Equation 1.107 relates the change in base current i_b to the corresponding change in collector current i_c, and the device has a small-signal input resistance given by

$$r_\pi = \frac{v_i}{i_b} \qquad (1.108)$$

Substitution of (1.107) in (1.108) gives

$$r_\pi = \frac{v_i}{i_c}\beta_0 \qquad (1.109)$$

and use of (1.90) in (1.109) gives

$$r_\pi = \frac{\beta_0}{g_m} \qquad (1.110)$$

Thus the small-signal input shunt resistance of a bipolar transistor depends on the current gain and is inversely proportional to I_C.

1.4.4 Output Resistance

In Section 1.3.2 the effect of changes in collector-emitter voltage V_{CE} on the large-signal characteristics of the transistor was described. It follows from that treatment that small changes ΔV_{CE} in V_{CE} produce corresponding changes ΔI_C in I_C, where

$$\Delta I_C = \frac{\partial I_C}{\partial V_{CE}} \Delta V_{CE} \qquad (1.111)$$

Substitution of (1.55) and (1.57) in (1.111) gives

$$\frac{\Delta V_{CE}}{\Delta I_C} = \frac{V_A}{I_C} = r_o \qquad (1.112)$$

where V_A is the Early voltage and r_o is the small-signal output resistance of the transistor. Since typical values of V_A are 50 to 100 V, corresponding values of r_o are 50 to 100 kΩ for $I_C = 1$mA. Note that r_o is inversely proportional to I_C, and thus r_o can be related to g_m, as are many of the other small-signal parameters.

$$r_o = \frac{1}{\eta g_m} \qquad (1.113)$$

where

$$\eta = \frac{kT}{qV_A} \qquad (1.114)$$

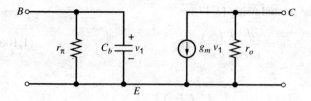

$$r_\pi = \frac{\beta}{g_m}, \quad r_o = \frac{1}{\eta g_m}, \quad g_m = \frac{qI_C}{kT}, \quad C_b = \tau_F g_m$$

Figure 1.18 Basic bipolar transistor small-signal equivalent circuit.

If $V_A = 100$ V, then $\eta = 2.6 \times 10^{-4}$ at 25°C. Note that $1/r_o$ is the slope of the output characteristics of Fig. 1.10.

1.4.5 Basic Small-Signal Model of the Bipolar Transistor

Combination of the above small-signal circuit elements yields the small-signal model of the bipolar transistor shown in Fig. 1.18. This is valid for both *npn* and *pnp* devices in the forward-active region and is called the *hybrid-π* model. Collector, base, and emitter nodes are labeled *C*, *B* and *E*, respectively. The elements in this circuit are present in the equivalent circuit of *any* bipolar transistor and are specified by relatively few parameters (β, τ_F, η, I_C). Note that in the evaluation of the small-signal parameters for *pnp* transistors, the *magnitude only* of I_C is used. In the following sections, further elements are added to this model to account for parasitics and second-order effects.

1.4.6 Collector-Base Resistance

Consider the effect of variations in V_{CE} on the minority charge in the base as illustrated in Fig. 1.9. An increase in V_{CE} causes an increase in the collector depletion-layer width and consequent reduction of base width. This causes a reduction in the total minority-carrier charge stored in the base and thus a reduction in base current I_B due to a reduction in I_{B1} given by (1.40). Since an increase ΔV_{CE} in V_{CE} causes a *decrease* ΔI_B in I_B, this effect can be modeled by inclusion of a resistor r_μ from collector to base of the model of Fig. 1.18. If V_{BE} is assumed held constant, the value of this resistor can be determined as follows.

$$r_\mu = \frac{\Delta V_{CE}}{\Delta I_{B1}} = \frac{\Delta V_{CE}}{\Delta I_C} \frac{\Delta I_C}{\Delta I_{B1}} \tag{1.115}$$

Substitution of (1.112) in (1.115) gives

$$r_\mu = r_o \frac{\Delta I_C}{\Delta I_{B1}} \tag{1.116}$$

If the base current I_B is composed entirely of component I_{B1}, then (1.107) can be used in (1.116) to give

$$r_\mu = \beta_0 r_o \tag{1.117}$$

This is a lower limit for r_μ. In practice, I_{B1} is typically less than 10 percent of I_B [component I_{B2} from (1.42) dominates] in integrated *npn* transistors, and since I_{B1} is very small, the change ΔI_{B1} in I_{B1} for a given ΔV_{CE} and ΔI_C is also very small. Thus a typical value for r_μ is greater than $10\beta_0 r_o$. For lateral *pnp* transistors, recombination in the base is more significant, and r_μ is in the range $2\beta_0 r_o$ to $5\beta_0 r_o$.

1.4.7 Parasitic Elements in the Small-Signal Model

The elements of the bipolar transistor small-signal equivalent circuit considered so far may be considered basic in the sense that they arise directly from essential processes in the device. However, technological limitations in the fabrication of transistors give rise to a number of parasitic elements that must be added to the equivalent circuit for most integrated-circuit transistors. A cross section of a typical *npn* transistor in a junction-isolated process is shown in Fig. 1.19. The means of fabricating such devices is described in Chapter 2.

As described in Section 1.2, all *pn* junctions have a voltage-dependent capacitance associated with the depletion region. In the cross section of Fig. 1.19, three depletion-region capacitances can be identified. The base-emitter junction has a depletion-region capacitance C_{je} and the base-collector and collector-substrate junctions have capacitances C_μ and C_{cs}, respectively. The base-emitter junction closely approximates an abrupt junction due to the steep rise of the doping density caused by the heavy doping in the emitter. Thus the variation of C_{je} with bias voltage is well approximated by (1.21). The collector-base junction behaves like a graded junction for small bias voltages since the doping density is a function of distance near the junction. However, for larger reverse-bias values (more than about a volt), the junction depletion region spreads into the collector, which is uniformly doped, and thus for devices with thick collectors the junction tends to behave like an abrupt junction with uniform doping. Many modern high-speed processes, however, have very thin collector regions (of the order of one micron), and the collector depletion region can extend all the way to the buried layer for quite small reverse-bias voltages. When this occurs, both the depletion region and the associated capacitance vary quite slowly with bias voltage. The collector-base capacitance C_μ thus tends to follow (1.22) for very small bias voltages and (1.21) for large bias voltages in thick-collector devices. In practice, measurements show that the variation of C_μ with bias voltage for most devices can be approximated by

$$C_\mu = \frac{C_{\mu 0}}{\left(1 - \dfrac{V}{\psi_o}\right)^n} \tag{1.117a}$$

where V is the forward bias on the junction and n is an exponent between about 0.2 and 0.5. The third parasitic capacitance in a monolithic *npn* transistor is the collector-substrate capacitance

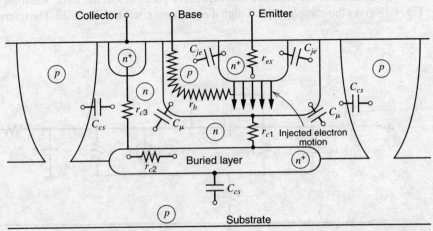

Figure 1.19 Integrated-circuit *npn* bipolar transistor structure showing parasitic elements. (Not to scale.)

C_{cs}, and for large reverse bias voltages this varies according to the abrupt junction equation (1.21) for junction-isolated devices. In the case of oxide-isolated devices, however, the deep p diffusions used to isolate the devices are replaced by oxide. The sidewall component of C_{cs} then consists of a fixed oxide capacitance. Equation 1.117a may then be used to model C_{cs}, but a value of n less than 0.5 gives the best approximation. In general, (1.117a) will be used to model all three parasitic capacitances with subscripts e, c, and s on n and ψ_0 used to differentiate emitter-base, collector-base, and collector-substrate capacitances, respectively. Typical zero-bias values of these parasitic capacitances for a minimum-size npn transistor in a modern oxide-isolated process are $C_{je0} \simeq 10$ fF, $C_{\mu 0} \simeq 10$ fF, and $C_{cs0} \simeq 20$ fF. Values for other devices are summarized in Chapter 2.

As described in Chapter 2, lateral pnp transistors have a parasitic capacitance C_{bs} from base to substrate in place of C_{cs}. Note that the substrate is always connected to the most negative voltage supply in the circuit in order to ensure that all isolation regions are separated by reverse-biased junctions. Thus the substrate is an ac ground, and all parasitic capacitance to the substrate is connected to ground in an equivalent circuit.

The final elements to be added to the small-signal model of the transistor are resistive parasitics. These are produced by the finite resistance of the silicon between the top contacts on the transistor and the active base region beneath the emitter. As shown in Fig. 1.19, there are significant resistances r_b and r_c in series with the base and collector contacts, respectively. There is also a resistance r_{ex} of several ohms in series with the emitter lead that can become important at high bias currents. (Note that the collector resistance r_c is actually composed of three parts labeled r_{c1}, r_{c2}, and r_{c3}.) Typical values of these parameters are $r_b = 50$ to $500 \,\Omega$, $r_{ex} = 1$ to $3 \,\Omega$, and $r_c = 20$ to $500 \,\Omega$. The value of r_b varies significantly with collector current because of *current crowding*.[15] This occurs at high collector currents where the dc base current produces a lateral voltage drop in the base that tends to forward bias the base-emitter junction preferentially around the edges of the emitter. Thus the transistor action tends to occur along the emitter periphery rather than under the emitter itself, and the distance from the base contact to the active base region is reduced. Consequently, the value of r_b is reduced, and in a typical npn transistor, r_b may decrease 50 percent as I_C increases from 0.1 mA to 10 mA.

The value of these parasitic resistances can be reduced by changes in the device structure. For example, a large-area transistor with multiple base and emitter stripes will have a smaller value of r_b. The value of r_c is reduced by inclusion of the low-resistance buried n^+ layer beneath the collector.

The addition of the resistive and capacitive parasitics to the basic small-signal circuit of Fig. 1.18 gives the complete small-signal equivalent circuit of Fig. 1.20. The internal base node

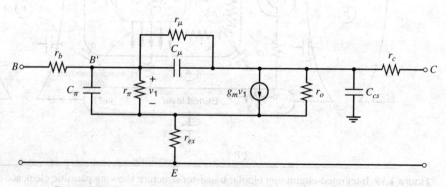

Figure 1.20 Complete bipolar transistor small-signal equivalent circuit.

is labeled B' to distinguish it from the external base contact B. The capacitance C_π contains the base-charging capacitance C_b and the emitter-base depletion layer capacitance C_{je}.

$$C_\pi = C_b + C_{je} \tag{1.118}$$

Note that the representation of parasitics in Fig. 1.20 is an approximation in that lumped elements have been used. In practice, as suggested by Fig. 1.19, C_μ is distributed across r_b and C_{cs} is distributed across r_c. This lumped representation is adequate for most purposes but can introduce errors at very high frequencies. It should also be noted that while the parasitic resistances of Fig. 1.20 can be very important at high bias currents or for high-frequency operation, they are usually omitted from the equivalent circuit for low-frequency calculations, particularly for collector bias currents less than 1 mA.

■ **EXAMPLE**

Derive the complete small-signal equivalent circuit for a bipolar transistor at $I_C = 1$ mA, $V_{CB} = 3$ V, and $V_{CS} = 5$ V. Device parameters are $C_{je0} = 10$ fF, $n_e = 0.5$, $\psi_{0e} = 0.9$ V, $C_{\mu 0} = 10$ fF, $n_c = 0.3$, $\psi_{0c} = 0.5$ V, $C_{cs0} = 20$ fF, $n_s = 0.3$, $\psi_{0s} = 0.65$ V, $\beta_0 = 100$, $\tau_F = 10$ ps, $V_A = 20$ V, $r_b = 300\ \Omega$, $r_c = 50\ \Omega$, $r_{ex} = 5\ \Omega$, $r_\mu = 10\ \beta_0 r_o$.

Since the base-emitter junction is forward biased, the value of C_{je} is difficult to determine for reasons described in Section 1.2.1. Either a value can be determined by computer or a reasonable estimation is to double C_{je0}. Using the latter approach, we estimate

$$C_{je} = 20 \text{ fF}$$

Using (1.117a) gives, for the collector-base capacitance,

$$C_\mu = \frac{C_{\mu 0}}{\left(1 + \dfrac{V_{CB}}{\psi_{0c}}\right)^{n_c}} = \frac{10}{\left(1 + \dfrac{3}{0.5}\right)^{0.3}} = 5.6 \text{ fF}$$

The collector-substrate capacitance can also be calculated using (1.117a)

$$C_{cs} = \frac{C_{cs0}}{\left(1 + \dfrac{V_{CS}}{\psi_{0s}}\right)^{n_s}} = \frac{20}{\left(1 + \dfrac{5}{0.65}\right)^{0.3}} = 10.5 \text{ fF}$$

From (1.91) the transconductance is

$$g_m = \frac{qI_C}{kT} = \frac{10^{-3}}{26 \times 10^{-3}} \text{A/V} = 38 \text{ mA/V}$$

From (1.104) the base-charging capacitance is

$$C_b = \tau_F g_m = 10 \times 10^{-12} \times 38 \times 10^{-3} \text{ F} = 0.38 \text{ pF}$$

The value of C_π from (1.118) is

$$C_\pi = 0.38 + 0.02 \text{ pF} = 0.4 \text{ pF}$$

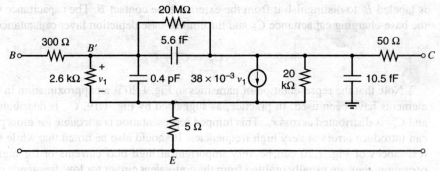

Figure 1.21 Complete small-signal equivalent circuit for a bipolar transistor at $I_C = 1$ mA, $V_{CB} = 3$ V, and $V_{CS} = 5$ V. Device parameters are $C_{je0} = 10$ fF, $n_e = 0.5$, $\psi_{0e} = 0.9$ V, $C_{\mu 0} = 10$ fF, $n_c = 0.3$, $\psi_{0c} = 0.5$ V, $C_{cs0} = 20$ fF, $n_s = 0.3$, $\psi_{0s} = 0.65$ V, $\beta_0 = 100$, $\tau_F = 10$ ps, $V_A = 20$ V, $r_b = 300\Omega$, $r_c = 50\ \Omega$, $r_{ex} = 5\ \Omega$, $r_\mu = 10\beta_0 r_0$.

The input resistance from (1.110) is

$$r_\pi = \frac{\beta_0}{g_m} = 100 \times 26\ \Omega = 2.6\ \text{k}\Omega$$

The output resistance from (1.112) is

$$r_o = \frac{20}{10^{-3}}\ \Omega = 20\ \text{k}\Omega$$

and thus the collector-base resistance is

$$r_\mu = 10\beta_0 r_o = 10 \times 100 \times 20\ \text{k}\Omega = 20\ \text{M}\Omega$$

■ The equivalent circuit with these parameter values is shown in Fig. 1.21.

1.4.8 Specification of Transistor Frequency Response

The high-frequency gain of the transistor is controlled by the capacitive elements in the equivalent circuit of Fig. 1.20. The frequency capability of the transistor is most often specified in practice by determining the frequency where the magnitude of the short-circuit, common-emitter current gain falls to unity. This is called the *transition frequency*, f_T, and is a measure of the maximum useful frequency of the transistor when it is used as an amplifier. The value of f_T can be measured as well as calculated, using the ac circuit of Fig. 1.22. A small-signal current i_i is applied to the base, and the output current i_o is measured with the collector short-circuited for ac signals. A small-signal equivalent circuit can be formed for this situation by using the equivalent circuit of Fig. 1.20 as shown in Fig. 1.23, where r_{ex} and r_μ have been neglected. If r_c is assumed small, then r_o and C_{cs} have no influence, and we have

$$v_1 \simeq \frac{r_\pi}{1 + r_\pi(C_\pi + C_\mu)s} i_i \tag{1.119}$$

If the current fed forward through C_μ is neglected,

$$i_o \simeq g_m v_1 \tag{1.120}$$

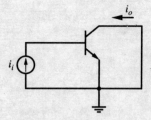

Figure 1.22 Schematic of ac circuit for measurement of f_T.

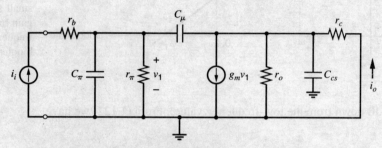

Figure 1.23 Small-signal equivalent circuit for the calculation of f_T.

Substitution of (1.119) in (1.120) gives

$$i_o \simeq i_i \frac{g_m r_\pi}{1 + r_\pi (C_\pi + C_\mu)s}$$

and thus

$$\frac{i_o}{i_i}(j\omega) = \frac{\beta_0}{1 + \beta_0 \dfrac{C_\pi + C_\mu}{g_m} j\omega} \qquad (1.121)$$

using (1.110).

Now if $i_o/i_i(j\omega)$ is written as $\beta(j\omega)$ (the high-frequency, small-signal current gain), then

$$\beta(j\omega) = \frac{\beta_0}{1 + \beta_0 \dfrac{C_\pi + C_\mu}{g_m} j\omega} \qquad (1.122)$$

At high frequencies the imaginary part of the denominator of (1.122) is dominant, and we can write

$$\beta(j\omega) \simeq \frac{g_m}{j\omega (C_\pi + C_\mu)} \qquad (1.123)$$

From (1.123), $|\beta(j\omega)| = 1$ when

$$\omega = \omega_T = \frac{g_m}{C_\pi + C_\mu} \qquad (1.124)$$

and thus

$$f_T = \frac{1}{2\pi} \frac{g_m}{C_\pi + C_\mu} \qquad (1.125)$$

The transistor behavior can be illustrated by plotting $|\beta(j\omega)|$ using (1.122) as shown in Fig. 1.24. The frequency ω_β is defined as the frequency where $|\beta(j\omega)|$ is equal to $\beta_0/\sqrt{2}$

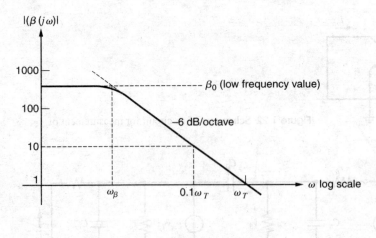

Figure 1.24 Magnitude of small-signal ac current gain $|\beta(j\omega)|$ versus frequency for a typical bipolar transistor.

(3 dB down from the low-frequency value). From (1.122) we have

$$\omega_\beta = \frac{1}{\beta_0}\frac{g_m}{C_\pi + C_\mu} = \frac{\omega_T}{\beta_0} \qquad (1.126)$$

From Fig. 1.24 it can be seen that ω_T can be determined by measuring $|\beta(j\omega)|$ at some frequency ω_x where $|\beta(j\omega)|$ is falling at 6 dB/octave and using

$$\omega_T = \omega_x|\beta(j\omega_x)| \qquad (1.127)$$

This is the method used in practice, since deviations from ideal behavior tend to occur as $|\beta(j\omega)|$ approaches unity. Thus $|\beta(j\omega)|$ is typically measured at some frequency where its magnitude is about 5 or 10, and (1.127) is used to determine ω_T.

It is interesting to examine the time constant, τ_T, associated with ω_T. This is defined as

$$\tau_T = \frac{1}{\omega_T} \qquad (1.128)$$

and use of (1.124) in (1.128) gives

$$\tau_T = \frac{C_\pi}{g_m} + \frac{C_\mu}{g_m} \qquad (1.129)$$

Substitution of (1.118) and (1.104) in (1.129) gives

$$\tau_T = \frac{C_b}{g_m} + \frac{C_{je}}{g_m} + \frac{C_\mu}{g_m} = \tau_F + \frac{C_{je}}{g_m} + \frac{C_\mu}{g_m} \qquad (1.130)$$

Equation 1.130 indicates that τ_T is dependent on I_C (through g_m) and approaches a constant value of τ_F at high collector bias currents. At low values of I_C, the terms involving C_{je} and C_μ dominate, and they cause τ_T to *rise* and f_T to *fall* as I_C is decreased. This behavior is illustrated in Fig. 1.25, which is a typical plot of f_T versus I_C for an integrated-circuit *npn* transistor. The decline in f_T at high collector currents is not predicted by this simple theory and is due to an increase in τ_F caused by high-level injection and Kirk effect at high currents. These are the same mechanisms that cause a decrease in β_F at high currents as described in Section 1.3.5.

1.4 Small-Signal Models of Bipolar Transistors

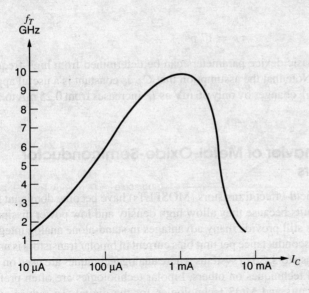

Figure 1.25 Typical curve of f_T versus I_C for an *npn* integrated-circuit transistor with 6 μm² emitter area in a high-speed process.

■ **EXAMPLE**

A bipolar transistor has a short-circuit, common-emitter current gain at 1 GHz of 8 with $I_C = 0.25$ mA and 9 with $I_C = 1$ mA. Assuming that high-level injection effects are negligible, calculate C_{je} and τ_F, assuming both are constant. The measured value of C_μ is 10 fF.

From the data, values of f_T are

$$f_{T1} = 8 \times 1 = 8 \text{ GHz} \quad \text{at} \quad I_C = 0.25 \text{ mA}$$
$$f_{T2} = 9 \times 1 = 9 \text{ GHz} \quad \text{at} \quad I_C = 1 \text{ mA}$$

Corresponding values of τ_T are

$$\tau_{T1} = \frac{1}{2\pi f_{T1}} = 19.9 \text{ ps}$$

$$\tau_{T2} = \frac{1}{2\pi f_{T2}} = 17.7 \text{ ps}$$

Using these data in (1.130), we have

$$19.9 \times 10^{-12} = \tau_F + 104(C_\mu + C_{je}) \tag{1.131}$$

at $I_C = 0.25$ mA. At $I_C = 1$ mA we have

$$17.7 \times 10^{-12} = \tau_F + 26(C_\mu + C_{je}) \tag{1.132}$$

Subtraction of (1.132) from (1.131) yields

$$C_\mu + C_{je} = 28.2 \text{ fF}$$

Since C_μ was measured as 10 fF, the value of C_{je} is given by

$$C_{je} \simeq 18.2 \text{ fF}$$

Substitution in (1.131) gives

$$\tau_F = 17 \text{ ps}$$

This is an example of how basic device parameters can be determined from high-frequency current-gain measurements. Note that the assumption that C_{je} is constant is a useful approximation in practice because V_{BE} changes by only 36 mV as I_C increases from 0.25 mA to 1 mA.

1.5 Large-Signal Behavior of Metal-Oxide-Semiconductor Field-Effect Transistors

Metal-oxide-semiconductor field-effect transistors (MOSFETs) have become dominant in the area of digital integrated circuits because they allow high density and low power dissipation. In contrast, bipolar transistors still provide many advantages in stand-alone analog integrated circuits. For example, the transconductance per unit bias current in bipolar transistors is usually much higher than in MOS transistors. So in systems where analog techniques are used on some integrated circuits and digital techniques on others, bipolar technologies are often preferred for the analog integrated circuits and MOS technologies for the digital. To reduce system cost and increase portability, both increased levels of integration and reduced power dissipation are required, forcing the associated analog circuits to use MOS-compatible technologies. One way to achieve these goals is to use a processing technology that provides both bipolar and MOS transistors, allowing great design flexibility. However, all-MOS processes are less expensive than combined bipolar and MOS processes. Therefore, economic considerations drive integrated-circuit manufacturers to use all-MOS processes in many practical cases. As a result, the study of the characteristics of MOS transistors that affect analog integrated-circuit design is important.

1.5.1 Transfer Characteristics of MOS Devices

A cross section of a typical enhancement-mode n-channel MOS (NMOS) transistor is shown in Fig. 1.26. Heavily doped n-type source and drain regions are fabricated in a p-type substrate (often called the body). A thin layer of silicon dioxide is grown over the substrate material and

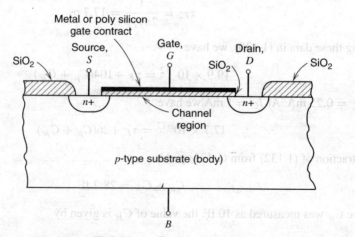

Figure 1.26 Typical enhancement-mode NMOS structure.

a conductive gate material (metal or polycrystalline silicon) covers the oxide between source and drain. Note that the gate is horizontal in Fig. 1.26, and we will use this orientation in all descriptions of the physical operation of MOS devices. In operation, the gate-source voltage modifies the conductance of the region under the gate, allowing the gate voltage to control the current flowing between source and drain. This control can be used to provide gain in analog circuits and switching characteristics in digital circuits.

The enhancement-mode NMOS device of Fig. 1.26 shows significant conduction between source and drain only when an n-type channel exists under the gate. This observation is the origin of the *n-channel* designation. The term *enhancement mode* refers to the fact that no conduction occurs for $V_{GS} = 0$. Thus, the channel must be *enhanced* to cause conduction. MOS devices can also be made by using an n-type substrate with a p-type conducting channel. Such devices are called enhancement-mode p-channel MOS (PMOS) transistors. In complementary MOS (CMOS) technology, both device types are present.

The derivation of the transfer characteristics of the enhancement-mode NMOS device of Fig. 1.26 begins by noting that with $V_{GS} = 0$, the source and drain regions are separated by back-to-back pn junctions. These junctions are formed between the n-type source and drain regions and the p-type substrate, resulting in an extremely high resistance (about $10^{12}\,\Omega$) between drain and source when the device is off.

Now consider the substrate, source, and drain grounded with a positive voltage V_{GS} applied to the gate as shown in Fig. 1.27. The gate and substrate then form the plates of a capacitor with the SiO_2 as a dielectric. Positive charge accumulates on the gate and negative charge in the substrate. Initially, the negative charge in the p-type substrate is manifested by the creation of a *depletion region* and the exclusion of holes under the gate as described in Section 1.2 for a pn-junction. The depletion region is shown in Fig. 1.27. The results of Section 1.2 can now be applied. Using (1.10), the depletion-layer width X under the oxide is

$$X = \left(\frac{2\epsilon\phi}{qN_A}\right)^{1/2} \tag{1.133}$$

where ϕ is the potential in the depletion layer at the oxide-silicon interface, N_A is the doping density (assumed constant) of the p-type substrate in atoms/cm^3, and ϵ is the permittivity of the silicon. The charge per area in this depletion region is

$$Q = qN_A X = \sqrt{2qN_A\epsilon\phi} \tag{1.134}$$

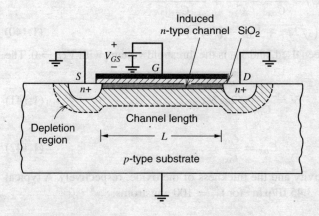

Figure 1.27 Idealized NMOS device cross section with positive V_{GS} applied, showing depletion regions and the induced channel.

When the surface potential in the silicon reaches a critical value equal to twice the Fermi level ϕ_f, a phenomenon known as *inversion* occurs.[16] The Fermi level ϕ_f is defined as

$$\phi_f = \frac{kT}{q} \ln\left[\frac{N_A}{n_i}\right] \tag{1.135}$$

where k is Boltzmann's constant. Also, n_i is the intrinsic carrier concentration, which is

$$n_i = \sqrt{N_c N_v} \exp\left(-\frac{E_g}{2kT}\right) \tag{1.136}$$

where E_g is the band gap of silicon at $T = 0°K$, N_c is the density of allowed states near the edge of the conduction band, and N_v is the density of allowed states near the edge of the valence band, respectively. The Fermi level ϕ_f is usually about 0.3 V. After the potential in the silicon reaches $2\phi_f$, further increases in gate voltage produce no further changes in the depletion-layer width but instead induce a thin layer of electrons in the depletion layer at the surface of the silicon directly under the oxide. Inversion produces a continuous n-type region with the source and drain regions and forms the conducting channel between source and drain. The conductivity of this channel can be modulated by increases or decreases in the gate-source voltage. In the presence of an inversion layer, and without substrate bias, the depletion region contains a fixed charge density

$$Q_{b0} = \sqrt{2qN_A \epsilon 2\phi_f} \tag{1.137}$$

If a substrate bias voltage V_{SB} (positive for n-channel devices) is applied between the source and substrate, the potential required to produce inversion becomes $(2\phi_f + V_{SB})$, and the charge density stored in the depletion region in general is

$$Q_b = \sqrt{2qN_A \epsilon (2\phi_f + V_{SB})} \tag{1.138}$$

The gate-source voltage V_{GS} required to produce an inversion layer is called the threshold voltage V_t and can now be calculated. This voltage consists of several components. First, a voltage $[2\phi_f + (Q_b/C_{ox})]$ is required to sustain the depletion-layer charge Q_b, where C_{ox} is the gate oxide capacitance per unit area. Second, a work-function difference ϕ_{ms} exists between the gate metal and the silicon. Third, positive charge density Q_{ss} always exists in the oxide at the silicon interface. This charge is caused by crystal discontinuities at the Si-SiO$_2$ interface and must be compensated by a gate-source voltage contribution of $-Q_{ss}/C_{ox}$. Thus we have a threshold voltage

$$V_t = \phi_{ms} + 2\phi_f + \frac{Q_b}{C_{ox}} - \frac{Q_{ss}}{C_{ox}} \tag{1.139}$$

$$= \phi_{ms} + 2\phi_f + \frac{Q_{b0}}{C_{ox}} - \frac{Q_{ss}}{C_{ox}} + \frac{Q_b - Q_{b0}}{C_{ox}}$$

$$= V_{t0} + \gamma\left(\sqrt{2\phi_f + V_{SB}} - \sqrt{2\phi_f}\right) \tag{1.140}$$

where (1.137) and (1.138) have been used, and V_{t0} is the threshold voltage with $V_{SB} = 0$. The parameter γ is defined as

$$\gamma = \frac{1}{C_{ox}} \sqrt{2q\epsilon N_A} \tag{1.141}$$

and

$$C_{ox} = \frac{\epsilon_{ox}}{t_{ox}} \tag{1.142}$$

where ϵ_{ox} and t_{ox} are the permittivity and the thickness of the oxide, respectively. A typical value of γ is 0.5 V$^{1/2}$, and $C_{ox} = 3.45$ fF/μm^2 for $t_{ox} = 100$ angstroms.

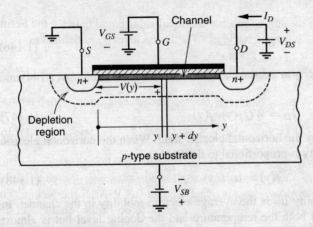

Figure 1.28 NMOS device with bias voltages applied.

In practice, the value of V_{t0} is usually adjusted in processing by implanting additional impurities into the channel region. Extra p-type impurities are implanted in the channel to set V_{t0} between 0.3 V and 1.5 V for n-channel enhancement devices. By implanting n-type impurities in the channel region, a conducting channel can be formed even for $V_{GS} = 0$, forming a *depletion* device with typical values of V_{t0} in the range -1 V to -4 V. If Q_i is the charge density due to the implant, then the threshold voltage given by (1.139) is shifted by approximately Q_i/C_{ox}.

The preceding equations can now be used to calculate the large-signal characteristics of an n-channel MOSFET. In this analysis, the source is assumed grounded and bias voltages V_{GS}, V_{DS}, and V_{SB} are applied as shown in Fig. 1.28. If $V_{GS} > V_t$, inversion occurs and a conducting channel exists. The channel conductivity is determined by the vertical electric field, which is controlled by the value of $(V_{GS} - V_t)$. If $V_{DS} = 0$, the current I_D that flows from drain to source is zero because the horizontal electric field is zero. Nonzero V_{DS} produces a horizontal electric field and causes current I_D to flow. The value of the current depends on both the horizontal and the vertical electric fields, explaining the term *field-effect* transistor. Positive voltage V_{DS} causes the reverse bias from the drain to the substrate to be larger than from the source to substrate, and thus the widest depletion region exists at the drain. For simplicity, however, we assume that the voltage drop along the channel itself is small so that the depletion-layer width is constant along the channel.

The drain current I_D is

$$I_D = \frac{dQ}{dt} \tag{1.143}$$

where dQ is the incremental channel charge at a distance y from the source in an incremental length dy of the channel, and dt is the time required for this charge to cross length dy. The charge dQ is

$$dQ = Q_I W dy \tag{1.144}$$

where W is the width of the device perpendicular to the plane of Fig. 1.28 and Q_I is the induced electron charge per unit area of the channel. At a distance y along the channel, the voltage with respect to the source is $V(y)$ and the gate-to-channel voltage at that point is $V_{GS} - V(y)$. We assume this voltage exceeds the threshold voltage V_t. Thus the induced electron charge per unit area in the channel is

$$Q_I(y) = C_{ox}[V_{GS} - V(y) - V_t] \tag{1.145}$$

Also,

$$dt = \frac{dy}{v_d(y)} \tag{1.146}$$

where v_d is the electron drift velocity at a distance y from the source. Combining (1.144) and (1.146) gives

$$I_D = WQ_I(y)v_d(y) \tag{1.147}$$

The drift velocity is determined by the horizontal electric field. When the horizontal electric field $\mathscr{E}(y)$ is small, the drift velocity is proportional to the field and

$$v_d(y) = \mu_n \mathscr{E}(y) \tag{1.148}$$

where the constant of proportionality μ_n is the average electron mobility in the channel. In practice, the mobility depends on both the temperature and the doping level but is almost constant for a wide range of normally used doping levels. Also, μ_n is sometimes called the surface mobility for electrons because the channel forms at the surface of the silicon. Typical values range from about 500 cm^2/(V-s) to about 700 cm^2/(V-s), which are much less than the mobility of electrons in the bulk of the silicon (about 1400 cm^2/V-s) because surface defects not present in the bulk impede the flow of electrons in MOS transistors.[17] The electric field $\mathscr{E}(y)$ is

$$\mathscr{E}(y) = \frac{dV}{dy} \tag{1.149}$$

where dV is the incremental voltage drop along the length of channel dy at a distance y from the source. Substituting (1.145), (1.148), and (1.149) into (1.147) gives

$$I_D = WC_{ox}[V_{GS} - V - V_t]\mu_n \frac{dV}{dy} \tag{1.150}$$

Separating variables and integrating gives

$$\int_0^L I_D \, dy = \int_0^{V_{DS}} W\mu_n C_{ox}(V_{GS} - V - V_t) \, dV \tag{1.151}$$

Carrying out this integration gives

$$I_D = \frac{k'}{2} \frac{W}{L} [2(V_{GS} - V_t)V_{DS} - V_{DS}^2] \tag{1.152}$$

where

$$k' = \mu_n C_{ox} = \frac{\mu_n \epsilon_{ox}}{t_{ox}} \tag{1.153}$$

When $V_{DS} \ll 2(V_{GS} - V_t)$, (1.152) predicts that I_D is approximately proportional to V_{DS}. This result is reasonable because the average horizontal electric field in this case is V_{DS}/L, and the average drift velocity of electrons is proportional to the average field when the field is small. Equation 1.152 is important and describes the I-V characteristics of an MOS transistor, assuming a continuous induced channel. A typical value of k' for $t_{ox} = 100$ angstroms is about 200 μA/V^2 for an *n*-channel device.

As the value of V_{DS} is increased, the induced conducting channel narrows at the drain end and (1.145) indicates that Q_I at the drain end approaches zero as V_{DS} approaches $(V_{GS} - V_t)$. That is, the channel is no longer connected to the drain when $V_{DS} > V_{GS} - V_t$. This phenomenon is called *pinch-off* and can be understood by writing a KVL equation around the transistor:

$$V_{DS} = V_{DG} + V_{GS} \tag{1.154}$$

Therefore, when $V_{DS} > V_{GS} - V_t$,

$$V_{DG} + V_{GS} > V_{GS} - V_t \qquad (1.155)$$

Rearranging (1.155) gives

$$V_{GD} < V_t \qquad (1.156)$$

Equation 1.156 shows that when drain-source voltage is greater than $(V_{GS} - V_t)$, the gate-drain voltage is less than a threshold, which means that the channel no longer exists at the drain. This result is reasonable because we know that the gate-to-channel voltage at the point where the channel disappears is equal to V_t by the definition of the threshold voltage. Therefore, at the point where the channel pinches off, the channel voltage is $(V_{GS} - V_t)$. As a result, the average horizontal electric field across the channel in pinch-off does not depend on the the drain-source voltage but instead on the voltage across the channel, which is $(V_{GS} - V_t)$. Therefore, (1.152) is no longer valid if $V_{DS} > V_{GS} - V_t$. The value of I_D in this region is obtained by substituting $V_{DS} = V_{GS} - V_t$ in (1.152), giving

$$I_D = \frac{k'}{2}\frac{W}{L}(V_{GS} - V_t)^2 \qquad (1.157)$$

Equation 1.157 predicts that the drain current is independent of V_{DS} in the pinch-off region. In practice, however, the drain current in the pinch-off region varies slightly as the drain voltage is varied. This effect is due to the presence of a depletion region between the physical pinch-off point in the channel at the drain end and the drain region itself. If this depletion-layer width is X_d, then the *effective* channel length is given by

$$L_{\text{eff}} = L - X_d \qquad (1.158)$$

If L_{eff} is used in place of L in (1.157), we obtain a more accurate formula for current in the pinch-off region

$$I_D = \frac{k'}{2}\frac{W}{L_{\text{eff}}}(V_{GS} - V_t)^2 \qquad (1.159)$$

Because X_d (and thus L_{eff}) are functions of the drain-source voltage in the pinch-off region, I_D varies with V_{DS}. This effect is called *channel-length modulation*. Using (1.158) and (1.159), we obtain

$$\frac{\partial I_D}{\partial V_{DS}} = -\frac{k'}{2}\frac{W}{L_{\text{eff}}^2}(V_{GS} - V_t)^2 \frac{dL_{\text{eff}}}{dV_{DS}} \qquad (1.160)$$

and thus

$$\frac{\partial I_D}{\partial V_{DS}} = \frac{I_D}{L_{\text{eff}}} \frac{dX_d}{dV_{DS}} \qquad (1.161)$$

This equation is analogous to (1.55) for bipolar transistors. Following a similar procedure, the Early voltage can be defined as

$$V_A = \frac{I_D}{\partial I_D/\partial V_{DS}} \qquad (1.162)$$

and thus

$$V_A = L_{\text{eff}} \left(\frac{dX_d}{dV_{DS}}\right)^{-1} \qquad (1.163)$$

For MOS transistors, a commonly used parameter for the characterization of channel-length modulation is the reciprocal of the Early voltage,

$$\lambda = \frac{1}{V_A} \qquad (1.164)$$

As in the bipolar case, the large-signal properties of the transistor can be approximated by assuming that λ and V_A are constants, independent of the bias conditions. Thus we can include the effect of channel-length modulation in the I-V characteristics by modifying (1.157) to

$$I_D = \frac{k'}{2}\frac{W}{L}(V_{GS} - V_t)^2 \left(1 + \frac{V_{DS}}{V_A}\right) = \frac{k'}{2}\frac{W}{L}(V_{GS} - V_t)^2(1 + \lambda V_{DS}) \qquad (1.165)$$

In practical MOS transistors, variation of X_d with voltage is complicated by the fact that the field distribution in the drain depletion region is not one-dimensional. As a result, the calculation of λ from the device structure is quite difficult,[18] and developing effective values of λ from experimental data is usually necessary. The parameter λ is inversely proportional to the effective channel length and a decreasing function of the doping level in the channel. Typical values of λ are in the range 0.05 V^{-1} to 0.005 V^{-1}.

Plots of I_D versus V_{DS} with V_{GS} as a parameter are shown in Fig. 1.29 for an NMOS transistor. The device operates in the pinch-off region when $V_{DS} > (V_{GS} - V_t)$. The pinch-off region for MOS devices is often called the *saturation* region. In saturation, the output characteristics are almost flat, which shows that the current depends mostly on the gate-source voltage and only to a small extent on the drain-source voltage. On the other hand, when $V_{DS} < (V_{GS} - V_t)$, the device operates in the *Ohmic* or *triode* region, where the device can be modeled as a nonlinear voltage-controlled resistor connected between the drain and source. The resistance of this resistor is *nonlinear* because the V_{DS}^2 term in (1.152) causes the resistance to depend on V_{DS}. Since this term is small when V_{DS} is small, however, the nonlinearity is also small when V_{DS} is small, and the triode region is also sometimes called the *linear* region. The boundary between the triode and saturation regions occurs when $V_{DS} = (V_{GS} - V_t)$. On this boundary, both (1.152) and (1.157) correctly predict I_D. Since $V_{DS} = (V_{GS} - V_t)$ along the boundary between triode and saturation, (1.157) shows that the boundary is $I_D = (k'/2)(W/L)V_{DS}^2$. This parabolic function of V_{DS} is shown in Fig. 1.29. For depletion n-channel MOS devices, V_t is negative, and I_D is nonzero even for $V_{GS} = 0$. For PMOS devices, all polarities of voltages and currents are reversed.

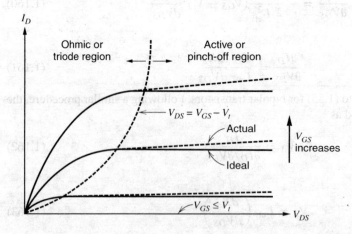

Figure 1.29 NMOS device characteristics.

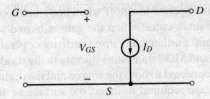

Figure 1.30 Large-signal model for the NMOS transistor.

The results derived above can be used to form a large-signal model of the NMOS transistor in saturation. The model topology is shown in Fig. 1.30, where I_D is given by (1.152) in the triode region and (1.157) in saturation, ignoring the effect of channel-length modulation. To include the effect of channel-length modulation, (1.159) or (1.165) should be used instead of (1.157) to find the drain current in saturation.

1.5.2 Comparison of Operating Regions of Bipolar and MOS Transistors

Notice that the meaning of the word *saturation* for MOS transistors is quite different than for bipolar transistors. Saturation in bipolar transistors refers to the region of operation where both junctions are forward biased and the collector-emitter voltage is approximately constant or saturated. On the other hand, saturation in MOS transistors refers to the region of operation where the channel is attached only to the source but not to the drain and the current is approximately constant or saturated. To avoid confusion, the term *active region* will be used in this book to describe the flat region of the MOS transistor characteristics, as shown in Fig. 1.29. This wording is selected to form a link between the operation of MOS and bipolar transistors. This link is summarized in the table of Fig. 1.31, which reviews the operating regions of *npn* bipolar and *n*-channel MOS transistors.

When the emitter junction is forward biased and the collector junction is reverse biased, bipolar transistors operate in the forward-active region. They operate in the reverse-active region when the collector junction is forward biased and the emitter junction is reverse biased. This distinction is important because integrated-circuit bipolar transistors are typically not symmetrical in practice; that is, the collector operates more efficiently as a collector of minority carriers than as an emitter. Similarly, the emitter operates more efficiently as an emitter of minority carriers than as a collector. One reason for this asymmetry is that the collector region surrounds the emitter region in integrated-circuit bipolar transistors, as shown in Fig. 1.19. A consequence of this asymmetry is that the current gain in the forward-active region β_F is usually much greater than the current gain in the reverse-active region β_R.

In contrast, the source and drain of MOS transistors are completely interchangeable based on the preceding description. (In practice, the symmetry is good but not perfect.) Therefore, distinguishing between the forward-active and reverse-active regions of operation of an MOS transistor is not necessary.

npn Bipolar Transistor			n-channel MOS Transistor		
Region	V_{BE}	V_{BC}	**Region**	V_{GS}	V_{GD}
Cutoff	$< V_{BE(on)}$	$< V_{BC(on)}$	Cutoff	$< V_t$	$< V_t$
Forward Active	$\geq V_{BE(on)}$	$< V_{BC(on)}$	Saturation(Active)	$\geq V_t$	$< V_t$
Reverse Active	$< V_{BE(on)}$	$\geq V_{BC(on)}$	Saturation(Active)	$< V_t$	$\geq V_t$
Saturation	$\geq V_{BE(on)}$	$\geq V_{BC(on)}$	Triode	$\geq V_t$	$\geq V_t$

Figure 1.31 Operating regions of *npn* bipolar and *n*-channel MOS transistors.

Figure 1.31 also shows that *npn* bipolar transistors operate in cutoff when both junctions are reversed biased. Similarly, MOS transistors operate in cutoff when the gate is biased so that inversion occurs at neither the source nor the drain. Furthermore, *npn* transistors operate in saturation when both junctions are forward biased, and MOS transistors operate in the triode region when the gate is biased so that the channel is connected to both the source and the drain. Therefore, this comparison leads us to view the voltage required to invert the surface of an MOS transistor as analogous to the voltage required to forward bias a *pn* junction in a bipolar transistor. To display this analogy, we will use the circuit symbols in Fig. 1.32a to represent MOS transistors. These symbols are intentionally chosen to appear similar to the symbols of the corresponding bipolar transistors. In bipolar-transistor symbols, the arrow at the emitter junction represents the direction of current flow when the emitter junction is forward biased. In MOS transistors, the *pn* junctions between the source and body and the drain and body are reverse biased for normal operation. Therefore, the arrows in Fig. 1.32a do not indicate *pn* junctions. Instead, they indicate the direction of current flow when the terminals are biased so that the terminal labeled as the drain operates as the drain and the terminal labeled as the source operates as the source. In NMOS transistors, the source is the source of electrons; therefore, the source operates at a lower voltage than the drain, and the current flows in a direction opposite that of the electrons in the channel. In PMOS transistors, the source is the source of holes; therefore, the source operates at a higher voltage than the drain, and the current flows in the same direction as the holes in the channel.

In CMOS technology, one device type is fabricated in the substrate, which is common to all devices, invariably connected to a dc power-supply voltage, and usually not shown on the circuit diagram. The other device type, however, is fabricated in separate isolation regions called *wells*, which may or may not be connected together and which may or may not be connected to a power-supply voltage. If these isolation regions are connected to the appropriate power supply, the symbols of Fig. 1.32a will be used, and the substrate connection will not

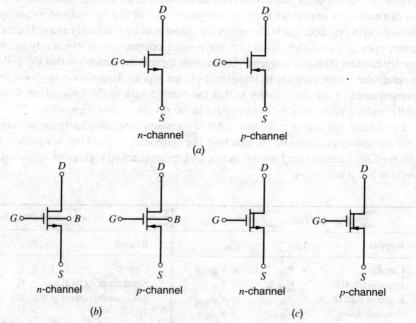

Figure 1.32 (a) NMOS and PMOS symbols used in CMOS circuits. (b) NMOS and PMOS symbols used when the substrate connection is nonstandard. (c) Depletion MOS device symbols.

be shown. On the other hand, if the individual isolation regions are connected elsewhere, the devices will be represented by the symbols of Fig. 1.32b, where the substrate is labeled B. Finally, symbols for depletion-mode devices, for which a channel forms for $V_{GS} = 0$, are shown in Fig. 1.32c.

1.5.3 Decomposition of Gate-Source Voltage

The gate-source voltage of a given MOS transistor is usually separated into two parts: the threshold, V_t, and the voltage over the threshold, $V_{GS} - V_t$. We will refer to this latter part of the gate-source voltage as the *overdrive*. This decomposition is used because these two components of the gate-source voltage have different properties. Assuming square-law behavior as in (1.157), the overdrive is

$$V_{ov} = V_{GS} - V_t = \sqrt{\frac{2I_D}{k'(W/L)}} \tag{1.166}$$

Since the transconductance parameter k' is proportional to mobility, and since mobility falls with increasing temperature, the overdrive rises with temperature. In contrast, the next section shows that the threshold falls with increasing temperature. Furthermore, (1.140) shows that the threshold depends on the source-body voltage, but not on the current; (1.166) shows that the overdrive depends directly on the current, but not on the source-body voltage.

1.5.4 Threshold Temperature Dependence

Assume that the source-body voltage is zero. Substituting (1.138) into (1.139) gives

$$V_t = \frac{\sqrt{2qN_A\epsilon(2\phi_f)}}{C_{ox}} + 2\phi_f + \phi_{ms} - \frac{Q_{ss}}{C_{ox}} \tag{1.167}$$

Assume that ϕ_{ms}, Q_{ss}, and C_{ox} are independent of temperature.[19] Then differentiating (1.167) gives

$$\frac{dV_t}{dT} = \frac{\sqrt{2qN_A\epsilon(2)}}{2C_{ox}\sqrt{\phi_f}}\frac{d\phi_f}{dT} + 2\frac{d\phi_f}{dT} = \frac{d\phi_f}{dT}\left[2 + \frac{1}{C_{ox}}\sqrt{\frac{qN_A\epsilon}{\phi_f}}\right] \tag{1.168}$$

Substituting (1.136) into (1.135) gives

$$\phi_f = \frac{kT}{q} \ln\left[\frac{N_A \exp\left(\frac{E_g}{2kT}\right)}{\sqrt{N_c N_v}}\right] \tag{1.169}$$

Assume both N_c and N_v are independent of temperature.[20] Then differentiating (1.169) gives

$$\frac{d\phi_f}{dT} = \frac{kT}{q}\left[-\frac{E_g}{2kT^2}\right] + \frac{k}{q}\ln\left[\frac{N_A \exp\left(\frac{E_g}{2kT}\right)}{\sqrt{N_c N_v}}\right] \tag{1.170}$$

Substituting (1.169) into (1.170) and simplifying gives

$$\frac{d\phi_f}{dT} = -\frac{E_g}{2qT} + \frac{\phi_f}{T} = -\frac{1}{T}\left[\frac{E_g}{2q} - \phi_f\right] \tag{1.171}$$

Substituting (1.141) and (1.171) into (1.168) gives

$$\frac{dV_t}{dT} = -\frac{1}{T}\left[\frac{E_g}{2q} - \phi_f\right]\left[2 + \frac{\gamma}{\sqrt{2\phi_f}}\right] \quad (1.172)$$

Equation 1.172 shows that the threshold voltage falls with increasing temperature if $\phi_f < E_g/(2q)$. The slope is usually in the range of -0.5 mV/°C to -4 mV/°C.[21]

■ **EXAMPLE**

Assume $T = 300°$K, $N_A = 10^{15}$ cm^{-3}, and $t_{ox} = 100$ Å. Find dV_t/dT.
From (1.135),

$$\phi_f = (25.8 \text{ mV}) \ln\left(\frac{10^{15} \text{cm}^{-3}}{1.45 \times 10^{10} \text{cm}^{-3}}\right) = 287 \text{ mV} \quad (1.173)$$

Also

$$\frac{E_g}{2q} = \frac{1.12 \text{ eV}}{2q} = 0.56 \text{ V} \quad (1.174)$$

Substituting (1.173) and (1.174) into (1.171) gives

$$\frac{d\phi_f}{dT} = -\frac{1}{300}(560 - 287) \frac{\text{mV}}{°\text{K}} = -0.91 \frac{\text{mV}}{°\text{K}} \quad (1.175)$$

From (1.142),

$$C_{ox} = \frac{3.9 \, (8.854 \times 10^{-14} \text{ F/cm})}{100 \times 10^{-8} \text{ cm}} = 3.45 \frac{\text{fF}}{\mu\text{m}^2} \quad (1.176)$$

Also,

$$\frac{\gamma}{\sqrt{2\phi_f}} = \frac{1}{C_{ox}} \sqrt{\frac{(2)(1.6 \times 10^{-19} \text{ C})(11.7)(8.854 \times 10^{-14} \text{ F/cm})(10^{15} \text{ cm}^{-3})}{(2)(0.287 \text{ V})}}$$

$$= \frac{2.4 \times 10^{-8} \text{ F/cm}^2}{3.45 \times 10^{-15} \text{ F/}\mu\text{m}^2} = \frac{2.4 \times 10^{-16} \text{ F/}\mu\text{m}^2}{3.45 \times 10^{-15} \text{ F/}\mu\text{m}^2} = 0.07 \quad (1.177)$$

Substituting (1.173) – (1.177) into (1.172) gives

$$\frac{dV_t}{dT} = \left(-0.91 \frac{\text{mV}}{°\text{K}}\right)(2 + 0.07) \simeq -1.9 \frac{\text{mV}}{°\text{K}} = -1.9 \frac{\text{mV}}{°\text{C}} \quad (1.178)$$

■

1.5.5 MOS Device Voltage Limitations

The main voltage limitations in MOS transistors are described next.[22,23] Some of these limitations have a strong dependence on the gate length L; others have little dependence on L. Also, some of the voltage limitations are inherently destructive; others cause no damage as long as overheating is avoided.

Junction Breakdown. For long channel lengths, the drain-depletion region has little effect on the channel, and the I_D-versus-V_{DS} curves closely follow the ideal curves of Fig. 1.29. For increasing V_{DS}, however, eventually the drain-substrate pn-junction breakdown voltage is exceeded, and the drain current increases abruptly by avalanche breakdown as described in Section 1.2.2. This phenomenon is not inherently destructive.

Punchthrough. If the depletion region around the drain in an MOS transistor touches the depletion region around the source before junction breakdown occurs, increasing the drain-source voltage increases the drain current by reducing the barrier to electron flow between the source and drain. This phenomenon is called *punchthrough*. Since it depends on the two depletion regions touching, it also depends on the gate length. Punchthrough is not inherently destructive and causes a more gradual increase in the drain current than is caused by avalanche breakdown. Punchthrough normally occurs below the surface of the silicon and is often prevented by an extra ion implantation below the surface to reduce the size of the depletion regions.

Hot Carriers. With sufficient horizontal or vertical electric fields, electrons or holes may reach sufficient velocities to be injected into the oxide, where most of them increase the gate current and some of them become trapped. Such carriers are called *hot* because the required velocity for injection into the oxide is usually greater than the random thermal velocity. Carriers trapped in the oxide shift the threshold voltage and may cause a transistor to remain on when it should turn off or vice versa. In this sense, injection of hot carriers into the oxide is a destructive process. This process is most likely to be problematic in short-channel technologies, where horizontal electric fields are likely to be high.

Oxide Breakdown. In addition to V_{DS} limitations, MOS devices must also be protected against excessive gate voltages. Typical gate oxides break down with an electric field of about 6×10^6 V/cm to 7×10^6 V/cm,[24,25] which corresponds to 6 to 7 V applied from gate to channel with an oxide thickness of 100 angstroms. Since this process depends on the vertical electrical field, it is independent of channel length. However, this process is destructive to the transistor, resulting in resistive connections between the gate and the channel. Oxide breakdown can be caused by static electricity and can be avoided by using pn diodes and resistors to limit the voltage range at sensitive nodes internal to the integrated circuit that connect to bonding pads.

1.6 Small-Signal Models of MOS Transistors

As mentioned in Section 1.5, MOS transistors are often used in analog circuits. To simplify the calculation of circuit gain and terminal impedances, *small-signal* models can be used. As in the case for bipolar transistors, a hierarchy of models with increasing complexity can be derived, and choosing the simplest model required to do a given analysis is important in practice.

Consider the MOS transistor in Fig. 1.33 with bias voltages V_{GS} and V_{DD} applied as shown. These bias voltages produce quiescent drain current I_D. If $V_{GS} > V_t$ and $V_{DD} > (V_{GS} - V_t)$, the device operates in the saturation or active region. A small-signal input voltage v_i is applied in series with V_{GS} and produces a small variation in drain current i_d. The total value of the drain current is $I_d = (I_D + i_d)$.

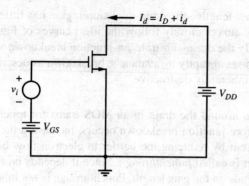

Figure 1.33 Schematic of an MOS transistor with biasing.

1.6.1 Transconductance

Assuming square-law operation, the transconductance from the gate can be determined from (1.165) by differentiating

$$g_m = \frac{\partial I_D}{\partial V_{GS}} = k'\frac{W}{L}(V_{GS} - V_t)(1 + \lambda V_{DS}) \qquad (1.179)$$

If $\lambda V_{DS} \ll 1$, (1.179) simplifies to

$$g_m = k'\frac{W}{L}(V_{GS} - V_t) = \sqrt{2k'\frac{W}{L}I_D} \qquad (1.180)$$

Unlike the bipolar transistor, the transconductance of the MOS transistor is proportional to the square root of the bias current and depends on device geometry (oxide thickness via k' and W/L). Another key difference between bipolar and MOS transistors can be seen by calculating the ratio of the transconductance to the current. Using (1.157) and (1.180) for MOS transistors shows that

$$\frac{g_m}{I_D} = \frac{2}{V_{GS} - V_t} = \frac{2}{V_{ov}} \qquad (1.181)$$

Also, for bipolar transistors, (1.91) shows that

$$\frac{g_m}{I_C} = \frac{q}{kT} = \frac{1}{V_T} \qquad (1.182)$$

At room temperature, the thermal voltage V_T is about equal to 26 mV. In contrast, the overdrive V_{ov} for MOS transistors in many applications is chosen to be approximately several hundred mV so that MOS transistors are fast enough for the given application. (Section 1.6.8 shows that the transition frequency f_T of an MOS transistor is proportional to the overdrive.) Under these conditions, the transconductance per given current is much higher for bipolar transistors than for MOS transistors. One of the key challenges in MOS analog circuit design is designing high-quality analog circuits with a low transconductance-to-current ratio.

The transconductance calculated in (1.180) is valid for small-signal analysis. To determine the limitation on the use of small-signal analysis, the change in the drain current resulting from a change in the gate-source voltage will be derived from a large-signal standpoint. The total drain current in Fig. 1.33 can be calculated using (1.157) as

$$I_d = \frac{k'}{2}\frac{W}{L}(V_{GS} + v_i - V_t)^2 = \frac{k'}{2}\frac{W}{L}\left[(V_{GS} - V_t)^2 + 2(V_{GS} - V_t)v_i + v_i^2\right] \qquad (1.183)$$

Substituting (1.157) in (1.183) gives

$$I_d = I_D + \frac{k'}{2}\frac{W}{L}\left[2(V_{GS} - V_t)v_i + v_i^2\right] \quad (1.184)$$

Rearranging (1.184) gives

$$i_d = I_d - I_D = k'\frac{W}{L}(V_{GS} - V_t)v_i\left[1 + \frac{v_i}{2(V_{GS} - V_t)}\right] \quad (1.185)$$

If the magnitude of the small-signal input $|v_i|$ is much less than twice the overdrive defined in (1.166), substituting (1.180) into (1.185) gives

$$i_d \simeq g_m v_i \quad (1.186)$$

In particular, if $|v_i| = |\Delta V_{GS}|$ is less than 20 percent of the overdrive, the small-signal analysis is accurate within about 10 percent.

1.6.2 Intrinsic Gate-Source and Gate-Drain Capacitance

If C_{ox} is the oxide capacitance per unit area from gate to channel, then the total capacitance under the gate is $C_{ox}WL$. This capacitance is intrinsic to the device operation and models the gate control of the channel conductance. In the triode region of device operation, the channel exists continuously from source to drain, and the gate-channel capacitance is usually lumped into two equal parts at the drain and source with

$$C_{gs} = C_{gd} = \frac{C_{ox}WL}{2} \quad (1.187)$$

In the saturation or active region, however, the channel pinches off before reaching the drain, and the drain voltage exerts little influence on either the channel or the gate charge. As a consequence, the intrinsic portion of C_{gd} is essentially zero in the saturation region. To calculate the value of the intrinsic part of C_{gs} in the saturation or active region, we must calculate the total charge Q_T stored in the channel. This calculation can be carried out by substituting (1.145) into (1.144) and integrating to obtain

$$Q_T = WC_{ox}\int_0^L [V_{GS} - V(y) - V_t]dy \quad (1.188)$$

Solving (1.150) for dy and substituting into (1.188) gives

$$Q_T = \frac{W^2 C_{ox}^2 \mu_n}{I_D}\int_0^{V_{GS}-V_t}(V_{GS} - V - V_t)^2 dV \quad (1.189)$$

where the limit $y = L$ corresponds to $V = (V_{GS} - V_t)$ in the saturation or active region. Solution of (1.189) and use of (1.153) and (1.157) gives

$$Q_T = \frac{2}{3}WLC_{ox}(V_{GS} - V_t) \quad (1.190)$$

Therefore, in the saturation or active region,

$$C_{gs} = \frac{\partial Q_T}{\partial V_{GS}} = \frac{2}{3}WLC_{ox} \quad (1.191)$$

and

$$C_{gd} = 0 \quad (1.192)$$

1.6.3 Input Resistance

The gate of an MOS transistor is insulated from the channel by the SiO$_2$ dielectric. As a result, the low-frequency gate current is essentially zero and the input resistance is essentially infinite. This characteristic is important in some circuits such as sample-and-hold amplifiers, where the gate of an MOS transistor can be connected to a capacitor to sense the voltage on the capacitor without leaking away the charge that causes that voltage. In contrast, bipolar transistors have small but nonzero base current and finite input resistance looking into the base, complicating the design of bipolar sample-and-hold amplifiers.

1.6.4 Output Resistance

In Section 1.5.1, the effect of changes in drain-source voltage on the large-signal characteristics of the MOS transistor was described. Increasing drain-source voltage in an n-channel MOS transistor increases the width of the depletion region around the drain and reduces the effective channel length of the device in the saturation or active region. This effect is called channel-length modulation and causes the drain current to increase when the drain-source voltage is increased. From that treatment, we can calculate the change in the drain current ΔI_D arising from changes in the drain-source voltage ΔV_{DS} as

$$\Delta I_D = \frac{\partial I_D}{\partial V_{DS}} \Delta V_{DS} \qquad (1.193)$$

Substitution of (1.161), (1.163), and (1.164) in (1.193) gives

$$\frac{\Delta V_{DS}}{\Delta I_D} = \frac{V_A}{I_D} = \frac{1}{\lambda I_D} = r_o \qquad (1.194)$$

where V_A is the Early voltage, λ is the channel-length modulation parameter, I_D is the drain current without channel-length modulation given by (1.157), and r_o is the small-signal output resistance of the transistor.

1.6.5 Basic Small-Signal Model of the MOS Transistor

Combination of the preceding small-signal circuit elements yields the small-signal model of the MOS transistor shown in Fig. 1.34. This model was derived for n-channel transistors in the saturation or active region and is called the *hybrid-π* model. Drain, gate, and source nodes are labeled D, G, and S, respectively. When the gate-source voltage is increased, the model predicts that the incremental current i_d flowing from drain to source increases. Since the dc drain current I_D also flows from drain to source in an n-channel transistor, increasing the gate-source voltage also increases the total drain current I_d. This result is reasonable physically because increasing the gate-source voltage in an n-channel transistor increases the channel conductivity and drain current.

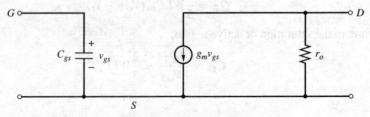

Figure 1.34 Basic small-signal model of an MOS transistor in the saturation or active region.

The model shown in Fig. 1.34 is also valid for *p*-channel devices. Therefore, the model again shows that increasing the gate-source voltage increases the incremental current i_d flowing from drain to source. Unlike in the *n*-channel case, however, the dc current I_D in a *p*-channel transistor flows from source to drain because the source acts as the source of holes. Therefore, the incremental drain current flows in a direction opposite to the dc drain current when the gate-source voltage increases, reducing the total drain current I_d. This result is reasonable physically because increasing the gate-source voltage in a *p*-channel transistor reduces the channel conductivity and drain current.

1.6.6 Body Transconductance

The drain current is a function of both the gate-source and body-source voltages. On the one hand, the gate-source voltage controls the vertical electric field, which controls the channel conductivity and therefore the drain current. On the other hand, the body-source voltage changes the threshold, which changes the drain current when the gate-source voltage is fixed. This effect stems from the influence of the substrate acting as a second gate and is called the *body effect*. Note that the body of an MOS transistor is usually connected to a constant power-supply voltage, which is a small-signal or ac ground. However, the source connection can have a significant ac voltage impressed on it, which changes the body-source voltage when the body voltage is fixed. Therefore, when the body-source voltage is not constant, two transconductance terms are required to model MOS transistors: one associated with the main gate and the other associated with the body or second gate.

Using (1.165), the transconductance from the body or second gate is

$$g_{mb} = \frac{\partial I_D}{\partial V_{BS}} = -k'\frac{W}{L}(V_{GS} - V_t)(1 + \lambda V_{DS})\frac{\partial V_t}{\partial V_{BS}} \quad (1.195)$$

From (1.140)

$$\frac{\partial V_t}{\partial V_{BS}} = -\frac{\gamma}{2\sqrt{2\phi_f + V_{SB}}} = -\chi \quad (1.196)$$

This equation defines a factor χ, which is the rate of change of threshold voltage with body bias voltage. Substitution of (1.141) in (1.196) and use of (1.20) gives

$$\chi = \frac{C_{js}}{C_{ox}} \quad (1.197)$$

where C_{js} is the capacitance per unit area of the depletion region under the channel, assuming a one-sided step junction with a built-in potential $\psi_0 = 2\phi_f$. Substitution of (1.196) in (1.195) gives

$$g_{mb} = \frac{\gamma k'(W/L)(V_{GS} - V_t)(1 + \lambda V_{DS})}{2\sqrt{2\phi_f + V_{SB}}} \quad (1.198)$$

If $\lambda V_{DS} \ll 1$, we have

$$g_{mb} = \frac{\gamma k'(W/L)(V_{GS} - V_t)}{2\sqrt{2\phi_f + V_{SB}}} = \gamma\sqrt{\frac{k'(W/L)I_D}{2(2\phi_f + V_{SB})}} \quad (1.199)$$

The ratio g_{mb}/g_m is an important quantity in practice. From (1.179) and (1.198), we find

$$\frac{g_{mb}}{g_m} = \frac{\gamma}{2\sqrt{2\phi_f + V_{SB}}} = \chi \quad (1.200)$$

1.6.7 Parasitic Elements in the Small-Signal Model

The factor χ is typically in the range 0.1 to 0.3; therefore, the transconductance from the main gate is typically a factor of about 3 to 10 times larger than the transconductance from the body or second gate.

The elements of the small-signal model for MOS transistors described above may be considered basic in the sense that they arise directly from essential processes in the device. As in the case of bipolar transistors, however, technological limitations in the fabrication of the devices give rise to a number of parasitic elements that must be added to the equivalent circuit for most integrated-circuit transistors. A cross section and top view of a typical n-channel MOS transistor are shown in Fig. 1.35. The means of fabricating such devices is described in Chapter 2.

All pn junctions in the MOS transistor should be reverse biased during normal operation, and each junction exhibits a voltage-dependent parasitic capacitance associated with its depletion region. The source-body and drain-body junction capacitances shown in Fig. 1.35a are C_{sb} and C_{db}, respectively. If the doping levels in the source, drain, and body regions are assumed to be constant, (1.21) can be used to express these capacitances as follows:

$$C_{sb} = \frac{C_{sb0}}{\left(1 + \frac{V_{SB}}{\psi_0}\right)^{1/2}} \quad (1.201)$$

$$C_{db} = \frac{C_{db0}}{\left(1 + \frac{V_{DB}}{\psi_0}\right)^{1/2}} \quad (1.202)$$

These capacitances are proportional to the source and drain region areas (including sidewalls). Since the channel is attached to the source in the saturation or active region, C_{sb} also includes

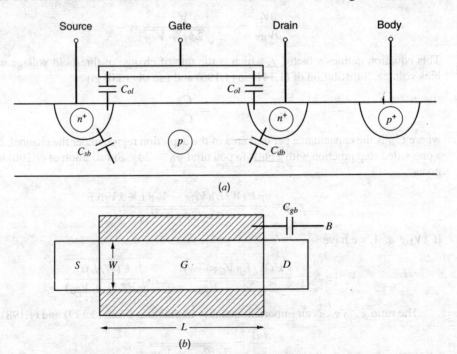

Figure 1.35 (a) Cross section and (b) top view of an n-channel MOS transistor.

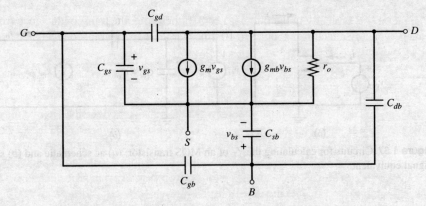

Figure 1.36 Small-signal MOS transistor equivalent circuit.

depletion-region capacitance from the induced channel to the body. A detailed analysis of the channel-body capacitance is given by Tsividis.[26]

In practice, C_{gs} and C_{gd}, given in (1.187) for the triode region of operation and in (1.191) and (1.192) for the saturation or active region, are increased due to parasitic oxide capacitances arising from gate overlap of the source and drain regions. These overlap capacitances C_{ol} are shown in Fig. 1.35a, and their values are calculated in Chapter 2.

Capacitance C_{gb} between gate and body or substrate models parasitic oxide capacitance between the gate-contact material and the substrate outside the active-device area. This capacitance is independent of the gate-body voltage and models coupling from polysilicon and metal interconnects to the underlying substrate, as shown by the shaded regions in the top view of Fig. 1.35b. Parasitic capacitance of this type underlies all polysilicon and metal traces on integrated circuits. Such parasitic capacitance should be taken into account when simulating and calculating high-frequency circuit and device performance. Typical values depend on oxide thicknesses. With a silicon dioxide thickness of 100 Å, the capacitance is about 3.45 fF per square micron. Fringing capacitance becomes important for lines narrower in width than several microns.

Parasitic resistance in series with the source and drain can be used to model the nonzero resistivity of the contacts and diffusion regions. In practice, these resistances are often ignored in hand calculations for simplicity but included in computer simulations. These parasitic resistances have an inverse dependence on channel width W. Typical values of these resistances are 50 Ω to 100 Ω for devices with W of about 1 μm. Similar parasitic resistances in series with the gate and body terminals are sometimes included but often ignored because very little current flows in these terminals, especially at low frequencies. The small-signal model including capacitive parasitics but ignoring resistive parasitics is shown in Fig. 1.36.

1.6.8 MOS Transistor Frequency Response

As for a bipolar transistor, the frequency capability of an MOS transistor is usually specified by finding the transition frequency f_T. For an MOS transistor, f_T is defined as the frequency where the magnitude of the short-circuit, common-source current gain falls to unity. Although the dc gate current of an MOS transistor is essentially zero, the high-frequency behavior of the transistor is controlled by the capacitive elements in the small-signal model, which cause the gate current to increase as frequency increases. To calculate f_T, consider the ac circuit of Fig. 1.37a and the small-signal equivalent of Fig. 1.37b. Since $v_{sb} = v_{ds} = 0$, g_{mb},

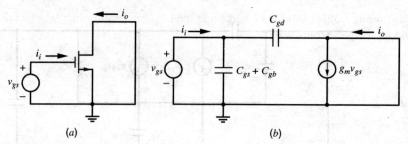

Figure 1.37 Circuits for calculating the f_T of an MOS transistor: (a) ac schematic and (b) small-signal equivalent.

r_o, C_{sb}, and C_{db} have no effect on the calculation and are ignored. The small-signal input current i_i is

$$i_i = s(C_{gs} + C_{gb} + C_{gd})v_{gs} \tag{1.203}$$

If the current fed forward through C_{gd} is neglected,

$$i_o \simeq g_m v_{gs} \tag{1.204}$$

Solving (1.203) for v_{gs} and substituting into (1.204) gives

$$\frac{i_o}{i_i} \simeq \frac{g_m}{s(C_{gs} + C_{gb} + C_{gd})} \tag{1.205}$$

To find the frequency response, we set $s = j\omega$. Then

$$\frac{i_o}{i_i} \simeq \frac{g_m}{j\omega(C_{gs} + C_{gb} + C_{gd})} \tag{1.206}$$

The magnitude of the small-signal current gain is unity when

$$\omega = \omega_T = \frac{g_m}{C_{gs} + C_{gb} + C_{gd}} \tag{1.207}$$

Therefore,

$$f_T = \frac{1}{2\pi}\omega_T = \frac{1}{2\pi}\frac{g_m}{C_{gs} + C_{gb} + C_{gd}} \tag{1.208}$$

Assume the intrinsic device capacitance C_{gs} is much greater than $(C_{gb} + C_{gd})$. Then substituting (1.180) and (1.191) into (1.208) gives

$$f_T = 1.5\frac{\mu_n}{2\pi L^2}(V_{GS} - V_t) \tag{1.209}$$

Comparison of this equation with the intrinsic f_T of a bipolar transistor when parasitic depletion-layer capacitance is neglected leads to an interesting result. From (1.128) and (1.130) with $\tau_F \gg (C_{je} + C_\mu)/g_m$,

$$f_T = \frac{1}{2\pi\tau_F} \tag{1.210}$$

Substituting from (1.99) for τ_F and using the Einstein relationship $D_n/\mu_n = kT/q = V_T$, we find for a bipolar transistor

$$f_T = 2\frac{\mu_n}{2\pi W_B^2}V_T \qquad (1.211)$$

The similarity in form between (1.211) and (1.209) is striking. In both cases, the intrinsic device f_T increases as the inverse square of the critical device dimension across which carriers are in transit. The voltage $V_T = 26$ mV is fixed for a bipolar transistor, but the f_T of an MOS transistor can be increased by operating at high values of $(V_{GS} - V_t)$. Note that the base width W_B in a bipolar transistor is a vertical dimension determined by diffusions or implants and can typically be made much smaller than the channel length L of an MOS transistor, which depends on surface geometry and photolithographic processes. Thus bipolar transistors generally have higher f_T than MOS transistors made with comparable processing. Finally, (1.209) was derived assuming that the MOS transistor exhibits square-law behavior as in (1.157). However, as described in Section 1.7, submicron MOS transistors depart significantly from square-law characteristics, and we find that for such devices f_T is proportional to L^{-1} rather than L^{-2}.

■ **EXAMPLE**

Derive the complete small-signal model for an NMOS transistor with $I_D = 100$ μA, $V_{SB} = 1$ V, $V_{DS} = 2$ V. Device parameters are $\phi_f = 0.3$ V, $W = 10$ μm, $L = 1$ μm, $\gamma = 0.5$ V$^{1/2}$, $k' = 200$ μA/V^2, $\lambda = 0.02$ V^{-1}, $t_{ox} = 100$ angstroms, $\psi_0 = 0.6$ V, $C_{sb0} = C_{db0} = 10$ fF. Overlap capacitance from gate to source and gate to drain is 1 fF. Assume $C_{gb} = 5$ fF.

From (1.166),

$$V_{ov} = V_{GS} - V_t = \sqrt{\frac{2I_D}{k'(W/L)}} = \sqrt{\frac{2 \times 100}{200 \times 10}} \simeq 0.316 \text{ V}$$

Since $V_{DS} > V_{ov}$, the transistor operates in the saturation or active region. From (1.180),

$$g_m = \sqrt{2k'\frac{W}{L}I_D} = \sqrt{2 \times 200 \times 10 \times 100} \text{ μA/V} \simeq 632 \text{ μA/V}$$

From (1.199),

$$g_{mb} = \gamma\sqrt{\frac{k'(W/L)I_D}{2(2\phi_f + V_{SB})}} = 0.5\sqrt{\frac{200 \times 10 \times 100}{2 \times 1.6}} \simeq 125 \text{ μA/V}$$

From (1.194),

$$r_o = \frac{1}{\lambda I_D} = \frac{1000}{0.02 \times 100} \text{ k}\Omega = 500 \text{ k}\Omega$$

Using (1.201) with $V_{SB} = 1$ V, we find

$$C_{sb} = \frac{10}{\left(1 + \frac{1}{0.6}\right)^{1/2}} \text{ fF} \simeq 6 \text{ fF}$$

The voltage from drain to body is

$$V_{DB} = V_{DS} + V_{SB} = 3 \text{ V}$$

and substitution in (1.202) gives

$$C_{db} = \frac{10}{\left(1 + \dfrac{3}{0.6}\right)^{1/2}} \text{ fF} \simeq 4 \text{ fF}$$

From (1.142), the oxide capacitance per unit area is

$$C_{ox} = \frac{3.9 \times 8.854 \times 10^{-14} \dfrac{\text{F}}{\text{cm}} \times \dfrac{100 \text{ cm}}{10^6 \text{ }\mu\text{m}}}{100 \text{ Å} \times \dfrac{10^6 \text{ }\mu\text{m}}{10^{10} \text{ Å}}} \simeq 3.45 \frac{\text{fF}}{\mu\text{m}^2}$$

The intrinsic portion of the gate-source capacitance can be calculated from (1.191), giving

$$C_{gs} \simeq \frac{2}{3} \times 10 \times 1 \times 3.45 \text{ fF} \simeq 23 \text{ fF}$$

The addition of overlap capacitance gives

$$C_{gs} \simeq 24 \text{ fF}$$

Finally, since the transistor operates in the saturation or active region, the gate-drain capacitance consists of only overlap capacitance and is

$$C_{gd} = 1 \text{ fF}$$

The complete small-signal equivalent circuit is shown in Fig. 1.38. The f_T of the device can be calculated from (1.208) as

$$f_T = \frac{1}{2\pi} \frac{g_m}{C_{gs} + C_{gb} + C_{gd}} = \frac{1}{2\pi} \times 632 \times 10^{-6} \times \frac{10^{15}}{24 + 5 + 1} \text{ Hz} = 3.4 \text{ GHz}$$

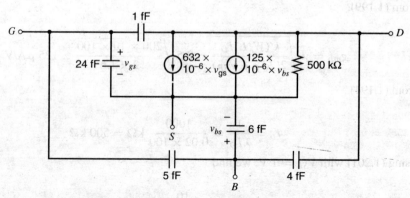

Figure 1.38 Complete small-signal equivalent circuit for an NMOS transistor with $I_D = 100 \text{ }\mu\text{A}$, $V_{SB} = 1 \text{ V}$, $V_{DS} = 2 \text{ V}$. Device parameters are $W = 10 \text{ }\mu\text{m}$, $L = 1 \text{ }\mu\text{m}$, $\gamma = 0.5 \text{ V}^{1/2}$, $k' = 200 \text{ }\mu\text{A/V}^2$, $\lambda = 0.02 \text{ V}^{-1}$, $t_{ox} = 100 \text{ Å}$, $\psi_0 = 0.6 \text{ V}$, $C_{sb0} = C_{db0} = 10 \text{ fF}$, $C_{gd} = 1 \text{ fF}$, and $C_{gb} = 5 \text{ fF}$.

1.7 Short-Channel Effects in MOS Transistors

The evolution of integrated-circuit processing techniques has led to continuing reductions in both the horizontal and vertical dimensions of the active devices. (The minimum allowed dimensions of passive devices have also decreased.) This trend is driven primarily by economics in that reducing dimensions increases the number of devices and circuits that can be processed at one time on a given wafer. A second benefit has been that the frequency capability of the active devices continues to increase, as intrinsic f_T values increase with smaller dimensions while parasitic capacitances decrease.

Vertical dimensions such as the base width of a bipolar transistor in production processes may now be on the order of 0.05 μm or less, whereas horizontal dimensions such as bipolar emitter width or MOS transistor gate length may be significantly less than 1 μm. Even with these small dimensions, the large-signal and small-signal models of bipolar transistors given in previous sections remain valid. However, significant short-channel effects become important in MOS transistors at channel lengths of about 1 μm or less and require modifications to the MOS models given previously. The primary effect is to modify the classical MOS square-law transfer characteristic in the saturation or active region to make the device voltage-to-current transfer characteristic more linear. However, even in processes with submicron capability, many of the MOS transistors in a given analog circuit may be deliberately designed to have channel lengths larger than the minimum and may be well approximated by the square-law model.

1.7.1 Velocity Saturation from the Horizontal Field

The most important short-channel effect in MOS transistors stems from velocity saturation of carriers in the channel.[27] When an MOS transistor operates in the triode region, the average horizontal electric field along the channel is V_{DS}/L. When V_{DS} is small and/or L is large, the horizontal field is low, and the linear relation between carrier velocity and field assumed in (1.148) is valid. At high fields, however, the carrier velocities approach the thermal velocities, and subsequently the slope of the carrier velocity decreases with increasing field. This effect is illustrated in Fig. 1.39, which shows typical measured electron drift velocity v_d versus horizontal electric field strength magnitude $\mathscr{E}$ in an NMOS surface channel. While the velocity at low field values is proportional to the field, the velocity at high field values approaches a constant called the *scattering-limited* velocity v_{scl}. A first-order analytical approximation to this curve is

$$v_d = \frac{\mu_n \mathscr{E}}{1 + \mathscr{E}/\mathscr{E}_c} \tag{1.212}$$

where $\mathscr{E}_c \simeq 1.5 \times 10^6$ V/m and $\mu_n \simeq 0.07$ m²/V-s is the low-field mobility close to the gate. Equation 1.212 is also plotted in Fig. 1.39. From (1.212), as $\mathscr{E} \to \infty$, $v_d \to v_{scl} = \mu_n \mathscr{E}_c$. At the critical field value $\mathscr{E}_c$, the carrier velocity is a factor of 2 less than the low-field formula would predict. In a device with a channel length $L = 0.5$ μm, we need a voltage drop of only 0.75 V along the channel to produce an average field equal to $\mathscr{E}_c$, and this condition is readily achieved in short-channel MOS transistors. Similar results are found for PMOS devices.

Substituting (1.212) and (1.149) into (1.147) and rearranging gives

$$I_D \left(1 + \frac{1}{\mathscr{E}_c} \frac{dV}{dy}\right) = W Q_I(y) \mu_n \frac{dV}{dy} \tag{1.213}$$

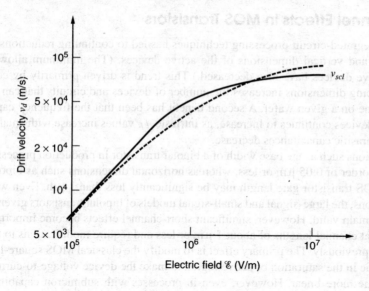

Figure 1.39 Typical measured electron drift velocity v_d versus horizontal electric field $\mathscr{E}$ in an MOS surface channel (solid plot). Also shown (dashed plot) is the analytical approximation of Eq. 1.212 with $\mathscr{E}_c = 1.5 \times 10^6$ V/m and $\mu_n = 0.07$ m²/V-s.

Note that as $\mathscr{E}_c \to \infty$ and velocity saturation becomes negligible, (1.213) approaches the original equation (1.147). Integrating (1.213) along the channel, we obtain

$$\int_0^L I_D \left(1 + \frac{1}{\mathscr{E}_c} \frac{dV}{dy}\right) dy = \int_0^{V_{DS}} W Q_I(y) \mu_n \, dV \tag{1.214}$$

and thus

$$I_D = \frac{\mu_n C_{ox}}{2\left(1 + \frac{V_{DS}}{\mathscr{E}_c L}\right)} \frac{W}{L} [2(V_{GS} - V_t)V_{DS} - V_{DS}^2] \tag{1.215}$$

In the limit as $\mathscr{E}_c \to \infty$, (1.215) is the same as (1.152), which gives the drain current in the triode region without velocity saturation. The quantity V_{DS}/L in (1.215) can be interpreted as the average horizontal electric field in the channel. If this field is comparable to $\mathscr{E}_c$, the drain current for a given V_{DS} is less than the simple expression (1.152) would predict.

Equation 1.215 is valid in the triode region. Let $V_{DS(act)}$ represent the maximum value of V_{DS} for which the transistor operates in the triode region, which is equivalent to the minimum value of V_{DS} for which the transistor operates in the active region. In the active region, the current should be independent of V_{DS} because channel-length modulation is not included here. Therefore, $V_{DS(act)}$ is the value of V_{DS} that sets $\partial I_D / \partial V_{DS} = 0$. From (1.215),

$$\frac{\partial I_D}{\partial V_{DS}} = \frac{k'}{2} \frac{W}{L} \left[\frac{\left(1 + \frac{V_{DS}}{\mathscr{E}_c L}\right)[2(V_{GS} - V_t) - 2V_{DS}] - \frac{[2(V_{GS} - V_t)V_{DS} - V_{DS}^2]}{\mathscr{E}_c L}}{\left(1 + \frac{V_{DS}}{\mathscr{E}_c L}\right)^2} \right] \tag{1.216}$$

where $k' = \mu_n C_{ox}$ as given by (1.153). To set $\partial I_D / \partial V_{DS} = 0$,

$$\left(1 + \frac{V_{DS}}{\mathscr{E}_c L}\right)[2(V_{GS} - V_t) - 2V_{DS}] - \frac{[2(V_{GS} - V_t)V_{DS} - V_{DS}^2]}{\mathscr{E}_c L} = 0 \tag{1.217}$$

Rearranging (1.217) gives

$$\frac{V_{DS}^2}{\mathscr{E}_c L} + 2V_{DS} - 2(V_{GS} - V_t) = 0 \tag{1.218}$$

1.7 Short-Channel Effects in MOS Transistors

Solving the quadratic equation gives

$$V_{DS(act)} = V_{DS} = -\mathscr{E}_c L \pm \mathscr{E}_c L \sqrt{1 + \frac{2(V_{GS} - V_t)}{\mathscr{E}_c L}} \quad (1.219)$$

Since the drain-source voltage must be greater than zero,

$$V_{DS(act)} = V_{DS} = \mathscr{E}_c L \left(\sqrt{1 + \frac{2(V_{GS} - V_t)}{\mathscr{E}_c L}} - 1 \right) \quad (1.220)$$

To determine $V_{DS(act)}$ without velocity-saturation effects, let $\mathscr{E}_c \to \infty$ so that the drift velocity is proportional to the electric field, and let $x = (V_{GS} - V_t)/(\mathscr{E}_c L)$. Then $x \to 0$, and a Taylor series can be used to show that

$$\sqrt{1 + 2x} = 1 + x - \frac{x^2}{2} + \cdots \quad (1.221)$$

Using (1.221) in (1.220) gives

$$V_{DS(act)} = (V_{GS} - V_t)\left(1 - \frac{V_{GS} - V_t}{2\mathscr{E}_c L} + \cdots\right) \quad (1.222)$$

When $\mathscr{E}_c \to \infty$, (1.222) shows that $V_{DS(act)} \to (V_{GS} - V_t)$, as expected.[28] This observation is confirmed by plotting the ratio of $V_{DS(act)}$ to the overdrive V_{ov} versus $\mathscr{E}_c L$ in Fig. 1.40. When $\mathscr{E}_c \to \infty$, $V_{DS(act)} \to V_{ov} = V_{GS} - V_t$, as predicted by (1.222). On the other hand, when $\mathscr{E}_c$ is small enough that velocity saturation is significant, Fig. 1.40 shows that $V_{DS(act)} < V_{ov}$.

To find the drain current in the active region with velocity saturation, substitute $V_{DS(act)}$ in (1.220) for V_{DS} in (1.215). After rearranging, the result is

$$I_D = \frac{\mu_n C_{ox}}{2} \frac{W}{L} [V_{DS(act)}]^2 \quad (1.223)$$

Equation 1.223 is in the same form as (1.157), where velocity saturation is neglected, except that $V_{DS(act)}$ is less than $(V_{GS} - V_t)$ when velocity saturation is significant, as shown in Fig. 1.40. Therefore, the current predicted by (1.157) overestimates the current that really flows when the carrier velocity saturates. To examine the limiting case when the velocity is completely

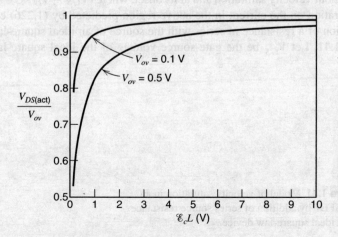

Figure 1.40 Ratio of the minimum drain-source voltage required for operation in the active region to the overdrive versus the product of the critical field and the channel length. When $\mathscr{E}_c \to \infty$, velocity saturation is not a factor, and $V_{DS(act)} \to V_{ov} = V_{GS} - V_t$, as expected. When velocity saturation is significant, $V_{DS(act)} < V_{ov}$.

saturated, let $\mathscr{E}_c \to 0$. Then (1.212) shows that the drift velocity approaches the scattering-limited velocity $v_d \to v_{scl} = \mu_n \mathscr{E}_c$. Substituting (1.220) into (1.223) gives

$$\lim_{\mathscr{E}_c \to 0} I_D = \mu_n C_{ox} W (V_{GS} - V_t) \mathscr{E}_c = W C_{ox} (V_{GS} - V_t) v_{scl} \qquad (1.224)$$

In contrast to the square-law behavior predicted by (1.157), (1.224) shows that the drain current is a *linear* function of the overdrive $(V_{GS} - V_t)$ when the carrier velocity saturates. Also, (1.224) shows that the drain current is independent of the channel length when the carrier velocity saturates. In this case, both the charge in the channel and the time required for the charge to cross the channel are proportional to L. Since the current is the ratio of the charge in the channel to the time required to cross the channel, the current does not depend on L as long as the channel length is short enough to produce an electric field that is high enough for velocity saturation to occur.[29] In contrast, when the carrier velocity is proportional to the electric field instead of being saturated, the time required for channel charge to cross the channel is proportional to L^2 because increasing L both reduces the carrier velocity and increases the distance between the source and the drain. Therefore, when velocity saturation is not significant, the drain current is inversely proportional to L, as we have come to expect through (1.157). Finally, (1.224) shows that the drain current in the active region is proportional to the scattering-limited velocity $v_{scl} = \mu_n \mathscr{E}_c$ when the velocity is saturated.

Substituting (1.222) into (1.223) gives

$$\begin{aligned} I_D &= \frac{\mu_n C_{ox}}{2} \frac{W}{L} (V_{GS} - V_t)^2 \left(1 - \frac{V_{GS} - V_t}{2 \mathscr{E}_c L} + \cdots \right)^2 \\ &= \frac{\mu_n C_{ox}}{2} \frac{W}{L} (V_{GS} - V_t)^2 \left(1 - \frac{x}{2} + \cdots \right)^2 \\ &= \frac{\mu_n C_{ox}}{2} \frac{W}{L} (V_{GS} - V_t)^2 (1 - x + \cdots) \\ &= \frac{\mu_n C_{ox}}{2} \frac{W}{L} (V_{GS} - V_t)^2 \left(1 - \frac{V_{GS} - V_t}{\mathscr{E}_c L} + \cdots \right) \end{aligned} \qquad (1.225)$$

where $x = (V_{GS} - V_t)/(\mathscr{E}_c L)$ as defined for (1.221). If $x \ll 1$, $(1 - x) \simeq 1/(1 + x)$, and

$$I_D \simeq \frac{\mu_n C_{ox}}{2 \left(1 + \dfrac{V_{GS} - V_t}{\mathscr{E}_c L}\right)} \frac{W}{L} (V_{GS} - V_t)^2 \qquad (1.226)$$

Equation 1.226 is valid without velocity saturation and at its onset, where $(V_{GS} - V_t) \ll \mathscr{E}_c L$. The effect of velocity saturation on the current in the active region predicted by (1.226) can be modeled with the addition of a resistance in series with the source of an ideal square-law device, as shown in Fig. 1.41. Let V'_{GS} be the gate-source voltage of the ideal square-law

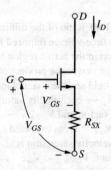

Figure 1.41 Model of velocity saturation in an MOSFET by addition of series source resistance to an ideal square-law device.

transistor. From (1.157),

$$I_D = \frac{\mu_n C_{ox}}{2} \frac{W}{L}(V'_{GS} - V_t)^2 \qquad (1.227)$$

Let V_{GS} be the sum of V'_{GS} and the voltage drop on R_{SX}. Then

$$V_{GS} = V'_{GS} + I_D R_{SX} \qquad (1.228)$$

This sum models the gate-source voltage of a real MOS transistor with velocity saturation. Substituting (1.228) into (1.227) gives

$$I_D = \frac{\mu_n C_{ox}}{2} \frac{W}{L}(V_{GS} - I_D R_{SX} - V_t)^2$$

$$I_D = \frac{\mu_n C_{ox}}{2} \frac{W}{L}\left((V_{GS} - V_t)^2 - 2(V_{GS} - V_t)I_D R_{SX} + (I_D R_{SX})^2\right) \qquad (1.229)$$

Rearranging (1.229) while ignoring the $(I_D R_{SX})^2$ term gives

$$I_D \simeq \frac{\mu_n C_{ox}}{2\left(1 + \mu_n C_{ox}\frac{W}{L} R_{SX}(V_{GS} - V_t)\right)} \frac{W}{L}(V_{GS} - V_t)^2 \qquad (1.230)$$

Equation 1.230 has the same form as (1.226) if we identify

$$\mu_n C_{ox} \frac{W}{L} R_{SX} = \frac{1}{\mathscr{E}_c L} \qquad (1.231)$$

Rearranging (1.231) gives

$$R_{SX} = \frac{1}{\mathscr{E}_c \mu_n C_{ox} W} \qquad (1.232)$$

Thus the influence of velocity saturation on the large-signal characteristics of an MOS transistor can be modeled to first order by a resistor R_{SX} in series with the source of an ideal square-law device. Note that R_{SX} varies inversely with W, as does the intrinsic physical series resistance due to the source and drain contact regions. Typically, R_{SX} is larger than the physical series resistance. For $W = 2$ µm, $k' = \mu_n C_{ox} = 200$ µA/V^2, and $\mathscr{E}_c = 1.5 \times 10^6$ V/m, we find $R_{SX} \simeq 1700\ \Omega$.

1.7.2 Transconductance and Transition Frequency

The values of all small-signal parameters can change significantly in the presence of short-channel effects.[30] One of the most important changes is to the transconductance. Substituting (1.220) into (1.223) and calculating $\partial I_D / \partial V_{GS}$ gives

$$g_m = \frac{\partial I_D}{\partial V_{GS}} = W C_{ox} v_{scl} \frac{\sqrt{1 + \frac{2(V_{GS} - V_t)}{\mathscr{E}_c L}} - 1}{\sqrt{1 + \frac{2(V_{GS} - V_t)}{\mathscr{E}_c L}}} \qquad (1.233)$$

where $v_{scl} = \mu_n \mathscr{E}_c$ as in Fig. 1.39. To determine g_m without velocity saturation, let $E_c \to \infty$ and $x = (V_{GS} - V_t)/(\mathscr{E}_c L)$. Then substituting (1.221) into (1.233) and rearranging gives

$$\lim_{\mathscr{E}_c \to \infty} g_m = k' \frac{W}{L}(V_{GS} - V_t) \qquad (1.234)$$

as predicted by (1.180). In this case, the transconductance increases when the overdrive increases or the channel length decreases. On the other hand, letting $\mathscr{E}_c \to 0$ to determine g_m when the velocity is saturated gives

$$\lim_{\mathscr{E}_c \to 0} g_m = W C_{ox} v_{scl} \tag{1.235}$$

Equation 1.235 shows that further decreases in L or increases in $(V_{GS} - V_t)$ do not change the transconductance when the velocity is saturated.

From (1.223) and (1.233), the ratio of the transconductance to the current can be calculated as

$$\frac{g_m}{I} = \frac{2}{(\mathscr{E}_c L)\sqrt{1 + \frac{2(V_{GS} - V_t)}{\mathscr{E}_c L}} \left(\sqrt{1 + \frac{2(V_{GS} - V_t)}{\mathscr{E}_c L}} - 1 \right)} \tag{1.236}$$

As $\mathscr{E}_c \to 0$, the velocity saturates and

$$\lim_{\mathscr{E}_c \to 0} \frac{g_m}{I} = \frac{1}{V_{GS} - V_t} \tag{1.237}$$

Comparing (1.237) to (1.181) shows that velocity saturation reduces the transconductance-to-current ratio for a given overdrive.

On the other hand, when $x = (V_{GS} - V_t)/(\mathscr{E}_c L) \ll 1$, substituting (1.221) into (1.236) gives

$$\frac{g_m}{I} \simeq \frac{2}{(V_{GS} - V_t)(1 + x)} \tag{1.238}$$

Therefore, as $\mathscr{E}_c \to \infty$, $x \to 0$, and (1.238) collapses to

$$\lim_{\mathscr{E}_c \to \infty} \frac{g_m}{I} = \frac{2}{V_{GS} - V_t} \tag{1.239}$$

as predicted by (1.181). Equation 1.238 shows that if $x < 0.1$, the error in using (1.181) to calculate the transconductance-to-current ratio is less than about 10 percent. Therefore, we will conclude that velocity-saturation effects are insignificant in hand calculations if

$$(V_{GS} - V_t) < 0.1(\mathscr{E}_c L) \tag{1.240}$$

Figure 1.42 plots the transconductance-to-current ratio versus the overdrive for three cases. The highest and lowest ratios come from (1.239) and (1.237), which correspond to asymptotes where velocity saturation is insignificant and dominant, respectively. In practice, the transition between these extreme cases is gradual and described by (1.236), which is plotted in Fig. 1.42 for an example where $\mathscr{E}_c = 1.5 \times 10^6$ V/m and $L = 0.5$ μm.

One reason the change in transconductance caused by velocity saturation is important is because it affects the transition frequency f_T. Assuming that $C_{gs} \gg C_{gb} + C_{gd}$, substituting (1.235) into (1.208) shows that

$$f_T = \frac{1}{2\pi} \frac{g_m}{C_{gs}} \propto \frac{W C_{ox} v_{scl}}{W L C_{ox}} \propto \frac{v_{scl}}{L} \tag{1.241}$$

One key point here is that the transition frequency is independent of the overdrive once velocity saturation is reached. In contrast, (1.209) shows that increasing $(V_{GS} - V_t)$ increases f_T before the velocity saturates. Also, (1.241) shows that the transition frequency is inversely proportional to the channel length when the velocity is saturated. In contrast, (1.209) predicts that f_T is

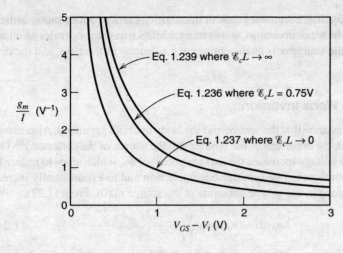

Figure 1.42
Transconductance-to-current ratio versus overdrive $(V_{GS} - V_t)$ where velocity saturation is insignificant $(\mathscr{E}_c L \to \infty)$, dominant $(\mathscr{E}_c L = 0)$, and of gradually increasing importance $(\mathscr{E}_c L = 0.75 \text{ V})$.

inversely proportional to the square of the channel length before the velocity saturates. As a result, velocity saturation reduces the speed improvement that can be achieved through reductions in the minimum channel length.

1.7.3 Mobility Degradation from the Vertical Field

Thus far, we have considered only the effects of the horizontal field due to the V_{DS} along the channel when considering velocity saturation. However, a vertical field originating from the gate voltage also exists and influences carrier velocity. A physical reason for this effect is that increasing the vertical electric field forces the carriers in the channel closer to the surface of the silicon, where surface imperfections impede their movement from the source to the drain, reducing mobility.[31] The vertical field at any point in the channel depends on the gate-channel voltage. Since the gate-channel voltage is not constant from the source to the drain, the effect of the vertical field on mobility should be included within the integration in (1.214) in principle.[32] For simplicity, however, this effect is often modeled after integration by changing the mobility in the previous equations to an effective mobility given by

$$\mu_{\text{eff}} = \frac{\mu_n}{1 + \theta(V_{GS} - V_t)} \quad (1.242)$$

where μ_n is the mobility with zero vertical field, and θ is inversely proportional to the oxide thickness. For $t_{ox} = 100 \text{ Å}$, θ is typically in the range from 0.1 V^{-1} to 0.4 V^{-1}.[33] In practice, θ is determined by a best fit to measured device characteristics.

1.8 Weak Inversion in MOS Transistors

The MOSFET analysis of Section 1.5 considered the normal region of operation for which a well-defined conducting channel exists under the gate. In this region of *strong inversion*, changes in the gate-source voltage are assumed to cause only changes in the channel charge and not in the depletion-region charge. In contrast, for gate-source voltages less than the extrapolated threshold voltage V_t but high enough to create a depletion region at the surface of the silicon, the device operates in *weak inversion*. In the weak-inversion region, the channel charge is much less than the charge in the depletion region, and the drain current arising from the drift of majority carriers is negligible. However, the total drain current in weak inversion is larger than that caused by drift because a gradient in minority-carrier concentration causes a diffusion current to flow. In weak inversion, an n-channel MOS transistor operates as an npn bipolar transistor, where the source acts as the emitter, the substrate as the base, and the drain as the collector.[34]

1.8.1 Drain Current in Weak Inversion

To analyze this situation, assume that the source and the body are both grounded. Also assume that $V_{DS} > 0$. (If $V_{DS} < 0$, the drain acts as the emitter and the source as the collector.)[35] Then increasing the gate-source voltage increases the surface potential ψ_s, which tends to reduce the reverse bias across the source-substrate (emitter-base) junction and to exponentially increase the concentration of electrons in the p-type substrate at the source $n_p(0)$. From (1.27),

$$n_p(0) = n_{po} \exp \frac{\psi_s}{V_T} \tag{1.243}$$

where n_{po} is the equilibrium concentration of electrons in the substrate (base). Similarly, the concentration of electrons in the substrate at the drain $n_p(L)$ is

$$n_p(L) = n_{po} \exp \frac{\psi_s - V_{DS}}{V_T} \tag{1.244}$$

From (1.31), the drain current due to the diffusion of electrons in the substrate is

$$I_D = qAD_n \frac{n_p(L) - n_p(0)}{L} \tag{1.245}$$

where D_n is the diffusion constant for electrons, and A is the cross-sectional area in which the diffusion current flows. The area A is the product of the transistor width W and the thickness X of the region in which I_D flows. Substituting (1.243) and (1.244) into (1.245) and rearranging gives

$$I_D = \frac{W}{L} q X D_n n_{po} \exp\left(\frac{\psi_s}{V_T}\right)\left[1 - \exp\left(-\frac{V_{DS}}{V_T}\right)\right] \tag{1.246}$$

In weak inversion, the surface potential is approximately a linear function of the gate-source voltage.[36] Assume that the charge stored at the oxide-silicon interface is independent of the surface potential. Then, in weak inversion, changes in the surface potential $\Delta\psi_s$ are controlled by changes in the gate-source voltage ΔV_{GS} through a voltage divider between the oxide capacitance C_{ox} and the depletion-region capacitance C_{js}. Therefore,

$$\frac{d\psi_s}{dV_{GS}} = \frac{C_{ox}}{C_{js} + C_{ox}} = \frac{1}{n} = \frac{1}{1+\chi} \tag{1.247}$$

in which $n = (1 + C_{js}/C_{ox})$ and $\chi = C_{js}/C_{ox}$, as defined in (1.197). Separating variables in (1.247) and integrating gives

$$\psi_s = \frac{V_{GS}}{n} + k_1 \tag{1.248}$$

where k_1 is a constant. Equation 1.248 is valid only when the transistor operates in weak inversion. When $V_{GS} = V_t$ with $V_{SB} = 0$, $\psi_s = 2\phi_f$ by definition of the threshold voltage. For $V_{GS} > V_t$, the inversion layer holds the surface potential nearly constant and (1.248) is not valid. Since (1.248) is valid only when $V_{GS} \leq V_t$, (1.248) is rewritten as follows:

$$\psi_s = \frac{V_{GS} - V_t}{n} + k_2 \tag{1.249}$$

where $k_2 = k_1 + V_t/n$. Substituting (1.249) into (1.246) gives

$$I_D = \frac{W}{L} q X D_n n_{po} \exp\left(\frac{k_2}{V_T}\right) \exp\left(\frac{V_{GS} - V_t}{nV_T}\right) \left[1 - \exp\left(-\frac{V_{DS}}{V_T}\right)\right] \quad (1.250)$$

Let

$$I_t = q X D_n n_{po} \exp\left(\frac{k_2}{V_T}\right) \quad (1.251)$$

represent the drain current with $V_{GS} = V_t$, $W/L = 1$, and $V_{DS} \gg V_T$. Then

$$I_D = \frac{W}{L} I_t \exp\left(\frac{V_{GS} - V_t}{nV_T}\right) \left[1 - \exp\left(-\frac{V_{DS}}{V_T}\right)\right] \quad (1.252)$$

Figure 1.43 plots the drain current versus the drain-source voltage for three values of the overdrive, with $W = 20$ μm, $L = 20$ μm, $n = 1.5$, and $I_t = 0.1$ μA. Notice that the drain current is almost constant when $V_{DS} > 3V_T$ because the last term in (1.252) approaches unity in this case. Therefore, unlike in strong inversion, the minimum drain-source voltage required to force the transistor to operate as a current source in weak inversion is independent of the overdrive.[37] Figure 1.43 and Equation 1.252 also show that the drain current is not zero when $V_{GS} \leq V_t$. To further illustrate this point, we show measured NMOS characteristics plotted on two different scales in Fig. 1.44. In Fig. 1.44a, we show $\sqrt{I_D}$ versus V_{GS} in the active region plotted on linear scales. For this device, $W = 20$ μm, $L = 20$ μm, and short-channel effects are negligible. (See Problem 1.21 for an example of a case in which short-channel effects are important.) The resulting straight line shows that the device characteristic is close to an ideal square law. Plots like the one in Fig. 1.44a are commonly used to obtain V_t by extrapolation (0.7 V in this case) and also k' from the slope of the curve (54 μA/V^2 in this case). Near the threshold voltage, the curve deviates from the straight line representing the square law. This region is weak inversion. The data are plotted a second time in Fig. 1.44b on log-linear scales. The straight line obtained for $V_{GS} < V_t$ fits (1.252) with $n = 1.5$. For $I_D < 10^{-12}$ A, the slope decreases because leakage currents are significant and do not follow (1.252).

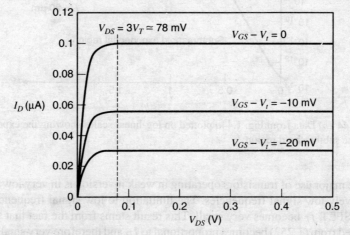

Figure 1.43 Drain current versus drain-source voltage in weak inversion.

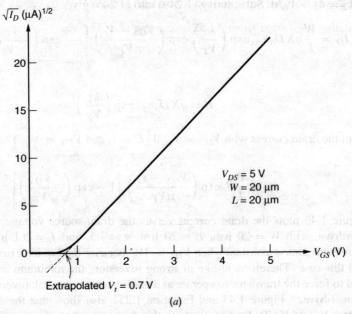

Figure 1.44 (a) Measured NMOS transfer characteristic in the active region plotted on linear scales as $\sqrt{I_D}$ versus V_{GS}, showing the square-law characteristic.

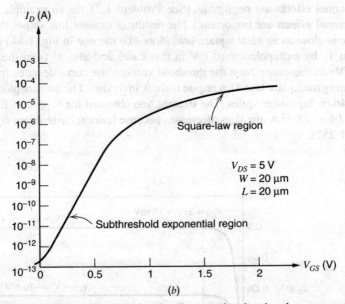

Figure 1.44 (b) Data from Fig. 1.44a plotted on log-linear scales showing the exponential characteristic in the subthreshold region.

The major use of transistors operating in weak inversion is in very low power applications at relatively low signal frequencies. The limitation to low signal frequencies occurs because the MOSFET f_T becomes very small. This result stems from the fact that the small-signal g_m calculated from (1.252) becomes proportional to I_D and therefore very small in weak inversion, as shown next.

1.8.2 Transconductance and Transition Frequency in Weak Inversion

Calculating $\partial I_D/\partial V_{GS}$ from (1.252) and using (1.247) gives

$$g_m = \frac{W}{L}\frac{I_t}{nV_T}\exp\left(\frac{V_{GS}-V_t}{nV_T}\right)\left[1-\exp\left(-\frac{V_{DS}}{V_T}\right)\right] = \frac{I_D}{nV_T} = \frac{I_D}{V_T}\frac{C_{ox}}{C_{js}+C_{ox}} \quad (1.253)$$

The transconductance of an MOS transistor operating in weak inversion is identical to that of a corresponding bipolar transistor, as shown in (1.182), except for the factor of $1/n = C_{ox}/(C_{js}+C_{ox})$. This factor stems from a voltage divider between the oxide and depletion capacitors in the MOS transistor, which models the indirect control of the gate on the surface potential.

From (1.253), the ratio of the transconductance to the current of an MOS transistor in weak inversion is

$$\frac{g_m}{I} = \frac{1}{nV_T} = \frac{1}{V_T}\frac{C_{ox}}{C_{js}+C_{ox}} \quad (1.254)$$

Equation 1.254 predicts that this ratio is independent of the overdrive. In contrast, (1.181) predicts that the ratio of transconductance to current is inversely proportional to the overdrive. Therefore, as the overdrive approaches zero, (1.181) predicts that this ratio becomes infinite. However, (1.181) is valid only when the transistor operates in strong inversion. To estimate the overdrive required to operate the transistor in strong inversion, we will equate the g_m/I ratios calculated in (1.254) and (1.181). The result is

$$V_{ov} = V_{GS} - V_t = 2nV_T \quad (1.255)$$

which is about 78 mV at room temperature with $n = 1.5$. Although this analysis implies that the transition from weak to strong inversion occurs abruptly, a nonzero transition width occurs in practice. Between weak and strong inversion, the transistor operates in a region of *moderate* inversion, where both diffusion and drift currents are significant.[38]

Figure 1.45 plots the transconductance-to-current ratio versus overdrive for an example case with $n = 1.5$. When the overdrive is negative but high enough to cause depletion at the surface, the transistor operates in weak inversion and the transconductance-to-current ratio is constant, as predicted by (1.254). When $V_{GS} - V_t = 0$, the surface potential is $2\psi_f$, which means that the surface concentration of electrons is equal to the bulk concentration of holes. This point is usually defined as the upper bound on the region of weak inversion. When

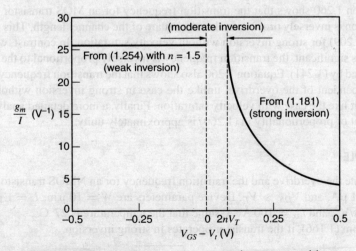

Figure 1.45 Transconductance-to-current ratio versus overdrive.

$V_{GS} - V_t > 2nV_T$, the transconductance-to-current ratio is given by (1.181), assuming that velocity saturation is negligible. If velocity saturation is significant, (1.236) should be used instead of (1.181) both to predict the transconductance-to-current ratio and to predict the overdrive required to operate in strong inversion. For $0 \leq V_{GS} - V_t \leq 2nV_T$, the transistor operates in moderate inversion. Because simple models for moderate inversion are not known in practice, we will ignore this region in the remainder of this book and assume that MOS transistors operate in weak inversion for overdrives less than the bound given in (1.255).

Equation 1.208 can be used to find the transition frequency. In weak inversion, $C_{gs} \simeq C_{gd} \simeq 0$ because the inversion layer contains little charge.[39] However, C_{gb} can be thought of as the series combination of the oxide and depletion capacitors. Therefore,

$$C_{gs} + C_{gb} + C_{gd} \simeq C_{gb} = WL \left(\frac{C_{ox} C_{js}}{C_{ox} + C_{js}} \right) \quad (1.256)$$

Substituting (1.253) and (1.256) into (1.208) gives

$$f_T = \frac{1}{2\pi} \omega_T = \frac{1}{2\pi} \frac{\frac{I_D}{V_T} \frac{C_{ox}}{C_{js} + C_{ox}}}{WL \frac{C_{ox} C_{js}}{C_{ox} + C_{js}}} = \frac{1}{2\pi} \frac{I_D}{V_T} \frac{1}{WLC_{js}} \quad (1.257)$$

Let I_M represent the maximum drain current that flows in the transistor in weak inversion. Then

$$I_M = \frac{W}{L} I_t \quad (1.258)$$

where I_t is given in (1.251). Multiplying numerator and denominator in (1.257) by I_M and using (1.258) gives

$$f_T = \frac{1}{2\pi} \frac{\frac{W}{L} I_t}{V_T} \frac{1}{WLC_{js}} \frac{I_D}{I_M} = \frac{1}{2\pi} \frac{I_t}{V_T} \frac{1}{C_{js}} \frac{1}{L^2} \frac{I_D}{I_M} \quad (1.259)$$

From (1.251), $I_t \propto D_n$. Using the Einstein relationship $D_n = \mu_n V_T$ gives

$$f_T \propto \frac{D_n}{L^2} \frac{I_D}{I_M} \propto \frac{\mu_n V_T}{L^2} \frac{I_D}{I_M} \quad (1.260)$$

Equation 1.260 shows that the transition frequency for an MOS transistor operating in weak inversion is inversely proportional to the square of the channel length. This result is consistent with (1.209) for strong inversion without velocity saturation. In contrast, when velocity saturation is significant, the transition frequency is inversely proportional to the channel length, as predicted by (1.241). Equation 1.260 also shows that the transition frequency in weak inversion is independent of the overdrive, unlike the case in strong inversion without velocity saturation, but like the case with velocity saturation. Finally, a more detailed analysis shows that the constant of proportionality in (1.260) is approximately unity.[39]

■ **EXAMPLE**

Calculate the overdrive and the transition frequency for an NMOS transistor with $I_D = 1$ μA, $I_t = 0.1$ μA, and $V_{DS} \gg V_T$. Device parameters are $W = 10$ μm, $L = 1$ μm, $n = 1.5$, $k' = 200$ μA/V², and $t_{ox} = 100$ Å. Assume that the temperature is 27°C.

From (1.166), if the transistor operates in strong inversion,

$$V_{ov} = V_{GS} - V_t = \sqrt{\frac{2I_D}{k'(W/L)}} = \sqrt{\frac{2 \times 1}{200 \times 10}} \simeq 32 \text{ mV}$$

Since the value of the overdrive calculated by (1.166) is less than $2nV_T \simeq 78$ mV, the overdrive calculated previously is not valid except to indicate that the transistor does not operate in strong inversion. From (1.252), the overdrive in weak inversion with $V_{DS} \gg V_T$ is

$$V_{ov} = nV_T \ln\left(\frac{I_D}{I_t}\frac{L}{W}\right) = (1.5)(26 \text{ mV})\ln\left(\frac{1}{0.1}\frac{1}{10}\right) = 0$$

From (1.253),

$$g_m = \frac{1 \text{ μA}}{1.5(26 \text{ mV})} \simeq 26 \frac{\text{μA}}{\text{V}}$$

From (1.247),

$$C_{js} = (n-1)C_{ox} = (0.5)C_{ox}$$

From (1.256),

$$C_{gs} + C_{gb} + C_{gd} \simeq C_{gb} = WL\frac{C_{ox}(0.5C_{ox})}{C_{ox} + 0.5C_{ox}} = WL\frac{C_{ox}}{3}$$

$$= \frac{10 \text{ μm}^2}{3}\frac{3.9 \times 8.854 \times 10^{-14}\frac{\text{F}}{\text{cm}} \times \frac{100 \text{ cm}}{10^6 \text{μm}}}{100 \text{ Å} \times \frac{10^6 \text{ μm}}{10^{10}\text{Å}}}$$

$$\simeq 11.5 \text{ fF}$$

From (1.208),

$$f_T = \frac{1}{2\pi}\omega_T = \frac{1}{2\pi}\frac{26 \text{ μA/V}}{11.5 \text{ fF}} \simeq 360 \text{ MHz}$$

Although 360 MHz may seem to be a high transition frequency at first glance, this result should be compared with the result of the example at the end of Section 1.6, where the same transistor operating in strong inversion with an overdrive of 316 mV had a transition frequency of 3.4 GHz.

■

1.9 Substrate Current Flow in MOS Transistors

In Section 1.3.4, the effects of avalanche breakdown on bipolar transistor characteristics were described. As the reverse-bias voltages on the device are increased, carriers traversing the depletion regions gain sufficient energy to create new electron-hole pairs in lattice collisions by a process known as *impact ionization*. Eventually, at sufficient bias voltages, the process results in large avalanche currents. For collector-base bias voltages well below the breakdown value, a small enhanced current flow may occur across the collector-base junction due to this process, with little apparent effect on the device characteristics.

Impact ionization also occurs in MOS transistors but has a significantly different effect on the device characteristics. This difference is because the channel electrons (for the NMOS case) create electron-hole pairs in lattice collisions in the drain depletion region, and some of the resulting holes then flow to the substrate, creating a substrate current. (The electrons created in the process flow out the drain terminal.) The carriers created by impact ionization are therefore not confined within the device as in a bipolar transistor. The effect of this phenomenon can be modeled by inclusion of a controlled current generator I_{DB} from drain to substrate, as shown

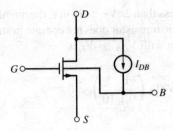

Figure 1.46 Representation of impact ionization in an MOSFET by a drain-substrate current generator.

in Fig. 1.46 for an NMOS device. The magnitude of this substrate current depends on the voltage across the drain depletion region (which determines the energy of the ionizing channel electrons) and also on the drain current (which is the rate at which the channel electrons enter the depletion region). Empirical investigation has shown that the current I_{DB} can be expressed as

$$I_{DB} = K_1(V_{DS} - V_{DS(\text{act})})I_D \exp\left(-\frac{K_2}{V_{DS} - V_{DS(\text{act})}}\right) \quad (1.261)$$

where K_1 and K_2 are process-dependent parameters and $V_{DS(\text{act})}$ is the minimum value of V_{DS} for which the transistor operates in the active region.[40] Typical values for NMOS devices are $K_1 = 5$ V^{-1} and $K_2 = 30$ V. The effect is generally much less significant in PMOS devices because the holes carrying the charge in the channel are much less efficient in creating electron-hole pairs than energetic electrons.

The major impact of this phenomenon on circuit performance is that it creates a parasitic resistance from drain to substrate. Because the common substrate terminal must always be connected to the most negative supply voltage in the circuit, the substrate of an NMOS device in a p-substrate process is an ac ground. Therefore, the parasitic resistance shunts the drain to ac ground and can be a limiting factor in many circuit designs. Differentiating (1.261), we find that the drain-substrate small-signal conductance is

$$g_{db} = \frac{\partial I_{DB}}{\partial V_D} = \frac{I_{DB}}{V_{DS} - V_{DS(\text{act})}}\left(\frac{K_2}{V_{DS} - V_{DS(\text{act})}} + 1\right) \simeq \frac{K_2 I_{DB}}{(V_{DS} - V_{DS(\text{act})})^2} \quad (1.262)$$

where the gate and the source are assumed to be held at fixed potentials.

■ **EXAMPLE**

Calculate $r_{db} = 1/g_{db}$ for $V_{DS} = 2$ V and 4 V, and compare with the device r_o. Assume $I_D = 100$ μA, $\lambda = 0.05$ V^{-1}, $V_{DS(\text{act})} = 0.3$ V, $K_1 = 5$ V^{-1}, and $K_2 = 30$ V.

For $V_{DS} = 2$ V, we have from (1.261)

$$I_{DB} = 5 \times 1.7 \times 100 \times 10^{-6} \times \exp\left(-\frac{30}{1.7}\right) \simeq 1.8 \times 10^{-11} \text{ A}$$

From (1.262),

$$g_{db} \simeq \frac{30 \times 1.8 \times 10^{-11}}{1.7^2} \simeq 1.9 \times 10^{-10} \frac{\text{A}}{\text{V}}$$

and thus

$$r_{db} = \frac{1}{g_{db}} \simeq 5.3 \times 10^9 \text{ } \Omega = 5.3 \text{ G}\Omega$$

This result is negligibly large compared with

$$r_o = \frac{1}{\lambda I_D} = \frac{1}{0.05 \times 100 \times 10^{-6}} = 200 \text{ k}\Omega$$

However, for $V_{DS} = 4$ V,

$$I_{DB} = 5 \times 3.7 \times 100 \times 10^{-6} \times \exp\left(-\frac{30}{3.7}\right) \simeq 5.6 \times 10^{-7} \text{ A}$$

The substrate leakage current is now about 0.5 percent of the drain current. More important, we find from (1.262)

$$g_{db} \simeq \frac{30 \times 5.6 \times 10^{-7}}{3.7^2} \simeq 1.2 \times 10^{-6} \frac{\text{A}}{\text{V}}$$

and thus

$$r_{db} = \frac{1}{g_{db}} \simeq 8.15 \times 10^5 \ \Omega = 815 \text{ k}\Omega$$

This parasitic resistor is now comparable to r_o and can have a dominant effect on high-output-impedance MOS current mirrors, as described in Chapter 4.

APPENDIX

A.1.1 SUMMARY OF ACTIVE-DEVICE PARAMETERS

(a) *npn* Bipolar Transistor Parameters

Quantity	Formula
Large-Signal Forward-Active Operation	
Collector current	$I_c = I_S \exp \dfrac{V_{be}}{V_T}$
Small-Signal Forward-Active Operation	
Transconductance	$g_m = \dfrac{qI_C}{kT} = \dfrac{I_C}{V_T}$
Transconductance-to-current ratio	$\dfrac{g_m}{I_C} = \dfrac{1}{V_T}$
Input resistance	$r_\pi = \dfrac{\beta_0}{g_m}$
Output resistance	$r_o = \dfrac{V_A}{I_C} = \dfrac{1}{\eta g_m}$
Collector-base resistance	$r_\mu = \beta_0 r_o$ to $5\beta_0 r_o$
Base-charging capacitance	$C_b = \tau_F g_m$
Base-emitter capacitance	$C_\pi = C_b + C_{je}$
Emitter-base junction depletion capacitance	$C_{je} \simeq 2C_{je0}$

(continued)

Quantity	Formula
Small-Signal Forward-Active Operation	
Collector-base junction capacitance	$C_\mu = \dfrac{C_{\mu 0}}{\left(1 - \dfrac{V_{BC}}{\psi_{0c}}\right)^{n_c}}$
Collector-substrate junction capacitance	$C_{cs} = \dfrac{C_{cs0}}{\left(1 - \dfrac{V_{SC}}{\psi_{0s}}\right)^{n_s}}$
Transition frequency	$f_T = \dfrac{1}{2\pi}\dfrac{g_m}{C_\pi + C_\mu}$
Effective transit time	$\tau_T = \dfrac{1}{2\pi f_T} = \tau_F + \dfrac{C_{je}}{g_m} + \dfrac{C_\mu}{g_m}$
Maximum gain	$g_m r_o = \dfrac{V_A}{V_T} = \dfrac{1}{\eta}$

(b) NMOS Transistor Parameters

Quantity	Formula
Large-Signal Operation	
Drain current (active region)	$I_d = \dfrac{\mu C_{ox}}{2}\dfrac{W}{L}(V_{gs} - V_t)^2$
Drain current (triode region)	$I_d = \dfrac{\mu C_{ox}}{2}\dfrac{W}{L}[2(V_{gs} - V_t)V_{ds} - V_{ds}^2]$
Threshold voltage	$V_t = V_{t0} + \gamma\left[\sqrt{2\phi_f + V_{sb}} - \sqrt{2\phi_f}\right]$
Threshold voltage parameter	$\gamma = \dfrac{1}{C_{ox}}\sqrt{2q\epsilon N_A}$
Oxide capacitance	$C_{ox} = \dfrac{\epsilon_{ox}}{t_{ox}} = 3.45\ \text{fF}/\mu\text{m}^2$ for $t_{ox} = 100\ \text{Å}$
Small-Signal Operation (Active Region)	
Top-gate transconductance	$g_m = \mu C_{ox}\dfrac{W}{L}(V_{GS} - V_t) = \sqrt{2I_D \mu C_{ox}\dfrac{W}{L}}$
Transconductance-to-current ratio	$\dfrac{g_m}{I_D} = \dfrac{2}{V_{GS} - V_t}$
Body-effect transconductance	$g_{mb} = \dfrac{\gamma}{2\sqrt{2\phi_f + V_{SB}}}g_m = \chi g_m$
Channel-length modulation parameter	$\lambda = \dfrac{1}{V_A} = \dfrac{1}{L_{\text{eff}}}\dfrac{dX_d}{dV_{DS}}$

(continued)

Quantity	Formula
Small-Signal Operation (Active Region)	
Output resistance	$r_o = \dfrac{1}{\lambda I_D} = \dfrac{L_{\text{eff}}}{I_D}\left(\dfrac{dX_d}{dV_{DS}}\right)^{-1}$
Effective channel length	$L_{\text{eff}} = L_{\text{drwn}} - 2L_d - X_d$
Maximum gain	$g_m r_o = \dfrac{1}{\lambda}\dfrac{2}{V_{GS}-V_t} = \dfrac{2V_A}{V_{GS}-V_t}$
Source-body depletion capacitance	$C_{sb} = \dfrac{C_{sb0}}{\left(1+\dfrac{V_{SB}}{\psi_0}\right)^{0.5}}$
Drain-body depletion capacitance	$C_{db} = \dfrac{C_{db0}}{\left(1+\dfrac{V_{DB}}{\psi_0}\right)^{0.5}}$
Gate-source capacitance	$C_{gs} = \dfrac{2}{3}WLC_{ox}$
Transition frequency	$f_T = \dfrac{g_m}{2\pi(C_{gs}+C_{gd}+C_{gb})}$

PROBLEMS

1.1(a) Calculate the built-in potential, depletion-layer depths, and maximum field in a plane-abrupt pn junction in silicon with doping densities $N_A = 8 \times 10^{15}$ atoms/cm^3 and $N_D = 10^{17}$ atoms/cm^3. Assume a reverse bias of 5 V.

(b) Repeat (a) for zero external bias and 0.3 V forward bias.

1.2 Calculate the zero-bias junction capacitance for the example in Problem 1.1, and also calculate the value at 3 V reverse bias and 0.5 V forward bias. Assume a junction area of 2×10^{-5} cm^2.

1.3 Calculate the breakdown voltage for the junction of Problem 1.1 if the critical field is $\mathscr{E}_{\text{crit}} = 2 \times 10^5$ V/cm.

1.4 If junction curvature causes the maximum field at a practical junction to be 1.5 times the theoretical value, calculate the doping density required to give a breakdown voltage of 200 V with an abrupt pn junction in silicon. Assume that one side of the junction is much more heavily doped than the other and $\mathscr{E}_{\text{crit}} = 3 \times 10^5$ V/cm.

1.5 If the collector doping density in a transistor is 6×10^{15} atoms/cm^3, and is much less than the base doping, find BV_{CEO} for $\beta_F = 200$ and $n = 4$. Use $\mathscr{E}_{\text{crit}} = 2 \times 10^5$ V/cm.

1.6 Repeat Problem 1.5 for a doping density of 10^{15} atoms/cm^3 and $\beta_F = 500$.

1.7(a) Sketch the I_C-V_{CE} characteristics in the forward-active region for an npn transistor with $\beta_F = 100$ (measured at low V_{CE}), $V_A = 50$ V, $BV_{CBO} = 120$ V, and $n = 4$. Use

$$I_C = \left(1 + \dfrac{V_{CE}}{V_A}\right)\dfrac{M\alpha_F}{1-M\alpha_F}I_B$$

where M is given by (1.78). Plot I_C from 0 to 10 mA and V_{CE} from 0 to 50 V. Use $I_B = 1\,\mu\text{A}, 10\,\mu\text{A}, 30\,\mu\text{A}$, and $60\,\mu\text{A}$.

(b) Repeat (a), but sketch V_{CE} from 0 to 10 V.

1.8 Derive and sketch the complete small-signal equivalent circuit for a bipolar transistor at $I_C = 0.2$ mA, $V_{CB} = 3$ V, $V_{CS} = 4$ V. Device parameters are $C_{je0} = 20$ fF, $C_{\mu 0} = 10$ fF, $C_{cs0} = 20$ fF, $\beta_0 = 100$, $\tau_F = 15$ ps, $\eta = 10^{-3}$, $r_b = 200\,\Omega$, $r_c = 100\,\Omega$, $r_{ex} = 4\,\Omega$, and $r_\mu = 5\beta_0 r_o$. Assume $\psi_0 = 0.55$ V for all junctions.

1.9 Repeat Problem 1.8 for $I_C = 1$ mA, $V_{CB} = 1$ V, and $V_{CS} = 2$ V.

1.10 Sketch the graph of small-signal, common-emitter current gain versus frequency on log scales from 0.1 MHz to 1000 MHz for the examples of

Problems 1.8 and 1.9. Calculate the f_T of the device in each case.

1.11 A lateral *pnp* transistor has an effective base width of 10 μm (1 μm = 10^{-4} cm).

(a) If the emitter-base depletion capacitance is 2 pF in the forward-bias region and is constant, calculate the device f_T at $I_C = -0.5$ mA. (Neglect C_μ.) Also, calculate the minority-carrier charge stored in the base of the transistor at this current level. *Data*: $D_p = 13$ cm²/s in silicon.

(b) If the collector-base depletion layer width changes 0.2 μm per volt of V_{CE}, calculate r_o for this transistor at $I_C = -0.5$ mA.

1.12 An integrated-circuit *npn* transistor has the following measured characteristics: $r_b = 100\ \Omega$, $r_c = 100\ \Omega$, $\beta_0 = 100$, $r_o = 50\ \mathrm{k}\Omega$ at $I_C = 1$ mA, $f_T = 600$ MHz with $I_C = 1$ mA and $V_{CB} = 10$ V, $f_T = 1$ GHz with $I_C = 10$ mA and $V_{CB} = 10$ V, $C_\mu = 0.15$ pF with $V_{CB} = 10$ V, and $C_{cs} = 1$ pF with $V_{CS} = 10$ V. Assume $\psi_0 = 0.55$ V for all junctions, and assume C_{je} is constant in the forward-bias region. Use $r_\mu = 5\beta_0 r_o$.

(a) Form the *complete* small-signal equivalent circuit for this transistor at $I_C = 0.1$ mA, 1 mA, and 5 mA with $V_{CB} = 2$ V and $V_{CS} = 15$ V.

(b) Sketch the graph of f_T versus I_C for this transistor on log scales from 1 μA to 10 mA with $V_{CB} = 2$ V.

1.13 If the area of the transistor in Problem 1.12 is effectively doubled by connecting two transistors in parallel, which model parameters in the small-signal equivalent circuit of the composite transistor would differ from those of the original device if the *total* collector current is unchanged? What is the relationship between the parameters of the composite and original devices?

1.14 An integrated *npn* transistor has the following characteristics: $\tau_F = 0.25$ ns, small-signal, short-circuit current gain is 9 with $I_C = 1$ mA at $f = 50$ MHz, $V_A = 40$ V, $\beta_0 = 100$, $r_b = 150\ \Omega$, $r_c = 150\ \Omega$, $C_\mu = 0.6$ pF, $C_{cs} = 2$ pF at the bias voltage used. Determine all elements in the small-signal equivalent circuit at $I_C = 2$ mA and sketch the circuit.

1.15 An NMOS transistor has parameters $W = 10$ μm, $L = 1$ μm, $k' = 194$ μA/V², $\lambda = 0.024$ V^{-1}, $t_{ox} = 80$ Å, $\phi_f = 0.3$ V, $V_{t0} = 0.6$ V, and $N_A = 5 \times 10^{15}$ atoms/cm³. Ignore velocity saturation effects.

(a) Sketch the I_D-V_{DS} characteristics for V_{DS} from 0 to 3 V and $V_{GS} = 0.5$ V, 1.5 V, and 3 V. Assume $V_{SB} = 0$.

(b) Sketch the I_D-V_{GS} characteristics for $V_{DS} = 2$ V as V_{GS} varies from 0 to 2 V with $V_{SB} = 0$, 0.5 V, and 1 V.

1.16 Derive and sketch the complete small-signal equivalent circuit for the device of Problem 1.15 with $V_{GS} = 1$ V, $V_{DS} = 2$ V, and $V_{SB} = 1$ V. Use $\psi_0 = 0.7$ V, $C_{sb0} = C_{db0} = 20$ fF, and $C_{gb} = 5$ fF. Overlap capacitance from gate to source and gate to drain is 2 fF.

1.17 Use the device data of Problems 1.15 and 1.16 to calculate the frequency of unity current gain of this transistor with $V_{DS} = 3$ V, $V_{SB} = 0$ V, $V_{GS} = 1$ V, 1.5 V, and 2 V.

1.18 Consider an NMOS transistor with $W = 2$ μm, $L = 0.5$ μm, $k' = 194$ μA/V², $\lambda = 0$, $V_{t0} = 0.6$ V, and $\mathscr{E}_c = 1.5 \times 10^6$ V/m. Compare the drain current predicted by the model of Fig. 1.41 to the drain current predicted by direct calculation using the equations including velocity saturation for V_{GS} from 0 to 3 V. Assume $V_{DS} = 3$ V and $V_{SB} = 0$. For what range of V_{GS} is the model of Fig. 1.41 accurate within 10 percent?

1.19 Examine the effect of velocity saturation on MOSFET characteristics by plotting $I_D - V_{DS}$ curves for $V_{GS} = 1$ V, 2 V, and 3 V, and $V_{DS} = 0$ to 3 V in the following cases, and by comparing the results with and without inclusion of velocity saturation effects. Assume $V_{SB} = 0$, $V_{t0} = 0.6$ V, $k' = 194$ μA/V², $\lambda = 0$, and $\mathscr{E}_c = 1.5 \times 10^6$ V/m.

(a) $W = 100$ μm and $L = 10$ μm.

(b) $W = 10$ μm and $L = 1$ μm.

(c) $W = 5$ μm and $L = 0.5$ μm.

1.20 Plot $\sqrt{I_D}$ versus V_{GS} for an *n*-channel MOSFET with $W = 1$ μm, $L = 1$ μm, $k' = 54$ μA/V², $\lambda = 0$, $V_{DS} = 5$ V, $V_{SB} = 0$, $V_{t0} = 0.7$ V, and $\mathscr{E}_c = 1.5 \times 10^6$ V/m. Ignore subthreshold conduction. Compare the plot with Fig. 1.44a and explain the main difference for large V_{GS}.

1.21 Calculate the transconductance of an *n*-channel MOSFET with $W = 10$ μm, $\mu_n = 450$ cm²/(V-s), and $\mathscr{E}_c = 1.5 \times 10^6$ V/m using channel lengths from 10 μm to 0.4 μm. Assume that $t_{ox} = L/50$ and that the device operates in the active region with $V_{GS} - V_t = 0.1$ V. Compare the result to a calculation that ignores velocity saturation. For what range of channel lengths is the model without velocity saturation accurate within 10 percent?

1.22 Calculate the transconductance of an *n*-channel MOSFET at $I_D = 10$ nA and $V_{DS} = 1$ V, assuming subthreshold operation and $n = 1.5$. Assuming $(C_{gs} + C_{gd} + C_{gb}) = 10$ fF, calculate the corresponding device f_T.

REFERENCES

1. P. E. Gray, D. DeWitt, A. R. Boothroyd, J. F. Gibbons. *Physical Electronics and Circuit Models of Transistors.* Wiley, New York, 1964, p. 20.

2. H. C. Poon and H. K. Gummel. "Modeling of Emitter Capacitance," *Proc. IEEE,* Vol. 57, pp. 2181–2182, December 1969.

3. B. R. Chawla and H. K. Gummel. "Transition Region Capacitance of Diffused pn Junctions," *IEEE Trans. Electron Devices,* Vol. ED-18, pp. 178–195, March 1971.

4. S. L. Miller. "Avalanche Breakdown in Germanium," *Phys. Rev.,* Vol. 99, p. 1234, 1955.

5. A. S. Grove. *Physics and Technology of Semiconductor Devices.* Wiley, New York, 1967, Ch. 6.

6. A. S. Grove. Op. cit., Ch. 4.

7. A. S. Grove. Op. cit., Ch. 7.

8. P. E. Gray et al. Op. cit., p. 10.

9. P. E. Gray et al. Op. cit., p. 129.

10. P. E. Gray et al. Op. cit., p. 180.

11. B. A. McDonald. "Avalanche Degradation of h_{FE}," *IEEE Trans. Electron Devices,* Vol. ED-17, pp. 871–878, October 1970.

12. H. DeMan. "The Influence of Heavy Doping on the Emitter Efficiency of a Bipolar Transistor," *IEEE Trans. Electron Devices,* Vol. ED-18, pp. 833–835, October 1971.

13. R. J. Whittier and D. A. Tremere. "Current Gain and Cutoff Frequency Falloff at High Currents," *IEEE Trans. Electron Devices,* Vol. ED-16, pp. 39–57, January 1969.

14. J. L. Moll and I. M. Ross. "The Dependence of Transistor Parameters on the Distribution of Base Layer Resistivity," *Proc. IRE,* Vol. 44, p. 72, 1956.

15. P. E. Gray et al. Op. cit., Ch. 8.

16. R. S. Muller and T. I. Kamins. *Device Electronics for Integrated Circuits.* Second Edition, Wiley, New York, 1986, p. 386.

17. Y. P. Tsividis. *Operation and Modeling of the MOS Transistor.* McGraw-Hill, New York, 1987, p. 141.

18. D. Frohman-Bentchkowsky and A. S. Grove. "Conductance of MOS Transistors in Saturation," *IEEE Trans. Electron Devices,* Vol. ED-16, pp. 108–113, January 1969.

19. S. M. Sze. *Physics of Semiconductor Devices.* Second Edition, Wiley, New York, 1981, pp. 451–452.

20. R. S. Muller and T. I. Kamins. Op. cit., p. 17.

21. Y. P. Tsividis. Op. cit., p. 148.

22. R. S. Muller and T. I. Kamins. Op. cit., pp. 490–496.

23. Y. P. Tsividis. Op. cit., pp. 150–151 and 198–200.

24. R. S. Muller and T. I. Kamins. Op. cit., p. 496.

25. Y. P. Tsividis. Op. cit., p. 151.

26. Y. P. Tsividis. Op. cit., pp. 310–328.

27. R. S. Muller and T. I. Kamins. Op. cit., p. 480.

28. R. S. Muller and T. I. Kamins. Op. cit., p. 482.

29. Y. P. Tsividis. Op. cit., p. 181.

30. Y. P. Tsividis. Op. cit., p. 294.

31. Y. P. Tsividis. Op. cit., p. 142.

32. R. S. Muller and T. I. Kamins. Op. cit., p. 484.

33. Y. P. Tsividis. Op. cit., p. 146.

34. S. M. Sze. Op. cit., p. 446.

35. Y. P. Tsividis. Op. cit., p. 136.

36. Y. P. Tsividis. Op. cit., p. 83.

37. Y. P. Tsividis. Op. cit., p. 139.

38. Y. P. Tsividis. Op. cit., p. 137.

39. Y. P. Tsividis. Op. cit., p. 324.

40. K. Y. Toh, P. K. Ko, and R. G. Meyer. "An Engineering Model for Short-Channel MOS Devices," *IEEE Journal of Solid-State Circuits,* Vol. 23, pp. 950–958, August 1988.

GENERAL REFERENCES

I. Getreu. *Modeling the Bipolar Transistor.* Tektronix Inc., 1976.

P. E. Gray and C. L. Searle, *Electronic Principles.* Wiley, New York, 1969.

R. S. Muller and T. I. Kamins. *Device Electronics for Integrated Circuits.* Wiley, New York, 1986.

Y. P. Tsividis. *Operation and Modeling of the MOS Transistor.* McGraw-Hill, New York, 1987.

CHAPTER 2

Bipolar, MOS, and BiCMOS Integrated-Circuit Technology

2.1 Introduction

For the designer and user of integrated circuits, a knowledge of the details of the fabrication process is important for two reasons. First, IC technology has become pervasive because it provides the economic advantage of the planar process for fabricating complex circuitry at low cost through batch processing. Thus a knowledge of the factors influencing the cost of fabrication of integrated circuits is essential for both the selection of a circuit approach to solve a given design problem by the designer and the selection of a particular circuit for fabrication as a custom integrated circuit by the user. Second, integrated-circuit technology presents a completely different set of cost constraints to the circuit designer from those encountered with discrete components. The optimum choice of a circuit approach to realize a specified circuit function requires an understanding of the degrees of freedom available with the technology and the nature of the devices that are most easily fabricated on the integrated-circuit chip.

At the present time, analog integrated circuits are designed and fabricated in bipolar technology, in MOS technology, and in technologies that combine both types of devices in one process. The necessity of combining complex digital functions on the same integrated circuit with analog functions has resulted in an increased use of digital MOS technologies for analog functions, particularly those functions such as analog-digital conversion required for interfaces between analog signals and digital systems. However, bipolar technology is now used and will continue to be used in a wide range of applications requiring high-current drive capability and the highest levels of precision analog performance.

In this chapter, we first enumerate the basic processes that are fundamental in the fabrication of bipolar and MOS integrated circuits: solid-state diffusion, lithography, epitaxial growth, ion implantation, selective oxidation, and polysilicon deposition. Next, we describe the sequence of steps that are used in the fabrication of bipolar integrated circuits and describe the properties of the passive and active devices that result from the process sequence. Also, we examine several modifications to the basic process. In the next subsection, we consider the sequence of steps in fabricating MOS integrated circuits and describe the types of devices resulting in that technology. This is followed by descriptions of BiCMOS technology, silicon-germanium heterojunction transistors, and interconnect materials under study to replace aluminum wires and silicon-dioxide dielectric. Next, we examine the factors affecting the manufacturing cost of monolithic circuits and, finally, present packaging considerations for integrated circuits.

2.2 Basic Processes in Integrated-Circuit Fabrication

The fabrication of integrated circuits and most modern discrete component transistors is based on a sequence of photomasking, diffusion, ion implantation, oxidation, and epitaxial growth steps applied to a slice of silicon starting material called a wafer.[1,2] Before beginning a description of the basic process steps, we will first review the effects produced on the electrical properties of silicon by the addition of impurity atoms.

2.2.1 Electrical Resistivity of Silicon

The addition of small concentrations of *n*-type or *p*-type impurities to a crystalline silicon sample has the effect of increasing the number of majority carriers (electrons for *n*-type, holes for *p*-type) and decreasing the number of minority carriers. The addition of impurities is called *doping* the sample. For practical concentrations of impurities, the density of majority carriers is approximately equal to the density of the impurity atoms in the crystal. Thus for *n*-type material,

$$n_n \simeq N_D \quad (2.1)$$

where n_n (cm^{-3}) is the equilibrium concentration of electrons and N_D (cm^{-3}) is the concentration of *n*-type donor impurity atoms. For *p*-type material,

$$p_p \simeq N_A \quad (2.2)$$

where p_p (cm^{-3}) is the equilibrium concentration of holes and N_A (cm^{-3}) is the concentration of *p*-type acceptor impurities. Any increase in the equilibrium concentration of one type of carrier in the crystal must result in a decrease in the equilibrium concentration of the other. This occurs because the holes and electrons recombine with each other at a rate that is proportional to the product of the concentration of holes and the concentration of electrons. Thus the number of recombinations per second, R, is given by

$$R = \gamma n p \quad (2.3)$$

where γ is a constant, and n and p are electron and hole concentrations, respectively, in the silicon sample. The generation of the hole-electron pairs is a thermal process that depends only on temperature; the rate of generation, G, is not dependent on impurity concentration. In equilibrium, R and G must be equal, so that

$$G = \text{constant} = R = \gamma n p \quad (2.4)$$

If no impurities are present, then

$$n = p = n_i(T) \quad (2.5)$$

where n_i (cm^{-3}) is the *intrinsic* concentration of carriers in a pure sample of silicon. Equations 2.4 and 2.5 establish that, for any impurity concentration, $\gamma n p = \text{constant} = \gamma n_i^2$, and thus

$$np = n_i^2(T) \quad (2.6)$$

Equation 2.6 shows that as the majority carrier concentration is increased by impurity doping, the minority carrier concentration is decreased by the same factor so that product np is constant in equilibrium. For impurity concentrations of practical interest, the majority carriers outnumber the minority carriers by many orders of magnitude.

The importance of minority- and majority-carrier concentrations in the operation of the transistor was described in Chapter 1. Another important effect of the addition of impurities is

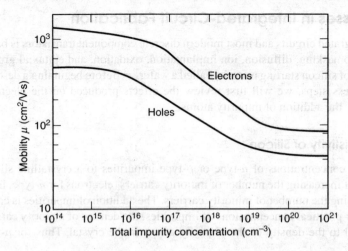

Figure 2.1 Hole and electron mobility as a function of doping in silicon.[3]

an increase in the ohmic conductivity of the material itself. This conductivity is given by

$$\sigma = q(\mu_n n + \mu_p p) \tag{2.7}$$

where μ_n (cm^2/V-s) is the electron mobility, μ_p (cm^2/V-s) is the hole mobility, and σ (Ω-cm)$^{-1}$ is the electrical conductivity. For an n-type sample, substitution of (2.1) and (2.6) in (2.7) gives

$$\sigma = q\left(\mu_n N_D + \mu_p \frac{n_i^2}{N_D}\right) \simeq q\mu_n N_D \tag{2.8}$$

For a p-type sample, substitution of (2.2) and (2.6) in (2.7) gives

$$\sigma = q\left(\mu_n \frac{n_i^2}{N_A} + \mu_p N_A\right) \simeq q\mu_p N_A \tag{2.9}$$

The mobility μ is different for holes and electrons and is also a function of the impurity concentration in the crystal for high impurity concentrations. Measured values of mobility in silicon as a function of impurity concentration are shown in Fig. 2.1. The resistivity ρ (Ω-cm) is usually specified in preference to the conductivity, and the resistivity of n- and p-type silicon as a function of impurity concentration is shown in Fig. 2.2. The conductivity and resistivity are related by the simple expression $\rho = 1/\sigma$.

2.2.2 Solid-State Diffusion

Solid-state diffusion of impurities in silicon is the movement, usually at high temperature, of impurity atoms from the surface of the silicon sample into the bulk material. During this high-temperature process, the impurity atoms replace silicon atoms in the lattice and are termed *substitutional impurities*. Since the doped silicon behaves electrically as p-type or n-type material depending on the type of impurity present, regions of p-type and n-type material can be formed by solid-state diffusion.

The nature of the diffusion process is illustrated by the conceptual example shown in Figs. 2.3 and 2.4. We assume that the silicon sample initially contains a uniform concentration of n-type impurity of 10^{15} atoms per cubic centimeter. Commonly used n-type impurities in silicon are phosphorus, arsenic, and antimony. We further assume that by some means we deposit atoms of p-type impurity on the top surface of the silicon sample. The most commonly used

2.2 Basic Processes in Integrated-Circuit Fabrication

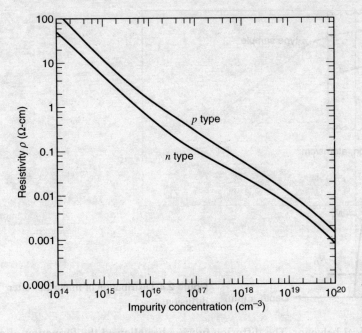

Figure 2.2 Resistivity of p- and n-type silicon as a function of impurity concentration.[4]

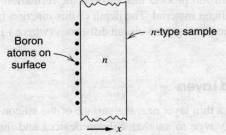

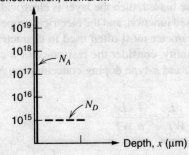

Figure 2.3 An n-type silicon sample with boron deposited on the surface.

p-type impurity in silicon device fabrication is boron. The distribution of impurities prior to the diffusion step is illustrated in Fig. 2.3. The initial placement of the impurity atoms on the surface of the silicon is called the *predeposition step* and can be accomplished by a number of different techniques.

If the sample is now subjected to a high temperature of about 1100°C for a time of about one hour, the impurities *diffuse* into the sample, as illustrated in Fig. 2.4. Within the silicon, the regions in which the p-type impurities outnumber the original n-type impurities display p-type electrical behavior, whereas the regions in which the n-type impurities are more numerous

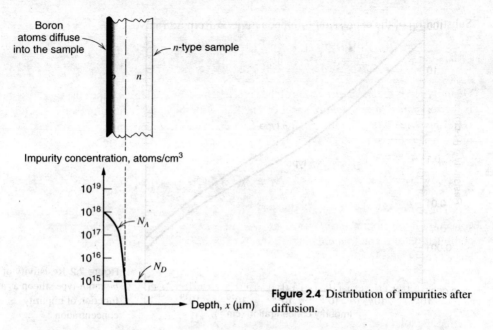

Figure 2.4 Distribution of impurities after diffusion.

display n-type electrical behavior. The diffusion process has allowed the formation of a *pn* junction within the continuous crystal of silicon material. The depth of this junction from the surface varies from 0.1 μm to 20 μm for silicon integrated-circuit diffusions (where 1 μm = 1 micrometer = 10^{-6} m).

2.2.3 Electrical Properties of Diffused Layers

The result of the diffusion process is often a thin layer near the surface of the silicon sample that has been converted from one impurity type to another. Silicon devices and integrated circuits are constructed primarily from these layers. From an electrical standpoint, if the *pn* junction formed by this diffusion is reverse biased, then the layer is electrically isolated from the underlying material by the reverse-biased junction, and the electrical properties of the layer itself can be measured. The electrical parameter most often used to characterize such layers is the *sheet resistance*. To define this quantity, consider the resistance of a uniformly doped sample of length L, width W, thickness T, and n-type doping concentration N_D, as shown in Fig. 2.5. The resistance is

$$R = \frac{\rho L}{WT} = \frac{1}{\sigma}\frac{L}{WT}$$

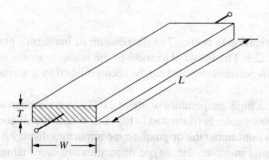

Figure 2.5 Rectangular sample for calculation of sheet resistance.

2.2 Basic Processes in Integrated-Circuit Fabrication

Substitution of the expression for conductivity σ from (2.8) gives

$$R = \left(\frac{1}{q\mu_n N_D}\right)\frac{L}{WT} = \frac{L}{W}\left(\frac{1}{q\mu_n N_D T}\right) = \frac{L}{W}R_\Box \qquad (2.10)$$

Quantity $R_\Box$ is the *sheet resistance* of the layer and has units of Ohms. Since the sheet resistance is the resistance of any *square* sheet of material with thickness T, its units are often given as *Ohms per square* ($\Omega/\Box$) rather than simply Ohms. The sheet resistance can be written in terms of the resistivity of the material, using (2.8), as

$$R_\Box = \frac{1}{q\mu_n N_D T} = \frac{\rho}{T} \qquad (2.11)$$

The diffused layer illustrated in Fig. 2.6 is similar to this case except that the impurity concentration is not uniform. However, we can consider the layer to be made up of a parallel combination of many thin conducting sheets. The conducting sheet of thickness dx at depth x has a conductance

$$dG = q\left(\frac{W}{L}\right)\mu_n N_D(x)\,dx \qquad (2.12)$$

To find the total conductance, we sum all the contributions.

$$G = \int_0^{x_j} q\frac{W}{L}\mu_n N_D(x)\,dx = \frac{W}{L}\int_0^{x_j} q\mu_n N_D(x)\,dx \qquad (2.13)$$

Inverting (2.13), we obtain

$$R = \frac{L}{W}\left[\frac{1}{\int_0^{x_j} q\mu_n N_D(x)\,dx}\right] \qquad (2.14)$$

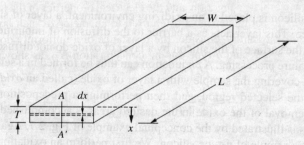

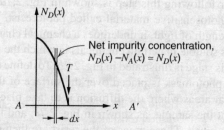

Figure 2.6 Calculation of the resistance of a diffused layer.

Comparison of (2.10) and (2.14) gives

$$R_\square = \left[\int_0^{x_j} q\mu_n N_D(x)\,dx\right]^{-1} \simeq \left[q\bar{\mu}_n \int_0^{x_j} N_D(x)\,dx\right]^{-1} \quad (2.15)$$

where $\bar{\mu}_n$ is the average mobility. Thus (2.10) can be used for diffused layers if the appropriate value of $R_\square$ is used. Equation 2.15 shows that the sheet resistance of the diffused layer depends on the total number of impurity atoms in the layer per unit area. The depth x_j in (2.13), (2.14), and (2.15) is actually the distance from the surface to the edge of the junction depletion layer, since the donor atoms within the depletion layer do not contribute to conduction. Sheet resistance is a useful parameter for the electrical characterization of diffusion processes and is a key parameter in the design of integrated resistors. The sheet resistance of a diffused layer is easily measured in the laboratory; the actual evaluation of (2.15) is seldom necessary.

■ **EXAMPLE**

Calculate the resistance of a layer with length 50 μm and width 5 μm in material of sheet resistance 200 Ω/□.

From (2.10)

$$R = \frac{50}{5} \times 200\,\Omega = 2\,\text{k}\Omega$$

Note that this region constitutes 10 squares in series, and R is thus 10 times the sheet resistance. ■

In order to use these diffusion process steps to fabricate useful devices, the diffusion must be restricted to a small region on the surface of the sample rather than the entire planar surface. This restriction is accomplished with photolithography.

2.2.4 Photolithography

When a sample of crystalline silicon is placed in an oxidizing environment, a layer of silicon dioxide will form at the surface. This layer acts as a barrier to the diffusion of impurities, so that impurities separated from the surface of the silicon by a layer of oxide do not diffuse into the silicon during high-temperature processing. A *pn* junction can thus be formed in a selected location on the sample by first covering the sample with a layer of oxide (called an *oxidation step*), removing the oxide in the selected region, and then performing a predeposition and diffusion step. The selective removal of the oxide in the desired areas is accomplished with photolithography. This process is illustrated by the conceptual example of Fig. 2.7. Again we assume the starting material is a sample of *n*-type silicon. We first perform an oxidation step in which a layer of silicon dioxide (SiO_2) is thermally grown on the top surface, usually of thickness of 0.2 μm to 1 μm. The wafer following this step is shown in Fig. 2.7a. Then the sample is coated with a thin layer of photosensitive material called photoresist. When this material is exposed to a particular wavelength of light, it undergoes a chemical change and, in the case of positive photoresist, becomes soluble in certain chemicals in which the unexposed photoresist is insoluble. The sample at this stage is illustrated in Fig. 2.7b. To define the desired diffusion areas on the silicon sample, a photomask is placed over the surface of the sample; this photomask is opaque except for clear areas where the diffusion is to take place. Light of the appropriate wavelength is directed at the sample, as shown in Fig. 2.7c, and falls on the photoresist only in the clear areas of the mask. These areas of the resist are then chemically

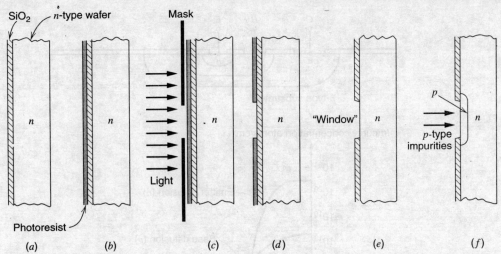

Figure 2.7 Conceptual example of the use of photolithography to form a *pn* junction diode. (*a*) Grow SiO_2. (*b*) Apply photoresist. (*c*) Expose through mask. (*d*) Develop photoresist. (*e*) Etch SiO_2 and remove photoresist. (*f*) Predeposit and diffuse impurities.

dissolved in the development step, as shown in Fig. 2.7*d*. The unexposed areas of the photoresist are impervious to the developer.

Since the objective is the formation of a region clear of SiO_2, the next step is the etching of the oxide. This step can be accomplished by dipping the sample in an etching solution, such as hydrofluoric acid, or by exposing it to an electrically produced plasma in a plasma etcher. In either case, the result is that in the regions where the photoresist has been removed, the oxide is etched away, leaving the bare silicon surface.

The remaining photoresist is next removed by a chemical stripping operation, leaving the sample with holes, or *windows*, in the oxide at the desired locations, as shown in Fig. 2.7*e*. The sample now undergoes a predeposition and diffusion step, resulting in the formation of *p*-type regions where the oxide had been removed, as shown in Fig. 2.7*f*. In some instances, the impurity to be locally added to the silicon surface is deposited by using ion implantation (see Section 2.2.6). This method of insertion can often take place through the silicon dioxide so that the oxide-etch step is unnecessary.

The minimum dimension of the diffused region that can be routinely formed with this technique in device production has decreased with time, and at present is approximately 0.1 μm × 0.1 μm. The number of such regions that can be fabricated simultaneously can be calculated by noting that the silicon sample used in the production of integrated circuits is a round slice, typically 4 inches to 12 inches in diameter and 250 μm thick. Thus the number of electrically independent *pn* junctions of dimension 0.1 μm × 0.1 μm spaced 0.1 μm apart that can be formed on one such wafer is on the order of 10^{12}. In actual integrated circuits, a number of masking and diffusion steps are used to form more complex structures such as transistors, but the key points are that photolithography is capable of defining a large number of devices on the surface of the sample and that all of these devices are batch fabricated at the same time. Thus the cost of the photomasking and diffusion steps applied to the wafer during the process is divided among the devices or circuits on the wafer. This ability to fabricate hundreds or thousands of devices at once is the key to the economic advantage of IC technology.

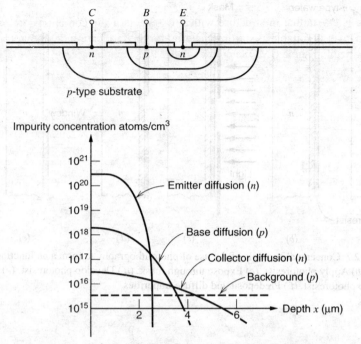

Figure 2.8 Triple-diffused transistor and resulting impurity profile.

2.2.5 Epitaxial Growth

Early planar transistors and the first integrated circuits used only photomasking and diffusion steps in the fabrication process. However, all-diffused integrated circuits had severe limitations compared with discrete component circuits. In a triple-diffused bipolar transistor, as illustrated in Fig. 2.8, the collector region is formed by an n-type diffusion into the p-type wafer. The drawbacks of this structure are that the series collector resistance is high and the collector-to-emitter breakdown voltage is low. The former occurs because the impurity concentration in the portion of the collector diffusion below the collector-base junction is low, giving the region high resistivity. The latter occurs because the concentration of impurities near the surface of the collector is relatively high, resulting in a low breakdown voltage between the collector and base diffusions at the surface, as described in Chapter 1. To overcome these drawbacks, the impurity concentration should be low at the collector-base junction for high breakdown voltage but high below the junction for low collector resistance. Such a concentration profile cannot be realized with diffusions alone, and the epitaxial growth technique was adopted as a result.

Epitaxial (epi) growth consists of formation of a layer of single-crystal silicon on the surface of the silicon sample so that the crystal structure of the silicon is continuous across the interface. The impurity concentration in the epi layer can be controlled independently and can be greater or smaller than in the substrate material. In addition, the epi layer is often of opposite impurity type from the substrate on which it is grown. The thickness of epi layers used in integrated-circuit fabrication varies from 1 μm to 20 μm, and the growth of the layer is accomplished by placing the wafer in an ambient atmosphere containing silicon tetrachloride ($SiCl_4$) or silane (SiH_4) at an elevated temperature. A chemical reaction takes place in which elemental silicon is deposited on the surface of the wafer, and the resulting surface layer of silicon is crystalline in structure with few defects if the conditions are carefully controlled. Such a layer is suitable as starting material for the fabrication of bipolar transistors. Epitaxy is also utilized in some CMOS and most BiCMOS technologies.

2.2.6 Ion Implantation

Ion implantation is a technique for directly inserting impurity atoms into a silicon wafer.[5,6] The wafer is placed in an evacuated chamber, and ions of the desired impurity species are directed at the sample at high velocity. These ions penetrate the surface of the silicon wafer to an average depth of from less than 0.1 μm to about 0.6 μm, depending on the velocity with which they strike the sample. The wafer is then held at a moderate temperature for a period of time (for example, 800°C for 10 minutes) in order to allow the ions to become mobile and fit into the crystal lattice. This is called an *anneal step* and is essential to allow repair of any crystal damage caused by the implantation. The principal advantages of ion implantation over conventional diffusion are (1) that small amounts of impurities can be reproducibly deposited, and (2) that the amount of impurity deposited per unit area can be precisely controlled. In addition, the deposition can be made with a high level of uniformity across the wafer. Another useful property of ion-implanted layers is that the peak of the impurity concentration profile can be made to occur below the surface of the silicon, unlike with diffused layers. This allows the fabrication of implanted bipolar structures with properties that are significantly better than those of diffused devices. This technique is also widely applied in MOS technology where small, well-controlled amounts of impurity are required at the silicon surface for adjustment of device thresholds, as described in Section 1.5.1.

2.2.7 Local Oxidation

In both MOS and bipolar technologies, the need often arises to fabricate regions of the silicon surface that are covered with relatively thin silicon dioxide, adjacent to areas covered by relatively thick oxide. Typically, the former regions constitute the active-device areas, whereas the latter constitute the regions that electrically isolate the devices from each other. A second requirement is that the transition from thick to thin regions must be accomplished without introducing a large vertical step in the surface geometry of the silicon, so that the metallization and other patterns that are later deposited can lie on a relatively planar surface. Local oxidation is used to achieve this result. The local oxidation process begins with a sample that already has a thin oxide grown on it, as shown in Fig. 2.9a. First a layer of silicon nitride (SiN) is deposited on the sample and subsequently removed with a masking step from all areas where thick oxide is to be grown, as shown in Fig. 2.9b. Silicon nitride acts as a barrier to oxygen atoms that might otherwise reach the Si-SiO$_2$ interface and cause further oxidation. Thus when a subsequent long, high-temperature oxidation step is carried out, a thick oxide is grown in the regions where there is no nitride, but no oxidation takes place under the nitride. The resulting geometry after nitride removal is shown in Fig. 2.9c. Note that the top surface of the silicon dioxide has a smooth transition from thick to thin areas and that the height of this transition is less than the oxide thickness difference because the oxidation in the thick oxide regions consumes some of the underlying silicon.

2.2.8 Polysilicon Deposition

Many process technologies utilize layers of polycrystalline silicon that are deposited during fabrication. After deposition of the polycrystalline silicon layer on the wafer, the desired features are defined by using a masking step and can serve as gate electrodes for silicon-gate MOS

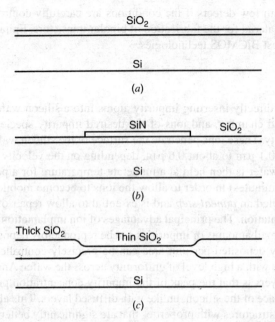

Figure 2.9 Local oxidation process. (*a*) Silicon sample prior to deposition of nitride. (*b*) After nitride deposition and definition. (*c*) After oxidation and nitride removal.

transistors, emitters of bipolar transistors, plates of capacitors, resistors, fuses, and interconnect layers. The sheet resistance of such layers can be controlled by the impurity added, much like bulk silicon, in a range from about 20 Ω/☐ up to very high values. The process that is used to deposit the layer is much like that used for epitaxy. However, since the deposition is usually over a layer of silicon dioxide, the layer does not form as a single-crystal extension of the underlying silicon but forms as a granular (or polysilicon) film. Some MOS technologies contain as many as three separate polysilicon layers, separated from one another by layers of SiO_2.

2.3 High-Voltage Bipolar Integrated-Circuit Fabrication

Integrated-circuit fabrication techniques have changed dramatically since the invention of the basic planar process. This change has been driven by developments in photolithography, processing techniques, and also the trend to reduce power-supply voltages in many systems. Developments in photolithography have reduced the minimum feature size attainable from tens of microns to the submicron level. The precise control allowed by ion implantation has resulted in this technique becoming the dominant means of predepositing impurity atoms. Finally, many circuits now operate from 3 V or 5 V power supplies instead of from the $\pm$ 15 V supplies used earlier to achieve high dynamic range in stand-alone integrated circuits, such as operational amplifiers. Reducing the operating voltages allows closer spacing between devices in an IC. It also allows shallower structures with higher frequency capability. These effects stem from the fact that the thickness of junction depletion layers is reduced by reducing operating voltages, as described in Chapter 1. Thus the highest-frequency IC processes are designed to operate from 5-V supplies or less and are generally not usable at higher supply voltages. In fact, a fundamental trade-off exists between the frequency capability of a process and its breakdown voltage.

In this section, we examine first the sequence of steps used in the fabrication of high-voltage bipolar integrated circuits using junction isolation. This was the original IC process

2.3 High-Voltage Bipolar Integrated-Circuit Fabrication

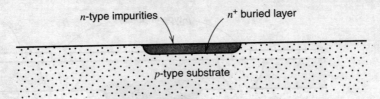

Figure 2.10 Buried-layer diffusion.

and is useful as a vehicle to illustrate the basic methods of IC fabrication. It is still used in various forms to fabricate high-voltage circuits.

The fabrication of a junction-isolated bipolar integrated circuit involves a sequence of from six to eight masking and diffusion steps. The starting material is a wafer of *p*-type silicon, usually 250 μm thick and with an impurity concentration of approximately 10^{16} atoms/cm^3. We will consider the sequence of diffusion steps required to form an *npn* integrated-circuit transistor. The first mask and diffusion step, illustrated in Fig. 2.10, forms a low-resistance *n*-type layer that will eventually become a low-resistance path for the collector current of the transistor. This step is called the *buried-layer diffusion*, and the layer itself is called the *buried layer*. The sheet resistance of the layer is in the range of 20 to 50 $\Omega/\square$, and the impurity used is usually arsenic or antimony because these impurities diffuse slowly and thus do not greatly redistribute during subsequent processing.

After the buried-layer step, the wafer is stripped of all oxide and an epi layer is grown, as shown in Fig. 2.11. The thickness of the layer and its *n*-type impurity concentration determine the collector-base breakdown voltage of the transistors in the circuit since this material forms the collector region of the transistor. For example, if the circuit is to operate at a power-supply voltage of 36 V, the devices generally are required to have BV_{CEO} breakdown voltages above this value. As described in Chapter 1, this implies that the plane breakdown voltage in the collector-base junction must be several times this value because of the effects of collector avalanche multiplication. For $BV_{CEO} = 36$ V, a collector-base plane breakdown voltage of approximately 90 V is required, which implies an impurity concentration in the collector of approximately 10^{15} atoms/cm^3 and a resistivity of 5 Ω-cm. The thickness of the epitaxial layer then must be large enough to accommodate the depletion layer associated with the collector-base junction. At 36 V, the results of Chapter 1 can be used to show that the depletion-layer thickness is approximately 6 μm. Since the buried layer diffuses outward approximately 8 μm during subsequent processing, and the base diffusion will be approximately 3 μm deep, a total epitaxial layer thickness of 17 μm is required for a 36-V circuit. For circuits with lower operating voltages, thinner and more heavily doped epitaxial layers are used to reduce the transistor collector series resistance, as will be shown later.

Following the epitaxial growth, an oxide layer is grown on the top surface of the epitaxial layer. A mask step and boron (*p*-type) predeposition and diffusion are performed, resulting in the structure shown in Fig. 2.12. The function of this diffusion is to isolate the collectors of the transistors from each other with reverse-biased *pn* junctions, and it is termed the *isolation*

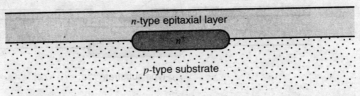

Figure 2.11 Bipolar integrated-circuit wafer following epitaxial growth.

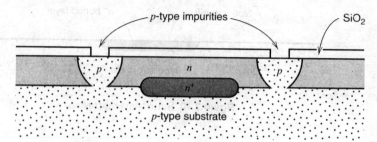

Figure 2.12 Structure following isolation diffusion.

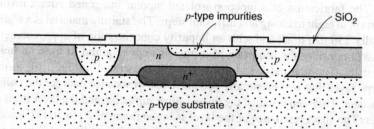

Figure 2.13 Structure following base diffusion.

diffusion. Because of the depth to which the diffusion must penetrate, this diffusion requires several hours in a diffusion furnace at temperatures of about 1200°C. The isolated diffused layer has a sheet resistance from 20 Ω/□ to 40 Ω/□.

The next steps are the base mask, base predeposition, and base diffusion, as shown in Fig. 2.13. The latter is usually a boron diffusion, and the resulting layer has a sheet resistance of from 100 Ω/□ to 300 Ω/□, and a depth of 1 μm to 3 μm at the end of the process. This diffusion forms not only the bases of the transistors, but also many of the resistors in the circuit, so that control of the sheet resistance is important.

Following the base diffusion, the emitters of the transistors are formed by a mask step, n-type predeposition, and diffusion, as shown in Fig. 2.14. The sheet resistance is between 2 Ω/□ and 10 Ω/□, and the depth is 0.5 μm to 2.5 μm after the diffusion. This diffusion step is also used to form a low-resistance region, which serves as the contact to the collector region. This is necessary because ohmic contact is difficult to accomplish between aluminum metallization and the high-resistivity epitaxial material directly. The next masking step, the contact mask, is used to open holes in the oxide over the emitter, the base, and the collector of the transistors so that electrical contact can be made to them. Contact windows are also opened

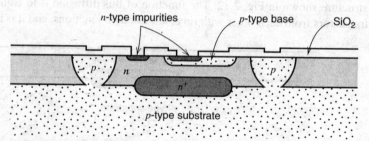

Figure 2.14 Structure following emitter diffusion.

2.3 High-Voltage Bipolar Integrated-Circuit Fabrication

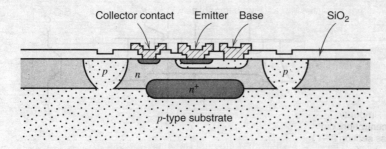

Figure 2.15 Final structure following contact mask and metallization.

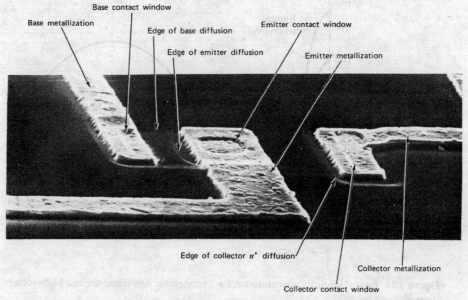

Figure 2.16 Scanning electron microscope photograph of *npn* transistor structure.

for the passive components on the chip. The entire wafer is then coated with a thin (about 1 μm) layer of aluminum that will interconnect the circuit elements. The actual interconnect pattern is defined by the last mask step, in which the aluminum is etched away in the areas where the photoresist is removed in the develop step. The final structure is shown in Fig. 2.15. A microscope photograph of an actual structure of the same type is shown in Fig. 2.16. The terraced effect on the surface of the device results from the fact that additional oxide is grown during each diffusion cycle, so that the oxide is thickest over the epitaxial region, where no oxide has been removed, is less thick over the base and isolation regions, which are both opened at the base mask step, and is thinnest over the emitter diffusion. A typical diffusion profile for a high-voltage, deep-diffused analog integrated circuit is shown in Fig. 2.17.

This sequence allows simultaneous fabrication of a large number (often thousands) of complex circuits on a single wafer. The wafer is then placed in an automatic tester, which checks the electrical characteristics of each circuit on the wafer and puts an ink dot on circuits that fail to meet specifications. The wafer is then broken up, by sawing or scribing and breaking, into individual circuits. The resulting silicon chips are called *dice*, and the singular is *die*. Each good die is then mounted in a package, ready for final testing.

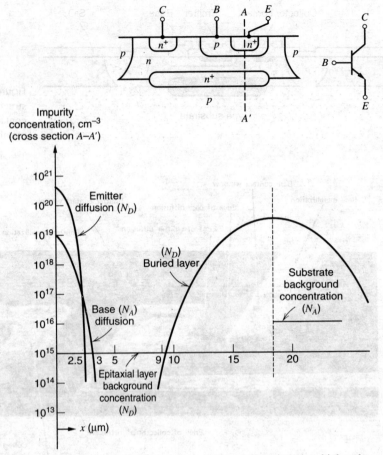

Figure 2.17 Typical impurity concentration for a monolithic *npn* transistor in a high-voltage, deep-diffused process.

2.4 Advanced Bipolar Integrated-Circuit Fabrication

A large fraction of bipolar analog integrated circuits currently manufactured uses the basic technology described in the previous section, or variations thereof. The fabrication sequence is relatively simple and low in cost. However, many of the circuit applications of commercial importance have demanded steadily increasing frequency response capability, which translates directly to a need for transistors of higher frequency-response capability in the technology. The higher speed requirement dictates a device structure with thinner base width to reduce base transit time and smaller dimensions overall to reduce parasitic capacitances. The smaller device dimensions require that the width of the junction depletion layers within the structure be reduced in proportion, which in turn requires the use of lower circuit operating voltages and higher impurity concentrations in the device structure. To meet this need, a class of bipolar fabrication technologies has evolved that, compared to the high-voltage process sequence described in the last section, use much thinner and more heavily doped epitaxial layers, selectively oxidized regions for isolation instead of diffused junctions, and a polysilicon layer as the source of dopant for the emitter. Because of the growing importance of this class of bipolar process, the sequence for such a process is described in this section.

The starting point for the process is similar to that for the conventional process, with a mask and implant step resulting in the formation of a heavily-doped n^+ buried layer in a *p*-type

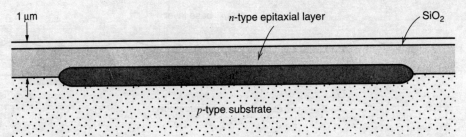

Figure 2.18 Device cross section following initial buried-layer mask, implant, and epitaxial-layer growth.

substrate. Following this step, a thin *n*-type epitaxial layer is grown, about 1 μm in thickness and about 0.5 Ω-cm in resistivity. The result after these steps is shown in cross section in Fig. 2.18.

Next, a selective oxidation step is carried out to form the regions that will isolate the transistor from its neighbors and also isolate the collector-contact region from the rest of the transistor. The oxidation step is as described in Section 2.2.7, except that prior to the actual growth of the thick SiO_2 layer, an etching step is performed to remove silicon material from the regions where oxide will be grown. If this is not done, the thick oxide growth results in elevated *humps* in the regions where the oxide is grown. The steps around these humps cause difficulty in coverage by subsequent layers of metal and polysilicon that will be deposited. The removal of some silicon material before oxide growth results in a nearly planar surface after the oxide is grown and removes the step coverage problem in subsequent processing. The resulting structure following this step is shown in Fig. 2.19. Note that the SiO_2 regions extend all the way down to the *p*-type substrate, electrically isolating the *n*-type epi regions from one another. These regions are often referred to as *moats*. Because growth of oxide layers thicker than a micron or so requires impractically long times, this method of isolation is practical only for very thin transistor structures.

Next, two mask and implant steps are performed. A heavy n^+ implant is made in the collector-contact region and diffused down to the buried layer, resulting in a low-resistance path to the collector. A second mask is performed to define the base region, and a thin-base *p*-type implant is performed. The resulting structure is shown in Fig. 2.20.

A major challenge in fabricating this type of device is the formation of very thin base and emitter structures, and then providing low-resistance ohmic contact to these regions. This is most often achieved using polysilicon as a doping source. An n^+ doped layer of polysilicon is deposited and masked to leave polysilicon only in the region directly over the emitter. During subsequent high-temperature processing steps, the dopant (usually arsenic) diffuses out of the polysilicon and into the crystalline silicon, forming a very thin, heavily doped emitter region. Following the poly deposition, a heavy *p*-type implant is performed, which results in a more

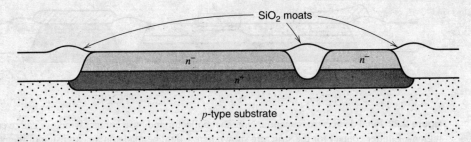

Figure 2.19 Device cross section following selective etch and oxidation to form thick-oxide moats.

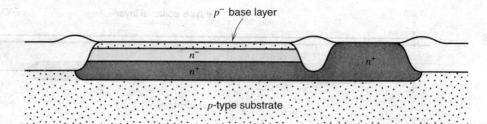

Figure 2.20 Device cross section following mask, implant, and diffusion of collector n^+ region, and mask and implant of base p-type region.

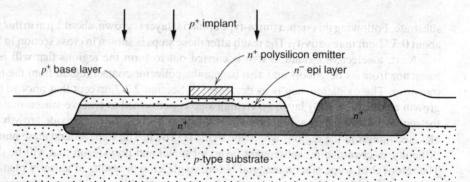

Figure 2.21 Device cross section following poly deposition and mask, base p-type implant, and thermal diffusion cycle.

heavily doped p-type layer at all points in the base region except directly under the polysilicon, where the polysilicon itself acts as a mask to prevent the boron atoms from reaching this part of the base region. The structure that results following this step is shown in Fig. 2.21.

This method of forming low-resistance regions to contact the base is called a *self-aligned structure* because the alignment of the base region with the emitter happens automatically and does not depend on mask alignment. Similar processing is used in MOS technology, described later in this chapter.

The final device structure after metallization is shown in Fig. 2.22. Since the moats are made of SiO_2, the metallization contact windows can overlap into them, a fact that dramatically reduces the minimum achievable dimensions of the base and collector regions. All exposed

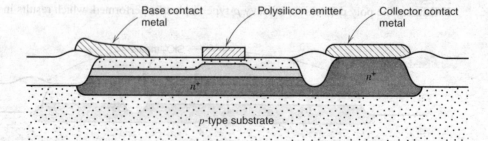

Figure 2.22 Final device cross section. Note that collector and base contact windows can overlap moat regions. Emitter contact for the structure shown here would be made on an extension of the polysilicon emitter out of the device active area, allowing the minimum possible emitter size.

2.5 Active Devices in Bipolar Analog Integrated Circuits

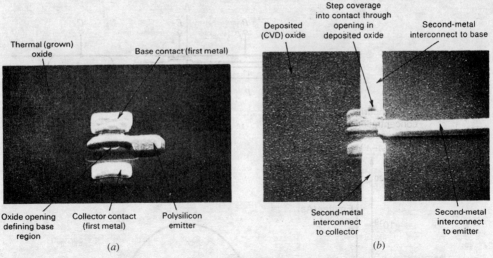

Figure 2.23 Scanning-electron-microscope photographs of a bipolar transistor in an advanced, polysilicon-emitter, oxide-isolated process. (*a*) After polysilicon emitter definition and first-metal contact to the base and collector. The polysilicon emitter is 1 μm wide. (*b*) After oxide deposition, contact etch, and second-metal interconnect. [QUBic process photograph courtesy of Signetics.]

silicon and polysilicon is covered with a highly conductive silicide (a compound of silicon and a refractory metal such as tungsten) to reduce series and contact resistance. For minimum-dimension transistors, the contact to the emitter is made by extending the polysilicon to a region outside the device active area and forming a metal contact to the polysilicon there. A photograph of such a device is shown in Fig. 2.23, and a typical impurity profile is shown in Fig. 2.24. The use of the remote emitter contact with polysilicon connection does add some series emitter resistance, so for larger device geometries or cases in which emitter resistance is critical, a larger emitter is used and the contact is placed directly on top of the polysilicon emitter itself. Production IC processes[7,8] based on technologies similar to the one just described yield bipolar transistors having f_T values well in excess of 10 GHz, compared to a typical value of 500 MHz for deep-diffused, high-voltage processes.

2.5 Active Devices in Bipolar Analog Integrated Circuits

The high-voltage IC fabrication process described previously is an outgrowth of the one used to make *npn* double-diffused discrete bipolar transistors, and as a result the process inherently produces double-diffused *npn* transistors of relatively high performance. The advanced technology process improves further on all aspects of device performance except for breakdown voltage. In addition to *npn* transistors, *pnp* transistors are also required in many analog circuits, and an important development in the evolution of analog IC technologies was the invention of device structures that allowed the standard technology to produce *pnp* transistors as well. In this section, we will explore the structure and properties of *npn*, lateral *pnp*, and substrate *pnp* transistors. We will draw examples primarily from the high-voltage technology. The available structures in the more advanced technology are similar, except that their frequency response is correspondingly higher. We will include representative device parameters from these newer technologies as well.

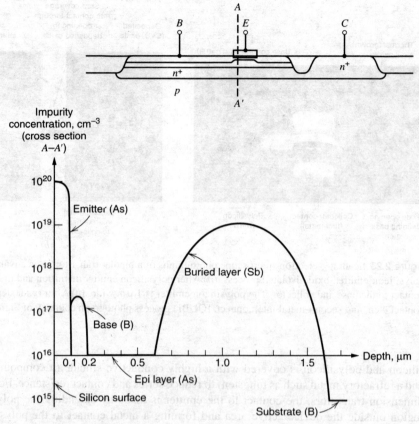

Figure 2.24 Typical impurity profile in a shallow oxide-isolated bipolar transistor.

2.5.1 Integrated-Circuit *npn* Transistors

The structure of a high-voltage, integrated-circuit *npn* transistor was described in the last section and is shown in plan view and cross section in Fig. 2.25. In the forward-active region of operation, the only electrically active portion of the structure that provides current gain is that portion of the base immediately under the emitter diffusion. The rest of the structure provides a top contact to the three transistor terminals and electrical isolation of the device from the rest of the devices on the same die. From an electrical standpoint, the principal effect of these regions is to contribute parasitic resistances and capacitances that must be included in the small-signal model for the complete device to provide an accurate representation of high-frequency behavior.

An important distinction between integrated-circuit design and discrete-component circuit design is that the IC designer has the capability to utilize a device geometry that is specifically optimized for the particular set of conditions found in the circuit. Thus the circuit-design problem involves a certain amount of device design as well. For example, the need often exists for a transistor with a high current-carrying capability to be used in the output stage of an amplifier. Such a device can be made by using a larger device geometry than the standard one, and the transistor then effectively consists of many standard devices connected in parallel. The larger geometry, however, will display larger base-emitter, collector-base, and collector-substrate capacitance than the standard device, and this must be taken into

2.5 Active Devices in Bipolar Analog Integrated Circuits

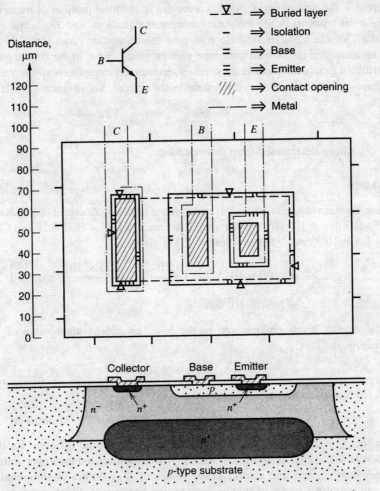

Figure 2.25 Integrated-circuit *npn* transistor. The mask layers are coded as shown.

account in analyzing the frequency response of the circuit. The circuit designer then must be able to determine the effect of changes in device geometry on device characteristics and to estimate the important device parameters when the device structure and doping levels are known. To illustrate this procedure, we will calculate the model parameters of the *npn* device shown in Fig. 2.25. This structure is typical of the devices used in circuits with a 5-Ω-cm, 17-μm epitaxial layer. The emitter diffusion is 20 μm × 25 μm, the base diffusion is 45 μm × 60 μm, and the base-isolation spacing is 25 μm. The overall device dimensions are 140 μm × 95 μm. Device geometries intended for lower epi resistivity and thickness can be much smaller; the base-isolation spacing is dictated by the side diffusion of the isolation region plus the depletion layers associated with the base-collector and collector-isolation junctions.

Saturation Current I_S. In Chapter 1, the saturation current of a graded-base transistor was shown to be

$$I_S = \frac{qA\,\overline{D}_n n_i^2}{Q_B} \tag{2.16}$$

where A is the emitter-base junction area, Q_B is the total number of impurity atoms per unit area in the base, n_i is the intrinsic carrier concentration, and $\overline{D}_n$ is the effective diffusion constant for electrons in the base region of the transistor. From Fig. 2.17, the quantity Q_B can be identified as the area under the concentration curve in the base region. This could be determined graphically but is most easily determined experimentally from measurements of the base-emitter voltage at a constant collector current. Substitution of (2.16) in (1.35) gives

$$\frac{Q_B}{\overline{D}_n} = A \frac{qn_i^2}{I_C} \exp \frac{V_{BE}}{V_T} \qquad (2.17)$$

and Q_B can be determined from this equation.

■ **EXAMPLE**

A base-emitter voltage of 550 mV is measured at a collector current of 10 μA on a test transistor with a 100 μm × 100 μm emitter area. Estimate Q_B if $T = 300°$K. From Chapter 1, we have $n_i = 1.5 \times 10^{10}$ cm^{-3}. Substitution in (2.17) gives

$$\frac{Q_B}{\overline{D}_n} = (100 \times 10^{-4})^2 \frac{1.6 \times 10^{-19} \times 2.25 \times 10^{20}}{10^{-5}} \exp(550/26)$$

$$= 5.54 \times 10^{11} \text{cm}^{-4}\text{s}$$

At the doping levels encountered in the base, an approximate value of $\overline{D}_n$, the electron diffusivity, is

$$\overline{D}_n = 13 \text{ cm}^2 \text{s}^{-1}$$

Thus for this example,

$$Q_B = 5.54 \times 10^{11} \times 13 \text{ cm}^{-2} = 7.2 \times 10^{12} \text{ atoms/cm}^2$$

Note that Q_B depends on the diffusion profiles and will be different for different types of processes. Generally speaking, fabrication processes intended for lower voltage operation use thinner base regions and display lower values of Q_B. Within one nominally fixed process, Q_B can vary by a factor of two or three to one because of diffusion process variations. The principal significance of the numerical value for Q_B is that it allows the calculation of the saturation current I_S for any device structure once the emitter-base junction area is known. ■

Series Base Resistance r_b. Because the base contact is physically removed from the active base region, a significant series ohmic resistance is observed between the contact and the active base. This resistance can have a significant effect on the high-frequency gain and on the noise performance of the device. As illustrated in Fig. 2.26a, this resistance consists of two parts. The first is the resistance r_{b1} of the path between the base contact and the edge of the emitter

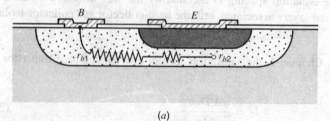

(a)

Figure 2.26 (a) Base resistance components for the *npn* transistor.

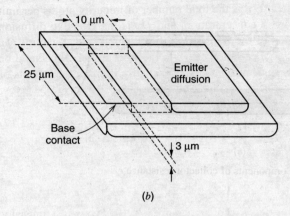

Figure 2.26 (b) Calculation of r_{b1}. The r_{b1} component of base resistance can be estimated by calculating the resistance of the rectangular block above.

diffusion. The second part r_{b2} is that resistance between the edge of the emitter and the site within the base region at which the current is actually flowing. The former component can be estimated by neglecting fringing and by assuming that this component of the resistance is that of a rectangle of material as shown in Fig. 2.26b. For a base sheet resistance of 100 $\Omega/\square$ and typical dimensions as shown in Fig. 2.26b, this would give a resistance of

$$r_{b1} = \frac{10 \ \mu m}{25 \ \mu m} 100 \ \Omega = 40 \ \Omega$$

The calculation of r_{b2} is complicated by several factors. First, the current flow in this region is not well modeled by a single resistor because the base resistance is distributed throughout the base region and two-dimensional effects are important. Second, at even moderate current levels, the effect of current crowding[9] in the base causes most of the carrier injection from the emitter into the base to occur near the periphery of the emitter diffusion. At higher current levels, essentially all of the injection takes place at the periphery and the effective value of r_b approaches r_{b1}. In this situation, the portion of the base directly beneath the emitter is not involved in transistor action. A typically observed variation of r_b with collector current for the *npn* geometry of Fig. 2.25 is shown in Fig. 2.27. In transistors designed for low-noise and/or high-frequency applications where low r_b is important, an effort is often made to maximize the periphery of the emitter that is adjacent to the base contact. At the same time, the emitter-base

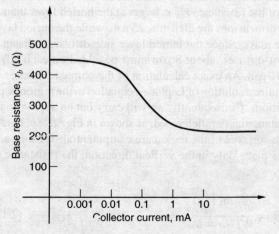

Figure 2.27 Typical variation of effective small-signal base resistance with collector current for integrated-circuit *npn* transistor.

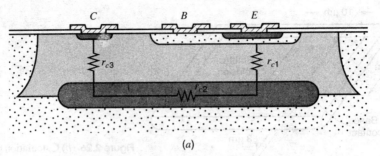

Figure 2.28 (a) Components of collector resistance r_c.

junction and collector-base junction areas must be kept small to minimize capacitance. In the case of high-frequency transistors, this usually dictates the use of an emitter geometry that consists of many narrow stripes with base contacts between them. The ease with which the designer can use such device geometries is an example of the flexibility allowed by monolithic IC construction.

Series Collector Resistance r_c. The series collector resistance is important both in high-frequency circuits and in low-frequency applications where low collector-emitter saturation voltage is required. Because of the complex three-dimensional shape of the collector region itself, only an approximate value for the collector resistance can be obtained by hand analysis. From Fig. 2.28, we see that the resistance consists of three parts: that from the collector-base junction under the emitter down to the buried layer, r_{c1}; that of the buried layer from the region under the emitter over to the region under the collector contact, r_{c2}; and finally, that portion from the buried layer up to the collector contact, r_{c3}. The small-signal series collector resistance in the forward-active region can be estimated by adding the resistance of these three paths.

■ **EXAMPLE**

Estimate the collector resistance of the transistor of Fig. 2.25, assuming the doping profile of Fig. 2.17. We first calculate the r_{c1} component. The thickness of the lightly doped epi layer between the collector-base junction and the buried layer is 6 μm. Assuming that the collector-base junction is at zero bias, the results of Chapter 1 can be used to show that the depletion layer is about 1 μm thick. Thus the undepleted epi material under the base is 5 μm thick.

The effective cross-sectional area of the resistance r_{c1} is larger at the buried layer than at the collector-base junction. The emitter dimensions are 20 μm × 25 μm, while the buried layer dimensions are 41 μm × 85 μm on the mask. Since the buried layer side-diffuses a distance roughly equal to the distance that it out-diffuses, about 8 μm must be added on each edge, giving an effective size of 57 μm × 101 μm. An exact calculation of the ohmic resistance of this three-dimensional region would require a solution of Laplace's equation in the region, with a rather complex set of boundary conditions. Consequently, we will carry out an approximate analysis by modeling the region as a rectangular parallelepiped, as shown in Fig. 2.28b. Under the assumptions that the top and bottom surfaces of the region are equipotential surfaces, and that the current flow in the region takes place only in the vertical direction, the resistance of the structure can be shown to be

$$R = \frac{\rho T}{WL} \frac{\ln\left(\frac{a}{b}\right)}{(a-b)} \qquad (2.18)$$

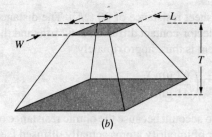

Figure 2.28 (*b*) Model for calculation of collector resistance.

where

T = thickness of the region
ρ = resistivity of the material
W, L = width, length of the top rectangle
a = ratio of the width of the bottom rectangle to the width of the top rectangle
b = ratio of the length of the bottom rectangle to the length of the top rectangle

Direct application of this expression to the case at hand would give an unrealistically low value of resistance, because the assumption of one-dimensional flow is seriously violated when the dimensions of the lower rectangle are much larger than those of the top rectangle. Equation 2.18 gives realistic results when the sides of the region form an angle of about 60° or less with the vertical. When the angle of the sides is increased beyond this point, the resistance does not decrease very much because of the long path for current flow between the top electrode and the remote regions of the bottom electrode. Thus the limits of the bottom electrode should be determined either by the edges of the buried layer or by the edges of the emitter plus a distance equal to about twice the vertical thickness T of the region, whichever is smaller. For the case of r_{c1},

$$T = 5 \text{ } \mu\text{m} = 5 \times 10^{-4} \text{ cm}$$
$$\rho = 5 \text{ } \Omega\text{-cm}$$

We assume that the effective emitter dimensions are those defined by the mask plus approximately 2 μm of side diffusion on each edge. Thus

$$W = 20 \text{ } \mu\text{m} + 4 \text{ } \mu\text{m} = 24 \times 10^{-4} \text{ cm}$$
$$L = 25 \text{ } \mu\text{m} + 4 \text{ } \mu\text{m} = 29 \times 10^{-4} \text{ cm}$$

For this case, the buried-layer edges are further away from the emitter edge than twice the thickness T on all four sides when side diffusion is taken into account. Thus the *effective* buried-layer dimensions that we use in (2.18) are

$$W_{BL} = W + 4T = 24 \text{ } \mu\text{m} + 20 \text{ } \mu\text{m} = 44 \text{ } \mu\text{m}$$
$$L_{BL} = L + 4T = 29 \text{ } \mu\text{m} + 20 \text{ } \mu\text{m} = 49 \text{ } \mu\text{m}$$

and

$$a = \frac{44 \text{ } \mu\text{m}}{24 \text{ } \mu\text{m}} = 1.83$$

$$b = \frac{49 \text{ } \mu\text{m}}{29 \text{ } \mu\text{m}} = 1.69$$

Thus from (2.18),

$$r_{c1} = \frac{(5)(5 \times 10^{-4})}{(24 \times 10^{-4})(29 \times 10^{-4})}(0.57) \text{ } \Omega = 204 \text{ } \Omega$$

We will now calculate r_{c2}, assuming a buried-layer sheet resistance of 20 $\Omega/\square$. The distance from the center of the emitter to the center of the collector-contact diffusion is 62 μm, and the width of the buried layer is 41 μm. The r_{c2} component is thus, approximately,

$$r_{c2} = (20\ \Omega/\square)\left(\frac{L}{W}\right) = 20\ \Omega/\square \left(\frac{62\ \mu m}{41\ \mu m}\right) = 30\ \Omega$$

Here the buried-layer side diffusion was not taken into account because the ohmic resistance of the buried layer is determined entirely by the number of impurity atoms actually diffused [see (2.15)] into the silicon, which is determined by the mask dimensions and the sheet resistance of the buried layer.

For the calculation of r_{c3}, the dimensions of the collector-contact n^+ diffusion are 18 μm × 49 μm, including side diffusion. The distance from the buried layer to the bottom of the n^+ diffusion is seen in Fig. 2.17 to be 6.5 μm, and thus $T = 6.5$ μm in this case. On the three sides of the collector n^+ diffusion that do not face the base region, the out-diffused buried layer extends only 4 μm outside the n^+ diffusion, and thus the effective dimension of the buried layer is determined by the actual buried-layer edge on these sides. On the side facing the base region, the effective edge of the buried layer is a distance $2T$, or 13 μm, away from the edge of the n^+ diffusion. The effective buried-layer dimensions for the calculation of r_{c3} are thus 35 μm × 57 μm. Using (2.18),

$$r_{c3} = \frac{(5)(6.5 \times 10^{-4})}{(18 \times 10^{-4})(49 \times 10^{-4})} 0.66 = 243\ \Omega$$

The total collector resistance is thus

$$r_c = r_{c1} + r_{c2} + r_{c3} = 531\ \Omega$$

The value actually observed in such devices is somewhat lower than this for three reasons. First, we have approximated the flow as one-dimensional, and it is actually three-dimensional. Second, for larger collector-base voltages, the collector-base depletion layer extends further into the epi, decreasing r_{c1}. Third, the value of r_c that is important is often that for a saturated device. In saturation, holes are injected into the epi region under the emitter by the forward-biased, collector-base function, and they modulate the conductivity of the region even at moderate current levels.[10] Thus the collector resistance one measures when the device is in saturation is closer to $(r_{c2} + r_{c3})$, or about 250 to 300 Ω. Thus r_c is smaller in saturation than in the forward-active region.

■

Collector-Base Capacitance. The collector-base capacitance is simply the capacitance of the collector-base junction including both the flat bottom portion of the junction and the side-walls. This junction is formed by the diffusion of boron into an n-type epitaxial material that we will assume has a resistivity of 5 Ω-cm, corresponding to an impurity concentration of 10^{15} atoms/cm^3. The uniformly doped epi layer is much more lightly doped than the p-diffused region, and as a result, this junction is well approximated by a step junction in which the depletion layer lies almost entirely in the epitaxial material. Under this assumption, the results of Chapter 1 regarding step junctions can be applied, and for convenience this relationship has been plotted in nomograph form in Fig. 2.29. This nomograph is a graphical representation of the relation

$$\frac{C_j}{A} = \sqrt{\frac{q\epsilon N_B}{2(\psi_0 + V_R)}} \qquad (2.19)$$

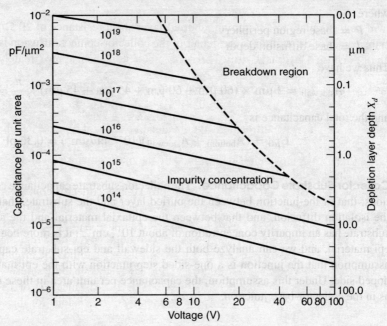

Figure 2.29 Capacitance and depletion-layer width of an abrupt *pn* junction as a function of applied voltage and doping concentration on the lightly doped side of the junction.[11]

where N_B is the doping density in the epi material and V_R is the reverse bias on the junction. The nomograph of Fig. 2.29 can also be used to determine the junction depletion-region width as a function of applied voltage, since this width is inversely proportional to the capacitance. The width in microns is given on the axis on the right side of the figure.

Note that the horizontal axis in Fig. 2.29 is the *total* junction potential, which is the applied potential plus the built-in voltage ψ_0. In order to use the curve, then, the built-in potential must be calculated. While this would be an involved calculation for a diffused junction, the built-in potential is actually only weakly dependent on the details of the diffusion profile and can be assumed to be about 0.55 V for the collector-base junction, 0.52 V for the collector-substrate junction, and about 0.7 V for the emitter-base junction.

■ **EXAMPLE**

Calculate the collector-base capacitance of the device of Fig. 2.25. The zero-bias capacitance per unit area of the collector-base junction can be found from Fig. 2.29 to be approximately 10^{-4} pF/μm^2. The total area of the collector-base junction is the sum of the area of the bottom of the base diffusion plus the base sidewall area. From Fig. 2.25, the bottom area is

$$A_{\text{bottom}} = 60 \ \mu\text{m} \times 45 \ \mu\text{m} = 2700 \ \mu\text{m}^2$$

The edges of the base region can be seen from Fig. 2.17 to have the shape similar to one-quarter of a cylinder. We will assume that the region is cylindrical in shape, which yields a sidewall area of

$$A_{\text{sidewall}} = P \times d \times \frac{\pi}{2}$$

where
- P = base region periphery
- d = base diffusion depth

Thus we have

$$A_{\text{sidewall}} = 3\ \mu\text{m} \times (60\ \mu\text{m} + 60\ \mu\text{m} + 45\ \mu\text{m} + 45\ \mu\text{m}) \times \frac{\pi}{2} = 989\ \mu\text{m}^2$$

and the total capacitance is

$$C_{\mu 0} = (A_{\text{bottom}} + A_{\text{sidewall}})(10^{-4}\ \text{pF}/\mu\text{m}^2) = 0.36\ \text{pF}$$

■

Collector-Substrate Capacitance. The collector-substrate capacitance consists of three portions: that of the junction between the buried layer and the substrate, that of the sidewall of the isolation diffusion, and that between the epitaxial material and the substrate. Since the substrate has an impurity concentration of about 10^{16} cm^{-3}, it is more heavily doped than the epi material, and we can analyze both the sidewall and epi-substrate capacitance under the assumption that the junction is a one-sided step junction with the epi material as the lightly doped side. Under this assumption, the capacitance per unit area in these regions is the same as in the collector-base junction.

■ **EXAMPLE**

Calculate the collector-substrate capacitance of the standard device of Fig. 2.25. The area of the collector-substrate sidewall is

$$A_{\text{sidewall}} = (17\ \mu\text{m})(140\ \mu\text{m} + 140\ \mu\text{m} + 95\ \mu\text{m} + 95\ \mu\text{m})\left(\frac{\pi}{2}\right) = 12{,}550\ \mu\text{m}^2$$

We will assume that the actual buried layer covers the area defined by the mask, indicated on Fig. 2.25 as an area of 41 μm × 85 μm, plus 8 μm of side-diffusion on each edge. This gives a total area of 57 μm × 101 μm. The area of the junction between the epi material and the substrate is the total area of the isolated region, minus that of the buried layer.

$$A_{\text{epi-substrate}} = (140\ \mu\text{m} \times 95\ \mu\text{m}) - (57\ \mu\text{m} \times 101\ \mu\text{m})$$
$$= 7543\ \mu\text{m}^2$$

The capacitances of the sidewall and epi-substrate junctions are, using a capacitance per unit area of 10^{-4} pF/μm^2

$$C_{cs0}\ (\text{sidewall}) = (12{,}550\ \mu\text{m}^2)(10^{-4}\ \text{pF}/\mu\text{m}^2) = 1.26\ \text{pF}$$
$$C_{cs0}(\text{epi-substrate}) = (7543\ \mu\text{m}^2)(10^{-4}\ \text{pF}/\mu\text{m}^2) = 0.754\ \text{pF}$$

For the junction between the buried layer and the substrate, the lightly doped side of the junction is the substrate. Assuming a substrate doping level of 10^{16} atoms/cm^3, and a built-in voltage of 0.52 V, we can calculate the zero-bias capacitance per unit area as 3.3×10^{-4} pF/μm^2. The area of the buried layer is

$$A_{BL} = 57\ \mu\text{m} \times 101\ \mu\text{m} = 5757\ \mu\text{m}^2$$

and the zero-bias capacitance from the buried layer to the substrate is thus

$$C_{cs0}(BL) = (5757\ \mu\text{m}^2)(3.3 \times 10^{-4}\ \text{pF}/\mu\text{m}^2) = 1.89\ \text{pF}$$

The total zero-bias, collector-substrate capacitance is thus

$$C_{cs0} = 1.26\ \text{pF} + 0.754\ \text{pF} + 1.89\ \text{pF} = 3.90\ \text{pF}$$

■

Emitter-Base Capacitance. The emitter-base junction of the transistor has a doping profile that is not well approximated by a step junction because the impurity concentration on both sides of the junction varies with distance in a rather complicated way. Furthermore, the sidewall capacitance per unit area is not constant but varies with distance from the surface because the base impurity concentration varies with distance. A precise evaluation of this capacitance can be carried out numerically, but a first-order estimate of the capacitance can be obtained by calculating the capacitance of an abrupt junction with an impurity concentration on the lightly doped side that is equal to the concentration in the base at the edge of the junction. The sidewall contribution is neglected.

■ **EXAMPLE**

Calculate the zero-bias, emitter-base junction capacitance of the standard device of Fig. 2.25.

We first estimate the impurity concentration at the emitter edge of the base region. From Fig. 2.17, it can be seen that this concentration is approximately 10^{17} atoms/cm^3. From the nomograph of Fig. 2.29, this abrupt junction would have a zero-bias capacitance per unit area of 10^{-3} pF/μm^2. Since the area of the bottom portion of the emitter-base junction is 25 μm $\times$ 20 μm, the capacitance of the bottom portion is

$$C_{\text{bottom}} = (500 \; \mu\text{m}^2)(10^{-3} \; \text{pF}/\mu\text{m}^2) = 0.5 \; \text{pF}$$

Again assuming a cylindrical cross section, the sidewall area is given by

$$A_{\text{sidewall}} = 2 \, (25 \; \mu\text{m} + 20 \; \mu\text{m}) \left(\frac{\pi}{2}\right)(2.5 \; \mu\text{m}) = 353 \; \mu\text{m}^2$$

Assuming that the capacitance per unit area of the sidewall is approximately the same as the bottom,

$$C_{\text{sidewall}} = (353 \; \mu\text{m}^2)(10^{-3} \; \text{pF}/\mu\text{m}^2) = 0.35 \; \text{pF}$$

The total emitter-base capacitance is

$$C_{je0} = 0.85 \; \text{pF}$$

■

Current Gain. As described in Chapter 1, the current gain of the transistor depends on minority-carrier lifetime in the base, which affects the base transport factor, and on the diffusion length in the emitter, which affects the emitter efficiency. In analog IC processing, the base minority-carrier lifetime is sufficiently long that the base transport factor is not a limiting factor in the forward current gain in *npn* transistors. Because the emitter region is heavily doped with phosphorus, the minority-carrier lifetime is degraded in this region, and current gain is limited primarily by emitter efficiency.[12] Because the doping level, and hence lifetime, vary with distance in the emitter, the calculation of emitter efficiency for the *npn* transistor is difficult, and measured parameters must be used. The room-temperature current gain typically lies between 200 and 1000 for these devices. The current gain falls with decreasing temperature, usually to a value of from 0.5 to 0.75 times the room temperature value at $-55°$C.

Summary of High-Voltage *npn* Device Parameters. A typical set of device parameters for the device of Fig. 2.25 is shown in Fig. 2.30. This transistor geometry is typical of that used for circuits that must operate at power supply voltages up to 40 V. For lower operating voltages,

Parameter	Typical Value, 5-Ω-cm, 17-μm epi 44-V Device	Typical Value, 1-Ω-cm, 10-μm epi 20-V Device
β_F	200	200
B_R	2	2
V_A	130 V	90 V
η	2×10^{-4}	2.8×10^{-4}
I_S	5×10^{-15} A	1.5×10^{-15} A
I_{CO}	10^{-10} A	10^{-10} A
BV_{CEO}	50 V	25 V
BV_{CBO}	90 V	50 V
BV_{EBO}	7 V	7 V
τ_F	0.35 ns	0.25 ns
τ_R	400 ns	200 ns
β_0	200	150
r_b	200 Ω	200 Ω
r_c (saturation)	200 Ω	75 Ω
r_{ex}	2 Ω	2 Ω
Base-emitter junction $\begin{cases} C_{je0} \\ \psi_{0e} \\ n_e \end{cases}$	1 pF 0.7 V 0.33	1.3 pF 0.7 V 0.33
Base-collector junction $\begin{cases} C_{\mu 0} \\ \psi_{0c} \\ n_c \end{cases}$	0.3 pF 0.55 V 0.5	0.6 pF 0.6 V 0.5
Collector-substrate junction $\begin{cases} C_{cs0} \\ \psi_{0s} \\ n_s \end{cases}$	3 pF 0.52 V 0.5	3 pF 0.58 V 0.5

Figure 2.30 Typical parameters for high-voltage integrated *npn* transistors with 500 μm² emitter area. The thick epi device is typical of those used in circuits operating at up to 44 V power-supply voltage, while the thinner device can operate up to about 20 V. While the geometry of the thin epi device is smaller, the collector-base capacitance is larger because of the heavier epi doping. The emitter-base capacitance is higher because the base is shallower, and the doping level in the base at the emitter-base junction is higher.

thinner epitaxial layers can be used, and smaller device geometries can be used as a result. Also shown in Fig. 2.30 are typical parameters for a device made with 1-Ω-cm epi material, which is 10 μm thick. Such a device is physically smaller and has a collector-emitter breakdown voltage of about 25 V.

Advanced-Technology Oxide-Isolated *npn* Bipolar Transistors. The structure of an advanced oxide-isolated, poly-emitter *npn* bipolar transistor is shown in plan view and cross section in Fig. 2.31. Typical parameters for such a device are listed in Fig. 2.32. Note the enormous reduction in device size, transit time, and parasitic capacitance compared to the high-voltage, deep-diffused process. These very small devices achieve optimum performance characteristics at relatively low bias currents. The value of β for such a device typically peaks at a collector current of about 50 μA. For these advanced-technology transistors, the use of ion implantation allows precise control of very shallow emitter (0.1 μm) and base (0.2 μm)

2.5 Active Devices in Bipolar Analog Integrated Circuits

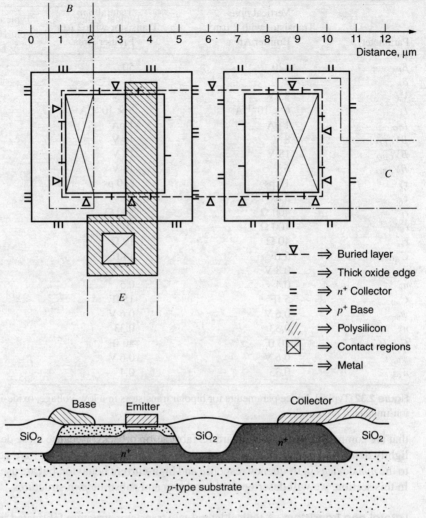

Figure 2.31 Plan view and cross section of a typical advanced-technology bipolar transistor. Note the much smaller dimensions compared with the high-voltage device.

regions. The resulting base width is of the order of 0.1 μm, and (1.99) predicts a base transit time about 25 times smaller than the deep-diffused device of Fig. 2.17. This is observed in practice, and the ion-implanted transistor has a peak f_T of about 13 GHz.

2.5.2 Integrated-Circuit *pnp* Transistors

As mentioned previously, the integrated-circuit bipolar fabrication process is an outgrowth of that used to build double-diffused epitaxial *npn* transistors, and the technology inherently produces *npn* transistors of high performance. However, *pnp* transistors of comparable performance are not easily produced in the same process, and the earliest analog integrated circuits used no *pnp* transistors. The lack of a complementary device for use in biasing, level shifting, and as load devices in amplifier stages proved to be a severe limitation on the performance attainable in analog circuits, leading to the development of several *pnp* transistor structures

Parameter	Vertical *npn* Transistor with 2 μm² Emitter Area	Lateral *pnp* Transistor with 2 μm² Emitter Area
β_F	120	50
β_R	2	3
V_A	35 V	30 V
I_S	6×10^{-18} A	6×10^{-18} A
I_{CO}	1 pA	1 pA
BV_{CEO}	8 V	14 V
BV_{CBO}	18 V	18 V
BV_{EBO}	6 V	18 V
τ_F	10 ps	650 ps
τ_R	5 ns	5 ns
r_b	400 Ω	200 Ω
r_c	100 Ω	20 Ω
r_{ex}	40 Ω	10 Ω
C_{je0}	5 fF	14 fF
ψ_{0e}	0.8 V	0.7 V
n_e	0.4	0.5
$C_{\mu 0}$	5 fF	15 fF
ψ_{0c}	0.6 V	0.6 V
n_c	0.33	0.33
C_{cs0} (C_{bs0})	20 fF	40 fF
ψ_{0s}	0.6 V	0.6 V
n_s	0.33	0.4

Figure 2.32 Typical device parameters for bipolar transistors in a low-voltage, oxide-isolated, ion-implanted process.

that are compatible with the standard IC fabrication process. Because these devices utilize the lightly doped *n*-type epitaxial material as the base of the transistor, they are generally inferior to the *npn* devices in frequency response and high-current behavior, but are useful nonetheless. In this section, we will describe the lateral *pnp* and substrate *pnp* structures.

Lateral *pnp* Transistors. A typical lateral *pnp* transistor structure fabricated in a high-voltage process is illustrated in Fig. 2.33a.[13] The emitter and collector are formed with the same diffusion that forms the base of the *npn* transistors. The collector is a *p*-type ring around the emitter, and the base contact is made in the *n*-type epi material *outside* the collector ring. The flow of minority carriers across the base is illustrated in Fig. 2.33b. Holes are injected from the emitter, flow parallel to the surface across the *n*-type base region, and ideally are collected by the *p*-type collector before reaching the base contact. Thus the transistor action is *lateral* rather than *vertical* as in the case for *npn* transistors. The principal drawback of the structure is the fact that the base region is more lightly doped than the collector. As a result, the collector-base depletion layer extends almost entirely into the base. The base region must then be made wide enough so that the depletion layer does not reach the emitter when the maximum collector-emitter voltage is applied. In a typical analog IC process, the width of this depletion layer is 6 μm to 8 μm when the collector-emitter voltage is in the 40-V range. Thus the minimum base width for such a device is about 8 μm, and the minimum base transit time can be estimated from (1.99) as

$$\tau_F = \frac{W_B^2}{2D_p} \tag{2.20}$$

2.5 Active Devices in Bipolar Analog Integrated Circuits

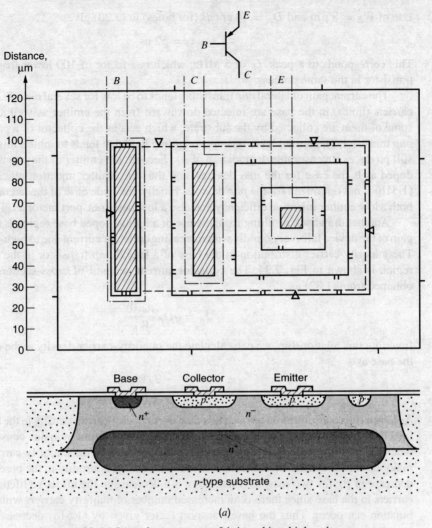

Figure 2.33 (a) Lateral *pnp* structure fabricated in a high-voltage process.

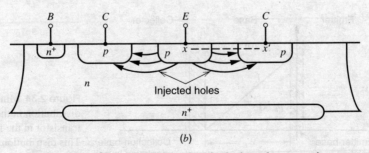

Figure 2.33 (b) Minority-carrier flow in the lateral *pnp* transistor.

Use of $W_B = 8$ μm and $D_p = 10$ cm²/s (for holes) in (2.20) gives

$$\tau_F = 32 \text{ ns}$$

This corresponds to a peak f_T of 5 MHz, which is a factor of 100 lower than a typical *npn* transistor in the same process.

The current gain of lateral *pnp* transistors tends to be low for several reasons. First, minority carriers (holes) in the base are injected downward from the emitter as well as laterally, and some of them are collected by the substrate, which acts as the collector of a parasitic vertical *pnp* transistor. The buried layer sets up a retarding field that tends to inhibit this process, but it still produces a measurable degradation of β_F. Second, the emitter of the *pnp* is not as heavily doped as is the case for the *npn* devices, and thus the emitter injection efficiency given by (1.51b) is not optimized for the *pnp* devices. Finally, the wide base of the lateral *pnp* results in both a low emitter injection efficiency and also a low base transport factor as given by (1.51a).

Another drawback resulting from the use of a lightly doped base region is that the current gain of the device falls very rapidly with increasing collector current due to high-level injection. The minority-carrier distribution in the base of a lateral *pnp* transistor in the forward-active region is shown in Fig. 2.34. The collector current per unit of cross-sectional area can be obtained from (1.32) as

$$J_p = qD_p \frac{p_n(0)}{W_B} \qquad (2.21)$$

Inverting this relationship, we can calculate the minority-carrier density at the emitter edge of the base as

$$p_n(0) = \frac{J_p W_B}{qD_p} \qquad (2.22)$$

As long as this concentration is much less than the majority-carrier density in the base, low-level injection conditions exist and the base minority-carrier lifetime remains constant. However, when the minority-carrier density becomes comparable with the majority-carrier density, the majority-carrier density must increase to maintain charge neutrality in the base. This causes a decrease in β_F for two reasons. First, there is a decrease in the effective lifetime of minority carriers in the base since there is an increased number of majority carriers with which recombination can occur. Thus the base transport factor given by (1.51a) decreases. Second, the increase in the majority-carrier density represents an effective increase in base doping density. This causes a decrease in emitter injection efficiency given by (1.51b). Both these mechanisms are also present in *npn* transistors, but occur at much higher current levels due to the higher doping density in the base of the *npn* transistor.

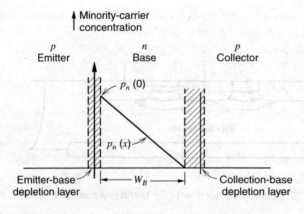

Figure 2.34 Minority-carrier distribution in the base of a lateral *pnp* transistor in the forward-active region. This distribution is that observed through section x-x' in Fig. 2.33b.

2.5 Active Devices in Bipolar Analog Integrated Circuits

The collector current at which these effects become significant can be calculated for a lateral *pnp* transistor by equating the minority-carrier concentration given by (2.22) to the equilibrium majority-carrier concentration. Thus

$$\frac{J_p W_B}{q D_p} = n_n \simeq N_D \tag{2.23}$$

where (2.1) has been substituted for n_n, and N_D is the donor density in the *pnp* base (*npn* collector). From (2.23), we can calculate the collector current for the onset of high-level injection in a *pnp* transistor as

$$I_C = \frac{q A N_D D_p}{W_B} \tag{2.24}$$

where A is the effective area of the emitter-base junction. Note that this current depends directly on the base doping density in the transistor, and since this is quite low in a lateral *pnp* transistor, the current density at which this fall-off begins is quite low.

Lateral *pnp* transistors are also widely used in shallow oxide-isolated bipolar IC technologies. The device structure used is essentially identical to that of Fig. 2.33, except that the device area is orders of magnitude smaller and the junction isolation is replaced by oxide isolation. Typical parameters for such a device are listed in Fig. 2.32. As in the case of *npn* transistors, we see dramatic reductions in device transit time and parasitic capacitance compared to the high-voltage, thick-epi process. The value of β for such a device typically peaks at a collector current of about 50 nA.

■ **EXAMPLE**

Calculate the collector current at which the current gain begins to fall for the *pnp* structure of Fig. 2.33a. The effective cross-sectional area A of the emitter is the sidewall area of the emitter, which is the *p*-type diffusion depth multiplied by the periphery of the emitter multiplied by $\pi/2$.

$$A = (3 \ \mu m)(30 \ \mu m + 30 \ \mu m + 30 \ \mu m + 30 \ \mu m)\left(\frac{\pi}{2}\right) = 565 \ \mu m^2 = 5.6 \times 10^{-6} \ cm^2$$

The majority-carrier density is 10^{15} atoms/cm^3 for an epi-layer resistivity of 5 Ω-cm. In addition, we can assume $W_B = 8 \ \mu m$ and $D_p = 10$ cm^2/s. Substitution of this data in (2.24) gives

$$I_C = 5.6 \times 10^{-6} \times 1.6 \times 10^{-19} \times 10^{15} \times 10 \frac{1}{8 \times 10^{-4}} A = 11.2 \ \mu A$$

■

The typical lateral *pnp* structure of Fig. 2.33a shows a low-current beta of approximately 30 to 50, which begins to decrease at a collector current of a few tens of microamperes, and has fallen to less than 10 at a collector current of 1 mA. A typical set of parameters for a structure of this type is shown in Fig. 2.35. Note that in the lateral *pnp* transistor, the substrate junction capacitance appears between the *base* and the substrate.

Substrate *pnp* Transistors. One reason for the poor high-current performance of the lateral *pnp* is the relatively small effective cross-sectional area of the emitter, which results from the lateral nature of the injection. A common application for a *pnp* transistor is in a Class-B output stage where the device is called on to operate at collector currents in the 10-mA range. A lateral *pnp* designed to do this would require a large amount of die area. In this application, a different structure is usually used in which the substrate itself is used as the collector instead of a diffused *p*-type region. Such a substrate *pnp* transistor in a high-voltage, thick-epi process is

Parameter		Typical Value, 5-Ω-cm, 17-μm epi 44-V Device	Typical Value, 1-Ω-cm, 10-μm epi 20-V Device
	β_F	50	20
	β_R	4	2
	V_A	50 V	50 V
	η	5×10^{-4}	5×10^{-4}
	I_S	2×10^{-15} A	2×10^{-15} A
	I_{CO}	10^{-10} A	5×10^{-9} A
	BV_{CEO}	60 V	30 V
	BV_{CBO}	90 V	50 V
	BV_{EBO}	90 V	50 V
	τ_F	30 ns	20 ns
	τ_R	3000 ns	2000 ns
	β_0	50	20
	r_b	300 Ω	150 Ω
	r_c	100 Ω	75 Ω
	r_{ex}	10 Ω	10 Ω
Base-emitter junction	C_{je0}	0.3 pF	0.6 pF
	ψ_{0e}	0.55 V	0.6 V
	n_e	0.5	0.5
Base-collector junction	$C_{\mu 0}$	1 pF	2 pF
	ψ_{0c}	0.55 V	0.6 V
	n_c	0.5	0.5
Base-substrate junction	C_{bs0}	3 pF	3.5 pF
	ψ_{0s}	0.52 V	0.58 V
	n_s	0.5	0.5

Figure 2.35 Typical parameters for lateral *pnp* transistors with 900 μm² emitter area in a high-voltage, thick-epi process.

shown in Fig. 2.36a. The *p*-type emitter diffusion for this particular substrate *pnp* geometry is rectangular with a rectangular hole in the middle. In this hole an n^+ region is formed with the *npn* emitter diffusion to provide a contact for the *n*-type base. Because of the lightly doped base material, the series base resistance can become quite large if the base contact is far removed from the active base region. In this particular structure, the n^+ base contact diffusion is actually allowed to come in contact with the *p*-type emitter diffusion, in order to get the low-resistance base contact diffusion as close as possible to the active base. The only drawback of this, in a substrate *pnp* structure, is that the emitter-base breakdown voltage is reduced to approximately 7 V. If larger emitter-base breakdown is required, then the *p*-emitter diffusion must be separated from the n^+ base contact diffusion by a distance of about 10 μm to 15 μm. Many variations exist on the substrate *pnp* geometry shown in Fig. 2.36a. They can also be realized in thin-epi, oxide-isolated processes.

The minority-carrier flow in the forward-active region is illustrated in Fig. 2.36b. The principal advantage of this device is that the current flow is vertical and the effective cross-sectional area of the emitter is much larger than in the case of the lateral *pnp* for the same overall device size. The device is restricted to use in emitter-follower configurations, however, since the collector is electrically identical with the substrate that must be tied to the most negative circuit potential. Other than the better current-handling capability, the properties of substrate *pnp* transistors are similar to those for lateral *pnp* transistors since the base width is similar

2.5 Active Devices in Bipolar Analog Integrated Circuits

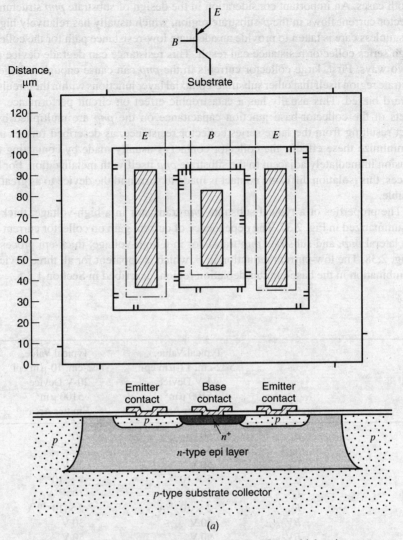

Figure 2.36 (a) Substrate *pnp* structure in a high-voltage, thick-epi process.

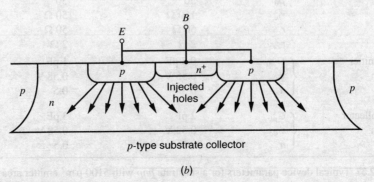

Figure 2.36 (b) Minority-carrier flow in the substrate *pnp* transistor.

in both cases. An important consideration in the design of substrate *pnp* structures is that the collector current flows in the *p*-substrate region, which usually has relatively high resistivity. Thus, unless care is taken to provide an adequate low-resistance path for the collector current, a high series collector resistance can result. This resistance can degrade device performance in two ways. First, large collector currents in the *pnp* can cause enough voltage drop in the substrate region itself that other substrate-epitaxial layer junctions within the circuit can become forward biased. This usually has a catastrophic effect on circuit performance. Second, the effects of the collector-base junction capacitance on the *pnp* are multiplied by the Miller effect resulting from the large series collector resistance, as described further in Chapter 7. To minimize these effects, the collector contact is usually made by contacting the isolation diffusion immediately adjacent to the substrate *pnp* itself with metallization. For high-current devices, this isolation diffusion contact is made to surround the device to as great an extent as possible.

The properties of a typical substrate *pnp* transistor in a high-voltage, thick-epi process are summarized in Fig. 2.37. The dependence of current gain on collector current for a typical *npn*, lateral *pnp*, and substrate *pnp* transistor in a high-voltage, thick-epi process are shown in Fig. 2.38. The low-current reduction in β, which is apparent for all three devices, is due to recombination in the base-emitter depletion region, described in Section 1.3.5.

	Parameter	Typical Value, 5-Ω-cm, 17-μm epi 44-V Device 5100 μm² Emitter Area	Typical Value, 1-Ω-cm, 10-μm epi 20-V Device 5100 μm² Emitter Area
	β_F	50	30
	β_R	4	2
	V_A	50 V	30 V
	η	5×10^{-4}	9×10^{-4}
	I_S	10^{-14} A	10^{-14} A
	I_{CO}	2×10^{-10} A	2×10^{-10} A
	BV_{CEO}	60 V	30 V
	BV_{CBO}	90 V	50 V
	BV_{EBO}	7 V or 90 V	7 V or 50 V
	τ_F	20 ns	14 ns
	τ_R	2000 ns	1000 ns
	β_0	50	30
	r_b	150 Ω	50 Ω
	r_c	50 Ω	50 Ω
	r_{ex}	2 Ω	2 Ω
Base-emitter junction	C_{je0}	0.5 pF	1 pF
	ψ_{0e}	0.55 V	0.58 V
	n_e	0.5	0.5
Base-collector junction	$C_{\mu 0}$	2 pF	3 pF
	ψ_{0c}	0.52 V	0.58 V
	n_c	0.5	0.5

Figure 2.37 Typical device parameters for a substrate *pnp* with 5100 μm² emitter area in a high-voltage, thick-epi process.

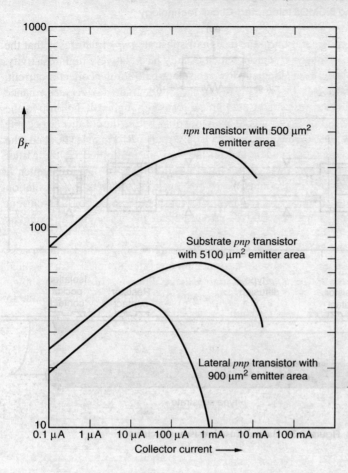

Figure 2.38 Current gain as a function of collector current for typical lateral *pnp*, substrate *pnp*, and *npn* transistor geometries in a high-voltage, thick-epi process.

2.6 Passive Components in Bipolar Integrated Circuits

In this section, we describe the structures available to the integrated-circuit designer for realization of resistance and capacitance. Resistor structures include base-diffused, emitter-diffused, ion-implanted, pinch, epitaxial, and pinched epitaxial resistors. Other resistor technologies, such as thin-film resistors, are considered in Section 2.7.3. Capacitance structures include MOS and junction capacitors. Inductors with values larger than a few nanohenries have not proven to be feasible in monolithic technology. However, such small inductors are useful in very high frequency integrated circuits.[14,15,16]

2.6.1 Diffused Resistors

In an earlier section of this chapter, the sheet resistance of a diffused layer was calculated. Integrated-circuit resistors are generally fabricated using one of the diffused or ion-implanted layers formed during the fabrication process, or in some cases a combination of two layers. The layers available for use as resistors include the base, the emitter, the epitaxial layer, the buried layer, the active-base region layer of a transistor, and the epitaxial layer pinched between the base diffusion and the *p*-type substrate. The choice of layer generally depends on the value, tolerance, and temperature coefficient of the resistor required.

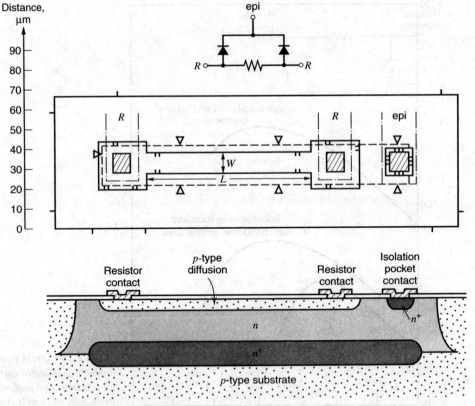

Figure 2.39 Base-diffused resistor structure.

Base and Emitter Diffused Resistors. The structure of a typical base-diffused resistor in a high-voltage process is shown in Fig. 2.39. The resistor is formed from the *p*-type base diffusion for the *npn* transistors and is situated in a separate isolation region. The epitaxial region into which the resistor structure is diffused must be biased in such a way that the *pn* junction between the resistor and the epi layer is always reverse biased. For this reason, a contact is made to the *n*-type epi region as shown in Fig. 2.39, and it is connected either to that end of the resistor that is most positive or to a potential that is more positive than either end of the resistor. The junction between these two regions contributes a parasitic capacitance between the resistor and the epi layer, and this capacitance is distributed along the length of the resistor. For most applications, this parasitic capacitance can be adequately modeled by separating it into two lumped portions and placing one lump at each end of the resistor as illustrated in Fig. 2.40.

The resistance of the structure shown in Fig. 2.39 is given by (2.10) as

$$R = \frac{L}{W} R_\square$$

where L is the resistor length and W is the width. The base sheet resistance $R_\square$ lies in the range 100 to 200 $\Omega/\square$, and thus resistances in the range 50 Ω to 50 kΩ are practical using the base diffusion. The resistance contributed by the *clubheads* at each end of the resistor can be significant, particularly for small values of L/W. The clubheads are required to allow space for ohmic contact to be made at the ends of the resistor.

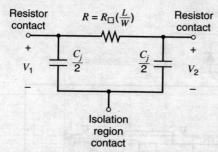

Figure 2.40 Lumped model for the base-diffused resistor.

Since minimization of die area is an important objective, the width of the resistor is kept as small as possible, the minimum practical width being limited to about 1 μm by photolithographic considerations. Both the tolerance on the resistor value and the precision with which two identical resistors can be matched can be improved by the use of wider geometries. However, for a given base sheet resistance and a given resistor value, the area occupied by the resistor increases as the *square* of its width. This can be seen from (2.10) since the ratio L/W is constant.

In shallow ion-implanted processes, the ion-implanted base can be used in the same way to form a resistor.

■ **EXAMPLE**

Calculate the resistance and parasitic capacitance of the base-diffused resistor structure shown in Fig. 2.39 for a base sheet resistance of 100 Ω/□, and an epi resistivity of 2.5 Ω-cm. Neglect end effects. The resistance is simply

$$R = 100 \ \Omega/\square \left(\frac{100 \ \mu m}{10 \ \mu m} \right) = 1 \ k\Omega$$

The capacitance is the total area of the resistor multiplied by the capacitance per unit area. The area of the resistor body is

$$A_1 = (10 \ \mu m)(100 \ \mu m) = 1000 \ \mu m^2$$

The area of the clubheads is

$$A_2 = 2 \ (30 \ \mu m \times 30 \ \mu m) = 1800 \ \mu m^2$$

The total zero-bias capacitance is, from Fig. 2.29,

$$C_{j0} = (10^{-4} \ pF/\mu m^2)(2800 \ \mu m^2) = 0.28 \ pF$$

As a first-order approximation, this capacitance can be divided into two parts, one placed at each end. Note that this capacitance will vary depending on the voltage at the clubhead with respect to the epitaxial pocket. ■

Emitter-diffused resistors are fabricated using geometries similar to the base resistor, but the emitter diffusion is used to form the actual resistor. Since the sheet resistance of this diffusion is in the 2 to 10 Ω/□ range, these resistors can be used to advantage where very low resistance values are required. In fact, they are widely used simply to provide a crossunder beneath an aluminum metallization interconnection. The parasitic capacitance can be

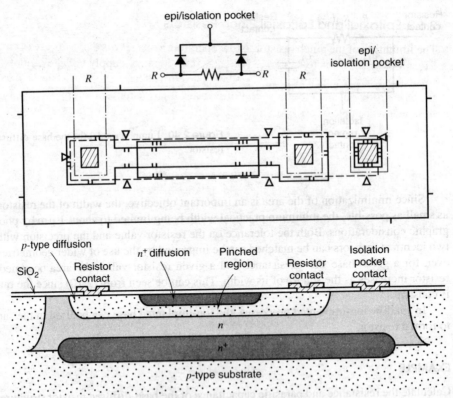

Figure 2.41 Pinch resistor structure.

calculated in a way similar to that for the base diffusion. However, these resistors have different temperature dependence from base-diffused resistors and the two types do not track with temperature.

Base Pinch Resistors. A third layer available for use as a resistor is the layer that forms the active base region in the *npn* transistor. This layer is *pinched* between the n^+ emitter and the *n*-type collector regions, giving rise to the term *pinch resistor*. The layer can be electrically isolated by reverse biasing the emitter-base and collector-base junctions, which is usually accomplished by connecting the *n*-type regions to the most positive end of the resistor. The structure of a typical pinch resistor is shown in Fig. 2.41; the n^+ diffusion overlaps the *p*-diffusion so that the n^+ region is electrically connected to the *n*-type epi region. The sheet resistance is in the 5 kΩ/□ to 15 kΩ/□ range. As a result, this resistor allows the fabrication of large values of resistance. Unfortunately, the sheet resistance undergoes the same process-related variations as does the Q_B of the transistor, which is approximately ± 50 percent. Also, because the material making up the resistor itself is relatively lightly doped, the resistance displays a relatively large variation with temperature. Another significant drawback is that the maximum voltage that can be applied across the resistor is limited to around 6 V because of the breakdown voltage between the emitter-diffused top layer and the base diffusion. Nonetheless, this type of resistor has found wide application where the large tolerance and low breakdown voltage are not significant drawbacks.

2.6.2 Epitaxial and Epitaxial Pinch Resistors

The limitation of the pinch resistor to low operating voltages disallows its use in circuits where a small bias current is to be derived directly from a power-supply voltage of more than about 7 V using a large-value resistor. The epitaxial layer itself has a sheet resistance much larger than the base diffusion, and the epi layer is often used as a resistor for this application. For example, the sheet resistance of a 17-μm thick, 5-Ω-cm epi layer can be calculated from (2.11) as

$$R_\Box = \frac{\rho_{epi}}{T} = \frac{5 \text{ }\Omega\text{-cm}}{(17 \text{ }\mu\text{m}) \times (10^{-4} \text{ cm/}\mu\text{m})} = 2.9 \text{ k}\Omega/\Box \qquad (2.25)$$

Large values of resistance can be realized in a small area using structures of the type shown in Fig. 2.42. Again, because of the light doping in the resistor body, these resistors display a rather large temperature coefficient. A still larger sheet resistance can be obtained by putting a *p*-type base diffusion over the top of an epitaxial resistor, as shown in Fig. 2.42. The depth of the *p*-type base and the thickness of the depletion region between the *p*-type base and the

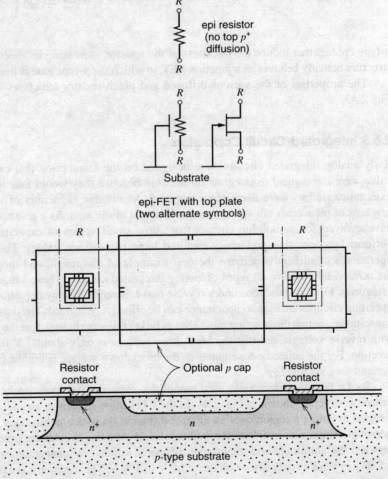

Figure 2.42 Epitaxial resistor structure. The *p*-cap diffusion is optional and forms an epitaxial pinch resistor.

Resistor Type	Sheet ρ $\Omega/\square$	Absolute Tolerance (%)	Matching Tolerance (%)	Temperature Coefficient
Base diffused	100 to 200	± 20	± 2 (5 μm wide) ± 0.2 (50 μm wide)	(+1500 to +2000) ppm/°C
Emitter diffused	2 to 10	± 20	± 2	+600 ppm/°C
Ion implanted	100 to 1000	± 3	± 1 (5 μm wide) ± 0.1 (50 μm wide)	Controllable to ± 100 ppm/°C
Base pinch	2k to 10k	± 50	± 10	+2500 ppm/°C
Epitaxial	2k to 5k	± 30	± 5	+3000 ppm/°C
Epitaxial pinch	4k to 10k	± 50	± 7	+3000 ppm/°C
Thin film	0.1k to 2k	± 5 to ± 20	± 0.2 to ± 2	(± 10 to ± 200) ppm/°C

Figure 2.43 Summary of resistor properties for different types of IC resistors.

n-type epi together reduce the thickness of the resistor, increasing its sheet resistance. Such a structure actually behaves as a junction FET, in which the p-type gate is tied to the substrate.[17]

The properties of the various diffused and pinch-resistor structures are summarized in Fig. 2.43.

2.6.3 Integrated-Circuit Capacitors

Early analog integrated circuits were designed on the assumption that capacitors of usable value were impractical to integrate on the chip because they would take too much area, and external capacitors were used where required. Monolithic capacitors of value larger than a few tens of picofarads are still expensive in terms of die area. As a result, design approaches have evolved for monolithic circuits that allow small values of capacitance to be used to perform functions that previously required large capacitance values. The compensation of operational amplifiers is perhaps the best example of this result, and monolithic capacitors are now widely used in all types of analog integrated circuits. These capacitors fall into two categories. First, pn junctions under reverse bias inherently display depletion capacitance, and in certain circumstances this capacitance can be effectively utilized. The drawbacks of junction capacitance are that the junction must always be kept reverse biased, that the capacitance varies with reverse voltage, and that the breakdown voltage is only about 7 V for the emitter-base junction. For the collector-base junction, the breakdown voltage is higher, but the capacitance per unit area is quite low.

By far the most commonly used monolithic capacitor in bipolar technology is the MOS capacitor structure shown in Fig. 2.44. In the fabrication sequence, an additional mask step is inserted to define a region over an emitter diffusion on which a thin layer of silicon dioxide is grown. Aluminum metallization is then placed over this thin oxide, producing a capacitor between the aluminum and the emitter diffusion, which has a capacitance of 0.3 fF/μm^2 to 0.5 fF/μm^2 and a breakdown voltage of 60 V to 100 V. This capacitor is extremely linear and has a low temperature coefficient. A sizable parasitic capacitance C_{ISO} is present between the n-type bottom plate and the substrate because of the depletion capacitance of the epi-substrate junction, but this parasitic is unimportant in many applications.

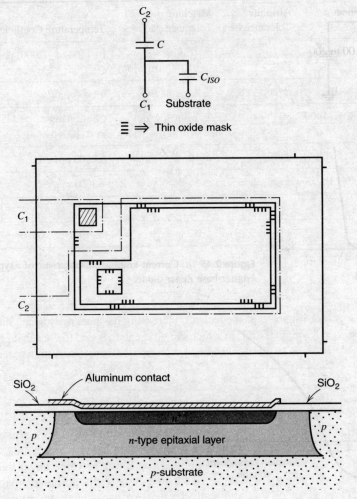

Figure 2.44 MOS capacitor structure.

2.6.4 Zener Diodes

As described in Chapter 1, the emitter-base junction of the *npn* transistor structure displays a reverse breakdown voltage of between 6 V and 8 V, depending on processing details. When the total supply voltage is more than this value, the reverse-biased, emitter-base junction is useful as a voltage reference for the stabilization of bias reference circuits, and for such functions as level shifting. The reverse bias *I-V* characteristic of a typical emitter-base junction is illustrated in Fig. 2.45a.

An important aspect of the behavior of this device is the temperature sensitivity of the breakdown voltage. The actual breakdown mechanism is dominated by quantum mechanical tunneling through the depletion layer when the breakdown voltage is below about 6 V; it is dominated by avalanche multiplication in the depletion layer at the larger breakdown voltages. Because these two mechanisms have opposite temperature coefficients of breakdown voltage, the actually observed breakdown voltage has a temperature coefficient that varies with the value of breakdown voltage itself, as shown in Fig. 2.45b.

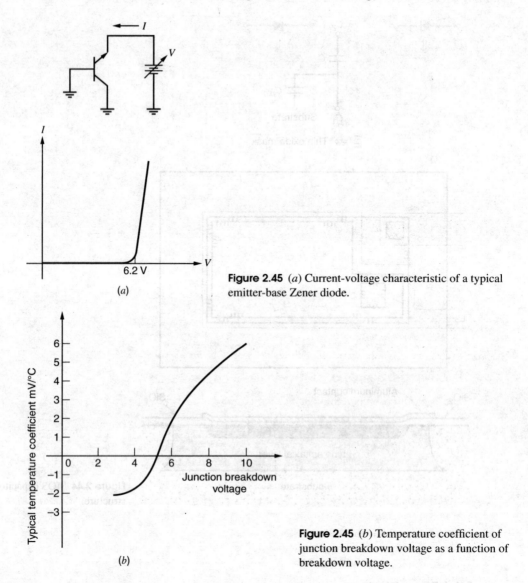

Figure 2.45 (*a*) Current-voltage characteristic of a typical emitter-base Zener diode.

Figure 2.45 (*b*) Temperature coefficient of junction breakdown voltage as a function of breakdown voltage.

2.6.5 Junction Diodes

Junction diodes can be formed by various connections of the *npn* and *pnp* transistor structures, as illustrated in Fig. 2.46. When the diode is forward biased in the diode connections *a*, *b*, and *d* of Fig. 2.46, the collector-base junction becomes forward biased as well. When this occurs, the collector-base junction injects holes into the epi region that can be collected by the reverse-biased, epi-isolation junction or by other devices in the same isolation region. A similar phenomenon occurs when a transistor enters saturation. As a result, substrate currents can flow that can cause voltage drops in the high-resistivity substrate material, and other epi-isolation junctions within the circuit can become inadvertently forward biased. Thus the diode connections of Fig. 2.46*c* are usually preferable since they keep the base-collector junction at zero bias. These connections have the additional advantage of resulting in the smallest amount of minority charge storage within the diode under forward-bias conditions.

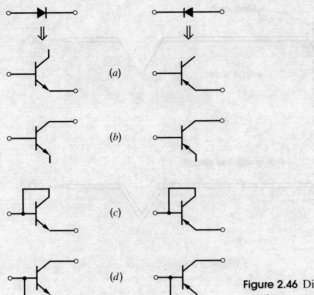

Figure 2.46 Diode connections for *npn* and *pnp* transistors.

2.7 Modifications to the Basic Bipolar Process

The basic high-voltage bipolar IC fabrication process described previously can be modified by the addition of extra processing steps to produce special devices or characteristics.

2.7.1 Dielectric Isolation

We first consider a special isolation technique—*dielectric isolation*—that has been used in digital and analog integrated circuits that must operate at very high speed and/or must operate in the presence of large amounts of radiation. The objective of the isolation technique is to electrically isolate the collectors of the devices from each other with a layer of silicon dioxide rather than with a *pn* junction. This layer has much lower capacitance per unit area than a *pn* junction, and as a result, the collector-substrate capacitance of the transistors is greatly reduced. Also, the reverse photocurrent that occurs with junction-isolated devices under intense radiation is eliminated.

The fabrication sequence used for dielectric isolation is illustrated in Figs. 2.47a–d. The starting material is a wafer of *n*-type material of resistivity appropriate for the collector region of the transistor. The first step is to etch grooves in the back side of the starting wafer, which will become the isolation regions in the finished circuit. These grooves are about 20 μm deep for typical analog circuit processing. This step, called *moat etch*, can be accomplished with a variety of techniques, including a preferential etch that allows precise definition of the depth of the moats. Next, an oxide is grown on the surface and a thick layer of polycrystalline silicon is deposited on the surface. This layer will be the mechanical support for the finished wafer and thus must be on the order of 200 μm thick. Next, the starting wafer is etched or ground from the top side until it is entirely removed except for the material left in the isolated islands between the moats, as illustrated in Fig. 2.47c. After the growth of an oxide, the wafer is ready for the rest of the standard process sequence. Note that the isolation of each device is accomplished by means of an oxide layer.

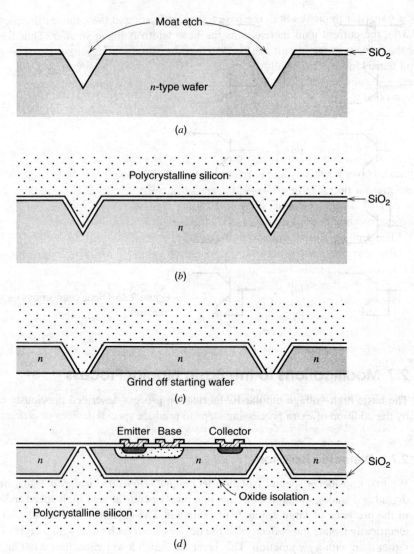

Figure 2.47 Fabrication steps in dielectric isolation. (a) Moat etch on bottom of starting wafer. (b) Deposit polycrystalline silicon support layer. (c) Grind off starting wafer and polish. (d) Carry out standard process, starting with base mask.

2.7.2 Compatible Processing for High-Performance Active Devices

Many specialized circuit applications require a particular type of active device other than the *npn* and *pnp* transistors that result from the standard process schedule. These include high-beta (*superbeta*) *npn* transistors for low-input-current amplifiers, MOSFETs for analog switching and low-input-current amplifiers, and high-speed *pnp* transistors for fast analog circuits. The fabrication of these devices generally requires the addition of one or more mask steps to the basic fabrication process. We now describe these special structures.

Superbeta Transistors. One approach to decreasing the input bias current in amplifiers is to increase the current gain of the input stage transistors.[18] Since a decrease in the base width of a transistor improves both the base transport factor and the emitter efficiency (see Section 1.3.1), the current gain increases as the base width is made smaller. Thus the current gain of the devices in the circuit can be increased by simply increasing the emitter diffusion time and narrowing the base width in the resulting devices. However, any increase in the current gain also causes a reduction in the breakdown voltage BV_{CEO} of the transistors. Section 1.3.4 shows that

$$BV_{CEO} = \frac{BV_{CBO}}{\sqrt[n]{\beta}} \qquad (2.26)$$

where BV_{CBO} is the plane breakdown voltage of the collector-base junction. Thus for a given epitaxial layer resistivity and corresponding collector-base breakdown voltage, an increase in beta gives a decrease in BV_{CEO}. As a result, using such a process modification to increase the beta of all the transistors in an operational amplifier is not possible because the modified transistors could not withstand the required operating voltage.

The problem of the trade-off between current gain and breakdown voltage can be avoided by fabricating two different types of devices on the same die. The standard device is similar to conventional transistors in structure. By inserting a second diffusion, however, high-beta devices also can be formed. A structure typical of such devices is shown in Fig. 2.48. These devices may be made by utilizing the same base diffusion for both devices and using separate emitter diffusions, or by using two different base diffusions and the same emitter diffusion. Both techniques are used. If the superbeta devices are used only as the input transistors in an operational amplifier, they are not required to have a breakdown voltage of more than about 1 V. Therefore, they can be diffused to extremely narrow base widths, giving current gain on the order of 2000 to 5000. At these base widths, the actual breakdown mechanism is often no longer collector multiplication at all but is due to the depletion layer of the collector-base junction depleting the whole base region and reaching the emitter-base depletion layer. This breakdown mechanism is called *punchthrough*. An application of these devices in op-amp design is described in Section 6.9.2.

MOS Transistors. MOS transistors are useful in bipolar integrated-circuit design because they provide high-performance analog switches and low-input-current amplifiers, and particularly because complex digital logic can be realized in a small area using MOS technology. The latter consideration is important since the partitioning of subsystems into analog and digital chips becomes more and more cumbersome as the complexity of the individual chips becomes greater.

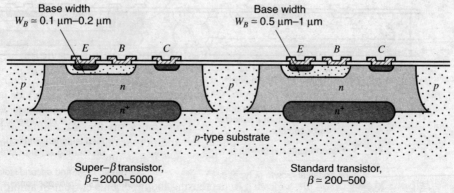

Figure 2.48 Superbeta device structure.

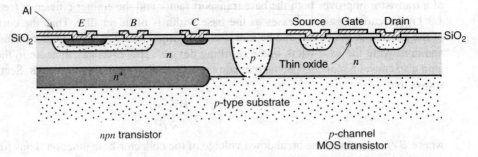

Figure 2.49 Compatible p-channel MOS transistor.

Metal-gate p-channel MOS transistors can be formed in a standard high-voltage bipolar analog IC process with one extra mask step.[19] If a capacitor mask is included in the original sequence, then no extra mask steps are required. As illustrated in Fig. 2.49, the source and drain are formed in the epi material using the base diffusion. The capacitor mask is used to define the oxide region over the channel and the aluminum metallization forms the metal gate.

A major development in IC processing in recent years has been the combination on the same chip of high-performance bipolar devices with CMOS devices in a BiCMOS process. This topic is considered in Section 2.11.

Double-Diffused pnp Transistors. The limited frequency response of the lateral pnp transistor places a limitation on the high-frequency performance attainable with certain types of analog circuits. While this problem can be circumvented by clever circuit design in many cases, the resulting circuit is often quite complex and costly. An alternative approach is to use a more complex process that produces a high-speed, double-diffused pnp transistor with properties comparable to those of the npn transistor.[20] The process usually utilizes three additional mask steps and diffusions: one to form a lightly doped p-type region, which will be the collector of the pnp; one n-type diffusion to form the base of the pnp; and one p-type diffusion to form the emitter of the pnp. A typical resulting structure is shown in Fig. 2.50. This process requires

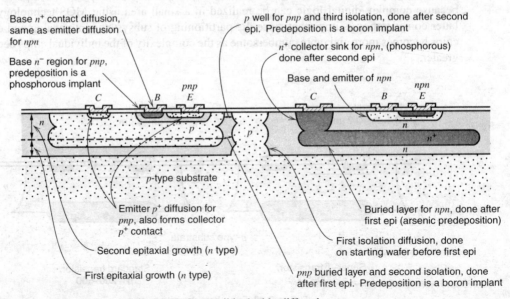

Figure 2.50 Compatible double-diffused pnp process.

2.8 MOS Integrated-Circuit Fabrication

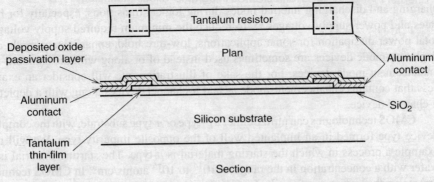

Figure 2.51 Typical thin-film resistor structure.

	Nichrome	Tantalum	Cermet (Cr-SiO)
Range of sheet resistance (Ω/□)	10 to 1000	10 to 1000	30 to 2500
Temperature coefficient (ppm/°C)	±10 to ±150	±5 to ±200	±50 to ±150

Figure 2.52 Properties of monolithic thin-film resistors.

ten masking steps and two epitaxial growth steps. Oxide isolation and poly-emitter technology have been incorporated into more advanced versions of this process.

2.7.3 High-Performance Passive Components

Diffused resistors have three drawbacks: They have high temperature coefficients, they have poor tolerance, and they are junction-isolated. The latter means that a parasitic capacitance is associated with each resistor, and exposure to radiation causes photocurrents to flow across the isolating junction. These drawbacks can be overcome by the use of thin-film resistors deposited on the top surface of the die over an insulating layer of oxide. After the resistor material itself is deposited, the individual resistors are defined in a conventional way using a masking step. They are then interconnected with the rest of the circuit using the standard aluminum interconnect process. The most common materials for the resistors are nichrome and tantalum, and a typical structure is shown in Fig. 2.51. The properties of the resulting resistors using these materials are summarized in Fig. 2.52.

2.8 MOS Integrated-Circuit Fabrication

Fabrication technologies for MOS integrated circuits span a considerably wider spectrum of complexity and performance than those for bipolar technology. CMOS technologies provide two basic types of transistors: enhancement-mode *n*-channel transistors (which have positive thresholds) and enhancement-mode *p*-channel transistors (which have negative thresholds). The magnitudes of the threshold voltages of these transistors are typically set to be 0.6 V to 0.8 V so that the drain current resulting from subthreshold conduction with zero gate-source voltage is very small. This property gives standard CMOS digital circuits high noise margins and essentially zero static power dissipation. However, such thresholds do not always minimize the *total* power dissipation because significant dynamic power is dissipated by charging and discharging internal nodes during logical transitions, especially for high clock rates and power-supply voltages.[21] To reduce the minimum required supply voltage and the total power dissipation for some applications, low-threshold, enhancement-mode devices or depletion-mode devices are sometimes used instead of or along with the standard-threshold, enhancement-mode devices. For the sake of illustration, we will consider an example process that contains enhancement-mode n- and p-channel devices along with a depletion-mode n-channel device.

CMOS technologies can utilize either a p-type or n-type substrate, with the complementary device type formed in an implanted well of the opposite impurity type. We will take as an example a process in which the starting material is p-type. The starting material is a silicon wafer with a concentration in the range of 10^{14} to 10^{15} atoms/cm^3. In CMOS technology, the first step is the formation of a *well* of opposite impurity-type material where the complementary device will be formed. In this case, the well is n-type and is formed by a masking operation and ion implantation of a donor species, typically phosphorus. Subsequent diffusion results in the structure shown in Fig. 2.53. The surface concentration in the well following diffusion is typically between 10^{15} and 10^{16} atoms/cm^3.

Next, a layer of silicon nitride is deposited and defined with a masking operation so that nitride is left only in the areas that are to become active devices. After this masking operation, additional ion implantations are carried out, which increase the surface concentrations in the areas that are not covered by nitride, called the *field regions*. This often involves an extra masking operation so that the surface concentration in the well and that in the substrate areas can be independently controlled by means of separate implants. This increase in surface concentration in the field is necessary because the field regions themselves are MOS transistors with very thick gate oxide. To properly isolate the active devices from one another, the field devices must have a threshold voltage high enough that they never turn on. This can be accomplished by increasing the surface concentration in the field regions. Following the field implants, a local oxidation is performed, which results in the structure shown in Fig. 2.54.

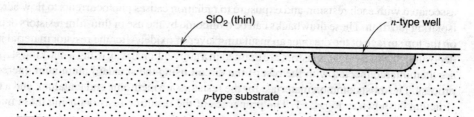

Figure 2.53 Cross section of sample following implantation and diffusion of the n-type well. Subsequent processing will result in formation of an n-channel device in the unimplanted p-type portions of the substrate and a p-type transistor in the n-type well region.

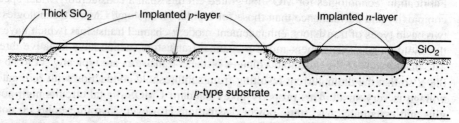

Figure 2.54 Cross section of the sample following field implant steps and field oxidation.

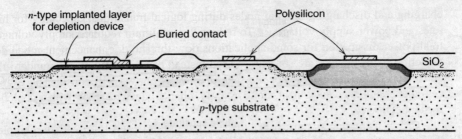

Figure 2.55 Cross section of the sample following deposition and definition of the polysilicon gate layer. Ion implantations have been performed in the thin-oxide regions to adjust the thresholds of the devices.

After field-oxide growth, the nitride is removed from the active areas, and implantation steps are carried out, which adjust the surface concentrations in what will become the channel of the MOS transistors. Equation 1.139, applied to the doping levels usually found in the active-device areas, gives an n-channel threshold of within a few hundred millivolts of zero, and p-channel threshold of about -2 V. To shift the magnitudes of the device threshold voltages to 0.6 V to 0.8 V, an implantation step that changes the impurity concentration at the surface in the channel regions of the two transistor types is usually included. This shift in threshold can sometimes be accomplished by using a single sheet implant over the entire wafer, which simultaneously shifts the thresholds of both types of devices. More typically, however, two separate masked implants are used, one for each device type. Also, if a depletion-mode n-channel device is included in the process, it is defined at this point by a masking operation and subsequent implant to shift the threshold of the selected devices to a negative value so that they are normally on.

Next, a layer of polysilicon is deposited, and the gates of the various devices are defined with a masking operation. The resulting structure is shown in Fig. 2.55. Silicon-gate MOS technology provides three materials that can be used for interconnection: polysilicon, diffusion, and several layers of metal. Unless special provision is made in the process, connections between polysilicon and diffusion layers require a metallization bridge, since the polysilicon layer acts as a mask for the diffused layers. To provide a direct electrical connection between polysilicon and diffusion layers, a buried contact can be included just prior to the polysilicon deposition. This masking operation opens a window in the silicon dioxide under the polysilicon, allowing it to touch the bare silicon surface when it is deposited, forming a direct polysilicon-silicon contact. The depletion device shown in Fig. 2.55 has such a buried contact connecting its source to its gate.

Next, a masking operation is performed such that photoresist covers the p-channel devices, and the wafer is etched to remove the oxide from the source and drain areas of the n-channel devices. Arsenic or phosphorus is then introduced into these areas, using either diffusion or ion implantation. After a short oxidation, the process is repeated for the p-channel source and drain areas, where boron is used. The resulting structure is shown in Fig. 2.56.

At this point in the process, a layer of silicon dioxide is usually deposited on the wafer, using chemical vapor deposition or some other similar technique. This layer is required to reduce the parasitic capacitance of the interconnect metallization and cannot be thermally grown because of the redistribution of the impurities within the device structures that would result during the growth. Following the oxide deposition, the contact windows are formed with a masking operation, and metallization is deposited and defined with a second masking operation. The final structure is shown in Fig. 2.57. A microscope photograph of such a device is shown in Fig. 2.58. Subsequent fabrication steps are as described in Section 2.3 for bipolar technology.

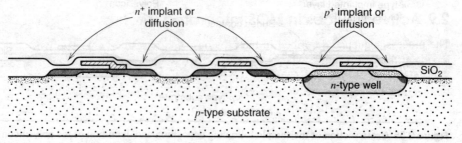

Figure 2.56 Cross section of the sample following the source drain masking and diffusion operations.

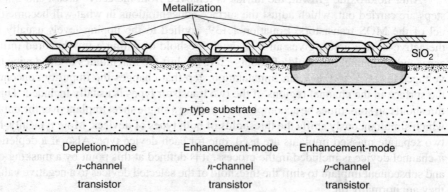

Figure 2.57 Cross section of the sample after final process step. The enhancement and depletion n-channel devices are distinguished from each other by the fact that the depletion device has received a channel implantation of donor impurities to lower its threshold voltage, usually to the range of -1.5 V to -3 V.

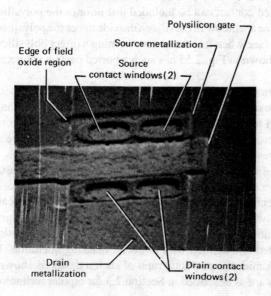

Figure 2.58 Photomicrograph of a silicon-gate MOS transistor. Visible in this picture are the polysilicon gate, field-oxide region boundary, source and drain metallization, and contact windows. In this particular device, the contact windows have been broken into two smaller rectangular openings rather than a single long one as shown in Fig. 2.59. Large contact windows are frequently implemented with an array of small openings so that all individual contact holes in the integrated circuit have the same nominal geometry. This results in better uniformity of the etch rate of the contact windows and better matching.

2.9 Active Devices in MOS Integrated Circuits

The process sequence described in the previous section results in a variety of device types having different threshold voltages, channel mobilities, and parasitic capacitances. In addition, the sequence allows the fabrication of a bipolar emitter follower, using the well as a base. In this section, we explore the properties of these different types of devices.

2.9.1 n-Channel Transistors

A typical layout of an *n*-channel MOS transistor is shown in Fig. 2.59. The electrically active portion of the device is the region under the gate; the remainder of the device area simply provides electrical contact to the terminals. As in the case of integrated bipolar transistors, these areas contribute additional parasitic capacitance and resistance.

In the case of MOS technology, the circuit designer has even greater flexibility than in the bipolar case to tailor the properties of each device to the role it is to play in the individual circuit application. Both the channel width (analogous to the emitter area in bipolar) and the channel length can be defined by the designer. The latter is analogous to the base width of a bipolar device, which is not under the control of the bipolar circuit designer since it is a process parameter and not a mask parameter. In contrast to a bipolar transistor, the transconductance

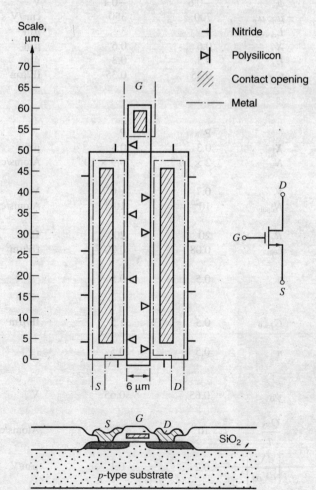

Figure 2.59 Example layout of an *n*-channel silicon-gate MOS transistor. The mask layers are coded as shown.

of an MOS device can be made to vary over a wide range at a fixed drain current by simply changing the device geometry. The same is true of the gate-source voltage. In making these design choices, the designer must be able to relate changes in device geometry to changes in the electrical properties of the device. To illustrate this procedure, we will calculate the model parameters of the device shown in Fig. 2.59. This device has a drawn channel length of 6 μm and channel width of 50 μm. We will assume the process has the parameters that are summarized in Table 2.1. This is typical of processes with minimum allowed gate lengths of 3 μm. Parameters for more advanced processes are given in Tables 2.2, 2.3, 2.4, 2.5, and 2.6.

Table 2.1 Summary of Process Parameters for a Typical Silicon-Gate n-Well CMOS Process with 3 μm Minimum Allowed Gate Length

Parameter	Symbol	Value n-Channel Transistor	Value p-Channel Transistor	Units
Substrate doping	N_A, N_D	1×10^{15}	1×10^{16}	Atoms/cm^3
Gate oxide thickness	t_{ox}	400	400	Å
Metal-silicon work function	ϕ_{ms}	−0.6	−0.1	V
Channel mobility	μ_n, μ_p	700	350	cm^2/V-s
Minimum drawn channel length	L_{drwn}	3	3	μm
Source, drain junction depth	X_j	0.6	0.6	μm
Source, drain side diffusion	L_d	0.3	0.3	μm
Overlap capacitance per unit gate width	C_{ol}	0.35	0.35	fF/μm
Threshold adjust implant (box dist)				
impurity type		P	P	
effective depth	X_i	0.3	0.3	μm
effective surface concentration	N_{si}	2×10^{16}	0.9×10^{16}	Atoms/cm^3
Nominal threshold voltage	V_t	0.7	−0.7	V
Polysilicon gate doping concentration	N_{dpoly}	10^{20}	10^{20}	Atoms/cm^3
Poly gate sheet resistance	R_s	20	20	Ω/□
Source, drain-bulk junction capacitances (zero bias)	C_{j0}	0.08	0.20	fF/μm^2
Source, drain-bulk junction capacitance grading coefficient	n	0.5	0.5	
Source, drain periphery capacitance (zero bias)	C_{jsw0}	0.5	1.5	fF/μm
Source, drain periphery capacitance grading coefficient	n	0.5	0.5	
Source, drain junction built-in potential	ψ_0	0.65	0.65	V
Surface-state density	$\dfrac{Q_{ss}}{q}$	10^{11}	10^{11}	Atoms/cm^2
Channel-length modulation parameter	$\left\vert\dfrac{dX_d}{dV_{DS}}\right\vert$	0.2	0.1	μm/V

2.9 Active Devices in MOS Integrated Circuits

Table 2.2 Summary of Process Parameters for a Typical Silicon-Gate n-Well CMOS Process with 1.5 μm Minimum Allowed Gate Length

Parameter	Symbol	Value n-Channel Transistor	Value p-Channel Transistor	Units
Substrate doping	N_A, N_D	2×10^{15}	1.5×10^{16}	Atoms/cm^3
Gate oxide thickness	t_{ox}	250	250	Å
Metal-silicon work function	ϕ_{ms}	-0.6	-0.1	V
Channel mobility	μ_n, μ_p	650	300	cm^2/V-s
Minimum drawn channel length	L_{drwn}	1.5	1.5	μm
Source, drain junction depth	X_j	0.35	0.4	μm
Source, drain side diffusion	L_d	0.2	0.3	μm
Overlap capacitance per unit gate width	C_{ol}	0.18	0.26	fF/μm
Threshold adjust implant (box dist)				
impurity type		P	P	
effective depth	X_i	0.3	0.3	μm
effective surface concentration	N_{si}	2×10^{16}	0.9×10^{16}	Atoms/cm^3
Nominal threshold voltage	V_t	0.7	-0.7	V
Polysilicon gate doping concentration	N_{dpoly}	10^{20}	10^{20}	Atoms/cm^3
Poly gate sheet resistance	R_s	20	20	Ω/□
Source, drain-bulk junction capacitances (zero bias)	C_{j0}	0.14	0.25	fF/μm^2
Source, drain-bulk junction capacitance grading coefficient	n	0.5	0.5	
Source, drain periphery capacitance (zero bias)	C_{jsw0}	0.8	1.8	fF/μm
Source, drain periphery capacitance grading coefficient	n	0.5	0.5	
Source, drain junction built-in potential	ψ_0	0.65	0.65	V
Surface-state density	$\dfrac{Q_{ss}}{q}$	10^{11}	10^{11}	Atoms/cm^2
Channel-length modulation parameter	$\left|\dfrac{dX_d}{dV_{DS}}\right|$	0.12	0.06	μm/V

Threshold Voltage. In Chapter 1, an MOS transistor was shown to have a threshold voltage of

$$V_t = \phi_{ms} + 2\phi_f + \frac{Q_b}{C_{ox}} - \frac{Q_{ss}}{C_{ox}} \tag{2.27}$$

where ϕ_{ms} is the metal-silicon work function, ϕ_f is the Fermi level in the bulk silicon, Q_b is the bulk depletion layer charge, C_{ox} is the oxide capacitance per unit area, and Q_{ss} is the

Table 2.3 Summary of Process Parameters for a Typical Silicon-Gate n-Well CMOS Process with 0.8 μm Minimum Allowed Gate Length

Parameter	Symbol	Value n-Channel Transistor	Value p-Channel Transistor	Units
Substrate doping	N_A, N_D	4×10^{15}	3×10^{16}	Atoms/cm^3
Gate oxide thickness	t_{ox}	150	150	Å
Metal-silicon work function	ϕ_{ms}	−0.6	−0.1	V
Channel mobility	μ_n, μ_p	550	250	cm^2/V-s
Minimum drawn channel length	L_{drwn}	0.8	0.8	μm
Source, drain junction depth	X_j	0.2	0.3	μm
Source, drain side diffusion	L_d	0.12	0.18	μm
Overlap capacitance per unit gate width	C_{ol}	0.12	0.18	fF/μm
Threshold adjust implant (box dist)				
impurity type		P	P	
effective depth	X_i	0.2	0.2	μm
effective surface concentration	N_{si}	3×10^{16}	2×10^{16}	Atoms/cm^3
Nominal threshold voltage	V_t	0.7	−0.7	V
Polysilicon gate doping concentration	N_{dpoly}	10^{20}	10^{20}	Atoms/cm^3
Poly gate sheet resistance	R_s	10	10	Ω/□
Source, drain-bulk junction capacitances (zero bias)	C_{j0}	0.18	0.30	fF/μm^2
Source, drain-bulk junction capacitance grading coefficient	n	0.5	0.5	
Source, drain periphery capacitance (zero bias)	C_{jsw0}	1.0	2.2	fF/μm
Source, drain periphery capacitance grading coefficient	n	0.5	0.5	
Source, drain junction built-in potential	ψ_0	0.65	0.65	V
Surface-state density	$\dfrac{Q_{ss}}{q}$	10^{11}	10^{11}	Atoms/cm^2
Channel-length modulation parameter	$\left|\dfrac{dX_d}{dV_{DS}}\right|$	0.08	0.04	μm/V

concentration of surface-state charge. An actual calculation of the threshold is illustrated in the following example.

Often the threshold voltage must be deduced from measurements, and a useful approach to doing this is to plot the square root of the drain current as a function of V_{GS}, as shown in Fig. 2.60. The threshold voltage can be determined as the extrapolation of the straight portion

2.9 Active Devices in MOS Integrated Circuits

Table 2.4 Summary of Process Parameters for a Typical Silicon-Gate n-Well CMOS Process with 0.4 μm Minimum Allowed Gate Length

Parameter	Symbol	Value n-Channel Transistor	Value p-Channel Transistor	Units
Substrate doping	N_A, N_D	5×10^{15}	4×10^{16}	Atoms/cm^3
Gate oxide thickness	t_{ox}	80	80	Å
Metal-silicon work function	ϕ_{ms}	−0.6	−0.1	V
Channel mobility	μ_n, μ_p	450	150	cm^2/V-s
Minimum drawn channel length	L_{drwn}	0.4	0.4	μm
Source, drain junction depth	X_j	0.15	0.18	μm
Source, drain side diffusion	L_d	0.09	0.09	μm
Overlap capacitance per unit gate width	C_{ol}	0.35	0.35	fF/μm
Threshold adjust implant (box dist)				
impurity type		P	P	
effective depth	X_i	0.16	0.16	μm
effective surface concentration	N_{si}	4×10^{16}	3×10^{16}	Atoms/cm^3
Nominal threshold voltage	V_t	0.6	−0.8	V
Polysilicon gate doping concentration	N_{dpoly}	10^{20}	10^{20}	Atoms/cm^3
Poly gate sheet resistance	R_s	5	5	Ω/□
Source, drain-bulk junction capacitances (zero bias)	C_{j0}	0.2	0.4	fF/μm^2
Source, drain-bulk junction capacitance grading coefficient	n	0.5	0.4	
Source, drain periphery capacitance (zero bias)	C_{jsw0}	1.2	2.4	fF/μm
Source, drain periphery capacitance grading coefficient	n	0.4	0.3	
Source, drain junction built-in potential	ψ_0	0.7	0.7	V
Surface-state density	$\dfrac{Q_{ss}}{q}$	10^{11}	10^{11}	Atoms/cm^2
Channel-length modulation parameter	$\left\vert \dfrac{dX_d}{dV_{DS}} \right\vert$	0.02	0.04	μm/V

of the curve to zero current. The slope of the curve also yields a direct measure of the quantity $\mu_n C_{ox} W/L_{eff}$ for the device at the particular drain-source voltage at which the measurement is made. The measured curve deviates from a straight line at low currents because of subthreshold conduction and at high currents because of mobility degradation in the channel as the carriers approach scattering-limited velocity.

Table 2.5 Summary of Process Parameters for a Typical CMOS Process with 0.2 μm Minimum Allowed Gate Length

Parameter	Symbol	Value n-Channel Transistor	Value p-Channel Transistor	Units
Substrate doping	N_A, N_D	8×10^{16}	8×10^{16}	Atoms/cm³
Gate oxide thickness	t_{ox}	42	42	Angstroms
Metal-silicon work function	ϕ_{ms}	−0.6	−0.1	V
Channel mobility	μ_n, μ_p	300	80	cm²/V-s
Minimum drawn channel length	L_{drwn}	0.2	0.2	μm
Source, drain junction depth	X_j	0.16	0.16	μm
Source, drain side diffusion	L_d	0.01	0.015	μm
Overlap capacitance per unit gate width	C_{ol}	0.36	0.33	fF/μm
Threshold adjust implant (box dist.)				
impurity type		P	P	
effective depth	X_i	0.12	0.12	μm
effective surface concentration	N_{si}	2×10^{17}	2×10^{17}	Atoms/cm³
Nominal threshold voltage	V_t	0.5	−0.45	V
Polysilicon gate doping concentration	N_{dpoly}	10^{20}	10^{20}	Atoms/cm³
Poly gate sheet resistance	R_s	7	7	Ω/□
Source, drain-bulk junction capacitances (zero bias)	C_{j0}	1.0	1.1	fF/μm²
Source, drain-bulk junction capacitance grading coefficient	n	0.36	0.45	
Source, drain periphery capacitance (zero bias)	C_{jsw0}	0.2	0.25	fF/μm
Source, drain periphery capacitance grading coefficient	n	0.2	0.24	
Source, drain junction built-in potential	ψ_0	0.68	0.74	V
Surface-state density	$\dfrac{Q_{ss}}{q}$	10^{11}	10^{11}	Atoms/cm²
Channel-length modulation parameter	$\left|\dfrac{dX_d}{dV_{DS}}\right|$	0.028	0.023	μm/V

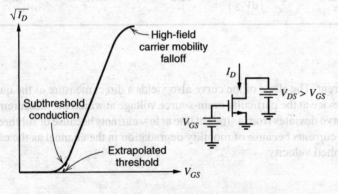

Figure 2.60 Typical experimental variation of drain current as a function of the square root of gate-source voltage in the active region.

2.9 Active Devices in MOS Integrated Circuits

Table 2.6 Summary of Process Parameters for a Typical CMOS Process with 0.1 μm Minimum Allowed Gate Length

Parameter	Symbol	Value n-Channel Transistor	Value p-Channel Transistor	Units
Substrate doping	N_A, N_D	1×10^{17}	1×10^{17}	Atoms/cm^3
Gate oxide thickness	t_{ox}	25	25	Angstroms
Gate leakage current density	J_G	1.2	0.4	nA/μm^2
Metal-silicon work function	ϕ_{ms}	−0.6	−0.1	V
Channel mobility	μ_n, μ_p	390	100	cm^2/V-s
Minimum drawn channel length	L_{drwn}	0.1	0.1	μm
Source, drain junction depth	X_j	0.15	0.16	μm
Source, drain side diffusion	L_d	0.005	0.005	μm
Overlap capacitance per unit gate width	C_{ol}	0.10	0.07	fF/μm
Threshold adjust implant (box dist.)				
impurity type		P	P	
effective depth	X_i	0.1	0.1	μm
effective surface concentration	N_{si}	5×10^{17}	5×10^{17}	Atoms/cm^3
Nominal threshold voltage	V_t	0.27	−0.28	V
Polysilicon gate doping concentration	N_{dpoly}	10^{20}	10^{20}	Atoms/cm^3
Poly gate sheet resistance	R_s	10	10	Ω/□
Source, drain-bulk junction capacitances (zero bias)	C_{j0}	1.0	1.1	fF/μm^2
Source, drain-bulk junction capacitance grading coefficient	n	0.25	0.35	
Source, drain periphery capacitance (zero bias)	C_{jsw0}	0.05	0.06	fF/μm
Source, drain periphery capacitance grading coefficient	n	0.05	0.05	
Source, drain junction built-in potential	ψ_0	0.6	0.65	V
Surface-state density	$\dfrac{Q_{ss}}{q}$	10^{11}	10^{11}	Atoms/cm^2
Channel-length modulation parameter	$\left|\dfrac{dX_d}{dV_{DS}}\right|$	0.06	0.05	μm/V

■ **EXAMPLE**

Calculate the zero-bias threshold voltage of the unimplanted and implanted NMOS transistors for the process given in Table 2.1.

Each of the four components in the threshold voltage expression (2.27) must be calculated. The first term is the metal-silicon work function. For an n-channel transistor with an n-type polysilicon gate electrode, this has a value equal to the difference in the Fermi potentials in the two regions, or approximately −0.6 V.

The second term in the threshold voltage equation represents the band bending in the semiconductor that is required to strongly invert the surface. To produce a surface concentration of electrons that is approximately equal to the bulk concentration of holes, the surface potential must be increased by approximately twice the bulk Fermi potential. The Fermi potential in the bulk is given by

$$\phi_f = \frac{kT}{q} \ln\left(\frac{N_A}{n_i}\right) \tag{2.28}$$

For the unimplanted transistor with the substrate doping given in Table 2.1, this value is 0.27 V. Thus the second term in (2.27) takes on a value of 0.54 V. The value of this term will be the same for the implanted transistor since we are defining the threshold voltage as the voltage that causes the surface concentration of electrons to be the same as that of holes in the bulk material beneath the channel implant. Thus the potential difference between the surface and the bulk silicon beneath the channel implant region that is required to bring this condition about is still given by (2.30), independent of the details of the channel implant.

The third term in (2.27) is related to the charge in the depletion layer under the channel. We first consider the unimplanted device. Using (1.137), with a value of N_A of 10^{15} atoms/cm^3,

$$Q_{b0} = \sqrt{2qN_A\epsilon 2\phi_f} = \sqrt{2(1.6 \times 10^{-19})(10^{15})(11.6 \times 8.86 \times 10^{-14})(0.54)}$$
$$= 1.34 \times 10^{-8} \text{ C/cm}^2 \tag{2.29}$$

Also, the capacitance per unit area of the 400-Å gate oxide is

$$C_{ox} = \frac{\epsilon_{ox}}{t_{ox}} = \frac{3.9 \times 8.86 \times 10^{-14} \text{ F/cm}}{400 \times 10^{-8} \text{ cm}} = 8.6 \times 10^{-8} \frac{\text{F}}{\text{cm}^2} = 0.86 \frac{\text{fF}}{\mu\text{m}^2} \tag{2.30}$$

The resulting magnitude of the third term is 0.16 V.

The fourth term in (2.27) is the threshold shift due to the surface-state charge. This positive charge has a value equal to the charge of one electron multiplied by the density of surface states, 10^{11} atoms/cm^2, from Table 2.1. The value of the surface-state charge term is then

$$\frac{Q_{ss}}{C_{ox}} = \frac{1.6 \times 10^{-19} \times 10^{11}}{8.6 \times 10^{-8}} = 0.19 \text{ V} \tag{2.31}$$

Using these calculations, the threshold voltage for the unimplanted transistor is

$$V_t = -0.6 \text{ V} + 0.54 \text{ V} + 0.16 \text{ V} - 0.19 \text{ V} = -0.09 \text{ V} \tag{2.32}$$

For the implanted transistor, the calculation of the threshold voltage is complicated by the fact that the depletion layer under the channel spans a region of nonuniform doping. A precise calculation of device threshold voltage would require consideration of this nonuniform profile. The threshold voltage can be approximated, however, by considering the implanted layer to be approximated by a box distribution of impurities with a depth X_i and a specified impurity concentration N_i. If the impurity profile resulting from the threshold-adjustment implant and subsequent process steps is sufficiently deep so that the channel-substrate depletion layer lies entirely within it, then the effect of the implant is simply to raise the effective substrate doping. For the implant specified in Table 2.1, the average doping in the layer is the sum of the implant doping and the background concentration, or 2.1×10^{16} atoms/cm^3. This increases the Q_{b0} term in the threshold voltage to 0.71 V and gives device threshold voltage of 0.47 V. The validity of the assumption regarding the boundary of the channel-substrate depletion layer can be checked by using Fig. 2.29. For a doping level of 2.1×10^{16} atoms/cm^3, a one-sided step junction displays a depletion region width of approximately 0.2 μm. Since the depth of the layer is 0.3 μm in this case, the assumption is valid.

Alternatively, if the implantation and subsequent diffusion had resulted in a layer that was very shallow, and was contained entirely within the depletion layer, the effect of the implanted layer would be simply to increase the effective value of Q_{ss} by an amount equal to the effective implant dose over and above that of the unimplanted transistor. The total active impurity dose for the implant given in Table 2.1 is the product of the depth and the impurity concentration, or 6×10^{11} atoms/cm^2. For this case, the increase in threshold voltage would have been 1.11 V, giving a threshold voltage of 1.02 V.

Body-Effect Parameter. For an unimplanted, uniform-channel transistor, the body-effect parameter is given by (1.141).

$$\gamma = \frac{1}{C_{ox}} \sqrt{2q\epsilon N_A} \qquad (2.33)$$

The application of this expression is illustrated in the following example.

■ **EXAMPLE**

Calculate the body-effect parameter for the unimplanted n-channel transistor in Table 2.1.
Utilizing in (2.33) the parameters given in Table 2.1, we obtain

$$\gamma = \frac{\sqrt{2(1.6 \times 10^{-19})(11.6 \times 8.86 \times 10^{-14})(10^{15})}}{8.6 \times 10^{-8}} = 0.21 \text{ V}^{1/2} \qquad (2.34)$$

The calculation of body effect in an implanted transistor is complicated by the fact that the channel is not uniformly doped and the preceding simple expression does not apply. The threshold voltage as a function of body bias voltage can be approximated again by considering the implanted layer to be approximated by a box distribution of impurity of depth X_i and concentration N_i. For small values of body bias where the channel-substrate depletion layer resides entirely within the implanted layer, the body effect is that corresponding to a transistor with channel doping $(N_i + N_A)$. For larger values of body bias for which the depletion layer extends into the substrate beyond the implanted distribution, the incremental body effect corresponds to a transistor with substrate doping N_A. A typical variation of threshold voltage as a function of substrate bias for this type of device is illustrated in Fig. 2.61.

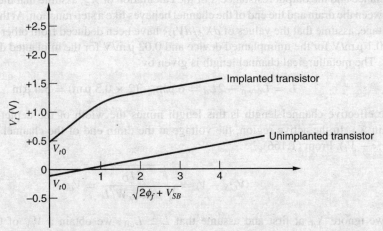

Figure 2.61 Typical variation of threshold voltage as a function of substrate bias for n-channel devices with uniform channel doping (no channel implant) and with nonuniform channel doping resulting from threshold adjustment channel implant.

Effective Channel Length. The gate dimension parallel to current flow that is actually drawn on the mask is called the drawn channel length L_{drwn}. This is the length referred to on circuit schematics. Because of exposure variations and other effects, the physical length of the polysilicon strip defining the gate may be somewhat larger or smaller than this value. The actual channel length of the device is the physical length of the polysilicon gate electrode minus the side or lateral diffusions of the source and the drain under the gate. This length will be termed the *metallurgical channel length* and is the distance between the metallurgical source and drain junctions. Assuming that the lateral diffusion of the source and drain are each equal to L_d, the metallurgical channel length is $L = (L_{\text{drwn}} - 2L_d)$.

When the transistor is biased in the active or saturation region, a depletion region exists between the drain region and the end of the channel. In Chapter 1, the width of this region was defined as X_d. Thus for a transistor operating in the active region, the actual effective channel length L_{eff} is given by

$$L_{\text{eff}} = L_{\text{drwn}} - 2L_d - X_d \tag{2.35}$$

A precise determination of X_d is complicated by the fact that the field distribution in the drain region is two-dimensional and quite complex. The drain depletion width X_d can be approximated by assuming that the electric field in the drain region is one-dimensional and that the depletion width is that of a one-sided step junction with an applied voltage of $V_{DS} - V_{ov}$, where $V_{ov} = V_{GS} - V_t$ is the potential at the drain end of the channel with respect to the source. This assumption is used in the following example.

As shown in Chapter 1, the small-signal output resistance of the transistor is inversely proportional to the effective channel length. Because the performance of analog circuits often depends strongly on the transistor small-signal output resistance, analog circuits often use channel lengths that are longer than the minimum channel length for digital circuits. This statement is particularly true for unimplanted transistors.

■ **EXAMPLE**

Estimate the effective channel length for the unimplanted and implanted transistors for the process shown in Table 2.1 and the device geometry shown in Fig. 2.59. Assume the device is biased at a drain-source voltage of 5 V and a drain current of 10 μA. Calculate the transconductance and the output resistance. For the calculation of X_d, assume that the depletion region between the drain and the end of the channel behaves like a step junction. At the given drain bias voltage, assume that the values of dX_d/dV_{DS} have been deduced from other measurements to be 0.1 μm/V for the unimplanted device and 0.02 μm/V for the implanted device.

The metallurgical channel length is given by

$$L = L_{\text{drwn}} - 2L_d = 6 \text{ μm} - (2 \times 0.3 \text{ μm}) = 5.4 \text{ μm} \tag{2.36}$$

The effective channel length is this length minus the width of the depletion region at the drain X_d. In the active region, the voltage at the drain end of the channel is approximately $(V_{GS} - V_t)$. From (1.166),

$$V_{GS} - V_t = \sqrt{\frac{2I_D}{\mu_n C_{ox} W/L}} = V_{ov} \tag{2.37}$$

If we ignore X_d at first and assume that $L \simeq L_{\text{eff}}$, we obtain a V_{ov} of 0.16 V using the data from Table 2.1. Thus the voltage across the drain depletion region is approximately 4.84 V. To estimate the depletion-region width, assume it is a one-sided step junction that mainly exists in the lightly doped side. Since the channel and the drain are both *n*-type regions, the built-in potential of the junction is near zero. The width of the depletion layer can be calculated using (1.14) or the nomograph in Fig. 2.29. Using (1.14), and assuming $N_D \gg N_A$,

$$X_d = \sqrt{\frac{2\epsilon(V_{DS} - V_{ov})}{qN_A}} \quad (2.38)$$

For the unimplanted device, this equation gives a depletion width of 2.4 μm. For the implanted device, the result is 0.5 μm, assuming an effective constant channel doping of 2.1×10^{16} atoms/cm^3. Thus the effective channel lengths of the two devices would be approximately 3.0 μm and 4.9 μm, respectively.

From (1.180), the device transconductance is given by

$$g_m = \sqrt{2\mu_n C_{ox}(W/L)I_D} \quad (2.39)$$

Assuming that $\mu_n = 700$ cm^2/V-s, we find

$$g_m = \sqrt{2(700)(8.6 \times 10^{-8})(50/3.0)(10 \times 10^{-6})} = 141 \text{ μA/V} \quad (2.40)$$

for the unimplanted transistor and

$$g_m = \sqrt{2(700)(8.6 \times 10^{-8})(50/4.9)(10 \times 10^{-6})} = 111 \text{ μA/V} \quad (2.41)$$

for the implanted transistor.

The output resistance can be calculated by using (1.163) and (1.194). For the unimplanted device,

$$r_o = \frac{L_{eff}}{I_D}\left(\frac{dX_d}{dV_{DS}}\right)^{-1} = \left(\frac{3.0 \text{ μm}}{10 \text{ μA}}\right)\frac{1}{0.1 \text{ μm/V}} = 3.0 \text{ M}\Omega \quad (2.42)$$

For the implanted device,

$$r_o = \left(\frac{4.9 \text{ μm}}{10 \text{ μA}}\right)\frac{1}{0.02 \text{ μm/V}} = 25 \text{ M}\Omega \quad (2.43)$$

Because the depletion region for unimplanted devices is much wider than for implanted devices, the channel length of unimplanted devices must be made longer than for implanted devices to achieve comparable punchthrough voltages and small-signal output resistances under identical bias conditions.

Effective Channel Width. The effective channel width of an MOS transistor is determined by the gate dimension parallel to the surface and perpendicular to the channel length over which the gate oxide is thin. Thick field oxide regions are grown at the edges of each transistor by using the local-oxidation process described in Sections 2.2.7 and 2.8. Before the field oxide is grown, nitride is deposited and patterned so that it remains only in areas that should become transistors. Therefore, the width of a nitride region corresponds to the the drawn width of a transistor. To minimize the width variation, the field oxide should grow only vertically; that is, the oxide thickness should increase only in regions where nitride does not cover the oxide. In practice, however, some lateral growth of oxide also occurs near the edges of the nitride during field-oxide growth. As a result, the edges of the field oxide are not vertical, as shown in Figures 2.9 and 2.54. This lateral growth of the oxide reduces the effective width of MOS transistors compared to their drawn widths. It is commonly referred to as the *bird's beak* because the gradually decreasing oxide thickness in the cross sections of Figures 2.9 and 2.54 resembles the corresponding portion of the profile of a bird.

As a result, both the effective lengths and the effective widths of transistors differ from the corresponding drawn dimensions. In analog design, the change in the effective length is usually much more important than the change in the effective width because transistors usually have drawn lengths much less than their drawn widths. As a result, the difference between the drawn and effective width is often ignored. However, this difference is sometimes important, especially when the matching between two ratioed transistors limits the accuracy of a given circuit. This topic is considered in Section 4.2.

Intrinsic Gate-Source Capacitance. As described in Chapter 1, the intrinsic gate-source capacitance of the transistor in the active region of operation is given by

$$C_{gs} = \frac{2}{3} W L_{\text{eff}} C_{ox} \tag{2.44}$$

The calculation of this parameter is illustrated in the next example.

Overlap Capacitance. Assuming that the source and drain regions each diffuse under the gate by L_d after implantation, the gate-source and gate-drain overlap capacitances are given by

$$C_{ol} = W L_d C_{ox} \tag{2.45}$$

This parasitic capacitance adds directly to the intrinsic gate-source capacitance. It constitutes the entire drain-gate capacitance in the active region of operation.

Junction Capacitances. Source-substrate and drain-substrate capacitances result from the junction-depletion capacitance between the source and drain diffusions and the substrate. A complicating factor in calculating these capacitances is the fact that the substrate doping around the source and drain regions is not constant. In the region of the periphery of the source and drain diffusions that border on the field regions, a relatively high surface concentration exists on the field side of the junction because of the field threshold adjustment implant. Although approximate calculations can be carried out, the zero-bias value and grading parameter of the periphery capacitance are often characterized experimentally by using test structures. The bulk-junction capacitance can be calculated directly by using (1.21) or can be read from the nomograph in Fig. 2.29.

An additional capacitance that must be accounted for is the depletion capacitance between the channel and the substrate under the gate, which we will term C_{cs}. Calculation of this capacitance is complicated by the fact that the channel-substrate voltage is not constant but varies along the channel. Also, the allocation of this capacitance to the source and drain varies with operating conditions in the same way as the allocation of C_{gs}. A reasonable approach is to develop an approximate total value for this junction capacitance under the gate and allocate it to source and drain in the same ratio as the gate capacitance is allocated. For example, in the active region, a capacitance of two-thirds of C_{cs} would appear in parallel with the source-substrate capacitance and none would appear in parallel with the drain-substrate capacitance.

■ **EXAMPLE**

Calculate the capacitances of an implanted device with the geometry shown in Fig. 2.59. Use the process parameters given in Table 2.1 and assume a drain-source voltage of 5 V, drain current of 10 μA, and no substrate bias voltage. Neglect the capacitance between the channel and the substrate. Assume that X_d is negligibly small.

From (2.44), the intrinsic gate-source capacitance is

$$C_{gs} = \frac{2}{3} W L_{\text{eff}} C_{ox} = \left(\frac{2}{3}\right) 50 \text{ μm} \times 5.4 \text{ μm} \times 0.86 \text{ fF/μm}^2 = 155 \text{ fF} \qquad (2.46)$$

From (2.45), the overlap capacitance is given by

$$C_{ol} = W L_d C_{ox} = 50 \text{ μm} \times 0.3 \text{ μm} \times 0.86 \text{ fF/μm}^2 = 12.9 \text{ fF} \qquad (2.47)$$

Thus the total gate-source capacitance is $(C_{gs} + C_{ol})$ or 168 fF. The gate-drain capacitance is equal to the overlap capacitance, or 12.9 fF.

The source- and drain-to-substrate capacitances consist of two portions. The periphery or sidewall part C_{jsw} is associated with that portion of the edge of the diffusion area that is adjacent to the field region. The second portion C_j is the depletion capacitance between the diffused junction and the bulk silicon under the source and drain. For the bias conditions given, the source-substrate junction operates with zero bias and the drain-substrate junction has a reverse bias of 5 V. Using Table 2.1, the periphery portion for the source-substrate capacitance is

$$C_{jsw}(\text{source}) = (50 \text{ μm} + 9 \text{ μm} + 9 \text{ μm})(0.5 \text{ fF/μm}) = 34 \text{ fF} \qquad (2.48)$$

Here, the perimeter is set equal to $W + 2L$ because that is the distance on the surface of the silicon around the part of the source and drain regions that border on field-oxide regions. Since the substrate doping is high along this perimeter to increase the magnitude of the threshold voltage in the field regions, the sidewall capacitance here is dominant. The bulk capacitance is simply the source-diffusion area multiplied by the capacitance per unit area from Table 2.1.

$$C_j(\text{source}) = (50 \text{ μm})(9 \text{ μm})(0.08 \text{ fF/μm}^2) = 36 \text{ fF} \qquad (2.49)$$

The total capacitance from source to bulk is the sum of these two, or

$$C_{sb} = 70 \text{ fF} \qquad (2.50)$$

For the geometry given for this example, the transistor is symmetrical, and the source and drain areas and peripheries are the same. From Table 2.1, both the bulk and periphery capacitances have a grading coefficient of 0.5. As a result, the drain-bulk capacitance is the same as the source-bulk capacitance modified to account for the 5 V reverse bias on the junction. Assuming $\psi_0 = 0.65$ V,

$$C_{db} = \frac{(70 \text{ fF})}{\sqrt{1 + V_{DB}/\psi_0}} = \frac{(70 \text{ fF})}{\sqrt{1 + 5/0.65}} = 24 \text{ fF} \qquad (2.51)$$

As the minimum channel length decreases, second-order effects cause the operation of short-channel MOS transistors to deviate significantly from the simple square-law models in Chapters 1 and 2.[22] Equations that include these second-order effects are complicated and make hand calculations difficult. Therefore, simple models and equations that ignore these effects are often used as a design aid and to develop intuition. SPICE simulations with highly accurate device models are used to verify circuit peformance and to refine a design.

For processes with minimum allowed channel length less than 0.2μm, the gate-oxide thickness can fall below 30 Angstroms (for example, see Table 2.6). With such thin gate oxide, enough carriers in the channel can tunnel through the gate oxide and create nonzero dc gate current that is sometimes important.[23] This current is referred to as gate-leakage current and is a complicated function of the operating point and oxide thickness.[24,25] The gate-leakage current I_G is the product of the gate-leakage-current density J_G and the gate area. SPICE models are available that include gate-leakage current and accurately predict short-channel-device operation.[26,27]

2.9.2 p-Channel Transistors

The *p*-channel transistor in most CMOS technologies displays dc and ac properties that are comparable to the *n*-channel transistor. One difference is that the transconductance parameter k' of a *p*-channel device is about one-half to one-third that of an *n*-channel device because holes have correspondingly lower mobility than electrons. As shown in (1.209), this difference in mobility also reduces the f_T of *p*-channel devices by the same factor. Another difference is that for a CMOS technology with a *p*-type substrate and *n*-type wells, the substrate terminal of the *p*-channel transistors can be electrically isolated since the devices are made in an implanted well. Good use can be made of this fact in analog circuits to alleviate the impact of the high body effect in these devices. For a CMOS process made on an *n*-type substrate with *p*-type wells, the *p*-channel devices are made in the substrate material, which is connected to the highest power-supply voltage, but the *n*-channel devices can have electrically isolated substrate terminals.

The calculation of device parameters for *p*-channel devices proceeds exactly as for *n*-channel devices. An important difference is the fact that for the *p*-channel transistors the threshold voltage that results if no threshold adjustment implant is used is relatively high, usually in the range of 1 to 3 V. This occurs because the polarities of the Q_{ss} term and the work-function term are such that they tend to increase the *p*-channel threshold voltages while decreasing the *n*-channel threshold voltages. Thus the *p*-type threshold adjustment implant is used to *reduce* the surface concentration by partially compensating the doping of the *n*-type well or substrate. Thus in contrast to the *n*-channel device, the *p*-channel transistor has an effective surface concentration in the channel that is lower than the bulk concentration, and as a result, often displays a smaller incremental body effect for low values of substrate bias and a larger incremental body effect for larger values of substrate bias.

2.9.3 Depletion Devices

The properties of depletion devices are similar to those of the enhancement device already considered, except that an implant has been added in the channel to make the threshold negative (for an *n*-channel device). In most respects a depletion device closely resembles an enhancement device with a voltage source in series with the gate lead of value $(V_{tD} - V_{tE})$, where V_{tD} is the threshold voltage of the depletion-mode transistor and V_{tE} is the threshold voltage of the enhancement-mode transistor. Depletion transistors are most frequently used with the gate tied to the source. Because the device is on with $V_{GS} = 0$, if it operates in the active region, it operates like a current source with a drain current of

$$I_{DSS} = I_D|_{V_{GS}=0} = \frac{\mu_n C_{ox}}{2} \frac{W}{L} V_{tD}^2 \qquad (2.52)$$

An important aspect of depletion-device performance is the variation of I_{DSS} with process variations. These variations stem primarily from the fact that the threshold voltage varies substantially from its nominal value due to processing variations. Since the transistor I_{DSS} varies as the square of the threshold voltage, large variations in I_{DSS} due to process variations often occur. Tolerances of ±40 percent or more from nominal due to process variations are common. Because I_{DSS} determines circuit bias current and power dissipation, the magnitude of this variation is an important factor. Another important aspect of the behavior of depletion devices stems from the body effect. Because the threshold voltage varies with body bias, a depletion device with $V_{GS} = 0$ and $v_{sb} \neq 0$ displays a finite conductance in the active region even if the effect of channel-length modulation is ignored. In turn, this finite conductance has a strong effect on the performance of analog circuits that use depletion devices as load elements.

2.9.4 Bipolar Transistors

Standard CMOS technologies include process steps that can be used to form a bipolar transistor whose collector is tied to the substrate. The substrate, in turn, is tied to one of the power supplies. Fig. 2.62a shows a cross section of such a device. The well region forms the base of the transistor, and the source/drain diffusion of the device in the well forms the emitter. Since the current flow through the base is perpendicular to the surface of the silicon, the device is a vertical bipolar transistor. It is a *pnp* transistor in processes that utilize *p*-type substrates as in Fig. 2.62a and an *npn* transistor in processes that use an *n*-type substrate. The device is particularly useful in band-gap references, described in Chapter 4, and in output stages, considered in Chapter 5. The performance of the device is a strong function of well depth and doping but is generally similar to the substrate *pnp* transistor in bipolar technology, described in Section 2.5.2.

The main limitation of such a vertical bipolar transistor is that its collector is the substrate and is connected to a power supply to keep the substrate *p-n* junctions reverse biased. Standard CMOS processes also provide another bipolar transistor for which the collector need not be connected to a power supply.[28] Figure 2.62b shows a cross section of such a device. As in the vertical transistor, the well region forms the base and a source/drain diffusion forms the emitter. In this case, however, another source/drain diffusion forms the collector C_1. Since the

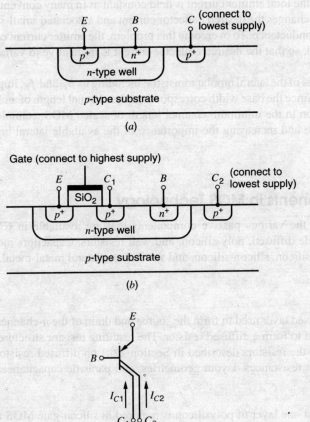

Figure 2.62 (a) Cross section of a vertical *pnp* transistor in an *n*-well CMOS process. (b) Cross section of lateral and vertical *pnp* transistors in an *n*-well CMOS process. (c) Schematic of the bipolar transistors in (b).

current flow through the base is parallel to the surface of the silicon, this device is a lateral bipolar transistor. Again, it is a *pnp* transistor in processes that utilize *n*-type wells and an *npn* transistor in processes that use *p*-type wells. The emitter and collector of this lateral device correspond to the source and drain of an MOS transistor. Since the goal here is to build a bipolar transistor, the MOS transistor is deliberately biased to operate in the cutoff region. In Fig. 2.62b, for example, the gate of the *p*-channel transistor must be connected to a voltage sufficient to bias it in the cutoff region. A key point here is that the base width of the lateral bipolar device corresponds to the channel length of the MOS device.

One limitation of this structure is that when a lateral bipolar transistor is intentionally formed, a vertical bipolar transistor is also formed. In Fig. 2.62b, the emitter and base connections of the vertical transistor are the same as for the lateral transistor, but the collector is the substrate, which is connected to the lowest supply voltage. When the emitter injects minority carriers into the base, some flow parallel to the surface and are collected by the collector of the lateral transistor C_1. However, others flow perpendicular to the surface and are collected by the substrate C_2. Figure 2.62c models this behavior by showing a transistor symbol with one emitter and one base but two collectors. The current I_{C1} is the collector current of the lateral transistor, and I_{C2} is the collector current of the vertical transistor. Although the base current is small because little recombination and reverse injection occur, the undesired current I_{C2} is comparable to the desired current I_{C1}. To minimize the ratio, the collector of the lateral transistor usually surrounds the emitter, and the emitter area as well as the lateral base width are minimized. Even with these techniques, however, the ratio of I_{C2}/I_{C1} is poorly controlled in practice.[28,29] If the total emitter current is held constant as in many conventional circuits, variation of I_{C2}/I_{C1} changes the desired collector current and associated small-signal parameters such as the transconductance. To overcome this problem, the emitter current can be adjusted by negative feedback so that the desired collector current is insensitive to variations in I_{C2}/I_{C1}.[30]

Some important properties of the lateral bipolar transistor, including its β_F and f_T, improve as the base width is reduced. Since the base width corresponds to the channel length of an MOS transistor, the steady reduction in the minimum channel length of scaled MOS technologies is improving the performance and increasing the importance of the available lateral bipolar transistor.

2.10 Passive Components in MOS Technology

In this section, we describe the various passive components that are available in CMOS technologies. Resistors include diffused, poly-silicon, and well resistors. Capacitors include poly-poly, metal-poly, metal-silicon, silicon-silicon, and vertical and lateral metal-metal.

2.10.1 Resistors

Diffused Resistors. The diffused layer used to form the source and drain of the *n*-channel and *p*-channel devices can be used to form a diffused resistor. The resulting resistor structure and properties are very similar to the resistors described in Section 2.6.1 on diffused resistors in bipolar technology. The sheet resistances, layout geometries, and parasitic capacitances are similar.

Polysilicon Resistors. At least one layer of polysilicon is required in silicon-gate MOS technologies to form the gates of the transistors, and this layer is often used to form resistors. The geometries employed are similar to those used for diffused resistors, and the resistor exhibits a

2.10 Passive Components in MOS Technology

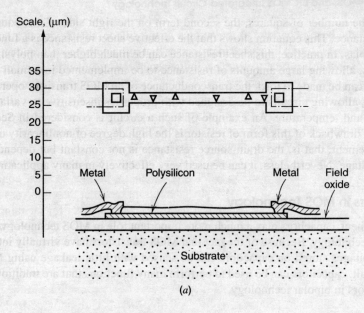

Figure 2.63 (a) Plan view and cross section of polysilicon resistor.

parasitic capacitance to the underlying layer much like a diffused resistor. In this case, however, the capacitance stems from the oxide layer under the polysilicon instead of from a reverse-biased *pn* junction. The nominal sheet resistance of most polysilicon layers that are utilized in MOS processes is on the order of 20 Ω/□ to 80 Ω/□ and typically displays a relatively large variation around the nominal value due to process variations. The matching properties of polysilicon resistors are similar to those of diffused resistors. A cross section and plan view of a typical polysilicon resistor are shown in Fig. 2.63a.

The sheet resistance of polysilicon can limit the speed of interconnections, especially in submicron technologies. To reduce the sheet resistance, a silicide layer is sometimes deposited on top of the polysilicon. Silicide is a compound of silicon and a metal, such as tungsten, that can withstand subsequent high-temperature processing with little movement. Silicide reduces the sheet resistance by about an order of magnitude. Also, it has little effect on the oxidation rate of polysilicon and is therefore compatible with conventional CMOS process technologies.[31] Finally, silicide can be used on the source/drain diffusions as well as on the polysilicon.

Well Resistors. In CMOS technologies the well region can be used as the body of a resistor. It is a relatively lightly doped region and when used as a resistor provides a sheet resistance on the order of 10 kΩ/□. Its properties and geometrical layout are much like the epitaxial resistor described in Section 2.6.2 and shown in Fig. 2.42. It displays large tolerance, high voltage coefficient, and high temperature coefficient relative to other types of resistors. Higher sheet resistance can be achieved by the addition of the pinching diffusion just as in the bipolar technology case.

MOS Devices as Resistors. The MOS transistor biased in the triode region can be used in many circuits to perform the function of a resistor. The drain-source resistance can be calculated by differentiating the equation for the drain current in the triode region with respect to the drain-source voltage. From (1.152),

$$R = \left(\frac{\partial I_D}{\partial V_{DS}}\right)^{-1} = \frac{L}{W} \frac{1}{k'(V_{GS} - V_t - V_{DS})} \quad (2.53)$$

Since L/W gives the number of squares, the second term on the right side of this equation gives the sheet resistance. This equation shows that the effective sheet resistance is a function of the applied gate bias. In practice, this sheet resistance can be much higher than polysilicon or diffused resistors, allowing large amounts of resistance to be implemented in a small area. Also, the resistance can be made to track the transconductance of an MOS transistor operating in the active region, allowing circuits to be designed with properties insensitive to variations in process, supply, and temperature. An example of such a circuit is considered in Section 9.4.3. The principal drawback of this form of resistor is the high degree of nonlinearity of the resulting resistor element; that is, the drain-source resistance is not constant but depends on the drain-source voltage. Nevertheless, it can be used very effectively in many applications.

2.10.2 Capacitors in MOS Technology

As a passive component, capacitors play a much more important role in MOS technology than they do in bipolar technology. Because of the fact that MOS transistors have virtually infinite input resistance, voltages stored on capacitors can be sensed with little leakage using MOS amplifiers. As a result, capacitors can be used to perform many functions that are traditionally performed by resistors in bipolar technology.

Poly-Poly Capacitors. Many MOS technologies that are used to implement analog functions have two layers of polysilicon. The additional layer provides an efficient capacitor structure, an extra layer of interconnect, and can also be used to implement floating-gate memory cells that are electrically programmable and optically erasable with UV light (EPROM). A typical poly-poly capacitor structure is shown in cross section and plan view in Fig. 2.63b. The plate separation is usually comparable to the gate oxide thickness of the MOS transistors.

An important aspect of the capacitor structure is the parasitic capacitance associated with each plate. The largest parasitic capacitance exists from the bottom plate to the underlying layer, which could be either the substrate or a well diffusion whose terminal is electrically isolated. This bottom-plate parasitic capacitance is proportional to the bottom-plate area and typically has a value from 10 to 30 percent of the capacitor itself.

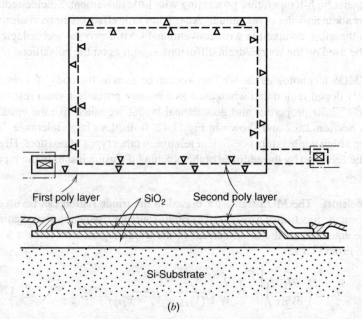

Figure 2.63 (b) Plan view and cross section of typical poly-poly capacitor.

The top-plate parasitic is contributed by the interconnect metallization or polysilicon that connects the top plate to the rest of the circuit, plus the parasitic capacitance of the transistor to which it is connected. In the structure shown in Fig. 2.63b, the drain-substrate capacitance of an associated MOS transistor contributes to the top-plate parasitic capacitance. The minimum value of this parasitic is technology dependent but is typically on the order of 5 fF to 50 fF.

Other important parameters of monolithic capacitor structures are the tolerance, voltage coefficient, and temperature coefficient of the capacitance value. The tolerance on the absolute value of the capacitor value is primarily a function of oxide-thickness variations and is usually in the 10 percent to 30 percent range. Within the same die, however, the matching of one capacitor to another identical structure is much more precise and can typically be in the range of 0.05 percent to 1 percent, depending on the geometry. Because the plates of the capacitor are a heavily doped semiconductor rather than an ideal conductor, some variation in surface potential relative to the bulk material of the plate occurs as voltage is applied to the capacitor.[32] This effect is analogous to the variation in surface potential that occurs in an MOS transistor when a voltage is applied to the gate. However, since the impurity concentration in the plate is usually relatively high, the variations in surface potential are small. The result of these surface potential variations is a slight variation in capacitance with applied voltage. Increasing the doping in the capacitor plates reduces the voltage coefficient. For the impurity concentrations that are typically used in polysilicon layers, the voltage coefficient is usually less than 50 ppm/V,[32,33] a level small enough to be neglected in most applications.

A variation in the capacitance value also occurs with temperature variations. This variation stems primarily from the temperature variation of the surface potential in the plates previously described.[32] Also, secondary effects include the temperature variation of the dielectric constant and the expansion and contraction of the dielectric. For heavily doped polysilicon plates, this temperature variation is usually less than 50 ppm/°C.[32,33]

MOS Transistors as Capacitors. The MOS transistor itself can be used as a capacitor when biased in the triode region, the gate forming one plate and the source, drain, and channel forming another. Unfortunately, because the underlying substrate is lightly doped, a large amount of surface potential variation occurs with changes in applied voltage and the capacitor displays a high voltage coefficient. In noncritical applications, however, it can be used effectively under two conditions. The circuit must be designed in such a way that the device is biased in the triode region when a high capacitance value is desired, and the high sheet resistance of the bottom plate formed by the channel must be taken into account.

Other Vertical Capacitor Structures. In processes with only one layer of polysilicon, alternative structures must be used to implement capacitive elements. One approach involves the insertion of an extra mask to reduce the thickness of the oxide on top of the polysilicon layer so that when the interconnect metallization is applied, a thin-oxide layer exists between the metal layer and the polysilicon layer in selected areas. Such a capacitor has properties that are similar to poly-poly capacitors.

Another capacitor implementation in single-layer polysilicon processing involves the insertion of an extra masking and diffusion operation such that a diffused layer with low sheet resistance can be formed underneath the polysilicon layer in a thin-oxide area. This is not possible in conventional silicon-gate processes because the polysilicon layer is deposited before the source-drain implants or diffusions are performed. The properties of such capacitors are similar to the poly-poly structure, except that the bottom-plate parasitic capacitance is that of a *pn* junction, which is voltage dependent and is usually larger than in the poly-poly case. Also, the bottom plate has a junction leakage current that is associated with it, which is important in some applications.

To avoid the need for extra processing steps, capacitors can also be constructed using the metal and poly layers with standard oxide thicknesses between layers. For example, in a process with one layer of polysilicon and two layers of metal, the top metal and the poly can be connected together to form one plate of a capacitor, and the bottom metal can be used to form the other plate. A key disadvantage of such structures, however, is that the capacitance per unit area is small because the oxide used to isolate one layer from another is thick. Therefore, such capacitors usually occupy large areas. Furthermore, the thickness of this oxide changes little as CMOS processes evolve with reduced minimum channel length. As a result, the area required by analog circuits using such capacitors undergoes a much smaller reduction than that of digital circuits in new technologies. This characteristic is important because reducing the area of an integrated circuit reduces its cost.

Lateral Capacitor Structures. To reduce the capacitor area, and to avoid the need for extra processing steps, lateral capacitors can be used.[34] A lateral capacitor can be formed in one layer of metal by separating one plate from another by spacing s, as shown in Fig. 2.64a. If w is the width of the metal and t is the metal thickness, the capacitance is $(wt\epsilon/s)$, where ϵ is the dielectric constant. As technologies evolve to reduced feature sizes, the minimum metal spacing shrinks but the thickness changes little; therefore, the die area required for a given lateral capacitance decreases in scaled technologies.[35] Note that the lateral capacitance is proportional to the perimeter of each plate that is adjacent to the other in a horizontal plane. Geometries to increase this perimeter in a given die area have been proposed.[35]

Lateral capacitors can be used in conjunction with vertical capacitors, as shown in Fig. 2.64b.[34] The key point here is that each metal layer is composed of multiple pieces, and each capacitor node is connected in an alternating manner to the pieces in each layer.

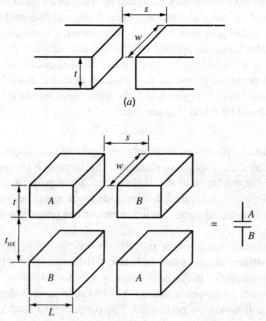

Figure 2.64 (a) Lateral capacitor in one level of metal. (b) Capacitor using two levels of metal in which both lateral and vertical capacitance contribute to the desired capacitance.

As a result, the total capacitance includes vertical and lateral components arising between all adjacent pieces. If the vertical and lateral dielectric constants are equal, the total capacitance is increased compared to the case in which the same die area is used to construct only a vertical capacitor when the minimum spacing $s < \sqrt{2t\,(t_{ox})}$, where t is the metal thickness and t_{ox} is the oxide thickness between metal layers. This concept can be extended to additional pieces in each layer and additional layers.

2.10.3 Latchup in CMOS Technology

The device structures that are present in standard CMOS technology inherently comprise a *pnpn* sandwich of layers. For example, consider the typical circuit shown in Fig. 2.65a. It uses one *n*-channel and one *p*-channel transistor and operates as an inverter if the two gates are connected together as the inverter input. Figure 2.65b shows the cross section in an *n*-well process. When the two MOS transistors are fabricated, two parasitic bipolar transistors are also formed: a lateral *npn* and a vertical *pnp*. In this example, the source of the *n*-channel transistor forms the emitter of the parasitic lateral *npn* transistor, the substrate forms the base, and the *n*-well forms the collector. The source of the *p*-channel transistor forms the emitter of a parasitic vertical *pnp* transistor, the *n*-well forms the base, and the *p*-type substrate forms the collector. The electrical connection of these bipolar transistors that results from the layout shown is illustrated in Fig. 2.65c. In normal operation, all the *pn* junctions in the structure are reverse biased. If the two bipolar transistors enter the active region for some reason, however, the circuit can display a large amount of positive feedback, causing both transistors to conduct heavily. This device structure is similar to that of a silicon-controlled rectifier (SCR), a widely used component in

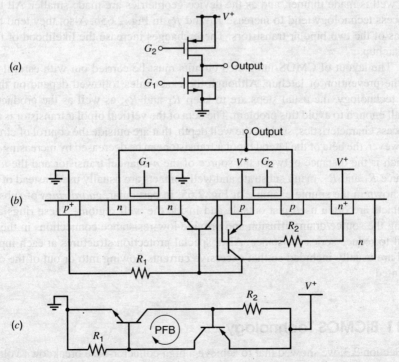

Figure 2.65 (*a*) Schematic of a typical CMOS device pair. (*b*) Cross section illustrating the parasitic bipolar transistors. (*c*) Schematic of the parasitic bipolar transistors.

power-control applications. In power-control applications, the property of the *pnpn* sandwich to remain in the *on* state with no externally supplied signal is a great advantage. However, the result of this behavior here is usually a destructive breakdown phenomenon called *latchup*.

The positive feedback loop is labeled in Fig. 2.65c. Feedback is studied in detail in Chapters 8 and 9. To explain why the feedback around this loop is positive, assume that both transistors are active and that the base current of the *npn* transistor increases by i for some reason. Then the collector current of the *npn* transistor increases by $\beta_{npn}i$. This current is pulled out of the base of the *pnp* transistor if R_2 is ignored. As a result, the current flowing out of the collector of the *pnp* transistor increases by $\beta_{npn}\beta_{pnp}i$. Finally, this current flows into the base of the *npn* transistor if R_1 is ignored. This analysis shows that the circuit generates a current that flows in the same direction as the initial disturbance; therefore, the feedback is positive. If the gain around the loop is more than unity, the response of the circuit to the initial disturbance continues to grow until one or both of the bipolar transistors saturate. In this case, a large current flows from the positive supply to ground until the power supply is turned off or the circuit burns out. This condition is called *latchup*. If R_1 and R_2 are large enough that base currents are large compared to the currents in these resistors, the gain around the loop is $\beta_{npn}\beta_{pnp}$. Therefore, latchup can occur if the product of the betas is greater than unity.

For latchup to occur, one of the junctions in the sandwich must become forward biased. In the configuration illustrated in Fig. 2.65, current must flow in one of the resistors between the emitter and the base of one of the two transistors in order for this to occur. This current can come from a variety of causes. Examples are an application of a voltage that is larger than the power-supply voltage to an input or output terminal, improper sequencing of the power supplies, the presence of large dc currents in the substrate or *p*- or *n*-well, or the flow of displacement current in the substrate or well due to fast-changing internal nodes. Latchup is more likely to occur in circuits as the substrate and well concentration is made lighter, as the well is made thinner, and as the device geometries are made smaller. All these trends in process technology tend to increase R_1 and R_2 in Fig. 2.65b. Also, they tend to increase the betas of the two bipolar transistors. These changes increase the likelihood of the occurrence of latchup.

The layout of CMOS-integrated circuits must be carried out with careful attention paid to the prevention of latchup. Although the exact rules followed depend on the specifics of the technology, the usual steps are to keep R_1 and R_2, as well as the product of the betas, small enough to avoid this problem. The beta of the vertical bipolar transistor is determined by process characteristics, such as the well depth, that are outside the control of circuit designers. However, the beta of the lateral bipolar transistor can be decreased by increasing its base width, which is the distance between the source of the *n*-channel transistor and the *n*-type well. To reduce R_1 and R_2, many substrate and well contacts are usually used instead of just one each, as shown in the simple example of Fig. 2.65. In particular, *guard rings* of substrate and well contacts are often used just outside and inside the well regions. These rings are formed by using the source/drain diffusion and provide low-resistance connections in the substrate and well to reduce series resistance. Also, special protection structures at each input and output pad are usually included so that excessive currents flowing into or out of the chip are safely shunted.

2.11 BiCMOS Technology

In Section 2.3, we showed that to achieve a high collector-base breakdown voltage in a bipolar transistor structure, a thick epitaxial layer is used (17 μm of 5 Ω-cm material for 36-V operation). This in turn requires a deep *p*-type diffusion to isolate transistors and other devices.

On the other hand, if a low breakdown voltage (say about 7 V to allow 5-V supply operation) can be tolerated, then a much more heavily doped (on the order of 0.5 Ω-cm) collector region can be used that is also much thinner (on the order of 1 μm). Under these conditions, the bipolar devices can be isolated by using the same local-oxidation technique used for CMOS, as described in Section 2.4. This approach has the advantage of greatly reducing the bipolar transistor collector-substrate parasitic capacitance because the heavily doped high-capacitance regions near the surface are now replaced by low-capacitance oxide isolation. The devices can also be packed much more densely on the chip. In addition, CMOS and bipolar fabrication technologies begin to look rather similar, and the combination of high-speed, shallow, ion-implanted bipolar transistors with CMOS devices in a BiCMOS technology becomes feasible (at the expense of several extra processing steps).[36] This technology has performance advantages in digital applications because the high current-drive capability of the bipolar transistors greatly facilitates driving large capacitive loads. Such processes are also attractive for analog applications because they allow the designer to take advantage of the unique characteristics of both types of devices.

We now describe the structure of a typical high-frequency, low-voltage, oxide-isolated BiCMOS process. A simplified cross section of a high-performance process[37] is shown in Fig. 2.66. The process begins with masking steps and the implantation of n-type antimony buried layers into a p-type substrate wherever an *npn* bipolar transistor or PMOS device is to be formed. A second implant of p-type boron impurities forms a p-well wherever an NMOS device is to be formed. This is followed by the growth of about 1 μm of n^- epi, which forms the collectors of the *npn* bipolar devices and the channel regions of the PMOS devices. During this and subsequent heat cycles, the more mobile boron atoms out-diffuse and the p-well extends to the surface, whereas the antimony buried layers remain essentially fixed.

A masking step defines regions where thick field oxide is to be grown and these regions are etched down into the epi layer. Field-oxide growth is then carried out, followed by a planarization step where the field oxide that has grown above the plane of the surface is etched back level with the other regions. This eliminates the lumpy surface shown in Fig. 2.57 and helps to overcome problems of ensuring reliable metal connections over the oxide steps (so-called *step coverage*). Finally, a series of masking steps and p- and n-type implants are carried out to form bipolar base and emitter regions, low-resistance bipolar collector contact, and source and drain regions for the MOSFETs. In this sequence, gate oxide is grown, polysilicon gates and emitters are formed, and threshold-adjusting implants are made for the MOS devices. Metal contacts are then made to the desired regions, and the chip is coated with a layer of deposited SiO_2. A second layer of metal interconnects is formed on top of this oxide with connections where necessary to the first layer of metal below. A further deposited layer of SiO_2 is then added with a third layer of metal interconnect and vias to give even more connection flexibility and thus to improve the density of the layout.

2.12 Heterojunction Bipolar Transistors

A *heterojunction* is a *pn* junction made of two different materials. Until this point, all the junctions we have considered have been *homojunctions* because the same material (silicon) has been used to form both the n-type and the p-type regions. In contrast, a junction between an n-type region of silicon and a p-type region of germanium or a compound of silicon and germanium forms a heterojunction.

In homojunction bipolar transistors, the emitter doping is selected to be much greater than the base doping to give an emitter injection efficiency γ of about unity, as shown by

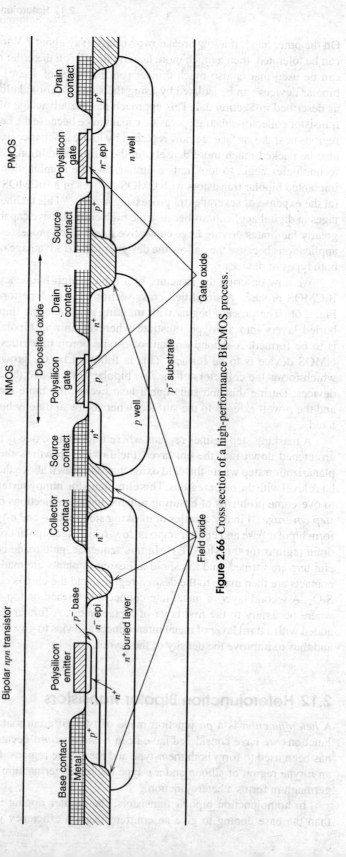

Figure 2.66 Cross section of a high-performance BiCMOS process.

(1.51b). As a result, the base is relatively lightly doped while the emitter is heavily doped in practice. Section 1.4.8 shows that the f_T of bipolar devices is limited in part by τ_F, which is the time required for minority carriers to cross the base. Maximizing f_T is important in some applications such as radio-frequency electronics. To increase f_T, the base width can be reduced. If the base doping is fixed to maintain a constant γ, however, this approach increases the base resistance r_b. In turn, this base resistance limits speed because it forms a time constant with capacitance attached to the base node. As a result, a tradeoff exists in standard bipolar technology between high f_T on the one hand and low r_b on the other, and both extremes limit the speed that can be attained in practice.

One way to overcome this tradeoff is to add some germanium to the base of bipolar transistors to form heterojunction transistors. The key idea is that the different materials on the two sides of the junction have different band gaps. In particular, the band gap of silicon is greater than for germanium, and forming a SiGe compound in the base reduces the band gap there. The relatively large band gap in the emitter can be used to increase the potential barrier to holes that can be injected from the base back to the emitter. Therefore, this structure does not require that the emitter doping be much greater than the base doping to give $\gamma \simeq 1$. As a result, the emitter doping can be decreased and the base doping can be increased in a heterojunction bipolar transistor compared to its homojunction counterpart. Increasing the base doping allows r_b to be constant even when the base width is reduced to increase f_T. Furthermore, this change also reduces the width of the base-collector depletion region in the base when the transistor operates in the forward active region, thus decreasing the effect of base-width modulation and increasing the early voltage V_A. Not only does increasing the base doping have a beneficial effect on performance, but also decreasing the emitter doping increases the width of the base-emitter, space-charge region in the emitter, reducing the C_{je} capacitance and further increasing the maximum speed.

The base region of the heterojunction bipolar transistors can be formed by growing a thin epitaxial layer of SiGe using ultra-high vacuum chemical vapor deposition (UHV/CVD).[38] Since this is an epi layer, it takes on the crystal structure of the silicon in the substrate. Because the lattice constant for germanium is greater than that for silicon, the SiGe layer forms under a compressive strain, limiting the concentration of germanium and the thickness of the layer to avoid defect formation after subsequent high-temperature processing used at the back end of conventional technologies.[39] In practice, with a base thickness of 0.1 μm, the concentration of germanium is limited to about 15 percent so that the layer is unconditionally stable.[40] With only a small concentration of germanium, the change in the band gap and the resulting shift in the potential barrier that limits reverse injection of holes into the emitter is small. However, the reverse injection is an exponential function of this barrier; therefore, even a small change in the barrier greatly reduces the reverse injection and results in these benefits.

In practice, the concentration of germanium in the base need not be constant. In particular, the UHV/CVD process is capable of increasing the concentration of germanium in the base from the emitter end to the collector end. This grading of the germanium concentration results in an electric field that helps electrons move across the base, further reducing τ_F and increasing f_T.

The heterojunction bipolar transistors described above can be included as the bipolar transistors in otherwise conventional BiCMOS processes. The key point is that the device processing sequence retains the well-established properties of silicon integrated-circuit processing because the average concentration of germanium in the base is small.[39] This characteristic is important because it allows the new processing steps to be included as a simple addition to an existing process, reducing the cost of the new technology. For example, a BiCMOS process with a minimum drawn CMOS channel length of 0.3 μm and heterojunction bipolar transistors with a f_T of 50 GHz has been reported.[40] The use of the heterojunction technology increases the f_T by about a factor of two compared to a comparable homojunction technology.

2.13 Interconnect Delay

As the minimum feature size allowed in integrated-circuit technologies is reduced, the maximum operating speed and bandwidth have steadily increased. This trend stems partly from the reductions in the minimum base width of bipolar transistors and the minimum channel length of MOS transistors, which in turn increase the f_T of these devices. While scaling has increased the speed of the transistors, however, it is also increasing the delay introduced by the interconnections to the point where it could soon limit the maximum speed of integrated circuits.[41] This delay is increasing as the minimum feature size is reduced because the width of metal lines and spacing between them are both being reduced to increase the allowed density of interconnections. Decreasing the width of the lines increases the number of squares for a fixed length, increasing the resistance. Decreasing the spacing between the lines increases the lateral capacitance between lines. The delay is proportional to the product of the resistance and capacitance. To reduce the delay, alternative materials are being studied for use in integrated circuits.

First, copper is replacing aluminum in metal layers because copper reduces the resistivity of the interconnection by about 40 percent and is less susceptible to electromigration and stress migration than aluminum. Electromigration and stress migration are processes in which the material of a conductor moves slightly while it conducts current and is under tension, respectively. These processes can cause open circuits to appear in metal interconnects and are important failure mechanisms in integrated circuits. Unfortunately, however, copper can not simply be substituted for aluminum with the same fabrication process. Two key problems are that copper diffuses through silicon and silicon dioxide more quickly than aluminum, and copper is difficult to plasma etch.[42] To overcome the diffusion problem, copper must be surrounded by a thin film of another metal that can endure high-temperature processing with little movement. To overcome the etch problem, a damascene process has been developed.[43] In this process, a layer of interconnection is formed by first depositing a layer of oxide. Then the interconnect pattern is etched into the oxide, and the wafer is uniformly coated by a thin diffusion-resistant layer and then copper. The wafer is then polished by a chemical-mechanical process until the surface of the oxide is reached, which leaves the copper in the cavities etched into the oxide. A key advantage of this process is that it results in a planar structure after each level of metalization.

Also, low-permittivity dielectrics are being studied to replace silicon dioxide to reduce the interconnect capacitance. The dielectric constant of silicon dioxide is 3.9 times more than for air. For relative dielectric constants between about 2.5 and 3.0, polymers have been studied. For relative dielectric constants below about 2.0, the proposed materials include foams and gels, which include air.[42] Other important requirements of low-permittivity dielectric materials include low leakage, high breakdown voltage, high thermal conductivity, stability under high-temperature processing, and adhesion to the metal layers.[41] The search for a replacement for silicon dioxide is difficult because it is an excellent dielectric in all these ways.

2.14 Economics of Integrated-Circuit Fabrication

The principal reason for the growing pervasiveness of integrated circuits in systems of all types is the reduction in cost attainable through integrated-circuit fabrication. Proper utilization of the technology to achieve this cost reduction requires an understanding of the factors influencing

2.14 Economics of Integrated-Circuit Fabrication

the cost of an integrated circuit in completed, packaged form. In this section, we consider these factors.

2.14.1 Yield Considerations in Integrated-Circuit Fabrication

As pointed out earlier in this chapter, integrated circuits are batch-fabricated on single wafers, each containing up to several thousand separate but identical circuits. At the end of the processing sequence, the individual circuits on the wafer are probed and tested prior to the breaking up of the wafer into individual dice. The percentage of the circuits that are electrically functional and within specifications at this point is termed the wafer-sort yield Y_{ws} and is usually in the range of 10 percent to 90 percent. The nonfunctional units can result from a number of factors, but one major source of yield loss is point defects of various kinds that occur during the photoresist and diffusion operations. These defects can result from mask defects, pinholes in the photoresist, airborne particles that fall on the surface of the wafer, crystalline defects in the epitaxial layer, and so on. If such a defect occurs in the active region on one of the transistors or resistors making up the circuit, a nonfunctional unit usually results. The frequency of occurrence per unit of wafer area of such defects is usually dependent primarily on the particular fabrication process used and not on the particular circuit being fabricated. Generally speaking, the more mask steps and diffusion operations that the wafer is subjected to, the higher will be the density of defects on the surface of the finished wafer.

The existence of these defects limits the size of the circuit that can be economically fabricated on a single die. Consider the two cases illustrated in Fig. 2.67, where two identical wafers with the same defect locations have been used to fabricate circuits of different area. Although the defect locations in both cases are the same, the wafer-sort yield of the large die would be zero. When the die size is cut to one-fourth of the original size, the wafer sort yield is 62 percent. This conceptual example illustrates the effect of die size on wafer-sort yield. Quantitatively, the expected yield for a given die size is a strong function of the complexity of the process, the nature of the individual steps in the process, and perhaps, most importantly, the maturity and degree of development of the process as a whole and the individual steps within it. Since the inception of the planar process, a steady reduction in defect densities has occurred as a result of improved lithography, increased use of low-temperature processing steps such as ion implantation, improved manufacturing environmental control, and so forth. Three typical curves derived from yield data on bipolar and MOS processes are shown in Fig. 2.68. These are representative of yields for processes ranging from a very complex process with many yield-reducing steps to a very simple process carried out in an advanced VLSI fabrication facility. Also, the yield curves can be raised or lowered by more conservative design rules, and other factors. Uncontrolled factors such as testing problems and design problems in the circuit can cause results for a particular integrated circuit to deviate widely from these curves, but still the overall trend is useful.

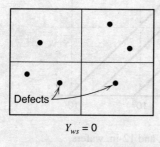

$Y_{ws} = 0$

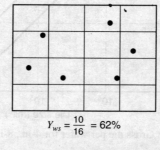

$Y_{ws} = \dfrac{10}{16} = 62\%$

Figure 2.67 Conceptual example of the effect of die size on yield.

158   Chapter 2 ▪ Bipolar, MOS, and BiCMOS Integrated-Circuit Technology

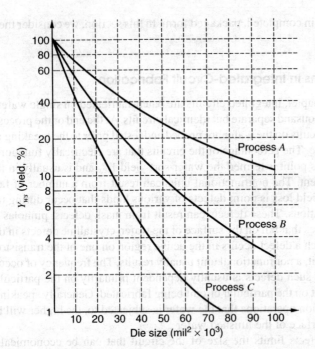

Figure 2.68 Typically observed yield versus die size for the three different processes, ranging from a very simple, well-developed process (curve A), to a very complex process with many yield-reducing steps (curve C).

In addition to affecting yield, the die size also affects the total number of dice that can be fabricated on a wafer of a given size. The total number of usable dice on the wafer, called the gross die per wafer N, is plotted in Fig. 2.69 as a function of die size for several wafer sizes. The product of the gross die per wafer and the wafer-sort yield gives the net good die per wafer, plotted in Fig. 2.70 for the yield curve of Fig. 2.68, assuming a 4-inch wafer.

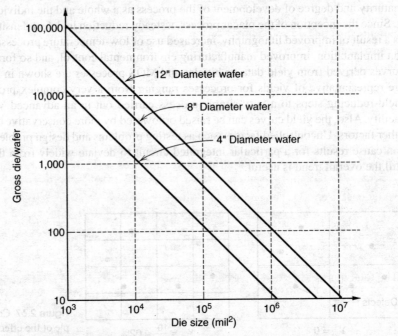

Figure 2.69 Gross die per wafer for 4-in., 8-in., and 12-in. wafers.

2.14 Economics of Integrated-Circuit Fabrication

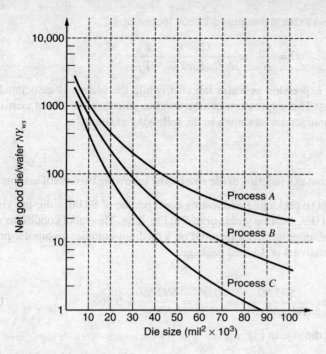

Figure 2.70 Net good die per wafer for the three processes in Fig. 2.66, assuming a 4-in. wafer. The same curve can be obtained approximately for other wafer sizes by simply scaling the vertical axis by a factor equal to the wafer area.

Once the wafer has undergone the wafer-probe test, it is separated into individual dice by sawing or scribing and breaking. The dice are visually inspected, sorted, and readied for assembly into packages. This step is termed *die fab*, and some loss of good dice occurs in the process. Of the original electrically good dice on the wafer, some will be lost in the die fab process due to breakage and scratching of the surface. The ratio of the electrically good dice following die fab to the number of electrically good dice on the wafer before die fab is called the *die fab yield* Y_{df}. The good dice are then inserted in a package, and the electrical connections to each die are made with bonding wires to the pins on the package. The packaged circuits then undergo a final test, and some loss of functional units usually occurs because of improper bonding and handling losses. The ratio of the number of good units at final test to the number of good dice into assembly is called the *final test yield* Y_{ft}.

2.14.2 Cost Considerations in Integrated-Circuit Fabrication

The principal direct costs to the manufacturer can be divided into two categories: those associated with fabricating and testing the wafer, called the *wafer fab cost* C_w, and those associated with packaging and final testing the individual dice, called the *packaging cost* C_p. If we consider the costs incurred by the complete fabrication of one wafer of dice, we first have the wafer cost itself C_w. The number of electrically good dice that are packaged from the wafer is $NY_{ws}Y_{df}$. The total cost C_t incurred once these units have been packaged and tested is

$$C_t = C_w + C_p NY_{ws} Y_{df} \tag{2.54}$$

The total number of good finished units N_g is

$$N_g = NY_{ws} Y_{df} Y_{ft} \tag{2.55}$$

Thus the cost per unit is

$$C = \frac{C_t}{N_g} = \frac{C_w}{NY_{ws}Y_{df}Y_{ft}} + \frac{C_p}{Y_{ft}} \quad (2.56)$$

The first term in the cost expression is wafer fab cost, while the second is associated with assembly and final testing. This expression can be used to calculate the direct cost of the finished product to the manufacturer as shown in the following example.

■ **EXAMPLE**

Plot the direct fabrication cost as a function of die size for the following two sets of assumptions.

(a) Wafer-fab cost of \$75.00, packaging and testing costs per die of \$0.06, a die-fab yield of 0.9, and a final-test yield of 0.9. Assume yield curve B in Fig. 2.68. This set of conditions might characterize an operational amplifier manufactured on a medium-complexity bipolar process and packaged in an inexpensive 8 or 14 lead package.

From (2.56),

$$C = \frac{\$75.00}{(NY_{ws})(0.81)} + \frac{0.06}{0.9} = \frac{\$92.59}{NY_{ws}} + 0.066 \quad (2.57)$$

This cost is plotted versus die size in Fig. 2.71a.

(b) A wafer-fab cost of \$100.00, packaging and testing costs of \$1.00, die-fab yield of 0.9, and final-test yield of 0.8. Assume yield curve A in Fig. 2.68. This might characterize a complex analog/digital integrated circuit, utilizing an advanced CMOS process and packaged in a large, multilead package. Again, from (2.56),

$$C = \frac{\$100.00}{(NY_{ws})(0.72)} + \frac{\$1.00}{0.8} = \frac{\$138.89}{NY_{ws}} + \$1.25 \quad (2.58)$$

■ This cost is plotted versus die size in Fig. 2.71b.

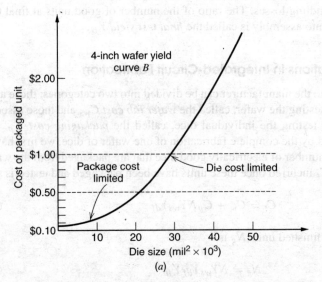

Figure 2.71 (a) Cost curve for example a.

2.14 Economics of Integrated-Circuit Fabrication

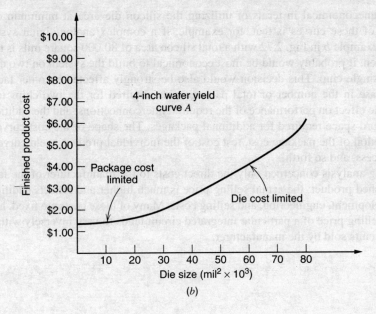

Figure 2.71 (b) Cost curve for example b.

This example shows that most of the cost comes from packaging and testing for small die sizes, whereas most of the cost comes from wafer-fab costs for large die sizes. This relationship is made clearer by considering the cost of the integrated circuit in terms of cost per unit area of silicon in the finished product, as illustrated in Fig. 2.72 for the examples previously given. These curves plot the ratio of the finished-product cost to the number of square mils of silicon on the die. The minimum cost per unit area of silicon results midway between the package-cost and die-cost limited regions for each example. Thus the fabrication of excessively large

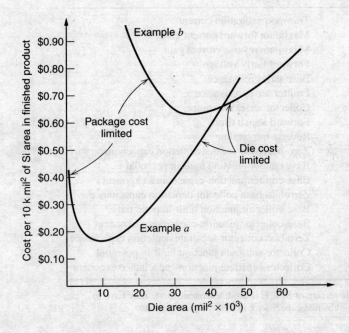

Figure 2.72 Cost of finished product in terms of cost per unit of silicon area for the two examples. Because the package and testing costs are lower in example a, the minimum cost point falls at a much smaller die size. The cost per unit of silicon area at large die sizes is smaller for example b because process A gives higher yield at large die sizes.

or small dice is uneconomical in terms of utilizing the silicon die area at minimum cost. The significance of these curves is that, for example, if a complex analog/digital system, characterized by example *b* in Fig. 2.72 with a total silicon area of 80,000 square mils is to be fabricated in silicon, it probably would be most economical to build the system on two chips rather than on a single chip. This decision would also be strongly affected by other factors such as the increase in the number of total package pins required for the two chips to be interconnected, the effect on performance of the required interconnections, and the additional printed circuit board space required for additional packages. The shape of the cost curves is also a strong function of the package cost, test cost of the individual product, yield curve for the particular process, and so forth.

The preceding analysis concerned only the direct costs to the manufacturer of the fabrication of the finished product; the actual selling price is much higher and reflects additional research and development, engineering, and selling costs. Many of these costs are fixed, however, so that the selling price of a particular integrated circuit tends to vary inversely with the quantity of the circuits sold by the manufacturer.

APPENDIX
A.2.1 SPICE MODEL-PARAMETER FILES

In this section, SPICE model-parameter symbols are compared with the symbols employed in the text for commonly used quantities.

Bipolar Transistor Parameters

SPICE Symbol	Text Symbol	Description
IS	I_S	Transport saturation current
BF	β_F	Maximum forward current gain
BR	β_R	Maximum reverse current gain
VAF	V_A	Forward Early voltage
RB	r_b	Base series resistance
RE	r_{ex}	Emitter series resistance
RC	r_c	Collector series resistance
TF	τ_F	Forward transit time
TR	τ_R	Reverse transit time
CJE	C_{je0}	Zero-bias base-emitter depletion capacitance
VJE	ψ_{0e}	Base-emitter junction built-in potential
MJE	n_e	Base-emitter junction-capacitance exponent
CJC	$C_{\mu 0}$	Zero-bias base-collector depletion capacitance
VJC	ψ_{0c}	Base-collector junction built-in potential
MJC	n_c	Base-collector junction-capacitance exponent
CJS	C_{CS0}	Zero-bias collector-substrate depletion capacitance
VJS	ψ_{0s}	Collector-substrate junction built-in potential
MJS	n_s	Collector-substrate junction-capacitance exponent

Note: Depending on which version of SPICE is used, a separate diode may have to be included to model base-substrate capacitance in a lateral *pnp* transistor.

MOSFET Parameters

SPICE Symbol	Text Symbol	Description
VTO	V_i	Threshold voltage with zero source-substrate voltage
KP	$k' = \mu C_{ox}$	Transconductance parameter
GAMMA	$\gamma = \dfrac{\sqrt{2q\epsilon N_A}}{C_{ox}}$	Threshold voltage parameter
PHI	$2\phi_f$	Surface potential
LAMBDA	$\lambda = \dfrac{1}{L_{\text{eff}}}\dfrac{dX_d}{dV_{DS}}$	Channel-length modulation parameter
CGSO	C_{ol}	Gate-source overlap capacitance per unit channel width
CGDO	C_{ol}	Gate-drain overlap capacitance per unit channel width
CJ	C_{j0}	Zero-bias junction capacitance per unit area from source and drain bottom to bulk (substrate)
MJ	n	Source-bulk and drain-bulk junction capacitance exponent (grading coefficient)
CJSW	C_{jsw0}	Zero-bias junction capacitance per unit junction perimeter from source and drain sidewall (periphery) to bulk
MJSW	n	Source-bulk and drain-bulk sidewall junction capacitance exponent
PB	ψ_0	Source-bulk and drain-bulk junction built-in potential
TOX	t_{ox}	Oxide thickness
NSUB	N_A, N_D	Substrate doping
NSS	Q_{ss}/q	Surface-state density
XJ	X_j	Source, drain junction depth
LD	L_d	Source, drain lateral diffusion

PROBLEMS

2.1 What impurity concentration corresponds to a 0.1 Ω-cm resistivity in p-type silicon? In n-type silicon?

2.2 What is the sheet resistance of a layer of 1 Ω-cm material that is 2 μm thick?

2.3 Consider a hypothetical layer of silicon that has an n-type impurity concentration of 10^{17} cm^{-3} at the top surface, and in which the impurity concentration decreases exponentially with distance into the silicon. Assume that the concentration has decreased to $1/e$ of its surface value at a depth of 0.5 μm, and that the impurity concentration in the sample before the insertion of the n-type impurities was 10^{15} cm^{-3} p-type. Determine the depth below the surface of the pn junction that results and determine the sheet resistance of the n-type layer. Assume a constant electron mobility of 1000 cm^2/V-s. Assume that the width of the depletion layer is negligible.

2.4 A diffused resistor has a length of 100 μm and a width of 5 μm. The sheet resistance of the base diffusion is 100 $\Omega/\square$ and the emitter diffusion is 5 $\Omega/\square$. The base pinched layer has a sheet resistance of 5 k$\Omega/\square$. Determine the resistance of the resistor if it is an emitter-diffused, base-diffused, or pinch resistor.

2.5 A base-emitter voltage of from 520 mV to 580 mV is measured on a test npn transistor structure with 20 μA collector current. The emitter dimensions on the test transistor are 100 μm $\times$ 100 μm. Determine the range of values of Q_B implied by this data. Use this information to calculate the range of values of sheet resistance that will be observed in the pinch resistors in the circuit. Assume a constant electron diffusivity, $\overline{D}_n$, of 13 cm^2/s, and a constant hole mobility of 150 cm^2/V-s. Assume that the width of the depletion layer is negligible.

2.6 Estimate the series base resistance, series collector resistance r_c, base-emitter capacitance, base-collector capacitance, and collector-substrate capacitance of the high-current npn transistor structure

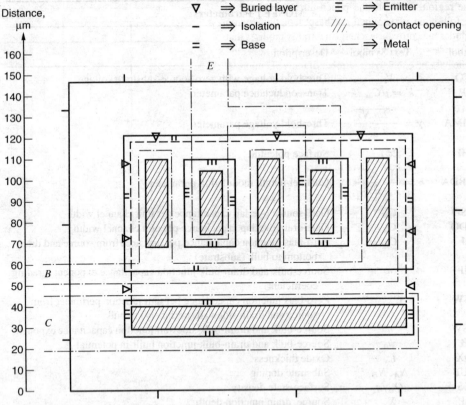

Figure 2.73 Device structure for Problem 2.6.

shown in Fig. 2.73. This structure is typical of those used as the output transistor in operational amplifiers that must supply up to about 20 mA. Assume a doping profile as shown in Fig. 2.17.

2.7 If the lateral *pnp* structure of Fig. 2.33a is fabricated with an epi layer resistivity of 0.5 Ω-cm, determine the value of collector current at which the current gain begins to fall off. Assume a diffusivity for holes of: $\overline{D}_p = 10$ cm²/s. Assume a base width of 5 μm.

2.8 The substrate *pnp* of Fig. 2.36a is to be used as a test device to monitor epitaxial layer thickness. Assume that the flow of minority carriers across the base is vertical, and that the width of the emitter-base and collector-base depletion layers is negligible. Assume that the epi layer resistivity is known to be 2 Ω-cm by independent measurement. The base-emitter voltage is observed to vary from 525 mV to 560 mV over several wafers at a collector current of 10 μA. What range of epitaxial layer thickness does this imply? What is the corresponding range of sheet resistance that will be observed in the epitaxial pinch resistors? Assume a hole diffusivity of 10 cm²/s, and an electron mobility of 800 cm²/V-s. Neglect the depletion layer thickness. Assume a junction depth of 3 μm for the base diffusion.

2.9 For the substrate *pnp* structure shown in Fig. 2.36a, calculate I_S, C_{je}, C_μ, and τ_F. Assume the doping profiles are as shown in Fig. 2.17.

2.10 Calculate the total parasitic junction capacitance associated with a 10-kΩ base-diffused resistor if the base sheet resistance is 100 Ω/□ and the resistor width is 6 μm. Repeat for a resistor width of 12 μm. Assume the doping profiles are as shown in Fig. 2.17. Assume the clubheads are 26 μm × 26 μm, and that the junction depth is 3 μm. Account for sidewall effects.

2.11 A base-emitter voltage of 480 mV is measured on a super-β test transistor with a 100 μm × 100 μm emitter area at a collector current of 10 μA. Calculate the Q_B and the sheet resistance of the base region. Estimate the punch-through voltage in the following way. When the base depletion region includes the entire base, charge neutrality requires that the number of ionized acceptors in the depletion region in the base be equal to the number of ionized donors in the depletion region on the collector side of the base. [See (1.2).] Therefore, when enough voltage is applied that the depletion region in the base region includes the whole base, the depletion region in the collector must include a number of ionized atoms equal to Q_B. Since the density of these atoms is known (equal to N_D), the width of the depletion layer in the collector region at punch-through can be determined. If we assume that the doping in the base N_A is much larger than that in the collector N_D, then (1.15) can be used to find the voltage that will result in this depletion layer width. Repeat this problem for the standard device, assuming a V_{BE} measured at 560 mV. Assume an electron diffusivity $\overline{D}_n$ of 13 cm²/s, and a hole mobility $\overline{\mu}_p$ of 150 cm²/V-s. Assume the epi doping is 10^{15} cm^{-3}. Use $\epsilon = 1.04 \times 10^{-12}$ F/cm for the permittivity of silicon. Also, assume ψ_o for the collector-base junction is 0.55 V.

2.12 An MOS transistor biased in the active region displays a drain current of 100 μA at a V_{GS} of 1.5 V and a drain current of 10 μA at a V_{GS} of 0.8 V. Determine the threshold voltage and $\mu_n C_{ox}(W/L)$. Neglect subthreshold conduction and assume that the mobility is constant.

2.13 Calculate the threshold voltage of the p-channel transistors for the process given in Table 2.1. First do the calculation for the unimplanted transistor, then for the case in which the device receives the channel implant specified. Note that this is a p-type implant, so that the effective surface concentration is the difference between the background substrate concentration and the effective concentration in the implant layer.

2.14 An n-channel implanted transistor from the process described in Table 2.1 displays a measured output resistance of 5 MΩ at a drain current of 10 μA, biased in the active region at a V_{DS} of 5 V. The drawn dimensions of the device are 100 μm by 7 μm. Find the output resistance of a second device on the same technology that has drawn dimensions of 50 μm by 12 μm and is operated at a drain current of 30 μA and a V_{DS} of 5 V.

2.15 Calculate the small-signal model parameters of the device shown in Fig. 2.74, including g_m, g_{mb}, r_o, C_{gs}, C_{gd}, C_{sb}, and C_{db}. Assume the transistor is biased at a drain-source voltage of 2 V and a drain current of 20 μA. Use the process parameters that are specified in Table 2.4. Assume $V_{SB} = 1$ V.

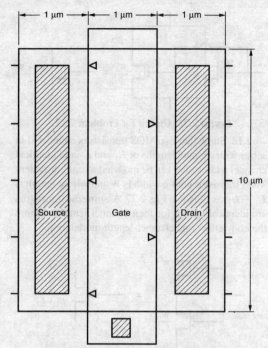

Figure 2.74 Transistor for Problem 2.15.

2.16 The transistor shown in Fig. 2.74 is connected in the circuit shown in Fig. 2.75. The gate is grounded, the substrate is connected to −1.5 V, and the drain is open circuited. An ideal current source is tied to the source, and this source has a value of zero for $t < 0$ and 10 μA for $t > 0$. The source and drain are at an initial voltage of +1.5 V at $t = 0$. Sketch the voltage at the source and drain from $t = 0$ until the drain voltage reaches −1.5 V. For simplicity, assume that the source-substrate and drain-substrate capacitances are constant at their zero-bias values. Assume the transistor has a threshold voltage of 0.6 V.

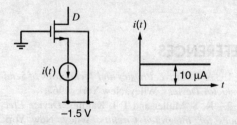

Figure 2.75 Circuit for Problem 2.16.

2.17 Show that two MOS transistors connected in parallel with channel widths of W_1 and W_2 and identical channel lengths of L can be modeled as one equivalent MOS transistor whose width is $W_1 + W_2$ and whose length is L, as shown in Fig. 2.76. Assume the transistors are identical except for their channel widths.

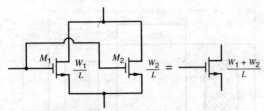

Figure 2.76 Circuit for Problem 2.17.

2.18 Show that two MOS transistors connected in series with channel lengths of L_1 and L_2 and identical channel widths of W can be modeled as one equivalent MOS transistor whose width is W and whose length is $L_1 + L_2$, as shown in Fig. 2.77. Assume the transistors are identical except for their channel lengths. Ignore the body effect and channel-length modulation.

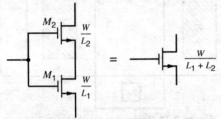

Figure 2.77 Circuit for Problem 2.18.

2.19 An integrated electronic subsystem is to be fabricated, which requires 40,000 square mils of silicon area. Determine whether the system should be put on one or two chips, assuming that the fabrication cost of the two chips is the only consideration. Assume that the wafer-fab cost is $100.00, the packaging and testing costs are $0.60, the die-fab yield is 0.9, and the final-test yield is 0.8. Assume the process used follows curve B in Fig. 2.68. Repeat the problem assuming yield curve A, and then yield curve C. Assume a 4-inch wafer.

2.20(a) A frequently used empirical approximation for the yield of an IC process as a function of die size is

$$Y_{ws} = \exp(-A/A_0)$$

where A is the die area and A_0 is a constant. Using Fig. 2.68, determine approximate values of A_0 for each of the three processes shown. Use the point on the curve at which the yield is e^{-1} to determine A_0. Plot the yield predicted by this expression and compare with the curves shown in Fig. 2.68.

(b) Use the expression derived in (a), together with the gross-die-per-wafer curves shown in Fig. 2.69, to develop an analytical expression for the cost of silicon per unit area as a function of die size, Y_{df}, Y_{ft}, C_p, and C_w for each of the three processes A, B, and C.

2.21 Determine the direct fabrication cost of an integrated circuit that is 150 mils on a side in size. Assume a wafer-fab cost of $130.00, a package and testing cost of $0.40, a die-fab yield of 0.8, and a final-test yield of 0.8. Work the problem for yield curves A, B, and C in Fig. 2.68. Assume a 4-inch wafer.

2.22 Calculate the small-signal model parameters g_m, r_o, C_{gs}, and C_{gd} for a NMOS transistor. Also calculate the gate-leakage current I_G. Assume the transistor operates in the active region with $I_D = 100$ μA, $V_{DS} = 1$ V, $V_{BS} = 0$ V, $W = 0.5$ μm, and $L = 0.1$ μm. Use the transistor model data in Table 2.6.

2.23 Calculate the small-signal model parameters g_m, r_o, C_{gs}, and C_{gd} for a NMOS transistor. Assume the transistor operates in the active region with $I_D = 100$ μA, $V_{DS} = 1$ V, $V_{BS} = 0$ V, $W = 0.9$ μm, and $L = 0.2$ μm. Use the transistor model data in Table 2.5.

REFERENCES

1. A. S. Grove. *Physics and Technology of Semiconductor Devices*. Wiley, New York, 1967.

2. R. S. Muller and T. I. Kamins. *Device Electronics for Integrated Circuits*. Wiley, New York, 1986.

3. E. M. Conwell. "Properties of Silicon and Germanium," *Proc. IRE*, Vol. 46, pp. 1281–1300, June 1958.

4. J. C. Irvin. "Resistivity of Bulk Silicon and of Diffused Layers in Silicon," *Bell System Tech. Journal*, Vol. 41, pp. 387–410, March 1962.

5. R. W. Russell and D. D. Culmer. "Ion-Implanted JFET-Bipolar Monolithic Analog Circuits," *Digest of Technical Papers, 1974 International Solid-State Circuits Conference*, Philadelphia, PA, pp. 140–141, February 1974.

6. D. J. Hamilton and W. G. Howard. *Basic Integrated Circuit Engineering*. McGraw-Hill, New York, 1975.

7. Y. Tamaki, T. Shiba, I. Ogiwara, T. Kure, K. Ohyu, and T. Nakamura. "Advanced Device Process Technology for 0.3 μm Self-Aligned Bipolar LSI," *Proceedings of the IEEE Bipolar Circuits and Technology Meeting*, pp. 166–168, September 1990.

8. M. Kurisu, Y. Sasyama, M. Ohuchi, A. Sawairi, M. Sigiyama, H. Takemura, and T. Tashiro.

"A Si Bipolar 21 GHz Static Frequency Divider," *Digest of Technical Papers, 1991 International Solid-State Circuits Conference*, pp. 158–159, February 1991.

9. R. M. Burger and R. P. Donovan. *Fundamentals of Silicon Integrated Device Technology*. Vol. 2, pp. 134–136. Prentice-Hall, Englewood Cliffs, NJ, 1968.

10. R. J. Whittier and D. A. Tremere. "Current Gain and Cutoff Frequency Falloff at High Currents," *IEEE Transactions Electron Devices*, Vol. ED-16, pp. 39–57, January 1969.

11. H. R. Camenzind. *Electronic Integrated Systems Design*. Van Nostrand Reinhold, New York, 1972. Copyright © 1972 Litton Educational Publishing, Inc. Reprinted by permission of Van Nostrand Reinhold Company.

12. H. J. DeMan. "The Influence of Heavy Doping on the Emitter Efficiency of a Bipolar Transistor," *IEEE Transactions on Electron Devices*, Vol. ED-18, pp. 833–835, October 1971.

13. H. C. Lin. *Integrated Electronics*. Holden-Day, San Francisco, 1967.

14. N. M. Nguyen and R. G. Meyer. "Si IC-Compatible Inductors and LC Passive Filters," *IEEE Journal of Solid-State Circuits*, Vol. 25, pp. 1028–1031, August 1990.

15. J. Y.-C. Chang, A. A. Abidi, and M. Gaitan. "Large Suspended Inductors on Silicon and Their Use in a 2 μm CMOS RF Amplifier," *IEEE Electron Device Letters*, Vol. 14, pp. 246–248, May 1993.

16. K. Negus, B. Koupal, J. Wholey, K. Carter, D. Millicker, C. Snapp, and N. Marion. "Highly Integrated Transmitter RFIC with Monolithic Narrowband Tuning for Digital Cellular Handsets," *Digest of Technical Papers, 1994 International Solid-State Circuits Conference*, San Francisco, CA, pp. 38–39, February 1994.

17. P. R. Gray and R. G. Meyer. *Analysis and Design of Analog Integrated Circuits*, Third Edition, Wiley, New York, 1993.

18. R. J. Widlar. "Design Techniques for Monolithic Operational Amplifiers," *IEEE Journal of Solid-State Circuits*, Vol. SC-4, pp. 184–191, August 1969.

19. K. R. Stafford, P. R. Gray, and R. A. Blanchard. "A Complete Monolithic Sample/Hold Amplifier," *IEEE Journal of Solid-State Circuits*, Vol. SC-9, pp. 381–387, December 1974.

20. P. C. Davis, S. F. Moyer, and V. R. Saari. "High Slew Rate Monolithic Operational Amplifier Using Compatible Complementary *pnp*'s," *IEEE Journal of Solid-State Circuits*, Vol. SC-9, pp. 340–346, December 1974.

21. A. P. Chandrakasan, S. Sheng, and R. W. Brodersen. "Low-Power CMOS Digital Design," *IEEE Journal of Solid-State Circuits*, Vol. 27, pp. 473–484, April 1992.

22. Y. P. Tsividis, *Operation and Modeling of the MOS Transistor*. McGraw-Hill, 1987.

23. S.-H. Lo, D. A. Buchanan, Y. Taur and W. Wang, "Quantum-Mechanical Modeling of Electron Tunneling from the Inversion Layer of Ultra-Thin-Oxide nMOSFET's," *IEEE Electron Device Letters*, pp. 209–211, May 1997.

24. K. F. Schuegraf, C. C. King, and C. Hu, "Ultra-thin Silicon Dioxide Leakage Current and Scaling Limits," *Symp. on VLSI Technology, Digest of Technical Papers*, pp. 18–19, 1992.

25. W-C. Lee, C. Hu, "Modeling Gate and Substrate Currents due to Conduction- and Valence-Band Electron and Hole Tunneling," *Symp. on VLSI Technology, Digest of Technical Papers*, pp. 198–199, 2000.

26. www-device.eecs.berkeley.edu/~bsim3/.

27. D. P. Foty, *MOSFET Modeling with SPICE*, Prentice-Hall, Upper Saddle River, NJ, 1997.

28. E. A. Vittoz. "MOS Transistors Operated in the Lateral Bipolar Mode and Their Application in CMOS Technology," *IEEE Journal of Solid-State Circuits*, Vol. SC-18, pp. 273–279, June 1983.

29. W. T. Holman and J. A. Connelly. "A Compact Low-Noise Operational Amplifier for a 1.2 μm Digital CMOS Technology," *IEEE Journal of Solid-State Circuits*, Vol. 30, pp. 710–714, June 1995.

30. C. A. Laber, C. F. Rahim, S. F. Dreyer, G. T. Uehara, P. T. Kwok, and P. R. Gray. "Design Considerations for a High-Performance 3-μm CMOS Analog Standard-Cell Library," *IEEE Journal of Solid-State Circuits*, Vol. SC-22, pp. 181–189, April 1987.

31. B. L. Crowder and S. Zirinsky. "1 μm MOSFET VLSI Technology: Part VII—Metal Silicide Interconnection Technology—A Future Perspective," *IEEE Journal of Solid-State Circuits*, Vol. SC-14, pp. 291–293, April 1979.

32. J. L. McCreary. "Matching Properties, and Voltage and Temperature Dependence of MOS Capacitors," *IEEE Journal of Solid-State Circuits*, Vol. SC-16, pp. 608–616, December 1981.

33. D. J. Allstot and W. C. Black, Jr. "Technological Design Considerations for Monolithic MOS Switched-Capacitor Filtering Systems," *Proceedings of the IEEE*, Vol. 71, pp. 967–986, August 1983.

34. O. E. Akcasu. "High Capacitance Structure in a Semiconductor Device," *U.S. Patent 5,208,725*, May 1993.

35. H. Samavati, A. Hajimiri, A. R. Shahani, G. N. Nasserbakht, and T. H. Lee. "Fractal Capacitors,"

IEEE Journal of Solid-State Circuits, Vol. 33, pp. 2035–2041, December 1998.

36. A. R. Alvarez. *BiCMOS Technology and Applications.* Kluwer Academic Publishers, Dordrecht, The Netherlands, 1989.

37. J. L. de Jong, R. Lane, B. van Schravendijk, and G. Conner. "Single Polysilicon Layer Advanced Super High-speed BiCMOS Technology," *Proceedings of the IEEE Bipolar Circuits and Technology Meeting,* pp. 182–185, September 1989.

38. D. L. Harame, J. H. Comfort, J. D. Cressler, E. F. Crabbé, J. Y.-C. Sun, B. S. Meyerson, and T. Tice. "Si/SiGe Epitaxial-Base Transistors–Part I: Materials, Physics, and Circuits," *IEEE Transactions on Electron Devices,* Vol. 42, pp. 455–468, March 1995.

39. J. D. Cressler, D. L. Harame, J. H. Comfort, J. M. C. Stork, B. S. Meyerson, and T. E. Tice. "Silicon-Germanium Heterojunction Bipolar Technology: The Next Leap in Silicon?" *Digest of Technical Papers, 1994 International Solid-State Circuits Conference,* San Francisco, CA, pp. 24–27, February 1994.

40. D. L. Harame, J. H. Comfort, J. D. Cressler, E. F. Crabbé, J. Y.-C. Sun, B. S. Meyerson, and T. Tice. "Si/SiGe Epitaxial-Base Transistors—Part II: Process Integration and Analog Applications," *IEEE Transactions on Electron Devices,* Vol. 42, pp. 469–482, March 1995.

41. M. T. Bohr. "Interconnect Scaling—The Real Limiter to High Performance ULSI," *Technical Digest, International Electron Devices Meeting,* pp. 241–244, December 1995.

42. C. S. Chang, K. A. Monnig, and M. Melliar-Smith. "Interconnection Challenges and the National Technology Roadmap for Semiconductors," *IEEE International Interconnect Technology Conference,* pp. 3–6, June 1998.

43. D. Edelstein, J. Heidenreich, R. Goldblatt, W. Cote, C. Uzoh, N. Lustig, P. Roper, T. McDevitt, W. Motsiff, A. Simon, J. Dukovic, R. Wachnik, H. Rathore, R. Schulz, L. Su, S. Luce, and J. Slattery. "Full Copper Wiring in a Sub-0.25 μm CMOS ULSI Technology," *IEEE International Electron Devices Meeting,* pp. 773–776, December 1997.

CHAPTER 3

Single-Transistor and Multiple-Transistor Amplifiers

The technology used to fabricate integrated circuits presents a unique set of component-cost constraints to the circuit designer. The most cost-effective circuit approach to accomplish a given function may be quite different when the realization of the circuit is to be in monolithic form as opposed to discrete transistors and passive elements.[1] As an illustration, consider the two realizations of a three-stage audio amplifier shown in Figs. 3.1 and 3.2. The first reflects a cost-effective solution in the context of discrete-component circuits, since passive components such as resistors and capacitors are less expensive than the active components, the transistors. Hence, the circuit contains a minimum number of transistors, and the interstage coupling is accomplished with capacitors. However, for the case of monolithic construction, a key determining factor in cost is the die area used. Capacitors of the values used in most discrete-component circuits are not feasible and would have to be external to the chip, increasing the pin count of the package, which increases cost. Therefore, a high premium is placed on eliminating large capacitors, and a dc-coupled circuit realization is very desirable. A second constraint is that the *cheapest* component that can be fabricated in the integrated circuit is the one that occupies the least area, usually a transistor. Thus a circuit realization that contains the minimum possible total resistance while using more active components may be optimum.[2,3] Furthermore, an important application of analog circuits is to provide interfaces between the real world and digital circuits. In building digital integrated circuits, CMOS technologies have become dominant because of their high densities and low power dissipations. To reduce the cost and increase the portability of mixed-analog-and-digital systems, both increased levels of integration and reduced power dissipations are required. As a result, we are interested in building analog interface circuits in CMOS technologies. The circuit of Fig. 3.2 reflects these constraints. It uses a CMOS technology and many more transistors than in Fig. 3.1, has less total resistance, and has no coupling capacitors. A differential pair is used to allow direct coupling between stages, while transistor current sources provide biasing without large amounts of resistance. In practice, feedback would be required around the amplifier shown in Fig. 3.2 but is not shown for simplicity. Feedback is described in Chapter 8.

The next three chapters analyze various circuit configurations encountered in linear integrated circuits. In discrete-component circuits, the number of transistors is usually minimized. The best way to analyze such circuits is usually to regard each individual transistor as a *stage* and to analyze the circuit as a collection of single-transistor stages. A typical monolithic circuit, however, contains a large number of transistors that perform many functions, both passive and active. Thus monolithic circuits are often regarded as a collection of *subcircuits* that perform specific functions, where the subcircuits may contain many transistors. In this chapter, we first consider the dc and low-frequency properties of the simplest subcircuits: common-emitter, common-base, and common-collector single-transistor amplifiers and their counterparts using

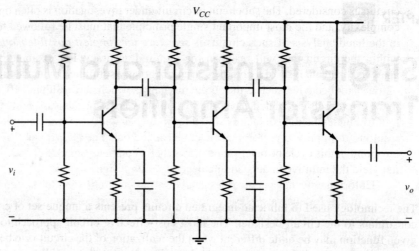

Figure 3.1 Typical discrete-component realization of an audio amplifier.

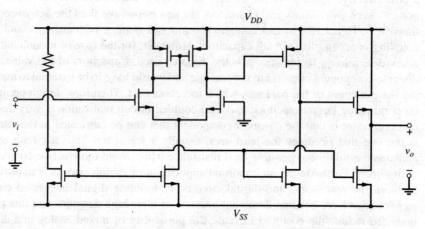

Figure 3.2 Typical CMOS integrated-circuit realization of an audio amplifier.

MOS transistors. We then consider some multi-transistor subcircuits that are useful as amplifying stages. The most widely used of these multi-transistor circuits are the differential pairs, which are analyzed extensively in this chapter.

3.1 Device Model Selection for Approximate Analysis of Analog Circuits

Much of this book is concerned with the salient performance characteristics of a variety of subcircuits commonly used in analog circuits and of complete functional blocks made up of these subcircuits. The aspects of the performance that are of interest include the dc currents and voltages within the circuit, the effect of mismatches in device characteristics on these voltages and currents, the small-signal, low-frequency input and output resistance, and the voltage gain of the circuit. In later chapters, the high-frequency, small-signal behavior of circuits is considered. The subcircuit or circuit under investigation is often one of considerable complexity, and the most important single principle that must be followed to achieve success in the hand analysis of such circuits is *selecting the simplest possible model* for the devices within the circuit that will result in the required accuracy. For example, in the case of dc analysis, hand analysis of a complex circuit is greatly simplified by neglecting certain aspects of transistor behavior, such as the output resistance, which may result in a 10 to 20 percent error in the dc currents calculated. The principal objective of hand analysis, however, is to obtain an intuitive understanding of factors affecting circuit behavior so that an iterative design procedure resulting in improved performance can be carried out. The performance of the circuit can at any point in this cycle be determined precisely by computer simulation, but this approach does not yield the intuitive understanding necessary for design.

Unfortunately, no specific rules can be formulated regarding the selection of the simplest device model for analysis. For example, in the dc analysis of bipolar biasing circuits, assuming constant base-emitter voltages and neglecting transistor output resistances often provides adequate accuracy. However, certain bias circuits depend on the nonlinear relation between the collector current and base-emitter voltage to control the bias current, and the assumption of a constant V_{BE} will result in gross errors in the analyses of these circuits. When analyzing the active-load stages in Chapter 4, the output resistance must be considered to obtain meaningful results. Therefore, a key step in every analysis is to inspect the circuit to determine what aspects of the behavior of the transistors strongly affect the performance of the circuit, and then simplify the model(s) to include only those aspects. This step in the procedure is emphasized in this and the following chapters.

3.2 Two-Port Modeling of Amplifiers

The most basic parameter of an amplifier is its gain. Since amplifiers may be connected to a wide variety of sources and loads, predicting the dependence of the gain on the source and load resistance is also important. One way to observe this dependence is to include these resistances in the amplifier analysis. However, this approach requires a completely new amplifier analysis each time the source or load resistance is changed. To simplify this procedure, amplifiers are often modeled as two-port equivalent networks. As shown in Fig. 3.3, two-port networks have four terminals and four port variables (a voltage and a current at each port). A pair of terminals is a port if the current that flows into one terminal is equal to the current that flows out of the other terminal. To model an amplifier, one port represents the amplifier input characteristics and the other represents the output. One variable at each port can be set independently. The other variable at each port is dependent on the two-port network and the independent variables. This dependence is expressed by two equations. We will focus here on the admittance-parameter equations, where the terminal currents are viewed as dependent variables controlled by the independent terminal voltages because we usually model transistors with voltage-controlled current sources. If the network is linear and

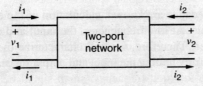

Figure 3.3 Two-port-network block diagram.

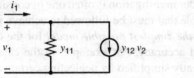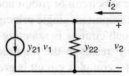

Figure 3.4 Admittance-parameter, two-port equivalent circuit.

contains no independent sources, the admittance-parameter equations are:

$$i_1 = y_{11}v_1 + y_{12}v_2 \tag{3.1}$$

$$i_2 = y_{21}v_1 + y_{22}v_2 \tag{3.2}$$

The voltages and currents in these equations are deliberately written as small-signal quantities because transistors behave in an approximately linear way only for small signals around a fixed operating point. An equivalent circuit for these equations is shown in Fig. 3.4. The parameters can be found and interpreted as follows:

$$y_{11} = \left.\frac{i_1}{v_1}\right|_{v_2=0} = \text{Input admittance with the output short-circuited} \tag{3.3}$$

$$y_{12} = \left.\frac{i_1}{v_2}\right|_{v_1=0} = \text{Reverse transconductance with the input short-circuited} \tag{3.4}$$

$$y_{21} = \left.\frac{i_2}{v_1}\right|_{v_2=0} = \text{Forward transconductance with the output short-circuited} \tag{3.5}$$

$$y_{22} = \left.\frac{i_2}{v_2}\right|_{v_1=0} = \text{Output admittance with the input short-circuited} \tag{3.6}$$

The y_{12} parameter represents feedback in the amplifier. When the signal propagates back from the output to the input as well as forward from the input to the output, the amplifier is said to be *bilateral*. In many practical cases, especially at low frequencies, this feedback is negligible and y_{12} is assumed to be zero. Then the amplifier is *unilateral* and characterized by the other three parameters. Since the model includes only one transconductance when $y_{12} = 0$, y_{21} is usually referred to simply as the *short-circuit transconductance*, which will be represented by G_m in this book. When an amplifier is unilateral, the calculation of y_{11} is simplified from that given in (3.3) because the connections at the output port do not affect the input admittance when $y_{12} = 0$.

Instead of calculating y_{11} and y_{22}, we will often calculate the reciprocals of these parameters, or the input and output impedances $Z_i = 1/y_{11}$ and $Z_o = 1/y_{22}$, as shown in the unilateral two-port model of Fig. 3.5a. Also, instead of calculating the short-circuit transconductance $G_m = y_{21}$, we will sometimes calculate the open-circuit voltage gain a_v. This substitution is justified by conversion of the Norton-equivalent output model shown in Fig. 3.5a to the Thévenin-equivalent output model shown in Fig. 3.5b. In general, finding any two of the three parameters including G_m, Z_o, and a_v specifies the third parameter because

$$a_v = \left.\frac{v_2}{v_1}\right|_{i_2=0} = -G_m Z_o \tag{3.7}$$

Once two of these parameters and the input impedance are known, calculation of the effects of loading at the input and output ports is possible. At low frequencies, the input and output impedances are usually dominated by resistances. Therefore, we will characterize the low-frequency behavior of many amplifiers in this book by finding the input and output resistances, R_i and R_o, as well as G_m or a_v.

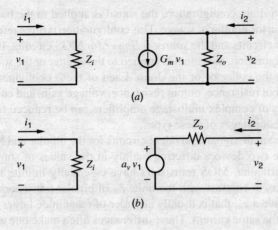

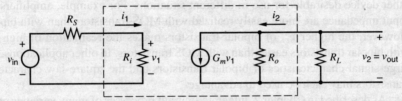

Figure 3.5 Unilateral two-port equivalent circuits with (a) Norton output model (b) Thévenin output model.

Figure 3.6 Example of loading at the input and output of an amplifier modeled by a two-port equivalent circuit.

■ EXAMPLE

A two-port model of a unilateral amplifier is shown in Fig. 3.6. Assume $R_i = 1\,\text{k}\Omega$, $R_o = 1\,\text{M}\Omega$, and $G_m = 1\,\text{mA/V}$. Let R_S and R_L represent the source resistance of the input generator and load resistance, respectively. Find the low-frequency gain $v_{\text{out}}/v_{\text{in}}$, assuming that the input is an ideal voltage source and the output is unloaded. Repeat, assuming that $R_S = 1\,\text{k}\Omega$ and $R_L = 1\,\text{M}\Omega$.

The open-circuit voltage gain of the two-port amplifier model by itself from v_1 to v_{out} is

$$\left.\frac{v_{\text{out}}}{v_1}\right|_{R_L \to \infty} = \left.\frac{v_2}{v_1}\right|_{i_2=0} = -G_m R_o = -(1\,\text{mA/V})(1000\,\text{k}\Omega) = -1000$$

Since the source and input resistances form a voltage divider, and since the output resistance appears in parallel with the load resistance, the overall gain from v_{in} to v_{out} is

$$\frac{v_{\text{out}}}{v_{\text{in}}} = \frac{v_1}{v_{\text{in}}}\frac{v_{\text{out}}}{v_1} = -\frac{R_i}{R_i + R_S} G_m (R_o \parallel R_L)$$

With an ideal voltage source at the input and no load at the output, $R_S = 0$, $R_L \to \infty$, and $v_{\text{out}}/v_{\text{in}} = -1000$. With $R_S = 1\,\text{k}\Omega$ and $R_L = 1\,\text{M}\Omega$, the gain is reduced by a factor of four to $v_{\text{out}}/v_{\text{in}} = -0.5(1\,\text{mA/V})(500\,\text{k}\Omega) = -250$.
■

3.3 Basic Single-Transistor Amplifier Stages

Bipolar and MOS transistors are capable of providing useful amplification in three different configurations. In the common-emitter or common-source configuration, the signal is applied to the base or gate of the transistor and the amplified output is taken from the collector or drain.

In the common-collector or common-drain configuration, the signal is applied to the base or gate and the output signal is taken from the emitter or source. This configuration is often referred to as the *emitter follower* for bipolar circuits and the *source follower* for MOS circuits. In the common-base or common-gate configuration, the signal is applied to the emitter or the source, and the output signal is taken from the collector or the drain. Each of these configurations provides a unique combination of input resistance, output resistance, voltage gain, and current gain. In many instances, the analysis of complex multistage amplifiers can be reduced to the analysis of a number of single-transistor stages of these types.

We showed in Chapter 1 that the small-signal equivalent circuits for the bipolar and MOS transistors are very similar, with the two devices differing mainly in the values of some of their small-signal parameters. In particular, MOS transistors have essentially infinite input resistance from the gate to the source, in contrast with the finite r_π of bipolar transistors. On the other hand, bipolar transistors have a g_m that is usually an order of magnitude larger than that of MOS transistors biased with the same current. These differences often make one or the other device desirable for use in different situations. For example, amplifiers with very high input impedance are more easily realized with MOS transistors than with bipolar transistors. However, the higher g_m of bipolar transistors makes the realization of high-gain amplifiers with bipolar transistors easier than with MOS transistors. In other applications, the exponential large-signal characteristics of bipolar transistors and the square-law characteristics of MOS transistors may each be used to advantage.

As described in Chapter 2, integrated-circuit processes of many varieties now exist. Examples include processes with bipolar or MOS transistors as the only active devices and combined bipolar and CMOS devices in BiCMOS processes. Because the more complex processes involve more masking steps and are thus somewhat more costly to produce, integrated-circuit designers generally use the simplest process available that allows the desired circuit specifications to be achieved. Therefore, designers must appreciate the similarities and differences between bipolar and MOS transistors so that appropriate choices of technology can be made.

3.3.1 Common-Emitter Configuration

The resistively loaded common-emitter (CE) amplifier configuration is shown in Fig. 3.7. The resistor R_C represents the collector load resistance. The short horizontal line labeled V_{CC} at the top of R_C implies that a voltage source of value V_{CC} is connected between that point and ground. This symbol will be used throughout the book. We first calculate the dc transfer characteristic of the amplifier as the input voltage is increased in the positive direction from

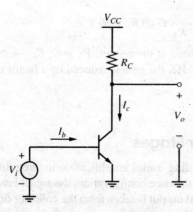

Figure 3.7 Resistively loaded common-emitter amplifier.

3.3 Basic Single-Transistor Amplifier Stages

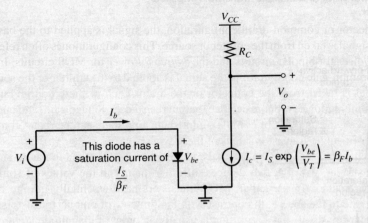

Figure 3.8 Large-signal equivalent circuit valid when the transistor is in the forward-active region. The saturation current of the equivalent base-emitter diode is I_S/β_F.

zero. We assume that the base of the transistor is driven by a voltage source of value V_i. When V_i is zero, the transistor operates in the cutoff state and no collector current flows other than the leakage current I_{CO}. As the input voltage is increased, the transistor enters the forward-active region, and the collector current is given by

$$I_c = I_S \exp \frac{V_i}{V_T} \qquad (3.8)$$

The equivalent circuit for the amplifier when the transistor operates in the forward-active region was derived in Chapter 1 and is repeated in Fig. 3.8. Because of the exponential relationship between I_c and V_{be}, the value of the collector current is very small until the input voltage reaches approximately 0.5 V. As long as the transistor operates in the forward-active region, the base current is equal to the collector current divided by β_F, or

$$I_b = \frac{I_c}{\beta_F} = \frac{I_S}{\beta_F} \exp \frac{V_i}{V_T} \qquad (3.9)$$

The output voltage is equal to the supply voltage, V_{CC}, minus the voltage drop across the collector resistor:

$$V_o = V_{CC} - I_c R_C = V_{CC} - R_C I_S \exp \frac{V_i}{V_T} \qquad (3.10)$$

When the output voltage approaches zero, the collector-base junction of the transistor becomes forward biased and the device enters saturation. Once the transistor becomes saturated, the output voltage and collector current take on nearly constant values:

$$V_o = V_{CE(\text{sat})} \qquad (3.11)$$

$$I_c = \frac{V_{CC} - V_{CE(\text{sat})}}{R_C} \qquad (3.12)$$

The base current, however, continues to increase with further increases in V_i. Therefore, the forward current gain I_c/I_b decreases from β_F as the transistor leaves the forward-active region of operation and moves into saturation. In practice, the current available from the signal source is limited. When the signal source can no longer increase the base current, V_i is maximum. The output voltage and the base current are plotted as a function of the input voltage in Fig. 3.9. Note that when the device operates in the forward-active region, small changes in the input voltage can give rise to large changes in the output voltage. The circuit thus provides *voltage gain*. We now proceed to calculate the voltage gain in the forward-active region.

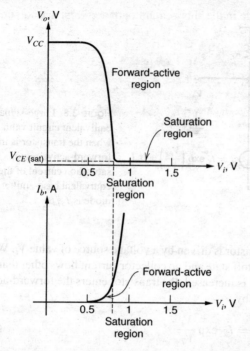

Figure 3.9 Output voltage and base current as a function of V_i for the common-emitter circuit.

While incremental performance parameters such as the voltage gain can be calculated from derivatives of the large-signal analysis, the calculations are simplified by using the small-signal hybrid-π model for the transistor developed in Chapter 1. The small-signal equivalent circuit for the common-emitter amplifier is shown in Fig. 3.10. Here we have neglected r_b, assuming that it is much smaller than r_π. We have also neglected r_μ. This equivalent circuit does not include the resistance of the load connected to the amplifier output. The collector resistor R_C is included because it is usually present in some form as a biasing element. Our objective is to characterize the amplifier alone so that the voltage gain can then be calculated under arbitrary conditions of loading at the input and output. Since the common-emitter amplifier is unilateral when r_μ is neglected, we will calculate the small-signal input resistance, transconductance, and output resistance of the circuit as explained in Section 3.2.

The input resistance is the Thévenin-equivalent resistance seen looking into the input. For the CE amplifier,

$$R_i = \frac{v_i}{i_i} = r_\pi = \frac{\beta_0}{g_m} \tag{3.13}$$

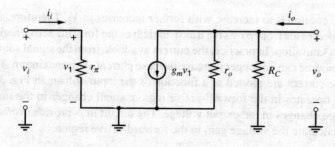

Figure 3.10 Small-signal equivalent circuit for the CE amplifier.

The transconductance G_m is the change in the short-circuit output current per unit change of input voltage and is given by

$$G_m = \left.\frac{i_o}{v_i}\right|_{v_o=0} = g_m \qquad (3.14)$$

Equation 3.14 shows that the transconductance of the CE amplifier is equal to the transconductance of the transistor. The output resistance is the Thévenin-equivalent resistance seen looking into the output with the input shorted, or

$$R_o = \left.\frac{v_o}{i_o}\right|_{v_i=0} = R_C \parallel r_o \qquad (3.15)$$

The *open-circuit*, or *unloaded*, voltage gain is

$$a_v = \left.\frac{v_o}{v_i}\right|_{i_o=0} = -g_m(r_o \parallel R_C) \qquad (3.16)$$

If the collector load resistor R_C is made very large, then a_v becomes

$$\lim_{R_C \to \infty} a_v = -g_m r_o = -\frac{I_C}{V_T}\frac{V_A}{I_C} = -\frac{V_A}{V_T} = -\frac{1}{\eta} \qquad (3.17)$$

where I_C is the dc collector current at the operating point, V_T is the thermal voltage, V_A is the Early voltage, and η is given in (1.114). This gain represents the maximum low-frequency voltage gain obtainable from the transistor. It is independent of the collector bias current for bipolar transistors, and the magnitude is approximately 5000 for typical *npn* devices.

Another parameter of interest is the *short-circuit current gain* a_i. This parameter is the ratio of i_o to i_i when the output is shorted. For the CE amplifier,

$$a_i = \left.\frac{i_o}{i_i}\right|_{v_o=0} = \frac{G_m v_i}{\frac{v_i}{R_i}} = g_m r_\pi = \beta_0 \qquad (3.18)$$

■ **EXAMPLE**

(a) Find the input resistance, output resistance, voltage gain, and current gain of the common-emitter amplifier in Fig. 3.11a. Assume that $I_C = 100$ μA, $\beta_0 = 100$, $r_b = 0$, and $r_o \to \infty$.

$$R_i = r_\pi = \frac{\beta_0}{g_m} \simeq \frac{100\,(26\text{ mV})}{100\text{ μA}} = 26\text{ k}\Omega$$

$$R_o = R_C = 5\text{ k}\Omega$$

$$a_v = -g_m R_C \simeq -\left(\frac{100\text{ μA}}{26\text{ mV}}\right)(5\text{ k}\Omega) \simeq -19.2$$

$$a_i = \beta_0 = 100$$

(b) Calculate the voltage gain of the circuit of Fig. 3.11b. Assume that V_{BIAS} is adjusted so that the dc collector current is maintained at 100 μA.

$$v_1 = v_s\left(\frac{R_i}{R_S + R_i}\right)$$

$$v_o = -G_m v_1(R_o \parallel R_L) = -G_m\left(\frac{R_i}{R_S + R_i}\right)(R_o \parallel R_L)v_s$$

$$\frac{v_o}{v_s} = -\left(\frac{1}{260\,\Omega}\right)\left(\frac{26\text{ k}\Omega}{26\text{ k}\Omega + 20\text{ k}\Omega}\right)\left[\frac{(10\text{ k}\Omega)(5\text{ k}\Omega)}{10\text{ k}\Omega + 5\text{ k}\Omega}\right] \simeq -7.25$$

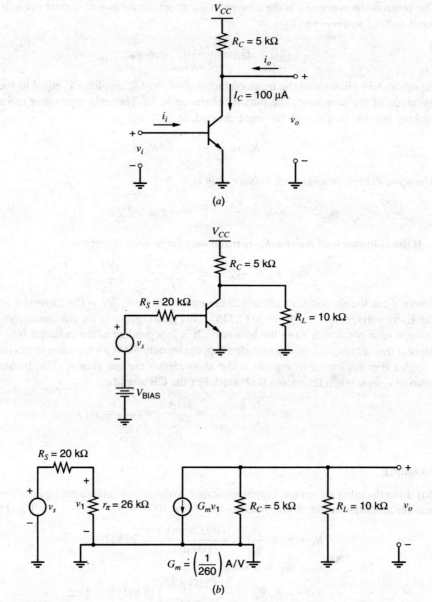

Figure 3.11 (a) Example amplifier circuit. (b) Circuit for calculation of voltage gain with typical source and load resistance values.

3.3.2 Common-Source Configuration

The resistively loaded common-source (CS) amplifier configuration is shown in Fig. 3.12a using an n-channel MOS transistor. The corresponding small-signal equivalent circuit is shown in Fig. 3.12b. As in the case of the bipolar transistor, the MOS transistor is cutoff for $V_i = 0$ and thus $I_d = 0$ and $V_o = V_{DD}$. As V_i is increased beyond the threshold voltage V_t, nonzero drain current flows and the transistor operates in the active region (which is often called saturation for MOS transistors) when $V_o > V_{GS} - V_t$. The large-signal model of Fig. 1.30 can then be

3.3 Basic Single-Transistor Amplifier Stages

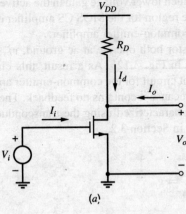

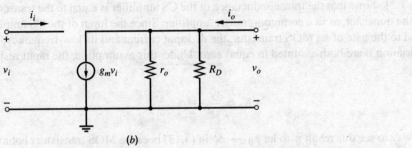

Figure 3.12 (a) Resistively loaded, common-source amplifier. (b) Small-signal equivalent circuit for the common-source amplifier.

used together with (1.157) to derive

$$V_o = V_{DD} - I_d R_D \quad (3.19)$$

$$= V_{DD} - \frac{\mu_n C_{ox}}{2} \frac{W}{L} R_D (V_i - V_t)^2 \quad (3.20)$$

The output voltage is equal to the drain-source voltage and decreases as the input increases. When $V_o < V_{GS} - V_t$, the transistor enters the triode region, where its output resistance becomes low and the small-signal voltage gain drops dramatically. In the triode region, the output voltage can be calculated by using (1.152) in (3.19). These results are illustrated in the plot of Fig. 3.13. The slope of this transfer characteristic at any operating point is the small-signal

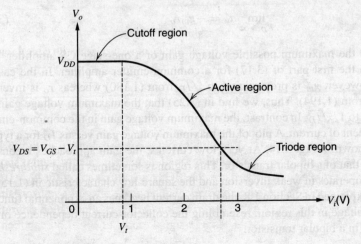

Figure 3.13 Output voltage versus input voltage for the common-source circuit.

voltage gain at that point. The MOS transistor has much lower voltage gain in the active region than does the bipolar transistor; therefore, the active region for the MOS CS amplifier extends over a much larger range of V_i than in the bipolar common-emitter amplifier.

Since the source and body of the MOS transistor both operate at ac ground, $v_{bs} = 0$ in Fig. 1.36; therefore, the g_{mb} generator is omitted in Fig. 3.12b. As a result, this circuit is topologically identical to the small-signal equivalent circuit for the common-emitter amplifier shown in Fig. 3.10. The CS amplifier is unilateral because it contains no feedback. Therefore, the low-frequency behavior of this circuit can be characterized using the transconductance, input resistance, and output resistance as described in Section 3.2.

The transconductance G_m is

$$G_m = \left.\frac{i_o}{v_i}\right|_{v_o=0} = g_m \qquad (3.21)$$

Equation 3.21 shows that the transconductance of the CS amplifier is equal to the transconductance of the transistor, as in a common-emitter amplifier. Since the input of the CS amplifier is connected to the gate of an MOS transistor, the dc input current and its low-frequency, small-signal variation i_i are both assumed to equal zero. Under this assumption, the input resistance R_i is

$$R_i = \frac{v_i}{i_i} \to \infty \qquad (3.22)$$

Another way to see this result is to let $\beta_0 \to \infty$ in (3.13) because MOS transistors behave like bipolar transistors with infinite β_0. The output resistance is the Thévenin-equivalent resistance seen looking into the output with the input shorted, or

$$R_o = \left.\frac{v_o}{i_o}\right|_{v_i=0} = R_D \parallel r_o \qquad (3.23)$$

The *open-circuit*, or *unloaded*, *voltage gain* is

$$a_v = \left.\frac{v_o}{v_i}\right|_{i_o=0} = -g_m(r_o \parallel R_D) \qquad (3.24)$$

If the drain load resistor R_D is replaced by a current source, $R_D \to \infty$ and a_v becomes

$$\lim_{R_D \to \infty} a_v = -g_m r_o \qquad (3.25)$$

Equation 3.25 gives the maximum possible voltage gain of a one-stage CS amplifier. This result is identical to the first part of (3.17) for a common-emitter amplifier. In the case of the CS amplifier, however, g_m is proportional to $\sqrt{I_D}$ from (1.180) whereas r_o is inversely proportional to I_D from (1.194). Thus, we find in (3.25) that the maximum voltage gain per stage is proportional to $1/\sqrt{I_D}$. In contrast, the maximum voltage gain in the common-emitter amplifier is independent of current. A plot of the maximum voltage gain versus I_D for a typical MOS transistor is shown in Fig. 3.14. At very low currents, the gain approaches a constant value comparable to that of a bipolar transistor. This region is sometimes called *subthreshold*, where the transistor operates in weak inversion and the square-law characteristic in (1.157) is no longer valid. As explained in Section 1.8, the drain current becomes an exponential function of the gate-source voltage in this region, resembling the collector-current dependence on the base-emitter voltage in a bipolar transistor.

3.3 Basic Single-Transistor Amplifier Stages

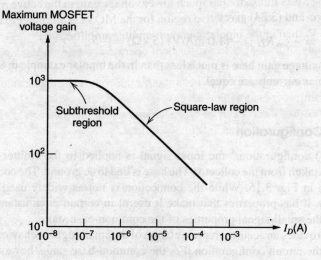

Figure 3.14 Typical variation of maximum MOSFET voltage gain with bias current.

Using (1.194), the limiting gain given by (3.25) can also be expressed as

$$\lim_{R_D \to \infty} a_v = -g_m r_o = -\frac{g_m}{I_D} I_D r_o = -\frac{g_m}{I_D} V_A \qquad (3.26)$$

In the square-law region in Fig. 3.14, substituting (1.181) into (3.26) gives

$$\lim_{R_D \to \infty} a_v = -\frac{V_A}{(V_{GS} - V_t)/2} = -\frac{2V_A}{V_{ov}} \qquad (3.27)$$

where $V_{ov} = V_{GS} - V_t$ is the gate *overdrive*. Since the gate overdrive is typically an order of magnitude larger than the thermal voltage V_T, the magnitude of the maximum gain predicted by (3.27) is usually much smaller than that predicted by (3.17) for the bipolar case. Substituting (1.163) into (3.27) gives

$$\lim_{R_D \to \infty} a_v = -\frac{2L_{\text{eff}}}{V_{GS} - V_t} \left(\frac{dX_d}{dV_{DS}} \right)^{-1} \qquad (3.28)$$

■ **EXAMPLE**

Find the voltage gain of the common-source amplifier of Fig 3.12a with $V_{DD} = 5$ V, $R_D = 5 \text{ k}\Omega, k' = \mu_n C_{ox} = 100 \text{ }\mu\text{A/V}^2, W = 50 \text{ }\mu\text{m}, L = 1 \text{ }\mu\text{m}, V_t = 0.8 \text{ V}, L_d = 0, X_d = 0,$ and $\lambda = 0$. Assume that the bias value of V_i is 1 V.

To determine whether the transistor operates in the active region, we first find the dc output voltage $V_O = V_{DS}$. If the transistor operates in the active region, (1.157) gives

$$I_D = \frac{k'}{2} \frac{W}{L} (V_{GS} - V_t)^2 = \frac{100}{2} \times 10^{-6} \times \frac{50}{1} (1 - 0.8)^2 = 100 \text{ }\mu\text{A}$$

Then

$$V_O = V_{DS} = V_{DD} - I_D R_D = 5 \text{ V} - (0.1 \text{ mA})(5 \text{ k}\Omega) = 4.5 \text{ V}$$

Since $V_{DS} = 4.5 \text{ V} > V_{GS} - V_t = 0.2 \text{ V}$, the transistor does operate in the active region, as assumed. Then from (1.180),

$$g_m = k' \frac{W}{L}(V_{GS} - V_t) = 100 \times 10^{-6} \times \frac{50}{1}(1 - 0.8) = 1000 \frac{\mu\text{A}}{\text{V}}$$

Then since $\lambda = 0$, $V_A \to \infty$ and (3.24) gives

$$a_v = -g_m R_D = -(1.0 \text{ mA/V})(5 \text{ k}\Omega) = -5$$

Note that the open-circuit voltage gain here is much less than in the bipolar example in Section 3.3.1 even though the dc bias currents are equal.

■

3.3.3 Common-Base Configuration

In the common-base (CB) configuration,[4] the input signal is applied to the emitter of the transistor, and the output is taken from the collector. The base is tied to ac ground. The common-base connection is shown in Fig. 3.15. While the connection is not as widely used as the common-emitter amplifier, it has properties that make it useful in certain circumstances. In this section, we calculate the small-signal properties of the common-base stage.

The hybrid-π model provides an accurate representation of the small-signal behavior of the transistor independent of the circuit configuration. For the common-base stage, however, the hybrid-π model is somewhat cumbersome because the dependent current source is connected between the input and output terminals.[4] The analysis of common-base stages can be simplified if the model is modified as shown in Fig. 3.16. The small-signal hybrid-π model is shown in Fig. 3.16a. First note that the dependent current source flows from the collector terminal to the emitter terminal. The circuit behavior is unchanged if we replace this single current source with two current sources of the same value, one going from the collector to the base and the other going from the base to the emitter, as shown in Fig. 3.16b. Since the currents fed into and removed from the base are equal, the equations that describe the operation of these circuits are identical. We next note that the controlled current source connecting the base and emitter is controlled by the voltage across its own terminals. Therefore, by the application of Ohm's law to this branch, this dependent current source can be replaced by a resistor of value $1/g_m$. This resistance appears in parallel with r_π, and the parallel combination of the two is called the emitter resistance r_e.

$$r_e = \cfrac{1}{g_m + \cfrac{1}{r_\pi}} = \cfrac{1}{g_m\left(1 + \cfrac{1}{\beta_0}\right)} = \frac{\alpha_0}{g_m} \qquad (3.29)$$

The new equivalent circuit is called the *T model* and is shown in Fig. 3.16c. It has terminal properties exactly equivalent to those of the hybrid-π model but is often more convenient to use for common-base calculations. For dc and low input frequencies, the capacitors C_π and C_μ appear as high-impedance elements and can be neglected. Assume at first that $r_b = 0$ and $r_o \to \infty$ so that the circuit is unilateral. When r_μ is also neglected, the model reduces to the simple form shown in Fig. 3.16d. Using the T model under these conditions, the small-signal

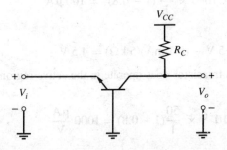

Figure 3.15 Typical common-base amplifier.

3.3 Basic Single-Transistor Amplifier Stages

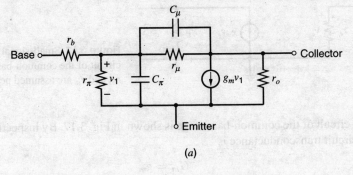

(a)

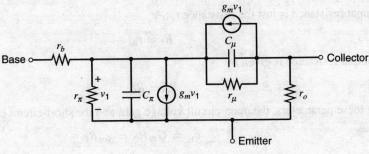

(b)

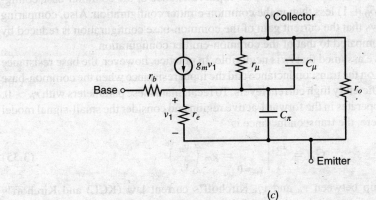

(c)

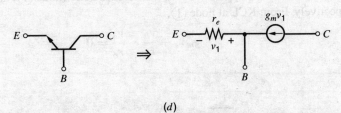

(d)

Figure 3.16 Generation of emitter-current-controlled T model from the hybrid-π. (a) Hybrid-π model. (b) The collector current source $g_m v_1$ is changed to two current sources in series, and the point between them attached to the base. This change does not affect the current flowing in the base. (c) The current source between base and emitter is converted to a resistor of value $1/g_m$. (d) T model for low frequencies, neglecting r_o, r_μ, and the charge-storage elements.

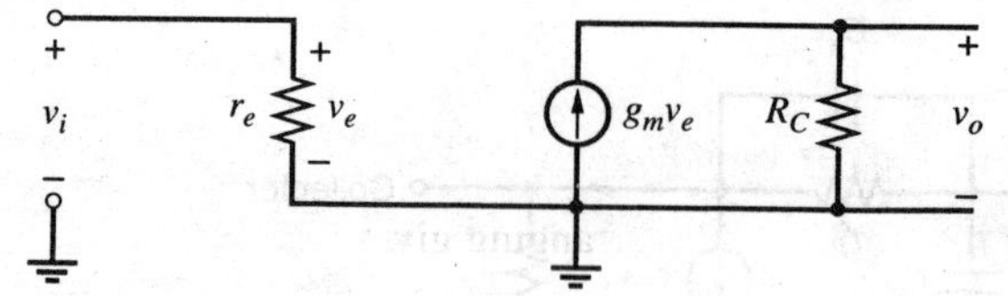

Figure 3.17 Small-signal equivalent circuit of the common-base stage; r_o, r_b, and r_μ are assumed negligible.

equivalent circuit of the common-base stage is shown in Fig. 3.17. By inspection of Fig. 3.17, the short-circuit transconductance is

$$G_m = g_m \tag{3.30}$$

The input resistance is just the resistance r_e:

$$R_i = r_e \tag{3.31}$$

The output resistance is given by

$$R_o = R_C \tag{3.32}$$

Using these parameters, the open-circuit voltage gain and the short-circuit current gain are

$$a_v = G_m R_o = g_m R_C \tag{3.33}$$

$$a_i = G_m R_i = g_m r_e = \alpha_0 \tag{3.34}$$

Comparing (3.31) and (3.13) shows that the input resistance of the common-base configuration is a factor of $(\beta_0 + 1)$ less than in the common-emitter configuration. Also, comparing (3.34) and (3.18) shows that the current gain of the common-base configuration is reduced by a factor of $(\beta_0 + 1)$ compared to that of the common-emitter configuration.

Until now, we have assumed that r_b is negligible. In practice, however, the base resistance has a significant effect on the transconductance and the input resistance when the common-base stage is operated at sufficiently high current levels. To recalculate these parameters with $r_b > 0$, assume the transistor operates in the forward-active region and consider the small-signal model shown in Fig. 3.18. Here, the transconductance is

$$G_m = \left.\frac{i_o}{v_i}\right|_{v_o=0} = g_m \left(\frac{v_e}{v_i}\right) \tag{3.35}$$

To find the relationship between v_e and v_i, Kirchoff's current law (KCL) and Kirchoff's voltage law (KVL) can be applied at the internal base node (node ①) and around the input loop, respectively. From KCL at node ①,

$$g_m v_e + \frac{v_b}{r_b} - \frac{v_e}{r_e} = 0 \tag{3.36}$$

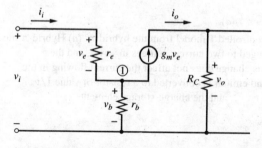

Figure 3.18 Small-signal model of the common-base stage with $r_b > 0$.

From KVL around the input loop,

$$v_i = v_e + v_b \tag{3.37}$$

Solving (3.37) for v_b, substituting into (3.36), and rearranging gives

$$\frac{v_i}{v_e} = 1 + \frac{g_m}{\beta_0}r_b = 1 + \frac{r_b}{r_\pi} \tag{3.38}$$

Substituting (3.38) into (3.35) gives

$$G_m = \frac{g_m}{1 + \dfrac{r_b}{r_\pi}} \tag{3.39}$$

Similarly, the input resistance in Fig. 3.18 is

$$R_i = \frac{v_i}{i_i} = \frac{v_i}{v_e/r_e} = r_e\left(\frac{v_i}{v_e}\right) \tag{3.40}$$

Substituting (3.38) into (3.40) gives

$$R_i = r_e\left(1 + \frac{r_b}{r_\pi}\right) = \frac{\alpha_0}{g_m}\left(1 + \frac{r_b}{r_\pi}\right) \tag{3.41}$$

Thus if the dc collector current is large enough that r_π is comparable with r_b, then the effects of base resistance must be included. For example, if $r_b = 100\,\Omega$ and $\beta_0 = 100$, then a collector current of 26 mA makes r_b and r_π equal.

The main motivation for using common-base stages is twofold. First, the collector-base capacitance does not cause high-frequency feedback from output to input as in the common-emitter amplifier. As described in Chapter 7, this change can be important in the design of high-frequency amplifiers. Second, as described in Chapter 4, the common-base amplifier can achieve much larger output resistance than the common-emitter stage in the limiting case where $R_C \to \infty$. As a result, the common-base configuration can be used as a current source whose current is nearly independent of the voltage across it.

3.3.4 Common-Gate Configuration

In the common-gate configuration, the input signal is applied to the source of the transistor, and the output is taken from the drain while the gate is connected to ac ground. This configuration is shown in Fig. 3.19, and its behavior is similar to that of a common-base stage.

As in the analysis of common-base amplifiers in Section 3.3.3, the analysis of common-gate amplifiers can be simplified if the model is changed from a hybrid-π configuration to a T model, as shown in Fig. 3.20. In Fig. 3.20a, the low-frequency hybrid-π model is shown. Note that both transconductance generators are now active. If the substrate or body connection is assumed to operate at ac ground, then $v_{bs} = v_{gs}$ because the gate also operates at ac ground.

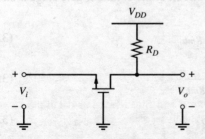

Figure 3.19 Common-gate configuration.

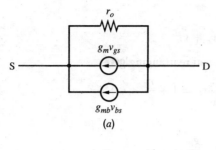
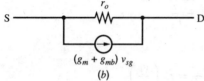
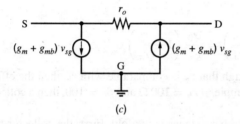
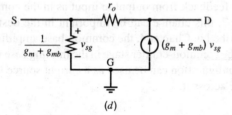

Figure 3.20 Conversion from hybrid-π to T model. (a) Low-frequency hybrid-π model. (b) The two dependent sources are combined. (c) The combined source is converted into two sources. (d) The current source between the source and gate is converted into a resistor.

Therefore, in Fig. 3.20b, the two dependent current sources are combined. In Fig. 3.20c, the combined current source from the source to the drain is replaced by two current sources: one from the source to the gate and the other from the gate to the drain. Since equal currents are pushed into and pulled out of the gate, the equations that describe the operation of the circuits in Figs. 3.20b and 3.20c are identical. Finally, because the current source from the source to the gate is controlled by the voltage across itself, it can be replaced by a resistor of value $1/(g_m + g_{mb})$, as in Fig. 3.20d.

If r_o is finite, the circuit of Fig. 3.20d is bilateral because of feedback provided through r_o. At first, we will assume that $r_o \to \infty$ so that the circuit is unilateral. Using the T model under these conditions, the small-signal equivalent circuit of the common-gate stage is shown in Fig. 3.21. By inspection of Fig. 3.21,

$$G_m = g_m + g_{mb} \tag{3.42}$$

$$R_i = \frac{1}{g_m + g_{mb}} \tag{3.43}$$

$$R_o = R_D \tag{3.44}$$

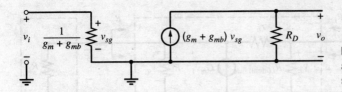

Figure 3.21 Small-signal equivalent circuit of the common-gate stage; r_o is assumed negligible.

Using these parameters, the open-circuit voltage gain and the short-circuit current gain are

$$a_v = G_m R_o = (g_m + g_{mb})R_D \qquad (3.45)$$

$$a_i = G_m R_i = 1 \qquad (3.46)$$

3.3.5 Common-Base and Common-Gate Configurations with Finite r_o

In calculating the expressions for G_m, R_i, and R_o of the common-base and common-gate amplifiers, we have neglected the effects of r_o. Since r_o is connected from each amplifier output back to its input, finite r_o causes each circuit to be bilateral, making the input resistance depend on the connection at the amplifier output. Let $R = R_C$ in Fig. 3.17 or $R = R_D$ in Fig. 3.21, depending on which circuit is under consideration. When R becomes large enough that it is comparable with r_o, r_o must be included in the small-signal model to accurately predict not only the input resistance, but also the output resistance. On the other hand, since the transconductance is calculated with the output shorted, the relationship between r_o and R has no effect on this calculation, and the effect of finite r_o on transconductance can be ignored if $r_o \gg 1/G_m$.

3.3.5.1 Common-Base and Common-Gate Input Resistance

Figure 3.22a shows a small-signal T model of a common-base or common-gate stage including finite r_o, where $R_{i(\text{ideal})}$ is given by (3.31) for a common-base amplifier or by (3.43) for a common-gate amplifier. Also, R represents R_C in Fig. 3.17 or R_D in Fig. 3.21. Connections to the load and the input source are shown in Fig. 3.22a to include their contributions to the input and output resistance, respectively. In Fig. 3.22a, the input resistance is $R_i = v_1/i_i$. To find the input resistance, a simplified equivalent circuit such as in Fig. 3.22b is often used. Here, a test voltage source v_t is used to drive the amplifier input, and the resulting test current i_t is calculated. KCL at the output node in Fig. 3.22b gives

$$\frac{v_o}{R \| R_L} + \frac{v_o - v_t}{r_o} = G_m v_t \qquad (3.47)$$

KCL at the input in Fig. 3.22b gives

$$i_t = \frac{v_t}{R_{i(\text{ideal})}} + \frac{v_t - v_o}{r_o} \qquad (3.48)$$

Solving (3.47) for v_o and substituting into (3.48) gives

$$\frac{i_t}{v_t} = \frac{1}{R_{i(\text{ideal})}} + \frac{1}{r_o}\left(1 - \frac{G_m + \dfrac{1}{r_o}}{\dfrac{1}{R \| R_L} + \dfrac{1}{r_o}}\right) \qquad (3.49)$$

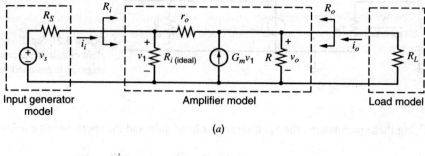

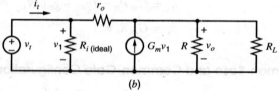

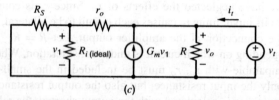

Figure 3.22 (a) Model of common-base and common-gate amplifiers with finite r_o, showing connections to the input source and load. (b) Equivalent circuit for calculation of R_i. (c) Equivalent circuit for calculation of R_o.

Rearranging (3.49) gives

$$R_i = \frac{v_t}{i_t} = \frac{r_o + R \parallel R_L}{1 - G_m(R \parallel R_L) + \frac{r_o + R \parallel R_L}{R_{i(\text{ideal})}}} \tag{3.50}$$

Common-Base Input Resistance. For the common-base amplifier, $G_m = g_m$ from (3.30), and $R_{i(\text{ideal})} = r_e = \alpha_0/g_m$ from (3.31). Substituting (3.30) and (3.31) into (3.50) with $R = R_C$ and rearranging gives

$$R_i = \frac{v_t}{i_t} = \frac{r_o + R_C \parallel R_L}{1 + \frac{g_m(R_C \parallel R_L)}{\beta_0} + \frac{g_m r_o}{\alpha_0}} = \frac{r_o + R_C \parallel R_L}{1 + \frac{g_m}{\beta_0}(R_C \parallel R_L + (\beta_0 + 1)r_o)} \tag{3.51}$$

From (3.51), when $(\beta_0 + 1)r_o \gg R_C \parallel R_L$,

$$R_i \simeq \frac{r_o + R_C \parallel R_L}{1 + \frac{g_m r_o}{\alpha_0}} \tag{3.52}$$

From (3.52), when $g_m r_o \gg \alpha_0$,

$$R_i \simeq \frac{\alpha_0}{g_m} + \frac{\alpha_0 (R_C \parallel R_L)}{g_m r_o} = r_e + \frac{\alpha_0 (R_C \parallel R_L)}{g_m r_o} \tag{3.53}$$

The first term on the right side of (3.53) is the same as in (3.31), where the common-base amplifier was unilateral because infinite r_o was assumed. The second term shows that the input resistance now depends on the connection to the output (because finite r_o provides feedback and makes the amplifier bilateral). The second term is about equal to the resistance at the amplifier output divided by the $G_m r_o$ product. When $r_o \gg (R_C \parallel R_L)$, the effect of the second term can be neglected.

Common-Gate Input Resistance. For the common-gate amplifier, $G_m = (g_m + g_{mb})$ from (3.42) and $R_{i(\text{ideal})} = 1/(g_m + g_{mb})$ from (3.43). Substituting (3.42) and (3.43) into (3.50) with $R = R_D$ and rearranging gives

$$R_i = \frac{v_t}{i_t} = \frac{r_o + R_D \parallel R_L}{1 + (g_m + g_{mb}) r_o} \tag{3.54}$$

When $(g_m + g_{mb}) r_o \gg 1$,

$$R_i \simeq \frac{1}{g_m + g_{mb}} + \frac{R_D \parallel R_L}{(g_m + g_{mb}) r_o} \tag{3.55}$$

The first term on the right side of (3.55) is the same as in (3.43), where the common-gate amplifier was unilateral because infinite r_o was assumed. The second term is about equal to the resistance at the amplifier output divided by the $G_m r_o$ product and shows the effect of finite r_o, which makes the circuit bilateral. When $r_o \gg (R_D \parallel R_L)$, the effect of the second term can be neglected. Neglecting the second term usually causes only a small error when R_D here or R_C in the common-base case is built as a physical resistor even if the amplifier is unloaded ($R_L \to \infty$). However, when R_D or R_C is replaced by a transistor current source, the effect of the second term can be significant. Chapter 4 describes techniques used to construct transistor current sources that can have very high equivalent resistance.

3.3.5.2 Common-Base and Common-Gate Output Resistance

The output resistance in Fig. 3.22a is $R_o = v_o/i_o$ with $v_s = 0$. For this calculation, consider the equivalent circuit shown in Fig. 3.22c, where $v_s = 0$. A test voltage v_t is used to drive the amplifier output, and the resulting test current i_t can be calculated. Since R appears in parallel with the amplifier output, the calculation will be done in two steps. First, the output resistance with $R \to \infty$ is calculated. Second, this result is placed in parallel with R to give the overall output resistance. From KCL at the input node in Fig. 3.22c,

$$\frac{v_1}{R_S} + \frac{v_1}{R_{i(\text{ideal})}} + \frac{v_1 - v_t}{r_o} = 0 \tag{3.56}$$

With $R \to \infty$, KCL at the output node gives

$$i_t = -G_m v_1 + \frac{v_t - v_1}{r_o} \tag{3.57}$$

Solving (3.56) for v_1 and substituting into (3.57) gives

$$\frac{i_t}{v_t} = \frac{1}{r_o} - \frac{1}{r_o} \left(\frac{G_m + \dfrac{1}{r_o}}{\dfrac{1}{R_S} + \dfrac{1}{R_{i(\text{ideal})}} + \dfrac{1}{r_o}} \right) \tag{3.58}$$

Rearranging (3.58) gives

$$\frac{v_t}{i_t} = \frac{r_o \left(\dfrac{1}{R_S} + \dfrac{1}{R_{i(\text{ideal})}} + \dfrac{1}{r_o} \right)}{\dfrac{1}{R_S} + \dfrac{1}{R_{i(\text{ideal})}} - G_m} \quad (3.59)$$

With finite R, the output resistance is

$$R_o = R \parallel \left(\frac{v_t}{i_t} \right) = R \parallel \left[\frac{r_o \left(\dfrac{1}{R_S} + \dfrac{1}{R_{i(\text{ideal})}} + \dfrac{1}{r_o} \right)}{\dfrac{1}{R_S} + \dfrac{1}{R_{i(\text{ideal})}} - G_m} \right] \quad (3.60)$$

Common-Base Output Resistance. For the common-base amplifier, $G_m = g_m$ from (3.30) and $R_{i(\text{ideal})} = r_e = \alpha_0/g_m$ from (3.31). Substituting (3.30) and (3.31) into (3.60) and rearranging gives

$$R_o = R \parallel \left[\frac{r_o + R_S \left(1 + \dfrac{g_m r_o}{\alpha_0} \right)}{1 + \dfrac{R_S}{r_\pi}} \right] \quad (3.61)$$

The term in brackets on the right side of (3.61) shows that the output resistance of the common-base amplifier depends on the resistance of the input source R_S when r_o is finite. For example, if the input comes from an ideal voltage source, $R_S = 0$ and

$$R_o = R \parallel r_o \quad (3.62)$$

On the other hand, if the input comes from an ideal current source, $R_S \to \infty$ and

$$R_o = R \parallel \left[\left(\frac{1 + g_m r_o}{\alpha_0} \right) r_\pi \right] \quad (3.63)$$

From (3.61), when $R_S \ll r_\pi$,

$$R_o \simeq R \parallel \left[r_o + R_S \left(\frac{1 + g_m r_o}{\alpha_0} \right) \right] \quad (3.64)$$

From (3.64), when $g_m r_o \gg \alpha_0$ and $g_m R_S \gg \alpha_0$,

$$R_o \simeq R \parallel \left(\frac{g_m r_o}{\alpha_0} R_S \right) \quad (3.65)$$

The term in parentheses in (3.65) is about equal to the input source resistance multiplied by the $G_m r_o$ product. Therefore, (3.65) and (3.53) together show that the common-base amplifier can be thought of as a resistance scaler, where the resistance is scaled up from the emitter to the collector and down from the collector to the emitter by a factor approximately equal to the $G_m r_o$ product in each case.

Common-Gate Output Resistance. For the common-gate amplifier, $G_m = (g_m + g_{mb})$ from (3.42) and $R_{i(\text{ideal})} = 1/(g_m + g_{mb})$ from (3.43). Substituting (3.42) and (3.43) into (3.60) and rearranging gives

$$R_o = R \parallel [r_o + R_S (1 + (g_m + g_{mb}) r_o)] \quad (3.66)$$

From (3.66), when $(g_m + g_{mb}) r_o \gg 1$ and $(g_m + g_{mb}) R_S \gg 1$,

$$R_o \simeq R \parallel ((g_m + g_{mb}) r_o R_S) \qquad (3.67)$$

The term in parentheses in (3.67) is equal to the input source resistance multiplied by the $G_m r_o$ product. Therefore, (3.67) and (3.55) together show that the common-gate amplifier is also a resistance scaler, where the resistance is scaled up from the source to the drain and down from the drain to the source by a factor approximately equal to the $G_m r_o$ product in each case.

3.3.6 Common-Collector Configuration (Emitter Follower)

The common-collector connection is shown in Fig. 3.23a. The distinguishing feature of this configuration is that the signal is applied to the base and the output is taken from the emitter.[4] From a large-signal standpoint, the output voltage is equal to the input voltage minus the base-emitter voltage. Since the base-emitter voltage is a logarithmic function of the collector current, the base-emitter voltage is almost constant even when the collector current varies. If the base-emitter voltage were exactly constant, the output voltage of the common-collector amplifier would be equal to the input voltage minus a constant offset, and the small-signal gain of the circuit would be unity. For this reason, the circuit is also known as an *emitter follower* because the emitter voltage follows the base voltage. In practice, the base-emitter voltage is not exactly constant if the collector current varies. For example, (1.82) shows that the base-emitter voltage must increase by about 18 mV to double the collector current and by about 60 mV to increase the collector current by a factor of 10 at room temperature. Furthermore, even if the collector current were exactly constant, the base-emitter voltage depends to some extent on the collector-emitter voltage if the Early voltage is finite. These effects are most easily studied using small-signal analysis.

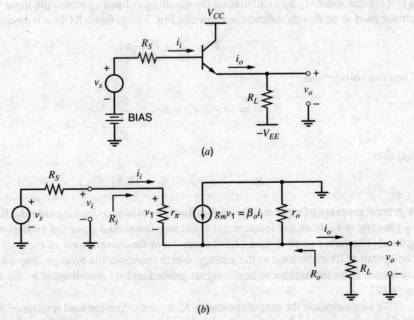

Figure 3.23 (a) Common-collector configuration. (b) Small-signal equivalent circuit of the emitter-follower circuit including R_L and R_S.

The appropriate small-signal transistor model is the hybrid-π, and the small-signal equivalent circuit is shown in Fig. 3.23b. When the input voltage v_s increases, the base-emitter voltage of the transistor increases, which increases the output current i_o. However, increasing i_o increases the output voltage v_o, which decreases the base-emitter voltage by negative feedback. Negative feedback is covered thoroughly in Chapter 8. The key point here is that the common-collector configuration is not unilateral. As a result, the input resistance depends on the load resistor R_L and the output resistance depends on the source resistance R_S. Therefore, the characterization of the emitter follower by the corresponding equivalent two-port network is not particularly useful for intuitive understanding. Instead, we will analyze the entire emitter-follower circuit of Fig. 3.23b, including both the source resistance R_S and the load resistor R_L. From KCL at the output node, we find

$$\frac{v_s - v_o}{R_S + r_\pi} + \beta_0 \left(\frac{v_s - v_o}{R_S + r_\pi} \right) - \frac{v_o}{R_L} - \frac{v_o}{r_o} = 0 \tag{3.68}$$

from which we find

$$\frac{v_o}{v_s} = \frac{1}{1 + \dfrac{R_S + r_\pi}{(\beta_0 + 1)(R_L \parallel r_o)}} \tag{3.69}$$

If the base resistance r_b is significant, it can simply be added to R_S in these expressions. The voltage gain is always less than unity and will be close to unity if $\beta_0 (R_L \parallel r_o) \gg (R_S + r_\pi)$. In most practical circuits, this condition holds. Note that because we have included the source resistance in this calculation, the value of v_o/v_s is not analogous to a_v calculated for the CE and CB stages. When $r_\pi \gg R_S$, $\beta_0 \gg 1$, and $r_o \gg R_L$, (3.69) can be approximated as

$$\frac{v_o}{v_s} \simeq \frac{g_m R_L}{1 + g_m R_L} \tag{3.70}$$

We calculate the input resistance R_i by removing the input source, driving the input with a test current source i_t, and calculating the resulting voltage v_t across the input terminals. The circuit used to do this calculation is shown in Fig. 3.24a. From KCL at the output node,

$$\frac{v_o}{R_L} + \frac{v_o}{r_o} = i_t + \beta_0 i_t \tag{3.71}$$

Then the voltage v_t is

$$v_t = i_t r_\pi + v_o = i_t r_\pi + \frac{i_t + \beta_0 i_t}{\dfrac{1}{R_L} + \dfrac{1}{r_o}} \tag{3.72}$$

and thus

$$R_i = \frac{v_t}{i_t} = r_\pi + (\beta_0 + 1)(R_L \parallel r_o) \tag{3.73}$$

A general property of emitter followers is that the resistance looking into the base is equal to r_π plus $(\beta_0 + 1)$ times the incremental resistance connected from the emitter to small-signal ground. The factor of $\beta_0 + 1$ in (3.73) stems from the current gain of the common-collector configuration from the base to the emitter, which increases the voltage drop on the resistance connected from the emitter to small-signal ground and its contribution to the test voltage v_t in (3.72).

We now calculate the output resistance R_o by removing the load resistance R_L and finding the Thevenin-equivalent resistance looking into the output terminals. We can do this by either inserting a test current and calculating the resulting voltage or applying a test voltage and

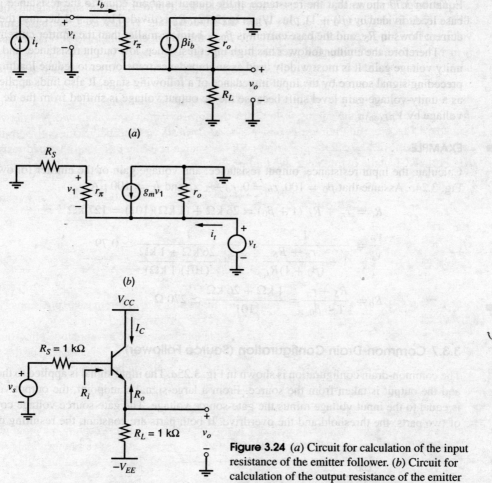

Figure 3.24 (a) Circuit for calculation of the input resistance of the emitter follower. (b) Circuit for calculation of the output resistance of the emitter follower. (c) Example emitter follower.

calculating the current. In this case, the calculation is simpler if a test voltage v_t is applied as shown in Fig. 3.24b. The voltage v_1 is given by

$$v_1 = -v_t \left(\frac{r_\pi}{r_\pi + R_S} \right) \tag{3.74}$$

The total output current i_t is thus

$$i_t = \frac{v_t}{r_\pi + R_S} + \frac{v_t}{r_o} + g_m v_t \left(\frac{r_\pi}{r_\pi + R_S} \right) \tag{3.75}$$

Therefore,

$$R_o = \frac{v_t}{i_t} = \left(\frac{r_\pi + R_S}{\beta_0 + 1} \right) \parallel r_o \tag{3.76}$$

If $\beta_0 \gg 1$ and $r_o \gg (1/g_m) + R_S/(\beta_0 + 1)$,

$$R_o \simeq \frac{1}{g_m} + \frac{R_S}{\beta_0 + 1} \tag{3.77}$$

Equation 3.77 shows that the resistance at the output is about equal to the resistance in the base lead, divided by $(\beta_0 + 1)$, plus $1/g_m$. In (3.77), R_S is divided by $\beta_0 + 1$ because the base current flows in R_S, and the base current is $\beta_0 + 1$ times smaller than the emitter current.

Therefore, the emitter follower has high input resistance, low output resistance, and near-unity voltage gain. It is most widely used as an impedance transformer to reduce loading of a preceding signal source by the input impedance of a following stage. It also finds application as a unity-voltage-gain level shift because the dc output voltage is shifted from the dc input voltage by $V_{BE(on)}$.

■ **EXAMPLE**

Calculate the input resistance, output resistance, and voltage gain of the emitter follower of Fig. 3.24c. Assume that $\beta_0 = 100$, $r_b = 0$, $r_o \to \infty$, and $I_C = 100\ \mu A$.

$$R_i = r_\pi + R_L(1 + \beta_0) = 26\ \text{k}\Omega + (1\ \text{k}\Omega)(101) = 127\ \text{k}\Omega$$

$$\frac{v_o}{v_s} = \frac{1}{1 + \dfrac{r_\pi + R_S}{(\beta_0 + 1)R_L}} = \frac{1}{1 + \dfrac{26\ \text{k}\Omega + 1\ \text{k}\Omega}{(101)(1\ \text{k}\Omega)}} \simeq 0.79$$

$$R_o = \frac{R_S + r_\pi}{1 + \beta_0} = \frac{1\ \text{k}\Omega + 26\ \text{k}\Omega}{101} \simeq 270\ \Omega$$

■

3.3.7 Common-Drain Configuration (Source Follower)

The common-drain configuration is shown in Fig. 3.25a. The input signal is applied to the gate and the output is taken from the source. From a large-signal standpoint, the output voltage is equal to the input voltage minus the gate-source voltage. The gate-source voltage consists of two parts: the threshold and the overdrive. If both parts are constant, the resulting output

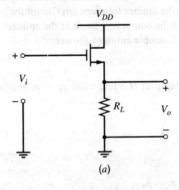

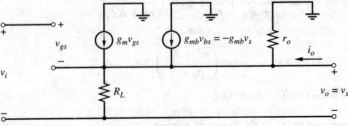

Figure 3.25 (a) Common-drain configuration. (b) Small-signal equivalent circuit of the common-drain configuration.

voltage is simply offset from the input, and the small-signal gain would be unity. Therefore, the source follows the gate, and the circuit is also known as a *source follower*. In practice, the body effect changes the threshold voltage, and the overdrive depends on the drain current, which changes as the output voltage changes unless $R_L \to \infty$. Furthermore, even if the current were exactly constant, the overdrive depends to some extent on the drain-source voltage unless the Early voltage is infinite. We will use small-signal analysis to study these effects.

The small-signal equivalent circuit is shown in Fig. 3.25b. Since the body terminal is not shown in Fig. 3.25a, we assume that the body is connected to the lowest supply voltage (ground here) to keep the source-body pn junction reverse biased. As a result, v_{bs} changes when the output changes because the source is connected to the output, and the g_{mb} generator is active in general.

From KVL around the input loop,

$$v_i = v_{gs} + v_o \tag{3.78}$$

With the output open circuited, $i_o = 0$, and KCL at the output node gives

$$g_m v_{gs} - g_{mb} v_o - \frac{v_o}{R_L} - \frac{v_o}{r_o} = 0 \tag{3.79}$$

Solving (3.78) for v_{gs}, substituting into (3.79), and rearranging gives

$$\left. \frac{v_o}{v_i} \right|_{i_o=0} = \frac{g_m}{g_m + g_{mb} + \frac{1}{R_L} + \frac{1}{r_o}} = \frac{g_m r_o}{1 + (g_m + g_{mb}) r_o + \frac{r_o}{R_L}} \tag{3.80}$$

If $R_L \to \infty$, (3.80) simplifies to

$$\lim_{R_L \to \infty} \left. \frac{v_o}{v_i} \right|_{i_o=0} = \frac{g_m r_o}{1 + (g_m + g_{mb}) r_o} \tag{3.81}$$

Equation 3.81 gives the open-circuit voltage gain of the source follower with the load resistor replaced by an ideal current source. If r_o is finite, this gain is less than unity even if the body effect is eliminated by connecting the source to the body to deactivate the g_{mb} generator. In this case, variation in the output voltage changes the drain-source voltage and the current through r_o. From a large-signal standpoint, solving (1.165) for $V_{GS} - V_t$ shows that the overdrive also depends on the drain-source voltage unless the channel-length modulation parameter λ is zero. This dependence causes the small-signal gain to be less than unity.

A significant difference between bipolar and MOS followers is apparent from (3.80). If $R_L \to \infty$ and $r_o \to \infty$,

$$\lim_{\substack{R_L \to \infty \\ r_o \to \infty}} \frac{v_o}{v_i} = \frac{g_m}{g_m + g_{mb}} = \frac{1}{1 + \chi} \tag{3.82}$$

Equation 3.82 shows that the source-follower gain is less than unity under these conditions and that the gain depends on $\chi = g_{mb}/g_m$, which is typically in the range of 0.1 to 0.3. In contrast, the gain of an emitter follower would be unity under these conditions. As a result, the source-follower gain is not as well specified as that of an emitter follower when body effect is a factor. Furthermore, (1.200) shows that χ depends on the source-body voltage which is equal to V_o when the body is connected to ground. Therefore, the gain calculated in (3.82) depends on the output voltage, causing distortion to arise for large-signal changes in the output as shown in Section 5.3.2. To overcome these limitations in practice, the type of source follower (n-channel or p-channel) can be chosen so that it can be fabricated in an isolated well. Then the well can be connected to the source of the transistor, setting $V_{SB} = 0$ and $v_{sb} = 0$. Unfortunately, the parasitic capacitance from the well to the substrate increases the capacitance attached to the source with this connection, reducing the bandwidth of the source follower. The frequency response of source followers is covered in Chapter 7.

The output resistance of the source follower can be calculated from Fig. 3.25b by setting $v_i = 0$ and driving the output with a voltage source v_o. Then $v_{gs} = -v_o$ and i_o is

$$i_o = \frac{v_o}{r_o} + \frac{v_o}{R_L} + g_m v_o + g_{mb} v_o \tag{3.83}$$

Rearranging (3.83) gives

$$R_o = \frac{v_o}{i_o} = \frac{1}{g_m + g_{mb} + \frac{1}{r_o} + \frac{1}{R_L}} \tag{3.84}$$

Equation 3.84 shows that the body effect reduces the output resistance, which is desirable because the source follower produces a voltage output. This beneficial effect stems from the nonzero small-signal current conducted by the g_{mb} generator in Fig. 3.25b, which increases the output current for a given change in the output voltage. As $r_o \rightarrow \infty$ and $R_L \rightarrow \infty$, this output resistance approaches $1/(g_m + g_{mb})$. The common-gate input resistance given in (3.54) approaches the same limiting value.

As with emitter followers, source followers are used as voltage buffers and level shifters. When used as a level shifter, they are more flexible than emitter followers because the dc value of V_{GS} can be altered by changing the W/L ratio.

3.3.8 Common-Emitter Amplifier with Emitter Degeneration

In the common-emitter amplifier considered earlier, the signal is applied to the base, the output is taken from the collector, and the emitter is attached to ac ground. In practice, however, the common-emitter circuit is often used with a nonzero resistance in series with the emitter as shown in Fig. 3.26a. The resistance has several effects, including reducing the transconductance, increasing the output resistance, and increasing the input resistance. These changes stem from negative feedback introduced by the emitter resistor R_E. When V_i increases, the base-emitter voltage increases, which increases the collector current. As a result, the voltage dropped across the emitter resistor increases, reducing the base-emitter voltage compared to the case where $R_E = 0$. Therefore, the presence of nonzero R_E reduces the base-emitter voltage through a negative-feedback process termed *emitter degeneration*. This circuit is examined from a feedback standpoint in Chapter 8.

In this section, we calculate the input resistance, output resistance, and transconductance of the emitter-degenerated, common-emitter amplifier. To find the input resistance and transconductance, consider the small-signal equivalent circuit shown in Fig. 3.26b, and focus on v_i, i_b, and i_o. From KCL at the emitter,

$$\frac{v_e}{R_E} + \frac{v_e + i_o R_C}{r_o} = (\beta_0 + 1) i_b \tag{3.85}$$

From KCL at the collector,

$$i_o + \frac{v_e + i_o R_C}{r_o} = \beta_0 i_b \tag{3.86}$$

From KVL around the input loop,

$$i_b = \frac{v_i - v_e}{r_\pi} \tag{3.87}$$

3.3 Basic Single-Transistor Amplifier Stages 197

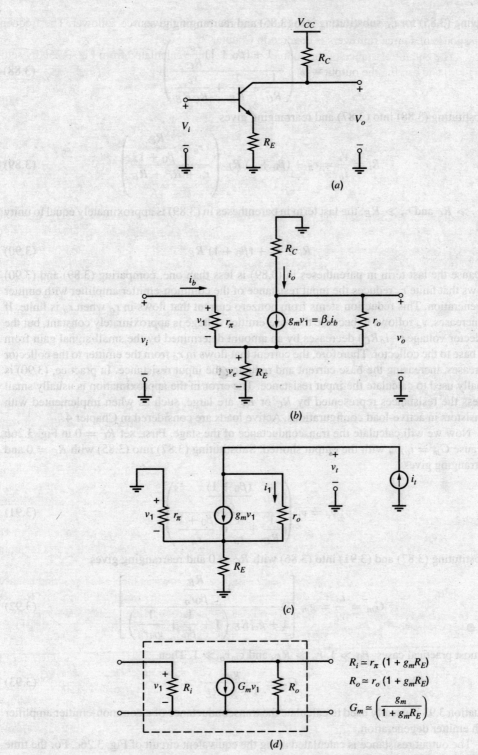

Figure 3.26 (a) Common-emitter amplifier with emitter degeneration. (b) Small-signal equivalent circuit for emitter-degenerated, common-emitter amplifier. (c) Circuit for calculation of output resistance. (d) Small-signal, two-port equivalent of emitter-degenerated CE amplifier.

Solving (3.85) for i_o, substituting into (3.86) and rearranging gives

$$v_e = i_b \left(\frac{1 + (\beta_0 + 1)\dfrac{r_o}{R_C}}{\dfrac{1}{R_C} + \dfrac{1}{R_E} + \dfrac{r_o}{R_C R_E}} \right) \quad (3.88)$$

Substituting (3.88) into (3.87) and rearranging gives

$$R_i = \frac{v_i}{i_b} = r_\pi + (\beta_0 + 1) R_E \left(\frac{r_o + \dfrac{R_C}{\beta_0 + 1}}{r_o + R_C + R_E} \right) \quad (3.89)$$

If $r_o \gg R_C$ and $r_o \gg R_E$, the last term in parentheses in (3.89) is approximately equal to unity and

$$R_i \simeq r_\pi + (\beta_0 + 1) R_E \quad (3.90)$$

Because the last term in parentheses in (3.89) is less than one, comparing (3.89) and (3.90) shows that finite r_o reduces the input resistance of the common-emitter amplifier with emitter degeneration. This reduction stems from nonzero current that flows in r_o when r_o is finite. If v_i increases, v_e follows v_i because the base-emitter voltage is approximately constant, but the collector voltage $(-i_o R_C)$ decreases by an amount determined by the small-signal gain from the base to the collector. Therefore, the current that flows in r_o from the emitter to the collector increases, increasing the base current and reducing the input resistance. In practice, (3.90) is usually used to calculate the input resistance. The error in the approximation is usually small unless the resistances represented by R_C or R_E are large, such as when implemented with transistors in active-load configurations. Active loads are considered in Chapter 4.

Now we will calculate the transconductance of the stage. First, set $R_C = 0$ in Fig. 3.26b because $G_m = i_o/v_i$ with the output shorted. Substituting (3.87) into (3.85) with $R_C = 0$ and rearranging gives

$$v_e = v_i \left(\frac{\dfrac{(\beta_0 + 1)}{r_\pi}}{\dfrac{1}{R_E} + \dfrac{1}{r_o} + \dfrac{\beta_0 + 1}{r_\pi}} \right) \quad (3.91)$$

Substituting (3.87) and (3.91) into (3.86) with $R_C = 0$ and rearranging gives

$$G_m = \frac{i_o}{v_i} = g_m \left[\frac{1 - \dfrac{R_E}{\beta_0 r_o}}{1 + g_m R_E \left(1 + \dfrac{1}{\beta_0} + \dfrac{1}{g_m r_o}\right)} \right] \quad (3.92)$$

In most practical cases, $\beta_0 \gg 1$, $r_o \gg R_E$, and $g_m r_o \gg 1$. Then

$$G_m \simeq \frac{g_m}{1 + g_m R_E} \quad (3.93)$$

Equation 3.93 is usually used to calculate the transconductance of a common-emitter amplifier with emitter degeneration.

The output resistance is calculated using the equivalent circuit of Fig. 3.26c. For the time being, assume that R_C is very large and can be neglected. The test current i_t flows in the parallel combination of r_π and R_E, so that

$$v_1 = -i_t(r_\pi \| R_E) \quad (3.94)$$

The current through r_o is

$$i_1 = i_t - g_m v_1 = i_t + i_t g_m (r_\pi \| R_E) \quad (3.95)$$

As a result, the voltage v_t is

$$v_t = -v_1 + i_1 r_o = i_t (r_\pi \| R_E) + i_t r_o [1 + g_m (r_\pi \| R_E)] \quad (3.96)$$

Thus

$$R_o = \frac{v_t}{i_t} = (r_\pi \| R_E) + r_o [1 + g_m (r_\pi \| R_E)] \quad (3.97)$$

In this equation, the first term is much smaller than the second. If the first term is neglected, we obtain,

$$R_o \simeq r_o \left(1 + g_m \frac{r_\pi R_E}{r_\pi + R_E}\right) = r_o \left(1 + \frac{g_m R_E}{1 + \frac{R_E}{r_\pi}}\right) = r_o \left(1 + \frac{g_m R_E}{1 + \frac{g_m R_E}{\beta_0}}\right) \quad (3.98)$$

If $g_m R_E \ll \beta_0$, then

$$R_o \simeq r_o (1 + g_m R_E) \quad (3.99)$$

Thus the output resistance is increased by a factor $(1 + g_m R_E)$. This fact makes the use of emitter degeneration desirable in transistor current sources. If the collector load resistor R_C is not large enough to neglect, it must be included in parallel with the expressions in (3.97)–(3.99). A small-signal equivalent circuit, neglecting R_C, is shown in Fig. 3.26d. On the other hand, if $g_m R_E \gg \beta_0$, (3.98) shows that

$$R_o \simeq r_o (1 + \beta_0) \quad (3.100)$$

The output resistance is finite even when $R_E \to \infty$ because nonzero test current flows in r_π when β_0 is finite.

3.3.9 Common-Source Amplifier with Source Degeneration

Source degeneration in MOS transistor amplifiers is not as widely used as emitter degeneration in bipolar circuits for at least two reasons. First, the transconductance of MOS transistors is normally much lower than that of bipolar transistors so that further reduction in transconductance is usually undesirable. Second, although degeneration increases the input resistance in the bipolar case, $R_i \to \infty$ even without degeneration in the MOS case. However, examining the effects of source degeneration is important in part because it is widely used to increase the output resistance of MOS current sources. Also, because small-geometry MOS transistors can be modeled as ideal square-law devices with added source resistors as shown in Section 1.7.1, we will consider the effects of source degeneration below.

A common-source amplifier with source degeneration is shown in Fig. 3.27. Its small-signal equivalent circuit is shown in Fig. 3.28. Because the input is connected to the gate of the MOS transistor, $R_i \to \infty$. To calculate the transconductance, set $R_D = 0$ because $G_m = i_o/v_i$ with the output shorted. Also, since a connection to the body is not shown in Fig. 3.27, we assume that the body is connected to the lowest power-supply voltage, which is ground. Therefore, the dc body voltage is constant and $v_b = 0$. From KCL at the source with $R_D = 0$,

$$\frac{v_s}{R_S} + \frac{v_s}{r_o} = g_m (v_i - v_s) + g_{mb} (0 - v_s) \quad (3.101)$$

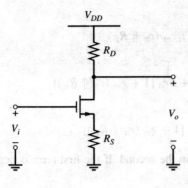

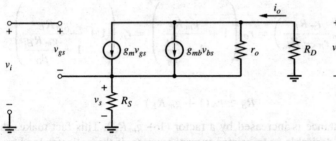

Figure 3.27 Common-source amplifier with source degeneration.

Figure 3.28 Small-signal equivalent of the source-degenerated, common-source amplifier.

From KCL at the drain with $R_D = 0$,

$$i_o + \frac{v_s}{r_o} = g_m (v_i - v_s) + g_{mb} (0 - v_s) \tag{3.102}$$

Solving (3.101) for v_s, substituting into (3.102), and rearranging gives

$$G_m = \frac{i_o}{v_i} = \frac{g_m}{1 + (g_m + g_{mb}) R_S + \dfrac{R_S}{r_o}} \tag{3.103}$$

If $r_o \gg R_S$,

$$G_m \simeq \frac{g_m}{1 + (g_m + g_{mb}) R_S} \tag{3.104}$$

For large R_S, (3.104) shows that the value of G_m approaches $1/[(1 + \chi) R_S]$. Even in this limiting case, the transconductance of the common-source amplifier with degeneration is dependent on an active-device parameter χ. Since χ is typically in the range of 0.1 to 0.3, the body effect causes the transconductance in this case to deviate from $1/R_S$ by about 10 to 20 percent. In contrast, (3.92) indicates that the value of G_m for a common-emitter amplifier with degeneration approaches $\beta_0/[(\beta_0 + 1) R_E]$ for large R_E, assuming that $r_o \gg R_E$ and $g_m r_o \gg 1$. If $\beta_0 > 100$, the transconductance of this bipolar amplifier is within 1 percent of $1/R_E$. Therefore, the transconductance of a common-source amplifier with degeneration is usually much more dependent on active-device parameters than in its bipolar counterpart.

The output resistance of the circuit can be calculated from the equivalent circuit of Fig. 3.29, where R_D is neglected. Since the entire test current flows in R_S,

$$v_s = i_t R_S \tag{3.105}$$

Then

$$v_t = v_s + i_1 r_o = v_s + r_o \left[i_t - g_m (0 - v_s) - g_{mb} (0 - v_s) \right] \tag{3.106}$$

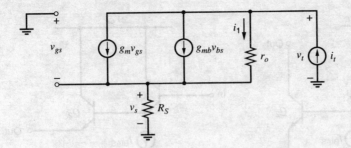

Figure 3.29 Circuit for calculation of output resistance.

Substituting (3.105) into (3.106) and rearranging gives

$$R_o = \frac{v_t}{i_t} = R_S + r_o\left[1 + (g_m + g_{mb})R_S\right] \qquad (3.107)$$

This equation shows that as R_S is made arbitrarily large, the value of R_o continues to increase. In contrast, (3.100) shows that R_o in the common-emitter amplifier with degeneration approaches a maximum value of about $(\beta_0 + 1)r_o$ as $R_E \to \infty$.

3.4 Multiple-Transistor Amplifier Stages

Most integrated-circuit amplifiers consist of a number of stages, each of which provides voltage gain, current gain, and/or impedance-level transformation from input to output. Such circuits can be analyzed by considering each transistor to be a *stage* and analyzing the circuit as a collection of individual transistors. However, certain combinations of transistors occur so frequently that these combinations are usually characterized as *subcircuits* and regarded as a single stage. The usefulness of these topologies varies considerably with the technology being used. For example, the Darlington two-transistor connection is widely used in bipolar integrated circuits to improve the effective current gain and input resistance of a single bipolar transistor. Since the current gain and input resistance are infinite with MOS transistors however, this connection finds little use in pure MOS integrated circuits. On the other hand, the cascode connection achieves a very high output resistance and is useful in both bipolar and MOS technologies.

3.4.1 The CC-CE, CC-CC and Darlington Configurations

The common-collector–common-emitter (CC-CE), common-collector–common-collector (CC-CC), and Darlington[5] configurations are all closely related. They incorporate an additional transistor to boost the current gain and input resistance of the basic bipolar transistor. The common-collector–common-emitter configuration is shown in Fig. 3.30a. The biasing current source I_{BIAS} is present to establish the quiescent dc operating current in the emitter-follower transistor Q_1; this current source may be absent in some cases or may be replaced by a resistor. The common-collector–common-collector configuration is illustrated in Fig. 3.30b. In both of these configurations, the effect of transistor Q_1 is to increase the current gain through the stage and to increase the input resistance. For the purpose of the low-frequency, small-signal analysis of circuits, the two transistors Q_1 and Q_2 can be thought of as a single composite transistor, as illustrated in Fig. 3.31. The small-signal equivalent circuit for this composite device is shown in Fig. 3.32, assuming that the effects of the r_o of Q_1 are negligible. We will now calculate effective values for the r_π, g_m, β_0, and r_o of the composite device, and we will designate these composite parameters with a superscript c. We will also denote the

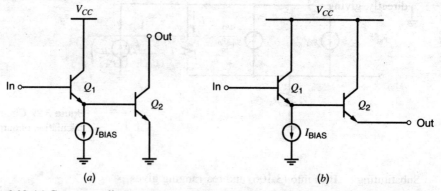

Figure 3.30 (a) Common-collector–common-emitter cascade. (b) Common-collector–common-collector cascade.

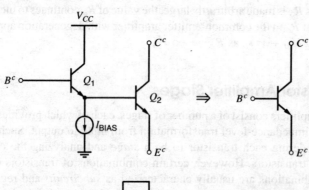

Figure 3.31 The composite transistor representation of the CC-CE and CC-CC connections.

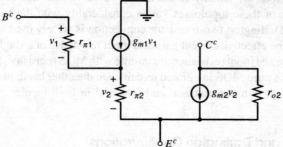

Figure 3.32 Small-signal equivalent circuit for the CC-CE and CC-CC connected transistors.

terminal voltages and currents of the composite device with a superscript c. We assume that β_0 is constant.

The effective value of r_π, r_π^c, is the resistance seen looking into the composite base B^c with the composite emitter E^c grounded. Referring to Fig. 3.32, we see that the resistance looking into the base of Q_2 with E^c grounded is simply $r_{\pi 2}$. Thus (3.73) for the input resistance of the emitter follower can be used. Substituting $r_{\pi 2}$ for R_L and allowing $r_o \to \infty$ gives

$$r_\pi^c = r_{\pi 1} + (\beta_0 + 1)\, r_{\pi 2} \qquad (3.108)$$

The effective transconductance of the configuration g_m^c is the change in the collector current of Q_2, i_c^c, for a unit change in v_{be}^c with C^c and E^c grounded. To calculate this transconductance, we first find the change in v_2 that occurs for a unit change in v_{be}^c. Equation 3.69 can be used directly, giving

$$\frac{v_2}{v_{be}^c} = \frac{1}{1 + \left(\dfrac{r_{\pi 1}}{(\beta_0 + 1) r_{\pi 2}}\right)} \quad (3.109)$$

Also

$$i_c^c = g_m^c v_{be}^c = g_{m2} v_2 = \frac{g_{m2} v_{be}^c}{1 + \left(\dfrac{r_{\pi 1}}{(\beta_0 + 1) r_{\pi 2}}\right)} \quad (3.110)$$

Thus

$$g_m^c = \frac{i_c^c}{v_{be}^c} = \frac{g_{m2}}{1 + \left(\dfrac{r_{\pi 1}}{(\beta_0 + 1) r_{\pi 2}}\right)} \quad (3.111)$$

For the special case in which the biasing current source I_{BIAS} is zero, the emitter current of Q_1 is equal to the base current of Q_2. Thus the ratio of $r_{\pi 1}$ to $r_{\pi 2}$ is $(\beta_0 + 1)$, and (3.111) reduces to

$$g_m^c = \frac{g_{m2}}{2} \quad (3.112)$$

The effective current gain β^c is the ratio

$$\beta^c = \frac{i_c^c}{i_b^c} = \frac{i_{c2}}{i_{b1}} \quad (3.113)$$

The emitter current of Q_1 is given by

$$i_{e1} = (\beta_0 + 1) i_{b1} \quad (3.114)$$

Since $i_{e1} = i_{b2}$,

$$i_{c2} = i_c^c = \beta_0 i_{b2} = \beta_0 (\beta_0 + 1) i_{b1} = \beta_0 (\beta_0 + 1) i_b^c \quad (3.115)$$

Therefore,

$$\beta^c = \beta_0 (\beta_0 + 1) \quad (3.116)$$

Equation 3.116 shows that the current gain of the composite transistor is approximately equal to β_0^2. Also, by inspection of Fig. 3.32, assuming r_μ is negligible, we have

$$r_o^c = r_{o2} \quad (3.117)$$

The small-signal, two-port network equivalent for the CC-CE connection is shown in Fig. 3.33, where the collector resistor R_C has not been included. This small-signal equivalent can be used to represent the small-signal operation of the composite device, simplifying the analysis of circuits containing this structure.

The Darlington configuration, illustrated in Fig. 3.34, is a composite two-transistor device in which the collectors are tied together and the emitter of the first device drives the base of the second. A biasing element of some sort is used to control the emitter current of Q_1. The result is a three-terminal composite transistor that can be used in place of a single transistor in common-emitter, common-base, and common-collector configurations. When used as an emitter follower, the device is identical to the CC-CC connection already described. When used as a common-emitter amplifier, the device is very similar to the CC-CE connection, except that the collector of Q_1 is connected to the output instead of to the power supply. One effect of this change is to reduce the effective output resistance of the device

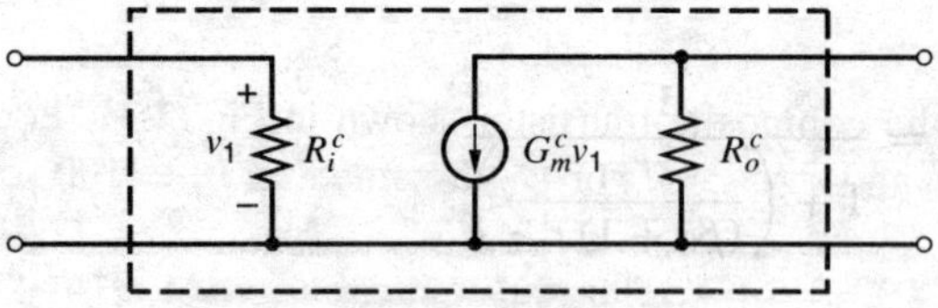

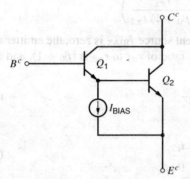

$$R_i^c = r_\pi^c = r_{\pi 1} + (\beta_0 + 1)\, r_{\pi 2}$$

$$G_m^c = g_m^c = \frac{g_{m2}}{1 + \left[\dfrac{r_{\pi 1}}{(\beta_0 + 1)\, r_{\pi 2}}\right]}$$

$$R_o^c = r_o^c = r_{o2}$$

Figure 3.33 Two-port representation, CC-CE connection.

Figure 3.34 The Darlington configuration.

because of feedback through the r_o of Q_1. Also, this change increases the input capacitance because of the connection of the collector-base capacitance of Q_1 from the input to the output. Because of these drawbacks, the CC-CE connection is normally preferable in integrated small-signal amplifiers. The term *Darlington* is often used to refer to both the CC-CE and CC-CC connections.

As mentioned previously, Darlington-type connections are used to boost the effective current gain of bipolar transistors and have no significant application in pure-MOS circuits. In BiCMOS technologies, however, a potentially useful connection is shown in Fig. 3.35, where an MOS transistor is used for Q_1. This configuration not only realizes the infinite input resistance and current gain of the MOS transistor, but also the large transconductance of the bipolar transistor.

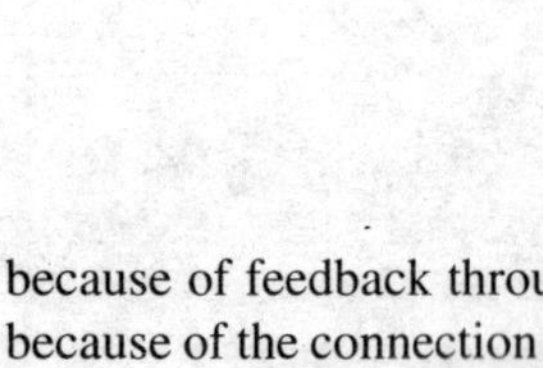

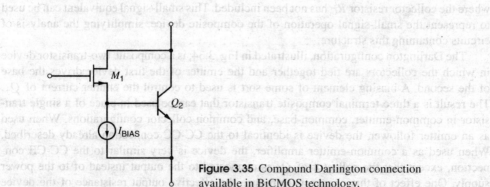

Figure 3.35 Compound Darlington connection available in BiCMOS technology.

■ **EXAMPLE**

Find the effective r_π^c, β^c, and g_m^c for the composite transistor shown in Fig. 3.31. For both devices, assume that $\beta_0 = 100$, $r_b = 0$, and $r_o \to \infty$. For Q_2, assume that $I_C = 100$ μA and that $I_{BIAS} = 10$ μA.

The base current of Q_2 is 100 μA/100 = 1 μA. Thus the emitter current of Q_1 is 11 μA. Then

$$r_{\pi 1} = \frac{\beta_0}{g_m} = \frac{100}{11 \text{ μA}/26 \text{ mV}} = 236 \text{ k}\Omega$$

$$g_{m1} = (2.36 \text{ k}\Omega)^{-1}$$

$$r_{\pi 2} = 26 \text{ k}\Omega$$

$$g_{m2} = (260 \text{ }\Omega)^{-1}$$

$$r_\pi^c = 236 \text{ k}\Omega + (101)(26 \text{ k}\Omega) = 2.8 \text{ M}\Omega$$

$$\beta^c = (101)(100) = 10,100$$

$$g_m^c = g_{m2}(0.916) = (283 \text{ }\Omega)^{-1}$$

Thus the composite transistor has much higher input resistance and current gain than a single transistor.
■

3.4.2 The Cascode Configuration

The cascode configuration was first invented for vacuum-tube circuits.[6,7] With vacuum tubes, the terminal that emits electrons is the *cathode*, the terminal that controls current flow is the *grid*, and the terminal that collects electrons is the *anode*. The *cascode* is a **casc**ade of common-cathode and common-grid stages joined at the an**ode** of the first stage and the cath**ode** of the second stage. The cascode configuration is important mostly because it increases output resistance and reduces unwanted capacitive feedback in amplifiers, allowing operation at higher frequencies than would otherwise be possible. The high output resistance attainable is particularly useful in desensitizing bias references from variations in power-supply voltage and in achieving large amounts of voltage gain. These applications are described further in Chapter 4. The topic of frequency response is covered in Chapter 7. Here, we will focus on the low-frequency, small-signal properties of the cascode configuration.

3.4.2.1 The Bipolar Cascode

In bipolar form, the cascode is a common-emitter–common-base (CE-CB) amplifier, as shown in Fig. 3.36. We will assume here that r_b in both devices is zero. Although the base resistances have a negligible effect on the low-frequency performance, the effects of nonzero r_b are important in the high-frequency performance of this combination. These effects are considered in Chapter 7.

The small-signal equivalent for the bipolar cascode circuit is shown in Fig. 3.37. Since we are considering the low-frequency performance, we neglect the capacitances in the model of each transistor. We will determine the input resistance, output resistance, and transconductance of the cascode circuit. By inspection of Fig. 3.37, the input resistance is simply

$$R_i = r_{\pi 1} \tag{3.118}$$

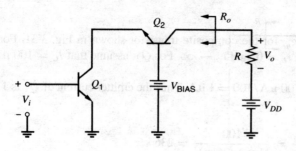

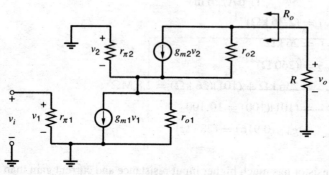

Figure 3.36 The cascode amplifier using bipolar transistors.

Figure 3.37 Small-signal equivalent circuit for the bipolar-transistor cascode connection.

Since the current gain from the emitter to the collector of Q_2 is nearly unity, the transconductance of the circuit from input to output is

$$G_m \simeq g_{m1} \tag{3.119}$$

The output resistance can be calculated by shorting the input v_i to ground and applying a test signal at the output. Then $v_1 = 0$ in Fig. 3.37 and the $g_{m1}v_1$ generator is inactive. The circuit is then identical to that of Fig. 3.26c for a bipolar transistor with emitter degeneration. Therefore, using (3.98) with $R_E = r_{o1}$ shows that the output resistance is

$$R_o \simeq r_{o2}\left(1 + \frac{g_{m2}r_{o1}}{1 + \frac{g_{m2}r_{o1}}{\beta_0}}\right) \tag{3.120}$$

If $g_{m2}r_{o1} \gg \beta_0$ and $\beta_0 \gg 1$,

$$R_o \simeq \beta_0 r_{o2} \tag{3.121}$$

Therefore, the CE-CB connection displays an output resistance that is larger by a factor of about β_0 than the CE stage alone. If this circuit is operated with a hypothetical collector load that has infinite incremental resistance, the voltage gain is

$$A_v = \frac{v_o}{v_i} = -G_m R_o \simeq -g_{m1}r_{o2}\beta_0 = -\frac{\beta_0}{\eta} \tag{3.122}$$

Thus the magnitude of the maximum available voltage gain is higher by a factor β_0 than for the case of a single transistor. For a typical *npn* transistor, the ratio of β_0/η is approximately 2×10^5. In this analysis, we have neglected r_μ. As described in Chapter 1, the value of r_μ for integrated-circuit *npn* transistors is usually much larger than $\beta_0 r_o$, and then r_μ has little effect on R_o. For lateral *pnp* transistors, however, r_μ is comparable with $\beta_0 r_o$ and decreases R_o somewhat.

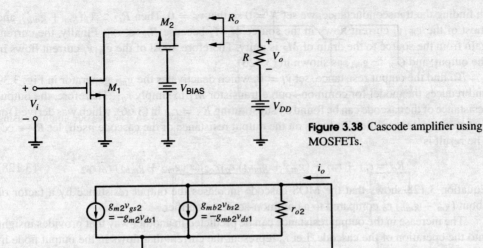

Figure 3.38 Cascode amplifier using MOSFETs.

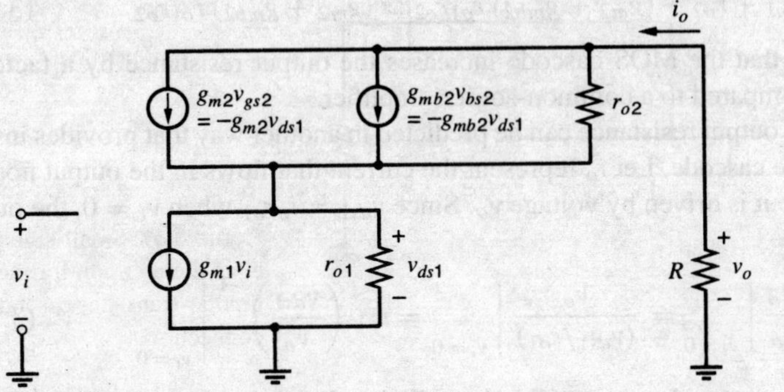

Figure 3.39 Small-signal equivalent circuit for the MOS-transistor cascode connection.

3.4.2.2 The MOS Cascode

In MOS form, the cascode is a common-source–common-gate (CS-CG) amplifier, as shown in Fig. 3.38. The small-signal equivalent circuit is shown in Fig. 3.39. Since the input is connected to the gate of M_1, the input resistance is

$$R_i \to \infty \qquad (3.123)$$

To find the transconductance, set $R = 0$ to short the output and calculate the current i_o. From KCL at the output,

$$i_o + g_{m2}v_{ds1} + g_{mb2}v_{ds1} + \frac{v_{ds1}}{r_{o2}} = 0 \qquad (3.124)$$

From KCL at the source of M_2,

$$g_{m1}v_i + g_{m2}v_{ds1} + g_{mb2}v_{ds1} + \frac{v_{ds1}}{r_{o1}} + \frac{v_{ds1}}{r_{o2}} = 0 \qquad (3.125)$$

Solving (3.125) for v_{ds1}, substituting into (3.124), and rearranging gives

$$G_m = \left.\frac{i_o}{v_i}\right|_{v_o=0} = g_{m1}\left(1 - \frac{1}{1 + (g_{m2} + g_{mb2})r_{o1} + \frac{r_{o1}}{r_{o2}}}\right) \simeq g_{m1} \qquad (3.126)$$

Equation 3.126 shows that the transconductance of the simple cascode is less than g_{m1}. If $(g_{m2} + g_{mb2})r_{o1} \gg 1$, however, the difference is small, and the main point here is that the cascode configuration has little effect on the transconductance. This result stems from the observation that R_{t2}, the resistance looking in the source of M_2, is much less than r_{o1}. From (3.54) and (3.55) with $R = R_D \parallel R_L$,

$$R_{i2} = \frac{r_{o2} + R}{1 + (g_{m2} + g_{mb2})r_{o2}} \simeq \frac{1}{g_{m2} + g_{mb2}} + \frac{R}{(g_{m2} + g_{mb2})r_{o2}} \qquad (3.127)$$

In finding the transconductance, we set $R = 0$ so that $v_o = 0$. Then $R_{i2} \simeq 1/(g_m + g_{mb})$, and most of the $g_{m1}v_i$ current flows in the source of M_2 because $R_{i2} \ll r_{o1}$. Finally, the current gain from the source to the drain of M_2 is unity. Therefore, most of the $g_{m1}v_i$ current flows in the output, and $G_m \simeq g_{m1}$, as shown in (3.126).

To find the output resistance, set $v_i = 0$, which deactivates the g_{m1} generator in Fig. 3.39 and reduces the model for common-source transistor M_1 to simply r_{o1}. Therefore, the output resistance of the cascode can be found by substituting $R_S = r_{o1}$ in (3.66), which was derived for a common-gate amplifier. To focus on the output resistance of the cascode itself, let $R \to \infty$. The result is

$$R_o = r_{o1} + r_{o2} + (g_{m2} + g_{mb2}) r_{o1} r_{o2} \simeq (g_{m2} + g_{mb2}) r_{o1} r_{o2} \qquad (3.128)$$

Equation 3.128 shows that the MOS cascode increases the output resistance by a factor of about $(g_m + g_{mb}) r_o$ compared to a common-source amplifier.

The increase in the output resistance can be predicted in another way that provides insight into the operation of the cascode. Let i_o represent the current that flows in the output node in Fig. 3.39 when the output is driven by voltage v_o. Since $v_{ds1} = i_o r_{o1}$ when $v_i = 0$, the output resistance is

$$R_o = \left.\frac{v_o}{i_o}\right|_{v_i=0} = \left.\frac{v_o}{(v_{ds1}/r_{o1})}\right|_{v_i=0} = r_{o1} \left.\left(\frac{v_{ds1}}{v_o}\right)^{-1}\right|_{v_i=0} \qquad (3.129)$$

To find the ratio v_{ds1}/v_o, consider the modified small-signal circuits shown in Fig. 3.40. In Fig. 3.40a, $R \to \infty$ so we can concentrate on the output resistance of the cascode circuit itself. Also, the $g_{m1}v_i$ generator is eliminated because $v_i = 0$, and the two generators $g_{m2}v_{ds1}$ and $g_{mb2}v_{ds1}$ have been combined into one equivalent generator $(g_{m2} + g_{mb2}) v_{ds1}$. In Fig. 3.40b,

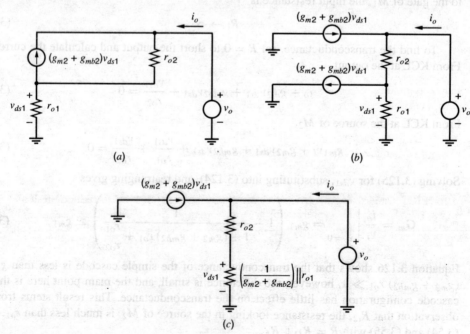

Figure 3.40 Construction of a cascode model to find v_{ds1}/v_o. (a) The dependent sources are combined. (b) The combined source is converted into two sources. (c) The current source between the source of M_2 and ground is converted into a resistor.

the $(g_{m2} + g_{mb2}) v_{ds1}$ generator from the source to the drain of M_2 has been replaced by two equal-valued generators: one from ground to the drain of M_2 and the other from the source of M_2 to ground. This replacement is similar to the substitution made in Fig. 3.20 to convert the hybrid-π model to a T model for a common-gate amplifier. Because the equations that describe the operation of the circuits in Fig. 3.40a and Fig. 3.40b are identical, the circuit in Fig. 3.40b is equivalent to that in Fig. 3.40a. Finally, in Fig. 3.40c, the current source from the source of M_2 to ground, which is controlled by the voltage across itself, is replaced by an equivalent resistor of value $1/(g_{m2} + g_{mb2})$. The current $(g_{m2} + g_{mb2}) v_{ds1}$ in Fig. 3.40c flows into the test source v_o. The two resistors in Fig. 3.40c form a voltage divider, giving

$$\frac{v_{ds1}}{v_o} = \frac{\left(\frac{1}{g_{m2} + g_{mb2}}\right) \| r_{o1}}{\left[\left(\frac{1}{g_{m2} + g_{mb2}}\right) \| r_{o1}\right] + r_{o2}} \simeq \frac{1}{(g_{m2} + g_{mb2}) r_{o2}} \quad (3.130)$$

Substituting (3.130) into (3.129) and rearranging gives the same result as in (3.128). In (3.130), the term $1/(g_{m2} + g_{mb2})$ represents the resistance looking into the source of the common-gate transistor M_2 when the output in Fig. 3.39 is voltage driven. The key point here is that the output resistance of the cascode can be increased by reducing the input resistance of the common-gate transistor under these conditions because this change reduces both v_{ds1} and i_o.

Unlike in the bipolar case, the maximum value of the output resistance in the MOS cascode does not saturate at a level determined by β_0; therefore, further increases in the output resistance can be obtained by using more than one level of cascoding. This approach is used in practice. Ultimately, the maximum output resistance is limited by impact ionization as described in Section 1.9 or by leakage current in the reverse-biased junction diode at the output. Also, the number of levels of cascoding is limited by the power-supply voltage and signal-swing requirements. Each additional level of cascoding places one more transistor in series with the input transistor between the power supply and ground. To operate all the transistors in the active region, the drain-source voltage of each transistor must be greater than its overdrive $V_{GS} - V_t$. Since the cascode transistors operate in series with the input transistor, additional levels of cascoding use some of the available power-supply voltage, reducing the amount by which the output can vary before pushing one or more transistors into the triode region. This topic is considered further in Chapter 4.

In BiCMOS technologies, cascodes are sometimes used with the MOS transistor M_2 in Fig. 3.38 replaced by a bipolar transistor, such as Q_2 in Fig. 3.36. This configuration has the infinite input resistance given by M_1. Also, the resistance looking into the emitter of the common-base stage Q_2 when the output is grounded is $R_{i2} \simeq 1/g_{m2}$ in this configuration. Since the transconductance for a given bias current of bipolar transistors is usually much greater than for MOS transistors, the BiCMOS configuration is often used to reduce the load resistance presented to M_1 and to improve the high-frequency properties of the cascode amplifier. The frequency response of a cascode amplifier is described in Chapter 7.

■ **EXAMPLE**

Calculate the transconductance and output resistance of the cascode circuit of Fig. 3.38. Assume that both transistors operate in the active region with $g_m = 1$ mA/V, $\chi = 0.1$, and $r_o = 20$ kΩ.

From (3.126),

$$G_m = \left(1 \frac{\text{mA}}{\text{V}}\right)\left(1 - \frac{1}{1 + (1.1)(20) + 1}\right) = 960 \frac{\mu\text{A}}{\text{V}}$$

From (3.128),

$$R_o = 20 \text{ k}\Omega + 20 \text{ k}\Omega + (1.1)(20)\,20 \text{ k}\Omega = 480 \text{ k}\Omega$$

The approximations in (3.126) and (3.128) give $G_m \simeq 1$ mA/V and $R_o \simeq 440$ kΩ. These approximations deviate from the exact results by about 4 percent and 8 percent, respectively, and are usually close enough for hand calculations.

3.4.3 The Active Cascode

As mentioned in the previous section, increasing the number of levels of cascoding increases the output resistance of MOS amplifiers. In practice, however, the power-supply voltage and signal-swing requirements limit the number of levels of cascoding that can be applied. One way to increase the output resistance of the MOS cascode circuit without increasing the number of levels of cascoding is to use the active-cascode circuit, as shown in Fig. 3.41.[8,9]

This circuit uses an amplifier in a negative feedback loop to control the voltage from the gate of M_2 to ground. If the amplifier gain a is infinite, the negative feedback loop adjusts the gate of M_2 until the voltage difference between the two amplifier inputs is zero. In other words, the drain-source voltage of M_1 is driven to equal V_{BIAS}. If the drain-source voltage of M_1 is constant, the change in the drain current in response to changes in the output voltage V_o is zero, and the output resistance is infinite. In practice, the amplifier gain a is finite, which means that the drain-source voltage of M_1 is not exactly constant and the output resistance is finite. The effect of negative feedback on output resistance is considered quantitatively in Chapter 8. In this section, we will derive the small-signal properties of the active-cascode circuit by comparing its small-signal model to that of the simple cascode described in the previous section.

Qualitatively, when the output voltage increases, the drain current of M_2 increases, which increases the drain current and drain-source voltage of M_1. This voltage increase is amplified by $-a$, causing the voltage from the gate of M_2 to ground to fall. The falling gate voltage of M_2 acts to reduce the change in its drain current, increasing the output resistance compared to a simple cascode, where the voltage from the gate of M_2 to ground is held constant.

Figure 3.42 shows the low-frequency, small-signal equivalent circuit. The body-effect transconductance generator for M_1 is inactive because $v_{bs1} = 0$. The gate-source voltage of M_2 is

$$v_{gs2} = v_{g2} - v_{s2} = v_{g2} - v_{ds1} = -(a)v_{ds1} - v_{ds1} = -(a+1)v_{ds1} \tag{3.131}$$

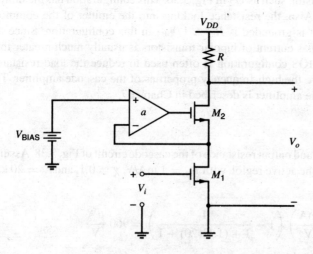

Figure 3.41 Active cascode amplifier using MOSFETs.

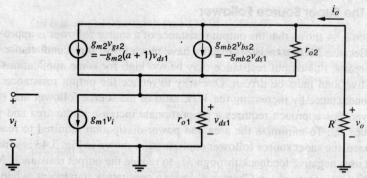

Figure 3.42 Small-signal equivalent circuit for the active-cascode connection with MOS transistors.

In contrast, $v_{gs2} = -v_{ds1}$ in a simple cascode because the voltage from the gate of M_2 to ground is constant in Fig. 3.38. Therefore, if $a > 0$, the factor $(a + 1)$ in (3.131) amplifies the gate-source voltage of M_2 compared to the case of a simple cascode. This amplification is central to the characteristics of the active-cascode circuit. Since the small-signal diagrams of the simple and active-cascode circuits are identical except for the value of v_{gs2}, and since v_{gs2} is only used to control the current flowing in the g_{m2} generator, the active-cascode circuit can be analyzed using the equations for the simple cascode with g_{m2} replaced by $(a + 1) g_{m2}$. In other words, the active cascode behaves as if it were a simple cascode with an enhanced value of g_{m2}.

To find the transconductance of the active cascode, $g_{m2}(a + 1)$ replaces g_{m2} in (3.126), giving

$$G_m = g_{m1}\left(1 - \frac{1}{1 + [g_{m2}(a+1) + g_{mb2}]r_{o1} + \frac{r_{o1}}{r_{o2}}}\right) \qquad (3.132)$$

Again, $G_m \simeq g_{m1}$ under most conditions; therefore, the active-cascode structure is generally not used to modify the transconductance.

The active cascode reduces R_{i2}, the resistance looking into the source of M_2, compared to the simple cascode, which reduces the v_{ds1}/v_o ratio given in (3.130) and increases the output resistance. Substituting (3.130) into (3.129) with $g_{m2}(a + 1)$ replacing g_{m2} gives

$$R_o = r_{o1} + r_{o2} + [g_{m2}(a+1) + g_{mb2}]r_{o1}r_{o2} \simeq [g_{m2}(a+1) + g_{mb2}]r_{o1}r_{o2} \qquad (3.133)$$

This result can also be derived by substituting $g_{m2}(a + 1)$ for g_{m2} in (3.128). Equation 3.133 shows that the active-cascode configuration increases the output resistance by a factor of about $[g_m(a + 1) + g_{mb}]r_o$ compared to a common-source amplifier.

A key limitation of the active-cascode circuit is that the output impedance is increased only at frequencies where the amplifier that drives the gate of M_2 provides some gain. In practice, the gain of this amplifier falls with increasing frequency, reducing the potential benefits of the active-cascode circuits in high-frequency applications. A potential problem with the active-cascode configuration is that the negative feedback loop through M_2 may not be stable in all cases.

3.4.4 The Super Source Follower

Equation 3.84 shows that the output resistance of a source follower is approximately $1/(g_m + g_{mb})$. Because MOS transistors usually have much lower transconductance than their bipolar counterparts, this output resistance may be too high for some applications, especially when a resistive load must be driven. One way to reduce the output resistance is to increase the transconductance by increasing the W/L ratio of the source follower and its dc bias current. However, this approach requires a proportionate increase in the area and power dissipation to reduce R_o. To minimize the area and power dissipation required to reach a given output resistance, the super source follower configuration shown in Fig. 3.43 is sometimes used. This circuit uses negative feedback through M_2 to reduce the output resistance. Negative feedback is studied quantitatively in Chapter 8. From a qualitative standpoint, when the input voltage is constant and the output voltage increases, the magnitude of the drain current of M_1 also increases, in turn increasing the gate-source voltage of M_2. As a result, the drain current of M_2 increases, reducing the output resistance by increasing the total current that flows into the output node under these conditions.

From a dc standpoint, the bias current in M_2 is the difference between I_1 and I_2; therefore, $I_1 > I_2$ is required for proper operation. This information can be used to find the small-signal parameters of both transistors. The small-signal equivalent circuit is shown in Fig. 3.44. The body-effect transconductance generator for M_2 is inactive because $v_{bs2} = 0$. Also, the polarities of the voltage-controlled current sources for n- and p-channel devices are identical. Finally, the output resistances of current sources I_1 and I_2 are represented by r_1 and r_2, respectively.

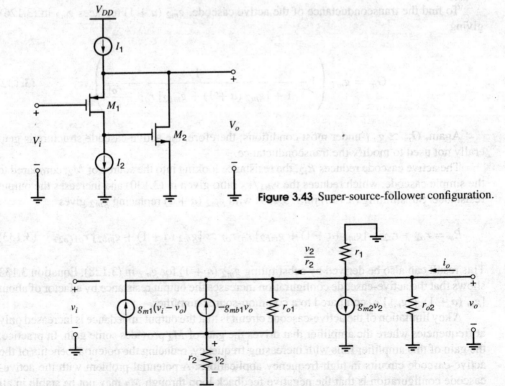

Figure 3.43 Super-source-follower configuration.

Figure 3.44 Small-signal equivalent circuit of the super-source follower.

If the current sources are ideal, $r_1 \to \infty$ and $r_2 \to \infty$. In practice, these resistances are large but finite. Techniques to build high-resistance current sources are considered in Chapter 4.

To find the output resistance, set $v_i = 0$ and calculate the current i_o that flows in the output node when the output is driven by a voltage v_o. From KCL at the output under these conditions,

$$i_o = \frac{v_o}{r_1} + \frac{v_o}{r_{o2}} + g_{m2}v_2 + \frac{v_2}{r_2} \tag{3.134}$$

From KCL at the drain of M_1 with $v_i = 0$,

$$\frac{v_2}{r_2} - g_{m1}v_o - g_{mb1}v_o + \frac{v_2 - v_o}{r_{o1}} = 0 \tag{3.135}$$

Solving (3.135) for v_2, substituting into (3.134), and rearranging gives

$$R_o = \left.\frac{v_o}{i_o}\right|_{v_i=0} = r_1 \parallel r_{o2} \parallel \left(\frac{r_{o1} + r_2}{[1 + (g_{m1} + g_{mb1})r_{o1}](1 + g_{m2}r_2)}\right) \tag{3.136}$$

Assume I_1 and I_2 are ideal current sources so that $r_1 \to \infty$ and $r_2 \to \infty$. If $r_{o2} \to \infty$, and if $(g_{m1} + g_{mb1})r_{o1} \gg 1$,

$$R_o \simeq \frac{1}{g_{m1} + g_{mb1}}\left(\frac{1}{g_{m2}r_{o1}}\right) \tag{3.137}$$

Comparing (3.84) and (3.137) shows that the negative feedback through M_2 reduces the output resistance by a factor of about $g_{m2}r_{o1}$.

Now we will calculate the open-circuit voltage gain of the super-source follower. With the output open circuited, KCL at the output node gives

$$\frac{v_o}{r_1} + \frac{v_o}{r_{o2}} + g_{m2}v_2 + \frac{v_2}{r_2} = 0 \tag{3.138}$$

From KCL at the drain of M_1,

$$\frac{v_2}{r_2} + g_{m1}(v_i - v_o) - g_{mb1}v_o + \frac{v_2 - v_o}{r_{o1}} = 0 \tag{3.139}$$

Solving (3.138) for v_2, substituting into (3.139), and rearranging gives

$$\left.\frac{v_o}{v_i}\right|_{i_o=0} = \frac{g_{m1}r_{o1}}{1 + (g_{m1} + g_{mb1})r_{o1} + \dfrac{(r_2 + r_{o1})}{(r_1 \parallel r_{o2})(1 + g_{m2}r_2)}} \tag{3.140}$$

With ideal current sources,

$$\lim_{\substack{r_1 \to \infty \\ r_2 \to \infty}} \left.\frac{v_o}{v_i}\right|_{i_o=0} = \frac{g_{m1}r_{o1}}{1 + (g_{m1} + g_{mb1})r_{o1} + \dfrac{1}{g_{m2}r_{o2}}} \tag{3.141}$$

Comparing (3.141) and (3.81) shows that the deviation of this gain from unity is greater than with a simple source follower. If $g_{m2}r_{o2} \gg 1$, however, the difference is small and the main conclusion is that the super-source-follower configuration has little effect on the open-circuit voltage gain.

As mentioned earlier, the super-source follower is sometimes used in MOS technologies to reduce the source-follower output resistance. It is also used in bipolar technologies in output stages to reduce the current conducted in a weak lateral *pnp* transistor. This application is described in Chapter 5. The main potential problem with the super-source-follower configuration is that the negative feedback loop through M_2 may not be stable in all cases, especially when driving a capacitive load. The stability of feedback amplifiers is considered in Chapter 9.

3.5 Differential Pairs

The differential pair is another example of a circuit that was first invented for use with vacuum tubes.[10] The original circuit uses two vacuum tubes whose cathodes are connected together. Modern differential pairs use bipolar or MOS transistors coupled at their emitters or sources, respectively, and are perhaps the most widely used two-transistor subcircuits in monolithic analog circuits. The usefulness of the differential pair stems from two key properties. First, cascades of differential pairs can be directly connected to one another without interstage coupling capacitors. Second, the differential pair is primarily sensitive to the difference between two input voltages, allowing a high degree of rejection of signals common to both inputs.[11,12] In this section, we consider the properties of emitter-coupled pairs of bipolar transistors and source-coupled pairs of MOS transistors in detail.

3.5.1 The dc Transfer Characteristic of an Emitter-Coupled Pair

The simplest form of an emitter-coupled pair is shown in Fig. 3.45. The biasing circuit in the lead connected to the emitters of Q_1 and Q_2 can be a transistor current source, which is called a *tail* current source, or a simple resistor. If a simple resistor R_{TAIL} is used alone, $I_{TAIL} = 0$ in Fig. 3.45. Otherwise, I_{TAIL} and R_{TAIL} together form a Norton-equivalent model of the tail current source.

The large-signal behavior of the emitter-coupled pair is important in part because it illustrates the limited range of input voltages over which the circuit behaves almost linearly. Also, the large-signal behavior shows that the amplitude of analog signals in bipolar circuits can be limited without pushing the transistors into saturation, where the response time would be increased because of excess charge storage in the base region. For simplicity in the analysis, we assume that the output resistance of the tail current source $R_{TAIL} \to \infty$, that the output resistance of each transistor $r_o \to \infty$, and that the base resistance of each transistor $r_b = 0$. These

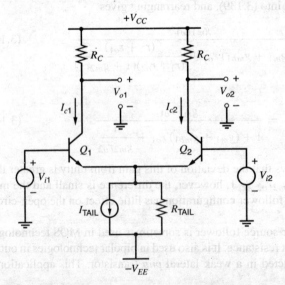

Figure 3.45 Emitter-coupled pair circuit diagram.

assumptions do not strongly affect the low-frequency, large-signal behavior of the circuit. From KVL around the input loop,

$$V_{i1} - V_{be1} + V_{be2} - V_{i2} = 0 \tag{3.142}$$

Assume the collector resistors are small enough that the transistors do not operate in saturation if $V_{i1} \leq V_{CC}$ and $V_{i2} \leq V_{CC}$. If $V_{be1} \gg V_T$ and $V_{be2} \gg V_T$, the Ebers-Moll equations show that

$$V_{be1} = V_T \ln \frac{I_{c1}}{I_{S1}} \tag{3.143}$$

$$V_{be2} = V_T \ln \frac{I_{c2}}{I_{S2}} \tag{3.144}$$

Assume the transistors are identical so that $I_{S1} = I_{S2}$. Then combining (3.142), (3.143), and (3.144), we find

$$\frac{I_{c1}}{I_{c2}} = \exp\left(\frac{V_{i1} - V_{i2}}{V_T}\right) = \exp\left(\frac{V_{id}}{V_T}\right) \tag{3.145}$$

where $V_{id} = V_{i1} - V_{i2}$. Since we have assumed that the transistors are identical, $\alpha_{F1} = \alpha_{F2} = \alpha_F$. Then KCL at the emitters of the transistors shows

$$-(I_{e1} + I_{e2}) = I_{\text{TAIL}} = \frac{I_{c1} + I_{c2}}{\alpha_F} \tag{3.146}$$

Combining (3.145) and (3.146), we find that

$$I_{c1} = \frac{\alpha_F I_{\text{TAIL}}}{1 + \exp\left(-\dfrac{V_{id}}{V_T}\right)} \tag{3.147}$$

$$I_{c2} = \frac{\alpha_F I_{\text{TAIL}}}{1 + \exp\left(\dfrac{V_{id}}{V_T}\right)} \tag{3.148}$$

These two currents are shown as a function of V_{id} in Fig. 3.46. When the magnitude of V_{id} is greater than about $3V_T$, which is approximately 78 mV at room temperature, the collector currents are almost independent of V_{id} because one of the transistors turns off and the other conducts all the current that flows. Furthermore, the circuit behaves in an approximately

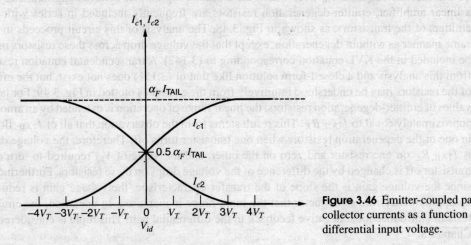

Figure 3.46 Emitter-coupled pair collector currents as a function of differential input voltage.

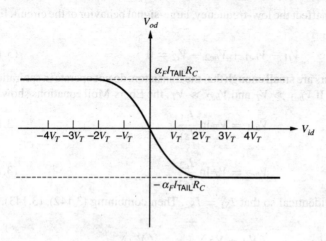

Figure 3.47 Emitter-coupled pair, differential output voltage as a function of differential input voltage.

linear fashion only when the magnitude of V_{id} is less than about V_T. We can now compute the output voltages as

$$V_{o1} = V_{CC} - I_{c1}R_C \tag{3.149}$$

$$V_{o2} = V_{CC} - I_{c2}R_C \tag{3.150}$$

The output signal of interest is often the difference between V_{o1} and V_{o2}, which we define as V_{od}. Then

$$V_{od} = V_{o1} - V_{o2} = \alpha_F I_{\text{TAIL}} R_C \tanh\left(\frac{-V_{id}}{2V_T}\right) \tag{3.151}$$

This function is plotted in Fig. 3.47. Here a significant advantage of differential amplifiers is apparent: When V_{id} is zero, V_{od} is zero if Q_1 and Q_2 are identical and if identical resistors are connected to the collectors of Q_1 and Q_2. This property allows direct coupling of cascaded stages without offsets.

3.5.2 The dc Transfer Characteristic with Emitter Degeneration

To increase the range of V_{id} over which the emitter-coupled pair behaves approximately as a linear amplifier, emitter-degeneration resistors are frequently included in series with the emitters of the transistors, as shown in Fig. 3.48. The analysis of this circuit proceeds in the same manner as without degeneration, except that the voltage drop across these resistors must be included in the KVL equation corresponding to (3.142). A transcendental equation results from this analysis and a closed-form solution like that of (3.151) does not exist, but the effect of the resistors may be understood intuitively from the examples plotted in Fig. 3.49. For large values of emitter-degeneration resistors, the linear range of operation is extended by an amount approximately equal to $I_{\text{TAIL}} R_E$. This result stems from the observation that all of I_{TAIL} flows in one of the degeneration resistors when one transistor turns off. Therefore, the voltage drop is $I_{\text{TAIL}} R_E$ on one resistor and zero on the other, and the value of V_{id} required to turn one transistor off is changed by the difference of the voltage drops on these resistors. Furthermore, since the voltage gain is the slope of the transfer characteristic, the voltage gain is reduced by approximately the same factor that the input range is increased. In operation, the emitter resistors introduce local negative feedback in the differential pair. This topic is considered in Chapter 8.

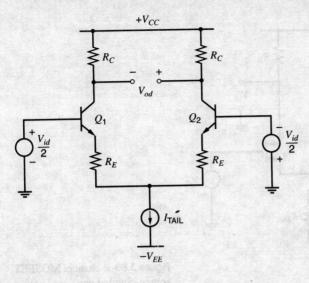

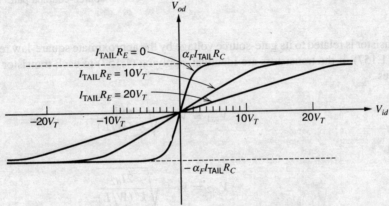

Figure 3.48 Circuit diagram of emitter-coupled pair with emitter degeneration.

Figure 3.49 Output voltage as a function of input voltage, emitter-coupled pair with emitter degeneration.

3.5.3 The dc Transfer Characteristic of a Source-Coupled Pair

Consider the *n*-channel MOS-transistor source-coupled pair shown in Fig. 3.50. The following analysis applies equally well to a corresponding *p*-channel source-coupled pair with appropriate sign changes. In monolithic form, a transistor current source, called a *tail* current source, is usually connected to the sources of M_1 and M_2. In that case, I_{TAIL} and R_{TAIL} together form a Norton-equivalent model of the tail current source.

For this large-signal analysis, we assume that the output resistance of the tail current source is $R_{TAIL} \to \infty$. Also, we assume that the output resistance of each transistor $r_o \to \infty$. Although these assumptions do not strongly affect the low-frequency, large-signal behavior of the circuit, they could have a significant impact on the small-signal behavior. Therefore, we will reconsider these assumptions when we analyze the circuit from a small-signal standpoint. From KVL around the input loop,

$$V_{i1} - V_{gs1} + V_{gs2} - V_{i2} = 0 \qquad (3.152)$$

We assume that the drain resistors are small enough that neither transistor operates in the triode region if $V_{i1} \leq V_{DD}$ and $V_{i2} \leq V_{DD}$. Furthermore, we assume that the drain current of each

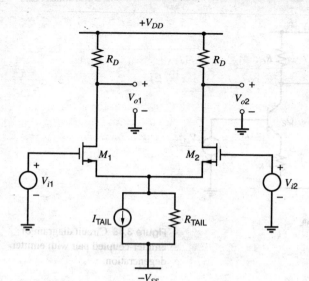

Figure 3.50 n-channel MOSFET source-coupled pair.

transistor is related to its gate-source voltage by the approximate square-law relationship given in (1.157). If the transistors are identical, applying (1.157) to each transistor and rearranging gives

$$V_{gs1} = V_t + \sqrt{\frac{2I_{d1}}{k'(W/L)}} \tag{3.153}$$

and

$$V_{gs2} = V_t + \sqrt{\frac{2I_{d2}}{k'(W/L)}} \tag{3.154}$$

Substituting (3.153) and (3.154) into (3.152) and rearranging gives

$$V_{id} = V_{i1} - V_{i2} = \frac{\sqrt{I_{d1}} - \sqrt{I_{d2}}}{\sqrt{\frac{k'}{2}\frac{W}{L}}} \tag{3.155}$$

From KCL at the source of M_1 and M_2,

$$I_{d1} + I_{d2} = I_{TAIL} \tag{3.156}$$

Solving (3.156) for I_{d2}, substituting into (3.155), rearranging, and using the quadratic formula gives

$$I_{d1} = \frac{I_{TAIL}}{2} \pm \frac{k'}{4}\frac{W}{L}V_{id}\sqrt{\frac{4I_{TAIL}}{k'(W/L)} - V_{id}^2} \tag{3.157}$$

Since $I_{d1} > I_{TAIL}/2$ when $V_{id} > 0$, the potential solution where the second term is subtracted from the first in (3.157) cannot occur in practice. Therefore,

$$I_{d1} = \frac{I_{TAIL}}{2} + \frac{k'}{4}\frac{W}{L}V_{id}\sqrt{\frac{4I_{TAIL}}{k'(W/L)} - V_{id}^2} \tag{3.158}$$

Substituting (3.158) into (3.156) gives

$$I_{d2} = \frac{I_{\text{TAIL}}}{2} - \frac{k'}{4}\frac{W}{L}V_{id}\sqrt{\frac{4I_{\text{TAIL}}}{k'(W/L)} - V_{id}^2} \qquad (3.159)$$

Equations 3.158 and 3.159 are valid when both transistors operate in the active or saturation region. Since we have assumed that neither transistor operates in the triode region, the limitation here stems from turning off one of the transistors. When M_1 turns off, $I_{d1} = 0$ and $I_{d2} = I_{\text{TAIL}}$. On the other hand, $I_{d1} = I_{\text{TAIL}}$ and $I_{d2} = 0$ when M_2 turns off. Substituting these values in (3.155) shows that both transistors operate in the active region if

$$|V_{id}| \leq \sqrt{\frac{2I_{\text{TAIL}}}{k'(W/L)}} \qquad (3.160)$$

Since $I_{d1} = I_{d2} = I_{\text{TAIL}}/2$ when $V_{id} = 0$, the range in (3.160) can be rewritten as

$$|V_{id}| \leq \sqrt{2} \cdot \left(\sqrt{\frac{2I_{d1}}{k'(W/L)}}\right)\bigg|_{V_{id}=0} = \sqrt{2}\,(V_{ov})|_{V_{id}=0} \qquad (3.161)$$

Equation 3.161 shows that the range of V_{id} for which both transistors operate in the active region is proportional to the overdrive calculated when $V_{id} = 0$. This result is illustrated in Fig. 3.51. The overdrive is an important quantity in MOS circuit design, affecting not only the input range of differential pairs, but also other characteristics including the speed, offset, and output swing of MOS amplifiers. Since the overdrive of an MOS transistor depends on its current and W/L ratio, the range of a source-coupled pair can be adjusted to suit a given application by adjusting the value of the tail current and/or the aspect ratio of the input devices. In contrast, the input range of the bipolar emitter-coupled pair is about $\pm 3V_T$, independent of bias current or device size. In fact, the source-coupled pair behaves somewhat like an emitter-coupled pair with emitter-degeneration resistors that can be selected to give a desired input voltage range.

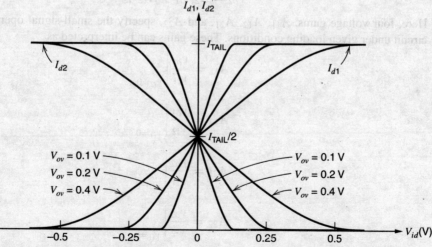

Figure 3.51 dc transfer characteristic of the MOS source-coupled pair. The parameter is the overdrive $V_{ov} = V_{GS} - V_t$ determined when $V_{id} = 0$.

In many practical cases, the key output of the differential pair is not I_{d1} or I_{d2} alone but the difference between these quantities. Subtracting (3.159) from (3.158) gives

$$\Delta I_d = I_{d1} - I_{d2} = \frac{k'}{2}\frac{W}{L} V_{id} \sqrt{\frac{4 I_{\text{TAIL}}}{k'(W/L)} - V_{id}^2} \qquad (3.162)$$

We can now compute the differential output voltage as

$$V_{od} = V_{o1} - V_{o2} = V_{DD} - I_{d1} R_D - V_{DD} + I_{d2} R_D = -(\Delta I_d) R_D \qquad (3.163)$$

Since $\Delta I_d = 0$ when $V_{id} = 0$, (3.163) shows that $V_{od} = 0$ when $V_{id} = 0$ if M_1 and M_2 are identical and if identical resistors are connected to the drains of M_1 and M_2. This property allows direct coupling of cascaded MOS differential pairs, as in the bipolar case.

3.5.4 Introduction to the Small-Signal Analysis of Differential Amplifiers

The features of interest in the performance of differential pairs are often the small-signal properties for dc differential input voltages near zero volts. In the next two sections, we assume that the dc differential input voltage is zero and calculate the small-signal parameters. If the parameters are constant, the small-signal model predicts that the circuit operation is linear. The results of the small-signal analysis are valid for signals that are small enough to cause insignificant nonlinearity.

In previous sections, we have considered amplifiers with two input terminals (V_i and ground) and two output terminals (V_o and ground). Small-signal analysis of such circuits leads to one equation for each circuit, such as

$$v_o = A v_i \qquad (3.164)$$

Here, A is the small-signal voltage gain under given loading conditions. In contrast, differential pairs have three input terminals (V_{i1}, V_{i2}, and ground) and three output terminals (V_{o1}, V_{o2}, and ground). Therefore, direct small-signal analysis of differential pairs leads to two equations for each circuit (one for each output), where each output depends on each input:

$$v_{o1} = A_{11} v_{i1} + A_{12} v_{i2} \qquad (3.165)$$

$$v_{o2} = A_{21} v_{i1} + A_{22} v_{i2} \qquad (3.166)$$

Here, four voltage gains, A_{11}, A_{12}, A_{21}, and A_{22}, specify the small-signal operation of the circuit under given loading conditions. These gains can be interpreted as

$$A_{11} = \left.\frac{v_{o1}}{v_{i1}}\right|_{v_{i2}=0} \qquad (3.167)$$

$$A_{12} = \left.\frac{v_{o1}}{v_{i2}}\right|_{v_{i1}=0} \qquad (3.168)$$

$$A_{21} = \left.\frac{v_{o2}}{v_{i1}}\right|_{v_{i2}=0} \qquad (3.169)$$

$$A_{22} = \left.\frac{v_{o2}}{v_{i2}}\right|_{v_{i1}=0} \qquad (3.170)$$

Although direct small-signal analysis of differential pairs can be used to calculate these four gain values in a straightforward way, the results are difficult to interpret because differential pairs usually are not used to react to v_{i1} or v_{i2} alone. Instead, differential pairs are used most often to sense the difference between the two inputs while trying to ignore the part of the two inputs that is common to each. Desired signals will be forced to appear as differences in differential circuits. In practice, undesired signals will also appear. For example, *mixed-signal* integrated circuits use both analog and digital signal processing, and the analog signals are vulnerable to corruption from noise generated by the digital circuits and transmitted through the common substrate. The hope in using differential circuits is that undesired signals will appear equally on both inputs and be rejected.

To highlight this behavior, we will define new differential and common-mode variables at the input and output as follows. The differential input, to which differential pairs are sensitive, is

$$v_{id} = v_{i1} - v_{i2} \tag{3.171}$$

The common-mode or average input, to which differential pairs are insensitive, is

$$v_{ic} = \frac{v_{i1} + v_{i2}}{2} \tag{3.172}$$

These equations can be inverted to give v_{i1} and v_{i2} in terms of v_{id} and v_{ic}:

$$v_{i1} = v_{ic} + \frac{v_{id}}{2} \tag{3.173}$$

$$v_{i2} = v_{ic} - \frac{v_{id}}{2} \tag{3.174}$$

The physical significance of these new variables can be understood by using (3.173) and (3.174) to redraw the input connections to a differential amplifier as shown in Fig. 3.52. The common-mode input is the input component that appears equally in v_{i1} and v_{i2}. The differential input is the input component that appears between v_{i1} and v_{i2}.

New output variables are defined in the same way. The differential output is

$$v_{od} = v_{o1} - v_{o2} \tag{3.175}$$

The common-mode or average output is

$$v_{oc} = \frac{v_{o1} + v_{o2}}{2} \tag{3.176}$$

Solving these equations for v_{o1} and v_{o2}, we obtain

$$v_{o1} = v_{oc} + \frac{v_{od}}{2} \tag{3.177}$$

$$v_{o2} = v_{oc} - \frac{v_{od}}{2} \tag{3.178}$$

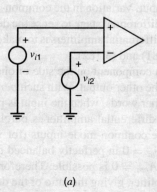

(a)

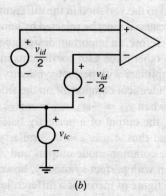

(b)

Figure 3.52 A differential amplifier with its inputs (*a*) shown as independent of each other and (*b*) redrawn in terms of the differential and common-mode components.

We have now defined two new input variables and two new output variables. By substituting the expressions for v_{i1}, v_{i2}, v_{o1}, and v_{o2} in terms of the new variables back into (3.165) and (3.166), we find

$$v_{od} = \left(\frac{A_{11} - A_{12} - A_{21} + A_{22}}{2}\right) v_{id} + (A_{11} + A_{12} - A_{21} - A_{22}) v_{ic} \quad (3.179)$$

$$v_{oc} = \left(\frac{A_{11} - A_{12} + A_{21} - A_{22}}{4}\right) v_{id} + \left(\frac{A_{11} + A_{12} + A_{21} + A_{22}}{2}\right) v_{ic} \quad (3.180)$$

Defining four new gain factors that are equal to the coefficients in these equations, (3.179) and (3.180) can be rewritten as

$$v_{od} = A_{dm} v_{id} + A_{cm-dm} v_{ic} \quad (3.181)$$
$$v_{oc} = A_{dm-cm} v_{id} + A_{cm} v_{ic} \quad (3.182)$$

The *differential-mode* gain A_{dm} is the change in the differential output per unit change in differential input:

$$A_{dm} = \left.\frac{v_{od}}{v_{id}}\right|_{v_{ic}=0} = \frac{A_{11} - A_{12} - A_{21} + A_{22}}{2} \quad (3.183)$$

The *common-mode* gain A_{cm} is the change in the common-mode output voltage per unit change in the common-mode input:

$$A_{cm} = \left.\frac{v_{oc}}{v_{ic}}\right|_{v_{id}=0} = \frac{A_{11} + A_{12} + A_{21} + A_{22}}{2} \quad (3.184)$$

The *differential-mode-to-common-mode* gain A_{dm-cm} is the change in the common-mode output voltage per unit change in the differential-mode input:

$$A_{dm-cm} = \left.\frac{v_{oc}}{v_{id}}\right|_{v_{ic}=0} = \frac{A_{11} - A_{12} + A_{21} - A_{22}}{4} \quad (3.185)$$

The *common-mode-to-differential-mode* gain A_{cm-dm} is the change in the differential-mode output voltage per unit change in the common-mode input:

$$A_{cm-dm} = \left.\frac{v_{od}}{v_{ic}}\right|_{v_{id}=0} = A_{11} + A_{12} - A_{21} - A_{22} \quad (3.186)$$

The purpose of a differential amplifier is to sense changes in its differential input while rejecting changes in its common-mode input. The desired output is differential, and its variation should be proportional to the variation in the differential input. Variation in the common-mode output is undesired because it must be rejected by another differential stage to sense the desired differential signal. Therefore, an important design goal in differential amplifiers is to make A_{dm} large compared to the other three gain coefficients in (3.181) and (3.182).

In differential amplifiers with perfect symmetry, each component on the side of one output corresponds to an identical component on the side of the other output. With such *perfectly balanced* amplifiers, when $v_{i1} = -v_{i2}$, $v_{o1} = -v_{o2}$. In other words, when the input is purely differential ($v_{ic} = 0$), the output of a perfectly balanced differential amplifier is purely differential ($v_{oc} = 0$), and thus $A_{dm-cm} = 0$. Similarly, pure common-mode inputs (for which $v_{id} = 0$) produce pure common-mode outputs and $A_{cm-dm} = 0$ in perfectly balanced differential amplifiers. Even with perfect symmetry, however, $A_{cm} \neq 0$ is possible. Therefore, the ratio A_{dm}/A_{cm} is one figure of merit for a differential amplifier, giving the ratio of the desired differential-mode gain to the undesired common-mode gain. In this book, we will define the magnitude of this ratio as the common-mode-rejection ratio, CMRR:

$$\text{CMRR} \equiv \left|\frac{A_{dm}}{A_{cm}}\right| \tag{3.187}$$

Furthermore, since differential amplifiers are not perfectly balanced in practice, $A_{dm-cm} \neq 0$ and $A_{cm-dm} \neq 0$. The ratios A_{dm}/A_{cm-dm} and A_{dm}/A_{dm-cm} are two other figures of merit that characterize the performance of differential amplifiers. Of these, the first is particularly important because ratio A_{dm}/A_{cm-dm} determines the extent to which the differential output is produced by the desired differential input instead of by the undesired common-mode input. This ratio is important because once a common-mode input is converted to a differential output, the result is treated as the *desired* signal by subsequent differential amplifiers. In fact, in multistage differential amplifiers, the common-mode-to-differential-mode gain of the first stage is usually an important factor in the overall CMRR. In Section 3.5.5, we consider perfectly balanced differential amplifiers from a small-signal standpoint; in Section 3.5.6.9, imperfectly balanced differential amplifiers from the same standpoint.

3.5.5 Small-Signal Characteristics of Balanced Differential Amplifiers

In this section, we will study perfectly balanced differential amplifiers. Therefore, $A_{cm-dm} = 0$ and $A_{dm-cm} = 0$ here, and our goal is to calculate A_{dm} and A_{cm}. Although calculating A_{dm} and A_{cm} from the entire small-signal equivalent circuit of a differential amplifier is possible, these calculations are greatly simplified by taking advantage of the symmetry that exists in perfectly balanced amplifiers. In general, we first find the response of a given circuit to pure differential and pure common-mode inputs separately. Then the results can be superposed to find the total solution. Since superposition is valid only for linear circuits, the following analysis is strictly valid only from a small-signal standpoint and approximately valid only for signals that cause negligible nonlinearity. In previous sections, we carried out large-signal analyses of differential pairs and assumed that the Norton-equivalent resistance of the tail current source was infinite. Since this resistance has a considerable effect on the small-signal behavior of differential pairs, however, we now assume that this resistance is finite.

Because the analysis here is virtually the same for both bipolar and MOS differential pairs, the two cases will be considered together. Consider the bipolar emitter-coupled pair of Fig. 3.45 and the MOS source-coupled pair of Fig. 3.50 from a small-signal standpoint. Then $V_{i1} = v_{i1}$ and $V_{i2} = v_{i2}$. These circuits are redrawn in Fig. 3.53a and Fig. 3.53b with the common-mode input voltages set to zero so we can consider the effect of the differential-mode input by itself. The small-signal equivalent circuit for both cases is shown in Fig. 3.54 with R used to replace R_C in Fig. 3.53a and R_D in 3.53b. Note that the small-signal equivalent circuit neglects finite r_o in both cases. Also, in the MOS case, nonzero g_{mb} is ignored and $r_\pi \to \infty$ because $\beta_0 \to \infty$.

Because the circuit in Fig. 3.54 is perfectly balanced, and because the inputs are driven by equal and opposite voltages, the voltage across R_{TAIL} does not vary at all. Another way to see this result is to view the two lower parts of the circuit as voltage followers. When one side pulls up, the other side pulls down, resulting in a constant voltage across the tail current source by superposition. Since the voltage across R_{TAIL} experiences no variation, the behavior of the small-signal circuit is unaffected by the placement of a short circuit across R_{TAIL}, as shown in Fig. 3.55. After placing this short circuit, we see that the two sides of the circuit are not only identical, but also independent because they are joined at a node that operates as a small-signal ground. Therefore, the response to small-signal differential inputs can be determined by analyzing one side of the original circuit with R_{TAIL} replaced by a short circuit.

224   Chapter 3 ■ Single-Transistor and Multiple-Transistor Amplifiers

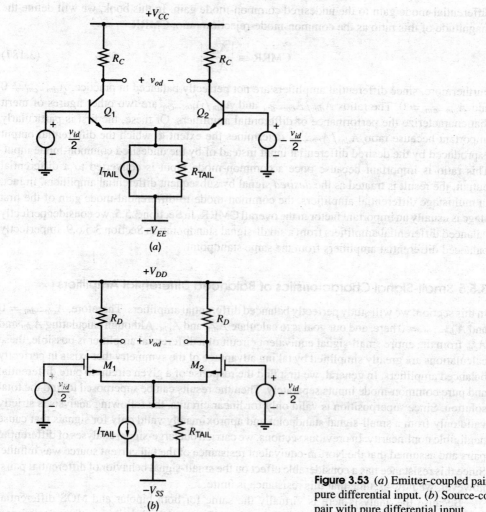

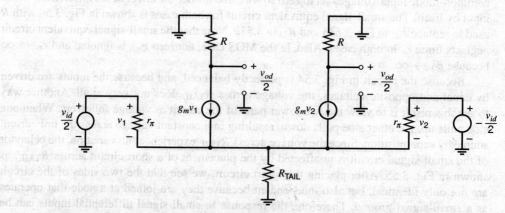

Figure 3.53 (*a*) Emitter-coupled pair with pure differential input. (*b*) Source-coupled pair with pure differential input.

Figure 3.54 Small-signal equivalent circuit for differential pair with pure differential-mode input.

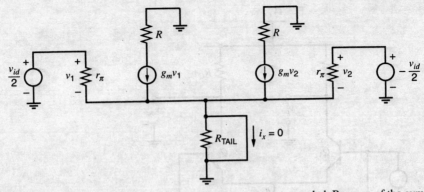

Figure 3.55 Differential-mode circuit with the tail current source grounded. Because of the symmetry of the circuit, $i_x = 0$.

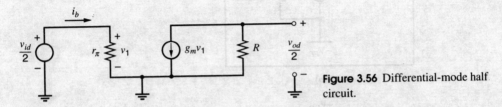

Figure 3.56 Differential-mode half circuit.

This simplified circuit, shown in Fig. 3.56, is called the *differential-mode* half circuit and is useful for analysis of both the low- and high-frequency performance of all types of differential amplifiers. By inspection of Fig. 3.56, we recognize this circuit as the small-signal equivalent of a common-emitter or common-source amplifier. Therefore,

$$\frac{v_{od}}{2} = -g_m R \frac{v_{id}}{2} \tag{3.188}$$

and

$$A_{dm} = \left. \frac{v_{od}}{v_{id}} \right|_{v_{ic}=0} = -g_m R \tag{3.189}$$

To include the output resistance of the transistor in the above analysis, R in (3.189) should be replaced by $R \parallel r_o$. Finally, note that neglecting g_{mb} from this analysis for MOS source-coupled pairs has no effect on the result because the voltage from the source to the body of the input transistors is the same as the voltage across the tail current source, which is constant with a pure differential input.

The circuits in Fig. 3.45 and Fig. 3.50 are now reconsidered from a small-signal, common-mode standpoint. Setting $V_{i1} = V_{i2} = v_{ic}$, the circuits are redrawn in Fig. 3.57a and Fig. 3.57b. The small-signal equivalent circuit is shown in Fig. 3.58, but with the modification that the resistor R_{TAIL} has been split into two parallel resistors, each of value twice the original. Also R has been used to replace R_C in Fig. 3.57a and R_D in 3.57b. Again r_o is neglected in both cases, and g_{mb} is neglected in the MOS case, where $r_\pi \to \infty$ because $\beta_0 \to \infty$.

Because the circuit in Fig. 3.58 is divided into two identical halves, and because each half is driven by the same voltage v_{ic}, no current i_x flows in the lead connecting the half circuits. The circuit behavior is thus unchanged when this lead is removed as shown in Fig. 3.59. As a result, we see that the two halves of the circuit in Fig. 3.58 are not only identical, but also independent because they are joined by a branch that conducts no small-signal current. Therefore, the response to small-signal, common-mode inputs can be determined by analyzing

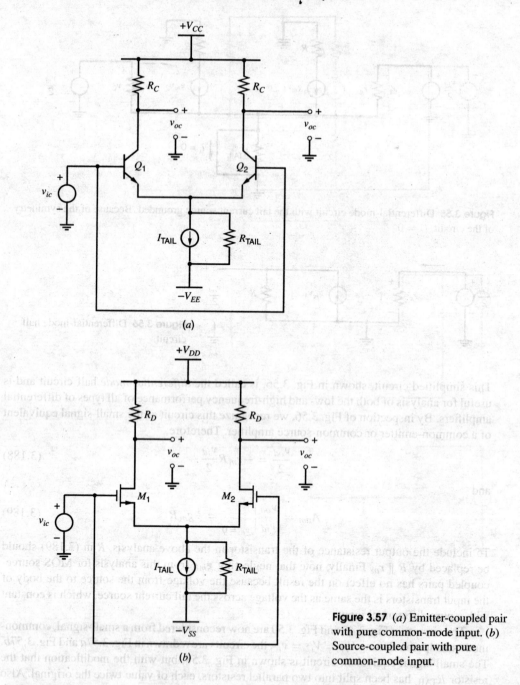

Figure 3.57 (a) Emitter-coupled pair with pure common-mode input. (b) Source-coupled pair with pure common-mode input.

one half of the original circuit with an open circuit replacing the branch that joins the two halves of the original circuit. This simplified circuit, shown in Fig. 3.60, is called the *common-mode half circuit*. By inspection of Fig. 3.60, we recognize this circuit as a common-emitter or common-source amplifier with degeneration. Then

$$v_{oc} = -G_m R v_{ic} \qquad (3.190)$$

3.5 Differential Pairs

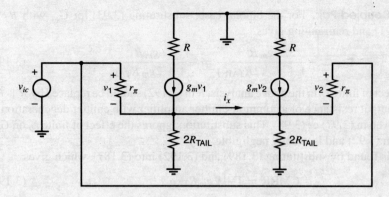

Figure 3.58 Small-signal equivalent circuit, pure common-mode input.

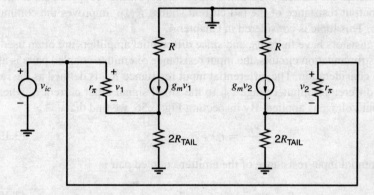

Figure 3.59 Modified common-mode equivalent circuit.

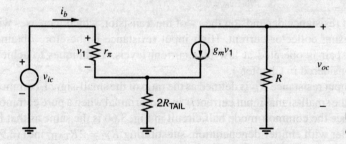

Figure 3.60 Common-mode half circuit.

and

$$A_{cm} = \left.\frac{v_{oc}}{v_{ic}}\right|_{v_{id}=0} = -G_m R \quad (3.191)$$

where G_m is the transconductance of a common-emitter or common-source amplifier with degeneration and will be considered quantitatively below. Since degeneration reduces the transconductance, and since degeneration occurs only in the common-mode case, (3.189) and (3.191) show that $|A_{dm}| > |A_{cm}|$; therefore, the differential pair is more sensitive to differential inputs than to common-mode inputs. In other words, the tail current source provides local negative feedback to common-mode inputs (or local common-mode feedback). Negative feedback is studied in Chapter 8.

Bipolar Emitter-Coupled Pair. For the bipolar case, substituting (3.93) for G_m with $R_E = 2R_{\text{TAIL}}$ into (3.191) and rearranging gives

$$A_{cm} \simeq -\frac{g_m R}{1 + g_m (2R_{\text{TAIL}})} = -\frac{g_m R}{1 + 2g_m R_{\text{TAIL}}} \tag{3.192}$$

To include the effect of finite r_o in the above analysis, R in (3.192) should be replaced by $R \parallel R_o$, where R_o is the output resistance of a common-emitter amplifier with emitter degeneration of $R_E = 2R_{\text{TAIL}}$, given in (3.97) or (3.98). This substitution ignores the effect of finite r_o on G_m, which is shown in (3.92) and is usually negligible.

The CMRR is found by substituting (3.189) and (3.192) into (3.187), which gives

$$\text{CMRR} = 1 + 2g_m R_{\text{TAIL}} \tag{3.193}$$

This expression applies to the particular case of a single-stage, emitter-coupled pair. It shows that increasing the output resistance of the tail current source R_{TAIL} improves the common-mode-rejection ratio. This topic is considered in Chapter 4.

Since bipolar transistors have finite β_0, and since differential amplifiers are often used as the input stage of instrumentation circuits, the input resistance of emitter-coupled pairs is also an important design consideration. The differential input resistance R_{id} is defined as the ratio of the small-signal differential input voltage v_{id} to the small-signal input current i_b when a pure differential input voltage is applied. By inspecting Fig. 3.56, we find that

$$\frac{v_{id}}{2} = i_b r_\pi \tag{3.194}$$

Therefore, the differential input resistance of the emitter-coupled pair is

$$R_{id} = \left.\frac{v_{id}}{i_b}\right|_{v_{ic}=0} = 2r_\pi \tag{3.195}$$

Thus the differential input resistance depends on the r_π of the transistor, which increases with increasing β_0 and decreasing collector current. High input resistance is therefore obtained when an emitter-coupled pair is operated at low bias current levels. Techniques to achieve small bias currents are considered in Chapter 4.

The common-mode input resistance R_{ic} is defined as the ratio of the small-signal, common-mode input voltage v_{ic} to the small-signal input current i_b in one terminal when a pure common-mode input is applied. Since the common-mode half circuit in Fig. 3.60 is the same as that for a common-emitter amplifier with emitter degeneration, substituting $R_E = 2R_{\text{TAIL}}$ into (3.90) gives R_{ic} as

$$R_{ic} = \left.\frac{v_{ic}}{i_b}\right|_{v_{id}=0} = r_\pi + (\beta_0 + 1)(2R_{\text{TAIL}}) \tag{3.196}$$

The small-signal input current that flows when both common-mode and differential-mode input voltages are applied can be found by superposition and is given by

$$i_{b1} = \frac{v_{id}}{R_{id}} + \frac{v_{ic}}{R_{ic}} \tag{3.197}$$

$$i_{b2} = -\frac{v_{id}}{R_{id}} + \frac{v_{ic}}{R_{ic}} \tag{3.198}$$

where i_{b1} and i_{b2} represent the base currents of Q_1 and Q_2, respectively.

The input resistance can be represented by the π equivalent circuit of Fig. 3.61a or by the T-equivalent circuit of Fig. 3.61b. For the π model, the common-mode input resistance is exactly R_{ic} independent of R_x. To make the differential-mode input resistance exactly R_{id}, the

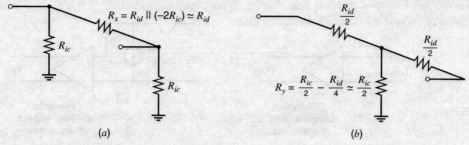

Figure 3.61 (a) General low-frequency, small-signal, π-equivalent input circuit for the differential amplifier. (b) T-equivalent input circuit.

value of R_x should be more than R_{id} to account for nonzero current in R_{ic}. On the other hand, for the T model, the differential-mode input resistance is exactly R_{id} independent of R_y, and the common-mode input resistance is R_{ic} if R_y is chosen to be less than $R_{ic}/2$ as shown. The approximations in Fig. 3.61 are valid if R_{ic} is much larger than R_{id}.

MOS Source-Coupled Pair. For the MOS case, substituting (3.104) for G_m with $g_{mb} = 0$ and $R_S = 2R_{\text{TAIL}}$ into (3.191) and rearranging gives

$$A_{cm} \simeq -\frac{g_m R}{1 + g_m (2R_{\text{TAIL}})} = -\frac{g_m R}{1 + 2g_m R_{\text{TAIL}}} \qquad (3.199)$$

Although (3.199) and the common-mode half circuit in Fig. 3.60 ignore the body-effect transconductance g_{mb}, the common-mode gain depends on g_{mb} in practice because the body effect changes the source-body voltage of the transistors in the differential pair. Since nonzero g_{mb} was included in the derivation of the transconductance of the common-source amplifier with degeneration, a simple way to include the body effect here is to allow nonzero g_{mb} when substituting (3.104) into (3.191). The result is

$$A_{cm} \simeq -\frac{g_m R}{1 + (g_m + g_{mb})(2R_{\text{TAIL}})} = -\frac{g_m R}{1 + 2(g_m + g_{mb})R_{\text{TAIL}}} \qquad (3.200)$$

To include the effect of finite r_o in the above analysis, R in (3.199) and (3.200) should be replaced by $R \parallel R_o$, where R_o is the output resistance of a common-source amplifier with source degeneration of $R_S = 2R_{\text{TAIL}}$, given in (3.107). This substitution ignores the effect of finite r_o on G_m, which is shown in (3.103) and is usually negligible.

The CMRR is found by substituting (3.189) and (3.200) into (3.187), which gives

$$\text{CMRR} \simeq 1 + 2(g_m + g_{mb})R_{\text{TAIL}} \qquad (3.201)$$

Equation 3.201 is valid for a single-stage, source-coupled pair and shows that increasing R_{TAIL} increases the CMRR. This topic is studied in Chapter 4.

3.5.6 Device Mismatch Effects in Differential Amplifiers

An important aspect of the performance of differential amplifiers is the minimum dc and ac differential voltages that can be detected. The presence of component mismatches within the amplifier itself and drifts of component values with temperature produce dc differential voltages at the output that are indistinguishable from the dc component of the signal being amplified. Also, such mismatches and drifts cause nonzero common-mode-to-differential-mode gain as well as nonzero differential-to-common-mode gain to arise. Nonzero A_{cm-dm} is especially important because it converts common-mode inputs to differential outputs, which

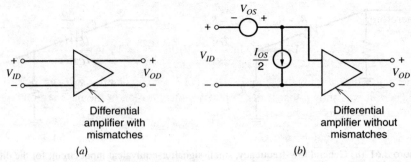

Figure 3.62 Equivalent input offset voltage (V_{OS}) and current (I_{OS}) for a differential amplifier. (*a*) Actual circuit containing mismatches. (*b*) Equivalent dc circuit with identically matched devices and the offset voltage and current referred to the input.

are treated as the desired signal by subsequent stages. In many analog systems, these types of errors pose the basic limitation on the resolution of the system, and hence consideration of mismatch-induced effects is often central to the design of analog circuits.

3.5.6.1 Input Offset Voltage and Current

For differential amplifiers, the effect of mismatches on dc performance is most conveniently represented by two quantities, the input offset voltage and the input offset current. These quantities represent the input-referred effect of all the component mismatches within the amplifier on its dc performance.[11,12] As illustrated in Fig. 3.62, the dc behavior of the amplifier containing the mismatches is identical to an ideal amplifier with no mismatches but with the input offset voltage source added in series with the input and the input offset current source in shunt across the input terminals. Both quantities are required to represent the effect of mismatch in general so that the model is valid for any source resistance. For example, if the input terminals are driven by an ideal voltage source with zero resistance, the input offset current does not contribute to the amplifier output, and the offset voltage generator is needed to model the effect of mismatch. On the other hand, if the input terminals are driven by an ideal current source with infinite resistance, the input offset voltage does not contribute to the amplifier output, and the offset current generator is needed to model the effect of mismatch. These quantities are usually a function of both temperature and common-mode input voltage. In the next several sections, we calculate the input offset voltage and current of the emitter-coupled pair and the source-coupled pair.

3.5.6.2 Input Offset Voltage of the Emitter-Coupled Pair

The predominant sources of offset error in the emitter-coupled pair are the mismatches in the base width, base doping level, and collector doping level of the transistors, mismatches in the effective emitter area of the transistors, and mismatches in the collector load resistors. To provide analytical results simple enough for intuitive interpretation, the analysis will be carried out assuming a uniform-base transistor. The results are similar for the nonuniform case, although the analytical procedure is more tedious. In most instances the dc base current is low enough that the dc voltage drop in r_b is negligible, so we neglect r_b.

Consider Fig. 3.45 with dc signals so that $V_{i1} = V_{I1}$, $V_{i2} = V_{I2}$, $V_{o1} = V_{O1}$, and $V_{o2} = V_{O2}$. Let $V_{ID} = V_{I1} - V_{I2}$. Also, assume that the collector resistors may not be identical. Let R_{C1} and R_{C2} represent the values of the resistors attached to Q_1 and Q_2, respectively. From KVL around the input loop,

$$V_{ID} - V_{BE1} + V_{BE2} = 0 \qquad (3.202)$$

Therefore,

$$V_{ID} = V_T \ln \frac{I_{C1}}{I_{S1}} - V_T \ln \frac{I_{C2}}{I_{S2}} = V_T \ln \frac{I_{C1}}{I_{C2}} \frac{I_{S2}}{I_{S1}} \qquad (3.203)$$

The factors determining the saturation current I_S of a bipolar transistor are described in Chapter 1. There it was shown that if the impurity concentration in the base region is uniform, these saturation currents can be written

$$I_{S1} = \frac{qn_i^2 \overline{D_n}}{N_A W_{B1}(V_{CB})} A_1 = \frac{qn_i^2 \overline{D_n}}{Q_{B1}(V_{CB})} A_1 \qquad (3.204)$$

$$I_{S2} = \frac{qn_i^2 \overline{D_n}}{N_A W_{B2}(V_{CB})} A_2 = \frac{qn_i^2 \overline{D_n}}{Q_{B2}(V_{CB})} A_2 \qquad (3.205)$$

where $W_B(V_{CB})$ is the base width as a function of V_{CB}, N_A is the acceptor density in the base, and A is the emitter area. We denote the product $N_A W_B(V_{CB})$ as $Q_B(V_{CB})$, the total base impurity doping per unit area.

The input offset voltage V_{OS} is equal to the value of $V_{ID} = V_{I1} - V_{I2}$ that must be applied to the input to drive the differential output voltage $V_{OD} = V_{O1} - V_{O2}$ to zero. For V_{OD} to be zero, $I_{C1} R_{C1} = I_{C2} R_{C2}$; therefore,

$$\frac{I_{C1}}{I_{C2}} = \frac{R_{C2}}{R_{C1}} \qquad (3.206)$$

Substituting (3.204), (3.205), and (3.206) into (3.203) gives

$$V_{OS} = V_T \ln \left[\left(\frac{R_{C2}}{R_{C1}} \right) \left(\frac{A_2}{A_1} \right) \left(\frac{Q_{B1}(V_{CB})}{Q_{B2}(V_{CB})} \right) \right] \qquad (3.207)$$

This expression relates the input offset voltage to the device parameters and R_C mismatch. Usually, however, the argument of the log function is very close to unity and the equation can be interpreted in a more intuitively satisfying way. In the following section we perform an approximate analysis, valid if the mismatches are small.

3.5.6.3 Offset Voltage of the Emitter-Coupled Pair: Approximate Analysis

In cases of practical interest involving offset voltages and currents, the mismatch between any two nominally matched circuit parameters is usually small compared with the absolute value of that parameter. This observation leads to a procedure by which the individual contributions to offset voltage can be considered separately and summed.

First, define new parameters to describe the mismatch in the components, using the relations

$$\Delta X = X_1 - X_2 \qquad (3.208)$$

$$X = \frac{X_1 + X_2}{2} \qquad (3.209)$$

Thus ΔX is the difference between two parameters, and X is the average of the two nominally matched parameters. Note that ΔX can be positive or negative. Next invert (3.208) and (3.209) to give

$$X_1 = X + \frac{\Delta X}{2} \qquad (3.210)$$

$$X_2 = X - \frac{\Delta X}{2} \qquad (3.211)$$

These relations can be applied to the collector resistances, the emitter areas, and the base doping parameters in (3.207) to give

$$V_{OS} = V_T \ln\left[\left(\frac{R_C - \frac{\Delta R_C}{2}}{R_C + \frac{\Delta R_C}{2}}\right)\left(\frac{A - \frac{\Delta A}{2}}{A + \frac{\Delta A}{2}}\right)\left(\frac{Q_B + \frac{\Delta Q_B}{2}}{Q_B - \frac{\Delta Q_B}{2}}\right)\right] \quad (3.212)$$

With the assumptions that $\Delta R_C \ll R_C$, $\Delta A \ll A$, and $\Delta Q_B \ll Q_B$, (3.212) can be simplified to

$$V_{OS} \simeq V_T \ln\left[\left(1 - \frac{\Delta R_C}{R_C}\right)\left(1 - \frac{\Delta A}{A}\right)\left(1 + \frac{\Delta Q_B}{Q_B}\right)\right]$$

$$\simeq V_T \left[\ln\left(1 - \frac{\Delta R_C}{R_C}\right) + \ln\left(1 - \frac{\Delta A}{A}\right) + \ln\left(1 + \frac{\Delta Q_B}{Q_B}\right)\right] \quad (3.213)$$

If $x \ll 1$, a Taylor series can be used to show that

$$\ln(1 + x) = x - \frac{x^2}{2} + \frac{x^3}{3} - \cdots \quad (3.214)$$

Applying (3.214) to each logarithm in (3.213) and ignoring terms higher than first order in the expansions gives

$$V_{OS} \simeq V_T \left(-\frac{\Delta R_C}{R_C} - \frac{\Delta A}{A} + \frac{\Delta Q_B}{Q_B}\right) \quad (3.215)$$

Thus, under the assumptions made, we have obtained an approximate expression for the input offset voltage, which is the linear superposition of the effects of the different components. It can be shown that this can always be done for small component mismatches. Note that the signs of the individual terms of (3.215) are not particularly significant, since the mismatch factors can be positive or negative depending on the direction of the random parameter variation. The worst-case offset occurs when the terms have signs such that the individual contributions add.

Equation 3.215 relates the offset voltage to mismatches in the resistors and in the structural parameters A and Q_B of the transistors. For the purpose of predicting the offset voltage from device parameters that are directly measurable electrically, we rewrite (3.215) to express the offset in terms of the resistor mismatch and the mismatch in the saturation currents of the transistors:

$$V_{OS} \simeq V_T \left(-\frac{\Delta R_C}{R_C} - \frac{\Delta I_S}{I_S}\right) \quad (3.216)$$

where

$$\frac{\Delta I_S}{I_S} = \frac{\Delta A}{A} - \frac{\Delta Q_B}{Q_B} \quad (3.217)$$

is the offset voltage contribution from the transistors themselves, as reflected in the mismatch in saturation current. Mismatch factors $\Delta R_C/R_C$ and $\Delta I_S/I_S$ are actually random parameters that take on a different value for each circuit fabricated, and the distribution of the observed values is described by a probability distribution. For large samples the distribution tends toward a normal, or Gaussian, distribution with zero mean. Typically observed standard deviations for the preceding mismatch parameters for small-area diffused devices are

$$\sigma_{\Delta R/R} = 0.01 \quad \sigma_{\Delta I_S/I_S} = 0.05 \quad (3.218)$$

In the Gaussian distribution, 68 percent of the samples have a value within $\pm \sigma$ of the mean value. If we assume that the mean value of the distribution is zero, then 68 percent of the resistor pairs in a large sample will match within 1 percent, and 68 percent of the transistor pairs will have saturation currents that match within 5 percent for the distributions described by (3.218). These values can be heavily influenced by device geometry and processing. If we pick one sample from each distribution so that the parameter mismatch is equal to the corresponding standard deviation, and if the mismatch factors are chosen in the direction so that they add, the resulting offset from (3.216) would be

$$V_{OS} \simeq (26\text{mV})(0.01 + 0.05) \simeq 1.5\text{mV} \tag{3.219}$$

Large ion-implanted devices with careful layout can achieve $V_{OS} \simeq 0.1$ mV. A parameter of more interest to the circuit designer than the offset of one sample is the standard deviation of the total offset voltage. Since the offset is the sum of two uncorrelated random parameters, the standard deviation of the sum is equal to the square root of the sum of the squares of the standard deviation of the two mismatch contributions, or

$$\sigma_{V_{OS}} = V_T \sqrt{(\sigma_{\Delta R/R})^2 + (\sigma_{\Delta I_S/I_S})^2} \tag{3.220}$$

The properties of the Gaussian distribution are summarized in Appendix A.3.1.

3.5.6.4 Offset Voltage Drift in the Emitter-Coupled Pair

When emitter-coupled pairs are used as low-level dc amplifiers where the offset voltage is critical, provision is sometimes made to manually adjust the input offset voltage to zero with an external potentiometer. When this adjustment is done, the important parameter becomes not the offset voltage itself, but the variation of this offset voltage with temperature, often referred to as *drift*. For most practical circuits, the sensitivity of the input offset voltage to temperature is not zero, and the wider the excursion of temperature experienced by the circuit, the more error the offset voltage drift will contribute. This parameter is easily calculated for the emitter-coupled pair by differentiating (3.207) as follows

$$\frac{dV_{OS}}{dT} = \frac{V_{OS}}{T} \tag{3.221}$$

using $V_T = kT/q$ and assuming the ratios in (3.207) are independent of temperature. Thus the drift and offset are proportional for the emitter-coupled pair. This relationship is observed experimentally. For example, an emitter-coupled pair with a measured offset voltage of 2 mV would display a drift of 2mV/300°K or 6.6μV/°C under the assumptions we have made.

Equation 3.221 appears to show that the drift also would be nulled by externally adjusting the offset to zero. This observation is only approximately true because of the way in which the nulling is accomplished.[13] Usually an external potentiometer is placed in parallel with a portion of one of the collector load resistors in the pair. The temperature coefficient of the nulling potentiometer generally does not match that of the diffused resistors, so a resistor-mismatch temperature coefficient is introduced that can make the drift worse than it was without nulling. Voltage drifts in the 1μV/°C range can be obtained with careful design.

3.5.6.5 Input Offset Current of the Emitter-Coupled Pair

The input offset current I_{OS} is measured with the inputs connected only to current sources and is the difference in the base currents that must be applied to drive the differential output voltage $V_{OD} = V_{O1} - V_{O2}$ to zero. Since the base current of each transistor is equal to the corresponding collector current divided by beta, the offset current is

$$I_{OS} = \frac{I_{C1}}{\beta_{F1}} - \frac{I_{C2}}{\beta_{F2}} \tag{3.222}$$

when $V_{OD} = 0$. As before, we can write

$$I_{C1} = I_C + \frac{\Delta I_C}{2} \qquad I_{C2} = I_C - \frac{\Delta I_C}{2} \qquad (3.223)$$

$$\beta_{F1} = \beta_F + \frac{\Delta \beta_F}{2} \qquad \beta_{F2} = \beta_F - \frac{\Delta \beta_F}{2} \qquad (3.224)$$

Inserting (3.223) and (3.224) into (3.222), the offset current becomes

$$I_{OS} = \left(\frac{I_C + \frac{\Delta I_C}{2}}{\beta_F + \frac{\Delta \beta_F}{2}} - \frac{I_C - \frac{\Delta I_C}{2}}{\beta_F - \frac{\Delta \beta_F}{2}} \right) \qquad (3.225)$$

Neglecting higher-order terms, this becomes

$$I_{OS} \simeq \frac{I_C}{\beta_F} \left(\frac{\Delta I_C}{I_C} - \frac{\Delta \beta_F}{\beta_F} \right) \qquad (3.226)$$

For V_{OD} to be zero, $I_{C1} R_{C1} = I_{C2} R_{C2}$; therefore, from (3.206), the mismatch in collector currents is

$$\frac{\Delta I_C}{I_C} = -\frac{\Delta R_C}{R_C} \qquad (3.227)$$

Equation 3.227 shows that the fractional mismatch in the collector currents must be equal in magnitude and opposite in polarity from the fractional mismatch in the collector resistors to force $V_{OD} = 0$. Substituting (3.227) into (3.226) gives

$$I_{OS} \simeq -\frac{I_C}{\beta_F} \left(\frac{\Delta R_C}{R_C} + \frac{\Delta \beta_F}{\beta_F} \right) \qquad (3.228)$$

A typically observed beta mismatch distribution displays a deviation of about 10 percent. Assuming a beta mismatch of 10 percent and a mismatch in collector resistors of 1 percent, we obtain

$$I_{OS} \simeq -\frac{I_C}{\beta_F} \left(\frac{\Delta R_C}{R_C} + \frac{\Delta \beta_F}{\beta_F} \right) = -\frac{I_C}{\beta_F}(0.11) = -0.11 \, (I_B) \qquad (3.229)$$

In many applications, the input offset current as well as the input current itself must be minimized. A good example is the input stage of operational amplifiers. Various circuit and technological approaches to reduce these currents are considered in Chapter 6.

3.5.6.6 Input Offset Voltage of the Source-Coupled Pair

As mentioned earlier in the chapter, MOS transistors inherently provide higher input resistance and lower input bias current than bipolar transistors when the MOS gate is used as the input. This observation also applies to differential-pair amplifiers. The input offset current of an MOS differential pair is the difference between the two gate currents and is essentially zero because the gates of the input transistors are connected to silicon dioxide, which is an insulator. However, MOS transistors exhibit lower transconductance than bipolar transistors at the same current, resulting in poorer input offset voltage and common-mode rejection ratio in MOS differential pairs than in the case of bipolar transistors. In this section we calculate the input offset voltage of the source-coupled MOSFET pair.

Consider Fig. 3.50 with dc signals so that $V_{i1} = V_{I1}$, $V_{i2} = V_{I2}$, $V_{o1} = V_{O1}$, and $V_{o2} = V_{O2}$. Let $V_{ID} = V_{I1} - V_{I2}$. Also, assume that the drain resistors may not be identical. Let R_{D1} and R_{D2} represent the values of the resistors attached to M_1 and M_2, respectively. KVL around the input loop gives

$$V_{ID} - V_{GS1} + V_{GS2} = 0 \qquad (3.230)$$

Solving (1.157) for the gate-source voltage and substituting into (3.230) gives

$$V_{ID} = V_{GS1} - V_{GS2}$$

$$= V_{t1} + \sqrt{\frac{2I_{D1}}{k'(W/L)_1}} - V_{t2} - \sqrt{\frac{2I_{D2}}{k'(W/L)_2}} \qquad (3.231)$$

As in the bipolar case, the input offset voltage V_{OS} is equal to the value of $V_{ID} = V_{I1} - V_{I2}$ that must be applied to the input to drive the differential output voltage $V_{OD} = V_{O1} - V_{O2}$ to zero. For V_{OD} to be zero, $I_{D1}R_{D1} = I_{D2}R_{D2}$; therefore,

$$V_{OS} = V_{t1} - V_{t2} + \sqrt{\frac{2I_{D1}}{k'(W/L)_1}} - \sqrt{\frac{2I_{D2}}{k'(W/L)_2}} \qquad (3.232)$$

subject to the constraint that $I_{D1}R_{D1} = I_{D2}R_{D2}$.

3.5.6.7 Offset Voltage of the Source-Coupled Pair: Approximate Analysis

The mismatch between any two nominally matched circuit parameters is usually small compared with the absolute value of that parameter in practice. As a result, (3.232) can be rewritten in a way that allows us to understand the contributions of each mismatch to the overall offset.

Defining difference and average quantities in the usual way, we have

$$\Delta I_D = I_{D1} - I_{D2} \qquad (3.233)$$

$$I_D = \frac{I_{D1} + I_{D2}}{2} \qquad (3.234)$$

$$\Delta(W/L) = (W/L)_1 - (W/L)_2 \qquad (3.235)$$

$$(W/L) = \frac{(W/L)_1 + (W/L)_2}{2} \qquad (3.236)$$

$$\Delta V_t = V_{t1} - V_{t2} \qquad (3.237)$$

$$V_t = \frac{V_{t1} + V_{t2}}{2} \qquad (3.238)$$

$$\Delta R_L = R_{L1} - R_{L2} \qquad (3.239)$$

$$R_L = \frac{R_{L1} + R_{L2}}{2} \qquad (3.240)$$

Rearranging (3.233) and (3.234) as well as (3.235) and (3.236) gives

$$I_{D1} = I_D + \frac{\Delta I_D}{2} \qquad I_{D2} = I_D - \frac{\Delta I_D}{2} \qquad (3.241)$$

$$(W/L)_1 = (W/L) + \frac{\Delta(W/L)}{2} \qquad (W/L)_2 = (W/L) - \frac{\Delta(W/L)}{2} \qquad (3.242)$$

Substituting (3.237), (3.241), and (3.242) into (3.232) gives

$$V_{OS} = \Delta V_t + \sqrt{\frac{2(I_D + \Delta I_D/2)}{k'[(W/L) + \Delta(W/L)/2]}} - \sqrt{\frac{2(I_D - \Delta I_D/2)}{k'[(W/L) - \Delta(W/L)/2]}} \qquad (3.243)$$

Rearranging (3.243) gives

$$V_{OS} = \Delta V_t + (V_{GS} - V_t)\left(\sqrt{\frac{1 + \Delta I_D/2I_D}{1 + \frac{\Delta(W/L)}{2(W/L)}}} - \sqrt{\frac{1 - \Delta I_D/2I_D}{1 - \frac{\Delta(W/L)}{2(W/L)}}}\right) \qquad (3.244)$$

If the mismatch terms are small, the argument of each square root in (3.244) is approximately unity. Using $\sqrt{x} \simeq (1+x)/2$ when $x \simeq 1$ for the argument of each square root in (3.244), we have

$$V_{OS} \simeq \Delta V_t + \frac{(V_{GS}-V_t)}{2} \left(\frac{1+\Delta I_D/2I_D}{1+\frac{\Delta(W/L)}{2(W/L)}} - \frac{1-\Delta I_D/2I_D}{1-\frac{\Delta(W/L)}{2(W/L)}} \right) \quad (3.245)$$

Carrying out the long divisions in (3.245) and ignoring terms higher than first order gives

$$V_{OS} \simeq \Delta V_t + \frac{(V_{GS}-V_t)}{2} \left(\frac{\Delta I_D}{I_D} - \frac{\Delta(W/L)}{(W/L)} \right) \quad (3.246)$$

When the differential input voltage is V_{OS}, the differential output voltage is zero; therefore, $I_{D1}R_{L1} = I_{D2}R_{L2}$, and

$$\frac{\Delta I_D}{I_D} = -\frac{\Delta R_L}{R_L} \quad (3.247)$$

In other words, the mismatch in the drain currents must be opposite of the mismatch of the load resistors to set $V_{OD} = 0$. Substituting (3.247) into (3.246) gives

$$V_{OS} = \Delta V_t + \frac{(V_{GS}-V_t)}{2} \left(-\frac{\Delta R_L}{R_L} - \frac{\Delta(W/L)}{(W/L)} \right) \quad (3.248)$$

The first term on the right side of (3.248) stems from threshold mismatch. This mismatch component is present in MOS devices but not in bipolar transistors. This component results in a constant offset component that is bias-current independent. Threshold mismatch is a strong function of process cleanliness and uniformity and can be substantially improved by the use of careful layout. Measurements indicate that large-geometry structures are capable of achieving threshold-mismatch distributions with standard deviations on the order of 2 mV in a modern silicon-gate MOS process. This offset component alone limits the minimum offset in the MOS case and is an order of magnitude larger than the total differential-pair offset in modern ion-implanted bipolar technologies.

The second term on the right side of (3.248) shows that another component of the offset scales with the overdrive $V_{ov} = (V_{GS} - V_t)$ and is related to a mismatch in the load elements or in the device W/L ratio. In the bipolar emitter-coupled pair offset, the corresponding mismatch terms were multiplied by V_T, typically a smaller number than $V_{ov}/2$. Thus source-coupled pairs of MOS transistors display higher input offset voltage than bipolar pairs for the same level of geometric mismatch or process gradient even when threshold mismatch is ignored. The key reason for this limitation is that the ratio of the transconductance to the bias current is much lower with MOS transistors than in the bipolar case. The quantities V_T in (3.216) and $(V_{GS} - V_t)/2 = V_{ov}/2$ in (3.248) are both equal to I_{BIAS}/g_m for the devices in question. This quantity is typically in the range 50 mV to 500 mV for MOS transistors instead of 26 mV for bipolar transistors.

3.5.6.8 Offset Voltage Drift in the Source-Coupled Pair

Offset voltage drift in MOSFET source-coupled pairs does not show the high correlation with offset voltage observed in bipolar pairs. The offset consists of several terms that have different temperature coefficients. Both V_t and V_{ov} have a strong temperature dependence, affecting V_{GS} in opposite directions. The temperature dependence of V_{ov} stems primarily from the mobility variation, which gives a negative temperature coefficient to the drain current, while the threshold voltage depends on the Fermi potential. As shown in Section 1.5.4, the latter decreases with temperature and contributes a positive temperature coefficient to the drain current. The drift due to the ΔV_t term in V_{OS} may be quite large if this term itself is large. These two effects can be made to cancel at one value of I_D, which is a useful phenomenon for temperature-stable biasing of single-ended amplifiers. In differential amplifiers, however, this phenomenon is not greatly useful because differential configurations already give first-order cancellation of V_{GS} temperature variations.

3.5.6.9 Small-Signal Characteristics of Unbalanced Differential Amplifiers[11]

As mentioned in Section 3.5.4, the common-mode-to-differential-mode gain and differential-mode-to-common-mode gain of unbalanced differential amplifiers are nonzero. The direct approach to calculation of these cross-gain terms requires analysis of the entire small-signal diagram. In perfectly balanced differential amplifiers, the cross-gain terms are zero, and the differential-mode and common-mode gains can be found by using two *independent* half circuits, as shown in Section 3.5.5. With imperfect matching, exact half-circuit analysis is still possible if the half circuits are *coupled* instead of independent. Furthermore, if the mismatches are small, a modified version of half-circuit analysis gives results that are approximately valid. This modified half-circuit analysis not only greatly simplifies the required calculations, but also gives insight about how to reduce A_{cm-dm} and A_{dm-cm} in practice.

First consider a pair of mismatched resistors R_1 and R_2 shown in Fig. 3.63. Assume that the branch currents are i_1 and i_2, respectively. From Ohm's law, the differential and common-mode voltages across the resistors can be written as

$$v_d = v_1 - v_2 = i_1 R_1 - i_2 R_2 \tag{3.249}$$

and

$$v_c = \frac{v_1 + v_2}{2} = \frac{i_1 R_1 + i_2 R_2}{2} \tag{3.250}$$

Define $i_d = i_1 - i_2$, $i_c = (i_1 + i_2)/2$, $\Delta R = R_1 - R_2$, and $R = (R_1 + R_2)/2$. Then (3.249) and (3.250) can be rewritten as

$$v_d = \left(i_c + \frac{i_d}{2}\right)\left(R + \frac{\Delta R}{2}\right) - \left(i_c - \frac{i_d}{2}\right)\left(R - \frac{\Delta R}{2}\right) = i_d R + i_c (\Delta R) \tag{3.251}$$

and

$$v_c = \frac{\left(i_c + \frac{i_d}{2}\right)\left(R + \frac{\Delta R}{2}\right) + \left(i_c - \frac{i_d}{2}\right)\left(R - \frac{\Delta R}{2}\right)}{2} = i_c R + \frac{i_d (\Delta R)}{4} \tag{3.252}$$

These equations can be used to draw differential and common-mode half circuits for the pair of mismatched resistors. Since the differential half circuit should give half the differential voltage dropped across the resistors, the two terms on the right-hand side of (3.251) are each divided by two and used to represent one component of a branch voltage of $v_d/2$. The differential half circuit is shown in Fig. 3.64a. The first component of the branch voltage is the voltage

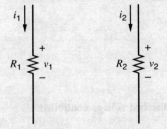

Figure 3.63 A pair of mismatched resistors.

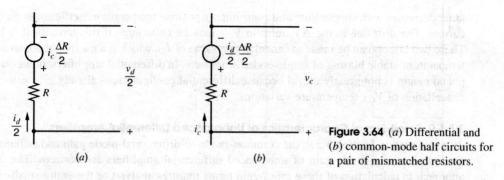

Figure 3.64 (a) Differential and (b) common-mode half circuits for a pair of mismatched resistors.

dropped across R and is half the differential current times the average resistor value. The second component is the voltage across the dependent voltage source controlled by the current flowing in the common-mode half circuit and is proportional to half the mismatch in the resistor values. The common-mode half circuit is constructed from (3.252) and is shown in Fig. 3.64b. Here the total branch voltage v_c is the sum of the voltages across a resistor and a dependent voltage source controlled by the current flowing in the differential half circuit. In the limiting case where $\Delta R = 0$, the voltage across each dependent source in Fig. 3.64 is zero, and each half circuit collapses to simply a resistor of value R. Therefore, the half circuits are independent in this case, as expected. In practice, however, $\Delta R \neq 0$, and Fig. 3.64 shows that the differential voltage depends not only on the differential current, but also on the common-mode current. Similarly, the common-mode voltage depends in part on the differential current. Thus the behavior of a pair of mismatched resistors can be represented exactly by using coupled half circuits.

Next consider a pair of mismatched voltage-controlled current sources as shown in Fig. 3.65. Assume that the control voltages are v_1 and v_2, respectively. Then the differential and common-mode currents can be written as

$$i_d = i_1 - i_2 = g_{m1}v_1 - g_{m2}v_2$$
$$= \left(g_m + \frac{\Delta g_m}{2}\right)\left(v_c + \frac{v_d}{2}\right) - \left(g_m - \frac{\Delta g_m}{2}\right)\left(v_c - \frac{v_d}{2}\right)$$
$$= g_m v_d + \Delta g_m v_c \tag{3.253}$$

and

$$i_c = \frac{i_1 + i_2}{2} = \frac{g_{m1}v_1 + g_{m2}v_2}{2}$$
$$= \frac{\left(g_m + \frac{\Delta g_m}{2}\right)\left(v_c + \frac{v_d}{2}\right) + \left(g_m - \frac{\Delta g_m}{2}\right)\left(v_c - \frac{v_d}{2}\right)}{2}$$
$$= g_m v_c + \frac{\Delta g_m v_d}{4} \tag{3.254}$$

where $v_d = v_1 - v_2$, $v_c = (v_1 + v_2)/2$, $\Delta g_m = g_{m1} - g_{m2}$, and $g_m = (g_{m1} + g_{m2})/2$.

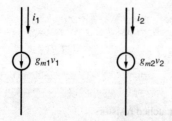

Figure 3.65 A pair of mismatched voltage-controlled current sources.

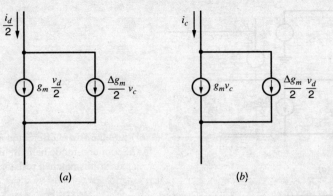

Figure 3.66 (a) Differential and (b) common-mode half circuits for a pair of mismatched voltage-controlled current sources.

The corresponding differential and common-mode half circuits each use two voltage-controlled current sources, as shown in Fig. 3.66. In each case, one dependent source is proportional to the average transconductance and the other to half the mismatch in the transconductances. With perfect matching, the mismatch terms are zero, and the two half circuits are independent. With imperfect matching, however, the mismatch terms are nonzero. In the differential half circuit, the mismatch current source is controlled by the common-mode control voltage. In the common-mode half circuit, the mismatch current source is controlled by half the differential control voltage. Thus, as for mismatched resistors, the behavior of a pair of mismatched voltage-controlled current sources can be represented exactly by using coupled half circuits.

With these concepts in mind, construction of the differential and common-mode half circuits of unbalanced differential amplifiers is straightforward. In the differential half circuit, mismatched resistors are replaced by the circuit shown in Fig. 3.64a, and mismatched voltage-controlled current sources are replaced by the circuit in Fig. 3.66a. Similarly, the circuits shown in Fig. 3.64b and Fig. 3.66b replace mismatched resistors and voltage-controlled current sources in the common-mode half circuit. Although mismatches change the differential and common-mode components of signals that appear at various points in the complete unbalanced amplifier, the differential components are still equal and opposite while the common-mode components are identical by definition. Therefore, small-signal short and open circuits induced by the differential and common-mode signals are unaffected by these replacements.

For example, the differential and common-mode half circuits of the unbalanced differential amplifier shown in Fig. 3.67 are shown in Fig. 3.68. KCL at the output of the differential half

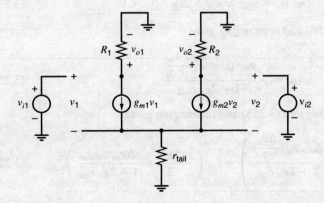

Figure 3.67 The small-signal diagram of an unbalanced differential amplifier.

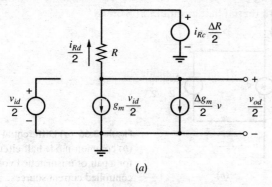

(a)

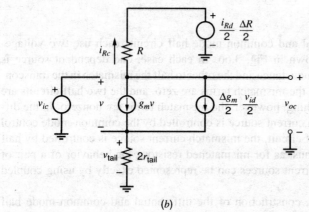

(b)

Figure 3.68 (a) Differential and (b) common-mode half circuits of the differential amplifier shown in Fig. 3.67.

circuit in Fig. 3.68a gives

$$\frac{i_{Rd}}{2} + g_m \frac{v_{id}}{2} + \frac{\Delta g_m}{2} v = 0 \quad (3.255)$$

KCL at the output of the common-mode half circuit in Fig. 3.68b gives

$$g_m v + \frac{\Delta g_m}{2} \frac{v_{id}}{2} + i_{Rc} = 0 \quad (3.256)$$

Also, KVL around the input loop in the common-mode half circuit gives

$$v = v_{ic} - v_{\text{tail}} = v_{ic} + 2i_{Rc} r_{\text{tail}} \quad (3.257)$$

Substituting (3.257) into (3.256) and rearranging gives

$$i_{Rc} = -\frac{g_m v_{ic} + \frac{\Delta g_m}{2} \frac{v_{id}}{2}}{1 + 2g_m r_{\text{tail}}} \quad (3.258)$$

Substituting (3.257) and (3.258) into (3.255) and rearranging gives

$$\frac{i_{Rd}}{2} = \frac{v_{id}}{2} \left(-g_m + \frac{\Delta g_m r_{\text{tail}} \frac{\Delta g_m}{2}}{1 + 2g_m r_{\text{tail}}} \right) + v_{ic} \left(-\frac{\Delta g_m}{2} + \frac{\Delta g_m r_{\text{tail}} g_m}{1 + 2g_m r_{\text{tail}}} \right) \quad (3.259)$$

From KVL in the R branch in the differential half circuit in Fig. 3.68a,

$$\frac{v_{od}}{2} = i_{Rc}\frac{\Delta R}{2} + \frac{i_{Rd}}{2}R \tag{3.260}$$

Substituting (3.258) and (3.259) into (3.260) and rearranging gives

$$v_{od} = A_{dm}v_{id} + A_{cm-dm}v_{ic} \tag{3.261}$$

where A_{dm} and A_{cm-dm} are

$$A_{dm} = \left.\frac{v_{od}}{v_{id}}\right|_{v_{ic}=0} = -g_m R + \frac{\Delta g_m r_{\text{tail}}\frac{\Delta g_m}{2}R - \frac{\Delta g_m}{2}\frac{\Delta R}{2}}{1 + 2g_m r_{\text{tail}}} \tag{3.262}$$

$$A_{cm-dm} = \left.\frac{v_{od}}{v_{ic}}\right|_{v_{id}=0} = -\left(\frac{g_m \Delta R + \Delta g_m R}{1 + 2g_m r_{\text{tail}}}\right) \tag{3.263}$$

From KVL in the R branch in the common-mode half circuit in Fig. 3.68b,

$$v_{oc} = \frac{i_{Rd}}{2}\frac{\Delta R}{2} + i_{Rc}R \tag{3.264}$$

Substituting (3.258) and (3.259) into (3.264) and rearranging gives

$$v_{oc} = A_{dm-cm}v_{id} + A_{cm}v_{ic} \tag{3.265}$$

where A_{dm-cm} and A_{cm} are

$$A_{dm-cm} = \left.\frac{v_{oc}}{v_{id}}\right|_{v_{ic}=0}$$

$$= -\frac{1}{4}\left[g_m \Delta R + \frac{\Delta g_m R - g_m \Delta R\left(2g_m r_{\text{tail}}\left(\frac{\Delta g_m}{2g_m}\right)^2\right)}{1 + 2g_m r_{\text{tail}}}\right] \tag{3.266}$$

$$A_{cm} = \left.\frac{v_{oc}}{v_{ic}}\right|_{v_{id}=0} = -\left(\frac{g_m R + \frac{\Delta g_m}{2}\frac{\Delta R}{2}}{1 + 2g_m r_{\text{tail}}}\right) \tag{3.267}$$

The calculations in (3.255) through (3.267) are based on the half circuits in Fig. 3.68 and give *exactly* the same results as an analysis of the entire differential amplifier shown in Fig. 3.67. Because the half circuits are coupled, however, exact half-circuit analysis requires the simultaneous consideration of both half circuits, which is about as complicated as the direct analysis of the entire original circuit.

In practice, the mismatch terms are usually a small fraction of the corresponding average values. As a result, the dominant contributions to the differential signals that control the mismatch generators in the common-mode half circuit stem from differential inputs. Similarly, the dominant part of the common-mode signals that control the mismatch generators in the differential half circuit arise from common-mode inputs. Therefore, we will assume that the signals controlling the mismatch generators can be found approximately by analyzing each half circuit independently without mismatch. The signals that control the mismatch generators in Fig. 3.68 are i_{Rc}, $i_{Rd}/2$, v, and $v_{id}/2$. We will find approximations to these quantities, $\hat{i}_{Rc}$, $\hat{i}_{Rd}/2$, $\hat{v}$, and $\hat{v}_{id}/2$ using the half circuits shown in Fig. 3.69, where the inputs are the same as in Fig. 3.68 but where the mismatch terms are set equal to zero. By ignoring the second-order

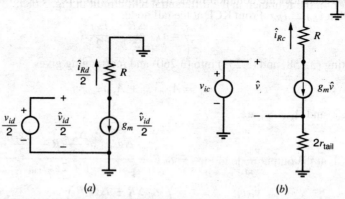

Figure 3.69 (a) Differential and (b) common-mode half circuits of the differential amplifier shown in Fig. 3.67 with mismatch terms set equal to zero.

interactions in which the mismatch generators influence the values of the control signals, this process greatly simplifies the required calculations, as shown next.

From inspection of the differential half circuit in Fig. 3.69a,

$$\frac{\hat{v}_{id}}{2} = \frac{v_{id}}{2} \tag{3.268}$$

and

$$\frac{\hat{i}_{Rd}}{2} = -g_m \frac{v_{id}}{2} \tag{3.269}$$

From the common-mode half circuit in Fig. 3.69b,

$$\hat{v} = v_{ic} - g_m \hat{v} \, (2r_{\text{tail}}) = \frac{v_{ic}}{1 + 2g_m r_{\text{tail}}} \tag{3.270}$$

Therefore,

$$\hat{i}_{Rc} = -\frac{g_m v_{ic}}{1 + 2g_m r_{\text{tail}}} \tag{3.271}$$

Now reconsider the differential half circuit with mismatch shown in Fig. 3.68a. Assume that $i_{Rc} \simeq \hat{i}_{Rc}$ and $v \simeq \hat{v}$. Then

$$\frac{v_{od}}{2} \simeq -\frac{\Delta R}{2}\left(\frac{g_m v_{ic}}{1 + 2g_m r_{\text{tail}}}\right) - g_m \frac{v_{id}}{2} R - \frac{\Delta g_m}{2} \frac{v_{ic}}{1 + 2g_m r_{\text{tail}}} R \tag{3.272}$$

From (3.272),

$$A_{dm} = \left.\frac{v_{od}}{v_{id}}\right|_{v_{ic}=0} \simeq -g_m R \tag{3.273}$$

and

$$A_{cm-dm} = \left.\frac{v_{od}}{v_{ic}}\right|_{v_{id}=0} \simeq -\left(\frac{g_m \Delta R + \Delta g_m R}{1 + 2g_m r_{\text{tail}}}\right) \tag{3.274}$$

Equation 3.274 shows that the ratio A_{dm}/A_{cm-dm} is approximately proportional to $1 + 2g_m r_{\text{tail}}$. Also, (3.274) agrees exactly with (3.263) in this case because the g_m generator in Fig. 3.68a is controlled by a purely differential signal. In other examples, the common-mode-to-differential-mode gain calculated in this way will be only approximately correct.

Now reconsider the common-mode half circuit with mismatch shown in Fig. 3.68b and assume that $i_{Rd} \simeq \hat{i}_{Rd}$. From KCL at the tail node,

$$v_{\text{tail}} \simeq \left(g_m v + \frac{\Delta g_m}{2} \frac{v_{id}}{2}\right) 2r_{\text{tail}} \tag{3.275}$$

Then

$$v = v_{ic} - v_{\text{tail}} \simeq \frac{v_{ic} - \frac{\Delta g_m}{2} \frac{v_{id}}{2}(2r_{\text{tail}})}{1 + 2g_m r_{\text{tail}}} \tag{3.276}$$

From KCL at the output node in Fig. 3.68b,

$$\frac{v_{oc} - \frac{i_{Rd}}{2} \frac{\Delta R}{2}}{R} + g_m v + \frac{\Delta g_m}{2} \frac{v_{id}}{2} = 0 \tag{3.277}$$

Assume that $i_{Rd} \simeq \hat{i}_{Rd}$. Substituting (3.269) and (3.276) into (3.277) and rearranging gives

$$v_{oc} \simeq -\frac{1}{4}\left(g_m \Delta R + \frac{\Delta g_m R}{1 + 2g_m r_{\text{tail}}}\right) v_{id} - \frac{g_m R}{1 + 2g_m r_{\text{tail}}} v_{ic} \tag{3.278}$$

From (3.278),

$$A_{dm-cm} = \left.\frac{v_{oc}}{v_{id}}\right|_{v_{ic}=0} \simeq -\frac{1}{4}\left(g_m \Delta R + \frac{\Delta g_m R}{1 + 2g_m r_{\text{tail}}}\right) \tag{3.279}$$

and

$$A_{cm} = \left.\frac{v_{oc}}{v_{ic}}\right|_{v_{id}=0} \simeq -\frac{g_m R}{1 + 2g_m r_{\text{tail}}} \tag{3.280}$$

These equations show that increasing the degeneration to common-mode inputs represented by the quantity $1 + 2g_m r_{\text{tail}}$ reduces the magnitude of A_{cm-dm}, A_{dm-cm}, and A_{cm}. As $r_{\text{tail}} \to \infty$ in this case, $A_{cm-dm} \to 0$ and $A_{cm} \to 0$. On the other hand, A_{dm-cm} does not approach zero when r_{tail} becomes infinite. Instead,

$$\lim_{r_{\text{tail}} \to \infty} A_{dm-cm} \simeq -\frac{g_m \Delta R}{4} \tag{3.281}$$

With finite and mismatched transistor output resistances, A_{cm-dm} also approaches a nonzero value as r_{tail} becomes infinite. Therefore, r_{tail} should be viewed as an important parameter because it reduces the sensitivity of differential pairs to common-mode inputs and helps reduce the effects of mismatch. However, even an ideal tail current source does not overcome all the problems introduced by mismatch. In Chapter 4, we will consider transistor current sources for which r_{tail} can be quite large.

■ **EXAMPLE**

Consider the unbalanced differential amplifier in Fig. 3.67. Assume that

$$g_{m1} = 1.001 \text{ mA/V} \quad g_{m2} = 0.999 \text{ mA/V}$$

$$R_1 = 101 \text{ k}\Omega \quad R_2 = 99 \text{ k}\Omega \quad r_{\text{tail}} = 1 \text{ M}\Omega$$

Find A_{dm}, A_{cm}, A_{cm-dm}, and A_{dm-cm}.

Calculating average and mismatch quantities gives

$$g_m = \frac{g_{m1} + g_{m2}}{2} = 1\frac{mA}{V} \qquad \Delta g_m = g_{m1} - g_{m2} = 0.002\frac{mA}{V}$$

$$R = \frac{R_1 + R_2}{2} = 100k\Omega \qquad \Delta R = R_1 - R_2 = 2k\Omega$$

From (3.269)

$$\frac{\hat{i}_{Rd}}{2} = -1\frac{mA}{V}\frac{v_{id}}{2} = -\frac{v_{id}}{2k\Omega}$$

From (3.271)

$$\hat{i}_{Rc} = -\frac{1\frac{mA}{V}v_{ic}}{1 + 2(1)(1000)} = -\frac{v_{ic}}{2001k\Omega}$$

From (3.273), (3.274), (3.279), and (3.280),

$$A_{dm} \simeq -1(100) = -100$$

$$A_{cm-dm} \simeq -\frac{1(2) + 0.002(100)}{1 + 2(1)(1000)} \simeq -0.0011$$

$$A_{dm-cm} \simeq -\frac{1}{4}\left(1(2) + \frac{0.002(100)}{1 + 2(1)(1000)}\right) \simeq -0.5$$

$$A_{cm} \simeq -\frac{1(100)}{1 + 2(1)(1000)} \simeq -0.05$$

APPENDIX

A.3.1 ELEMENTARY STATISTICS AND THE GAUSSIAN DISTRIBUTION

From the standpoint of a circuit designer, many circuit parameters are best regarded as random variables whose behavior is described by a probability distribution. This view is particularly important in the case of a parameter such as offset voltage. Even though the offset may be zero with perfectly matched components, random variations in resistors and transistors cause a spread of offset voltage around the mean value, and the size of this spread determines the fraction of circuits that meet a given offset specification.

Several factors cause the parameters of an integrated circuit to show random variations. One of these factors is the randomness of the edge definition when regions are defined to form resistors and active devices. In addition, random variations across the wafer in the diffusion of impurities can be a significant factor. These processes usually give rise to a *Gaussian* distribution (sometimes called a *normal* distribution) of the parameters. A Gaussian distribution of a parameter x is specified by a probability density function $p(x)$ given by

$$p(x) = \frac{1}{\sqrt{2\pi}\sigma}\exp\left[-\frac{(x-m)^2}{2\sigma^2}\right] \qquad (3.282)$$

where σ is the standard deviation of the distribution and m is the mean or average value of x. The significance of this function is that, for one particular circuit chosen at random from a large collection of circuits, the probability of the parameter having values between x and $(x + dx)$ is given by $p(x)dx$, which is the *area under the curve $p(x)$ in the range x to $(x + dx)$*. For

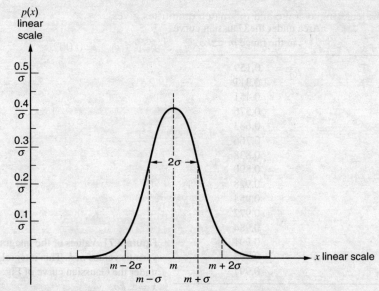

Figure 3.70 Probability density function $p(x)$ for a Gaussian distribution with mean value m and standard deviation σ. $p(x) = \exp[-(x-m)^2/(2\sigma^2)]/(\sqrt{2\pi}\sigma)$.

example, the probability that x has a value less than X is obtained by integrating (3.282) to give

$$P(x < X) = \int_{-\infty}^{X} p(x)\, dx \tag{3.283}$$

$$= \int_{-\infty}^{X} \frac{1}{\sqrt{2\pi}\sigma} \exp\left[-\frac{(x-m)^2}{2\sigma^2}\right] dx \tag{3.284}$$

In a large sample, the *fraction of circuits* where x is less than X will be given by the probability $P(x < X)$, and thus this quantity has real practical significance. The probability density function $p(x)$ in (3.282) is sketched in Fig. 3.70 and shows a characteristic bell shape. The peak value of the distribution occurs when $x = m$, where m is the mean value of x. The standard deviation σ is a measure of the *spread* of the distribution, and large values of σ give rise to a broad distribution. The distribution extends over $-\infty < x < \infty$, as shown by (3.282), but most of the area under the curve is found in the range $x = m \pm 3\sigma$, as will be seen in the following analysis.

The development thus far has shown that the probability of the parameter x having values in a certain range is just equal to the area under the curve of Fig. 3.70 in that range. Since x must lie somewhere in the range $\pm \infty$, the total area under the curve must be unity, and integration of (3.282) will show that this is so. The most common specification of interest to circuit designers is the fraction of a large sample of circuits that lies inside a band around the mean. For example, if a circuit has a gain x that has a Gaussian distribution with mean value 100, what fraction of circuits have gain values in the range 90 to 110? This fraction can be found by evaluating the probability that x takes on values in the range $x = m \pm 10$ where $m = 100$. This probability could be found from (3.282) if σ is known by integrating as follows:

$$P(m - 10 < x < m + 10) = \int_{m-10}^{m+10} \frac{1}{\sqrt{2\pi}\sigma} \exp\left[-\frac{(x-m)^2}{2\sigma^2}\right] dx \tag{3.285}$$

This equation gives the area under the Gaussian curve in the range $x = m \pm 10$.

k	Area under the Gaussian curve in the range $m \pm k\sigma$
0.2	0.159
0.4	0.311
0.6	0.451
0.8	0.576
1.0	0.683
1.2	0.766
1.4	0.838
1.6	0.890
1.8	0.928
2.0	0.954
2.2	0.972
2.4	0.984
2.6	0.991
2.8	0.995
3.0	0.997

Figure 3.71 Values of the integral in (3.286) for various values of k. This integral gives the area under the Gaussian curve of Fig. 3.70 in the range $x = \pm k\sigma$.

To simplify calculations of the kind described above, values of the integral in (3.285) have been calculated and tabulated. To make the tables general, the range of integration is normalized to σ to give

$$P(m - k\sigma < x < m + k\sigma) = \int_{m-k\sigma}^{m+k\sigma} \frac{1}{\sqrt{2\pi}\sigma} \exp\left[-\frac{(x-m)^2}{2\sigma^2}\right] dx \quad (3.286)$$

Values of this integral for various values of k are tabulated in Fig. 3.71. This table shows that $P = 0.683$ for $k = 1$ and thus 68.3 percent of a large sample of a Gaussian distribution lies within a range $x = m \pm \sigma$. For $k = 3$, the value of $P = 0.997$ and thus 99.7 percent of a large sample lies within a range $x = m \pm 3\sigma$.

Circuit parameters such as offset or gain often can be expressed as a linear combination of other parameters as shown in (3.216) and (3.248) for offset voltage. If all the parameters are independent random variables with Gaussian distributions, the standard deviations and means can be related as follows. Assume that the random variable x can be expressed in terms of random variables a, b, and c using

$$x = a + b - c \quad (3.287)$$

Then it can be shown that

$$m_x = m_a + m_b - m_c \quad (3.288)$$

$$\sigma_x^2 = \sigma_a^2 + \sigma_b^2 + \sigma_c^2 \quad (3.289)$$

where m_x is the mean value of x and σ_x is the standard deviation of x. Equation 3.289 shows that the square of the standard deviation of x is the sum of the square of the standard deviations of a, b, and c. This result extends to any number of variables.

These results were created in the context of the random variations found in circuit parameters. The Gaussian distribution is also useful in the treatment of random noise, as described in Chapter 11.

■ **EXAMPLE**

The offset voltage of a circuit has a mean value of $m = 0$ and a standard deviation of $\sigma = 2$ mV. What fraction of circuits will have offsets with magnitudes less than 4 mV?

A range of offset of ± 4 mV corresponds to $\pm 2\sigma$. From Fig. 3.71, we find that the area under the Gaussian curve in this range is 0.954, and thus 95.4 percent of circuits will have offsets with magnitudes less than 4 mV.

PROBLEMS

For the *npn* bipolar transistors in these problems, use the high-voltage bipolar device parameters given in Fig. 2.30, unless otherwise specified.

3.1 Determine the input resistance, transconductance, and output resistance of the CE amplifier of Fig. 3.7 if $R_C = 10\,\text{k}\Omega$ and $I_C = 250\,\mu\text{A}$. Assume that $r_b = 0$.

3.2 A CE transistor is to be used in the amplifier of Fig. 3.72 with a source resistance R_S and collector resistor R_C. First, find the overall small-signal gain v_o/v_i as a function of R_S, R_C, β_0, V_A, and the collector current I_C. Next, determine the value of dc collector bias current I_C that maximizes the small-signal voltage gain. Explain qualitatively why the gain falls at very high and very low collector currents. Do not neglect r_o in this problem. What is the voltage gain at the optimum I_C? Assume that $r_b = 0$.

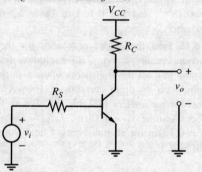

Figure 3.72 Circuit for Problem 3.2.

3.3 Assume that $R_S = R_C = 100\,\text{k}\Omega$ in Problem 3.2, and calculate the optimum I_C. What is the dc voltage drop across R_C? What is the voltage gain?

3.4 For the common-source amplifier of Fig. 3.12, calculate the small-signal voltage gain and the bias values of V_i and V_o at the edge of the triode region. Also calculate the bias values of V_i and V_o where the small-signal voltage gain is unity with the transistor operating in the active region. What is the maximum voltage gain of this stage? Assume $V_{DD} = 3\text{V}$, $R_D = 5\,\text{k}\Omega$, $\mu_n C_{ox} = 200\,\mu\text{A/V}^2$, $W = 10\,\mu\text{m}$, $L = 1\,\mu\text{m}$, $V_t = 0.6\,\text{V}$, and $\lambda = 0$. Check your answer with SPICE.

3.5 Determine the input resistance, transconductance, and output resistance of the CB amplifier of Fig. 3.15 if $I_C = 250\,\mu\text{A}$ and $R_C = 20\,\text{k}\Omega$. Neglect r_b and r_o.

3.6 Assume that R_C is made large compared with r_o in the CB amplifier of Fig. 3.15. Use the equivalent circuit of Fig. 3.17 and add r_o between the input (emitter terminal) and the output (collector terminal) to calculate the output resistance when

(a) The amplifier is driven by an ideal current source.

(b) The amplifier is driven by an ideal voltage source. Neglect r_b.

3.7 Determine the input resistance of the CG amplifier of Fig. 3.19 if the transistor operates in the active region with $I_D = 100\,\mu\text{A}$. Let $R_D = 10\,\text{k}\Omega$, $\mu_n C_{ox} = 200\,\mu\text{A/V}^2$, $\lambda = 0.01\,\text{V}^{-1}$, $W = 400\,\mu\text{m}$, and $L = 1\,\mu\text{m}$. Ignore the body effect. Repeat with $R_D = 1\,\text{M}\Omega$. If the 100 μA current flows through R_D in this case, a power-supply voltage of at least 100 V would be required. To overcome this problem, assume that an ideal 100-μA current source is placed in parallel with R_D here.

3.8 Determine the input resistance, voltage gain v_o/v_s, and output resistance of the CC amplifier of Fig. 3.23a if $R_S = 2\,\text{k}\Omega$, $R_L = 500\,\Omega$, and $I_C = 1\,\text{mA}$. Neglect r_b and r_o. Do not include R_S in calculating the input resistance. In calculating the output resistance, however, include R_L. Include both R_S and R_L in the gain calculation.

3.9 For the common-drain amplifier of Fig. 3.73, assume $W/L = 10$ and $\lambda = 0$. Use Table 2.2 for other parameters. Find the dc output voltage V_O and the small-signal gain v_o/v_i under the following conditions:

(a) Ignoring the body effect and with $R \to \infty$.
(b) Including the body effect and with $R \to \infty$.
(c) Including the body effect and with $R = 100\,\text{k}\Omega$.
(d) Including the body effect and with $R = 10\,\text{k}\Omega$.

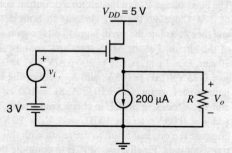

Figure 3.73 Circuit for Problem 3.9.

3.10 Determine the dc collector currents in Q_1 and Q_2, and then the input resistance and voltage gain for the Darlington emitter follower of Fig. 3.74. Neglect r_μ, r_b, and r_o. Assume that $V_{BE(on)} = 0.7$ V. Check your answer with SPICE and also use SPICE to determine the output resistance of the stage.

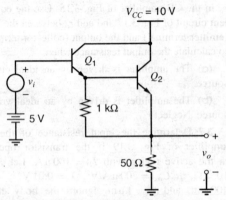

Figure 3.74 Circuit for Problem 3.10.

3.11 Calculate the output resistance r_o^c of the common-emitter Darlington transistor of Fig. 3.75 as a function of I_{BIAS}. Do not neglect either r_{o1} or r_{o2} in this calculation, but you may neglect r_b and r_μ. If $I_{C2} = 1$mA, what is r_o^c for $I_{BIAS} = 1$mA? For $I_{BIAS} = 0$?

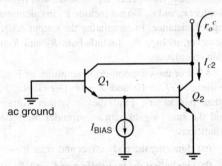

Figure 3.75 Circuit for Problem 3.11.

3.12 A BiCMOS Darlington is shown in Fig. 3.76. The bias voltage V_B is adjusted for a dc output voltage of 2 V. Calculate the bias currents in both devices and then calculate the small-signal voltage gain v_o/v_i of the circuit. For the MOS transistor, assume $W = 10$ μm, $L = 1$ μm, $\mu_n C_{ox} = 200$ μA/V^2, $V_t = 0.6$ V, $\gamma = 0.25$ V$^{1/2}$, $\phi_f = 0.3$ V, and $\lambda = 0$. For the bipolar transistor, assume $I_S = 10^{-16}$ A, $\beta_F = 100$, $r_b = 0$, and $V_A \to \infty$. Use SPICE to check your result. Then add $\lambda = 0.05$ V^{-1}, $r_b = 100\Omega$, and $V_A = 20$ V and compare the original result to the result with this new transistor data. Finally, use SPICE to compute the dc transfer characteristic of the circuit.

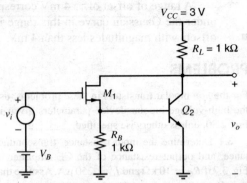

Figure 3.76 BiCMOS Darlington circuit for Problem 3.12.

3.13 Determine the input resistance, transconductance, output resistance, and maximum open-circuit voltage gain for the CE-CB circuit of Fig. 3.36 if $I_{C1} = I_{C2} = 200$ μA.

3.14 Determine the input resistance, transconductance, output resistance, and maximum open-circuit voltage gain for the CS-CG circuit of Fig. 3.38 if $I_{D1} = I_{D2} = 250$ μA. Assume $W/L = 100$, $\lambda = 0.1$V^{-1}, and $\chi = 0.1$. Use Table 2.2 for other parameters.

3.15 Find the output resistance for the active-cascode circuit of Fig. 3.77 excluding resistor R. Assume that all the transistors operate in the active region with dc drain currents of 100 μA. Use the transistor parameters in Table 2.4. Ignore the body effect. Assume $W = 10$ μm, $L_{drwn} = 0.4$ μm, and $X_d = 0.1$ μm for all transistors. Check your answer with SPICE.

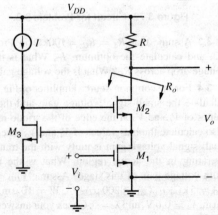

Figure 3.77 Active-cascode circuit for Problem 3.15.

3.16 Find the short-circuit transconductance of the super-source follower shown in Fig. 3.43. Assume $I_1 = 200$ μA, $I_2 = 100$ μA, $W_1 = 30$ μm, and $W_2 = 10$ μm. Also, assume that both transistors

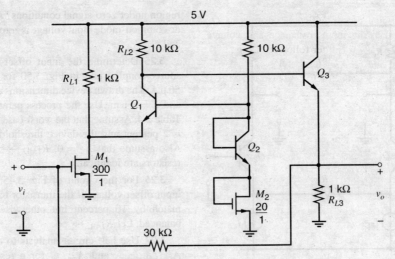

Figure 3.78 BiCMOS amplifier for Problem 3.17.

operate in the active region, and ignore the body effect. Use the transistor parameters in Table 2.4. Assume $L_{\text{drwn}} = 0.4\,\mu\text{m}$ and $X_d = 0.1\,\mu\text{m}$ for all transistors.

3.17 A BiCMOS amplifier is shown in Fig. 3.78. Calculate the small-signal voltage gain v_o/v_i. Assume $I_S = 10^{-16}\,\text{A}$, $\beta_F = 100$, $r_b = 0$, $V_A \to \infty$, $\mu_n C_{ox} = 200\,\mu\text{A/V}^2$, $V_t = 0.6\,\text{V}$, and $\lambda = 0$. Check your answer with SPICE and then use SPICE to investigate the effects of velocity saturation by including source degeneration in the MOS transistors as shown in Fig. 1.41 using $\mathscr{E}_c = 1.5 \times 10^6\,\text{V/m}$.

3.18 Determine the differential-mode gain, common-mode gain, differential-mode input resistance, and common-mode input resistance for the circuit of Fig. 3.45 with $I_{\text{TAIL}} = 20\,\mu\text{A}$, $R_{\text{TAIL}} = 10\,\text{M}\Omega$, $R_C = 100\,\text{k}\Omega$, and $V_{EE} = V_{CC} = 5\,\text{V}$. Neglect r_b, r_o, and r_μ. Calculate the CMRR. Check with SPICE and use SPICE to investigate the effects of adding nonzero r_b and finite V_A as given in Fig. 2.30.

3.19 Repeat Problem 3.18, but with the addition of emitter-degeneration resistors of value $4\,\text{k}\Omega$ each.

3.20 Determine the overall input resistance, voltage gain, and output resistance of the CC-CB connection of Fig. 3.79. Neglect r_o, r_μ, and r_b. Note that the addition of a 10-kΩ resistor in the collector of Q_1 would not change the results, so that the results of the emitter-coupled pair analysis can be used.

3.21 Consider the circuit of Fig. 3.80 except replace both *npn* transistors with *n*-channel MOS transistors. Neglect the body effect, and assume $\lambda = 0$. Use half-circuit concepts to determine the differential-mode and common-mode gain of this modified circuit.

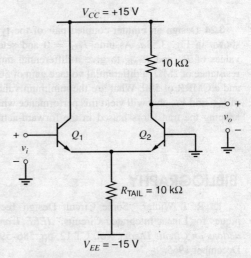

Figure 3.79 Circuit for Problem 3.20.

3.22 Use half-circuit concepts to determine the differential-mode and common-mode gain of the circuit shown in Fig. 3.80. Neglect r_o, r_μ, and r_b. Calculate the differential-mode and common-mode input resistance.

3.23 Determine the required bias current and device sizes to design a source-coupled pair to have the following two characteristics. First, the small-signal transconductance with zero differential input voltage should be 1.0 mA/V. Second, a differential input voltage of 0.2 V should result in a differential output current of 85 percent of the maximum value. Assume that the devices are *n*-channel transistors that are made with the technology summarized in Table 2.4. Use a drawn device channel length of 1 μm. Neglect channel-length modulation, and assume $X_d = 0$.

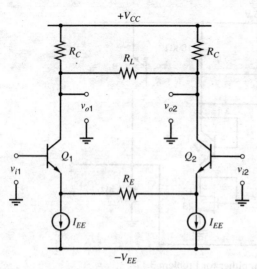

Figure 3.80 Circuit for Problem 3.21.

3.24 Design an emitter-coupled pair of the type shown in Fig. 3.53a. Assume $I_{TAIL} = 0$ and select values of R_C and R_{TAIL} to give a differential input resistance of $2 M\Omega$, a differential voltage gain of 500, and a CMRR of 500. What are the minimum values of V_{CC} and V_{EE} that will yield this performance while keeping the transistors biased in the forward-active region under zero-signal conditions? Assume that the dc common-mode input voltage is zero. Neglect r_b, r_μ, and r_o.

3.25 Determine the input offset voltage of the source-coupled pair in Fig. 3.50 for which $I_{TAIL} = 50\,\mu A$. The drawn device dimensions are $W = 10\,\mu m$ and $L = 1\,\mu m$. Use the process parameters given in Table 2.4. Assume that the worst-case W/L mismatch is 2 percent and the device thresholds are identical. Also assume that $X_d = 0$, $R_{TAIL} \to \infty$ and the load resistors are identical.

3.26 For the circuit of Fig. 3.45, determine the input offset voltage if the transistor base widths mismatch by 10 percent but otherwise the circuit is balanced. Let $R_{TAIL} \to \infty$.

3.27 Use half-circuit analysis to determine A_{dm}, A_{cm}, A_{cm-dm}, and A_{dm-cm} for a resistively loaded differential pair with mismatched resistive loads, R_1 and R_2. Assume that $R_1 = 10.1\,k\Omega$ and $R_2 = 9.9\,k\Omega$. Also assume that $g_{m1} = g_{m2} = 1\,mA/V$, $r_{o1} \to \infty$, and $r_{o2} \to \infty$. Finally, assume that the equivalent resistance of the tail current source $r_{tail} = 1\,M\Omega$.

3.28 Repeat Problem 3.27 but with matched loads and mismatched transistor output resistances. Assume $R_1 = R_2 = 10\,k\Omega$, $r_{o1} = 505\,k\Omega$, and $r_{o2} = 495\,k\Omega$. What happens when $r_{tail} \to \infty$?

BIBLIOGRAPHY

1. R. J. Widlar. "Some Circuit Design Techniques for Linear Integrated Circuits," *IEEE Transactions on Circuit Theory,* Vol. CT-12, pp. 586–590, December 1965.

2. H. R. Camenzind and A. B. Grebene. "An Outline of Design Techniques for Linear Integrated Circuits," *IEEE Journal of Solid-State Circuits,* Vol. SC-4, pp. 110–122, June 1969.

3. J. Giles. *Fairchild Semiconductor Linear Integrated Circuits Applications Handbook.* Fairchild Semiconductor, 1967.

4. C. L. Searle, A. R. Boothroyd, E. J. Angelo, P. E. Gray, and D. O. Pederson. *Elementary Circuit Properties of Transistors.* Chapter 7, Wiley, New York, 1964.

5. S. Darlington. "Semiconductor Signal Translating Device," U.S. Patent 2,663,806, May 1952.

6. F. V. Hunt and R. W. Hickman. "On Electronic Voltage Stabilizers," *Review of Scientific Instruments,* Vol. 10, pp. 6–21, January 1939.

7. H. Wallman, A. B. Macnee, and C. P. Gadsden. "A Low-Noise Amplifier," *Proceedings of the I.R.E.,* Vol. 36, pp. 700–708, June 1948.

8. B. J. Hosticka. "Improvement of the Gain of MOS Amplifiers," *IEEE Journal of Solid-State Circuits,* Vol. SC-14, pp. 1111–1114, December 1979.

9. E. Säckinger and W. Guggenbühl. "A High-Swing, High-Impedance MOS Cascode Circuit," *IEEE Journal of Solid-State Circuits,* Vol. 25, pp. 289–298, February 1990.

10. A. D. Blumlein. "Improvements in or Relating to Thermionic Valve Amplifying Circuit Arrangements," British Patent 482,740, July, 1936.

11. R. D. Middlebrook. *Differential Amplifiers.* Wiley, New York, 1963.

12. L. J. Giacoletto. *Differential Amplifiers.* Wiley, New York, 1970.

13. G. Erdi. "A Low-Drift, Low-Noise Monolithic Operational Amplifier for Low Level Signal Processing," *Fairchild Semiconductor Applications Brief.* No. 136, July 1969.

CHAPTER 4

Current Mirrors, Active Loads, and References

4.1 Introduction

Current mirrors made by using active devices have come to be widely used in analog integrated circuits both as biasing elements and as load devices for amplifier stages. The use of current mirrors in biasing can result in superior insensitivity of circuit performance to variations in power supply and temperature. Current mirrors are frequently more economical than resistors in terms of the die area required to provide bias current of a certain value, particularly when the required value of bias current is small. When used as a load element in transistor amplifiers, the high incremental resistance of the current mirror results in high voltage gain at low power-supply voltages.

The first section of this chapter describes the general properties of current mirrors and compares various bipolar and MOS mirrors to each other using these properties. The next section deals with the use of current mirrors as load elements in amplifier stages. The last section shows how current mirrors are used to construct references that are insensitive to variations in supply and temperature. Finally, the appendix analyzes the effects of device mismatch.

4.2 Current Mirrors

4.2.1 General Properties

A current mirror is an element with at least three terminals, as shown in Fig. 4.1. The common terminal is connected to a power supply, and the input current source is connected to the input terminal. Ideally, the output current is equal to the input current multiplied by a desired current gain. If the gain is unity, the input current is reflected to the output, leading to the name *current mirror*. Under ideal conditions, the current-mirror gain is independent of input frequency, and the output current is independent of the voltage between the output and common terminals. Furthermore, the voltage between the input and common terminals is ideally zero because this condition allows the entire supply voltage to appear across the input current source, simplifying its transistor-level design. More than one input and/or output terminals are sometimes used.

In practice, real transistor-level current mirrors suffer many deviations from this ideal behavior. For example, the gain of a real current mirror is never independent of the input frequency. The topic of frequency response is covered in Chapter 7, and mainly dc and low-frequency ac signals are considered in the rest of this chapter. Deviations from ideality that will be considered in this chapter are listed below.

1. One of the most important deviations from ideality is the variation of the current-mirror output current with changes in voltage at the output terminal. This effect is characterized

252 Chapter 4 ▪ Current Mirrors, Active Loads, and References

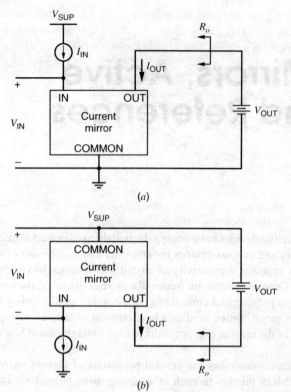

Figure 4.1 Current-mirror block diagrams referenced to (*a*) ground and (*b*) the positive supply.

by the small-signal output resistance, R_o, of the current mirror. A Norton-equivalent model of the output of the current mirror includes R_o in parallel with a current source controlled by the input current. The output resistance directly affects the performance of many circuits that use current mirrors. For example, the common-mode rejection ratio of the differential amplifier depends directly on this resistance, as does the gain of the active-load circuits. Increasing the output resistance reduces the dependence of the output current on the output voltage and is therefore desirable. Generally speaking, the output resistance increases in practical circuits when the output current decreases. Unfortunately, decreasing the output current also decreases the maximum operating speed. Therefore, when comparing the output resistance of two current mirrors, they should be compared at identical output currents.

2. Another important error source is the gain error, which is the deviation of the gain of a current mirror from its ideal value. The gain error is separated into two parts: (1) the systematic gain error and (2) the random gain error. The systematic gain error, ϵ, is the gain error that arises even when all matched elements in the mirror are perfectly matched and will be calculated for each of the current mirrors presented in this section. The random gain error is the gain error caused by unintended mismatches between matched elements.

3. When the input current source is connected to the input terminal of a real current mirror, it creates a positive voltage drop, V_{IN}, that reduces the voltage available across the input current source. Minimizing V_{IN} is important because it simplifies the design of the input current source, especially in low-supply applications. To reduce V_{IN}, current mirrors sometimes have more than one input terminal. In that case, we will calculate an input voltage for each input terminal. An example is the MOS high-swing cascode current mirror considered in Section 4.2.5.

4. A positive output voltage, V_{OUT}, is required in practice to make the output current depend mainly on the input current. This characteristic is summarized by the minimum voltage across the output branch, $V_{OUT(min)}$, that allows the output device(s) to operate in the active region. Minimizing $V_{OUT(min)}$ maximizes the range of output voltages for which the current-mirror output resistance is almost constant, which is important in applications where current mirrors are used as active loads in amplifiers (especially with low power-supply voltages). This topic is covered in Section 4.3. When current mirrors have more than one output terminal, each output must be biased above its $V_{OUT(min)}$ to make the corresponding output current depend mainly on the input current.

In later sections, the performance of various current mirrors will be compared to each other through these four parameters: R_o, ϵ, V_{IN}, and $V_{OUT(min)}$.

4.2.2 Simple Current Mirror

4.2.2.1 Bipolar

The simplest form of a current mirror consists of two transistors. Fig. 4.2 shows a bipolar version of this mirror. Transistor Q_1 is diode connected, forcing its collector-base voltage to zero. In this mode, the collector-base junction is off in the sense that no injection takes place there, and Q_1 operates in the forward-active region. Assume that Q_2 also operates in the forward-active region and that both transistors have infinite output resistance. Then I_{OUT} is controlled by V_{BE2}, which is equal to V_{BE1} by KVL. A KVL equation is at the heart of the operation of all current mirrors. Neglecting junction leakage currents,

$$V_{BE2} = V_T \ln \frac{I_{C2}}{I_{S2}} = V_{BE1} = V_T \ln \frac{I_{C1}}{I_{S1}} \tag{4.1}$$

where $V_T = kT/q$ is the thermal voltage and I_{S1} and I_{S2} are the transistor saturation currents. From (4.1),

$$I_{C2} = \frac{I_{S2}}{I_{S1}} I_{C1} \tag{4.2}$$

If the transistors are identical, $I_{S1} = I_{S2}$ and (4.2) shows that the current flowing in the collector of Q_1 is mirrored to the collector of Q_2. KCL at the collector of Q_1 yields

$$I_{IN} - I_{C1} - \frac{I_{C1}}{\beta_F} - \frac{I_{C2}}{\beta_F} = 0 \tag{4.3}$$

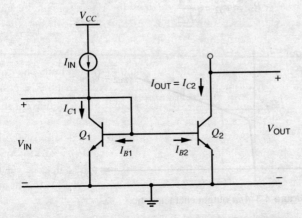

Figure 4.2 A simple bipolar current mirror.

Therefore, with identical transistors,

$$I_{OUT} = I_{C2} = I_{C1} = \frac{I_{IN}}{1 + \dfrac{2}{\beta_F}} \quad (4.4)$$

If β_F is large, the base currents are small and

$$I_{OUT} = I_{C1} \simeq I_{IN} \quad (4.5)$$

Thus for identical devices Q_1 and Q_2, the gain of the current mirror is approximately unity. This result holds for both dc and low-frequency ac currents. Above the 3-dB frequency of the mirror, however, the base current increases noticeably because the impedance of the base-emitter capacitance decreases, reducing the gain of the current mirror. Frequency response is studied in Chapter 7. The rest of this section considers dc currents only.

In practice, the devices need not be identical. Then from (4.2) and (4.3),

$$I_{OUT} = \frac{I_{S2}}{I_{S1}} I_{C1} = \left(\frac{I_{S2}}{I_{S1}} I_{IN}\right) \left(\frac{1}{1 + \dfrac{1 + (I_{S2}/I_{S1})}{\beta_F}}\right) \quad (4.6)$$

When $I_{S2} = I_{S1}$, (4.6) is the same as (4.4). Since the saturation current of a bipolar transistor is proportional to its emitter area, the first term in (4.6) shows that the gain of the current mirror can be larger or smaller than unity because the emitter areas can be ratioed. If the desired current-mirror gain is a rational number, M/N, the area ratio is usually set by connecting M identical devices called *units* in parallel to form Q_2 and N units in parallel to form Q_1 to minimize mismatch arising from lithographic effects in forming the emitter regions. However, area ratios greater than about five to one consume a large die area dominated by the area of the larger of the two devices. Thus other methods described in later sections are preferred for the generation of large current ratios. The last term in (4.6) accounts for error introduced by finite β_F. Increasing I_{S2}/I_{S1} increases the magnitude of this error by increasing the base current of Q_2 compared to that of Q_1.

In writing (4.1) and (4.2), we assumed that the collector currents of the transistors are independent of their collector-emitter voltages. If a transistor is biased in the forward-active region, its collector current actually increases slowly with increasing collector-emitter voltage. Fig. 4.3 shows an output characteristic for Q_2. The output resistance of the current mirror at any given operating point is the reciprocal of the slope of the output characteristic at that point. In the forward-active region,

$$R_o = r_{o2} = \frac{V_A}{I_{C2}} \quad (4.7)$$

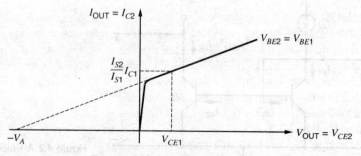

Figure 4.3 *npn* output characteristic.

The point where $V_{CE2} = V_{CE1}$ and $V_{BE2} = V_{BE1}$ is labeled on the characteristic. Because the collector current is controlled by the base-emitter and collector-emitter voltages, $I_{C2} = (I_{S2}/I_{S1})I_{C1}$ at this point. If the slope of the characteristic in saturation is constant, the variation in I_{C2} for changes in V_{CE2} can be predicted by a straight line that goes through the labeled point. As described in Chapter 1, extrapolation of the output characteristic in the forward-active region back to the V_{CE2} axis gives an intercept at $-V_A$, where V_A is the Early voltage. If $V_A \gg V_{CE1}$, the slope of the straight line is about equal to $(I_{S2}/I_{S1})(I_{C1}/V_A)$. Therefore,

$$I_{OUT} = \frac{I_{S2}}{I_{S1}} I_{C1} \left(1 + \frac{V_{CE2} - V_{CE1}}{V_A}\right) = \frac{\frac{I_{S2}}{I_{S1}} I_{IN} \left(1 + \frac{V_{CE2} - V_{CE1}}{V_A}\right)}{1 + \frac{1 + (I_{S2}/I_{S1})}{\beta_F}} \quad (4.8)$$

Since the ideal gain of the current mirror is I_{S2}/I_{S1}, the systematic gain error, ϵ, of the current mirror can be calculated from (4.8).

$$\epsilon = \left(\frac{1 + \frac{V_{CE2} - V_{CE1}}{V_A}}{1 + \frac{1 + (I_{S2}/I_{S1})}{\beta_F}}\right) - 1 \simeq \frac{V_{CE2} - V_{CE1}}{V_A} - \frac{1 + (I_{S2}/I_{S1})}{\beta_F} \quad (4.9)$$

The first term in (4.9) stems from finite output resistance and the second term from finite β_F. If $V_{CE2} > V_{CE1}$, the polarities of the two terms are opposite. Since the two terms are independent, however, cancellation is unlikely in practice. The first term dominates when the difference in the collector-emitter voltages and β_F are large. For example, with identical transistors and $V_A = 130$ V, if the collector-emitter voltage of Q_1 is held at $V_{BE(on)}$, and if the collector-emitter voltage of Q_2 is 30 V, then the systematic gain error $(30 - 0.6)/130 - 2/200 \simeq 0.22$. Thus for a circuit operating at a power-supply voltage of 30 V, the current-mirror currents can differ by more than 20 percent from those values calculated by assuming that the transistor output resistance and β_F are infinite. Although the first term in (4.9) stems from finite output resistance, it does not depend on r_{o2} directly but instead on the collector-emitter and Early voltages. The Early voltage is independent of the bias current, and

$$V_{IN} = V_{CE1} = V_{BE1} = V_{BE(on)} \quad (4.10)$$

Since $V_{BE(on)}$ is proportional to the natural logarithm of the collector current, V_{IN} changes little with changes in bias current. Therefore, changing the bias current in a current mirror changes systematic gain error mainly through changes in V_{CE2}.

Finally, the minimum output voltage required to keep Q_2 in the forward-active region is

$$V_{OUT(min)} = V_{CE2(sat)} \quad (4.11)$$

4.2.2.2 MOS

Figure 4.4 shows an MOS version of the simple current mirror. The drain-gate voltage of M_1 is zero; therefore, the channel does not exist at the drain, and the transistor operates in the saturation or active region if the threshold is positive. Although the principle of operation for MOS transistors does not involve forward biasing any diodes, M_1 is said to be *diode connected* in an analogy to the bipolar case. Assume that M_2 also operates in the active region and that both transistors have infinite output resistance. Then I_{D2} is controlled by V_{GS2}, which is equal to V_{GS1} by KVL. A KVL equation is at the heart of the operation of all current mirrors. As described in Section 1.5.3, the gate-source voltage of a given MOS transistor is usually

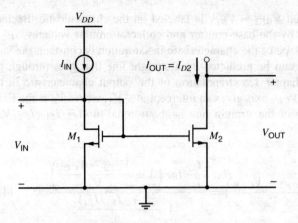

Figure 4.4 A simple MOS current mirror.

separated into two parts: the threshold V_t and the overdrive V_{ov}. Assuming square-law behavior as in (1.157), the overdrive for M_2 is

$$V_{ov2} = V_{GS2} - V_t = \sqrt{\frac{2I_{D2}}{k'(W/L)_2}} \tag{4.12}$$

Since the transconductance parameter k' is proportional to mobility, and since mobility falls with increasing temperature, the overdrive rises with temperature. In contrast, Section 1.5.4 shows that the threshold falls with increasing temperature. From KVL and (1.157),

$$V_{GS2} = V_t + \sqrt{\frac{2I_{D2}}{k'(W/L)_2}} = V_{GS1} = V_t + \sqrt{\frac{2I_{D1}}{k'(W/L)_1}} \tag{4.13}$$

Equation 4.13 shows that the overdrive of M_2 is equal to that of M_1.

$$V_{ov2} = V_{ov1} = V_{ov} \tag{4.14}$$

If the transistors are identical, $(W/L)_2 = (W/L)_1$, and therefore

$$I_{OUT} = I_{D2} = I_{D1} \tag{4.15}$$

Equation 4.15 shows that the current that flows in the drain of M_1 is mirrored to the drain of M_2. Since $\beta_F \to \infty$ for MOS transistors, (4.15) and KCL at the drain of M_1 yield

$$I_{OUT} = I_{D1} = I_{IN} \tag{4.16}$$

Thus for identical devices operating in the active region with infinite output resistance, the gain of the current mirror is unity. This result holds when the gate currents are zero; that is, (4.16) is at least approximately correct for dc and low-frequency ac currents. As the input frequency increases, however, the gate currents of M_1 and M_2 increase because each transistor has a nonzero gate-source capacitance. The part of the input current that flows into the gate leads does not flow into the drain of M_1 and is not mirrored to M_2; therefore, the gain of the current mirror decreases as the frequency of the input current increases. The rest of this section considers dc currents only.

In practice, the devices need not be identical. Then from (4.13) and (4.16),

$$I_{OUT} = \frac{(W/L)_2}{(W/L)_1} I_{D1} = \frac{(W/L)_2}{(W/L)_1} I_{IN} \tag{4.17}$$

Equation 4.17 shows that the gain of the current mirror can be larger or smaller than unity because the transistor sizes can be ratioed. To ratio the transistor sizes, either the widths or the lengths can be made unequal in principle. In practice, however, the lengths of M_1 and M_2 are rarely made unequal. The lengths that enter into (4.17) are the effective channel lengths given by (2.35). Equation 2.35 shows that the effective channel length of a given transistor differs from its drawn length by offset terms stemming from the depletion region at the drain and lateral diffusion at the drain and source. Since the offset terms are independent of the drawn length, a ratio of two effective channel lengths is equal to the drawn ratio only if the drawn lengths are identical. As a result, a ratio of unequal channel lengths depends on process parameters that may not be well controlled in practice. Similarly, Section 2.9.1 shows that the effective width of a given transistor differs from the drawn width because of lateral oxidation resulting in a *bird's beak*. Therefore, a ratio of unequal channel widths will also be process dependent. In many applications, however, the shortest channel length allowed in a given technology is selected for most transistors to maximize speed and minimize area. In contrast, the drawn channel widths are usually many times larger than the minimum dimensions allowed in a given technology. Therefore, to minimize the effect of the offset terms when the current-mirror gain is designed to differ from unity, the widths are ratioed rather than the lengths in most practical cases. If the desired current-mirror gain is a rational number, M/N, the ratio is usually set by connecting M identical devices called *units* in parallel to form M_2 and N units in parallel to form M_1 to minimize mismatch arising from lithographic effects in forming the gate regions. As in the bipolar case, ratios greater than about five to one consume a large die area dominated by the area of the larger of the two devices. Thus other methods described in later sections are preferred for the generation of large current ratios.

In writing (4.13) and (4.15), we assumed that the drain currents of the transistors are independent of their drain-source voltages. If a transistor is biased in the active region, its drain current actually increases slowly with increasing drain-source voltage. Figure 4.5 shows an output characteristic for M_2. The output resistance of the current mirror at any given operating point is the reciprocal of the slope of the output characteristic at that point. In the active region,

$$R_o = r_{o2} = \frac{V_A}{I_{D2}} = \frac{1}{\lambda I_{D2}} \qquad (4.18)$$

The point where $V_{DS2} = V_{DS1}$ and $V_{GS2} = V_{GS1}$ is labeled on the characteristic. Because the drain current is controlled by the gate-source and drain-source voltages, $I_{D2} = [(W/L)_2/(W/L)_1]I_{D1}$ at this point. If the slope of the characteristic in saturation is constant, the variation in I_{D2} for changes in V_{DS2} can be predicted by a straight line that goes through the

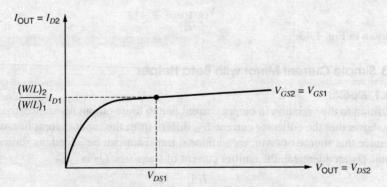

Figure 4.5 Output characteristic of simple MOS current mirror.

labeled point. As described in Chapter 1, extrapolation of the output characteristic in the active region back to the V_{DS2} axis gives an intercept at $-V_A = -1/\lambda$, where V_A is the Early voltage. If $V_A \gg V_{DS1}$, the slope of the straight line is about equal to $[(W/L)_2/(W/L)_1][I_{D1}/V_A]$. Therefore,

$$I_{OUT} = \frac{(W/L)_2}{(W/L)_1} I_{IN} \left(1 + \frac{V_{DS2} - V_{DS1}}{V_A}\right) \tag{4.19}$$

Since the ideal gain of the current mirror is $(W/L)_2/(W/L)_1$, the systematic gain error, ϵ, of the current mirror can be calculated from (4.19).

$$\epsilon = \frac{V_{DS2} - V_{DS1}}{V_A} \tag{4.20}$$

For example, if the drain-source voltage of M_1 is held at 1.2 V, and if the drain-source voltage of M_2 is 5 V, then the systematic gain error is $(5 - 1.2)/10 \simeq 0.38$ with $V_A = 10$ V. Thus for a circuit operating at a power-supply voltage of 5 V, the current-mirror currents can differ by more than 35 percent from those values calculated by assuming that the transistor output resistance is infinite. Although ϵ stems from finite output resistance, it does not depend on r_{o2} directly but instead on the drain-source and Early voltages. Since the Early voltage is independent of the bias current, this observation shows that changing the input bias current in a current mirror changes systematic gain error mainly through changes to the drain-source voltages.

For the simple MOS current mirror, the input voltage is

$$V_{IN} = V_{GS1} = V_t + V_{ov1} = V_t + V_{ov} \tag{4.21}$$

With square-law behavior, the overdrive in (4.21) is proportional to the square root of the input current. In contrast, (4.10) shows that the entire V_{IN} in a simple bipolar mirror is proportional to the natural logarithm of the input current. Therefore, for a given change in the input current, the variation in V_{IN} in a simple MOS current mirror is generally larger than in its bipolar counterpart.

Finally, the minimum output voltage required to keep M_2 in the active region is

$$V_{OUT(min)} = V_{ov2} = V_{ov} = \sqrt{\frac{2I_{OUT}}{k'(W/L)_2}} \tag{4.22}$$

Equation 4.22 predicts that $V_{OUT(min)}$ depends on the transistor geometry and can be made arbitrarily small in a simple MOS mirror, unlike in the bipolar case. However, if the overdrive predicted by (4.22) is less than $2nV_T$, where n is defined in (1.247) and V_T is a thermal voltage, the result is invalid except to indicate that the transistors operate in weak inversion. At room temperature with $n = 1.5$, $2nV_T \simeq 78$ mV. If the transistors operate in weak inversion,

$$V_{OUT(min)} \simeq 3V_T \tag{4.23}$$

as shown in Fig. 1.43.[1]

4.2.3 Simple Current Mirror with Beta Helper

4.2.3.1 Bipolar

In addition to the variation in output current due to finite output resistance, the second term in (4.9) shows that the collector current I_{C2} differs from the input current because of finite β_F. To reduce this source of error, an additional transistor can be added, as shown in Fig. 4.6. If Q_1 and Q_3 are identical, the emitter current of transistor Q_2 is

$$I_{E2} = -\frac{I_{C1}}{\beta_F} - \frac{I_{C3}}{\beta_F} = -\frac{2}{\beta_F} I_{C1} \tag{4.24}$$

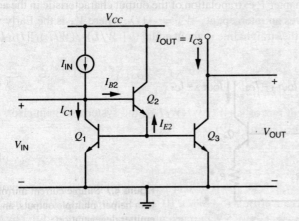

Figure 4.6 Simple current mirror with beta helper.

where I_E, I_C, and I_B are defined as positive when flowing into the transistor, and where we have neglected the effects of finite output resistance. The base current of transistor Q_2 is equal to

$$I_{B2} = -\frac{I_{E2}}{\beta_F + 1} = \frac{2}{\beta_F(\beta_F + 1)} I_{C1} \tag{4.25}$$

Finally, KCL at the collector of Q_1 gives

$$I_{IN} - I_{C1} - \frac{2}{\beta_F(\beta_F + 1)} I_{C1} = 0 \tag{4.26}$$

Since I_{C1} and I_{C3} are equal when Q_1 and Q_3 are identical,

$$I_{OUT} = I_{C3} = \frac{I_{IN}}{1 + \dfrac{2}{\beta_F(\beta_F + 1)}} \simeq I_{IN}\left(1 - \frac{2}{\beta_F(\beta_F + 1)}\right) \tag{4.27}$$

Equation 4.27 shows that the systematic gain error from finite β_F has been reduced by a factor of $[\beta_F + 1]$, which is the current gain of emitter follower Q_2. As a result, Q_2 is often referred to as a *beta helper*.

Although the beta helper has little effect on the output resistance and the minimum output voltage of the current mirror, it increases the input voltage by the base-emitter voltage of Q_2:

$$V_{IN} = V_{BE1(on)} + V_{BE2(on)} \tag{4.28}$$

If multiple emitter followers are cascaded to further reduce the gain error arising from finite β_F, V_{IN} increases by an extra base-emitter voltage for each additional emitter follower, posing one limit to the use of cascaded emitter followers.

Current mirrors often use a beta helper when they are constructed with *pnp* transistors because the value of β_F for *pnp* transistors is usually less than for *npn* transistors. Another application of the beta-helper configuration is in current mirrors with multiple outputs. An example with two independent outputs is shown in Fig. 4.7. At first, ignore Q_2 and imagine that Q_1 is simply diode connected. Also, let $R_1 = R_3 = R_4 = 0$ here. (The effects of nonzero resistances will be considered in Section 4.2.4.) Then the gain from the input to each output is primarily determined by the area ratios I_{S3}/I_{S1} and I_{S4}/I_{S1}. Because the bases of three instead of two transistors are connected together, the total base current is increased here, which increases the gain error from the input to either output arising from finite β_F. Furthermore, the

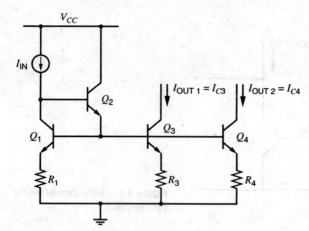

Figure 4.7 Simple current mirror with beta helper, multiple outputs, and emitter degeneration.

gain errors worsen as the number of independent outputs increases. Since the beta helper, Q_2, reduces the gain error from the input to each output by a factor of $[\beta_F + 1]$, it is often used in bipolar current mirrors with multiple outputs.

4.2.3.2 MOS

Since $\beta_F \rightarrow \infty$ for an MOS transistor, beta helpers are not used in simple MOS current mirrors to reduce the systematic gain error. However, a beta-helper configuration can increase the bandwidth of MOS and bipolar current mirrors.

4.2.4 Simple Current Mirror with Degeneration

4.2.4.1 Bipolar

The performance of the simple bipolar transistor current mirror of Fig. 4.6 can be improved by the addition of emitter degeneration as shown in Fig. 4.7 for a current mirror with two independent outputs. The purpose of the emitter resistors is twofold. First, Appendix A.4.1 shows that the matching between I_{IN} and outputs I_{C3} and I_{C4} can be greatly improved by using emitter degeneration. Second, as shown in Section 3.3.8, the use of emitter degeneration boosts the output resistance of each output of the current mirror. Transistors Q_1 and Q_2 combine to present a very low resistance at the bases of Q_3 and Q_4. Therefore, from (3.99), the small-signal output resistance seen at the collectors of Q_3 and Q_4 is

$$R_o \simeq r_o(1 + g_m R_E) \tag{4.29}$$

if $r_\pi \gg R_E$. Taking Q_3 as an example and using $g_{m3} = I_{C3}/V_T$, we find

$$R_o \simeq r_{o3}\left(1 + \frac{I_{C3} R_3}{V_T}\right) \tag{4.30}$$

This increase in the output resistance for a given output current also decreases the component of systematic gain error that stems from finite output resistance by the same factor. From (4.9) and (4.30) with infinite β_F,

$$\epsilon \simeq \frac{V_{CE2} - V_{CE1}}{V_A \left(1 + \dfrac{I_{C3} R_3}{V_T}\right)} \tag{4.31}$$

The quantity $I_{C3}R_3$ is just the dc voltage drop across R_3. If this quantity is 260 mV, for example, then R_o is about $10r_o$ at room temperature, and ϵ is reduced by a factor of about eleven. Unfortunately, this improvement in R_o is limited by corresponding increases in the input and minimum output voltages of the mirror:

$$V_{IN} \simeq V_{BE1(on)} + V_{BE2(on)} + I_{IN}R_1 \qquad (4.32)$$

and

$$V_{OUT(min)} = V_{CE3(sat)} + I_{C3}R_3 \qquad (4.33)$$

The emitter areas of Q_1, Q_3, and Q_4 may be matched or ratioed. For example, if we want $I_{OUT1} = I_{IN}$ and $I_{OUT2} = 2I_{IN}$, we would make Q_3 identical to Q_1, and Q_4 consist of two copies of Q_1 connected in parallel so that $I_{S4} = 2I_{S1}$. In addition, we could make $R_3 = R_1$, and R_4 consist of two copies of R_1 connected in parallel so that $R_4 = R_1/2$. Note that all the dc voltage drops across R_1, R_3, and R_4 would then be equal. Using KVL around the loop including Q_1 and Q_4 and neglecting base currents, we find

$$I_{C1}R_1 + V_T \ln \frac{I_{C1}}{I_{S1}} = I_{C4}R_4 + V_T \ln \frac{I_{C4}}{I_{S4}} \qquad (4.34)$$

from which

$$I_{OUT2} = I_{C4} = \frac{1}{R_4}\left(I_{IN}R_1 + V_T \ln \frac{I_{IN}}{I_{C4}} \frac{I_{S4}}{I_{S1}}\right) \qquad (4.35)$$

Since $I_{S4} = 2I_{S1}$, the solution to (4.35) is

$$I_{OUT2} = \frac{R_1}{R_4} I_{IN} = 2I_{IN} \qquad (4.36)$$

because the last term in (4.35) goes to zero. If we make the voltage drops $I_{IN}R_1$ and $I_{C4}R_4$ much greater than V_T, the current-mirror gain to the Q_4 output is determined primarily by the resistor ratio R_4/R_1, and only to a secondary extent by the emitter area ratio, because the natural log term in (4.35) varies slowly with its argument.

4.2.4.2 MOS
Source degeneration is rarely used in MOS current mirrors because, in effect, MOS transistors are inherently controlled resistors. Thus, matching in MOS current mirrors is improved simply by increasing the gate areas of the transistors.[2,3,4] Furthermore, the output resistance can be increased by increasing the channel length. To increase the output resistance while keeping the current and $V_{GS} - V_t$ constant, the W/L ratio must be held constant. Therefore, the channel width must be increased as much as the length, and the price paid for the improved output resistance is that increased chip area is consumed by the current mirror.

4.2.5 Cascode Current Mirror

4.2.5.1 Bipolar
Section 3.4.2 shows that the cascode connection achieves a very high output resistance. Since this is a desirable characteristic for a current mirror, exploring the use of cascodes for high-performance current mirrors is natural. A bipolar-transistor current mirror based on the cascode connection is shown in Fig. 4.8. Transistors Q_3 and Q_1 form a simple current mirror, and emitter resistances can be added to improve the matching. Transistor Q_2 acts as the common-base part of the cascode and transfers the collector current of Q_1 to the output while presenting a high

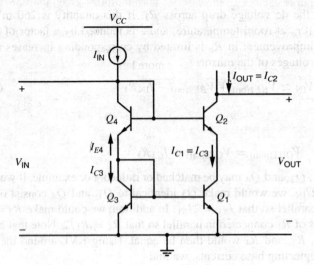

Figure 4.8 Cascode current mirror with bipolar transistors.

output resistance. Transistor Q_4 acts as a diode level shifter and biases the base of Q_2 so that Q_1 operates in the forward-active region with $V_{CE1} \simeq V_{CE3} = V_{BE3(on)}$. If we assume that the small-signal resistances of diodes Q_3 and Q_4 are small, a direct application of (3.98) with $R_E = r_{o1}$ concludes that

$$R_o = r_{o2}\left(1 + \frac{g_{m2}r_{o1}}{1 + \frac{g_{m2}r_{o1}}{\beta_0}}\right) \simeq \beta_0 r_{o2} \qquad (4.37)$$

because $g_{m2}r_{o1} \simeq g_{m1}r_{o1} \gg \beta_0$. This calculation assumes that almost all of the small-signal current that flows into the collector of Q_2 flows out its base because the small-signal resistance connected to the emitter of Q_2 is much greater than that connected to its base. A key problem with this calculation, however, is that it ignores the effect of the simple current mirror formed by Q_3 and Q_1. Let i_{b2} and i_{e2} represent increases in the base and emitter currents flowing out of Q_2 caused by increasing output voltage. Then the simple mirror forces $i_{e2} \simeq i_{b2}$. As a result, the variation in the collector current of Q_2 splits into two equal parts and half flows in $r_{\pi 2}$. A small-signal analysis shows that R_o in (4.37) is reduced by half to

$$R_o \simeq \frac{\beta_0 r_{o2}}{2} \qquad (4.38)$$

Thus, the cascode configuration boosts the output resistance by approximately $\beta_0/2$. For $\beta_0 = 100$, $V_A = 130$ V, and $I_{C2} = 1$ mA,

$$R_o = \frac{\beta_0 V_A}{2I_{C2}} = \frac{100(130)}{2 \text{ mA}} = 6.5 \text{ M}\Omega \qquad (4.39)$$

In this calculation of output resistance, we have neglected the effects of r_μ. Although this assumption is easy to justify in the case of the simple current mirror, it must be reexamined here because the output resistance is so high. The collector-base resistance r_μ results from modulation of the base-recombination current as a consequence of the Early effect, as described in Chapter 1. For a transistor whose base current is composed entirely of base-recombination current, the percentage change in base current when V_{CE} is changed at a constant V_{BE} would equal that of the collector current, and r_μ would be equal to $\beta_0 r_o$. In this case, the effect of

r_μ would be to reduce the output resistance of the cascode current mirror given in (4.38) by a factor of 1.5.

In actual integrated-circuit *npn* transistors, however, only a small percentage of the base current results from recombination in the base. Since only this component is modulated by the Early effect, the observed values of r_μ are a factor of 10 or more larger than $\beta_0 r_o$. Therefore, r_μ has a negligible effect here with *npn* transistors. On the other hand, for lateral *pnp* transistors, the feedback resistance r_μ is much smaller than for *npn* transistors because most of the base current results from base-region recombination. The actual value of this resistance depends on a number of process and device-geometry variables, but observed values range from 2 to 5 times $\beta_0 r_o$. Therefore, for a cascode current mirror constructed with lateral *pnp* transistors, the effect of r_μ on the output resistance can be significant. Furthermore, when considering current mirrors that give output resistances higher than $\beta_0 r_o$, the effects of r_μ must be considered.

In the cascode current mirror, the base of Q_1 is connected to a low-resistance point because Q_3 is diode connected. As a result, feedback from $r_{\mu 1}$ is greatly attenuated and has negligible effect on the output resistance. On the other hand, if the resistance from the base of Q_1 to ground is increased while all other parameters are held constant, local feedback from $r_{\mu 1}$ significantly affects the base-emitter voltage of Q_1 and reduces the output resistance. In the limit where the resistance from the base of Q_1 to ground becomes infinite, Q_1 acts as if it were diode connected. Local feedback is considered in Chapter 8.

The input voltage of the cascode current mirror is

$$V_{IN} = V_{BE3} + V_{BE4} = 2V_{BE(on)} \tag{4.40}$$

Although V_{IN} is higher here than in (4.10) for a simple current mirror, the increase becomes a limitation only if the power-supply voltage is reduced to nearly two diode drops.

The minimum output voltage for which the output resistance is given by (4.38) must allow both Q_1 and Q_2 to be biased in the forward-active region. Since $V_{CE1} \simeq V_{CE3} = V_{BE(on)}$,

$$V_{OUT(min)} = V_{CE1} + V_{CE2(sat)} \simeq V_{BE(on)} + V_{CE2(sat)} \tag{4.41}$$

Comparing (4.41) and (4.11) shows that the minimum output voltage for a cascode current mirror is higher than for a simple current mirror by a diode drop. This increase poses an important limitation on the minimum supply voltage when the current mirror is used as an active load for an amplifier.

Since $V_{CE1} \simeq V_{CE3}$, $I_{C1} \simeq I_{C3}$, and the systematic gain error arising from finite transistor output resistance is almost zero. A key limitation of the cascode current mirror, however, is that the systematic gain error arising from finite β_F is worse than for a simple current mirror. From KCL at the collector of Q_3,

$$-I_{E4} = I_{C3} + \frac{2I_{C3}}{\beta_F} \tag{4.42}$$

From KCL at the collector of Q_4,

$$I_{IN} = -I_{E4} + \frac{I_{C2}}{\beta_F} \tag{4.43}$$

The collector current of Q_2 is

$$I_{C2} = \frac{\beta_F}{\beta_F + 1} I_{C3} \tag{4.44}$$

Substituting (4.42) and (4.44) into (4.43) gives

$$I_{IN} = I_{C3} + \frac{2I_{C3}}{\beta_F} + \frac{I_{C3}}{\beta_F + 1} \tag{4.45}$$

Rearranging (4.45) to find I_{C3} and substituting back into (4.44) gives

$$I_{OUT} = I_{C2} = \left(\frac{\beta_F}{\beta_F + 1}\right)\left(\frac{I_{IN}}{1 + \frac{2}{\beta_F} + \frac{1}{\beta_F + 1}}\right) \quad (4.46)$$

Equation 4.46 can be rearranged to give

$$I_{OUT} = I_{IN}\left(1 - \frac{4\beta_F + 2}{\beta_F^2 + 4\beta_F + 2}\right) \quad (4.47)$$

Equation 4.47 shows that the systematic gain error is

$$\epsilon = -\frac{4\beta_F + 2}{\beta_F^2 + 4\beta_F + 2} \quad (4.48)$$

When $\beta_F \gg 1$, (4.48) simplifies to

$$\epsilon \simeq -\frac{4}{\beta_F + 4} \quad (4.49)$$

In contrast, the systematic gain error stemming from finite β_F in a simple current mirror with identical transistors is about $-2/\beta_F$, which is less in magnitude than (4.49) predicts for a cascode current mirror if $\beta_F > 4$. This limitation of a cascode current mirror is overcome by the Wilson current mirror described in Section 4.2.6.

4.2.5.2 MOS

The cascode current mirror is widely used in MOS technology, where it does not suffer from finite β_F effects. Figure 4.9 shows the simplest form. From (3.107), the small-signal output resistance is

$$R_o = r_{o2}[1 + (g_{m2} + g_{mb2})r_{o1}] + r_{o1} \quad (4.50)$$

As shown in the previous section, the bipolar cascode current mirror cannot realize an output resistance larger than $\beta_0 r_o/2$ because β_0 is finite and nonzero small-signal base current flows in the cascode transistor. In contrast, the MOS cascode is capable of realizing arbitrarily high output resistance by increasing the number of stacked cascode devices because $\beta_0 \to \infty$ for MOS transistors. However, the MOS substrate leakage current described in Section 1.9 can create a resistive shunt to ground from the output node, which can dominate the output resistance for $V_{OUT} > V_{OUT(min)}$.[5]

■ **EXAMPLE**

Find the output resistance of the double-cascode current mirror shown in Fig. 4.10. Assume all the transistors operate in the active region with $I_D = 10$ μA, $V_A = 50$ V, and $g_m r_o = 50$. Neglect body effect.

The output resistance of each transistor is

$$r_o = \frac{V_A}{I_D} = \frac{50 \text{ V}}{10 \text{ μA}} = 5 \text{ M}\Omega$$

From (4.50), looking into the drain of M_2:

$$R_{o2} = r_{o2}(1 + g_{m2}r_{o1}) + r_{o1} \quad (4.51)$$

4.2 Current Mirrors   265

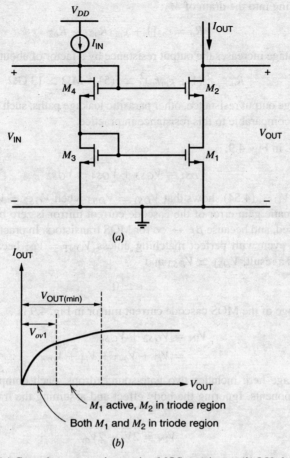

Figure 4.9 (a) Cascode current mirror using MOS transistors. (b) I-V characteristic.

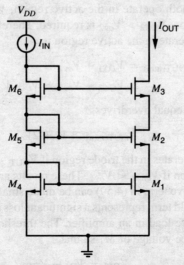

Figure 4.10 Example of a double-cascode current mirror.

Similarly, looking into the drain of M_3:

$$R_o = r_{o3}[1 + g_{m3}R_{o2}] + R_{o2} \tag{4.52}$$

Each cascode stage increases the output resistance by a factor of about $(1 + g_m r_o)$. Therefore,

$$R_o \simeq r_o(1 + g_m r_o)^2 \simeq 5(51)^2 \text{ M}\Omega \simeq 13 \text{ G}\Omega \tag{4.53}$$

With such a large output resistance, other parasitic leakage paths, such as the substrate leakage path, could be comparable to this resistance in practice.

From KVL in Fig. 4.9,

$$V_{DS1} = V_{GS3} + V_{GS4} - V_{GS2} \tag{4.54}$$

Since $V_{DS3} = V_{GS3}$, (4.54) shows that $V_{DS1} = V_{DS3}$ when $V_{GS2} = V_{GS4}$. Under this condition, the systematic gain error of the cascode current mirror is zero because M_1 and M_3 are identically biased, and because $\beta_F \to \infty$ for MOS transistors. In practice, V_{GS2} is not exactly equal to V_{GS4} even with perfect matching unless $V_{OUT} = V_{IN}$ because of channel-length modulation. As a result, $V_{DS1} \simeq V_{DS3}$ and

$$\epsilon \simeq 0 \tag{4.55}$$

The input voltage of the MOS cascode current mirror in Fig. 4.9 is

$$\begin{aligned} V_{IN} &= V_{GS3} + V_{GS4} \\ &= V_{t3} + V_{ov3} + V_{t4} + V_{ov4} \end{aligned} \tag{4.56}$$

The input voltage here includes two gate-source drops, each composed of threshold and overdrive components. Ignoring the body effect and assuming the transistors all have equal overdrives,

$$V_{IN} = 2V_t + 2V_{ov} \tag{4.57}$$

Also, adding extra cascode levels increases the input voltage by another threshold and another overdrive component for each additional cascode. Furthermore, the body effect increases the threshold of all transistors with $V_{SB} > 0$. Together, these facts increase the difficulty of designing the input current source for low power-supply voltages.

When M_1 and M_2 both operate in the active region, $V_{DS1} \simeq V_{DS3} = V_{GS3}$. For M_2 to operate in the active region, $V_{DS2} > V_{ov2}$ is required. Therefore, the minimum output voltage for which M_1 and M_2 operate in the active region is

$$\begin{aligned} V_{OUT(min)} &= V_{DS1} + V_{ov2} \\ &\simeq V_{GS3} + V_{ov2} = V_t + V_{ov3} + V_{ov2} \end{aligned} \tag{4.58}$$

If the transistors all have equal overdrives,

$$V_{OUT(min)} \simeq V_t + 2V_{ov} \tag{4.59}$$

On the other hand, M_2 operates in the triode region if $V_{OUT} < V_{OUT(min)}$, and both M_1 and M_2 operate in the triode region if $V_{OUT} < V_{ov1}$. These results are shown graphically in Fig. 4.9b.

Although the overdrive term in (4.59) can be made small by using large values of W for a given current, the threshold term represents a significant loss of voltage swing when the current mirror is used as an active load in an amplifier. The threshold term in (4.59) stems from the biasing of the drain-source voltage of M_1 so that

$$V_{DS1} = V_{IN} - V_{GS2} \tag{4.60}$$

Ignoring the body effect and assuming that M_1-M_4 all operate in the active region with equal overdrives,

$$V_{DS1} = V_t + V_{ov} \qquad (4.61)$$

Therefore, the drain-source voltage of M_1 is a threshold larger than necessary to operate M_1 in the active region. To reduce V_{DS1}, the voltage from the gate of M_2 to ground can be level shifted down by a threshold as shown in Fig. 4.11a. In practice, a source follower is used to implement the level shift, as shown in Fig. 4.11b.[6] Transistor M_5 acts as the source follower and is biased by the output of the simple current mirror M_3 and M_6. Because the gate-source voltage of M_5 is greater than its threshold by the overdrive, however, the drain-source voltage of M_1 would be zero with equal thresholds and overdrives on all transistors. To bias M_1 at the

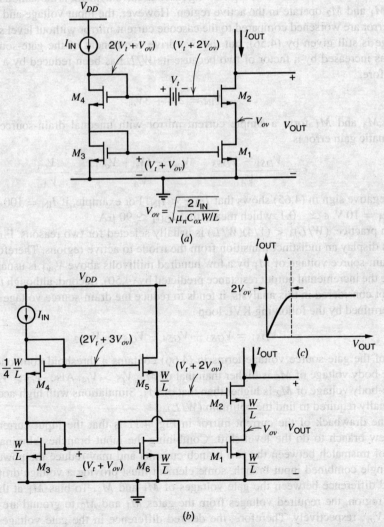

Figure 4.11 (a) MOS cascode current mirror with improved biasing for maximum voltage swing. (b) Practical implementation. (c) I-V characteristic.

boundary between the active and triode regions,

$$V_{DS1} = V_{ov} \tag{4.62}$$

is required. Therefore, the overdrive on M_4 is doubled by reducing its W/L by a factor of four to satisfy (4.62). As a result, the threshold term in (4.59) is eliminated and

$$V_{OUT(min)} \simeq 2V_{ov} \tag{4.63}$$

Because the minimum output voltage does not contain a threshold component, the range of output voltages for which M_1 and M_2 both operate in the active region is significantly improved. Therefore, the current mirror in Fig. 4.11 places much less restriction on the range of output voltages that can be achieved in an amplifier using this current mirror as an active load than the mirror in Fig. 4.9. For this reason, the mirror in Fig. 4.11 is called a *high-swing* cascode current mirror. This type of level shifting to reduce $V_{OUT(min)}$ can also be applied to bipolar circuits.

The output resistance of the high-swing cascode current mirror is the same as in (4.50) when both M_1 and M_2 operate in the active region. However, the input voltage and the systematic gain error are worsened compared to the cascode current mirror without level shift. The input voltage is still given by (4.56), but the overdrive component of the gate-source voltage of M_4 has increased by a factor of two because its W/L has been reduced by a factor of four. Therefore,

$$V_{IN} = 2V_t + 3V_{ov} \tag{4.64}$$

Since M_3 and M_1 form a simple current mirror with unequal drain-source voltages, the systematic gain error is

$$\epsilon = \frac{V_{DS1} - V_{DS3}}{V_A} \simeq \frac{V_{ov1} - (V_t + V_{ov1})}{V_A} = -\frac{V_t}{V_A} \tag{4.65}$$

The negative sign in (4.65) shows that $I_{OUT} < I_{IN}$. For example, if $I_{IN} = 100$ μA, $V_t = 1$ V, and $V_A = 10$ V, $\epsilon \simeq -0.1$, which means that $I_{OUT} \simeq 90$ μA.

In practice, $(W/L)_4 < (1/4)(W/L)$ is usually selected for two reasons. First, MOS transistors display an indistinct transition from the triode to active regions. Therefore, increasing the drain-source voltage of M_1 by a few hundred millivolts above V_{ov1} is usually required to realize the incremental output resistance predicted by (4.50). Second, although the body effect was not considered in this analysis, it tends to reduce the drain-source voltage on M_1, which is determined by the following KVL loop

$$V_{DS1} = V_{GS3} + V_{GS4} - V_{GS5} - V_{GS2} \tag{4.66}$$

Each of the gate-source voltage terms in (4.66) contains a threshold component. Since the source-body voltage of M_5 is higher than that of M_4, $V_{t5} > V_{t4}$. Also, $V_{t2} > V_{t3}$ because the source-body voltage of M_2 is higher than that of M_3. Simulations with high-accuracy models are usually required to find the optimum $(W/L)_4$.

One drawback of the current mirror in Fig. 4.11 is that the input current is mirrored to a new branch to do the level shift. Combining the input branches eliminates the possibility of mismatch between the two branch currents and may reduce the power dissipation. In a single combined input branch, some element must provide a voltage drop equal to the desired difference between the gate voltages of M_1 and M_2. To bias M_1 at the edge of the active region, the required voltages from the gates M_1 and M_2 to ground are $V_t + V_{ov}$ and $V_t + 2V_{ov}$, respectively. Therefore, the desired difference in the gate voltages is V_{ov}. This voltage difference can be developed across the drain to the source of a transistor deliberately operated in the triode region, as shown in Fig. 4.12a.[7] Since M_6 is diode connected, it operates

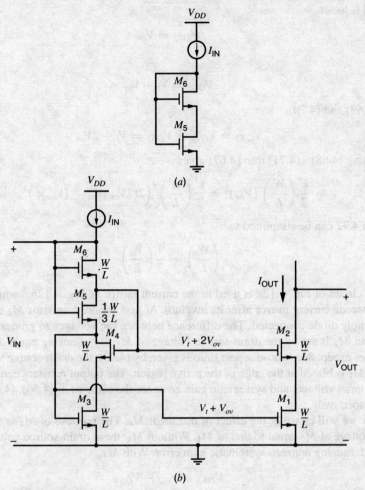

Figure 4.12 (a) Circuit that forces M_5 to operate in the triode region. (b) Sooch cascode current mirror using the circuit in (a).

in the active region as long as the input current and threshold are positive. However, since the gate-source voltage of M_6 is equal to the gate-drain voltage of M_5, a channel exists at the drain of M_5 when it exists at the source of M_6. In other words, M_6 forces M_5 to operate in the triode region.

To use the circuit in Fig. 4.12a in a current mirror, we would like to choose the aspect ratios of the transistors so that the drain-source voltage of M_5 is V_{ov}. Since M_6 operates in the active region,

$$I_{IN} = \frac{k'}{2}\left(\frac{W}{L}\right)_6 (V_{GS6} - V_t)^2 \qquad (4.67)$$

Since M_5 operates in the triode region,

$$I_{IN} = \frac{k'}{2}\left(\frac{W}{L}\right)_5 \left(2(V_{GS5} - V_t)V_{DS5} - (V_{DS5})^2\right) \qquad (4.68)$$

The goal is to set

$$V_{DS5} = V_{ov} \quad (4.69)$$

when

$$V_{GS6} = V_t + V_{ov} \quad (4.70)$$

From (4.69) and (4.70),

$$V_{GS5} = V_{GS6} + V_{DS5} = V_t + 2V_{ov} \quad (4.71)$$

Substituting (4.68) - (4.71) into (4.67) gives

$$\frac{k'}{2}\left(\frac{W}{L}\right)_6 (V_{ov})^2 = \frac{k'}{2}\left(\frac{W}{L}\right)_5 \left(2(2V_{ov})V_{ov} - (V_{ov})^2\right) \quad (4.72)$$

Equation 4.72 can be simplified to

$$\left(\frac{W}{L}\right)_5 = \frac{1}{3}\left(\frac{W}{L}\right)_6 \quad (4.73)$$

The circuit of Fig. 4.12a is used in the current mirror of Fig. 4.12b,[7] which is called the *Sooch* cascode current mirror after its inventor. At first, ignore transistor M_4 and assume that M_3 is simply diode connected. The difference between the voltages to ground from the gates of M_1 and M_2 is set by the drain-source voltage of M_5. By choosing equal aspect ratios for all devices except M_5, whose aspect ratio is given by (4.73), the drain-source voltage of M_5 is V_{ov} and M_1 is biased at the edge of the active region. The output resistance, minimum output voltage, input voltage, and systematic gain error are the same as in (4.50), (4.63), (4.64), and (4.65) respectively.

Now we will consider the effect of transistor M_4. The purpose of M_4 is to set the drain-source voltage of M_3 equal to that of M_1. Without M_4, these drain-source voltages differ by a threshold, causing nonzero systematic gain error. With M_4,

$$V_{DS3} = V_{G2} - V_{GS4} \quad (4.74)$$

where

$$V_{G2} = V_{GS3} + V_{DS5} \quad (4.75)$$

Ignoring channel-length modulation,

$$V_{G2} = (V_t + V_{ov}) + V_{ov} = V_t + 2V_{ov} \quad (4.76)$$

Ignoring the body effect and assuming that M_4 operates in the active region,

$$V_{GS4} = V_t + V_{ov} \quad (4.77)$$

Then substituting (4.76) and (4.77) into (4.74) gives

$$V_{DS3} = V_{ov} \quad (4.78)$$

If M_2 also operates in the active region under these conditions, $V_{DS3} = V_{DS1}$. As a result, the systematic gain error is

$$\epsilon = 0 \quad (4.79)$$

Therefore, the purpose of M_4 is to equalize the drain-source voltages of M_3 and M_1 to reduce the systematic gain error.

For M_4 to operate in the active region, $V_{DS4} > V_{ov}$ is required. Since

$$V_{DS4} = V_{GS3} - V_{DS3} = (V_t + V_{ov}) - V_{ov} = V_t \tag{4.80}$$

Equation 4.80 shows that M_4 operates in the active region if $V_t > V_{ov}$. Although this condition is usually satisfied, a low threshold and/or high overdrive may cause M_4 to operate in the triode region. If this happens, the gate-source voltage of M_4 depends strongly on its drain-source voltage, increasing the systematic gain error. Since increasing temperature causes the threshold to decrease, but the overdrive to increase, checking the region of operation of M_4 in simulation at the maximum expected operating temperature is important in practice.

The main limitation of the high-swing cascode current mirrors just presented, is that the input voltage is large. In Fig. 4.11, the input voltage is the sum of the gate-source voltages of M_3 and M_4 and is given by (4.64) ignoring body effect. In Fig. 4.12, the input voltage is

$$\begin{aligned} V_{IN} &= V_{GS3} + V_{DS5} + V_{GS6} \\ &= V_t + V_{ov} + V_{ov} + V_t + V_{ov} \\ &= 2V_t + 3V_{ov} \end{aligned} \tag{4.81}$$

Equation 4.81 shows that the input voltage of the high-swing cascode current mirror in Fig. 4.12 is the same as in (4.64) for Fig. 4.11. The large input voltages may limit the minimum power-supply voltage because a transistor-level implementation of the input current source requires some nonzero drop for proper operation. With threshold voltages of about 1 V, the cascode current mirrors in Figs. 4.11 and 4.12 can operate properly for power-supply voltages greater than about 3 V. Below about 2 V, however, reduced thresholds or a new configuration is required. Reducing the magnitude of the threshold for all transistors increases the difficulty in turning off transistors that are used as switches. This problem can be overcome by using low-threshold devices in the current mirror and high-threshold devices as switches, but this solution increases process complexity and cost. Therefore, circuit techniques to reduce the input voltage are important to minimize cost.

To reduce the input voltage, the input branch can be split into two branches, as shown in Fig. 4.13. If M_1 and M_2 are biased in the active region, the output resistance is still given by (4.50). Also, the minimum output voltage for which (4.50) applies is still given by (4.63). Furthermore, if M_4 operates in the active region, the drain-source voltage of M_3 is equal to that of M_1, and the systematic gain error is still zero as in (4.79).

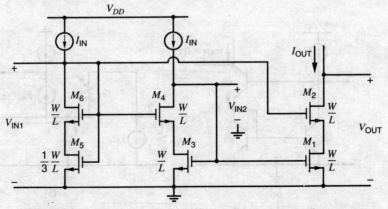

Figure 4.13 MOS high-swing current mirror with two input branches.

Since the mirror in Fig. 4.13 has two input branches, an input voltage can be calculated for each:

$$V_{IN1} = V_{DS5} + V_{GS6} = V_t + 2V_{ov} \qquad (4.82)$$

$$V_{IN2} = V_{GS3} = V_t + V_{ov} \qquad (4.83)$$

Both V_{IN1} and V_{IN2} are less than the input voltage given in (4.64) for Fig. 4.12b by more than a threshold, allowing the input current sources to operate properly with power-supply voltages greater than about 2 V, assuming thresholds of about 1 V.

Finally, in Fig. 4.13, the drain-source voltage of M_5 is only used to bias the source of M_6. Therefore, M_5 and M_6 can be collapsed into one diode-connected transistor whose source is grounded. Call this replacement transistor M_7. The aspect ratio of M_7 should be a factor of four smaller than the aspect ratios of M_1-M_4 to maintain the bias conditions as in Fig. 4.13. In practice, the aspect ratio of M_7 is further reduced to bias M_1 past the edge of the active region and to overcome a mismatch in the thresholds of M_7 and M_2 caused by body effect.

4.2.6 Wilson Current Mirror

4.2.6.1 Bipolar

The main limitation of the bipolar cascode current mirror is that the systematic gain error stemming from finite β_F was large, as given in (4.49). To overcome this limitation, the Wilson current mirror can be used as shown in Fig. 4.14a.[8] This circuit uses negative feedback through Q_1, activating Q_3 to reduce the base-current error and raise the output resistance. (See Chapter 8.)

From a qualitative standpoint, the difference between the input current and I_{C3} flows into the base of Q_2. This base current is multiplied by $(\beta_F + 1)$ and flows in the diode-connected transistor Q_1, which causes current of the same magnitude to flow in Q_3. A feedback path is thus formed that regulates I_{C3} so that it is nearly equal to the input current, reducing the systematic gain error caused by finite β_F. Similarly, when the output voltage increases, the collector current of Q_2 also increases, in turn increasing the collector current of Q_1. As a result, the collector current of Q_3 increases, which reduces the base current of Q_2. The decrease in the base current of Q_2 caused by negative feedback reduces the original change in the collector current of Q_2 and increases the output resistance.

To find the output resistance of the Wilson current mirror when all transistors operate in the active region, we will analyze the small-signal model shown in Fig. 4.14b, in which a

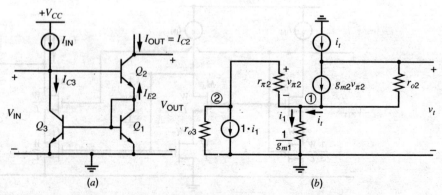

Figure 4.14 (a) Bipolar Wilson current mirror. (b) Small-signal model.

test current source i_t is applied at the output. Transistors Q_1 and Q_3 form a simple current mirror. Since Q_1 is diode connected, the small-signal resistance from the base of Q_1 to ground is $(1/g_{m1})||r_{\pi1}||r_{\pi3}||r_{o1}$. Assume that an unknown current i_1 flows in this resistance. When $g_{m1}r_{\pi1} \gg 1$, $g_{m1}r_{\pi3} \gg 1$, and $g_{m1}r_{o1} \gg 1$, this resistance is approximately equal to $1/g_{m1}$. Transistor Q_3 could be modeled as a voltage-controlled current source of value $g_{m3}v_{\pi3}$ in parallel with r_{o3}. Since $v_{\pi3} = v_{\pi1} \simeq i_1/g_{m1}$, the voltage-controlled current source in the model for Q_3 can be replaced by a current-controlled current source of value $(g_{m3}/g_{m1})(i_1) = 1(i_1)$, as shown in Fig. 4.14b. This model represents the behavior of the simple current mirror directly: The input current i_1 is mirrored to the output by the current-controlled current source.

Using this model, the resulting voltage v_t is

$$v_t = \frac{i_1}{g_{m1}} + (i_t - g_{m2}v_{\pi2})r_{o2} \quad (4.84)$$

To find the relationship between i_1 and $v_{\pi2}$, note that the voltage across r_{o3} is $(i_1/g_{m1} + v_{\pi2})$ and use KCL at node ② in Fig. 4.14b to show that

$$\frac{v_{\pi2}}{r_{\pi2}} + i_1 + \frac{\frac{i_1}{g_{m1}} + v_{\pi2}}{r_{o3}} = 0 \quad (4.85)$$

Rearranging (4.85) gives

$$v_{\pi2} = -i_1 r_{\pi2} \left(\frac{1 + \dfrac{1}{g_{m1}r_{o3}}}{1 + \dfrac{r_{\pi2}}{r_{o3}}} \right) \quad (4.86)$$

To find the relationship between i_1 and i_t, use KCL at node ① in Fig. 4.14b to show that

$$i_t = i_1 - \frac{v_{\pi2}}{r_{\pi2}} \quad (4.87)$$

Substituting (4.86) into (4.87) and rearranging gives

$$i_1 = \frac{i_t}{1 + \left(\dfrac{1 + \dfrac{1}{g_{m1}r_{o3}}}{1 + \dfrac{r_{\pi2}}{r_{o3}}} \right)} \quad (4.88)$$

Substituting (4.88) into (4.86) and rearranging gives

$$v_{\pi2} = -i_t r_{\pi2} \left(\frac{1 + \dfrac{1}{g_{m1}r_{o3}}}{2 + \dfrac{r_{\pi2}}{r_{o3}} + \dfrac{1}{g_{m1}r_{o3}}} \right) \quad (4.89)$$

Substituting (4.88) and (4.89) into (4.84) and rearranging gives

$$R_o = \frac{v_t}{i_t} = \frac{1}{g_{m1} \left[1 + \left(\dfrac{1 + \dfrac{1}{g_{m1}r_{o3}}}{1 + \dfrac{r_{\pi2}}{r_{o3}}} \right) \right]} + r_{o2} + \frac{g_{m2}r_{\pi2}r_{o2}\left(1 + \dfrac{1}{g_{m1}r_{o3}}\right)}{2 + \dfrac{r_{\pi2}}{r_{o3}} + \dfrac{1}{g_{m1}r_{o3}}} \quad (4.90)$$

If $r_{o3} \to \infty$, the small-signal current that flows in the collector of Q_3 is equal to i_1 and (4.90) reduces to

$$R_o = \frac{1}{g_{m1}(2)} + r_{o2} + \frac{g_{m2}r_{\pi 2}r_{o2}}{2} \simeq \frac{\beta_0 r_{o2}}{2} \qquad (4.91)$$

This result is the same as (4.38) for the cascode current mirror. In the cascode current mirror, the small-signal current that flows in the base of Q_2 is mirrored through Q_3 to Q_1 so that the small-signal base and emitter currents leaving Q_2 are approximately equal. On the other hand, in the Wilson current mirror, the small-signal current that flows in the emitter of Q_2 is mirrored through Q_1 to Q_3 and then flows in the base of Q_2. Although the cause and effect relationship here is opposite of that in a cascode current mirror, the output resistance is unchanged because the small-signal base and emitter currents leaving Q_2 are still forced to be equal. Therefore, the small-signal collector current of Q_2 that flows because of changes in the output voltage still splits into two equal parts with half flowing in $r_{\pi 2}$.

For the purpose of dc analysis, we assume that $V_A \to \infty$ and that the transistors are identical. Then the input voltage is

$$V_{IN} = V_{CE3} = V_{BE1} + V_{BE2} = 2V_{BE(\text{on})} \qquad (4.92)$$

which is the same as in (4.40) for a cascode current mirror. Also, the minimum output voltage for which both transistors in the output branch operate in the forward-active region is

$$V_{OUT(\text{min})} = V_{CE1} + V_{CE2(\text{sat})} = V_{BE(\text{on})} + V_{CE2(\text{sat})} \qquad (4.93)$$

The result in (4.93) is the same as in (4.41) for a cascode current mirror.

To find the systematic gain error, start with KCL at the collector of Q_1 to show that

$$-I_{E2} = I_{C1} + I_{B1} + I_{B3} = I_{C1}\left(1 + \frac{1}{\beta_F}\right) + \frac{I_{C3}}{\beta_F} \qquad (4.94)$$

Since we assumed that the transistors are identical and $V_A \to \infty$,

$$I_{C3} = I_{C1} \qquad (4.95)$$

Substituting (4.95) into (4.94) gives

$$-I_{E2} = I_{C1}\left(1 + \frac{2}{\beta_F}\right) \qquad (4.96)$$

Using (4.96), the collector current of Q_2 is then

$$I_{C2} = -I_{E2}\left(\frac{\beta_F}{1+\beta_F}\right) = I_{C1}\left(1 + \frac{2}{\beta_F}\right)\left(\frac{\beta_F}{1+\beta_F}\right) \qquad (4.97)$$

Rearranging (4.97) we obtain

$$I_{C1} = I_{C2}\left[\frac{1}{\left(1+\dfrac{2}{\beta_F}\right)\left(\dfrac{\beta_F}{1+\beta_F}\right)}\right] \qquad (4.98)$$

From KCL at the base of Q_2,

$$I_{C3} = I_{IN} - \frac{I_{C2}}{\beta_F} \qquad (4.99)$$

Inserting (4.98) and (4.99) into (4.95), we find that

$$I_{OUT} = I_{C2} = I_{IN}\left(1 - \frac{2}{\beta_F^2 + 2\beta_F + 2}\right) = \frac{I_{IN}}{1 + \frac{2}{\beta_F(\beta_F + 2)}} \quad (4.100)$$

In the configuration shown in Fig. 4.14a, the systematic gain error arising from finite output resistance is not zero because Q_3 and Q_1 operate with collector-emitter voltages that differ by the base-emitter voltage of Q_2. With finite V_A and finite β_F,

$$\begin{aligned} I_{OUT} &\simeq I_{IN}\left(1 - \frac{2}{\beta_F^2 + 2\beta_F + 2}\right)\left(1 + \frac{V_{CE1} - V_{CE3}}{V_A}\right) \\ &\simeq I_{IN}\left(1 - \frac{2}{\beta_F^2 + 2\beta_F + 2}\right)\left(1 - \frac{V_{BE2}}{V_A}\right) \end{aligned} \quad (4.101)$$

Therefore, the systematic gain error is

$$\epsilon \simeq -\left(\frac{2}{\beta_F^2 + 2\beta_F + 2} + \frac{V_{BE2}}{V_A}\right) \quad (4.102)$$

Comparing (4.102) to (4.49) shows two key points. First, the systematic gain error arising from finite β_F in a Wilson current mirror is much less than in a cascode current mirror. Second, the systematic gain error arising from finite output resistance is worse in the Wilson current mirror shown in Fig. 4.14a than in the cascode current mirror shown in Fig. 4.9. However, this limitation is not fundamental because it can be overcome by introducing a new diode-connected transistor between the collector of Q_3 and the base of Q_2 to equalize the collector-emitter voltages of Q_3 and Q_1.

4.2.6.2 MOS

Wilson current mirrors are also used in MOS technology, as shown in Fig. 4.15. Ignoring M_4, the circuit operation is essentially identical to the bipolar case with $\beta_F \to \infty$. One way to calculate the output resistance is to let $r_{\pi 2} \to \infty$ in (4.90), which gives

$$R_o = \frac{1}{g_{m1}} + r_{o2} + g_{m2}r_{o2}\left(1 + \frac{1}{g_{m1}r_{o3}}\right)r_{o3} \simeq (1 + g_{m2}r_{o3})r_{o2} \quad (4.103)$$

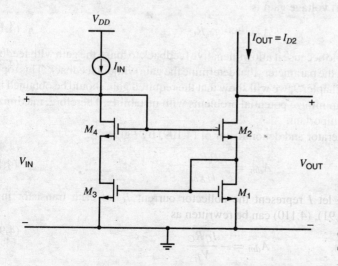

Figure 4.15 Improved MOS Wilson current mirror with an additional device such that the drain voltages of M_1 and M_3 are equal.

Since the calculation in (4.103) is based on the small-signal model for the bipolar Wilson current mirror in Fig. 4.14b, it ignores the body effect in transistor M_2. Repeating the analysis with a body-effect generator in parallel with r_{o2} gives

$$R_o \simeq (2 + g_{m2}r_{o3})r_{o2} \qquad (4.104)$$

The body effect on M_2 has little effect on (4.104) because M_1 is diode connected and therefore the voltage from the source of M_2 to ground is almost constant.

Although $\beta_F \to \infty$ for MOS transistors, the systematic gain error is not zero without M_4 because the drain-source voltage of M_3 differs from that of M_1 by the gate-source voltage of M_2. Therefore, without M_4,

$$\epsilon = \frac{V_{DS1} - V_{DS3}}{V_A} = -\frac{V_{GS2}}{V_A} \qquad (4.105)$$

Transistor M_4 is inserted in series with M_3 to equalize the drain-source voltages of M_3 and M_1 so that

$$\epsilon \simeq 0 \qquad (4.106)$$

With M_4, the output resistance is still given by (4.104) if all transistors operate in the active region. Also, insertion of M_4 does not change either the minimum output voltage for which (4.104) applies or the input voltage. Ignoring body effect and assuming equal overdrives on all transistors, the minimum output voltage is

$$V_{OUT(min)} = V_{GS1} + V_{ov2} = V_t + 2V_{ov} \qquad (4.107)$$

Under the same conditions, the input voltage is

$$V_{IN} = V_{GS1} + V_{GS2} = 2V_t + 2V_{ov} \qquad (4.108)$$

4.3 Active Loads

4.3.1 Motivation

In differential amplifiers of the type described in Chapter 3, resistors are used as the load elements. For example, consider the differential amplifier shown in Fig. 3.45. For this circuit, the differential-mode (dm) voltage gain is

$$A_{dm} = -g_m R_C \qquad (4.109)$$

Large gain is often desirable because it allows negative feedback to make the gain with feedback insensitive to variations in the parameters that determine the gain without feedback. This topic is covered in Chapter 8. In Chapter 9, we will show that the required gain should be obtained in as few stages as possible to minimize potential problems with instability. Therefore, maximizing the gain of each stage is important.

Multiplying the numerator and denominator of (4.109) by I gives

$$A_{dm} = -\frac{I(R_C)}{I/g_m} \qquad (4.110)$$

With bipolar transistors, let I represent the collector current I_C of each transistor in the differential pair. From (1.91), (4.110) can be rewritten as

$$A_{dm} = -\frac{I_C R_C}{V_T} \qquad (4.111)$$

4.3 Active Loads

To achieve large voltage gain, (4.111) shows that the $I_C R_C$ product must be made large, which in turn requires a large power-supply voltage. Furthermore, large values of resistance are required when low current is used to limit the power dissipation. As a result, the required die area for the resistors can be large.

A similar situation occurs in MOS amplifiers with resistive loads. Let I represent the drain current I_D of each transistor in the differential pair, and let the resistive loads be R_D. From (1.157) and (1.180), (4.110) can be rewritten as

$$A_{dm} = -\frac{I_D R_D}{(V_{GS} - V_t)/2} = -\frac{2 I_D R_D}{V_{ov}} \qquad (4.112)$$

Equation 4.112 shows that the $I_D R_D$ product must be increased to increase the gain with constant overdrive. As a result, a large power supply is usually required for large gain, and large resistance is usually required to limit power dissipation. Also, since the overdrive is usually much larger than the thermal voltage, comparing (4.111) and (4.112) shows that the gain of an MOS differential pair is usually much less than the gain of its bipolar counterpart with equal resistive drops. This result stems from the observation that bipolar transistors provide much more transconductance for a given current than MOS transistors provide.

If the power-supply voltage is only slightly larger than the drop on the resistors, the range of common-mode input voltages for which the input transistors would operate in the active region would be severely restricted in both bipolar and MOS amplifiers. To overcome this problem and provide large gain without large power-supply voltages or resistances, the r_o of a transistor can be used as a load element.[9] Since the load element in such a circuit is a transistor instead of a resistor, the load element is said to be *active* instead of passive.

4.3.2 Common-Emitter–Common-Source Amplifier with Complementary Load

A common-emitter amplifier with *pnp* current-mirror load is shown in Fig. 4.16a. The common-source counterpart with a *p*-channel MOS current-mirror load is shown in Fig. 4.16b. In both cases, there are two output variables: the output voltage, V_{out}, and the output current, I_{out}. The relationship between these variables is governed by both the input transistor and the load transistor. From the standpoint of the input transistor T_1,

$$I_{\text{out}} = I_{c1} \quad \text{or} \quad I_{\text{out}} = I_{d1} \qquad (4.113)$$

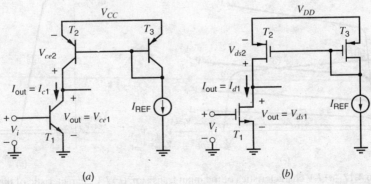

Figure 4.16 (a) Common-emitter amplifier with active load. (b) Common-source amplifier with active load.

and

$$V_{\text{out}} = V_{ce1} \quad \text{or} \quad V_{\text{out}} = V_{ds1} \tag{4.114}$$

Equations 4.113 and 4.114 show that the output I-V characteristics of T_1 can be used directly in the analysis of the relationship between the output variables. Since the input voltage is the base-emitter voltage of T_1 in Fig. 4.16a and the gate-source voltage of T_1 in Fig. 4.16b, the input voltage is the parameter that determines the particular curve in the family of output characteristics under consideration at any point, as shown in Fig. 4.17a.

In contrast, the base-emitter or gate-source voltage of the load transistor T_2 is fixed by diode-connected transistor T_3. Therefore, only one curve in the family of output I-V characteristics needs to be considered for the load transistor, as shown in Fig. 4.17b. From the standpoint of the load transistor,

$$I_{\text{out}} = -I_{c2} \quad \text{or} \quad I_{\text{out}} = -I_{d2} \tag{4.115}$$

and

$$V_{\text{out}} = V_{CC} + V_{ce2} \quad \text{or} \quad V_{\text{out}} = V_{DD} + V_{ds2} \tag{4.116}$$

Equation 4.115 shows that the output characteristic of the load transistor should be mirrored along the horizontal axis to plot in the same quadrant as the output characteristics of the input

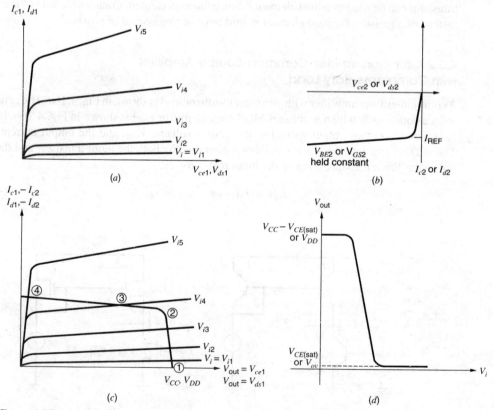

Figure 4.17 (a) I-V characteristics of the input transistor. (b) I-V characteristic of the active load. (c) I-V characteristics with load characteristic superimposed. (d) dc transfer characteristic of common-emitter or common-source amplifier with current-mirror load.

transistor. Equation 4.116 shows that the load curve should be shifted to the right by an amount equal to the power-supply voltage.

We now consider the dc transfer characteristic of the circuits. Initially, assume that $V_i = 0$. Then the input transistor is turned off, and the load is saturated in the bipolar case and linear in the MOS case, corresponding to point ① in Fig. 4.17c. As V_i is increased, the input transistor eventually begins to conduct current but the load remains saturated or linear until point ② is reached. Here the load enters the active region and a small further increase in V_i moves the operating point through point ③ to point ④, where the input transistor saturates in the bipolar case or enters the linear region in the MOS case. The change in V_i required to move from point ② to point ④ is small because the slopes of the output I-V characteristics in the active region are small for both transistors. The transfer curve (V_{out} as a function of V_i) is sketched in Fig. 4.17d.

A key point of this analysis is that the slope of the output characteristic is not constant, which is important because the slope is the gain of the amplifier. Since the gain of the amplifier depends on the input voltage, the amplifier is nonlinear in general, causing distortion to appear in the amplifier output. For low V_i, the output is high and the gain is low because the load transistor does not operate in the active region. Similarly, for large V_i, the output is low and the gain is low because the input transistor does not operate in the active region. To minimize distortion while providing gain, the amplifier should be operated in the intermediate region of V_i, where all transistors operate in the active region. The range of outputs for which all transistors operate in the active region should be maximized to use the power-supply voltage to the maximum extent. The active loads in Fig. 4.16 maintain high incremental output resistance as long as the drop across the load is more than $V_{\text{OUT(min)}}$ of the current mirror, which is $|V_{CE2(\text{sat})}|$ in the bipolar case and $|V_{ov2}|$ in the MOS case here. Therefore, minimizing $V_{\text{OUT(min)}}$ of the mirror maximizes the range of outputs over which the amplifier provides high and nearly constant gain. In contrast, an ideal passive load requires a large voltage drop to give high gain, as shown in (4.111) and (4.112). As a result, the range of outputs for which the gain is high and nearly constant is much less than with an active load.

The gain at any output voltage can be found by finding the slope in Fig. 4.17d. In general, this procedure requires writing equations for the various curves in all of Fig. 4.17. Although this process is required to study the nonlinear behavior of the circuits, it is so complicated analytically that it is difficult to carry out for more than just a couple of transistors at a time. Furthermore, after completing such a large-signal analysis, the results are often so complicated that the effects of the key parameters are difficult to understand, increasing the difficulty of designing with these results. Since we are ultimately interested in being able to analyze and design circuits with a large number of transistors, we will concentrate on the small-signal analysis, which is much simpler to carry out and interpret than the large-signal analysis. Unfortunately, the small-signal analysis provides no information about nonlinearity because it assumes that all transistor parameters are constant.

The primary characteristics of interest in the small-signal analysis here are the voltage gain and output resistance when both devices operate in the active region. The small-signal equivalent circuit is shown in Fig. 4.18. It is drawn for the bipolar case but applies for the MOS case as well when $r_{\pi 1} \to \infty$ and $r_{\pi 2} \to \infty$ because $\beta_0 \to \infty$. Since I_{REF} in Fig. 4.16 is assumed constant, the large-signal base-emitter or gate-source voltage of the load transistor is constant. Therefore, the small-signal base-emitter or gate-source voltage of the load transistor, v_2, is zero. As a result, the small-signal voltage-controlled current $g_{m2}v_2 = 0$. To find the output resistance of the amplifier, we set the input to zero. Therefore, $v_1 = 0$ and $g_{m1}v_1 = 0$, and the output resistance is

$$R_o = r_{o1} \| r_{o2} \qquad (4.117)$$

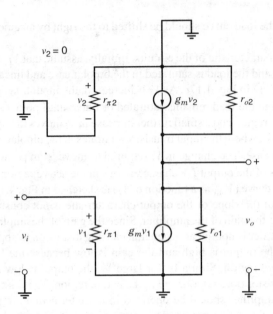

Figure 4.18 Small-signal equivalent circuit for common-emitter amplifier with active load.

Equation 4.117 together with (1.112) and (1.194) show that the output resistance is inversely proportional to the current in both the bipolar and MOS cases.

Since $v_2 = 0$, $g_{m1}v_1$ flows in $r_{o1}\|r_{o2}$ and

$$A_v = -g_{m1}(r_{o1}\|r_{o2}) \tag{4.118}$$

Substituting (1.91) and (1.112) into (4.118) gives for the bipolar case,

$$A_v = -\frac{1}{\dfrac{V_T}{V_{A1}} + \dfrac{V_T}{V_{A2}}} \tag{4.119}$$

Equation 4.119 shows that the gain is independent of the current in the bipolar case because the transconductance is proportional to the current while the output resistance is inversely proportional to the current. Typical values for this voltage gain are in the 1000 to 2000 range. Therefore, the actively loaded bipolar stage provides very high voltage gain.

In contrast, (1.180) shows that the transconductance is proportional to the square root of the current in the MOS case assuming square-law operation. Therefore, the gain in (4.118) is inversely proportional to the square root of the current. With channel lengths less than 1 μm, however, the drain current is almost linearly related to the gate-source voltage, as shown in (1.224). Therefore, the transconductance is almost constant, and the gain is inversely proportional to the current with very short channel lengths. Furthermore, typical values for the voltage gain in the MOS case are between 10 and 100, which is much less than with bipolar transistors.

4.3.3 Common-Emitter–Common-Source Amplifier with Depletion Load

Actively loaded gain stages using MOS transistors can be realized in processes that include only n-channel or only p-channel transistors if depletion devices are available. A depletion transistor is useful as a load element because it behaves like a current source when the transistor operates in the active region with the gate shorted to the source.

4.3 Active Loads

The *I-V* characteristic of an *n*-channel MOS depletion-load transistor is illustrated in Fig. 4.19. Neglecting body effect, the device exhibits a very high output resistance (equal to the device r_o) as long as the device operates in the active region. When the body effect is included, the resistance seen across the device drops to approximately $1/g_{mb}$. A complete gain stage is shown in Fig. 4.20 together with its dc transfer characteristic. The small-signal equivalent model when both transistors operate in the active region is shown in Fig. 4.21. From this circuit, we find that the gain is

$$\frac{v_o}{v_i} = -g_{m1}\left(r_{o1}\|r_{o2}\|\frac{1}{g_{mb2}}\right) \simeq -\frac{g_{m1}}{g_{mb2}} \qquad (4.120)$$

For a common-source amplifier with a depletion load, rearranging (4.120) and using (1.180) and (1.200) gives

$$\frac{v_o}{v_i} \simeq -\frac{g_{m1}}{\frac{g_{mb2}}{g_{m2}}g_{m2}} = -\frac{1}{\chi}\sqrt{\frac{(W/L)_1}{(W/L)_2}} \qquad (4.121)$$

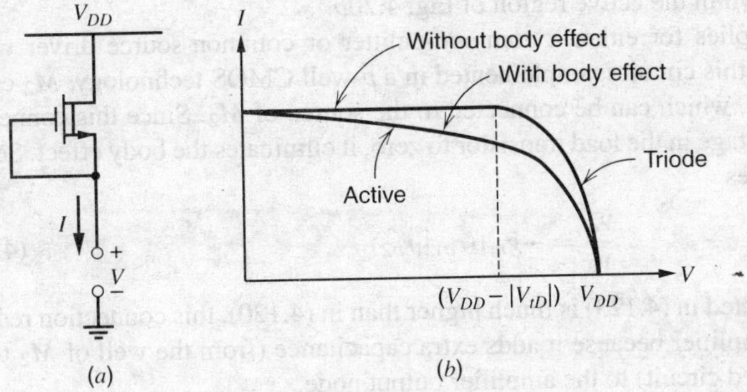

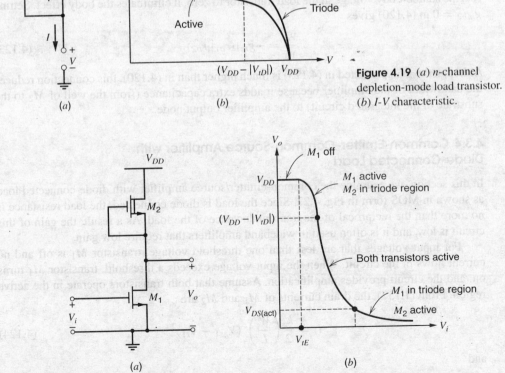

Figure 4.19 (*a*) *n*-channel depletion-mode load transistor. (*b*) *I-V* characteristic.

Figure 4.20 (*a*) Common-source amplifier with depletion-mode transistor load. (*b*) dc transfer characteristic.

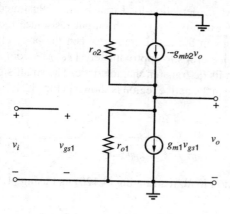

Figure 4.21 Small-signal equivalent circuit of the common-source amplifier with depletion load, including the body effect in the load and the channel-length modulation in the load and the common-source device.

From (1.196) and (1.141),

$$\frac{1}{\chi} = 2\sqrt{2\phi_f}C_{ox}\sqrt{\frac{1 + V_{SB}/(2\phi_f)}{2q\epsilon N_A}} \tag{4.122}$$

Since χ depends on $V_o = V_{SB}$, the incremental voltage gain varies with output voltage, giving the slope variation shown in the active region of Fig. 4.20b.

Equation 4.120 applies for either a common-emitter or common-source driver with a depletion MOS load. If this circuit is implemented in a p-well CMOS technology, M_2 can be built in an isolated well, which can be connected to the source of M_2. Since this connection sets the source-body voltage in the load transistor to zero, it eliminates the body effect. Setting $g_{mb2} = 0$ in (4.120) gives

$$\frac{v_o}{v_i} = -g_{m1}(r_{o1}||r_{o2}) \tag{4.123}$$

Although the gain predicted in (4.123) is much higher than in (4.120), this connection reduces the bandwidth of the amplifier because it adds extra capacitance (from the well of M_2 to the substrate of the integrated circuit) to the amplifier output node.

4.3.4 Common-Emitter–Common-Source Amplifier with Diode-Connected Load

In this section, we examine the common-emitter/source amplifier with diode-connected load as shown in MOS form in Fig. 4.22. Since the load is diode connected, the load resistance is no more than the reciprocal of the transconductance of the load. As a result, the gain of this circuit is low, and it is often used in wideband amplifiers that require low gain.

For input voltages that are less than one threshold voltage, transistor M_1 is off and no current flows in the circuit. When the input voltage exceeds a threshold, transistor M_1 turns on, and the circuit provides amplification. Assume that both transistors operate in the active region. From (1.157), the drain currents of M_1 and M_2 are

$$I_1 = \frac{k'}{2}\left(\frac{W}{L}\right)_1 (V_{gs1} - V_{t1})^2 \tag{4.124}$$

and

$$I_2 = \frac{k'}{2}\left(\frac{W}{L}\right)_2 (V_{gs2} - V_{t2})^2 \tag{4.125}$$

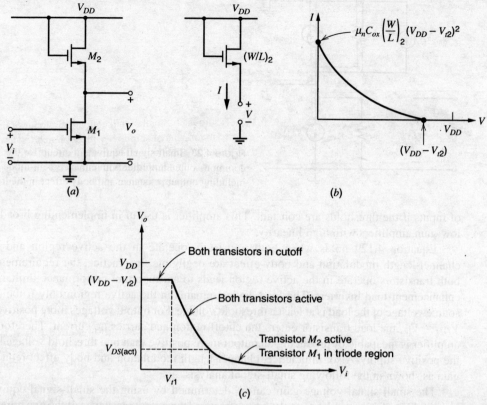

Figure 4.22 (a) Common-source amplifier with enhancement-mode load. (b) I-V characteristic of load transistor. (c) Transfer characteristic of the circuit.

From KVL in Fig. 4.22,

$$V_o = V_{DD} - V_{gs2} \tag{4.126}$$

Solving (4.125) for V_{gs2} and substituting into (4.126) gives

$$V_o = V_{DD} - V_{t2} - \sqrt{\frac{2I_2}{k'(W/L)_2}} \tag{4.127}$$

Since $I_2 = I_1$, (4.127) can be rewritten as

$$V_o = V_{DD} - V_{t2} - \sqrt{\frac{2I_1}{k'(W/L)_2}} \tag{4.128}$$

Substituting (4.124) into (4.128) with $V_{gs1} = V_i$ gives

$$V_o = V_{DD} - V_{t2} - \sqrt{\frac{(W/L)_1}{(W/L)_2}}(V_i - V_{t1}) \tag{4.129}$$

Equation 4.129 shows that the slope of the transfer characteristic is the square root of the aspect ratios, assuming that the thresholds are constant. Since the slope of the transfer characteristic is the gain of the amplifier, the gain is constant and the amplifier is linear for a wide range

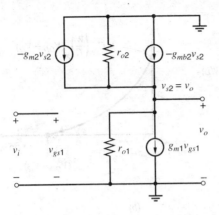

Figure 4.23 Small-signal equivalent circuit for the common-source amplifier with enhancement-mode load, including output resistance and body effect in the load.

of inputs if the thresholds are constant. This amplifier is useful in implementing broadband, low-gain amplifiers with high linearity.

Equation 4.129 holds when both transistors operate in the active region and when channel-length modulation and body effect are negligible. In practice, the requirement that both transistors operate in the active region leads to an important performance limitation in enhancement-load inverters. The load device remains in the active region only if the drain-source voltage of the load is at least a threshold voltage. For output voltages more positive than $V_{DD} - V_{t2}$, the load transistor enters the cutoff region and carries no current. Therefore, the amplifier is incapable of producing an output more positive than one threshold voltage below the positive supply. Also, in practice, channel-length modulation and body effect reduce the gain as shown in the following small-signal analysis.

The small-signal voltage gain can be determined by using the small-signal equivalent circuit of Fig. 4.23, in which both the body effect and the output resistance of the two transistors have been included. From KCL at the output node,

$$g_{m1}v_i + \frac{v_o}{r_{o1}} + \frac{v_o}{r_{o2}} + g_{m2}v_o + g_{mb2}v_o = 0 \qquad (4.130)$$

Rearranging (4.130) gives

$$\frac{v_o}{v_i} = -g_{m1}\left(\frac{1}{g_{m2}} \middle\| \frac{1}{g_{mb2}} \middle\| r_{o1} \middle\| r_{o2}\right)$$

$$= -\frac{g_{m1}}{g_{m2}}\left(\frac{1}{1 + \frac{g_{mb2}}{g_{m2}} + \frac{1}{g_{m2}r_{o1}} + \frac{1}{g_{m2}r_{o2}}}\right) \qquad (4.131)$$

If $g_{m2}/g_{mb2} \gg 1$, $g_{m2}r_{o1} \gg 1$, and $g_{m2}r_{o2} \gg 1$,

$$\frac{v_o}{v_i} \simeq -\frac{g_{m1}}{g_{m2}} = -\sqrt{\frac{(W/L)_1}{(W/L)_2}} \qquad (4.132)$$

as in (4.129). For practical device geometries, this relationship limits the maximum voltage gain to values on the order of 10 to 20.

The bipolar counterpart of the circuit in Fig. 4.22 is a common-emitter amplifier with a diode-connected load. The magnitude of its gain would be approximately equal to the ratio of the transconductances, which would be unity. However, the current that would flow in this circuit would be extremely large for inputs greater than $V_{be(on)}$ because the collector current in a bipolar transistor is an exponential function of its base-emitter voltage. To limit the current but maintain unity gain, equal-value resistors can be placed in series with the emitter of each transistor. Alternatively, the input transistors can be replaced by a differential pair, where the current is limited by the tail current source. In this case, emitter degeneration is used in the differential pair to increase the range of inputs for which all transistors operate in the active region, as in Fig. 3.49. In contrast, source degeneration is rarely used in MOS differential pairs because their transconductance and linear range can be controlled through the device aspect ratios.

4.3.5 Differential Pair with Current-Mirror Load

4.3.5.1 Large-Signal Analysis

A straightforward application of the active-load concept to the differential pair would yield the circuit shown in Fig. 4.24a. Assume at first that all n-channel transistors are identical

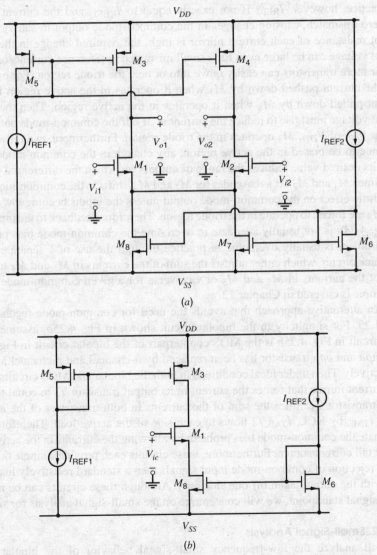

Figure 4.24 (a) Differential pair with active load. (b) Common-mode half circuit for differential pair with active load.

and that all *p*-channel transistors are identical. Then the differential-mode half circuit for this differential pair is just a common-source amplifier with an active-load, as in Fig. 4.16*b*. Thus the differential-mode voltage gain is large when all the transistors are biased in the active region. The circuit as it stands, however, has the drawback that the quiescent value of the common-mode output voltage is very sensitive to changes in the drain currents of M_3, M_4, M_7, and M_8. As a result, some transistors may operate in or near the triode region, reducing the differential gain or the range of outputs for which the differential gain is high.

This fact is illustrated by the dc common-mode half-circuit shown in Fig. 4.24*b*. In the common-mode half circuit, the combination of M_1, M_6, and M_8 form a cascode current mirror, which is connected to the simple current mirror formed by M_3 and M_5. If all transistors operate in the active region, M_3 pushes down a current about equal to I_{REF1}, and M_8 pulls down a current about equal to I_{REF2}. KCL requires that the current in M_3 must be equal to the current in M_8. If $I_{REF2} = I_{REF1}$, KCL can be satisfied while all transistors operate in the active region. In practice, however, I_{REF2} is not exactly equal to I_{REF1}, and the current mirrors contain nonzero mismatch, causing changes in the common-mode output to satisfy KCL. Since the output resistance of each current mirror is high, the required change in the common-mode output voltage can be large even for a small mismatch in reference currents or transistors, and one or more transistors can easily move into or near the triode region. For example, suppose that the current pushed down by M_3 when it operates in the active region is more than the current pulled down by M_8 when it operates in the active region. Then the common-mode output voltage must rise to reduce the current in M_3. If the common-mode output voltage rises within V_{ov3} of V_{DD}, M_3 operates in the triode region. Furthermore, even if all the transistors continue to be biased in the active region, any change in the common-mode output voltage from its desired value reduces the range of outputs for which the differential gain is high.

Since M_1 and M_2 act as cascodes for M_7 and M_8, shifts in the common-mode input voltage have little effect on the common-mode output unless the inputs become low enough that M_7 and M_8 are forced to operate in the triode region. Therefore, feedback to the inputs of the circuit in Fig. 4.24*a* is not usually adequate to overcome the common-mode bias problem. Instead, this problem is usually overcome in practice through the use of a separate common-mode feedback circuit, which either adjusts the sum of the currents in M_3 and M_4 to be equal to the sum of the currents in M_7 and M_8 or vice versa for a given common-mode output voltage. This topic is covered in Chapter 12.

An alternative approach that avoids the need for common-mode feedback is shown in Fig. 4.25. For simplicity in the bipolar circuit shown in Fig. 4.25*a*, assume that $\beta_F \to \infty$. The circuit in Fig. 4.25*b* is the MOS counterpart of the bipolar circuit in Fig. 4.25*a* because each *npn* and *pnp* transistor has been replaced by *n*-channel and *p*-channel MOS transistors, respectively. Then under ideal conditions in both the bipolar and MOS circuits, the active load is a current mirror that forces the current in its output transistor T_4 to equal the current in its input transistor T_3. Since the sum of the currents in both transistors of the active load must equal I_{TAIL} by KCL, $I_{TAIL}/2$ flows in each side of the active load. Therefore, these circuits eliminate the common-mode bias problem by allowing the currents in the active load to be set by the tail current source. Furthermore, these circuits each provide a single output with much better rejection of common-mode input signals than a standard resistively loaded differential pair with the output taken off one side only. Although these circuits can be analyzed from a large-signal standpoint, we will concentrate on the small-signal analysis for simplicity.

4.3.5.2 Small-Signal Analysis

We will analyze the low-frequency small-signal behavior of the bipolar circuit shown in Fig. 4.25*a* because these results cover both the bipolar and MOS cases by letting $\beta_0 \to \infty$ and $r_\pi \to \infty$. Key parameters of interest in this circuit include the small-signal

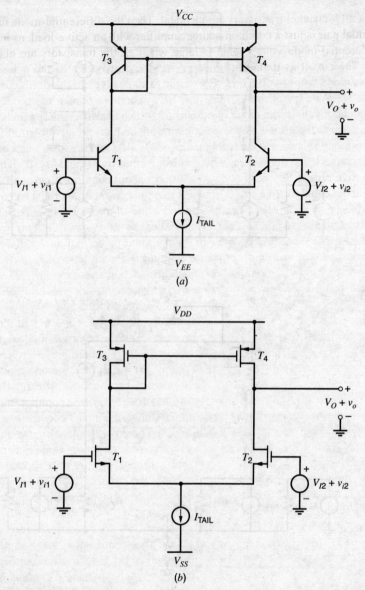

Figure 4.25 (a) Emitter-coupled pair with current-mirror load. (b) Source-coupled pair with current-mirror load (MOS counterpart).

transconductance and output resistance. (The product of these two quantities gives the small-signal voltage gain with no load.) Since only one transistor in the active load is diode connected, the circuit is not symmetrical and a half-circuit approach is not useful. Therefore, we will analyze the small-signal model of this circuit directly. Assume that all transistors operate in the active region with $r_\mu \to \infty$ and $r_b = 0$. Let r_tail represent the output resistance of the tail current source I_TAIL. The resulting small-signal circuit is shown in Fig. 4.26a.

Since T_3 and T_4 form a current mirror, we expect the mirror output current to be approximately equal to the mirror input current. Therefore, we will write

$$g_{m4}v_3 = i_3(1 - \epsilon_m) \qquad (4.133)$$

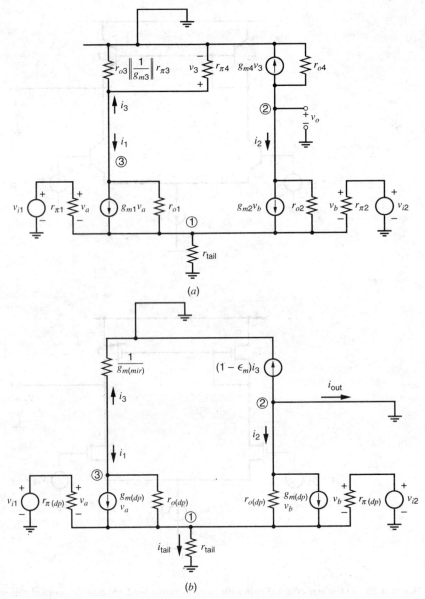

Figure 4.26 (a) Small-signal equivalent circuit, differential pair with current-mirror load. (b) Simplified drawing of small-signal model of differential pair with current-mirror load.

where ϵ_m is the systematic gain error of the current mirror calculated from small-signal parameters. Let r_3 represent the total resistance connected between the base or gate of T_3 and the power supply. Then r_3 is the parallel combination of $1/g_{m3}, r_{\pi 3}, r_{\pi 4}$, and r_{o3}. Under the simplifying assumptions that $\beta_0 \gg 1$ and $g_m r_o \gg 1$, this parallel combination is approximately equal to $1/g_{m3}$. Then the drop across $r_{\pi 4}$ is

$$v_3 = i_3 r_3 \simeq \frac{i_3}{g_{m3}} \qquad (4.134)$$

We will also assume that the two transistors in the differential pair match perfectly and operate with equal dc currents, as do the two transistors in the current-mirror load. Then $g_{m(dp)} = g_{m1}$

$= g_{m2}$, $g_{m(mir)} = g_{m3} = g_{m4}$, $r_{\pi(dp)} = r_{\pi 1} = r_{\pi 2}$, $r_{\pi(mir)} = r_{\pi 3} = r_{\pi 4}$, $r_{o(dp)} = r_{o1} = r_{o2}$, and $r_{o(mir)} = r_{o3} = r_{o4}$. From (4.134), the resulting voltage-controlled current $g_{m4}v_3$ is

$$g_{m4}v_3 = g_{m(mir)}v_3 \simeq g_{m(mir)}\frac{i_3}{g_{m(mir)}} = i_3 \tag{4.135}$$

Equations 4.133 and 4.135 show that $\epsilon_m \simeq 0$ and thus the active load acts as a current mirror in a small-signal sense, as expected. Using (4.133), the small-signal circuit is redrawn in Fig. 4.26b with the output grounded to find the transconductance. Note that r_{o4} is omitted because it is attached to a small-signal ground on both ends.

From KCL at node ①,

$$(v_{i1} - v_1 + v_{i2} - v_1)\left(\frac{1}{r_{\pi(dp)}} + g_{m(dp)}\right) + \frac{v_3 - v_1}{r_{o(dp)}} - \frac{v_1}{r_{o(dp)} \| r_{\text{tail}}} = 0 \tag{4.136}$$

where v_1 and v_3 are the voltages to ground from nodes ① and ③. To complete an exact small-signal analysis, KCL equations could also be written at nodes ② and ③, and these KCL equations plus (4.136) could be solved simultaneously. However, this procedure is complicated algebraically and leads to an equation that is difficult to interpret. To simplify the analysis, we will assume at first that $r_{\text{tail}} \to \infty$ and $r_{o(dp)} \to \infty$ since the transistors are primarily controlled by their base-emitter or gate-source voltages. Then from (4.136)

$$v_1 = \frac{v_{i1} + v_{i2}}{2} = v_{ic} \tag{4.137}$$

where v_{ic} is the common-mode component of the input. Let $v_{id} = v_{i1} - v_{i2}$ represent the differential-mode component of the input. Then $v_{i1} = v_{ic} + v_{id}/2$ and $v_{i2} = v_{ic} - v_{id}/2$, and the small-signal collector or drain currents

$$i_1 = g_{m(dp)}(v_{i1} - v_1) = \frac{g_{m(dp)}v_{id}}{2} \tag{4.138}$$

and

$$i_2 = g_{m(dp)}(v_{i2} - v_1) = -\frac{g_{m(dp)}v_{id}}{2} \tag{4.139}$$

With a resistive load and a single-ended output, only i_2 flows in the output. Therefore, the transconductance for a differential-mode (dm) input with a passive load is

$$G_m[dm] = \left.\frac{i_{\text{out}}}{v_{id}}\right|_{v_{\text{out}}=0} = -\frac{i_2}{v_{id}} = \frac{g_{m(dp)}}{2} \tag{4.140}$$

On the other hand, with the active loads in Fig. 4.25, not only i_2 but also most of i_3 flows in the output because of the action of the current mirror, as shown by (4.135). Therefore, the output current in Fig. 4.26b is

$$i_{\text{out}} = -(1 - \epsilon_m)i_3 - i_2 \tag{4.141}$$

Assume at first that the current mirror is ideal so that $\epsilon_m = 0$. Then since $i_3 = -i_1$, substituting (4.138) and (4.139) in (4.141) gives

$$i_{\text{out}} = g_{m(dp)}v_{id} \tag{4.142}$$

Therefore, with an active load,

$$G_m[dm] = \left.\frac{i_{\text{out}}}{v_{id}}\right|_{v_{\text{out}}=0} = g_{m(dp)} \tag{4.143}$$

Equation 4.143 applies for both the bipolar and MOS amplifiers shown in Fig. 4.25. Comparing (4.140) and (4.143) shows that the current-mirror load doubles the differential transconductance compared to the passive-load case. This result stems from the fact that the current mirror creates a second signal path to the output. (The first path is through the differential pair.) Although frequency response is not analyzed in this chapter, note that the two signal paths usually have different frequency responses, which is often important in high-speed applications.

The key assumptions that led to (4.142) and (4.143) are that the current mirror is ideal so $\epsilon_m = 0$ and that $r_{tail} \to \infty$ and $r_{o(dp)} \to \infty$. Under these assumptions, the output current is independent of the common-mode input. In practice, none of these assumptions is exactly true, and the output current depends on the common-mode input. However, this dependence is small because the active load greatly enhances the common-mode rejection ratio of this stage, as shown in Section 4.3.5.3.

Another important parameter of the differential pair with active load is the output resistance. The output resistance is calculated using the circuit of Fig. 4.27, in which a test voltage source v_t is applied at the output while the inputs are connected to small-signal ground. The resulting current i_t has four components. The current in r_{o4} is

$$i_{t1} = \frac{v_t}{r_{o4}} \tag{4.144}$$

The resistance in the emitter or source lead of T_2 is r_{tail} in parallel with the resistance seen looking into the emitter or source of T_1, which is approximately $1/g_{m1}$. Thus, using (3.99) for a transistor with degeneration, we find that the effective output resistance looking into the collector or drain of T_2 is

$$R_{o2} \simeq r_{o2}\left(1 + g_{m2}\frac{1}{g_{m1}}\right) = 2r_{o2} \tag{4.145}$$

Hence

$$i_{t2} + i_{t4} \simeq \frac{v_t}{2r_{o2}} \tag{4.146}$$

If $r_{tail} \gg 1/g_{m1}$, this current flows into the emitter or source of T_1, and is mirrored to the output with a gain of approximately unity to produce

$$i_{t3} \simeq i_{t2} + i_{t4} \simeq \frac{v_t}{2r_{o2}} \tag{4.147}$$

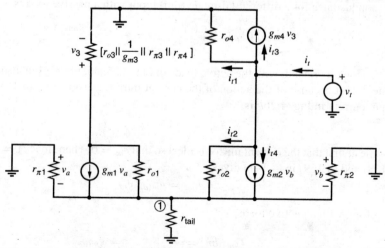

Figure 4.27 Circuit for calculation of the output resistance of the differential pair with current-mirror load.

Thus

$$i_t = i_{t1} + i_{t2} + i_{t3} + i_{t4} \simeq v_t \left(\frac{1}{r_{o4}} + \frac{1}{r_{o2}} \right) \quad (4.148)$$

Since $r_{o2} = r_{o(dp)}$ and $r_{o4} = r_{o(mir)}$,

$$R_o = \left. \frac{v_t}{i_t} \right|_{\substack{v_{i1}=0 \\ v_{i2}=0}} \simeq \frac{1}{\frac{1}{r_{o(dp)}} + \frac{1}{r_{o(mir)}}} = r_{o(dp)} \| r_{o(mir)} \quad (4.149)$$

The result in (4.149) applies for both the bipolar and MOS amplifiers shown in Fig. 4.25. In multistage bipolar amplifiers, the low-frequency gain of the loaded circuit is likely to be reduced by the input resistance of the next stage because the output resistance is high. In contrast, low-frequency loading is probably not an issue in multistage MOS amplifiers because the next stage has infinite input resistance if the input is the gate of an MOS transistor.

Finally, although the source-coupled pair has infinite input resistance, the emitter-coupled pair has finite input resistance because β_0 is finite. If the effects of the r_o of T_2 and T_4 are neglected, the differential input resistance of the actively loaded emitter-coupled pair is simply $2r_{\pi(dp)}$ as in the resistively loaded case. In practice, however, the asymmetry of the circuit together with the high voltage gain cause feedback to occur through the output resistance of T_2 to node ①. This feedback causes the input resistance to differ slightly from $2r_{\pi(dp)}$.

In summary, the actively loaded differential pair is capable of providing differential-to-single-ended conversion, that is, the conversion from a differential voltage to a voltage referenced to the ground potential. The high output resistance of the circuit requires that the next stage must have high input resistance if the large gain is to be realized. A small-signal two-port equivalent circuit for the stage is shown in Fig. 4.28.

4.3.5.3 Common-Mode Rejection Ratio

In addition to providing high voltage gain, the circuits in Fig. 4.25 provide conversion from a differential input signal to an output signal that is referenced to ground. Such a conversion is required in all differential-input, single-ended output amplifiers.

The simplest differential-to-single-ended converter is a resistively loaded differential pair in which the output is taken from only one side, as shown in Fig. 4.29a. In this case, $A_{dm} > 0$, $A_{cm} < 0$, and the output is

$$v_o = -\frac{v_{od}}{2} + v_{oc} = -\frac{A_{dm} v_{id}}{2} + A_{cm} v_{ic} \quad (4.150)$$

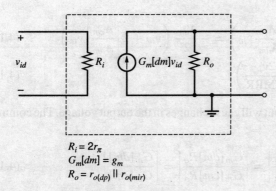

$R_i = 2r_\pi$
$G_m[dm] = g_m$
$R_o = r_{o(dp)} \| r_{o(mir)}$

Figure 4.28 Two-port representation of small-signal properties of differential pair with current-mirror load. The effects of asymmetrical input resistance have been neglected.

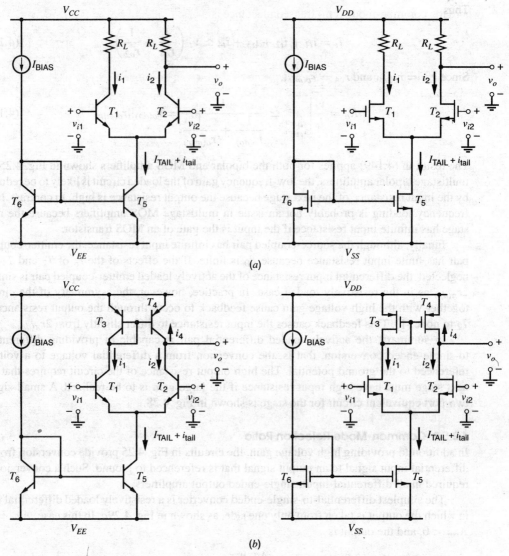

Figure 4.29 Differential-to-single-ended conversion using (a) resistively loaded differential pairs and (b) actively loaded differential pairs.

$$v_o = -\frac{A_{dm}}{2}\left(v_{id} - \frac{2A_{cm}}{A_{dm}}v_{ic}\right) = -\frac{A_{dm}}{2}\left(v_{id} + 2\left|\frac{A_{cm}}{A_{dm}}\right|v_{ic}\right) \quad (4.151)$$

$$= -\frac{A_{dm}}{2}\left(v_{id} + \frac{2v_{ic}}{\text{CMRR}}\right) \quad (4.152)$$

Thus, common-mode signals at the input will cause changes in the output voltage. The common-mode rejection ratio (CMRR) is

$$\text{CMRR} = \left|\frac{A_{dm}}{A_{cm}}\right| = \left|\frac{G_m[dm]R_o}{G_m[cm]R_o}\right| = \left|\frac{G_m[dm]}{G_m[cm]}\right| \quad (4.153)$$

where the common-mode (cm) transconductance is

$$G_m[cm] = \left. \frac{i_{out}}{v_{ic}} \right|_{v_{out}=0} \quad (4.154)$$

Since the circuits in Fig. 4.29a are symmetrical, a common-mode half circuit can be used to find $G_m[cm]$. The common-mode half circuit is a common-emitter/source amplifier with degeneration. From (3.93) and (3.104),

$$G_m[cm] = -\frac{i_2}{v_{ic}} \simeq -\frac{g_{m(dp)}}{1 + g_{m(dp)}(2r_{tail})} \quad (4.155)$$

where $g_{m(dp)} = g_{m1} = g_{m2}$ and r_{tail} represents the output resistance of the tail current source T_5. The negative sign appears in (4.155) because the output current is defined as positive when it flows from the output terminal into the small-signal ground to be consistent with the differential case, as in Fig. 4.26b. Equation 4.155 applies for both the bipolar and MOS cases if the base current is ignored in the bipolar case, the body effect is ignored in the MOS case, and r_{o1} and r_{o2} are ignored in both cases. Substituting (4.140) and (4.155) into (4.153) gives

$$CMRR = \frac{1 + 2g_{m(dp)}r_{tail}}{2} \simeq g_{m(dp)}r_{tail} = g_{m1}r_{o5} \quad (4.156)$$

Equation 4.156 shows that the common-mode rejection ratio here is about half that in (3.193) because the outputs in Fig. 4.29 are taken only from one side of each differential pair instead of from both sides, reducing the differential-mode gain by a factor of two. The result in (4.156) applies for both the bipolar and MOS amplifiers shown in Fig. 4.29a. Because $g_m r_o$ is much higher for bipolar transistors than MOS transistors, the CMRR of a bipolar differential pair with resistive load is much higher than that of its MOS counterpart.

On the other hand, the active-load stages shown in Fig 4.29b have common-mode rejection ratios much superior to those of the corresponding circuits in Fig. 4.29a. Assume that the outputs in Fig. 4.29b are connected to small-signal ground to allow calculation of the common-mode transconductance. The small-signal model is the same as shown in Fig. 4.26b with $v_{i1} = v_{i2} = v_{ic}$. For simplicity, let $\beta_0 \to \infty$ and $r_\pi \to \infty$ at first. As with a resistive load, changes in the common-mode input will cause changes in the tail bias current i_{tail} because the output resistance of T_5 is finite. If we assume that the currents in the differential-pair transistors are controlled only by the base-emitter or gate-source voltages, the change in the current in T_1 and T_2 is

$$i_1 = i_2 = \frac{i_{tail}}{2} \quad (4.157)$$

If $\epsilon_m = 0$, the gain of the current mirror is unity. Then substituting (4.157) into (4.141) with $i_3 = -i_1$ gives

$$i_{out} = -i_3 - i_2 = i_1 - i_2 = 0 \quad (4.158)$$

As a result,

$$G_m[cm] = \left. \frac{i_{out}}{v_{ic}} \right|_{v_{out}=0} = 0 \quad (4.159)$$

Therefore,

$$CMRR \to \infty \quad (4.160)$$

The common-mode rejection ratio in (4.160) is infinite because the change in the current in T_4 cancels that in T_2 even when r_{tail} is finite under these assumptions.

The key assumptions that led to (4.160) are that $r_{o(dp)} \to \infty$ so $i_1 = i_2$ and that the current mirror is ideal so $\epsilon_m = 0$. In practice, the currents in the differential-pair transistors are not only controlled by their base-emitter or gate-source voltages, but also to some extent by their collector-emitter or drain-source voltages. As a result, i_1 is not exactly equal to i_2 because of finite $r_{o(dp)}$ in T_1 and T_2. Furthermore, the gain of the current mirror is not exactly unity, which means that ϵ_m is not exactly zero in practice because of finite $r_{o(mir)}$ in T_3 and T_4. Finite β_0 also affects the systematic gain error of the current mirror when bipolar transistors are used. For these reasons, the common-mode rejection ratio is finite in practice. However, the use of the active load greatly improves the common-mode rejection ratio compared to the resistive load case, as we will show next.

Suppose that

$$i_1 = i_2(1 - \epsilon_d) \tag{4.161}$$

where ϵ_d can be thought of as the gain error in the differential pair. Substituting (4.161) into (4.141) with $i_3 = -i_1$ gives

$$i_{\text{out}} = i_1(1 - \epsilon_m) - i_2 = i_2((1 - \epsilon_d)(1 - \epsilon_m) - 1) \tag{4.162}$$

Rearranging (4.162) gives

$$i_{\text{out}} = -i_2(\epsilon_d + \epsilon_m - \epsilon_d \epsilon_m) \tag{4.163}$$

If $\epsilon_d \ll 1$ and $\epsilon_m \ll 1$, the product term $\epsilon_d \epsilon_m$ is a second-order error and can be neglected. Therefore,

$$i_{\text{out}} \simeq -i_2(\epsilon_d + \epsilon_m) \tag{4.164}$$

Substituting (4.164) into (4.154) gives

$$G_m[cm] \simeq -\left(\frac{i_2}{v_{ic}}\right)(\epsilon_d + \epsilon_m) \tag{4.165}$$

Equation 4.165 applies for the active-load circuits shown in Fig. 4.29b; however, the first term has approximately the same value as in the passive-load case. Therefore, we will substitute (4.155) into (4.165), which gives

$$G_m[cm] \simeq -\left(\frac{g_{m(dp)}}{1 + g_{m(dp)}(2r_{\text{tail}})}\right)(\epsilon_d + \epsilon_m) \tag{4.166}$$

Substituting (4.166) and (4.143) into (4.153) gives

$$\text{CMRR} = \left|\frac{G_m[dm]}{G_m[cm]}\right| \simeq \frac{1 + 2g_{m(dp)}r_{\text{tail}}}{(\epsilon_d + \epsilon_m)} \tag{4.167}$$

Comparing (4.167) and (4.156) shows that the active load improves the common-mode rejection ratio by a factor of $2/(\epsilon_d + \epsilon_m)$. The factor of 2 in the numerator of this expression stems from the increase in the differential transconductance, and the denominator stems from the decrease in the common-mode transconductance.

To find ϵ_d, we will refer to Fig. 4.26b with $v_{i1} = v_{i2} = v_{ic}$. First, we write

$$i_1 = g_{m(dp)}(v_{ic} - v_1) + \frac{v_3 - v_1}{r_{o(dp)}} \tag{4.168}$$

and

$$i_2 = g_{m(dp)}(v_{ic} - v_1) - \frac{v_1}{r_{o(dp)}} \tag{4.169}$$

Substituting (4.134) and $i_3 = -i_1$ into (4.168) gives

$$i_1 \simeq g_{m(dp)}(v_{ic} - v_1) - \frac{v_1}{r_{o(dp)}} - \frac{i_1}{g_{m(mir)}r_{o(dp)}} \tag{4.170}$$

Equation 4.170 can be rearranged to give

$$\left(\frac{1+g_{m(mir)}r_{o(dp)}}{g_{m(mir)}r_{o(dp)}}\right)i_1 \simeq g_{m(dp)}(v_{ic}-v_1) - \frac{v_1}{r_{o(dp)}} \qquad (4.171)$$

Substituting (4.169) into (4.171) gives

$$i_1 \simeq \left(\frac{g_{m(mir)}r_{o(dp)}}{1+g_{m(mir)}r_{o(dp)}}\right)i_2 \qquad (4.172)$$

Substituting (4.161) into (4.172) gives

$$\epsilon_d \simeq \frac{1}{1+g_{m(mir)}r_{o(dp)}} \qquad (4.173)$$

To find ϵ_m, we will again refer to Fig. 4.26b with $v_{i1}=v_{i2}=v_{ic}$. In writing (4.134), we assumed that $r_3 \simeq 1/g_{m3}$. We will now reconsider this assumption and write

$$r_3 = \frac{1}{g_{m3}}||r_{\pi 3}||r_{\pi 4}||r_{o3} \qquad (4.174)$$

We will still assume that the two transistors in the differential pair match perfectly and operate with equal dc currents, as do the two transistors in the active load. Then (4.174) can be rewritten as

$$r_3 = \frac{r_{\pi(mir)}r_{o(mir)}}{r_{\pi(mir)}+2r_{o(mir)}+g_{m(mir)}r_{\pi(mir)}r_{o(mir)}} \qquad (4.175)$$

Substituting (4.175) into (4.135) gives

$$g_{m4}v_3 = g_{m4}i_3 r_3 = \frac{g_{m(mir)}r_{\pi(mir)}r_{o(mir)}i_3}{r_{\pi(mir)}+2r_{o(mir)}+g_{m(mir)}r_{\pi(mir)}r_{o(mir)}} \qquad (4.176)$$

Substituting (4.133) into (4.176) gives

$$\epsilon_m = \frac{r_{\pi(mir)}+2r_{o(mir)}}{r_{\pi(mir)}+2r_{o(mir)}+g_{m(mir)}r_{\pi(mir)}r_{o(mir)}} \qquad (4.177)$$

For bipolar transistors, r_π is usually much less than r_o; therefore,

$$\epsilon_m[bip] = \frac{2+\dfrac{r_{\pi(mir)}}{r_{o(mir)}}}{2+\dfrac{r_{\pi(mir)}}{r_{o(mir)}}+g_{m(mir)}r_{\pi(mir)}} \simeq \frac{1}{1+\dfrac{g_{m(mir)}r_{\pi(mir)}}{2}} = \frac{1}{1+\dfrac{\beta_0}{2}} \qquad (4.178)$$

Since $r_\pi \to \infty$ for MOS transistors,

$$\epsilon_m[MOS] = \frac{1}{1+g_{m(mir)}r_{o(mir)}} \qquad (4.179)$$

For the bipolar circuit in Fig. 4.29b, substituting (4.173) and (4.178) into (4.167) gives

$$\text{CMRR} \simeq \frac{1+2g_{m(dp)}r_{\text{tail}}}{\left(\dfrac{1}{1+g_{m(mir)}r_{o(dp)}}+\dfrac{1}{1+\dfrac{g_{m(mir)}r_{\pi(mir)}}{2}}\right)} \qquad (4.180)$$

If $(g_{m(mir)}r_{o(dp)}) \gg 1$ and $(g_{m(mir)}r_{\pi(mir)}/2) \gg 1$, (4.180) can be simplified to give

$$\begin{aligned} \text{CMRR} &\simeq (1 + 2g_{m(dp)}r_{\text{tail}})g_{m(mir)}\left(r_{o(dp)}\|\frac{r_{\pi(mir)}}{2}\right) \\ &\simeq (2g_{m(dp)}r_{\text{tail}})g_{m(mir)}\left(r_{o(dp)}\|\frac{r_{\pi(mir)}}{2}\right) \end{aligned} \quad (4.181)$$

Comparing (4.181) and (4.156) shows that the active load increases the common-mode rejection ratio by a factor of about $2g_{m(mir)}\left(r_{o(dp)}\|(r_{\pi(mir)}/2)\right)$ for the bipolar circuit in Fig. 4.29b compared to its passive-load counterpart in Fig. 4.29a.

On the other hand, for the MOS circuit in Fig. 4.29b, substituting (4.173) and (4.179) into (4.167) gives

$$\text{CMRR} \simeq \frac{1 + 2g_{m(dp)}r_{\text{tail}}}{\left(\dfrac{1}{1 + g_{m(mir)}r_{o(dp)}} + \dfrac{1}{1 + g_{m(mir)}r_{o(mir)}}\right)} \quad (4.182)$$

If $(g_{m(mir)}r_{o(dp)}) \gg 1$ and $(g_{m(mir)}r_{o(mir)}) \gg 1$, (4.182) can be simplified to give

$$\begin{aligned} \text{CMRR} &\simeq (1 + 2g_{m(dp)}r_{\text{tail}})g_{m(mir)}(r_{o(dp)}\|r_{o(mir)}) \\ &\simeq (2g_{m(dp)}r_{\text{tail}})g_{m(mir)}(r_{o(dp)}\|r_{o(mir)}) \end{aligned} \quad (4.183)$$

Comparing (4.183) and (4.156) shows that the active load increases the common-mode rejection ratio by a factor of about $2g_{m(mir)}(r_{o(dp)}\|r_{o(mir)})$ for the MOS circuit in Fig. 4.29b compared to its passive-load counterpart in Fig. 4.29a.

For these calculations, perfect matching was assumed so that $g_{m1} = g_{m2}$, $g_{m3} = g_{m4}$, $r_{o1} = r_{o2}$, and $r_{o3} = r_{o4}$. In practice, however, nonzero mismatch occurs. With mismatch in a MOS differential pair using a current-mirror load, the differential-mode transconductance is

$$G_m[dm] \simeq g_{m1-2}\left[\frac{1 - \left(\dfrac{\Delta g_{m1-2}}{2g_{m1-2}}\right)^2}{1 + \left(\dfrac{\Delta g_{m3-4}}{2g_{m3-4}}\right)}\right] \quad (4.184)$$

where $\Delta g_{m1-2} = g_{m1} - g_{m2}$, $g_{m1-2} = (g_{m1} + g_{m2})/2$, $\Delta g_{m3-4} = g_{m3} - g_{m4}$, and $g_{m3-4} = (g_{m3} + g_{m4})/2$. See Problem 4.18. The approximation in (4.184) is valid to the extent that $g_m r_o \gg 1$ for each transistor and $(g_{m1} + g_{m2})r_{\text{tail}} \gg 1$ for the tail current source. Equation 4.184 shows that the mismatch between g_{m1} and g_{m2} has only a minor effect on $G_m[dm]$. This result stems from the fact that the small-signal voltage across the tail current source, v_{tail}, is zero with a purely differential input only when $g_{m1} = g_{m2}$, assuming $r_{o1} \to \infty$ and $r_{o2} \to \infty$. For example, increasing g_{m1} compared to g_{m2} tends to increase the small-signal drain current i_1 if v_{gs1} is constant. However, this change also increases v_{tail}, which reduces v_{gs1} for a fixed v_{id}. The combination of these two effects causes i_1 to be insensitive to $g_{m1} - g_{m2}$. On the other hand, mismatch between g_{m3} and g_{m4} directly modifies the contribution of i_1 through the current mirror to the output current. Therefore, (4.184) shows that $G_m[dm]$ is most sensitive to the mismatch between g_{m3} and g_{m4}.

With mismatch in a MOS differential pair using a current-mirror load, the common-mode transconductance is

$$G_m[cm] \simeq -\frac{1}{2r_{\text{tail}}}(\epsilon_d + \epsilon_m) \quad (4.185)$$

where ϵ_d is the gain error in the source-coupled pair with a pure common-mode input defined in (4.161) and ϵ_m is the gain error in the current mirror defined in (4.133). From Problem 4.19,

$$\epsilon_d \simeq \frac{1}{g_{m3}r_{o(dp)}} - \frac{\Delta g_{m1-2}}{g_{m1-2}}\left(1 + \frac{2r_{\text{tail}}}{r_{o(dp)}}\right) - \frac{2r_{\text{tail}}}{r_{o(dp)}}\frac{\Delta r_{o(dp)}}{r_{o(dp)}} \quad (4.186)$$

Each term in (4.186) corresponds to one source of gain error by itself, and interactions between terms are ignored. The first term in (4.186) is consistent with (4.173) when $g_{m3}r_{o(dp)} \gg 1$ and stems from the observation that the drain of T_1 is not connected to a small-signal ground during the calculation of $G_m[dm]$, unlike the drain of T_2. The second term in (4.186) stems from mismatch between g_{m1} and g_{m2} alone. The third term in (4.186) stems from the mismatch between r_{o1} and r_{o2} alone. The contribution of this mismatch to $G_m[cm]$ is significant because the action of the current mirror nearly cancels the contributions of the input g_m generators to $G_m[cm]$ under ideal conditions, causing $G_m[cm] \simeq 0$ in (4.166). In contrast, $G_m[dm]$ is insensitive to the mismatch between r_{o1} and r_{o2} because the dominant contributions to $G_m[dm]$ arising from the input g_m generators do not cancel at the output. From Problem 4.19,

$$\epsilon_m = \frac{1}{1 + g_{m3}r_{o3}} + \frac{(g_{m3} - g_{m4})r_{o3}}{1 + g_{m3}r_{o3}} \simeq \frac{1}{g_{m3}r_{o3}} + \frac{\Delta g_{m3-4}}{g_{m3-4}} \quad (4.187)$$

Each term in (4.187) corresponds to one source of gain error by itself, and interactions between terms are ignored. The first term in (4.187), which is consistent with (4.179) when $g_{m3}r_{o3} \gg 1$, stems from the observation that the small-signal input resistance of the current mirror is not exactly $1/g_{m3}$ but $(1/g_{m3})||r_{o3}$. The second term in (4.187) stems from mismatch between g_{m3} and g_{m4} alone. The CMRR with mismatch is the ratio of $G_m[dm]$ in (4.184) to $G_m[cm]$ in (4.185), using (4.186) and (4.187) for ϵ_d and ϵ_m, respectively. Since the common-mode transconductance is very small without mismatch (as a result of the behavior of the current-mirror load), mismatch usually reduces the CMRR by increasing $|G_m[cm]|$.

4.4 Voltage and Current References

4.4.1 Low-Current Biasing

4.4.1.1 Bipolar Widlar Current Source

In ideal operational amplifiers, the current is zero in each of the two input leads. However, the input current is not zero in real op amps with bipolar input transistors because β_F is finite. Since the op-amp inputs are usually connected to a differential pair, the tail current must usually be very small in such op amps to keep the input current small. Typically, the tail current is on the order of 5 μA. Bias currents of this magnitude are also required in a variety of other applications, especially where minimizing power dissipation is important. The simple current mirrors shown in Fig. 4.30 are usually not optimum for such small currents. For example, using a simple bipolar current mirror as in Fig. 4.30a and assuming a maximum practical emitter area ratio between transistors of ten to one, the mirror would need an input current of 50 μA for an output current of 5 μA. If the power-supply voltage in Fig. 4.30a is 5 V, and if $V_{BE(\text{on})} = 0.7$ V, $R = 86$ kΩ would be required. Resistors of this magnitude are costly in terms of die area. Currents of such low magnitude can be obtained with moderate values of resistance, however, by modifying the simple current mirror so that the transistors operate with unequal base-emitter voltages. In the Widlar current source of Fig. 4.31a, resistor R_2 is inserted in series with the emitter of Q_2, and transistors Q_1 and Q_2 operate with unequal base emitter voltages if $R_2 \neq 0$.[10,11] This circuit is referred to as a current source rather than a current mirror because the output current in Fig. 4.31a is much less dependent on the input

298 Chapter 4 ■ Current Mirrors, Active Loads, and References

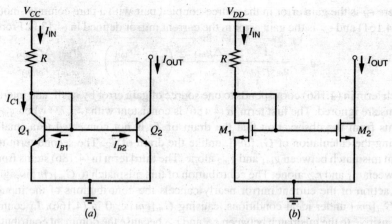

Figure 4.30 Simple two-transistor current mirrors where the input current is set by the supply voltage and a resistor using (*a*) bipolar and (*b*) MOS transistors.

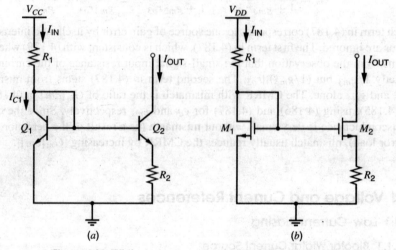

Figure 4.31 Widlar current sources: (*a*) bipolar and (*b*) MOS.

current and the power-supply voltage than in the simple current mirror of Fig. 4.30*a*, as shown in Section 4.4.2. We will now calculate the output current of the Widlar current source.

If $I_{IN} > 0$, Q_1 operates in the forward-active region because it is diode connected. Assume that Q_2 also operates in the forward active region. KVL around the base-emitter loop gives

$$V_{BE1} - V_{BE2} - \frac{\beta_F + 1}{\beta_F} I_{OUT} R_2 = 0 \qquad (4.188)$$

If we assume that $V_{BE1} = V_{BE2} = V_{BE(on)} = 0.7$ V in (4.188), we would predict that $I_{OUT} = 0$. Although I_{OUT} is small in practice, it is greater than zero under the usual bias conditions, which means that the standard assumption about $V_{BE(on)}$ is invalid here. In contrast, the standard assumption is usually valid in calculating I_{IN} because small variations in V_{BE1} have little effect on I_{IN} if $V_{CC} \gg V_{BE1}$. When one base-emitter voltage is subtracted from another, however, small differences between them are important. If $V_A \to \infty$, (4.188) can be rewritten using (1.35) as

$$V_T \ln \frac{I_{C1}}{I_{S1}} - V_T \ln \frac{I_{OUT}}{I_{S2}} - \frac{\beta_F + 1}{\beta_F} I_{OUT} R_2 = 0 \tag{4.189}$$

If $\beta_F \to \infty$, (4.189) simplifies to

$$V_T \ln \frac{I_{IN}}{I_{S1}} - V_T \ln \frac{I_{OUT}}{I_{S2}} - I_{OUT} R_2 = 0 \tag{4.190}$$

For identical transistors, I_{S1} and I_{S2} are equal, and (4.190) becomes

$$V_T \ln \frac{I_{IN}}{I_{OUT}} = I_{OUT} R_2 \tag{4.191}$$

This transcendental equation can be solved by trial and error to find I_{OUT} if R_2 and I_{IN} are known, as in typical analysis problems. Because the logarithm function compresses changes in its argument, attention can be focused on the linear term in (4.191), simplifying convergence of the trial-and-error process. In design problems, however, the desired I_{IN} and I_{OUT} are usually known, and (4.191) provides the required value of R_2.

- **EXAMPLE**

In the circuit of Fig. 4.31a, determine the proper value of R_2 to give $I_{OUT} = 5$ μA. Assume that $V_{CC} = 5$ V, $R_1 = 4.3$ kΩ, $V_{BE(on)} = 0.7$ V, and $\beta_F \to \infty$.

$$I_{IN} = \frac{5 \text{ V} - 0.7 \text{ V}}{4.3 \text{ k}\Omega} = 1 \text{ mA}$$

$$V_T \ln \frac{I_{IN}}{I_{OUT}} = 26 \text{ mV} \ln \left(\frac{1 \text{ mA}}{5 \text{ μA}}\right) = 137 \text{ mV}$$

Thus from (4.191)

$$I_{OUT} R_2 = 137 \text{ mV}$$

and

$$R_2 = \frac{137 \text{ mV}}{5 \text{ μA}} = 27.4 \text{ k}\Omega$$

- The total resistance in the circuit is 31.7 kΩ.

- **EXAMPLE**

In the circuit of Fig. 4.31a, assume that $I_{IN} = 1$ mA, $R_2 = 5$ kΩ, and $\beta_F \to \infty$. Find I_{OUT}. From (4.191),

$$V_T \ln \frac{1 \text{ mA}}{I_{OUT}} - 5 \text{ k}\Omega (I_{OUT}) = 0$$

Try

$$I_{OUT} = 15 \text{ μA}$$

$$108 \text{ mV} - 75 \text{ mV} \neq 0$$

The linear term $I_{OUT} R_2$ is too small; therefore, $I_{OUT} > 15$ μA should be tried. Try

$$I_{OUT} = 20 \text{ μA}$$

$$101.7 \text{ mV} - 100 \text{ mV} \simeq 0$$

Therefore, the output current is close to 20 μA. Notice that while the linear term increased by 25 mV from the first to the second trial, the logarithm term decreased by only about 6 mV because the logarithm function compresses changes in its argument.

4.4.1.2 MOS Widlar Current Source

The Widlar configuration can also be used in MOS technology, as shown in Fig. 4.31b.

If $I_{IN} > 0$, M_1 operates in the active region because it is diode connected. Assume that M_2 also operates in the forward active region. KVL around the gate-source loop gives

$$V_{GS1} - V_{GS2} - I_{OUT} R_2 = 0 \tag{4.192}$$

If we ignore the body effect, the threshold components of the gate-source voltages cancel and (4.192) simplifies to

$$I_{OUT} R_2 + V_{ov2} - V_{ov1} = 0 \tag{4.193}$$

If the transistors operate in strong inversion and $V_A \to \infty$,

$$I_{OUT} R_2 + \sqrt{\frac{2 I_{OUT}}{k'(W/L)_2}} - V_{ov1} = 0 \tag{4.194}$$

This quadratic equation can be solved for $\sqrt{I_{OUT}}$.

$$\sqrt{I_{OUT}} = \frac{-\sqrt{\frac{2}{k'(W/L)_2}} \pm \sqrt{\frac{2}{k'(W/L)_2} + 4R_2 V_{ov1}}}{2R_2} \tag{4.195}$$

where $V_{ov1} = \sqrt{2I_{IN}/[k'(W/L)_1]}$. From (1.157),

$$\sqrt{I_{OUT}} = \sqrt{\frac{k'(W/L)_2}{2}} (V_{GS2} - V_t) \tag{4.196}$$

Equation 4.196 applies only when M_2 operates in the active region, which means that $V_{GS2} > V_t$. As a result, $\sqrt{I_{OUT}} > 0$ and the potential solution where the second term in the numerator of (4.195) is subtracted from the first, cannot occur in practice. Therefore,

$$\sqrt{I_{OUT}} = \frac{-\sqrt{\frac{2}{k'(W/L)_2}} + \sqrt{\frac{2}{k'(W/L)_2} + 4R_2 V_{ov1}}}{2R_2} \tag{4.197}$$

Equation 4.197 shows that a closed-form solution for the output current can be written for a Widlar current source that uses MOS transistors operating in strong inversion, unlike the bipolar case where trial and error is required to find I_{OUT}.

■ **EXAMPLE**

In Fig. 4.31b, find I_{OUT} if $I_{IN} = 100$ μA, $R_2 = 4$ kΩ, $k' = 200$ μA/V², and $(W/L)_1 = (W/L)_2 = 25$. Assume the temperature is 27°C and that $n = 1.5$ in (1.247). Then $R_2 = 0.004$ MΩ, $V_{ov1} = \sqrt{200/(200 \times 25)}$ V $= 0.2$ V,

$$\sqrt{I_{OUT}} = \frac{-\sqrt{\frac{2}{200(25)}} + \sqrt{\frac{2}{200(25)} + 4(0.004)(0.2)}}{2(0.004)} \sqrt{\mu A} = 5\sqrt{\mu A}$$

and $I_{OUT} = 25\ \mu A$. Also,

$$V_{ov2} = V_{ov1} - I_{OUT} R_2 = 0.2 - 25 \times 0.004 = 0.1\ \text{V} > 2nV_T \simeq 78\ \text{mV}$$

- Therefore, both transistors operate in strong inversion, as assumed.

4.4.1.3 Bipolar Peaking Current Source

The Widlar source described in Section 4.4.1.1 allows currents in the microamp range to be realized with moderate values of resistance. Biasing integrated-circuit stages with currents on the order of nanoamps is often desirable. To reach such low currents with moderate values of resistance, the circuit shown in Fig. 4.32 can be used.[12,13,14] Neglecting base currents, we have

$$V_{BE1} - I_{IN} R = V_{BE2} \tag{4.198}$$

If $V_A \rightarrow \infty$, (4.198) can be rewritten using (1.35) as

$$V_T \ln \frac{I_{IN}}{I_{S1}} - V_T \ln \frac{I_{OUT}}{I_{S2}} = I_{IN} R \tag{4.199}$$

If Q_1 and Q_2 are identical, (4.199) can be rewritten as

$$I_{OUT} = I_{IN} \exp\left(-\frac{I_{IN} R}{V_T}\right) \tag{4.200}$$

Equation 4.200 is useful for analysis of a given circuit. For design with identical Q_1 and Q_2, (4.199) can be rewritten as

$$R = \frac{V_T}{I_{IN}} \ln \frac{I_{IN}}{I_{OUT}} \tag{4.201}$$

For example, for $I_{IN} = 10\ \mu A$ and $I_{OUT} = 100\ \text{nA}$, (4.201) can be used to show that $R \simeq 12\ k\Omega$.

A plot of I_{OUT} versus I_{IN} from (4.200) is shown in Fig. 4.33. When the input current is small, the voltage drop on the resistor is small, and $V_{BE2} \simeq V_{BE1}$ so $I_{OUT} \simeq I_{IN}$. As the input current increases, V_{BE1} increases in proportion to the logarithm of the input current while the drop on the resistor increases linearly with the input current. As a result, increases in the input current eventually cause the base-emitter voltage of Q_2 to decrease. The output current

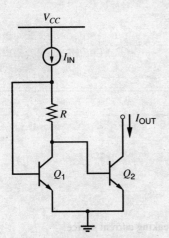

Figure 4.32 Bipolar peaking current source.

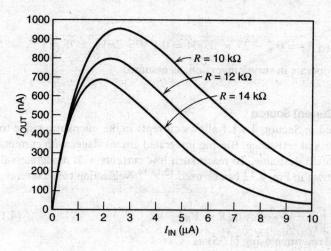

Figure 4.33 Transfer characteristics of the bipolar peaking current source with $T = 27°C$.

reaches a maximum when V_{BE2} is maximum. The name *peaking current source* stems from this behavior, and the location and magnitude of the peak both depend on R.

4.4.1.4 MOS Peaking Current Source

The peaking-current configuration can also be used in MOS technology, as shown in Fig. 4.34. If I_{IN} is small and positive, the voltage drop on R is small and M_1 operates in the active region. Assume that M_2 also operates in the active region. KVL around the gate-source loop gives

$$V_{GS1} - I_{IN}R - V_{GS2} = 0 \tag{4.202}$$

Since the sources of M_1 and M_2 are connected together, the thresholds cancel and (4.202) simplifies to

$$V_{ov2} = V_{ov1} - I_{IN}R \tag{4.203}$$

From (1.157),

$$I_{OUT} = \frac{k'(W/L)_2}{2}(V_{ov2})^2 = \frac{k'(W/L)_2}{2}(V_{ov1} - I_{IN}R)^2 \tag{4.204}$$

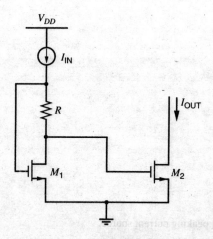

Figure 4.34 MOS peaking current source.

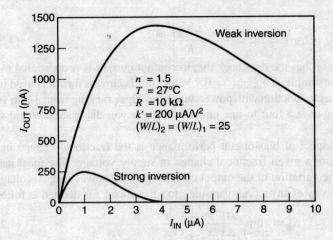

Figure 4.35 Transfer characteristics of the MOS peaking current source assuming both transistors operate in weak inversion or in strong inversion.

where $V_{ov1} = \sqrt{2I_{IN}/[k'(W/L)_1]}$. Equation 4.204 assumes that the transistors operate in strong inversion. In practice, the input current is usually small enough that the overdrive of M_1 is less than $2nV_T$, where n is defined in (1.247) and V_T is a thermal voltage. Equation 4.203 shows that the overdrive of M_2 is even smaller than that of M_1. Therefore, both transistors usually operate in weak inversion, where the drain current is an exponential function of the gate-source voltage as shown in (1.252). If $V_{DS1} > 3V_T$, applying (1.252) to M_1 and substituting into (4.202) gives

$$V_{GS2} - V_t \simeq nV_T \ln\left(\frac{I_{IN}}{(W/L)_1 I_t}\right) - I_{IN} R \qquad (4.205)$$

Then if the transistors are identical and $V_{DS2} > 3V_T$, substituting (4.205) into (1.252) gives

$$I_{OUT} \simeq \frac{W}{L} I_t \exp\left(\frac{V_{GS2} - V_t}{nV_T}\right) \simeq I_{IN} \exp\left(-\frac{I_{IN} R}{nV_T}\right) \qquad (4.206)$$

where I_t is given by (1.251) and represents the drain current of M_2 with $V_{GS2} = V_t$, $W/L = 1$, and $V_{DS} \gg V_T$. Comparing (4.206) with (4.200) shows that the output current in an MOS peaking current source where both transistors operate in weak inversion is the same as in the bipolar case except that $1.3 \leq n \leq 1.5$ in the MOS case and $n = 1$ in the bipolar case.

Plots of (4.206) and (4.204) are shown in Fig. 4.35 for $n = 1.5$, $T = 27°C$, $R = 10$ kΩ, $k' = 200$ μA/V^2, and $(W/L)_2 = (W/L)_1 = 25$. In both cases, when the input current is small, the voltage drop on the resistor is small, and $I_{OUT} \simeq I_{IN}$. As the input current increases, V_{GS1} increases more slowly than the drop on the resistor. As a result, increases in the input current eventually cause the gate-source voltage of M_2 to decrease. The output current reaches a maximum when V_{GS2} is maximum. As in the bipolar case, the name *peaking current source* stems from this behavior, and the location and magnitude of the peak both depend on R. Because the overdrives on both transistors are usually very small, the strong-inversion equation (4.204) usually underestimates the output current.

4.4.2 Supply-Insensitive Biasing

Consider the simple current mirror of Fig. 4.30a, where the input current source has been replaced by a resistor. Ignoring the effects of finite β_F and V_A, (4.5) shows that the output current is

$$I_{OUT} \simeq I_{IN} = \frac{V_{CC} - V_{BE(on)}}{R} \qquad (4.207)$$

If $V_{CC} \gg V_{BE(on)}$, this circuit has the drawback that the output current is proportional to the power-supply voltage. For example, if $V_{BE(on)} = 0.7$ V, and if this current mirror is used in an operational amplifier that has to function with power-supply voltages ranging from 3 V to 10 V, the bias current would vary over a four-to-one range, and the power dissipation would vary over a thirteen-to-one range.

One measure of this aspect of bias-circuit performance is the fractional change in the bias current that results from a given fractional change in supply voltage. The most useful parameter for describing the variation of the output current with the power-supply voltage is the sensitivity S. The sensitivity of any circuit variable y to a parameter x is defined as follows:

$$S_x^y = \lim_{\Delta x \to 0} \frac{\Delta y/y}{\Delta x/x} = \frac{x}{y}\frac{\partial y}{\partial x} \qquad (4.208)$$

Applying (4.208) to find the sensitivity of the output current to small variations in the power-supply voltage gives

$$S_{V_{SUP}}^{I_{OUT}} = \frac{V_{SUP}}{I_{OUT}} \frac{\partial I_{OUT}}{\partial V_{SUP}} \qquad (4.209)$$

The supply voltage V_{SUP} is usually called V_{CC} in bipolar circuits and V_{DD} in MOS circuits. If $V_{CC} \gg V_{BE(on)}$ in Fig. 4.30a, and if $V_{DD} \gg V_{GS1}$ in Fig. 4.30b,

$$S_{V_{SUP}}^{I_{OUT}} \simeq 1 \qquad (4.210)$$

Equation 4.210 shows that the output currents in the simple current mirrors in Fig. 4.30 depend strongly on the power-supply voltages. Therefore, this configuration should not be used when supply insensitivity is important.

4.4.2.1 Widlar Current Sources

For the case of the bipolar Widlar source in Fig. 4.31a, the output current is given implicitly by (4.191). To determine the sensitivity of I_{OUT} to the power-supply voltage, this equation is differentiated with respect to V_{CC}:

$$V_T \frac{\partial}{\partial V_{CC}} \ln \frac{I_{IN}}{I_{OUT}} = R_2 \frac{\partial I_{OUT}}{\partial V_{CC}} \qquad (4.211)$$

Differentiating yields

$$V_T \left(\frac{I_{OUT}}{I_{IN}}\right)\left(\frac{1}{I_{OUT}}\frac{\partial I_{IN}}{\partial V_{CC}} - \frac{I_{IN}}{I_{OUT}^2}\frac{\partial I_{OUT}}{\partial V_{CC}}\right) = R_2 \frac{\partial I_{OUT}}{\partial V_{CC}} \qquad (4.212)$$

Solving this equation for $\partial I_{OUT}/\partial V_{CC}$, we obtain

$$\frac{\partial I_{OUT}}{\partial V_{CC}} = \left(\frac{1}{1 + \frac{I_{OUT}R_2}{V_T}}\right) \frac{I_{OUT}}{I_{IN}} \frac{\partial I_{IN}}{\partial V_{CC}} \qquad (4.213)$$

Substituting (4.213) into (4.209) gives

$$S_{V_{CC}}^{I_{OUT}} = \left(\frac{1}{1 + \frac{I_{OUT}R_2}{V_T}}\right) \frac{V_{CC}}{I_{IN}} \frac{\partial I_{IN}}{\partial V_{CC}} = \left(\frac{1}{1 + \frac{I_{OUT}R_2}{V_T}}\right) S_{V_{CC}}^{I_{IN}} \qquad (4.214)$$

If $V_{CC} \gg V_{BE(on)}$, $I_{IN} \simeq V_{CC}/R_1$ and the sensitivity of I_{IN} to V_{CC} is approximately unity, as in the simple current mirror of Fig. 4.30a. For the example in Section 4.4.1.1 where $I_{IN} = 1$ mA, $I_{OUT} = 5$ μA, and $R_2 = 27.4$ kΩ, (4.214) gives

$$S_{V_{CC}}^{I_{OUT}} = \frac{V_{CC}}{I_{OUT}} \frac{\partial I_{OUT}}{\partial V_{CC}} \simeq \frac{1}{1 + \dfrac{137 \text{ mV}}{26 \text{ mV}}} \simeq 0.16 \quad (4.215)$$

Thus for this case, a 10 percent power-supply voltage change results in only 1.6 percent change in I_{OUT}.

For the case of the MOS Widlar source in Fig. 4.31b, the output current is given by (4.197). Differentiating with respect to V_{DD} gives

$$\frac{1}{2\sqrt{I_{OUT}}} \frac{\partial I_{OUT}}{\partial V_{DD}} = \frac{1}{4R_2} \frac{1}{\sqrt{\dfrac{2}{k'(W/L)_2} + 4R_2 V_{ov1}}} 4R_2 \frac{\partial V_{ov1}}{\partial V_{DD}} \quad (4.216)$$

where

$$\frac{\partial V_{ov1}}{\partial V_{DD}} = \sqrt{\frac{2}{k'(W/L)_1}} \frac{1}{2\sqrt{I_{IN}}} \frac{\partial I_{IN}}{\partial V_{DD}} = \frac{V_{ov1}}{2I_{IN}} \frac{\partial I_{IN}}{\partial V_{DD}} \quad (4.217)$$

Substituting (4.216) and (4.217) into (4.209) gives

$$S_{V_{DD}}^{I_{OUT}} = \frac{V_{ov1}}{\sqrt{V_{ov2}^2 + 4I_{OUT} R_2 V_{ov1}}} S_{V_{DD}}^{I_{IN}} \quad (4.218)$$

Since I_{OUT} is usually much less than I_{IN}, V_{ov2} is usually small and $I_{OUT} R_2 \simeq V_{ov1}$ and (4.218) simplifies to

$$S_{V_{DD}}^{I_{OUT}} \simeq \frac{V_{ov1}}{\sqrt{4V_{ov1}^2}} S_{V_{DD}}^{I_{IN}} = 0.5 S_{V_{DD}}^{I_{IN}} \quad (4.219)$$

If $V_{DD} \gg V_{GS1}$, $I_{IN} \simeq V_{DD}/R_1$ and the sensitivity of I_{IN} to V_{DD} is approximately unity, as in the simple current mirror of Fig. 4.30b. Thus for this case, a 10 percent power-supply voltage change results in a 5 percent change in I_{OUT}.

4.4.2.2 Current Sources Using Other Voltage Standards

The level of power-supply independence provided by the bipolar and MOS Widlar current sources is not adequate for many types of analog circuits. Much lower sensitivity can be obtained by causing bias currents in the circuit to depend on a voltage standard other than the supply voltage. Bias reference circuits can be classified according to the voltage standard by which the bias currents are established. The most convenient standards are the base-emitter or threshold voltage of a transistor, the thermal voltage, or the breakdown voltage of a reverse-biased *pn* junction (a Zener diode). Each of these standards can be used to reduce supply sensitivity, but the drawback of the first three standards is that the reference voltage is quite temperature dependent. Both the base-emitter and threshold voltages have negative temperature coefficients of magnitude 1 to 2 mV/°C, and the thermal voltage has a positive temperature coefficient of $k/q \simeq 86$ μV/°C. The Zener diode has the disadvantage that at least 7 to 10 V of supply voltage are required because standard integrated-circuit processes produce a minimum breakdown voltage of about 6 V across the most highly doped junctions (usually *npn* transistor emitter-base junctions). Furthermore, *pn* junctions produce large amounts of voltage

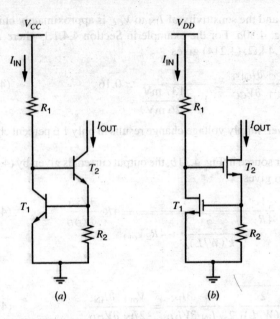

Figure 4.36 (a) Base-emitter referenced current source. (b) Threshold referenced current source.

noise under the reverse-breakdown conditions encountered in a bias reference circuit. Noise in avalanche breakdown is considered further in Chapter 11.

We now consider bias reference circuits based on the base-emitter or gate-source voltage. The circuit in simplest form in bipolar technology is shown in Fig. 4.36a. This circuit is similar to a Wilson current mirror where the diode-connected transistor is replaced by a resistor. For the input current to flow in T_1, transistor T_2 must supply enough current into R_2 so that the base-emitter voltage of T_1 is

$$V_{BE1} = V_T \ln \frac{I_{IN}}{I_{S1}} \qquad (4.220)$$

If we neglect base currents, I_{OUT} is equal to the current flowing through R_2. Since the voltage drop on R_2 is V_{BE1}, the output current is proportional to this base-emitter voltage. Thus, neglecting base currents, we have

$$I_{OUT} = \frac{V_{BE1}}{R_2} = \frac{V_T}{R_2} \ln \frac{I_{IN}}{I_{S1}} \qquad (4.221)$$

Differentiating (4.221) and substituting into (4.209) gives

$$S^{I_{OUT}}_{V_{CC}} = \frac{V_T}{I_{OUT} R_2} S^{I_{IN}}_{V_{CC}} = \frac{V_T}{V_{BE(on)}} S^{I_{IN}}_{V_{CC}} \qquad (4.222)$$

If $V_{CC} \gg 2V_{BE(on)}$, $I_{IN} \simeq V_{CC}/R_1$ and the sensitivity of I_{IN} to V_{CC} is approximately unity. With $V_{BE(on)} = 0.7$ V,

$$S^{I_{OUT}}_{V_{CC}} = \frac{0.026 \text{ V}}{0.7 \text{ V}} \simeq 0.037 \qquad (4.223)$$

Thus for this case, a 10 percent power-supply voltage change results in a 0.37 percent change in I_{OUT}. The result is significantly better than for a bipolar Widlar current source.

4.4 Voltage and Current References

The MOS counterpart of the base-emitter reference is shown in Fig. 4.36b. Here

$$I_{OUT} = \frac{V_{GS1}}{R_2} = \frac{V_t + V_{ov1}}{R_2} = \frac{V_t + \sqrt{\frac{2I_{IN}}{k'(W/L)_1}}}{R_2} \tag{4.224}$$

The case of primary interest is when the overdrive of T_1 is small compared to the threshold voltage. This case can be achieved in practice by choosing sufficiently low input current and large $(W/L)_1$. In this case, the output current is determined mainly by the threshold voltage and R_2. Therefore, this circuit is known as a *threshold-referenced* bias circuit. Differentiating (4.224) with respect to V_{DD} and substituting into (4.209) gives

$$S_{V_{DD}}^{I_{OUT}} = \frac{V_{ov1}}{2I_{OUT}R_2} S_{V_{DD}}^{I_{IN}} = \frac{V_{ov1}}{2V_{GS1}} S_{V_{DD}}^{I_{IN}} \tag{4.225}$$

For example, if $V_t = 1$ V, $V_{ov1} = 0.1$ V, and $S_{V_{DD}}^{I_{IN}} \simeq 1$

$$S_{V_{DD}}^{I_{OUT}} \simeq \frac{0.1}{2(1.1)} \simeq 0.045 \tag{4.226}$$

These circuits are not fully supply independent because the base-emitter or gate-source voltages of T_1 change slightly with power-supply voltage. This change occurs because the collector or drain current of T_1 is approximately proportional to the supply voltage. The resulting supply sensitivity is often a problem in bias circuits whose input current is derived from a resistor connected to the supply terminal, since this configuration causes the currents in some portion of the circuit to change with the supply voltage.

4.4.2.3 Self-Biasing

Power-supply sensitivity can be greatly reduced by the use of the so-called *bootstrap* bias technique, also referred to as *self-biasing*. Instead of developing the input current by connecting a resistor to the supply, the input current is made to depend directly on the output current of the current source itself. The concept is illustrated in block-diagram form in Fig. 4.37a. Assuming that the feedback loop formed by this connection has a stable operating point, the currents flowing in the circuit are much less sensitive to power-supply voltage than in the resistively biased case. The two key variables here are the input current, I_{IN}, and the output current, I_{OUT}. The relationship between these variables is governed by both the current source and the current

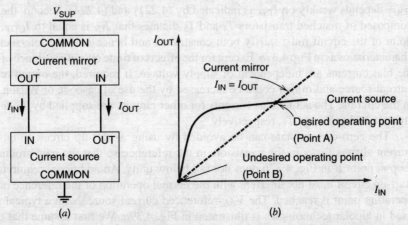

Figure 4.37 (a) Block diagram of a self-biased reference. (b) Determination of operating point.

mirror. From the standpoint of the current source, the output current is almost independent of the input current for a wide range of input currents as shown in Fig. 4.37b. From the standpoint of the current mirror, I_{IN} is set equal to I_{OUT}, assuming that the gain of the current mirror is unity. The operating point of the circuit must satisfy both constraints and hence is at the intersection of the two characteristics. In the plot of Fig. 4.37b, two intersections or potential operating points are shown. Point A is the desired operating point, and point B is an undesired operating point because $I_{OUT} = I_{IN} = 0$.

If the output current in Fig. 4.37a increases for any reason, the current mirror increases the input current by the same amount because the gain of the current mirror is assumed to be unity. As a result, the current source increases the output current by an amount that depends on the gain of the current source. Therefore, the loop responds to an initial change in the output current by further changing the output current in a direction that reinforces the initial change. In other words, the connection of a current source and a current mirror as shown in Fig. 4.37a forms a positive feedback loop, and the gain around the loop is the gain of the current source. In Chapter 9, we will show that circuits with positive feedback are stable if the gain around the loop is less than unity. At point A, the gain around the loop is quite small because the output current of the current source is insensitive to changes in the input current around point A. On the other hand, at point B, the gain around the feedback loop is deliberately made greater than unity so that the two characteristics shown in Fig. 4.37b intersect at a point away from the origin. As a result, this simplified analysis shows that point B is an unstable operating point in principle, and the circuit would ideally tend to drive itself out of this state.

In practice, however, point B is frequently a stable operating point because the currents in the transistors at this point are very small, often in the picoampere range. At such low current levels, leakage currents and other effects reduce the current gain of both bipolar and MOS transistors, usually causing the gain around the loop to be less than unity. As a result, actual circuits of this type are usually unable to drive themselves out of the zero-current state. Thus, unless precautions are taken, the circuit may operate in the zero-current condition. For these reasons, self-biased circuits often have a stable state in which zero current flows in the circuit even when the power-supply voltage is nonzero. This situation is analogous to a gasoline engine that is not running even though it has a full tank of fuel. An electrical or mechanical device is required to start the engine. Similarly, a start-up circuit is usually required to prevent the self-biased circuit from remaining in the zero-current state.

The application of this technique to the V_{BE}-referenced current source is illustrated in Fig. 4.38a, and the threshold-referenced MOS counterpart is shown in Fig. 4.38b. We assume for simplicity that $V_A \to \infty$. The circuit composed of T_1, T_2, and R dictates that the current I_{OUT} depends weakly on I_{IN}, as indicated by (4.221) and (4.224). Second, the current mirror composed of matched transistors T_4 and T_5 dictates that I_{IN} is equal to I_{OUT}. The operating point of the circuit must satisfy both constraints and hence is at the intersection of the two characteristics as in Fig. 4.37b. Except for the effects of finite output resistance of the transistors, the bias currents are independent of supply voltage. If required, the output resistance of the current source and mirror could be increased by the use of cascode or Wilson configurations in the circuits. The actual bias currents for other circuits are supplied by T_6 and/or T_3, which are matched to T_5 and T_1, respectively.

The zero-current state can be avoided by using a start-up circuit to ensure that some current always flows in the transistors in the reference so that the gain around the feedback loop at point B in Fig. 4.37b does not fall below unity. An additional requirement is that the start-up circuit must not interfere with the normal operation of the reference once the desired operating point is reached. The V_{BE}-referenced current source with a typical start-up circuit used in bipolar technologies is illustrated in Fig. 4.39a. We first assume that the circuit is in the undesired zero-current state. If this were true, the base-emitter voltage of T_1 would be

4.4 Voltage and Current References

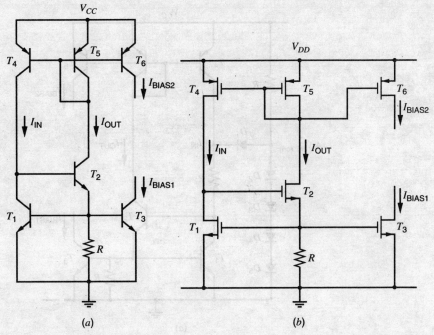

Figure 4.38 (a) Self-biasing V_{BE} reference. (b) Self-biasing V_t reference.

zero. The base-emitter voltage T_2 would be tens of millivolts above ground, determined by the leakage currents in the circuit. However, the voltage on the left-hand end of D_1 is four diode drops above ground, so that a voltage of at least three diode drops would appear across R_x, and a current would flow through R_x into the T_1-T_2 combination. This action would cause current to flow in T_4 and T_5, avoiding the zero-current state.

The bias reference circuit then drives itself toward the desired stable state, and we require that the start-up circuit not affect the steady-state current values. This can be accomplished by causing R_x to be large enough that when the steady-state current is established in T_1, the voltage drop across R_x is large enough to reverse bias D_1. In the steady state, the collector-emitter voltage of T_1 is two diode drops above ground, and the left-hand end of D_1 is four diode drops above ground. Thus if we make $I_{IN}R_x$ equal to two diode drops, D_1 will have zero voltage across it in the steady state. As a result, the start-up circuit composed of R_s, D_2-D_5, and D_1 is, in effect, disconnected from the circuit for steady-state operation.

Floating diodes are not usually available in MOS technologies. The threshold-referenced current source with a typical start-up circuit used in MOS technologies is illustrated in Fig. 4.39b. If the circuit is in the undesired zero-current state, the gate-source voltage of T_1 would be less than a threshold voltage. As a result, T_7 is off and T_8 operates in the triode region, pulling the gate-source voltage of T_9 up to V_{DD}. Therefore, T_9 is on and pulls down on the gates of T_4 and T_5. This action causes current to flow in T_4 and T_5, avoiding the zero-current state.

In steady state, the gate-source voltage of T_7 rises to $I_{OUT}R$, which turns on T_7 and reduces the gate-source voltage of T_9. In other words, T_7 and T_8 form a CMOS inverter whose output falls when the reference circuit turns on. Since the start-up circuit should not interfere with normal operation of the reference in steady state, the inverter output should fall low enough to turn T_9 off in steady state. Therefore, the gate-source voltage of T_9 must fall below a threshold voltage when the inverter input rises from zero to $I_{OUT}R$. In practice,

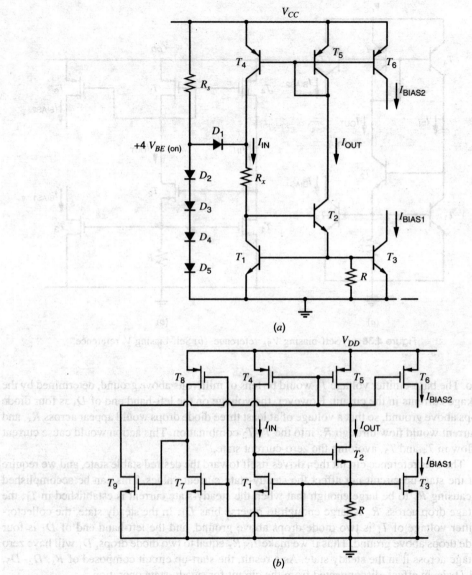

Figure 4.39 (a) Self-biasing V_{BE} reference with start-up circuit. (b) Self-biasing V_t reference with start-up circuit.

this requirement is satisfied by choosing the aspect ratio of T_7 to be much larger than that of T_8.

Another important aspect of the performance of biasing circuits is their dependence on temperature. This variation is most conveniently expressed in terms of the fractional change in output current per degree centigrade of temperature variation, which we call the fractional temperature coefficient TC_F:

$$TC_F = \frac{1}{I_{OUT}} \frac{\partial I_{OUT}}{\partial T} \qquad (4.227)$$

4.4 Voltage and Current References

For the V_{BE}-referenced circuit of Fig. 4.38a,

$$I_{OUT} = \frac{V_{BE1}}{R} \tag{4.228}$$

$$\frac{\partial I_{OUT}}{\partial T} = \frac{1}{R}\frac{\partial V_{BE1}}{\partial T} - \frac{V_{BE1}}{R^2}\frac{\partial R}{\partial T} \tag{4.229}$$

$$= I_{OUT}\left(\frac{1}{V_{BE1}}\frac{\partial V_{BE1}}{\partial T} - \frac{1}{R}\frac{\partial R}{\partial T}\right) \tag{4.230}$$

Therefore,

$$TC_F = \frac{1}{I_{OUT}}\frac{\partial I_{OUT}}{\partial T} = \frac{1}{V_{BE1}}\frac{\partial V_{BE1}}{\partial T} - \frac{1}{R}\frac{\partial R}{\partial T} \tag{4.231}$$

Thus the temperature dependence of the output current is related to the difference between the resistor temperature coefficient and that of the base-emitter junction. Since the former has a positive and the latter a negative coefficient, the net TC_F is quite large.

■ **EXAMPLE**

Design a bias reference as shown in Fig. 4.38a to produce 100 μA output current. Find the TC_F. Assume that for T_1, $I_S = 10^{-14}$ A. Assume that $\partial V_{BE}/\partial T = -2$ mV/°C and that $(1/R)(\partial R/\partial T) = +1500$ ppm/°C.

The current in T_1 will be equal to I_{OUT}, so that

$$V_{BE1} = V_T \ln \frac{100 \text{ μA}}{10^{-14} \text{ A}} = 598 \text{ mV}$$

Thus from (4.228),

$$R = \frac{598 \text{ mV}}{0.1 \text{ mA}} = 5.98 \text{ k}\Omega$$

From (4.231),

$$TC_F \simeq \frac{-2 \text{ mV/°C}}{598 \text{ mV}} - 1.5 \times 10^{-3} \simeq -3.3 \times 10^{-3} - 1.5 \times 10^{-3}$$

and thus

$$TC_F \simeq -4.8 \times 10^{-3}/°C = -4800 \text{ ppm/°C}$$

■ The term ppm is an abbreviation for parts per million and implies a multiplier of 10^{-6}.

For the threshold-referenced circuit of Fig. 4.38b,

$$I_{OUT} = \frac{V_{GS1}}{R} \simeq \frac{V_t}{R} \tag{4.232}$$

Differentiating (4.232) and substituting into (4.227) gives

$$TC_F = \frac{1}{I_{OUT}}\frac{\partial I_{OUT}}{\partial T} \simeq \frac{1}{V_t}\frac{\partial V_t}{\partial T} - \frac{1}{R}\frac{\partial R}{\partial T} \tag{4.233}$$

Since the threshold voltage of an MOS transistor and the base-emitter voltage of a bipolar transistor both change at about -2 mV/°C, (4.233) and (4.231) show that the temperature dependence of the threshold-referenced current source in Fig. 4.38b is about the same as the V_{BE}-referenced current source in Fig. 4.38a.

V_{BE}-referenced bias circuits are also used in CMOS technology. An example is shown in Fig. 4.40, where the *pnp* transistor is the parasitic device inherent in *p*-substrate CMOS

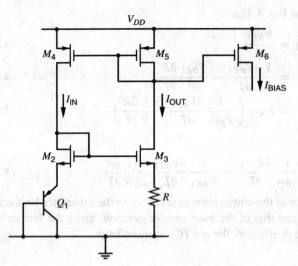

Figure 4.40 Example of a V_{BE}-referenced self-biased circuit in CMOS technology.

technologies. A corresponding circuit utilizing *npn* transistors can be used in *n*-substrate CMOS technologies. The feedback circuit formed by M_2, M_3, M_4, and M_5 forces the current in transistor Q_1 to be the same as in resistor R. Assuming matched devices, $V_{GS2} = V_{GS3}$ and thus

$$I_{OUT} = \frac{V_{BE1}}{R} \qquad (4.234)$$

An alternate source for the voltage reference is the thermal voltage V_T. The difference in junction potential between two junctions operated at different current densities can be shown to be proportional to V_T. This voltage difference must be converted to a current to provide the bias current. For the Widlar source shown in Fig. 4.31a, (4.190) shows that the voltage across the resistor R_2 is

$$I_{OUT} R_2 = V_T \ln \frac{I_{IN}}{I_{OUT}} \frac{I_{S2}}{I_{S1}} \qquad (4.235)$$

Thus if the ratio of the input to the output current is held constant, the voltage across R_2 is indeed proportional to V_T. This fact is utilized in the self-biased circuit of Fig. 4.41. Here Q_3 and Q_4 are selected to have equal areas. Therefore, if we assume that $\beta_F \to \infty$ and $V_A \to \infty$, the current mirror formed by Q_3 and Q_4 forces the collector current of Q_1 to equal that of Q_2. Figure 4.41 also shows that Q_2 has two emitters and Q_1 has one emitter, which indicates that the emitter area of Q_2 is twice that of Q_1 in this example. This selection is made to force the gain around the positive feedback loop at point B in Fig. 4.37b to be more than unity so that point B is an unstable operating point and a stable nonzero operating point exists at point A. As in other self-biased circuits, a start-up circuit is required to make sure that enough current flows to force operation at point A in Fig. 4.37b in practice. Under these conditions, $I_{S2} = 2I_{S1}$ and the voltage across R_2 is

$$I_{OUT} R_2 = V_T \ln \frac{I_{IN}}{I_{OUT}} \frac{I_{S2}}{I_{S1}} = V_T \ln 2 \qquad (4.236)$$

Therefore, the output current is

$$I_{OUT} = \frac{V_T}{R_2} \ln 2 \qquad (4.237)$$

4.4 Voltage and Current References

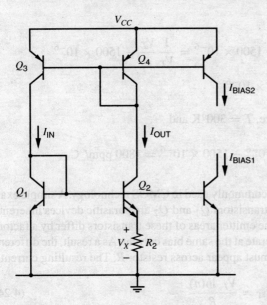

Figure 4.41 Bias current source using the thermal voltage.

The temperature variation of the output current can be calculated as follows. From (4.237)

$$\frac{\partial I_{OUT}}{\partial T} = (\ln 2) \frac{R_2 \frac{\partial V_T}{\partial T} - V_T \frac{\partial R_2}{\partial T}}{R_2^2}$$

$$= \frac{V_T}{R_2}(\ln 2)\left(\frac{1}{V_T}\frac{\partial V_T}{\partial T} - \frac{1}{R_2}\frac{\partial R_2}{\partial T}\right) \qquad (4.238)$$

Substituting (4.237) in (4.238) gives

$$TC_F = \frac{1}{I_{OUT}}\frac{\partial I_{OUT}}{\partial T} = \frac{1}{V_T}\frac{\partial V_T}{\partial T} - \frac{1}{R_2}\frac{\partial R_2}{\partial T} \qquad (4.239)$$

This circuit produces much smaller temperature coefficient of the output current than the V_{BE} reference because the fractional sensitivities of both V_T and that of a diffused resistor R_2 are positive and tend to cancel in (4.239). We have chosen a transistor area ratio of two to one as an example. In practice, this ratio is often chosen to minimize the total area required for the transistors and for resistor R_2.

■ **EXAMPLE**

Design a bias reference of the type shown in Fig. 4.41 to produce an output current of 100 μA. Find the TC_F of I_{OUT}. Assume the resistor temperature coefficient $(1/R)(\partial R/\partial T) = +1500$ ppm/°C.

From (4.237)

$$R_2 = \frac{V_T(\ln 2)}{I_{OUT}} = \frac{(26 \text{ mV})(\ln 2)}{100 \text{ μA}} \simeq 180 \text{ Ω}$$

From (4.239)

$$\frac{1}{I_{\text{OUT}}} \frac{\partial I_{\text{OUT}}}{\partial T} = \frac{1}{V_T} \frac{\partial V_T}{\partial T} - 1500 \times 10^{-6} = \frac{1}{V_T} \frac{V_T}{T} - 1500 \times 10^{-6}$$

$$= \frac{1}{T} - 1500 \times 10^{-6}$$

Assuming operation at room temperature, $T = 300°\text{K}$ and

$$\frac{1}{I_{\text{OUT}}} \frac{\partial I_{\text{OUT}}}{\partial T} \simeq 3300 \times 10^{-6} - 1500 \times 10^{-6} = 1800 \text{ ppm/°C}$$

V_T-referenced bias circuits are also commonly used in CMOS technology. A simple example is shown in Fig. 4.42, where bipolar transistors Q_1 and Q_2 are parasitic devices inherent in p-substrate CMOS technologies. Here the emitter areas of these transistors differ by a factor n, and the feedback loop forces them to operate at the same bias current. As a result, the difference between the two base-emitter voltages must appear across resistor R. The resulting current is

$$I_{\text{OUT}} = \frac{V_T \ln(n)}{R} \tag{4.240}$$

In the circuit of Fig. 4.42, small differences in the gate-source voltages of M_3 and M_4 result in large variations in the output current because the voltage drop across R is only on the order of 100 mV. Such gate-source voltage differences can result from device mismatches or from channel-length modulation in M_3 and M_4 because they have different drain-source voltages. Practical implementations of this circuit typically utilize large geometry devices for M_3 and M_4 to minimize offsets and cascode or Wilson current sources to minimize channel-length modulation effects. A typical example of a practical circuit is shown in Fig. 4.43. In general, cascoding is often used to improve the performance of reference circuits in all technologies. The main limitation of the application of cascoding is that it increases the minimum required power-supply voltage to operate all transistors in the active region.

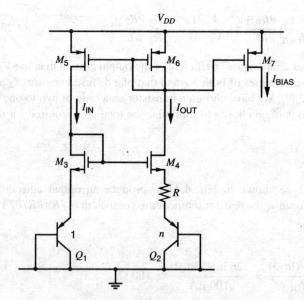

Figure 4.42 Example of a CMOS V_T-referenced self-biased circuit.

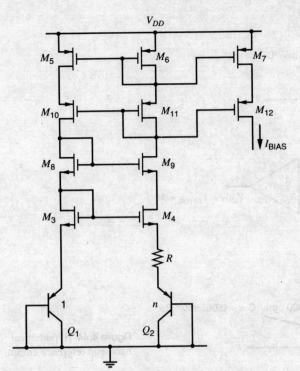

Figure 4.43 Example of a CMOS V_T-referenced self-biased reference circuit with cascoded devices to improve power-supply rejection and initial accuracy.

4.4.3 Temperature-Insensitive Biasing

As illustrated by the examples in Section 4.4.2, the base-emitter-voltage- and thermal-voltage-referenced circuits have rather high temperature coefficients of output current. Although the temperature sensitivity is reduced considerably in the thermal-voltage circuit, even its temperature coefficient is not low enough for many applications. Thus we are led to examine other possibilities for the realization of a biasing circuit with low temperature coefficient.

4.4.3.1 Band-Gap-Referenced Bias Circuits in Bipolar Technology

Since the bias sources referenced to $V_{BE(\text{on})}$ and V_T have opposite TC_F, the possibility exists for referencing the output current to a composite voltage that is a weighted sum of $V_{BE(\text{on})}$ and V_T. By proper weighting, zero temperature coefficient should be attainable.

In the biasing sources described so far, we have concentrated on the problem of obtaining a *current* with low temperature coefficient. In practice, requirements often arise for low-temperature-coefficient voltage bias or reference *voltages*. The voltage reference for a voltage regulator is a good example. The design of these two types of circuits is similar except that in the case of the current source, a temperature coefficient must be intentionally introduced into the voltage reference to compensate for the temperature coefficient of the resistor that will define the current. In the following description of the band-gap reference, we assume for simplicity that the objective is a *voltage* source of low temperature coefficient.

First consider the hypothetical circuit of Fig. 4.44. An output voltage is developed that is equal to $V_{BE(\text{on})}$ plus a constant M times V_T. To determine the required value for M, we must determine the temperature coefficient of $V_{BE(\text{on})}$. Neglecting base current,

$$V_{BE(\text{on})} = V_T \ln \frac{I_1}{I_S} \qquad (4.241)$$

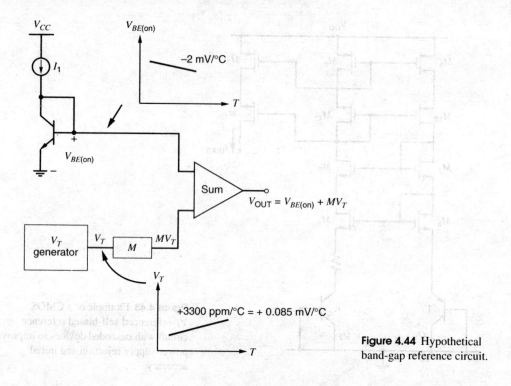

Figure 4.44 Hypothetical band-gap reference circuit.

As shown in Chapter 1, the saturation current I_S can be related to the device structure by

$$I_S = \frac{qAn_i^2 \overline{D}_n}{Q_B} = Bn_i^2 \overline{D}_n = B'n_i^2 T\overline{\mu}_n \qquad (4.242)$$

where n_i is the intrinsic minority-carrier concentration, Q_B is the total base doping per unit area, $\overline{\mu}_n$ is the average electron mobility in the base, A is the emitter-base junction area, and T is the temperature. Here, the constants B and B' involve only temperature-independent quantities. The Einstein relation $\mu_n = (q/kT)D_n$ was used to write I_S in terms of $\overline{\mu}_n$ and n_i^2. The quantities in (4.242) that are temperature dependent are given by[15]

$$\overline{\mu}_n = CT^{-n} \qquad (4.243)$$

$$n_i^2 = DT^3 \exp\left(-\frac{V_{G0}}{V_T}\right) \qquad (4.244)$$

where V_{G0} is the band-gap voltage of silicon extrapolated to $0°K$. Here again C and D are temperature-independent quantities whose exact values are unimportant in the analysis. The exponent n in the expression for base-region electron mobility $\overline{\mu}_n$ is dependent on the doping level in the base. Combining (4.241), (4.242), (4.243), and (4.244) yields

$$V_{BE(on)} = V_T \ln\left(I_1 T^{-\gamma} E \exp\frac{V_{G0}}{V_T}\right) \qquad (4.245)$$

where E is another temperature-independent constant and

$$\gamma = 4 - n \qquad (4.246)$$

In actual band-gap circuits, the current I_1 is not constant but varies with temperature. We assume for the time being that this temperature variation is known and that it can be written in the form

$$I_1 = GT^\alpha \tag{4.247}$$

where G is another temperature-independent constant. Combining (4.245) and (4.247) gives

$$V_{BE(on)} = V_{G0} - V_T[(\gamma - \alpha)\ln T - \ln(EG)] \tag{4.248}$$

From Fig. 4.44, the output voltage is

$$V_{OUT} = V_{BE(on)} + MV_T \tag{4.249}$$

Substitution of (4.248) into (4.249) gives

$$V_{OUT} = V_{G0} - V_T(\gamma - \alpha)\ln T + V_T[M + \ln(EG)] \tag{4.250}$$

This expression gives the output voltage as a function of temperature in terms of the circuit parameters G, α, and M, and the device parameters E and γ. Our objective is to make V_{OUT} independent of temperature. To this end, we take the derivative of V_{OUT} with respect to temperature to find the required values of G, γ, and M to give zero TC_F. Differentiating (4.250) gives

$$0 = \left.\frac{dV_{OUT}}{dT}\right|_{T=T_0} = \frac{V_{T0}}{T_0}[M + \ln(EG)] - \frac{V_{T0}}{T_0}(\gamma - \alpha)\ln T_0 - \frac{V_{T0}}{T_0}(\gamma - \alpha) \tag{4.251}$$

where T_0 is the temperature at which the TC_F of the output is zero and V_{T0} is the thermal voltage V_T evaluated at T_0. Equation 4.251 can be rearranged to give

$$[M + \ln(EG)] = (\gamma - \alpha)\ln T_0 + (\gamma - \alpha) \tag{4.252}$$

This equation gives the required values of circuit parameters M, α, and G in terms of the device parameters E and γ. In principle, these values could be calculated directly from (4.252). However, further insight is gained by back-substituting (4.252) into (4.250). The result is

$$V_{OUT} = V_{G0} + V_T(\gamma - \alpha)\left(1 + \ln\frac{T_0}{T}\right) \tag{4.253}$$

Thus the temperature dependence of the output voltage is entirely described by the single parameter T_0, which in turn is determined by the constants M, E, and G.

Using (4.253), the output voltage at the zero TC_F temperature ($T = T_0$) is given by

$$V_{OUT}|_{T=T_0} = V_{G0} + V_{T0}(\gamma - \alpha) \tag{4.254}$$

For example, to achieve zero TC_F at 27°C, assuming that $\gamma = 3.2$ and $\alpha = 1$,

$$V_{OUT}|_{T=T_0=25°C} = V_{G0} + 2.2V_{T0} \tag{4.255}$$

The band-gap voltage of silicon is $V_{G0} = 1.205$ V so that

$$V_{OUT}|_{T=T_0=25°C} = 1.205 \text{ V} + (2.2)(0.026 \text{ V}) = 1.262 \text{ V} \tag{4.256}$$

Therefore, the output voltage for zero temperature coefficient is close to the band-gap voltage of silicon, which explains the name given to these bias circuits.

Differentiating (4.253) with respect to temperature yields

$$\frac{dV_{OUT}}{dT} = \frac{1}{T}\left[V_T(\gamma - \alpha)\left(1 + \ln\frac{T_0}{T}\right)\right] - \frac{V_T}{T}(\gamma - \alpha)$$

$$= (\gamma - \alpha)\frac{V_T}{T}\left(\ln\frac{T_0}{T}\right) \tag{4.257}$$

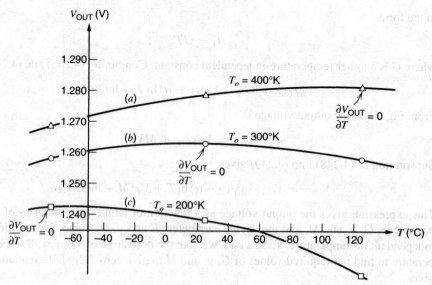

Figure 4.45 Variation of band-gap reference output voltage with temperature.

Equation 4.257 gives the slope of the output as a function of temperature. A typical family of output-voltage-variation characteristics is shown in Fig. 4.45 for different values of T_0 for the special case in which $\alpha = 0$ and I_1 is temperature independent. The slope of each curve is zero at $T = T_0$. When $T < T_0$, the slope is positive because the argument of the logarithm in (4.257) is more than unity. Similarly, the slope is negative when $T > T_0$. For values of T near T_0,

$$\ln\frac{T_0}{T} = \ln\left(1 + \frac{T_0 - T}{T}\right) \simeq \frac{T_0 - T}{T} \qquad (4.258)$$

and we have

$$\frac{dV_{OUT}}{dT} \simeq (\gamma - \alpha)\frac{V_T}{T}\left(\frac{T_0 - T}{T}\right) \qquad (4.259)$$

As shown by (4.257) and (4.259), the temperature coefficient of the output is zero only at one temperature $T = T_0$. This result stems from the addition of a weighted thermal voltage to a base-emitter voltage as in Fig. 4.44. Since the temperature coefficient of base-emitter voltage is not exactly constant, the gain M can be chosen to set the temperature coefficient of the output to zero only at one temperature. In other words, the thermal voltage generator is used to cancel the linear dependence of the base-emitter voltage with temperature. After this cancellation, the changing outputs in Fig. 4.45 stem from the nonlinear dependence of the base-emitter voltage with temperature. Band-gap references that compensate for this nonlinearity are said to be *curvature compensated*.[16,17,18]

- **EXAMPLE**

A band-gap reference is designed to give a nominal output voltage of 1.262 V, which gives zero TC_F at 27°C. Because of component variations, the actual room temperature output voltage is 1.280 V. Find the temperature of actual zero TC_F of V_{OUT}. Also, write an equation for V_{OUT} as a function of temperature, and calculate the TC_F at room temperature. Assume that $\gamma = 3.2$ and $\alpha = 1$.

From (4.253) at $T = 27°C = 300°K$,

$$1.280 \text{ V} = 1.205 + (0.026 \text{ V})(2.2)\left(1 + \ln \frac{T_0}{300°K}\right)$$

and thus

$$T_0 = 300°K \left(\exp \frac{18 \text{ mV}}{57 \text{ mV}}\right) = 411°K$$

Therefore, the TC_F will be zero at $T_0 = 411°K = 138°C$, and we can express V_{OUT} as

$$V_{OUT} = 1.205 \text{ V} + 57 \text{ mV}\left(1 + \ln \frac{411°K}{T}\right)$$

From (4.259) with $T = 300°K$ and $T_0 = 411°K$,

$$\frac{dV_{OUT}}{dT} \simeq (2.2)\frac{26 \text{ mV}}{300°K}\left(\frac{411-300}{300}\right) \simeq 70 \text{ }\mu\text{V}/°K = 70 \text{ }\mu\text{V}/°C$$

Therefore, the TC_F at room temperature is

$$TC_F = \frac{1}{V_{OUT}}\frac{dV_{OUT}}{dT} \simeq \frac{70 \text{ }\mu\text{V}/°C}{1.280 \text{ V}} \simeq 55 \text{ ppm}/°C$$

■

To reduce the TC_F, the constant M in (4.249)–(4.252) is often trimmed at one temperature so that the band-gap output is set to a desired target voltage.[19] In principle, the target voltage is given by (4.253). In practice, however, significant inaccuracy in (4.253) stems from an approximation in (4.244).[20] As a result, the target voltage is usually determined experimentally by measuring the TC_F directly for several samples of each band-gap reference in a given process.[21,22] This procedure reduces the TC_F at the reference temperature to a level of about 10 ppm/°C.

A key parameter of interest in reference sources is the variation of the output that is encountered over the entire temperature range. Since the TC_F expresses the temperature sensitivity only at one temperature, a different parameter must be used to characterize the behavior of the circuit over a broad temperature range. An effective temperature coefficient can be defined for a voltage reference as

$$TC_{F(\text{eff})} = \frac{1}{V_{OUT}}\left(\frac{V_{MAX} - V_{MIN}}{T_{MAX} - T_{MIN}}\right) \quad (4.260)$$

where V_{MAX} and V_{MIN} are the largest and smallest output voltages observed over the temperature range, and $T_{MAX} - T_{MIN}$ is the temperature excursion. V_{OUT} is the nominal output voltage. By this standard, $TC_{F(\text{eff})}$ over the -55 to $125°C$ range for case (b) of Fig. 4.45 is 44 ppm/°C. If the temperature range is restricted to 0 to 70°C, $TC_{F(\text{eff})}$ improves to 17 ppm/°C. Thus over a restricted temperature range, this reference is comparable with the standard cell in temperature stability once the zero TC_F temperature has been set at room temperature. Saturated standard cells (precision batteries) have a TC_F of about ± 30 ppm/°C.

Practical realizations of band-gap references in bipolar technologies take on several forms.[15,19,23] One such circuit is illustrated in Fig. 4.46a.[15] This circuit uses a feedback loop to establish an operating point in the circuit such that the output voltage is equal to a $V_{BE(\text{on})}$ plus a voltage proportional to the difference between two base-emitter voltages. The operation of the feedback loop is best understood by reference to Fig. 4.46b, in which a portion of the circuit is shown. We first consider the variation of the output voltage V_2 as the input voltage V_1 is varied from zero in the positive direction. Initially, with $V_1 = 0$, devices Q_1 and Q_2 do not

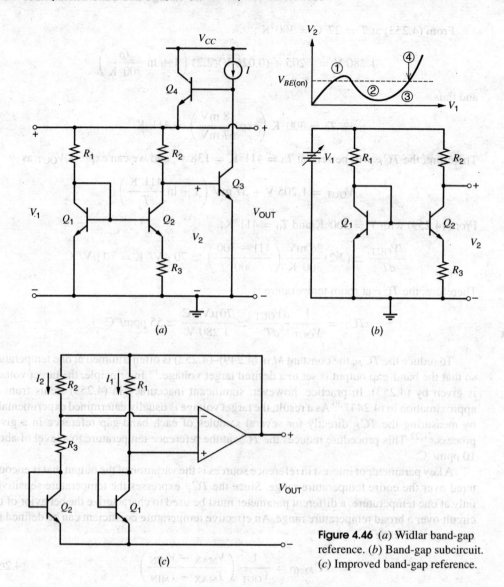

Figure 4.46 (a) Widlar band-gap reference. (b) Band-gap subcircuit. (c) Improved band-gap reference.

conduct and $V_2 = 0$. As V_1 is increased, Q_1 and Q_2 do not conduct significant current until the input voltage reaches about 0.6 V. When $V_1 < 0.6$ V, $V_2 = V_1$ since the voltage drop on R_2 is zero. When V_1 exceeds 0.6 V, however, Q_1 begins to conduct current, corresponding to point ① in Fig. 4.46b. The magnitude of the current in Q_1 is roughly equal to $(V_1 - 0.6 \text{ V})/R_1$. For small values of this current, Q_1 and Q_2 carry the same current because the drop across R_3 is negligible at low currents. Since R_2 is much larger than R_1, the voltage drop across R_2 is much larger than $(V_1 - 0.6 \text{ V})$, and transistor Q_2 saturates, corresponding to point ② in Fig. 4.46b. Because of the presence of R_3, the collector current that *would* flow in Q_2 if it were in the forward-active region has an approximately logarithmic dependence on V_1, exactly as in the Widlar source. Thus as V_1 is further increased, a point is reached at which Q_2 comes out of saturation because V_1 increases faster than the voltage drop across R_2. This point is labeled point ③ in Fig. 4.46b.

Now consider the complete circuit of Fig. 4.46a. If transistor Q_3 is initially turned off, transistor Q_4 will drive V_1 in the positive direction. This process will continue until enough voltage is developed at the base of Q_3 to produce a collector current in Q_3 approximately equal to I. Thus the circuit stabilizes with voltage V_2 equal to one diode drop, the base-emitter voltage of Q_3, which can occur at point ① or point ④ in Fig. 4.46b. Appropriate start-up circuitry must be included to ensure operation at point ④.

Assuming that the circuit has reached a stable operating point at point ④, the output voltage V_{OUT} is the sum of the base-emitter voltage of Q_3 and the voltage drop across R_2. The drop across R_2 is equal to the voltage drop across R_3 multiplied by R_2/R_3 because the collector current of Q_2 is approximately equal to its emitter current. The voltage drop across R_3 is equal to the difference in base-emitter voltages of Q_1 and Q_2. The ratio of currents in Q_1 and Q_2 is set by the ratio of R_2 to R_1.

A drawback of this reference is that the current I is set by the power supply and may vary with power-supply variations. A self-biased band-gap reference circuit is shown in Fig. 4.46c. Assume that a stable operating point exists for this circuit and that the op amp is ideal. Then the differential input voltage of the op amp must be zero and the voltage drops across resistors R_1 and R_2 are equal. Thus the ratio of R_2 to R_1 determines the ratio of I_1 to I_2. These two currents are the collector currents of the two diode-connected transistors Q_2 and Q_1, assuming base currents are negligible. The voltage across R_3 is

$$V_{R3} = \Delta V_{BE} = V_{BE1} - V_{BE2} = V_T \ln \frac{I_1}{I_2} \frac{I_{S2}}{I_{S1}} = V_T \ln \frac{R_2}{R_1} \frac{I_{S2}}{I_{S1}} \quad (4.261)$$

Since the same current that flows in R_3 also flows in R_2, the voltage across R_2 must be

$$V_{R2} = \frac{R_2}{R_3} V_{R3} = \frac{R_2}{R_3} \Delta V_{BE} = \frac{R_2}{R_3} V_T \ln \frac{R_2}{R_1} \frac{I_{S2}}{I_{S1}} \quad (4.262)$$

This equation shows that the voltage across R_2 is proportional to absolute temperature (PTAT) because of the temperature dependence of the thermal voltage. Since the op amp forces the voltages across R_1 and R_2 to be equal, the currents I_1 and I_2 are both proportional to temperature if the resistors have zero temperature coefficient. Thus for this reference, $\alpha = 1$ in (4.247). The output voltage is the sum of the voltage across Q_2, R_3, and R_2:

$$V_{OUT} = V_{BE2} + V_{R3} + V_{R2} = V_{BE2} + \left(1 + \frac{R_2}{R_3}\right) \Delta V_{BE}$$

$$= V_{BE2} + \left(1 + \frac{R_2}{R_3}\right) V_T \ln \frac{R_2}{R_1} \frac{I_{S2}}{I_{S1}} = V_{BE2} + M V_T \quad (4.263)$$

The circuit thus behaves as a band-gap reference, with the value of M set by the ratios of R_2/R_3, R_2/R_1, and I_{S2}/I_{S1}.

4.4.3.2 Band-Gap-Referenced Bias Circuits in CMOS Technology

Band-gap-referenced biasing also can be implemented using the parasitic bipolar devices inherent in CMOS technology. For example, in a n-well process, substrate pnp transistors can be used to replace the npn transistors in Fig. 4.46c, as shown in Fig. 4.47. Assume that the CMOS op amp has infinite gain but nonzero input-referred offset voltage V_{OS}. (The input-referred offset voltage of an op amp is defined as the differential input voltage required to drive the output to zero.) Because of the threshold mismatch and the low transconductance per current of CMOS transistors, the offset of op amps in CMOS technologies is usually larger than in bipolar technologies. With the offset voltage, the voltage across R_3 is

$$V_{R3} = V_{EB1} - V_{EB2} + V_{OS} = \Delta V_{EB} + V_{OS} \quad (4.264)$$

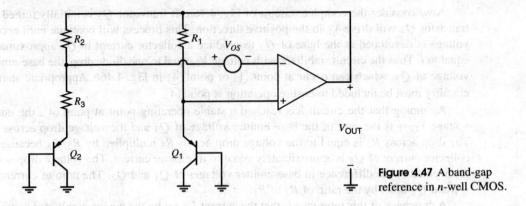

Figure 4.47 A band-gap reference in n-well CMOS.

The emitter-base voltages are used here because the base-emitter voltages of the *pnp* transistors operating in the forward-active region are negative. Then the voltage across R_2 is

$$V_{R2} = \frac{R_2}{R_3}V_{R3} = \frac{R_2}{R_3}(V_{EB1} - V_{EB2} + V_{OS}) = \frac{R_2}{R_3}(\Delta V_{EB} + V_{OS}) \quad (4.265)$$

and the output voltage is [18]

$$V_{OUT} = V_{EB2} + V_{R3} + V_{R2}$$
$$= V_{EB2} + \left(1 + \frac{R_2}{R_3}\right)(\Delta V_{EB} + V_{OS}) \quad (4.266)$$

Since the difference in the base-emitter voltages is proportional to the thermal voltage, comparing (4.266) with $V_{OS} = 0$ to (4.249) shows that the gain M here is proportional to $(1 + R_2/R_3)$. Rearranging (4.266) gives

$$V_{OUT} = V_{EB2} + \left(1 + \frac{R_2}{R_3}\right)(\Delta V_{EB}) + V_{OS(out)} \quad (4.267)$$

where the output-referred offset is

$$V_{OS(out)} = \left(1 + \frac{R_2}{R_3}\right)V_{OS} \quad (4.268)$$

Equations 4.267 and 4.268 show that the output contains an offset voltage that is a factor of $(1 + R_2/R_3)$ times bigger than the input-referred offset voltage. Therefore, the same gain that is applied to the difference in the base-emitter voltages is also applied to the input-referred offset voltage.

Assume that the offset voltage is independent of temperature. To set TC_F of the output equal to zero, the gain must be changed so that temperature coefficients of the V_{EB} and ΔV_{EB} terms cancel. Since the offset is assumed to be temperature independent, this cancellation occurs when the output is equal to the target, where zero offset was assumed, plus the output-referred offset. If the gain is trimmed at $T = T_0$ to set the output to a target voltage assuming the offset is zero, (4.267) shows that the gain is too small if the offset voltage is positive and that the gain is too big if the offset voltage is negative. Since the gain is applied to the PTAT term, the resulting slope of the output versus temperature is negative when the offset is positive and the gain is too small. On the other hand, this slope is positive when the offset is negative and the gain is too big.

4.4 Voltage and Current References

We will now calculate the magnitude of the slope of the output versus temperature at $T = T_0$. With zero offset and a target that assumes zero offset, trimming R_2 and/or R_3 to set the output in (4.267) to the target forces the slope of the ΔV_{EB} term to cancel the slope of the V_{EB} term. With nonzero offset but the same target, the factor $(1 + R_2/R_3)$ differs from its ideal value after trimming by $-V_{OS(out)}/\Delta V_{EB}$. Since this error is multiplied by ΔV_{EB} in (4.267), the resulting slope of the output versus temperature is

$$\left.\frac{dV_{OUT}}{dT}\right|_{T=T_0} = -\left(\frac{V_{OS(out)}}{\Delta V_{EB}}\right)\frac{d\Delta V_{EB}}{dT} \quad (4.269)$$

Since ΔV_{EB} is proportional to the thermal voltage V_T,

$$\Delta V_{EB} = H V_T \quad (4.270)$$

where H is a temperature-independent constant. Substituting (4.270) into (4.269) gives

$$\left.\frac{dV_{OUT}}{dT}\right|_{T=T_0} = -\frac{V_{OS(out)}}{HV_T}\frac{HV_T}{T}\bigg|_{T=T_0} = -\frac{V_{OS(out)}}{T_0} \quad (4.271)$$

Therefore, when the gain is trimmed at one temperature to set the band-gap output to a desired target voltage, variation in the op-amp offset causes variation in the output temperature coefficient. In practice, the op-amp offset is usually the largest source of nonzero temperature coefficient.[18] Equation 4.271 shows that the temperature coefficient at $T = T_0$ is proportional to the output-referred offset under these circumstances. Furthermore, (4.268) shows that the output-referred offset is equal to the gain that is applied to the ΔV_{EB} term times the input-referred offset. Therefore, minimizing this gain minimizes the variation in the temperature coefficient at the output. Since the reference output at $T = T_0$ for zero TC_F is approximately equal to the band-gap voltage, the required gain can be minimized by maximizing the ΔV_{EB} term.

To maximize the ΔV_{EB} term, designers generally push a large current into a small transistor and a small current into a large transistor, as shown in Fig. 4.48. Ignoring base currents,

$$\Delta V_{EB} = V_{EB1} - V_{EB2} = V_T \ln\left(\frac{I_1}{I_2}\frac{I_{S2}}{I_{S1}}\right) \quad (4.272)$$

Equation 4.272 shows that maximizing the product of the ratios I_1/I_2 and I_{S2}/I_{S1} maximizes ΔV_{EB}. In Fig. 4.48, $I_1 > I_2$ is emphasized by drawing the symbol for I_1 larger than the symbol for I_2. Similarly, the emitter area of Q_2 is larger than that of Q_1 to make $I_{S2} > I_{S1}$,

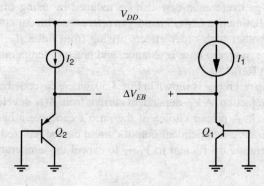

Figure 4.48 A circuit that increases ΔV_{EB} by increasing I_1/I_2 and I_{S2}/I_{S1}.

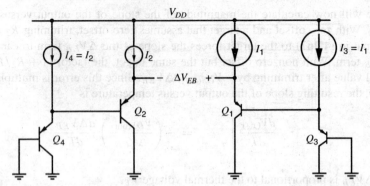

Figure 4.49 A circuit that cascades emitter followers to double ΔV_{EB} if $I_{S3} = I_{S1}$ and $I_{S4} = I_{S2}$.

and this relationship is shown by drawing the symbol of Q_2 larger than the symbol of Q_1.[24] In practice, these ratios are often each set to be about equal to ten, and the resulting $\Delta V_{EB} \simeq 120$ mV at room temperature. Because the logarithm function compresses its argument, however, a limitation of this approach arises. For example, if the argument is increased by a factor of ten, ΔV_{EB} increases by only $V_T \ln(10) \simeq 60$ mV. Therefore, to double ΔV_{EB} to 240 mV, $(I_1/I_2)(I_{S2}/I_{S1})$ must be increased by a factor of 100 to 10,000. On the other hand, if Q_1 and the transistors that form I_2 are minimum-sized devices when $(I_1/I_2)(I_{S2}/I_{S1}) = 100$, the required die area would be dominated by the biggest devices (Q_2 and/or the transistors that form I_1). Therefore, increasing $(I_1/I_2)(I_{S2}/I_{S1})$ from 100 to 10,000 would increase the die area by about a factor of 100 but only double ΔV_{EB}.

To overcome this limitation, stages that each contribute to ΔV_{EB} can be cascaded.[25] For example, consider Fig. 4.49, where two emitter-follower stages are cascaded. Here

$$\Delta V_{EB} = V_{EB3} - V_{EB4} + V_{EB1} - V_{EB2} \quad (4.273)$$

Assume the new devices in Fig. 4.49 are identical to the corresponding original devices in Fig. 4.48 so that $I_3 = I_1$, $I_4 = I_2$, $I_{S3} = I_{S1}$, and $I_{S4} = I_{S2}$. Then ignoring base currents

$$\Delta V_{EB} = 2(V_{EB1} - V_{EB2}) = 2V_T \ln\left(\frac{I_1}{I_2}\frac{I_{S2}}{I_{S1}}\right) \quad (4.274)$$

Thus cascading two identical emitter followers doubles ΔV_{EB} while only doubling the required die area.

The effect of the offset in a band-gap reference can also be reduced by using offset cancellation. An example of offset cancellation in a CMOS band-gap reference with curvature correction in addition to an analysis of other high-order effects arising from finite β_F, β_F mismatch, β_F variation with temperature, nonzero base resistance, and nonzero temperature coefficient in the resistors is presented in Ref. 18.

A high-performance CMOS band-gap reference is shown in Fig. 4.50, where cascoded current mirrors are used to improve supply rejection. A V_T-dependent current from M_{11} develops a V_T-dependent voltage across resistor xR. A proper choice of the ratio x can give a band-gap voltage at V_{OUT}. If desired, a temperature-independent output current can be realized by choosing x to give an appropriate temperature coefficient to V_{OUT} to cancel the temperature coefficient of resistor R_2.

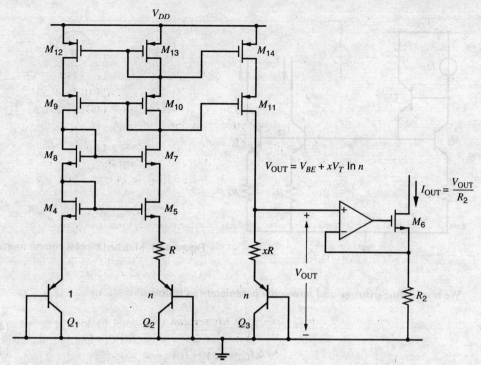

Figure 4.50 Example of a V_{BE}-referenced self-biased reference circuit in CMOS technology.

APPENDIX

A.4.1 MATCHING CONSIDERATIONS IN CURRENT MIRRORS

In many types of circuits, an objective of current-mirror design is generation of two or more current sources whose values are identical. This objective is particularly important in the design of digital-to-analog converters, operational amplifiers, and instrumentation amplifiers. We first examine the factors affecting matching in active current mirrors in bipolar technologies and then in MOS technologies.

A.4.1.1 BIPOLAR

Consider the bipolar current mirror with two outputs in Fig. 4.51. If the resistors and transistors are identical and the collector voltages are the same, the collector currents will match precisely. However, mismatch in the transistor parameters α_F and I_S and in the emitter resistors will cause the currents to be unequal. For Q_3,

$$V_T \ln \frac{I_{C3}}{I_{S3}} + \frac{I_{C3}}{\alpha_{F3}} R_3 = V_B \tag{4.275}$$

For Q_4,

$$V_T \ln \frac{I_{C4}}{I_{S4}} + \frac{I_{C4}}{\alpha_{F4}} R_4 = V_B \tag{4.276}$$

Subtraction of these two equations gives

$$V_T \ln \frac{I_{C3}}{I_{C4}} - V_T \ln \frac{I_{S3}}{I_{S4}} + \frac{I_{C3}}{\alpha_{F3}} R_3 - \frac{I_{C4}}{\alpha_{F4}} R_4 = 0 \tag{4.277}$$

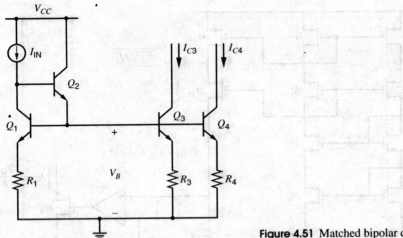

Figure 4.51 Matched bipolar current sources.

We now define *average* and *mismatch* parameters as follows:

$$I_C = \frac{I_{C3} + I_{C4}}{2} \tag{4.278}$$

$$\Delta I_C = I_{C3} - I_{C4} \tag{4.279}$$

$$I_S = \frac{I_{S3} + I_{S4}}{2} \tag{4.280}$$

$$\Delta I_S = I_{S3} - I_{S4} \tag{4.281}$$

$$R = \frac{R_3 + R_4}{2} \tag{4.282}$$

$$\Delta R = R_3 - R_4 \tag{4.283}$$

$$\alpha_F = \frac{\alpha_{F3} + \alpha_{F4}}{2} \tag{4.284}$$

$$\Delta \alpha_F = \alpha_{F3} - \alpha_{F4} \tag{4.285}$$

These relations can be inverted to give the original parameters in terms of the average and mismatch parameters. For example,

$$I_{C3} = I_C + \frac{\Delta I_C}{2} \tag{4.286}$$

$$I_{C4} = I_C - \frac{\Delta I_C}{2} \tag{4.287}$$

This set of equations for the various parameters is now substituted into (4.277). The result is

$$V_T \ln\left(\frac{I_C + \frac{\Delta I_C}{2}}{I_C - \frac{\Delta I_C}{2}}\right) - V_T \ln\left(\frac{I_S + \frac{\Delta I_S}{2}}{I_S - \frac{\Delta I_S}{2}}\right)$$
$$+ \frac{\left(I_C + \frac{\Delta I_C}{2}\right)\left(R + \frac{\Delta R}{2}\right)}{\alpha_F + \frac{\Delta \alpha_F}{2}} - \frac{\left(I_C - \frac{\Delta I_C}{2}\right)\left(R - \frac{\Delta R}{2}\right)}{\alpha_F - \frac{\Delta \alpha_F}{2}} = 0 \tag{4.288}$$

The first term in this equation can be rewritten as

$$V_T \ln\left(\frac{I_C + \frac{\Delta I_C}{2}}{I_C - \frac{\Delta I_C}{2}}\right) = V_T \ln\left(\frac{1 + \frac{\Delta I_C}{2I_C}}{1 - \frac{\Delta I_C}{2I_C}}\right) \tag{4.289}$$

If $\Delta I_C/2I_C \ll 1$, this term can be rewritten as

$$V_T \ln\left(\frac{1 + \frac{\Delta I_C}{2I_C}}{1 - \frac{\Delta I_C}{2I_C}}\right) \simeq V_T \ln\left[\left(1 + \frac{\Delta I_C}{2I_C}\right)\left(1 + \frac{\Delta I_C}{2I_C}\right)\right] \tag{4.290}$$

$$\simeq V_T \ln\left[1 + \frac{\Delta I_C}{I_C} + \left(\frac{\Delta I_C}{2I_C}\right)^2\right] \tag{4.291}$$

$$\simeq V_T \ln\left(1 + \frac{\Delta I_C}{I_C}\right) \tag{4.292}$$

where the squared term is neglected. The logarithm function has the infinite series

$$\ln(1 + x) = x - \frac{x^2}{2} + \cdots \tag{4.293}$$

If $x \ll 1$,

$$\ln(1 + x) \simeq x \tag{4.294}$$

To simplify (4.292) when $\Delta I_C/I_C \ll 1$, let $x = \Delta I_C/I_C$. Then

$$V_T \ln\left(\frac{I_C + \frac{\Delta I_C}{2}}{I_C - \frac{\Delta I_C}{2}}\right) \simeq V_T \frac{\Delta I_C}{I_C} \tag{4.295}$$

Applying the same approximations to the other terms in (4.288), we obtain

$$\frac{\Delta I_C}{I_C} \simeq \left(\frac{1}{1 + \frac{g_m R}{\alpha_F}}\right)\frac{\Delta I_S}{I_S} + \frac{\frac{g_m R}{\alpha_F}}{1 + \frac{g_m R}{\alpha_F}}\left(-\frac{\Delta R}{R} + \frac{\Delta \alpha_F}{\alpha_F}\right) \tag{4.296}$$

We will consider two important limiting cases. First, since $g_m = I_C/V_T$, when $g_m R \ll 1$, the voltage drop on an emitter resistor is much smaller than the thermal voltage. In this case, the second term in (4.296) is small and the mismatch is mainly determined by the transistor I_S mismatch in the first term. Observed mismatches in I_S typically range from ± 10 to ± 1 percent depending on geometry. Second, when $g_m R \gg 1$, the voltage drop on an emitter resistor is much larger than the thermal voltage. In this case, the first term in (4.296) is small and the mismatch is mainly determined by the resistor mismatch and transistor α_F mismatch in the second term. Resistor mismatch typically ranges from ± 2 to ± 0.1 percent depending on geometry, and α_F matching is in the ± 0.1 percent range for *npn* transistors. Thus for *npn* current sources, the use of emitter resistors offers significantly improved current matching. On the other hand, for *pnp* current sources, the α_F mismatch is larger due to the lower β_F, typically around ± 1 percent. Therefore, the advantage of emitter degeneration is less significant with *pnp* than *npn* current sources.

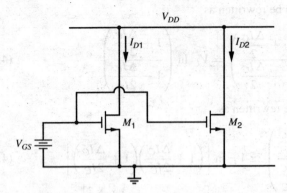

Figure 4.52 Matched MOS current sources.

A.4.1.2 MOS

Matched current sources are often required in MOS analog integrated circuits. The factors affecting this mismatch can be calculated using the circuit of Fig. 4.52. The two transistors M_1 and M_2 will have mismatches in their W/L ratios and threshold voltages. The drain currents are given by

$$I_{D1} = \frac{1}{2}\mu_n C_{ox} \left(\frac{W}{L}\right)_1 (V_{GS} - V_{t1})^2 \tag{4.297}$$

$$I_{D2} = \frac{1}{2}\mu_n C_{ox} \left(\frac{W}{L}\right)_2 (V_{GS} - V_{t2})^2 \tag{4.298}$$

Defining average and mismatch quantities, we have

$$I_D = \frac{I_{D1} + I_{D2}}{2} \tag{4.299}$$

$$\Delta I_D = I_{D1} - I_{D2} \tag{4.300}$$

$$\frac{W}{L} = \frac{1}{2}\left[\left(\frac{W}{L}\right)_1 + \left(\frac{W}{L}\right)_2\right] \tag{4.301}$$

$$\Delta \frac{W}{L} = \left(\frac{W}{L}\right)_1 - \left(\frac{W}{L}\right)_2 \tag{4.302}$$

$$V_t = \frac{V_{t1} + V_{t2}}{2} \tag{4.303}$$

$$\Delta V_t = V_{t1} - V_{t2} \tag{4.304}$$

Substituting these expressions into (4.297) and (4.298) and neglecting high-order terms, we obtain

$$\frac{\Delta I_D}{I_D} = \frac{\Delta \frac{W}{L}}{\frac{W}{L}} - \frac{\Delta V_t}{(V_{GS} - V_t)/2} \tag{4.305}$$

The current mismatch consists of two components. The first is geometry dependent and contributes a fractional current mismatch that is independent of bias point. The second is dependent on threshold voltage mismatch and increases as the overdrive $(V_{GS} - V_t)$ is reduced. This change occurs because as the overdrive is reduced, the fixed threshold mismatch progressively becomes a larger fraction of the total gate drive that is applied to the transistors and therefore contributes a progressively larger percentage error to the current mismatch. In practice, these

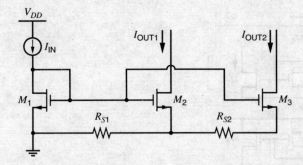

Figure 4.53 Current mirror with two outputs used to compare voltage- and current-routing techniques.

observations are important because they affect the techniques used to distribute bias signals on integrated circuits.

Consider the current mirror shown in Fig. 4.53, which has one input and two outputs. At first, assume that $R_{S1} = R_{S2} = 0$. Also, assume that the input current is generated by a circuit with desirable properties. For example, a self-biased band-gap reference might be used to make I_{IN} insensitive to changes in the power supply and temperature. Finally, assume that each output current is used to provide the required bias in one analog circuit on the integrated circuit (IC). For example, M_2 and M_3 could each act as the tail current source of a differential pair.

One way to build the circuit in Fig. 4.53 is to place M_1 on the IC near the input current source I_{IN}, while M_2 and M_3 are placed near the circuits that they bias, respectively. Since the gate-source voltage of M_1 must be routed to M_2 and M_3 here, this case is referred to as an example of the *voltage routing* of bias signals. An advantage of this approach is that by routing only two nodes (the gate and the source of M_1) around the IC, any number of output currents can be produced. Furthermore the gains from the input to each output of the current mirror are not affected by the number of outputs in MOS technologies because $\beta_F \to \infty$. (In bipolar technologies, β_F is finite, and the gain error increases as the number of outputs increase, but a beta-helper configuration can be used to reduce such errors as described in Section 4.2.3.)

Unfortunately, voltage routing has two important disadvantages. First, the input and output transistors in the current mirror may be separated by distances that are large compared to the size of the IC, increasing the potential mismatches in (4.305). In particular, the threshold voltage typically displays considerable gradient with distance across a wafer. Therefore, when the devices are physically separated by large distances, large current mismatch can result from biasing current sources sharing the same gate-source bias, especially when the overdrive is small. The second disadvantage of voltage routing is that the output currents are sensitive to variations in the supply resistances R_{S1} and R_{S2}. Although these resistances were assumed to be zero above, they are nonzero in practice because of imperfect conduction in the interconnect layers and their contacts. Since I_{OUT2} flows in R_{S2} and $(I_{OUT1} + I_{OUT2})$ flows in R_{S1}, nonzero resistances cause $V_{GS2} < V_{GS1}$ and $V_{GS3} < V_{GS1}$ when $I_{OUT1} > 0$ and $I_{OUT2} > 0$. Therefore, with perfectly matched transistors, the output currents are less than the input current, and the errors in the output currents can be predicted by an analysis similar to that presented in Section 4.4.1 for Widlar current sources. The key point here is that R_{S1} and R_{S2} increase as the distances between the input and output transistors increase, increasing the errors in the output currents. As a result of both of these disadvantages, the output currents may have considerable variation from one IC to another with voltage routing, increasing the difficulty of designing the circuits biased by M_2 and M_3 to meet the required specifications even if I_{IN} is precisely controlled.

To overcome these problems, the circuit in Fig. 4.53 can be built so that M_1–M_3 are close together physically, and the current outputs I_{OUT1} and I_{OUT2} are routed as required on the IC. This case is referred to as an example of the *current routing* of bias signals. Current routing reduces the problems with mismatch and supply resistance by reducing the distances

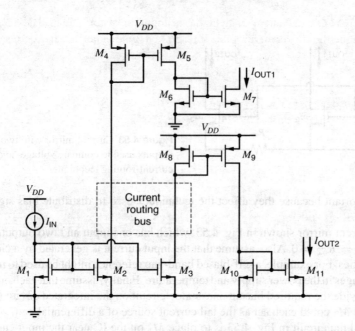

Figure 4.54 Bias-distribution circuit using both current routing and voltage routing.

between the input and output transistors in the current mirror in Fig. 4.53 compared to voltage routing. One disadvantage of current routing is that it requires one node to be routed for each bias signal. Therefore, when the number of bias outputs is large, the die area required for the interconnect to distribute the bias currents can be much larger than that required with voltage routing. Another disadvantage of current routing is that it can increase the parasitic capacitance on the drains of M_2 and M_3. If these nodes are connected to circuits that process high-frequency signals, increased parasitic capacitance can reduce performance in some ways. For example, if M_2 and M_3 act as the tail current sources of differential pairs, increased parasitic capacitance will increase the common-mode gain and reduce the common-mode rejection ratio of each differential pair at high frequencies.

In practice, many ICs use a combination of current- and voltage-routing techniques. For example, Fig. 4.54 shows a circuit with five current mirrors, where the input and output currents are still referenced as in Fig. 4.53. If the current routing bus in Fig. 4.54 travels over a large distance, the parasitic capacitances on the drains of M_2 and M_3 may be large. However, the parasitic capacitances on the drains of M_7 and M_{11} are minimized by using voltage routing within each current mirror. Although simple current mirrors are shown in Fig. 4.54, cascoding is often used in practice to reduce gain errors stemming from a finite Early voltage. In ICs using both current and voltage routing, currents are routed globally and voltages locally, where the difference between global and local routing depends on distance. When the distance is large enough to significantly worsen mismatch or supply-resistance effects, the routing is *global*. Otherwise, it is *local*. An effective combination of these bias distribution techniques is to divide an IC into blocks, where bias currents are routed between blocks and bias voltages within the blocks.

A.4.2 INPUT OFFSET VOLTAGE OF DIFFERENTIAL PAIR WITH ACTIVE LOAD

A.4.2.1 BIPOLAR

For the resistively loaded emitter-coupled pair, we showed in Chapter 3 that the input offset voltage arises primarily from mismatches in I_S in the input transistors and from mismatches in the collector load resistors. In the active-load case, the input offset voltage results from nonzero base current of the load devices and mismatches in the input transistors and load devices. Refer to Fig. 4.25a. Assume the inputs are grounded. If the matching is perfect and if $\beta_F \to \infty$ in T_3 and T_4,

$$V_{OUT} = V_{CC} - |V_{BE3}| \tag{4.306}$$

Equation 4.306 holds because only this output voltage forces $V_{CE3} = V_{CE4}$, where $I_{C1} = I_{C2}$ and $V_{BE1} = V_{BE2}$, which is required by KVL when $V_{I1} = V_{I2}$.

The differential input required to drive the output to the value given by (4.306) is the input-referred offset voltage. With finite β_F in the active-load transistors and/or device mismatch, the offset is usually nonzero. In the active-load, KVL shows that

$$V_{BE3} = V_{BE4} \tag{4.307}$$

Solving (1.58) for V_{BE3} and V_{BE4} and substituting in (4.307) gives

$$\frac{I_{C3}}{I_{S3}}\left(\frac{1}{1+\dfrac{V_{CE3}}{V_{A3}}}\right) = \frac{I_{C4}}{I_{S4}}\left(\frac{1}{1+\dfrac{V_{CE4}}{V_{A4}}}\right) \tag{4.308}$$

Assume that the Early voltages of T_3 and T_4 are identical. Since $V_{CE3} = V_{CE4}$ when (4.306) is satisfied, (4.308) can be simplified to

$$I_{C4} = I_{C3}\left(\frac{I_{S4}}{I_{S3}}\right) \tag{4.309}$$

Since $I_{C2} = -I_{C4}$, (4.309) can be written as

$$I_{C2} = -I_{C3}\left(\frac{I_{S4}}{I_{S3}}\right) \tag{4.310}$$

From KCL at the collector of T_3,

$$I_{C1} = -I_{C3}\left[1 + \left(\frac{2}{\beta_F}\right)\right] \tag{4.311}$$

where β_F is the ratio of the collector to base current in the active-load devices. From KVL in the input loop,

$$V_{ID} = V_{I1} - V_{I2} = V_{BE1} - V_{BE2} \tag{4.312}$$

Then the input offset voltage, V_{OS}, is the value of V_{ID} for which the output voltage is given by (4.306). If the Early voltages of T_1 and T_2 are identical, solving (1.58) for V_{BE1} and V_{BE2} and substituting into (4.312) gives

$$V_{OS} = V_{ID} = V_T \ln\left(\frac{I_{C1}}{I_{C2}}\frac{I_{S2}}{I_{S1}}\right) \tag{4.313}$$

because $V_{CE1} = V_{CE2}$ when (4.306) is satisfied. Substituting (4.310) and (4.311) in (4.313) gives

$$V_{OS} = V_T \ln\left[\frac{I_{S3}}{I_{S4}}\frac{I_{S2}}{I_{S1}}\left(1 + \frac{2}{\beta_F}\right)\right] \tag{4.314}$$

If the mismatches are small, this expression can be approximated as

$$V_{OS} \simeq V_T \left(\frac{\Delta I_{SP}}{I_{SP}} - \frac{\Delta I_{SN}}{I_{SN}} + \frac{2}{\beta_F}\right) \tag{4.315}$$

using the technique described in Section 3.5.6.3, where

$$\Delta I_{SP} = I_{S3} - I_{S4} \tag{4.316}$$

$$I_{SP} = \frac{I_{S3} + I_{S4}}{2} \tag{4.317}$$

$$\Delta I_{SN} = I_{S1} - I_{S2} \tag{4.318}$$

$$I_{SN} = \frac{I_{S1} + I_{S2}}{2} \tag{4.319}$$

In the derivation of (4.315), we assumed that the Early voltages of matched transistors are identical. In practice, mismatch in Early voltages also contributes to the offset, but the effect is usually negligible when the transistors are biased with collector-emitter voltages much less than their Early voltages.

Assuming a worst-case value for $\Delta I_S/I_S$ of ± 5 percent and a *pnp* beta of 20, the worst-case offset voltage is

$$V_{OS} \simeq V_T(0.05 + 0.05 + 0.1) = 0.2 V_T \simeq 5 \text{ mV} \tag{4.320}$$

To find the worst-case offset, we have added the mismatch terms for the *pnp* and *npn* transistors in (4.320) instead of subtracting them as shown in in (4.315) because the mismatch terms are random and independent of each other in practice. Therefore, the polarity of the mismatch terms is unknown in general. Comparing (4.320) to (3.219) shows that the actively loaded differential pair has significantly higher offset than the resistively loaded case under similar conditions. The offset arising here from mismatch in the load devices can be reduced by inserting resistors in series with the emitters of T_3 and T_4 as shown in Section A.4.1. To reduce the offset arising from finite β_F in the load devices, the current mirror in the load can use a beta helper transistor as described in Section 4.2.3.

A.4.2.2 MOS

The offset in the CMOS differential pair with active load shown in Fig. 4.25b is similar to the bipolar case. If the matching is perfect with the inputs grounded,

$$V_{OUT} = V_{DD} - |V_{GS3}| \tag{4.321}$$

Equation 4.321 holds because only this output voltage forces $V_{DS3} = V_{DS4}$, where $I_1 = I_2$ and $V_{GS1} = V_{GS2}$, which is required by KVL when $V_{I1} = V_{I2}$.

The differential input required to drive the output to the value given by (4.321) is the input-referred offset voltage. With device mismatch, the offset is usually nonzero.

$$V_{ID} = V_{GS1} - V_{GS2} = V_{t1} + V_{ov1} - V_{t2} - V_{ov2} \tag{4.322}$$

Assume that the Early voltages of T_1 and T_2 are identical. Since $V_{DS1} = V_{DS2} = V_{DSN}$ when $V_{ID} = V_{OS}$, applying (1.165) to V_{ov1} and V_{ov2} in (4.322) gives

$$V_{OS} = V_{t1} - V_{t2} + \sqrt{\frac{1}{1 + \lambda_N V_{DSN}}} \left(\sqrt{\frac{2I_1}{k'(W/L)_1}} - \sqrt{\frac{2I_2}{k'(W/L)_2}} \right) \tag{4.323}$$

If the mismatches are small, this expression can be approximated as

$$V_{OS} \simeq V_{t1} - V_{t2} + \frac{V_{ovN}}{2} \left(\frac{\Delta I_N}{I_N} - \frac{\Delta(W/L)_N}{(W/L)_N} \right) \tag{4.324}$$

using the technique described in Section 3.5.6.7, where

$$V_{ovN} = \sqrt{\frac{2I_N}{k'(W/L)_N(1 + \lambda_N V_{DSN})}} \tag{4.325}$$

$$\Delta I_N = I_1 - I_2 \qquad (4.326)$$

$$I_N = \frac{I_1 + I_2}{2} \qquad (4.327)$$

$$\Delta(W/L)_N = (W/L)_1 - (W/L)_2 \qquad (4.328)$$

$$(W/L)_N = \frac{(W/L)_1 + (W/L)_2}{2} \qquad (4.329)$$

Since $I_1 = -I_3$ and $I_2 = -I_4$

$$\frac{\Delta I_N}{I_N} = \frac{\Delta I_P}{I_P} \qquad (4.330)$$

where

$$\Delta I_P = I_3 - I_4 \qquad (4.331)$$

$$I_P = \frac{I_3 + I_4}{2} \qquad (4.332)$$

To find $\Delta I_P/I_P$, we will use KVL in the gate-source loop in the load as follows

$$0 = V_{GS3} - V_{GS4} = V_{t3} + V_{ov3} - V_{t4} - V_{ov4} \qquad (4.333)$$

Since T_3 and T_4 are p-channel transistors, their overdrives are negative. Assume that the Early voltages of T_3 and T_4 are identical. Since $V_{DS3} = V_{DS4} = V_{DSP}$ when $V_{ID} = V_{OS}$ (4.333) can be rewritten as

$$0 = V_{t3} - V_{t4} - \sqrt{\frac{1}{1+|\lambda_P V_{DSP}|}}\left(\sqrt{\frac{2|I_3|}{k'(W/L)_3}} - \sqrt{\frac{2|I_4|}{k'(W/L)_4}}\right) \qquad (4.334)$$

In (4.334), absolute value functions have been used so that the arguments of the square-root functions are positive. If the mismatches are small, this expression can be approximated as

$$\frac{\Delta I_P}{I_P} \simeq \frac{V_{t3} - V_{t4}}{\frac{|V_{ovP}|}{2}} + \frac{\Delta(W/L)_P}{(W/L)_P} \qquad (4.335)$$

using the technique described in Section 3.5.6.7, where

$$|V_{ovP}| = \sqrt{\frac{2|I_P|}{k'(W/L)_P(1+|\lambda_P V_{DSP}|)}} \qquad (4.336)$$

$$\Delta(W/L)_P = (W/L)_3 - (W/L)_4 \qquad (4.337)$$

$$(W/L)_P = \frac{(W/L)_3 + (W/L)_4}{2} \qquad (4.338)$$

Substituting (4.335) and (4.330) into (4.324) gives

$$V_{OS} \simeq V_{t1} - V_{t2} + \frac{V_{ovN}}{2}\left(\frac{V_{t3} - V_{t4}}{\frac{|V_{ovP}|}{2}} + \frac{\Delta(W/L)_P}{(W/L)_P} - \frac{\Delta(W/L)_N}{(W/L)_N}\right) \qquad (4.339)$$

Comparing (4.339) to (4.315) shows that the MOS differential pair with active load includes terms to account for threshold mismatch but excludes a term to account for finite beta in the active load because $\beta_F \to \infty$ in MOS transistors.

■ **EXAMPLE**

Find the input-referred offset voltage of the circuit in Fig. 4.25b using the transistor parameters in the table below.

Transistor	V_t (V)	W (μm)	L (μm)	k' (μA/V^2)
T_1	0.705	49	1	100
T_2	0.695	51	1	100
T_3	-0.698	103	1	50
T_4	-0.702	97	1	50

Assume that $I_{TAIL} = 200$ μA and that $\lambda_N V_{DSN} \ll 1$ and $|\lambda_P V_{DSP}| \ll 1$. From (4.327) and KCL,

$$I_N = \frac{I_1 + I_2}{2} = \frac{I_{TAIL}}{2} = 100 \text{ μA} \quad (4.340)$$

Substituting (4.340) and (4.329) into (4.325) gives

$$V_{ovN} \simeq \sqrt{\frac{200}{100(49+51)/2}} \text{ V} = 0.2 \text{ V} \quad (4.341)$$

Similarly, from (4.332) and KCL

$$I_P = \frac{I_3 + I_4}{2} = -\frac{I_{TAIL}}{2} = -100 \text{ μA} \quad (4.342)$$

Substituting (4.342) and (4.338) into (4.336) gives

$$|V_{ovP}| \simeq \sqrt{\frac{200}{50(103+97)/2}} \text{ V} = 0.2 \text{ V} \quad (4.343)$$

Substituting (4.337) and (4.328) into (4.339) gives

$$V_{OS} \simeq 0.705 \text{ V} - 0.695 \text{ V}$$
$$+ 0.1 \left(\frac{-0.698 + 0.702}{0.1} + \frac{103 - 97}{(103+97)/2} - \frac{49 - 51}{(49+51)/2} \right) \text{ V}$$
$$\simeq 0.01 \text{ V} + 0.1(0.04 + 0.06 + 0.04) \text{ V} = 0.024 \text{ V} \quad (4.344)$$

In this example, the mismatches have been chosen so that the individual contributions to the offset add constructively to give the worst-case offset.

■

PROBLEMS

For the bipolar transistors in these problems, use the high-voltage device parameters given in Fig. 2.30 and Fig. 2.35, unless otherwise specified. Assume that $r_b = 0$ and $r_\mu \to \infty$ in all problems. Assume all bipolar transistors operate in the forward-active region, and neglect base currents in bias calculations unless otherwise specified.

4.1 Determine the output current and output resistance of the bipolar current mirror shown in Fig. 4.55. Find the output current if $V_{OUT} = 1$ V, 5 V, and 20 V. Ignore the effects of nonzero base currents. Compare your answer with a SPICE simulation.

4.2 Repeat Problem 4.1 including the effects of nonzero base currents.

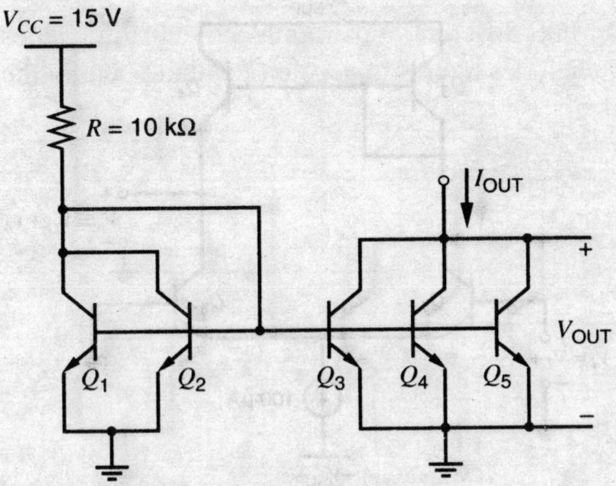

$V_{CC} = 15$ V

$R = 10$ kΩ

I_{OUT}

Q_1 Q_2 Q_3 Q_4 Q_5

V_{OUT}

Figure 4.55 Circuit for Problem 4.1.

4.3 Design a simple MOS current mirror of the type shown in Fig. 4.4 to meet the following constraints:

(a) Transistor M_2 must operate in the active region for values of V_{OUT} to within 0.2 V of ground.

(b) The output current must be 50 μA.

(c) The output current must change less than 2 percent for a change in output voltage of 1 V.

Make M_1 and M_2 identical. You are to minimize the total device area within the given constraints. Here the device area will be taken to be the total gate area ($W \times L$ product). Assume $X_d = 0$ and take other device data from Table 2.4.

4.4 Calculate an analytical expression for the small-signal output resistance R_o of the bipolar cascode current mirror of Fig. 4.8. Assume that the input current source is not ideal and that the nonideality is modeled by placing a resistor R_1 in parallel with I_{IN}. Show that for large R_1, the output resistance approaches $\beta_0 r_o / 2$. Calculate the value of R_o if $V_{CC} = 5$ V, $I_{IN} = 0$, and $R_1 = 10$ kΩ, and estimate the value of V_{OUT} below which R_o will begin to decrease substantially. Use SPICE to check your calculations and also to investigate the β_F sensitivity by varying β_F by -50 percent and examining I_{OUT}. Use SPICE to plot the large-signal I_{OUT}-V_{OUT} characteristic.

4.5 Calculate the output resistance of the circuit of Fig. 4.9, assuming that $I_{IN} = 100$ μA and the devices have drawn dimensions of 100 μm/1 μm. Use the process parameters given in Table 2.4, and assume for all devices that $X_d = 0$. Also, ignore the body effect for simplicity. Compare your answer with a SPICE simulation and also use SPICE to plot the I_{OUT}-V_{OUT} characteristic for V_{OUT} from 0 to 3 V.

4.6 Using the data given in the example of Section 1.9, include the effects of substrate leakage in the calculation of the output resistance for the circuit of Problem 4.5. Let $V_{OUT} = 2$ V and 3 V.

4.7 Design the circuit of Fig. 4.11b to satisfy the constraints in Problem 4.3 except the output resistance objective is that the output current change less than 0.02 percent for a 1 V change in the output voltage. Ignore the body effect for simplicity. Make all devices identical except for M_4. Use SPICE to check your design and also to plot the I_{OUT}-V_{OUT} characteristic for V_{OUT} from 0 to 3 V.

4.8 For the circuit of Fig. 4.56, assume that $(W/L)_8 = (W/L)$. Ignoring the body effect, find $(W/L)_6$ and $(W/L)_7$ so that $V_{DS6} = V_{DS7} = V_{ov8}$. Draw the schematic of a double-cascode current mirror that uses the circuit of Fig. 4.56 to bias both cascode devices in the output branch. For this current mirror, calculate the output resistance, the minimum output voltage for which all three transistors in the output branch operate in the active region, the total voltage across all the devices in the input branch, and the systematic gain error.

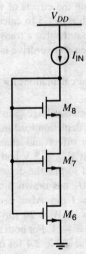

Figure 4.56 Circuit for Problem 4.8.

4.9 Calculate the output resistance of the Wilson current mirror shown in Fig. 4.57. What is the percentage change in I_{OUT} for a 5-V change in V_{OUT}? Compare your answer with a SPICE simulation using a full device model. Use SPICE to check the β_F sensitivity by varying β_F by -50 percent and examining I_{OUT}. Also, use SPICE to plot the large-signal I_{OUT}-V_{OUT} characteristic for V_{OUT} from 0 to 15 V.

4.10 Calculate the small-signal voltage gain of the common-source amplifier with active load in Fig. 4.16b. Assume that $V_{DD} = 3$ V and that all the transistors operate in the active region. Do the calculations for values of I_{REF} of 1 mA, 100 μA, 10 μA, and 1 μA.

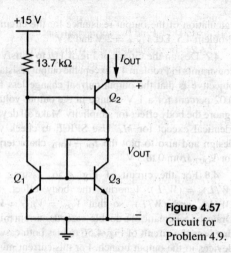

Figure 4.57 Circuit for Problem 4.9.

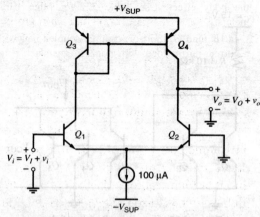

Figure 4.58 Circuit for Problem 4.12.

Assume that the drawn dimensions of each transistor are $W = 100$ μm and $L = 1$ μm. Assume $X_d = 0$ and use Table 2.4 for other parameters.

(a) At first, assume the transistors operate in strong inversion in all cases.

(b) Repeat part (a) including the effects of weak inversion by using (1.253) with $n = 1.5$ to calculate the transconductance of M_1. Assume that a transistor operates in weak inversion when its overdrive is less than $2nV_T$, as given in (1.255).

(c) Use SPICE to check your calculations for both parts (a) and (b).

4.11 Calculate the small-signal voltage gain of a common-source amplifier with depletion load in Fig. 4.20, including both the body effect and channel-length modulation. Assume that $V_{DD} = 3$ V and that the dc input voltage is adjusted so that the dc output voltage is 1 V. Assume that M_1 has drawn dimensions of $W = 100$ μm and $L = 1$ μm. Also, assume that M_2 has drawn dimensions of $W = 20$ μm and $L = 1$ μm. For M_2, assume $V_{t0} = -1$ V. For both transistors, assume that $X_d = 0$. Use Table 2.4 for other parameters of both transistors.

4.12 Determine the unloaded voltage gain v_o/v_i and output resistance for the circuit of Fig. 4.58. Check with SPICE and also use SPICE to plot out the large-signal V_O-V_I transfer characteristic for $V_{SUP} = 2.5$ V. Use SPICE to determine the CMRR if the current-source output resistance is 1 MΩ.

4.13 Repeat Problem 4.12, but now assuming that 2-kΩ resistors are inserted in series with the emitters of Q_3 and Q_4.

4.14 Repeat Problem 4.12 except replace Q_1 and Q_2 with n-channel MOS transistors M_1 and M_2. Also, replace Q_3 and Q_4 with p-channel MOS transistors M_3 and M_4. Assume $W_n = 50$ μm and $W_p = 100$ μm.

For all transistors, assume $L_{drwn} = 1$ μm and $X_d = 0$. Use Table 2.3 for other parameters.

4.15 Repeat Problem 4.14, but now assuming that 2 kΩ resistors are inserted in series with the sources of M_3 and M_4. Ignore the body effect.

4.16 Determine the unloaded voltage gain v_o/v_i and output resistance for the circuit of Fig. 4.59. Neglect r_μ. Verify with SPICE and also use SPICE to plot the large-signal V_O-V_I transfer characteristic for $V_{SUP} = 2.5$ V.

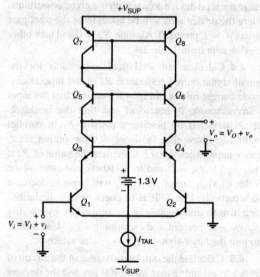

Figure 4.59 Cascode active-load circuit for Problem 4.16.

4.17 Repeat Problem 4.16 except replace the npn and pnp transistors with n-channel and p-channel MOS transistors, respectively. Assume $W_n = 50$ μm and $W_p = 100$ μm. For all transistors, assume $L_{drwn} = 1$ μm and $X_d = 0$. Let $I_{TAIL} = 100$ μA. Ignore the body effect. Use Table 2.3 for other parameters.

4.18 Find $G_m[dm]$ of a source-coupled pair with a current-mirror load with nonzero mismatch (Fig. 4.29b) and show that it is approximately given by (4.184). Calculate the value of $G_m[dm]$ using the following data:

	T_1	T_2	T_3	T_4	T_5
g_m (mA/V)	1.05	0.95	1.1	0.9	2.0
r_o (MΩ)	0.95	1.05	1.0	1.0	0.5

Compare your answer with a SPICE simulation. Also, compare your answer to the result that would apply without mismatch.

4.19 Although $G_m[cm]$ of a differential pair with a current-mirror load can be calculated exactly from a small-signal diagram where mismatch is allowed, the calculation is complicated because the mismatch terms interact, and the results are difficult to interpret. In practice, the mismatch terms are often a small fraction of the corresponding average values, and the interactions between mismatch terms are often negligible. Using the following steps as a guide, calculate an approximation to $G_m[cm]$ including the effects of mismatch.

(a) Derive the ratio i_2/v_{ic} included in (4.165) and show that this ratio is approximately $1/2r_{tail}$ as shown in (4.185) if $\epsilon_d \ll 2$, $g_{m2}r_{o2} \gg 1$, and $2g_{m2}r_{tail} \gg 1$.

(b) Use (4.173) to calculate ϵ_d with perfect matching, where ϵ_d represents the gain error of the differential pair with a pure common-mode input and is defined in (4.161).

(c) Calculate ϵ_d if $1/g_{m3} = 0$ and if the only mismatch is $g_{m1} \neq g_{m2}$.

(d) Calculate ϵ_d if $1/g_{m3} = 0$ and if the only mismatch is $r_{o1} \neq r_{o2}$.

(e) Now estimate the total ϵ_d including mismatch by adding the values calculated in parts (b), (c), and (d). Show that the result agrees with (4.186) if $g_{m3}r_{o(dp)} \gg 1$.

(f) Calculate ϵ_m, which represents the gain error of the current mirror and is defined in (4.133). Show that the result agrees with (4.187).

(g) Calculate the value of $G_m[cm]$ using (4.185) and the CMRR for the data given in Problem 4.18. Compare your answer with a SPICE simulation. Also, compare your answer to the result that would apply without mismatch.

4.20 Design a Widlar current source using *npn* transistors that produces a 10-μA output current. Use Fig. 4.31a with identical transistors, $V_{CC} = 30$ V, and $R_1 = 30$ kΩ. Find the output resistance.

4.21 In the design of a Widlar current source of Fig. 4.31a to produce a specified output current, two resistors must be selected. Resistor R_1 sets I_{IN}, and the emitter resistor R_2 sets I_{OUT}. Assuming a supply voltage of V_{CC} and a desired output current I_{OUT}, determine the values of the two resistors so that the total resistance in the circuit is minimized. Your answer should be given as expressions for R_1 and R_2 in terms of V_{CC} and I_{OUT}. What values would these expressions give for Problem 4.20? Are these values practical?

4.22 Determine the output current in the circuit of Fig. 4.60.

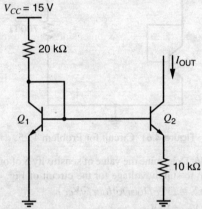

Figure 4.60 Circuit for Problem 4.22.

4.23 Design a MOS Widlar current source using the circuit shown in Fig. 4.31b to meet the following constraints with $V_{DD} = 3$ V:

(a) The input current should be 100 μA, and the output current should be 10 μA.

(b) $V_{ov1} = 0.2$ V.

(c) Transistor M_2 must operate in the active region if the voltage from the drain of M_2 to ground is at least 0.2 V.

(d) The output resistance should be 50 MΩ.

Ignore the body effect. Assume $L_{drwn} = 1$ μm and $X_d = L_d = 0$. Use Table 2.4 for other parameters.

4.24 Design the MOS peaking current source in Fig. 4.34 so that $I_{OUT} = 0.1$ μA.

(a) First, let $I_{IN} = 1$ μA and find the required value of R.

(b) Second, let $R = 10$ kΩ and find the required I_{IN}.

In both cases, assume that both transistors are identical and operate in weak inversion with $I_t = 0.1$ μA and $n = 1.5$. Also, find the minimum W/L

in both cases, assuming that $V_{GS} - V_t < 0$ is required to operate a transistor in weak inversion as shown in Fig. 1.45.

4.25 Determine the output current and output resistance of the circuit shown in Fig. 4.61.

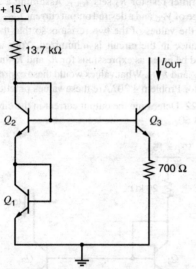

Figure 4.61 Circuit for Problem 4.25.

4.26 Determine the value of sensitivity S of output current to supply voltage for the circuit of Fig. 4.62, where $S = (V_{CC}/I_{OUT})(\partial I_{OUT}/\partial V_{CC})$.

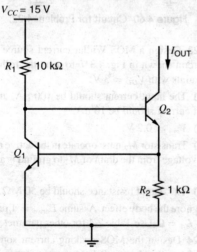

Figure 4.62 Circuit for Problem 4.26.

4.27 In the analysis of the hypothetical reference of Fig. 4.44, the current I_1 was assumed proportional to temperature. Assume instead that this current is derived from a diffused resistor, and thus has a TC_F of -1500 ppm/°C. Determine the new value of V_{OUT} required to achieve zero TC_F at 25°C. Neglect base current.

4.28 The circuit of Fig. 4.46c is to be used as a band-gap reference. If the op amp is ideal, its differential input voltage and current are both zero and

$$V_{OUT} = (V_{BE1} + I_1 R_1) = (V_{BE1} + I_2 R_2)$$
$$V_{OUT} = V_{BE1} + R_2 \left(\frac{V_{BE1} - V_{BE2}}{R_3} \right)$$

Assume that I_1 is to be made equal to 200 μA, and that $(V_{BE1} - V_{BE2})$ is to be made equal to 100 mV. Determine R_1, R_2, and R_3 to realize zero TC_F of V_{OUT} at 25°C. Neglect base currents.

4.29 A band-gap reference like that of Fig. 4.47 is designed to have nominally zero TC_F at 25°C. Due to process variations, the saturation current I_S of the transistors is actually twice the nominal value. Assume $V_{OS} = 0$. What is dV_{OUT}/dT at 25°C? Neglect base currents.

4.30 Repeat Problem 4.29 assuming that the values of I_S, R_2, and R_1 are nominal but that R_3 is 1 percent low. Assume $V_{BE(on)} = 0.6$ V.

4.31 Simulate the band-gap reference from Problem 4.29 on SPICE. Assume that the amplifier is just a voltage-controlled voltage source with an open-loop gain of 10,000 and that the resistor values are independent of temperature. Also assume that $I_{S1} = 1.25 \times 10^{-17}$ A and $I_{S2} = 1 \times 10^{-16}$ A. In SPICE, adjust the closed-loop gain of the amplifier (by choosing suitable resistor values) so that the output TC_F is zero at 25°C. What is the resulting target value of V_{OUT}? Now double I_{S1} and I_{S2}. Use SPICE to adjust the gain so that V_{OUT} is equal to the target at 25°C. Find the new dV_{OUT}/dT at 25°C with SPICE. Compare this result with the calculations from Problem 4.29.

4.32 A band-gap reference circuit is shown in Fig. 4.63. Assume that $\beta_F \to \infty$, $V_A \to \infty$, $I_{S1} = 1 \times 10^{-15}$ A, and $I_{S2} = 8 \times 10^{-15}$ A. Assume the op amp is ideal except for a possibly nonzero offset voltage V_{OS}, which is modeled by a voltage source in Fig. 4.63.

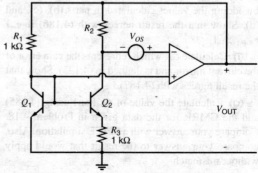

Figure 4.63 Band-gap reference circuit for Problem 4.32.

(a) Suppose that R_2 is trimmed to set V_{OUT} equal to the target voltage for which $dV_{OUT}/dT = 0$ at $T = 25°C$ when $V_{OS} = 0$. Find dV_{OUT}/dT at $T = 25°C$ when $V_{OS} = 30$ mV.

(b) Under the conditions in part (a), is dV_{OUT}/dT positive or negative? Explain.

4.33 For the circuit of Fig. 4.64, find the value of W/L for which $dV_{GS}/dT = 0$ at $25°C$. Assume that the threshold voltage falls 2 mV for each $1°C$ increase in temperature. Also, assume that the mobility temperature dependence is given by (4.243) with $n = 1.5$. Finally, use Table 2.4 for other parameters at $25°C$, and let $I = 200$ μA.

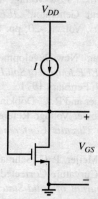

Figure 4.64 Circuit for Problem 4.33.

4.34 The circuit of Fig. 4.65 produces a supply-insensitive current. Calculate the ratio of small-signal variations in I_{BIAS} to small-signal variations in V_{DD} at low frequencies. Ignore the body effect but include finite transistor r_o in this calculation.

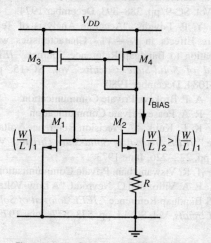

Figure 4.65 Circuit for Problem 4.34.

4.35 Calculate the bias current of the circuit shown in Fig. 4.65 as a function of R, $\mu_n C_{ox}$, $(W/L)_1$, and $(W/L)_2$. Comment on the temperature behavior of the bias current. For simplicity, assume that $X_d = L_d = 0$ and ignore the body effect. Assume M_4 is identical to M_3.

4.36 For the bias circuit shown in Fig. 4.66, determine the bias current. Assume that $X_d = L_d = 0$. Neglect base currents and the body effect. Comment on the temperature dependence of the bias current. Assume a channel mobility and oxide thickness from Table 2.4. Compare your calculations to a SPICE simulation using a full circuit model from Table 2.4, and also use SPICE to determine the supply-voltage sensitivity of I_{BIAS}.

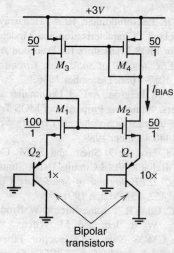

Figure 4.66 Circuit for Problem 4.36.

4.37 A pair of bipolar current sources is to be designed to produce output currents that match with ±1 percent. If resistors display a worst-case mismatch of ±0.5 percent, and transistors a worst-case V_{BE} mismatch of 2 mV, how much voltage must be dropped across the emitter resistors?

4.38 Determine the worst-case input offset voltage for the circuit of Fig. 4.58. Assume the worst-case I_S mismatches in the transistors are ±5 percent and $\beta_F = 15$ for the pnp transistors. Assume the dc output voltage is $V_{SUP} - |V_{BE(on)}|$.

4.39 Repeat Problem 4.38 but replace the bipolar transistors with MOS transistors as in Problem 4.14. Assume the worst-case W/L mismatches in the transistors are ±5 percent and the worst-case V_t mismatches are ±10 mV. Assume the dc output voltage to ground is $V_{SUP} - |V_{GS3}|$. Also, assume that $(W/L)_1 + (W/L)_2 = 20$ and $(W/L)_3 + (W/L)_4 = 60$. Use Table 2.4 to calculate the transconductance parameters.

4.40 Repeat Problem 4.38, but assume that 2-kΩ resistors are placed in series with the emitters of Q_3 and Q_4. Assume the worst-case resistor mismatch is ±0.5 percent and the worst-case *pnp* β_F mismatch is ±10 percent.

REFERENCES

1. Y. P. Tsividis. *Operation and Modeling of the MOS Transistor,* McGraw Hill, New York, 1987, p. 139.

2. J.-B. Shyu, G. C. Temes, and F. Krummenacher. "Random Error Effects in Matched MOS Capacitors and Current Sources," *IEEE Journal of Solid-State Circuits,* Vol. SC-19, pp. 948–955, December 1984.

3. K. R. Lakshmikumar, R. A. Hadaway, and M. A. Copeland. "Characterisation and Modeling of Mismatch in MOS Transistors for Precision Analog Design," *IEEE Journal of Solid-State Circuits,* Vol. SC-21, pp. 1057–1066, December 1986.

4. M. J. M. Pelgrom, A. C. J. Duinmaijer, and A. P. G. Welbers. "Matching Properties of MOS Transistors," *IEEE Journal of Solid-State Circuits,* Vol. 24, pp. 1433–1440, October 1989.

5. W.-J. Hsu, B. J. Sheu, and S. M. Gowda. "Design of Reliable VLSI Circuits Using Simulation Techniques," *IEEE Journal of Solid-State Circuits,* Vol. 26, pp. 452–457, March 1991.

6. T. C. Choi, R. T. Kaneshiro, R. W. Brodersen, P. R. Gray, W. B. Jett, and M. Wilcox. "High-Frequency CMOS Switched-Capacitor Filters for Communications Application," *IEEE Journal of Solid-State Circuits,* Vol. SC-18, pp. 652–664, December 1983.

7. N. S. Sooch. "MOS Cascode Current Mirror," U.S. Patent 4,550,284, October 1985.

8. G. R. Wilson. "A Monolithic Junction FET—n-p-n Operational Amplifier," *IEEE Journal of Solid-State Circuits,* Vol. SC-3, pp. 341–348, December 1968.

9. D. Fullagar. "A New-Performance Monolithic Operational Amplifier," *Fairchild Semiconductor Applications Brief.* May 1968.

10. R. J. Widlar. "Some Circuit Design Techniques for Linear Integrated Circuits," *IEEE Transactions on Circuit Theory,* Vol. CT-12, pp. 586–590, December 1965.

11. R. J. Widlar. "Design Techniques for Monolithic Operational Amplifiers," *IEEE Journal of Solid-State Circuits,* Vol. SC-4, pp. 184–191, August 1969.

12. M. Nagata. "Constant Current Circuits," Japanese Patent 628,228, May 6, 1971.

13. T. M. Frederiksen. "Constant Current Source," U.S. Patent 3,659,121, April 25, 1972.

14. C. Y. Kwok. "Low-Voltage Peaking Complementary Current Generator," *IEEE Journal of Solid-State Circuits,* Vol. SC-20, pp. 816–818, June 1985.

15. R. J. Widlar. "New Developments in IC Voltage Regulators," *IEEE Journal of Solid-State Circuits,* Vol. SC-6, pp. 2–7, February 1971.

16. C. R. Palmer and R. C. Dobkin. "A Curvature Corrected Micropower Voltage Reference," *International Solid-State Circuits Conference,* pp. 58–59, February 1981.

17. G. C. M. Meijer, P. C. Schmale, and K. van Zalinge. "A New Curvature-Corrected Bandgap Reference," *IEEE Journal of Solid-State Circuits,* Vol. SC-17, pp. 1139–1143, December 1982.

18. B.-S. Song and P. R. Gray. "A Precision Curvature-Compensated CMOS Bandgap Reference," *IEEE Journal of Solid-State Circuits,* Vol. SC-18, pp. 634–643, December 1983.

19. A. P. Brokaw. "A Simple Three-Terminal IC Bandgap Reference," *IEEE Journal of Solid-State Circuits,* Vol. SC-9, pp. 388–393, December 1974.

20. Y. P. Tsividis. "Accurate Analysis of Temperature Effects in $I_C - V_{BE}$ Characteristics with Application to Bandgap Reference Sources," *IEEE Journal of Solid-State Circuits,* Vol. SC-15, pp. 1076–1084, December 1980.

21. A. P. Brokaw. Private Communication.

22. R. A. Pease. Private Communication.

23. K. E. Kuijk. "A Precision Reference Voltage Source," *IEEE Journal of Solid-State Circuits,* Vol. SC-8, pp. 222–226, June 1973.

24. T. R. Viswanathan. Private Communication.

25. E. A. Vittoz and O. Neyroud. "A Low-Voltage CMOS Bandgap Reference," *IEEE Journal of Solid-State Circuits,* Vol. SC-14, pp. 573–577, June 1979.

CHAPTER 5

Output Stages

5.1 Introduction

The output stage of an amplifier must satisfy a number of special requirements. One of the most important requirements is to deliver a specified amount of signal power to a load with acceptably low levels of signal distortion. Another common objective of output-stage design is to minimize the output impedance so that the voltage gain is relatively unaffected by the value of load impedance. A well-designed output stage should achieve these performance specifications while consuming low quiescent power and, in addition, should not be a major limitation on the frequency response of the amplifier.

In this chapter, several output-stage configurations will be considered to satisfy the above requirements. The simplest output-stage configurations are the emitter and source followers. More complex output stages employing multiple output devices are also treated, and comparisons are made of power-output capability and efficiency.

Because of their excellent current-handling capability, bipolar transistors are the preferred devices for use in output stages. Although parasitic bipolar transistors can be used in some CMOS output stages, output stages in CMOS technologies are usually constructed without bipolar transistors and are also described in this chapter.

5.2 The Emitter Follower as an Output Stage

An emitter-follower output stage is shown in Fig. 5.1. To simplify the analysis, positive and negative bias supplies of equal magnitude V_{CC} are assumed, although these supplies may have different values in practice. When output voltage V_o is zero, output current I_o is also zero. The emitter-follower output device Q_1 is biased to a quiescent current I_Q by current source Q_2. The output stage is driven by voltage V_i, which has a quiescent dc value of V_{be1} for $V_o = 0$ V. The bias components R_1, R_3, and Q_3 can be those used to bias other stages in the circuit. Since the quiescent current I_Q in Q_2 will usually be larger than the reference current I_R, resistor R_2 is usually smaller than R_1 to accommodate this difference.

This circuit topology can also be implemented in CMOS technologies using an MOS current source for bias and the parasitic bipolar-transistor emitter follower available in standard CMOS processes. Because any large current flow to the substrate can initiate the *pnpn* latch-up phenomenon described in Chapter 2, however, this configuration should be used carefully in CMOS technologies with lightly doped substrates. Extensive substrate taps in the vicinity of the emitter follower are essential to collect the substrate current flow.

5.2.1 Transfer Characteristics of the Emitter-Follower

The circuit of Fig. 5.1 must handle large signal amplitudes; that is, the current and voltage swings resulting from the presence of signals may be a large fraction of the bias values. As

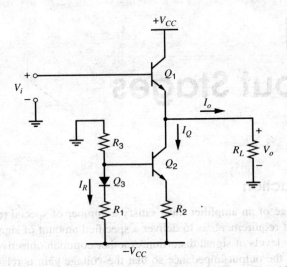

Figure 5.1 Emitter-follower output stage with current-mirror bias.

a result, the small-signal analyses that have been used extensively up to this point must be used with care in this situation. For this reason, we first determine the dc transfer characteristic of the emitter follower. This characteristic allows calculation of the gain of the circuit and also gives important information on the *linearity* and thus on the *distortion* performance of the stage.

Consider the circuit of Fig. 5.1. The large-signal transfer characteristic can be derived as follows:

$$V_i = V_{be1} + V_o \tag{5.1}$$

In this case, the base-emitter voltage V_{be1} of Q_1 cannot be assumed constant but must be expressed in terms of the collector current I_{c1} of Q_1 and the saturation current I_S. If the load resistance R_L is small compared with the output resistance of the transistors,

$$V_{be1} = \frac{kT}{q} \ln\left(\frac{I_{c1}}{I_S}\right) \tag{5.2}$$

if Q_1 is in the forward-active region. Also

$$I_{c1} = I_Q + \frac{V_o}{R_L} \tag{5.3}$$

if Q_2 is in the forward-active region and β_F is assumed large. Substitution of (5.3) and (5.2) in (5.1) gives

$$V_i = \frac{kT}{q} \ln\left(\frac{I_Q + \frac{V_o}{R_L}}{I_S}\right) + V_o \tag{5.4}$$

Equation 5.4 is a nonlinear equation relating V_o and V_i if both Q_1 and Q_2 are in the forward-active region.

The transfer characteristic from (5.4) has been plotted in Fig. 5.2. First, consider the case where R_L is large, which is labeled R_{L1}. In this case, the first term on the right-hand side of (5.4), which represents the base-emitter voltage V_{be1} of Q_1, is almost constant as V_o changes. This result stems from the observation that the current in the load is small for a large R_L;

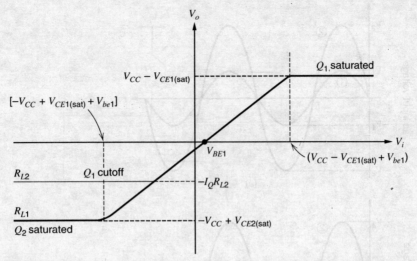

Figure 5.2 Transfer characteristic of the circuit of Fig. 5.1 for a low (R_{L2}) and a high (R_{L1}) value of load resistance.

therefore, the current in Q_1 and V_{be1} are both almost constant as V_o changes in this case. As a result, the center part of the transfer characteristic for $R_L = R_{L1}$ is nearly a straight line with unity slope that is offset on the V_i axis by V_{BE1}, the quiescent value of V_{be1}. This near-linear region depends on both Q_1 and Q_2 being in the forward-active region. However, as V_i is made large positive or negative, one of these devices *saturates* and the transfer characteristic abruptly changes slope.

Consider V_i made large and positive. Output voltage V_o follows V_i until $V_o = V_{CC} - V_{CE1(\text{sat})}$ at which point Q_1 saturates. The collector-base junction of Q_1 is then forward biased and large currents flow from base to collector. In practice, the transistor base resistance (and any source resistance present) limit the current in the forward-biased collector-base junction and prevent the voltage at the internal transistor base from rising appreciably higher. Further increases in V_i thus produce little change in V_o and the characteristic flattens out, as shown in Fig. 5.2. The value of V_i required to cause this behavior is slightly larger than the supply voltage because V_{be1} is larger than the saturation voltage $V_{CE(\text{sat})}$. Consequently, the preceding stage often limits the maximum positive output voltage in a practical circuit because a voltage larger than V_{CC} usually cannot be generated at the base of the output stage. (The portion of the curve for V_i large positive where Q_1 is saturated actually has a positive slope if the effect of the collector series resistance r_c of Q_1 is included. In any event, this portion of the transfer characteristic must be avoided, because the saturation of Q_1 results in large nonlinearity and a major reduction of power gain.)

Now consider V_i made large and negative. The output voltage follows the input until $V_o = -V_{CC} + V_{CE2(\text{sat})}$, at which point Q_2 saturates. (The voltage drop across R_2 is assumed small and is neglected. It could be lumped in with the saturation voltage $V_{CE2(\text{sat})}$ of Q_2 if necessary.) When Q_2 saturates, another discontinuity in the transfer curve occurs, and the slope abruptly decreases. For acceptable distortion performance in the circuit, the voltage swing must be limited to the region between these two break points. As mentioned above, the driver stage supplying V_i usually cannot produce values of V_i that have a magnitude exceeding V_{CC} (if it is connected to the same supply voltages) and the driver itself then sets the upper limit.

Next consider the case where R_L in Fig. 5.1 has a relatively small value. Then when V_o is made large and negative, the first term in (5.4) can become large. In particular, this term

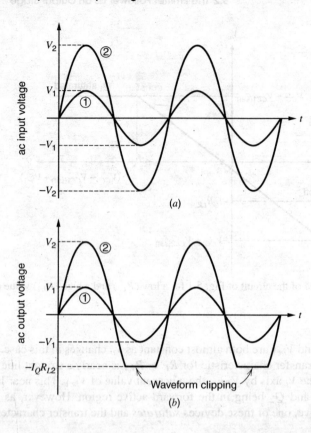

Figure 5.3 (a) ac input signals applied to the circuit of Fig. 5.1. (b) ac output waveforms corresponding to the inputs in (a) with $R_L = R_{L2}$.

approaches minus infinity when V_o approaches the critical value

$$V_o = -I_Q R_L \tag{5.5}$$

In this situation, the current drawn from the load ($-V_o/R_L$) is equal to the current I_Q, and device Q_1 cuts off, leaving Q_2 to draw the current I_Q from the load. Further decreases in V_i produce no change in V_o, and the transfer characteristic is the one labeled R_{L2} in Fig. 5.2. The transfer characteristic for positive V_i is similar for both cases.

For the case $R_L = R_{L2}$, the stage will produce severe waveform distortion if V_i is a sinusoid with amplitude exceeding $I_Q R_{L2}$. Consider the two sinusoidal waveforms in Fig. 5.3a. Waveform ① has an amplitude $V_1 < I_Q R_{L2}$ and waveform ② has an amplitude $V_2 > I_Q R_{L2}$. If these signals are applied as inputs at V_i in Fig. 5.1 (together with a bias voltage), the output waveforms that result are as shown in Fig. 5.3b for $R_L = R_{L2}$. For the smaller input signal, the circuit acts as a near-linear amplifier and the output is near sinusoidal. The output waveform distortion, which is apparent for the larger input, is termed "clipping" and must be avoided in normal operation of the circuit as a linear output stage. For a given I_Q and R_L, the onset of clipping limits the maximum signal that can be handled. Note that if $I_Q R_L$ is larger than V_{CC}, the situation shown for $R_L = R_{L1}$ in Fig. 5.2 holds, and the output voltage can swing almost to the positive and negative supply voltages before excessive distortion occurs.

5.2.2 Power Output and Efficiency

Further insight into the operation of the circuit of Fig. 5.1 can be obtained from Fig. 5.4 where three different load lines are drawn on the $I_c - V_{ce}$ characteristics of Q_1. The equation for the

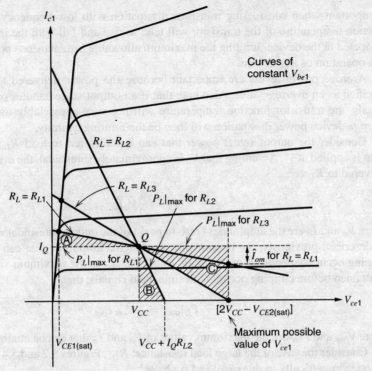

Figure 5.4 Load lines in the $I_{c1} - V_{ce1}$ plane for emitter follower Q_1 of Fig. 5.1.

load lines can be written from Fig. 5.1 and is

$$V_{ce1} = V_{CC} - (I_{c1} - I_Q)R_L \qquad (5.6)$$

when both Q_1 and Q_2 are in the forward-active region. The values of V_{ce1} and I_{c1} are related by (5.6) for any value of V_i and the line includes the quiescent point Q, where $I_{c1} = I_Q$ and $V_{ce1} = V_{CC}$. Equation 5.6 is plotted in Fig. 5.4 for load resistances R_{L1}, R_{L2}, and R_{L3} and the device operating point moves up and down these lines as V_i varies. As V_i increases and V_{ce1} decreases, Q_1 eventually saturates, as was illustrated in Fig. 5.2. As V_i decreases and V_{ce1} increases, there are two possibilities as described above. If R_L is large (R_{L1}), V_o decreases and V_{ce1} increases until Q_2 saturates. Thus the maximum possible value that V_{ce1} can attain is $[2V_{CC} - V_{CE2(sat)}]$ and this value is marked on Fig. 5.4. However, if R_L is small (R_{L2}), the maximum negative value of V_o as illustrated in Fig. 5.2 is $-I_Q R_{L2}$ and the maximum possible value of V_{ce1} is $(V_{CC} + I_Q R_{L2})$.

Thus far no mention has been made of the maximum voltage limitations of the output stage. As described in Chapter 1, avalanche breakdown of a bipolar transistor occurs for $V_{ce} = BV_{CEO}$ in the common-emitter configuration, which is the worst case for breakdown voltage. In a conservative design, the value of V_{ce} in the circuit of Fig. 5.1 should always be less than BV_{CEO} by an appropriate safety margin. In the preceding analysis, the maximum value that V_{ce1} can attain in this circuit for *any* load resistance was calculated as approximately $2V_{CC}$, and thus BV_{CEO} must be greater than this value.

Consider now the power relationships in the circuit. When sinusoidal signals are present, the power dissipated in various elements varies with time. We are concerned both with the *instantaneous* power dissipated and with the *average* power dissipated. Instantaneous power is important when considering transistor dissipation with low-frequency or dc signals. The junction temperature of the transistor will tend to rise and fall with the instantaneous power dissipated in the device, limiting the maximum allowable instantaneous power dissipation for safe operation of any device.

Average power levels are important because the power delivered to a load is usually specified as an *average* value. Also note that if an output stage handles only high-frequency signals, the transistor junction temperature will not vary appreciably over a cycle and the *average* device power dissipation will then be the limiting quantity.

Consider the output signal power that can be delivered to load R_L when a *sinusoidal* input is applied at V_i. Assuming that V_o is approximately sinusoidal, the *average* output power delivered to R_L is

$$P_L = \frac{1}{2}\hat{V}_o \hat{I}_o \tag{5.7}$$

where $\hat{V}_o$ and $\hat{I}_o$ are the amplitudes (zero to peak) of the output sinusoidal voltage and current. As described previously, the maximum output signal amplitude that can be attained before clipping occurs depends on the value of R_L. If $P_L|_{\max}$ is the maximum value of P_L that can be attained before clipping occurs with sinusoidal signals, then

$$P_L|_{\max} = \frac{1}{2}\hat{V}_{om}\hat{I}_{om} \tag{5.7a}$$

where $\hat{V}_{om}$ and $\hat{I}_{om}$ are the maximum values of $\hat{V}_o$ and $\hat{I}_o$ that can be attained before clipping.

Consider the case of the large load resistance, R_{L1}. Figures 5.2 and 5.4 show that clipping occurs symmetrically in this case, and we have

$$\hat{V}_{om} = V_{CC} - V_{CE(\text{sat})} \tag{5.8}$$

assuming equal saturation voltages in Q_1 and Q_2. The corresponding sinusoidal output current amplitude is $\hat{I}_{om} = \hat{V}_{om}/R_{L1}$. The maximum average power that can be delivered to R_{L1} is calculated by substituting these values in (5.7a). This value of power can be interpreted geometrically as the area of the triangle A in Fig. 5.4 since the base of the triangle equals $\hat{V}_{om}$ and its height is $\hat{I}_{om}$. As R_{L1} is increased, the maximum average output power that can be delivered diminishes because the triangle becomes smaller. The maximum output voltage amplitude remains essentially the same but the current amplitude decreases as R_{L1} increases.

If $R_L = R_{L2}$ in Fig. 5.4, the maximum output voltage swing before clipping occurs is

$$\hat{V}_{om} = I_Q R_{L2} \tag{5.9}$$

The corresponding current amplitude is $\hat{I}_{om} = I_Q$. Using (5.7a), the maximum average output power $P_L|_{\max}$ that can be delivered is given by the area of triangle B, shown in Fig. 5.4. As R_{L2} is decreased, the maximum average power that can be delivered is diminished.

An examination of Fig. 5.4 shows that the power-output capability of the stage is maximized for $R_L = R_{L3}$, which can be calculated from (5.6) and Fig. 5.4 as

$$R_{L3} = \frac{V_{CC} - V_{CE(\text{sat})}}{I_Q} \tag{5.10}$$

This load line gives the triangle of largest area (C) and thus the largest average output power. In this case, $\hat{V}_{om} = [V_{CC} - V_{CE(\text{sat})}]$ and $\hat{I}_{om} = I_Q$. Using (5.7a), we have

$$P_L|_{\max} = \frac{1}{2}\hat{V}_{om}\hat{I}_{om} = \frac{1}{2}[V_{CC} - V_{CE(\text{sat})}]I_Q \tag{5.11}$$

To calculate the *efficiency* of the circuit, the power drawn from the supply voltages must now be calculated. The current drawn from the positive supply is the collector current of Q_1, which is assumed sinusoidal with an average value I_Q. The current flowing in the negative supply is constant and equal to I_Q (neglecting bias current I_R). Since the supply voltages are constant, the *average power* drawn from the supplies is constant and *independent* of the presence of sinusoidal signals in the circuit. The total power drawn from the two supplies is thus

$$P_{\text{supply}} = 2V_{CC}I_Q \tag{5.12}$$

The *power conversion efficiency* (η_C) of the circuit at an arbitrary output power level is defined as the ratio of the average power delivered to the load to the average power drawn from the supplies.

$$\eta_C = \frac{P_L}{P_{\text{supply}}} \tag{5.13}$$

Since the power drawn from the supplies is constant in this circuit, the efficiency increases as the output power increases. Also, since the previous analysis shows that the power-output capability of the circuit depends on the value of R_L, the efficiency also depends on R_L. The best efficiency occurs for $R_L = R_{L3}$ since this value gives maximum average power output. If $R_L = R_{L3}$ and $\hat{V}_o = \hat{V}_{om}$, then substitution of (5.11) and (5.12) in (5.13) gives for the maximum possible efficiency

$$\eta_{\max} = \frac{1}{4}\left(1 - \frac{V_{CE(\text{sat})}}{V_{CC}}\right) \tag{5.14}$$

Thus if $V_{CE(\text{sat})} \ll V_{CC}$, the maximum efficiency of the stage is 1/4 or 25 percent.

Another important aspect of circuit performance is the power dissipated in the active device. The current and voltage waveforms in Q_1 at maximum signal swing and with $R_L = R_{L3}$ are shown in Fig. 5.5 (assuming $V_{CE(\text{sat})} \simeq 0$ for simplicity) together with their product, which

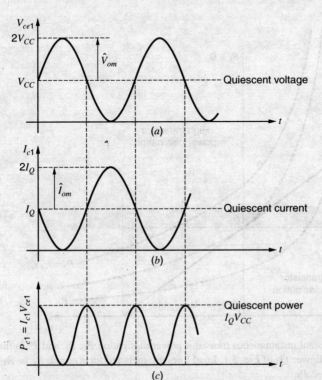

Figure 5.5 Waveforms for the transistor Q_1 of Figure 5.1 at full output with $R_L = R_{L3}$. (*a*) Collector-emitter voltage waveform. (*b*) Collector current waveform. (*c*) Collector power dissipation waveform.

is the *instantaneous* power dissipation in the transistor. The curve of instantaneous power dissipation in Q_1 as a function of time varies at twice the signal frequency and has an average value of one-half the quiescent value. This result can be shown analytically as follows. The instantaneous power dissipation in Q_1 is

$$P_{c1} = V_{ce1} I_{c1} \tag{5.15}$$

At maximum signal swing with a sinusoidal signal, P_{c1} can be expressed as (from Fig. 5.5)

$$P_{c1} = V_{CC}(1 + \sin \omega t) I_Q (1 - \sin \omega t) = \frac{V_{CC} I_Q}{2}(1 + \cos 2\omega t) \tag{5.15a}$$

The average value of P_{c1} from (5.15a) is $V_{CC} I_Q / 2$. Thus at maximum output the *average* power dissipated in Q_1 is half its quiescent value, and the average device temperature when delivering power with $R_L = R_{L3}$ is *less than* its quiescent value.

Further information on the power dissipated in Q_1 can be obtained by plotting curves of constant device dissipation in the $I_c - V_{ce}$ plane. Equation 5.15 indicates that such curves are hyperbolas, which are plotted in Fig. 5.6 for constant transistor instantaneous power dissipation values P_1, P_2, and P_3 (where $P_1 < P_2 < P_3$). The power hyperbola of value P_2 passes through the quiescent point Q, and the equation of this curve can be calculated from (5.15) as

$$I_{c1} = \frac{P_2}{V_{ce1}} \tag{5.16}$$

The slope of the curve is

$$\frac{dI_{c1}}{dV_{ce1}} = -\frac{P_2}{V_{ce1}^2}$$

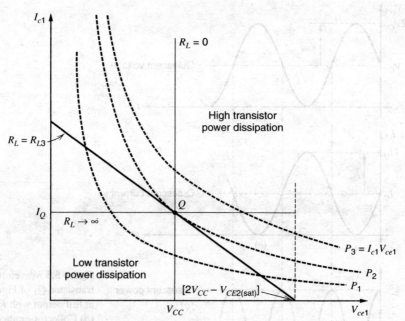

Figure 5.6 Hyperbolas of constant instantaneous transistor power dissipation P_1, P_2, and P_3 in the $I_{c1} - V_{ce1}$ plane for emitter follower Q_1 of Fig. 5.1. Load lines are included for $R_L = 0$, $R_L = R_{L3}$, and $R_L \to \infty$. Note that $P_1 < P_2 < P_3$.

and substitution of (5.16) in this equation gives

$$\frac{dI_{c1}}{dV_{ce1}} = -\frac{I_{c1}}{V_{ce1}} \tag{5.17}$$

At the quiescent point Q, we have $I_{c1} = I_Q$ and $V_{ce1} = V_{CC}$. Thus the slope is

$$\left.\frac{dI_{c1}}{dV_{ce1}}\right|_Q = -\frac{I_Q}{V_{CC}} \tag{5.18}$$

From (5.6), the slope of the load line with $R_L = R_{L3}$ is $-(1/R_{L3})$. Using (5.10) for R_{L3} gives

$$-\frac{1}{R_{L3}} \simeq -\frac{I_Q}{V_{CC}} \tag{5.19}$$

Comparing (5.18) with (5.19) shows that the load line with $R_L = R_{L3}$ is tangent to the power hyperbola passing through the quiescent point, since both curves have the same slope at that point. This result is illustrated in Fig. 5.6. As the operating point leaves the quiescent point and moves on the load line with $R_L = R_{L3}$, the load line then intersects constant-power hyperbolas representing lower power values; therefore, the instantaneous device power dissipation *decreases*. This point of view is consistent with the power waveform shown in Fig. 5.5.

The load line for $R_L \to \infty$ (open-circuit load) is also shown in Fig. 5.6. In that case, the transistor collector current does not vary over a period but is constant. For values of V_{ce1} greater than the quiescent value, the instantaneous device power dissipation increases. The maximum possible value of V_{ce1} is $(2V_{CC} - V_{CE2(\text{sat})})$. At this value, the instantaneous power dissipation in Q_1 is approximately $2V_{CC}I_Q$ if $V_{CE2(\text{sat})} \ll V_{CC}$. This dissipation is twice the quiescent value of $V_{CC}I_Q$, and this possibility should be taken into account when considering the power-handling requirements of Q_1. At the other extreme of the swing where $V_{ce1} \simeq 0$, the power dissipation in Q_2 is also $2V_{CC}I_Q$.

A situation that is potentially even more damaging can occur if the load is short circuited. In that case, the load line is vertical through the quiescent point, as shown in Fig. 5.6. With large input signals, the collector current (and thus the device power dissipation) of Q_1 can become quite large. The limit on collector current is set by the ability of the driver to supply the base current to Q_1, and also by the high-current fall-off in β_F of Q_1, described in Chapter 1. In practice, these limits may be sufficient to prevent burnout of Q_1, but current-limiting circuitry may be necessary. An example of such protection is given in Section 5.4.6.

A useful general result can be derived from the calculations above involving load lines and constant-power hyperbolas. Figure 5.6 shows that the maximum instantaneous device power dissipation for $R_L = R_{L3}$ occurs at the quiescent point Q (since $P_1 < P_2 < P_3$), which is the *midpoint* of the load line if $V_{CE2(\text{sat})} \ll V_{CC}$. (The midpoint of the load line is assumed to be midway between its intersections with the I_c and V_{ce} axes.) It can be seen from (5.17) that *any* load line tangent to a power hyperbola makes contact with the hyperbola at the midpoint of the load line. Consequently, the midpoint is the point of maximum instantaneous device power dissipation with *any* load line. For example, in Fig. 5.4 with $R_L = R_{L2}$, the maximum instantaneous device power dissipation occurs at the midpoint of the load line where $V_{ce1} = \frac{1}{2}(V_{CC} + I_Q R_{L2})$.

An output stage of the type described in this section, where the output device always conducts appreciable current, is called a *Class A* output stage. This type of operation can be realized with different transistor configurations but always has a maximum efficiency of 25 percent.

Finally, the emitter follower in this analysis was assumed to have a current source I_Q in its emitter as a bias element. In practice, the current source is sometimes replaced by a resistor connected to the negative supply, and such a configuration will give some deviations from the above calculations. In particular, the output power available from the circuit will be reduced.

EXAMPLE

An output stage such as shown in Fig. 5.1 has the following parameters: $V_{CC} = 10$ V, $R_3 = 5$ kΩ, $R_1 = R_2 = 0$, $V_{CE(\text{sat})} = 0.2$ V, $R_L = 1$ kΩ. Assume that the dc input voltage is adjusted so that the dc output voltage is 0 volts.

(a) Calculate the maximum average output power that can be delivered to R_L before clipping occurs, and the corresponding efficiency. What is the maximum possible efficiency with this stage and what value of R_L is required to achieve this efficiency? Assume the signals are sinusoidal.

(b) Calculate the maximum possible instantaneous power dissipation in Q_1. Also calculate the average power dissipation in Q_1 when $\hat{V}_o = 1.5$ V and the output voltage is sinusoidal.

The solution proceeds as follows.

(a) The bias current I_Q is first calculated.

$$I_Q = I_R = \frac{V_{CC} - V_{BE3}}{R_3} = \frac{10 - 0.7}{5} \text{ mA} = 1.86 \text{ mA}$$

The product, $I_Q R_L$, is given by

$$I_Q R_L = 1.86 \times 1 = 1.86 \text{ V}$$

Since the dc output voltage is assumed to be 0 volts, and since $I_Q R_L$ is less than V_{CC}, the maximum sinusoidal output voltage swing is limited to 1.86 V by clipping on negative voltage swings and the situation corresponds to $R_L = R_{L2}$ in Fig. 5.4. The maximum output voltage and current swings are thus $\hat{V}_{om} = 1.86$ V and $\hat{I}_{om} = 1.86$ mA. The maximum average output power available from the circuit for sinusoidal signals can be calculated from (5.7a) as

$$P_L|_{\max} = \frac{1}{2}\hat{V}_{om}\hat{I}_{om} = \frac{1}{2} \times 1.86 \times 1.86 \text{ mW} = 1.73 \text{ mW}$$

The power drawn from the supplies is calculated from (5.12) as

$$P_{\text{supply}} = 2V_{CC}I_Q = 2 \times 10 \times 1.86 \text{ mW} = 37.2 \text{ mW}$$

The efficiency of the circuit at the output power level calculated above can be determined from (5.13)

$$\eta_C = \frac{P_L|_{\max}}{P_{\text{supply}}} = \frac{1.73}{37.2} = 0.047$$

The efficiency of 4.7 percent is quite low and is due to the limitation on the negative voltage swing.

The maximum possible efficiency with this stage occurs for $R_L = R_{L3}$ in Fig. 5.4, and R_{L3} is given by (5.10) as

$$R_{L3} = \frac{V_{CC} - V_{CE(\text{sat})}}{I_Q} = \frac{10 - 0.2}{1.86} \text{ k}\Omega = 5.27 \text{ k}\Omega$$

In this instance, the maximum average power that can be delivered to the load before clipping occurs is found from (5.11) as

$$P_L|_{\max} = \frac{1}{2}[V_{CC} - V_{CE(\text{sat})}]I_Q = \frac{1}{2}(10 - 0.2)1.86 \text{ mW} = 9.11 \text{ mW}$$

The corresponding efficiency using (5.14) is

$$\eta_{max} = \frac{1}{4}\left[1 - \frac{V_{CE(sat)}}{V_{CC}}\right] = \frac{1}{4}\left(1 - \frac{0.2}{10}\right) = 0.245$$

This result is close to the theoretical maximum of 25 percent.

(b) The maximum possible instantaneous power dissipation in Q_1 occurs at the midpoint of the load line. Reference to Fig. 5.4 and the load line $R_L = R_{L2}$ shows that this occurs for

$$V_{ce1} = \frac{1}{2}(V_{CC} + I_Q R_L) = \frac{1}{2}(10 + 1.86) = 5.93 \text{ V}$$

The corresponding collector current in Q_1 is $I_{c1} = 5.93/R_L = 5.93$ mA since $R_L = 1$ kΩ. Thus the maximum instantaneous power dissipation in Q_1 is

$$P_{c1} = I_{c1} V_{ce1} = 35.2 \text{ mW}$$

This power dissipation occurs for $V_{ce1} = 5.93$ V, which represents a signal swing beyond the linear limits of the circuit [clipping occurs when the output voltage reaches -1.86 V as calculated in (a)]. However, this condition could easily occur if the circuit is overdriven by a large input signal.

The average power dissipation in Q_1 can be calculated by noting that for sinusoidal signals, the average power drawn from the two supplies is constant and independent of the presence of signals. Since the power input to the circuit from the supplies is constant, the *total* average power dissipated in Q_1, Q_2, and R_L must be constant and independent of the presence of sinusoidal signals. The average power dissipated in Q_2 is constant because I_Q is constant, and thus the average power dissipated in Q_1 and R_L together is constant. Thus as $\hat{V}_o$ is increased, the average power dissipated in Q_1 *decreases* by the same amount as the average power in R_L *increases*. With no input signal, the quiescent power dissipated in Q_1 is

$$P_{CQ} = V_{CC} I_Q = 10 \times 1.86 \text{ mW} = 18.6 \text{ mW}$$

For $\hat{V}_o = 1.5$ V, the average power delivered to the load is

$$P_L = \frac{1}{2}\frac{\hat{V}_o^2}{R_L} = \frac{1}{2}\frac{2.25}{1} \text{ mW} = 1.13 \text{ mW}$$

Thus the *average* power dissipated in Q_1 when $\hat{V}_o = 1.5$ V with a sinusoidal signal is

$$P_{av} = P_{CQ} - P_L = 17.5 \text{ mW}$$

■

5.2.3 Emitter-Follower Drive Requirements

The calculations above have been concerned with the performance of the emitter-follower output stage when driven by a sinusoidal input voltage. The stage preceding the output stage is called the driver stage, and in practice it may introduce additional limitations on the circuit performance. For example, it was shown that to drive the output voltage V_o of the emitter follower to its maximum positive value required an input voltage slightly greater than the supply voltage. Since the driver stage is connected to the same supplies as the output stage in most cases, the driver stage generally cannot produce voltages greater than the supply, further reducing the possible output voltage swing.

The above limitations stem from the observation that the emitter follower has a voltage gain of unity and thus the driver stage must handle the same voltage swing as the output. However, the driver can be a much *lower power* stage than the output stage because the current it must

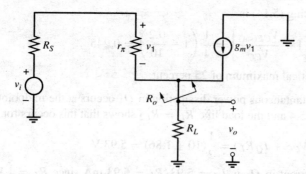

Figure 5.7 Low-frequency, small-signal equivalent circuit for the emitter follower of Fig. 5.1.

deliver is the base current of the emitter follower, which is about $1/\beta_F$ times the emitter current. Consequently, the driver bias current can be much lower than the output-stage bias current, and a smaller geometry can be used for the driver device. Although it has only unity voltage gain, the emitter follower has substantial *power gain*, which is a requirement of any output stage.

5.2.4 Small-Signal Properties of the Emitter Follower

A simplified low-frequency, small-signal equivalent circuit of the emitter follower of Fig. 5.1 is shown in Fig. 5.7. As described in Chapter 7, the emitter follower is an extremely wideband circuit and rarely is a source of frequency limitation in the small-signal gain of an amplifier. Thus the equivalent circuit of Fig. 5.7 is useful over a wide frequency range and an analysis of this circuit shows that the voltage gain A_v and the output resistance R_o can be expressed approximately for $\beta_0 \gg 1$ as

$$A_v = \frac{v_o}{v_i} \simeq \frac{R_L}{R_L + \dfrac{1}{g_m} + \dfrac{R_S}{\beta_0}} \tag{5.20}$$

$$R_o = \frac{1}{g_m} + \frac{R_S}{\beta_0} \tag{5.21}$$

These quantities are small-signal quantities, and since $g_m = qI_C/kT$ is a function of bias point, both A_v and R_o are functions of I_C. Since the emitter follower is being considered here for use as an output stage where the signal swing may be large, (5.20) and (5.21) must be applied with caution. However, for small to moderate signal swings, these equations may be used to estimate the average gain and output resistance of the stage if quiescent bias values are used for transistor parameters in the equations. Equation 5.20 can also be used as a means of estimating the nonlinearity[1] in the stage by recognizing that it gives the *incremental slope* of the large-signal characteristic of Fig. 5.2 at any point. If this equation is evaluated at the extremes of the signal swing, an estimate of the curvature of the characteristic is obtained, as illustrated in the following example.

■ **EXAMPLE**

Calculate the incremental slope of the transfer characteristic of the circuit of Fig. 5.1 as the quiescent point and at the extremes of the signal swing with a peak sinusoidal output of 0.6 V. Use data as in the previous example and assume that $R_S = 0$.

From (5.20) the small-signal gain with $R_S = 0$ is

$$A_v = \frac{R_L}{R_L + \dfrac{1}{g_m}} \quad (5.22)$$

Since $I_Q = 1.86$ mA, $1/g_m = 14\ \Omega$ at the quiescent point and the quiescent gain is

$$A_{vQ} = \frac{1000}{1000 + 14} = 0.9862$$

Since the output voltage swing is 0.6 V, the output current swing is

$$\hat{I}_o = \frac{\hat{V}_o}{R_L} = \frac{0.6}{1000} = 0.6\ \text{mA}$$

Thus at the positive signal peak, the transistor collector current is

$$I_Q + \hat{I}_o = 1.86 + 0.6 = 2.46\ \text{mA}$$

At this current, $1/g_m = 10.6\ \Omega$ and use of (5.22) gives the small-signal gain as

$$A_v^+ = \frac{1000}{1010.6} = 0.9895$$

This gain is 0.3 percent more than the quiescent value. At the negative signal peak, the transistor collector current is

$$I_Q - \hat{I}_o = 1.86 - 0.6 = 1.26\ \text{mA}$$

At this current, $1/g_m = 20.6\ \Omega$ and use of (5.22) gives the small-signal gain as

$$A_v^- = \frac{1000}{1020.6} = 0.9798$$

This gain is 0.7 percent less than the quiescent value. Although the collector-current signal amplitude is one-third of the bias current in this example, the small-signal gain variation is extremely small. This circuit thus has a high degree of linearity. Since the nonlinearity is small, the resulting distortion can be determined from the three values of the small-signal gain calculated in this example. See Problem 5.8.

∎

5.3 The Source Follower as an Output Stage

The small-signal properties of the source follower are calculated in Chapter 3. Since this circuit has low output resistance, it is often used as an output stage. The large signal properties of the source follower are considered next.

5.3.1 Transfer Characteristics of the Source Follower

A source-follower output stage is shown in Fig. 5.8 with equal magnitude positive and negative power supplies for simplicity. The large-signal transfer characteristic can be derived as follows:

$$V_i = V_o + V_{gs1} = V_o + V_{t1} + V_{ov1} \quad (5.23)$$

If the threshold and overdrive terms are exactly constant, the output voltage follows the input voltage with a constant difference. In practice, however, the body effect changes the threshold voltage. Also, the overdrive is not constant mainly because the drain current is not constant.

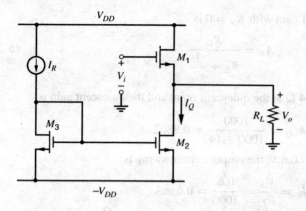

Figure 5.8 Source-follower output stage with current-mirror bias.

Substituting (1.140) with $V_{sb} = V_o + V_{DD}$ and (1.166) with $I_{d1} = I_Q + V_o/R_L$ into (5.23) gives

$$V_i = V_o + V_{t0} + \gamma\left(\sqrt{2\phi_f + V_o + V_{DD}} - \sqrt{2\phi_f}\right) + \sqrt{\dfrac{2\left(I_Q + \dfrac{V_o}{R_L}\right)}{k'(W/L)_1}} \quad (5.24)$$

This equation is valid provided that M_1 and M_2 operate in the active region with output resistances much larger than R_L.

The transfer characteristic is plotted in Fig. 5.9. It intersects the x axis at the input-referred offset voltage, which is

$$V_i|_{V_o=0} = V_{t0} + \gamma\left(\sqrt{2\phi_f + V_{DD}} - \sqrt{2\phi_f'}\right) + \sqrt{\dfrac{2I_Q}{k'(W/L)_1}} \quad (5.25)$$

The slope at this point is the incremental gain calculated in (3.80). With $r_o \to \infty$, the slope is

$$\dfrac{v_o}{v_i} = \dfrac{g_m R_L}{1 + (g_m + g_{mb})R_L} \quad (5.26)$$

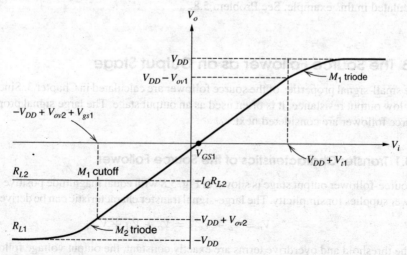

Figure 5.9 Transfer characteristic of the circuit of Fig. 5.8 for a low (R_{L2}) and a high (R_{L1}) value of load resistance.

When $R_L \to \infty$,

$$\frac{v_o}{v_i} = \frac{g_m}{g_m + g_{mb}} = \frac{1}{1+\chi} \tag{5.27}$$

Since χ is typically in the range of 0.1 to 0.3, the slope typically ranges from about 0.7 to 0.9. In contrast, the slope of the emitter-follower transfer characteristic is unity under these conditions. Furthermore, (1.200) shows that χ depends on the source-body voltage, which is $V_o + V_{DD}$ in Fig. 5.8. Therefore, the slope calculated in (5.27) changes as V_o changes even when M_1 operates in the active region, causing distortion. Fig. 5.9 ignores this variation in the slope, but it is considered in Section 5.3.2.

When the output voltage rises within an overdrive of the positive supply, M_1 enters the triode region, dramatically reducing the slope of the transfer characteristic. The overdrive at this break point is calculated using the total drain current, which exceeds I_Q if R_L is finite. Therefore, the location of the break point depends on R_L, but this effect is not shown in Fig. 5.9. Unlike in the emitter follower, the output can pull up asymptotically to the supply voltage with unlimited input voltage. In practice, however, the input must be larger than V_{DD} by at least the threshold voltage to bias M_1 in the triode region. If the input voltage is limited to V_{DD}, M_1 never reaches the triode region.

For negative input voltages, the minimum value of the output voltage depends on R_L, as in the emitter follower. If $I_Q R_L > V_{DD}$, the slope of the transfer characteristic is approximately constant until M_2 enters the triode region, which happens when

$$V_o = -V_{DD} + V_{ov2} = -V_{DD} + \sqrt{\frac{2I_Q}{k'(W/L)_2}} \tag{5.28}$$

This case is labeled as R_{L1} in Fig. 5.9. On the other hand, if $I_Q R_L < V_{DD}$, the slope is almost constant until M_1 turns off. The corresponding minimum output voltage is

$$V_o = -I_Q R_L \tag{5.29}$$

This case is labeled as R_{L2} in Fig. 5.9. From a design standpoint, $I_Q R_L$ is usually set larger than V_{DD}, so that the situation shown for $R_L = R_{L1}$ in Fig. 5.9 holds. Under this condition, the output voltage can swing almost to the positive and negative supply voltages before excessive distortion occurs.

5.3.2 Distortion in the Source Follower

The transfer function of the source-follower stage was calculated in (5.23), where V_i is expressed as a function of V_o. The calculation of signal distortion from a nonlinear transfer function will now be illustrated using the source-follower stage as an example.

Using a Taylor series, the input voltage can be written as

$$V_i = V_I + v_i = \sum_{n=0}^{\infty} \frac{f^{(n)}(V_o = V_O)(V_o - V_O)^n}{n!} \tag{5.30}$$

where $f^{(n)}$ represents the nth derivative of f. Since $v_o = V_o - V_O$, (5.30) can be rewritten as

$$V_i = V_I + v_i = \sum_{n=0}^{\infty} b_n (v_o)^n \tag{5.31}$$

where $b_n = f^{(n)}(V_o = V_O)/(n!)$. For simplicity, assume that $R_L \to \infty$. From (1.140) and (5.23),

$$V_i = f(V_o) = V_o + V_{t0} + \gamma\left(\sqrt{V_o + V_{DD} + 2\phi_f} - \sqrt{2\phi_f}\right) + V_{ov1} \quad (5.32)$$

Then

$$f'(V_o) = 1 + \frac{\gamma}{2}(V_o + V_{DD} + 2\phi_f)^{-1/2} \quad (5.33)$$

$$f''(V_o) = -\frac{\gamma}{4}(V_o + V_{DD} + 2\phi_f)^{-3/2} \quad (5.34)$$

$$f'''(V_o) = \frac{3\gamma}{8}(V_o + V_{DD} + 2\phi_f)^{-5/2} \quad (5.35)$$

Therefore,

$$b_0 = f(V_o = V_O) = V_O + V_{t0} + \gamma\left(\sqrt{V_O + V_{DD} + 2\phi_f} - \sqrt{2\phi_f}\right) + V_{ov1} \quad (5.36)$$

$$b_1 = f'(V_o = V_O) = 1 + \frac{\gamma}{2}(V_O + V_{DD} + 2\phi_f)^{-1/2} \quad (5.37)$$

$$b_2 = \frac{f''(V_o = V_O)}{2} = -\frac{\gamma}{8}(V_O + V_{DD} + 2\phi_f)^{-3/2} \quad (5.38)$$

$$b_3 = \frac{f'''(V_o = V_O)}{3!} = \frac{\gamma}{16}(V_O + V_{DD} + 2\phi_f)^{-5/2} \quad (5.39)$$

Since the constant b_0 is the dc input voltage V_I, (5.31) can be rewritten as

$$v_i = \sum_{n=1}^{\infty} b_n(v_o)^n = b_1 v_o + b_2 v_o^2 + b_3 v_o^3 + \ldots \quad (5.40)$$

To find the distortion, we would like to rearrange this equation into the following form

$$v_o = \sum_{n=1}^{\infty} a_n(v_i)^n = a_1 v_i + a_2 v_i^2 + a_3 v_i^3 + \ldots \quad (5.41)$$

Substituting (5.41) into (5.40) gives

$$v_i = b_1(a_1 v_i + a_2 v_i^2 + a_3 v_i^3 + \ldots) + b_2(a_1 v_i + a_2 v_i^2 + a_3 v_i^3 + \ldots)^2$$
$$+ b_3(a_1 v_i + a_2 v_i^2 + a_3 v_i^3 + \ldots)^3 + \ldots$$
$$= b_1 a_1 v_i + (b_1 a_2 + b_2 a_1^2)v_i^2 + (b_1 a_3 + 2b_2 a_1 a_2 + b_3 a_1^3)v_i^3 + \ldots \quad (5.42)$$

Matching coefficients in (5.42) shows that

$$1 = b_1 a_1 \quad (5.43)$$
$$0 = b_1 a_2 + b_2 a_1^2 \quad (5.44)$$
$$0 = b_1 a_3 + 2b_2 a_1 a_2 + b_3 a_1^3 \quad (5.45)$$

From (5.43),

$$a_1 = \frac{1}{b_1} \quad (5.46)$$

Substituting (5.46) into (5.44) and rearranging gives

$$a_2 = -\frac{b_2}{b_1^3} \quad (5.47)$$

5.3 The Source Follower as an Output Stage

Substituting (5.46) and (5.47) into (5.45) and rearranging gives

$$a_3 = \frac{2b_2^2}{b_1^5} - \frac{b_3}{b_1^4} \tag{5.48}$$

For the source follower, substituting (5.37) into (5.46) gives

$$a_1 = \frac{1}{1 + \frac{\gamma}{2}(V_O + V_{DD} + 2\phi_f)^{-1/2}} \tag{5.49}$$

Substituting (5.37) and (5.38) into (5.47) and rearranging gives

$$a_2 = \frac{\frac{\gamma}{8}(V_O + V_{DD} + 2\phi_f)^{-3/2}}{\left(1 + \frac{\gamma}{2}(V_O + V_{DD} + 2\phi_f)^{-1/2}\right)^3} \tag{5.50}$$

Substituting (5.37), (5.38), and (5.39) into (5.48) and rearranging gives

$$a_3 = -\frac{\frac{\gamma}{16}(V_O + V_{DD} + 2\phi_f)^{-5/2}}{\left(1 + \frac{\gamma}{2}(V_O + V_{DD} + 2\phi_f)^{-1/2}\right)^5} \tag{5.51}$$

Equations 5.41, 5.49, 5.50, and 5.51 can be used to calculate the distortion of the source-follower stage. For small values of v_i such that $a_2 v_i^2 \ll a_1 v_i$, the first term on the right-hand side of (5.41) dominates and the circuit is essentially linear. However, as v_i becomes comparable to a_1/a_2, other terms become significant and distortion products are generated, as is illustrated next. A common method of describing the nonlinearity of an amplifier is the specification of *harmonic distortion*, which is defined for a single sinusoidal input applied to the amplifier. Thus let

$$v_i = \hat{v}_i \sin \omega t \tag{5.52}$$

Substituting (5.52) into (5.41) gives

$$\begin{aligned}v_o &= a_1 \hat{v}_i \sin \omega t + a_2 \hat{v}_i^2 \sin^2 \omega t + a_3 \hat{v}_i^3 \sin^3 \omega t + \ldots \\ &= a_1 \hat{v}_i \sin \omega t + \frac{a_2 \hat{v}_i^2}{2}(1 - \cos 2\omega t) + \frac{a_3 \hat{v}_i^3}{4}(3 \sin \omega t - \sin 3\omega t) + \ldots\end{aligned} \tag{5.53}$$

Equation 5.53 shows that the output voltage contains frequency components at the fundamental frequency, ω (the input frequency), and also at harmonic frequencies 2ω, 3ω, and so on. The latter terms represent *distortion products* that are not present in the input signal. Second-harmonic distortion HD_2 is defined as the ratio of the amplitude of the output-signal component at frequency 2ω to the amplitude of the first harmonic (or fundamental) at frequency ω. For small distortion, the term $(3/4)a_3\hat{v}_i^3 \sin \omega t$ in (5.53) is small compared to $a_1 \hat{v}_i \sin \omega t$, and the amplitude of the fundamental is approximately $a_1 \hat{v}_i$. Again for small distortion, higher-order terms in (5.53) may be neglected and

$$HD_2 = \frac{a_2 \hat{v}_i^2}{2} \frac{1}{a_1 \hat{v}_i} = \frac{1}{2} \frac{a_2}{a_1} \hat{v}_i \tag{5.54}$$

Under these assumptions, HD_2 varies *linearly* with the peak signal level $\hat{v}_i$. The value of HD_2 can be expressed in terms of known parameters by substituting (5.49) and (5.50) in (5.54) to give

$$HD_2 = \frac{\gamma}{16} \frac{(V_O + V_{DD} + 2\phi_f)^{-3/2}(\hat{v}_i)}{\left(1 + \frac{\gamma}{2}(V_O + V_{DD} + 2\phi_f)^{-1/2}\right)^2} \tag{5.55}$$

If $\gamma \ll 2\sqrt{V_O + V_{DD} + 2\phi_f}$, then

$$HD_2 \simeq \frac{\gamma}{16}(V_O + V_{DD} + 2\phi_f)^{-3/2}\hat{v}_i \qquad (5.56)$$

This equation shows that the second-harmonic distortion can be reduced by increasing the dc output voltage V_O. This result is reasonable because this distortion stems from the body effect. Therefore, increasing V_O decreases the variation of the source-body voltage compared to its dc value caused by an input with fixed peak amplitude.[2] Equation 5.56 also shows that the second-harmonic distortion is approximately proportional to γ, neglecting the effect of γ on V_O.

Similarly, third-harmonic distortion HD_3 is defined as the ratio of the output signal component at frequency 3ω to the first harmonic. From (5.53) and assuming small distortion,

$$HD_3 = \frac{a_3 \hat{v}_i^3}{4} \frac{1}{a_1 \hat{v}_i} = \frac{1}{4}\frac{a_3}{a_1}\hat{v}_i^2 \qquad (5.57)$$

Under these assumptions, HD_3 varies as the *square* of the signal amplitude. The value of HD_3 can be expressed in terms of known parameters by substituting (5.49) and (5.51) in (5.57) to give

$$HD_3 = -\frac{\gamma}{64} \frac{(V_O + V_{DD} + 2\phi_f)^{-5/2}(\hat{v}_i^2)}{\left(1 + \frac{\gamma}{2}(V_O + V_{DD} + 2\phi_f)^{-1/2}\right)^4} \qquad (5.58)$$

Since the distortion calculated above stems from the body effect, it can be eliminated by placing the source follower in an isolated well and connecting the source to the well. However, this approach specifies the type of source-follower transistor because it must be opposite the type of the doping in the well. Also, this approach adds the well-substrate parasitic capacitance to the output load of the source follower, reducing its bandwidth.

■ **EXAMPLE**

Calculate second- and third-harmonic distortion in the circuit of Fig. 5.8 for a peak sinusoidal input voltage $\hat{v}_i = 0.5$ V. Assume that $V_I = 0$, $V_{DD} = 2.5$ V, $I_Q = 1$ mA, and $R_L \to \infty$. Also, assume that $(W/L)_1 = 1000$, $k' = 200$ μA/V^2, $V_{t0} = 0.7$ V, $\phi_f = 0.3$ V, and $\gamma = 0.5$ V$^{1/2}$.

First, the dc output voltage V_O is

$$V_O = V_I - V_{t0} - \gamma\left(\sqrt{V_O + V_{DD} + 2\phi_f} - \sqrt{2\phi_f}\right) - V_{ov1} \qquad (5.59)$$

Rearranging (5.59) gives

$$(V_O + V_{DD} + 2\phi_f) + \gamma\sqrt{V_O + V_{DD} + 2\phi_f}$$
$$-V_I + V_{ov1} + V_{t0} - \gamma\sqrt{2\phi_f} - V_{DD} - 2\phi_f = 0 \qquad (5.60)$$

This quadratic equation can be solved for $\sqrt{V_O + V_{DD} + 2\phi_f}$. Since the result must be positive,

$$\sqrt{V_O + V_{DD} + 2\phi_f} =$$
$$-\frac{\gamma}{2} + \sqrt{\left(\frac{\gamma}{2}\right)^2 + V_I - V_{ov1} - V_{t0} + \gamma\sqrt{2\phi_f} + V_{DD} + 2\phi_f} \qquad (5.61)$$

Squaring both sides and rearranging gives

$$V_O = -V_{DD} - 2\phi_f + \left(-\frac{\gamma}{2} + \sqrt{\left(\frac{\gamma}{2}\right)^2 + V_I - V_{ov1} - V_{t0} + \gamma\sqrt{2\phi_f} + V_{DD} + 2\phi_f}\right)^2 \quad (5.62)$$

In this example,

$$V_{ov1} = \sqrt{\frac{2I_Q}{k'(W/L)_1}} = \sqrt{\frac{2(1000)}{200(1000)}} \text{ V} = 0.1 \text{ V} \quad (5.63)$$

Since $V_{DD} + 2\phi_f = 3.1$ V and $V_I - V_{ov1} - V_{t0} = -0.8$ V,

$$V_O = -3.1 \text{ V} + \left(-0.25 + \sqrt{(0.25)^2 - 0.8 + 0.5\sqrt{0.6} + 3.1}\right)^2 \text{ V}$$
$$= -1.117 \text{ V}$$

Therefore,

$$V_O + V_{DD} + 2\phi_f = (-1.117 + 2.5 + 0.6) \text{ V} = 1.983 \text{ V}$$

From (5.55),

$$HD_2 = \frac{0.5}{16} \frac{(1.983)^{-3/2}(0.5)}{\left(1 + \frac{0.5}{2}(1.983)^{-1/2}\right)^2} = 0.0040 \quad (5.64)$$

From (5.58),

$$HD_3 = -\frac{0.5}{64} \frac{(1.983)^{-5/2}(0.5)^2}{\left(1 + \frac{0.5}{2}(1.983)^{-1/2}\right)^4} = -1.8 \times 10^{-4} \quad (5.65)$$

Thus the second-harmonic distortion is 0.40 percent and the third-harmonic distortion is 0.018 percent. In practice, the second-harmonic distortion is usually dominant.[3]

5.4 Class B Push–Pull Output Stage [4,5]

The major disadvantage of Class A output stages is that large power dissipation occurs even for no ac input. In many applications of power amplifiers, the circuit may spend long periods of time in a *standby* condition with no input signal, or with intermittent inputs as in the case of voice signals. Power dissipated in these standby periods is wasted, which is important for two reasons. First, in battery-operated equipment, supply power must be conserved to extend the battery life. Second, any power wasted in the circuit is dissipated in the active devices, increasing their operating temperatures and thus the chance of failure. Furthermore, the power dissipated in the devices affects the physical size of device required, and larger devices are more expensive in terms of silicon area.

A Class B output stage alleviates this problem by having essentially zero power dissipation with zero input signal. *Two* active devices are used to deliver the power instead of one, and each device conducts for alternate half cycles. This behavior is the origin of the name *push–pull*. Another advantage of Class B output stages is that the efficiency is much higher than for a Class A output stage (ideally 78.6 percent at full output power).

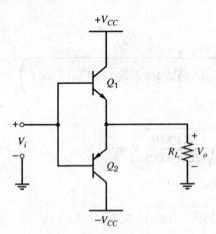

Figure 5.10 Simple integrated-circuit Class B output stage.

A typical integrated-circuit realization of the Class B output stage is shown in Fig. 5.10 in bipolar technology. This circuit uses both *pnp* and *npn* devices and is known as a *complementary* output stage. The *pnp* transistor is usually a substrate *pnp*. Note that the load resistance R_L is connected to the emitters of the active devices; therefore, the devices act as emitter followers.

5.4.1 Transfer Characteristic of the Class B Stage

The transfer characteristic of the circuit of Fig. 5.10 is shown in Fig. 5.11. For V_i equal to zero, V_o is also zero and both devices are off with $V_{be} = 0$. As V_i is made positive, the base-emitter voltage of Q_1 increases until it reaches the value $V_{BE(\text{on})}$, when appreciable current will start to flow in Q_1. At this point, V_o is still approximately zero, but further increases in V_i will cause similar increases in V_o because Q_1 acts as an emitter follower. When $V_i > 0$, Q_2 is off with a *reverse* bias of $V_{BE(\text{on})}$ across its base-emitter junction. As V_i is made even more positive, Q_1 eventually saturates (for $V_i = V_{CC} + V_{be1} - V_{CE1(\text{sat})}$), and the characteristic flattens out as for the conventional emitter follower considered earlier.

As V_i is taken negative from $V_i = 0$, a similar characteristic is obtained except that Q_2 now acts as an emitter follower for V_i more negative than $-V_{BE(\text{on})}$. In this region, Q_1 is held in the off condition with a reverse bias of $V_{BE(\text{on})}$ across its base-emitter junction.

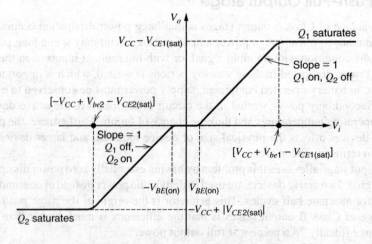

Figure 5.11 Transfer characteristic of the Class B output stage of Fig. 5.10.

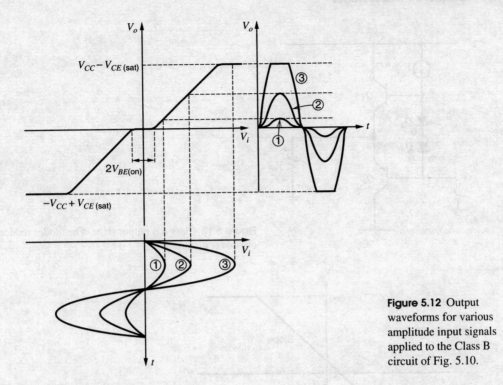

Figure 5.12 Output waveforms for various amplitude input signals applied to the Class B circuit of Fig. 5.10.

The characteristic of Fig. 5.11 shows a notch (or *deadband*) of $2V_{BE(on)}$ in V_i centered around $V_i = 0$. This deadband is common in Class B output stages and gives rise to *crossover* distortion, which is illustrated in Fig. 5.12, where the output waveforms from the circuit are shown for various amplitude input sinusoidal signals. In this circuit, the distortion is high for small input signals with amplitudes somewhat larger than $V_{BE(on)}$. The effect of this source of distortion diminishes as the input signal becomes larger and the deadband represents a smaller fraction of the signal amplitude. Eventually, for very large signals, saturation of Q_1 and Q_2 occurs and distortion rises sharply again due to clipping. This behavior is characteristic of Class B output stages and is why distortion figures are often quoted for both low and high output power operation.

The crossover distortion described above can be reduced by using *Class AB* operation of the circuit. In this scheme, the active devices are biased so that each conducts a small quiescent current for $V_i = 0$. Such biasing can be achieved as shown in Fig. 5.13, where the current source I_Q forces bias current in diodes Q_3 and Q_4. Since the diodes are connected in parallel with the base-emitter junctions of Q_1 and Q_2, the output transistors are biased on with a current that is dependent on the area ratios of Q_1, Q_2, Q_3, and Q_4. A typical transfer characteristic for this circuit is shown in Fig. 5.14, and the deadband has been effectively eliminated. The remaining nonlinearities due to crossover in conduction from Q_1 to Q_2 can be reduced by using negative feedback, as described in Chapter 8.

The operation of the circuit in Fig. 5.13 is quite similar to that of Fig. 5.11. As V_i is taken negative from its quiescent value, emitter follower Q_2 forces V_o to follow. The load current flows through Q_2, whose base-emitter voltage will increase slightly. Since the diodes maintain a constant total bias voltage across the base-emitter junctions of Q_1 and Q_2, the base-emitter voltage of Q_1 will decrease by the same amount that Q_2 increased. Thus during the negative output voltage excursion, Q_1 stays on but conducts little current and plays no part in delivering

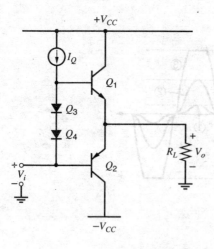

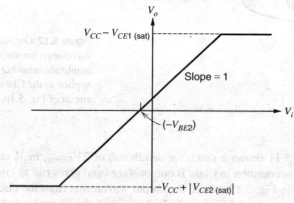

Figure 5.13 Class AB output stage. The diodes reduce crossover distortion.

Figure 5.14 Transfer characteristic of the circuit of Fig. 5.13.

output power. For V_i taken positive, the opposite occurs, and Q_1 acts as the emitter follower delivering current to R_L with Q_2 conducting only a very small current. In this case, the current source I_Q supplies the base-current drive to Q_1.

In the derivation of the characteristics of Figures 5.11 and 5.14, we assumed that the magnitude of the input voltage V_i was unlimited. In the characteristic of Fig. 5.11, the magnitude of V_i required to cause saturation of Q_1 or Q_2 exceeds the supply voltage V_{CC}. However, as in the case of the single emitter follower described earlier, practical driver stages generally cannot produce values of V_i exceeding V_{CC} if they are connected to the same supply voltages as the output stage. For example, the current source I_Q in Fig. 5.13 is usually realized with a *pnp* transistor and thus the voltage at the base of Q_1 cannot exceed $(V_{CC} - V_{CE(\text{sat})})$, at which point saturation of the current-source transistor occurs. Consequently, the positive and negative limits of V_o where clipping occurs are generally somewhat less than shown in Fig. 5.11 and Fig. 5.14, and the limitation usually occurs in the driver stage. This point will be investigated further when practical output stages are considered in later sections.

5.4.2 Power Output and Efficiency of the Class B Stage

The method of operation of a Class B stage can be further appreciated by plotting the collector current waveforms in the two devices, as in Fig. 5.15, where crossover distortion is ignored. Note that each transistor conducts current to R_L for half a cycle.

5.4 Class B Push-Pull Output Stage

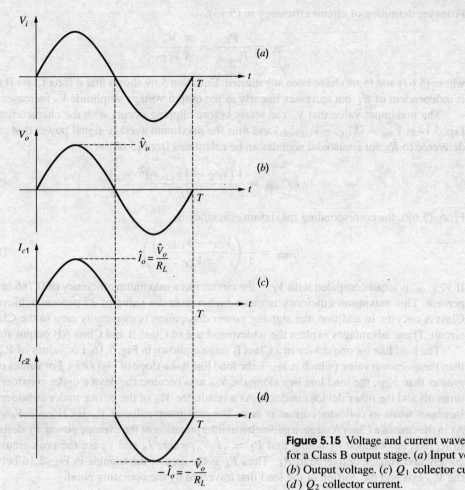

Figure 5.15 Voltage and current waveforms for a Class B output stage. (*a*) Input voltage. (*b*) Output voltage. (*c*) Q_1 collector current. (*d*) Q_2 collector current.

The collector current waveforms of Fig. 5.15 also represent the waveforms of the current drawn from the two supplies. If the waveforms are assumed to be half-sinusoids, then the average current drawn from the $+V_{CC}$ supply is

$$I_{\text{supply}} = \frac{1}{T}\int_0^T I_{c1}(t)dt = \frac{1}{T}\int_0^{T/2} \frac{\hat{V}_o}{R_L}\sin\left(\frac{2\pi t}{T}\right)dt = \frac{1}{\pi}\frac{\hat{V}_o}{R_L} = \frac{1}{\pi}\hat{I}_o \quad (5.66)$$

where T is the period of the input signal. Also, $\hat{V}_o$ and $\hat{I}_o$ are the zero-to-peak amplitudes of the output sinusoidal voltage and current. Since each supply delivers the same current magnitude, the total average power drawn from the two supplies is

$$P_{\text{supply}} = 2V_{CC}I_{\text{supply}} = \frac{2}{\pi}\frac{V_{CC}}{R_L}\hat{V}_o \quad (5.67)$$

where (5.66) has been substituted. Unlike in the Class A case, the average power drawn from the supplies *does vary* with signal level for a Class B stage, and is directly proportional to $\hat{V}_o$.

The average power delivered to R_L is given by

$$P_L = \frac{1}{2}\frac{\hat{V}_o^2}{R_L} \quad (5.68)$$

From the definition of circuit efficiency in (5.13),

$$\eta_C = \frac{P_L}{P_{\text{supply}}} = \frac{\pi}{4} \frac{\hat{V}_o}{V_{CC}} \tag{5.69}$$

where (5.67) and (5.68) have been substituted. Equation 5.69 shows that η for a Class B stage is independent of R_L but increases linearly as the output voltage amplitude $\hat{V}_o$ increases.

The maximum value that $\hat{V}_o$ can attain before clipping occurs with the characteristic of Fig. 5.14 is $\hat{V}_{om} = (V_{CC} - V_{CE(\text{sat})})$ and thus the maximum average signal power that can be delivered to R_L for sinusoidal signals can be calculated from (5.68) as

$$P_L|_{\max} = \frac{1}{2} \frac{[V_{CC} - V_{CE(\text{sat})}]^2}{R_L} \tag{5.70}$$

From (5.69), the corresponding maximum efficiency is

$$\eta_{\max} = \frac{\pi}{4} \left(\frac{V_{CC} - V_{CE(\text{sat})}}{V_{CC}} \right) \tag{5.71}$$

If $V_{CE(\text{sat})}$ is small compared with V_{CC}, the circuit has a maximum efficiency of 0.786 or 78.6 percent. This maximum efficiency is much higher than the value of 25 percent achieved in Class A circuits. In addition, the standby power dissipation is essentially zero in the Class B circuit. These advantages explain the widespread use of Class B and Class AB output stages.

The load line for one device in a Class B stage is shown in Fig. 5.16. For values of V_{ce} less than the quiescent value (which is V_{CC}), the load line has a slope of $(-1/R_L)$. For values of V_{ce} greater than V_{CC}, the load line lies along the V_{ce} axis because the device under consideration turns off and the other device conducts. As a result, the V_{ce} of the device under consideration increases while its collector current is zero. The maximum value of V_{ce} is $(2V_{CC} - V_{CE(\text{sat})})$. As in the case of a Class A stage, a geometrical interpretation of the average power P_L delivered to R_L can be obtained by noting that $P_L = \frac{1}{2}\hat{I}_o\hat{V}_o$, where $\hat{I}_o$ and $\hat{V}_o$ are the peak sinusoidal current and voltage delivered to R_L. Thus P_L is the area of the triangle in Fig. 5.16 between the V_{ce} axis and the portion of the load line traversed by the operating point.

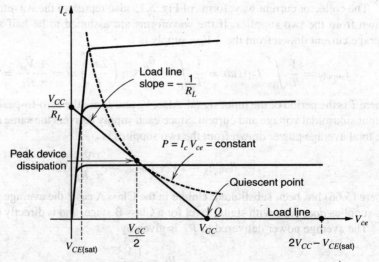

Figure 5.16 Load line for one device in a Class B stage.

Consider the instantaneous power dissipated in one device:

$$P_c = V_{ce} I_c \tag{5.72}$$

But

$$V_{ce} = V_{CC} - I_c R_L \tag{5.73}$$

Substitution of (5.73) in (5.72) gives

$$P_c = I_c(V_{CC} - I_c R_L) = I_c V_{CC} - I_c^2 R_L \tag{5.74}$$

Differentiation of (5.74) shows that P_c reaches a peak for

$$I_c = \frac{V_{CC}}{2R_L} \tag{5.75}$$

This peak lies on the load line midway between the I_c and V_{ce} axis intercepts and agrees with the result derived earlier for the Class A stage. As in that case, the load line in Fig. 5.16 is tangent to a power hyperbola at the point of peak dissipation. Thus, in a Class B stage, maximum instantaneous device dissipation occurs for an output voltage equal to about half the maximum swing. Since the quiescent device power dissipation is zero, the operating temperature of a Class B device always *increases* when nonzero signal is applied.

The instantaneous device power dissipation as a function of time is shown in Fig. 5.17, where collector current, collector-emitter voltage, and their product are displayed for one device in a Class B stage at maximum output. (Crossover distortion is ignored, and $V_{CE(sat)} = 0$ is assumed.) When the device conducts, the power dissipation varies at twice the signal frequency. The device power dissipation is zero for the half cycle when the device is cut off. For an open-circuited load, the load line in Fig. 5.16 lies along the V_{ce} axis and the device has zero

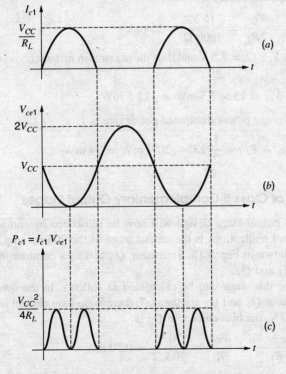

Figure 5.17 Waveforms at maximum output for one device in a Class B stage.
(a) Collector current waveform.
(b) Collector voltage waveform.
(c) Collector power dissipation waveform.

EXAMPLE

A Class B stage of the type shown in Fig. 5.10 drives a load $R_L = 500\ \Omega$. If the positive and negative supplies have magnitudes of 15 V, calculate the maximum average power that is delivered to R_L for $\hat{V}_o = 14.4\ \text{V}$, the corresponding efficiency, and the maximum instantaneous device dissipation. Assume that V_o is sinusoidal.

From (5.66), the average of supply current is

$$I_{\text{supply}} = \frac{1}{\pi}\frac{\hat{V}_o}{R_L} = \frac{1}{\pi}\frac{14.4}{500} = 9.17\ \text{mA}$$

Use of (5.67) gives the average power drawn from the supplies as

$$P_{\text{supply}} = I_{\text{supply}} \times 2V_{CC} = 9.17 \times 30\ \text{mW} = 275\ \text{mW}$$

From (5.68), the average power delivered to R_L is

$$P_L = \frac{1}{2}\frac{\hat{V}_o^2}{R_L} = \frac{1}{2}\frac{14.4^2}{500} = 207\ \text{mW}$$

From (5.13), the corresponding efficiency is

$$\eta_C = \frac{P_L}{P_{\text{supply}}} = \frac{207}{275} = 75.3\ \text{percent}$$

This result is close to the theoretical maximum of 78.6 percent. From (5.75), the maximum instantaneous device power dissipation occurs when

$$I_c = \frac{V_{CC}}{2R_L} = \frac{15\ \text{V}}{1000\ \Omega} = 15\ \text{mA}$$

The corresponding value of V_{ce} is $V_{CC}/2 = 7.5\ \text{V}$ and thus the maximum instantaneous device dissipation is

$$P_c = I_c V_{ce} = 15 \times 7.5\ \text{mW} = 112.5\ \text{mW}$$

By conservation of power, the *average* power dissipated per device is

$$P_{\text{av}} = \frac{1}{2}(P_{\text{supply}} - P_L) = \frac{1}{2}(275 - 207)\ \text{mW} = 34\ \text{mW}$$

5.4.3 Practical Realizations of Class B Complementary Output Stages[6]

The practical aspects of Class B output-stage design will now be illustrated by considering two examples. One of the simplest realizations is the output stage of the 709 op amp, and a simplified schematic of this is shown in Fig. 5.18. Transistor Q_3 acts as a common-emitter driver stage for output devices Q_1 and Q_2.

The transfer characteristic of this stage can be calculated as follows. In the quiescent condition, $V_o = 0$ and $V_1 = 0$. Since Q_1 and Q_2 are then off, there is no base current in these devices. Therefore, for $V_{CC} = 10\ \text{V}$, the bias current in Q_3 is

$$I_{C3} = \frac{V_{CC} - V_1}{R_1} = \frac{V_{CC}}{R_1} = \frac{10\ \text{V}}{20\ \text{k}\Omega} = 0.50\ \text{mA}$$

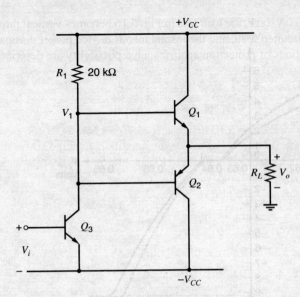

Figure 5.18 Simplified schematic of the output stage of the 709 op amp.

The limiting values that V_o can take are determined by the *driver stage*. When V_i is taken large positive, V_1 decreases until Q_3 saturates, at which point the negative voltage limit V_o^- is reached:

$$V_o^- = -V_{CC} + V_{CE3(sat)} - V_{be2} \quad (5.76)$$

For values of V_1 between $(-V_{CC} + V_{CE3(sat)})$ and $(-V_{BE(on)})$, both Q_3 and Q_2 are in the forward-active region, and V_o follows V_1 with Q_2 acting as an emitter follower.

As V_i is taken negative, the current in Q_3 decreases and V_1 rises, turning Q_1 on. The positive voltage limit V_o^+ is reached when Q_3 cuts off and the base of Q_1 is simply fed from the positive supply via R_1. Then

$$V_{CC} = I_{b1} R_1 + V_{be1} + V_o^+ \quad (5.77)$$

If β_{F1} is large then

$$V_o^+ = I_{c1} R_L = \beta_{F1} I_{b1} R_L$$

where β_{F1} is the current gain of Q_1. Thus

$$I_{b1} = \frac{V_o^+}{\beta_{F1} R_L} \quad (5.78)$$

Substituting (5.78) in (5.77) and rearranging gives

$$V_o^+ = \frac{V_{CC} - V_{be1}}{1 + \dfrac{R_1}{\beta_{F1} R_L}} \quad (5.79)$$

For $R_L = 10 \text{ k}\Omega$ and $\beta_{F1} = 100$, (5.79) gives

$$V_o^+ = 0.98(V_{CC} - V_{be1})$$

In this case, the limit on V_o is similar for positive and negative swings. However, if $R_L = 1 \text{ k}\Omega$ and $\beta_{F1} = 100$, (5.79) gives

$$V_o^+ = 0.83(V_{CC} - V_{be1})$$

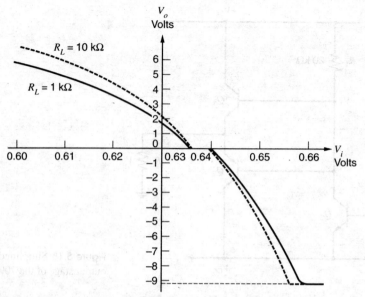

Figure 5.19 SPICE-generated transfer characteristic for the circuit of Fig. 5.18 with $V_{CC} = 10$ V and $R_L = 1$ kΩ and 10 kΩ.

For this lower value of R_L, the maximum positive value of V_o is reduced, and clipping on a sine wave occurs first for V_o going positive.

Computer-generated transfer curves using SPICE for this circuit with $V_{CC} = 10$ V are shown in Fig. 5.19 for $R_L = 1$ kΩ and $R_L = 10$ kΩ. ($\beta_F = 100$ is assumed for all devices.) The reduced positive voltage capability for $R_L = 1$ kΩ is apparent, as is the deadband present in the transfer characteristic. The curvature in the characteristic is due to the exponential nonlinearity of the driver Q_3. In practice, the transfer characteristic may be even more nonlinear than shown in Fig. 5.19 because β_F for the *npn* transistor Q_1 is generally larger than β_F for the *pnp* transistor Q_2, causing the positive and negative sections of the characteristic to differ significantly. This behavior can be seen by calculating the small-signal gain $\Delta V_o/\Delta V_i$ for positive and negative V_o. In the actual 709 integrated circuit, negative feedback is applied around this output stage to reduce these nonlinearities in the transfer characteristic.

A second example of a practical Class B output stage is shown Fig. 5.20, where SPICE-calculated bias currents are included. This circuit is a simplified schematic of the 741 op amp output circuitry. The output devices Q_{14} and Q_{20} are biased to a collector current of about 0.17 mA by the diodes Q_{18} and Q_{19}. The value of the bias current in Q_{14} and Q_{20} depends on the effective area ratio between diodes Q_{18} and Q_{19} and the output devices. (Q_{18} and Q_{19} are implemented with transistors in practice.) The output stage is driven by lateral *pnp* emitter-follower Q_{23}, which is driven by common-emitter stage Q_{17} biased to 0.68 mA by current source Q_{13B}.

The diodes in Fig. 5.20 essentially eliminate crossover distortion in the circuit, which can be seen in the SPICE-generated transfer characteristic of Fig. 5.21. The linearity of this stage is further improved by the fact that the output devices are driven from a low resistance provided by the emitter follower Q_{23}. Consequently, differences in β_F between Q_{14} and Q_{20} produce little effect on the transfer characteristic because small-signal gain $\Delta V_o/\Delta V_1 \simeq 1$ for *any* practical value of β_F with either Q_{14} or Q_{20} conducting.

The limits on the output voltage swing shown in Fig. 5.21 can be determined as follows. As V_i is taken positive, the voltage V_1 at the base of Q_{23} goes negative, and voltages V_2 and

5.4 Class B Push-Pull Output Stage

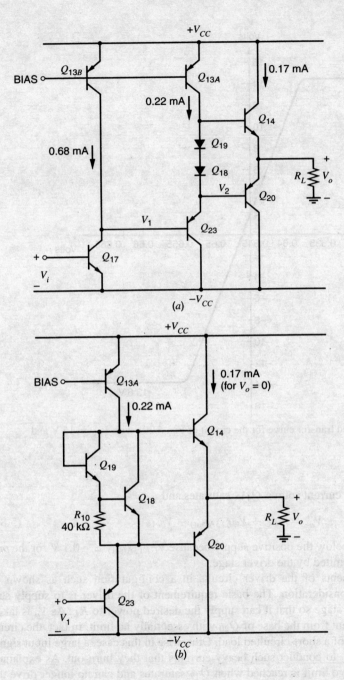

Figure 5.20 (a) Simplified schematic of the 741 op amp output stage. (b) Schematic of the 741 output stage showing the detail of Q_{18} and Q_{19}.

V_o follow with Q_{20} drawing current from R_L. When Q_{17} saturates, the output voltage limit for negative excursions is reached at

$$V_o^- = -V_{CC} + V_{CE17(\text{sat})} - V_{be23} - V_{be20} \qquad (5.80)$$

This limit is about 1.4 V more positive than the negative supply. Thus V_o^- is limited by saturation in Q_{17}, which is the stage *preceding* driver stage Q_{23}.

As V_i is taken negative from its quiescent value (where $V_o = 0$), voltage V_1 rises and voltages V_2 and V_o follow with Q_{14} delivering current to the load. The positive output voltage

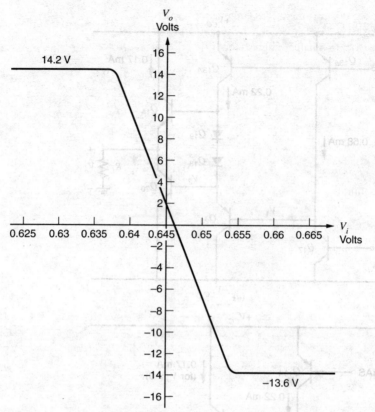

Figure 5.21 SPICE-generated transfer curve for the circuit of Fig. 5.20a with $V_{CC} = 15$ V and $R_L = 1 \text{ k}\Omega$.

limit V_o^+ is reached when current source Q_{13A} saturates and

$$V_o^+ = V_{CC} + V_{CE13A(\text{sat})} - V_{be14} \qquad (5.81)$$

This limit is about 0.8 V below the positive supply because $V_{CE13A(\text{sat})} \simeq -0.1$ V for the *pnp* device. Thus V_o^+ is also limited by the driver stage.

The power requirements of the driver circuits in a configuration such as shown in Fig. 5.20 require some consideration. The basic requirement of the driver is to supply sufficient drive to the output stage so that it can supply the desired power to R_L. As V_o is taken negative, Q_{23} draws current from the base of Q_{20} with essentially no limit. In fact, the circuit must be protected in case of a short-circuited load. Otherwise in this case, a large input signal could cause Q_{23} and Q_{20} to conduct such heavy currents that they burn out. As explained before, the negative voltage limit is reached when Q_{17} saturates and can no longer drive the base of Q_{23} negative.

As V_o is taken positive (by V_i going negative and V_1 going positive), Q_{23} conducts less, and current source Q_{13A} supplies base current to Q_{14}. The maximum output current is limited by the current of 0.22 mA available for driving Q_{14}. As V_1, V_2, and V_o go positive, the current in Q_{14} increases, and the current in Q_{13A} is progressively diverted to the base of Q_{14}. The maximum possible output current delivered by Q_{14} is thus

$$I_o = \beta_{F14} \times 0.22 \text{ mA}$$

If $\beta_{F14} = 100$, the maximum output current is 22 mA. The driver stage may thus limit the maximum positive current available from the output stage. However, this output current level is only reached if R_L is small enough so that Q_{13A} does not saturate on the positive voltage excursion.

The stage preceding the driver in this circuit is Q_{17}. As mentioned above, the negative voltage limit of V_o is reached when Q_{17} saturates. The bias current of 0.68 mA in Q_{17} is much greater than the base current of Q_{23}, and thus Q_{23} produces very little loading on Q_{17}. Consequently, voltage V_1 at the base of Q_{23} can be driven to within $V_{CE(\text{sat})}$ of either supply voltage with only a very small fractional change in the collector current of Q_{17}.

Finally, we will now examine the detail of the fabrication of diodes Q_{18} and Q_{19} in the 741. The actual circuit is shown in Fig. 5.20b with the output protection circuitry omitted. Diode Q_{19} conducts only a current equal to the base current of Q_{18} plus the bleed current in pinch resistor R_{10}. Transistor Q_{18} thus conducts most of the bias current of current source Q_{13A}. This arrangement is used for two reasons. First, the basic aim of achieving a voltage drop equal to two base-emitter voltages is achieved. Since Q_{18} and Q_{19} have common collectors, however, they can be placed in the same isolation region, reducing die area. Second, since Q_{19} conducts only a small current, the bias voltage produced by Q_{18} and Q_{19} across the bases of Q_{14} and Q_{20} is less than would result from a connection as shown in Fig. 5.20a. This observation is important because output transistors Q_{14} and Q_{20} generally have emitter areas larger than the standard device geometry (typically four times larger or more) so that they can maintain high β_F while conducting large output currents. Thus in the circuit of Fig. 5.20a, the bias current in Q_{14} and Q_{20} would be about four times the current in Q_{18} and Q_{19}, which would be excessive in a 741-type circuit. However, the circuit of Fig. 5.20b can be designed to bias Q_{14} and Q_{20} to a current comparable to the current in the diodes, even though the output devices have a large area. The basic reason for this result is that the small bias current in Q_{19} in Fig. 5.20b gives it a smaller base-emitter voltage than for the same device in Fig. 5.20a, reducing the total bias voltage between the bases of Q_{14} and Q_{20}.

The results described above can be illustrated quantitatively by calculating the bias currents in Q_{14} and Q_{20} of Fig. 5.20b. From KVL,

$$V_{BE19} + V_{BE18} = V_{BE14} + |V_{BE20}|$$

and thus

$$V_T \ln \frac{I_{C19}}{I_{S19}} + V_T \ln \frac{I_{C18}}{I_{S18}} = V_T \ln \frac{I_{C14}}{I_{S14}} + V_T \ln \left| \frac{I_{C20}}{I_{S20}} \right| \tag{5.82}$$

If we assume that the circuit is biased for $V_o = 0$ V and also that $\beta_{F14} \gg 1$ and $\beta_{F20} \gg 1$, then $|I_{C14}| = |I_{C20}|$ and (5.82) becomes

$$\frac{I_{C19} I_{C18}}{I_{S18} I_{S19}} = \frac{I_{C14}^2}{I_{S14} I_{S20}}$$

from which

$$I_{C14} = -I_{C20} = \sqrt{I_{C19} I_{C18}} \sqrt{\frac{I_{S14} I_{S20}}{I_{S18} I_{S19}}} \tag{5.83}$$

Equation 5.83 may be used to calculate the output bias current in circuits of the type shown in Fig. 5.20b. The output-stage bias current from (5.83) is proportional to $\sqrt{I_{C18}}$ and $\sqrt{I_{C19}}$. For this specific example, the collector current in Q_{19} is approximately equal to the current in R_{10} if β_F is large and thus

$$I_{C19} \simeq \frac{V_{BE18}}{R_{10}} \simeq \frac{0.6}{40} \text{ mA} = 15 \text{ μA}$$

If the base currents of Q_{14} and Q_{20} are neglected, the collector current of Q_{18} is

$$I_{C18} \simeq |I_{C13A}| - I_{C19} = (220 - 15)\ \mu A = 205\ \mu A$$

To calculate the output-stage bias currents from (5.83), values for the various reverse saturation currents are required. These values depend on the particular IC process used, but typical values are $I_{S18} = I_{S19} = 2 \times 10^{-15}$ A, $I_{S14} = 4I_{S18} = 8 \times 10^{-15}$ A, and $I_{S20} = 4 \times 10^{-15}$ A. Substitution of these data in (5.83) gives $I_{C14} = -I_{C20} = 0.16$ mA.

■ **EXAMPLE**

For the output stage of Fig. 5.20a, calculate bias currents in all devices for $V_o = +10$ V. Assume that $V_{CC} = 15$ V, $R_L = 2$ kΩ, and $\beta_F = 100$. For simplicity, assume all devices have equal area and for each device

$$|I_C| = 10^{-14} \exp\left|\frac{V_{be}}{V_T}\right| \tag{5.84}$$

Assuming that Q_{14} supplies the load current for positive output voltages, we have

$$I_{c14} = \frac{V_o}{R_L} = \frac{10\ V}{2\ k\Omega} = 5\ mA$$

Substitution in (5.84) and rearranging gives

$$V_{be14} = (26\ mV) \ln\left(\frac{5 \times 10^{-3}}{10^{-14}}\right) = 700\ mV$$

Also

$$I_{b14} = \frac{I_{c14}}{\beta_{F14}} = \frac{5\ mA}{100} = 0.05\ mA$$

Thus

$$I_{c19} \simeq I_{c18} \simeq -I_{c23} = (0.22 - 0.05)\ mA = 0.17\ mA$$

Substitution in (5.84) and rearranging gives

$$V_{be19} = V_{be18} = -V_{be23} = (26\ mV)\ln\left(\frac{0.17 \times 10^{-3}}{10^{-14}}\right) = 613\ mV$$

Thus

$$V_{be20} = -(V_{be19} + V_{be18} - V_{be14}) = -525\ mV$$

Use of (5.84) gives

$$I_{c20} = -5.9\ \mu A$$

and the collector current in Q_{20} is quite small as predicted. Finally

$$I_{c17} = 0.68\ mA - \frac{I_{c23}}{\beta_{F23}} = \left(0.68 - \frac{0.17}{100}\right) mA = 0.68\ mA$$

and also

$$V_2 = V_o - |V_{be20}| = (10 - 0.525)\ V = 9.475\ V$$

and

$$V_1 = V_2 - |V_{be23}| = (9.475 - 0.613)\ V = 8.862\ V$$

■

5.4.4 All-*npn* Class B Output Stage[7,8,9]

The Class B circuits described above are adequate for many integrated-circuit applications where the output power to be delivered to the load is of the order of several hundred milliwatts or less. However, if output-power levels of several watts or more are required, these circuits are inadequate because the substrate *pnp* transistors used in the output stage have a limited current-carrying capability. This limit stems from the fact that the doping levels in the emitter, base, and collector of these devices are not optimized for *pnp* structures because the *npn* devices in the circuit have conflicting requirements.

A circuit design that uses high-power *npn* transistors in both halves of a Class B configuration is shown in Fig. 5.22. In this circuit, common-emitter transistor Q_1 delivers power to the load during the negative half-cycle, and emitter follower Q_2 delivers power during the positive half-cycle.

To examine the operation of this circuit, consider V_i taken negative from its quiescent value so that Q_1 is off and $I_{c1} = 0$. Then diodes D_1 and D_2 must both be off and all of the collector current of Q_3 is delivered to the base of Q_2. The output voltage then has its maximum positive value V_o^+. If R_L is big enough, Q_3 saturates and

$$V_o^+ = V_{CC} - |V_{CE3(\text{sat})}| - V_{be2} \tag{5.85}$$

To attain this maximum positive value, transistor Q_3 must saturate in this extreme condition. In contrast, Q_2 in this circuit cannot saturate because the collector of Q_2 is connected to the positive supply and the base voltage of Q_2 cannot exceed the positive supply voltage. The condition for Q_3 to be saturated is that the nominal collector bias current I_{Q3} in Q_3 (when Q_3 is *not* saturated) should be larger than the required base current of Q_2 when $V_o = V_o^+$. Thus we require

$$I_{Q3} > I_{b2} \tag{5.86}$$

Since Q_2 supplies the current to R_L for $V_o > 0$, we have

$$V_o^+ = -I_{e2}R_L = (\beta_2 + 1)I_{b2}R_L \tag{5.87}$$

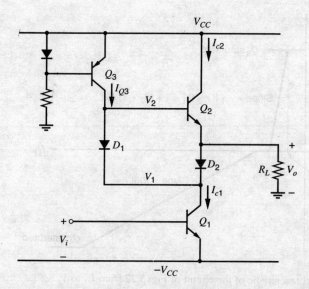

Figure 5.22 All-*npn* Class B output stage.

Substitution of (5.87) and (5.85) in (5.86) gives the requirement on the bias current of Q_3 as

$$I_{Q3} > \frac{V_{CC} - V_{CE3(\text{sat})} - V_{be2}}{(\beta_2 + 1)R_L} \tag{5.88}$$

Equation 5.88 also applies to the circuit of Fig. 5.20a. It gives limits on I_{C3}, β_2, and R_L for V_o to be able to swing close to the positive supply. If I_{Q3} is less than the value given by (5.88), V_o will begin clipping at a positive value *less than* that given by (5.85) and Q_3 will never saturate.

Now consider V_i made positive to turn Q_1 on and produce nonzero I_{c1}. Since the base of Q_2 is more positive than its emitter, diode D_1 will turn on in preference to D_2, which will be off with zero volts across its junction. The current I_{c1} will flow through D_1 and will be drawn from Q_3, which is assumed saturated at first. As I_{c1} increases, Q_3 will eventually come out of saturation, and voltage V_2 at the base of Q_2 will then be pulled down. Since Q_2 acts as an emitter follower, V_o will follow V_2 down. This behavior occurs during the positive half of the cycle, and Q_1 acts as a driver with Q_2 as the output device.

When V_o is reduced to 0 V, the load current is zero and $I_{c2} = 0$. This point corresponds to $I_{c1} = |I_{C3}|$, and all of the bias current in Q_3 passes through D_1 to Q_1. If I_{c1} is increased further, V_o stays constant at 0 V while V_2 is reduced to 0 V also. Therefore, V_1 is negative by an amount equal to the diode voltage drop of D_1, and thus power diode D_2 turns on. Since the current in D_1 is essentially fixed by Q_3, further increases in I_{c1} cause increasing current to flow through D_2. The negative half of the cycle consists of Q_1 acting as the output device and feeding R_L through D_2. The maximum negative voltage occurs when Q_1 saturates and is

$$V_o^- = -V_{CC} + V_{CE1(\text{sat})} + V_{d2} \tag{5.89}$$

where V_{d2} is the forward voltage drop across D_2.

The sequence just described gives rise to a highly nonlinear transfer characteristic, as shown in Fig. 5.23, where V_o is plotted as a function of I_{c1} for convenience. When V_o is positive, the current I_{c1} feeds into the base of Q_2 and the small-signal gain is

$$\frac{\Delta V_o}{\Delta I_{c1}} \simeq \frac{\Delta V_2}{\Delta I_{c1}} = r_{o1} \parallel r_{o3} \parallel [r_{\pi 2} + (\beta_2 + 1)R_L]$$

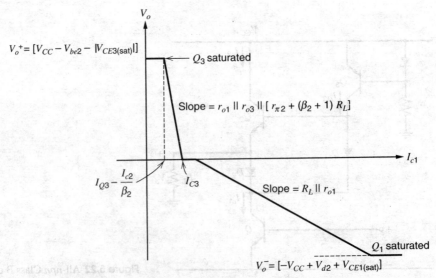

Figure 5.23 Transfer characteristic of the circuit of Fig. 5.22 from I_{c1} to V_o.

where the impedance of D_1 is assumed negligible. That is, the impedance at the base of Q_2 is equal to the parallel combination of the output resistances of Q_1 and Q_3 and the input resistance of emitter follower Q_2.

When V_o in Fig. 5.22 is negative, I_{c1} feeds R_L directly and the small-signal gain is

$$\frac{\Delta V_o}{\Delta I_{c1}} \simeq r_{o1} \parallel R_L$$

where the impedance of D_2 is assumed negligible.

Note the small deadband in Fig. 5.23 where diode D_2 turns on. This deadband can be eliminated by adding a second diode in series with D_1. In practice, negative feedback must be used around this circuit to linearize the transfer characteristic, and such feedback will reduce any crossover effects. The transfer characteristic of the circuit from V_i to V_o is even more nonlinear than shown in Fig. 5.23 because it includes the exponential nonlinearity of Q_1.

In integrated-circuit fabrication of the circuit of Fig. 5.22, devices Q_1 and Q_2 are identical large-power transistors. In high-power circuits (delivering several watts or more), they may occupy 50 percent of the whole die. Diode D_2 is a large-power diode that also occupies considerable area. These features are illustrated in Fig. 5.24, which is a die photo of the 791 high-power op amp. This circuit can dissipate 10 W of power and can deliver 15 W of output power into an 8-Ω load. The large power transistors in the output stage can be seen on the right-hand side of the die.

Finally, the power and efficiency results derived previously for the complementary Class B stage apply equally to the all-*npn* Class B stage if allowance is made for the voltage drop in D_2. Thus the ideal maximum efficiency is 79 percent.

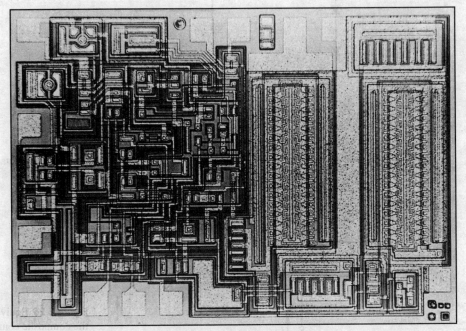

Figure 5.24 Die photo of the 791 high-power op amp.

5.4.5 Quasi-Complementary Output Stages[10]

The all-*npn* stage described above is one solution to the problem of the limited power-handling capability of the substrate *pnp*. Another solution is shown in Fig. 5.25, where a *composite pnp* has been made from a lateral *pnp* Q_3 and a high-power *npn* transistor Q_4. This circuit is called a quasi-complementary output stage.

The operation of the circuit of Fig. 5.25 is almost identical to that of Fig. 5.20. The pair Q_3-Q_4 is equivalent to a *pnp* transistor as shown in Fig. 5.26, and the collector current of Q_3 is

$$I_{C3} = -I_S \exp\left(-\frac{V_{BE}}{V_T}\right) \tag{5.90}$$

The composite collector current I_C is the emitter current of Q_4, which is

$$I_C = (\beta_{F4} + 1)I_{C3} = -(\beta_{F4} + 1)I_S \exp\left(-\frac{V_{BE}}{V_T}\right) \tag{5.91}$$

The composite device thus shows the standard relationship between I_C and V_{BE} for a *pnp* transistor. However, most of the current is carried by the high-power *npn* transistor. Note that the magnitude of the saturation voltage of the composite device is $(|V_{CE3(\text{sat})}| + V_{BE4})$. This

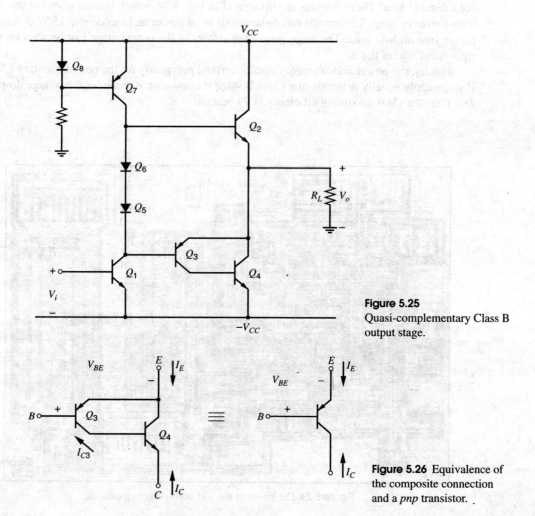

Figure 5.25 Quasi-complementary Class B output stage.

Figure 5.26 Equivalence of the composite connection and a *pnp* transistor.

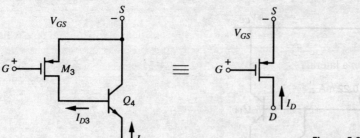

Figure 5.27 Compound high-current PMOS connection.

magnitude is higher than normal because saturation occurs when Q_3 saturates, and V_{BE4} must be added to this voltage.

The major problem with the configuration of Fig. 5.25 is potential instability of the local feedback loop formed by Q_3 and Q_4, particularly with capacitive loads on the amplifier. The stability of feedback loops is considered in Chapter 9.

The quasi-complementary Class B stage can also be effectively implemented in BiCMOS technology. In the circuit of Fig. 5.25, the compound bipolar device Q_3-Q_4 can be replaced by the MOS-bipolar combination[11] of Fig. 5.27, where Q_4 is a large-area high-current bipolar device. The overall transfer characteristic is

$$I_D = (\beta_{F4} + 1)I_{D3} = -(\beta_{F4} + 1)\frac{\mu_n C_{ox}}{2}\left(\frac{W}{L}\right)_3 (V_{GS} - V_t)^2 \quad (5.92)$$

Equation 5.92 shows that the composite PMOS device appears to have a W/L ratio $(\beta_{F4} + 1)$ times larger than the physical PMOS device M_3. With this circuit, one of the diodes Q_5 or Q_6 in Fig. 5.25 would now be replaced by a diode-connected PMOS transistor to set up a temperature-stable standby current in the output stage. A bias bleed resistor can be connected from the base to the emitter of Q_4 to optimize the bias current in M_3 and to speed up the turn-off of Q_4 in high-frequency applications by allowing reverse base current to remove the base charge. Such a resistor can also be connected from the base to the emitter of Q_4 in Fig. 5.25.

5.4.6 Overload Protection

The most common type of overload protection in integrated-circuit output stages is short-circuit current protection. As an example, consider the 741 output stage shown in Fig. 5.28 with partial short-circuit protection included. Initially assume that $R_6 = 0$ and ignore Q_{15}. The maximum positive drive delivered to the output stage occurs for V_i large positive. If $R_L = 0$ then V_o is held at zero volts and V_a in Fig. 5.28 is equal to V_{be14}. Thus, as V_i is taken positive, Q_{23}, Q_{18}, and Q_{19} will cut off and all of the current of Q_{13A} is fed to Q_{14}. If this is a high-β_F device, then the output current can become destructively large:

$$I_{c14} = \beta_{F14}|I_{C13A}| \quad (5.93)$$

If

$$\beta_{F14} = 500$$

then

$$I_{c14} = 500 \times 0.22 = 110 \text{ mA}$$

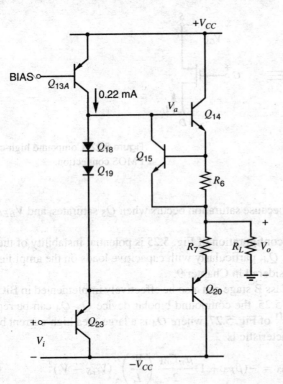

Figure 5.28 Schematic of the 741 op amp showing partial short-circuit protection.

If $V_{CC} = 15$ V, this current level gives a power dissipation in Q_{14} of

$$P_{c14} = V_{ce}I_c = 15 \times 110 \text{ mW} = 1.65 \text{ W}$$

which is sufficient to destroy the device. Thus the current under short-circuit conditions must be limited, and this objective is achieved using R_6 and Q_{15} for positive V_o.

The short-circuit protection operates by sensing the output current with resistor R_6 of about 25 Ω. The voltage developed across R_6 is the base-emitter voltage of Q_{15}, which is normally off. When the current through R_6 reaches about 20 mA (the maximum safe level), Q_{15} begins to conduct appreciably and diverts any further drive away from the base of Q_{14}. The drive current is thus harmlessly passed to the output instead of being multiplied by the β_F of Q_{14}.

The operation of this circuit can be seen by calculating the transfer characteristic of Q_{14} when driving a short-circuit load. This can be done using Fig. 5.29.

$$I_i = I_{b14} + I_{c15} \tag{5.94}$$

$$I_{c15} = I_{S15} \exp \frac{V_{be15}}{V_T} \tag{5.95}$$

Also

$$V_{be15} \simeq I_{c14}R \tag{5.96}$$

From (5.94)

$$I_{b14} = I_i - I_{c15}$$

But

$$I_{c14} = \beta_{F14}I_{b14} = \beta_{F14}(I_i - I_{c15}) \tag{5.97}$$

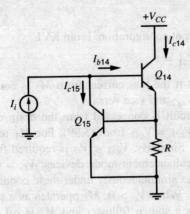

Figure 5.29 Equivalent circuit for the calculation of the effect of Q_{15} on the transfer characteristic of Q_{14} in Fig. 5.28 when $R_L = 0$.

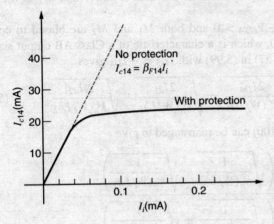

Figure 5.30 Transfer characteristic of the circuit of Fig. 5.29 with and without protection transistor Q_{15} ($\beta_{F14} = 500$).

Substitution of (5.95) and (5.96) in (5.97) gives

$$I_{c14} + \beta_{F14} I_{S15} \exp \frac{I_{c14} R}{V_T} = \beta_{F14} I_i \qquad (5.98)$$

The second term on the left side of (5.98) stems from Q_{15}. If this term is negligible, then $I_{c14} = \beta_{F14} I_i$ as expected. The transfer characteristic of the stage is plotted from (5.98) in Fig. 5.30, using $\beta_{F14} = 500$, $I_{S15} = 10^{-14}$ A, and $R = 25\ \Omega$. For a maximum drive of $I_i = 0.22$ mA, the value of I_{c14} is effectively limited to 24 mA. For values of I_{c14} below 20 mA, Q_{15} has little effect on circuit operation.

Similar protection for negative output voltages in Fig. 5.29 is achieved by sensing the voltage across R_7 and diverting the base drive away from one of the preceding stages.

5.5 CMOS Class AB Output Stages

The classical Class AB topology of Fig. 5.13 can also be implemented in standard CMOS technology. However, the output swing of the resulting circuit is usually much worse than in the bipolar case. Although the swing can be improved using a common-source configuration, this circuit suffers from poor control of the quiescent current in the output devices. These issues are described below.

5.5.1 Common-Drain Configuration

Figure 5.31 shows the common-drain Class AB output configuration. From KVL,

$$V_{SG5} + V_{GS4} = V_{gs1} + V_{sg2} \tag{5.99}$$

Ignoring the body effect, $V_{SG5} + V_{GS4}$ is constant if the bias current from M_3 is constant. Under these conditions, increasing V_{gs1} decreases V_{sg2} and vice versa.

For simplicity, assume at first that a short circuit is connected from the drain of M_4 to the drain of M_5. Then $V_{SG5} + V_{GS4} = 0$ and $V_{gs1} = V_{gs2}$ from (5.99). For M_1 to conduct nonzero drain current, $V_{gs1} > V_{t1}$ is required. Similarly, $V_{gs2} < V_{t2}$ is required for M_2 to conduct nonzero drain current. With standard enhancement-mode devices, $V_{t1} > 0$ and $V_{t2} < 0$. Therefore, M_1 and M_2 do not both conduct simultaneously under these conditions, which is a characteristic of a Class B output stage. When $V_o > 0$, M_1 operates as a source follower and M_2 is off. Similarly, M_2 operates as a source follower and M_1 is off when $V_o < 0$.

In Fig. 5.31, however, $V_{SG5} + V_{GS4} > 0$ and both M_1 and M_2 are biased to conduct nonzero drain current when $V_o = 0$, which is a characteristic of a Class AB output stage. If $V_{t1} = V_{t4}$ and $V_{t2} = V_{t5}$, using (1.157) in (5.99) with $I_{D5} = -I_{D4}$ gives

$$\sqrt{\frac{2I_{D4}}{k'_p(W/L)_5}} + \sqrt{\frac{2I_{D4}}{k'_n(W/L)_4}} = \sqrt{\frac{2I_{d1}}{k'_n(W/L)_1}} + \sqrt{\frac{2|I_{d2}|}{k'_p(W/L)_2}} \tag{5.100}$$

If $V_o = 0$, then $I_{d2} = -I_{d1}$ and (5.100) can be rearranged to give

$$I_{D1} = I_{D4} \frac{\left(\sqrt{\frac{1}{k'_n(W/L)_4}} + \sqrt{\frac{1}{k'_p(W/L)_5}}\right)^2}{\left(\sqrt{\frac{1}{k'_n(W/L)_1}} + \sqrt{\frac{1}{k'_p(W/L)_2}}\right)^2} \tag{5.101}$$

where $I_{D1} = I_{d1}$ with $V_o = 0$. The key point of this equation is that the quiescent current in the output transistors is well controlled with respect to the bias current that flows in the diode-connected transistors, as in the bipolar case.

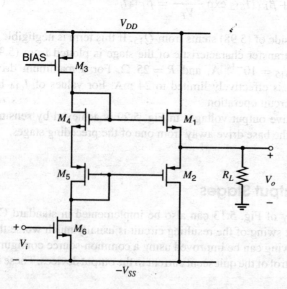

Figure 5.31 Complementary source-follower CMOS output stage based on traditional bipolar implementation.

An important problem with this circuit is that its output swing can be much less than the corresponding bipolar circuit with equal supply voltages. For $V_o > 0$, $V_{gs1} > V_{t1}$ and M_1 acts as a source follower. Therefore,

$$V_o = V_{DD} - V_{sd3} - V_{gs1} \tag{5.102}$$

The minimum V_{sd3} required to operate M_3 as a current source is $|V_{ov3}| = |V_{GS3} - V_{t3}|$. From (5.102), the maximum output voltage V_o^+ is

$$V_o^+ = V_{DD} - |V_{ov3}| - V_{gs1} \tag{5.103}$$

The minimum output voltage can be found by similar reasoning. (See Problem 5.21.) Although (5.103) appears to be quite similar to (5.81) if V_{CC} in Fig. 5.20 is equal to V_{DD} in Fig. 5.31, the limit in (5.103) is usually much less than in (5.81) for three reasons. First, the gate-source voltage includes a threshold component that is absent in the base-emitter voltage. Second, the body effect increases the threshold voltage V_{t1} as V_o increases. Finally, the overdrive part of the gate-source voltage rises more steeply with increasing current than the entire base-emitter voltage because the overdrive is proportional to the square root of the current and the base-emitter voltage is proportional to the logarithm of the current. In practice, the output voltage swing can be increased by increasing the W/L ratios of the output devices to reduce their overdrives. However, the required transistor sizes are sometimes so large that the parasitic capacitances associated with the output devices can dominate the overall performance at high frequencies. Thus the circuit of Fig. 5.31 is generally limited to much smaller currents than its bipolar equivalent.

■ **EXAMPLE**

An output stage such as shown in Fig. 5.31 is required to produce a maximum output voltage of 0.7 V with $R_L = 35\ \Omega$ and $V_{DD} = V_{SS} = 1.5$ V. Using the transistor parameters in Table 2.4, find the required W/L of M_1. Assume $|V_{ov3}| = 100$ mV, and ignore the body effect.
From (5.103),

$$V_{gs1} = V_{DD} - |V_{ov3}| - V_o^+ = (1.5 - 0.1 - 0.7)\ \text{V} = 0.7\ \text{V}$$

Since Table 2.4 gives $V_{t1} = 0.6$ V,

$$V_{ov1} = V_{gs1} - V_{t1} = (0.7 - 0.6)\ \text{V} = 0.1\ \text{V}$$

With $V_o = 0.7$ V, the current in the load is $(0.7\ \text{V})/(35\ \Omega) = 20$ mA. If $I_{d2} = 0$ under these conditions, $I_{d1} = 20$ mA. Rearranging (1.157) gives

$$\left(\frac{W}{L}\right)_1 = \frac{2I}{k'_n V_{ov1}^2} = \frac{2(20000)}{194(0.1)^2} \simeq 20,000 \tag{5.104}$$

■ which is a very large transistor.

5.5.2 Common-Source Configuration with Error Amplifiers

Another alternative is the use of quasi-complementary configurations. In this case, a common-source transistor together with an error amplifier replaces an output source-follower device. A circuit with this substitution for both output transistors is shown conceptually in Fig. 5.32.[12,13,14] The combination of the error amplifier and the common-source device mimics the behavior of a source follower with high dc transconductance. The function of the amplifiers is to sense the voltage difference between the input and the output of the stage and drive the gates of the output transistors so as to make the difference as small as possible. This

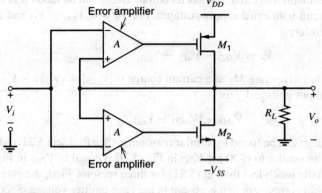

Figure 5.32 A complementary Class AB output stage using embedded common-source output devices.

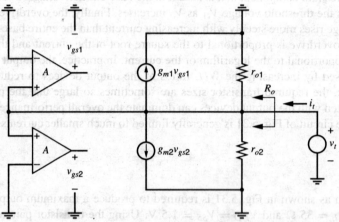

Figure 5.33 Small-signal model of the output stage in Fig. 5.32 used to find R_o.

operation can be viewed as negative feedback. A key advantage of the use of negative feedback here is that it reduces the output resistance. Since negative feedback is covered in Chapter 8, we will analyze this structure with straightforward circuit analysis.

To find the output resistance, consider the small-signal model of this output stage shown in Fig. 5.33. The current i_t is

$$i_t = \frac{v_t}{r_{o1}} + \frac{v_t}{r_{o2}} + g_{m1}Av_t + g_{m2}Av_t \qquad (5.105)$$

Rearranging this equation to solve for v_t/i_t gives

$$R_o = \frac{v_t}{i_t}\bigg|_{v_i=0} = \frac{1}{(g_{m1}+g_{m2})A} \parallel r_{o1} \parallel r_{o2} \qquad (5.106)$$

This equation shows that increasing the gain A of the error amplifiers reduces R_o and that R_o is much less than the drain-source resistance of M_1 or M_2 because of the negative feedback.

To find the transfer characteristic, consider the dc model of the output stage shown in Fig. 5.34. The model includes the input-referred offset voltages of the error amplifiers as voltage sources. Assume $k'_p(W/L)_1 = k'_n(W/L)_2 = k'(W/L)$ and $-V_{t1} = V_{t2} = V_t$. Also assume that the error amplifiers are designed so that $-I_{D1} = I_{D2} = I_Q$ when $V_i = 0$, $V_{OSP} = 0$, and $V_{OSN} = 0$. Under these conditions, $V_o = 0$ and

$$V_{gs1} = -V_t - V_{ov} \qquad (5.107)$$

$$V_{gs2} = V_t + V_{ov} \qquad (5.108)$$

5.5 CMOS Class AB Output Stages

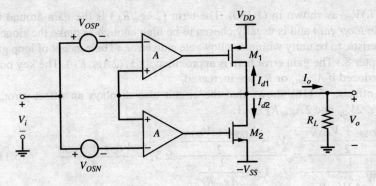

Figure 5.34 A dc model of the output stage in Fig. 5.32 used to find the transfer characteristic.

where

$$V_{ov} = \sqrt{\frac{2I_Q}{k'(W/L)}} \quad (5.109)$$

With nonzero input and offsets, the output may not be zero. As a result, the differential input to the top error amplifier changes from zero to $V_o - (V_i - V_{OSP})$. Similarly, the differential input to the bottom error amplifier changes from zero to $V_o - (V_i - V_{OSN})$. Assuming that the output of each error amplifier changes by its gain A times the change in its input,

$$V_{gs1} = -V_t - V_{ov} + A[V_o - (V_i - V_{OSP})] \quad (5.110)$$

$$V_{gs2} = V_t + V_{ov} + A[V_o - (V_i - V_{OSN})] \quad (5.111)$$

If M_1 and M_2 operate in the active region,

$$I_{d1} = -\frac{k'_p}{2}\left(\frac{W}{L}\right)_1 (V_{gs1} - V_{t1})^2 = -\frac{k'}{2}\frac{W}{L}(V_{gs1} + V_t)^2 \quad (5.112)$$

$$I_{d2} = \frac{k'_n}{2}\left(\frac{W}{L}\right)_2 (V_{gs2} - V_{t2})^2 = \frac{k'}{2}\frac{W}{L}(V_{gs2} - V_t)^2 \quad (5.113)$$

Also,

$$I_o = \frac{V_o}{R_L} \quad (5.114)$$

From KCL at the output,

$$I_o + I_{d1} + I_{d2} = 0 \quad (5.115)$$

Substituting (5.110)–(5.114) into (5.115) and rearranging gives

$$V_o = \frac{V_i - \dfrac{V_{OSP} + V_{OSN}}{2}}{1 + \dfrac{1}{k'\dfrac{W}{L}A[2V_{ov} - A(V_{OSP} - V_{OSN})]R_L}} \quad (5.116)$$

If $V_{OSP} = V_{OSN} = 0$,

$$V_o = \frac{V_i}{1 + \dfrac{1}{k'\dfrac{W}{L}A2V_{ov}R_L}} = \frac{V_i}{1 + \dfrac{1}{2Ag_mR_L}} \simeq V_i\left(1 - \frac{1}{2Ag_mR_L}\right) \quad (5.117)$$

where $g_m = k'(W/L)V_{ov}$ as shown in (1.180). The term $(2Ag_mR_L)$ is the gain around the feedback loop or the *loop gain* and is usually chosen to be high enough to make the slope of the transfer characteristic to be unity within an allowable gain error. (The concept of loop gain is described in Chapter 8.) The gain error here is approximately $1/(2Ag_mR_L)$. The key point is that the error is reduced if A, g_m, or R_L are increased.

With nonzero offsets, (5.116) shows that the circuit also displays an offset error. If $A(V_{OSP} - V_{OSN}) \ll 2V_{ov}$ and $2Ag_mR_L \gg 1$,

$$V_o \simeq \frac{V_i - \frac{V_{OSP} + V_{OSN}}{2}}{1 + \frac{1}{k'\frac{W}{L}A2V_{ov}R_L}} = \frac{V_i - \frac{V_{OSP} + V_{OSN}}{2}}{1 + \frac{1}{2Ag_mR_L}} \simeq V_i - \frac{V_{OSP} + V_{OSN}}{2} \quad (5.118)$$

Therefore, the input offset voltage of the buffer is about $-(V_{OSP} + V_{OSN})/2$.

Equation 5.116 is valid as long as both M_1 and M_2 operate in the active region. If the magnitude of the output voltage is large enough, however, one of the two output transistors turns off. For example, when V_i increases, V_o also increases but the gain is slightly less than unity. As a result, the differential inputs to the error amplifiers both decrease, decreasing V_{gs1} and V_{gs2}. In turn, these changes increase $|I_{d1}|$ but reduce I_{d2}, and M_2 turns off for large enough V_i. To find the portion of the transfer characteristic with M_1 in the active region but M_2 off, the above analysis can be repeated with $I_{d2} = 0$. See Problem 5.23.

The primary motivation for using the quasi-complementary configuration is to increase the output swing. If the output transistors are not allowed to operate in the triode region, the output voltage can pull within an overdrive of either supply. This result is an improvement compared to the limit given in (5.103) for the common-drain output stage mostly because the threshold voltages of the output transistors do not limit the output swing in the common-source configuration.

Although quasi-complementary circuits improve the output swing, they suffer from two main problems. First, the error amplifiers must have large bandwidth to prevent crossover distortion problems for high input frequencies. Unfortunately, increasing the bandwidth of the error amplifiers worsens the stability margins, especially in the presence of large capacitive loads. As a result, these circuits present difficult design problems in compensation. The topics of stability and compensation are covered in Chapter 9. Second, nonzero offset voltages in the error amplifiers change the quiescent current flowing in the output transistors. From a design standpoint, the quiescent current is chosen to be barely high enough to limit crossover distortion to an acceptable level. Although further increases in the quiescent current reduce the crossover distortion, such increases also increase the power dissipation and reduce the output swing. Therefore, proper control of the quiescent current with nonzero offsets is also a key design constraint.

One way to control the quiescent current is to sense and feedback a copy of the current.[12] This method is not considered further here. Another way to limit the variation in the quiescent current is to design the error amplifiers to have low gain.[13,14] The concept is that the quiescent current is controlled by the gate-source voltages on the output transistors, which in turn are controlled by the outputs of the error amplifiers. Therefore, reducing the error-amplifier gain reduces the variation of gate-source voltages and the quiescent current for a given variation in the offset voltages.

To study this situation quantitatively, define the quiescent current in the output devices as the common-mode component of the current flowing from V_{DD} to $-V_{SS}$ with $V_i = 0$. Then

$$I_Q = \frac{I_{D2} - I_{D1}}{2} \quad (5.119)$$

Subtraction is used in the above equation because the drain current of each transistor is defined as positive when it flows into the transistor. Substituting (5.110)–(5.113) into (5.119) gives

$$I_Q = \frac{k'}{4}\frac{W}{L}\left((V_{ov} + A[V_o + V_{OSN}])^2 + (-V_{ov} + A[V_o + V_{OSP}])^2\right) \quad (5.120)$$

Since $V_o = 0$ if $V_{OSP} = V_{OSN} = 0$, (5.120) shows that

$$I_Q\Big|_{\substack{V_{OSP}=0\\V_{OSN}=0}} = \frac{k'}{4}\frac{W}{L}\left((V_{ov})^2 + (-V_{ov})^2\right) = \frac{k'}{2}\frac{W}{L}(V_{ov})^2 \quad (5.121)$$

From (5.118) with $V_i = 0$,

$$V_o + V_{OSP} \simeq \frac{V_{OSP} - V_{OSN}}{2} \quad (5.122)$$

$$V_o + V_{OSN} \simeq -\frac{V_{OSP} - V_{OSN}}{2} \quad (5.123)$$

Substituting (5.122) and (5.123) into (5.120) gives

$$I_Q = \frac{k'}{2}\frac{W}{L}\left(V_{ov} - A\left[\frac{V_{OSP} - V_{OSN}}{2}\right]\right)^2 \quad (5.124)$$

Define ΔI_Q as the change in I_Q caused by nonzero offsets; that is,

$$\Delta I_Q = I_Q\Big|_{\substack{V_{OSP}=0\\V_{OSN}=0}} - I_Q \quad (5.125)$$

Substituting (5.124) and (5.121) into (5.125) gives

$$\Delta I_Q = \frac{k'}{2}\frac{W}{L}A(V_{OSP} - V_{OSN})\left[V_{ov} - A\left(\frac{V_{OSP} - V_{OSN}}{4}\right)\right] \quad (5.126)$$

To evaluate the magnitude of ΔI_Q, we will compare it to the quiescent current with zero offsets by dividing (5.126) by (5.121). The result is

$$\frac{\Delta I_Q}{I_Q\Big|_{\substack{V_{OSP}=0\\V_{OSN}=0}}} = A\left(\frac{V_{OSP} - V_{OSN}}{V_{ov}}\right)\left[1 - A\left(\frac{V_{OSP} - V_{OSN}}{4V_{ov}}\right)\right] \quad (5.127)$$

If $A(V_{OSP} - V_{OSN}) \ll 4V_{ov}$,

$$\frac{\Delta I_Q}{I_Q\Big|_{\substack{V_{OSP}=0\\V_{OSN}=0}}} \simeq A\left(\frac{V_{OSP} - V_{OSN}}{V_{ov}}\right) \quad (5.128)$$

Therefore, to keep the fractional change in the quiescent current less than a given amount, the maximum error-amplifier gain is

$$A < \left(\frac{V_{ov}}{V_{OSP} - V_{OSN}}\right)\left(\frac{\Delta I_Q}{I_Q\Big|_{\substack{V_{OSP}=0\\V_{OSN}=0}}}\right) \quad (5.129)$$

For example, if $V_{ov} = 200$ mV, $V_{OSP} - V_{OSN} = 5$ mV, and up to 20 percent variation in the quiescent current is allowed, (5.129) shows that the error amplifier gain should be less than about 8.[13,14]

Figure 5.35 shows a schematic of the top error amplifier and M_1 from Fig. 5.32.[14] A complementary structure used to drive M_2 is not shown. The difference between V_i and V_o is sensed by the differential pair M_{11} and M_{12}, which is biased by the tail current source I_{TAIL}. The load of the differential pair consists of two parts: current mirror M_{13} and M_{14} and common-drain transistors M_{15} and M_{16}. The purpose of the common-drain transistors is to reduce the

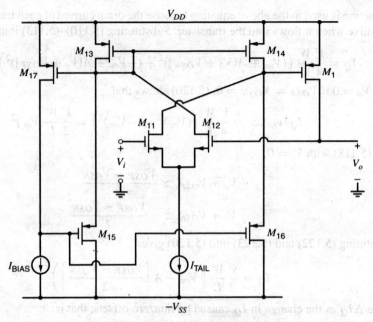

Figure 5.35 Schematic of the top error amplifier and output transistor M_1.

output resistance of the error amplifier to set its gain to a well-defined low value. The gates of the common-drain transistors are biased by a negative feedback loop including M_{13}, M_{17}, I_{BIAS}, and M_{15}. This circuit adjusts the voltage at the gate of M_{15} so that M_{17} operates in the active region and conducts I_{BIAS}. Although negative feedback is studied in Chapter 8, the basic idea can be understood here as follows. If $|I_{D17}|$ is less than I_{BIAS}, current source I_{BIAS} pulls the gate voltage of M_{15} down. Since M_{15} operates as a source follower, the source of M_{15} is pulled down, increasing $|I_{D13}|$. Because M_{13} and M_{17} together form a current mirror, $|I_{D17}|$ also increases until $|I_{D17}| = I_{BIAS}$. Similar reasoning shows that this equality is established when $|I_{D17}|$ is initially greater than I_{BIAS}. If M_{15} and M_{17} are enhancement-mode devices, M_{17} operates in the active region because $V_{GD17} = V_{SG15} = |V_{t15}| + |V_{ov15}| > 0 > V_{t17}$; therefore, the channel does not exist at the drain of M_{17}.

Since M_{13} and M_{17} form a current mirror,

$$I_{D13} = I_{D17}\frac{(W/L)_{13}}{(W/L)_{17}} = -I_{BIAS}\frac{(W/L)_{13}}{(W/L)_{17}} \quad (5.130)$$

Since M_{13} and M_{14} also form a current mirror, and since $(W/L)_{14} = (W/L)_{13}$,

$$I_{D14} = I_{D13} = -I_{BIAS}\frac{(W/L)_{13}}{(W/L)_{17}} \quad (5.131)$$

With $V_i = V_o$,

$$I_{D11} = I_{D12} = \frac{I_{TAIL}}{2} \quad (5.132)$$

From KCL,

$$I_{D16} = I_{D14} + I_{D11} \quad (5.133)$$

Substituting (5.131) and (5.132) into (5.133) gives

$$I_{D16} = -I_{BIAS}\frac{(W/L)_{13}}{(W/L)_{17}} + \frac{I_{TAIL}}{2} \quad (5.134)$$

Since $I_{D15} = I_{D16}$ when $I_{D11} = I_{D12}$,

$$V_{SD14} = V_{SD13} = V_{SG13} = V_{SG1} \quad (5.135)$$

Therefore, ignoring channel-length modulation,

$$I_{D1} = I_{D13}\frac{(W/L)_1}{(W/L)_{13}} \quad (5.136)$$

Substituting (5.130) into (5.136) and rearranging gives

$$I_{D1} = -I_{BIAS}\frac{(W/L)_1}{(W/L)_{17}} \quad (5.137)$$

This equation shows that the drain current in M_1 is controlled by I_{BIAS} and a ratio of transistor sizes if the offset voltage of the error amplifier is zero so that $V_o = 0$ when $V_i = 0$. In practice, $(W/L)_1 \gg (W/L)_{17}$ so that little power is dissipated in the bias circuits.

Another design consideration comes out of these equations. To keep the gain of the error amplifier low under all conditions, M_{16} must never cut off. Therefore, from (5.133), $|I_{D14}|$ should be greater than the maximum value of I_{D11}. Since the maximum value of I_{D11} is I_{TAIL}, (5.131) and (5.133) show that

$$|I_{D14}| = I_{BIAS}\frac{(W/L)_{13}}{(W/L)_{17}} > I_{TAIL}. \quad (5.138)$$

To find the gain of the error amplifier, the key observation is that the small-signal resistance looking into the source of of M_{15} is zero, ignoring channel-length modulation. This result stems from the operation of the same negative feedback loop that biases the gate of M_{15}. If the small-signal voltage at the source of M_{15} changes, the negative feedback loop works to undo the change. For example, suppose that the source voltage of M_{15} increases. This change reduces the gate voltage of M_{15} because M_{17} operates as a common-source amplifier. Then the source voltage of M_{15} falls because M_{15} operates as a source follower. Ignoring channel-length modulation, the source voltage of M_{15} must be held exactly constant because $i_{d17} = 0$ if I_{BIAS} is constant. Therefore, the small-signal resistance looking into the source of M_{15} is zero. As a result, none of the small-signal drain current from M_{12} flows in M_{13}. Instead, it all flows in the source of M_{15}. To calculate the transconductance of the error amplifier, the output at the drain of M_{16} is connected to a small-signal ground. Since M_{15} and M_{16} share the same gate connection, and since their sources each operate at small-signal grounds, the small-signal current in M_{15} is copied to M_{16}. Therefore, the entire small-signal current from the differential pair flows at the error-amplifier output and the transconductance is

$$G_m = g_{m11} = g_{m12} \quad (5.139)$$

Ignoring channel-length modulation, the output resistance is set by common-drain transistor M_{16}. From (3.84),

$$R_o = \frac{1}{g_{m16} + g_{mb16}} \quad (5.140)$$

Therefore, the gain of the error amplifier is

$$A = G_m R_o = \frac{g_{m11}}{g_{m16} + g_{mb16}} \quad (5.141)$$

5.5.3 Alternative Configurations

The main potential advantage of the common-source output stage described in the last section is that it can increase the output swing compared to the common-drain case. However, the common-source configuration suffers from an increase in harmonic distortion, especially at high frequencies, for two main reasons. First, the bandwidth of the error amplifiers is usually limited, to avoid stability problems. Second, the gain of these amplifiers is limited to establish adequate control on the quiescent output current.

5.5.3.1 Combined Common-Drain Common-Source Configuration

One way to overcome this problem is to use a combination of the common-drain and common-source configurations shown in Fig. 5.31 and Fig. 5.32.[15] The combined schematic is shown in Fig. 5.36. The key aspect of this circuit is that it uses two buffers connected to the output: a Class AB common-drain buffer (M_1-M_2) and a Class B quasi-complementary common-source buffer (M_{11}-M_{12} and the error amplifiers A). The common-source buffer is dominant when the output swing is maximum, but off with zero output. On the other hand, the common-drain buffer controls the quiescent output current and improves the frequency response, as described next.

From KVL,

$$V_o = V_1 + V_{gs4} - V_{gs1} \tag{5.142}$$

If $V_{t1} = V_{t4}$, this equation can be rewritten as

$$V_o = V_1 + V_{ov4} - V_{ov1} \tag{5.143}$$

Therefore, when V_i is adjusted so that $V_1 = 0$, M_1-M_5 force $V_o = 0$ if $V_{ov1} = V_{ov4}$. From (1.166), $V_{ov1} = V_{ov4}$ if

$$\frac{I_{D1}}{(W/L)_1} = \frac{I_{D4}}{(W/L)_4} \tag{5.144}$$

Substituting (5.101) into (5.144) and rearranging shows that $V_{ov1} = V_{ov4}$ if

$$\frac{(W/L)_1}{(W/L)_2} = \frac{(W/L)_4}{(W/L)_5} \tag{5.145}$$

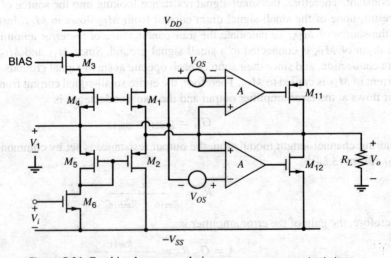

Figure 5.36 Combined common-drain, common-source output stage.

We will assume that this condition holds so that $V_o = 0$ when $V_1 = 0$. In this case, M_{11} and M_{12} are designed to be cut off. This characteristic stems from small offsets designed into the error amplifiers. These offsets are shown as voltage sources V_{OS} in Fig. 5.36 and can be introduced by intentionally mismatching the input differential pair in each error amplifier. With $V_1 = V_o = 0$ and $V_{OS} > 0$, the error amplifiers are designed to give $V_{gs11} > V_{t11}$ and $V_{gs12} < V_{t12}$ so that M_{11} and M_{12} are off. As a result, the quiescent output current is controlled by the common-drain stage and its biasing circuit M_1-M_5, as shown in (5.101).

As V_i decreases from the value that forces $V_1 = 0$, V_1 increases and V_o follows but with less than unity gain if R_L is finite. Therefore $V_1 - V_o$ increases, and both V_{gs11} and V_{gs12} decrease, eventually turning on M_{11} but keeping M_{12} off. After M_{11} turns on, both I_{d1} and $|I_{d11}|$ increase as V_o rises until $V_{gs1} - V_{t1}$ reaches its maximum value. Such a maximum occurs when the output swing from the common-source stage is greater than that of the common-drain stage. As V_o rises beyond this point, $|I_{d11}|$ increases but I_{d1} decreases, and the common-source stage becomes dominant.

From (5.103), the output swing allowed by the common-drain stage in Fig. 5.31 is limited in part by V_{gs1}, which includes a threshold component. On the other hand, the output swing limitation of common-source stage in Fig. 5.32 does not include a threshold term, and this circuit can swing within an overdrive of the positive supply. So with proper design, we expect the common-source stage to have a larger output swing than the common-drain stage. When the two circuits are combined as in Fig. 5.36, however, the output swing is limited by the driver stage that produces V_1. Define V_1^+ as the maximum value of V_1 for which M_3 operates in the active region. Then

$$V_1^+ = V_{DD} - |V_{ov3}| - V_{gs4} \quad (5.146)$$

Since V_1 is the input to the common-source stage, the maximum output V_o^+ can be designed to be

$$V_o^+ \simeq V_1^+ = V_{DD} - |V_{ov3}| - V_{gs4} \quad (5.147)$$

Comparing (5.147) with (5.103) shows that the positive swing of the combined output stage in Fig. 5.36 exceeds that of the common-drain output stage in Fig. 5.31 provided that $V_{gs4} < V_{gs1}$ when $V_o = V_o^+$. This condition is usually satisfied because $I_{d4} \ll I_{d1}$ when the output is maximum with finite R_L. Therefore, the circuit in Fig. 5.36 can be designed to increase the output swing.

Since the common-source stage in Fig. 5.36 is not responsible for establishing the quiescent output current, the gain of the error amplifiers in Fig. 5.36 is not limited as in (5.129). In practice, the error amplifiers are often designed as one-stage amplifiers with a gain related to the product of g_m and r_o that can be achieved in a given technology. This increase in gain reduces the harmonic distortion because it reduces the error between the input and the output of the common-source stage.

Finally, we will consider the frequency response of the circuit in Fig. 5.36 qualitatively. The circuit has two paths from V_1 to the output. The path through the common-source transistors M_{11} and M_{12} may be slow because of the need to limit the bandwidth of the error amplifiers to guarantee that the circuit is stable. (Stability is studied in Chapter 9.) On the other hand, the path through the common-drain transistors M_1 and M_2 is fast because source followers are high-bandwidth circuits, as shown in Chapter 7. Since the circuit sums the current from the common-drain and common-source stages in the load to produce the output voltage, the fast path will dominate for high-frequency signals. This technique is called *feedforward,* and other instances are described in Chapter 9. It causes the circuit to take on the characteristics of the source followers for high frequencies, reducing the phase shift that would otherwise be introduced by the slow error amplifiers. As a result, the design required

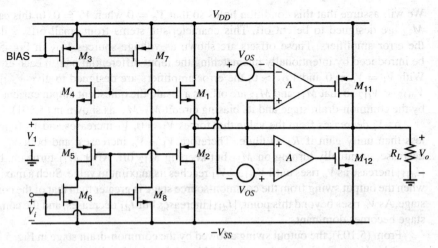

Figure 5.37 Combined common-drain, common-source output stage with improved swing.

to guarantee stability is simplified,[15] and the harmonic distortion for high-frequency signals is reduced.

5.5.3.2 Combined Common-Drain Common-Source Configuration with High Swing

Although the swing of the circuit in Fig. 5.36 is improved compared to the circuit in Fig. 5.31, it can be improved even further. As shown in (5.147), the main limitation to the positive swing in Fig. 5.36 stems from V_{gs4}. This voltage includes a threshold component, which increases with increasing V_1 and V_o because of the body effect. Similarly, the negative swing is limited by V_{gs5}, whose threshold component increases in magnitude as V_1 and V_o decrease. In practice, these terms alone reduce the available output swing by about 1.5 V to 2 V.

Figure 5.37 shows a circuit that overcomes this limitation.[16] The circuit is the same as in Fig. 5.36 except that one extra branch is included. The new branch consists of transistors M_7 and M_8 and operates in parallel with the branch containing M_3-M_6 to produce voltage V_1. The swing of V_1 in Fig. 5.36 is limited by the threshold voltages of M_4 and M_5 as described above. In contrast, the new branch in Fig. 5.37 can drive V_1 within an overdrive of either supply while M_7 and M_8 operate in the active region. Since the output swing in Fig. 5.36 is limited by the swing of V_1, improving the swing of V_1 as in Fig. 5.37 also improves the output swing.

5.5.3.3 Parallel Common-Source Configuration

Another circuit that overcomes the problem described in the introduction of Section 5.5.3 is shown in Fig. 5.38.[17] Like the circuit in Fig. 5.37, this circuit combines two buffers in parallel at the output. The error amplifiers with gain A_1 along with M_1 and M_2 form one buffer, which controls the operation of the output stage with $V_i = 0$. The error amplifiers with gain A_2 together with M_{11} and M_{12} form the other buffer, which dominates the operation of the output stage for large-magnitude output voltages.

This behavior stems from small offset voltages intentionally built into the A_1 amplifiers. These offsets are shown as voltage sources V_{OS} in Fig. 5.38 and are introduced in practice by intentionally mismatching the input differential pairs in the A_1 amplifiers. At first, assume that these offsets have little effect on the drain currents of M_1 and M_2 because A_1 is intentionally chosen to be small. Therefore, M_1 and M_2 operate in the active region when $V_i = 0$, and the buffer that includes these transistors operates in a Class AB mode. On the other hand, the offsets force M_{11} and M_{12} to operate in cutoff when $V_i = 0$ because the gates of these transistors

5.5 CMOS Class AB Output Stages

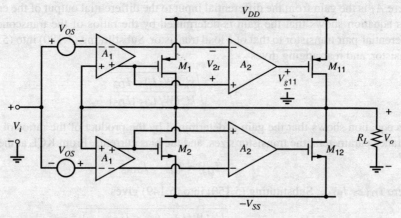

Figure 5.38 Output stage with a Class AB common-source buffer and a Class B common-source buffer.

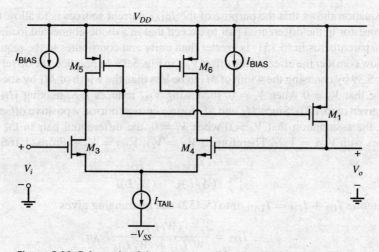

Figure 5.39 Schematic of the top A_1 amplifier and output transistor M_1.

are driven by the outputs of the A_2 amplifiers, which in turn are driven by the differential outputs of the A_1 amplifiers. In particular, the product $V_{OS}A_1A_2$ is chosen by design to be big enough to force $V_{gs11} > V_{t11}$ and $V_{gs12} < V_{t12}$ so that M_{11} and M_{12} are off when $V_i = 0$. As a result, the quiescent output current of this output stage is controlled by M_1, M_2, and the A_1 amplifiers.

Figure 5.39 shows a schematic of the top A_1 amplifier and M_1 from Fig. 5.38.[17] A complementary configuration is used for the bottom A_1 amplifier but is not shown for simplicity. The difference between V_i and V_o is sensed by the differential pair M_3 and M_4. The load of the differential pair consists of two parts: diode-connected transistors M_5 and M_6 and current sources I_{BIAS}, which are implemented by the outputs of p-channel current mirrors in practice. The purpose of the diode-connected transistors is to limit the gain of the error amplifier to a small value so that the output quiescent current is well controlled. Ignoring channel-length modulation, the gain of this error amplifier is

$$A_1 = \frac{g_{m3}}{g_{m5}} \qquad (5.148)$$

where A_1 is the gain from the differential input to the differential output of the error amplifier. This equation shows that the gain is determined by the ratios of the transconductance of a differential-pair transistor to that of a load transistor. Substituting (1.180) into (5.148) for each transistor, and rearranging gives

$$A_1 = \sqrt{\frac{k'_n}{k'_p} \frac{(W/L)_3}{(W/L)_5} \frac{I_{D3}}{|I_{D5}|}} \qquad (5.149)$$

This equation shows that the gain is determined by the product of the ratios of the transconductance parameters, the transistor sizes, and the bias currents. From KCL at the drain of M_5,

$$-I_{D5} = I_{D3} - I_{\text{BIAS}} \qquad (5.150)$$

where $I_{D3} > I_{\text{BIAS}}$. Substituting (5.150) into (5.149) gives

$$A_1 = \sqrt{\frac{k'_n}{k'_p} \frac{(W/L)_3}{(W/L)_5} \left(\frac{I_{D3}}{I_{D3} - I_{\text{BIAS}}}\right)} \qquad (5.151)$$

This equation shows that the purpose of the I_{BIAS} current sources is to allow the bias current in a transistor in the differential pair to exceed that in a diode-connected load. As a result, the term in parentheses in (5.151) is greater than unity and contributes to the required gain.

Now consider the effect of the offset V_{OS} in Fig. 5.38. In practice, the offset is implemented in Fig. 5.39 by choosing the width of M_3 to be less than the width of M_4 by about 20 percent.[17] Assume that $V_o = 0$ when $V_i = 0$. Increasing V_{OS} reduces I_{D3}, making $|I_{D5}|$ less than the value given in (5.150). Since M_5 and M_1 form a current mirror, a positive offset reduces $|I_{D1}|$. Under the assumption that $V_o = 0$ when $V_i = 0$, the differential pair in the error amplifier operates with $V_{gs3} = V_{gs4}$. Therefore, if $V_{t3} = V_{t4}$, $V_{ov3} = V_{ov4}$. From (1.166),

$$\frac{I_{D3}}{(W/L)_3} = \frac{I_{D4}}{(W/L)_4} \qquad (5.152)$$

Substituting $I_{D3} + I_{D4} = I_{\text{TAIL}}$ into (5.152) and rearranging gives

$$I_{D3} = \frac{(W/L)_3}{(W/L)_3 + (W/L)_4} I_{\text{TAIL}} \qquad (5.153)$$

when $V_o = V_i = 0$. Similarly, a positive offset in the bottom A_1 amplifier as labeled in Fig. 5.38 reduces I_{D2}. If the tail and bias current in the top A_1 amplifier are identical to the corresponding values in the bottom A_1 amplifier, and if the fractional mismatches in the differential pairs are identical, the reductions in $|I_{D1}|$ and I_{D2} caused by the mismatches intentionally introduced into the differential pairs are equal and $V_o = 0$ when $V_i = 0$ is assumed. In other words, the offset of the entire output stage shown in Fig. 5.38 is zero even with nonzero V_{OS} in the error amplifiers.

Because a design goal is to bias M_1 in the active region when $V_i = 0$, the offset must be chosen to be small enough that $I_{D3} > 0$ when $V_o = V_i = 0$. Also, the error amplifiers contain some unintentional mismatches stemming from random effects in practice. Because another design goal is to bias M_{11} in cutoff, the random component must not be allowed to be larger than the systematic offset in magnitude and opposite in polarity. Therefore, V_{OS} is chosen to be bigger than the expected random offset.

Since the quiescent output current is controlled by the A_1 error amplifiers along with M_1 and M_2, the gain of the A_2 error amplifiers driving M_{11} and M_{12} need not be small. With large A_2, M_{11} or M_{12} becomes the dominant output device for large output magnitudes if the aspect ratios of M_{11} and M_{12} are at least as big as M_1 and M_2, respectively. Furthermore, increasing A_2 has the advantage of increasing the loop gain when M_{11} or M_{12} turns on. This loop gain

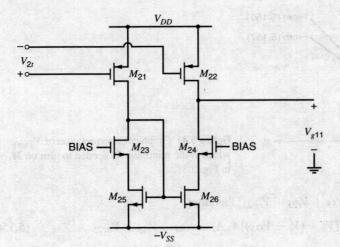

Figure 5.40 Schematic of the top A_2 amplifier.

is related to the product of A_2, g_{m11} or g_{m12}, and R_L. Increasing the loop gain reduces the error between the input and output, as shown in Chapter 8. If M_{11} conducts, increasing A_2 allows the output stage to drive reduced loads with constant g_{m11} and error. If the load is fixed, increasing A_2 allows the transconductance to be reduced, which in turn allows $(W/L)_{11}$ to be reduced. One potential concern here is that reducing the transistor sizes also reduces the range of outputs for which the devices operate in the active region. If M_{11} operates in the triode region, its drain-source resistance r_{o11} is finite, and the loop gain is proportional to the product $A_2 g_{m11}(r_{o11} \parallel R_L)$. Thus, operation of M_{11} in the triode region increases the error by reducing g_{m11} and r_{o11}. However, increasing A_2 compensates for this effect. Therefore, a key advantage of the output-stage configuration shown in Fig. 5.38 is that it allows the output swing with a given level of nonlinearity to be increased by allowing the dominant transistor to operate in the triode region.

The A_1 and A_2 amplifiers together form the two-stage error amplifiers that drive M_{11} and M_{12}. The A_2 amplifiers operate on the differential outputs of the A_1 amplifiers. Since the common-mode components of the outputs of the A_1 amplifiers are well controlled by diode-connected loads, differential pairs are not required at the inputs of the A_2 amplifiers. Figure 5.40 shows a schematic of the top A_2 amplifier.[17] A complementary configuration used for the bottom A_2 amplifier is not shown for simplicity. The inputs are applied to the gates of common-source transistors M_{21} and M_{22}. The cascode current mirror M_{23}-M_{26} then converts the differential signal into a single-ended output. Because the output resistance of the cascode-current mirror is large compared to the output resistance of the common-source transistor M_{22},

$$A_2 \simeq g_{m22} r_{o22} \qquad (5.154)$$

and $A_2 \simeq 70$ in Ref.17.

To determine the range of input voltages for which M_{11} and M_{12} are both off, let V_{g11} represent the voltage from the gate of M_{11} to ground. Assuming that the gains A_1 and A_2 are constant,

$$V_{g11} = [V_o - (V_i - V_{OS})]A_1 A_2 + K \qquad (5.155)$$

where K is a constant. If the differential input voltage to the top A_2 amplifier shown in Fig. 5.40 is zero, $V_{g11} = -V_{SS} + V_{t25} + V_{ov25}$ so that $I_{d21} = I_{d22}$. Substituting this boundary condition

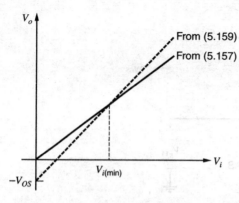

Figure 5.41 Graphical interpretation of $V_{i(\min)}$, which is the minimum V_i needed to turn on M_{11} in Fig. 5.38.

into (5.155) gives $K = -V_{SS} + V_{t25} + V_{ov25}$; therefore,

$$V_{g11} = [V_o - (V_i - V_{OS})]A_1 A_2 - V_{SS} + V_{t25} + V_{ov25} \quad (5.156)$$

Also, (5.117) with $g_m = g_{m1} = g_{m2}$ gives V_o in terms of V_i for the output stage in Fig. 5.38 as long as the random offset is negligible, both M_1 and M_2 operate in the active region, and M_{11} and M_{12} are off. Since the gates of M_1 and M_2 are each driven by only one output of the corresponding A_1 amplifiers, $A = A_1/2$. With these substitutions, (5.117) gives

$$V_o = \frac{V_i}{1 + \dfrac{1}{A_1 g_{m1} R_L}} \quad (5.157)$$

To turn M_{11} on, $V_{g11} < V_{DD} - |V_{t11}|$. Substituting this condition and (5.157) into (5.156) gives

$$V_{i(\min)} = V_{OS}(1 + A_1 g_{m1} R_L) - \frac{(V_{DD} + V_{SS} - |V_{t11}| - V_{t25} - V_{ov25})(1 + A_1 g_{m1} R_L)}{A_1 A_2} \quad (5.158)$$

where $V_{i(\min)}$ is the minimum value of V_i for which M_{11} conducts nonzero drain current. To interpret this result, let $A_2 \to \infty$. Then to turn M_{11} on, the required differential input voltage of the top A_2 amplifier $V_{2t} = 0$. Therefore, the required differential input of the top A_1 amplifier is zero; that is,

$$V_o = V_i - V_{OS} \quad (5.159)$$

This equation and (5.157) are both plotted in Fig. 5.41. As the input voltage increases, the output voltage follows with a slope less than unity if R_L is finite, as shown by the solid plot. Therefore, as V_i increases, $V_i - V_o$ also increases. To turn M_{11} barely on, this difference must be equal to V_{OS} so that the circuit operates at the intersection of the two lines in Fig. 5.41. Substituting (5.157) into (5.159) gives

$$V_{i(\min)} = V_{OS}(1 + A_1 g_{m1} R_L) \quad (5.160)$$

This equation agrees with the result that would be obtained by allowing $A_2 \to \infty$ in (5.158). The term in parentheses in (5.160) is equal to the reciprocal of the difference in the slopes of the two lines in Fig. 5.41.

■ **EXAMPLE**

Find the minimum input voltage in Fig. 5.38 for which M_{11} turns on, assuming at first that $A_2 \to \infty$ and then that $A_2 = 70$. Let $V_{OS} = 10$ mV, $A_1 = 8$, $g_{m1} = 5$ mA/V, $R_L = 60$ Ω, $V_{DD} = V_{SS} = 2.5$ V, $V_{t11} = -0.7$ V, $V_{t25} = 0.7$ V, and $V_{ov25} = 0.1$ V.

From (5.160),
$$V_{i(\min)} = 10\text{ mV}[1 + 8(0.005)(60)] = 34\text{ mV}$$
when $A_2 \to \infty$. On the other hand, when $A_2 = 70$, (5.158) shows that
$$V_{i(\min)} = 10\text{ mV}[1 + 8(0.005)(60)] - \frac{3.5[1 + 8(0.005)(60)]}{8(70)} \simeq 0.13\text{ mV}$$

This example shows that the minimum input voltage required to turn on M_{11} is reduced from the value given in (5.160) when A_2 is finite, because the fractional term in (5.158) is positive. ∎

The key point of this analysis is that M_{11} and M_{12} in Fig. 5.38 remain off for only a small range of input voltages. Therefore, the nonlinearity introduced by turning on M_{11} or M_{12} occurs when $|V_i|$ is small. As a result, this circuit is well suited for the ISDN (Integrated Service Digital Network) line-driving application for which it was designed because the required four-level output code does not include zero, avoiding distortion that would be introduced by turning M_{11} and M_{12} on or off[17] if a zero-level output pulse were required.

PROBLEMS

5.1 A circuit as shown in Fig. 5.1 has $V_{CC} = 15\text{V}$, $R_1 = R_2 = 0$, $R_3 = 5\text{ k}\Omega$, $R_L = 2\text{k}\Omega$, $V_{CE(\text{sat})} = 0.2\text{ V}$, and $V_{BE(\text{on})} = 0.7\text{ V}$. All device areas are equal.

(a) Sketch the transfer characteristic from V_i to V_o.

(b) Repeat (a) if $R_L = 10\text{ k}\Omega$.

(c) Sketch the waveform of V_o if a sinusoidal input voltage with an amplitude (zero to peak) of 10 V is applied at V_i in (a) and (b) above.

(d) Use SPICE to verify (a), (b), and (c) and also to determine second and third harmonic distortion in V_o for the conditions in (c).

5.2 (a) Prove that any load line tangent to a power hyperbola makes contact with the hyperbola at the midpoint of the load line.

(b) Calculate the maximum possible instantaneous power dissipation in Q_1 for the circuit of Problem 5.1 with $R_L = 2\text{ k}\Omega$ and $R_L = 10\text{ k}\Omega$.

(c) Calculate the *average* power dissipated in Q_1 for the circuit of Problem 5.1 with $R_L = 2\text{ k}\Omega$ and $R_L = 10\text{ k}\Omega$. Assume that V_o is sinusoidal with an amplitude equal to the maximum possible before clipping occurs.

5.3 (a) For the circuit of Problem 5.1, sketch load lines in the I_c-V_{ce} plane for $R_L = 2\text{ k}\Omega$ and $R_L = 10\text{ k}\Omega$.

(b) Calculate the maximum average sinusoidal output power that can be delivered to R_L (both values) before clipping occurs in (a) above. Sketch corresponding waveforms for I_{c1}, V_{ce1}, and P_{c1}.

(c) Calculate the circuit efficiency for each value of R_L in (b). (Neglect power dissipated in Q_3 and R_3.)

(d) Select R_L for maximum efficiency in this circuit and calculate the corresponding average output power with sinusoidal signals.

5.4 If $\beta_F = 50$ for Q_1 in Problem 5.1, calculate the average signal power delivered to Q_1 by its driver stage if V_o is sinusoidal with an amplitude equal to the maximum possible before clipping occurs. Repeat for $R_L = 10\text{ k}\Omega$. Thus calculate the power gain of the circuit.

5.5 Calculate the incremental slope of the transfer characteristic of the circuit of Problem 5.1 at the quiescent point and at the extremes of the signal swing with a peak sinusoidal output of 1 V and $R_L = 10\text{ k}\Omega$.

5.6 (a) For the circuit of Problem 5.1, draw load lines in the I_c-V_{ce} plane for $R_L = 0$ and $R_L \to \infty$. Use an I_c scale from 0 to 30 mA. Also draw constant power hyperbolas for $P_c = 0.1$ W, 0.2 W, and 0.3 W. What is the maximum possible instantaneous power dissipation in Q_1 for the above values of R_L? Assume that the driver stage can supply a maximum base current to Q_1 of 0.3 mA and $\beta_F = 100$ for Q_1.

(b) If the maximum allowable instantaneous power dissipation in Q_1 is 0.2 W, calculate the minimum allowable value of R_L. (A graphical solution is the easiest.)

5.7 Calculate the incremental slope of the transfer characteristic of the circuit of Fig. 5.8 at the quiescent point and at the extremes of the signal swing with $v_i = \hat{v}_i \sin \omega t$ and $\hat{v}_i = 0.5$ V:

(a) Let $A_v^+ = A_v$ when v_i is maximum.

(b) Let $A_{vQ} = A_v$ when $v_i = 0$.

(c) Let $A_v^- = A_v$ when v_i is minimum.

Assume that $V_I = 0$, $V_{DD} = 2.5$ V, $I_Q = 1$ mA, and $R_L \to \infty$. Also, assume that $(W/L)_1 = 1000$, $k' = 200$ μA/V^2, $V_{t0} = 0.7$ V, $\phi_f = 0.3$ V, and $\gamma = 0.5$ V$^{1/2}$.

5.8 When the distortion is small, the second and third harmonic-distortion terms of an amplifier can be calculated from the small-signal gains at the quiescent and extreme operating points. Starting with the power series given in (5.41),

(a) Calculate an expression for the small-signal gain $A_v = dv_o/dv_i$.

(b) Let $v_i = \hat{v}_i \sin \omega t$ as in (5.52), and derive expressions for A_v^+, A_{vQ}, and A_v^- as defined in Problem 5.7.

(c) Define two normalized differential gain error terms as

(i) $E^+ = (A_v^+ - A_{vQ})/A_{vQ}$

(ii) $E^- = (A_v^- - A_{vQ})/A_{vQ}$

and calculate expressions for $(E^+ + E^-)$ and $(E^+ - E^-)$.

(d) Compare the results of part (c) with (5.54) and (5.57) to calculate HD_2 and HD_3 in terms of E^+ and E^-.

(e) Use the results of part (d) and Problem 5.7 to calculate HD_2 and HD_3 for the circuit of Fig. 5.8 under the conditions given in Problem 5.7. Compare the results to the results of the Example in Section 5.3.2.

5.9 The circuit of Fig. 5.10 has $V_{CC} = 15$ V, $R_L = 2$ kΩ, $V_{BE(on)} = 0.6$ V, and $V_{CE(sat)} = 0.2$ V.

(a) Sketch the transfer characteristic from V_i to V_o assuming that the transistors turn on abruptly for $V_{be} = V_{BE(on)}$.

(b) Sketch the output voltage waveform and the collector current waveform in each device for a sinusoidal input voltage of amplitude 1 V, 10 V, 20 V.

(c) Check (a) and (b) using SPICE with $I_S = 10^{-16}$ A, $\beta_F = 100$, $r_b = 100$ Ω, and $r_c = 20$ Ω for each device. Use SPICE to determine second and third harmonic distortion in V_o for the conditions in (b).

5.10 Calculate second-harmonic distortion in the common-source amplifier with a depletion load shown in Fig. 4.20a for a peak sinusoidal input voltage $\hat{v}_i = 0.01$ V and $V_{DD} = 3$ V. Assume that the dc input voltage is adjusted so that the dc output voltage is 1 V. For simplicity, assume that the two transistors have identical parameters except for unequal threshold voltages. Let $W/L = 100$, $k' = 200$ μA/V^2, and $\lambda = 0$. Assume $V_{t1}|_{V_{SB1}=0} = 0.6$ V, $V_{t2}|_{V_{SB2}=0} = -0.6$ V, $\phi_f = 0.3$ V, and $\gamma = 0.5$ V$^{1/2}$. Use SPICE to verify the result.

5.11 For the circuit of Fig. 5.10, assume that $V_{CC} = 15$ V, $R_L = 1$ kΩ, and $V_{CE(sat)} = 0.2$ V. Assume that there is sufficient sinusoidal input voltage available at V_i to drive V_o to its limits of clipping. Calculate the maximum average power that can be delivered to R_L before clipping occurs, the corresponding efficiency, and the maximum instantaneous device dissipation. Neglect crossover distortion.

5.12 For the circuit of Problem 5.10, calculate and sketch the waveforms of I_{c1}, V_{ce1}, and P_{c1} for device Q_1 over one cycle. Do this for output voltage amplitudes (zero to peak) of 11.5 V, 6 V, and 3 V. Neglect crossover distortion and assume sinusoidal signals.

5.13 In the circuit of Fig. 5.13, $V_{CC} = 12$ V, $I_Q = 0.1$ mA, $R_L = 1$ kΩ, and for all devices $I_S = 10^{-15}$ A, $\beta_F = 150$. Calculate the value of V_i and the current in each device for $V_o = 0$, ±5 V, and ±10 V. Then sketch the transfer characteristic from $V_o = 10$ V to $V_o = -10$ V.

5.14 For the output stage of Fig. 5.18, assume that $V_{CC} = 15$ V and for all devices $V_{CE(sat)} = 0.2$ V, $V_{BE(on)} = 0.7$ V, and $\beta_F = 50$.

(a) Calculate the maximum positive and negative limits of V_o for $R_L = 10$ kΩ and $R_L = 1$ kΩ.

(b) Calculate the maximum average power that can be delivered to R_L before clipping occurs for $R_L = 10$ kΩ and $R_L = 1$ kΩ. Calculate the corresponding circuit efficiency (for the output devices only) and the average power dissipated per output device. Neglect crossover distortion and assume sinusoidal signals.

5.15 For the output stage of Fig. 5.20a, assume that $V_{CC} = 15$ V, $\beta_F(pnp) = 50$, $\beta_F(npn) = 200$, and for all devices $V_{BE(on)} = 0.7$ V, $V_{CE(sat)} = 0.2$ V, $I_S = 10^{-14}$ A. Assume that the magnitude of the collector current in Q_{13A} is 0.2 mA.

(a) Calculate the maximum positive and negative limits of V_o for $R_L = 10$ kΩ, $R_L = 1$ kΩ, and $R_L = 100$ Ω.

(b) Calculate the maximum average power that can be delivered to $R_L = 1$ kΩ before clipping occurs, and the corresponding circuit efficiency (for the output devices only). Also calculate the peak instantaneous power dissipation in each output device. Assume sinusoidal signals.

5.16 (a) For the circuit of Problem 5.15, calculate the maximum possible average output power than can be delivered to a load R_L if the instantaneous power dissipation per device must be less than 100 mW. Also

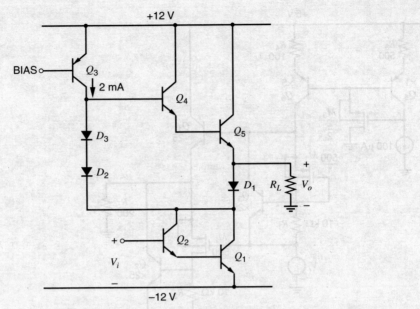

Figure 5.42 All-*npn* Darlington output stage.

specify the corresponding value of R_L and the circuit efficiency (for the output devices only). Assume sinusoidal signals.

(b) Repeat (a) if the maximum instantaneous power dissipation per device is 200 mW.

5.17 For the circuit of Problem 5.15, calculate bias currents in $Q_{23}, Q_{20}, Q_{19}, Q_{18}$, and Q_{14} for $V_o = -10$ V with $R_L = 10\,\text{k}\Omega$. Use $I_S = 10^{-14}$ A for all devices.

5.18 For the circuit of Fig. 5.25, assume that $V_{CC} = 15\,\text{V}, \beta_F(pnp) = 30, \beta_F(npn) = 150, I_S(npn) = 10^{-14}$ A, $I_S(pnp) = 10^{-15}$ A, and for all devices $V_{BE(on)} = 0.7\,\text{V}, V_{CE(sat)} = 0.2\,\text{V}$. Assume that Q_5 and Q_6 are *npn* devices and the collector current in Q_7 is 0.15 mA.

(a) Calculate the maximum positive and negative limits of V_o for $R_L = 1\,\text{k}\Omega$.

(b) Calculate quiescent currents in Q_1-Q_7 for $V_o = 0\,\text{V}$.

(c) Calculate the maximum average output power (sine wave) that can be delivered to R_L if the maximum instantaneous dissipation in any device is 100 mW. Calculate the corresponding value of R_L, and the peak currents in Q_3 and Q_4.

5.19 An all-*npn* Darlington output stage is shown in Fig. 5.42. For all devices $V_{BE(on)} = 0.7\,\text{V}, V_{CE(sat)} = 0.2\,\text{V}, \beta_F = 100$. The magnitude of the collector current in Q_3 is 2 mA.

(a) If $R_L = 8\,\Omega$, calculate the maximum positive and negative limits of V_o.

(b) Calculate the power dissipated in the circuit for $V_o = 0\,\text{V}$.

(c) Calculate the maximum average power that can be delivered to $R_L = 8\,\Omega$ before clipping occurs and the corresponding efficiency of the *complete* circuit. Also calculate the maximum instantaneous power dissipated in each output transistor. Assume that feedback is used around the circuit so that V_o is approximately sinusoidal.

(d) Use SPICE to plot the dc transfer characteristic from V_i to V_o as V_o is varied over the complete output voltage range with $R_L = 8\,\Omega$. For Q_1, Q_5, and D_1 assume $I_S = 10^{-15}$ A, $r_b = 1\,\Omega, r_c = 0.2\,\Omega, \beta_F = 100$, and $V_A = 30\,\text{V}$. Assume Q_4, Q_2, D_2, and D_3 are 1/100 the size of the large devices. For Q_3 assume $r_c = 50\,\Omega$ and $V_A = 30\,\text{V}$.

5.20 A BiCMOS Class AB output stage is shown in Fig. 5.43. Device parameters are $\beta_F(npn) = 80, \beta_F(pnp) = 20, V_{BE(on)} = 0.8\,\text{V}, \mu_p C_{ox} = 26\,\mu\text{A/V}^2$, and $V_t = -0.7\,\text{V}$.

(a) Calculate bias currents in all devices for $V_o = 0$.

(b) Calculate the positive and negative limits of V_o for $R_L = 200\,\Omega$. Thus calculate the maximum average power that can be delivered to R_L before clipping occurs.

(c) Use SPICE to check (a) and also to plot the complete dc transfer characteristic of the circuit from V_i to V_o. Also plot the waveforms of I_{c1}, I_{c2}, and I_{d2} for a sinusoidal output voltage at V_o of 2 V and then

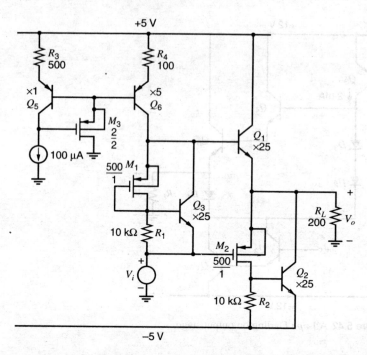

Figure 5.43 BiCMOS Class AB output stage.

4 V zero-to-peak. In the simulation, assume bipolar parameters as in Fig. 2.32 and MOS parameters as in Table 2.3 (apart from the values of β_F and $\mu_p C_{ox}$ given above).

5.21 Find the minimum output voltage for the circuit in Fig. 5.31.

5.22 Design a CMOS output stage based on the circuit of Fig. 5.31 to deliver ± 1 V before clipping at V_o with $R_L = 1$ kΩ and $V_{DD} = V_{SS} = 2.5$ V. Use 10 μA bias current in M_3 and 100 μA idling current in M_1 and M_2. Set $(W/L)_3 = 50/1$ and $(W/L)_6 = 25/1$. Specify the W/L for M_1-M_6 that minimizes the total chip area. Use the transistor parameters in Table 2.3 except assume that $L_{eff} = L_{drwn}$ for simplicity. The minimum channel length is 1 μm. Assume the body of each n-channel transistor is connected to $-V_{SS}$, and the body of each p-channel transistor is connected to V_{DD}. Use SPICE to verify your design by plotting the V_o versus V_i characteristic.

5.23 For the circuit in Fig. 5.34, assume that the input voltage V_i is high enough that M_1 operates in the active region but M_2 is cut off. Using the same assumptions as in the derivation of (5.116), show that V_o is related to V_i by the following expression

$$V_o = V_i + \frac{V_{ov}}{A} - V_{OSP} + \frac{1}{k'\dfrac{W}{L}A^2 R_L}$$

$$- \frac{1}{k'\dfrac{W}{L}A^2}\sqrt{\frac{1}{R_L^2} + 2k'\frac{W}{L}\frac{A^2}{R_L}\left(V_i + \frac{V_{ov}}{A} - V_{OSP}\right)}$$

5.24 Using a circuit that is the complement of the one in Fig. 5.35, draw the schematic for the bottom error amplifier and output transistor M_2, which are shown in block diagram form in Fig. 5.32. In the error amplifier, label the transistors as M_{21}-M_{27}, where M_{21} is the complement of M_{11}, M_{22} is the complement of M_{12}, etc. Also, label the current sources complementary to I_{BIAS} and I_{TAIL} as I_{BIASP} and I_{TAILP}, respectively.

5.25 Using the schematics from Fig. 5.35 and Problem 5.24, design the output stage shown in Fig. 5.32 to satisfy the following requirements.

(a) $V_{DD} = V_{SS} = 2.5$ V.

(b) The standby power dissipation should be no more than 70 mW.

(c) $R_L = 100$ Ω.

(d) The maximum allowed gain error with zero offsets and all transistors operating in the active region is 1 percent.

(e) In Fig. 5.35, $(W/L)_{17} = (W/L)_1/100$ and $(W/L)_{13} = (W/L)_{14} = (W/L)_1/10$. Similarly, in Problem 5.24, $(W/L)_{27} = (W/L)_2/100$ and $(W/L)_{23} = (W/L)_{24} = (W/L)_2/10$.

(f) To control the quiescent current, the maximum allowed error-amplifier gain is 5.

(g) Your solution is allowed to use four ideal current sources, two in each error amplifier. To allow these ideal current sources to be replaced by real transistors in a design step not required in this problem, the voltage across each ideal current source in Fig. 5.35 must be at least 0.5 V when $V_o \geq 0$. Similarly, the voltage across each ideal current source in the complementary circuit from Problem 5.24 must be at least 0.5 V when $V_o \leq 0$.

(h) To guarantee that M_{16} and M_{26} do not cut off under these conditions, assume $I_{\text{TAIL}} = 5I_{\text{BIAS}}$ in Fig. 5.35 and $I_{\text{TAIL}P} = 5I_{\text{BIAS}P}$ in Problem 5.24.

(i) For both n- and p-channel transistors, assume that $\lambda = 0$, $L_d = X_d = 0$, and ignore the body effect. Use $L_{\text{drwn}} = 1$ μm for all transistors, and use Table 2.3 for other transistor parameters.

(j) The distortion of the output stage should be minimized under the above conditions.

Verify your design using SPICE.

REFERENCES

1. E. M. Cherry and D. E. Hooper. *Amplifying Devices and Low-Pass Amplifier Design.* Wiley, New York, 1968, Ch. 9.

2. E. Fong and R. Zeman. "Analysis of Harmonic Distortion in Single-Channel MOS Integrated Circuits," *IEEE Journal of Solid-State Circuits,* Vol. SC-17, pp. 83–86, February 1982.

3. Y. P. Tsividis and D. L. Fraser, Jr. "Harmonic Distortion in Single-Channel MOS Integrated Circuits," *IEEE Journal of Solid-State Circuits,* Vol. SC-16, pp. 694–702, December 1981.

4. E. M. Cherry and D. E. Hooper. Op. cit., Ch. 16.

5. J. Millman and C. C. Halkias. *Integrated Electronics.* McGraw-Hill, New York, 1972, Ch. 18.

6. A. B. Grebene. *Analog Integrated Circuit Design.* Van Nostrand Reinhold, New York, 1972, Ch. 5–6.

7. T. M. Frederiksen and J. E. Solomon. "A High-Performance 3-Watt Monolithic Class B Power Amplifier," *IEEE Journal of Solid-State Circuits,* Vol. SC-3, pp. 152–160, June 1968.

8. P. R. Gray. "A 15-W Monolithic Power Operational Amplifier," *IEEE Journal of Solid-State Circuits,* Vol. SC-7, pp. 474–480, December 1972.

9. R. G. Meyer and W. D. Mack. "A Wideband Class AB Monolithic Power Amplifier," *IEEE Journal of Solid-State Circuits,* Vol. 24, pp. 7–12, February 1989.

10. E. L. Long and T. M. Frederiksen. "High-Gain 15-W Monolithic Power Amplifier with Internal Fault Protection," *IEEE Journal of Solid-State Circuits,* Vol. SC-6, pp. 35–44, February 1971.

11. K. Tsugaru, Y. Sugimoto, M. Noda, T. Ito, and Y. Suwa. "A Single-Power-Supply 10-b Video BiCMOS Sample-and-Hold IC," *IEEE Journal of Solid-State Circuits,* Vol. 25, pp. 653–659, June 1990.

12. K. E. Brehmer and J. B. Wieser. "Large-Swing CMOS Power Amplifier," *IEEE Journal of Solid-State Circuits,* Vol. SC-18, pp. 624–629, December 1983.

13. B. K. Ahuja, P. R. Gray, W. M. Baxter, and G. T. Uehara. "A Programmable CMOS Dual Channel Interface Processor for Telecommunications Applications," *IEEE Journal of Solid-State Circuits,* Vol. SC-19, pp. 892–899, December 1984.

14. H. Khorramabadi. "A CMOS Line Driver with 80 dB Linearity for ISDN Applications," *IEEE Journal of Solid-State Circuits,* Vol. 27, pp. 539–544, April 1992.

15. J. A. Fisher. "A High-Performance CMOS Power Amplifier," *IEEE Journal of Solid-State Circuits,* Vol. SC-20, pp. 1200–1205, December 1985.

16. K. Nagaraj. "Large-Swing CMOS Buffer Amplifier," *IEEE Journal of Solid-State Circuits,* Vol. 24, pp. 181–183, February 1989.

17. H. Khorramabadi, J. Anidjar, and T. R. Peterson. "A Highly Efficient CMOS Line Driver with 80-dB Linearity for ISDN U-Interface Applications," *IEEE Journal of Solid-State Circuits,* Vol. 27, pp. 1723–1729, December 1992.

CHAPTER 6
Operational Amplifiers with Single-Ended Outputs

In the previous three chapters, the most important circuit building blocks utilized in analog integrated circuits (ICs) have been studied. Most analog ICs consist primarily of these basic circuits connected in such a way as to perform the desired function. Although the variety of standard and special-purpose custom ICs is almost limitless, a few standard circuits stand out as perhaps having the widest application in systems of various kinds. These include operational amplifiers, voltage regulators, and analog-to-digital (A/D) and digital-to-analog (D/A) converters. In this chapter, we will consider monolithic operational amplifiers (op amps) with single-ended outputs, both as an example of the utilization of the previously described circuit building blocks and as an introduction to the design and application of this important class of analog circuit. Op amps with fully differential outputs are considered in Chapter 12, and voltage-regulator circuits are considered in Chapter 8. The design of A/D and D/A converters is not covered, but it involves application of the circuit techniques described throughout the book.

An ideal op amp with a single-ended output has a differential input, infinite voltage gain, infinite input resistance, and zero output resistance. A conceptual schematic diagram is shown in Fig. 6.1. While actual op amps do not have these ideal characteristics, their performance is usually sufficiently good that the circuit behavior closely approximates that of an ideal op amp in most applications.

In op-amp design, bipolar transistors offer many advantages over their CMOS counterparts, such as higher transconductance for a given current, higher gain ($g_m r_o$), higher speed, lower input-referred offset voltage and lower input-referred noise voltage. (The topic of noise is considered in Chapter 11.) As a result, op amps made from bipolar transistors offer the best performance in many cases, including for example dc-coupled, low-offset, low-drift applications. For these reasons, bipolar op amps became commercially significant first and still usually offer superior analog performance. However, CMOS technologies have become dominant in building the digital portions of signal-processing systems because CMOS digital circuits are smaller and dissipate less power than their bipolar counterparts. Since these systems often operate on signals that originate in analog form, analog circuits such as op amps are required to interface to the digital CMOS circuits. To reduce system cost and increase portability, analog and digital circuits are now often integrated together, providing a strong economic incentive to use CMOS op amps.

In this chapter, we first explore several applications of op amps to illustrate their versatility in analog circuit and system design. CMOS op amps are considered next. Then a general-purpose bipolar monolithic op amp, the NE5234, is analyzed, and the ways in which the performance of the circuit deviates from ideality are described. Design considerations for improving the various aspects of monolithic op-amp low-frequency performance are described. The high-frequency and transient response of op amps are covered in Chapters 7 and 9.

6.1 Applications of Operational Amplifiers

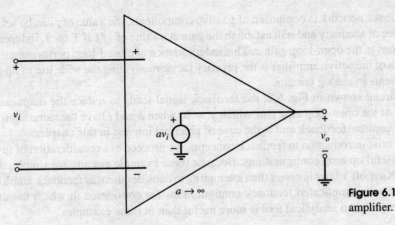

Figure 6.1 Ideal operational amplifier.

6.1 Applications of Operational Amplifiers

6.1.1 Basic Feedback Concepts

Virtually all op-amp applications rely on the principles of *feedback*. The topic of feedback amplifiers is covered in detail in Chapter 8; we now consider a few basic concepts necessary for an understanding of op-amp circuits. A generalized feedback amplifier is shown in Fig. 6.2. The block labeled a is called the forward or basic amplifier, and the block labeled f is called the feedback network. The gain of the basic amplifier when the feedback network is not present is called the *open-loop gain*, a, of the amplifier. The function of the feedback network is to sense the output signal S_o and develop a feedback signal S_{fb}, which is equal to fS_o, where f is usually less than unity. This feedback signal is subtracted from the input signal S_i, and the difference S_ϵ is applied to the basic amplifier. The gain of the system when the feedback network is present is called the *closed-loop gain*. For the basic amplifier we have

$$S_o = aS_\epsilon = a(S_i - S_{fb}) = a(S_i - fS_o) \tag{6.1}$$

and thus

$$\frac{S_o}{S_i} = \frac{a}{1+af} = \frac{1}{f}\left(\frac{af}{1+af}\right) = \frac{1}{f}\left(\frac{T}{1+T}\right) \tag{6.2}$$

where $T = af$ is called the *loop gain*. When T becomes large compared to unity, the closed-loop gain becomes

$$\lim_{T \to \infty} \frac{S_o}{S_i} = \frac{1}{f} \tag{6.3}$$

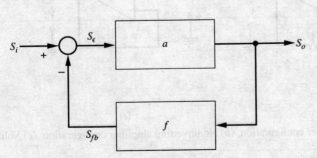

Figure 6.2 A conceptual feedback amplifier.

Since the feedback network is composed of passive components, the value of f can be set to an arbitrary degree of accuracy and will establish the gain at a value of $1/f$ if $T \gg 1$, independent of any variations in the open-loop gain a. This independence of closed-loop performance from the parameters of the active amplifier is the primary factor motivating the wide use of op amps as active elements in analog circuits.

For the circuit shown in Fig. 6.2, the feedback signal tends to *reduce* the magnitude of S_ϵ below that of the open-loop case (for which $f = 0$) when a and f have the same sign. This case is called *negative* feedback and is the case of practical interest in this chapter.

With this brief introduction to feedback concepts, we proceed to a consideration of several examples of useful op-amp configurations. Because these example circuits are simple, direct analysis with Kirchoff's laws is easier than attempting to consider them as feedback amplifiers. In Chapter 8, more complicated feedback configurations are considered in which the use of feedback concepts as an analytical tool is more useful than in these examples.

6.1.2 Inverting Amplifier

The inverting amplifier connection is shown in Fig. 6.3a.[1,2,3] We assume that the op-amp input resistance is infinite, and that the output resistance is zero as shown in Fig. 6.1. From KCL at node X,

$$\frac{V_s - V_i}{R_1} + \frac{V_o - V_i}{R_2} = 0 \tag{6.4}$$

Since R_2 is connected between the amplifier output and the inverting input, the feedback is negative. Therefore, V_i would be driven to zero with infinite open-loop gain. On the other hand, with finite open-loop gain a,

$$V_i = \frac{-V_o}{a} \tag{6.5}$$

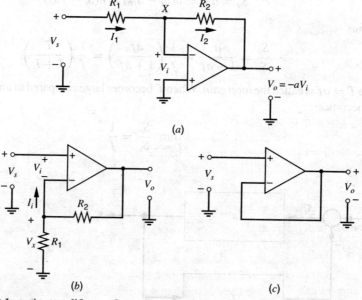

Figure 6.3 (*a*) Inverting amplifier configuration. (*b*) Noninverting amplifier configuration. (*c*) Voltage-follower configuration.

Substituting (6.5) into (6.4) and rearranging gives

$$\frac{V_o}{V_s} = -\frac{R_2}{R_1} \left[\frac{1}{1 + \frac{1}{a}\left(1 + \frac{R_2}{R_1}\right)} \right] \quad (6.6)$$

If the gain of the op amp is large enough that

$$a\left(\frac{R_1}{R_1 + R_2}\right) \gg 1 \quad (6.7)$$

then the closed-loop gain is

$$\frac{V_o}{V_s} \simeq -\frac{R_2}{R_1} \quad (6.8)$$

When the inequality in (6.7) holds, (6.8) shows that the closed-loop gain depends primarily on the external passive components R_1 and R_2. Since these components can be selected with arbitrary accuracy, a high degree of precision can be obtained in closed-loop performance independent of variations in the active device (op-amp) parameters. For example, if the op-amp gain were to change from 5×10^4 to 10^5, this 100 percent increase in gain would have almost no observable effect on closed-loop performance provided that (6.7) is valid.

■ **EXAMPLE**

Calculate the gain of the circuit Fig. 6.3a, for $a = 10^4$ and $a = 10^5$, and $R_1 = 1$ kΩ, $R_2 = 10$ kΩ.

From (6.6) with $a = 10^4$,

$$A = \frac{V_o}{V_s} = -10 \left(\frac{1}{1 + \frac{11}{10^4}} \right) = -9.9890 \quad (6.9a)$$

From (6.6) with $a = 10^5$,

$$A = \frac{V_o}{V_s} = -10 \left(\frac{1}{1 + \frac{11}{10^5}} \right) = -9.99890 \quad (6.9b)$$

■

The large gain of op amps allows the approximate analysis of circuits like that of Fig. 6.3a to be performed by the use of *summing-point constraints*.[1] If the op amp is connected in a negative-feedback circuit, and if the gain of the op amp is very large, then for a finite value of output voltage the input voltage must approach zero since

$$V_i = -\frac{V_o}{a} \quad (6.10)$$

Thus one can analyze such circuits approximately by assuming a priori that the op-amp input voltage is driven to zero. An implicit assumption in doing so is that the feedback is negative, and that the circuit has a *stable* operating point at which (6.10) is valid.

The assumption that $V_i = 0$ is called a *summing-point constraint*. A second constraint is that no current can flow into the op-amp input terminals, since no voltage exists across the input resistance of the op amp if $V_i = 0$. This summing-point approach allows an intuitive understanding of the operation of the inverting amplifier configuration of Fig. 6.3a. Since the inverting input terminal is forced to ground potential, the resistor R_1 serves to convert the voltage V_s to an input current of value V_s/R_1. This current cannot flow in the input terminal of an ideal op amp; therefore, it flows through R_2, producing a voltage drop of $V_s R_2/R_1$. Because the op-amp input terminal operates at ground potential, the input resistance of the overall circuit as seen by V_s is equal to R_1. Since the inverting input of the amplifier is forced to ground potential by the negative feedback, it is sometimes called a *virtual* ground.

6.1.3 Noninverting Amplifier

The noninverting amplifier is shown in Fig. 6.3b.[1,2,3] Using Fig. 6.1, assume that no current flows into the inverting op-amp input terminal. If the open-loop gain is a, $V_i = V_o/a$ and

$$V_x = V_o \left(\frac{R_1}{R_1 + R_2} \right) = V_s - \frac{V_o}{a} \tag{6.11}$$

Rearranging (6.11) gives

$$\frac{V_o}{V_s} = \left(1 + \frac{R_2}{R_1} \right) \frac{\frac{aR_1}{R_1 + R_2}}{1 + \frac{aR_1}{R_1 + R_2}} \simeq \left(1 + \frac{R_2}{R_1} \right) \tag{6.12}$$

The approximation in (6.12) is valid to the extent that $aR_1/(R_1 + R_2) \gg 1$.

In contrast to the inverting case, this circuit displays a very high input resistance as seen by V_s because of the type of feedback used. (See Chapter 8.) Also unlike the inverting case, the noninverting connection causes the common-mode input voltage of the op amp to be equal to V_s. An important variation of this connection is the voltage follower, in which $R_1 \to \infty$ and $R_2 = 0$. This circuit is shown in Fig. 6.3c, and its gain is close to unity if $a \gg 1$.

6.1.4 Differential Amplifier

The differential amplifier is used to amplify the *difference* between two voltages. The circuit is shown in Fig. 6.4.[1,2] For this circuit, $I_{i1} = 0$ and thus resistors R_1 and R_2 form a voltage

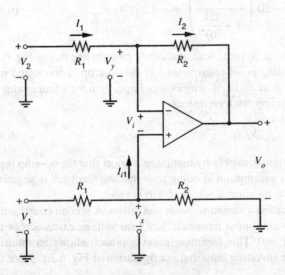

Figure 6.4 Differential amplifier configuration.

divider. Voltage V_x is then given by

$$V_x = V_1 \left(\frac{R_2}{R_1 + R_2} \right) \tag{6.13}$$

The current I_1 is

$$I_1 = \left(\frac{V_2 - V_y}{R_1} \right) = I_2 \tag{6.14}$$

The output voltage is given by

$$V_o = V_y - I_2 R_2 \tag{6.15}$$

If the open-loop gain is infinite, the summing-point constraint that $V_i = 0$ is valid and forces $V_y = V_x$. Substituting $V_y = V_x$, (6.13), and (6.14) into (6.15) and rearranging gives

$$V_o = \frac{R_2}{R_1}(V_1 - V_2) \tag{6.16}$$

The circuit thus amplifies the *difference* voltage $(V_1 - V_2)$.

Differential amplifiers are often required to detect and amplify small differences between two sizable voltages. For example, a typical application is the measurement of the difference voltage between the two arms of a Wheatstone bridge. As in the case of the noninverting amplifier, the op amp of Fig. 6.4 experiences a common-mode input that is almost equal to the common-mode voltage $(V_1 + V_2)/2$ applied to the input terminals when $R_2 \gg R_1$.

6.1.5 Nonlinear Analog Operations

By including nonlinear elements in the feedback network, op amps can be used to perform nonlinear operations on one or more analog signals. The logarithmic amplifier, shown in Fig. 6.5, is an example of such an application. Log amplifiers find wide application in instrumentation systems where signals of very large dynamic range must be sensed and recorded. The operation of this circuit can again be understood by application of the summing-point constraints. Because the input voltage of the op amp must be zero, the resistor R serves to convert the input voltage V_s into a current. This same current must then flow into the collector of the transistor. Thus the circuit forces the collector current of the transistor to be proportional to the input voltage. Furthermore, the transistor operates in the forward-active region because $V_{CB} \simeq 0$. Since the base-emitter voltage of a bipolar transistor in the forward-active region is logarithmically related to the collector current, and since the output voltage is just the emitter-base voltage of the transistor, a logarithmic transfer characteristic is produced. In terms of equations,

$$I_1 = \frac{V_s}{R} = I_c = I_S \left[\exp\left(\frac{V_{be}}{V_T}\right) - 1 \right] \simeq I_S \exp\left(\frac{V_{be}}{V_T}\right) \tag{6.17}$$

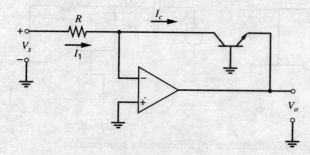

Figure 6.5 Logarithmic amplifier configuration.

and
$$V_o = -V_{be} \tag{6.18}$$

Thus
$$V_o = -V_T \ln\left(\frac{V_s}{I_S R}\right) \tag{6.19}$$

The log amplifier is only one example of a wide variety of op-amp applications in which a nonlinear feedback element is used to develop a nonlinear transfer characteristic. For example, two log amplifiers can be used to develop the logarithm of two different signals. These voltages can be summed, and then the exponential function of the result can be developed using an inverting amplifier connection with R_1 replaced with a diode. The result is an analog multiplier. Other nonlinear operations such as limiting, rectification, peak detection, squaring, square rooting, raising to a power, and division can be performed in conceptually similar ways.

6.1.6 Integrator, Differentiator

The integrator and differentiator circuits, shown in Fig. 6.6, are examples of using op amps with reactive elements in the feedback network to realize a desired frequency response or time-domain response.[1,2] In the case of the integrator, the resistor R is used to develop a current I_1 that is proportional to the input voltage. This current flows into the capacitor C, whose voltage is proportional to the integral of the current I_1 with respect to time. Since the output voltage is equal to the negative of the capacitor voltage, the output is proportional to the integral of the input voltage with respect to time. In terms of equations,

$$I_1 = \frac{V_s}{R} = I_2 \tag{6.20}$$

and
$$V_o = -\frac{1}{C}\int_0^t I_2 d\tau + V_o(0) \tag{6.21}$$

Combining (6.20) and (6.21) yields
$$V_o(t) = -\frac{1}{RC}\int_0^t V_s(\tau)d\tau + V_o(0) \tag{6.22}$$

The performance limitations of real op amps limit the range of V_o and the rate of change of V_o for which this relationship is maintained.

In the case of the differentiator, the capacitor C is connected between V_s and the inverting op-amp input. The current through the capacitor is proportional to the time derivative of the

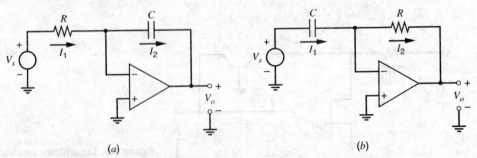

Figure 6.6 (a) Integrator configuration. (b) Differentiator configuration.

voltage across it, which is equal to the input voltage. This current flows through the feedback resistor R, producing a voltage at the output proportional to the capacitor current, which is proportional to the time rate of change of the input voltage. In terms of equations,

$$I_1 = C\frac{dV_s}{dt} = I_2 \tag{6.23}$$

$$V_o = -RI_2 = -RC\frac{dV_s}{dt} \tag{6.24}$$

6.1.7 Internal Amplifiers

The performance objectives for op amps to be used within a monolithic analog subsystem are often quite different from those of general-purpose op amps that use external feedback elements. In a monolithic analog subsystem, only a few of the amplifiers must drive a signal off-chip where the capacitive and resistive loads are significant and variable. These amplifiers will be termed *output buffers,* and the amplifiers whose outputs do not go off chip will be termed *internal amplifiers.* Perhaps the most important difference is that the load an internal amplifier has to drive is well defined and is often purely capacitive with a value of a few picofarads. In contrast, stand-alone general-purpose amplifiers usually must be designed to achieve a certain level of performance that is independent of changes in capacitive loads up to several hundred picofarads and resistive loads down to 2 kΩ or less.

6.1.7.1 Switched-Capacitor Amplifier

In MOS technologies, capacitors instead of resistors are often used as passive elements in feedback amplifiers in part because capacitors are often the best available passive components. Also, capacitors can store charge proportional to analog signals of interest, and MOS transistors can act as switches to connect to the capacitors without offset and with little leakage, allowing the discrete-time signal processing of analog quantities. The topic of MOS switched-capacitor amplifiers is one important application of internal amplifiers. Here, we introduce the application to help explain the construction of MOS op amps.

Figure 6.7 shows the schematic of an inverting amplifier with capacitive feedback in contrast to the resistive feedback shown in Fig. 6.3a. If the op amp is ideal, the gain from a change in the input ΔV_s to a change in the output ΔV_o is still given by the ratio of the impedance of the feedback element C_2 to the impedance of the input element C_1, or

$$\frac{\Delta V_o}{\Delta V_s} = -\frac{\frac{1}{\omega C_2}}{\frac{1}{\omega C_1}} = -\frac{C_1}{C_2} \tag{6.25}$$

Unlike the case when resistors are used as passive elements, however, this circuit does not provide dc bias for the inverting op-amp input because the impedances of both capacitors are

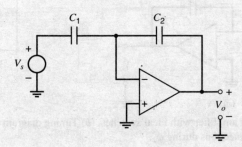

Figure 6.7 Inverting amplifier configuration with capacitive feedback without dc bias for the inverting op-amp input.

infinite at dc. To overcome this problem, switches controlled by a two-phase nonoverlapping clock are used to control the operation of the above circuit. The resulting circuit is known as a *switched-capacitor amplifier*.

Figure 6.8a shows the schematic of a switched-capacitor amplifier. Each switch in the schematic is controlled by one of two clock phases ϕ_1 and ϕ_2, and the timing diagram is shown

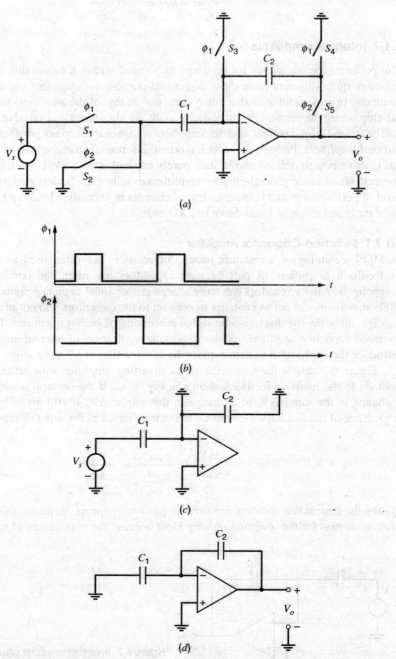

Figure 6.8 (a) Schematic of a switched-capacitor amplifier with ideal switches. (b) Timing diagram of clock signals. (c) Connections during ϕ_1. (d) Connections during ϕ_2.

in Fig. 6.8b. We will assume that each switch is closed when its controlling clock signal is high, and open when its clock signal is low. Since ϕ_1 and ϕ_2 are never both high at the same time, the switches controlled by one clock phase are never closed at the same time as the switches controlled by the other clock phase. Because of this property, the clock signals in Fig. 6.8b are known as *nonoverlapping*. For example, the left side of C_1 is connected to the input V_s through switch S_1 when ϕ_1 is high and to ground through switch S_2 when ϕ_2 is high, but this node is never simultaneously connected to both the input and ground.

To simplify the description of the operation of switched-capacitor circuits, they are often redrawn twice, once for each nonoverlapping clock phase. Figures 6.8c and 6.8d show the connections when ϕ_1 is high (*during ϕ_1*) and when ϕ_2 is high (*during ϕ_2*), respectively. Switches that are closed are assumed to be short circuits, and switches that are open are assumed to be open circuits.

To find the output for a given input, an analysis based on charge conservation is used. After switch S_3 opens, the charge on the plates of the capacitors that connect to the op-amp input node is conserved until this switch closes again. This property stems partly from the fact that the passive elements connected to the op-amp input node are capacitors, which conduct zero dc current. Also, if the op amp is constructed with an MOS differential input pair, the op-amp input is connected to the gate of one transistor in the differential pair, and the gate conducts zero dc current. Finally, if we assume that switch S_3 conducts zero current when it is open, the charge stored cannot leak away while S_3 is open.

Since both sides of C_2 are grounded in Fig. 6.8c, the charge stored on the plates of the capacitors that connect to the the op-amp input node during ϕ_1 is

$$Q_1 = (0 - V_s)C_1 + (0)C_2 = (0 - V_s)C_1 \tag{6.26}$$

Because the input voltage is sampled or stored onto capacitor C_1 during ϕ_1, this phase is known as the *input sample* phase. If the op amp is ideal, the voltage V_i from the inverting op-amp input to ground is driven to zero by negative feedback during ϕ_2. Therefore, the charge stored during ϕ_2 is

$$Q_2 = (0)C_1 + (0 - V_o)C_2 = (0 - V_o)C_2 \tag{6.27}$$

Because the sampled charge appears on C_1 during ϕ_1 and on C_2 during ϕ_2, ϕ_2 is known as the *charge-transfer* phase. By charge conservation, $Q_2 = Q_1$; therefore,

$$\frac{V_o}{V_s} = \frac{C_1}{C_2} \tag{6.28}$$

In (6.28), V_s represents the input voltage at the end of ϕ_1, and V_o represents the output voltage at the end of ϕ_2. The shape of the output voltage waveform is not predicted by (6.28) and depends on the rates at which the capacitors are charged and discharged. In practice, these rates depend on the bandwidth of the op amp and the resistances of the closed switches. The result in (6.28) is valid as long as the op amp is ideal, the input V_s is dc, and the intervals over which ϕ_1 and ϕ_2 are high are long enough to completely charge and discharge the associated capacitors. The ratio of V_o/V_s in (6.28) is positive because if $V_s > 0$ during ϕ_1, the voltage applied between the left side of C_1 and ground decreases from a positive value to zero during ϕ_2. Therefore, this negative change in the applied voltage during ϕ_2 is multiplied by a negative closed-loop amplifier gain, giving a positive ratio in (6.28).

MOS technologies are well suited to building switched-capacitor circuits for two key reasons. First, the dc current that flows into the input terminals of MOS op amps is zero as long as the inputs are connected only to the gates of MOS transistors. In contrast, bipolar op amps have nonzero dc input currents that stem from finite β_F in bipolar transistors. Second, the switches in Fig. 6.8 can be implemented without offset by using MOS transistors, as shown

410   Chapter 6 ■ Operational Amplifiers with Single-Ended Outputs

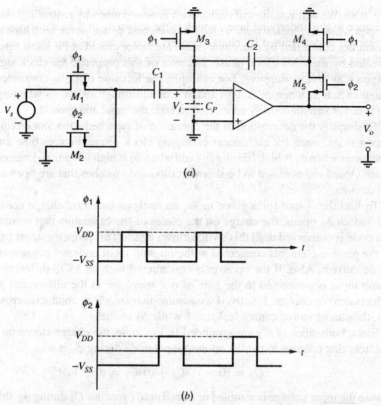

Figure 6.9 (a) Schematic of a switched-capacitor amplifier with *n*-channel MOS transistors used as switches. (b) Clock waveforms with labeled voltages.

in Fig. 6.9a. The arrows indicating the source terminals of M_1-M_5 are arbitrarily chosen in the sense that the source and drain terminals are interchangeable. (Since the source is the source of electrons in *n*-channel transistors, the source-to-ground voltage is lower than the drain-to-ground voltage.) Assume that the clock voltages alternate between $-V_{SS}$ and V_{DD} as shown in Fig. 6.9b. Also assume that all node voltages are no lower than $-V_{SS}$ and no higher than V_{DD}. Finally, assume that the transistor threshold voltages are positive. Under these conditions, each transistor turns off when its gate is low. Furthermore, each transistor turns on when its gate is high as long as its source operates at least a threshold below V_{DD}. If V_s is a dc signal, all drain currents approach zero as the capacitors become charged. As the drain currents approach zero, the MOS transistors that are on operate in the triode region, where the drain-source voltage is zero when the drain current is zero. Therefore, the input voltage V_s is sampled onto C_1 with zero offset inserted by the MOS transistors operating as switches. This property of MOS transistors is important in the implementation of switched-capacitor circuits. In contrast, bipolar transistors operating as switches do not give zero collector-emitter voltage with zero collector current.

The switched-capacitor amplifier shown in Fig. 6.9a is important in practice mainly because the gain of this circuit has little dependence on the various parasitic capacitances that are present on all the nodes in the circuit. These undesired capacitances stem in part from the drain-body and source-body junction capacitances of each transistor. Also, the op-amp input capacitance contributes to the parasitic capacitance on the op-amp input node. Furthermore, as described in Section 2.10.2, the bottom plates of capacitors C_1 and C_2 exhibit at least some capacitance to the underlying layer, which is the substrate or a well diffusion. Since the gain of the circuit is determined by charge conservation at the op-amp input node, the presence of parasitic capacitance from any node except the op-amp input to any node with a constant voltage to ground makes no difference to the accuracy of the circuit. (Such parasitics do reduce the maximum clock rate, however.) On the other hand, parasitic capacitance on the op-amp input node does affect the accuracy of the circuit gain, but the error is inversely proportional to the op-amp gain, as we will now show.

Let C_P represent the total parasitic capacitance from the op-amp input to all nodes with constant voltage to ground. If the op-amp gain is a, the voltage from the inverting op-amp input to ground during ϕ_2 is given by (6.10). Therefore, with finite op-amp gain, C_1 and C_P are not completely discharged during ϕ_2. Under these conditions, the charge stored on the op-amp input node during ϕ_2 becomes

$$Q_2 = \left(-\frac{V_o}{a}\right) C_1 + \left(-\frac{V_o}{a}\right) C_P + \left(-\frac{V_o}{a} - V_o\right) C_2 \qquad (6.29)$$

When the op-amp gain becomes infinite, (6.29) collapses to (6.27), as expected. Setting Q_2 in (6.29) equal to Q_1 in (6.26) by charge conservation gives

$$\frac{V_o}{V_s} = \frac{C_1}{C_2} \left[\frac{1}{1 + \frac{1}{a}\left(\frac{C_1 + C_2 + C_P}{C_2}\right)} \right] \qquad (6.30)$$

This closed-loop gain can be written as

$$\frac{V_o}{V_s} = \frac{C_1}{C_2}(1 - \epsilon) \qquad (6.31)$$

where ϵ is a gain error given by

$$\epsilon = \frac{1}{1 + a\left(\frac{C_2}{C_1 + C_2 + C_P}\right)} \qquad (6.32)$$

As $a \to \infty$, $\epsilon \to 0$, and the gain of the switched-capacitor amplifier approaches C_1/C_2 as predicted in (6.28). As a result, the circuit gain is said to be *parasitic insensitive* to an extent that depends on the op-amp gain.

One important parameter of the switched-capacitor amplifier shown in Fig. 6.9a is the minimum clock period. This period is divided into two main parts, one for each clock phase. The duration of ϕ_2 must be long enough for the op-amp output to reach and stay within a given level of accuracy. This time is defined as the op-amp settling time and depends on the switch resistances, the circuit capacitances, and the op-amp properties. The settling time is usually determined by SPICE simulations. Such simulations should be run for both clock phases because the op-amp output voltage during ϕ_1 is not well controlled in practice. If the op amp is ideal, this output voltage is zero. With nonzero offset voltage, however, the output voltage will be driven to a nonzero value during ϕ_1 that depends on both the offset voltage and the op-amp gain. If the op-amp gain is large, the offset can easily be large enough to force the output voltage to clip near one of the supplies. The output voltage at the end of ϕ_1 can be thought of as an initial condition for the circuit during ϕ_2. If the initial condition and the desired final output at the end of ϕ_2 happen to have far different values, the time required for the output to reach a given level of accuracy during ϕ_2 can be increased. Furthermore, nonzero offset can increase the time required for the op-amp output voltage to reach a constant level during ϕ_1. Although the circuit output voltage defined in (6.30) only appears during ϕ_2, failure to reach a constant output voltage during ϕ_1 causes the initial condition defined above to vary, depending on the value of V_o at the end of the preceding ϕ_2. As a result of this memory effect, the circuit can behave as a filter (which is possibly nonlinear), weighting together the results of more than one previous input sample to determine any given output. The key point is that an increase in the minimum duration of either phase requires an increase in the minimum clock period. This effect should be included in simulation by intentionally simulating with nonzero offset voltages.

One way to reduce the effect of nonzero offset voltage on the minimum clock period is to include a reset switch at the op-amp output. If the op amp has a single-ended output as shown in Fig. 6.9a, such a switch can be connected between the op-amp output and a bias point between the two supplies. On the other hand, if the op amp has differential outputs as described in Chapter 12, such a reset switch can be connected between the two op-amp outputs. In either case, the reset switch would be turned on during ϕ_1 and off during ϕ_2. The main value of such a reset switch is that it can reduce both the maximum output voltage produced by a nonzero offset voltage during ϕ_1 and the time required to reach this value, in turn reducing the effect of nonzero offset on the minimum durations of both phases.

The accuracy of the switched-capacitor amplifier shown in Fig. 6.9a is limited by several other factors that we will now consider briefly. First, even with an ideal op amp, the gain depends on the ratio of capacitors C_1/C_2, which is not controlled perfectly in practice because of random-mismatch effects. Second, op-amp offset limits the minimum signal that can be distinguished from the offset in the switched-capacitor amplifier. As shown in Section 3.5.6, the input-referred offset of CMOS differential pairs is usually worse than for bipolar differential pairs. This property extends to op amps and stems partly from the reduced transconductance-to-current ratio of MOS transistors compared to their bipolar counterparts and partly from the threshold mismatch term, which appears only in the MOS case. Third, the charge-conservation equation and accuracy of the switched-capacitor amplifier are affected by the charge stored under the gates of some of the MOS transistors acting as switches in Fig. 6.9a. For example, some of the charge stored under the gate of transistor M_3 in Fig. 6.9a is injected onto the op-amp input node after M_3 is turned off. Techniques to overcome these limitations are often used in practice but are not considered here.

6.1.7.2 Switched-Capacitor Integrator

Another application of an internal op amp, a switched-capacitor integrator, is illustrated in its simplest form in Fig. 6.10a. This circuit is widely utilized as the basic element of monolithic switched-capacitor filters for two main reasons. First, the frequency response of the integrator is insensitive to the various parasitic capacitances that are present on all nodes in the circuit.[4,5] Second, using switched-capacitor integrators as the basic elements, the synthesis of desired filter frequency responses is relatively straightforward. In this section, we will analyze the frequency response of the switched-capacitor integrator.

The integrator consists of an op amp, a sampling capacitor C_S, an integrating capacitor C_I, and four MOS transistor switches. The load capacitance shown represents the sampling capacitor of the following integrator, plus any parasitic capacitances that may be present. Typical values for the sampling, integrating, and load capacitances are labeled in Fig. 6.10a.

Figure 6.10b shows the timing diagram of two nonoverlapping clock signals, ϕ_1 and ϕ_2, that control the operation of the circuit as well as typical input and output waveforms. During the interval when clock phase ϕ_1 is high, transistors M_1 and M_3 operate in the triode region and serve to charge the sampling capacitor to a voltage that is equal to the input voltage. Subsequently, clock signal ϕ_1 falls. Then clock signal ϕ_2 rises, causing transistors M_2 and M_4 to turn on and the sampling capacitor to be connected between the inverting op-amp input, which is sometimes called the *summing node*, and ground. If the op amp is ideal, the resulting

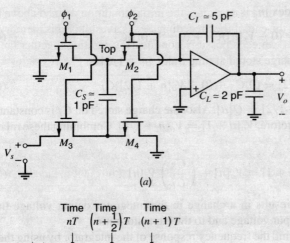

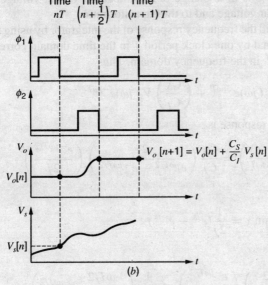

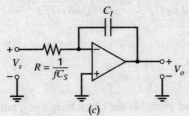

Figure 6.10 (a) Schematic of a switched-capacitor integrator. (b) Timing diagram. (c) Continuous-time equivalent circuit for input frequencies much less than the clock frequency.

change in the summing-node voltage causes the op-amp output to move so that the summing-node voltage is driven back to ground. After the transient has gone to completion, the voltage across C_S is driven to zero.

To find the relationship between the input and output, a charge-conservation analysis is used. After transistor M_1 opens in Fig. 6.10a, the charge on the plates of the capacitors connected to node *Top* and the inverting op-amp input is conserved until M_1 closes again. Define time points [n] and [n + 1/2] as the time indexes at which ϕ_1 and ϕ_2 first fall in Fig. 6.10b, respectively. Point [n + 1] is defined as the next time index at which ϕ_1 falls. The points [n] and [n + 1] are separated by one clock period T. If the switches and the op amp are ideal, the charge stored at time index $[n]$ is

$$Q[n] = (0 - V_s[n])C_S + (0 - V_o[n])C_I \qquad (6.33)$$

Under the same conditions, the charge stored at time index $[n + 1/2]$ is

$$Q[n + 1/2] = (0)C_S + (0 - V_o[n + 1/2])C_I \qquad (6.34)$$

From charge conservation, $Q[n + 1/2] = Q[n]$. Also, the charge stored on C_I is constant during ϕ_1 under these conditions; therefore, $V_o[n + 1] = V_o[n + 1/2]$. Combining these relations gives

$$V_o[n+1] = V_o[n] + \left(\frac{C_S}{C_I}\right) V_s[n] \qquad (6.35)$$

Thus, one complete clock cycle results in a change in the integrator output voltage that is proportional to the value of the input voltage and to the capacitor ratio.

Equation 6.35 can be used to find the frequency response of the integrator by using the fact that the operation of delaying a signal by one clock period T in the time domain corresponds to multiplication by the factor $e^{-j\omega T}$ in the frequency domain. Thus

$$V_o(j\omega) = V_o(j\omega)e^{-j\omega T} + \left(\frac{C_S}{C_I}\right) V_s(j\omega)e^{-j\omega T} \qquad (6.36)$$

Therefore, the integrator frequency response is

$$\frac{V_o}{V_s}(j\omega) = -\frac{C_S}{C_I}\left(\frac{1}{1-e^{j\omega T}}\right) = \frac{C_S}{C_I}\left(\frac{2j}{e^{j\omega T/2} - e^{-j\omega T/2}}\right)\left(\frac{e^{-j\omega T/2}}{2j}\right) \qquad (6.37)$$

Using the identity

$$\sin x = \frac{1}{2j}(e^{jx} - e^{-jx}) \qquad (6.38)$$

in (6.37) with $x = \omega T/2$ we find

$$\frac{V_o}{V_s}(j\omega) = \frac{C_S}{C_I}\left(\frac{\omega T/2}{\sin \omega T/2}\right)\left(\frac{e^{-j\omega T/2}}{2j\omega T/2}\right) = \frac{1}{j\omega}\underbrace{\left(\frac{\omega T/2}{\sin \omega T/2} e^{-j\omega T/2}\right)}_{\omega_o} \qquad (6.39)$$

where

$$\omega_o = \frac{C_S}{TC_I} = \frac{fC_S}{C_I} = \frac{1}{\tau} \qquad (6.40)$$

where τ is the time constant of the integrator. Here f is the clock frequency, equal to $1/T$. For input frequencies that are much less than the clock frequency, the quantity ωT is much less than unity, and the right-most term in parentheses in (6.39) reduces to unity. The remaining term is simply the frequency response of an analog integrator, as desired. In practical designs, the excess phase and magnitude error contributed by the term in parentheses in (6.39) must often be taken into account. From a conceptual standpoint, however, the circuit can be thought of as providing an analog integration of the signal. Note that the time constant of the integrator is the same as would occur if the sampling capacitor and switches were replaced by a continuous-value resistor of value $(1/fC_S)$. This equivalence is illustrated in Fig. 6.10c.

A key advantage of a switched-capacitor integrator compared to its continuous-time counterpart is that the time constant of the switched-capacitor integrator can be much better controlled in practice. The time constant of a continuous-time integrator depends on the product of a resistance and a capacitance as in (6.22). In monolithic technologies, resistance and capacitance values do not track each other. Therefore, the time constant of a continuous-time integrator is not well controlled over variations in process, supply, and temperature in general. However, the time constant of a switched-capacitor integrator is determined by the *ratio* of two capacitor values and the clock frequency, as in (6.40). If the two capacitors have the same properties, the ratio is well controlled even when the absolute values are poorly controlled. Since the clock frequency can be precisely determined by a crystal-controlled clock generator, the time constant of a switched-capacitor integrator can be well controlled in monolithic technologies.

A key requirement for the op amps in Figs. 6.9a and 6.10a is that dc currents at the input terminals must be extremely small to minimize the loss of charge over the time when the above analyses assumed that charge was conserved. Therefore, switched-capacitor amplifiers and integrators are ideally suited to the use of op amps with MOS transistors in the input stage.

6.2 Deviations from Ideality in Real Operational Amplifiers

Real op amps deviate from ideal behavior in significant ways. The main effects of these deviations are to limit the frequency range of the signals that can be accurately amplified, to place a lower limit on the magnitude of dc signals that can be detected, and to place an upper limit on the magnitudes of the impedance of the passive elements that can be used in the feedback network with the amplifier. This section summarizes the most important deviations from ideality and their effects in applications.

6.2.1 Input Bias Current

An input stage for a bipolar transistor op amp is shown in Fig. 6.11. Here Q_1 and Q_2 are the input transistors of the amplifier. The base currents of Q_1 and Q_2 flow into the amplifier input terminals, and the input bias current is defined as the average of the two input currents:

$$I_{\text{BIAS}} = \frac{I_{B1} + I_{B2}}{2} \quad (6.41)$$

Nonzero bias current violates the assumption made in summing-point analysis that the current into the input terminals is zero. Typical magnitudes for the bias current are 10 to 100 nA

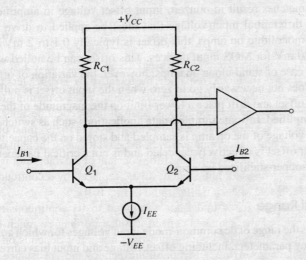

Figure 6.11 Typical op-amp input stage.

for bipolar input devices and less than 0.001 pA for MOS input devices. In dc inverting and noninverting amplifiers, this bias current can cause undesired voltage drops in the resistors forming the feedback network, with the result that a residual dc voltage appears at the output when the amplifier is ideal in all other ways and the external input voltage is zero. In integrator circuits, the input bias current is indistinguishable from the current being integrated and causes the output voltage to change at a constant rate even when V_s is zero. To the extent that the currents are equal in the two input leads, however, their effects can be canceled in some applications by including a balancing resistor in series with one of the input leads so that the same resistance is seen looking away from each op-amp input. For example, the differential amplifier of Fig. 6.4 produces zero output with $V_1 = V_2 = 0$ if identical currents flow in both op-amp input leads. In practice, however, the two input currents are not exactly equal because of random mismatches, causing nonzero output in Fig. 6.4 when $V_1 = V_2 = 0$.

6.2.2 Input Offset Current

For the emitter-coupled pair shown in Fig. 6.11, the two input bias currents will be equal only if the two transistors have equal betas. Geometrically identical devices on the same IC die typically display beta mismatches that are described by a normal distribution with a standard deviation of a few percent of the mean value. Since this mismatch in the two currents varies randomly from circuit to circuit, it cannot be compensated by a fixed resistor. This aspect of op-amp performance is characterized by the input offset current, defined as

$$I_{OS} = I_{B1} - I_{B2} \tag{6.42}$$

Consider the differential amplifier in Fig. 6.4. Repeating the analysis in Section 6.1.4 for that circuit with nonzero I_{OS} and $V_1 = V_2 = 0$ gives a dc output voltage of

$$V_O = (I_{B2} - I_{B1})R_2 = -I_{OS}R_2 \tag{6.43}$$

If $I_{OS} = 0$, then $V_O = 0$ here. This equation shows that the error in the dc output voltage is proportional to both the input offset current and the feedback resistance under these conditions. The key point is that the size of the feedback resistance is limited by the maximum offset current that can arise and by the allowed error in the dc output voltage in practice. See Problem 6.6.

6.2.3 Input Offset Voltage

As described in Chapter 3, mismatches result in nonzero input offset voltage in amplifiers. The input offset voltage is the differential input voltage that must be applied to drive the output to zero. For untrimmed monolithic op amps, this offset is typically 0.1 to 2 mV for bipolar input devices and 1 to 20 mV for MOS input devices. This offset can be nulled with an external potentiometer in the case of stand-alone op amps; however, the variation of offset with temperature (called *drift*) does not necessarily go to zero when the input offset is nulled. In dc amplifier applications, the offset and drift place a lower limit on the magnitude of the dc voltage that can be accurately amplified. In some sampled-data applications such as switched-capacitor filters, the input offset voltage of the op amp is sampled and stored on the capacitors every clock cycle. Thus the input offset is effectively canceled and is not a critical parameter. This same principle is used in chopper-stabilized op amps.

6.2.4 Common-Mode Input Range

The common-mode input range is the range of dc common-mode input voltages for which an op amp behaves normally with its key parameters, including offset voltage and input bias current, within specifications. Many years ago, op amps were usually designed to use large equal-and-opposite power-supply voltages. For example, the 741 op amp, described in previous editions of this book, often operated with supply voltages of ± 15 V, with a corresponding common-mode input range of about ± 13 V. In contrast, modern op amps often operate between ground and one positive power-supply voltage of 3 V or less. If the common-mode input range were limited to be 2 V above ground and 2 V below the positive supply in this case, the input stage in such op amps would not operate properly for any common-mode input voltage. Although reducing the gap between the common-mode input range and the supplies overcomes this problem, the inverting amplifier configuration shown in Fig. 6.3a cannot be used without modification when the op amp operates from a single nonzero supply voltage unless the common-mode input range is extended to include ground. In practice, including both power-supply voltages in the common-mode input range is often important. For example, in the noninverting amplifier and voltage follower shown in Fig. 6.3b and Fig. 6.3c, respectively, the op-amp common-mode input voltage is approximately equal to V_s. Extending the common-mode input range to include both supplies avoids an unnecessary limit to op-amp performance in these configurations. Section 6.8 describes an example op amp with this property, which is called a *rail-to-rail common-mode input range*.

6.2.5 Common-Mode Rejection Ratio (CMRR)

If an op amp has a differential input and a single-ended output, its small-signal output voltage can be described in terms of its differential and common-mode input voltages (v_{id} and v_{ic}) by the following equation

$$v_o = A_{dm} v_{id} + A_{cm} v_{ic} \qquad (6.44)$$

where A_{dm} is the differential-mode gain and A_{cm} is the common-mode gain. As defined in (3.187), the common-mode rejection ratio of the op amp is

$$\text{CMRR} = \left| \frac{A_{dm}}{A_{cm}} \right| \qquad (6.45)$$

From an applications standpoint, the CMRR can be regarded as the change in input offset voltage that results from a unit change in common-mode input voltage. For example, assume that we apply zero common-mode input voltage to the amplifier and then apply just enough differential voltage to the input to drive the output voltage to zero. The dc voltage we have applied is just the input offset voltage V_{OS}. If we keep the applied differential voltage constant and increase the common-mode input voltage by an amount ΔV_{ic}, the output voltage will change, by an amount

$$v_o = \Delta V_o = A_{cm} \Delta V_{ic} = A_{cm} v_{ic} \qquad (6.46)$$

To drive the output voltage back to zero, we will have to change the differential input voltage by an amount

$$v_{id} = \Delta V_{id} = \frac{\Delta V_o}{A_{dm}} = \frac{A_{cm} \Delta V_{ic}}{A_{dm}} \qquad (6.47)$$

Thus we can regard the effect of finite CMRR as causing a change in the input offset voltage whenever the common-mode input voltage is changed. Using (6.45) and (6.47), we obtain

$$\text{CMRR} = \left| \frac{A_{dm}}{A_{cm}} \right| = \left(\left. \frac{\Delta V_{id}}{\Delta V_{ic}} \right|_{V_o=0} \right)^{-1} = \left(\frac{\Delta V_{OS}}{\Delta V_{ic}} \right)^{-1} \simeq \left(\left. \frac{\partial V_{OS}}{\partial V_{ic}} \right|_{V_o=0} \right)^{-1} \qquad (6.48)$$

In circuits such as the differential amplifier of Fig. 6.11, an offset voltage is produced that is a function of the common-mode signal input, producing a voltage at the output that is indistinguishable from the desired signal. For a common-mode rejection ratio of 10^4 (or 80 dB), (6.48) shows that a 10-V common-mode signal produces a 1-mV change in the input offset voltage.

6.2.6 Power-Supply Rejection Ratio (PSRR)

In (6.44), we assumed that the power-supply voltages are constant so that the op-amp output voltage depends only on the differential and common-mode input voltages provided to the op amp. In practice, however, the power-supply voltages are not exactly constant, and variations in the power-supply voltages contribute to the op-amp output. Figure 6.12 shows a block diagram of an op amp with varying power-supply voltages. The small-signal variation on the positive and negative power supplies is v_{dd} and v_{ss}, respectively. If $v_{ic} = 0$ is assumed for simplicity, the resulting small-signal op-amp output voltage is

$$v_o = A_{dm} v_{id} + A^+ v_{dd} + A^- v_{ss} \qquad (6.49)$$

where A^+ and A^- are the small-signal gains from the positive and negative power-supplies to the output, respectively. Since op amps should be sensitive to changes in their differential-mode input voltage but insensitive to changes in their supply voltages, this equation is rewritten below in a form that simplifies comparison of these gains:

$$\begin{aligned} v_o &= A_{dm} \left(v_{id} + \frac{A^+}{A_{dm}} v_{dd} + \frac{A^-}{A_{dm}} v_{ss} \right) \\ &= A_{dm} \left(v_{id} + \frac{v_{dd}}{\text{PSRR}^+} + \frac{v_{ss}}{\text{PSRR}^-} \right) \end{aligned} \qquad (6.50)$$

where

$$\text{PSRR}^+ = \frac{A_{dm}}{A^+} \quad \text{and} \quad \text{PSRR}^- = \frac{A_{dm}}{A^-} \qquad (6.51)$$

Figure 6.13 shows one way to interpret (6.50), where the diagram in Fig. 6.12 is redrawn using an op amp with constant power supplies. To set the output in Fig. 6.13 equal to that in Fig. 6.12, the power-supply variations from Fig. 6.12 are included as equivalent differential inputs

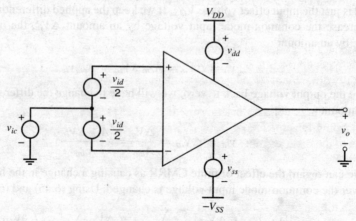

Figure 6.12 Block diagram of an op amp with varying power-supply voltages.

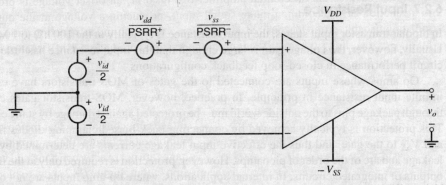

Figure 6.13 Block diagram of an op amp with supply variations modeled in the input differential loop and with $v_{ic} = 0$.

in Fig. 6.13. Equation 6.50 and Fig. 6.13 show that the power-supply rejection ratios should be maximized to minimize the undesired contributions to the op-amp output voltage. In practice, the power-supply rejection ratios are functions of frequency and often decrease for increasing frequency.

Power-supply rejection ratio has become an increasingly important parameter in MOS amplifier design as the level of integration increases. With small-scale integration, few transistors could be integrated on one integrated circuit. Therefore, analog and digital functions were isolated from each other on separate chips, avoiding some coupling from the digital circuits to the analog supplies. Also, such separation provides an opportunity to filter interference generated by the digital circuits at the printed-circuit-board level with external capacitors connected in parallel with the supplies. With large-scale integration, however, many transistors can be integrated on one integrated circuit. Integrating analog and digital functions on the same chip reduces cost but increases the coupling from the digital circuits to the analog supplies. In principle, monolithic filter capacitors can be used to reduce the resulting supply variations; however, the required areas of such capacitors are large in practice. For example, if the oxide thickness is 100 Å, the capacitance per unit area is 3.45 fF/μm^2. For a capacitor of 0.01 μF (a commonly used value to filter supplies on printed-circuit boards), the required area is 1.7 mm^2. Since many integrated circuits occupy areas less than 100 mm^2, this single capacitor would account for a significant fraction of the cost of many integrated circuits.

To reduce the cost, instead of concentrating only on reducing supply variations through filtering, another option is to build circuits with low sensitivities to power-supply variations. The use of fully differential circuit techniques has emerged as an important tool in this effort. Fully differential circuits represent all signals of interest as differences between two corresponding quantities such as voltages or currents. If two identical signal paths are used to determine corresponding quantities, and if the coupling from supply variations to one quantity is the same as to the other quantity, the difference can be independent of the supply variations and the coupling in principle. In practice, mismatches cause differences in the two signal paths, and the coupling may not be identical, causing imperfect cancellation. Also, if the power-supply noise is large enough, nonlinearity may result and limit the extent of the cancellation. Although the op amps considered in this chapter have differential inputs, they are not fully differential because their outputs are single-ended. Fully differential op amps are considered in Chapter 12.

6.2.7 Input Resistance

In bipolar transistor input stages, the input resistance is typically in the 100 kΩ to 1 MΩ range. Usually, however, the voltage gain is large enough that this input resistance has little effect on circuit performance in closed-loop feedback configurations.

Op amps whose inputs are connected to the gates of MOS transistors have essentially infinite input resistance in principle. In practice, however, MOS-transistor gates connected through package pins to the outside world must be protected against damage by static electricity. This protection is typically achieved by connecting back-biased clamping diodes from V_{DD} and V_{SS} to the gate, and thus the effective input leakage currents are determined by junction leakage and are of the order of picoamps. However, protection is required only at the inputs and outputs of integrated circuits. In internal applications, where op-amp inputs are not connected to the external pins of an integrated circuit, protection is not required and op amps with MOS transistor gates as inputs do realize ultra-high input resistance.

6.2.8 Output Resistance

General-purpose bipolar op amps usually use a buffer as an output stage, which typically produces an output resistance on the order of 40 Ω to 100 Ω. On the other hand, in MOS technologies, internal op amps usually do not have to drive resistive loads. Therefore, internal MOS op amps usually do not use a buffer output stage, and the resulting output resistance can be much larger than in the bipolar case. In both cases, however, the output resistance does not strongly affect the closed-loop performance except as it affects stability under large capacitive loading, and in the case of power op amps that must drive a small load resistance.

6.2.9 Frequency Response

Because of the capacitances associated with devices in the op amp, the voltage gain decreases at high frequencies. This fall-off must usually be controlled by the addition of extra capacitance, called *compensation capacitance*, to ensure that the circuit does not oscillate when connected in a feedback loop. (See Chapter 9.) This aspect of op-amp behavior is characterized by the unity-gain bandwidth, which is the frequency at which the magnitude of the open-loop voltage gain is equal to unity. For general-purpose amplifiers, this frequency is typically in the 1 to 100 MHz range. This topic is considered in detail in Chapters 7 and 9.

A second aspect of op-amp high-frequency behavior is a limitation of the rate at which the output voltage can change under large-signal conditions. This limitation stems from the limited current available within the circuit to charge the compensation capacitor. This maximum rate, called the slew rate, is described more extensively in Chapter 9.

6.2.10 Operational-Amplifier Equivalent Circuit

The effect of some of these deviations from ideality on the low-frequency performance of an op amp in a particular application can be calculated using the equivalent circuit shown in Fig. 6.14. (This model does not include the effects of finite PSRR or CMRR.) Here, the two current sources labeled I_{BIAS} represent the *average* value of dc current flowing into the input terminals. The polarity of these current sources shown in Fig. 6.14 applies for an *npn* transistor input stage. The current source labeled I_{OS} represents the *difference* between the currents flowing into the amplifier terminals. For example, if a particular circuit displayed a current of 1.5 μA flowing into the noninverting input terminal and a current of 1 μA flowing into the inverting input terminal, then the value of I_{BIAS} in Fig. 6.14 would be 1.25 μA, and the value of I_{OS} would be 0.5 μA.

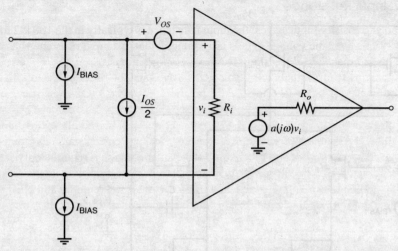

Figure 6.14 Equivalent circuit for the operational amplifier including input offset voltage and current, input and output resistance, and voltage gain.

6.3 Basic Two-Stage MOS Operational Amplifiers

Figure 6.15 shows a schematic of a basic two-stage CMOS op amp.[6,7,8] A differential input stage drives an active load followed by a second gain stage. An output stage is usually not used but may be added for driving heavy loads off-chip. This circuit configuration provides good common-mode range, output swing, voltage gain, and CMRR in a simple circuit that can be compensated with a single capacitor. The circuit is redrawn in Fig. 6.16, where the ideal current sources are replaced with transistor current mirrors. In this section, we will analyze the various performance parameters of this CMOS op-amp circuit.

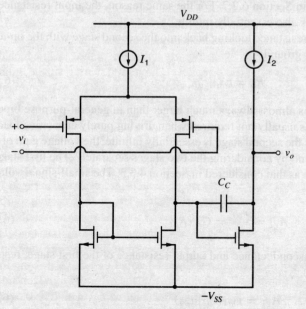

Figure 6.15 Basic two-stage CMOS operational amplifier.

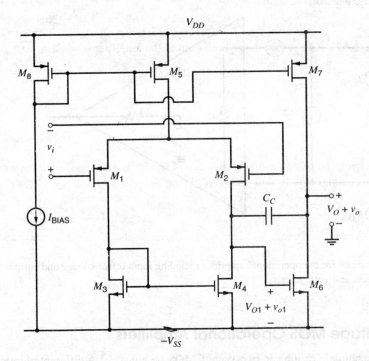

Figure 6.16 More detailed schematic diagram of a typical two-stage CMOS operational amplifier.

6.3.1 Input Resistance, Output Resistance, and Open-Circuit Voltage Gain

The first stage in Fig. 6.16 consists of a *p*-channel differential pair M_1-M_2 with an *n*-channel current mirror load M_3-M_4 and a *p*-channel tail current source M_5. The second stage consists of an *n*-channel common-source amplifier M_6 with a *p*-channel current-source load M_7. Because the op-amp inputs are connected to the gates of MOS transistors, the input resistance is essentially infinite when the op amp is used in internal applications, which do not require the protection diodes described in Section 6.2.7. For the same reason, the input resistance of the second stage of the op amp is also essentially infinite.

The output resistance is the resistance looking back into the second stage with the op-amp inputs connected to small-signal ground:

$$R_o = r_{o6} || r_{o7} \tag{6.52}$$

Although this output resistance is almost always much larger than in general-purpose bipolar op amps, low output resistance is usually not required when driving purely capacitive loads.

Since the input resistance of the second stage is essentially infinite, the voltage gain of the amplifier in Fig. 6.16 can be found by considering the two stages separately. The first stage is precisely the same configuration as that considered in Section 4.3.5. The small-signal voltage gain is

$$A_{v1} = \frac{v_{o1}}{v_i} = G_{m1} R_{o1} \tag{6.53}$$

where G_{m1} and R_{o1} are the transconductance and output resistance of the first stage, respectively. From (4.143) and (4.149),

$$A_{v1} = g_{m1}(r_{o2} || r_{o4}) \tag{6.54}$$

Similarly, the second-stage voltage gain is

$$A_{v2} = -g_{m6} R_o \tag{6.55}$$

where R_o is given in (6.52). As a result, the overall gain of the amplifier is

$$A_v = A_{v1} A_{v2} = -g_{m1}(r_{o2}||r_{o4})g_{m6}(r_{o6}||r_{o7}) \tag{6.56}$$

This equation shows that the overall gain is related to the quantity $(g_m r_o)^2$. Recall from (3.27) that

$$g_m r_o = \frac{2V_A}{V_{ov}} \tag{6.57}$$

Therefore, the overall voltage gain is a strong function of the Early voltage (which is proportional to the effective channel length) and the overdrive (which is set by the bias conditions).

■ **EXAMPLE**

Calculate the gain of the op amp in Fig. 6.16 assuming that it uses the 0.8-μm process technology described in Table 2.3. Also, assume that $L_{\text{eff}} = 0.8$ μm and $|V_{ov}| = |V_{GS} - V_t| = 0.2$ V for all devices.

Let I_{D2}, I_{D4}, I_{D6}, and I_{D7} represent the bias currents flowing into the drains of M_2, M_4, M_6, and M_7, respectively. Since $I_{D4} = -I_{D2}$ and $I_{D7} = -I_{D6}$, (6.56) shows that

$$A_v = -g_{m1} \left(\frac{\frac{|V_{A2}|}{|I_{D2}|} \frac{V_{A4}}{|I_{D2}|}}{\frac{|V_{A2}|}{|I_{D2}|} + \frac{V_{A4}}{|I_{D2}|}} \right) g_{m6} \left(\frac{\frac{V_{A6}}{I_{D6}} \frac{|V_{A7}|}{I_{D6}}}{\frac{V_{A6}}{I_{D6}} + \frac{|V_{A7}|}{I_{D6}}} \right)$$

$$= -\frac{g_{m1}}{|I_{D2}|} \frac{g_{m6}}{I_{D6}} \left(\frac{|V_{A2}|V_{A4}}{|V_{A2}| + V_{A4}} \right) \left(\frac{V_{A6}|V_{A7}|}{V_{A6} + |V_{A7}|} \right) \tag{6.58}$$

where the absolute-value function has been used so that each quantity in (6.58) is positive. From (1.181),

$$A_v = -\frac{2}{|V_{ov1}|} \frac{2}{V_{ov6}} \left(\frac{|V_{A2}|V_{A4}}{|V_{A2}| + V_{A4}} \right) \left(\frac{V_{A6}|V_{A7}|}{V_{A6} + |V_{A7}|} \right) \tag{6.59}$$

because $I_{D1} = I_{D2}$ with zero differential input. From (1.163),

$$V_A = L_{\text{eff}} \left(\frac{dX_d}{dV_{DS}} \right)^{-1} \tag{6.60}$$

Substituting (6.60) into (6.59) with the given data and dX_d/dV_{DS} from Table 2.3 gives

$$A_v = -\frac{2}{0.2} \frac{2}{0.2} \left(\frac{\frac{0.8}{0.04} \times \frac{0.8}{0.08}}{\frac{0.8}{0.04} + \frac{0.8}{0.08}} \right)^2 \simeq -4400$$

■

The overall gain can be increased by either increasing the channel lengths of the devices to increase the Early voltages or by reducing the bias current to reduce the overdrives.

6.3.2 Output Swing

The output swing is defined to be the range of output voltages $V_o = V_O + v_o$ for which all transistors operate in the active region so that the gain calculated in (6.56) is approximately constant. From inspection of Fig. 6.16, M_6 operates in the triode region if the output voltage is less than $V_{ov6} - V_{SS}$. Similarly, M_7 operates in the triode region if the output voltage is more than $V_{DD} - |V_{ov7}|$. Therefore, the output swing is

$$V_{ov6} - V_{SS} \leq V_o \leq V_{DD} - |V_{ov7}| \tag{6.61}$$

This inequality shows that the op amp can provide high gain while its output voltage swings within one overdrive of each supply. Beyond these limits, one of the output transistors enters the triode region, where the overall gain of the amplifier would be greatly diminished. As a result, the output swing can be increased by reducing the overdrives of the output transistors.

6.3.3 Input Offset Voltage

In Sections 3.5.6 and 6.2.3, the input offset voltage of a differential amplifier was defined as the differential input voltage for which the differential output voltage is zero. Because the op amp in Fig. 6.16 has a single-ended output, this definition must be modified here. Referring to the voltage between the output node and ground as the *output voltage*, the most straightforward modification is to define the input offset voltage of the op amp as the differential input voltage for which the op-amp output voltage is zero. This definition is reasonable if $V_{DD} = V_{SS}$ because setting the output voltage to zero maximizes the allowed variation in the output voltage before one transistor operates in the triode region provided that $V_{ov6} = |V_{ov7}|$. If $V_{DD} \neq V_{SS}$, however, the output voltage should be set midway between the supply voltages to maximize the output swing. Therefore, we will define the input offset voltage of op amps with differential inputs and single-ended outputs as the differential input voltage for which the dc output voltage is midway between the supplies.

The offset voltage of an op amp is composed of two components: the systematic offset and the random offset. The former results from the design of the circuit and is present even when all the matched devices in the circuit are identical. The latter results from mismatches in supposedly identical pairs of devices.

Systematic Offset Voltage. In bipolar technologies, the gain of each stage in an op amp can be quite high (on the order of 500) because the $g_m r_o$ product is usually greater than 1000. As a result, the input-referred offset voltage of a bipolar op amp usually depends mainly on the design of the first stage. In MOS technologies, however, the $g_m r_o$ product is usually between about 20 and 100, reducing the gain per stage and sometimes causing the offset of the second stage to play an important role in determining the op-amp offset voltage.

To study the systematic offset, Fig. 6.17 shows the op amp of Fig. 6.16 split into two separate stages. If the inputs of the first stage are grounded, and if the matching is perfect, then the dc drain-source voltage of M_4 must be equal to the dc drain-source voltage of M_3. This result stems from the observation that if $V_{DS3} = V_{DS4}$, then $V_{DS1} = V_{DS2}$ and $I_{D1} = I_{D2} = I_{D5}/2$. Therefore, with $V_{DS3} = V_{DS4}$, $I_{D3} = I_{D4} = -I_{D5}/2$. As a result, V_{DS3} must be equal to V_{DS4} because this operating point is the only point for which the current flowing out of the drain of M_2 is equal to the current flowing into the drain of M_4. For example, increasing the drain-source voltage of M_4 would increase the current flowing into the drain of M_4 but decrease the current flowing out of the drain of M_2 because of the effects of channel-length modulation. Therefore, the dc drain-source voltages of M_3 and M_4 must be equal under these conditions.

On the other hand, the value of the gate-source voltage of M_6 required to set the amplifier output voltage midway between the supplies may differ from the dc output voltage of the first stage. If the first stage gain is 50, for example, each 50-mV difference in these voltages results in 1 mV of input-referred systematic offset. Ignoring channel-length modulation in M_5 and M_7, the current in these transistors is independent of their drain-source voltages if they operate

6.3 Basic Two-Stage MOS Operational Amplifiers

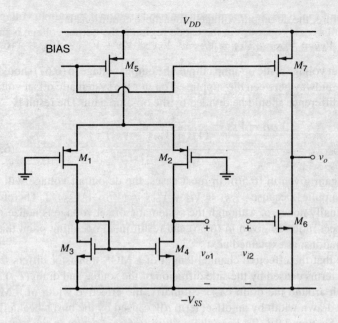

Figure 6.17 Two-stage amplifier with first and second stages disconnected to show the effect of interstage coupling on input-referred offset voltage.

in the active region. To set the output voltage midway between the supplies, the gate-source voltage of M_6 should be chosen so that the drain current of M_6 is equal to the drain current of M_7 while both transistors operate in the active region. When the input of the second stage is connected to the output of the first stage, $V_{GS6} = V_{DS4}$. With perfect matching and zero input voltages, $V_{DS4} = V_{DS3} = V_{GS3}$ and $V_{t3} = V_{t4} = V_{t6}$. Therefore,

$$V_{ov3} = V_{ov4} = V_{ov6} \quad (6.62)$$

is required. Substituting (1.166) into (6.62) gives

$$\frac{I_{D3}}{(W/L)_3} = \frac{I_{D4}}{(W/L)_4} = \frac{I_{D6}}{(W/L)_6} \quad (6.63)$$

In other words, requiring that the transistors have equal overdrives is equivalent to requiring that they have equal drain-current-to-W/L ratios (or *current densities*). Since $I_{D3} = I_{D4} = |I_{D5}|/2$ and $I_{D6} = |I_{D7}|$,

$$\frac{|I_{D5}|}{2(W/L)_3} = \frac{|I_{D5}|}{2(W/L)_4} = \frac{|I_{D7}|}{(W/L)_6} \quad (6.64)$$

Since M_5 and M_7 have equal gate-source voltages,

$$\frac{I_{D5}}{I_{D7}} = \frac{(W/L)_5}{(W/L)_7} \quad (6.65)$$

Substituting (6.65) into (6.64) gives

$$\frac{(W/L)_3}{(W/L)_6} = \frac{(W/L)_4}{(W/L)_6} = \frac{1}{2}\frac{(W/L)_5}{(W/L)_7} \quad (6.66)$$

With the aspect ratios chosen to satisfy (6.66), M_3, M_4, and M_6 operate with equal current densities. In the active region, the current density of a device depends not only on its gate-source voltage, but also on its drain-source voltage to some extent. Since the gate-source voltages and current densities of M_3, M_4, and M_6 are equal, the drain-source voltages of these transistors must also be equal. Therefore, the dc output voltage under these conditions is

$$V_O = V_{DS6} - V_{SS} = V_{DS3} - V_{SS} = V_{GS3} - V_{SS} = V_{t3} + V_{ov3} - V_{SS} \quad (6.67)$$

To find the systematic offset voltage at the op-amp output, the output voltage in (6.67) should be subtracted from a voltage midway between the supplies. To refer the systematic offset voltage to the op-amp input, this difference should be divided by the op-amp gain. The result is

$$V_{OS(sys)} = \frac{\dfrac{V_{DD} - V_{SS}}{2} - (V_{t3} + V_{ov3} - V_{SS})}{A_v} \quad (6.68)$$

where A_v is the op-amp gain given in (6.56). In most cases, the dc output voltage will not be midway between the supplies because $V_{GS3} = V_{t3} + V_{ov3} \neq (V_{DD} + V_{SS})/2$. Therefore, the systematic offset is usually nonzero. Although the systematic offset voltage is nonzero in general, the choice of aspect ratios as given in (6.66) can result in an operating point that is insensitive to process variations, as explained next.

Equation 2.35 shows that the effective channel length of a MOS transistor differs from its drawn length by offset terms caused by the side diffusion of the source and drain (L_d) and the depletion region width around the drain (X_d). Similarly, the effective width of a MOS transistor differs from the drawn width by an offset term dW caused by the bird's-beak effect in the oxide described in Section 2.9.1. To keep the ratio in (6.66) constant in the presence of process-induced variations in L_d, X_d, and dW, the drawn channel lengths and widths of the ratioed transistors can each be chosen to be identical. In this case, the ratio in (6.66) can be set equal to any rational number J/K by connecting J identical devices called *n-channel units* in parallel to form M_3 and M_4 while K n-channel units in parallel form M_6. Then if M_5 is constructed of $2J$ identical devices called *p-channel units*, M_7 should be constructed from K p-channel units. In practice for matched devices, the channel lengths are almost never ratioed directly because the use of small channel lengths for high-speed operation would result in a large sensitivity to process variations. On the other hand, the channel widths of matched devices are sometimes ratioed directly when the width is large enough to make the resulting sensitivity to process variations insignificant.

A key point of this analysis is that the use of identical channel lengths for M_3, M_4, and M_6 conflicts with a combination of other requirements. First, for stability reasons described in Chapter 9, M_6 should have a large transconductance and thus a short channel length. Second, for low noise and random input offset voltage, M_3 and M_4 should have a small transconductance and therefore a long channel length. Noise is considered in Chapter 11, and random input offset voltage is considered next.

Random Input Offset Voltage. As described in Section 3.5.6, source-coupled pairs generally display a higher random offset than their bipolar counterparts. Ignoring the contribution of the second stage in the op amp to the input-referred random offset, a straightforward analysis for the offset voltage of the circuit of Fig. 6.16, which is analogous to the analysis leading to (3.248), gives

$$V_{OS} \simeq \Delta V_{t(1-2)} + \Delta V_{t(3-4)} \left(\frac{g_{m3}}{g_{m1}} \right)$$

$$+ \frac{V_{ov(1-2)}}{2} \left[\frac{\Delta\left(\dfrac{W}{L}\right)_{(3-4)}}{\left(\dfrac{W}{L}\right)_{(3-4)}} - \frac{\Delta\left(\dfrac{W}{L}\right)_{(1-2)}}{\left(\dfrac{W}{L}\right)_{(1-2)}} \right] \quad (6.69)$$

The first term represents the threshold mismatch of the input transistors. The second is the threshold mismatch of the current-mirror-load devices and is minimized by choosing the W/L ratio of the load devices so that their transconductance is small compared to that of the input transistors. For this reason, selecting a longer channel length for M_3 and M_4 than for M_1 and M_2 reduces the random input offset voltage. The third term represents the effects of W/L mismatches in the input transistors and loads and is minimized by operating the input transistors at low values of overdrive, typically on the order of 50 to 200 mV.

6.3.4 Common-Mode Rejection Ratio

For the op amp in Fig. 6.16, (6.45) gives

$$\text{CMRR} = \left| \frac{A_{dm}}{A_{cm}} \right| = \left| \frac{\frac{v_o}{v_{o1}} \frac{v_{o1}}{v_{id}}}{\frac{v_o}{v_{o1}} \frac{v_{o1}}{v_{ic}}} \right| = \text{CMRR}_1 \qquad (6.70)$$

where CMRR_1 is the common-mode rejection ratio of the first stage. The second stage does not contribute to the common-mode rejection ratio of the op amp because the second stage has a single-ended input and a single-ended output. In (4.182) and (4.183), the common-mode rejection ratio of a stage with a differential pair and a current-mirror load was calculated assuming perfect matching. Applying (4.183) here gives

$$\text{CMRR} \simeq (2g_{m(dp)} r_{\text{tail}}) g_{m(mir)} (r_{o(dp)} || r_{o(mir)}) \qquad (6.71)$$

where $g_{m(dp)}$ and $r_{o(dp)}$ are the transconductance and output resistance of M_1 and M_2, $g_{m(mir)}$ and $r_{o(mir)}$ are the transconductance and output resistance of M_3 and M_4, and r_{tail} is the output resistance of M_5. By a process similar to the derivation of (6.59), this equation can be simplified to give

$$\text{CMRR} \simeq \left| \frac{2}{V_{ov(dp)}} \frac{2}{V_{ov(mir)}} \left(\frac{V_{A(dp)} V_{A(mir)}}{|V_{A(dp)}| + |V_{A(mir)}|} \right) \right| \qquad (6.72)$$

where $V_{ov(dp)}$ and $V_{A(dp)}$ are the overdrive and Early voltage of the differential pair, and $V_{ov(mir)}$ and $V_{A(mir)}$ are the overdrive and Early voltage of the mirror. Equation 6.72 shows that the common-mode rejection ratio of the op amp can be increased by reducing the overdrive voltages.

Another way to increase the common-mode rejection ratio is to replace the simple current mirror M_5 and M_8 with one of the high-output-resistance current mirrors considered in Chapter 4. Unfortunately, such a replacement would also worsen the common-mode input range.

6.3.5 Common-Mode Input Range

The common-mode input range of the op amp in Fig. 6.16 is the range of dc common-mode input voltages for which all transistors in the first stage operate in the active region. To operate in the active region, the gate-drain voltages of n-channel transistors must be less than their thresholds so that their channels do not exist at their drains. Similarly, p-channel transistors operate in the active region only if their gate-drain voltages are more than their thresholds, again so that their channels do not exist at their drains. With a pure common-mode input V_{IC} applied to the inputs of the op amp in Fig. 6.16,

$$V_{DS4} = V_{DS3} = V_{GS3} = V_{t3} + V_{ov3} \qquad (6.73)$$

The gate-drain voltage of M_1 and M_2 is

$$V_{GD1} = V_{GD2} = V_{IC} - V_{t3} - V_{ov3} + V_{SS} \quad (6.74)$$

When V_{IC} is reduced to the point at which $V_{GD1} = V_{GD2} = V_{t1} = V_{t2}$, M_1 and M_2 operate at the edge between the the triode and active regions. This point defines the lower end of the common-mode range, which is

$$V_{IC} > V_{t1} + V_{t3} + V_{ov3} - V_{SS} \quad (6.75)$$

If V_{IC} is too high, however, M_5 operates in the triode region. The drain-source voltage of M_5 is

$$V_{DS5} = V_{IC} - V_{GS1} - V_{DD} = V_{IC} - V_{t1} - V_{ov1} - V_{DD} \quad (6.76)$$

From the standpoint of drain-source voltages, n-channel transistors operate in the active region only if their drain-source voltage is more than their overdrive. On the other hand, p-channel transistors operate in the active region only if their drain-source voltage is less than their overdrive. Therefore, the upper end of the common-mode input range is

$$V_{IC} < V_{t1} + V_{ov1} + V_{ov5} + V_{DD} \quad (6.77)$$

Since M_1 and M_5 are p-channel transistors, their overdrives are negative; that is, their gate-source voltages must be less than their thresholds for the channel to exist at the source. Furthermore, if M_1 is an enhancement-mode device, its threshold is negative because it is p-type. Under this assumption, the common-mode range limits in (6.75) and (6.77) can be rewritten as

$$V_{t3} - |V_{t1}| + V_{ov3} - V_{SS} < V_{IC} < V_{DD} - |V_{t1}| - |V_{ov1}| - |V_{ov5}| \quad (6.78)$$

This inequality shows that the magnitudes of the overdrive terms should be minimized to maximize the common-mode range. Also, the body effect on the input transistors can be used to increase the range. If the bodies of M_1 and M_2 are connected to V_{DD} as implied in Fig. 6.16, the source-body voltage of these transistors is low when V_{IC} is high. Therefore, the upper limit in (6.78) can be found approximately by using the zero-bias value of V_{t1}. On the other hand, when V_{IC} decreases, the source-body voltage of M_1 and M_2 becomes more negative, widening the depletion region around the source and making the threshold voltage of these transistors more negative. Therefore, the body effect can be used to include the negative supply in the common-mode range.

■ **EXAMPLE**

For the two-stage CMOS op amp in Fig. 6.16, choose the device sizes to give a dc voltage gain greater than 5000 and a peak output swing of at least 1 V. Use the 0.4 μm CMOS model parameters in Table 2.4. Use bias currents of $|I_{D1}| = |I_{D2}| = 100$ μA, and $I_{D6} = 400$ μA. Assume $V_{DD} = V_{SS} = 1.65$ V $\pm$ 0.15 V. Assume perfect matching and that all transistors operate in the active (or saturation) region with dc voltages $V_{IC} = 0$ (where V_{IC} is the common-mode input voltage), $V_I = 0$ and $V_O \simeq 0$. Ignore the body effect.

To simplify the design, a drawn channel length $L = 1$ μm will be used for all transistors. This selection avoids short-channel effects that degrade output resistance and cause transistor operation to deviate from the square-law equations.

Since the peak output swing should be 1 V and the magnitude of each supply is at least 1.5 V, (6.61) shows that

$$V_{ov6} = |V_{ov7}| \leq 0.5 \text{ V}$$

To maximize the transition frequency f_T of each device subject to this constraint, we will choose $V_{ov6} = |V_{ov7}| = 0.5$ V. Using (1.157) with $|I_{D7}| = I_{D6} = 400$ μA gives

$$\left(\frac{W}{L}\right)_7 = \frac{2|I_{D7}|}{k'_p(V_{ov7})^2} = \frac{2(400)}{64.7(-0.5)^2} \simeq 50$$

and

$$\left(\frac{W}{L}\right)_6 = \frac{2I_{D6}}{k'_n(V_{ov6})^2} = \frac{2(400)}{194(0.5)^2} \simeq 16$$

Since $V_{ov5} = V_{ov7}$ by KVL and $I_{D1} + I_{D2} = I_{D7}/2$,

$$\left(\frac{W}{L}\right)_5 = \frac{1}{2}\left(\frac{W}{L}\right)_7 \simeq 25$$

From (6.66),

$$\left(\frac{W}{L}\right)_3 = \left(\frac{W}{L}\right)_4 = \frac{1}{2}\frac{(W/L)_5}{(W/L)_7}(W/L)_6 \simeq \frac{1}{2}\left(\frac{25}{50}\right)16 = 4$$

Since the common-mode input range should include $V_{IC} = 0$, the allowed overdrives on M_1 and M_2 are limited by (6.77), and rearranging this equation with $V_{IC} = 0$ and $V_{DD} = 1.5$ V gives

$$V_{ov1} > V_{IC} - V_{t1} - V_{ov5} - V_{DD} = 0 - (-0.8 \text{ V}) - (-0.5 \text{ V}) - 1.5 \text{ V} = -0.2 \text{ V}$$

Therefore,

$$\left(\frac{W}{L}\right)_1 = \left(\frac{W}{L}\right)_2 > \frac{2|I_{D1}|}{k'_p(V_{ov1})^2} = \frac{2(100)}{64.7(-0.2)^2} \simeq 77$$

From (6.59) and (6.60) with $L_{\text{eff}} \simeq L_{\text{drwn}} - 2L_d$, and using data from Table 2.4,

$$A_v = -\frac{2}{|V_{ov1}|}\frac{2}{V_{ov6}}\left(\frac{|V_{A2}|V_{A4}}{|V_{A2}| + V_{A4}}\right)\left(\frac{V_{A6}|V_{A7}|}{V_{A6} + |V_{A7}|}\right)$$

$$= -\frac{2}{0.2}\frac{2}{0.5}\left(\frac{\frac{0.82}{0.04} \times \frac{0.82}{0.02}}{\frac{0.82}{0.04} + \frac{0.82}{0.02}}\right)^2 \simeq -7500$$

This calculation assumes that dX_d/dV_{DS} and L_{eff} are constant for each type of transistor, allowing us to use constant Early voltages. In practice, however, dX_d/dV_{DS} and L_{eff} both depend on the operating point, and accurate values of the Early voltages are rarely available to circuit designers when channel lengths are less than about 1.5 μm. As a result, circuit simulations are an important part of the design process. SPICE simulation of the op amp under the conditions described above gives a gain of about 6200, which shows that the hand calculations are accurate within about 20 percent.

∎

6.3.6 Power-Supply Rejection Ratio (PSRR)

To calculate the PSRR from the V_{dd} supply for the op amp in Fig. 6.16, we will divide the small-signal gain $A^+ = v_o/v_{dd}$ into the gain from the input. For this calculation, assume that the V_{ss} supply voltage is constant and that both op-amp inputs in Fig. 6.16 are connected to small-signal grounds. The current in M_8 is equal to I_{BIAS}. If this current is constant, the gate-source voltage of M_8 must be constant because M_8 is diode connected. Therefore, $v_{gs8} = v_{gs5} = v_{gs7} = 0$, and the g_m generators in M_5 and M_7 are inactive. As a result, if $r_{o5} = r_{tail} \to \infty$ and $r_{o7} \to \infty$, $v_o/v_{dd} = 0$. To find the gain with finite r_{tail} and r_{o7}, consider the small-signal diagrams shown in Fig. 6.18, where the g_m generators for M_5 and M_7 are omitted because they are inactive. In Fig. 6.18a, the output is defined as v_{oa}, and the v_{dd} supply variation is set equal to zero at the point where r_{tail} is connected. In Fig. 6.18b, the output is defined as v_{ob}, and the v_{dd} supply variation is set equal to zero at the point where r_{o7} is connected. We will find v_{oa} and v_{ob} separately and use superposition to find the total gain $v_o/v_{dd} = (v_{oa} + v_{ob})/v_{dd}$.

In Fig. 6.18a, the first stage experiences no variation, and $v_{gs6} = 0$. Therefore, g_{m6} is inactive, and the output stage appears as a simple voltage divider to the supply variation. Since the dc drain current in M_6 is equal and opposite to that in M_7,

$$\frac{v_{oa}}{v_{dd}} = \frac{r_{o6}}{r_{o6} + r_{o7}} = \frac{\frac{V_{A6}}{I_{D6}}}{\frac{V_{A6}}{I_{D6}} + \frac{|V_{A7}|}{I_{D6}}} = \frac{V_{A6}}{V_{A6} + |V_{A7}|} \quad (6.79)$$

In Fig. 6.18b,

$$\frac{v_{ob}}{v_{dd}} = \frac{v_{gs6}}{v_{dd}} \frac{v_{ob}}{v_{gs6}} \quad (6.80)$$

where the first term on the right side represents the gain of the first stage, and the second term represents the gain of the second stage. The v_{dd} input to the first stage in Fig. 6.18b is applied between the top of r_{tail} and ground while the gates of M_1 and M_2 are grounded. This situation is equivalent to grounding the top of r_{tail} and applying a voltage of $-v_{dd}$ between the gates of M_1 and M_2 and ground. In other words, the v_{dd} input in Fig. 6.18b appears as a common-mode

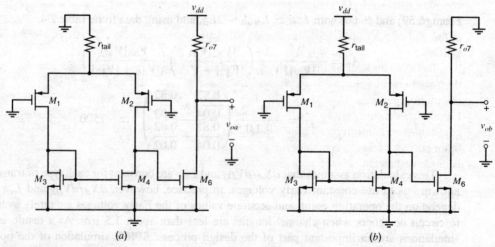

Figure 6.18 Small-signal diagrams of the two-stage op amp used to calculate the coupling from v_{dd} to the output (a) through the second stage and (b) through the first stage.

input $v_{ic} = -v_{dd}$ to the first stage. Therefore, the gain of the first stage can be expressed as

$$\frac{v_{gs6}}{v_{dd}} = -\frac{v_{gs6}}{v_{ic}} = -G_m[cm]R_{o1} \qquad (6.81)$$

where $G_m[cm]$ is the common-mode transconductance of the first stage and R_{o1} is the output resistance of the first stage. Substituting (4.149), (4.166), (4.173), and (4.179) into (6.81) gives

$$\frac{v_{gs6}}{v_{dd}} \simeq \frac{g_{m(dp)}(r_{o(dp)}||r_{o(mir)})}{1 + 2g_{m(dp)}r_{tail}} \left(\frac{1}{1 + g_{m(mir)}r_{o(dp)}} + \frac{1}{1 + g_{m(mir)}r_{o(mir)}} \right) \qquad (6.82)$$

If $2g_{m(dp)}r_{tail} \gg 1$, $g_{m(mir)}r_{o(dp)} \gg 1$, and $g_{m(mir)}r_{o(mir)} \gg 1$,

$$\frac{v_{gs6}}{v_{dd}} \simeq \frac{r_{o(dp)}||r_{o(mir)}}{2r_{tail}g_{m(mir)}(r_{o(dp)}||r_{o(mir)})} = \frac{1}{2g_{m(mir)}r_{tail}} \qquad (6.83)$$

Substituting (6.83), (6.55), and (6.52) into (6.80) gives

$$\frac{v_{ob}}{v_{dd}} \simeq -\frac{g_{m6}(r_{o6}||r_{o7})}{2g_{m(mir)}r_{tail}} \qquad (6.84)$$

If $V_{ov3} = V_{ov6}$ as in (6.62) to control the systematic offset, $g_{m6}/g_{m(mir)} = I_{D6}/I_{D3}$. Since the dc drain current in M_6 is equal and opposite to that in M_7,

$$\frac{v_{ob}}{v_{dd}} \simeq -\frac{I_{D6}}{2I_{D3}} \left(\frac{\frac{V_{A6}}{I_{D6}} \frac{|V_{A7}|}{I_{D6}}}{\frac{V_{A6}}{I_{D6}} + \frac{|V_{A7}|}{I_{D6}}} \right) \frac{|I_{D5}|}{|V_{A5}|} = -\frac{|I_{D5}|}{2I_{D3}} \left(\frac{V_{A6}}{V_{A6} + |V_{A7}|} \right) \frac{|V_{A7}|}{|V_{A5}|}$$

$$= -\frac{V_{A6}}{V_{A6} + V_{A7}} \qquad (6.85)$$

because $V_{A5} = V_{A7}$ and $|I_{D5}| = 2I_{D3}$. Combining (6.79) and (6.85) gives

$$A^+ = \frac{v_o}{v_{dd}} = \frac{v_{oa} + v_{ob}}{v_{dd}} \simeq 0 \qquad (6.86)$$

Therefore, from (6.51), PSRR$^+ \to \infty$ for low frequencies with perfect matching because the coupling from v_{dd} to the output through the first stage cancels that through the second stage. In practice, mismatch can increase the common-mode transconductance of the first stage as shown at the end of Section 4.3.5.3, disrupting this cancellation and decreasing the low-frequency PSRR$^+$.

To calculate the PSRR from the V_{ss} supply for the op amp in Fig. 6.16, we will calculate the small-signal gain $A^- = v_o/v_{ss}$ and then normalize to the gain from the input. For this calculation, assume that the V_{dd} supply voltage is constant and that both op-amp inputs in Fig. 6.16 are connected to small-signal grounds. Under these conditions, M_1 and M_2 act as common-gate amplifiers, attempting to keep the bias current in M_3 and M_4 constant. If the drain current of M_3 is constant, the gate-source voltage of M_3 must be constant because M_3 is diode connected. Therefore, $v_{gs3} = 0$. Since $v_{ds3} = v_{gs3}$, and since $v_{ds4} = v_{ds3}$ under these conditions, $v_{ds4} = v_{gs6} = 0$. Therefore, g_{m6} is inactive, and the output stage appears as a simple voltage divider to the supply variation. Since the drain current in M_6 is equal and opposite to that in M_7,

$$A^- = \frac{v_o}{v_{ss}} = \frac{r_{o7}}{r_{o6} + r_{o7}} = \frac{\frac{|V_{A7}|}{I_{D6}}}{\frac{V_{A6}}{I_{D6}} + \frac{|V_{A7}|}{I_{D6}}} = \frac{|V_{A7}|}{V_{A6} + |V_{A7}|} \qquad (6.87)$$

Substituting (6.59) and (6.87) into (6.51) gives

$$\text{PSRR}^- = \frac{A_{dm}}{A^-} = \frac{\dfrac{v_o}{v_{id}}}{\dfrac{v_o}{v_{ss}}} = -\frac{2}{|V_{ov1}|}\frac{2}{V_{ov6}}\left(\frac{|V_{A2}|V_{A4}}{|V_{A2}|+V_{A4}}\right)V_{A6} \quad (6.88)$$

This equation gives the low-frequency supply rejection from the negative supply. This rejection worsens as frequency increases. The topic of frequency response is covered in detail in Chapters 7 and 9, but the essence of this behavior can be understood without a complete frequency-response analysis. As the applied frequency increases, the impedance of the compensation capacitor C_C in Fig. 6.16 decreases, effectively shorting the gate of M_6 to its drain for high-frequency ac signals. If the gate-source voltage on M_6 is constant, the variation on the negative supply is fed directly to the output at high frequencies. Therefore, $A^- \simeq 1$ at frequencies high enough to short circuit C_C, assuming that $C_C \gg C_L$, where C_L is the load capacitance of the op amp connected between the op-amp output and ground. The same phenomenon causes the gains A_{dm} and A^+ to decrease as frequency increases, so that the PSRR$^+$ remains relatively constant with increasing frequency. Since A^- increases to unity as A_{dm} decreases, however, PSRR$^-$ decreases and reaches unity at the frequency where $|A_{dm}| = 1$.

Power-Supply Rejection and Supply Capacitance. Another important contribution to nonzero gain between the power supplies and the op-amp output is termed supply capacitance.[9,10] This phenomenon manifests itself as a capacitive coupling between one or both of the power supplies and the op-amp input leads when the op amp is connected with capacitive feedback C_I as shown in Fig. 6.19. For simplicity, assume that the op-amp open-loop gain is infinite. If the supply-coupling capacitance is C_{sup}, the gain from C_{sup} to the op-amp output is $-C_{\text{sup}}/C_I$. Figure 6.19 shows two possible sources of supply capacitance, which are the gate-drain and gate-source capacitance of M_1. Four important ways in which supply capacitance can occur are described below.

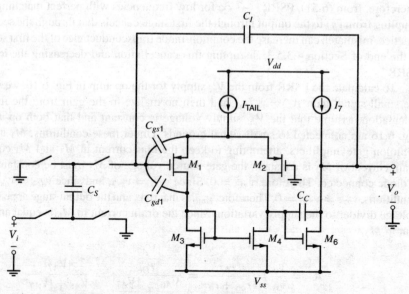

Figure 6.19 Supply capacitance in a two-stage MOS amplifier with capacitive feedback.

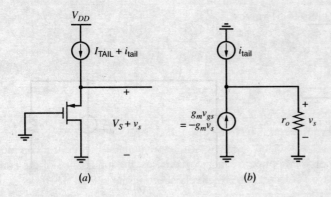

Figure 6.20 (a) Source follower and (b) small-signal diagram to calculate the dependence of v_s on i_{tail}.

1. If the drain current of M_3 is constant, a variation on V_{ss} causes the voltage from the drain of M_1 to ground to vary to hold the gate-source voltage of M_3 constant. This variation couples to the summing node through the gate-drain capacitance of M_1; that is, the supply capacitance $C_{\text{sup}} = C_{gd1}$. This problem is usually overcome by the use of cascode transistors in series with the drains of the input transistors.

2. A variation on V_{dd} or V_{ss} causes the current flowing in the tail current source to vary. To understand the effect of this bias-current variation, consider Fig. 6.20a, which shows a p-channel source follower whose biasing current source $I_{\text{tail}} = I_{\text{TAIL}} + i_{\text{tail}}$ is not constant. The source follower models the behavior of M_1 in Fig. 6.19 from the standpoint of variation in I_{tail} for two reasons. First, the voltage from the gate of M_1 to ground is held to small-signal ground by negative feedback. Second, MOS transistors that operate in the active region are controlled mainly by their gate-source voltages. For simplicity, ignore the body effect because it is not needed to demonstrate the problem here. The small-signal diagram of the source follower is shown in Fig. 6.20b. From KCL at the source,

$$i_{\text{tail}} = g_m v_s + \frac{v_s}{r_o} \tag{6.89}$$

Rearranging this equation gives

$$v_s = \frac{i_{\text{tail}} r_o}{1 + g_m r_o} \simeq \frac{i_{\text{tail}}}{g_m} \tag{6.90}$$

Therefore, nonzero i_{tail} arising from supply variations causes nonzero v_s in the source follower. Similarly, in Fig. 6.19, the voltage from the source of M_1 to ground varies with I_{tail}, and this variation couples to the summing node through the gate-source capacitance of M_1; that is, the supply capacitance $C_{\text{sup}} = C_{gs1}$. A supply-independent bias reference is usually used to overcome this problem.

3. If the substrate terminal of the input transistors is connected to a supply or a supply-related voltage, then the substrate bias changes as the supply voltage changes. In turn, substrate bias variation changes the threshold through the body effect, which changes the gate-source voltage. Again, this mechanism can be studied with the help of a source follower, as shown in Fig. 6.21a. Here I_{TAIL} is assumed to be constant, but $V_{dd} = V_{DD} + v_{dd}$ is assumed to vary. The small-signal diagram is shown in Fig. 6.21b. From KCL at the source,

$$g_{mb} v_{dd} = g_m v_s + g_{mb} v_s + \frac{v_s}{r_o} \tag{6.91}$$

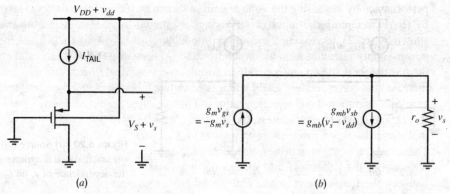

Figure 6.21 (a) Source follower and (b) small-signal diagram to calculate the dependence of v_s on v_{dd} through the body effect.

Rearranging this equation gives

$$v_s = \frac{g_{mb} r_o}{1 + (g_m + g_{mb}) r_o} v_{dd} \simeq \frac{g_{mb}}{g_m + g_{mb}} v_{dd} \qquad (6.92)$$

Therefore, nonzero v_{dd} in the source follower in Fig. 6.21 causes nonzero v_s. Similarly, in Fig. 6.19, the voltage from the source of M_1 to ground varies with V_{dd}, and this variation couples to the summing node through the gate-source capacitance of M_1; that is, the supply capacitance $C_{\text{sup}} = C_{gs1}$.

A solution to this problem is to place the input transistors in a well and connect the well to the sources of the input transistors to eliminate the body effect. This solution has two potential disadvantages. First, it disallows the use of the body effect on the input devices to increase the common-mode input range of the op amp, as described in Section 6.3.5. Second, this solution dictates the polarity of the input transistors in a given process. For example, in a p-well process, the input devices must be n-channel devices to place them in a p well. This requirement might conflict with the polarity of the input transistors that would be chosen for other reasons. For example, p-channel input transistors would be used to minimize the input-referred flicker noise. (See Chapter 11.)

4. Interconnect crossovers in the op-amp and system layout can produce undesired capacitive coupling between the supplies and the summing node. In this case, the supply capacitance is a parasitic or undesired capacitance. This problem is usually overcome with careful layout. In particular, one important layout technique is to shield the op-amp inputs with metal lines connected to ground.

The result of supply capacitance can be quite poor power-supply rejection in switched-capacitor filters and other sampled-data analog circuits that use capacitive feedback. In addition to the solutions to this problem mentioned above, another solution is to use fully differential op amps, which have two outputs. The output voltage of interest is the voltage difference between these outputs. Fully differential op amps, which are considered in Chapter 12, overcome the supply capacitance problem to the extent that the coupling from a given supply to one output is the same as to the other output.

6.3.7 Effect of Overdrive Voltages

The overdrive of a MOS transistor can be reduced by reducing the ratio of its drain current to its W/L. Reducing the overdrive voltages in the op amp in Fig. 6.16 improves the op-amp performance by increasing the voltage gain as shown by (6.59), increasing the swing as shown by (6.61), reducing the input offset voltage as shown by (6.69), increasing the CMRR as shown by (6.72), increasing the common-mode range as shown by (6.78), and increasing the power-supply rejection ratio as shown by (6.88). These observations are valid provided that the transistors in the op amp operate in strong inversion. Also, increasing the channel lengths increases the corresponding Early voltages as shown by (1.163), and thereby increases the op-amp gain, common-mode rejection ratio, and power-supply rejection ratio as shown by (6.59), (6.72), and (6.88). Unfortunately, the transition frequency of MOS transistors is proportional to the overdrive and inversely proportional to the square of the channel length from (1.209). Therefore, reducing overdrives and increasing the channel lengths degrades the frequency response of the transistors and in turn the amplifier. Thus we find a fundamental trade-off between the frequency response and the other measures of performance in CMOS op-amp design.

6.3.8 Layout Considerations

A basic objective in op-amp design is to minimize the mismatch between the two signal paths in the input differential pair so that common-mode input signals are rejected to the greatest possible extent. Mismatch affects the performance of the differential pair not only at dc, where it causes nonzero offset voltage, but also at high frequencies where it reduces the common-mode and power-supply rejection ratios.

Figure 6.22a shows a possible layout of a differential pair. Five nodes are labeled: two gates, two drains, and one source. Connections to each region are omitted for simplicity. The

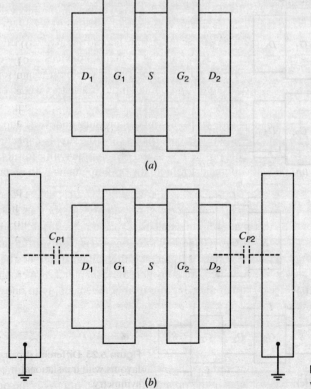

Figure 6.22 Differential-pair layouts with mirror symmetry.

sources of the two transistors are connected to each other by merging the two sources together into one diffusion region. Although such a layout saves area and minimizes undesired capacitance connected to the sources, this layout is not optimum from the standpoint of matching in part because it is sensitive to alignment shifts between masks that define various layers in an integrated circuit. The key problem is that the layout in Fig. 6.22a uses only *mirror* symmetry in the sense that each transistor is a mirror image of the other. For example, suppose that two additional grounded segments of metal are added to the layout to produce the layout shown in Fig. 6.22b. In exactly these locations, the parasitic capacitance C_{P1} from D_1 to ground is equal to the parasitic capacitance C_{P2} from D_2 to ground. However, if the mask that defines the metal shifts to the right compared to the mask that defines the diffusion, C_{P1} increases but C_{P2} decreases, creating mismatch. In practice, balancing the parasitics in a way that is insensitive to alignment shifts is most important in amplifiers that have both differential inputs and differential outputs. Such amplifiers are considered in Chapter 12.

Figure 6.23 shows layouts that overcome these problems. In Fig. 6.23a-b, the transistors are drawn using translational symmetry; that is, each transistor is a copy of the other without rotation. Another option is shown in Fig. 6.23c, where each transistor has been split into two pieces. To maintain the same width/length ratio as in the previous drawings, the width of each transistor in Fig. 6.23c has been reduced by a factor of two. This structure has both translational and mirror symmetry. Structures with translational symmetry are insensitive to alignment shifts.

One limitation of these layouts is that they are sensitive to process gradients perpendicular to the line of symmetry. For example, in Fig. 6.23c, suppose the oxide thickness increases from

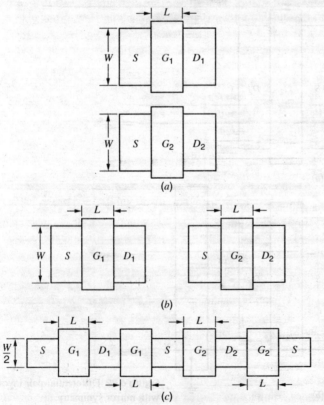

Figure 6.23 Differential-pair layouts with translational symmetry.

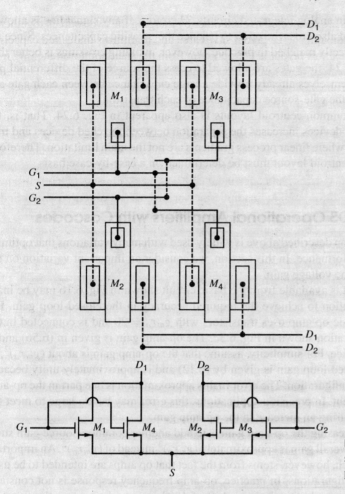

Figure 6.24 Common-centroid structure for the MOS differential pair.

left to right. Then the gate stripes connected to G_1 have thinner oxide than those connected to G_2, causing the transistors to have unequal thresholds and unequal transconductance parameters. The effects of process-related gradients across the die can be partially alleviated by use of common-centroid geometries. Figure 6.24 shows a common-centroid layout of the differential pair. Each side of the differential pair is split into two components that are cross connected in the layout. In a geometric sense, the centroid of both composite devices lies at center of the structure. Because any gradient can be decomposed into horizontal and vertical components, this layout overcomes the effect of linear process gradients in any direction.

Figure 6.24 also shows two layers of metal used to cross connect the devices. One layer of metal is drawn with solid lines. The other is drawn with dashed lines. The two layers connect at intersections only where dots are drawn. The interconnect drawn here is shift insensitive and balanced in the sense that any signal line that crosses a node on one side of the differential pair also crosses the corresponding node on the other side. This balance helps to keep undesired signals in common-mode form so they can be rejected by the differential pair. Finally, the only line that crosses the metal connected to the two input gates is the metal line to the sources of the differential pair. This characteristic is important because such crossings create a small parasitic capacitance between the two layers that cross, and can allow undesired signals to couple to the op-amp inputs. Since op amps are designed to have high gain, op-amp inputs are the most sensitive nodes in analog integrated circuits. Therefore, if any signal line is allowed to cross one gate, it should also cross the other to balance the parasitic capacitances. Since the parasitics may not be perfectly matched in practice, however, avoiding crossings is better than balancing them. In Fig. 6.24, the gates are allowed to cross the source of the differential pair because the transistors themselves already provide a large capacitance between each gate and the source in the form of the gate-source capacitance of each transistor.

A disadvantage of common-centroid layouts is also apparent in Fig. 6.24. That is, the need to cross-connect the devices increases the separation between matched devices and may worsen matching in cases where linear process gradients are not the main limitation. Therefore, the value of a common-centroid layout must be determined on a case-by-case basis.

6.4 Two-Stage MOS Operational Amplifiers with Cascodes

The basic two-stage op amp described above is widely used with many variations that optimize certain aspects of the performance. In this section, we consider an important variation on the basic circuit to increase the voltage gain.

The voltage gain that is available from the basic circuit shown in Fig. 6.16 may be inadequate in a given application to achieve the required accuracy in the closed-loop gain. For example, suppose the basic op amp uses transistors with $g_m r_o = 20$ and is connected in the voltage-follower configuration shown in Fig. 6.3c. The op-amp gain is given in (6.56) and is less than $(g_m r_o)^2$ in practice. For simplicity, assume that the op-amp gain is about $(g_m r_o)^2$, or 400 in this case. The closed-loop gain is given by (6.12) and is approximately unity because $R_2 = 0$ in the follower configuration. The error in this approximation is one part in the op-amp gain or at least 0.25 percent. In precision applications, this error may be too large to meet the given specifications, requiring an increase in the op-amp gain.

One approach to increasing the op-amp gain is to add another common-source gain stage to the op amp so that the overall gain is approximately $(g_m r_o)^3$ instead of $(g_m r_o)^2$. An important problem with this approach, however, stems from the fact that op amps are intended to be used in negative-feedback configurations. In practice, op-amp frequency response is not constant. If the op amp introduces an additional phase shift of $180°$ at some frequency, the negative feedback that was intended becomes positive feedback at that frequency, and the op amp may be unstable. The topics of frequency response and stability are covered in detail in Chapters 7 and 9, respectively. The key point here is that if an op amp is unstable in a given feedback configuration, it does not act as an amplifier but as a latch or oscillator. To avoid this problem, op amps are usually designed with no more than two gain stages because each stage contains a node for which the impedance to ground is high and as a result contributes a significant pole to the op-amp transfer function. Since the phase shift from one pole approaches $-90°$ asymptotically, an op amp with no more than two poles cannot provide the $180°$ phase shift that is required to convert negative feedback into positive feedback.

To increase the voltage gain without adding another common-source gain stage, common-gate transistors can be added. Together with a common-source transistor, a common-gate transistor forms a cascode that increases the output resistance and gain of the stage while contributing a less significant pole to the amplifier transfer function than would be contributed by another common-source stage. Figure 6.25 illustrates the use of cascodes to increase the voltage gain of a two-stage amplifier. Here, a series connection of two transistors, one in the common-source connection and one in the common-gate connection, replace each common-source transistor in the first stage. Therefore, M_1 and M_{1A} in Fig. 6.25 replace M_1 in Fig. 6.16. Similarly, M_2 and M_{2A} in Fig. 6.25 replace M_2 in Fig. 6.16. Transistor M_9 and current source I_C have also been added to bias the gates of M_{1A} and M_{2A}. In practice, the

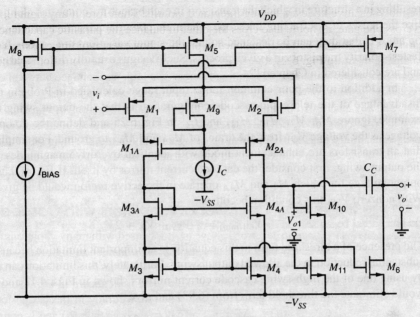

Figure 6.25 Two-stage amplifier with cascoded first stage.

W/L of M_9 is chosen so that M_1 and M_2 are operated barely in the active region. The effect of these replacements is to increase the unloaded output impedance of the differential pair by a factor that is approximately equal to $g_m r_o$ of the cascode device.

If the current mirror M_3-M_4 were not also cascoded, however, the output resistance of the first stage including the current-mirror load would be limited by the mirror. To overcome this limitation, a cascode current mirror shown in Fig. 4.9 is used instead. As a result, the stage gain and output resistance including the load are increased by approximately a factor $g_m r_o$. In this circuit, M_{10} and M_{11} are included to level shift the output of the first stage down by V_{GS10} so that the second stage input is driven by a signal whose dc level is V_{GS3} above $-V_{SS}$. If the aspect ratios are chosen to satisfy (6.66), and if each common-gate transistor is identical to its common-source counterpart, the systematic offset voltage with this level shift is given by (6.68), where the op-amp gain here is on the order of $(g_m r_o)^3$ instead of $(g_m r_o)^2$ in Fig. 6.16. One disadvantage of this circuit is a substantial reduction in the common-mode input range. (See Problem 6.16.) To overcome this problem, cascoding could be added instead to the second stage. In that case, however, the output swing of the op amp would be degraded by the cascodes.

6.5 MOS Telescopic-Cascode Operational Amplifiers

As mentioned in the previous section, cascode configurations may be used to increase the voltage gain of CMOS transistor amplifier stages. In many applications, the stage gain can be increased enough so that a single common-source–common-gate stage can provide enough voltage gain to meet the accuracy requirements. The first stage of Fig. 6.25 is sometimes used by itself as an op amp and provides a gain comparable to the gain of the two-stage op-amp in Fig. 6.16. This structure has been called a *telescopic-cascode* op amp[11] because the cascodes are connected between the power supplies in series with the transistors in the differential pair, resulting in a structure in which the transistors in each branch are connected along a straight line like the lenses of a refracting telescope. The main potential advantage of telescopic cascode op amps is that they can be designed so that the signal variations are entirely handled by the fastest-polarity transistors in a given process. Such designs use fully differential configurations and are considered in Chapter 12.

In addition to the poor common-mode input range calculated in Problem 6.16, another disadvantage of the telescopic-cascode configuration is that the output swing is small. For example, ignore M_6, M_7, M_{10}, M_{11}, and C_C in Fig. 6.25 and define the dc op-amp output voltage as the voltage V_{O1} from the drains of M_{2A} and M_{4A} to ground. For simplicity, assume that all transistors are enhancement mode with identical overdrive magnitudes. To calculate the output swing, first consider the cascode current mirror by itself. If $V_{SS} = 0$, the minimum output voltage for which M_4 and M_{4A} operate in the active region would be given by (4.59). With nonzero V_{SS}, this condition becomes

$$V_{O1(\min)} = -V_{SS} + V_{tn} + 2V_{ov} \tag{6.93}$$

The presence of a threshold term in this equation is an important limitation because it causes a substantial reduction in the allowed output swing. Fortunately, this limitation can be overcome by using one of the high-swing cascode current mirrors shown in Figs. 4.11 and 4.12, which would eliminate the threshold term from (6.93) and give

$$V_{O1(\min)} = -V_{SS} + 2V_{ov} \tag{6.94}$$

With this change, we can see that to achieve a gain comparable to $(g_m r_o)^2$ in one stage, the swing is limited at best to two overdrives away from the supply. In contrast, the basic two-stage op amp in Fig. 6.16 gives about the same gain but allows the output to swing within one overdrive of each supply, as shown by (6.61).

To find the maximum output voltage swing of the telescopic-cascode op amp, consider the cascoded differential pair shown in Fig. 6.25. Assume that the common-mode input from the gates of M_1 and M_2 to ground is V_{IC}. The voltage from the source of M_1 and M_2 to ground is

$$V_S = V_{IC} + |V_{tp}| + |V_{ov}| \tag{6.95}$$

To operate M_5 in the active region, its source-drain voltage should be at least $|V_{ov}|$. Therefore,

$$V_{DD} - V_S \geq |V_{ov}| \tag{6.96}$$

Substituting (6.95) into (6.96) and rearranging gives

$$V_{IC} \leq V_{DD} - |V_{tp}| - 2|V_{ov}| \tag{6.97}$$

If we assume that M_9 and I_C are chosen to operate M_1 and M_2 at the edge of the active region, the maximum output voltage for which M_2 and M_{2A} operate in the active region is

$$V_{O1(\max)} = V_S - 2|V_{ov}| \tag{6.98}$$

Substituting (6.95) into (6.98) gives

$$V_{O1(\max)} = V_{IC} + |V_{tp}| - |V_{ov}| \tag{6.99}$$

This equation shows another limitation of the telescopic-cascode op amp from the standpoint of output swing; that is, the maximum output voltage depends on the common-mode input. However, this limitation as well as the limitation on the common-mode input range calculated in Problem 6.16 can be overcome in switched-capacitor circuits. Such circuits allow the op-amp common-mode input voltage to be set to a level that is independent of all other common-mode voltages on the same integrated circuit. This property holds because the only coupling of signals to the op-amp inputs is through capacitors, which conduct zero dc current even with a nonzero dc voltage drop. Assuming that the op-amp inputs are biased to the maximum common-mode input voltage for which M_5 operates in the active region, the maximum output voltage can be found by substituting the maximum V_{IC} from (6.97) into (6.99), which gives

$$V_{O1(\text{max})} = V_{DD} - 3|V_{ov}| \tag{6.100}$$

This equation shows that the maximum output voltage of a telescopic op amp that consists of the first stage of Fig. 6.25 with optimum common-mode input biasing is three overdrives less than the positive supply. This result stems from the observation that three transistors (M_5, M_2, and M_{2A}) are connected between V_{DD} and the output. In contrast, (6.94) shows that the minimum output swing is limited by two overdrives.

To determine the minimum required supply voltage difference, we will subtract (6.94) from (6.100), which gives

$$V_{O1(\text{max})} - V_{O1(\text{min})} = V_{DD} - (-V_{SS}) - 5|V_{ov}| \tag{6.101}$$

assuming that the magnitudes of all the overdrives are all equal. Rearranging this equation gives

$$V_{DD} - (-V_{SS}) = V_{O1(\text{max})} - V_{O1(\text{min})} + 5|V_{ov}| \tag{6.102}$$

This equation shows that the minimum difference between the supply voltages in a telescopic cascode op amp must be at least equal to the peak-to-peak output signal swing plus five overdrive voltages to operate all transistors in the active region. For example, with a peak-to-peak output swing of 1 V and $|V_{ov}| = 100$ mV for each transistor, the minimum difference between the supply voltages is 1.5 V.

In practice, this calculation has two main limitations. First, if transistors are deliberately biased at the edge of the active region, a small change in the process, supply, or temperature may cause one or more transistors to operate in the triode region, reducing the output resistance and gain of the op amp. To avoid this problem, transistors in practical op amps are usually biased so that the magnitude of the drain-source voltage of each transistor is more than the corresponding overdrive by a margin of typically at least one hundred millivolts. The margin allowed for each transistor directly adds to the minimum required supply voltage difference. Second, this calculation determines the supply requirements only for transistors between the output node and each supply, and other branches may require a larger supply difference than given in (6.102). For example, consider the path from one supply to the other through M_5, M_1, M_{1A}, M_{3A}, and M_3 in Fig. 6.25. Ignore the body effect for simplicity. Since M_{3A} and M_3 are diode connected, the drain-source voltage of each is $V_t + V_{ov}$. Furthermore, if the W/L of M_{3A} is reduced by a factor of four to build the high-swing cascode current mirror shown in Fig. 4.11, the voltage drop from the drain of M_{3A} to the source of M_3 is $2V_t + 3V_{ov}$. If the other three transistors in this path are biased so that $|V_{DS}| = |V_{ov}|$, the required supply difference for all the transistors in this path to operate in the active region is $2V_t + 6|V_{ov}|$. This requirement exceeds the requirement given in (6.102) if $2V_t + |V_{ov}|$ is more than the peak-to-peak output swing. However, this result does not pose a fundamental limitation to the minimum required power-supply voltage because low-threshold devices are sometimes available.

The minimum supply voltage difference in (6.102) includes five overdrive terms. In contrast, the corresponding equation for the op amp in Fig. 6.16 would include only two overdrive terms, one for M_6 and the other for M_7. The presence of the three extra overdrive terms increases the minimum required supply difference or reduces the allowed overdrives for a given supply difference. The extra three overdrive terms in (6.102) stem from the two cascode devices and the tail current source. The overdrives from the cascodes should be viewed as the cost of using cascodes; however, the overdrive of the tail current source can be eliminated from the minimum required supply difference using the circuit described in the next section.

6.6 MOS Folded-Cascode Operational Amplifiers

Figure 6.26 shows two cascode circuits where $V_{DD} = 0$ for simplicity. In Fig. 6.26a, both M_1 and M_{1A} are p-channel devices. In Fig. 6.26b, M_1 is still a p-channel device but M_{1A} is now an n-channel device. In both cases, however, M_1 is connected in a common-source configuration, and M_{1A} is connected in a common-gate configuration. Small-signal variations in the drain current of M_1 are conducted primarily through M_{1A} in both cases because I_{BIAS} is a constant current source. Therefore, both circuits are examples of cascodes. The cascode in Fig. 6.26b is said to be *folded* in the sense that it reverses the direction of the signal flow back toward ground. This reversal has two main advantages when used with a differential pair. First, it increases the output swing. Second, it increases the common-mode input range.

Figure 6.27 shows a simplified schematic of a circuit that applies the folded-cascode structure to both sides of a differential pair. As in Fig. 6.26b, M_1 and M_{1A} form one cascode structure in Fig. 6.27. Also, M_2 and M_{2A} form another. The current mirror converts the differential signal into a single-ended output by sending variations in the drain current of M_{1A} to the output. The resulting op amp is called a *folded-cascode op amp*.[7,12] A complete schematic is shown in Fig. 6.28. Bias is realized by making the currents in current sources M_{11} and M_{12} larger than $|I_{D5}|/2$. Thus

$$I_{D1A} = I_{D2A} = I_{D11} - \frac{|I_{D5}|}{2} = I_{D12} - \frac{|I_{D5}|}{2} = I_{BIAS} - \frac{I_{TAIL}}{2} \qquad (6.103)$$

Compared to the other op-amp configurations we have considered, the folded-cascode configuration improves the common-mode input range. The upper end of the range is the same as in the basic two-stage op amp and the telescopic cascode op amp. On the other hand, the lower end of the range can be reduced significantly compared to both of the other configurations if V_{BIAS2} is adjusted so that M_{11} and M_{12} operate at the edge of the active region. Under this condition, the bias voltage from the drain of M_1 to $-V_{SS}$ is V_{ov11}, which can be much less than in the other configurations. See (6.75), (6.77), and Problem 6.18.

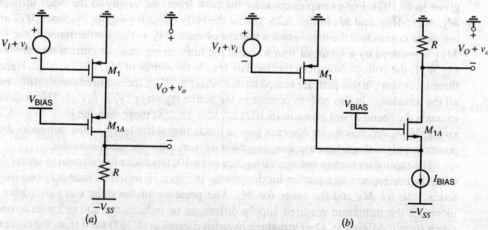

Figure 6.26 (a) Standard cascode configuration. (b) Folded-cascode configuration.

6.6 MOS Folded-Cascode Operational Amplifiers

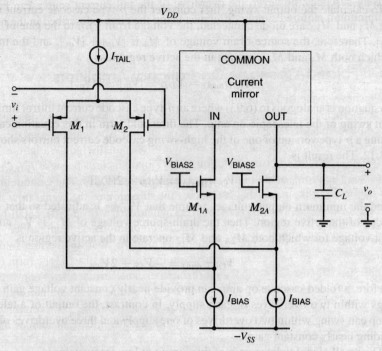

Figure 6.27 Simplified schematic of a folded-cascode op amp.

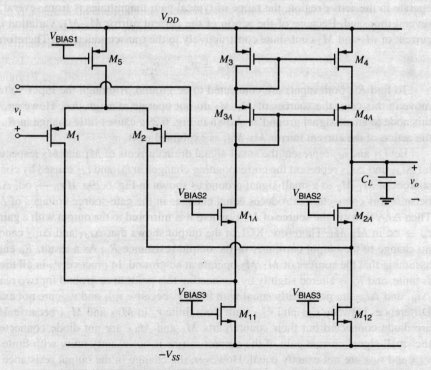

Figure 6.28 More detailed schematic of a folded-cascode op amp.

To calculate the output swing, first consider the *p*-type cascode current mirror by itself. Since M_3 and M_{3A} are diode connected, the voltage from V_{DD} to the gate of M_{4A} is $2|V_{tp}| + 2|V_{ov}|$. Therefore, the source-drain voltage of M_4 is $|V_{tp}| + |V_{ov}|$, and the maximum output for which both M_4 and M_{4A} operate in the active region is

$$V_{\text{OUT(max)}} = V_{DD} - |V_{tp}| - 2|V_{ov}| \tag{6.104}$$

This equation is analogous to (6.93) where an *n*-type cascode current mirror limits the minimum output swing of the telescopic op amp. The threshold term in this equation can be eliminated by using a *p*-type version of one of the high-swing cascode current mirrors shown in Figs. 4.11 and 4.12. The result is

$$V_{\text{OUT(max)}} = V_{DD} - 2|V_{ov}| \tag{6.105}$$

To find the minimum output voltage, assume that V_{BIAS2} is adjusted so that M_{12} operates at the edge of the active region. Then the drain-source voltage of M_{12} is V_{ov} and the minimum output voltage for which both M_{2A} and M_{12} operate in the active region is

$$V_{\text{OUT(min)}} = -V_{SS} + 2V_{ov} \tag{6.106}$$

Therefore, a folded-cascode op amp can provide nearly constant voltage gain while its output swings within two overdrives of each supply. In contrast, the output of a telescopic-cascode op amp can swing within two overdrives of one supply and three overdrives of the other while providing nearly constant gain.

The small-signal voltage gain of this circuit at low frequencies is

$$A_v = G_m R_o \tag{6.107}$$

where G_m is the transconductance and R_o is the output resistance. When all the transistors operate in the active region, the range of typical gain magnitudes is from several hundred to several thousand. Because of the action of the current mirror M_3-M_4, variation in the drain current of M_1 and M_2 contribute constructively to the transconductance. Therefore,

$$G_m = g_{m1} = g_{m2} \tag{6.108}$$

To find R_o, both inputs are connected to ac ground. Although the input voltages do not move in this case, the sources of M_1-M_2 do not operate at ac ground. However, connecting this node to small-signal ground as shown in Fig. 6.29a causes little change in R_o because of the action of the current mirror M_3-M_4, as explained next.

Let i_{d1} and i_{d2} represent the small-signal drain currents of M_1 and M_2 respectively. Also, let Δi_{d1} and Δi_{d2} represent the corresponding changes in i_{d1} and i_{d2} caused by connecting the sources of M_1-M_2 to a small-signal ground as shown in Fig. 6.29a. If $r_o \to \infty$, $\Delta i_{d1} = \Delta i_{d2}$ because this connection introduces equal changes in the gate-source voltages of M_1 and M_2. Then Δi_{d1} flows in the source of M_{1A}, where it is mirrored to the output with a gain of unity if $r_o \to \infty$ in M_3-M_4. Therefore, KCL at the output shows that Δi_{d1} and Δi_{d2} cancel, causing no change to the output current i_x or the output resistance R_o. As a result, R_o can be found, assuming that the sources of M_1-M_2 operate at ac ground. In practice, r_o in all the transistors is finite, and R_o is altered slightly by connecting this point to ac ground for two reasons. First, Δi_{d1} and Δi_{d2} are not exactly equal with finite r_o because v_{ds1} and v_{ds2} are not exactly equal. Differences between v_{ds1} and v_{ds2} stem from finite r_o in M_{1A} and M_{2A} because M_3 and M_{3A} are diode connected but their counterparts M_4 and M_{4A} are not diode connected. Second, the small-signal current gain of the current mirror is not exactly unity with finite r_o because v_{ds3} and v_{ds4} are not exactly equal. However, the change in the output resistance introduced by these considerations is usually negligible. (These effects are related to the explanation of

6.6 MOS Folded-Cascode Operational Amplifiers

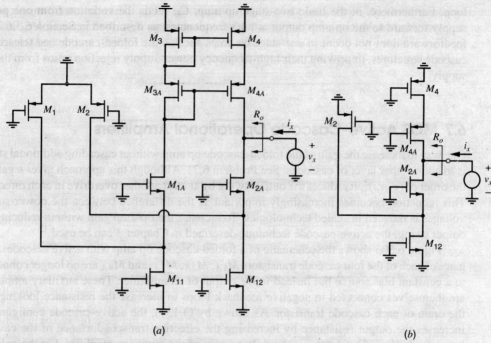

Figure 6.29 (a) Test voltage source applied to the output to calculate the output resistance. (b) Simplified circuit.

how a current-mirror load increases the common-mode rejection ratio of a differential pair presented in Section 4.3.5.3.)

With the sources of M_1-M_2 connected to ac ground, the drain current of M_1 is constant. Furthermore, the Thévenin equivalent resistances presented to the gates of M_4 and M_{4A} are small because M_3 and M_{3A} are diode connected. So little error is introduced by assuming that the gates of M_4 and M_{4A} are connected to small-signal ground. Therefore, the calculation of R_o can be carried out using the circuit of Fig. 6.29b. By inspection,

$$R_o = (R_{\text{out}}|M2A) \parallel (R_{\text{out}}|M4A) \tag{6.109}$$

The output resistance of transistor current sources with nonzero source resistance was considered in Chapter 4. The result is the same as for a common-source amplifier with source degeneration. The incremental resistance in the source of M_{2A} is the r_o of M_2 in parallel with the r_o of M_{12} while the incremental resistance in the source of M_{4A} is the r_o of M_4. From (3.107),

$$R_{\text{out}}|M2A = (r_{o2} \| r_{o12}) + r_{o2A}[1 + (g_{m2A} + g_{mb2A})(r_{o2} \| r_{o12})]$$
$$\simeq [g_{m2A}(r_{o2} \| r_{o12})]r_{o2A} \tag{6.110}$$

and

$$R_{\text{out}}|M4A = r_{o4} + r_{o4A}[1 + (g_{m4A} + g_{mb4A})(r_{o4})]$$
$$\simeq (g_{m4A}r_{o4})r_{o4A} \tag{6.111}$$

An important advantage of this circuit is that the load capacitance C_L performs the compensation function (see Chapter 9). Thus no additional capacitance (such as C_C in previous circuits) need be added to keep the amplifier from oscillating when connected in a feedback loop. Furthermore, in the basic two-stage op amp, C_C feeds the variation from one power supply forward to the op-amp output at high frequencies, as described in Section 6.3.6. This feedforward does not occur in one-stage op amps such as the folded-cascode and telescopic-cascode structures, improving their high-frequency power-supply rejection ratios from the V_{ss} supply.

6.7 MOS Active-Cascode Operational Amplifiers

One way to increase the gain of the folded-cascode op amp without cascading additional stages is to add another layer of cascodes. See Problem 6.21. Although this approach gives a gain on the order of $(g_m r_o)^3$, it reduces the output swing by at least another overdrive in each direction. This reduction becomes increasingly important as the difference between the power-supply voltages is reduced in scaled technologies. To increase the op-amp gain without reducing the output swing, the active-cascode technique described in Chapter 3 can be used.[13]

Figure 6.30a shows the schematic of a folded-cascode op amp with active cascodes. The gates of each of the four cascode transistors M_{1A}, M_{2A}, M_{3A}, and M_{4A} are no longer connected to a constant bias source but instead to the output of an amplifier. These auxiliary amplifiers are themselves connected in negative feedback loops to increase the resistance looking into the drain of each cascode transistor. As shown by (3.133), the active-cascode configuration increases the output resistance by increasing the effective transconductance of the cascode transistor by $(a + 1)$, where a is the voltage gain of the auxiliary amplifier. Let the gains of the auxiliary amplifiers driving M_{3A} and M_{4A} be A_1. Applying (3.133) to (6.111) to find the

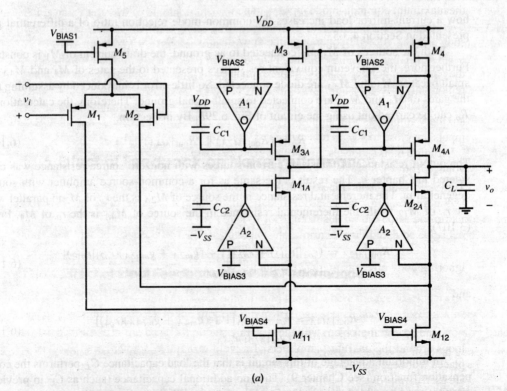

Figure 6.30 (a) Folded-cascode op amp with active-cascode gain-enhancement auxiliary amplifiers.

output resistance looking into the drain of M_{4A} gives

$$R_{\text{out}}|_{M4A} = r_{o4} + r_{o4A}\{1 + [g_{m4A}(A_1 + 1) + g_{mb4A}](r_{o4})\}$$
$$\simeq (A_1 + 1)(g_{m4A}r_{o4})r_{o4A} \tag{6.112}$$

Let the gains of the auxiliary amplifiers driving M_{1A} and M_{2A} be A_2. Applying (3.133) to (6.110) to find the output resistance looking into the drain of M_{2A} gives

$$R_{\text{out}}|_{M2A} = (r_{o2}||r_{o12}) + r_{o2A}\{1 + [g_{m2A}(A_2 + 1) + g_{mb2A}](r_{o2}||r_{o12})\}$$
$$\simeq (A_2 + 1)[g_{m2A}(r_{o2} \| r_{o12})]r_{o2A} \tag{6.113}$$

To find the overall op-amp gain, (6.112) and (6.113) can be substituted into (6.109) and the result in (6.107). This analysis shows that the gain enhancement in the folded-cascode op amp does not rely on the use of auxiliary amplifiers driving the gates of M_{1A} and M_{3A}. However, these auxiliary amplifiers are included in Fig. 6.30a because they reduce the systematic offset of the folded-cascode op amp. Also, using identical auxiliary amplifiers to drive the gates of both M_{1A} and M_{2A} balances the two signal paths until the differential signal is converted into single-ended form by the current mirror.

In Fig. 6.30a, the auxiliary amplifiers with gain A_1 drive the gates of M_{3A} and M_{4A} so that the voltages from the drains of M_3 and M_4 to ground are approximately equal to V_{BIAS2}. For simplicity, assume that the overdrive voltages for all p- and n-channel transistors operating in the active region are V_{ovp} and V_{ovn}, respectively. Also assume that all n-channel transistors have positive thresholds and all p-channel transistors have negative thresholds. To maximize the positive output swing of the folded-cascode amplifier, the voltage drop from V_{DD} to V_{BIAS2} is chosen to be about $|V_{ovp}|$. Therefore, the A_1 amplifiers must operate with a high common-mode input voltage. If these amplifiers use a p-channel differential input pair, the maximum common-mode input voltage would be no more than $V_{DD} - |V_{tp}| - 2|V_{ovp}|$. To overcome this limitation, the A_1 amplifiers use an n-channel differential pair M_{21} and M_{22} as shown in Fig. 6.30b. In operation, the dc voltage from V_{DD} to the output of the A_1 amplifiers is about $|V_{tp}| + 2|V_{ovp}|$ so that the source-drain voltages of M_3 and M_4 are $|V_{ovp}|$. Therefore, the gate-drain voltage of M_{22} is approximately $|V_{tp}| + |V_{ovp}|$, and M_{22} operates in the active region only if its threshold (with the body effect) is greater than this value.

A similar argument can be made to explain the use of p-channel differential pairs in the auxiliary amplifiers with gain A_2. The schematic is shown in Fig. 6.30c, and the common-mode inputs are close to V_{SS} in this case.

Figure 6.30d shows a circuit that produces the bias voltages needed in Fig. 6.30a–c. The voltage from V_{DD} to V_{BIAS1} is $|V_{tp}| + |V_{ovp}|$, and the voltage from V_{BIAS4} to V_{SS} is $V_{tn} + V_{ovn}$. Transistor M_{105} forces M_{106} to operate in the triode region, and the voltage from V_{DD} to V_{BIAS2} is at least $|V_{ovp}|$ if

$$\left(\frac{W}{L}\right)_{106} \leq \frac{1}{3}\left(\frac{W}{L}\right)_{105} \tag{6.114}$$

ignoring body effect as in (4.73). Similarly, the voltage from V_{BIAS3} to V_{SS} is at least V_{ovn} if

$$\left(\frac{W}{L}\right)_{114} \leq \frac{1}{3}\left(\frac{W}{L}\right)_{113} \tag{6.115}$$

One potential problem with the structure shown in Fig. 6.30a is instability in the feedback loops around the auxiliary amplifiers. To avoid instability, a compensation capacitor can be placed from each auxiliary-amplifier output to a small-signal ground. Since the A_1 amplifiers are used to improve the performance of a p-channel current mirror, where signals are referenced to V_{DD}, compensation capacitors for the A_1 amplifiers C_{C1} are connected to V_{DD}. Similarly,

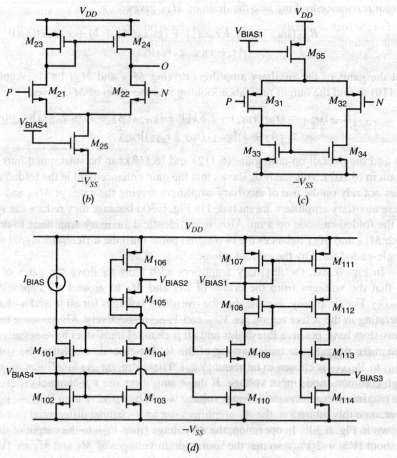

Figure 6.30 (b) Auxiliary amplifier with gain A_1. (c) Auxiliary amplifier with gain A_2. (d) Bias circuit.

compensation capacitors for the A_2 amplifiers C_{C2} are connected to V_{SS}. The need for such capacitors stems from the observation that the capacitance looking into the gates of M_{1A}, M_{2A}, M_{3A}, and M_{4A} can be quite small because the gate-source capacitances of these transistors are *bootstrapped*. This expression means that the source of each of these transistors follows its gate when the corresponding drain current is constant. If the gate-source voltages are exactly constant, zero ac current flows into the gate-source capacitances, and the capacitances looking into the gates of the cascode transistors are independent of their gate-source capacitances. In practice, the gate-source voltages are not exactly constant because of variations in the drain currents caused by variations in the differential input voltage of the folded-cascode op amp, but the bootstrapping effect is significant. As a result, the load capacitances of the auxiliary amplifiers are dominated by parasitics that may vary considerably over variations in processing unless a capacitor is added at the output of each auxiliary amplifier. The issue of stability in feedback amplifiers is considered in detail in Chapter 9.

6.8 Bipolar Operational Amplifiers

A basic topology for the input stage of bipolar-transistor op amps is shown in Fig. 6.31a. As in the case of the basic MOS op amp, the input stage consists of a differential pair, a tail current source, and a current-mirror load. The common-mode input range of op amps with this input stage is the range of dc common-mode input voltages for which all transistors in Fig. 6.31a operate in the forward-active region. Assume that a pure common-mode voltage V_{IC} is applied from the op-amp inputs to ground. Also, assume that Q_1 is identical to Q_2, and Q_3 is identical to Q_4. Then the collector-emitter voltage of Q_5 is

$$V_{CE5} = V_{IC} - V_{BE1} - V_{CC} \tag{6.116}$$

This equation shows that increasing V_{IC} increases V_{CE5}. Since Q_5 is a *pnp* transistor, it operates in the active region when $V_{CE5} < V_{CE5(sat)}$. Therefore, (6.116) can be rearranged to give the following upper limit for the common-mode input range

$$V_{IC} < V_{CC} + V_{BE1} + V_{CE5(sat)} \tag{6.117}$$

For example, if $V_{BE1} = -0.7$ V and $V_{CE5(sat)} = -0.1$ V, $V_{IC} < V_{CC} - 0.8$ V is required to operate Q_5 in the forward-active region.

To find the lower limit for the common-mode range, ignore base currents for simplicity. Then with a pure common-mode input,

$$V_{CE4} = V_{CE3} = V_{BE3} \tag{6.118}$$

As a result, the collector-emitter voltages of Q_1 and Q_2 are

$$V_{CE1} = V_{CE2} = -V_{EE} + V_{BE3} - (V_{IC} - V_{BE1}) \tag{6.119}$$

Because Q_1 and Q_2 are *pnp* transistors, they operate in the forward-active region when $V_{CE1} < V_{CE1(sat)}$ and $V_{CE2} < V_{CE2(sat)}$. Therefore, the lower end of the common-mode range is

$$V_{IC} > -V_{EE} + V_{BE3} + V_{BE1} - V_{CE1(sat)} \tag{6.120}$$

For example, if $V_{BE3} = 0.7$ V, $V_{BE1} = -0.7$ V, and $V_{CE5(sat)} = -0.1$ V, $V_{IC} > -V_{EE} + 0.1$ V is required to operate Q_1 and Q_2 in the forward-active region.

This inequality shows that the common-mode input range of the op amp in Fig. 6.31a almost includes $-V_{EE}$. A simple way to see this result is to let $V_{IC} = -V_{EE}$ in Fig. 6.31a. Then the collector-base voltages of Q_1 and Q_2 are

$$V_{CB1} = V_{CB2} = V_{BE3} \tag{6.121}$$

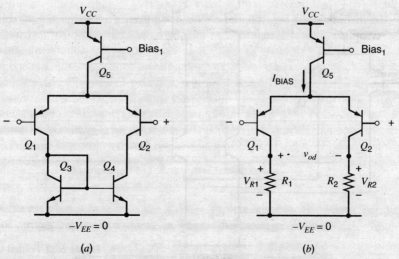

Figure 6.31 (a) Input stage in a basic bipolar operational amplifier. (b) Input stage with resistive loads to include $-V_{EE}$ in the input common-mode range. Assume $R_1 = R_2$.

Since Q_3 is diode connected and conducts current from its collector to its emitter, V_{BE3} is big enough to forward bias the base-emitter junction of Q_3. Therefore, V_{BE3} is also big enough to forward bias the collector-base junctions of Q_1 and Q_2, pushing these transistors into saturation.

To extend the common-mode input range to include $-V_{EE}$ (which is important in single-supply applications with $V_{EE} = 0$), the V_{BE3} term in (6.120) and (6.121) can be reduced by replacing the Q_3-Q_4 current mirror by resistive loads,[14] as shown in Fig. 6.31b. The voltage drops across the resistors V_{R1} and V_{R2} are chosen to be too small to forward bias the collector-base junctions of Q_1 and Q_2 when $V_{IC} = -V_{EE}$. In practice, these drops are usually set to about 200 mV with a pure common-mode input so that the common-mode input range extends a few hundred millivolts below $-V_{EE}$. One example in which this property is useful is in the inverting amplifier configuration shown in Fig. 6.3a when $V_{EE} = 0$. Since the configuration is inverting, $V_s < 0$ causes $V_o > 0$, and single-supply op amps can have high gain with positive outputs and $V_{CC} > 0$. However, if the op amp has zero offset and its gain a is finite, (6.5) shows that $V_i < 0$ when $V_o > 0$. As a result, the op-amp common-mode input voltage is negative, requiring that the common-mode input range extend below ground to handle this case.

The main drawback of the circuit in Fig. 6.31b is that its gain is low.[15] If the matching is perfect and $V_{R1} = V_{R2} = 0.2$ V,

$$\frac{v_{od}}{v_{id}} = g_{m1}R_1 = \frac{I_{BIAS}}{2V_T}R_1 = \frac{V_{R1}}{V_T} = \frac{0.2}{0.026} \simeq 7.7 \qquad (6.122)$$

To increase the gain, the input stage can use the folded-cascode structure shown in Fig. 6.32. Transistors Q_3 and Q_4 are connected in a common-base configuration. Combined with the input differential pair, this circuit forms a folded-cascode amplifier. The gain is

$$\frac{v_{od}}{v_{id}} = g_{m1}\rho_1 R_{o1} \qquad (6.123)$$

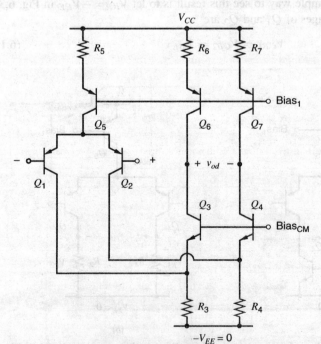

Figure 6.32 Folded-cascode input stage.

where ρ_1 represents the fraction of the small-signal collector current in Q_1 that flows into the emitter of Q_3 and R_{o1} represents the output resistance of this stage. To find ρ_1, the small-signal resistance looking into the emitter of Q_3 is needed. Since Q_3 operates as a common-base amplifier, this resistance can be found using (3.53) and is approximately equal to r_{e3}, ignoring the last term in (3.53) for simplicity. Then

$$\rho_1 \simeq \frac{R_3}{R_3 + r_{e3}} \tag{6.124}$$

Assume that Bias$_1$ and Bias$_{CM}$ are adjusted so that Q_6, Q_7, Q_9, and Q_{10} operate in the forward-active region. Then the output resistance of the stage is

$$R_{o1} = R_{\text{up1}} \| R_{\text{down1}} \tag{6.125}$$

where

$$R_{\text{up1}} \simeq r_{o6}(1 + g_{m6} R_6) \tag{6.126}$$

and

$$R_{\text{down1}} \simeq r_{o3}(1 + g_{m3} R_3) \tag{6.127}$$

assuming $g_{m6} R_6 \ll \beta_{pnp}$, $g_{m3} R_3 \ll \beta_{npn}$, and the resistance looking back into the collector of Q_1 is much greater than R_3. Although the folded-cascode circuit in Fig. 6.32 increases the stage gain significantly and allows proper operation for common-mode inputs slightly below $-V_{EE}$, it does not allow operation with common-mode inputs up to V_{CC} as shown by (6.117).

To extend the common-mode input range to V_{CC}, an npn differential pair can be added, as shown in Fig. 6.33.[16] Assume that the dc voltage drops on R_5-R_{10} are 200 mV each. Then

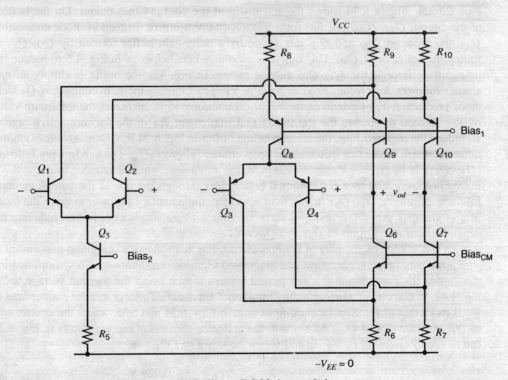

Figure 6.33 Rail-to-rail folded-cascode input stage.

the *pnp* differential pair operates normally for common-mode inputs slightly below ground to about 1 V below V_{CC}. Similarly, the *npn* differential pair operates normally for common-mode inputs slightly above V_{CC} to about 1 V above ground. Therefore, if $V_{CC} < 2$ V, neither differential pair operates normally when $V_{IC} = V_{CC}/2$ because both Q_5 and Q_8 operate in saturation under these conditions. On the other hand, if $V_{CC} \geq 2$ V, at least one of the two differential pairs operates normally for any common-mode input voltage from just below $-V_{EE}$ to just above V_{CC}. In other words, the stage has a rail-to-rail common-mode input range for $V_{CC} \geq 2$ V.

However, an important disadvantage of this input stage is that its small-signal transconductance is not constant. The stage transconductance is the sum of the transconductances from both differential pairs. In the center of the common-mode range, both differential pairs operate normally, and the stage transconductance is high. On the other hand, when the common-mode input voltage is within about 1 V of either supply, the tail current source in one of the differential pairs saturates, reducing that current as well as the transconductances of that differential pair and the input stage. This variation in the stage transconductance causes variation in the open-loop gain and frequency response of the op amp, compromising its performance when it is designed to be stable in a closed-loop configuration. (The topic of stability is covered in Chapter 9.) This disadvantage is overcome in the NE5234 op amp, which is described below.[17,18] This stand-alone op amp is widely used, and its popularity stems from the fact that it can operate with $V_{EE} = 0$ and V_{CC} as low as 2 V, giving constant input transconductance over a common-mode input range slightly beyond the power-supply voltages (or *rails*) and almost a rail-to-rail output swing.

6.8.1 The dc Analysis of the NE5234 Operational Amplifier

Bias Circuit. Figure 6.34 shows the schematic of the NE5234 bias circuit. On the bottom of the circuit, Q_{49}, Q_{60}, and R_{60} form a Widlar current mirror. Instead of diode connecting Q_{49}, the bases of Q_{49} and Q_{60} are driven by a unity-gain buffer formed by Q_{50}-Q_{53} to follow the collector of Q_{49}. This buffer is used as a beta helper as in Fig. 4.6 to reduce the base-current error in the Q_{49}-Q_{60} current mirror. In Fig. 4.6, the buffer is simply an *npn* emitter follower. As a result, $V_{CE1} = V_{BE1} + V_{BE2} = 2V_{BE(\text{on})}$ there. In contrast, if Q_1 were diode connected, V_{CE1} would equal $V_{BE(\text{on})}$. Increasing V_{CE1} increases the minimum value of V_{CC} needed to operate the transistors that implement I_{IN} in the forward-active region. To reduce the required V_{CC}, the corresponding buffer in Fig. 6.34 is a *complementary emitter follower*, which means that it consists of a *pnp* emitter follower (Q_{50}-Q_{51}) and an *npn* follower (Q_{52}-Q_{53}). If $V_{BE51} = -V_{BE52}$, $V_{CE49} = V_{BE49} = V_{BE(\text{on})}$, which is reduced by about 0.7 V in practice compared to a conventional beta helper through the use of the complementary follower. Similarly, Q_{54}-Q_{57} in Fig. 6.34 act as a complementary follower to drive the bases of Q_{47} and Q_{58}. Also, Q_{48} and Q_{59} serve as common-base stages or cascodes, reducing the dependence of the currents in Q_{47} and Q_{60} on V_{CC}.

To concentrate on the core of the bias circuit, Fig. 6.35 shows a simplified bias circuit in which the beta helpers, cascodes, and associated elements are removed. This simplification shows that the core circuit is a self-biased current source using the thermal voltage, as in Fig. 4.41. In that circuit, corresponding transistors are matched except that the emitter area of Q_2 is twice that of Q_1. Similar conditions hold in Figs. 6.34 and 6.35, where the emitter area of Q_{60} is twice that of Q_{49}. As a result, the collector currents of the transistors in Fig. 6.35 can be found assuming $\beta_F \to \infty$ and $V_A \to \infty$ using (4.237)

$$|I_{C47}| = I_{C49} = |I_{C58}| = I_{C60} = \frac{V_T}{R_{60}}\ln 2 = \frac{0.026 \text{ V}}{3 \text{ k}\Omega}\ln 2 = 6 \text{ }\mu\text{A} \qquad (6.128)$$

6.8 Bipolar Operational Amplifiers

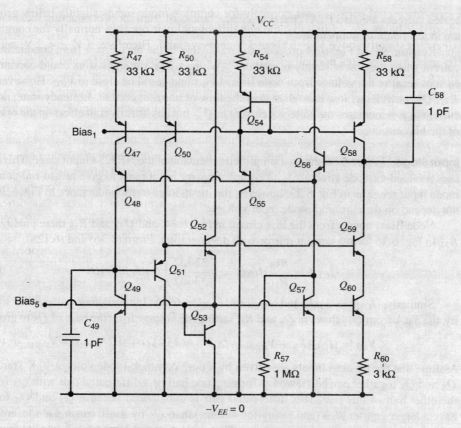

Figure 6.34 Schematic of the NE5234 bias circuit.

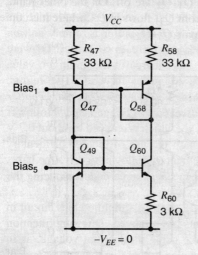

Figure 6.35 Simplified schematic of the NE5234 bias circuit.

If R_{60} is constant, these currents are proportional to absolute temperature (PTAT). Although β_F and V_A are finite in practice, this equation also gives the corresponding currents in Fig. 6.34 because that circuit uses the beta helpers and cascodes mentioned above. The bias circuit produces dc voltages at two nodes, Bias_1 and Bias_5, that are used to establish scaled copies of the 6 μA current in *pnp* and *npn* transistors in the op amp, respectively. Since these scaled currents are also PTAT, the transconductances of transistors conducting these currents are independent of temperature.

Resistor R_{57} in Fig. 6.34 prevents the transistors in the bias circuit from conducting zero current when the power supply is applied.[18] Without R_{57}, the transistors could conduct zero current because the voltage from node Bias$_1$ to ground could be close to V_{CC}. However, with R_{57}, Q_{55} pulls Bias$_1$ low enough to start the flow of nonzero current. In steady state, nonzero current in R_{57} increases the collector current in Q_{56} but has little overall effect on the operation of the bias circuit.

Input Stage. Figure 6.36 shows a simplified schematic of the NE5234 input stage. This circuit uses a folded-cascode structure with complementary input pairs to give rail-to-rail common-mode input range as in Fig. 6.33; however, the input-stage transconductance in Fig. 6.36 does not depend on the common-mode input voltage.

Node Bias$_1$ comes from the bias circuit in Fig. 6.34, and Q_{58} and R_{58} there plus Q_{11} and R_{11} in Fig. 6.36 form a current mirror with degeneration. From (4.36) and (6.128),

$$|I_{C11}| = \frac{R_{58}}{R_{11}}|I_{C58}| = \left(\frac{33 \text{ k}\Omega}{33 \text{ k}\Omega}\right) 6 \text{ }\mu\text{A} = 6 \text{ }\mu\text{A} \qquad (6.129)$$

Similarly, $|I_{C12}| = 3$ μA, and $|I_{C13}| = |I_{C14}| = 6$ μA. Ignoring base currents for simplicity, the 3 μA from Q_{12} flows in Q_8 and R_8, setting the voltage from the base of Q_5 to ground to

$$V_{B5} = |I_{C12}|R_8 + V_{BE8(\text{on})} \simeq 3 \text{ }\mu\text{A}(33 \text{ k}\Omega) + 0.7 \text{ V} = 0.8 \text{ V} \qquad (6.130)$$

Assume that the op-amp inputs are driven by a pure common-mode voltage V_{IC}. Transistors Q_3 and Q_4 together can be viewed as forming one half of a differential pair with Q_5 forming the other half of this pair. This differential pair is unbalanced because Q_3 and Q_4 together have a larger emitter area (and saturation current) than Q_5 by itself, causing a nonzero offset voltage. Ignore this offset for simplicity. Then if $V_{IC} \ll 0.8$ V, the 6 μA current from Q_{11} flows in the *pnp* input pair Q_3-Q_4, and Q_5-Q_7 plus Q_1-Q_2 are off. On the other hand, if $V_{IC} \gg 0.8$ V, the *pnp* pair is off and the 6 μA current from Q_{11} flows in Q_5, where it is copied by the Q_7-Q_6 current mirror to activate the *npn* input pair Q_1-Q_2.

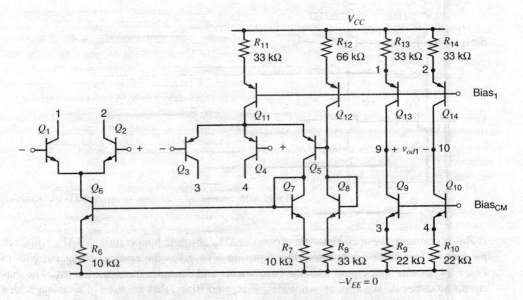

Figure 6.36 Simplified schematic of the NE5234 input stage.

In both extreme cases, the transconductance of the input stage is constant because it is set by the 6 μA current. In the transition region around 0.8 V, both input pairs Q_1-Q_2 and Q_3-Q_4 are active. The width of this region is about $\pm\, 3V_T$ or $\pm\, 78$ mV. In this region, some of the 6 μA from Q_{11} flows in Q_3-Q_4, and the rest flows in Q_5, where it ultimately biases Q_1-Q_2 through Q_6. Therefore, the sum of the collector currents in Q_1-Q_4 is a constant of 6 μA ignoring base currents, base-width modulation, and mismatch. Since the transconductance of each input pair is proportional to its bias current, and since the total input-stage transconductance is the sum of the transconductances of both input pairs, the input-stage transconductance does not depend on V_{IC}.[17]

To obtain high gain in the first stage, transistors Q_9, Q_{10}, Q_{13}, and Q_{14} must operate in the forward-active region so that the output resistance of the stage is high. Since Bias_1 is set so that $|I_{C13}| = |I_{C14}| = 6$ μA, Bias_{CM} should be adjusted so that I_{C9} and I_{C10} also equal 6 μA. However, if the currents in Q_9 and Q_{10} are simply adjusted to this constant value, the common-mode output voltage at nodes 9 and 10 would be sensitive to small changes in transistors Q_9, Q_{10}, Q_{13}, and Q_{14}. For example, suppose that Q_9 and Q_{10} conduct 6 μA as expected but Q_{13} and Q_{14} conduct 6.1 μA each because of a slight increase in their emitter areas. Then the common-mode output voltage would rise until KCL is satisfied at nodes 9 and 10, forcing transistors Q_{13} and Q_{14} to operate at the edge of saturation to reduce their currents to 6 μA. Similarly, if Q_9 and Q_{10} pull slightly more current than pushed by Q_{13} and Q_{14}, the common-mode output output voltage would fall until KCL is satisfied. The key point is that the common-mode output voltage of the input stage is not well controlled by the circuit in Fig. 6.36. This problem is overcome in the second stage of the amplifier, which adjusts Bias_{CM} to set the common-mode output voltage of the first stage so that Q_9, Q_{10}, Q_{13}, and Q_{14} operate in the forward-active region.

Second Stage. Figure 6.37 shows a schematic of the NE5234 second stage. Nodes 9 and 10 are the outputs of the first stage and the inputs of the second stage. Capacitors C_{21} and C_{22} are used for frequency *compensation*; that is, they control the frequency response so that the op amp is stable when connected in a negative feedback loop. The topic of compensation is covered in Chapter 9. Capacitor C_{22} is connected to the op-amp output, which is generated in the output stage. Both capacitors are ignored here. Emitter followers Q_{21} and Q_{22} reduce the loading of the second stage on the first. Ignore Q_{23} and Q_{24} at first because they are normally off. Transistors Q_{25}-Q_{28} form a differential pair in which each side of the pair is split into two transistors. This differential pair amplifies the differential output of the first stage v_{od1}. Splitting the pair allows it to produce two outputs that are in phase with respect to each other, one at node 25 and the other at node 26, as required by the output stage.

From a dc standpoint, (4.36) and (6.128) can be applied to find

$$|I_{C15}| = \frac{R_{58}}{R_{15}}|I_{C58}| = \left(\frac{33\text{ k}\Omega}{66\text{ k}\Omega}\right) 6 \text{ μA} = 3 \text{ μA} \tag{6.131}$$

Similarly, $|I_{C16}| = |I_{C19}| = 4$ μA, $|I_{C17}| = |I_{C18}| = 21$ μA, and $|I_{C20}| = 6.6$ μA. For simplicity, assume that $|I_{C16}|$ and $|I_{C19}|$ set $V_{EB21} = V_{EB22} = 0.7$ V. Ignoring base currents, $|I_{C15}|$ flows in Schottky diode D_1. The voltage across the diode is

$$V_{D1} = V_T \ln\frac{|I_{C15}|}{I_{S(D1)}} \tag{6.132}$$

where $I_{S(D1)}$ is the saturation current of D_1. Assume $I_{S(D1)} = 6 \times 10^{-13}$ A. Then

$$V_{D1} = 0.026 \ln\left(\frac{3 \times 10^{-6}}{6 \times 10^{-13}}\right) \text{V} = 0.4 \text{ V} \tag{6.133}$$

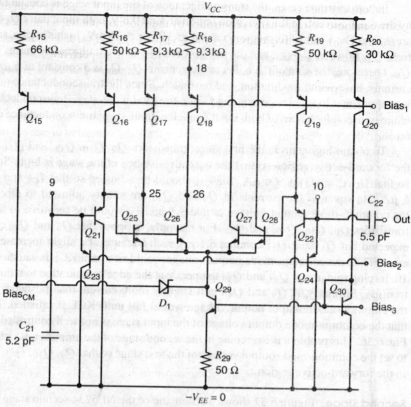

Figure 6.37 Schematic of the NE5234 second stage. Node Out is the output of the third stage (shown in Fig. 6.39).

Resistor R_{29} along with transistors Q_{29}-Q_{30} form a Widlar current mirror. If $\beta_F \to \infty$, (4.190) gives

$$V_T \ln \frac{I_{C30}}{I_{S30}} - V_T \ln \frac{I_{C29}}{I_{S29}} - I_{C29} R_{29} = 0 \tag{6.134}$$

In the NE5234 op amp, $I_{S29}/I_{S30} = 7$ and

$$V_T \ln \frac{7 I_{C30}}{I_{C29}} = I_{C29} R_{29} \tag{6.135}$$

Since $I_{C30} = |I_{C20}| = 6.6\ \mu\text{A}$, a trial-and-error solution of this equation gives $I_{C29} = 42\ \mu\text{A}$. Therefore, the current available for Q_{25}-Q_{28} is $I_{C29} - I_{D1} = 39\ \mu\text{A}$. Let V_9 and V_{10} represent the voltages from nodes 9 and 10 to ground, respectively, and assume $V_9 = V_{10}$. Then $I_{C25} = I_{C26} = I_{C27} = I_{C28} = 39/4\ \mu\text{A} \simeq 10\ \mu\text{A}$. For simplicity, assume that these currents set $V_{BE25} = V_{BE26} = V_{BE27} = V_{BE28} = 0.7$ V.

Let V_{cmout1} represent the common-mode output voltage of the first stage, including both dc and ac components. It is the average of the voltages from nodes 9 and 10 to ground:

$$V_{cmout1} = \frac{1}{2}(V_9 + V_{10}) \tag{6.136}$$

Similarly, let V_{biascm} represent the voltage from node Bias_{CM} to ground. Elements in the first and second stages work together to set both of these voltages, as explained below.

Consider the first stage, where V_{biascm} is an input and V_{cmout1} is an output. The relationship between these voltages here is controlled by the common-emitter amplifiers with degeneration (Q_9 and Q_{10}) and active loads (Q_{13} and Q_{14}). Let v_{biascm}, v_9, and v_{10} represent the changes in the voltages from nodes $Bias_{CM}$, 9, and 10 to ground, respectively. Let the small-signal gain v_9/v_{biascm} equal A. Then $A = v_{10}/v_{biascm}$ from symmetry. In other words, A is the common-mode gain from $Bias_{CM}$ to the first-stage output. The polarity of A is negative because pulling up on $Bias_{CM}$ increases the collector currents in Q_9 and Q_{10}, reducing the common-mode output voltage. The magnitude of A is large, but its magnitude is not calculated here because it is not important to a first-order analysis that finds the operating point. For simplicity, assume $A \to -\infty$ when Q_9, Q_{10}, Q_{13}, and Q_{14} operate in the forward-active region.

Assume the op-amp common-mode input voltage V_{IC} is much less than the 0.8 V threshold given in (6.130). Then Q_1 and Q_2 are off. Since $|I_{C13}| = |I_{C14}| = 6$ μA, the dc voltages across R_{13} and R_{14} are

$$V_{R13} = V_{R14} = 6 \text{ μA}(33 \text{ k}\Omega) = 0.2 \text{ V} \tag{6.137}$$

If $V_{CE13(sat)} = V_{CE14(sat)} = -0.1$ V, Q_{13} and Q_{14} operate in the forward-active region if $V_{cmout1} < V_{CC} - 0.3$ V.

To satisfy KCL at nodes 9 and 10, Q_9 and Q_{10} must adapt to pull the same current pushed by Q_{13} and Q_{14}, which is 6 μA under nominal conditions. Assume that this current sets $V_{BE9} = V_{BE10} = 0.7$ V. Also, since $V_{IC} \ll 0.8$ V is assumed, $|I_{C3}| = |I_{C4}| = |I_{C11}|/2 = 3$ μA, and the dc currents in R_9 and R_{10} are

$$I_{R9} = I_{R10} = I_{C9} + |I_{C3}| = I_{C10} + |I_{C4}| = 9 \text{ μA} \tag{6.138}$$

As a result, the dc voltages across these resistors are

$$V_{R9} = V_{R10} = 9 \text{ μA}(22 \text{ k}\Omega) = 0.2 \text{ V} \tag{6.139}$$

Assume $V_{CE9(sat)} = V_{CE10(sat)} = 0.1$ V. Then Q_9 and Q_{10} operate in the forward active region if $V_{cmout1} > 0.3$ V and V_{biascm} is

$$V_{biascm} = V_{R9} + V_{BE9} = V_{R10} + V_{BE10} = 0.9 \text{ V} \tag{6.140}$$

This behavior is summarized in Fig. 6.38, which shows two plots of V_{cmout1} versus V_{biascm}. One plot is labeled *Amplifier Characteristic*. For $V_{biascm} < 0.9$ V, it shows that V_{cmout1} is constant at $V_{CC} - 0.3$ V because Q_{13} and Q_{14} saturate. For $V_{biascm} > 0.9$ V, this plot shows that V_{cmout1} is constant at 0.3 V because Q_9 and Q_{10} saturate. For $V_{biascm} = 0.9$ V, V_{cmout1} falls from $V_{CC} - 0.3$ V to 0.3 V because $A \to -\infty$ is assumed above.

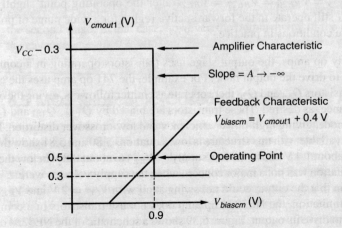

Figure 6.38 Plots of V_{cmout1} versus V_{biascm}. The amplifier characteristic comes from stage 1, and the feedback characteristic comes from stage 2.

Now consider the second stage, where V_{cmout1} is an input and V_{biascm} is an output. Assume that the first-stage common-mode output voltage increases by v_{cmout1}. Then the voltages from the emitters of Q_{21} and Q_{22} to ground also increase by approximately v_{cmout1} because these transistors act as emitter followers. These changes cause the voltage from the collector of Q_{29} to ground to rise by approximately v_{cmout1} because the base-emitter voltages of Q_{25}-Q_{28} are almost constant. This result stems from the fact that the combination of Q_{25}-Q_{26} on the one hand and Q_{27}-Q_{28} on the other hand forms a differential pair. Since the input to this differential pair is a pure common-mode signal under the conditions given above, the collector currents of the individual transistors are approximately constant assuming that the tail current source Q_{29} has high output resistance. Furthermore, from a dc standpoint, the emitter followers Q_{21} and Q_{22} level shift V_9 and V_{10} up by 0.7 V, and the split differential pair Q_{25}-Q_{28} level shifts the emitter follower outputs back down by the same amount. Then because the voltage across Schottky diode D_1 is constant as shown by (6.133), the voltage from Bias$_{CM}$ to ground is the first-stage common-mode output voltage shifted up by 0.4 V. Since the collector voltage of Q_{29} rises by v_{cmout1} in this example, the voltage from Bias$_{CM}$ to ground also rises by the same amount. As a result, the voltage from node Bias$_{CM}$ to ground is

$$V_{biascm} = V_{cmout1} + 0.4 \text{ V} \tag{6.141}$$

This equation is plotted in Fig. 6.38. It is labeled as the *Feedback Characteristic* because the role of the second stage is to sense, level shift, and feed the first-stage common-mode output voltage back to the first stage to control the common-mode operating point. This loop is an example of negative feedback because increasing V_{cmout1} increases V_{biascm}, which then reduces V_{cmout1} through the action of the common-mode amplifiers with gain A in the first stage described above. Although this negative feedback loop operates on the first-stage common-mode output voltage, it does not operate on the first-stage differential output. For example, suppose that V_9 increases incrementally and V_{10} decreases by the same amount. Then the voltage from node 29 to ground is constant because Q_{25} and Q_{26} pull up while Q_{27} and Q_{28} pull down with equal strength. As a result, V_{biascm} is constant, and the loop is inactive. Therefore, the loop under consideration is called a *common-mode feedback loop*. The topic of common-mode feedback is covered in detail in Chapter 12.

Figure 6.38 presents a graphical analysis to explain how the bias point is set here. Two variables, V_{biascm} and V_{cmout1}, are central to this analysis. These variables are related to each other in two ways, by the amplifier and feedback characteristics described above. Since both characteristics must be satisfied, the circuits operate at their intersection, where the average value of $V_{biascm} = 0.9$ V and the average value of $V_{cmout1} = 0.5$ V. In principle, finite A and $V_{EB21} = V_{EB22} \neq V_{BE25} = V_{BE26} = V_{BE27} = V_{BE28}$ alter the operating point slightly, but Q_9, Q_{10}, Q_{13}, and Q_{14} still operate in the forward-active region for a wide range of process, supply, and temperature conditions in practice.

Output Stage. In many op amps, the output stage uses transistors operating in a common-collector configuration to drive the output node. For example, the 741 op amp uses the circuit shown in Fig. 5.20. Transistors Q_{14} and Q_{20} there operate as emitter followers, giving the output stage low output resistance, as desired. These transistors are biased by Q_{13A}, Q_{18}, and Q_{19} to operate in a Class AB mode, giving high-power efficiency and low-crossover distortion. However, the output swing available with this structure is low. Equations 5.80 and 5.81 show that the output can swing from about 1.4 V above the low supply ($-V_{EE}$) to about 0.8 V below the high supply (V_{CC}). This limitation was not a major concern when the supply voltages were $\pm$ 15 V. However, it would mean that the output could not swing at all with $V_{CC} = 2$ V and $V_{EE} = 0$.

To overcome this limitation, the NE5234 op amp does not use transistors in a common-collector configuration to drive its output. Figure 6.39 shows a schematic of the NE5234 output

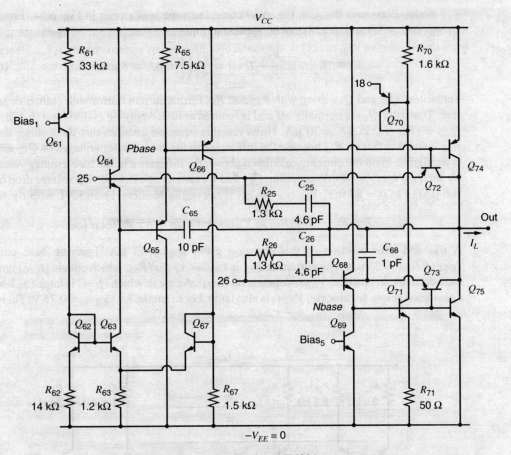

Figure 6.39 Schematic of the NE5234 output stage.

stage. Nodes 25 and 26 are the inputs of this stage and outputs of the second stage. Capacitors C_{25} and C_{26} along with resistors R_{25} and R_{26} are for frequency compensation and are ignored here. The output is driven by common-emitter transistors Q_{74} and Q_{75}. Although the resulting output resistance of the stage is high, the output resistance is reduced dramatically by negative feedback around the op amp. Negative feedback is covered in Chapter 8. The main reason to use such an output stage is that it can drive the output node to within $V_{CE75(sat)}$ or about 0.1 V of the low supply and to within $|V_{CE74(sat)}|$ or about 0.1 V of the high supply. In other words, this output stage gives almost rail-to-rail output swing.

A design goal is for Q_{74} to be able to push up to 10 mA into a load connected to the output. Similarly, Q_{75} should be able to pull the same current out of the load.[18] The bias currents of Q_{25} and Q_{26} in Fig. 6.37 are about 10 μA each. To limit the current required to flow into or out of nodes 25 and 26 to this amount, the output stage should provide a current gain of at least 1000. To meet this requirement when Q_{75} pulls current from the load, emitter follower Q_{68} is used. The current gain from node 26 to the output is approximately $\beta_{F68}\beta_{F75} = \beta_{F(npn)}^2$, which is greater than 1000 because the minimum value of $\beta_{F(npn)}$ is about 40 in this process. On the other hand, when Q_{74} pushes current into the load, the current gain might be less than 1000 if Q_{74} were driven by only a *pnp* emitter follower because the minimum value of $\beta_{F(pnp)}$ is about 10 in this process. To overcome this problem, the buffer that drives the base of Q_{74} consists of a complementary emitter follower Q_{64}-Q_{65}, giving a current gain from node 25 to the output of approximately $\beta_{F64}\beta_{F65}\beta_{F74} = \beta_{F(pnp)}^2\beta_{F(npn)}$.

Nodes Bias$_1$ and Bias$_5$ in Fig. 6.39 come from the bias circuit in Fig. 6.34. From a dc standpoint, (4.36) and (6.128) can be applied to find

$$|I_{C61}| = \frac{R_{58}}{R_{61}}|I_{C58}| = \frac{33 \text{ k}\Omega}{33 \text{ k}\Omega} 6 \text{ }\mu\text{A} = 6 \text{ }\mu\text{A} \tag{6.142}$$

Transistors Q_{62} and Q_{63} along with R_{62} and R_{63} form a current mirror with emitter degeneration. Transistor Q_{67} is normally off and is ignored at first. Applying (4.136) and (6.128) gives $I_{C63} = (R_{62}/R_{63})6 \text{ }\mu\text{A} = 70 \text{ }\mu\text{A}$. However, this equation assumes that the voltage drop on R_{62} is equal to that on R_{63} because the difference in the base-emitter voltages of Q_{62} and Q_{63} is negligible. With the currents calculated above, the difference in the base-emitter voltages is $V_T \ln(I_{C62}/I_{C63}) = 64$ mV assuming Q_{62} and Q_{63} are identical. Since the voltage drop on R_{62} is 6 μA(14 kΩ) = 84 mV, $V_{BE62} - V_{BE63}$ is not negligible here. From KVL with $\beta_F \rightarrow \infty$,

$$I_{C62}R_{62} - I_{C63}R_{63} = V_{BE63} - V_{BE62} = V_T \ln(I_{C63}/I_{C62}) \tag{6.143}$$

A trial-and-error solution of this equation gives $I_{C63} = 33$ μA. Ignoring base currents, $I_{C64} = I_{C63} = 33$ μA. The current in R_{65} is $I_{R65} = V_{EB74}/R_{65}$, which equals $|I_{C65}|$ ignoring base currents. In practice, V_{EB74} depends on $|I_{C74}|$. A case in which $|I_{C74}|$ is large (> 1 mA) is considered below. In this case, V_{EB74} is also large. For example, let $V_{EB74} = 0.75$ V. Therefore,

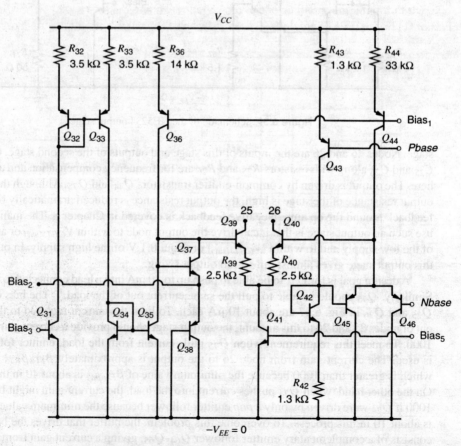

Figure 6.40 Schematic of the bias circuit for the output stage.

$|I_{C65}| = I_{R65} = 100$ μA. Since the base-emitter voltage of Q_{69} is equal to that of Q_{49} in Fig. 6.34, $I_{C69} = I_{C49} = 6$ μA. Ignoring base currents, $I_{C68} = I_{C69} = 6$ μA. However, nonzero base currents often have a significant effect in the output stage, especially on I_{C64}, $|I_{C65}|$, and I_{C68}, which are recalculated below after $|I_{C74}|$ and I_{C75} are found. Transistors Q_{70}, Q_{72}, and Q_{73} are normally off and ignored at first.

The dc currents in the other transistors in the output stage are set by the bias circuit for the output stage, which is shown in Fig. 6.40. Nodes *Pbase* and *Nbase* are outputs of the output stage and inputs to this bias circuit. Node $Bias_2$ is an output that becomes an input to the second stage in Fig. 6.37 and is ignored at first. Node $Bias_3$ comes from the second stage. Since $V_{BE31} = V_{BE30}$, $I_{C31} = I_{C30} = 6.6$ μA. This current is mirrored through Q_{32}-Q_{33} and then Q_{34}-Q_{35} so that $I_{C35} = 6.6$ μA ignoring base currents and base-width modulation. Applying (4.136) and (6.128) gives $|I_{C44}| = 6$ μA. Similarly, $|I_{C36}| = (R_{58}/R_{36})6$ μA $= 14$ μA. Then $I_{C37} = I_{C38} = |I_{C36}| - I_{C35} = 7.4$ μA. Node $Bias_5$ comes from the bias circuit in Fig. 6.34. Since $V_{BE41} = V_{BE49}$, and since $I_{S41}/I_{S49} = 3$ in the NE5234 op amp, $I_{C41} = 3I_{C49} = 18$ μA.

To focus on the interaction between the circuits of Figs. 6.39 and 6.40, Fig. 6.41 shows a simplified schematic of the NE5234 output stage with its bias circuit. For simplicity here, the transistors that are normally off are omitted, and the bias currents calculated above are set by current sources. As in many output stages, this stage operates in a Class AB mode. However, the basic relationship between the collector currents of the two output transistors differs here compared to that in many op amps, as explained below.

In op amps with common-collector output drivers, the classical Class AB biasing technique sets the product of the collector currents of the two output transistors to a constant. For example,

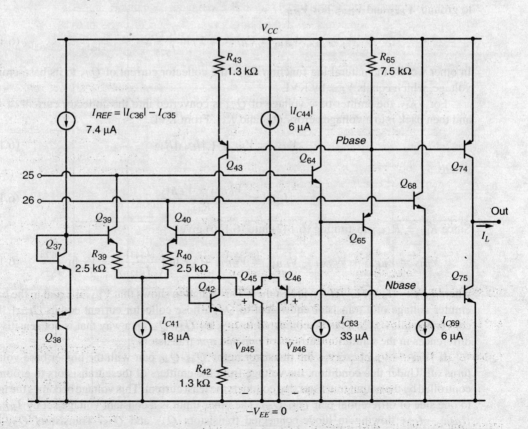

Figure 6.41 Simplified schematic of the NE5234 output stage with its bias circuit.

in the 741 op amp, Q_{14} and Q_{20} in Fig. 5.20 are the output transistors. These transistors are biased by Q_{18} and Q_{19} through a KVL loop, as shown in (5.82). Manipulating this equation gives

$$I_{C14}|I_{C20}| = I_{C18}I_{C19}\left(\frac{I_{S14}|I_{S20}|}{I_{S18}I_{S19}}\right) = \text{constant} \tag{6.144}$$

As a result, neither I_{C14} nor $|I_{C20}|$ goes to zero in principle even when the other is large. However, in practice, the KVL loop includes terms omitted from (5.82) for simplicity. For example, voltage drops appear on nonzero resistances in the base and emitter regions of the driving transistor. These extra drops increase the $|V_{BE}|$ of the driving transistor as the magnitude of the load current increases, turning the inactive transistor off at some point. As a result, the time required to turn this transistor back on increases the delay in driving the output in the other direction, worsening crossover distortion.[19]

To overcome this problem, the output transistors in the NE5234 op amp are biased so that neither turns off even for extreme signal swings.[19,20] The circuit compares the collector currents of the two output transistors and controls the smaller current through negative feedback when the driving current is large. The collector currents can be sensed by placing elements in series with the collectors or emitters of the output transistors; however, the voltage drops across such elements would not be zero and would reduce the output swing. To maximize the output swing, the collector currents are sensed through the base-emitter voltages of the output transistors. These voltages are manipulated and sent to differential pair Q_{45}-Q_{46} for comparison. This differential pair operates on the voltages from the bases of these transistors to ground, V_{B45} and V_{B46}. For V_{B46},

$$V_{B46} = V_{BE75} = V_T \ln \frac{I_{C75}}{I_{S75}} \tag{6.145}$$

In other words, the natural log function maps the collector current of Q_{75} to its base-emitter voltage, which equals V_{B46} by KVL.

For V_{B45}, the emitter-base voltage of Q_{74} is converted into the collector current of Q_{43} and then back into a voltage across Q_{42} and R_{42}. From KVL,

$$V_{B45} = V_{BE42} + |I_{C43}|R_{42} \tag{6.146}$$

where

$$|I_{C43}| = \frac{V_{EB74} - V_{EB43}}{R_{43}} \tag{6.147}$$

Since $R_{42} = R_{43}$, substituting (6.147) into (6.146) gives

$$V_{B45} = V_{BE42} + V_{EB74} - V_{EB43} = V_{EB74} + V_T \ln\left(\frac{|I_{S43}|}{I_{S42}}\right) = V_T \ln \frac{|I_{C74}|}{I_{S75}} \tag{6.148}$$

where $I_{S75} = |I_{S74}I_{S42}/|I_{S43}||$ is assumed. This equation shows that V_{B45} is equal to the base-emitter voltage of a transistor equivalent to Q_{75} whose collector current equals $|I_{C74}|$. This conversion allows $|I_{C74}|$ to be compared to I_{C75} by Q_{45}-Q_{46} in a way that is not sensitive to differences in the saturation currents of pnp and npn transistors.

If $|V_{B45} - V_{B46}| > 3V_T$, the transistor in the Q_{45}-Q_{46} pair with the higher base voltage turns off. Under this condition, the voltage from the emitters of these transistors to ground is controlled by the output transistor that conducts the least current. This voltage becomes the input to one side of differential pair Q_{39}-Q_{40}. The other input is a constant voltage set by $I_{REF} = |I_{C36}| - I_{C35}$ flowing in diode-connected transistors Q_{37} and Q_{38}. Transistors Q_{39}-Q_{40} form the core of a differential amplifier, and this amplifier operates in a negative feedback loop. For example, assume that Q_{75} conducts a large current to pull the op-amp output low. Then the loop controls the collector current of Q_{74}. If the voltage from the base of Q_{40} rises for some reason, I_{C40} is increased and I_{C39} is reduced. As a result, the voltage from node 25 to ground increases, which increases the voltage from node $Pbase$ to ground because Q_{64} and Q_{65} are emitter followers. This change reduces V_{B45} because common-emitter amplifier Q_{43} provides inverting gain at low frequencies. Finally, reducing V_{B45} reduces the voltage from the base of Q_{40} to ground because V_{EB45} is almost constant. In other words, the loop responds to a change in the voltage at the base of Q_{40} by driving this voltage in the opposite direction of the original change. Similar reasoning applies if Q_{74} conducts a large current to pull the op-amp output high. Therefore, the loop has negative gain, which means that it is a negative feedback loop, as stated above. Negative feedback is covered in Chapter 8. The key point here is that if the magnitude of the gain around this loop is high enough, the loop drives the voltage from the base of Q_{40} to ground to equal the voltage from the base of Q_{39} to ground. This result can be viewed as an example of the summing-point constraints used to analyze ideal op-amp circuits such as those in Fig. 6.3.

When Q_{45} and Q_{46} are both on, the relationship between the collector currents of Q_{74} and Q_{75} is determined by the interaction of one KCL equation and two KVL equations in Fig. 6.41. First, from KCL

$$|I_{C45}| + |I_{C46}| = |I_{C44}| \tag{6.149}$$

Also, from KVL,

$$V_{BE75} + V_{EB46} - V_{EB45} - V_{B45} = 0 \tag{6.150}$$

Finally, from KVL,

$$V_{BE75} + V_{EB46} - V_{BE40} - I_{R40}R_{40} + I_{R39}R_{39} + V_{BE39} - V_{BE37} - V_{EB38} = 0 \tag{6.151}$$

where I_{R39} and I_{R40} are the currents in R_{39} and R_{40}, respectively. Since the voltage from the base of Q_{40} is driven to equal the voltage from the base of Q_{39} by negative feedback,

$$V_{BE40} + I_{R40}R_{40} = V_{BE39} + I_{R39}R_{39} \tag{6.152}$$

and (6.151) reduces to

$$V_{BE75} + V_{EB46} - V_{BE37} - V_{EB38} = 0 \tag{6.153}$$

Substituting (1.82) for each term in this equation gives

$$V_T \ln \frac{I_{C75}}{I_{S75}} + V_T \ln \frac{|I_{C46}|}{|I_{S46}|} - V_T \ln \frac{I_{C37}}{I_{S37}} - V_T \ln \frac{|I_{C38}|}{|I_{S38}|} = 0 \tag{6.154}$$

Ignoring base currents, $I_{C37} = |I_{C38}| = I_{REF} = |I_{C36}| - I_{C35}$. Therefore,

$$\left(\frac{I_{C75}}{I_{REF}}\right)\frac{I_{S37}}{I_{S75}} = \left(\frac{I_{REF}}{|I_{C46}|}\right)\frac{|I_{S46}|}{|I_{S38}|} \tag{6.155}$$

Also, substituting (1.82) for the first three terms in (6.150) and (6.148) for the last term gives

$$V_T \ln \frac{I_{C75}}{I_{S75}} + V_T \ln \frac{|I_{C46}|}{|I_{S46}|} - V_T \ln \frac{|I_{C45}|}{|I_{S45}|} - V_T \ln \frac{|I_{C74}|}{I_{S75}} = 0 \quad (6.156)$$

Assuming Q_{45} and Q_{46} are identical, this equation becomes

$$\frac{I_{C75}}{|I_{C74}|} = \frac{|I_{C45}|}{|I_{C46}|} \quad (6.157)$$

Substituting (6.149) into (6.157) gives

$$\frac{I_{C75}}{|I_{C74}|} = \frac{|I_{C44}| - |I_{C46}|}{|I_{C46}|} \quad (6.158)$$

Solving for $|I_{C46}|$ gives

$$|I_{C46}| = \frac{|I_{C44}||I_{C74}|}{I_{C75} + |I_{C74}|} \quad (6.159)$$

Substituting this equation into (6.155) gives

$$\frac{I_{C75}|I_{C74}|}{I_{C75} + |I_{C74}|} = \frac{(I_{REF})^2}{|I_{C44}|} \frac{I_{S75}}{I_{S37}} \frac{|I_{S46}|}{|I_{S38}|} \quad (6.160)$$

In the NE5234 op amp, $I_{REF} = 7.4$ μA and $|I_{C44}| = 6$ μA as calculated above. Also, $I_{S75}/I_{S37} = 10$ and $|I_{S46}|/|I_{S38}| = 2$. When the load current $I_L = 0$, (6.160) gives

$$I_{C75} = |I_{C74}| = 2\frac{(7.4 \text{ μA})^2}{6 \text{ μA}} 20 = 360 \text{ μA} \quad (6.161)$$

When I_L is big and negative, it flows in Q_{75}, and V_{BE75} is big enough to turn off Q_{46}. Therefore, the negative feedback loop in Fig. 6.41 sets the collector current in Q_{74}. To find $|I_{C74}|$, the limit of (6.160) with $I_{C75} \to \infty$ gives

$$\lim_{I_{C75} \to \infty} \left(\frac{I_{C75}|I_{C74}|}{I_{C75} + I_{C74}} \right) = |I_{C74}| = \frac{(I_{REF})^2}{|I_{C44}|} \frac{I_{S75}}{I_{S37}} \frac{|I_{S46}|}{|I_{S38}|} = \frac{(7.4 \text{ μA})^2}{6 \text{ μA}} 20 = 180 \text{ μA} \quad (6.162)$$

Similarly, when I_L is big and positive, it comes from Q_{74}, and Q_{45} turns off. Then

$$\lim_{|I_{C74}| \to \infty} \left(\frac{I_{C75}|I_{C74}|}{I_{C75} + I_{C74}} \right) = I_{C75} = \frac{(I_{REF})^2}{|I_{C44}|} \frac{I_{S75}}{I_{S37}} \frac{|I_{S46}|}{|I_{S38}|} = 180 \text{ μA} \quad (6.163)$$

These equations give the currents flowing in the inactive output transistor when the other transistor drives a load current large enough to turn either Q_{45} or Q_{46} off. These currents are half the bias current that flows in both output transistors when the load current is zero.

The output either sources or sinks current, depending on the output voltage and load. As a result, the properties of the output stage are dependent on the particular value of output voltage and current. For example, assume that the load current $I_L = 1$ mA and that this current flows *out* of the output terminal. Further assume that the load resistance is 2 kΩ. From KCL at the output,

$$|I_{C74}| = I_{C75} + I_L = I_{C75} + 1 \text{ mA} \quad (6.164)$$

Substituting this equation into (6.160) and solving for I_{C75} in the NE5234 op amp gives $I_{C75} = 210$ μA. Therefore, $|I_{C74}| = 1.2$ mA. If the load current increases to 1.1 mA, I_{C75} decreases by only a few μA and is still about 210 μA while $|I_{C74}|$ becomes about 1.3 mA. In other words, the current in Q_{75} is almost constant under these conditions because its current is regulated by the loop described above.

Now that the collector currents of Q_{74} and Q_{75} have been calculated, we are able to calculate the collector currents in Q_{42}, Q_{43}, Q_{45}, and Q_{46}. Ignoring base currents, KVL across the emitter-base junctions of Q_{43} and Q_{74} gives

$$|I_{C43}|R_{43} = V_T \ln \frac{|I_{C74}|}{|I_{C43}|} \frac{|I_{S43}|}{|I_{S74}|} \tag{6.165}$$

The ratio of the emitter area of Q_{43} to that of Q_{74} is 3/32. Therefore,

$$|I_{C43}|(1.3 \text{ k}\Omega) = V_T \ln \frac{1200 \text{ }\mu\text{A}}{|I_{C43}|} \frac{3}{32} \tag{6.166}$$

Solving this equation by trial and error as in a Widlar current mirror gives $|I_{C43}| = 28$ µA. As a result, $I_{C42} = 28$ µA, and the differential input voltage to the Q_{45}-Q_{46} differential pair is

$$\begin{aligned} V_{B45} - V_{B46} &= V_{BE42} + I_{C42}R_{42} - V_{BE75} \\ &= V_T \ln \frac{I_{C42}}{I_{C75}} \frac{I_{S75}}{I_{S42}} + I_{C42}R_{42} \\ &= (26 \text{ mV}) \ln \left(\frac{28}{210} \cdot \frac{10}{1} \right) + (28 \text{ }\mu\text{A})(1.3 \text{ k}\Omega) = 44 \text{ mV} \end{aligned} \tag{6.167}$$

because the ratio of the emitter area of Q_{75} to that of Q_{42} is 10. From (3.147) and (3.148) with $\alpha_F = 1$,

$$|I_{C45}| = \frac{|I_{C44}|}{1 + \exp \frac{V_{B45} - V_{B46}}{V_T}} = \frac{6 \text{ }\mu\text{A}}{1 + \exp \frac{44}{26}} = 0.93 \text{ }\mu\text{A} \tag{6.168}$$

and

$$|I_{C46}| = \frac{|I_{C44}|}{1 + \exp \frac{V_{B46} - V_{B45}}{V_T}} = \frac{6 \text{ }\mu\text{A}}{1 + \exp \frac{-44}{26}} = 5.1 \text{ }\mu\text{A} \tag{6.169}$$

where $|I_{C46}| > |I_{C45}|$ because $V_{B45} > V_{B46}$.

Finally, with $|I_{C74}| = 1.2$ mA and $I_{C75} = 210$ µA, we will recalculate the collector currents of Q_{64}, Q_{65}, and Q_{68}, taking nonzero base currents into account. Assume $\beta_{F(npn)} = 40$ and $\beta_{F(pnp)} = 10$. Then $|I_{B74}| = 1.2$ mA$/10 = 120$ µA. Therefore,

$$|I_{C65}| = \left(\frac{\beta_{F(pnp)}}{\beta_{F(pnp)} + 1} \right) (I_{R65} + |I_{B74}|) = \left(\frac{10}{11} \right) (100 + 120) \text{ }\mu\text{A} = 200 \text{ }\mu\text{A} \tag{6.170}$$

This equation ignores the base current in Q_{66}, which is reasonable because the emitter area of Q_{66} is 32 times smaller than that of Q_{74}. Therefore, $|I_{B65}| = 20$ µA and

$$I_{C64} = \left(\frac{\beta_{F(npn)}}{\beta_{F(npn)} + 1} \right) (I_{C63} - |I_{B65}|) = \left(\frac{40}{41} \right) (33 - 20) \text{ }\mu\text{A} \simeq 13 \text{ }\mu\text{A} \tag{6.171}$$

Similarly, $I_{B75} = 210$ µA$/40 = 5.3$ µA and

$$I_{C68} = \left(\frac{\beta_{F(npn)}}{\beta_{F(npn)} + 1} \right) (I_{C69} + I_{B75}) = \left(\frac{40}{41} \right) (6 + 5.3) \text{ }\mu\text{A} \simeq 11 \text{ }\mu\text{A} \tag{6.172}$$

ignoring the base current in Q_{71}, which is ten times smaller than Q_{75}.

Figure 6.42 shows the relative emitter areas of the transistors in the NE5234 op amp, the dc collector currents calculated above, and the dc collector currents predicted by SPICE simulations under the conditions described above. The calculated collector currents of Q_{55}

Trans.	Rel. Area	I_C (μA) (Calc.)	I_C (μA) (Sim.)	Trans.	Rel. Area	I_C (μA) (Calc.)	I_C (μA) (Sim.)
Q_1	2	0	0	Q_{37}	1	7.4	7.29
Q_{1d}	4	0	0	Q_{38}	1	−7.4	−6.79
Q_2	2	0	0	Q_{39}	1	9.0	8.33
Q_{2d}	4	0	0	Q_{40}	1	9.0	8.48
Q_3	2	−3.0	−2.65	Q_{41}	3	18	17.2
Q_{3d}	4	0	0	Q_{42}	1	28	27.3
Q_4	2	−3.0	−2.63	Q_{43}	3	−28	−27.9
Q_{4d}	4	0	0	Q_{44}	1	−6.0	−5.79
Q_5	1	0	0	Q_{45}	2	−0.93	−0.939
Q_6	1	0	0	Q_{46}	2	−5.1	−4.15
Q_7	1	0	0	Q_{47}	1	−6.0	−5.76
Q_8	1	3.0	3.06	Q_{48}	1	−6.0	−5.25
Q_9	2	6.0	6.16	Q_{49}	1	6.0	5.74
Q_{10}	2	6.0	6.17	Q_{50}	1	−6.0	−5.80
Q_{11}	1	−6.0	−5.81	Q_{51}	1	−6.0	−5.01
Q_{12}	1	−3.0	−3.14	Q_{52}	1	6.0	6.73
Q_{13}	1	−6.0	−5.84	Q_{53}	1	6.0	5.74
Q_{14}	1	−6.0	−5.84	Q_{54}	1	−6.0	−5.76
Q_{15}	1	−3.0	−3.13	Q_{55}	1	−6.0	−15.9
Q_{16}	1	−4.0	−4.01	Q_{56}	1	6.0	5.01
Q_{17}	1	−21	−17.6	Q_{57}	1	6.0	5.85
Q_{18}	1	−21	−17.7	Q_{58}	1	−6.0	−5.76
Q_{19}	1	−4.0	−4.01	Q_{59}	1	6.0	5.64
Q_{20}	1	−6.6	−6.35	Q_{60}	2	6.0	5.78
Q_{21}	1	−4.0	−3.25	Q_{61}	1	−6.0	−5.83
Q_{22}	1	−4.0	−3.38	Q_{62}	1	6.0	5.08
Q_{23}	1	0	0	Q_{63}	1	33	25.3
Q_{24}	1	0	0	Q_{64}	1	13	5.04
Q_{25}	1	10	9.13	Q_{65}	8	−200	−216
Q_{26}	1	10	8.94	Q_{66}	1	−37	−43.2
Q_{27}	1	10	6.19	Q_{67}	1	0	0
Q_{28}	1	10	6.19	Q_{68}	1	11	12.0
Q_{29}	7	42	34.0	Q_{69}	1	6.0	5.75
Q_{30}	1	6.6	5.23	Q_{70}	1	0	0
Q_{31}	1	6.6	5.50	Q_{71}	1	21	25.9
Q_{32}	1	−6.6	−4.59	Q_{72}	3	0	0
Q_{33}	1	−6.6	−4.81	Q_{73}	1	0	0
Q_{34}	1	6.6	4.58	Q_{74}	32	−1200	−1260
Q_{35}	1	6.6	4.70	Q_{75}	10	210	264
Q_{36}	1	−14	−12.4				

Figure 6.42 Relative emitter areas, calculated collector currents, and simulated collector currents of transistors in the NE5234 op amp with $V_{IC} \ll 0.8$ V and $I_L = 1$ mA.

and Q_{64} have large errors. Transistor Q_{55} is an emitter follower that drives the node Bias₁ in Fig. 6.34. The calculation of I_{C55} ignores the base currents in all the transistors with bases connected to Bias₁. This calculation is not repeated here because it makes little difference to the parameters calculated in this book. Similarly, Q_{64} is an emitter follower that drives emitter follower Q_{65} in Fig. 6.39. The error in I_{C64} stems from the calculation of I_{C62}. Ignoring base currents I_{C62} is estimated as 6 μA, and (6.143) is solved by trial and error to estimate I_{C63} as

33 μA. However, the combined base current of Q_{62} and Q_{63} is then $(39 \mu A)/\beta_{F(npn)}$, which is about 1 μA with $\beta_{F(npn)} = 40$. As a result, I_{C62} is about 5 μA instead of 6 μA. When I_{C63} is recalculated by trial and error from (6.143) with the new value of I_{C62}, the result is $I_{C63} = 24$ μA. Although the change from 33 μA to 24 μA does not seem large, the collector current of Q_{64} is significantly reduced by this change because it is determined by the difference between two currents. From KCL,

$$I_{C64} = \left(\frac{\beta_{F(npn)}}{\beta_{F(npn)} + 1} \right) (I_{C63} - |I_{B65}|) = \left(\frac{40}{41} \right) \left(24 - \frac{200}{10} \right) \mu A \simeq 4 \mu A \quad (6.173)$$

instead of 13 μA as calculated in (6.171).

Of course, improved estimates of I_{C55} and I_{C64} could have been calculated in the first place by including base currents in the original calculations. However, that approach would have added a significant level of complexity to those calculations. In practice, the first level of analysis usually makes broad simplifying assumptions. For example, base currents are normally ignored in dc analysis. Then the results are compared to SPICE simulations, and the comparison may lead to some further calculations including previously ignored parameters, exactly as done here. This approach has the advantage of simplicity. Simplifying analysis is important in practice because design is the inverse of analysis. Simplifying assumptions allow designers to improve designs by quickly repeating analysis under modified conditions. This approach allows designers to focus on parameters that limit performance at every step in the design process.

6.8.2 Transistors that Are Normally Off [18]

Consider Fig. 6.36, and assume that the common-mode input voltage is set at a level at which Q_1, Q_2, and Q_6 operate in the forward-active region. If the differential op-amp input voltage increases, the collector current of Q_2 increases and that of Q_1 decreases. Therefore, the voltage from node 2 to ground falls and the voltage from node 1 to ground rises. The key point is that when Q_1 and Q_2 operate in the forward-active region, they introduce 180° of phase shift from their bases to their collectors. Now assume that the common-mode input voltage is raised above V_{CC} by more than a few hundred millivolts so that Q_1 and Q_2 saturate. Under this condition, the collector-base junctions of these transistors become forward biased, keeping the collector-base voltages approximately constant. Therefore, increasing the differential op-amp input voltage increases the voltage from node 2 to ground and decreases the voltage from node 1 to ground. In other words, operating the transistors in saturation eliminates the 180° of phase shift normally introduced. Reversing the polarity of the gain through the input transistors reverses the polarity of the op-amp gain, converting negative feedback to positive feedback and potentially causing instability. A similar problem happens when the common-mode input voltage is lowered below $-V_{EE}$ until Q_3 and Q_4 saturate.

This problem occurs at common-mode input voltages beyond the common-mode range. In principle, this problem can be avoided by specifying that the op amp must not operate with common-mode input voltages that saturate any of the input transistors. However, preventing the polarity reversal through circuit design extends the applications in which the op amp can be used.

Figure 6.43 shows a schematic of part of the NE5234 input stage with transistors Q_{1d}, Q_{2d}, Q_{3d}, and Q_{4d}, which were omitted from Fig. 6.36 for simplicity. The base-emitter junctions of these transistors are short circuited; therefore, each of these transistors operates as a diode between its collector and base. When Q_1 and Q_2 saturate, their collector-base junctions become forward biased. The base of Q_{1d} is connected to the base of Q_1. The base of Q_{2d} is connected to the base of Q_2. The collectors of Q_{1d} and Q_{2d} are connected to the collectors of Q_2 and Q_1, respectively. Therefore, with perfect matching and a pure common-mode input voltage,

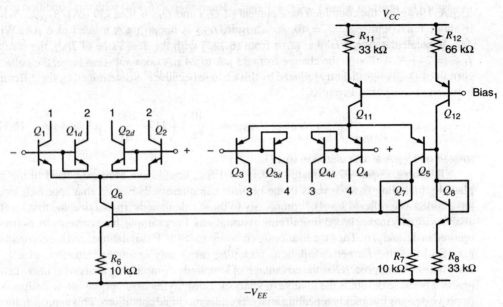

Figure 6.43 Partial schematic of the NE5234 input stage with diodes in both differential pairs. Transistors Q_9, Q_{10}, Q_{13}, and Q_{14} and their degeneration resistors have been omitted for simplicity.

$V_{BC1} = V_{BC1d} = V_{BC2} = V_{BC2d}$, and the collector-base junctions of these four transistors are equally biased. To prevent the polarity reversal described above, Q_{1d} and Q_{2d} have larger areas than Q_1 and Q_2. (In the NE5234 op amp, the area ratio is 2.) Then when the collector-base junctions become forward biased, the collector-base junctions of Q_{1d} and Q_{2d} conduct more current than those of Q_1 and Q_2. Therefore, changes in the noninverting op-amp input voltage exert more influence on node 1 than on node 2, and changes in the inverting op-amp input voltage exert more influence on node 2 than on node 1. As a result, when the differential op-amp input voltage increases, the voltage from node 1 to ground rises, and the voltage from node 2 to ground falls, as in normal operation. Similar statements apply to the *pnp* differential pair. This circuit allows the common-mode input voltage to range from about 700 mV above V_{CC} to about 700 mV below $-V_{EE}$. In the top and bottom few hundred millivolts of this range, transistors in the input stage operate in saturation, and some op-amp specifications are not met. However, the polarity of the op-amp gain is not reversed in this range. Beyond this range, damage to transistors Q_{1d}, Q_{2d}, Q_{3d}, and Q_{4d} could occur. Resistors in series with each op-amp input can be added to avoid damage by limiting the current that flows in the diodes when they turn on.[14]

Transistors Q_{23} and Q_{24} in Fig. 6.37 are normally off. Their bases are connected to node Bias$_2$, which is an output of the bias circuit in Fig. 6.40. The voltage from Bias$_2$ to ground is $V_{EB38} \simeq 0.7$ V. Therefore, if the voltage from node 9 to ground or node 10 to ground rises to about 1.4 V when a large differential voltage is provided to the op-amp inputs, Q_{23} or Q_{24} turn on to prevent further increases in these voltages. This limitation is important to avoid saturating Q_{16} and Q_{19}, reducing the delay required to properly handle a sudden reduction in the magnitude of the differential op-amp input voltage.

Transistors Q_{67} and Q_{70} in Fig. 6.39 are normally off. These transistors turn on to limit $|I_L|$ to prevent destruction of Q_{74} and Q_{75}. For example, if Q_{67} is off, its base-emitter voltage is

$$V_{BE67} = |I_{C66}|R_{67} - I_{C63}R_{63} \tag{6.174}$$

Transistor Q_{67} turns on when $V_{BE67} = V_{BE(\text{on})}$. Rearranging (6.174) with this condition gives

$$|I_{C66}| = \frac{V_{BE(\text{on})}}{R_{67}} + I_{C63}\left(\frac{R_{63}}{R_{67}}\right) = \frac{0.7\text{ V}}{1.5\text{ k}\Omega} + 33\text{ }\mu\text{A}\left(\frac{1.2\text{ k}\Omega}{1.5\text{ k}\Omega}\right) = 490\text{ }\mu\text{A} \quad (6.175)$$

Since the emitter area of Q_{74} is thirty-two times larger than that of Q_{66}, Q_{67} turns on when

$$|I_{C74}| = 32(490\text{ }\mu\text{A}) = 16\text{ mA} \quad (6.176)$$

Once Q_{67} turns on, it raises the voltage across R_{63}, decreasing I_{C63}. This change reduces the ability of Q_{63} to pull down on the base of Q_{65}, which then limits the ability of Q_{65} to pull down on the base of Q_{74}. As a result, the maximum $|I_{C74}|$ is limited to about 16 mA.

Similar reasoning applies to limiting the current in Q_{75}. The current is sensed by Q_{71}. When I_{C71} is large enough, Q_{70} turns on and pulls down on the emitter of Q_{18} in Fig. 6.37. This change reduces $|I_{C18}|$, limiting the current that it can provide to pull up on node 26, thereby limiting I_{C75}. See Problem 6.25.

Transistors Q_{72} and Q_{73} in Fig. 6.39 are normally off. Their purpose is to limit the extent to which Q_{74} and Q_{75} can saturate.[21] This limitation is important because allowing either output transistor to thoroughly saturate fills its base region with minority carriers and increases the delay required to drive the output in the other direction, worsening crossover distortion.

If Q_{74} begins to saturate, its collector-base junction becomes forward biased. As a result, the collector-base junction of Q_{72} also becomes forward biased. Since the voltage from node 25 to ground is level shifted down about 0.7 V by Q_{64} and up about the same amount by Q_{65}, $V_{EB72} \simeq 0$. Therefore, Q_{72} operates in the reverse-active mode when Q_{74} begins to saturate. In this mode, the current in Q_{72} flows into its collector, out its emitter, and into node 25. This current raises the voltage from node 25 to ground, which reduces the output voltage and the extent to which Q_{74} saturates. Similarly, when Q_{75} begins to saturate, its base-collector junction becomes forward biased, which forward biases the base-collector junction of Q_{73}. This transistor then operates in the reverse-active region, pulling current out of node 26 to reduce the voltage from node 26 to ground and limit the extent of the saturation in Q_{75}.

6.8.3 Small-Signal Analysis of the NE5234 Operational Amplifier

Our next objective is to determine the small-signal properties of the amplifier. We will break the circuit up into its three stages—the input stage, second stage, and output stage—and analyze each. In this section, we will assume that $\beta_{0(npn)} = 40$, $\beta_{0(pnp)} = 10$, $V_{A(npn)} = 30$ V, and $V_{A(pnp)} = 20$ V unless otherwise stated. These numbers are estimates of the minimum values that occur in practice.

Input Stage. The input stage of the NE5234 op amp is a fully differential circuit with two emitter-coupled pairs, and the input resistance depends on which pair or pairs are conducting. Assume that $V_{IC} \ll 0.8$ V. Then the *npn* pair Q_1-Q_2 is off, and Q_{11} biases the *pnp* pair Q_3-Q_4. Figure 6.44 shows the differential-mode half circuit under this condition. The load resistance is $R_{in2}/2$, which is half the input resistance of the second stage. Since the emitter of Q_3 operates at a small-signal ground, the input resistance of the differential-mode half circuit is $r_{\pi 3}$, as shown in Fig. 3.56. From (3.195), the differential input resistance is

$$R_{id} = 2r_{\pi 3} \quad (6.177)$$

Since $|I_{C11}| = 6$ μA, $|I_{C3}| = |I_{C4}| = 3$ μA, and

$$R_{id} = 2\left(\frac{\beta_{0(pnp)}}{g_{m3}}\right) = 2\left(\frac{\beta_{0(pnp)}}{|I_{C3}|}\right)V_T = 2\left(\frac{10}{3\text{ }\mu\text{A}}\right)26\text{ mV} = 170\text{ k}\Omega \quad (6.178)$$

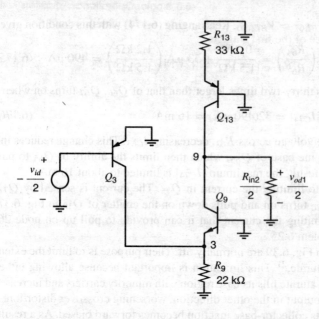

Figure 6.44 Differential-mode half circuit of the input stage with $V_{IC} \ll 0.8$ V.

With a pure differential-mode input, the bases of Q_9 and Q_{13} also operate as small-signal grounds, as shown in Fig. 6.44. As a result, Q_9 and Q_{13} are common-base amplifiers. From the term in brackets in (3.61), the output resistance looking up into the collector of Q_{13} is

$$R_{up1} = \frac{r_{o13} + R_{13}\left(1 + \frac{g_{m13} r_{o13}(\beta_{0(pnp)} + 1)}{\beta_{0(pnp)}}\right)}{1 + \frac{g_{m13} R_{13}}{g_{m13} r_{\pi 13}}} \simeq \frac{r_{o13}\left(1 + g_{m13} R_{13} \frac{\beta_{0(pnp)} + 1}{\beta_{0(pnp)}}\right)}{1 + \frac{g_{m13} R_{13}}{\beta_{0(pnp)}}}$$

(6.179)

because $R_{13} \ll r_{o13}$. Then,

$$R_{up1} = \frac{\frac{20\text{ V}}{6\text{ μA}}\left[1 + \left(\frac{6\text{ μA}}{26\text{ mV}}\right) 33\text{ kΩ} \left(\frac{11}{10}\right)\right]}{1 + \frac{\left(\frac{6\text{ μA}}{26\text{ mV}}\right) 33\text{ kΩ}}{10}} = 18\text{ MΩ}$$

(6.180)

Similarly, if the resistance looking back into the collector of Q_3 is much greater than R_9, the output resistance looking down into the collector of Q_9 is

$$R_{down1} \simeq \frac{r_{o9}\left(1 + g_{m9} R_9 \frac{\beta_{0(npn)} + 1}{\beta_{0(npn)}}\right)}{1 + \frac{g_{m9} R_9}{\beta_{0(npn)}}} \simeq \frac{r_{o9}(1 + g_{m9} R_9)}{1 + \frac{g_{m9} R_9}{\beta_{0(npn)}}}$$

(6.181)

because $R_9 \ll r_{o9}$ and $\beta_{0(npn)} + 1 \simeq \beta_{0(npn)}$. Then,

$$R_{down1} \simeq \frac{\frac{30\text{ V}}{6\text{ μA}}\left[1 + \left(\frac{6\text{ μA}}{26\text{ mV}}\right) 22\text{ kΩ}\right]}{1 + \frac{\left(\frac{6\text{ μA}}{26\text{ mV}}\right) 22\text{ kΩ}}{40}} = 27\text{ MΩ}$$

(6.182)

Therefore, the output resistance of the first stage is

$$R_{o1} = R_{up1} \| R_{down1} = 18 \text{ M}\Omega \| 27 \text{ M}\Omega = 11 \text{ M}\Omega \qquad (6.183)$$

The transconductance of the input stage is

$$G_{m1} = g_{m3} \rho_3 \qquad (6.184)$$

where g_{m3} is the transconductance of Q_3 and ρ_3 represents the fraction of the small-signal collector current in Q_3 that flows into the emitter of Q_9. As in (6.124),

$$\rho_3 \simeq \frac{R_9}{R_9 + r_{e9}} = \frac{R_9}{R_9 + \left(\frac{\beta_{0(npn)}}{\beta_{0(npn)}+1}\right)\frac{V_T}{I_{C9}}} = \frac{22 \text{ k}\Omega}{22 \text{ k}\Omega + \left(\frac{40}{41}\right)\frac{26 \text{ mV}}{6 \text{ }\mu\text{A}}} = 0.84 \qquad (6.185)$$

Therefore,

$$G_{m1} = \left(\frac{|I_{C3}|}{V_T}\right) \rho_3 = \left(\frac{3 \text{ }\mu\text{A}}{26 \text{ mV}}\right) 0.84 = 97 \frac{\mu\text{A}}{\text{V}} \qquad (6.186)$$

As pointed out in Section 6.8.1, the transconductance of the input stage does not depend on the common-mode input voltage.

Second Stage. Since nodes 9 and 10 move equally but in opposite directions with a pure differential input, the emitters of Q_{25}-Q_{28} operate at a small-signal ground. Therefore, the resistance looking in the second stage (Fig. 6.37) is

$$R_{i2} = 2 \left[r_{\pi 21} + (\beta_{0(pnp)} + 1)(r_{\pi 25} \| r_{\pi 26}) \right] \qquad (6.187)$$

assuming the resistance looking into the collector of Q_{16} is much bigger than $r_{\pi 25} \| r_{\pi 26}$. Since $r_{\pi 25} = r_{\pi 26} = \beta_{0(npn)}/g_{m25}$,

$$R_{i2} = 2 \left[\frac{\beta_{0(pnp)}}{g_{m21}} + \frac{\beta_{0(pnp)} + 1}{2} \left(\frac{\beta_{0(npn)}}{g_{m25}} \right) \right]$$

$$= 2 \left[\frac{10(26 \text{ mV})}{4 \text{ }\mu\text{A}} + \frac{11}{2} \left(\frac{40(26 \text{ mV})}{10 \text{ }\mu\text{A}} \right) \right] = 1.3 \text{ M}\Omega \qquad (6.188)$$

To find the stage output resistance, we first calculate the resistance looking up into the collector of Q_{17} or Q_{18}, which is R_{up2}. Since these transistors can be viewed as operating as common-base amplifiers, the term in brackets in (3.61) can be used to give

$$R_{up2} = \frac{r_{o17} + R_{17}\left(1 + \frac{g_{m17} r_{o17}(\beta_{0(pnp)} + 1)}{\beta_{0(pnp)}}\right)}{1 + \frac{g_{m17} R_{17}}{g_{m17} r_{\pi 17}}} \simeq \frac{r_{o17}\left(1 + g_{m17} R_{17} \frac{\beta_{0(pnp)} + 1}{\beta_{0(pnp)}}\right)}{1 + \frac{g_{m17} R_{17}}{\beta_{0(pnp)}}} \qquad (6.189)$$

because $R_{17} \ll r_{o17}$. Then,

$$R_{up2} = \frac{\frac{20 \text{ V}}{21 \text{ }\mu\text{A}}\left[1 + \left(\frac{21 \text{ }\mu\text{A}}{26 \text{ mV}}\right) 9.3 \text{ k}\Omega \left(\frac{11}{10}\right)\right]}{1 + \frac{\left(\frac{21 \text{ }\mu\text{A}}{26 \text{ mV}}\right) 9.3 \text{ k}\Omega}{10}} = 5.0 \text{ M}\Omega \qquad (6.190)$$

Now looking down into the collector of Q_{25} or Q_{26}, we see

$$R_{down2} = r_{o25}(1 + g_{m25}R_{E25}) \tag{6.191}$$

where R_{E25} is the small-signal resistance looking away from the emitter of Q_{25}. Assuming that the resistance looking into the collector of Q_{29} is high enough to be ignored, R_{E25} is the resistance looking into the emitters of Q_{26}-Q_{28}. Since the collectors of Q_{27} and Q_{28} are connected to a small-signal ground (V_{CC}), the resistance looking into each of their emitters is just $r_{e27} = r_{e28}$ from (3.53) with $R_C = 0$. However, for Q_{26}, the calculation is more complicated than for Q_{27} or Q_{28} because the resistance connected to the collector of Q_{26} can be large. Let $R_{in3(26)}$ represent the input resistance of the third stage at node 26. Then from (3.53),

$$R_{E25} = \left[r_{e26} + \left(\frac{\beta_{0(npn)}}{\beta_{0(npn)} + 1} \right) \frac{(R_{up2} || R_{in3(26)})}{g_{m26}r_{o26}} \right] || r_{e27} || r_{e28} \tag{6.192}$$

If $R_{in3(26)}/(g_{m26}r_{o26}) \ll r_{e26}$, which turns out to be true with the small value of $\beta_{0(npn)}$ assumed here, then

$$R_{E25} \simeq \left(\frac{\beta_{0(npn)}}{\beta_{0(npn)} + 1} \right) \left[\frac{1}{g_{m26}} || \frac{1}{g_{m27}} || \frac{1}{g_{m28}} \right]$$

$$= \left(\frac{40}{41} \right) \left[\frac{26 \text{ mV}}{10 \text{ μA}} || \frac{26 \text{ mV}}{10 \text{ μA}} || \frac{26 \text{ mV}}{10 \text{ μA}} \right] = 0.85 \text{ k}\Omega \tag{6.193}$$

Substituting this result into (6.191) gives

$$R_{down2} = \frac{30 \text{ V}}{10 \text{ μA}} \left[1 + \left(\frac{10 \text{ μA}}{26 \text{ mV}} \right) 0.85 \text{ k}\Omega \right] = 4.0 \text{ M}\Omega \tag{6.194}$$

As a result, the output resistance of the second stage looking back into node 25 is

$$R_{o2} = R_{up2} || R_{down2} = 5.0 \text{ M}\Omega || 4.0 \text{ M}\Omega = 2.2 \text{ M}\Omega \tag{6.195}$$

Let $R_{in3(25)}$ represent the input resistance of the third stage at node 25. If $R_{in3(25)} \simeq R_{in3(26)}$, then (6.195) also gives the output resistance of the second stage looking back into node 26.

Figure 6.45 shows a small-signal model of the second stage for finding its transconductance. The resistors labeled R_{up2} represent the small-signal resistance looking into the collector of Q_{17} or Q_{18} in Fig. 6.37. As shown in (6.190), these resistances are 5 MΩ each. The resistances labeled $R_{in3(25)}$ and $R_{in3(26)}$ are the input resistances of the third stage from nodes 25 and 26, respectively. Let v_9, v_{10}, and v_{25} represent the small-signal voltages to ground from nodes 9, 10, and 25, respectively. The emitters of Q_{25}-Q_{28} are floating in Fig. 6.45 because Q_{29} in Fig. 6.37 operates as a current source with an output resistance that is much higher than the resistances looking into the emitters of Q_{25}-Q_{28}. Let a_{v21} and a_{v22} represent the small-signal

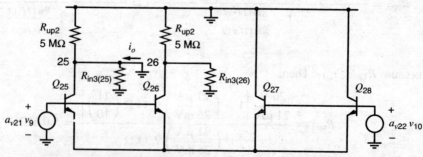

Figure 6.45 Small-signal model of second stage to find its transconductance.

gains of emitter followers Q_{21} and Q_{22} in Fig. 6.37, respectively. Each emitter follower drives a load of $R_L = r_{\pi 25} || r_{\pi 26} = r_{\pi 27} || r_{\pi 28} = r_{\pi 25}/2$. From (3.69) with $R_S = 0$,

$$a_{v21} = a_{v22} = \cfrac{1}{1 + \cfrac{r_{\pi 21}}{(\beta_{0(pnp)} + 1)[(r_{\pi 25}/2) || r_{o21}]}}$$

$$= \cfrac{1}{1 + \cfrac{\cfrac{1}{10(26\text{ mV})}}{\cfrac{4\ \mu A}{(11)\left[\left(\cfrac{40(26\text{ mV})}{2(10\ \mu A)}\right) || \left(\cfrac{20\text{ V}}{4\ \mu A}\right)\right]}}} = 0.90 \qquad (6.196)$$

The transconductance of the second stage is

$$G_{m2} = \left.\frac{i_o}{v_9 - v_{10}}\right|_{v_{25}=0} \qquad (6.197)$$

Because the common-mode output voltage of the first stage is set to a constant as shown in Fig. 6.38, $v_9 = v_{od1}/2$ and $v_{10} = -v_{od1}/2$, where v_{od1} is the small-signal differential output voltage of the first stage. For simplicity in finding the stage transconductance, assume that $v_{10} = 0$. This change introduces a nonzero common-mode input voltage to the second stage. However, this common-mode input causes a negligible change in G_{m2} because the tail current source Q_{29} has high output resistance. Setting $v_{10} = 0$ also halves the differential input voltage to the second stage. However, since the small-signal model is linear, halving the input voltage halves i_o but does not change G_{m2}. As a result,

$$G_{m2} = \left.\frac{i_o}{v_9}\right|_{v_{25}=0} \qquad (6.198)$$

Since the circuit is linear from a small-signal standpoint, superposition can be used to find i_o. Figure 6.46 shows the circuit configuration to find the first component of i_o. The bases of Q_{25} and Q_{26} are driven separately, and the base of Q_{26} is connected to a small-signal ground here. The resistances connected to node 25 are removed because zero small-signal current flows in them. To find i_{o1}, Q_{25} is viewed as a common-emitter amplifier with degeneration. The small-signal resistance looking away from the emitter of Q_{25} is $R_{E25} = 0.85\text{ k}\Omega$ as shown in (6.193). From (3.93),

$$i_{o1} = \frac{g_{m25}}{1 + g_{m25}R_{E25}}a_{v21}v_9 \qquad (6.199)$$

Figure 6.47 shows the circuit configuration to find the second component of i_o. Here the base of Q_{25} is connected to small-signal ground. To find i_{c26}, Q_{26} is viewed as another

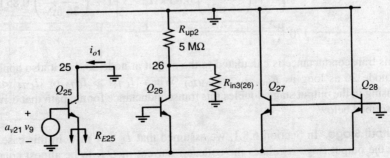

Figure 6.46 Modified small-signal model to find the first component of i_o.

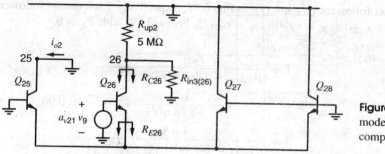

Figure 6.47 Small-signal model to find the second component of i_o.

common-emitter amplifier with degeneration. The small-signal resistance looking away from the emitter of Q_{26} is

$$R_{E26} = r_{e25}||r_{e27}||r_{e28} = r_{e25}/3 = 0.85 \text{ k}\Omega \qquad (6.200)$$

From (3.99), the small-signal resistance looking down into the collector of Q_{26} is

$$R_{C26} = r_{o26}(1 + g_{m26}R_{E26}) = \frac{30 \text{ V}}{10 \text{ }\mu\text{A}}\left[1 + \left(\frac{10 \text{ }\mu\text{A}}{26 \text{ mV}}\right)0.85 \text{ k}\Omega\right] = 4.0 \text{ M}\Omega \qquad (6.201)$$

To find the small-signal collector current in Q_{26}, (3.93) can be used along with a fraction to account for the current divider at the collector of Q_{26}:

$$i_{c26} \simeq \frac{g_{m26}}{1 + g_{m26}R_{E26}}a_{v21}v_9\left(\frac{R_{C26}}{R_{C26} + R_{up2}||R_{in3(26)}}\right) \simeq \frac{g_{m26}}{1 + g_{m26}R_{E26}}a_{v21}v_9 \qquad (6.202)$$

where $R_{in3(26)} \ll R_{C26}$ is assumed in the last approximation. Since $\beta_{0(npn)}r_{o26} \gg R_{E26}$ and $g_{m26}r_{o26} \gg 1$, this equation ignores the terms involving r_o in (3.92). Also, the $1/\beta_0$ term there is ignored here as well as in the rest of the calculation of G_{m2} for simplicity.

Since Q_{25}, Q_{27}, and Q_{28} are identical and their bases and collectors are all connected to small-signal grounds, i_{c26} splits into three equal parts. Therefore, ignoring base currents, i_{o2} is

$$i_{o2} \simeq -i_{c26}/3 \qquad (6.203)$$

Since $i_o = i_{o1} + i_{o2}$,

$$G_{m2} \simeq a_{v21}\left(\frac{g_{m25}}{1 + g_{m25}R_{E25}} - \frac{g_{m26}}{3(1 + g_{m26}R_{E26})}\right)$$

$$= 0.9\left\{\frac{\frac{10 \text{ }\mu\text{A}}{26 \text{ mV}}}{1 + \left(\frac{10 \text{ }\mu\text{A}}{26 \text{ mV}}\right)0.85 \text{ k}\Omega} - \frac{\frac{10 \text{ }\mu\text{A}}{26 \text{ mV}}}{3\left[1 + \left(\frac{10 \text{ }\mu\text{A}}{26 \text{ mV}}\right)0.85 \text{ k}\Omega\right]}\right\}$$

$$\simeq 170 \frac{\mu\text{A}}{\text{V}} \qquad (6.204)$$

This transconductance is calculated to the output at node 25, but it also applies to the output at node 26 as long as $R_{in3(25)} \simeq R_{in3(26)}$. When $|I_{C74}| \gg I_{C75}$ or $|I_{C74}| \ll I_{C75}$, feedback biasing in the output stage doubles this transconductance for the path that drives the output, as described below.

Output Stage. In Section 6.8.1, we assumed that $I_L = 1$ mA. In this case, the bias circuit for the output stage regulates the collector current in Q_{75} to be almost constant. Since Q_{75} is driven by emitter follower Q_{68}, the base current for Q_{68} must be approximately constant

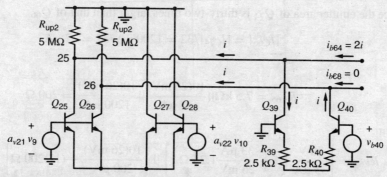

Figure 6.48 Small-signal model of the interaction between the second stage and differential pair Q_{39}-Q_{40} in the bias circuit for the third stage when I_{C75} is almost constant.

under this condition. To explain this result, consider Fig. 6.48, which shows a small-signal model of the key part of the second stage plus differential pair Q_{39}-Q_{40} from the bias circuit that controls the output stage. The outputs of this model are nodes 25 and 26, which drive the bases of Q_{64} and Q_{68} in Fig. 6.39.

First consider the second stage. The two outputs are driven by matched circuits in the second stage, and the loading placed on these two nodes in the third stage is almost identical. Therefore, the small-signal currents drawn into nodes 25 and 26 from the second stage are almost identical and labeled as i in Fig. 6.48.

Now consider differential pair Q_{39}-Q_{40}. This pair forms the input of a differential amplifier that operates in a negative feedback loop and keeps I_{C75} almost constant when $|I_{C74}| \gg I_{C75}$, as described in Section 6.8.1. The base of Q_{39} is connected to a small-signal ground in Fig. 6.48 because the collector currents of Q_{37} and Q_{38} in Fig. 6.40 are constant. On the other hand, the base of Q_{40} is driven by v_{b40}, which is the small-signal voltage from the base of Q_{40} to ground. Since Q_{75} is driven by Q_{68}, keeping I_{C75} constant means that the small-signal base current of Q_{68}, i_{b68}, is approximately zero. Therefore, Q_{40} must inject a small-signal current of i into node 26 under these conditions, as shown in Fig. 6.48. Since the total current in the differential pair is constant, Q_{39} pulls the same current i out of node 25. This result doubles the small-signal base current of Q_{64}, i_{b64}, to $2i$, effectively doubling G_{m2} to node 25 and setting G_{m2} to node 26 to approximately zero when $|I_{C74}| \gg I_{C75}$. Therefore, in the analysis of the output stage below, we will focus on the path from node 25 to the output. In the opposite case (when $|I_{C74}| \ll I_{C75}$), the path from node 26 to the output is dominant. When $|I_{C74}| \simeq I_{C75}$, both paths must be considered.

Ignoring the resistance looking down into the collector of Q_{39}, the input resistance of the third stage (Fig. 6.39) at node 25 is

$$R_{i3(25)} = r_{\pi 64} + (\beta_{0(npn)} + 1)R_{E64} \qquad (6.205)$$

In this equation, R_{E64} is the incremental resistance from the emitter of Q_{64} to small-signal ground, which is

$$R_{E64} = [r_{o63}(1 + g_{m63}R_{63})] \| [r_{\pi 65} + (\beta_{0(pnp)} + 1)R_{E65}] \qquad (6.206)$$

Similarly, R_{E65} is the incremental resistance from the emitter of Q_{65} to small-signal ground. Ignoring the resistance looking into the base of Q_{43} in Fig. 6.40, R_{E65} is

$$R_{E65} = R_{65} \| r_{\pi 66} \| r_{\pi 74} = R_{65} \| \frac{\beta_{0(pnp)} V_T}{|I_{C66}|} \| \frac{\beta_{0(pnp)} V_T}{|I_{C74}|} \qquad (6.207)$$

Since the emitter area of Q_{74} is thirty-two times larger than that of Q_{66},

$$|I_{C66}| = |I_{C74}|/32 = 1200\ \mu A/32 = 37\ \mu A \tag{6.208}$$

So

$$R_{E65} = 7.5\ k\Omega \Big|\Big| \frac{10(26\ mV)}{37\ \mu A} \Big|\Big| \frac{10(26\ mV)}{1200\ \mu A} = 200\ \Omega \tag{6.209}$$

Then

$$R_{E64} = \left[\frac{30\ V}{33\ \mu A}\left(1 + \frac{33\ \mu A}{26\ mV} 1.2\ k\Omega\right)\right] \Big|\Big| \left[\frac{10(26\ mV)}{200\ \mu A} + (11)200\ \Omega\right] = 3.5\ k\Omega \tag{6.210}$$

Finally, from (6.205) with $I_{C64} = 4\ \mu A$ as calculated in (6.173),

$$R_{i3(25)} = \frac{40(26\ mV)}{4\ \mu A} + (41)3.5\ k\Omega = 404\ k\Omega \tag{6.211}$$

The output resistance of the stage is

$$R_{o3} = r_{o74} || r_{o75} = \frac{20\ V}{1200\ \mu A} \Big|\Big| \frac{30\ V}{210\ \mu A} = 15\ k\Omega \tag{6.212}$$

The transconductance of the stage is

$$G_{m3} = a_{v64} a_{v65} g_{m74} \tag{6.213}$$

where a_{v64}, and a_{v65} are the gains of emitter followers Q_{64} and Q_{65}, respectively. Since $r_{o64} \gg R_{E64}$, (3.69) gives

$$a_{v64} = \frac{1}{1 + \dfrac{r_{\pi 64}}{(\beta_{0(npn)}+1)R_{E64}}} = \frac{1}{1 + \dfrac{40(26\ mV)}{\dfrac{4\ \mu A}{41(3.5\ k\Omega)}}} = 0.36 \tag{6.214}$$

Since $r_{o65} \gg R_{E65}$, (3.69) gives

$$a_{v65} = \frac{1}{1 + \dfrac{r_{\pi 65}}{(\beta_{0(pnp)}+1)R_{E65}}} = \frac{1}{1 + \dfrac{10(26\ mV)}{\dfrac{200\ \mu A}{11(200\ \Omega)}}} = 0.63 \tag{6.215}$$

Therefore,

$$G_{m3} = 0.36(0.63)\left(\frac{1200\ \mu A}{26\ mV}\right) = \frac{1}{96\ \Omega} \tag{6.216}$$

Overall Gain. The voltage gain of the input stage when loaded by the second stage is

$$a_{v1} = G_{m1}\left(R_{o1} || \frac{R_{i2}}{2}\right) = 97\ \frac{\mu A}{V}\left(11\ M\Omega || \frac{1.3\ M\Omega}{2}\right) = 60 \tag{6.217}$$

The voltage gain of the second stage when loaded by the third stage is

$$a_{v2} = 2G_{m2}(R_{o2} || R_{i3(25)}) = 2 \times 170\ \frac{\mu A}{V}(2.2\ M\Omega || 404\ k\Omega) = 120 \tag{6.218}$$

In this equation, G_{m2} is multiplied by two because of the effect of feedback biasing explained above. The voltage gain of the third stage when loaded by a 2 kΩ resistor conducting a current of 1 mA is

$$a_{v3} = G_{m3}(R_{o3}||R_L) = \frac{1}{96\,\Omega}(15\text{ k}\Omega||2\text{ k}\Omega) = 18 \quad (6.219)$$

Therefore, the overall gain of the NE5234 op amp is

$$a_v = a_{v1}a_{v2}a_{v3} = 130{,}000 \quad (6.220)$$

This gain is an estimate of the minimum gain of the op amp because the load is usually larger than 2 kΩ and because the values of Early voltages and betas used are the minimum values expected to occur when building many op amps. In practice, the gain is usually much higher than this value. See Problem 6.28.

6.8.4 Calculation of the Input Offset Voltage and Current of the NE5234

Two important aspects of the performance of op amps are the input offset voltage and current. Since the offsets are indistinguishable from the dc component in the signal of interest, these deviations from ideality limit the ability of the circuit to amplify small dc signals accurately. Furthermore, calculation of these parameters is fundamentally different from calculation of other parameters such as voltage gain. Offset voltage and current are random parameters with mean values that are usually near zero. These offsets arise from randomly occurring mismatches between pairs of matched elements in the input stage of the circuit, and are best described by a probability distribution with (ideally) zero mean and some standard deviation. The parameters of interest from the designer's standpoint are the standard deviations of the offset voltage and current distributions, which determine the limits that can be placed on the offset voltage and current of production units being tested while maintaining an acceptable yield. From the user's standpoint, the details of the distribution are of little significance except as they affect the specified maximum offset voltage chosen by the designer. The information available to the designer at the design stage is the distribution of mismatches in resistor values, transistor saturation currents, and transistor betas. Given these distributions, the task is to design an input stage with the minimum offset while meeting the other circuit requirements. We will calculate the offset voltage and current of the NE5234 op amp under the assumption that the various device mismatches are described by normal or Gaussian distributions.

In this analysis, we will assume that the voltage and current gains of the input stage are large enough that other stages have negligible effect on the input offset voltage and current of the op amp. Provided that the mismatches are small enough that they only slightly perturb the currents in the circuit, we can simplify the problem by considering each device pair independently. Figure 6.49 shows a simplified schematic of the input stage of the NE5234. This figure assumes that the common-mode input voltage to the op amp is low enough that Q_3 and Q_4 are active but Q_1 and Q_2 are off. The collector currents of Q_3 and Q_4 are modeled here by current sources $|I_{C3}|$ and $|I_{C4}|$. The first problem is to determine the difference between $|I_{C3}|$ and $|I_{C4}|$ required to set $I_{C9} - I_{C10} = 0$ in the presence of mismatches between Q_9 and Q_{10} as well as between R_9 and R_{10}.

From KVL,

$$V_B = V_T \ln\left(\frac{I_{C9}}{I_{S9}}\right) + \left(\frac{I_{C9}}{\alpha_{F9}} + |I_{C3}|\right) R_9 \quad (6.221)$$

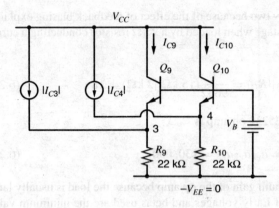

Figure 6.49 A model of the NE5234 input stage for finding $|I_{C3}| - |I_{C4}|$ to set $I_{C9} - I_{C10} = 0$.

and

$$V_B = V_T \ln\left(\frac{I_{C10}}{I_{S10}}\right) + \left(\frac{I_{C10}}{\alpha_{F10}} + |I_{C4}|\right) R_{10} \quad (6.222)$$

Subtracting these equations gives

$$0 = V_T \ln\left(\frac{I_{C9}}{I_{C10}}\right) - V_T \ln\left(\frac{I_{S9}}{I_{S10}}\right) + \frac{I_{C9}}{\alpha_{F9}} R_9 - \frac{I_{C10}}{\alpha_{F10}} R_{10} + |I_{C3}|R_9 - |I_{C4}|R_{10} \quad (6.223)$$

To simplify the first two terms on the right side of this equation, use standard definitions of average and mismatch quantities as in Section A.4.1.1: $I_{C9,10} = (I_{C9} + I_{C10})/2$, $\Delta I_{C9,10} = I_{C9} - I_{C10}$, $I_{S9,10} = (I_{S9} + I_{S10})/2$, and $\Delta I_{S9,10} = I_{S9} - I_{S10}$. Also, assume that $\Delta I_{C9,10}/2I_{C9,10} \ll 1$, $\Delta I_{S9,10}/2I_{S9,10} \ll 1$, and $\ln(1+x) \simeq x$ for $x \ll 1$. Then

$$V_T \ln\left(\frac{I_{C9}}{I_{C10}}\right) - V_T \ln\left(\frac{I_{S9}}{I_{S10}}\right) \simeq V_T \left(\frac{\Delta I_{C9,10}}{I_{C9,10}} - \frac{\Delta I_{S9,10}}{I_{S9,10}}\right) \quad (6.224)$$

With corresponding definitions and assumptions for other parameters, the next two terms in (6.223) become

$$\frac{I_{C9}}{\alpha_{F9}} R_9 - \frac{I_{C10}}{\alpha_{F10}} R_{10} \simeq \frac{I_{C9,10} R_{9,10}}{\alpha_{F9,10}} \left(\frac{\Delta I_{C9,10}}{I_{C9,10}} + \frac{\Delta R_{9,10}}{R_{9,10}} - \frac{\Delta \alpha_{F9,10}}{\alpha_{F9,10}}\right) \quad (6.225)$$

Similarly, the last two terms in (6.223) become

$$|I_{C3}|R_9 - |I_{C4}|R_{10} \simeq |I_{C3,4}|R_{9,10}\left(\frac{\Delta |I_{C3,4}|}{|I_{C3,4}|} + \frac{\Delta R_{9,10}}{R_{9,10}}\right) \quad (6.226)$$

Substituting (6.224), (6.225), and (6.226) into (6.223) and finding $\Delta|I_{C3,4}|/|I_{C3,4}|$ needed to set $\Delta I_{C9,10}/I_{C9,10} = 0$ gives

$$\frac{\Delta |I_{C3,4}|}{|I_{C3,4}|} \simeq \frac{1}{g_{m3,4} R_{9,10}} \frac{\Delta I_{S9,10}}{I_{S9,10}} - \left(1 + \frac{g_{m9,10}}{\alpha_{F9,10} g_{m3,4}}\right) \frac{\Delta R_{9,10}}{R_{9,10}} + \frac{g_{m9,10}}{\alpha_{F9,10} g_{m3,4}} \frac{\Delta \alpha_{F9,10}}{\alpha_{F9,10}} \quad (6.227)$$

Now consider Fig. 6.50, which shows a slightly more complicated model of the NE5234 input stage. This figure includes $|I_{C13}|$ and $|I_{C14}|$ to model the currents produced by Q_{13}, Q_{14}, R_{13}, and R_{14}. These elements form the outputs of a pair of current mirrors, and the effect of

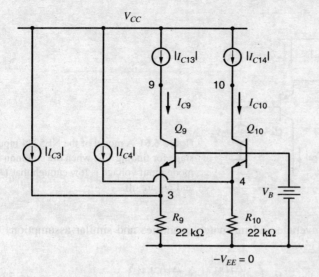

Figure 6.50 A model of the NE5234 input stage for finding $|I_{C3}| - |I_{C4}|$ to set $I_{C9} - I_{C10} = |I_{C13}| - |I_{C14}|$.

mismatch in these elements is considered in Section A.4.1.1. From (4.296),

$$\frac{\Delta|I_{C13,14}|}{|I_{C13,14}|} \simeq \left(\frac{1}{1 + \dfrac{g_{m13,14}R_{13,14}}{\alpha_{F13,14}}}\right) \frac{\Delta|I_{S13,14}|}{|I_{S13,14}|}$$

$$+ \frac{\dfrac{g_{m13,14}R_{13,14}}{\alpha_{F13,14}}}{1 + \dfrac{g_{m13,14}R_{13,14}}{\alpha_{F13,14}}} \left(-\frac{\Delta R_{13,14}}{R_{13,14}} + \frac{\Delta \alpha_{F13,14}}{\alpha_{F13,14}}\right) \quad (6.228)$$

To include these mismatches in the stage offsets, we substitute (6.224), (6.225), and (6.226) into (6.223) and find $\Delta|I_{C3,4}|/|I_{C3,4}|$ needed to set $\Delta I_{C9,10}/I_{C9,10} = \Delta|I_{C13,14}|/|I_{C13,14}|$, which causes the differential output voltage of the stage to be zero and gives

$$\frac{\Delta|I_{C3,4}|}{|I_{C3,4}|} \simeq \frac{1}{g_{m3,4}R_{9,10}} \frac{\Delta I_{S9,10}}{I_{S9,10}} - \left(1 + \frac{g_{m9,10}}{\alpha_{F9,10}g_{m3,4}}\right) \frac{\Delta R_{9,10}}{R_{9,10}} + \frac{g_{m9,10}}{\alpha_{F9,10}g_{m3,4}} \frac{\Delta \alpha_{F9,10}}{\alpha_{F9,10}}$$

$$- \left(\frac{1 + \dfrac{g_{m9,10}R_{9,10}}{\alpha_{F9,10}}}{g_{m3,4}R_{9,10}}\right)\left(\frac{\dfrac{g_{m13,14}R_{13,14}}{\alpha_{F13,14}}}{1 + \dfrac{g_{m13,14}R_{13,14}}{\alpha_{F13,14}}}\right)\left(-\frac{\Delta R_{13,14}}{R_{13,14}} + \frac{\Delta \alpha_{F13,14}}{\alpha_{F13,14}}\right)$$

$$- \left(\frac{1 + \dfrac{g_{m9,10}R_{9,10}}{\alpha_{F9,10}}}{g_{m3,4}R_{9,10}}\right)\left(\frac{1}{1 + \dfrac{g_{m13,14}R_{13,14}}{\alpha_{F13,14}}}\right) \frac{\Delta|I_{S13,14}|}{|I_{S13,14}|} \quad (6.229)$$

Now the problem is to refer $\Delta|I_{C3,4}|/|I_{C3,4}|$ to the input of the op amp. Fig. 6.51 shows a model of the input stage used to find the input offset voltage. From KVL,

$$V_4 - V_3 = V_{EB3} - V_{EB4} = V_T \ln\left(\frac{|I_{S4}|}{|I_{S3}|}\frac{|I_{C3}|}{|I_{C4}|}\right) \quad (6.230)$$

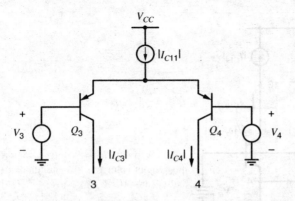

Figure 6.51 A model of the NE5234 input stage for finding V_{OS} when the common-mode input voltage is low enough that Q_1 and Q_2 are off.

Using standard definitions of average and mismatch quantities and similar assumptions to those made to simplify (6.223) gives

$$V_4 - V_3 \simeq V_T \left(-\frac{\Delta |I_{S3,4}|}{|I_{S3,4}|} + \frac{\Delta |I_{C3,4}|}{|I_{C3,4}|} \right) \tag{6.231}$$

Then the input offset voltage, V_{OS}, is the differential input voltage, $V_4 - V_3$, for which the differential output voltage of the stage is zero. This condition is satisfied when $\Delta|I_{C3,4}|/|I_{C3,4}|$ is given by (6.229). Substituting (6.229) into (6.231) with $g_{m3,4} = 3$ µA/26 mV, $g_{m9,10} = g_{m13,14} = 6$ µA/26 mV, $R_{9,10} = 22$ kΩ, $R_{13,14} = 33$ kΩ, $\alpha_{F9,10} = 40/41$, and $\alpha_{F13,14} = 10/11$ gives

$$V_{OS} \simeq V_T \left[-\left(\frac{\Delta|I_{S3,4}|}{|I_{S3,4}|}\right) + 0.39 \left(\frac{\Delta I_{S9,10}}{I_{S9,10}}\right) - 3.1 \left(\frac{\Delta R_{9,10}}{R_{9,10}}\right) + 2.1 \left(\frac{\Delta \alpha_{F9,10}}{\alpha_{F9,10}}\right) \right.$$
$$\left. - 2.2 \left(-\frac{\Delta R_{13,14}}{R_{13,14}} + \frac{\Delta \alpha_{F13,14}}{\alpha_{F13,14}}\right) - 0.26 \left(\frac{\Delta|I_{S13,14}|}{|I_{S13,14}|}\right) \right] \tag{6.232}$$

For a given set of mismatches, this expression gives the input offset voltage. However, the information of interest to the designer is the distribution of observed offset voltages over a large number of samples, and the information available is the distribution of the mismatch factors. If each of the quantities

$$\frac{\Delta|I_{S3,4}|}{|I_{S3,4}|} \quad \frac{\Delta I_{S9,10}}{I_{S9,10}} \quad \frac{\Delta R_{9,10}}{R_{9,10}} \quad \frac{\Delta \alpha_{F9,10}}{\alpha_{F9,10}} \quad \frac{\Delta R_{13,14}}{R_{13,14}} \quad \frac{\Delta \alpha_{F13,14}}{\alpha_{F13,14}} \quad \frac{\Delta|I_{S13,14}|}{|I_{S13,14}|}$$

are regarded as independent random variables with normal distributions, the standard deviation of the sum is given by

$$\sigma_{\text{sum}} = \sqrt{\sum_n \sigma_n^2} \tag{6.233}$$

as described in Appendix A3.1. Thus the standard deviation of the distribution for V_{OS} is calculated by taking the square root of the sum of the squares of the individual contributions.

Assuming that the standard deviation of resistor matching is 1 percent, that of I_S matching is 5 percent, and that of beta matching is 10 percent, we obtain

$$\frac{\sigma_{V_{OS}}}{V_T} = \sqrt{(0.050)^2 + (0.020)^2 + (0.031)^2 + (0.0050)^2 + (0.022)^2 + (0.020)^2 + (0.013)^2}$$
$$\sigma_{V_{OS}} = V_T(0.070) = 1.8 \text{ mV} \tag{6.234}$$

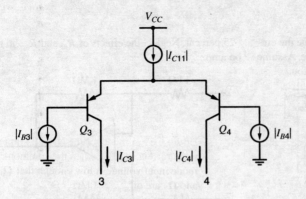

Figure 6.52 A model of the NE5234 input stage for finding I_{OS} when the common-mode input voltage is low enough that Q_1 and Q_2 are off.

The largest single offset contribution is thus the saturation current mismatch of the input devices Q_3 and Q_4. If the offset voltage has a standard deviation of 1.8 mV, then the fraction, Y, of all devices fabricated that will have an offset voltage of less than the NE5234 specification of 5 mV can be found using (3.285) and is

$$Y = \int_{-5}^{5} \frac{1}{1.8\sqrt{2\pi}} \exp\left[-\frac{x^2}{2(1.8)^2}\right] dx \quad (6.235)$$

Evaluating this integral with the aid of Fig. 3.71 gives $Y = 0.995$. Thus a 0.5 percent yield loss will be suffered from offset voltage variations with the 5-mV offset specification.

Figure 6.52 shows a model of the input stage used to find the input offset *current*. The difference of the magnitudes of the base currents is

$$|I_{B4}| - |I_{B3}| = \frac{|I_{C4}|}{\beta_{F4}} - \frac{|I_{C3}|}{\beta_{F3}} = \frac{|I_{C3,4}|}{\beta_{F3,4}}\left(\frac{1 - \frac{\Delta|I_{C3,4}|}{2|I_{C3,4}|}}{1 - \frac{\Delta\beta_{F3,4}}{2\beta_{F3,4}}} - \frac{1 + \frac{\Delta|I_{C3,4}|}{2|I_{C3,4}|}}{1 + \frac{\Delta\beta_{F3,4}}{2\beta_{F3,4}}}\right)$$

$$\simeq |I_{B3,4}|\left(\frac{\Delta\beta_{F3,4}}{\beta_{F3,4}} - \frac{\Delta|I_{C3,4}|}{|I_{C3,4}|}\right) \quad (6.236)$$

Then the input offset current, I_{OS}, is the differential input current, $|I_{B4}| - |I_{B3}|$, for which the differential output voltage of the stage is zero. This condition is satisfied when $\Delta|I_{C3,4}|/|I_{C3,4}|$ is given by (6.229). Substituting (6.229) into (6.236) with the same parameter values as in (6.232) gives

$$I_{OS} \simeq |I_{B3,4}|\left[\left(\frac{\Delta\beta_{F3,4}}{\beta_{F3,4}}\right) - 0.39\left(\frac{\Delta I_{S9,10}}{I_{S9,10}}\right) + 3.1\left(\frac{\Delta R_{9,10}}{R_{9,10}}\right) - 2.1\left(\frac{\Delta\alpha_{F9,10}}{\alpha_{F9,10}}\right)\right.$$
$$\left. + 2.2\left(-\frac{\Delta R_{13,14}}{R_{13,14}} + \frac{\Delta\alpha_{F13,14}}{\alpha_{F13,14}}\right) + 0.26\left(\frac{\Delta|I_{S13,14}|}{|I_{S13,14}|}\right)\right] \quad (6.237)$$

Under the same conditions used to calculate $\sigma_{V_{OS}}$, we obtain

$$\frac{\sigma_{I_{OS}}}{|I_{B3,4}|} = \sqrt{(0.10)^2 + (0.020)^2 + (0.031)^2 + (0.0050)^2 + (0.022)^2 + (0.020)^2 + (0.013)^2}$$

$$\sigma_{I_{OS}} = \frac{3\ \mu A}{10}(0.11) = 33\text{ nA} \quad (6.238)$$

and is dominated by the beta mismatch between Q_3 and Q_4.

PROBLEMS

6.1 For the circuit of Fig. 6.53, determine the output current as a function of the input voltage. Assume that the transistor operates in the active region.

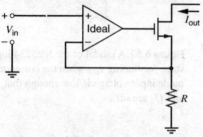

Figure 6.53 Circuit for Problem 6.1.

6.2 Determine the output voltage as a function of the input voltage for the circuit of Fig. 6.54. Assume the op amp is ideal.

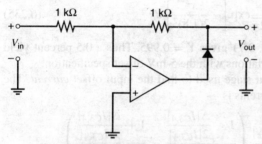

Figure 6.54 Circuit for Problem 6.2.

6.3 In the circuit of Fig. 6.55, determine the correct value of R_x so that the output voltage is zero when the input voltage is zero. Assume a nonzero input bias current, but zero input offset current and input offset voltage.

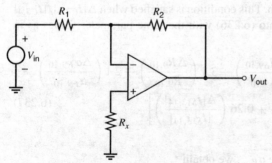

Figure 6.55 Circuit for Problem 6.3.

6.4 The differential instrumentation amplifier shown in Fig. 6.56 must have a voltage gain of 10^3 with an accuracy of 0.1 percent. What is the minimum required open-loop gain of the op amp? Assume the op amp open-loop gain has a tolerance of +100 percent, −25 percent. Neglect the effects of R_{in} and R_{out} in the op amp.

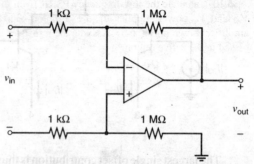

Figure 6.56 Circuit for Problem 6.4.

6.5 Once the offset voltage of the differential amplifier in Problem 6.4 is adjusted to zero, the input-referred offset voltage must remain less than 5 mV in magnitude for common-mode input voltages between ± 10 V. What is the minimum CMRR allowable for the amplifier to meet this requirement?

6.6 Consider the differential amplifier shown in Fig. 6.4. Choose values of R_1 and R_2 for which the gain is equal to −20 and the magnitude of the dc output voltage is less than or equal to −20 mV with $V_1 = V_2 = 0$. Assume that the op amp is ideal except that $|I_{OS}| = 100$ nA.

6.7 Suppose an op amp with $PSRR^+ = 10$ is connected in the voltage-follower configuration shown in Fig. 6.3c. The input V_S is set to zero, but a low-frequency ac signal with peak magnitude $v_{sup} = 10$ mV is superimposed on the positive power supply. Calculate the peak magnitude of the output voltage.

6.8 In the switched-capacitor amplifier of Fig. 6.9a, assume that the source of M_4 is connected to V_S instead of to ground. Calculate the output voltage that appears during ϕ_2 for a given V_S. Assume the op amp is ideal except that it has a finite gain a and a nonzero input capacitance C_P. Assume ideal MOS switches with zero on-resistance and infinite off-resistance.

6.9(a) Calculate and sketch the output voltage waveform of the switched-capacitor integrator of Fig. 6.10a from $t = 0$ to $t = 20$ μs assuming a fixed $V_s = 1$ V and a clock rate of 1 MHz. Assume an ideal MOS op amp with infinite gain and zero output rise time. Assume ideal MOS switches with zero on-resistance and infinite off-resistance.

(b) Compare the result for (a) with the output waveform of the continuous-time equivalent circuit of Fig. 6.10c.

(c) Investigate the effect on the output voltage waveform in (a) of a finite voltage gain of 1000 in the MOS op amp.

6.10 Calculate the low-frequency PSRR from the V_{dd} and V_{ss} power supplies for the common-source amplifier shown in Fig. 6.57. Assume the transistor is biased in the active region.

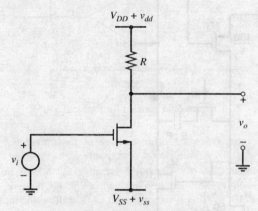

Figure 6.57 Circuit for Problem 6.10.

6.11(a) Equation 6.69 gives the random input offset voltage of the op amp in Fig. 6.16. Explain the polarity of each term in (6.69) by assuming that the matching is perfect except for the term under consideration. Keep in mind that the overdrive is negative for p-channel transistors. Therefore, (6.69) predicts that the offset stemming from $W_3 > W_4$ is negative.

(b) Repeat (a) for an op amp that uses an n-channel differential pair and a p-channel current-mirror load. Equation 6.69 still applies in this case.

6.12 Draw a two-stage op amp similar to the op amp in Fig. 6.16 except reverse the polarity of every transistor. For example, the resulting op amp should have an n-channel input pair. Calculate the following parameters: (a) low-frequency voltage gain, (b) output swing, (c) systematic input offset voltage assuming (6.66) is satisfied, (d) common-mode rejection ratio, (e) common-mode input range, and (f) low-frequency power-supply rejection ratio from both supplies.

6.13(a) Calculate the random input offset voltage for the op amp in Fig. 6.16. Assume the matching is perfect except that $V_{t3} - V_{t4} = 20$mV. Also assume that all transistors have equal W/L and operate in the active region. Ignore short-channel effects and use the data in Table 2.4.

(b) Repeat (a) for an op amp that uses an n-channel differential pair and a p-channel current-mirror load.

(c) Which of these two configurations gives lower input offset voltage? Explain.

6.14 List and explain at least three reasons to select a two-stage op amp with an n-channel input pair instead of with a p-channel input pair for a given application.

6.15 Calculate bias currents and the low-frequency small-signal voltage gain for the CMOS op amp of Fig. 6.58. Use the parameters given in Table 2.4, and assume that $X_d = 0.1$ μm and $dX_d/dV_{DS} = 0.04$ μm/V for all the transistors at the operating point. Calculate the input common-mode range assuming that the wells of M_1 and M_2 are connected to their common-source point. Calculate the low-frequency gain from each supply to the output. Check these calculations with SPICE simulations.

6.16 Draw a telescopic-cascode op amp similar to the first stage in Fig. 6.25 except use an n-channel input pair and a high-swing p-type cascode current-mirror load. Calculate the maximum output swing in terms of the common-mode input voltage. Determine the optimum common-mode input voltage for maximizing the output swing, and calculate the swing with this common-mode input voltage under the same assumptions given in Problem 6.16.

6.17 Calculate the common-mode input range of the op amp in Fig. 6.25. Assume that all the transistors are enhancement-mode devices with $|V_t| = 1$ V, and ignore the body effect. Also assume that the biasing is arranged so that $|V_{ov}| = 0.2$ V for each transistor except M_9. Finally, assume that M_1 and M_2 are biased at the edge of the active region by M_9 and I_C.

6.18 Calculate the common-mode input range of the folded-cascode op amp in Fig. 6.28. Assume that all the transistors are enhancement-mode devices with $|V_t| = 0.8$ V, and ignore the body effect. Also assume that the biasing is arranged so that $|V_{ov}| = 0.2$ V for each transistor. Finally, assume that M_{11} and M_{12} are biased at the edge of the active region.

6.19 Design a CMOS op amp based on the folded-cascode architecture of Fig. 6.28 using supply voltages of ± 1.5 V. Use the bias circuit of Fig. 4.42 (with M_3 and M_4 cascoded) to generate the bias current I_{BIAS}. Then design an extension to this bias circuit that produces the bias voltages V_{BIAS1}, V_{BIAS2}, and V_{BIAS3} based on I_{BIAS}. The output current-drive capability is to be ± 100 μA, the output voltage-swing capability 1.5 Volts peak-peak, and the input common-mode range should extend from 0.5 V to the negative supply. Matching requirements dictate a minimum effective channel length of 1 μm. To make the gain insensitive to small shifts in the operating point, design the circuit so that the magnitude of the drain-source voltage for each transistor operating in the active region exceeds the magnitude of its overdrive by at least 100 mV.

484 Chapter 6 ■ Operational Amplifiers with Single-Ended Outputs

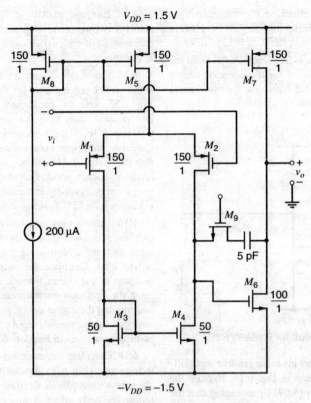

Figure 6.58 Circuit for Problem 6.15.

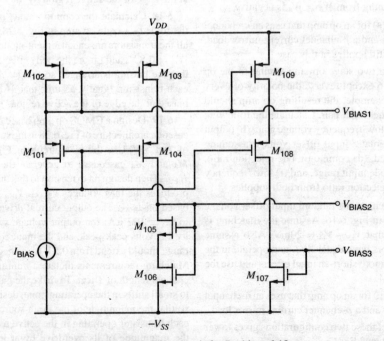

Figure 6.59 Bias circuit for Problem 6.19.

Specify all device geometries and bias currents. The process is the *n*-well process with parameters given in Table 2.4. Assume $X_d = 0$ and $\gamma = 0.25\text{V}^{1/2}$ for both *n*- and *p*-channel transistors, but ignore the body effect in the hand calculations. Use SPICE to verify and refine your design as well as to determine the gain.

6.20 Find the low-frequency voltage gain from variation on each power supply to the op-amp output in Fig. 6.28. Assume that the bias voltages V_{BIAS1}, V_{BIAS2}, and V_{BIAS3} are produced by the circuit shown in Fig. 6.59, where M_{106} is the only transistor that operates in the triode region. Assume that the W/L ratios are chosen so that all transistors in the op amp operate in the active region. Compare your calculations to SPICE simulations. Use models and supply voltages of your choice provided that the above conditions are satisfied.

6.21 Draw the schematic of a folded-cascode op amp similar to the op amp in Fig. 6.28 except with two layers of both *n*- and *p*-type cascodes. Choose a current mirror that maximizes the output swing. Assume that all transistors have equal overdrive magnitudes except where changes are needed to maximize the output swing. Use the models in Table 2.4 and ignore the body effect. Specify the W/L ratios in multiples of $(W/L)_1$.

6.22 For the folded-active-cascode op amp in Fig. 6.30, choose the device sizes to give a peak-peak output swing of at least 2.5 V. Use the 0.4 μm CMOS model parameters in Table 2.4 except let $\gamma = 0.25 \text{ V}^{1/2}$ and $X_d = 0$ for all transistors and $V_{t0} = 0.7$ V and -0.7 V for *n*-channel and *p*-channel transistors, respectively. Assume the drawn channel length is $L = 1$ μm is used for all transistors to simplify the design. With $I_{BIAS} = 25$ μA, the bias currents should be $|I_{D5}| = I_{D11} = I_{D12} = I_{D25} = |I_{D35}| = 200$ μA. Assume $V_{DD} = V_{SS} = 1.65$ V and that the matching is perfect. When the dc input voltage $V_I = 0$, assume that all transistors except M_{106} and M_{114} operate in the active region with equal overdrive magnitudes. To make the gain insensitive to small shifts in the operating point, design the circuit so that the magnitude of the drain-source voltage for each transistor operating in the active region exceeds the magnitude of its overdrive by at least 100 mV. Ignore the body effect in the hand calculations. Use SPICE to verify your design, to choose the widths of M_{106} and M_{114}, and to determine the gain. Also, use SPICE to determine the gain if $V_{t0} = 0.6V$ for *n*-channel transistors and $V_{t0} = -0.8$ V for *p*-channel transistors, as given in Table 2.4. Explain the resulting change in the op-amp again.

6.23 Suppose that the peak-peak output swing requirement in Problem 6.22 is reduced while the other conditions are held constant. This change allows the overdrive magnitude to be increased. Which transistor in the bias circuit of Fig. 6.30d enters the triode region first if the overdrive magnitudes are increased uniformly? Explain. Exclude M_{106} and M_{114} from consideration because they are deliberately operated in the triode region. How can the bias circuit be redesigned to increase the allowed overdrive magnitude while operating all transistors except M_{106} and M_{114} in the active region? What are the disadvantages of the modified bias circuit?

6.24(a) Figure 6.60a shows a folded version of the op amp in Fig. 6.15. A differential interstage level-shifting network composed of voltage sources *V*

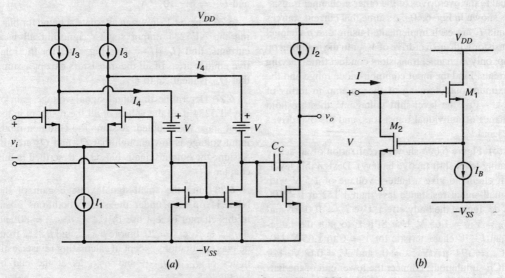

Figure 6.60 Circuits for Problem 6.24.

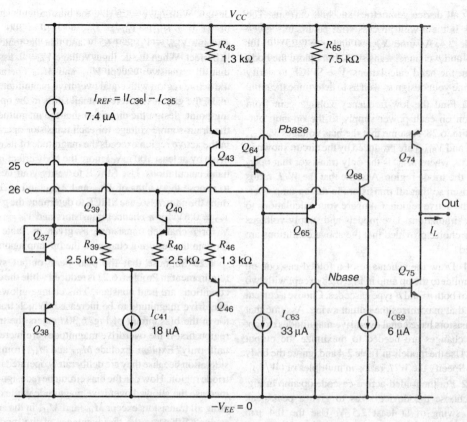

Figure 6.61 Circuit for Problem 6.26.

has been inserted between the first and second stages. Assume that current source I_1 is implemented using an n-channel transistor with overdrive of V_{ovn}, which is equal to the overdrives of the other n-channel transistors shown in Fig. 6.60a. Assume that current sources I_2 and I_3 are each implemented using one p-channel transistor with an overdrive of V_{ovp}. In the resulting op amp, only n-channel transistors conduct time-varying currents. Find the input common-mode range and the maximum output swing of the op-amp in terms of V_{DD}, $-V_{SS}$, the level-shift voltage V, the threshold voltages of individual transistors, and the overdrives V_{ovn} and V_{ovp}.

(b) Figure 6.60b shows a realization of an active floating level shift (active battery). Design this level-shift circuit to give a battery voltage of 1.5 V with a small-signal resistance less than 1 kΩ at I = 100 μA dc. Ignore the body effect. Use $I_B = 100$ μA and $V_{DD} = V_{SS} = 1.65$ V. Use SPICE to plot the large-signal I − V characteristic for V = 0 to 1.65 V. Use $\mu_n C_{ox} = 194$ μA/V^2, $\lambda = 0$, and $V_t = 0.6$ V. For SPICE simulations, connect the lower end of the battery to the negative supply.

6.25 Assume that Q_{75} does not saturate and calculate the maximum value of I_{C75} in the NE5234 op amp. Assume $V_{EB70(on)} = 0.7$ V, $I_{S71} = 6 \times 10^{-18}$ A and $I_{S75} = 6 \times 10^{-17}$ A.

6.26 Fig. 6.61 shows an alternate scheme for biasing the NE5234 output stage.[17] Ignoring all base currents, find $|I_{C74}|/I_{C75}$. Use Fig. 6.42 for the relative emitter areas of all the transistors except assume that Q_{46} is identical to Q_{43}.

6.27 Determine the small-signal voltage gain of the NE5234 if all the values of all the resistors in the input stage are doubled. Assume the common-mode input voltage is low enough that Q_1 and Q_2 are off. Assume the output stage is biased as described in this chapter.

6.28 Find the small-signal voltage gain of the NE5234 op amp under the same conditions given in this chapter except use $\beta_{F(npn)} = \beta_{0(npn)} = 80$ and $\beta_{F(pnp)} = \beta_{0(pnp)} = 20$. Ignore $\beta_{F(npn)}$ and $\beta_{F(pnp)}$ from dc bias calculations except in calculating revised estimates of I_{C64}, $|I_{C65}|$, and I_{C68}, as at the end of Section 6.8.1.

6.29 In Fig. 6.31b, resistive loads were used to extend the common-mode input range to include $-V_{EE}$. Fig. 6.62 shows another circuit with this characteristic.[22] Find the common-mode range of this circuit. Assume that $V_{BE(on)} = 0.7$ V and $V_{CE(sat)} = 0.1$ V for the *npn* transistors. Also, assume that $V_{BE(on)} = -0.7$ V and $V_{CE(sat)} = -0.1$ V for the *pnp* transistors. Ignore base currents for simplicity.

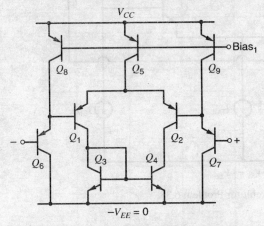

Figure 6.62 Circuit for Problem 6.29.

6.30 Repeat Problem 6.29 except replace each *pnp* transistor with a *p*-channel MOS transistor and replace each *npn* transistor with an *n*-channel MOS transistor. Assume $V_t = 0.5$ V and $V_{ov} = 0.1$ V for all *n*-channel MOS transistors. Also, assume $V_t = -0.5$ V and $V_{ov} = -0.1$ V for all *p*-channel MOS transistors.

6.31(a) Suppose that the *npn* and *pnp* transistors in the input stage of the NE5234 op amp are changed into *n*-channel and *p*-channel MOS transistors, respectively. If transistors conducting nonzero current operate in strong inversion, the transconductance of the revised circuit depends on the common-mode input voltage, V_{IC}. Explain why.

(b) Fig. 6.63 shows one circuit that overcomes this problem.[23] Let R represent the gains of the M_8-M_7 and the M_{11}-M_{12} current mirrors. When $V_{B1} \ll V_{IC} \ll V_{DD} - V_{B2}$, M_5-M_8 as well as M_{11} and M_{12} are off. Therefore, the tail current for M_1-M_2 is I_{B1}, and the tail current for M_3-M_4 is I_{B2}. On the other hand, when $0 \le V_{IC} \ll V_{B1}$, M_1 and M_2 are off, and the tail current in M_3 and M_4 becomes $RI_{B1} + I_{B2}$. Similarly, when $V_{DD} \ge V_{IC} \gg V_{DD} - V_{B2}$, M_3 and M_4 are off, and the tail current in M_1 and M_2 becomes $I_{B1} + RI_{B2}$. Find the value of R for which the transconductance of the input stage when only one differential pair is active is the same as when both pairs are active. Assume that $I_{B1} = I_{B2}$ and $\mu_n C_{ox,n}(W/L)_n = \mu_p C_{ox,p}(W/L)_p$ for M_1-M_4. Also, assume that M_9, M_{10}, M_{13}, and M_{14} operate in saturation.

6.32 Find the minimum value of V_{CC} for proper operation of the NE5234 op amp. For simplicity, assume $V_{EE} = 0$, $|V_{BE(on)}| = 0.7$ V, and $|V_{CE(sat)}| = 0.1$ V. Also, ignore base currents. Assume the bias circuits in Figures 6.34 and 6.40 operate properly even though some transistors in these circuits may operate in or near saturation with low V_{CC}.

6.33 In BiCMOS technology, MOS source followers can be used to drive a bipolar differential pair to reduce the average current flowing in the stage input leads. See Fig. 6.64. Calculate the input-referred

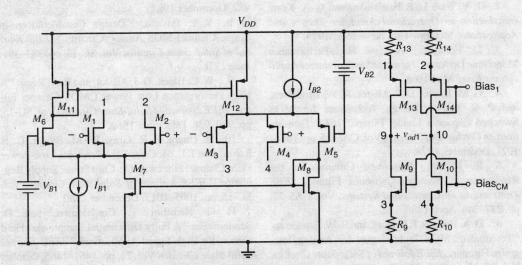

Figure 6.63 Circuit for Problem 6.31.

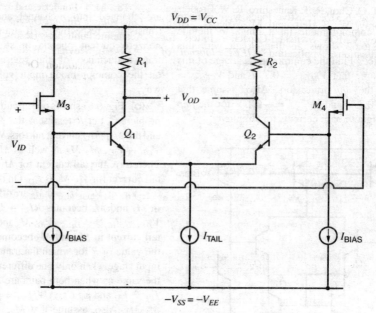

Figure 6.64 Circuit for Problem 6.33.

random offset voltage of this structure. Assume nonzero mismatch occurs in the following parameters: the thresholds of the MOS transistors, the W/L of the MOS transistors, the bias currents, the saturation currents of the bipolar transistors, and the load resistors. Ignore the body effect.

REFERENCES

1. G. E. Tobey, J. G. Graeme, and L. P. Huelsman. *Operational Amplifiers.* McGraw-Hill, New York, 1971.

2. J. V. Wait, L. P. Huelsman, and G. A. Korn. *Introduction to Operational Amplifier Theory and Applications.* McGraw-Hill, New York, 1975.

3. D. Fullagar. "A New High-Performance Monolithic Op Amp," *Fairchild Semiconductor Applications Brief,* May 1968.

4. G. M. Jacobs, D. J. Allstot, R. W. Brodersen, and P. R. Gray, "Design Techniques for MOS Switched Capacitor Ladder Filters," *IEEE Transactions on Circuits and Systems,* Vol. CAS-25, pp. 1014–1021, December 1978.

5. K. Martin, "Improved Circuits for the Realization of Switched-Capacitor Filters," *IEEE Transactions on Circuits and Systems,* Vol. CAS-27, pp. 237–244, April 1980.

6. D. A. Hodges, P. R. Gray, and R. W. Brodersen. "Potential of MOS Technologies for Analog Integrated Circuits," *IEEE Journal of Solid-State Circuits,* Vol. SC-13, pp. 285–294, June 1978.

7. P. R. Gray and R. G. Meyer. "MOS Operational Amplifier Design-A Tutorial Overview," *IEEE Journal of Solid-State Circuits,* Vol. SC-17, pp. 969–982, December 1982.

8. Y. P. Tsividis. "Design Considerations in Single-Channel MOS Analog Circuits," *IEEE Journal of Solid- State Circuits,* Vol. SC-13, pp. 383–391, June 1978.

9. W. C. Black, D. J. Allstot, and R. A. Reed. "A High Performance Low Power CMOS Channel Filter," *IEEE Journal of Solid-State Circuits,* Vol. SC-15, pp. 929–938, December 1980.

10. H. Ohara, P. R. Gray, W. M. Baxter, C. F. Rahim, and J. L. McCreary. "A Precision Low-Power PCM Channel Filter with On-Chip Power Supply Regulation," *IEEE Journal of Solid-State Circuits,* Vol. SC-15, pp. 1005–1013, December 1980.

11. G. Nicollini, P. Confalonieri, and D. Senderowicz. "A Fully Differential Sample-and-Hold Circuit for High-Speed Applications," *IEEE Journal Solid-State Circuits*, Vol. 24, pp. 1461–1465, October 1989.

12. T. C. Choi, R. T. Kaneshiro, R. W. Brodersen, P. R. Gray, W. B. Jett, and M. Wilcox. "High-Frequency CMOS Switched-Capacitor Filters for Communications Application," *IEEE Journal of Solid-State Circuits*, Vol. SC-18, pp. 652–664, December 1983.

13. K. Bult and G. J. G. M. Geelen. "A Fast-Settling CMOS Op Amp for SC Circuits with 90-dB DC Gain," *IEEE Journal of Solid-State Circuits*, Vol. 25, pp. 1379–1384, December 1990.

14. R. J. Widlar. "Low Voltage Techniques," *IEEE Journal of Solid-State Circuits*, Vol. SC-13, pp. 838–846, Dec. 1978.

15. D. F. Bowers and S. A. Wurcer. "Recent Developments in Bipolar Operational Amplifiers," *Proceedings of the IEEE Bipolar/BiCMOS Circuits and Technology Meeting*, pp. 38–45, 1999.

16. R. A. Blauschild. "Differential Amplifier with Rail-to-Rail Capability," U.S. Patent 4,532,479, July 1985.

17. J. H. Huijsing and D. Linebarger. "Low-Voltage Operational Amplifier with Rail-to-Rail Input and Output Ranges," *IEEE Journal of Solid-State Circuits*, Vol. SC-20, pp. 1144–1150, Dec. 1985.

18. M. J. Fonderie and J. H. Huijsing. *Design of Low-Voltage Bipolar Operational Amplifiers*, Kluwer Academic Publishers, Boston, 1993.

19. E. Seevinck, W. de Jager, and P. Buitendijk. "A Low-Distortion Output Stage with Improved Stability for Monolithic Power Amplifiers," *IEEE Journal of Solid-State Circuits*, Vol. 23, pp. 794–801, June 1988.

20. J. H. Huijsing. *Operational Amplifiers Theory and Design*, Kluwer Academic Publishers, Boston, 2001.

21. J. Fonderie, M. M. Maris, E. J. Schnitger, and J. H. Huijsing. "1-V Operational Amplifier with Rail-to-Rail Input and Output Ranges," *IEEE Journal of Solid-State Circuits*, Vol. 24, pp. 1551–1159, Dec. 1989.

22. K. Fukahori, Y. Nishikawa, and A. R. Hamade. "A High Precision Micropower Operational Amplifier," *IEEE Journal of Solid-State Circuits*, Vol. SC-14, pp. 1048–1058, December 1979.

23. R. Hogervorst, J. P. Tero, R. G. H. Eschauzier, and J. H. Huijsing. "A Compact Power-Efficient 3 V CMOS Rail-to-Rail Input/Output Operational Amplifier for VLSI Cell Libraries," *IEEE Journal of Solid-State Circuits*, Vol. 29, pp. 1505–1513, Dec. 1994.

CHAPTER 7

Frequency Response of Integrated Circuits

7.1 Introduction

The analysis of integrated-circuit behavior in previous chapters was concerned with low-frequency performance, and the effects of parasitic capacitance in transistors were not considered. However, as the frequency of the signal being processed by a circuit increases, the capacitive elements in the circuit eventually become important.

In this chapter, the small-signal behavior of integrated circuits at high frequencies is considered. The frequency response of single-stage amplifiers is treated first, followed by an analysis of multistage amplifiers. Finally, the frequency response of the NE5234 operational amplifier is considered, and those parts of the circuit that limit its frequency response are identified.

7.2 Single-Stage Amplifiers

The basic topology of the small-signal equivalent circuits of bipolar and MOS single-stage amplifiers are similar. Therefore in the following sections, the frequency-response analysis for each type of single-stage circuit is initially carried out using a general small-signal model that applies to both types of transistors, and the general results are then applied to each type of transistor. The general small-signal transistor model is shown in Fig. 7.1. Table 7.1 lists the parameters of this small-signal model and the corresponding parameters that transform it into a bipolar or MOS model. For example, C_{in} in the general model becomes C_π in the bipolar model and C_{gs} in the MOS model. However, some device-specific small-signal elements are not included in the general model. For example, the g_{mb} generator and capacitors C_{sb} and C_{gb} in the MOS models are not incorporated in the general model. The effect of such device-specific elements will be handled separately in the bipolar and MOS sections.

The common-emitter and common-source stages are analyzed in the sections below on differential amplifiers.

7.2.1 Single-Stage Voltage Amplifiers and the Miller Effect

Single-transistor voltage-amplifier stages are widely used in integrated circuits. Figures 7.2a and 7.2b show the ac schematics for common-emitter and common-source amplifiers with resistive loads, respectively. Resistance R_S is the source resistance, and R_L is the load resistance. A simple linear model that can be applied to both of these circuits is shown in Fig. 7.2c. The elements in the dashed box form the general small-signal transistor model from Fig. 7.1 without r_o. We will assume that the output resistance of the transistor r_o is much larger than R_L. Since these resistors are connected in parallel in the small-signal circuit, r_o can be neglected. An approximate analysis of this circuit can be made using the *Miller-effect* approximation.

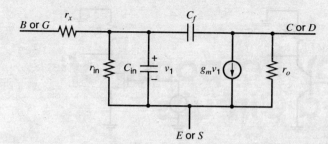

Figure 7.1 A general small-signal transistor model.

This analysis is done by considering the input impedance seen looking across the plane AA in Fig. 7.2c. To find this impedance, we calculate the current i_1 produced by the voltage v_1.

$$i_1 = (v_1 - v_o)sC_f \tag{7.1}$$

KCL at the output node gives

$$g_m v_1 + \frac{v_o}{R_L} + (v_o - v_1)sC_f = 0 \tag{7.2}$$

From (7.2), the voltage gain A_v from v_1 to v_o can be expressed as

$$A_v(s) = \frac{v_o}{v_1} = -g_m R_L \left(\frac{1 - s\dfrac{C_f}{g_m}}{1 + sR_L C_f} \right) \tag{7.3}$$

Using $v_o = A_v(s)v_1$ from (7.3) in (7.1) gives

$$i_1 = [1 - A_v(s)]sC_f v_1 \tag{7.4}$$

Equation 7.4 indicates that the admittance seen looking across the plane AA has a value $[1 - A_v(s)]sC_f$. This modification to the admittance sC_f stems from the voltage gain across C_f and is referred to as the Miller effect. Unfortunately, this admittance is complicated, due to the frequency dependence of $A_v(s)$. Replacing the voltage gain $A_v(s)$ in (7.4) with its low-frequency value $A_{v0} = A_v(0)$, (7.4) indicates that a capacitance of value

$$C_M = (1 - A_{v0})C_f \tag{7.5}$$

is seen looking across plane AA. The use of the low-frequency voltage gain here is called the *Miller approximation*, and C_M is called the *Miller capacitance*. From (7.3), $A_{v0} = A_v(0) = -g_m R_L$; therefore, (7.5) can be written as

$$C_M = (1 + g_m R_L)C_f \tag{7.6}$$

The Miller capacitance is often much larger than C_f because usually $g_m R_L \gg 1$.

Table 7.1 Small-Signal Model Elements

General Model	Bipolar Model	MOS Model
r_x	r_b	0
r_{in}	r_π	∞
C_{in}	C_π	C_{gs}
C_f	C_μ	C_{gd}
r_o	r_o	r_o

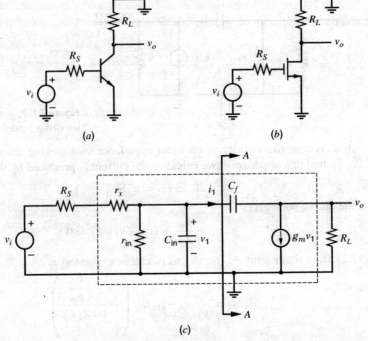

(a) (b) (c)

Figure 7.2 (a) An ac schematic of a common-emitter amplifier. (b) An ac schematic of a common-source amplifier. (c) A general model for both amplifiers.

We can now form a new equivalent circuit that is useful for calculating the *forward transmission* and input impedance of the circuit. This is shown in Fig. 7.3 using the Miller-effect approximation. Note that this equivalent circuit is *not* useful for calculating high-frequency reverse transmission or output impedance. From this circuit, we can see that at high frequencies the input impedance will eventually approach r_x.

The physical origin of the Miller capacitance is found in the voltage gain of the circuit. At low frequencies, a small input voltage v_1 produces a large output voltage $v_o = A_{v0}v_1 = -g_m R_L v_1$ of *opposite polarity*. Thus the voltage across C_f in Fig. 7.2c is $(1 + g_m R_L)v_1$ and a correspondingly large current i_1 flows in this capacitor. The voltage across C_M in Fig. 7.3 is only v_1, but C_M is larger than C_f by the factor $(1 + g_m R_L)$; therefore, C_M conducts the same current as C_f.

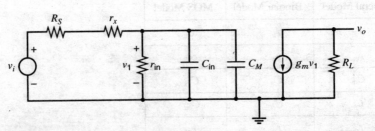

Figure 7.3 Equivalent circuit for Fig. 7.2c using the Miller approximation.

7.2 Single-Stage Amplifiers

In Fig. 7.3, the Miller capacitance adds directly to C_{in} and thus reduces the bandwidth of the amplifier, which can be seen by calculating the gain of the amplifier as follows:

$$v_1 = \frac{\frac{r_{in}}{1 + sr_{in}C_t}}{\frac{r_{in}}{1 + sr_{in}C_t} + R_S + r_x} v_i \quad (7.7)$$

$$v_o = -g_m R_L v_1 \quad (7.8)$$

where

$$C_t = C_M + C_{in} \quad (7.9)$$

Substitution of (7.7) in (7.8) gives the gain

$$A(s) = \frac{v_o}{v_i} = -g_m R_L \frac{r_{in}}{R_S + r_x + r_{in}} \frac{1}{1 + sC_t \frac{(R_S + r_x)r_{in}}{R_S + r_x + r_{in}}} \quad (7.10a)$$

$$= K \frac{1}{1 - \frac{s}{p_1}} \quad (7.10b)$$

where K is the low-frequency voltage gain and p_1 is the pole of the circuit. Comparing (7.10a) and (7.10b) shows that

$$K = -g_m R_L \frac{r_{in}}{R_S + r_x + r_{in}} \quad (7.11a)$$

$$p_1 = -\frac{R_S + r_x + r_{in}}{(R_S + r_x)r_{in}} \cdot \frac{1}{C_t} = -\frac{1}{[(R_S + r_x)||r_{in}]C_t}$$

$$= -\frac{1}{[(R_S + r_x)||r_{in}] \cdot [C_{in} + C_f(1 + g_m R_L)]} \quad (7.11b)$$

This analysis indicates that the circuit has a single pole, and setting $s = j\omega$ in (7.10b) shows that the voltage gain is 3 dB below its low-frequency value at a frequency

$$\omega_{-3dB} = |p_1| = \frac{R_S + r_x + r_{in}}{(R_S + r_x)r_{in}} \cdot \frac{1}{C_t} = \frac{1}{[(R_S + r_x)||r_{in}] \cdot [C_{in} + C_f(1 + g_m R_L)]} \quad (7.12)$$

As C_t, R_L, or R_S increase, the -3-dB frequency of the amplifier is reduced.

The exact gain expression for this circuit can be found by analyzing the equivalent circuit shown in Fig. 7.4. The poles from an exact analysis can be compared to the pole found using the Miller effect. In Fig. 7.4, a Norton equivalent is used at the input where

$$R = (R_S + r_x)||r_{in} \quad (7.13)$$

$$i_i = \frac{v_i}{R_S + r_x} \quad (7.14)$$

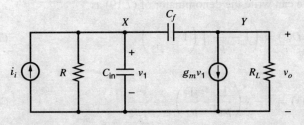

Figure 7.4 Figure 7.2c redrawn using a Norton equivalent circuit at the input.

KCL at node X gives

$$i_i = \frac{v_1}{R} + v_1 s C_{in} + (v_1 - v_o) s C_f \qquad (7.15)$$

KCL at node Y gives

$$g_m v_1 + \frac{v_o}{R_L} + (v_o - v_1) s C_f = 0 \qquad (7.16a)$$

Equation 7.16a can be written as

$$v_1 (g_m - s C_f) = -v_o \left(\frac{1}{R_L} + s C_f \right) \qquad (7.16b)$$

and thus

$$v_1 = -v_o \frac{\frac{1}{R_L} + s C_f}{g_m - s C_f} \qquad (7.17)$$

Substitution of (7.17) in (7.15) gives

$$i_i = -\left(\frac{1}{R} + s C_{in} + s C_f \right) \frac{\frac{1}{R_L} + s C_f}{g_m - s C_f} v_o - s C_f v_o$$

and the transfer function can be calculated as

$$\frac{v_o}{i_i} = -\frac{R R_L (g_m - s C_f)}{1 + s(C_f R_L + C_f R + C_{in} R + g_m R_L R C_f) + s^2 R_L R C_f C_{in}} \qquad (7.18)$$

Substitution of i_i from (7.14) in (7.18) gives

$$\frac{v_o}{v_i} = -\frac{g_m R_L R}{R_S + r_x} \frac{1 - s \frac{C_f}{g_m}}{1 + s(C_f R_L + C_f R + C_{in} R + g_m R_L R C_f) + s^2 R_L R C_f C_{in}} \qquad (7.19)$$

Substitution of R from (7.13) into (7.19) gives, for the low-frequency gain,

$$\left. \frac{v_o}{v_i} \right|_{\omega=0} = -g_m R_L \frac{r_{in}}{R_S + r_x + r_{in}} \qquad (7.20)$$

as obtained in (7.10).

Equation 7.19 shows that the transfer function v_o/v_i has a positive real zero with magnitude g_m/C_f. This zero stems from the transmission of the signal directly through C_f to the output. The effect of this zero is small except at very high frequencies, and it will be neglected here. However, this positive real zero can be important in the stability analysis of some operational amplifiers, and it will be considered in detail in Chapter 9. The denominator of (7.19) shows that the transfer function has two poles, which are usually real and widely separated in practice. If the poles are at p_1 and p_2, we can write the denominator of (7.19) as

$$D(s) = \left(1 - \frac{s}{p_1} \right) \left(1 - \frac{s}{p_2} \right) \qquad (7.21)$$

and thus

$$D(s) = 1 - s \left(\frac{1}{p_1} + \frac{1}{p_2} \right) + \frac{s^2}{p_1 p_2} \qquad (7.22)$$

We now assume that the poles are real and widely separated, and we let the lower frequency pole be p_1 (the dominant pole) and the higher frequency pole be p_2 (the nondominant pole). Then $|p_2| \gg |p_1|$ and (7.22) becomes

$$D(s) \approx 1 - \frac{s}{p_1} + \frac{s^2}{p_1 p_2} \qquad (7.23)$$

If the coefficient of s in (7.23) is compared with that in (7.19), we can identify

$$p_1 = -\frac{1}{C_{in}R + C_f(R + g_m R_L R + R_L)}$$

$$= -\frac{1}{R\left[C_{in} + C_f\left(1 + g_m R_L + \frac{R_L}{R}\right)\right]} \qquad (7.24)$$

If the value of R from (7.13) is substituted in (7.24), then the dominant pole is

$$p_1 = -\frac{R_S + r_x + r_{in}}{(R_S + r_x)r_{in}} \frac{1}{\left[C_{in} + C_f\left(1 + g_m R_L + \frac{R_L}{R}\right)\right]}$$

$$= -\frac{1}{[(R_S + r_x)\|r_{in}] \cdot \left[C_{in} + C_f\left(1 + g_m R_L + \frac{R_L}{R}\right)\right]} \qquad (7.25)$$

This value of p_1 is almost identical to that given in (7.11b) by the Miller approximation. The only difference between these equations is in the last term in the denominator of (7.25), R_L/R, and this term is usually small compared to the $(1 + g_m R_L)$ term. This result shows that the Miller-effect calculation is nearly equivalent to calculating the dominant pole of the amplifier and neglecting higher frequency poles. The Miller approximation gives a good estimate of ω_{-3dB} in many circuits.

Let us now calculate the nondominant pole by equating the coefficient of s^2 in (7.23) with that in (7.19), giving

$$p_2 = \frac{1}{p_1} \frac{1}{R_L R C_f C_{in}} \qquad (7.26)$$

Substitution of p_1 from (7.24) in (7.26) gives

$$p_2 = -\left(\frac{1}{R_L C_f} + \frac{1}{R C_{in}} + \frac{1}{R_L C_{in}} + \frac{g_m}{C_{in}}\right) \qquad (7.27)$$

The results in this section were derived using a general small-signal model. The general model parameters and the corresponding parameters for a bipolar and MOS transistor are listed in Table 7.1. By substituting values from Table 7.1, the general results of this section will be extended to the bipolar common-emitter and MOS common-source amplifiers, which appear in the half-circuits for differential amplifiers in the following sections.

7.2.1.1 The Bipolar Differential Amplifier: Differential-Mode Gain

A basic building block of analog bipolar integrated circuits is the differential stage shown in Fig. 7.5. For small-signal differential inputs at v_i, the node E is a virtual ground, and we can form the differential-mode (DM) ac half-circuit of Fig. 7.6a. The gain of this common-emitter circuit is equal to the DM gain of the full circuit. The circuit analysis that follows applies to this DM half-circuit as well as any single-stage common-emitter amplifier of the form shown in Fig. 7.6a. The small-signal equivalent circuit of Fig. 7.6a is shown in Fig. 7.6b

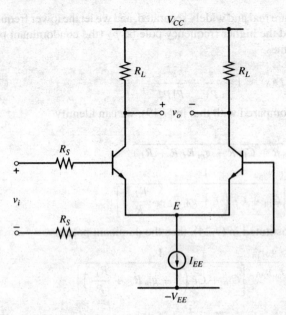

Figure 7.5 Bipolar differential amplifier circuit.

and, for compactness, the factor of 1/2 has been omitted from the input and output voltages. This change does not alter the analysis in any way. Also, for simplicity, the collector-substrate capacitance of the transistor has been omitted. Since this capacitance would be connected in parallel with R_L, its effect could be included in the following analysis by replacing R_L with Z_L, where Z_L equals R_L in parallel with C_{cs}.

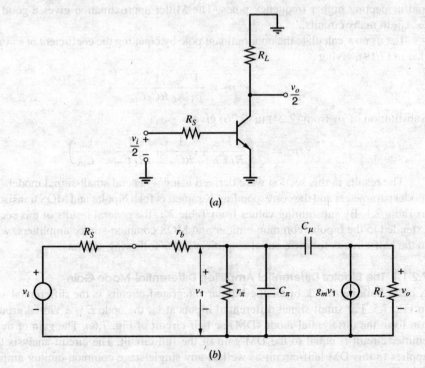

Figure 7.6 (a) Differential-mode ac half-circuit for Fig. 7.5. (b) Small-signal equivalent circuit for (a).

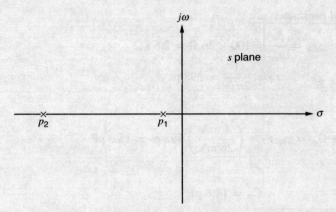

Figure 7.7 Typical pole positions for the circuit in Fig. 7.4.

The small-signal circuit in Fig. 7.2c becomes the circuit in Fig. 7.6b when the bipolar model parameters in the second column of Table 7.1 are substituted for the general model parameters in Fig. 7.2c. Therefore, the analysis results from the previous section can be used here. Substituting the values in the second column of Table 7.1 into (7.19), the voltage gain is given by

$$\frac{v_o}{v_i} = -\frac{\dfrac{g_m R_L R}{R_S + r_b}\left(1 - s\dfrac{C_\mu}{g_m}\right)}{1 + s(C_\mu R_L + C_\mu R + C_\pi R + g_m R_L R C_\mu) + s^2 R_L R C_\mu C_\pi} \quad (7.28)$$

where $R = (R_S + r_b)||r_\pi$. Using (7.25), the dominant pole is given by

$$p_1 = -\frac{1}{[(R_S + r_b)||r_\pi]\left[C_\pi + C_\mu\left(1 + g_m R_L + \dfrac{R_L}{R}\right)\right]} \quad (7.29)$$

Calculation of p_1 using (7.11b), which is based on the Miller-effect approximation, gives

$$p_1 = -\frac{1}{[(R_S + r_b)||r_\pi]\left[C_\pi + C_\mu(1 + g_m R_L)\right]} = -\frac{1}{[(R_S + r_b)||r_\pi](C_\pi + C_M)} \quad (7.30)$$

where

$$C_M = C_\mu(1 + g_m R_L) \quad (7.31)$$

is the Miller capacitance. Equation 7.30 gives virtually the same p_1 as (7.29) if $R_L/R \ll (1 + g_m R_L)$, which is usually true. This result shows that the Miller approximation is useful for finding the dominant pole. From (7.27), the nondominant pole is given by

$$p_2 = -\left(\frac{1}{R_L C_\mu} + \frac{1}{R C_\pi} + \frac{1}{R_L C_\pi} + \frac{g_m}{C_\pi}\right) \quad (7.32)$$

The last term of (7.32) is $g_m/C_\pi > g_m/(C_\pi + C_\mu) = \omega_T$ and thus $|p_2| > \omega_T$. (Recall that ω_T is the transition frequency for the transistor, as defined in Chapter 1.) Consequently, $|p_2|$ is a very high frequency; therefore, $|p_1|$ is almost always much less than $|p_2|$, as assumed. In the s plane, the poles of the amplifier are thus widely separated, as shown in Fig. 7.7.

■ **EXAMPLE**

Using the Miller approximation, calculate the -3-dB frequency of a common-emitter transistor stage using the following parameters:

$$R_S = 1\ \text{k}\Omega \quad r_b = 200\ \Omega \quad I_C = 1\ \text{mA} \quad \beta_0 = 100$$
$$f_T = 400\ \text{MHz (at } I_C = 1\ \text{mA)} \quad C_\mu = 0.5\ \text{pF} \quad R_L = 5\ \text{k}\Omega$$

The transistor small-signal parameters are

$$r_\pi = \frac{\beta_0}{g_m} = 100 \times 26\ \Omega = 2.6\ \text{k}\Omega$$

$$\tau_T = \frac{1}{2\pi f_T} = 398\ \text{ps}$$

Using (1.129) gives

$$C_\pi + C_\mu = g_m\,\tau_T = \left(\frac{1\ \text{mA}}{26\ \text{mV}}\right) 398\ \text{ps} = 15.3\ \text{pF}$$

Thus

$$C_\pi = 14.8\ \text{pF}$$

Substitution of data in (7.31) gives for the Miller capacitance

$$C_M = (1 + g_m R_L)C_\mu = \left(1 + \frac{1\ \text{mA}}{26\ \text{mV}}(5\ \text{k}\Omega)\right)(0.5\ \text{pF}) = 96.7\ \text{pF}$$

This term is much greater than C_π and dominates the frequency response. Substitution of values in (7.30) gives

$$f_{-3\text{dB}} = \frac{|p_1|}{2\pi} = \frac{1}{2\pi}\frac{1000 + 200 + 2600}{(1000 + 200)2600}\frac{10^{12}}{14.8 + 96.7} = 1.74\ \text{MHz}$$

For comparison, using (7.29) gives $|p_1| = 10.7$ Mrad/s and $f_{-3\text{dB}} = 1.70$ MHz, which is in close agreement with the value using the Miller effect. The low-frequency gain can be calculated from (7.28) as

$$\left.\frac{v_o}{v_i}\right|_{\omega=0} = -g_m R_L \frac{r_\pi}{R_S + r_b + r_\pi} = -\frac{5000}{26}\frac{2.6}{5 + 0.2 + 2.6} = -64.1$$

The gain magnitude at low frequency is thus 36.1 dB and the gain versus frequency on log scales is plotted in Fig. 7.8 for frequencies below and slightly above $|p_1|$.

■

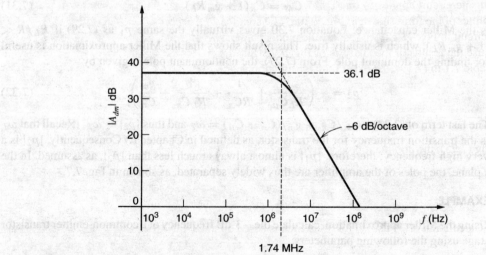

Figure 7.8 Gain magnitude versus frequency for the circuit in Fig. 7.3 using typical bipolar transistor data.

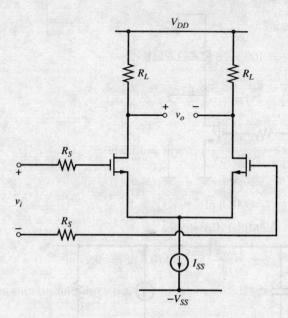

Figure 7.9 MOS differential amplifier circuit.

7.2.1.2 The MOS Differential Amplifier: Differential-Mode Gain

A MOS differential amplifier with resistive loads is shown in Fig. 7.9. The differential-mode (DM) ac half-circuit and the corresponding small-signal circuit are shown in Figs. 7.10a and 7.10b, respectively. For compactness, the factor of 1/2 has been omitted from the input and output voltages in Fig. 7.10b, but this change does not alter the analysis in any way. This circuit is a common-source amplifier. The g_{mb} generator and source-body capacitance C_{sb} are not shown; they have no effect because $v_{bs} = 0$ here. For simplicity, the drain-body capacitance C_{db} of the transistor has been omitted. Since this capacitance would be connected in parallel with R_L, its effect could be handled in the following analysis by replacing R_L with Z_L, where Z_L equals R_L in parallel with C_{db}. For simplicity, the gate-body capacitance C_{gb} is ignored. It could be included by simply adding it to C_{gs} since C_{gb} appears in parallel with C_{gs} in the common-source amplifier. However, usually $C_{gs} \gg C_{gb}$, so $C_{gs} + C_{gb} \approx C_{gs}$. The analysis of this ac circuit applies to this DM half-circuit as well as any single-stage common-source amplifier of the form shown in Fig. 7.10a. The small-signal circuit in Fig. 7.10b is the same as the circuit in Fig. 7.2c if we rename the model parameters as listed in Table 7.1. Therefore, we can use the results of the analysis of Fig. 7.2c. Substituting the values from the third column in Table 7.1 in (7.19), the exact transfer function is given by

$$\frac{v_o}{v_i} = -\frac{g_m R_L \left(1 - s\dfrac{C_{gd}}{g_m}\right)}{1 + s(C_{gd}R_L + C_{gd}R_S + C_{gs}R_S + g_m R_L R_S C_{gd}) + s^2 R_L R_S C_{gd} C_{gs}} \quad (7.33)$$

Using (7.25), the dominant pole is given by

$$p_1 = -\frac{1}{R_S \left[C_{gs} + C_{gd}(1 + g_m R_L + \dfrac{R_L}{R_S})\right]} \quad (7.34)$$

Calculation of p_1 using (7.11b), which is based on the Miller-effect approximation, gives

$$p_1 \approx -\frac{1}{R_S[C_{gs} + C_{gd}(1 + g_m R_L)]} = -\frac{1}{R_S(C_{gs} + C_M)} \quad (7.35)$$

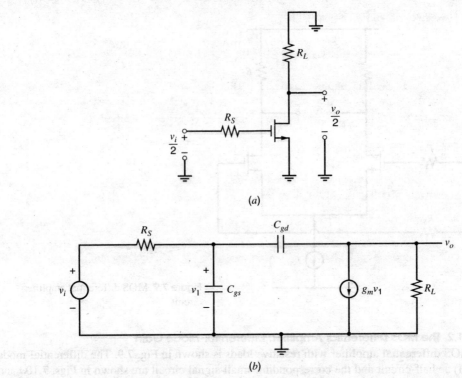

Figure 7.10 (a) Differential-mode ac half-circuit for Fig. 7.9. (b) Small-signal equivalent circuit for (a).

where

$$C_M = C_{gd}(1 + g_m R_L) \quad (7.36)$$

is the Miller capacitance. Equation 7.35 gives nearly the same value for p_1 as (7.34) if $R_L/R_S \ll (1 + g_m R_L)$, which shows that the Miller approximation is useful for finding the dominant pole. From (7.27), the nondominant pole is given by

$$p_2 = -\left(\frac{1}{R_L C_{gd}} + \frac{1}{R_S C_{gs}} + \frac{1}{R_L C_{gs}} + \frac{g_m}{C_{gs}}\right) \quad (7.37)$$

The last term of (7.37) is $g_m/C_{gs} > g_m/(C_{gs} + C_{gd} + C_{gb}) = \omega_T$ and thus $|p_2| > \omega_T$. (Recall that ω_T is the transition frequency for the transistor, as defined in Chapter 1.) Consequently, $|p_2|$ is a very high frequency; therefore, $|p_1|$ is almost always much less than $|p_2|$. In the s plane, the poles of the amplifier are thus widely separated, as shown in Fig. 7.7.

■ **EXAMPLE**

Using the Miller approximation, calculate the -3-dB frequency of a common-source transistor stage using the following parameters:

$$R_S = 1\,\text{k}\Omega \quad I_D = 1\,\text{mA} \quad k'\frac{W}{L} = 100\,\frac{\text{mA}}{\text{V}^2}$$

$$f_T = 400\,\text{MHz (at } I_D = 1\,\text{mA)} \quad C_{gd} = 0.5\,\text{pF} \quad C_{gb} = 0 \quad R_L = 5\,\text{k}\Omega$$

The small-signal transconductance is

$$g_m = \sqrt{2\left(100\frac{\text{mA}}{\text{V}^2}\right)(1 \text{ mA})} = 14.1\frac{\text{mA}}{\text{V}}$$

Using (1.207) from Chapter 1 and $C_{gb} = 0$ gives

$$C_{gs} + C_{gd} = \frac{g_m}{\omega_T} = \frac{14.1 \text{ mA/V}}{2\pi(400 \text{ MHz})} = 5.6 \text{ pF}$$

Thus

$$C_{gs} = 5.6 \text{ pF} - C_{gd} = 5.1 \text{ pF}$$

Substitution of data in (7.36) gives for the Miller capacitance

$$C_M = (1 + g_m R_L)C_{gd} = \left[1 + \left(14.1\frac{\text{mA}}{\text{V}}\right)(5 \text{ k}\Omega)\right](0.5 \text{ pF}) = 35.7 \text{ pF}$$

This capacitance is much greater than C_{gs} and dominates the frequency response. Substitution of values in (7.35) gives

$$f_{-3\text{dB}} = \frac{|p_1|}{2\pi} = \frac{1}{2\pi}\frac{1}{(1000 \text{ }\Omega)(5.1 \text{ pF} + 35.7 \text{ pF})} = 3.9 \text{ MHz}$$

For comparison, using (7.34) gives $|p_1| = 23.1$ Mrad/s and $f_{-3\text{dB}} = 3.7$ MHz, which is close to the result using the Miller effect. The low-frequency gain can be calculated from (7.33) as

$$\left.\frac{v_o}{v_i}\right|_{\omega=0} = -g_m R_L = -\left(14.1\frac{\text{mA}}{\text{V}}\right)(5000 \text{ }\Omega) = -70.5$$

■

7.2.2 Frequency Response of the Common-Mode Gain for a Differential Amplifier

In Chapter 3, the importance of the common-mode (CM) gain of a differential amplifier was described. It was shown that low values of CM gain are desirable so that the circuit can reject undesired signals that are applied equally to both inputs. Because undesired CM signals may have high-frequency components, the frequency response of the CM gain is important. The CM frequency response of the differential circuits in Figures 7.5 and 7.9 can be calculated from the CM half-circuits shown in Figs. 7.11a and 7.11b. In Fig. 7.11, R_T and C_T are the equivalent output resistance and capacitance of the tail current source. Since impedances common to the two devices are doubled in the CM half-circuit, R_T and C_T become $2R_T$ and $C_T/2$, respectively. A general small-signal equivalent circuit for Fig. 7.11a-b is shown in Fig. 7.11c. The parallel combination of $2R_T$ and $C_T/2$ will be referred to as Z_T.

The complete analysis of Fig. 7.11c is quite complex. However, the important aspects of the frequency response can be calculated by making some approximations. Consider the time constant $R_T C_T$. The resistance R_T is the output resistance of the current source and is usually greater than or equal to the r_o of a transistor. Let us assume that this resistance is on the order of 1 MΩ. The capacitor C_T includes C_{cs} of the bipolar current-source transistor or C_{db} of the current-source transistor plus C_{sb} of the input transistors in the MOS circuit. Typically, C_T is 1 pF or less. Using $R_T = 1$ MΩ and $C_T = 1$ pF, the time constant $R_T C_T$ is 1 μs, and the break frequency corresponding to this time constant is $1/(2\pi R_T C_T) = 166$ kHz. Below this frequency the impedance Z_T is dominated by R_T, and above this frequency C_T dominates. Thus as the frequency of operation is increased, the impedance Z_T will exhibit frequency variation before the rest of the circuit. We now calculate the frequency response assuming

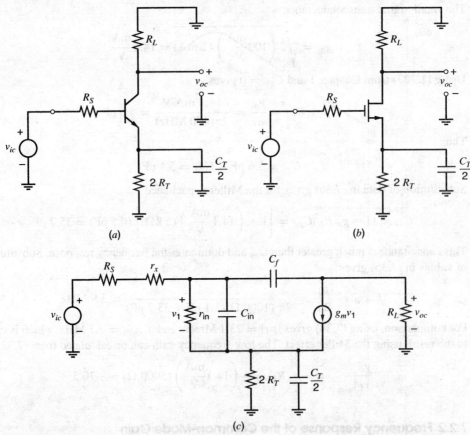

Figure 7.11 (a) Common-mode ac half-circuit for Fig. 7.5. (b) Common-mode ac half-circuit for Fig. 7.9. (c) A general model for both half-circuits.

that C_T is the only significant capacitance. Since the impedance Z_T is high, almost all of v_{ic} appears across Z_T if R_S is small. Therefore, we can approximate the CM gain as

$$A_{cm} = \frac{v_{oc}}{v_{ic}} \approx -\frac{R_L}{Z_T} \qquad (7.38)$$

where

$$Z_T = \frac{2R_T}{1 + sC_T R_T} \qquad (7.39)$$

Substitution of (7.39) in (7.38) gives

$$A_{cm}(s) = \frac{v_{oc}}{v_{ic}}(s) \approx -\frac{R_L}{2R_T}(1 + sC_T R_T) \qquad (7.40)$$

Equation 7.40 shows that the CM gain expression contains a zero, which causes the CM gain to rise at 6 dB/octave above a frequency $\omega = 1/R_T C_T$. This behavior is undesirable because the CM gain should ideally be as small as possible. The increase in CM gain cannot continue indefinitely, however, because the other capacitors in the circuit of Fig. 7.11c eventually become important. The other capacitors cause the CM gain to fall at very high frequencies, and this behavior is shown in the plot of CM gain versus frequency in Fig. 7.12a.

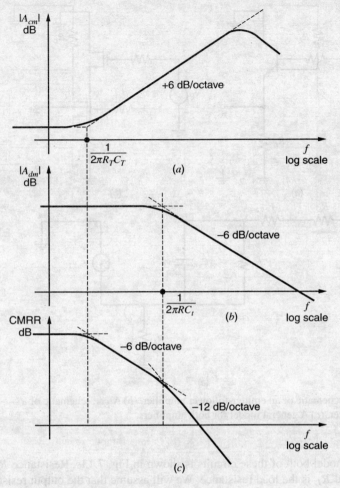

Figure 7.12 Variation with frequency of the gain parameters for the differential amplifier in Fig. 7.5 or Fig. 7.9. (a) Common-mode gain. (b) Differential-mode gain. (c) Common-mode rejection ratio.

The differential-mode (DM) gain A_{dm} of the circuit of Fig. 7.5 or Fig. 7.9 is plotted versus frequency in Fig. 7.12b using (7.10). As described earlier, $|A_{dm}|$ begins to fall off at a frequency given by $f = 1/2\pi RC_t$, where $R = (R_S + r_x)||r_{in}$ and $C_t = C_{in} + C_M$. As pointed out in Chapter 3, an important differential amplifier parameter is the common-mode rejection ratio (CMRR) defined as

$$\text{CMRR} = \frac{|A_{dm}|}{|A_{cm}|} \tag{7.41}$$

The CMRR is plotted as a function of frequency in Fig. 7.12c by simply taking the magnitude of the ratio of the DM and CM gains. This quantity begins to decrease at frequency $f = 1/2\pi R_T C_T$ when $|A_{cm}|$ begins to increase. The rate of decrease of CMRR further increases when $|A_{dm}|$ begins to fall with increasing frequency. Thus differential amplifiers are far less able to reject CM signals as the frequency of those signals increases.

7.2.3 Frequency Response of Voltage Buffers

Single-stage voltage buffers are often used in integrated circuits. The ac circuits for bipolar and MOS voltage buffers are shown in Figs. 7.13a and 7.13b, respectively. A small-signal model

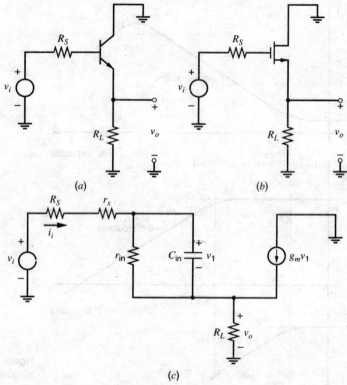

Figure 7.13 (a) An ac schematic of an emitter-follower amplifier. (b) An ac schematic of a source-follower amplifier. (c) A general model for both amplifiers.

that can be used to model both of these circuits is shown in Fig. 7.13c. Resistance R_S is the source resistance, and R_L is the load resistance. We will assume that the output resistance of the transistor r_o is much larger than R_L. Since these resistors are in parallel in the small-signal circuit, r_o can be neglected. The series input resistance in the transistor model and the source resistance are in series and can be lumped together as $R'_S = R_S + r_x$. For simplicity, the effect of capacitor C_f in Fig. 7.1 is initially neglected, a reasonable approximation if R'_S is small. The effect of C_f is to form a low-pass circuit with R'_S and to cause the gain to decrease at very high frequencies. From Fig. 7.13c,

$$v_i = i_i R'_S + v_1 + v_o \tag{7.42}$$

$$i_i = \frac{v_1}{z_{in}} \tag{7.43}$$

$$z_{in} = \frac{r_{in}}{1 + sC_{in}r_{in}} \tag{7.44}$$

$$i_i + g_m v_1 = \frac{v_o}{R_L} \tag{7.45}$$

Using (7.43) and (7.44) in (7.45) gives

$$\frac{v_1}{r_{in}}(1 + sC_{in}r_{in}) + g_m v_1 = \frac{v_o}{R_L}$$

and thus

$$v_1 = \frac{v_o}{R_L} \frac{1}{g_m + \frac{1}{r_{in}}(1 + sC_{in}r_{in})} \quad (7.46)$$

Using (7.46) and (7.43) in (7.42) gives

$$v_i = \left(\frac{R'_S}{z_{in}} + 1\right) \frac{v_o}{R_L} \frac{1}{g_m + \frac{1}{r_{in}}(1 + sC_{in}r_{in})} + v_o$$

Collecting terms in this equation, we find

$$\frac{v_o}{v_i} = \frac{g_m R_L + \frac{R_L}{r_{in}}}{1 + g_m R_L + \frac{R'_S + R_L}{r_{in}}} \left[\frac{1 - \frac{s}{z_1}}{1 - \frac{s}{p_1}}\right] \quad (7.47)$$

where

$$z_1 = -\frac{g_m + \frac{1}{r_{in}}}{C_{in}} \quad (7.48)$$

$$p_1 = -\frac{1}{R_1 C_{in}} \quad (7.49)$$

with

$$R_1 = r_{in} \| \frac{R'_S + R_L}{1 + g_m R_L} \quad (7.50)$$

Equation 7.47 shows that, as expected, the low-frequency voltage gain is about unity if $g_m R_L \gg 1$ and $g_m R_L \gg (R'_S + R_L)/r_{in}$. The high-frequency gain is controlled by the presence of a pole at p_1 and a zero at z_1.

7.2.3.1 Frequency Response of the Emitter Follower

The small-signal circuit for the emitter follower in Fig. 7.13a is shown in Fig. 7.14. We initially ignore C_μ, as in the general analysis in Section 7.2.3. The transfer function for the emitter follower can be found by substituting the appropriate values in Table 7.1 into (7.47) through

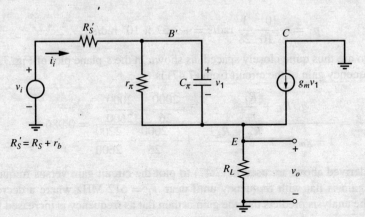

Figure 7.14 Small-signal model for the emitter follower in Fig. 7.13a.

(7.50). If $g_m R_L \gg 1$ and $g_m R_L \gg (R'_S + R_L)/r_\pi$, where $R'_S = R_S + r_b$, the low-frequency gain is about unity. The zero and pole are given by

$$z_1 = -\frac{g_m + \frac{1}{r_\pi}}{C_\pi} \approx -\frac{g_m}{C_\pi} \approx -\omega_T \qquad (7.51)$$

$$p_1 = -\frac{1}{C_\pi R_1} \qquad (7.52)$$

where

$$R_1 = r_\pi \| \frac{R'_S + R_L}{1 + g_m R_L} \qquad (7.53)$$

Typically, the zero has a magnitude that is slightly larger than the pole, and both are approximately equal to ω_T of the device. In particular, if $g_m R_L \gg 1$ and $R'_S \ll R_L$ in (7.53), then $R_1 \approx 1/g_m$ and (7.52) gives $p_1 \approx -g_m/C_\pi \approx -\omega_T$. However, if R_S is large, then R'_S in (7.53) becomes large compared to R_L, and the pole magnitude will be significantly less than ω_T.

■ **EXAMPLE**

Calculate the transfer function for an emitter follower with $C_\pi = 10$ pF, $C_\mu = 0$, $R_L = 2$ kΩ, $R_S = 50$ Ω, $r_b = 150$ Ω, $\beta = 100$, and $I_C = 1$ mA.

From the data, $g_m = 38$ mA/V, $r_\pi = 2.6$ kΩ, and $R'_S = R_S + r_b = 200$ Ω. Since $C_\mu = 0$, ω_T of the device is

$$\omega_T = \frac{g_m}{C_\pi} = \frac{1 \text{ mA}}{26 \text{ mV}} \frac{1}{10 \text{ pF}} = 3.85 \times 10^9 \text{ rad/s} \qquad (7.54)$$

and thus $f_T = 612$ MHz. From (7.51) and (7.54), the zero of the transfer function is

$$z_1 \approx -\omega_T = -3.85 \times 10^9 \text{ rad/s}$$

From (7.53)

$$R_1 = 2.6 \text{ k}\Omega \left\| \left(\frac{200 + 2000}{1 + \frac{2000}{26}} \right) \Omega \approx 28 \text{ }\Omega \right.$$

Using (7.52), the pole is

$$p_1 = -\frac{10^{12}}{10} \frac{1}{28} \text{ rad/s} = -3.57 \times 10^9 \text{ rad/s}$$

The pole and zero are thus quite closely spaced, as shown in the s-plane plot of Fig. 7.15a. The low-frequency gain of the circuit from (7.47) is

$$\frac{v_o}{v_i} = \frac{g_m R_L + \frac{R_L}{r_\pi}}{1 + g_m R_L + \frac{R'_S + R_L}{r_\pi}} = \frac{\frac{2000}{26} + \frac{2000}{2600}}{1 + \frac{2000}{26} + \frac{2200}{2600}} = 0.986$$

The parameters derived above are used in (7.47) to plot the circuit gain versus frequency in Fig. 7.15b. The gain is flat with frequency until near $f_T = 612$ MHz where a decrease of 0.4 dB occurs. The analysis predicts that the gain is then flat as frequency is increased further.

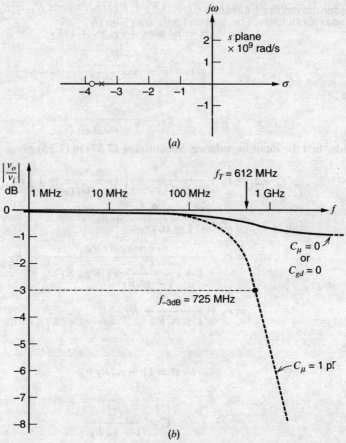

Figure 7.15 (a) Pole-zero plot for a voltage buffer. (b) Voltage gain versus frequency for the voltage buffer.

By inspecting Fig. 7.14 we can see that the high-frequency gain is asymptotic to $R_L/(R_L + R'_S)$ since C_π becomes a short circuit. This forces $v_1 = 0$ and thus the controlled current $g_m v_1$ is also zero. If a value of $C_\mu = 1$ pF is included in the equivalent circuit, the more realistic dashed frequency response of Fig. 7.15b is obtained. Since the collector is grounded, C_μ is connected from B' to ground and thus high-frequency signals are attenuated by voltage division between R'_S and C_μ. As a result, the circuit has a -3-dB frequency of 725 MHz due to the low-pass action of R'_S and C_μ. However, the bandwidth of the emitter follower is still quite large, and bandwidths of the order of the f_T of the device can be obtained in practice.

The preceding considerations have shown that large bandwidths are available from the emitter-follower circuit. One of the primary uses of an emitter follower is as a voltage buffer circuit due to its high input impedance and low output impedance. The behavior of these terminal impedances as a function of frequency is thus significant, especially when driving large loads, and will now be examined.

In Chapter 3, the terminal impedances of the emitter follower were calculated using a circuit similar to that of Fig. 7.14b except that C_π was not included. The results obtained there can be used here if r_π is replaced by z_π, which is a parallel combination of r_π and C_π. In the low-frequency calculation, β_0 was used as a symbol for $g_m r_\pi$ and thus is now replaced by $g_m z_\pi$. Using these substitutions in (3.73) and (3.76), including r_b and letting $r_o \to \infty$, we obtain for the emitter follower

$$z_i = r_b + z_\pi + (g_m z_\pi + 1)R_L \tag{7.55}$$

$$z_o = \frac{z_\pi + R_S + r_b}{1 + g_m z_\pi} \tag{7.56}$$

where

$$z_\pi = \frac{r_\pi}{1 + sC_\pi r_\pi} \tag{7.57}$$

Consider first the input impedance. Substituting (7.57) in (7.55) gives

$$z_i = r_b + \frac{r_\pi}{1 + sC_\pi r_\pi} + \left(\frac{g_m r_\pi}{1 + sC_\pi r_\pi} + 1\right) R_L$$

$$= r_b + \frac{(1 + g_m R_L) r_\pi}{1 + sC_\pi r_\pi} + R_L \tag{7.58}$$

$$= r_b + \frac{(1 + g_m R_L) r_\pi}{1 + s \dfrac{C_\pi}{1 + g_m R_L}(1 + g_m R_L) r_\pi} + R_L$$

$$= r_b + \frac{R}{1 + sCR} + R_L \tag{7.59}$$

where

$$R = (1 + g_m R_L) r_\pi \tag{7.59a}$$

and

$$C = \frac{C_\pi}{1 + g_m R_L} \tag{7.59b}$$

Thus z_i can be represented as a parallel R-C circuit in series with r_b and R_L as shown in Fig. 7.16. The effective input capacitance is $C_\pi/(1 + g_m R_L)$ and is much less than C_π for typical values of $g_m R_L$. The collector-base capacitance C_μ may dominate the input capacitance and can be added to this circuit from B' to ground. Thus, at high frequencies, the input impedance of the emitter follower becomes capacitive and its magnitude decreases.

The emitter-follower high-frequency output impedance can be calculated by substituting (7.57) in (7.56). Before proceeding, we will examine (7.57) to determine the high and low

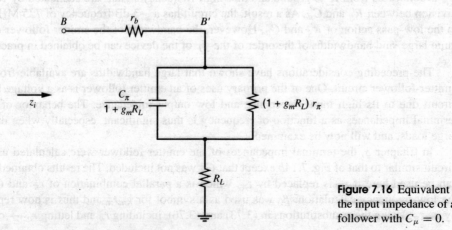

Figure 7.16 Equivalent circuit for the input impedance of an emitter follower with $C_\mu = 0$.

frequency limits on $|z_o|$. At low frequencies, $z_\pi = r_\pi$ and

$$z_o|_{\omega=0} \approx \frac{1}{g_m} + \frac{R_S + r_b}{\beta_0} \qquad (7.60)$$

At high frequencies, $z_\pi \to 0$ because C_π becomes a short circuit and thus

$$z_o|_{\omega=\infty} = R_S + r_b \qquad (7.61)$$

Thus z_o is resistive at very low and very high frequencies and its behavior in between depends on parameter values. At very low collector currents, $1/g_m$ is large. If $1/g_m > (R_S + r_b)$, a comparison of (7.60) and (7.61) shows that $|z_o|$ decreases as frequency increases and the output impedance appears capacitive. However, at collector currents of more than several hundred micro amperes, we usually find that $1/g_m < (R_S + r_b)$. Then $|z_o|$ increases with frequency, which represents inductive behavior that can have a major influence on the circuit behavior, particularly when driving capacitive loads. If $1/g_m = (R_S + r_b)$ then the output impedance is resistive and independent of frequency over a wide bandwidth. To maintain this condition over variations in process, supply, and temperature, practical design goals are $R_S \approx 1/g_m$ and $r_b \ll R_S$.

Assuming that the collector bias current is such that z_o is inductive, we can postulate an equivalent circuit for z_o as shown in Fig. 7.17. At low frequencies the inductor is a short circuit and

$$z_o|_{\omega=0} = R_1 \| R_2 \qquad (7.62)$$

At high frequencies the inductor is an open circuit and

$$z_o|_{\omega=\infty} = R_2 \qquad (7.63)$$

If we assume that $z_o|_{\omega=0} \ll z_o|_{\omega=\infty}$ then $R_1 \ll R_2$, and we can simplify (7.62) to

$$z_o|_{\omega=0} \approx R_1 \qquad (7.64)$$

The impedance of the circuit of Fig. 7.17 can be expressed as

$$z_o = \frac{(R_1 + sL)R_2}{R_1 + R_2 + sL} \approx \frac{(R_1 + sL)R_2}{R_2 + sL} \qquad (7.65)$$

assuming that $R_1 \ll R_2$.

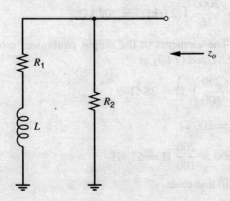

Figure 7.17 Equivalent circuit for the output impedance of an emitter follower at moderate current levels.

The complete emitter-follower output impedance can be calculated by substituting (7.57) in (7.56) with $R'_S = r_b + R_S$, which gives

$$z_o = \frac{\dfrac{r_\pi}{1 + sC_\pi r_\pi} + R'_S}{1 + \dfrac{g_m r_\pi}{1 + sC_\pi r_\pi}} = \frac{r_\pi + R'_S + sC_\pi r_\pi R'_S}{\beta_0 + 1 + sC_\pi r_\pi}$$

$$\approx \frac{\left(\dfrac{1}{g_m} + \dfrac{R'_S}{\beta_0} + sC_\pi r_\pi \dfrac{R'_S}{\beta_0}\right) R'_S}{R'_S + sC_\pi r_\pi \dfrac{R'_S}{\beta_0}} \qquad (7.66)$$

where $\beta_0 \gg 1$ is assumed.

Comparing (7.66) with (7.65) shows that, under the assumptions made in this analysis, the emitter-follower output impedance can be represented by the circuit of Fig. 7.17 with

$$R_1 = \frac{1}{g_m} + \frac{R'_S}{\beta_0} \qquad (7.67)$$

$$R_2 = R'_S \qquad (7.68)$$

$$L = C_\pi r_\pi \frac{R'_S}{\beta_0} \qquad (7.69)$$

The effect of C_μ was neglected in this calculation, which is a reasonable approximation for low to moderate values of R'_S.

The preceding calculations have shown that the input and output impedances of the emitter follower are frequency dependent. One consequence of this dependence is that the variation of the terminal impedances with frequency may limit the useful bandwidth of the circuit.

■ **EXAMPLE**

Calculate the elements in the equivalent circuits for input and output impedance of the emitter follower in the previous example. In Fig. 7.16 the input capacitance can be calculated from (7.59)

$$\frac{C_\pi}{1 + g_m R_L} = \frac{10}{1 + \dfrac{2000}{26}} \text{ pF} = 0.13 \text{ pF}$$

The resistance in shunt with this capacitance is

$$(1 + g_m R_L) r_\pi = \left(1 + \frac{2000}{26}\right)(2.6 \text{ k}\Omega) = 202 \text{ k}\Omega$$

In addition, $r_b = 150\ \Omega$ and $R_L = 2\ \text{k}\Omega$. The elements in the output equivalent circuit of Fig. 7.17 can be calculated from (7.67), (7.68), and (7.69) as

$$R_1 = \left(26 + \frac{200}{100}\right) \Omega = 28\ \Omega$$

$$R_2 = 200\ \Omega$$

$$L = 10^{-11} \times 2600 \times \frac{200}{100}\ \text{H} = 52\ \text{nH}$$

■ Note that the assumption $R_1 \ll R_2$ is valid in this case.

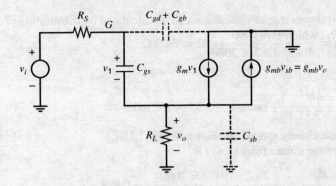

Figure 7.18 Small-signal model for the source follower in Fig. 7.13b.

7.2.3.2 Frequency Response of the Source Follower

The small-signal circuit for the source follower in Fig. 7.13b is shown in Fig. 7.18. Here, C_{gd}, C_{gb}, and C_{sb} are ignored initially. One key difference between Figures 7.18 and 7.13c is the g_{mb} generator. Since the current through the g_{mb} generator is controlled by the voltage across it, this generator can be replaced with a resistor of value $1/g_{mb}$ from v_o to ground, which is in parallel with R_L. Therefore, the total effective load resistance is $R'_L = R_L \| (1/g_{mb})$. The transfer function for the source follower can be found by substituting the appropriate values from Table 7.1 into (7.47), (7.48), and (7.49). If $g_m R'_L \gg 1$, the low-frequency gain is about unity. The zero and pole are given by

$$z_1 = -\frac{g_m}{C_{gs}} \approx -\omega_T \tag{7.70}$$

$$p_1 = -\frac{1}{C_{gs} R_1} \tag{7.71}$$

where

$$R_1 = \frac{R_S + R'_L}{1 + g_m R'_L} \tag{7.72}$$

Typically, the zero has a magnitude that is slightly larger than the pole. If $g_m R'_L \gg 1$ and $R_S \ll R'_L$ in (7.72), then $R_1 \approx 1/g_m$ and (7.71) gives $p_1 \approx -g_m/C_{gs} \approx -\omega_T$. However, if R_S in (7.72) becomes large compared to R_L or if $g_m R'_L$ is not much larger than one, the pole magnitude will be significantly less than ω_T.

EXAMPLE

Calculate the transfer function for a source follower with $C_{gs} = 7.33$ pF, $k'W/L = 100$ mA/V^2, $R_L = 2$ kΩ, $R_S = 190$ Ω, and $I_D = 4$ mA. Ignore body effect, and let $C_{gd} = 0$, $C_{gb} = 0$, and $C_{sb} = 0$.

From the data, $g_m = \sqrt{2(100)4}$ mA/V $= 28.2$ mA/V. Ignoring body effect, we have $R'_L = R_L \| (1/g_{mb}) = R_L$. Since $C_{gd} = 0$, ω_T of the device is

$$\omega_T = \frac{g_m}{C_{gs}} = \frac{28.2 \frac{\text{mA}}{\text{V}}}{7.33 \text{ pF}} = 3.85 \times 10^9 \text{ rad/s} \tag{7.73}$$

and thus $f_T = 612$ MHz. From (7.70) and (7.73), the zero of the transfer function is

$$z_1 = -\frac{g_m}{C_{gs}} = -3.85 \times 10^9 \text{ rad/s}$$

From (7.72)

$$R_1 = \frac{190 + 2000}{1 + 0.0282 \times 2000} \, \Omega = 38.2 \, \Omega$$

The pole from (7.71) is

$$p_1 = -\frac{10^{12}}{7.33} \frac{1}{38.2} \text{ rad/s} = -3.57 \times 10^9 \text{ rad/s}$$

The pole and zero are thus quite closely spaced, as shown in Fig. 7.15a.

The low-frequency gain of the circuit from (7.67) is

$$\frac{v_o}{v_i} = \frac{g_m R'_L}{1 + g_m R'_L} = \frac{28.2 \times 10^{-3} \times 2000}{1 + 28.2 \times 10^{-3} \times 2000} = 0.983$$

The parameters derived above can be used in (7.47) to plot the circuit gain versus frequency, and this plot is similar to the magnitude plot shown in Fig. 7.15b for $C_{gd} = 0$. The gain is flat with frequency until near $f_T = 612$ MHz, where a decrease of about 0.4 dB occurs. The analysis predicts that the gain is then flat as frequency is increased further because the input signal is simply fed forward to R'_L via C_{gs} at very high frequencies. In practice, capacitors C_{gd}, C_{gb}, and C_{sb} that were assumed to be zero in this example cause the gain to roll off at high frequencies, as will be demonstrated in the next example.

When capacitors C_{gd}, C_{gb}, and C_{sb} that were ignored above are included in the analysis, the voltage-gain expression becomes more complicated than (7.47). An exact analysis, with all the capacitors included, follows the same steps as the analysis of Fig. 7.13c and yields

$$\frac{v_o}{v_i} = \frac{g_m R'_L}{1 + g_m R'_L} \frac{1 + s\frac{C_{gs}}{g_m}}{1 + as + bs^2} \quad (7.74a)$$

with

$$a = \frac{R'_L(C_{gs} + C_{sb}) + R_S(C_{gs} + C'_{gd}) + R_S g_m R'_L C'_{gd}}{1 + g_m R'_L}$$

$$\approx \frac{R'_L(C_{gs} + C_{sb}) + R_S C_{gs} + R_S g_m R'_L C'_{gd}}{1 + g_m R'_L} \quad (7.74b)$$

$$b = \frac{R_S R'_L [C_{sb}(C_{gs} + C'_{gd}) + C_{gs} C'_{gd}]}{1 + g_m R'_L} \approx \frac{R_S R'_L [C_{sb} C_{gs} + C_{gs} C'_{gd}]}{1 + g_m R'_L} \quad (7.74c)$$

where $C'_{gd} = C_{gd} + C_{gb}$. The approximations for a and b use $C_{gs} + C'_{gd} \approx C_{gs}$ since $C_{gs} \gg C'_{gd}$. This exact transfer function has a zero at $-g_m/C_{gs}$, in agreement with (7.70), and two poles. If C'_{gd} and C_{sb} are set to zero, (7.74) has one pole as given by (7.71). Making approximations in (7.74) that lead to simple, useful expressions for the poles is difficult since a dominant real pole does not always exist. In fact, the poles can be complex.

■ **EXAMPLE**

Calculate the transfer function for a source follower with $C_{gs} = 7.33$ pF, $C_{gd} = 0.1$ pF, $C_{gb} = 0.05$ pF, $C_{sb} = 0.5$ pF, $k'W/L = 100$ mA/V², $R_L = 2$ kΩ, $R_S = 190$ Ω, and $I_D = 4$ mA. Ignore body effect.

These data are the same as in the last example, except we now have non-zero C_{gd}, C_{gb}, and C_{sb}. From the data, $g_m = 28.2$ mA/V and $C'_{gd} = C_{gd} + C_{gb} = 0.15$ pF. Also, ignoring body effect, we have $R'_L = R_L \| (1/g_{mb}) = R_L$. The low-frequency gain of the circuit, as calculated in the previous example, is 0.983. From (7.74a), the zero is

$$z = -\frac{g_m}{C_{gs}} = -\frac{28.2 \frac{\text{mA}}{\text{V}}}{7.33 \text{ pF}} = -3.85 \times 10^9 \text{ rad/s}$$

From (7.74b) and (7.74c), the coefficients of the denominator of the transfer function are

$$a \approx \frac{2k(7.33p + 0.5p) + 190(7.33p) + 190(0.0282)(2k)(0.15p)}{1 + (0.0282)(2k)} \text{ s} = 0.324 \text{ ns}$$

$$b \approx \frac{190(2k)[(0.5p)(7.33p) + (7.33p)(0.15p)]}{1 + (0.0282)(2k)} \text{ s}^2 = 0.0315 \text{ (ns)}^2$$

Using the quadratic formula to solve for the poles, we find that the poles are

$$p_{1,2} = -5.1 \times 10^9 \pm j2.3 \times 10^9 \text{ rad/s}$$

The poles and zero are fairly close together, as shown in Fig. 7.19a. The gain magnitude and phase are plotted in Fig. 7.19. The -3-dB bandwidth is 1.6 GHz. Because the two poles and zero are fairly close together, the gain versus frequency approximates a one-pole roll-off.

High-frequency input signals are attenuated by capacitors C_{gd} and C_{gb} that connect between the gate and ground, causing the gain to roll off and approach zero as $\omega \to \infty$. However, the bandwidth of the source follower is still quite large, and bandwidths of the order of the f_T of the device can be obtained in practice.

■

If the source follower drives a load capacitance in parallel with the load resistance, its value can be added to C_{sb} in (7.74). Such a load capacitance is present whenever the source-follower transistor is fabricated in a well and its source is connected to its well to avoid body effect. The well-body capacitance can be large and may significantly affect the 3-dB bandwidth of the circuit.

In Section 7.2.3.1, calculations were carried out for the input and output impedances of the emitter follower. Since the equivalent circuit for the MOSFET is similar to that for the

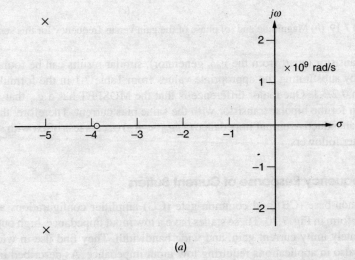

Figure 7.19 (a) Pole-zero plot for the source-follower example using (7.74).

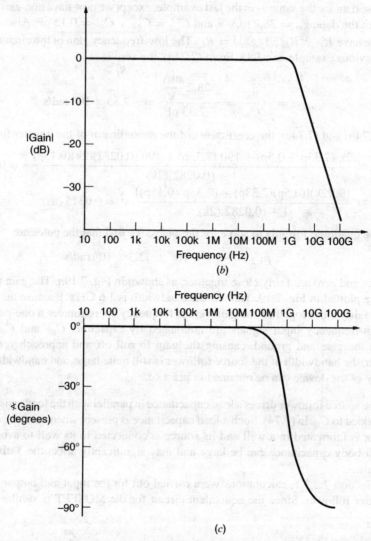

Figure 7.19 (b) Magnitude and (c) phase of the gain versus frequency for this source follower.

bipolar transistor (apart from the g_{mb} generator), similar results can be found for the source follower by substituting the appropriate values from Table 7.1 in the formulas for z_i and z_o in Section 7.2.3.1. One major difference is that the MOSFET has a g_m that is usually much lower than for the bipolar transistor with the same bias current. Therefore, the condition that produces an inductive output impedance ($1/g_m < R_S$) occurs less often with source followers than emitter followers.

7.2.4 Frequency Response of Current Buffers

The common-base (CB) and common-gate (CG) amplifier configurations are shown in ac schematic form in Fig. 7.20. These stages have a low input impedance, high output impedance, approximately unity current gain, and wide bandwidth. They find use in wideband applications and also in applications requiring low input impedance. As described in Chapter 1, the bipolar transistor breakdown voltage is maximum in this configuration. The combination of

7.2 Single-Stage Amplifiers

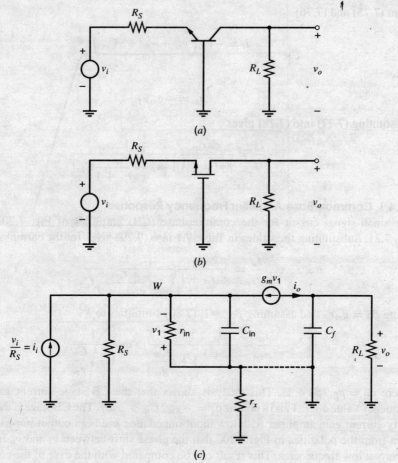

Figure 7.20 (a) An ac schematic of a common-base amplifier. (b) An ac schematic of a common-gate amplifier. (c) A general model for both amplifiers.

this property and the wideband property make the CB stages useful in high-voltage wideband output stages driving oscilloscope deflection plates.

A small-signal equivalent circuit that can model both the CB and CG stage by using a general small-signal model is shown in Fig. 7.20c. The input voltage source and source resistance are represented by a Norton equivalent. Resistance R_S is neglected in the following analysis since the input impedance of the amplifier is quite low. Another good approximation if r_x is small is that C_f simply shunts R_L, as shown in Fig. 7.20c. In this analysis, r_o will be neglected and the output signal i_o will be taken as the output of the g_m controlled source. The approximate output voltage v_o is obtained by assuming i_o flows in the parallel combination of R_L and C_f.

The analysis of the circuit of Fig. 7.20c proceeds by applying KCL at node W. Neglecting R_S,

$$i_i + \frac{v_1}{z_{in}} + g_m v_1 = 0 \qquad (7.75)$$

where

$$z_{in} = \frac{r_{in}}{1 + sC_{in}r_{in}} \qquad (7.76)$$

From (7.75) and (7.76)

$$i_i = -v_1 \left(g_m + \frac{1}{r_{in}} + sC_{in} \right) \quad (7.77)$$

Now

$$i_o = -g_m v_1 \quad (7.78)$$

Substituting (7.77) into (7.78) gives

$$\frac{i_o}{i_i} = \frac{g_m r_{in}}{g_m r_{in} + 1} \frac{1}{1 + s \dfrac{r_{in}}{g_m r_{in} + 1} C_{in}} \quad (7.79)$$

7.2.4.1 Common-Base Amplifier Frequency Response

The small-signal circuit for the common-base (CB) amplifier of Fig. 7.20a is shown in Fig. 7.21. Substituting the values in Table 7.1 into (7.79) gives for the current gain

$$\frac{i_o}{i_i} = \frac{g_m r_\pi}{g_m r_\pi + 1} \frac{1}{1 + s \dfrac{r_\pi}{g_m r_\pi + 1} C_\pi} \quad (7.80)$$

Using $\beta_0 = g_m r_\pi$ and assuming $\beta_0 \gg 1$, (7.80) simplifies to

$$\frac{i_o}{i_i} \approx \frac{\beta_0}{\beta_0 + 1} \frac{1}{1 + s \dfrac{C_\pi}{g_m}} = \alpha_0 \frac{1}{1 + s \dfrac{C_\pi}{g_m}} \quad (7.81)$$

where $\alpha_0 = \beta_0/(\beta_0 + 1)$. This analysis shows that the CB stage current gain has a low-frequency value $\alpha_0 \approx 1$ and a pole at $p_1 = -g_m/C_\pi \approx -\omega_T$. The CB stage is thus a wide-band unity-current gain amplifier with low input impedance and high output impedance. It can be seen from the polarities in Fig. 7.20a that the phase shift between v_i and v_o in the CB stage is zero at low frequencies. This result can be compared with the case of the common-emitter stage of Fig. 7.2a, which has $180°$ phase shift between v_i and v_o at low frequencies.

If the desired output is the current flowing through R_L, then C_μ and R_L form a current divider from i_o to this desired current (under the assumption that C_μ is in parallel with R_L because r_b is very small). When included in the analysis, this current divider introduces an additional pole $p_2 = -1/R_L C_\mu$ in the transfer function.

Comparing Fig. 7.20a with Fig. 7.13a shows that the input impedance of the common-base stage is the same as the output impedance of the emitter follower with $R_S = 0$. Thus the

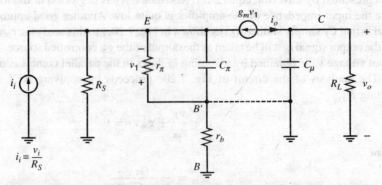

Figure 7.21 Small-signal model for the common-base circuit in Fig. 7.20a.

common-base stage input impedance is low at low frequencies and becomes inductive at high frequencies for collector bias currents of several hundred microamperes or more. As shown in Chapter 3, the output resistance of the common-base stage at low frequencies with large R_S is approximately $\beta_0 r_o$, which is extremely large. At high frequencies the output impedance is capacitive and is dominated by C_μ (and C_{cs} for npn transistors).

Unlike the common-emitter stage where C_μ is Miller multiplied, the common-base stage does not contain a feedback capacitance from collector to emitter to cause the Miller effect. As a consequence, the effect of large values of R_L on the frequency response of the common-base stage is much less than in the common-emitter stage.

7.2.4.2 Common-Gate Amplifier Frequency Response

The small-signal circuit for the common-gate (CG) amplifier of Fig. 7.20b is shown in Fig. 7.22. One element in this circuit that does not appear in the general model in Fig. 7.20c is the g_{mb} generator. Since $v_{bs} = v_{gs}$ and the g_m and g_{mb} generators are in parallel here, these controlled sources can be combined. Also, capacitors C_{gb}, C_{db}, and C_{sb} are not included in the general model. Here, C_{gb} is shorted and can be ignored. Capacitance C_{db} is in parallel with R_L and therefore can be ignored if the output variable of interest is the current i_o. Capacitance C_{sb} appears in parallel with C_{gs} in the small-signal circuit since the body and gate both connect to small-signal ground. Using the combined transconductance and input capacitance and values from Table 7.1 in (7.79) gives

$$\frac{i_o}{i_i} = \frac{1}{1 + s\dfrac{C_{gs} + C_{sb}}{g_m + g_{mb}}} \qquad (7.82)$$

From this equation, the current gain of the common-gate stage has a low-frequency value of unity and a pole at $p_1 = -(g_m + g_{mb})/(C_{gs} + C_{sb})$. If $C_{gs} \gg C_{sb}$, then $|p_1| \approx (g_m + g_{mb})/C_{gs} > g_m/C_{gs} \approx \omega_T$. The CG stage is thus a wideband unity-current-gain amplifier with low input impedance and high output impedance. It can be seen from the polarities in Fig. 7.20b that there is zero phase shift between v_i and v_o in the CG stage at low frequencies. This phase shift can be compared with the case of the common-source stage of Fig. 7.2b, which has 180° phase shift between v_i and v_o at low frequencies.

If the desired output is the current flowing through R_L, then C_{db}, C_{gd}, and R_L form a current divider from i_o to this desired current. When included in the analysis, this current divider introduces an additional pole $p_2 = -1/R_L(C_{db} + C_{gd})$ in the transfer function.

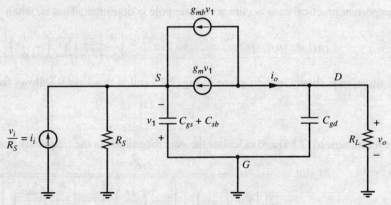

Figure 7.22 Small-signal model for the common-gate circuit in Fig. 7.20b.

Unlike the common-source stage where C_{gd} is Miller multiplied, the common-gate stage does not contain a feedback capacitance from drain to source to cause the Miller effect. As a consequence, the effect of large values of R_L on the frequency response of the common-gate stage is much less than in the common-source stage.

7.3 Multistage Amplifier Frequency Response

The above analysis of the frequency behavior of single-stage circuits indicates the complexity that can arise even with simple circuits. The complete analysis of the frequency response of multistage circuits with many capacitive elements rapidly becomes very difficult and the answers become so complicated that little use can be made of the results. For this reason approximate methods of analysis have been developed to aid in the circuit design phase, and computer simulation is used to verify the final design. One such method of analysis is the *zero-value time constant* analysis that will now be described. First some ideas regarding dominant poles are developed.

7.3.1 Dominant-Pole Approximation

For any electronic circuit we can derive a transfer function $A(s)$ by small-signal analysis to give

$$A(s) = \frac{N(s)}{D(s)} = \frac{a_0 + a_1 s + a_2 s^2 + \cdots + a_m s^m}{1 + b_1 s + b_2 s^2 + \cdots + b_n s^n} \tag{7.83}$$

where $a_0, a_1, \ldots, a_m$, and $b_1, b_2, \ldots, b_n$ are constants. Very often the transfer function contains poles only (or the zeros are unimportant). In this case we can factor the denominator of (7.83) to give

$$A(s) = \frac{K}{\left(1 - \dfrac{s}{p_1}\right)\left(1 - \dfrac{s}{p_2}\right) \cdots \left(1 - \dfrac{s}{p_n}\right)} \tag{7.84}$$

where K is a constant and $p_1, p_2, \ldots, p_n$ are the poles of the transfer function.

It is apparent from (7.84) that

$$b_1 = \sum_{i=1}^{n} \left(-\frac{1}{p_i}\right) \tag{7.85}$$

An important practical case occurs when one pole is dominant. That is, when

$$|p_1| \ll |p_2|, |p_3|, \ldots \qquad \text{so that} \qquad \left|\frac{1}{p_1}\right| \gg \left|\sum_{i=2}^{n}\left(-\frac{1}{p_i}\right)\right|$$

This situation is shown in the s plane in Fig. 7.23 and in this case it follows from (7.85) that

$$b_1 \simeq \left|\frac{1}{p_1}\right| \tag{7.86}$$

If we return now to (7.84) and calculate the gain magnitude in the frequency domain, we obtain

$$|A(j\omega)| = \frac{K}{\sqrt{\left[1 + \left(\dfrac{\omega}{p_1}\right)^2\right]\left[1 + \left(\dfrac{\omega}{p_2}\right)^2\right] \cdots \left[1 + \left(\dfrac{\omega}{p_n}\right)^2\right]}} \tag{7.87}$$

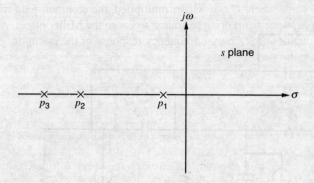

Figure 7.23 Pole diagram for a circuit with a dominant pole.

If a dominant pole exists, then (7.87) can be approximated by

$$|A(j\omega)| \simeq \frac{K}{\sqrt{1 + \left(\dfrac{\omega}{p_1}\right)^2}} \qquad (7.88)$$

This approximation will be quite accurate at least until $\omega \simeq |p_1|$, and thus (7.88) will accurately predict the -3-dB frequency and we can write

$$\omega_{-3\text{dB}} \simeq |p_1| \qquad (7.89)$$

Use of (7.86) in (7.89) gives

$$\omega_{-3\text{dB}} \simeq \frac{1}{b_1} \qquad (7.90)$$

for a dominant-pole situation.

7.3.2 Zero-Value Time Constant Analysis

This is an approximate method of analysis that allows an estimate to be made of the dominant pole (and thus the -3-dB frequency) of complex circuits. Considerable saving in computational effort is achieved because a full analysis of the circuit is not required. The method will be developed by considering a practical example.

Consider the equivalent circuit shown in Fig. 7.24. This is a single-stage bipolar transistor amplifier with resistive source and load impedances. The feedback capacitance is split into two parts (C_x and C_μ) as shown. This is a slightly better approximation to the actual situation than the single collector-base capacitor we have been using, but is rarely used in hand calculations because of the analysis complexity. For purposes of analysis, the capacitor voltages v_1, v_2, and v_3 are chosen as variables. The external input v_i is removed and the circuit excited with three independent current sources i_1, i_2, and i_3 across the capacitors, as shown in Fig. 7.24. We can show that with this choice of variables the circuit equations are of the form

$$i_1 = (g_{11} + sC_\pi)v_1 + g_{12}v_2 + g_{13}v_3 \qquad (7.91)$$

$$i_2 = g_{21}v_1 + (g_{22} + sC_\mu)v_2 + g_{23}v_3 \qquad (7.92)$$

$$i_3 = g_{31}v_1 + g_{32}v_2 + (g_{33} + sC_x)v_3 \qquad (7.93)$$

where the g terms are conductances. Note that the terms involving s contributed by the capacitors are associated only with their respective capacitor voltage variables and only appear on the diagonal of the system determinant.

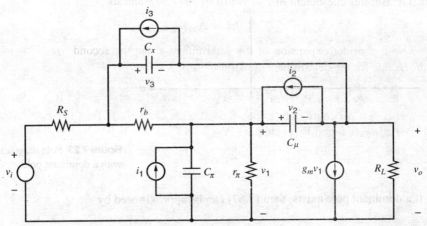

Figure 7.24 Small-signal equivalent circuit of a common-emitter stage with internal feedback capacitors C_μ and C_x.

The poles of the circuit transfer function are the zeros of the determinant Δ of the circuit equations, which can be written in the form

$$\Delta(s) = K_3 s^3 + K_2 s^2 + K_1 s + K_0 \tag{7.94}$$

where the coefficients K are composed of terms from the above equations. For example, K_3 is the sum of the coefficients of all terms involving s^3 in the expansion of the determinant. Equation 7.94 can be expressed as

$$\Delta(s) = K_0(1 + b_1 s + b_2 s^2 + b_3 s^3) \tag{7.95}$$

where this form corresponds to (7.83). Note that this is a third-order determinant because there are three capacitors in the circuit. The term K_0 in (7.94) is the value of $\Delta(s)$ if all capacitors are zero ($C_x = C_\mu = C_\pi = 0$). This can be seen from (7.91), (7.92), and (7.93). Thus

$$K_0 = \Delta|_{C_\pi = C_\mu = C_x = 0}$$

and it is useful to define

$$K_0 \triangleq \Delta_0 \tag{7.96}$$

Consider now the term $K_1 s$ in (7.94). This is the sum of all the terms involving s that are obtained when the system determinant is evaluated. However, from (7.91) to (7.93) it is apparent that s only occurs when associated with a capacitance. Thus the term $K_1 s$ can be written as

$$K_1 s = h_1 s C_\pi + h_2 s C_\mu + h_3 s C_x \tag{7.97}$$

where the h terms are constants. The term h_1 can be evaluated by expanding the determinant of (7.91) to (7.93) about the first row:

$$\Delta(s) = (g_{11} + sC_\pi)\Delta_{11} + g_{12}\Delta_{12} + g_{13}\Delta_{13} \tag{7.98}$$

where Δ_{11}, Δ_{12}, and Δ_{13} are cofactors of the determinant. Inspection of (7.91), (7.92), and (7.93) shows that C_π occurs only in the first term of (7.98). Thus the coefficient of $C_\pi s$ in (7.98) is found by evaluating Δ_{11} with $C_\mu = C_x = 0$, which will eliminate the other capacitive terms in Δ_{11}. But this coefficient of $C_\pi s$ is just h_1 in (7.97), and so

$$h_1 = \Delta_{11}|_{C_\mu = C_x = 0} \tag{7.99}$$

Now consider expansion of the determinant about the second row. This must give the same value for the determinant, and thus

$$\Delta(s) = g_{21}\Delta_{21} + (g_{22} + sC_\mu)\Delta_{22} + g_{23}\Delta_{23} \tag{7.100}$$

In this case C_μ occurs only in the second term of (7.100). Thus the coefficient of $C_\mu s$ in this equation is found by evaluating Δ_{22} with $C_\pi = C_x = 0$, which will eliminate the other capacitive terms. This coefficient of $C_\mu s$ is just h_2 in (7.97), and thus

$$h_2 = \Delta_{22}|_{C_\pi = C_x = 0} \tag{7.101}$$

Similarly by expanding about the third row it follows that

$$h_3 = \Delta_{33}|_{C_\pi = C_\mu = 0} \tag{7.102}$$

Combining (7.97) with (7.99), (7.101), and (7.102) gives

$$K_1 = (\Delta_{11}|_{C_\mu = C_x = 0} \times C_\pi) + (\Delta_{22}|_{C_\pi = C_x = 0} \times C_\mu) + (\Delta_{33}|_{C_\pi = C_\mu = 0} \times C_x) \tag{7.103}$$

and

$$b_1 = \frac{K_1}{K_0} = \frac{\Delta_{11}|_{C_\mu = C_x = 0}}{\Delta_0} \times C_\pi + \frac{\Delta_{22}|_{C_\pi = C_x = 0}}{\Delta_0} \times C_\mu$$

$$+ \frac{\Delta_{33}|_{C_\mu = C_\pi = 0}}{\Delta_0} \times C_x \tag{7.104}$$

where the boundary conditions on the determinants are the same as in (7.103). Now consider putting $i_2 = i_3 = 0$ in Fig. 7.24. Solving (7.91) to (7.93) for v_1 gives

$$v_1 = \frac{\Delta_{11} i_1}{\Delta(s)}$$

and thus

$$\frac{v_1}{i_1} = \frac{\Delta_{11}}{\Delta(s)} \tag{7.105}$$

Equation 7.105 is an expression for the driving-point impedance at the C_π node pair. Thus

$$\frac{\Delta_{11}|_{C_\mu = C_x = 0}}{\Delta_0}$$

is the driving-point *resistance* at the C_π node pair with *all* capacitors equal to zero because

$$\frac{\Delta_{11}|_{C_\mu = C_x = 0}}{\Delta_0} = \frac{\Delta_{11}}{\Delta}\bigg|_{C_\mu = C_x = C_\pi = 0} \tag{7.106}$$

We now define

$$R_{\pi 0} = \frac{\Delta_{11}}{\Delta_0}\bigg|_{C_\mu = C_x = 0} \tag{7.107}$$

Similarly,

$$\frac{\Delta_{22}|_{C_\pi = C_x = 0}}{\Delta_0}$$

is the driving-point *resistance* at the C_μ node pair with all capacitors put equal to zero and is represented by $R_{\mu 0}$. Thus we can write from (7.104)

$$b_1 = R_{\pi 0} C_\pi + R_{\mu 0} C_\mu + R_{x 0} C_x \qquad (7.108)$$

The time constants in (7.108) are called *zero-value time constants* because all capacitors are set equal to zero to perform the calculation. Although derived in terms of a specific example, this result is true in any circuit for which the various assumptions made in this analysis are valid. In its most general form, (7.108) becomes

$$b_1 = \Sigma T_0 \qquad (7.109)$$

where ΣT_0 is the sum of the zero-value time constants.

We showed previously that if there are no dominant zeros in the circuit transfer function, and if there is a dominant pole p_1, then

$$\omega_{-3\text{dB}} \approx |p_1| \qquad (7.110)$$

Using (7.90), (7.109), and (7.110) we can write

$$\omega_{-3\text{dB}} \approx |p_1| \approx \frac{1}{b_1} = \frac{1}{\Sigma T_0} \qquad (7.111)$$

For example, consider the circuit of Fig. 7.24. By inspection,

$$R_{\pi 0} = r_\pi \| (R_S + r_b) \qquad (7.112)$$

In order to calculate $R_{\mu 0}$ it is necessary to write some simple circuit equations. We apply a test current i at the C_μ terminals as shown in Fig. 7.25 and calculate the resulting v.

$$v_1 = R_{\pi 0} i \qquad (7.113)$$
$$v_o = -(i + g_m v_1) R_L \qquad (7.114)$$

Substituting (7.113) in (7.114) gives

$$v_o = -(i + g_m R_{\pi 0} i) R_L \qquad (7.115)$$

Now

$$R_{\mu 0} = \frac{v}{i}$$

and

$$R_{\mu 0} = \frac{v_1 - v_o}{i} \qquad (7.116)$$

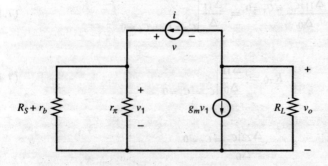

Figure 7.25 Equivalent circuit for the calculation of $R_{\mu 0}$ for Fig. 7.24.

Substitution of (7.113) and (7.115) in (7.116) gives

$$R_{\mu 0} = R_{\pi 0} + R_L + g_m R_L R_{\pi 0} \tag{7.117}$$

R_{x0} can be calculated in a similar fashion, and it is apparent that $R_{x0} \simeq R_{\mu 0}$ if $r_b \ll r_\pi$. This justifies the common practice of lumping C_x in with C_μ if r_b is small. Assuming that this is done, (7.111) gives, for the -3-dB frequency,

$$\omega_{-3\text{dB}} = \frac{1}{R_{\pi 0} C_\pi + R_{\mu 0} C_\mu} \tag{7.118}$$

Using (7.117) in (7.118) gives

$$\omega_{-3\text{dB}} = \frac{1}{R_{\pi 0} \left\{ C_\pi + C_\mu \left[(1 + g_m R_L) + \dfrac{R_L}{R_{\pi 0}} \right] \right\}} \tag{7.119}$$

Equation 7.119 is identical with the result obtained in (7.29) by exact analysis. [Recall that $(R_S + r_b) \| r_\pi$ in (7.29) is the same as $R_{\pi 0}$ in (7.119).] However, the zero-value time-constant analysis gives the result with *much less* effort. It does *not* give any information on the nondominant pole.

As a further illustration of the uses and limitations of the zero-value time-constant approach, consider the emitter-follower circuit of Fig. 7.13a, where only the capacitance C_π has been included. The value of $R_{\pi 0}$ can be calculated by inserting a current source i as shown in Fig. 7.26 and calculating the resulting voltage v_1:

$$i = \frac{v_1}{r_\pi} + \frac{v_1 + v_o}{R_S + r_b} \tag{7.120}$$

$$\frac{v_1}{r_\pi} - i + g_m v_1 = \frac{v_o}{R_L} \tag{7.121}$$

Substituting (7.121) in (7.120) gives

$$i = \frac{v_1}{r_\pi} + \frac{v_1}{R_S + r_b} + \frac{R_L}{R_S + r_b} \left(\frac{v_1}{r_\pi} + g_m v_1 - i \right)$$

and this equation can be expressed as

$$i = \frac{v_1}{r_\pi} + v_1 \frac{1 + g_m R_L}{R_S + r_b + R_L}$$

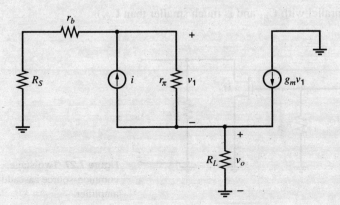

Figure 7.26 Equivalent circuit for the calculation of $R_{\pi 0}$ for the emitter follower.

Finally, $R_{\pi 0}$ can be calculated as

$$R_{\pi 0} = \frac{v_1}{i} = r_\pi \left\| \frac{R_S + r_b + R_L}{1 + g_m R_L} \right. \qquad (7.122)$$

Thus the dominant pole of the emitter follower is at

$$|p| = \frac{1}{R_{\pi 0} C_\pi} \qquad (7.123)$$

This is in agreement with the result obtained in (7.52) by exact analysis and requires less effort. However, the zero-value time-constant approach tells us nothing of the zero that exact analysis showed. Because of the dominant zero, the dominant-pole magnitude is *not* the −3-dB frequency in this case. This shows that care must be exercised in interpreting the results of zero-value time-constant analysis. However, it *is* a useful technique, and with experience the designer can recognize circuits that are likely to contain dominant zeros. *Such circuits usually have a capacitive path directly coupling input and output as C_π does in the emitter follower.*

7.3.3 Cascade Voltage-Amplifier Frequency Response

The real advantages of the zero-value time-constant approach appear when circuits containing more than one device are analyzed. For example, consider the two-stage common-source amplifier shown in Fig. 7.27. This circuit could be a drawing of a single-ended amplifier or the differential half-circuit of a fully differential amplifier. Exact analysis of this circuit to find the −3-dB frequency is extremely arduous, but the zero-value time-constant analysis is quite straightforward, as shown below. To show typical numerical calculations, specific parameter values are assumed. In the example below, as in others in this chapter, parasitic capacitance associated with resistors is neglected. This approximation is often reasonable for monolithic resistors of several thousand ohms or less, but should be checked in each case.

The zero-value time-constant analysis for Fig. 7.27 is carried out in the following example. Analysis of a two-stage common-emitter amplifier would follow similar steps.

■ **EXAMPLE**

Calculate the −3-dB frequency of the circuit of Fig. 7.27 assuming the following parameter values:

$$R_S = 10 \text{ k}\Omega \qquad R_{L1} = 10 \text{ k}\Omega \qquad R_{L2} = 5 \text{ k}\Omega$$
$$C_{gs1} = 5 \text{ pF} \qquad C_{gs2} = 10 \text{ pF} \qquad C_{gd1} = C_{gd2} = 1 \text{ pF}$$
$$C_{db1} = C_{db2} = 2 \text{ pF} \qquad g_{m1} = 3 \text{ mA/V} \qquad g_{m2} = 6 \text{ mA/V}$$

Ignore C_{gb} (which is in parallel with C_{gs} and is much smaller than C_{gs}).

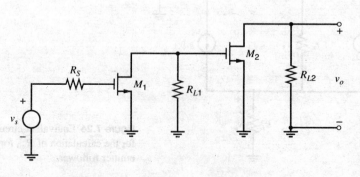

Figure 7.27 Two-stage, common-source cascade amplifier.

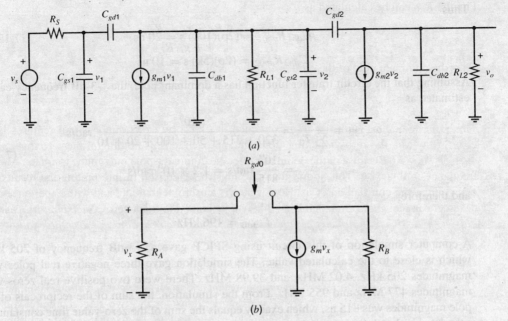

Figure 7.28 Small-signal equivalent circuit of Fig. 7.27.

The small-signal equivalent circuit of Fig. 7.27 is shown in Fig. 7.28a. The zero-value time constants for this circuit are determined by calculating the resistance seen by each capacitor across its own terminals. However, significant effort can be saved by recognizing that some capacitors in the circuit are in similar configurations and the same formula can be applied to them. For example, consider C_{gd1} and C_{gd2}. The resistance seen by either C_{gd} capacitor can be found by calculating the resistance R_{gd0} in the circuit in Fig. 7.28b. This circuit is the same as the circuit in Fig. 7.25 if we let $R_A = R_{\pi 0}$ and $R_B = R_L$. Therefore, the resistance R_{gd0} in Fig. 7.28b is given by substituting R_A for $R_{\pi 0}$ and R_B for R_L in (7.117):

$$R_{gd0} = R_A + R_B + g_m R_B R_A \tag{7.124}$$

This equation can be used to find the zero-value time constants for both gate-drain capacitors:

$$C_{gd1} R_{gd01} = C_{gd1}(R_S + R_{L1} + g_{m1} R_{L1} R_S)$$
$$= (1p)[10k + 10k + (3 \times 10^{-3})(10k)(10k)] \text{ s} = 320 \text{ ns}$$
$$C_{gd2} R_{gd02} = C_{gd2}(R_{L1} + R_{L2} + g_{m2} R_{L2} R_{L1})$$
$$= (1p)[10k + 5k + (6 \times 10^{-3})(10k)(5k)] \text{ s} = 315 \text{ ns}$$

The value of R_{gs0} for each device can be found by inspection

$$R_{gs01} = R_S$$
$$R_{gs02} = R_{L1}$$

The corresponding time constants are

$$C_{gs1} R_{gs01} = C_{gs1} R_S = (5p)(10k) \text{ s} = 50 \text{ ns}$$
$$C_{gs2} R_{gs02} = C_{gs2} R_{L1} = (10p)(10k) \text{ s} = 100 \text{ ns}$$

Also, the value for R_{db0} for each device can also be found without computation

$$R_{db01} = R_{L1}$$
$$R_{db02} = R_{L2}$$

Thus

$$C_{db1}R_{db01} = (2p)(10k) \text{ s} = 20 \text{ ns}$$
$$C_{db2}R_{db02} = (2p)(5k) \text{ s} = 10 \text{ ns}$$

Assuming that the circuit transfer function has a dominant pole, the -3-dB frequency can be estimated as

$$\omega_{-3dB} = \frac{1}{\sum T_0} = \frac{10^9}{320 + 315 + 50 + 100 + 20 + 10} \text{ rad/s}$$
$$= \frac{10^9}{815} \text{ rad/s} = 1.2 \times 10^6 \text{ rad/s} \quad (7.125)$$

and therefore

$$f_{-3dB} = 196 \text{ kHz}$$

A computer simulation of this circuit using SPICE gave a -3-dB frequency of 205 kHz, which is close to the calculated value. The simulation gave three negative real poles with magnitudes 205 kHz, 4.02 MHz, and 39.98 MHz. There were two positive real zeros with magnitudes 477 MHz and 955 MHz. From the simulation, the sum of the reciprocals of the pole magnitudes was 815 ns, which exactly equals the sum of the zero-value time constants as calculated by hand.

An exact analysis of Fig. 7.28a would first apply KCL at three nodes and produce a transfer function with a third-order denominator, in which some coefficients would consist of a sum of many products of small-signal model parameters. Many simplifying approximations would be needed to give useful design equations. As the circuit complexity increases, the number of equations increases and the order of the denominator increases, eventually making exact analysis by hand impractical and making time-constant analysis quite attractive.

■

The foregoing analytical result was obtained with relatively small effort and the calculation has focused on the contributions to the -3-dB frequency from the various capacitors in the circuit. In this example, as is usually the case in a cascade of this kind, the time constants associated with the gate-drain capacitances are the major contributors to the -3-dB frequency of the circuit. *One of the major benefits of the zero-value time-constant analysis is the information it gives on the circuit elements that most affect the -3-dB frequency of the circuit.*

In the preceding calculation, we assumed that the circuit of Fig. 7.28 had a dominant pole. The significance of this assumption will now be examined in more detail. For purposes of illustration, assume that capacitors C_{gd1}, C_{gd2}, C_{db1}, and C_{db2} in Fig. 7.28 are zero and that $R_S = R_{L1} = R_{L2}$ and $C_{gs1} = C_{gs2}$. Then the circuit has two identical stages, and each will contribute a pole with the same magnitude; that is, a dominant-pole situation does not exist because the circuit has two identical poles. However, inclusion of nonzero C_{gd1} and C_{gd2} tends to cause these poles to split apart and produces a dominant-pole situation. (See Chapter 9.) For this reason, most practical circuits of this kind do have a dominant pole, and the zero-value time-constant analysis gives a good estimate of ω_{-3dB}. Even if the circuit has two identical poles, however, the zero-value time-constant analysis is still useful. Equations 7.85 and 7.109 are valid in general, and thus

$$\Sigma T_0 = \sum_{i=1}^{n} \left(-\frac{1}{p_i}\right) \quad (7.126)$$

is always true. That is, the sum of the zero-value time constants equals the sum of the negative reciprocals of all the poles whether or not a dominant pole exists. Consider a circuit with two identical negative real poles with magnitudes ω_x. Then the gain magnitude of the circuit is

$$|G(j\omega)| = \frac{G_0}{1 + \left(\dfrac{\omega}{\omega_x}\right)^2} \tag{7.127}$$

The -3-dB frequency of this circuit is the frequency where $|G(j\omega)| = G_0/\sqrt{2}$, which can be shown to be

$$\omega_{-3\text{dB}} = \omega_x \sqrt{\sqrt{2} - 1} = 0.64\omega_x \tag{7.128}$$

The zero-value time-constant approach predicts

$$\Sigma T_0 = \frac{2}{\omega_x}$$

and thus

$$\omega_{-3\text{dB}} = \frac{1}{\Sigma T_0} = 0.5\omega_x \tag{7.129}$$

Even in this extreme case, the prediction is only 22 percent in error and gives a pessimistic estimate.

7.3.4 Cascode Frequency Response

The cascode connection is a multiple-device configuration that is useful in high-frequency applications. An ac schematic of a bipolar version, shown in Fig. 7.29a, consists of a

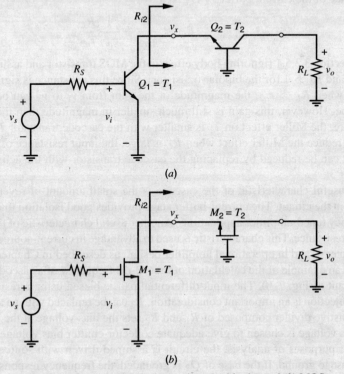

Figure 7.29 Cascode circuit connections (a) bipolar and (b) MOS.

common-emitter stage driving a common-base stage. An MOS version, shown in Fig. 7.29b, consists of a common-source stage driving a common-gate stage. In both circuits, transistor T_2 operates as a current buffer. Therefore, the voltage gain of the cascode circuit is approximately

$$\frac{v_o}{v_i} \approx -g_{m1} R_L \tag{7.130}$$

assuming that the output resistance of the cascode circuit is large compared to R_L. This result is the same as the voltage gain for a common-emitter or common-source stage without the current buffer T_2. The cascode derives its advantage at high frequencies from the fact that the load for transistor T_1 is the low input impedance of the current buffer. This impedance at low frequencies was shown in Sections 3.3.3 and 3.3.4 to be

$$R_{i2} \approx \frac{1}{g_{m2}} \tag{7.131}$$

if $r_{o2} \to \infty$, ignoring body effect in the MOS transistor and assuming $r_b/(\beta_0 + 1) \ll 1/g_{m2}$ and $\beta_0 \gg 1$ for the bipolar transistor. If transistors T_1 and T_2 have equal bias currents and device dimensions, then $g_{m1} = g_{m2}$. Since the load resistance seen by T_1 is about $1/g_{m2}$, the magnitude of the voltage gain from v_i to v_x is about unity. Thus the influence of the Miller effect on T_1 is minimal, even for fairly large values of R_L. Since the current-buffer stage T_2 has a wide bandwidth (see Section 7.2.4), the cascode circuit overall has good high-frequency performance when compared to a single common-emitter or common-source stage, especially for large R_L. (See Problems 7.29 and 7.30.)

If the assumption that $r_{o2} \to \infty$ is removed, the magnitude of the voltage gain from v_i to v_x can be larger than one. For example, R_L might be the output resistance of a cascoded current source in an amplifier stage. In this case, R_L is large compared to r_{o2}, and the input resistance of the current buffer T_2 is given by

$$R_{i2} \approx \frac{1}{g_{m2}} + \frac{1}{g_{m2}} \frac{R_L}{r_{o2}} \tag{7.132}$$

from Section 3.3.5.1 [ignoring body effect in the MOS transistor and assuming $r_b/(\beta_0 + 1) \ll 1/g_{m2}$ and $\beta_0 \gg 1$ for the bipolar transistor]. Since this resistance is significantly bigger than $1/g_{m2}$ when $R_L \gg r_{o2}$, the magnitude of the gain from v_i to v_x can be significantly larger than one. However, this gain is still much smaller in magnitude than the gain from v_i to v_o; therefore, the Miller effect on T_1 is smaller with the cascode transistor T_2 than without it. To further reduce the Miller effect when R_L is large, the input resistance of the current buffer in (7.132) can be reduced by replacing the cascode transistor with the active cascode shown in Chapter 3.

A useful characteristic of the cascode is the small amount of reverse transmission that occurs in the circuit. The current-buffer stage provides good isolation that is required in high-frequency tuned-amplifier applications. Another useful characteristic of the cascode is its high output resistance. This characteristic is used to advantage in current-source design, as described in Chapter 4, and in operational amplifier design, as described in Chapter 6.

As an example of the calculation of the -3-dB frequency of a cascode amplifier, consider the circuit of Fig. 7.30. The input differential pair is biased using a resistor R_3. If common-mode rejection is an important consideration, R_3 can be replaced with an active current source. The resistive divider composed of R_1 and R_2 sets the bias voltage at the bases of Q_3 and Q_4, and this voltage is chosen to give adequate collector-emitter bias voltage for each device.

For purposes of analysis, the circuit is assumed driven with source resistance R_S from each base to ground. If the base of Q_2 is grounded, the frequency response of the circuit is not greatly affected if R_S is small. The circuit of Fig. 7.30 can be analyzed using the ac differential

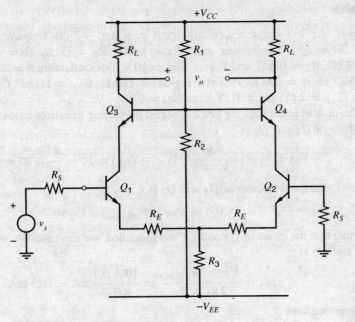

Figure 7.30 Cascode differential amplifier.

half-circuit of Fig. 7.31a. Note that in forming the differential half-circuit, the common-base point of Q_3 and Q_4 is assumed to be a virtual ground for differential signals. The frequency response $(v_o/v_s)(j\omega)$ of the circuit of Fig. 7.31a will be the same as that of Fig. 7.30 if R_3 in Fig. 7.30 is large enough to give a reasonable value of common-mode rejection. The small-signal equivalent circuit of Fig. 7.31a is shown in Fig. 7.31b.

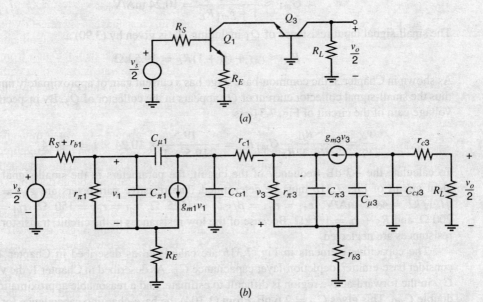

Figure 7.31 (a) The ac differential half-circuit of Fig. 7.30. (b) Small-signal equivalent of the circuit in (a).

■ **EXAMPLE**

Calculate the low-frequency, small-signal gain and -3-dB frequency of the circuit of Fig. 7.30 using the following data: $R_S = 1$ kΩ, $R_E = 75$ Ω, $R_3 = 4$ kΩ, $R_L = 1$ kΩ, $R_1 = 4$ kΩ, $R_2 = 10$ kΩ, and $V_{CC} = V_{EE} = 10$ V. Device data are $\beta = 200$, $V_{BE(on)} = 0.7$ V, $\tau_F = 0.25$ ns, $r_b = 200$ Ω, r_c(active region) = 150 Ω, $C_{je0} = 1.3$ pF, $C_{\mu 0} = 0.6$ pF, $\psi_{0c} = 0.6$ V, $C_{cs0} = 2$ pF, $\psi_{0s} = 0.58$ V, and $n_s = 0.5$.

The dc bias conditions are first calculated neglecting transistor base currents. The voltage at the base of Q_3 and Q_4 is

$$V_{B3} = V_{CC} - \frac{R_1}{R_1 + R_2}(V_{CC} + V_{EE}) = 10 - \frac{4}{14} \times 20 = 4.3 \text{ V}$$

The voltage at the collectors of Q_1 and Q_2 is

$$V_{C1} = V_{B3} - V_{BE3(on)} = 3.6 \text{ V}$$

Assuming that the bases of Q_1 and Q_2 are grounded, we can calculate the collector currents of Q_1 and Q_2 as

$$I_{C1} = \frac{V_{EE} - V_{BE(on)}}{2R_3 + R_E} = \frac{10 - 0.7}{8.075} \text{mA} = 1.15 \text{ mA}$$

Therefore, we have

$$I_{C1} = I_{C2} = I_{C3} = I_{C4} = 1.15 \text{ mA}$$

The dc analysis is completed by noting that the voltage at the collectors of Q_3 and Q_4 is

$$V_{C3} = V_{CC} - I_{C3}R_L = 10 \text{ V} - 1.15 \text{ V} = 8.85 \text{ V}$$

The low-frequency gain can be calculated from the ac differential half-circuit of Fig. 7.31a, using the results derived in Chapter 3 for a stage with emitter resistance. If we neglect base resistance, the small-signal transconductance of Q_1 including R_E is given by (3.93) as

$$G_{m1} \approx \frac{g_{m1}}{1 + g_{m1}R_E} = 10.24 \text{ mA/V}$$

The small-signal input resistance of Q_1 including R_E is given by (3.90) as

$$R_{i1} \approx r_{\pi 1} + (\beta + 1)R_E = 19.5 \text{ k}\Omega$$

As shown in Chapter 3, the common-base stage has a current gain of approximately unity, and thus the small-signal collector current of Q_1 appears in the collector of Q_3. By inspection, the voltage gain of the circuit of Fig. 7.31a is

$$\frac{v_o}{v_s} = -\frac{R_{i1}}{R_{i1} + R_S}G_{m1}R_L = -\frac{19.5}{19.5 + 1} \times 10.24 \times 1 = -9.74$$

To calculate the -3-dB frequency of the circuit, the parameters in the small-signal equivalent circuit of Fig. 7.31b must be determined. The resistive parameters are $g_{m1} = g_{m3} = qI_{C1}/kT = 44.2$ mA/V, $r_{\pi 1} = r_{\pi 3} = \beta/g_{m1} = 4525$ Ω, $r_{c1} = r_{c3} = 150$ Ω, $r_{b1} = r_{b3} = 200$ Ω, and $R_S + r_b = 1.2$ kΩ. Because of the low resistances in the circuit, transistor output resistances are neglected.

The capacitive elements in Fig. 7.31b are calculated as described in Chapter 1. First consider base-emitter, depletion-layer capacitance C_{je}. As described in Chapter 1, the value of C_{je} in the forward-active region is difficult to estimate, and a reasonable approximation is to double C_{je0}. This gives $C_{je} = 2.6$ pF. From (1.104) the base-charging capacitance for Q_1 is

$$C_{b1} = \tau_F g_{m1} = 0.25 \times 10^{-9} \times 44.2 \times 10^{-3} \text{ F} = 11.1 \text{ pF}$$

Use of (1.118) gives

$$C_{\pi 1} = C_{b1} + C_{je1} = 13.7 \text{ pF}$$

Since the collector currents of Q_1 and Q_3 are equal, $C_{\pi 3} = C_{\pi 1} = 13.7$ pF.

The collector-base capacitance $C_{\mu 1}$ of Q_1 can be calculated using (1.117a) and noting that the collector-base bias voltage of Q_1 is $V_{CB1} = 3.6$ V. Thus

$$C_{\mu 1} = \frac{C_{\mu 0}}{\sqrt{1 + \frac{V_{CB}}{\psi_{0c}}}} = \frac{0.6}{\sqrt{1 + \frac{3.6}{0.6}}} \text{ pF} = 0.23 \text{ pF}$$

The collector-substrate capacitance of Q_1 can also be calculated using (1.117a) with a collector-substrate voltage of $V_{CS} = V_{C1} + V_{EE} = 13.6$ V. (The substrate is assumed connected to the negative supply voltage.) Thus we have

$$C_{cs1} = \frac{C_{cs0}}{\sqrt{1 + \frac{V_{CS}}{\psi_{0s}}}} = \frac{2}{\sqrt{1 + \frac{13.6}{0.58}}} \text{ pF} = 0.40 \text{ pF}$$

Similar calculations show that the parameters of Q_3 are $C_{\mu 3} = 0.20$ pF and $C_{cs3} = 0.35$ pF.

The -3-dB frequency of the circuit can now be estimated by calculating the zero-value time constants for the circuit. First consider $C_{\pi 1}$. The resistance seen across its terminals is given by (7.122), which was derived for the emitter follower. The presence of resistance in series with the collector of Q_1 makes no difference to the calculation because of the infinite impedance of the current generator $g_{m1} v_1$. Thus from (7.122)

$$R_{\pi 01} = r_{\pi 1} \| \frac{R_S + r_{b1} + R_E}{1 + g_{m1} R_E} = \left(4525 \| \frac{1000 + 200 + 75}{1 + 44.2 \times 0.075} \right) \Omega = (4525 \| 295) \Omega = 277 \Omega$$

Note that the effect of R_E is to reduce $R_{\pi 01}$, which increases the bandwidth of the circuit by reducing the zero-value time constant associated with $C_{\pi 1}$. This time constant has a value

$$C_{\pi 1} R_{\pi 01} = 13.7 \times 0.277 \text{ ns} = 3.79 \text{ ns}$$

The collector-substrate capacitance of Q_1 sees a resistance equal to r_{c1} plus the common-base stage input resistance, which is

$$R_{i3} = \frac{1}{g_{m3}} + \frac{r_{b3}}{\beta + 1} = 23.6 \, \Omega$$

and thus C_{cs1} sees a resistance

$$R_{cs\,01} = R_{i3} + r_{c1} = 174 \, \Omega$$

The zero-value time constant is

$$C_{cs1} R_{cs\,01} = 0.4 \times 0.174 \text{ ns} = 0.07 \text{ ns}$$

The zero-value time constant associated with $C_{\mu 1}$ of Q_1 can be determined by calculating the resistance $R_{\mu 01}$ seen across the terminals of $C_{\mu 1}$ using the equivalent circuit of Fig. 7.32a. To simplify the analysis, the circuit in Fig. 7.32a is transformed into the circuit of Fig. 7.32b, where the transistor with emitter degeneration is represented by parameters R_{i1} and G_{m1}, which were defined previously. The circuit of Fig. 7.32b is in the form of a common-emitter stage as shown in Fig. 7.25, and the formula derived for that case can be used now. Thus from (7.117)

$$R_{\mu 01} = R_1 + R_{L1} + G_{m1} R_{L1} R_1 \qquad (7.133)$$

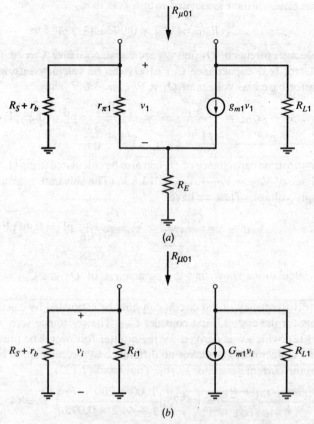

Figure 7.32 (a) Circuit for the calculation of $R_{\mu 01}$ for Q_1. (b) Equivalent circuit for the circuit in (a).

where

$$R_1 = R_{i1}||(R_S + r_b) = (19.5||1.2)\,\text{k}\Omega = 1.13\,\text{k}\Omega$$

The load resistance R_{L1} is just r_{c1} plus the input resistance of Q_3. Using the previously calculated values, we obtain

$$R_{L1} = 174\,\Omega$$

Substituting into (7.133) gives

$$R_{\mu 01} = [1.13 + 0.17 + (10.24 \times 1.13 \times 0.17)]\,\text{k}\Omega = 3.27\,\text{k}\Omega$$

The zero-value time constant associated with $C_{\mu 1}$ is thus

$$C_{\mu 1} R_{\mu 01} = 0.23 \times 3.27\,\text{ns} = 0.75\,\text{ns}$$

Because the input impedance of the common-base stage is small, the contribution of $C_{\mu 1}$ to the sum of the zero value time constants is much smaller than that due to $C_{\pi 1}$.

The time constant associated with $C_{\pi 3}$ of Q_3 can be calculated by recognizing that (7.122) derived for the emitter follower also applies here. The effective source resistance R_S is zero as the base is grounded, and the effective emitter resistance R_L is infinite because the collector of Q_1 is connected to the emitter of Q_3. Thus (7.122) gives

$$R_{\pi 03} = r_{\pi 3}||\frac{1}{g_{m3}} = 22.6\,\Omega$$

The zero-value time constant associated with $C_{\pi 3}$ is thus

$$C_{\pi 3} R_{\pi 03} = 13.7 \times 0.0023 \text{ ns} = 0.32 \text{ ns}$$

The time constant associated with collector-base capacitance $C_{\mu 3}$ of Q_3 can be calculated using (7.133) with G_{m1} equal to zero since the effective value of R_E is infinite in this case. In (7.133) the effective value of R_1 is just r_b and thus

$$R_{\mu 03} = r_b + R_{L3}$$

where

$$R_{L3} = r_{c3} + R_L$$

and R_{L3} is the load resistance seen by Q_3. Thus

$$R_{\mu 03} = [200 + 150 + 1000] \, \Omega = 1.35 \text{ k}\Omega$$

and the time constant is

$$C_{\mu 3} R_{\mu 03} = 0.2 \times 1.35 \text{ ns} = 0.27 \text{ ns}$$

Finally, the collector-substrate capacitance of Q_3 sees a resistance

$$R_{cs03} = r_{c3} + R_L = 1.15 \text{ k}\Omega$$

and

$$C_{cs3} R_{cs03} = 0.35 \times 1.15 \text{ ns} = 0.4 \text{ ns}$$

The sum of the zero-value time constants is thus

$$\Sigma T_0 = (3.79 + 0.07 + 0.75 + 0.32 + 0.27 + 0.4) \text{ ns} = 5.60 \text{ ns}$$

The -3-dB frequency is estimated as

$$f_{-3\text{dB}} = \frac{1}{2\pi \Sigma T_0} = 28.4 \text{ MHz}$$

Computer simulation of this circuit using SPICE gave a -3-dB frequency of 34.7 MHz. The computer simulation showed six poles, of which the first two were negative real poles with magnitudes 35.8 MHz and 253 MHz. The zero-value time constant analysis has thus given a reasonable estimate of the -3-dB frequency and has also shown that the major limitation on the circuit frequency response comes from $C_{\pi 1}$ of Q_1. The circuit can thus be broadbanded even further by increasing resistance R_E in the emitter of Q_1, since the calculation of $R_{\pi 01}$ showed that increasing R_E will reduce the value of $R_{\pi 01}$. Note that this change will reduce the gain of the circuit.

■

Further useful information regarding the circuit frequency response can be obtained from the previous calculations by recognizing that Q_3 in Fig. 7.31 effectively isolates $C_{\mu 3}$ and C_{cs3} from the rest of the circuit. In fact, if r_{b3} is zero then these two capacitors are connected in parallel across the output and will contribute a separate pole to the transfer function. The magnitude of this pole can be estimated by summing zero-value time constants for $C_{\mu 3}$ and C_{cs3} alone to give $\Sigma T_0 = 0.67$ ns. This time constant corresponds to a pole with magnitude $1/(2\pi \Sigma T_0) = 237$ MHz, which is very close to the second pole calculated by the computer. The dominant pole would then be estimated by summing the rest of the time constants to give $\Sigma T_0 = 4.93$ ns, which corresponds to a pole with magnitude 32.3 MHz and is also close to the computer-calculated value. This technique can be used any time a high degree of isolation exists between various portions of a circuit. To estimate the dominant pole of a given section, the zero-value time constants may be summed for that section.

In this example, the bandwidth of the differential-mode gain was estimated by computing the zero-value time constants for the differential half-circuit. The bandwidth of the common-mode gain could be estimated by computing the zero-value time constants for the common-mode half-circuit.

7.3.5 Frequency Response of a Current Mirror Loading a Differential Pair

A CMOS differential pair with current-mirror load is shown in Fig. 7.33a. The current mirror here introduces a pole and a zero that are not widely separated. To show this result, consider the simplified small-signal circuit in Fig. 7.33b for finding the transconductance $G_m = i_o/v_{id}$ with $v_o = 0$. The circuit has been simplified by letting $r_o \to \infty$ for all transistors and ignoring

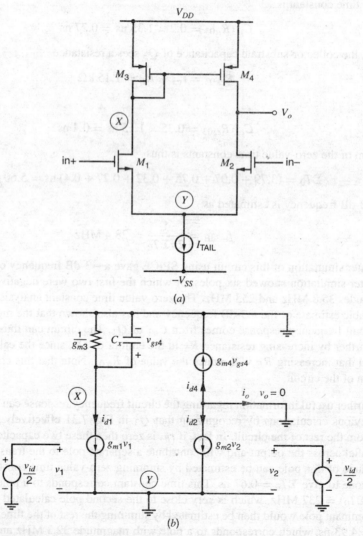

Figure 7.33 (a) Differential pair with current-mirror load and (b) a simplified small-signal model.

all capacitors except C_x. Here C_x models the total capacitance from node X to ground, which consists of C_{gs3}, C_{gs4}, and other smaller capacitances. With a purely differential input, node Y is an ac ground. The zero-value time constant associated with C_x is $T_0 = C_x/g_{m3}$; therefore,

$$p = -\frac{g_{m3}}{C_x} \quad (7.134)$$

An exact analysis of this circuit yields a transfer function with a pole given by (7.134) and a zero at $z = -2g_{m3}/C_x$. (See Problem 7.48.) The magnitudes of the pole and zero are separated by one octave. The magnitude and phase responses for $G_m(s)$ are plotted in Fig. 7.34. The phase shift from this pole-zero pair is between 0 and -19.4 degrees.

The frequency-response plot can be explained as follows. The drain currents in the differential pair are $i_{d1} = g_{m1}v_{id}/2$ and $i_{d2} = -g_{m1}v_{id}/2$. At low frequencies, M_3 and M_4 mirror i_{d1}, giving for the output current

$$i_o = -i_{d2} - i_{d4} = -i_{d2} + i_{d1} = g_{m1}v_{id} \quad (7.135)$$

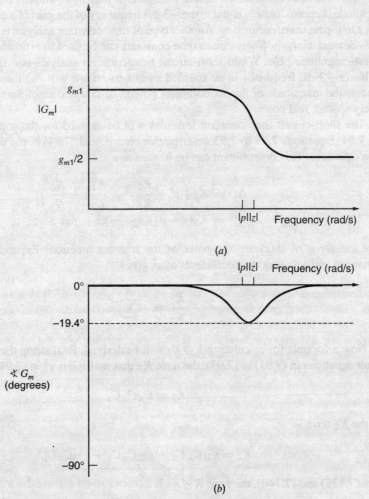

Figure 7.34 (a) Magnitude and (b) phase versus frequency of the transconductance $G_m = i_o/v_{id}$ for the circuit in Fig. 7.33.

At high frequencies ($\omega \to \infty$), C_x becomes a short, so $v_{gs4} \to 0$ and $i_{d4} \to 0$. Therefore,

$$i_o = -i_{d2} - i_{d4} = -i_{d2} - 0 = \frac{g_{m1} v_{id}}{2} \tag{7.136}$$

These equations show that the transconductance falls from g_{m1} at low frequencies to $g_{m1}/2$ at high frequencies. This result stems from the observation that the current mirror does not contribute to the output current when C_x becomes a short. The change in the transconductance occurs between frequencies $|p|$ and $|z|$. This analysis shows that the pole and zero are both important in this circuit. Since $C_x \approx C_{gs3} + C_{gs4} = 2C_{gs3}$, (7.134) gives $|p| \approx g_{m3}/2C_{gs3} \approx \omega_{T(M_3)}/2$. Therefore, the pole-zero pair has an effect only at very high frequencies, and the effect of this pair is much less than that of either an isolated pole or zero.

Analysis of a bipolar version of the circuit in Fig. 7.33a gives a similar result.

7.3.6 Short-Circuit Time Constants

Zero-value time-constant analysis (which is sometimes called open-circuit time-constant analysis) can be used to estimate the smallest-magnitude pole of an amplifier. This estimated pole magnitude is approximately equal to the -3-dB frequency of the gain of a dc-coupled amplifier with a low-pass transfer function. Another type of time-constant analysis is called *short-circuit time-constant* analysis. Short-circuit time constants can be used to estimate the location of the largest-magnitude pole. While short-circuit time-constant analysis is often used to estimate the lower -3-dB frequency in ac-coupled amplifiers,[1,2] we will use these time constants to estimate the magnitude of the nondominant pole in dc-coupled amplifiers that have only two widely spaced, real poles.

The short-circuit time-constant formulas will be derived for the small-signal circuit in Fig. 7.24. Equations 7.91 to 7.93 describe this circuit, and (7.94) is the determinant $\Delta(s)$ of these equations. This determinant can be written as

$$\begin{aligned}\Delta(s) &= K_3 \left(s^3 + \frac{K_2}{K_3}s^2 + \frac{K_1}{K_3}s + \frac{K_0}{K_3} \right) \\ &= K_3(s - p_1)(s - p_2)(s - p_3)\end{aligned} \tag{7.137}$$

since the zeros of $\Delta(s)$ are the poles of the transfer function. Expanding the right-most expression and equating the coefficients of s^2 gives

$$\frac{K_2}{K_3} = -\sum_{i=1}^{3} p_i \tag{7.138}$$

Now a formula for calculating K_2/K_3 will be derived. Evaluating the determinant of the circuit equations in (7.91) to (7.93), the term K_3 that multiplies s^3 in (7.94) is given by

$$K_3 = C_\pi C_\mu C_x \tag{7.139}$$

and the K_2 term is

$$K_2 = g_{11} C_\mu C_x + g_{22} C_\pi C_x + g_{33} C_\pi C_\mu \tag{7.140}$$

From (7.139) and (7.140), the ratio K_2/K_3 is

$$\frac{K_2}{K_3} = \frac{g_{11}}{C_\pi} + \frac{g_{22}}{C_\mu} + \frac{g_{33}}{C_x} = \frac{1}{r_{11}C_\pi} + \frac{1}{r_{22}C_\mu} + \frac{1}{r_{33}C_x} \tag{7.141}$$

where $r_{ii} = 1/g_{ii}$. Now, let us examine each term in the right-most expression. The first term is $1/(r_{11}C_\pi)$. Using (7.91), $r_{11} = 1/g_{11}$ can be found as

$$r_{11} = \frac{1}{g_{11}} = \frac{v_1}{i_1}\bigg|_{v_2=v_3=0,\ C_\pi=0} \quad (7.142)$$

That is, r_{11} is the resistance in parallel with C_π, computed with C_π removed from the circuit and with the other capacitors C_μ and C_x shorted. (Note that shorting C_μ makes $v_2 = 0$; shorting C_x makes $v_3 = 0$.) Therefore, the product $r_{11}C_\pi$ is called the *short-circuit time constant* for capacitor C_π. Similarly, from (7.92)

$$r_{22} = \frac{1}{g_{22}} = \frac{v_2}{i_2}\bigg|_{v_1=v_3=0,\ C_\mu=0} \quad (7.143)$$

This r_{22} is the resistance in parallel with C_μ, computed with C_μ removed from the circuit and with the other capacitors C_π and C_x shorted. Finally, from (7.93)

$$r_{33} = \frac{1}{g_{33}} = \frac{v_3}{i_3}\bigg|_{v_1=v_2=0,\ C_x=0} \quad (7.144)$$

So r_{33} is the resistance in parallel with C_x, computed with C_x removed from the circuit and with the other capacitors C_π and C_μ shorted. Using (7.142) to (7.144), (7.141) can be rewritten as

$$\frac{K_2}{K_3} = \sum_{i=1}^{3} \frac{1}{\tau_{si}} \quad (7.145)$$

where τ_{si} is the short-circuit time constant associated with the ith capacitor in the circuit. The short-circuit time constant for the ith capacitor is found by multiplying its capacitance by the driving-point resistance in parallel with the ith capacitor, computed with all other capacitors shorted.

Combining (7.138) and (7.145) gives

$$\sum_{i=1}^{3} p_i = -\sum_{i=1}^{3} \frac{1}{\tau_{si}} \quad (7.146)$$

This key equation relates the sum of the poles and the sum of the reciprocals of the short-circuit time constants. This relationship holds true for any small-signal circuit that consists of resistors, capacitors, and controlled sources, assuming that the circuit has no loops of capacitors.[1]

If a circuit has n poles and pole p_n has a magnitude that is much larger than the magnitude of every other pole, then a general version of (7.146) can be written as

$$p_n \approx \sum_{i=1}^{n} p_i = -\sum \frac{1}{\tau_{si}} \quad (7.147)$$

For a circuit that has only two widely spaced, real poles, (7.147) simplifies to

$$p_2 \approx -\sum \frac{1}{\tau_{si}} \quad (7.148)$$

This simple relationship allows the magnitude of the nondominant pole to be readily estimated from the short-circuit time constants.

■ **EXAMPLE**

Estimate the nondominant pole magnitude for the circuit in Fig. 7.35 with

$$R_S = 10\ \text{k}\Omega \quad R_L = 10\ \text{k}\Omega$$
$$C_{gs} = 1\ \text{pF} \quad C_f = 20\ \text{pF} \quad g_m = 3\ \text{mA/V}$$

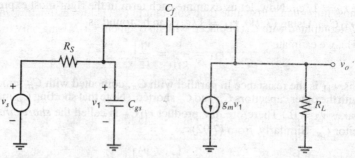

Figure 7.35 Example circuit for calculating short-circuit time constants.

This circuit has two poles because it has two independent capacitors. First, we will calculate the short-circuit time constant τ_{s1} for C_f. With C_{gs} shorted and the independent source set to zero, we have $v_1 = 0$, so the current through the dependent source is also zero. Therefore, the resistance seen by C_f is just R_L, and the corresponding time constant is

$$\tau_{s1} = C_f R_L = (20 \text{ pF})(10 \text{ k}\Omega) = 200 \text{ ns}$$

To find the short-circuit time constant for C_{gs}, we short C_f and find the resistance seen by C_{gs}. With C_f shorted, the dependent source is controlled by the voltage across it; therefore, it acts as a resistance of $1/g_m$. This resistance is in parallel with R_L and R_S, so the short-circuit time constant for C_{gs} is

$$\tau_{s2} = C_{gs}[R_S || 1/g_m || R_L] = (1 \text{ pF})[10 \text{ k}\Omega || 333 \text{ }\Omega || 10 \text{ k}\Omega] = 0.312 \text{ ns} \quad (7.149)$$

From (7.148),

$$p_2 \approx -\sum \frac{1}{\tau_{si}} = -\left(\frac{1}{200 \text{ ns}} + \frac{1}{0.312 \text{ ns}}\right) = -3.21 \text{ Grad/s} \quad (7.150)$$

if the poles are real and widely spaced. An exact analysis of this circuit gives $p_2 = -3.20$ Grad/s, which is very close to the estimate above, and $p_1 = -156$ krad/s.

Using zero-value time constants, the dominant-pole magnitude can be estimated. From (7.124), the zero-value time constant for C_f is

$$C_f R_{f0} = C_f (R_S + R_L + g_m R_L R_S) \quad (7.151)$$

$$= (20\text{p})[10\text{k} + 10\text{k} + (3 \times 10^{-3})(10\text{k})(10\text{k})] \text{ s} = 6.4 \text{ μs}$$

The zero-value time constant for C_{gs} is simply

$$C_{gs} R_S = (1 \text{ pF})(10 \text{ k}\Omega) = 10 \text{ ns}$$

Therefore, (7.111) gives

$$p_1 \approx -\frac{1}{6.4 \text{ μs} + 10 \text{ ns}} = -156 \text{ krad/s}$$

Exact analysis of this circuit gives $p_1 = -156$ krad/s, which is the same as the estimate above. This example demonstrates that for circuits with two widely spaced, real poles, time constant analyses can give accurate estimates of the magnitudes of the dominant and nondominant poles.

In this case of widely spaced poles, note that p_1 and p_2 can be accurately estimated using only one time constant for each pole. Since C_f is large compared to C_{gs} and the resistance C_f sees when computing its zero-value time constant is large compared to the resistance seen by C_{gs}, a reasonable conclusion is that C_{gs} has negligible effect near the frequency $|p_1|$. Therefore, the most important zero-value time constant in (7.111) for estimating p_1 is the time constant for C_f (computed with C_{gs} replaced with an open circuit). Using only that time constant from (7.151), we estimate

$$p_1 \approx -\frac{1}{6.4 \text{ μs}} = -156 \text{ krad/s}$$

which is a very accurate estimate of p_1.

If the real poles are widely separated and the dominant pole was set by C_f, a reasonable assumption is that C_f is a short circuit near the frequency $|p_2|$. Therefore, the important short-circuit time constant in (7.150) is the time constant for C_{gs}, computed with C_f shorted. Using only that time constant from (7.149) in (7.148) gives

$$p_2 \approx -\frac{1}{0.312 \text{ ns}} = -3.21 \text{ Grad/s}$$

which is an accurate estimate of p_2.

This last set of calculations shows that if two real poles are widely spaced and if one capacitor is primarily responsible for p_1 and another capacitor is responsible for p_2, we need only compute one zero-value time constant to estimate p_1 and one short-circuit time constant to estimate p_2.

7.4 Analysis of the Frequency Response of the NE5234 Op Amp

Up to this point, the analysis of the frequency response of integrated circuits has been limited to fairly simple configurations. The reason for this is apparent in previous sections, where the large amount of calculation required to estimate the dominant pole of some simple circuits was illustrated. A complete frequency analysis by hand of a large integrated circuit is thus out of the question. However, a circuit designer often needs insight into the frequency response of large circuits such as the NE5234 op amp, and, by making some sensible approximations, the methods of analysis described previously can be used to provide such information. We will now illustrate this point by analyzing the frequency response of the NE5234.

7.4.1 High-Frequency Equivalent Circuit of the NE5234

Schematics of the stages in the NE5234 are shown Chapter 6. Its frequency response is dominated by the 5.5-pF integrated capacitor C_{22} in Fig. 6.37, which is a compensation capacitor designed to prevent the circuit from oscillating when connected in a feedback loop. The choice of C_{22} and its function are described in Chapter 9.

Since the NE5234 contains more than seventy interconnected transistors, a complete analysis is not attempted even using zero-value time-constant techniques. In order to obtain an estimate of the frequency response of this circuit, the circuit designer must be able to recognize those parts of the circuit that have little or no influence on the frequency response and to discard them from the analysis. As a general rule, elements involved in the bias circuit can often be eliminated. This approach leads to the ac schematic of Fig. 7.36, which is adequate for an approximate calculation of the high-frequency behavior of the NE5234 assuming that the common-mode input voltage, V_{IC}, is low enough that Q_1 and Q_2 in Fig. 6.36 are off.

All bias elements have been eliminated. When the magnitude of the load current is large, either Q_{74} or Q_{75} in the output stage will be regulated by the circuit in Fig. 6.41 to conduct a current that is almost constant as shown by (6.162) and (6.163), depending on the polarity of the load current. The frequency response of the circuit will be slightly different in these two cases.

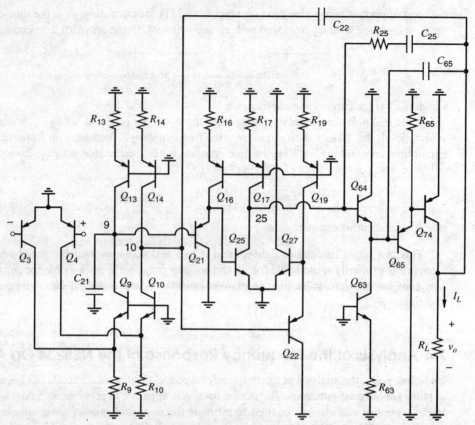

Figure 7.36 An ac schematic of the high-frequency gain path of the NE5234 assuming that V_{IC} is low enough that Q_1 and Q_2 in Fig. 6.36 are off.

Here, the dc load current is assumed to be $I_L = 1$ mA as in the calculations in Chapter 6. Therefore, Q_{75} conducts a nearly constant current and is omitted in Fig. 7.36 along with the circuits that control it for simplicity. From the standpoint of finding the op-amp bandwidth, the major effect of these circuits is to double the transconductance of the second stage in the NE5234, as explained with Fig. 6.48 and included in the calculation below. This simplification is the major approximation in Fig. 7.36, and computer simulation shows that it is a reasonable approximation.

As mentioned above, the frequency response of the NE5234 is dominated by C_{22}, and the -3-dB frequency can be estimated by considering the effect of this capacitance alone. However, as described in Chapter 9, the presence of poles other than the dominant one has a crucial effect on the behavior of the circuit when feedback is applied. The magnitude of the nondominant poles is thus of considerable interest, and their magnitudes can be determined by simulation.

7.4.2 Calculation of the −3-dB Frequency of the NE5234

The -3-dB frequency of the circuit of Fig. 7.36 can be estimated by calculating the zero-value resistance $R_{c22,0}$ seen by C_{22}. This capacitor is connected between the input of the second stage at node 10 and the op-amp output. Resistance $R_{c22,0}$ can be calculated from the equivalent circuit of Fig. 7.37. Here, R_{o1} is the output resistance of the first stage of the NE5234, and R_{i2}, G_{m2}, and R_{o2} are the input resistance, transconductance, and output resistance of the

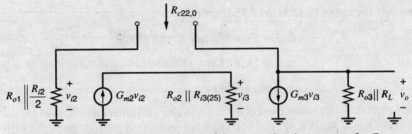

Figure 7.37 Circuit for the calculation of the zero-value time constant for C_{22}.

second stage, respectively. Similarly, $R_{i3(25)}$, G_{m3}, and R_{o3}, are the input resistance from node 25, transconductance, and output resistance of the third stage. Finally, R_L is the load resistance connected to the op-amp output. These quantities were calculated or given in Chapter 6 as follows:

$$R_{o1} = 11 \text{ M}\Omega \qquad R_{i3(25)} = 404 \text{ k}\Omega$$
$$R_{i2}/2 = 650 \text{ k}\Omega \qquad G_{m3} = 1/96 \text{ A/V}$$
$$G_{m2} = 2 \times 170 \text{ } \mu\text{A/V} \qquad R_{o3} = 15 \text{ k}\Omega$$
$$R_{o2} = 2.2 \text{ M}\Omega \qquad R_L = 2 \text{ k}\Omega$$

The circuit of Fig. 7.37 is similar to the circuit in Fig. 7.25 except that the output voltage-controlled current source, $G_{m3}v_{i3}$, is not controlled directly by the input voltage, v_{i2}. Instead, v_{i2} controls the voltage-controlled current source in the model of the second stage, $G_{m2}v_{i2}$. The current from this source flows in $R_{o2}||R_{i3(25)}$, generating the input voltage of the third stage, v_{i3}, which controls $G_{m3}v_{i3}$. As a result, the effective transconductance from the input to the output in Fig. 7.37 is $G_{m(\text{eff})} = G_{m2}(R_{o2}||R_{i3(25)})G_{m3} = 2 \times 170 \times (2.2||0.404)/(96 \Omega) = 1.2 \text{ A/V}$. Then $R_{c22,0}$ can be found using this effective transconductance in (7.117)

$$R_{c22,0} = R_{o1} \left\| \frac{R_{i2}}{2} + R_{o3}||R_L + G_{m(\text{eff})} \left(R_{o1} \left\| \frac{R_{i2}}{2} \right. \right) (R_{o3}||R_L) \right.$$

$$= [11||0.65 + 0.015||0.002 + (1.2 \times (11||0.65) \times (15||2) \times 10^3)] \text{ M}\Omega$$

$$= 1.3 \times 10^9 \text{ } \Omega \tag{7.152}$$

This extremely large resistance when combined with $C_{22} = 5.5$ pF gives a time constant

$$C_{22}R_{c22,0} = 5.5 \times 10^{-12} \times 1.3 \times 10^9 \text{ s} = 7.2 \times 10^{-3} \text{ s}$$

This totally dominates the sum of the zero-value time constants and gives a -3-dB frequency of

$$f_{-3\text{dB}} = \frac{1}{2\pi C_{22}R_{c22,0}} = 22 \text{ Hz}$$

A computer simulation of the complete NE5234 gave $f_{-3\text{dB}} = 21$ Hz.

An alternative means of calculating the effect of the frequency compensation is using the Miller effect as described in Section 7.2.1. The compensation capacitor is connected from node 10 to the output, and the voltage gain between these two points can be calculated from the equivalent circuit of Fig. 7.37:

$$A_v = \frac{v_o}{v_{i2}} = -G_{m(\text{eff})}(R_{o3}||R_L) \tag{7.153}$$

From (7.5) the Miller capacitance seen at node 10 is

$$C_M = (1 - A_v)C_{22} \tag{7.154}$$

and substitution of (7.153) in (7.154) gives

$$C_M = [1 + G_{m(\text{eff})}(R_{o3}\|R_L)]C_{22} \qquad (7.155)$$
$$= [1 + (1.2 \times (15\|2) \times 10^3)] \times 5.5 \text{ pF}$$
$$= 12{,}000 \text{ pF}$$

This extremely large effective capacitance swamps all other capacitances and, when combined with resistance $R_{o1}\|(R_{i2}/2) = 610 \text{ k}\Omega$ from node 10 to ground, gives a -3-dB frequency for the circuit of

$$f_{-3\text{dB}} = \frac{1}{2\pi C_M \left(R_{o1} \left\| \frac{R_{i2}}{2}\right.\right)}$$
$$= \frac{1}{2\pi \times 12{,}000 \times 10^{-12} \times 610 \times 10^3} \text{ Hz}$$
$$= 22 \text{ Hz}$$

This is the same value as predicted by the zero-value time-constant approach.

7.4.3 Nondominant Poles of the NE5234

The foregoing calculations have shown that the 5.5-pF compensation capacitor produces a dominant pole in the NE5234 with a magnitude of about $2\pi(22)$ rad/sec. From the complexity of the circuit, it is evident that there will be a large number of poles with larger magnitudes. Although calculation of the locations of these higher frequency poles is quite difficult, computer simulations show that they contribute negative phase shift at the unity-gain frequency of the amplifier.

7.5 Relation Between Frequency Response and Time Response

In this chapter the effect of increasing signal frequency on circuit performance has been illustrated by considering the circuit response to a sinusoidal input signal. In practice, however, an amplifier may be required to amplify nonsinusoidal signals such as pulse trains or square waves. In addition, such signals are often used in testing circuit frequency response. The response of a circuit to such input signals is thus of some interest and will now be calculated.

Initially we consider a circuit whose small-signal transfer function can be approximated by a single-pole expression

$$\frac{v_o}{v_i}(s) = \frac{K}{1 - \dfrac{s}{p_1}} \qquad (7.156)$$

where K is the low-frequency gain and p_1 is the pole of the transfer function. As described earlier, the -3-dB frequency of this circuit for sinusoidal signals is $\omega_{-3\text{dB}} = -p_1$. Now consider a small input voltage step of amplitude v_a applied to the circuit. If we assume that the circuit responds linearly, we can use (7.156) to calculate the circuit response using $v_i(s) = v_a/s$. Thus

$$v_o(s) = \frac{K v_a}{s} \frac{1}{1 - \dfrac{s}{p_1}} = K v_a \left(\frac{1}{s} - \frac{1}{s - p_1}\right)$$

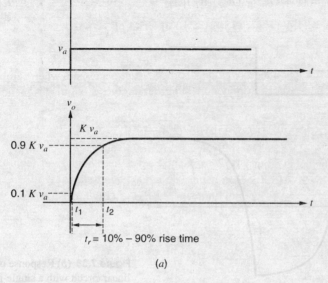

Figure 7.38 (a) Step response of a linear circuit with gain K and a single-pole transfer function.

and the circuit response to a step input is

$$v_o(t) = Kv_a(1 - e^{p_1 t}) \tag{7.157}$$

The output voltage thus approaches Kv_a and the time constant of the exponential in (7.157) is $-1/p_1$. Equation 7.157 is sketched in Fig. 7.38a together with v_i. The rise time of the output is usually specified by the time taken to go from 10 percent to 90 percent of the final value. From (7.157) we have

$$0.1 Kv_a = Kv_a(1 - e^{p_1 t_1}) \tag{7.158}$$

$$0.9 Kv_a = Kv_a(1 - e^{p_1 t_2}) \tag{7.159}$$

From (7.158) and (7.159) we obtain, for the 10 percent to 90 percent rise time,

$$t_r = t_2 - t_1 = -\frac{1}{p_1}\ln 9 = \frac{2.2}{\omega_{-3\text{dB}}} = \frac{0.35}{f_{-3\text{dB}}} \tag{7.160}$$

This equation shows that the pulse rise time is directly related to the -3-dB frequency of the circuit. For example, if $f_{-3\text{dB}} = 10$ MHz, then (7.160) predicts $t_r = 35$ ns. If a square wave is applied to a circuit with a single-pole transfer function, the response is as shown in Fig. 7.38b. The edges of the square wave are rounded as described above for a single pulse.

The calculations in this section have shown the relation between frequency response and time response for small signals applied to a circuit with a single-pole transfer function. For circuits with multiple-pole transfer functions, the same general trends apply, but the pulse response may differ greatly from that shown in Fig. 7.38. In particular, if the circuit transfer function contains complex poles leading to a frequency response with a high-frequency peak (see Chapter 9), then the pulse response will exhibit overshoot[3] and damped sinusoidal oscillation as shown in Fig. 7.39. Such a response is usually undesirable in pulse amplifiers.

Finally, it should be pointed out that all the foregoing results were derived on the assumption that the applied signals were small in the sense that the amplifier acted linearly. If the applied pulse is large enough to cause nonlinear operation of the circuit, the pulse response may differ significantly from that predicted here. This point is discussed further in Chapter 9.

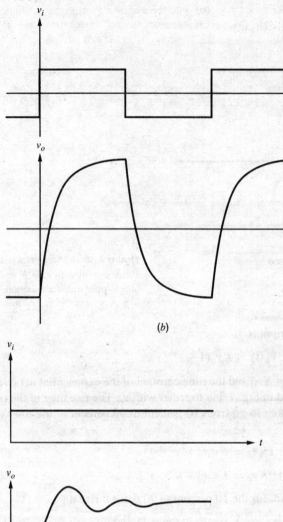

Figure 7.38 (*b*) Response of a linear circuit with a single-pole when a square-wave input is applied.

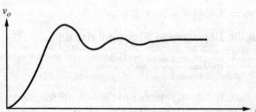

Figure 7.39 Typical step response of a linear circuit whose transfer function contains complex poles.

PROBLEMS

7.1(a) Use the Miller approximation to calculate the -3-dB frequency of the small-signal voltage gain of a common-emitter transistor stage as shown in Fig. 7.2a using $R_S = 5$ kΩ, $R_L = 3$ kΩ, and the following transistor parameters: $r_b = 300$ Ω, $I_C = 0.5$ mA, $\beta = 200$, $f_T = 500$ MHz (at $I_C = 0.5$ mA), $C_\mu = 0.3$ pF, $C_{cs} = 0$, and $V_A = \infty$.

(b) Calculate the nondominant pole magnitude for the circuit in (a). Compare your answer with a SPICE simulation.

7.2 Repeat Problem 7.1 for the MOS common-source stage shown in Fig. 7.2b using $R_S = 10$ kΩ, $R_L = 5$ kΩ, $I_D = 0.5$ mA, and the following NMOS transistor data: NMOS: $W = 100$ μm, $L_{drwn} = 2$ μm, $L_d = 0.2$ μm, $X_d = 0$, $\lambda = 0$, $k'_n = 60$ μA/V^2, $\gamma = 0$, $C_{sb} = C_{db} = 0$, $C_{ox} = 0.7$ fF/(μm^2), and $C_{gd} = 14$ fF.

7.3 Calculate an expression for the output impedance of the circuit in Problem 7.1 as seen by R_L and form an equivalent circuit. Plot the magnitude of this impedance on log scales from $f = 1$ kHz to $f = 100$ MHz.

7.4 Repeat Problem 7.3 for $R_S = 0$ and $R_S = \infty$.

7.5 Repeat Problem 7.3 for the MOS circuit in Problem 7.2.

7.6 A bipolar differential amplifier as shown in Fig. 7.5 has $I_{EE} = 1$ mA. The resistor values and transistor data are as given in Problem 7.1. If the tail current source has an associated resistance $R_T = 300$ kΩ and capacitance $C_T = 2$ pF as defined in Fig. 7.11a, calculate the CM and DM gain and CMRR as a function of frequency. Sketch the magnitude of these quantities in decibels from $f = 10$ kHz to $f = 20$ MHz, using a log frequency scale. Compare your answer with a SPICE simulation.

7.7 A MOS differential amplifier is shown in Fig. 7.9. For this circuit, carry out the calculations in Problem 7.6. Use $I_{SS} = 1$ mA, the values of $R_T = 300$ kΩ and $C_T = 2$ pF as defined in Fig. 7.11b, and the transistor data in Problem 7.2.

7.8 A lateral *pnp* emitter follower has $R_S = 250$ Ω, $r_b = 200$ Ω, $\beta = 50$, $I_C = -300$ μA, $f_T = 4$ MHz, $R_E = 4$ kΩ, $C_\mu = 0$, and $r_o = \infty$. Calculate the small-signal voltage gain as a function of frequency. Sketch the magnitude of the voltage gain in decibels from $f = 10$ kHz to $f = 20$ MHz, using a log frequency scale.

7.9 Calculate the values of the elements in the small-signal equivalent circuits for the input and output impedances of the emitter follower of Problem 7.8. Sketch the magnitudes of these impedances as a function of frequency from $f = 10$ kHz to $f = 20$ MHz, using log scales. Use SPICE to determine the small-signal step response of the circuit for a resistive load of 1 kΩ and then a capacitive load of 400 pF. Use a 1-mV input pulse amplitude with zero rise time. Comment on the shape of the time-domain responses. (Bias the circuit with an ideal 300-μA current source connected to the emitter for the capacitive load test.)

7.10 For the source follower in Fig. 7.13b, find the low-frequency gain and plot the magnitude and phase of its voltage gain versus frequency from $f = 10$ kHz to $f = 20$ GHz, using log scales. Compare your plot with a SPICE simulation. Use the transistor data given in Problem 7.2 with a resistive load of 1 kΩ and then a capacitive load of 400 pF. In both cases, take $I_D = 0.5$ mA. Use a 1-mV input pulse amplitude with zero rise time. Comment on the shape of the time-domain responses. (Bias the circuit with an ideal 0.5 mA current source connected to the source for the capacitive load test.)

7.11(a) Find expressions for R_1, R_2, and L in the output impedance model for a MOS source follower assuming $R_S \gg 1/g_m$, $\gamma \neq 0$, and $v_{sb} = v_o$.

(b) Plot the magnitude of the output impedance versus frequency from $f = 10$ kHz to $f = 10$ GHz, using log scales, when $R_S = 1$ MΩ, $g_m = 0.3$ mA/V, and $\gamma = 0$.

7.12 A common-base stage has the following parameters: $I_C = 0.5$ mA, $C_\pi = 10$ pF, $C_\mu = 0.3$ pF, $r_b = 200$ Ω, $\beta = 100$, $r_o = \infty$, $R_L = 0$, and $R_S = \infty$.

(a) Calculate an expression for the small-signal current gain of the stage as a function of frequency and thus determine the frequency where the current gain is 3 dB below its low-frequency value.

(b) Calculate the values of the elements in the small-signal equivalent circuits for the input and output impedances of the stage and sketch the magnitudes of these impedances from $f = 100$ kHz to $f = 100$ MHz using log scales.

7.13 Repeat Problem 7.12 for a NMOS common-gate stage using $R_L = 0$ and $R_S = \infty$. Use $I_D = 0.5$ mA and the MOS transistor data in Problem 7.2. Plot the impedance magnitudes from $f = 100$ kHz to $f = 100$ GHz.

7.14 The ac schematic of a common-emitter stage is shown in Fig. 7.2a. Calculate the low-frequency small-signal voltage gain v_o/v_i and use the zero-value time-constant method to estimate the -3-dB frequency for $R_S = 10$ kΩ and $R_L = 5$ kΩ. Data: $\beta = 200$, $f_T = 600$ MHz (at $I_C = 1$ mA), $C_\mu = 0.2$ pF, $C_{je} = 2$ pF, $C_{cs} = 1$ pF, $r_b = 0$, $r_o = \infty$, and $I_C = 1$ mA.

7.15 Repeat Problem 7.14 if an emitter degeneration resistor of value 100 Ω is included in the circuit.

7.16 Repeat Problem 7.14 if a resistor of value 20 kΩ is connected between collector and base of the transistor.

7.17 Repeat the calculations in Problem 7.14 for the common-source stage in Fig. 7.2b. Take $V_{DB} = 7.5$ V and $I_D = 1$ mA. Use the same transistor data and resistor values as in Problem 7.2 with the following exceptions:

1. C_{ox} and C_{gd} are not given, but $f_T = 3$ GHz.

2. C_{db} is not equal to 0. Calculate the zero-bias drain-bulk capacitance as $C_{db0} = A_D(C_{j0}) + P_D(C_{jsw0})$, where $A_D = (5 \ \mu\text{m})W$, and use $P_D = W$. Let $C_{j0} = 0.4$ fF/(μm^2) and $C_{jsw0} = 0.4$ fF/μm. Then use (1.202) with $\psi_0 = 0.6$ V to calculate C_{db}.

7.18 Repeat Problem 7.14 using a NMOS transistor in place of the bipolar transistor. Use $I_D = 0.5$ mA and the transistor data in Problem 7.2.

7.19 Repeat Problem 7.18 if a 500 Ω source-degeneration resistor is included in the circuit.

7.20 Repeat Problem 7.18 if a resistor of value 50 kΩ is connected between drain and gate of the transistor.

7.21 A Darlington stage and a common-collector–common-emitter cascade are shown schematically in Fig. 7.40, where $R_S = 100$ kΩ and $R_L = 3$ kΩ.

(a) Calculate the low-frequency small-signal voltage gain v_o/v_i for each circuit.

(b) Use the zero-value time-constant method to calculate the -3-dB frequency of the gain of each circuit. Data: $\beta = 100$, $f_T = 500$ MHz at $I_C = 1$ mA, $C_\mu = 0.4$ pF, $C_{je} = 2$ pF, $C_{cs} = 1$ pF, $r_b = 0$, $r_o = \infty$, $I_{C1} = 10$ µA, and $I_{C2} = 1$ mA. (Values of C_μ, C_{cs}, and C_{je} are at the bias point.)

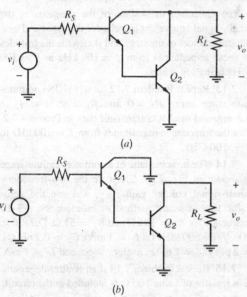

Figure 7.40 The ac schematics of (a) Darlington stage and (b) common-collector–common-emitter stage.

7.22 Repeat Problem 7.21 if the input signal is a current source of value i_i applied at the base of Q_1. (That is, i_i replaces the voltage source v_i and resistance R_S.) The transfer function is then a transresistance v_o/i_i.

7.23 Repeat Problem 7.21 if a bleed resistor of 15 kΩ is added from the emitter of Q_1 to ground, which increases the collector bias current in Q_1 to 50 µA.

7.24 Replace the bipolar transistors in Fig. 7.40 with NMOS transistors. Repeat the calculations in Problem 7.21, using $R_S = 100$ kΩ, $R_L = 3$ kΩ, and the NMOS transistor model data in Problem 7.2, but use $C_{db} = 200$ fF and $C_{sb} = 180$ fF here. Take $I_{D1} = 50$ µA and $I_{D2} = 1$ mA.

7.25 An amplifier stage is shown in Fig. 7.41 where bias current I_B is adjusted so that $V_O = 0$ V dc. Take $V_{\text{SUPPLY}} = 10$ V.

(a) Calculate the low-frequency, small-signal transresistance v_o/i_i and use the zero-value time-constant method to estimate the -3-dB frequency. Data: npn: $\beta = 100$, $f_T = 500$ MHz at $I_C = 1$ mA, $C_{\mu 0} = 0.7$ pF, $C_{je} = 3$ pF (at the bias point), $C_{cs0} = 2$ pF, $r_b = 0$, and $V_A = 120$ V. Assume $n = 0.5$ and $\psi_0 = 0.55$ V for all junctions. pnp: $\beta = 50$, $f_T = 4$ MHz at $I_C = -0.5$ mA, $C_{\mu 0} = 1.0$ pF, $C_{je} = 3$ pF (at the bias point), $C_{bs0} = 2$ pF, $r_b = 0$, and $|V_A| = 50$ V. Assume $n = 0.5$ and $\psi_0 = 0.55$ V for all junctions.

(b) Repeat (a) if a 20-pF capacitor is connected from collector to base of Q_1.

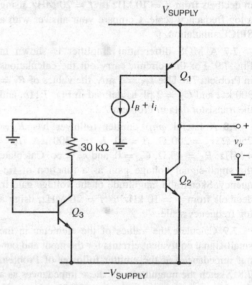

Figure 7.41 Amplifier stage.

7.26 Repeat Problem 7.25 with the following changes:

1. Replace Q_1 with a p-channel MOS transistor, M_1. Replace Q_2 and Q_3 with n-channel MOS transistors, M_2 and M_3.

2. Add a resistor of value $1/g_{m1}$ from the gate to the source of M_1.

3. Take $V_{\text{SUPPLY}} = 2.5$ V.

4. Use the formula for C_{db0} given in Problem 7.17.

5. For all transistors: $L_{\text{drwn}} = 2$ µm, $L_d = 0.2$ µm, $X_d = 1$ µm, and $\gamma = 0$. $W_1 = 200$ µm and $W_2 = W_3 = 100$ µm. Use (1.201) and (1.202) with $\psi_0 = 0.6$ V for the junction capacitances. Use the equations in Problem 7.17 for C_{db0}. NMOS data: $V_{tn} = 1$ V, $k'_n = 60$ µA/V^2, $\lambda_n = 1/(100$ V$)$, $C_{ox} = 0.7$ fF/(µm^2), $C_{j0} = 0.4$ fF/(µm^2), and $C_{jsw0} = 0.4$ fF/µm. PMOS data: $V_{tp} = -1$ V, $k'_p = 20$ µA/V^2, $|\lambda_p| = 1/(50$ V$)$, $C_{ox} = 0.7$ fF/(µm^2), $C_{j0} = 0.2$ fF/(µm^2), and $C_{jsw0} = 0.2$ fF/µm.

7.27 A differential circuit employing active loads is shown in Fig. 7.42. Bias voltage V_B is adjusted

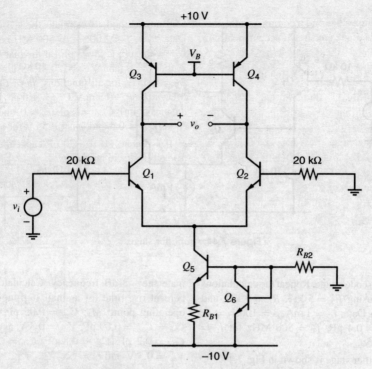

Figure 7.42 Differential circuit with active loads.

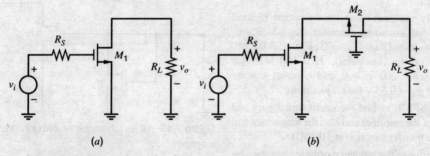

Figure 7.43 An ac schematic of (a) common-source stage and (b) cascode stage.

so that the collectors of Q_1 and Q_2 are at +5 V dc. Biasing resistors are $R_{B1} = 10$ kΩ and $R_{B2} = 20$ kΩ. Calculate the low-frequency, small-signal voltage gain v_o/v_i, and use the zero-value time-constant method in the DM half-circuit to estimate the −3-dB frequency of the DM gain. Use the device data in Problem 7.25.

7.28 Repeat Problem 7.27, replacing the bipolar transistors with MOS transistors. Assume that the values of R_{B1} and R_{B2} set $I_{D5} = 1$ mA. Use $W_1 = W_2 = W_5 = W_6 = 100$ μm, $W_3 = W_4 = 50$ μm, and $L_{drwn} = 2$ μm. See Problem 7.26 for all other MOS transistor data.

7.29 The ac schematics of a common-source stage and a common-source–common-gate (cascode) stage are shown in Fig. 7.43 with $R_S = 10$ kΩ and $R_L = 20$ kΩ. Using the transistor and operating-point data in Problem 7.2:

(a) Calculate the low-frequency, small-signal voltage gain v_o/v_i for each circuit.

(b) Use the zero-value time-constant method to calculate and compare the −3-dB frequencies of the gain of the two circuits.

(c) Estimate the 10 to 90 percent rise time for each circuit for a small step input and sketch the output voltage waveform over 0 to 300 ns for a 1-mV step input.

7.30 Replace the NMOS transistors in Fig. 7.43 with npn transistors. The resulting ac schematics are of a common-emitter stage and a common-emitter–

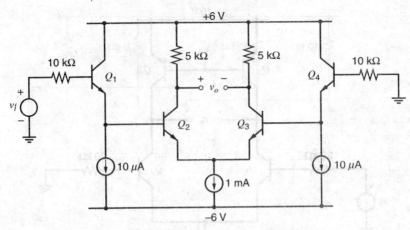

Figure 7.44 Amplifier stage.

common-base (cascode) stage. Repeat the calculations in Problem 7.29 using $R_S = 5$ kΩ, $R_L = 3$ kΩ, and the following data. Data: $I_C = 1$ mA, $\beta = 100$, $r_b = 0$, $C_{cs} = 1$ pF, $C_\mu = 0.4$ pF, $f_T = 500$ MHz (at $I_C = 1$ mA), and $r_o = \infty$.

7.31 An amplifier stage is shown in Fig. 7.44.

(a) Calculate the low-frequency, small-signal voltage gain v_o/v_i.

(b) Apply the zero-value time-constant method to the DM half-circuit to calculate the -3-dB frequency of the gain. Data: $C_{cs0} = 2$ pF, $C_{\mu 0} = 0.5$ pF, $C_{je} = 4$ pF (at the bias point), $f_T = 500$ MHz (at $I_C = 2$ mA), $\beta = 200$, $r_b = 0$, and $r_o = \infty$. Assume $n = 0.5$ and $\psi_0 = 0.55$ V for all junctions.

(c) Use SPICE to find the small-signal gain and bandwidth of the amplifier and also the magnitude and phase of the transfer function at 100 MHz.

(d) Investigate the influence of base resistance by repeating (c) with $r_b = 200$ Ω and comparing the results.

7.32 Repeat Problem 7.31 with n-channel MOS transistors replacing all bipolar transistors. Assume $W = 100$ μm, $L_{drwn} = 2$ μm, $L_d = 0.2$ μm, $X_d = 0$, $\lambda = 0$, $k'_n = 60$ μA/V^2, $\gamma = 0$, $V_t = 1$ V, $C_{ox} = 0.7$ fF/μm^2, $C_{ol} = 0.15$ fF/μm, $C_{j0} = 0.4$ fF/μm^2, and $C_{jsw0} = 0.4$ fF/μm. Use (1.201) and (1.202) with $\psi_0 = 0.6$ V for the junctions. Use the information in Problem 7.17 to calculate $C_{db0} = C_{sb0}$. Skip part (d) in Problem 7.31.

7.33 The ac schematic of a wideband MOS current amplifier is shown in Fig. 7.45. The W/L of M_2 is four times that of M_1 and corresponding bias currents are $I_{D1} = 0.5$ mA and $I_{D2} = 2$ mA. Calculate the low-frequency, small-signal current gain i_o/i_i and use the zero-value time-constant method to estimate the -3-dB frequency. Calculate the 10 to 90 percent rise time for a small step input. Data at the operating point: M_1: $C_{gd} = 0.05$ pF, $C_{gs} = 0.2$ pF, $C_{sb} = C_{db} = 0.09$ pF, $V_{ov} = 0.3$ V, and $r_o = \infty$. M_2: $C_{gd} = 0.2$ pF, $C_{gs} = 0.8$ pF, $C_{sb} = C_{db} = 0.36$ pF, $V_{ov} = 0.3$ V, and $r_o = \infty$.

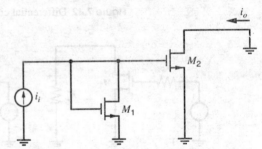

Figure 7.45 An ac schematic of a MOS current amplifier.

7.34 Replace the MOS transistors in the amplifier in Fig. 7.45 with bipolar npn transistors. The emitter area of Q_2 is four times that of Q_1 and corresponding bias currents are $I_{C1} = 1$ mA and $I_{C2} = 4$ mA. Repeat the calculations in Problem 7.33 using the following data. Data at the operating point: Q_1: $\beta = 200$, $\tau_F = 0.2$ ns, $C_\mu = 0.2$ pF, $C_{je} = 1$ pF, $C_{cs} = 1$ pF, $r_b = 0$, and $r_o = \infty$. Q_2: $\beta = 200$, $\tau_F = 0.2$ ns, $C_\mu = 0.8$ pF, $C_{je} = 4$ pF, $C_{cs} = 4$ pF, $r_b = 0$, and $r_o = \infty$.

7.35 A two-stage amplifier is shown in Fig. 7.46. Calculate the low-frequency, small-signal gain and use the zero-value time-constant method to estimate the -3-dB frequency. Calculate the 10 to 90 percent rise time for a small step input. Use SPICE to determine the -3-dB frequency and also the frequency where the phase shift in the transfer function is $-135°$ beyond the low-frequency value. Data: $\beta = 200$,

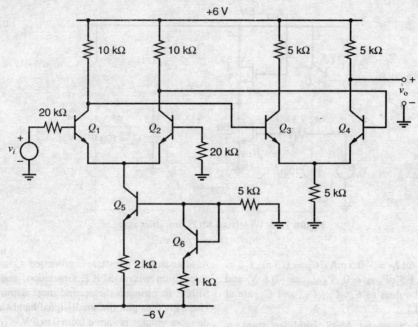

Figure 7.46 Two-stage amplifier.

$f_T = 600$ MHz at $I_C = 1$ mA, $C_\mu = 0.2$ pF, $C_{je} = 2$ pF, $C_{cs} = 1$ pF, $r_b = 0$, $V_{BE(on)} = 0.6$ V, and $r_o = \infty$. (Values of C_μ, C_{cs}, and C_{je} are at the bias point.)

7.36 A two-stage bipolar amplifier is shown in Fig. 7.47. Calculate the low-frequency, small-signal voltage gain v_o/v_i and use the zero-value time-constant method to estimate the -3-dB frequency. Use a DM half-circuit for the differential pair. Use SPICE to estimate the first and second most dominant poles of the circuit. Data: npn: $\beta = 200$, $f_T = 400$ MHz (at $I_C = 1$ mA), $C_\mu = 0.3$ pF, $C_{je} = 3$ pF, $C_{cs} = 1.5$ pF, $r_b = 0$, $V_{BE(on)} = 0.6$ V, and $r_o = \infty$. pnp: $\beta = 100$,

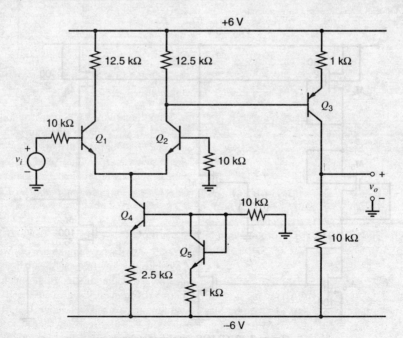

Figure 7.47 Two-stage amplifier with *pnp* second stage.

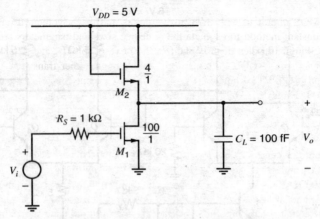

Figure 7.48 Wideband MOS amplifier stage.

$f_T = 6$ MHz (at $I_C = -0.5$ mA), $C_\mu = 0.3$ pF, $C_{je} = 3$ pF, $C_{bs} = 1.5$ pF, $r_b = 0$, $V_{BE(on)} = -0.6$ V, and $r_o = \infty$. (All values of C_μ, C_{cs}, C_{bs}, and C_{je} are at the bias point.)

7.37(a) A wideband MOS amplifier stage is shown in Fig. 7.48. Calculate the small-signal, low-frequency gain and use the zero-value time-constant method to estimate the −3-dB bandwidth. Use $\mu_n C_{ox} = 60$ μA/V^2, $t_{ox} = 20$ nm, $C_{ol} = 0.3$ fF/(μm of gate width), $\psi_o = 0.6$ V, $V_t = 0.7$ V, $\gamma = 0.4$ V$^{1/2}$, $\lambda = 0$, and $V_O = 2.5$ V dc. For C_{db} and C_{sb}, use $C_{db0} = C_{sb0} = 0.8$ fF/(μm of gate width).

Assume that the substrate is grounded. Compare your calculation with a SPICE simulation, and also use SPICE to estimate the second most dominant pole. Use SPICE to plot the small-signal bandwidth as the dc input voltage is varied from 0 to 5 V.

(b) Calculate the small-signal gain and −3-dB bandwidth including short-channel effects with $\mathscr{E}_c = 1.5 \times 10^6$ V/m. Assume the same bias currents as in (a) and model MOSFET short-channel effects with a resistor in series with the source. Connect the device capacitances to the lower end of the added source resistor.

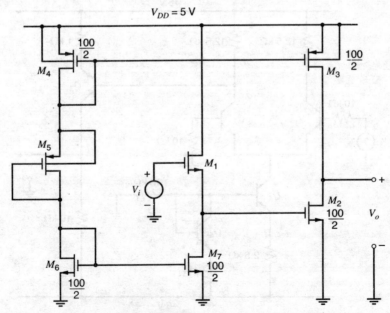

Figure 7.49 CMOS amplifier stage.

7.38 For the BiCMOS circuit of Fig. 3.78, use the zero-value time-constant method to estimate the first and second most dominant poles of the circuit. Assume an input voltage drive. Use bipolar transistor data from Fig. 2.32 and MOSFET parameters $C_{gd} = 90$ fF, $C_{sb} = C_{db} = 200$ fF, and $C_{gs} = 200$ fF at the bias point. Further assume $\mu_n C_{ox} = 40\ \mu\text{A/V}^2$, $V_t = 0.8$ V, $\lambda = 0$, and $\gamma = 0$. Use SPICE to check your result.

7.39 A CMOS amplifier stage is shown in Fig. 7.49. Select W/L for M_1 and M_5 to give $V_i = V_O = 2.5$ V dc and $|I_D| = 100\ \mu\text{A}$ bias in all devices. The minimum value of L and W is 2 μm. Calculate the small-signal, low-frequency gain and the -3-dB frequency of the stage. Verify with SPICE. Use device data from Problem 7.37(a) with $|\lambda| = 0.03$ V^{-1}, $\mu_p C_{ox} = 30\ \mu\text{A/V}^2$, and $V_{tp} = -0.7$ V.

7.40 Use the zero-value time-constant method to estimate the small-signal dominant pole for the current gain of the MOS cascode current mirror of Fig. 4.9. Assume an input ac current source in parallel with I_{IN} and a zero load impedance with $V_{\text{out}} = V_{GS3} + V_{GS4}$. The bias current $I_{\text{IN}} = 100\ \mu\text{A}$. Compare your answer with the f_T value of the devices. Device parameters are $\mu_n C_{ox} = 60\ \mu\text{A/V}^2$, $\gamma = 0$, $\lambda = 0$, $V_t = 0.7$ V, $W = 10\ \mu\text{m}$, $L_{\text{eff}} = 1\ \mu\text{m}$, $C_{gs} = 20$ fF, $C_{gd} = 3$ fF, $C_{sb} = C_{db} = 10$ fF at the bias point. Compare your answer with a SPICE simulation of the bandwidth of the circuit and use SPICE to find the bandwidth for $I_{\text{IN}} = 50\ \mu\text{A}$ and $I_{\text{IN}} = 200\ \mu\text{A}$.

7.41 Repeat Problem 7.40 including short-channel effects with $\mathscr{E}_c = 1.5 \times 10^6$ V/m.

7.42 A MOS cascode stage is shown in Fig. 7.43b. Replace the load resistor with a load capacitor $C_L = 2$ pF. Assume the total capacitance that connects to the drain of M_1 can be modeled by a capacitor $C_p = 0.2$ pF from that drain to ac ground. Ignore all other capacitors. Therefore, the gain for this circuit has only two poles. For both transistors, take $I_D = 100\ \mu\text{A}$, $W = 20\ \mu\text{m}$, $L_{\text{eff}} = 0.5\ \mu\text{m}$, $k' = 180\ \mu\text{A/V}^2$, and $\lambda = 0.04$ V^{-1}.

(a) Use zero-value time constants to estimate the dominant pole.

(b) Use short-circuit time constants to estimate the nondominant pole.

(c) Compare your answers with a SPICE simulation.

7.43 Use the short-circuit time-constant method to estimate the nondominant pole that originates at the drain nodes of M_1 and M_2 in the CMOS folded cascode of Fig. 6.28. Assume the gates of M_{1A}, M_{2A}, M_5, M_{11}, and M_{12} are biased from low-impedance points and that a voltage drive is applied at v_i. All device sizes and parameters are as given in Problem 7.40 except for M_{11} and M_{12}, which have twice the width of the other transistors. All bias currents are 100 μA except for M_5, M_{11}, and M_{12}, which have $|I_D| = 200\ \mu\text{A}$. Assume V_{GD} of M_{11} and M_{12} is zero volts. Use $C_L = 1$ pF and $\mu_p C_{ox} = 30\ \mu\text{A/V}^2$. How much phase shift is contributed to the amplifier transfer function by this nondominant pole at the amplifier unity-gain frequency? Check your calculations with SPICE simulations.

7.44(a) For the common-emitter amplifier in Problem 7.1, use zero-value time constants to estimate the dominant pole and short-circuit time constants to estimate the nondominant pole.

(b) Compare your estimates with a SPICE simulation.

7.45(a) Use zero-value time constants to estimate the dominant pole and short-circuit time constants to estimate the nondominant pole for the common-source amplifier in Problem 7.2.

(b) Compare your answers with a SPICE simulation.

7.46(a) An integrator is shown in Fig. 7.50. Use zero-value time constants to estimate the dominant pole and short-circuit time constants to estimate the nondominant pole for this circuit. Use $R = 20$ kΩ, $C = 50$ pF, $C_{\text{in}} = 0.2$ pF, $a_v = 1000$, and $R_o = 5$ kΩ.

(b) Compare your answers with a SPICE simulation.

7.47 Add a 0.5 pF load capacitor from the output to ground to the integrator in Fig. 7.50. When this capacitor is added, the circuit has a loop of three capacitors. Direct application of the short-circuit time-constant method here gives zero for each short-circuit time-constant. (Verify this.)

The problem here is that three short-circuit time constants are being calculated for the three capacitors, and the sum of the reciprocal of these time constants equals the sum of three poles, as shown in (7.146). However, this circuit has only two poles because only two of the capacitor voltages are independent.

An alternative approach to estimating the nondominant pole is to calculate the zero-value time constants and determine if one zero-value time constant is much larger than the others. If so, the capacitor associated with the largest zero-value time constant is shorted, and one time constant for the remaining capacitors, which are in parallel, is the short-circuit time constant. Carry out these steps, and compare the estimated nondominant pole with a SPICE simulation.

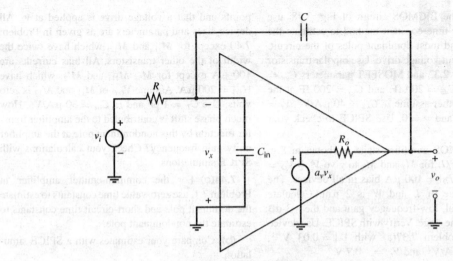

Figure 7.50 An integrator stage.

7.48 Find an expression for $G_m(s) = i_o(s)/v_{id}(s)$ for the circuit in Fig. 7.33 and verify the equations for the pole and zero given in Section 7.3.5.

7.49 Calculate the pole and zero associated with the current-mirror load in Fig. 7.33 if $I_{D3} = -200\,\mu\text{A}$, $|V_{ov3}| = 0.2$ V, and $C_x = 0.1$ pF.

REFERENCES

1. P. E. Gray and C. L. Searle. *Electronic Principles: Physics, Models, and Circuits*. Wiley, New York, 1969.

2. A. S. Sedra and K. C. Smith. *Microelectronic Circuits*. Oxford University Press, New York, 1998.

3. K. Ogata. *Modern Control Engineering*. Second Edition, Prentice-Hall, Englewood Cliffs, N.J., 1990.

CHAPTER 8

Feedback

Negative feedback is widely used in amplifier design because it produces several important benefits. One of the most significant is that negative feedback stabilizes the gain of the amplifier against parameter changes in the active devices due to supply voltage variation, temperature changes, or device aging. A second benefit is that negative feedback allows the designer to modify the input and output impedances of the circuit in any desired fashion. Another significant benefit of negative feedback is the reduction in signal waveform distortion that it produces, and for this reason almost all high-quality audio amplifiers employ negative feedback around the power output stage. Finally, negative feedback can produce an increase in the bandwidth of circuits and is widely used in broadband amplifiers.

However, the benefits of negative feedback listed above are accompanied by two disadvantages. First, the gain of the circuit is reduced in almost direct proportion to the other benefits achieved. Thus, it is often necessary to make up the decrease in gain by adding extra amplifier stages with a consequent increase in hardware cost. The second potential problem associated with the use of feedback is the tendency for oscillation to occur in the circuit, and careful attention by the designer is often required to overcome this problem.

In this chapter, the various benefits of negative feedback are considered, together with a systematic classification of feedback configurations. Two different methods for analyzing feedback circuits are presented. The problem of feedback-induced oscillation and its solution are considered in Chapter 9.

8.1 Ideal Feedback Equation

Consider the idealized feedback configuration of Fig. 8.1. In this figure S_i and S_o are input and output signals that may be voltages or currents. The feedback network (which is usually linear and passive) has a transfer function f and feeds back a signal S_{fb} to the input. At the input, signal S_{fb} is subtracted from input signal S_i at the input differencing node. Error signal S_ϵ is the difference between S_i and S_{fb}, and S_ϵ is fed to the basic amplifier with transfer function a. Note that another common convention is to assume that S_i and S_{fb} are added together in an input *summing* node, and this leads to some sign changes in the analysis. It should be pointed out that negative-feedback amplifiers in practice have an input differencing node and thus the convention assumed here is more convenient for amplifier analysis.

From Fig. 8.1

$$S_o = aS_\epsilon \qquad (8.1)$$

assuming that the feedback network does not load the basic amplifier. Also

$$S_{fb} = fS_o \qquad (8.2)$$

$$S_\epsilon = S_i - S_{fb} \qquad (8.3)$$

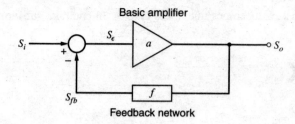

Figure 8.1 Ideal feedback configuration.

Substituting (8.2) in (8.3) gives

$$S_\epsilon = S_i - fS_o \tag{8.4}$$

Substituting (8.4) in (8.1) gives

$$S_o = aS_i - afS_o$$

and thus

$$\frac{S_o}{S_i} = A = \frac{a}{1 + af} \tag{8.5}$$

Equation 8.5 is the fundamental equation for negative feedback circuits where A is the overall gain with feedback applied. (A is often called the *closed-loop gain*.)

It is useful to define a quantity T called the *loop gain* such that

$$T = af \tag{8.6}$$

and

$$\frac{S_o}{S_i} = A = \frac{a}{1 + T} \tag{8.7}$$

T is the total gain around the feedback loop. If $T \gg 1$, then, from (8.5), gain A is given by

$$A \simeq \frac{1}{f} \tag{8.8}$$

That is, for large values of loop gain T, the overall amplifier gain is determined by the feedback transfer function f. Since the feedback network is usually formed from stable, passive elements, the value of f is well defined and so is the overall amplifier gain.

The feedback loop operates by forcing S_{fb} to be nearly equal to S_i. This is achieved by amplifying the difference $S_\epsilon = S_i - S_{fb}$, and the feedback loop then effectively minimizes error signal S_ϵ. This can be seen by substituting (8.5) in (8.4) to obtain

$$S_\epsilon = S_i - f\frac{aS_i}{1 + af}$$

and this leads to

$$\frac{S_\epsilon}{S_i} = \frac{1}{1 + af} = \frac{1}{1 + T} \tag{8.9}$$

As T becomes much greater than 1, S_ϵ becomes much less than S_i. In addition, substituting (8.5) in (8.2) gives

$$S_{fb} = fS_i \frac{a}{1+af}$$

and thus

$$\frac{S_{fb}}{S_i} = \frac{T}{1+T} \qquad (8.10)$$

If $T \gg 1$, then S_{fb} is approximately equal to S_i. That is, feedback signal S_{fb} is a replica of the input signal. Since S_{fb} and S_o are directly related by (8.2), it follows that if $|f| < 1$, then S_o is an amplified replica of S_i. This is the aim of a feedback amplifier.

8.2 Gain Sensitivity

In most practical situations, gain a of the basic amplifier is not well defined. It is dependent on temperature, active-device operating conditions, and transistor parameters. As mentioned previously, the negative-feedback loop reduces variations in overall amplifier gain due to variations in a. This effect may be examined by differentiating (8.5) to obtain

$$\frac{dA}{da} = \frac{(1+af) - af}{(1+af)^2}$$

and this reduces to

$$\frac{dA}{da} = \frac{1}{(1+af)^2} \qquad (8.11)$$

If a changes by δa, then A changes by δA where

$$\delta A = \frac{\delta a}{(1+af)^2}$$

The fractional change in A is

$$\frac{\delta A}{A} = \frac{1+af}{a} \frac{\delta a}{(1+af)^2}$$

This can be expressed as

$$\frac{\delta A}{A} = \frac{\dfrac{\delta a}{a}}{1+af} = \frac{\dfrac{\delta a}{a}}{1+T} \qquad (8.12)$$

Equation 8.12 shows that the fractional change in A is reduced by $(1+T)$ compared to the fractional change in a. For example, if $T = 100$ and a changes by 10 percent due to temperature change, then the overall gain A changes by only 0.1 percent using (8.12).

8.3 Effect of Negative Feedback on Distortion

The foregoing results show that even if the basic-amplifier gain a changes, the negative feedback keeps overall gain A approximately constant. This suggests that feedback should be effective in reducing distortion because distortion is caused by changes in the slope of the

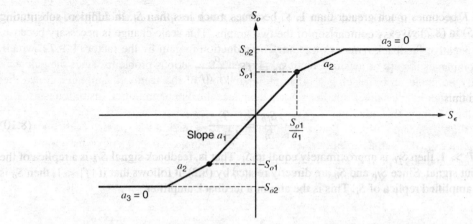

Figure 8.2 Basic-amplifier transfer characteristic.

basic-amplifier transfer characteristic. The feedback should tend to reduce the effect of these slope changes since A is relatively independent of a. This is explained next.

Suppose the basic amplifier has a transfer characteristic with a nonlinearity as shown in Fig. 8.2. It is assumed that two regions exist, each with constant but different slopes a_1 and a_2. When feedback is applied, the overall gain will still be given by (8.5) but the appropriate value of a must be used, depending on which region of Fig. 8.2 is being traversed. Thus the *overall* transfer characteristic with feedback applied will also have two regions of different slope, as shown in Fig. 8.3. However, slopes A_1 and A_2 are almost equal because of the effect of the negative feedback. This can be seen by substituting in (8.5) to give

$$A_1 = \frac{a_1}{1 + a_1 f} \simeq \frac{1}{f} \tag{8.13}$$

$$A_2 = \frac{a_2}{1 + a_2 f} \simeq \frac{1}{f} \tag{8.14}$$

Thus the transfer characteristic of the feedback amplifier of Fig. 8.3 shows much less nonlinearity than the original basic-amplifier characteristics of Fig. 8.2.

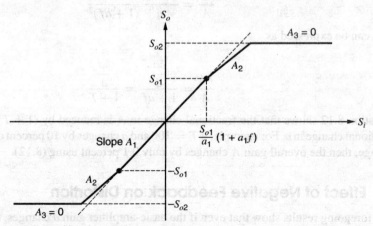

Figure 8.3 Feedback-amplifier transfer characteristic corresponding to the basic-amplifier characteristic of Fig. 8.2.

Note that the horizontal scale in Fig. 8.3 has been *compressed* as compared to Fig. 8.2 in order to allow easy comparison of the two graphs. This scale change is necessary because the negative feedback *reduces the gain*. The reduction in gain by the factor $(1 + T)$, which accompanies the use of negative feedback, presents few serious problems, since the gain can easily be made up by placing a preamplifier in front of the feedback amplifier. Since the preamplifier handles much smaller signals than does the output amplifier, distortion is usually not a problem in that amplifier.

One further point that should be made about Figs. 8.2 and 8.3 is that both show hard saturation of the output amplifier (i.e., the output becomes independent of the input) at an output signal level of S_{o2}. Since the incremental slope $a_3 = 0$ in that region, negative feedback cannot improve the situation as $A_3 = 0$ also, using (8.5).

8.4 Feedback Configurations

The treatment in the previous sections was based on the idealized configuration shown in Fig. 8.1. Practical feedback amplifiers are composed of circuits that have current or voltage signals as inputs and produce output currents or voltages. In order to pursue feedback amplifier design at a practical level, it is necessary to specify the details of the feedback sampling process and the circuits used to realize this operation. There are four basic feedback amplifier connections. These are specified according to whether the output signal S_o is a current or a voltage and whether the feedback signal S_{fb} is a current or a voltage. It is apparent that four combinations exist and these are now considered.

8.4.1 Series-Shunt Feedback

Suppose it is required to design a feedback amplifier that stabilizes a voltage transfer function. That is, a given input voltage should produce a well-defined proportional output voltage. This will require sampling the output voltage and feeding back a proportional voltage for comparison with the incoming voltage. This situation is shown schematically in Fig. 8.4. The basic amplifier has gain a, and the feedback network is a two-port with transfer function f that *shunts* the output of the basic amplifier to sample v_o. Ideally, the impedance $z_{22f} = \infty$, and the feedback network does not load the basic amplifier. The feedback voltage v_{fb} is connected in *series* with the input to allow comparison with v_i and, ideally, $z_{11f} = 0$. The signal v_ϵ is the *difference* between v_i and v_{fb} and is fed to the basic amplifier. The basic amplifier and feedback circuits are assumed *unilateral* in that the basic amplifier transmits only from v_ϵ to v_o and the feedback network transmits only from v_o to v_{fb}. This point will be taken up later.

This feedback is called *series-shunt* feedback because the feedback network is connected in *series* with the input and *shunts* the output.

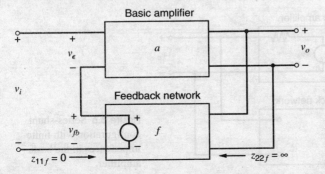

Figure 8.4 Series-shunt feedback configuration.

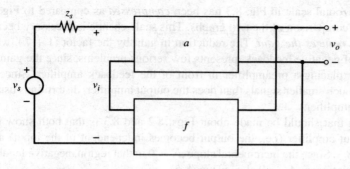

Figure 8.5 Series-shunt configuration fed from a finite source impedance.

From Fig. 8.4

$$v_o = av_\epsilon \quad (8.15)$$
$$v_{fb} = fv_o \quad (8.16)$$
$$v_\epsilon = v_i - v_{fb} \quad (8.17)$$

From (8.15), (8.16), and (8.17),

$$\frac{v_o}{v_i} = \frac{a}{1 + af} \quad (8.18)$$

Thus the ideal feedback equation applies. Equation 8.18 indicates that the transfer function that is stabilized is v_o/v_i, as desired. If the circuit is fed from a high source impedance as shown in Fig. 8.5, the ratio v_o/v_i is still stabilized [and given by (8.18)], *but* now v_i is given by

$$v_i = \frac{Z_i}{Z_i + z_s}v_s \quad (8.19)$$

where Z_i is the input impedance seen by v_i. If $z_s \approx Z_i$, then v_i depends on Z_i, which is *not* usually well defined since it often depends on active-device parameters. Thus the overall gain v_o/v_s will not be stabilized. Consequently, the full benefits of gain stabilization are achieved for a series-shunt feedback amplifier when the source impedance is low compared to the input impedance of the closed-loop amplifier. The ideal driving source is a voltage source.

Consider now the effect of series-shunt feedback on the terminal impedances of the amplifier. Assume the basic amplifier has input and output impedances z_i and z_o as shown in Fig. 8.6. Again assume the feedback network is ideal and feeds back a voltage fv_o as shown. Both networks are unilateral. The applied voltage v_i produces input current i_i and output voltage v_o. From Fig. 8.6

$$v_o = av_\epsilon \quad (8.20)$$
$$v_i = v_\epsilon + fv_o \quad (8.21)$$

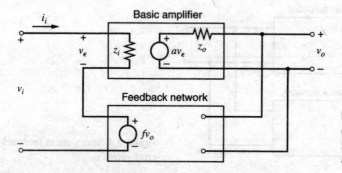

Figure 8.6 Series-shunt configuration with finite impedances in the basic amplifier.

8.4 Feedback Configurations

Substituting (8.20) in (8.21) gives

$$v_i = v_\epsilon + afv_\epsilon = v_\epsilon(1 + af) \quad (8.22)$$

Also

$$i_i = \frac{v_\epsilon}{z_i} \quad (8.23)$$

Substituting (8.22) in (8.23) gives

$$i_i = \frac{v_i}{z_i} \frac{1}{1 + af} \quad (8.24)$$

Thus, from (8.24), input impedance Z_i with feedback applied is

$$Z_i = \frac{v_i}{i_i} = (1 + T)z_i \quad (8.25)$$

Series feedback at the input *always* raises the input impedance by $(1 + T)$.

The effect of series-shunt feedback on the output impedance can be calculated using the circuit of Fig. 8.7. The input voltage is removed (the input is shorted) and a voltage v applied at the output. From Fig. 8.7

$$v_\epsilon + fv = 0 \quad (8.26)$$

$$i = \frac{v - av_\epsilon}{z_o} \quad (8.27)$$

Substituting (8.26) in (8.27) gives

$$i = \frac{v + afv}{z_o} \quad (8.28)$$

From (8.28) output impedance Z_o with feedback applied is

$$Z_o = \frac{v}{i} = \frac{z_o}{1 + T} \quad (8.29)$$

Shunt feedback at the output *always* lowers the output impedance by $(1 + T)$. This makes the output a better voltage source so that series-shunt feedback produces a *good voltage amplifier*. It stabilizes v_o/v_i, raises Z_i, and lowers Z_o.

The original series-shunt feedback amplifier of Fig. 8.6 can now be represented as shown in Fig. 8.8a using (8.18), (8.25), and (8.29). As the forward gain a approaches infinity, the equivalent circuit approaches that of Fig. 8.8b, which is an ideal voltage amplifier.

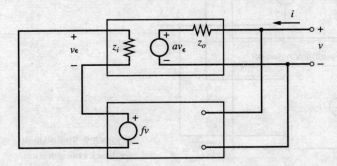

Figure 8.7 Circuit for the calculation of the output impedance of the series-shunt feedback configuration.

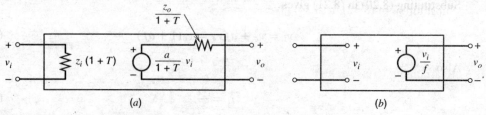

Figure 8.8 (a) Equivalent circuit of a series-shunt feedback amplifier. (b) Equivalent circuit of a series-shunt feedback amplifier for $a \to \infty$.

8.4.2 Shunt-Shunt Feedback

This configuration is shown in Fig. 8.9. The feedback network again shunts the output of the basic amplifier and samples v_o and, ideally, $z_{22f} = \infty$ as before. However, the feedback network now *shunts* the input of the main amplifier as well and feeds back a proportional current fv_o. Ideally, $z_{11f} = \infty$ so that the feedback network does not produce any shunt loading on the amplifier input. Since the feedback signal is a current it is more convenient to deal with an error *current* i_ϵ at the input. The input signal in this case is ideally a current i_i and this is assumed. From Fig. 8.9

$$a = \frac{v_o}{i_\epsilon} \tag{8.30}$$

where a is a *transresistance*,

$$f = \frac{i_{fb}}{v_o} \tag{8.31}$$

where f is a *transconductance*, and

$$v_o = a i_\epsilon \tag{8.32}$$

$$i_\epsilon = i_i - i_{fb} \tag{8.33}$$

Substitution of i_{fb} from (8.31) in (8.33) gives

$$i_\epsilon = i_i - f v_o \tag{8.34}$$

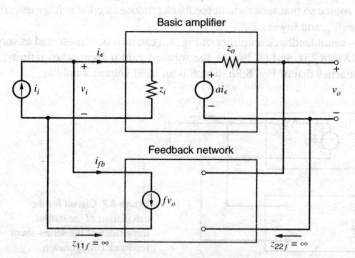

Figure 8.9 Shunt-shunt feedback configuration.

Substitution of (8.32) in (8.34) gives

$$\frac{v_o}{a} = i_i - fv_o$$

Rearranging terms we find

$$\frac{v_o}{i_i} = \frac{a}{1+af} = A \qquad (8.35)$$

Again the ideal feedback equation applies. Note that although a and f have dimensions of resistance and conductance, the loop gain $T = af$ is *dimensionless*. This is always true.

In this configuration, if the source impedance z_s is finite, a division of input current i_i occurs between z_s and the amplifier input, and the ratio v_o/i_i will not be as well defined as (8.35) suggests. The full benefits of negative feedback for a shunt-shunt feedback amplifier are thus obtained for $z_s \gg Z_i$, which approaches a current-source drive.

The input impedance of the circuit of Fig. 8.9 can be calculated using (8.32) and (8.35) to give

$$i_\epsilon = \frac{i_i}{1+af} \qquad (8.36)$$

The input impedance Z_i with feedback is

$$Z_i = \frac{v_i}{i_i} \qquad (8.37)$$

Substituting (8.36) in (8.37) gives

$$Z_i = \frac{v_i}{i_\epsilon} \frac{1}{1+af} = \frac{z_i}{1+T} \qquad (8.38)$$

Thus shunt feedback at the input *reduces* the amplifier input impedance by $(1+T)$. This is always true.

It is easily shown that the output impedance in this case is

$$Z_o = \frac{z_o}{1+T} \qquad (8.39)$$

as before, for shunt feedback at the output.

Shunt-shunt feedback has made this amplifier a good *transresistance* amplifier. The transfer function v_o/i_i has been stabilized and both Z_i and Z_o are lowered.

The original shunt-shunt feedback amplifier of Fig. 8.9 can now be represented as shown in Fig. 8.10a using (8.35), (8.38), and (8.39). As forward gain a approaches infinity, the equivalent circuit approaches that of Fig. 8.10b, which is an ideal transresistance amplifier.

8.4.3 Shunt-Series Feedback

The shunt-series configuration is shown in Fig. 8.11. The feedback network samples i_o and feeds back a proportional current $i_{fb} = fi_o$. Since the desired output signal is a current i_o, it is more convenient to represent the output of the basic amplifier with a Norton equivalent. In this case both a and f are dimensionless current ratios, and the ideal input source is a current

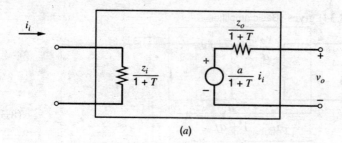

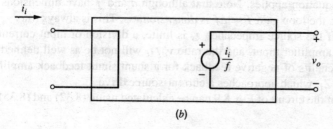

Figure 8.10 (a) Equivalent circuit of a shunt-shunt feedback amplifier. (b) Equivalent circuit of a shunt-shunt feedback amplifier for $a \to \infty$.

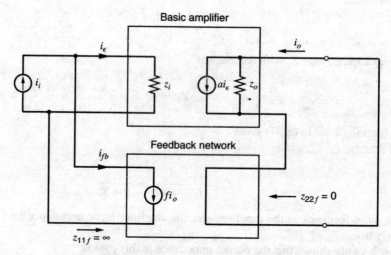

Figure 8.11 Shunt-series feedback configuration.

source i_i. It can be shown that

$$\frac{i_o}{i_i} = \frac{a}{1 + af} \tag{8.40}$$

$$Z_i = \frac{z_i}{1 + T} \tag{8.41}$$

$$Z_o = z_o(1 + T) \tag{8.42}$$

This amplifier is a good *current* amplifier and has stable current gain i_o/i_i, low Z_i, and high Z_o.

8.4.4 Series-Series Feedback

The series-series configuration is shown in Fig. 8.12. The feedback network samples i_o and feeds back a proportional voltage v_{fb} in *series* with the input. The forward gain a is a transconductance and f is a transresistance, and the ideal driving source is a voltage source v_i. It can

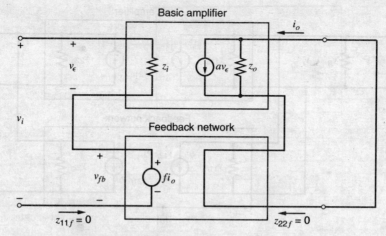

Figure 8.12 Series-series feedback configuration.

be shown that

$$\frac{i_o}{v_i} = \frac{a}{1+af} \tag{8.43}$$

$$Z_i = z_i(1+T) \tag{8.44}$$

$$Z_o = z_o(1+T) \tag{8.45}$$

This amplifier is a good transconductance amplifier and has a stabilized gain i_o/v_i, as well as high Z_i and Z_o.

8.5 Practical Configurations and the Effect of Loading

In practical feedback amplifiers, the feedback network causes loading at the input and output of the basic amplifier, and the division into basic amplifier and feedback network is not as obvious as the above treatment implies. In such cases, the circuit can always be analyzed by writing circuit equations for the whole amplifier and solving for the transfer function and terminal impedances. However, this procedure becomes very tedious and difficult in most practical cases, and the equations so complex that one loses sight of the important aspects of circuit performance. Thus it is profitable to identify a basic amplifier and feedback network in such cases and then to use the ideal feedback equations derived above. In general it will be necessary to include the loading effect of the feedback network on the basic amplifier, and methods of including this loading in the calculations are now considered. The *method* will be developed through the use of two-port representations of the circuits involved, although this method of representation is not necessary for practical calculations, as we will see.

8.5.1 Shunt-Shunt Feedback

Consider the shunt-shunt feedback amplifier of Fig. 8.9. The effect of nonideal networks may be included as shown in Fig. 8.13a, where finite input and output admittances are assumed in both forward and feedback paths, as well as reverse transmission in each. Finite source and load admittances y_S and y_L are assumed. The most convenient two-port representation in this case is the short-circuit admittance parameters or y parameters,[1] as used in Fig. 8.13a. The reason for this is that the basic amplifier and the feedback network are connected in parallel at input and output, and thus have identical *voltages* at their terminals. The y parameters specify the response of a network by expressing the terminal currents in terms of the terminal voltages, and

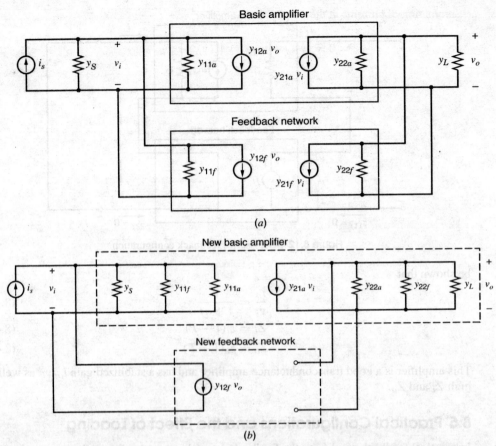

Figure 8.13 (a) Shunt-shunt feedback configuration using the y-parameter representation. (b) Circuit of (a) redrawn with generators $y_{21f}v_i$ and $y_{12a}v_o$ omitted.

this results in very simple calculations when two networks have identical terminal voltages. This will be evident in the circuit calculations to follow. The y-parameter representation is illustrated in Fig. 8.14.

From Fig. 8.13a, at the input

$$i_s = (y_S + y_{11a} + y_{11f})v_i + (y_{12a} + y_{12f})v_o \qquad (8.46)$$

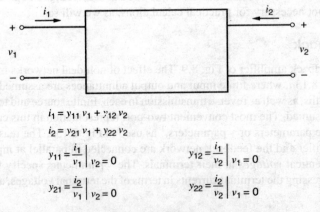

$i_1 = y_{11}v_1 + y_{12}v_2$
$i_2 = y_{21}v_1 + y_{22}v_2$

$y_{11} = \dfrac{i_1}{v_1}\bigg|_{v_2=0} \qquad y_{12} = \dfrac{i_1}{v_2}\bigg|_{v_1=0}$

$y_{21} = \dfrac{i_2}{v_1}\bigg|_{v_2=0} \qquad y_{22} = \dfrac{i_2}{v_2}\bigg|_{v_1=0}$

Figure 8.14 The y-parameter representation of a two-port.

Summation of currents at the output gives

$$0 = (y_{21a} + y_{21f})v_i + (y_L + y_{22a} + y_{22f})v_o \tag{8.47}$$

It is useful to define

$$y_i = y_S + y_{11a} + y_{11f} \tag{8.48}$$
$$y_o = y_L + y_{22a} + y_{22f} \tag{8.49}$$

Solving (8.46) and (8.47) by using (8.48) and (8.49) gives

$$\frac{v_o}{i_s} = \frac{-(y_{21a} + y_{21f})}{y_i y_o - (y_{21a} + y_{21f})(y_{12a} + y_{12f})} \tag{8.50}$$

The equation can be put in the form of the ideal feedback equation of (8.35) by dividing by $y_i y_o$ to give

$$\frac{v_o}{i_s} = \frac{\dfrac{-(y_{21a} + y_{21f})}{y_i y_o}}{1 + \dfrac{-(y_{21a} + y_{21f})}{y_i y_o}(y_{12a} + y_{12f})} \tag{8.51}$$

Comparing (8.51) with (8.35) gives

$$a = -\frac{y_{21a} + y_{21f}}{y_i y_o} \tag{8.52}$$

$$f = y_{12a} + y_{12f} \tag{8.53}$$

At this point, a number of approximations can be made that greatly simplify the calculations. First, we assume that the signal transmitted by the basic amplifier is much greater than the signal fed forward by the feedback network. Since the former has gain (usually large) while the latter has loss, this is almost invariably a valid assumption. This means that

$$|y_{21a}| \gg |y_{21f}| \tag{8.54}$$

Second, we assume that the signal fed back by the feedback network is much greater than the signal fed back through the basic amplifier. Since most active devices have very small reverse transmission, the basic amplifier has a similar characteristic, and this assumption is almost invariably quite accurate. This assumption means that

$$|y_{12a}| \ll |y_{12f}| \tag{8.55}$$

Using (8.54) and (8.55) in (8.51) gives

$$\frac{v_o}{i_s} = A \simeq \frac{\dfrac{-y_{21a}}{y_i y_o}}{1 + \left(\dfrac{-y_{21a}}{y_i y_o}\right) y_{12f}} \tag{8.56}$$

Comparing (8.56) with (8.35) gives

$$a = -\frac{y_{21a}}{y_i y_o} \tag{8.57}$$

$$f = y_{12f} \tag{8.58}$$

A circuit representation of (8.57) and (8.58) can be found as follows. Equations 8.54 and 8.55 mean that in Fig. 8.13a the feedback generator of the basic amplifier and the forward-transmission generator of the feedback network may be neglected. If this is done the circuit may be redrawn as in Fig. 8.13b, where the terminal admittances y_{11f} and y_{22f} of the feedback network have been absorbed into the basic amplifier, together with source and load impedances y_S and y_L. The new basic amplifier thus *includes the loading effect* of the original feedback network, and the new feedback network is an ideal one as used in Fig. 8.9. If the transfer function of the basic amplifier of Fig. 8.13b is calculated (by first removing the feedback network), the result given in (8.57) is obtained. Similarly, the transfer function of the feedback network of Fig. 8.13b is given by (8.58). Thus Fig. 8.13b is a *circuit representation* of (8.57) and (8.58).

Since Fig. 8.13b has a direct correspondence with Fig. 8.9, all the results derived in Section 8.4.2 for Fig. 8.9 can now be used. The loading effect of the feedback network on the basic amplifier is now included by simply shunting input and output with y_{11f} and y_{22f}, respectively. As shown in Fig. 8.14, these terminal admittances of the feedback network are calculated with the other port of the network short-circuited. In practice, loading term y_{11f} is simply obtained by shorting the output node of the amplifier and calculating the feedback circuit input admittance. Similarly, term y_{22f} is calculated by shorting the input node in the amplifier and calculating the feedback circuit output admittance. The feedback transfer function f given by (8.58) is the short-circuit reverse transfer admittance of the feedback network and is defined in Fig. 8.14. This is readily calculated in practice and is often obtained by inspection. Note that the use of y parameters in further calculations is *not* necessary. Once the circuit of Fig. 8.13b is established, any convenient network analysis method may be used to calculate gain a of the basic amplifier. We have simply used the two-port representation as a general means of illustrating how loading effects may be included in the calculations.

For example, consider the common shunt-shunt feedback circuit using an op amp as shown in Fig. 8.15a. The equivalent circuit is shown in Fig. 8.15b and is redrawn in 8.15c to allow for loading of the feedback network on the basic amplifier. The y parameters of the feedback network can be found from Fig. 8.15d.

$$y_{11f} = \left.\frac{i_1}{v_1}\right|_{v_2=0} = \frac{1}{R_F} \tag{8.59}$$

$$y_{22f} = \left.\frac{i_2}{v_2}\right|_{v_1=0} = \frac{1}{R_F} \tag{8.60}$$

$$y_{12f} = \left.\frac{i_1}{v_2}\right|_{v_1=0} = -\frac{1}{R_F} = f \tag{8.61}$$

Using (8.54), we neglect y_{21f}.

The basic-amplifier gain a can be calculated from Fig. 8.15c by putting $i_{fb} = 0$ to give

$$v_1 = \frac{z_i R_F}{z_i + R_F} i_i \tag{8.62}$$

$$v_o = -\frac{R}{R + z_o} a_v v_1 \tag{8.63}$$

where

$$R = R_F \| R_L \tag{8.64}$$

8.5 Practical Configurations and the Effect of Loading

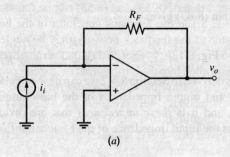

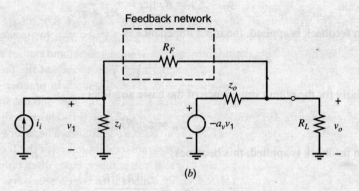

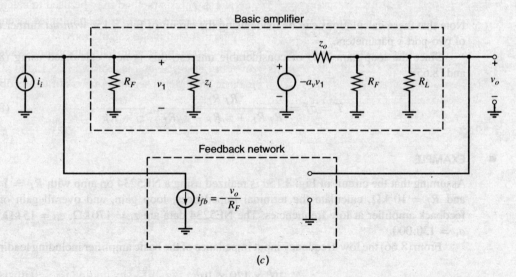

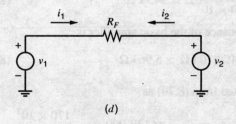

Figure 8.15 (a) Shunt-shunt feedback circuit using an op amp as the gain element. (b) Equivalent circuit of (a) including load resistance R_L. (c) Division of the circuit in (b) into forward and feedback paths. (d) Circuit for the calculation of the y parameters of the feedback network of the circuit in (b).

Substituting (8.62) in (8.63) gives

$$\frac{v_o}{i_i} = a = -\frac{R}{R + z_o} a_v \frac{z_i R_F}{z_i + R_F} \qquad (8.65)$$

Using the formulas derived in Section 8.4.2 we can now calculate all parameters of the feedback circuit. The input and output impedances of the basic amplifier now *include* the effect of feedback loading, and it is *these impedances* that are divided by $(1 + T)$ as described in Section 8.4.2. Thus the input impedance of the basic amplifier of Fig. 8.15c is

$$z_{ia} = R_F || z_i = \frac{R_F z_i}{R_F + z_i} \qquad (8.66)$$

When feedback is applied, the input impedance is

$$Z_i = \frac{z_{ia}}{1 + T} \qquad (8.67)$$

Similarly for the output impedance of the basic amplifier

$$z_{oa} = z_o || R_F || R_L \qquad (8.68)$$

When feedback is applied, this becomes

$$Z_o = \frac{z_o || R_F || R_L}{1 + T} \qquad (8.69)$$

Note that these calculations can be made using the circuit of Fig. 8.15c *without* further need of two-port y parameters.

Since the loop gain T is of considerable interest, this is now calculated using (8.61) and (8.65):

$$T = af = \frac{R_F R_L}{R_F R_L + z_o R_F + z_o R_L} a_v \frac{z_i}{z_i + R_F} \qquad (8.70)$$

■ **EXAMPLE**

Assuming that the circuit of Fig. 8.15a is realized using a NE5234 op amp with $R_F = 1$ MΩ and $R_L = 10$ kΩ, calculate the terminal impedances, loop gain, and overall gain of the feedback amplifier at low frequencies. The NE5234 data are $z_i = 170$ kΩ, $z_o = 15$ kΩ, and $a_v = 130{,}000$.

From (8.66) the low-frequency input impedance of the basic amplifier including loading is

$$z_{ia} = \frac{10^6 \times 170 \times 10^3}{10^6 + 170 \times 10^3} \Omega = 145 \text{ k}\Omega \qquad (8.71)$$

From (8.68), the low-frequency output impedance of the basic amplifier is

$$z_{oa} = 15 \text{ k}\Omega || 1 \text{ M}\Omega || 10 \text{ k}\Omega \simeq 5.96 \text{ k}\Omega \qquad (8.72)$$

The low-frequency loop gain can be calculated from (8.70) as

$$T = \frac{10^6 \times 10^4}{10^6 \times 10^4 + 15 \times 10^3 \times 10^6 + 15 \times 10^3 \times 10^4} \times 130{,}000 \times \frac{170 \times 10^3}{170 \times 10^3 + 10^6}$$
$$= 7510 \qquad (8.73)$$

The loop gain in this case is large. Note that a finite source resistance at the input could reduce this significantly.

The input impedance with feedback applied is found by substituting (8.71) and (8.73) in (8.67) to give

$$Z_i = \frac{145 \times 10^3}{7511} \, \Omega = 19.3 \, \Omega$$

The output impedance with feedback applied is found by substituting (8.72) and (8.73) in (8.69) to give

$$Z_o = \frac{5.96 \times 10^3}{7511} \, \Omega = 0.794 \, \Omega$$

In practice, second-order effects in the circuit may result in a larger value of Z_o.

The overall transfer function with feedback can be found approximately from (8.8) as

$$\frac{v_o}{i_i} = A \simeq \frac{1}{f} \qquad (8.74)$$

Using (8.61) in (8.74) gives

$$\frac{v_o}{i_i} = A \simeq -R_F$$

Substituting for R_F we obtain

$$\frac{v_o}{i_i} = A \simeq -1 \, \mathrm{M}\Omega \qquad (8.75)$$

A more exact value of A can be calculated from (8.5). Since the loop gain is large in this case, it is useful to transform (8.5) as follows:

$$A = \frac{1}{f} \frac{1}{1 + \dfrac{1}{af}} \qquad (8.76)$$

$$= \frac{1}{f} \frac{1}{1 + \dfrac{1}{T}} \qquad (8.77)$$

Since T is high in this example, A differs little from $1/f$. Substituting $T = 7510$ and $1/f = -1 \, \mathrm{M}\Omega$ in (8.77), we obtain

$$A = -999{,}867 \, \Omega \qquad (8.78)$$

■ For most practical purposes, (8.75) is sufficiently accurate.

8.5.2 Series-Series Feedback

Consider the series-series feedback connection of Fig. 8.12. The effect of nonideal networks can be calculated using the representation of Fig. 8.16a. In this case the most convenient two-port representation is the use of the open-circuit impedance parameters or z parameters because the basic amplifier and the feedback network are now connected in series at input and output and thus have identical *currents* at their terminals. As shown in Fig. 8.17, the z

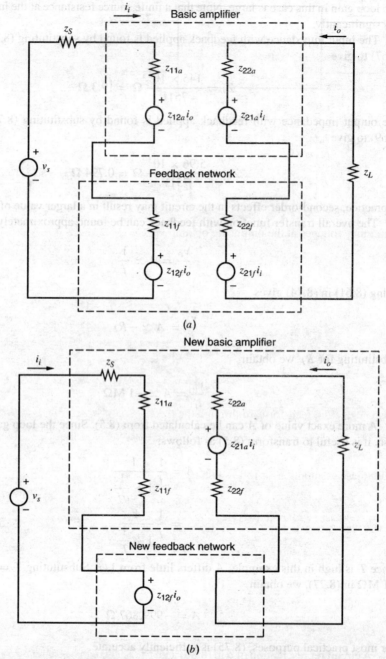

Figure 8.16 (a) Series-series feedback configuration using the z-parameter representation. (b) Circuit of (a) redrawn with generators $z_{21f}i_i$ and $z_{12a}i_o$ omitted.

parameters specify the network by expressing terminal voltages in terms of terminal currents, and this results in simple calculations when the two networks have common terminal currents. The calculation in this case proceeds as the exact dual of that in Section 8.5.1. From Fig. 8.16, summation of voltages at the input gives

$$v_s = (z_S + z_{11a} + z_{11f})i_i + (z_{12a} + z_{12f})i_o \qquad (8.79)$$

8.5 Practical Configurations and the Effect of Loading

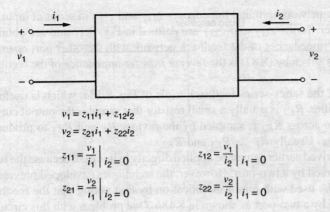

$$v_1 = z_{11}i_1 + z_{12}i_2$$
$$v_2 = z_{21}i_1 + z_{22}i_2$$

$$z_{11} = \left.\frac{v_1}{i_1}\right|_{i_2=0} \qquad z_{12} = \left.\frac{v_1}{i_2}\right|_{i_1=0}$$

$$z_{21} = \left.\frac{v_2}{i_1}\right|_{i_2=0} \qquad z_{22} = \left.\frac{v_2}{i_2}\right|_{i_1=0}$$

Figure 8.17 The z-parameter representation of a two-port.

Summing voltages at the output we obtain

$$0 = (z_{21a} + z_{21f})i_i + (z_L + z_{22a} + z_{22f})i_o \tag{8.80}$$

It is useful to define

$$z_i = z_S + z_{11a} + z_{11f} \tag{8.81}$$
$$z_o = z_L + z_{22a} + z_{22f} \tag{8.82}$$

Again neglecting reverse transmission through the basic amplifier, we assume that

$$|z_{12a}| \ll |z_{12f}| \tag{8.83}$$

Also neglecting feed-forward through the feedback network, we can write

$$|z_{21a}| \gg |z_{21f}| \tag{8.84}$$

With these assumptions it follows that

$$\frac{i_o}{v_s} = A \simeq \frac{\dfrac{-z_{21a}}{z_i z_o}}{1 + \left(\dfrac{-z_{21a}}{z_i z_o}\right) z_{12f}} = \frac{a}{1 + af} \tag{8.85}$$

where

$$a = -\frac{z_{21a}}{z_i z_o} \tag{8.86}$$

$$f = z_{12f} \tag{8.87}$$

A circuit representation of a in (8.86) and f in (8.87) can be found by removing generators $z_{21f}i_i$ and $z_{12a}i_o$ from Fig. 8.16a in accord with (8.83) and (8.84). This gives the approximate representation of Fig. 8.16b, where the new basic amplifier includes the loading effect of the original feedback network. The new feedback network is an ideal one as used in Fig. 8.12. The transfer function of the basic amplifier of Fig. 8.16b is the same as in (8.86), and the transfer function of the feedback network of Fig. 8.16b is given by (8.87). Thus Fig. 8.16b is a circuit representation of (8.86) and (8.87).

Since Fig. 8.16b has a direct correspondence with Fig. 8.12, all the results of Section 8.4.4 can now be used. The loading effect of the feedback network on the basic amplifier is included by connecting the feedback-network terminal impedances z_{11f} and z_{22f} in series at input and output of the basic amplifier. Terms z_{11f} and z_{22f} are defined in Fig. 8.17 and are obtained by calculating the terminal impedances of the feedback network with the other port *open circuited*. Feedback function f given by (8.87) is the *reverse transfer impedance* of the feedback network.

Consider, for example, the series-series feedback triple of Fig. 8.18*a*, which is useful as a wideband feedback amplifier. R_{E2} is usually a small resistor that samples the output current i_o, and the resulting voltage across R_{E2} is sampled by the divider R_F and R_{E1} to produce a feedback voltage across R_{E1}. Usually $R_F \gg R_{E1}$ and R_{E2}.

The two-port theory derived earlier cannot be applied directly in this case because the basic amplifier cannot be represented by a two-port. However, the techniques developed previously using two-port theory can be used with minor modification by first noting that the feedback network can be represented by a two-port as shown in 8.18*b*. One problem with this circuit is that the feedback generator $z_{12f}i_{e3}$ is in the emitter of Q_1 and not in the input lead where it can be compared directly with v_s. This problem can be overcome by considering the small-signal equivalent of the input portion of this circuit as shown in Fig. 8.19. For this circuit

$$v_s = i_i z_S + v_{be} + i_{e1} z_{11f} + z_{12f} i_{e3} \tag{8.88}$$

Using

$$i_{e3} = \frac{i_o}{\alpha_3} \tag{8.89}$$

in (8.88) gives

$$v_s - z_{12f} \frac{i_o}{\alpha_3} = i_i z_S + v_{be} + i_{e1} z_{11f} \tag{8.90}$$

where the quantities in these equations are *small-signal* quantities. Equation 8.90 shows that the feedback voltage generator $z_{12f}(i_o/\alpha_3)$ can be *moved back* in series with the input lead; if this was done, exactly the same equation would result. (See Fig. 8.18*c*.) Note that the common-base current gain α_3 of Q_3 appears in this feedback expression because the output current is sampled by R_{E2} in the *emitter* of Q_3 in order to feed back a correcting signal to the input. This problem is common to most circuits employing series feedback at the output, and the α_3 of Q_3 is *outside* the feedback loop. There are many applications where this is not a problem, since $\alpha \simeq 1$. However, if high gain precision is required, variations in α_3 can cause difficulties.

The z parameters of the feedback network can be determined from Fig. 8.20 as

$$z_{12f} = \left.\frac{v_1}{i_2}\right|_{i_1=0} = \frac{R_{E1} R_{E2}}{R_{E1} + R_{E2} + R_F} \tag{8.91}$$

$$z_{22f} = \left.\frac{v_2}{i_2}\right|_{i_1=0} = R_{E2} \| (R_{E1} + R_F) \tag{8.92}$$

$$z_{11f} = \left.\frac{v_1}{i_1}\right|_{i_2=0} = R_{E1} \| (R_F + R_{E2}) \tag{8.93}$$

Using (8.84), we neglect z_{21f}.

From the foregoing results we can redraw the circuit of Fig. 8.18*b* as shown in Fig. 8.18*c*. As in previous calculations, the signal fed forward via the feedback network (in this case $z_{21f} i_{e1}$) is neglected. The feedback voltage generator is placed in series with the input lead

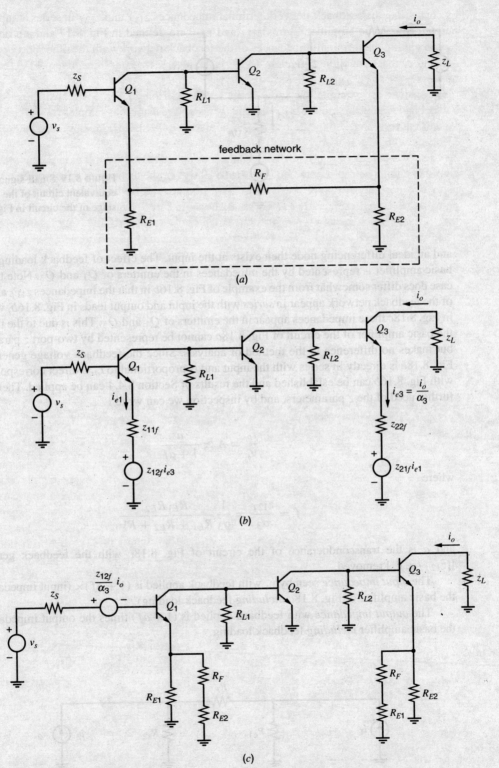

Figure 8.18 (a) Series-series feedback triple. (b) Circuit of (a) redrawn using a two-port z-parameter representation of the feedback network. (c) Approximate representation of the circuit in (b).

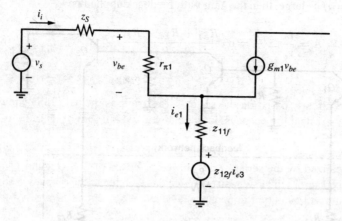

Figure 8.19 Small-signal equivalent circuit of the input stage of the circuit in Fig. 8.18b.

and an ideal differencing node then exists at the input. The effect of feedback loading on the basic amplifier is represented by the impedances in the emitters of Q_1 and Q_3. Note that this case does differ somewhat from the example of Fig. 8.16b in that the impedances z_{11f} and z_{22f} of the feedback network appear in *series* with the input and output leads in Fig. 8.16b, whereas in Fig. 8.18c these impedances appear in the emitters of Q_1 and Q_3. This is due to the fact that the basic amplifier of the circuit of Fig. 8.18a cannot be represented by two-port z parameters but makes no difference to the method of analysis. Since the feedback voltage generator in Fig. 8.18c is directly in series with the input and is proportional to i_o, a direct correspondence with Fig. 8.16b can be established and the results of Section 8.4.4 can be applied. There is no further need of the z parameters, and by inspection we can write

$$\frac{i_o}{v_s} = A \simeq \frac{a}{1 + af} \tag{8.94}$$

where

$$f = \frac{z_{12f}}{\alpha_3} = \frac{1}{\alpha_3} \frac{R_{E1} R_{E2}}{R_{E1} + R_{E2} + R_F} \tag{8.95}$$

and a is the transconductance of the circuit of Fig. 8.18c with the feedback generator $[(z_{12f}/\alpha_3)i_o]$ removed.

The *input impedance* seen by v_s with feedback applied is $(1 + af) \times$ (input impedance of the basic amplifier of Fig. 8.18c *including* feedback loading).

The *output impedance* with feedback applied is $(1 + af)$ times the output impedance of the basic amplifier *including* feedback loading.

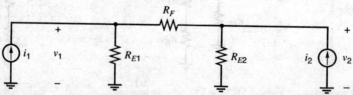

Figure 8.20 Circuit for the calculation of the z parameters of the feedback network of the circuit in Fig 8.18a.

If the loop gain $T = af$ is large, then the gain with feedback applied is

$$A = \frac{i_o}{v_s} \simeq \frac{1}{f} = \alpha_3 \frac{R_{E1} + R_{E2} + R_F}{R_{E1} R_{E2}} \qquad (8.96)$$

■ **EXAMPLE**

A commercial integrated circuit[2] based on the series-series triple is the MC 1553 shown in Fig. 8.21a. Calculate the terminal impedances, loop gain, and overall gain of this amplifier at low frequencies.

The MC 1553 is a wideband amplifier with a bandwidth of 50 MHz at a voltage gain of 50. The circuit gain is realized by the series-series triple composed of Q_1, Q_2, and Q_3. The output voltage is developed across the load resistor R_C and is then fed to the output via emitter follower Q_4, which ensures a low output impedance. The rest of the circuit is largely for bias

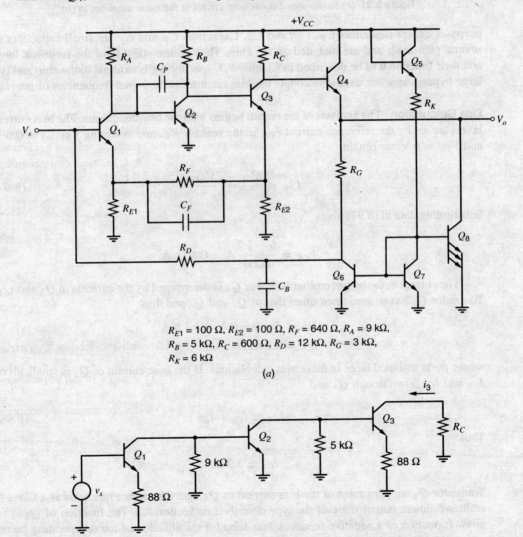

$R_{E1} = 100\,\Omega$, $R_{E2} = 100\,\Omega$, $R_F = 640\,\Omega$, $R_A = 9\,\text{k}\Omega$,
$R_B = 5\,\text{k}\Omega$, $R_C = 600\,\Omega$, $R_D = 12\,\text{k}\Omega$, $R_G = 3\,\text{k}\Omega$,
$R_K = 6\,\text{k}\Omega$

(a)

(b)

Figure 8.21 (a) Circuit of the MC 1553 wideband integrated circuit. (b) Basic amplifier of the series-series triple in (a).

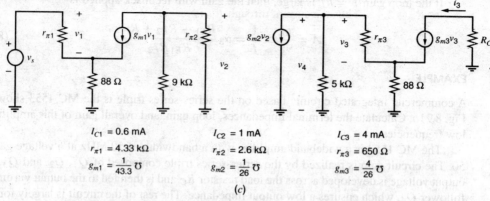

$I_{C1} = 0.6$ mA $I_{C2} = 1$ mA $I_{C3} = 4$ mA
$r_{\pi 1} = 4.33$ kΩ $r_{\pi 2} = 2.6$ kΩ $r_{\pi 3} = 650$ Ω
$g_{m1} = \frac{1}{43.3}$ ℧ $g_{m2} = \frac{1}{26}$ ℧ $g_{m3} = \frac{4}{26}$ ℧

(c)

Figure 8.21 (c) Small-signal equivalent circuit of the basic amplifier in (b).

purposes except capacitors C_P, C_F, and C_B. Capacitors C_P and C_F are small capacitors of several picofarads and are included on the chip. They ensure stability of the feedback loop, and their function will be described in Chapter 9. Capacitor C_B is external to the chip and is a large bypass capacitor used to decouple the bias circuitry at the signal frequencies of interest.

Bias Calculation. The analysis of the circuit begins with the bias conditions. The bias current levels are set by the reference current I_{RK} in the resistor R_K, and assuming $V_{BE(on)} = 0.6$ V and $V_{CC} = 6$ V, we obtain

$$I_{RK} = \frac{V_{CC} - 2V_{BE(on)}}{R_K} \qquad (8.97)$$

Substituting data in (8.97) gives

$$I_{RK} = \frac{6 - 1.2}{6000} \text{A} = 0.80 \text{ mA}$$

The current in the output emitter follower Q_4 is determined by the currents in Q_6 and Q_8. Transistor Q_8 has an area three times that of Q_7 and Q_6 and thus

$$I_{C8} = 3 \times 0.8 \text{ mA} = 2.4 \text{ mA}$$
$$I_{C6} = 0.8 \text{ mA}$$

where β_F is assumed large in these bias calculations. If the base current of Q_1 is small, all of I_{C6} and I_{C8} flow through Q_4 and

$$I_{C4} \simeq I_{C6} + I_{C8} \qquad (8.98)$$

Thus

$$I_{C4} \simeq 3.2 \text{ mA}$$

Transistor Q_8 supplies most of the bias current to Q_4, and this device functions as a Class A emitter-follower output stage of the type described in Section 5.2. The function of Q_6 is to allow formation of a negative-feedback bias loop for stabilization of the dc operating point, and resistor R_G is chosen to cause sufficient dc voltage drop to allow connection of R_D back to the base of Q_1. Transistors Q_1, Q_2, Q_3, and Q_4 are then connected in a negative-feedback bias loop and the dc conditions can be ascertained approximately as follows.

If we assume that Q_2 is on, the voltage at the collector of Q_1 is about 0.6 V and the voltage across R_A is 5.4 V. Thus the current through R_A is

$$I_{RA} = \frac{5.4}{R_A}$$

$$= \frac{5.4}{9000} = 0.6 \text{ mA} \tag{8.99}$$

If β_F is assumed high, it follows that

$$I_{C1} \simeq I_{RA} = 0.6 \text{ mA} \tag{8.100}$$

Since the voltage across R_{E1} is small, the voltage at the base of Q_1 is approximately 0.6 V, and if the base current of Q_1 is small, this is also the voltage at the collector of Q_6 since any voltage across R_D will be small. The dc output voltage can be written

$$V_O = V_{C6} + I_{C6}R_G \tag{8.101}$$

Substitution of data gives

$$V_O = (0.6 + 0.8 \times 3) \text{ V} = 3 \text{ V}$$

The voltage at the base of Q_4 (collector of Q_3) is V_{BE} above V_O and is thus 3.6 V. The collector current of Q_3 is

$$I_{C3} \simeq \frac{V_{CC} - V_{C3}}{R_C} \tag{8.102}$$

Substitution of parameter values gives

$$I_{C3} \simeq \frac{6 - 3.6}{600} \text{ A} = 4 \text{ mA}$$

The voltage at the base of Q_3 (collector of Q_2) is

$$V_{B3} \simeq -I_{E3}R_{E2} + V_{BE(\text{on})} \tag{8.103}$$

Thus

$$V_{B3} = V_{C2} \simeq (4 \times 0.1 + 0.6) \text{ V} = 1 \text{ V}$$

I_{C2} may be calculated from

$$I_{C2} \simeq \frac{V_{CC} - V_{C2}}{R_B} \tag{8.104}$$

and substitution of parameter values gives

$$I_{C2} \simeq \frac{6 - 1}{5000} \text{ A} = 1 \text{ mA}$$

The ac Calculation. The ac analysis can now proceed using the methods previously developed in this chapter. For purposes of ac analysis, the feedback triple composed of Q_1, Q_2, and Q_3 in Fig. 8.21a is identical to the circuit in Fig. 8.18a, and the results derived previously for the latter circuit are directly applicable to the triple in Fig. 8.21a. To obtain the voltage gain of the circuit of Fig. 8.21a, we simply multiply the transconductance of the triple by the load resistor R_C, since the gain of the emitter follower Q_4 is almost exactly unity. Note that resistor R_D is assumed grounded for ac signals by the large capacitor C_B, and thus has no influence on the ac circuit operation, except for a shunting effect at the input that will be discussed later. From (8.95) the feedback factor f of the series-series triple of Fig. 8.21a is

$$f = \frac{1}{0.99} \frac{100 \times 100}{100 + 100 + 640} \, \Omega = 12.0 \, \Omega \tag{8.105}$$

where $\beta_0 = 100$ has been assumed.

If the loop gain is large, the transconductance of the triple of Fig. 8.21a can be calculated from (8.96) as

$$\frac{i_{o3}}{v_s} \simeq \frac{1}{f} = \frac{1}{12} \text{ A/V} \tag{8.106}$$

where i_{o3} is the small-signal collector current in Q_3 in Fig. 8.21a. If the input impedance of the emitter follower Q_4 is large, the load resistance seen by Q_3 is $R_C = 600 \, \Omega$ and the voltage gain of the circuit is

$$\frac{v_o}{v_s} = -\frac{i_{o3}}{v_s} \times R_C \tag{8.107}$$

Substituting (8.106) in (8.107) gives

$$\frac{v_o}{v_s} = -50.0 \tag{8.108}$$

Consider now the loop gain of the circuit of Fig. 8.21a. This can be calculated by using the basic-amplifier representation of Fig. 8.18c to calculate the forward gain a. Fig. 8.18c is redrawn in Fig. 8.21b using data from this example, assuming that $z_S = 0$ and omitting the feedback generator. The small-signal, low-frequency equivalent circuit is shown in Fig. 8.21c assuming $\beta_0 = 100$, and it is a straightforward calculation to show that the gain of the basic amplifier is

$$a = \frac{i_3}{v_s} = 20.3 \text{ A/V} \tag{8.109}$$

Combination of (8.105) and (8.109) gives

$$T = af = 12 \times 20.3 = 243.6 \tag{8.110}$$

The transconductance of the triple can now be calculated more accurately from (8.94) as

$$\frac{i_{o3}}{v_s} = \frac{a}{1+T} = \frac{20.3}{244.6} \text{ A/V} = 0.083 \text{ A/V} \tag{8.111}$$

Substitution of (8.111) in (8.107) gives for the overall voltage gain

$$\frac{v_o}{v_s} = -\frac{i_{o3}}{v_s} R_C = -0.083 \times 600 = -49.8 \tag{8.112}$$

This is close to the approximate value given by (8.108).

The input resistance of the basic amplifier is readily determined from Fig. 8.21c to be

$$r_{ia} = 13.2 \text{ k}\Omega \tag{8.113}$$

The input resistance when feedback is applied is

$$R_i = r_{ia}(1+T) \tag{8.114}$$

8.5 Practical Configurations and the Effect of Loading

Substituting (8.113) and (8.110) in (8.114) gives

$$R_i = 13.2 \times 244.6 \text{ k}\Omega = 3.23 \text{ M}\Omega \tag{8.115}$$

As expected, series feedback at the input results in a high input resistance. In this example, however, the bias resistor R_D directly shunts the input for ac signals and is *outside* the feedback loop. Since $R_D = 12$ kΩ and is much less than R_i, the resistor R_D determines the input resistance for this circuit.

Finally, the output resistance of the circuit is of some interest. The output resistance of the triple can be calculated from Fig. 8.21c by including output resistance r_o in the model for Q_3. The resistance obtained is then multiplied by $(1 + T)$ and the resulting value is much greater than the collector load resistor of Q_3, which is $R_C = 600$ Ω. The output resistance of the full circuit is thus essentially the output resistance of emitter follower Q_4 fed from a 600-Ω source resistance, and this is

$$R_o = \frac{1}{g_{m4}} + \frac{R_C}{\beta_4} = \left(\frac{26}{3.2} + \frac{600}{100}\right) \Omega = 14 \ \Omega \tag{8.116}$$

8.5.3 Series-Shunt Feedback

Series-shunt feedback is shown schematically in Fig. 8.4. The basic amplifier and the feedback network have the same input current and the same output voltage. A two-port representation that uses input current and output voltage as the independent variables is the hybrid h-parameter representation shown in Fig. 8.22. The h parameters can be used to represent nonideal circuits in a series-shunt feedback as shown in Fig. 8.23a. Summation of voltages at the input of this figure gives

$$v_s = (z_S + h_{11a} + h_{11f})i_i + (h_{12a} + h_{12f})v_o \tag{8.117}$$

Summing currents at the output yields

$$0 = (h_{21a} + h_{21f})i_i + (y_L + h_{22a} + h_{22f})v_o \tag{8.118}$$

We now define

$$z_i = z_S + h_{11a} + h_{11f} \tag{8.119}$$

$$y_o = y_L + h_{22a} + h_{22f} \tag{8.120}$$

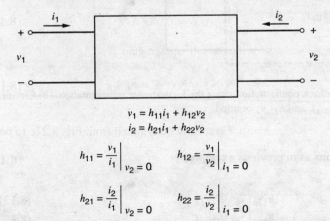

Figure 8.22 The h-parameter representation of a two-port.

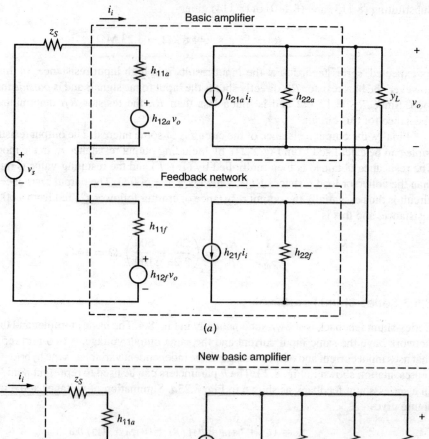

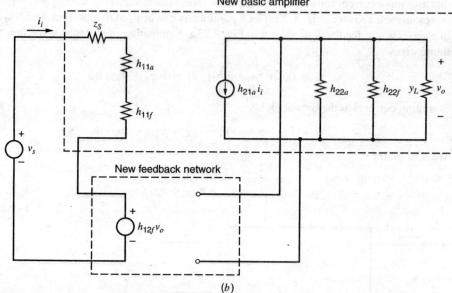

Figure 8.23 (a) Series-shunt feedback configuration using the h-parameter representation. (b) Circuit of (a) redrawn with generators $h_{21f}i_i$ and $h_{12a}v_o$ omitted.

and make the same assumptions as in previous examples:

$$|h_{12a}| \ll |h_{12f}| \qquad (8.121)$$
$$|h_{21a}| \gg |h_{21f}| \qquad (8.122)$$

8.5 Practical Configurations and the Effect of Loading

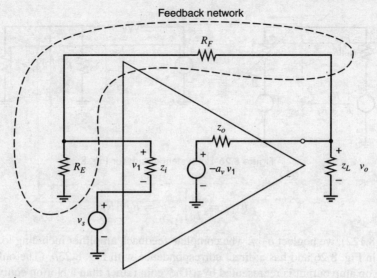

Figure 8.24 Series-shunt feedback circuit using an op amp as the gain element.

It can then be shown that

$$\frac{v_o}{v_s} = A \simeq \frac{-\dfrac{h_{21a}}{z_i y_o}}{1 + \left(-\dfrac{h_{21a}}{z_i y_o}\right) h_{12f}} = \frac{a}{1 + af} \qquad (8.123)$$

where

$$a = -\frac{h_{21a}}{z_i y_o} \qquad (8.124)$$

$$f = h_{12f} \qquad (8.125)$$

A circuit representation of a in (8.124) and f in (8.125) can be found by removing the generators $h_{12a}v_o$ and $h_{21f}i_i$ from Fig. 8.23a as suggested by the approximations of (8.121) and (8.122). This gives the approximate representation of Fig. 8.23b, where the new basic amplifier includes the loading effect of the original feedback network. As in previous examples, the circuit of Fig. 8.23b is a circuit representation of (8.123), (8.124), and (8.125) and has the form of an ideal feedback loop. Thus all the equations of Section 8.4.1 can be applied to the circuit.

For example, consider the common series-shunt op amp circuit of Fig. 8.24, which fits exactly the model described above. We first determine the h parameters of the feedback network from Fig. 8.25:

$$h_{22f} = \left.\frac{i_2}{v_2}\right|_{i_1=0} = \frac{1}{R_F + R_E} \qquad (8.126)$$

$$h_{12f} = \left.\frac{v_1}{v_2}\right|_{i_1=0} = \frac{R_E}{R_E + R_F} \qquad (8.127)$$

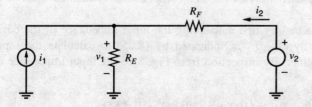

Figure 8.25 Circuit for the determination of the h parameters of the feedback network in Fig. 8.24.

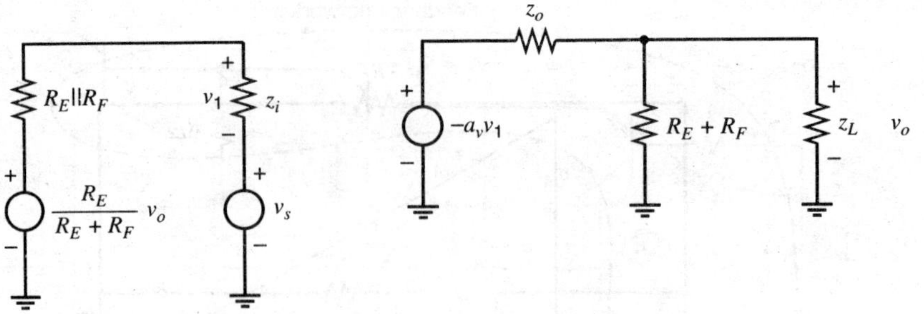

Figure 8.26 Equivalent circuit for Fig. 8.24.

$$h_{11f} = \left.\frac{v_1}{i_1}\right|_{v_2=0} = R_E\|R_F \qquad (8.128)$$

Using (8.122), we neglect h_{21f}. The complete feedback amplifier including loading effects is shown in Fig. 8.26 and has a direct correspondence with Fig. 8.23b. (The only difference is that the op amp output is represented by a Thévenin rather than a Norton equivalent.)

The gain a of the basic amplifier can be calculated from Fig. 8.26 by initially disregarding the feedback generator to give

$$a = \frac{z_i}{z_i + R} a_v \frac{z_{LX}}{z_{LX} + z_o} \qquad (8.129)$$

where

$$R = R_E\|R_F \qquad (8.130)$$
$$z_{LX} = z_L\|(R_E + R_F) \qquad (8.131)$$

Also

$$f = \frac{R_E}{R_E + R_F} \qquad (8.132)$$

Thus the overall gain of the feedback circuit is

$$A = \frac{v_o}{v_s} = \frac{a}{1 + af} \qquad (8.133)$$

and A can be evaluated using (8.129) and (8.132).

■ **EXAMPLE**

Assume that the circuit of Fig. 8.24 is realized, using a differential amplifier with low-frequency parameters $z_i = 100$ kΩ, $z_o = 10$ kΩ, and $a_v = 3000$. Calculate the input impedance of the feedback amplifier at low frequencies if $R_E = 5$ kΩ, $R_F = 20$ kΩ, and $z_L = 10$ kΩ. Note that z_o in this case is not small as is usually the case for an op amp, and this situation can arise in some applications.

This problem is best approached by first calculating the input impedance of the basic amplifier and then multiplying by $(1 + T)$ as indicated by (8.25) to calculate the input impedance of the feedback amplifier. By inspection from Fig. 8.26 the input impedance of the basic amplifier is

$$z_{ia} = z_i + R_E\|R_F = (100 + 5\|20) \text{ k}\Omega = 104 \text{ k}\Omega$$

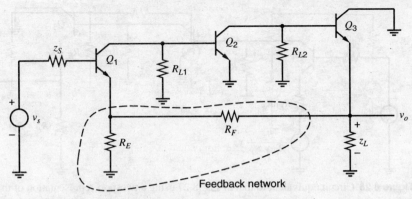

Figure 8.27 Series-shunt feedback circuit.

The parallel combination of z_L and $(R_F + R_E)$ in Fig. 8.27 is

$$z_{LX} = \frac{10 \times 25}{35} = 7.14 \text{ k}\Omega$$

Substitution in (8.129) gives, for the gain of the basic amplifier of Fig. 8.26,

$$a = \frac{100}{100 + 4} \times 3000 \times \frac{7.14}{7.14 + 10} = 1202$$

From (8.127) the feedback factor f for this circuit is

$$f = \frac{5}{5 + 20} = 0.2$$

and thus the loop gain is

$$T = af = 1202 \times 0.2 = 240$$

The input impedance of the feedback amplifier is thus

$$Z_i = z_{ia}(1 + T) = 104 \times 241 \text{ k}\Omega = 25 \text{ M}\Omega$$

In this example the loading effect produced by $(R_F + R_E)$ on the output has a significant effect on the gain a of the basic amplifier and thus on the input impedance of the feedback amplifier.

As another example of a series-shunt feedback circuit, consider the series-series triple of Fig. 8.18a but assume that the output signal is taken as the voltage at the emitter of Q_3, as shown in Fig. 8.27. This is an example of how the same circuit can realize two different feedback functions if the output is taken from different nodes. As in the case of Fig. 8.18a, the basic amplifier of Fig. 8.27 cannot be represented as a two-port. However, the feedback network *can* be represented as a two-port and the appropriate parameters are the h parameters as shown in Fig. 8.28. The h parameters for this feedback network are given by (8.126), (8.127), and (8.128) and h_{21f} is neglected. The analysis of Fig. 8.28 then proceeds in the usual manner.

8.5.4 Shunt-Series Feedback

Shunt-series feedback is shown schematically in Fig. 8.11. In this case the basic amplifier and the feedback network have common input voltages and output currents, and hybrid g parameters as defined in Fig. 8.29 are best suited for use in this case. The feedback circuit is

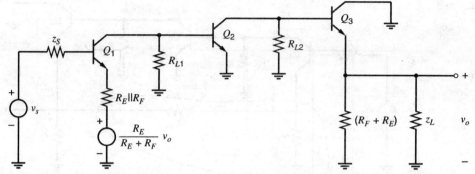

Figure 8.28 Circuit equivalent to that of Fig. 8.27 using a two-port representation of the feedback network.

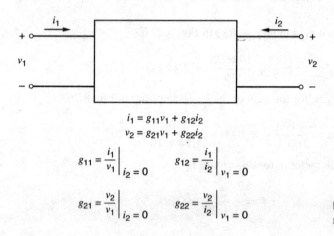

Figure 8.29 The g-parameter representation of a two-port.

shown in Fig. 8.30a, and, at the input, we find that

$$i_s = (y_S + g_{11a} + g_{11f})v_i + (g_{12a} + g_{12f})i_o \qquad (8.134)$$

At the output

$$0 = (g_{21a} + g_{21f})v_i + (z_L + g_{22a} + g_{22f})i_o \qquad (8.135)$$

Defining

$$y_i = y_S + g_{11a} + g_{11f} \qquad (8.136)$$
$$z_o = z_L + g_{22a} + g_{22f} \qquad (8.137)$$

and making the assumptions

$$|g_{12a}| \ll |g_{12f}| \qquad (8.138)$$
$$|g_{21a}| \gg |g_{21f}| \qquad (8.139)$$

we find

$$\frac{i_o}{i_s} = A \simeq \frac{-\dfrac{g_{21a}}{y_i z_o}}{1 + \left(-\dfrac{g_{21a}}{y_i z_o}\right) g_{12f}} = \frac{a}{1 + af} \qquad (8.140)$$

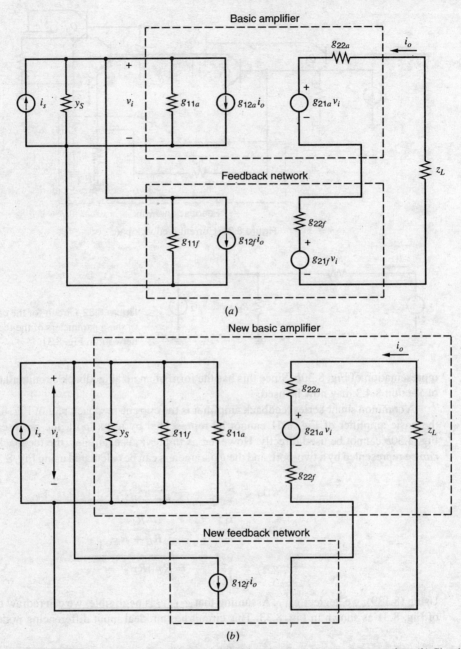

Figure 8.30 (a) Shunt-series feedback configuration using the g-parameter representation. (b) Circuit of (a) redrawn with generators $g_{21f}v_i$ and $g_{12a}i_o$ omitted.

where

$$a = -\frac{g_{21a}}{y_i z_o} \qquad (8.141)$$

$$f = g_{12f} \qquad (8.142)$$

Following the procedure for the previous examples, we can find a circuit representation for this case by eliminating the generators $g_{21f}v_i$ and $g_{12a}i_o$ to obtain the approximate

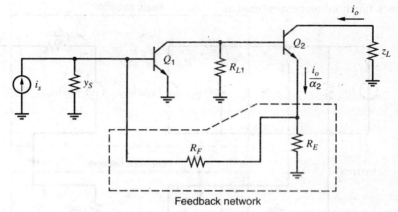

Figure 8.31 Current feedback pair.

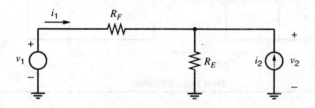

Figure 8.32 Circuit for the calculation of the g parameters of the feedback network in Fig. 8.31.

representation of Fig. 8.30b. Since this has the form of an ideal feedback circuit, all the results of Section 8.4.3 may now be used.

A common shunt-series feedback amplifier is the current-feedback pair of Fig. 8.31. Since the basic amplifier of Fig. 8.31 cannot be represented by a two-port, the representation of Fig. 8.30b cannot be used directly. However, as in previous examples, the feedback network *can* be represented by a two-port, and the g parameters can be calculated using Fig. 8.32 to give

$$g_{11f} = \left.\frac{i_1}{v_1}\right|_{i_2=0} = \frac{1}{R_F + R_E} \tag{8.143}$$

$$g_{12f} = \left.\frac{i_1}{i_2}\right|_{v_1=0} = -\frac{R_E}{R_E + R_F} \tag{8.144}$$

$$g_{22f} = \left.\frac{v_2}{i_2}\right|_{v_1=0} = R_E \| R_F \tag{8.145}$$

Using (8.139), we neglect g_{21f}. Assuming that $g_{21f} v_i$ is negligible, we can redraw the circuit of Fig. 8.31 as shown in Fig. 8.33. This circuit has an ideal input differencing node, and the

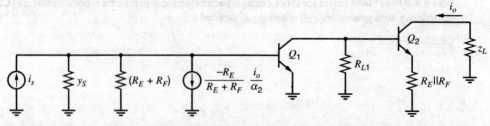

Figure 8.33 Circuit equivalent to that of Fig. 8.31 using a two-port representation of the feedback network.

feedback function can be identified as

$$f = -\frac{R_E}{R_E + R_F}\frac{1}{\alpha_2} \qquad (8.146)$$

The gain a of the basic amplifier is determined by calculating the current gain of the circuit of Fig. 8.33 with the feedback generator removed. The overall current gain with feedback applied can then be calculated from (8.40).

8.5.5 Summary

The results derived above regarding practical feedback circuits and the effect of feedback loading can be summarized as follows.

First, input and output variables must be identified and the feedback identified as shunt or series at input and output.

The feedback function f is found by the following procedure. If the feedback is *shunt* at the *input*, short the input feedback node to ground and calculate the feedback *current*. If the feedback is *series* at the *input*, open circuit the input feedback node and calculate the feedback *voltage*. In both these cases, if the feedback is *shunt* at the *output*, drive the feedback network with a voltage source. If the feedback is *series* at the *output*, drive with a current source.

The effect of feedback loading on the basic amplifier is found as follows. If the feedback is *shunt* at the *input*, *short* the input feedback node to ground to find the feedback loading on the *output*. If the feedback is *series* at the input, *open* circuit the input feedback node to calculate *output* feedback loading. Similarly, if the feedback is *shunt* or *series* at the output, then *short* or *open* the output feedback node to calculate feedback loading on the input.

These results along with other information are summarized in Table 8.1.

8.6 Single-Stage Feedback

The considerations of feedback circuits in this chapter have been mainly directed toward the general case of circuits with multiple stages in the basic amplifier. However, in dealing with some of these circuits (such as the series-series triple of Fig. 8.18a), equivalent circuits were derived in which one or more stages contained an emitter resistor. (See Fig. 8.18c.) Such a stage represents *in itself* a feedback circuit as will be shown. Thus the circuit of Fig. 8.18c contains feedback loops within a feedback loop, and this has a direct effect on the amplifier performance. For example, the emitter of Q_3 in Fig. 8.18c has a linearizing effect on Q_3, and so does the overall feedback when the loop is closed. It is thus important to calculate the effects of both *local* and *overall* feedback loops. The term *local* feedback is often used instead of *single-stage* feedback. Local feedback is often used in isolated stages as well as being found inside overall feedback loops. In this section, the low-frequency characteristics of two basic single-stage feedback circuits will be analyzed.

8.6.1 Local Series-Series Feedback

A local series-series feedback stage in bipolar technology (emitter degeneration) is shown in Fig. 8.34a. The characteristics of this circuit were considered previously in Section 3.3.8,

Table 8.1 Summary of Feedback Configurations

Feedback Configuration	Two-Port Parameter Representation	Output Variable	Input Variable	Transfer Function Stabilized	Z_i	Z_o	To Calculate Feedback Loading At Input	To Calculate Feedback Loading At Output	To Calculate Feedback Function f
Shunt-shunt	y	v_o	i_s	$\dfrac{v_o}{i_s}$ Transresistance	Low	Low	Short circuit output feedback node	Short circuit input feedback node	Drive the feedback network with a voltage at its output and calculate the current flowing into a short at its input
Shunt-series	g	i_o	i_s	$\dfrac{i_o}{i_s}$ Current gain	Low	High	Open circuit output feedback node	Short circuit input feedback node	Drive the feedback network with a current and calculate the current flowing into a short
Series-shunt	h	v_o	v_s	$\dfrac{v_o}{v_s}$ Voltage gain	High	Low	Short circuit output feedback node	Open circuit input feedback node	Drive the feedback network with a voltage and calculate the voltage produced into an open circuit
Series-series	z	i_o	v_s	$\dfrac{i_o}{v_s}$ Transconductance	High	High	Open circuit output feedback node	Open circuit input feedback node	Drive the feedback network with a current and calculate the voltage produced into an open circuit

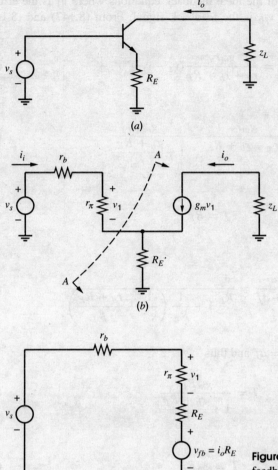

Figure 8.34 (a) Single-stage series-series feedback circuit. (b) Low-frequency equivalent circuit of (a). (c) Circuit equivalent to (b) using a Thévenin equivalent across the plane AA.

and it will be considered again here from the feedback viewpoint. Similar results are found for MOS transistors with source degeneration.

The circuit of Fig. 8.34a can be recognized as a degenerate series-series feedback configuration as described in Sections 8.4.4 and 8.5.2. Instead of attempting to use the generalized forms of those sections, we can perform a more straightforward calculation in this simple case by working directly from the low-frequency, small-signal equivalent circuit shown in Fig. 8.34b. For simplicity, source impedance is assumed zero but can be lumped in with r_b if desired. If a Thévenin equivalent across the plane AA is calculated, Fig. 8.34b can be redrawn as in Fig. 8.34c. The existence of an input differencing node is apparent, and quantity $i_o R_E$ is identified as the feedback voltage v_{fb}. Writing equations for Fig. 8.34c we find

$$v_1 = \frac{r_\pi}{r_\pi + r_b + R_E}(v_s - v_{fb}) \qquad (8.147)$$

$$i_o = g_m v_1 \qquad (8.148)$$

$$v_{fb} = i_o R_E \qquad (8.149)$$

These equations are of the form of the ideal feedback equations where v_1 is the error voltage, i_o is the output signal, and v_{fb} is the feedback signal. From (8.147) and (8.148) we know that

$$i_o = \frac{g_m r_\pi}{r_\pi + r_b + R_E}(v_s - v_{fb}) \tag{8.150}$$

and thus we can identify

$$a = \frac{g_m r_\pi}{r_\pi + r_b + R_E} = \frac{g_m}{1 + \dfrac{r_b + R_E}{r_\pi}} \tag{8.151}$$

From (8.149)

$$f = R_E \tag{8.152}$$

Thus for the complete circuit

$$\frac{i_o}{v_s} = A = \frac{a}{1+af} = \frac{1}{R_E}\frac{1}{1 + \dfrac{1}{R_E}\left(\dfrac{1}{g_m} + \dfrac{r_b + R_E}{\beta_0}\right)} \tag{8.153}$$

and $A \simeq 1/R_E$ for large loop gain.

The loop gain is given by $T = af$ and thus

$$T = \frac{g_m R_E}{1 + \dfrac{r_b + R_E}{r_\pi}} \tag{8.154}$$

If $(r_b + R_E) \ll r_\pi$ we find

$$T \simeq g_m R_E \tag{8.155}$$

The input resistance of the circuit is given by

$$\text{Input resistance} = (1 + T) \times (\text{input resistance with } v_{fb} = 0)$$
$$= (1 + T)(r_b + r_\pi + R_E) \tag{8.156}$$

Using (8.154) in (8.156) gives

$$\text{Input resistance} = r_b + R_E + r_\pi(1 + g_m R_E) \tag{8.157}$$
$$= r_b + r_\pi + (\beta_0 + 1)R_E \tag{8.158}$$

We can also show that if the output resistance r_o of the transistor is included, the output resistance of the circuit is given by

$$\text{Output resistance} \simeq r_o\left(1 + \frac{g_m R_E}{1 + \dfrac{r_b + R_E}{r_\pi}}\right) \tag{8.159}$$

Both input and output resistance are increased by the application of emitter feedback, as expected.

■ **EXAMPLE**

Calculate the low-frequency parameters of the series-series feedback stage represented by Q_3 in Fig. 8.21b. The relevant parameters are as follows:

$$R_E = 88\,\Omega \qquad r_\pi = 650\,\Omega \qquad g_m = \frac{4}{26}\,\text{A/V} \qquad \beta_0 = 100 \qquad r_b = 0$$

The loading produced by Q_3 at the collector of Q_2 is given by the input resistance expression of (8.158) and is

$$R_{i3} = (650 + 101 \times 88)\,\Omega = 9.54\,\text{k}\Omega \tag{8.160}$$

The output resistance seen at the collector of Q_3 can be calculated from (8.159) using $r_b = 5\,\text{k}\Omega$ to allow for the finite source resistance in Fig. 8.21b. If we assume that $r_o = 25\,\text{k}\Omega$ at $I_C = 4$ mA for Q_3, then from (8.159) we find that

$$R_{o3} = 25\left(1 + \frac{\frac{4}{26} \cdot 88}{1 + \frac{5088}{650}}\right)\,\text{k}\Omega = 63\,\text{k}\Omega \tag{8.161}$$

In the example of Fig. 8.21b, the above output resistance would be multiplied by the loop gain of the series-series triple.

Finally, when ac voltage v_4 at the collector of Q_2 in Fig. 8.21c is determined, output current i_3 in Q_3 can be calculated using (8.153):

$$\frac{i_3}{v_4} = \frac{1}{88}\frac{1}{1 + \frac{1}{88}\left(\frac{26}{4} + \frac{88}{100}\right)}\,\text{A/V} = \frac{1}{95.4}\,\text{A/V} \tag{8.162}$$

Note that since the voltage v_4 exists at the base of Q_3, the effective source resistance in the above calculation is zero.

■

8.6.2 Local Series-Shunt Feedback

Another example of a local feedback stage is the common-drain stage shown in Fig. 8.35a. This circuit is a series-shunt feedback configuration. The small-signal model is shown in Fig. 8.35b. The current through the g_{mb} controlled source is controlled by the voltage across it; therefore, it can be replaced by a resistor of value $1/g_{mb}$. Using this transformation, the small-signal model in Fig. 8.35b is redrawn in Fig. 8.35c, where

$$R'_L = R_L \| r_o \| \frac{1}{g_{mb}} \tag{8.163}$$

The feedback network is taken to be R'_L. Using Fig. 8.35d, the h two-port parameters for the feedback network are

$$h_{11f} = \left.\frac{v_1}{i_1}\right|_{v_2=0} = 0 \tag{8.164}$$

$$h_{22f} = \left.\frac{i_2}{v_2}\right|_{i_1=0} = \frac{1}{R'_L} \tag{8.165}$$

592   Chapter 8 ■ Feedback

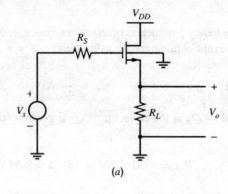

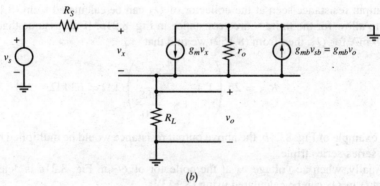

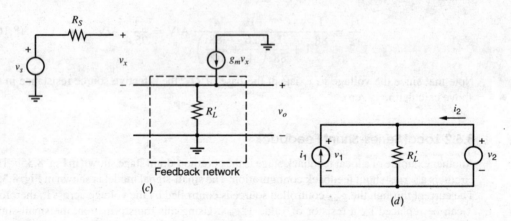

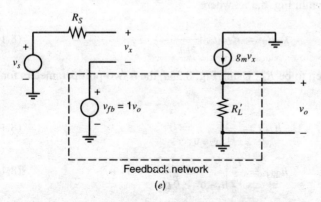

Figure 8.35 (a) A source follower driving a resistive load. (b) A small-signal model for the circuit in (a). (c) A simplified small-signal model. (d) Circuit for finding the two-port parameters for the feedback network. (e) The circuit in (c) with the feedback network replaced by a two-port model.

and the feedback factor is

$$f = h_{12f} = \left.\frac{v_1}{v_2}\right|_{i_1=0} = 1 \qquad (8.166)$$

Here we neglect the forward transmission through the feedback network, h_{21f}, as was done in Section 8.5.3.

Figure 8.35b is redrawn in Fig. 8.35e with the feedback network replaced by a two-port model. The gain of the basic amplifier a is found by setting the feedback to zero in Fig. 8.35e

$$a = \left.\frac{v_o}{v_s}\right|_{v_{fb}=0} = g_m R'_L \qquad (8.167)$$

Using (8.166) and (8.167), the closed-loop gain is

$$A = \frac{a}{1+af} = \frac{g_m R'_L}{1 + g_m R'_L} \qquad (8.168)$$

This closed-loop gain is always less than one and approaches unity as $g_m R'_L \to \infty$.

From Fig. 8.35e, the output resistance without feedback is R'_L, and the closed-loop output resistance is

$$R_o = \frac{R'_L}{1+af} = \frac{R'_L}{1 + g_m R'_L} = \frac{1}{\frac{1}{R'_L} + g_m} = \frac{1}{\frac{1}{R_L} + \frac{1}{r_o} + g_{mb} + g_m} \qquad (8.169)$$

where (8.163) is used in the right-most expression. The output resistance is reduced because of the feedback. The input resistance is infinite because the resistance from the gate to the source in the model in Fig. 8.35 is infinite.

8.7 The Voltage Regulator as a Feedback Circuit

As an example of a practical feedback circuit, the operation of a voltage regulator will be examined. This section is introduced for the dual purpose of illustrating the use of feedback in practice and for describing the elements of voltage regulator design.

Voltage regulators are widely used components that accept a poorly specified (possibly fluctuating) dc input voltage and produce from it a constant, well-specified output voltage that can then be used as a supply voltage for other circuits.[3] In this way, fluctuations in the supply voltage are essentially eliminated, and this usually results in improved performance for circuits powered from such a supply.

A common type of voltage regulator is the *series* regulator shown schematically in Fig. 8.36. The name *series* comes from the fact that the output voltage is controlled by a power transistor in series with the output. This is the last stage of a high-gain voltage amplifier, as shown in Fig. 8.36.

Many of the techniques discussed in previous chapters are utilized in the design of circuits of this kind. A stable reference voltage V_R can be generated using a Zener diode or a bandgap reference, as described in Chapter 4. This is then fed to the noninverting input of the high-gain amplifier, where it is compared with a sample of the output taken by resistors R_1 and R_2. This is recognizable as a series-shunt feedback arrangement, and using (8.132) we find that for large loop gain

$$V_O = V_R \frac{R_1 + R_2}{R_2} \qquad (8.170)$$

The output voltage can be varied by changing ratio R_1/R_2.

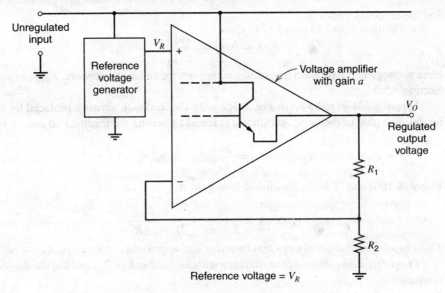

Figure 8.36 Schematic of a series voltage regulator.

The characteristics required in the amplifier of Fig. 8.36 are those of a good op amp, as described in Chapter 6. In particular, low drift and offset are essential so that the output voltage V_O is as stable as possible. Note that the series-shunt feedback circuit will present a high input impedance to the reference generator, which is desirable to minimize loading effects. In addition, a very low output impedance will be produced at V_O, which is exactly the requirement for a good voltage source. If the effects of feedback loading are neglected (usually a good assumption in such circuits), the low-frequency output resistance of the regulator is given by (8.29) as

$$R_o = \frac{r_{oa}}{1+T} \qquad (8.171)$$

where

$$T = a\frac{R_2}{R_1 + R_2} \qquad (8.172)$$

r_{oa} = output resistance of the amplifier without feedback

a = magnitude of the forward gain of the regulator amplifier

If the output voltage of the regulator is varied by changing ratio R_1/R_2, then (8.171) and (8.172) indicate that T and thus R_o also change. Assuming that V_R is constant and $T \gg 1$, we can describe this behavior by substituting (8.170) and (8.172) in (8.171) to give

$$R_o = \frac{r_{oa}}{aV_R}V_O \qquad (8.173)$$

which shows R_o to be a function of V_O if a, V_R, and r_{oa} are fixed. If the output current drawn from the regulator changes by ΔI_O, then V_O changes by ΔV_O, where

$$\Delta V_O = R_o \Delta I_O \qquad (8.174)$$

Substitution of (8.173) in (8.174) gives

$$\frac{\Delta V_O}{V_O} = \frac{r_{oa}}{aV_R}\Delta I_O \qquad (8.175)$$

This equation allows calculation of the *load regulation* of the regulator. This is a widely used specification, which gives the percentage change in V_O for a specified change in I_O and should be as small as possible.

Another common regulator specification is the *line regulation*, which is the percentage change in output voltage for a specified change in input voltage. Since V_O is directly proportional to V_R, the line regulation is determined by the change in reference voltage V_R with changes in input voltage and depends on the particular reference circuit used.

As an example of a practical regulator, consider the circuit diagram of the 723 monolithic voltage regulator shown in Fig. 8.37. The correspondence to Fig. 8.36 can be recognized, with the portion of Fig. 8.37 to the right of the dashed line being the voltage amplifier with feedback. The divider resistors R_1 and R_2 in Fig. 8.36 are labeled R_A and R_B in Fig. 8.37 and are external to the chip. The output power transistor Q_{15} is on the chip and is Darlington connected with Q_{14} for high gain. Differential pair Q_{11} and Q_{12}, together with active load Q_8, contribute most of the gain of the amplifier. Resistor R_C couples the reference voltage to the amplifier and C_2 is an external capacitor, which is needed to prevent oscillation in the high-gain feedback loop. Its function is discussed in Chapter 9.

■ **EXAMPLE**

Calculate the bias conditions and load regulation of the 723. Assume the total supply voltage is 15 V.

The bias calculation begins at the left-hand side of Fig. 8.37. Current source I_1 models a transistor current source that uses a junction field-effect transistor (JFET)[4] that behaves like an *n*-channel MOS transistor with a negative threshold voltage. Diodes D_1 and D_2 are Zener diodes.[4] When operating in reverse breakdown, the voltage across a Zener diode is nearly constant as described in Chapter 2.

The Zener diode D_1 produces a voltage drop of about 6.2 V, which sets up a reference current in Q_2:

$$\begin{aligned}I_{C2} &= -\frac{6.2 - |V_{BE2}|}{R_1 + R_2} \\ &= -\frac{6.2 - 0.6}{16{,}000}\text{ A} \\ &= -348\ \mu\text{A}\end{aligned} \qquad (8.176)$$

Note that I_{C2} is almost independent of supply voltage because it is dependent only on the Zener diode voltage.

The voltage across R_1 and Q_2 establishes the currents in current sources Q_3, Q_7, and Q_8.

$$I_{C7} = I_{C8} = -174\ \mu\text{A} \qquad (8.177)$$
$$I_{C3} = -10.5\ \mu\text{A} \qquad (8.178)$$

Current source Q_3 establishes the operating current in the voltage reference circuit composed of transistors Q_4, Q_5, Q_6, resistors R_6, R_7, R_8, and Zener diode D_2. This circuit can be recognized as a variation of the Wilson current source described in Chapter 4, and

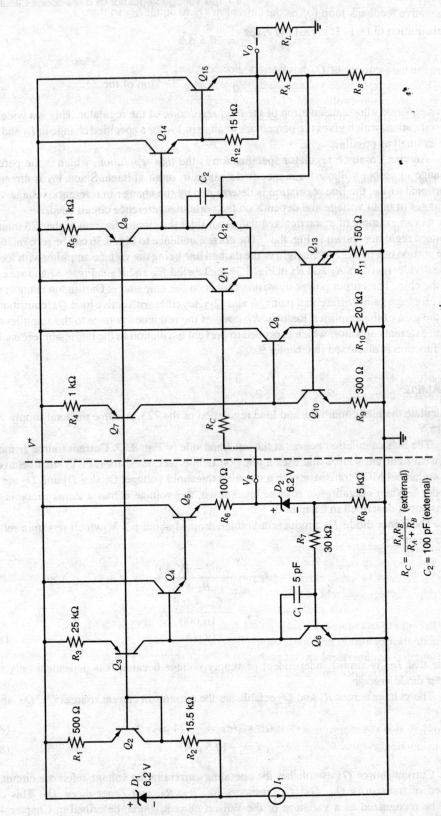

Figure 8.37 Circuit diagram of the 723 monolithic voltage regulator.

the negative feedback loop forces the current in Q_6 to equal I_{C3} so that

$$I_{C6} = 10.5 \ \mu A \tag{8.179}$$

where the base current of Q_4 has been neglected.

The output reference voltage V_R is composed of the sum of the Zener diode voltage D_2 plus the base-emitter voltage of Q_6, giving a reference voltage of about 6.8 V. The current in the Zener is established by V_{BE6} and R_8 giving

$$I_{D2} = \frac{V_{BE6}}{R_8} = \frac{600}{5} \ \mu A = 120 \ \mu A \tag{8.180}$$

The Darlington pair Q_4, Q_5 helps give a large loop gain that results in a very low output impedance at the voltage reference node. Resistor R_6 limits the current and protects Q_5 in case of an accidental grounding of the voltage reference node. Resistor R_7 and capacitor C_1 form the high-frequency compensation required to prevent oscillation in the feedback loop. Note that the feedback is shunt at the output node. Any changes in reference-node voltage (due to loading for example) are detected at the base of Q_6, amplified, and fed to the base of Q_4 and thus back to the output where the original change is opposed.

The biasing of the amplifier is achieved via current sources Q_7 and Q_8. The current in Q_7 also appears in Q_{10} (neglecting the base current of Q_9). Transistor Q_{13} has an area twice that of Q_{10} and one half the emitter resistance. Thus

$$I_{C13} = 2I_{C10} = 2I_{C7} = 348 \ \mu A \tag{8.181}$$

Transistor Q_9 provides current gain to minimize the effect of base current in Q_{10} and Q_{13}. This beta-helper current mirror was described in Chapter 4.

The bias current in each half of the differential pair Q_{11}, Q_{12} is thus

$$I_{C11} = I_{C12} = \frac{1}{2} I_{C13} = 174 \ \mu A \tag{8.182}$$

Since Q_8 and R_5 are identical to Q_7 and R_4, the current in Q_8 is given by

$$I_{C8} = -174 \ \mu A \tag{8.183}$$

Transistor Q_8 functions as an active load for Q_{12}, and, since the collector currents in these two devices are nominally equal, the input offset voltage for the differential pair is nominally zero.

The current in output power transistor Q_{15} depends on the load resistance but can go as high as 150 mA before a current-limit circuit (not shown) prevents further increase. Resistor R_{12} provides bleed current so that Q_{14} always has at least 0.04 mA of bias current, even when the current in Q_{15} is low and/or its current gain is large.

In order to calculate the load regulation of the 723, (8.175) indicates that it is necessary to calculate the open-loop gain and output resistance of the regulator amplifier. For this purpose, a differential ac equivalent circuit of this amplifier is shown in Fig. 8.38. Load resistance R_{L12} is the output resistance presented by Q_8, which is

$$R_{L12} \simeq r_{o8}(1 + g_{m8} R_5) \tag{8.184}$$

Assuming that the magnitude of the Early voltage of Q_8 is 100 V and $I_{C8} = -174 \ \mu A$, we can calculate the value of R_{L12} as

$$R_{L12} = \frac{100}{0.174} \left(1 + \frac{0.174}{26} 1000 \right) \ k\Omega = 4.42 \ M\Omega \tag{8.185}$$

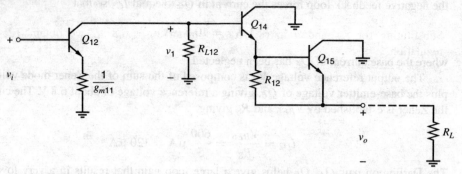

Figure 8.38 An ac equivalent circuit of the regulator amplifier of the 723 voltage regulator.

Since $g_{m11} = g_{m12}$, the impedance in the emitter of Q_{12} halves the transconductance and gives an effective output resistance of

$$R_{o12} = \left(1 + g_{m12}\frac{1}{g_{m11}}\right) r_{o12} \tag{8.186}$$

where r_{o12} is the output resistance of Q_{12} alone and is 575 kΩ if the Early voltage is 100 V. Thus

$$R_{o12} = 1.15 \text{ M}\Omega \tag{8.187}$$

The external load resistance R_L determines the load current and thus the bias currents in Q_{14} and Q_{15}. However, R_L is not included in the small-signal calculation of output resistance because this quantity is the resistance seen by R_L looking back into the circuit. Thus, for purposes of calculating the ac output resistance, R_L may be assumed infinite and the output Darlington pair then produces no loading at the collector of Q_{12}. The voltage gain of the circuit may then be calculated as

$$a = \left|\frac{v_o}{v_i}\right| = \left|\frac{v_1}{v_i}\right| = \frac{g_{m12}}{2}(R_{o12}\|R_{L12})$$

$$= \frac{0.174}{26}\frac{1}{2}(1.15\|4.42) \times 10^6$$

$$= 3054 \tag{8.188}$$

The output resistance r_{oa} of the circuit of Fig. 8.38 is the output resistance of a Darlington emitter follower. If R_{12} is assumed large compared with $r_{\pi 15}$ then

$$r_{oa} = \frac{1}{g_{m15}} + \frac{1}{\beta_{0(15)}}\left(\frac{1}{g_{m14}} + \frac{R_S}{\beta_{0(14)}}\right) \tag{8.189}$$

where

$$R_S = R_{o12}\|R_{L12} = 913 \text{ k}\Omega \tag{8.190}$$

If we assume collector bias currents of 20 mA in Q_{15} and 0.5 mA in Q_{14} together with $\beta_{0(15)} = \beta_{0(14)} = 100$, then substitution in (8.189) gives

$$r_{oa} = \left[1.3 + \frac{1}{100}\left(52 + \frac{913,000}{100}\right)\right] \Omega$$

$$= (1.3 + 92) \Omega$$

$$= 93 \Omega \tag{8.191}$$

Substituting for r_{oa} and a in (8.175) and using $V_R = 6.8$ V, we obtain, for the load regulation,

$$\frac{\Delta V_O}{V_O} = \frac{93}{3054 \times 6.8} \Delta I_O$$
$$= 4.5 \times 10^{-3} \Delta I_O \qquad (8.192)$$

where ΔI_O is in Amps.

If ΔI_O is 50 mA, then (8.192) gives

$$\frac{\Delta V_O}{V_O} = 2 \times 10^{-4} = 0.02 \text{ percent} \qquad (8.193)$$

This answer is close to the value of 0.03 percent given on the specification sheet. Note the extremely small percentage change in output voltage for a 50-mA change in load current.

■

8.8 Feedback Circuit Analysis Using Return Ratio

The feedback analysis presented so far has used two-ports to manipulate a feedback circuit into unilateral forward amplifier and feedback networks. Since real feedback circuits have bilateral feedback networks and possibly bilateral amplifiers, some work is required to find the amplifier a and feedback f networks. The correct input and output variables and the type of feedback must be identified, and the correct two-port representation (y, z, h, or g) must be used. After this work, the correspondence between the original circuit and modified two-ports is small, which can make these techniques difficult to use.

Alternatively, a feedback circuit can be analyzed in a way that does not use two-ports. This alternative analysis, which is often easier than two-port analysis, is called return-ratio analysis.[5,6,7] Here, the closed-loop properties of a feedback circuit are described in terms of the *return ratio* for a dependent source in the small-signal model of an active device. The return ratio for a dependent source in a feedback loop is found by the following procedure:

1. Set all independent sources to zero.
2. Disconnect the dependent source from the rest of the circuit, which introduces a break in the feedback loop.
3. On the side of the break that is not connected to the dependent source, connect an independent test source s_t of the same sign and type as the dependent source.
4. Find the return signal s_r generated by the dependent source.

Then the return ratio ($\mathcal{R}$) for the dependent source is $\mathcal{R} = -s_r/s_t$, where the variable s represents either a current or a voltage.

Figure 8.39a shows a negative feedback amplifier that includes a dependent voltage source. Figure 8.39b shows how the circuit is modified to find the return ratio. The dependent source is disconnected from the rest of the circuit by breaking the connections at the two X's marked in Fig. 8.39a. A test signal v_t is connected on the side of the break that is not connected to the controlled source. The return signal v_r is measured at the open circuit across the controlled source to find the return ratio $\mathcal{R} = -v_r/v_t$.

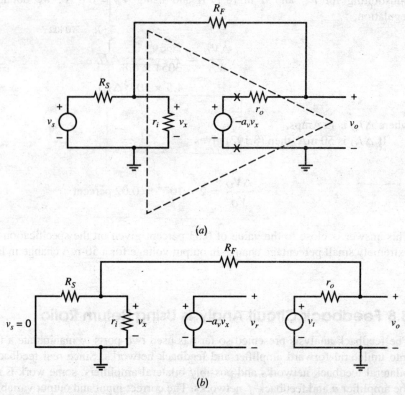

Figure 8.39 (a) A feedback circuit that gives an inverting voltage gain. The "X" marks indicate where the loop will be broken. (b) The circuit in (a) modified to find the return ratio for the dependent voltage source.

■ **EXAMPLE**

Find the return ratio for the circuit in Fig. 8.39b.

The return ratio can be found with little computation because the resistors form a voltage divider

$$v_x = \frac{R_S \| r_i}{R_S \| r_i + R_F + r_o} v_t \qquad (8.194)$$

The return voltage v_r is

$$v_r = -a_v v_x \qquad (8.195)$$

Combining these equations gives

$$\mathcal{R} = -\frac{v_r}{v_t} = \frac{R_S \| r_i}{R_S \| r_i + R_F + r_o} a_v \qquad (8.196)$$

■

Figure 8.40a is a single-stage feedback circuit. Its small-signal model shown in Fig. 8.40b includes a dependent current source. Figure 8.40c illustrates how the return ratio is found in this case. Here, the dependent source is disconnected from the rest of the circuit, and a test current source i_t is connected on the side of the break that is not connected to the dependent source. A short circuit is applied across the dependent current source to provide a path for the return current i_r to flow. The return ratio is computed as $\mathcal{R} = -i_r/i_t$.

8.8 Feedback Circuit Analysis Using Return Ratio 601

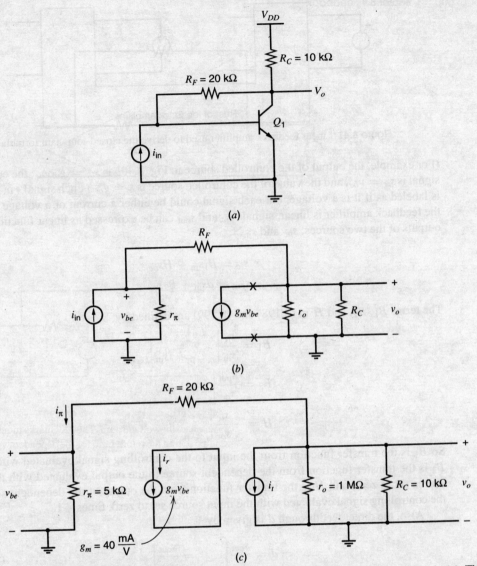

Figure 8.40 (a) A transresistance feedback circuit. (b) The small-signal model for the circuit in (a). The "X" marks indicate where the loop will be broken. (c) The circuit in (b) modified to find the return ratio for the dependent current source.

8.8.1 Closed-Loop Gain Using Return Ratio

A formula for the closed-loop gain of a feedback amplifier in terms of the return ratio will now be derived. Consider a feedback amplifier as shown in Fig. 8.41. The feedback amplifier consists of linear elements: passive components, controlled sources, and transistor small-signal models. A controlled source with value k that is part of the small-signal model of an active device is shown explicitly. The output of the controlled source is s_{oc}, and the controlling signal is s_{ic}. The equation that describes the controlled source is

$$s_{oc} = k s_{ic} \tag{8.197}$$

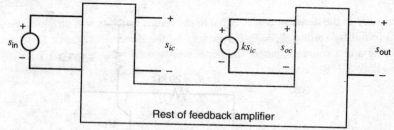

Figure 8.41 Linear feedback amplifier used to derive the closed-loop gain formula.

(For example, the output of the controlled source in Fig. 8.40b is $s_{oc} = g_m v_{be}$; the controlling signal is $s_{ic} = v_{be}$, and the value of the controlled source is $k = g_m$.) Each signal s in the figure is labeled as if it is a voltage, but each signal could be either a current or a voltage. Because the feedback amplifier is linear, signals s_{ic} and s_{out} can be expressed as linear functions of the outputs of the two sources, s_{oc} and s_{in},

$$s_{ic} = B_1 s_{in} - H s_{oc} \qquad (8.198)$$

$$s_{out} = d s_{in} + B_2 s_{oc} \qquad (8.199)$$

The terms B_1, B_2, and H in (8.198) and (8.199) are defined by

$$B_1 = \left.\frac{s_{ic}}{s_{in}}\right|_{s_{oc}=0} = \left.\frac{s_{ic}}{s_{in}}\right|_{k=0} \qquad (8.200a)$$

$$B_2 = \left.\frac{s_{out}}{s_{oc}}\right|_{s_{in}=0} \qquad (8.200b)$$

$$H = -\left.\frac{s_{ic}}{s_{oc}}\right|_{s_{in}=0} \qquad (8.200c)$$

So B_1 is the transfer function from the input to the controlling signal evaluated with $k = 0$, B_2 is the transfer function from the dependent source to the output evaluated with the input source set to zero, and H is the transfer function from the output of the dependent source to the controlling signal evaluated with the input source set to zero, times -1.

Also, the direct feedthrough d is given by

$$d = \left.\frac{s_{out}}{s_{in}}\right|_{s_{oc}=0} = \left.\frac{s_{out}}{s_{in}}\right|_{k=0} \qquad (8.200d)$$

which is the transfer function from the input to the output evaluated with $k = 0$. The calculation of d usually involves signal transfer through passive components that provide a signal path directly from the input to output, a path that goes around rather than through the controlled source k.

Equations 8.197, 8.198, and 8.199 can be solved for the closed-loop gain. Substituting (8.197) in (8.198) and rearranging gives

$$s_{ic} = \frac{B_1}{1 + kH} s_{in} \qquad (8.201)$$

Substituting (8.197) in (8.199), then substituting (8.201) in the resulting equation and rearranging terms gives the closed-loop gain A

$$A = \frac{s_{out}}{s_{in}} = \frac{B_1 k B_2}{1 + kH} + d \qquad (8.202)$$

The term kH in the denominator is equal to the return ratio, as will be shown next. The return ratio is found by setting $s_{in} = 0$, disconnecting the dependent source from the circuit, and connecting a test source s_t where the dependent source was connected. After these changes, $s_{oc} = s_t$ and (8.198) becomes

$$s_{ic} = -Hs_t \tag{8.203}$$

Then the output of the dependent source is the return signal $s_r = ks_{ic} = -kHs_t$. Therefore

$$\mathcal{R} = -\frac{s_r}{s_t} = kH \tag{8.204}$$

So the closed-loop gain in (8.202) can be rewritten as

$$A = \frac{s_{out}}{s_{in}} = \frac{B_1 k B_2}{1 + \mathcal{R}} + d \tag{8.205a}$$

or

$$A = \frac{s_{out}}{s_{in}} = \frac{g}{1 + \mathcal{R}} + d \tag{8.205b}$$

where

$$g = B_1 k B_2 \tag{8.206}$$

Here g is the gain from s_{in} to s_{out} if $H = 0$ and $d = 0$, and d is the direct signal feedthrough, which is the value of A when the controlled source is set to zero ($k = 0$).

The closed-loop gain formula in (8.205a) requires calculations of four terms: B_1, B_2, d, and $\mathcal{R}$. That equation can be manipulated into a more convenient form with only three terms. Combining terms in (8.205b) using a common denominator $1 + \mathcal{R}$ gives

$$A = \frac{g + d(1 + \mathcal{R})}{1 + \mathcal{R}} = \frac{g + d\mathcal{R}}{1 + \mathcal{R}} + \frac{d}{1 + \mathcal{R}} = \frac{\left(\frac{g}{\mathcal{R}} + d\right)\mathcal{R}}{1 + \mathcal{R}} + \frac{d}{1 + \mathcal{R}} \tag{8.207}$$

Defining

$$A_\infty = \frac{g}{\mathcal{R}} + d \tag{8.208}$$

allows (8.207) to be rewritten as

$$A = A_\infty \frac{\mathcal{R}}{1 + \mathcal{R}} + \frac{d}{1 + \mathcal{R}} \tag{8.209}$$

This is a useful expression for the closed-loop gain. Here, if $\mathcal{R} \to \infty$, then $A = A_\infty$ because $\mathcal{R}/(1 + \mathcal{R}) \to 1$ and $d/(1 + \mathcal{R}) \to 0$. So A_∞ is the closed-loop gain when the feedback circuit is ideal (that is, when $\mathcal{R} \to \infty$).

A block-diagram representation of (8.209) is shown in Fig. 8.42. The gain around the feedback loop is $\mathcal{R}$, and the effective forward gain in the loop is $\mathcal{R}A_\infty$. A key difference between the two-port and return-ratio analyses can be seen by comparing Figs. 8.1 and 8.42. In the two-port analysis, all forward signal transfer through the amplifier and the feedback network is lumped into a. In the return-ratio analysis, there are two forward signal paths: one

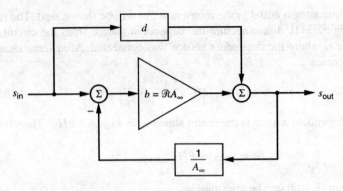

Figure 8.42 A block diagram for the closed-loop gain formula in (8.209).

path (d) for the feedforward through the feedback network and another path ($\mathcal{R} A_\infty$) for the effective forward gain.

Typically, A_∞ is determined by a passive feedback network and is equal to $1/f$ from two-port analysis. The value of A_∞ can be found readily since $A_\infty = A$ when $k \to \infty$. Letting $k \to \infty$ causes $\mathcal{R} = kH \to \infty$. (Here we assumed $k > 0$. If $k < 0$ in a negative feedback circuit, then $\mathcal{R} \to \infty$ when $k \to -\infty$.) When $k \to \infty$, the controlling signal s_{ic} for the dependent source must be zero if the output of the dependent source is finite. The controlled source output will be finite if the feedback is negative. These facts can be used to find A_∞ with little computation in many circuits, as is demonstrated in the next example.

■ **EXAMPLE**

Compute the closed-loop gain for the circuit of Fig. 8.40 using (8.209). Use the component values given in the figure.

To use (8.209), A_∞, $\mathcal{R}$, and d are needed. Here, the only controlled source is the g_m source, so $k = g_m$. Also, $s_{in} = i_{in}$, $s_{out} = v_o$, and $s_{ic} = v_{be}$. To find A_∞, let $g_m \to \infty$, which forces the controlling voltage v_{be} to equal zero, assuming that the output current from the g_m generator is finite. With $v_{be} = 0$, no current flows through r_π, and therefore the input current i_{in} flows through R_F to produce v_o. Thus

$$A_\infty = \left.\frac{v_o}{i_{in}}\right|_{g_m = \infty} = -R_F = -20 \text{ k}\Omega. \tag{8.210}$$

Next, d is found by setting $k = g_m = 0$ and computing the transfer function from input to output

$$d = \left.\frac{v_o}{i_{in}}\right|_{g_m = 0} = (r_o \| R_C) \cdot \frac{r_\pi}{r_\pi + R_F + r_o \| R_C}$$

$$= (1 \text{ M}\Omega \| 10 \text{ k}\Omega) \cdot \frac{5 \text{ k}\Omega}{5 \text{ k}\Omega + 10 \text{ k}\Omega + 1 \text{ M}\Omega \| 100 \text{ k}\Omega} \tag{8.211}$$

$$= 1.4 \text{ k}\Omega$$

Finally, the return ratio for the g_m generator can be calculated using Fig. 8.40c. Applying a current-divider formula gives the current i_π through r_π as

$$i_\pi = -\frac{r_o \| R_C}{r_o \| R_C + R_F + r_\pi} i_t \tag{8.212}$$

The return current is

$$i_r = g_m v_{be} = g_m r_\pi i_\pi \tag{8.213}$$

8.8 Feedback Circuit Analysis Using Return Ratio

Combining these equations gives

$$\mathcal{R} = -\frac{i_r}{i_t} = g_m r_\pi \cdot \frac{r_o \| R_C}{r_o \| R_C + R_F + r_\pi} \quad (8.214)$$

$$= (40 \text{ mA/V}) \cdot 5 \text{ k}\Omega \cdot \frac{1 \text{ M}\Omega \| 10 \text{ k}\Omega}{1 \text{ M}\Omega \| 10 \text{ k}\Omega + 20 \text{ k}\Omega + 5 \text{ k}\Omega} = 56.7 \quad (8.215)$$

Then, using (8.209),

$$A = A_\infty \frac{\mathcal{R}}{1+\mathcal{R}} + \frac{d}{1+\mathcal{R}} = -20 \text{ k}\Omega \frac{56.7}{1+56.7} + \frac{1.4 \text{ k}\Omega}{1+56.7} = -19.6 \text{ k}\Omega \quad (8.216)$$

In (8.209), the second term that includes d can be neglected whenever $|d| \ll |A_\infty \mathcal{R}|$. This condition usually holds at low frequencies because d is the forward signal transfer through a passive network, while $|A_\infty \mathcal{R}|$ is large because it includes the gain through the active device(s). For example, ignoring the $d/(1+\mathcal{R})$ term in (8.216) gives $A \approx -19.7$ kΩ, which is close to the exact value. As the frequency increases, however, the gain provided by the transistors falls. As a result, d may become significant at high frequencies.

The effective forward gain $A_\infty \mathcal{R}$ can be computed after A_∞ and $\mathcal{R}$ have been found. Alternatively, this effective forward gain can be found directly from the feedback circuit.[8] Call this forward gain b, so

$$b = A_\infty \cdot \mathcal{R} \quad (8.217)$$

Then the closed-loop gain in (8.209) can be written as

$$A = \frac{b}{1+\mathcal{R}} + \frac{d}{1+\mathcal{R}} \quad (8.218)$$

Using (8.206), (8.208), and $\mathcal{R} = kH$, b can be expressed as

$$b = A_\infty \cdot \mathcal{R} = \left(\frac{g}{\mathcal{R}} + d\right) \cdot \mathcal{R} = (B_1 k B_2 + d\mathcal{R})$$

$$= (B_1 k B_2 + dkH) = \left[B_1 + \frac{dH}{B_2}\right] k B_2 \quad (8.219)$$

This final expression breaks b into parts that can be found by analyzing the feedback circuit. In (8.200b), B_2 is defined as the transfer function from the output of the controlled source s_{oc} to s_{out} evaluated with $s_{in} = 0$. The term in brackets in (8.219) is equal to the transfer function from s_{in} to s_{ic} when $s_{out} = 0$, as will be shown next.

If $s_{out} = 0$, then (8.199) simplifies to

$$B_2 s_{oc} = -d s_{in} \quad (8.220)$$

Substituting (8.220) into (8.198) gives

$$s_{ic} = B_1 s_{in} + \frac{dH}{B_2} s_{in} \quad (8.221)$$

Therefore

$$\left.\frac{s_{ic}}{s_{in}}\right|_{s_{out}=0} = B_1 + \frac{dH}{B_2} \quad (8.222)$$

606   Chapter 8 ■ Feedback

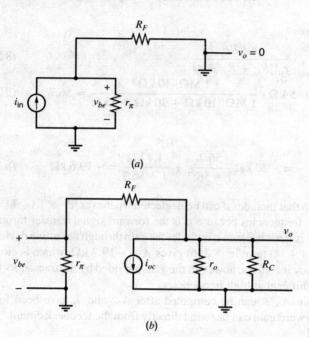

Figure 8.43 Circuits for finding the effective forward gain b. (a) The circuit for the input side. (b) The circuit for the output side.

which is the expression in brackets in (8.219). Substituting (8.222) and (8.200b) into (8.219) gives

$$b = \left.\frac{s_{ic}}{s_{in}}\right|_{s_{out}=0} \cdot k \cdot \left.\frac{s_{out}}{s_{oc}}\right|_{s_{in}=0} \qquad (8.223)$$

The effective forward gain b can be found using this formula.

■ **EXAMPLE**

Compute the effective forward gain $b = A_\infty \mathcal{R}$ for the circuit in Fig. 8.40.

As in the previous example, $k = g_m$, $s_{in} = i_{in}$, $s_{out} = v_o$, $s_{ic} = v_{be}$, and $s_{oc} = i_{oc} = g_m v_{be}$. To compute the first term in (8.223), the output v_o must be set to zero by shorting the output to ground. The resulting circuit is shown in Fig. 8.43a. The calculation gives

$$\left.\frac{s_{ic}}{s_{in}}\right|_{s_{out}=0} = \left.\frac{v_{be}}{i_{in}}\right|_{v_o=0} = r_\pi || R_F = 5\,\text{k}\Omega || 20\,\text{k}\Omega = 4.0\,\text{k}\Omega \qquad (8.224)$$

The last term in (8.223) is found by setting the input i_{in} to zero. This input current can be set to zero by replacing the source i_{in} with an open circuit, as shown in Fig. 8.43b. Treating the g_m generator as an independent source with value i_{oc} for this calculation, the result is

$$\left.\frac{s_{out}}{s_{oc}}\right|_{s_{in}=0} = \left.\frac{v_o}{i_{oc}}\right|_{i_{in}=0} = -[r_o || R_c || (R_F + r_\pi)]$$

$$= -[1\,\text{M}\Omega || 10\,\text{k}\Omega || (20\,\text{k}\Omega + 5\,\text{k}\Omega)] = -7.09\,\text{k}\Omega \qquad (8.225)$$

Substituting (8.224) and (8.225) into (8.223) gives

$$b = 4.0\,\text{k}\Omega(g_m)(-7.09\,\text{k}\Omega) = 4.0\,\text{k}\Omega(40\,\text{mA/V})(-7.09\,\text{k}\Omega) = -1134\,\text{k}\Omega$$

For comparison, we can find b using (8.217) and the values of A_∞ and $\mathcal{R}$ computed in the previous example:

$$b = A_\infty \mathcal{R} = -20 \text{ k}\Omega \, (56.7) = -1134 \text{ k}\Omega$$

- Both calculations give the same value for the effective forward gain b.

8.8.2 Closed-Loop Impedance Formula Using Return Ratio

Feedback affects the input and output impedance of a circuit. In this section, a useful expression for the impedance at any port in a feedback circuit in terms of return ratio[9] is derived. Consider the feedback circuit shown in Fig. 8.44. This feedback amplifier consists of linear elements: passive components, controlled sources, and transistor small-signal models. A controlled source k that is part of the small-signal model of an active device is shown explicitly. The derivation is carried out for the impedance Z_port looking into an arbitrary port that is labeled as *port X* in Fig. 8.44a. The port impedance can be found by driving the port by an independent current source as shown in Fig. 8.44b and computing $Z_\text{port} = v_x/i_x$. Since the circuit in Fig. 8.44b is linear, the signals s_{ic} and v_x are linear functions of the signals i_x and s_y applied to the ports labeled X and Y. Therefore, we can write

$$v_x = a_1 i_x + a_2 s_y \tag{8.226}$$

$$s_{ic} = a_3 i_x + a_4 s_y \tag{8.227}$$

From (8.226), the impedance looking into the port when $k = 0$ is

$$Z_\text{port}(k=0) = \left.\frac{v_x}{i_x}\right|_{k=0} = \left.\frac{v_x}{i_x}\right|_{s_y=0} = a_1 \tag{8.228}$$

Next we compute two return ratios for the controlled source k under different conditions. Both are used in the final formula for the closed-loop impedance. The first return ratio is found

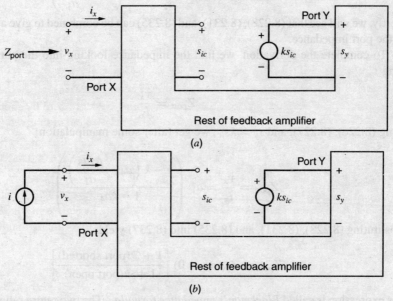

Figure 8.44 (*a*) The linear feedback circuit used to derive Blackman's impedance formula with respect to port X. (*b*) The circuit with port X driven by an independent current source.

with the port open. With port X open, $i_x = 0$. The return ratio is found by disconnecting the controlled source from the circuit and connecting a test source s_t where the dependent source was connected. With these changes to the circuit, $s_y = s_t$ and (8.227) becomes

$$s_{ic} = a_4 s_t \tag{8.229}$$

The output of the controlled source is the return signal

$$s_r = k s_{ic} \tag{8.230}$$

From the last two equations, we find

$$\mathcal{R}(\text{port open}) = -\frac{s_r}{s_t} = -k a_4 \tag{8.231}$$

The other return ratio is found with the port shorted. With port X shorted, the voltage v_x is zero. To find the return ratio, we disconnect the controlled source and connect test source s_t where the dependent source was connected. With these changes, (8.226) gives

$$i_x = -\frac{a_2}{a_1} s_t \tag{8.232}$$

Substituting (8.232) into (8.227), using $s_y = s_t$, and rearranging terms gives

$$s_{ic} = \left(a_4 - \frac{a_2 a_3}{a_1} \right) s_t \tag{8.233}$$

The return signal is

$$s_r = k s_{ic} \tag{8.234}$$

Combining these last two equations gives the return ratio with the port shorted

$$\mathcal{R}(\text{port shorted}) = -\frac{s_r}{s_t} = -k \left(a_4 - \frac{a_2 a_3}{a_1} \right) \tag{8.235}$$

Shortly, we will see that (8.228), (8.231), and (8.235) can be combined to give a useful formula for the port impedance.

To complete the derivation, we find the impedance looking into the port in Fig. 8.44b using

$$Z_{\text{port}} = \frac{v_x}{i_x} \tag{8.236}$$

Using (8.226), (8.227), and $s_y = k s_{ic}$, we get (after some manipulation)

$$Z_{\text{port}} = \frac{v_x}{i_x} = a_1 \left(\frac{1 - k \left(a_4 - \dfrac{a_2 a_3}{a_1} \right)}{1 - k a_4} \right) \tag{8.237}$$

Substituting (8.228), (8.231), and (8.235) into (8.237) yields

$$Z_{\text{port}} = Z_{\text{port}}(k=0) \left[\frac{1 + \mathcal{R}(\text{port shorted})}{1 + \mathcal{R}(\text{port open})} \right] \tag{8.238}$$

This expression is called *Blackman's impedance formula*.[9] The two return ratios, with the port open and shorted, are computed with respect to the same controlled source k. Equation 8.238

can be used to compute the impedance at any port, including the input and output ports. A key advantage of this formula is that it applies to any feedback circuit, regardless of the type of feedback. Usually, one of the two return ratios in (8.238) is zero, and in those cases Blackman's formula shows that feedback either increases or decreases the impedance by a factor $(1 + \mathcal{R})$.

■ **EXAMPLE**

Use Blackman's formula to find the output resistance for the feedback circuit in Fig. 8.40.
From Blackman's formula, the output resistance is given by

$$R_{out} = R_{out}(g_m = 0) \left[\frac{1 + \mathcal{R}(\text{output port shorted})}{1 + \mathcal{R}(\text{output port open})} \right] \quad (8.239)$$

Shorting the output port in Fig. 8.40c causes $v_{be} = 0$ so $i_r = g_m v_{be} = 0$; therefore, $\mathcal{R}(\text{output port shorted}) = 0$. The $\mathcal{R}(\text{output port open})$ is the same return ratio that was computed in (8.215), so $\mathcal{R}(\text{output port open}) = 56.7$. The only remaining value to be computed is the resistance at the output port when $g_m = 0$:

$$R_{out}(g_m = 0) = r_o || R_C || (R_F + r_\pi) = 1 \, M\Omega || 10 \, k\Omega || (20 \, k\Omega + 5 \, k\Omega)$$
$$= 7.1 \, k\Omega \quad (8.240)$$

Substituting into (8.239) yields

$$R_{out} = 7.1 \, k\Omega \left[\frac{1 + 0}{1 + 56.7} \right] = 120 \, \Omega \quad (8.241)$$

The negative feedback reduces the output resistance, which is desirable because the output is a voltage, and a low output resistance is desired in series with a voltage source.

■ **EXAMPLE**

Find the output resistance for the MOS *super-source follower* shown in Fig. 8.45a. Ignore body effect here to simplify the analysis.

The super-source follower uses feedback to reduce the output impedance. Ideal current sources I_1 and I_2 bias the transistors and are shown rather than transistor current sources to simplify the circuit. With current source I_1 forcing the current in M_1 to be constant, M_2 provides the output current when driving a load. There is feedback from v_{out} to v_{gs2} through M_1. The small-signal model for this circuit is shown in Fig. 8.45b. In this circuit, either g_{m1} or g_{m2} could be chosen as k. Here, we will use $k = g_{m2}$. In all the following calculations, the input source v_{in} is set to zero. First, the output resistance when $g_{m2} = 0$ is

$$R_{out}(g_{m2} = 0) = r_{o2} \quad (8.242)$$

This result may seem surprising at first, since the output is connected to the source of M_1, which is usually a low-impedance point. However, an ideal current source, which is a small-signal open circuit, is connected to the drain of M_1. Therefore the current in the g_{m1} generator flows only in r_{o1}, so M_1 has no effect on the output resistance when $g_{m2} = 0$.

The return ratio for the g_{m2} source with the output port open is found to be

$$\mathcal{R}(\text{output open}) = g_{m2} r_{o2} (1 + g_{m1} r_{o1}) \quad (8.243)$$

The return ratio with the output port shorted is

$$\mathcal{R}(\text{output shorted}) = 0 \quad (8.244)$$

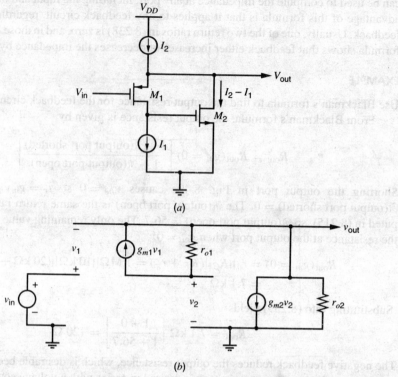

Figure 8.45 (a) The super-source-follower circuit. (b) The circuit with each transistor replaced by its small-signal model.

because shorting the output port forces $v_{out} = 0$ and $v_1 = -v_{out} = 0$. Hence no current flows in M_1, so $v_2 = 0$ and therefore the return ratio is zero. Substituting the last three equations into (8.238) gives the closed-loop output resistance

$$R_{out} = r_{o2} \left[\frac{1+0}{1+g_{m2}r_{o2}(1+g_{m1}r_{o1})} \right] \tag{8.245}$$

Assuming $g_m r_o \gg 1$, then

$$R_{out} \approx \frac{r_{o2}}{g_{m2}r_{o2}g_{m1}r_{o1}} = \frac{1}{g_{m2}g_{m1}r_{o1}} \tag{8.246}$$

which is much lower than the output resistance of a conventional source follower, which is about $1/g_m$. This result agrees with (3.137), which was derived without the use of feedback principles. Although g_{mb1} appears in (3.137), it does not appear in (8.246) because the body effect is ignored here.

The next example demonstrates an unusual case where neither return ratio vanishes in Blackman's impedance formula.

■ **EXAMPLE**

In the Wilson current source in Fig. 8.46a, assume that the three bipolar transistors are identical with $\beta_0 \gg 1$ and are biased in the forward-active region. Find the output resistance.

Blackman's impedance formula can be used here because there is a feedback loop formed by the current mirror Q_1-Q_3 with Q_2. With all transistors forward active and $\beta_F \gg 1$,

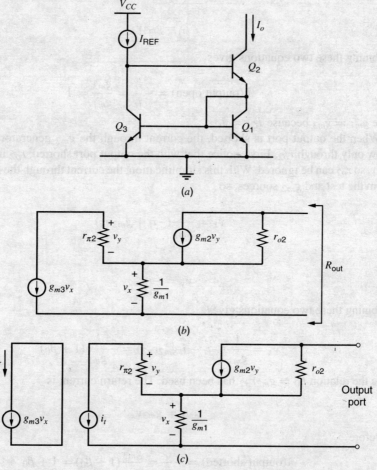

Figure 8.46 (a) The Wilson current mirror. (b) The circuit with each transistor replaced by its small-signal model. (c) The small-signal model modified for calculation of $\mathcal{R}$ for g_{m1}.

$I_{C1} = I_{C2} = I_{C3} = I_{REF}$. (The output is connected to other circuitry that is not shown, so I_{C2} is nonzero.) The small-signal model is shown in Fig. 8.46b, where diode-connected Q_1 is modeled by a resistor of value $1/g_{m1}$. Resistor $r_{\pi 3}$, which is in parallel with $1/g_{m1}$, is ignored (since $r_{\pi 3} = \beta_0/g_{m3} = \beta_0/g_{m1} \gg 1/g_{m1}$). Also r_{o3} is ignored, assuming that it is much larger than the resistance looking into the base of Q_2. Selecting $k = g_{m3}$ and calculating the first term in (8.238) gives

$$R_{\text{out}}(g_{m3} = 0) = r_{o2} + \frac{1}{g_{m1}} \approx r_{o2} \quad (8.247)$$

since setting $g_{m3} = 0$ forces the current through $r_{\pi 2}$ to be zero. Therefore, the voltage across $r_{\pi 2}$ is zero, which causes the current through the g_{m2} source to be zero.

The return ratios in Blackman's formula can be found using the circuit shown in Fig. 8.46c. First, let us find $\mathcal{R}$(output port open). When the output port is open, the current in the g_{m2} generator can only flow through the parallel resistor r_{o2}, so the currents through $r_{\pi 2}$ and $1/g_{m1}$ are equal and are supplied by the test source i_t. Therefore,

$$v_x = -i_t \frac{1}{g_{m1}} \quad (8.248)$$

Also

$$i_r = g_{m3}v_x \qquad (8.249)$$

Combining these two equations gives

$$\mathcal{R}(\text{output open}) = -\frac{i_r}{i_t} = \frac{g_{m3}}{g_{m1}} = 1 \qquad (8.250)$$

where $g_{m1} = g_{m3}$ because $I_{C1} = I_{C3}$.

When the output port is shorted, the current through the g_{m2} generator is not restricted to flow only through r_{o2}. First, notice that with the output port shorted, r_{o2} is in parallel with $1/g_{m1}$, so r_{o2} can be ignored. With this simplification, the current through the resistance $1/g_{m1}$ is from the test and g_{m2} sources, so

$$v_x = \frac{1}{g_{m1}}(-i_t + g_{m2}v_y) \qquad (8.251)$$

Now

$$v_y = -i_t r_{\pi 2} \qquad (8.252)$$

Combining these two equations gives

$$v_x = \frac{1}{g_{m1}}(-i_t - i_t g_{m2} r_{\pi 2}) = -\frac{i_t}{g_{m1}}(1 + \beta_0) \qquad (8.253)$$

where the relation $\beta_0 = g_{m2}r_{\pi 2}$ has been used. The return current is

$$i_r = g_{m3}v_x \qquad (8.254)$$

Therefore

$$\mathcal{R}(\text{output shorted}) = -\frac{i_r}{i_t} = \frac{g_{m3}}{g_{m1}}(1 + \beta_0) = 1 + \beta_0 \qquad (8.255)$$

Substituting in Blackman's formula gives

$$R_{\text{out}}(\text{closed loop}) = r_{o2} \cdot \frac{1 + (\beta_0 + 1)}{1 + 1} = \frac{r_{o2}(\beta_0 + 2)}{2} \approx \frac{\beta_0 r_{o2}}{2} \qquad (8.256)$$

This approximate result agrees with (4.91), which was derived without the use of Blackman's formula.

8.8.3 Summary—Return-Ratio Analysis

Return-ratio analysis is an alternative approach to feedback circuit analysis that does not use two-ports. The loop transmission is measured by the return ratio $\mathcal{R}$. The return ratio is a different measure of loop transmission than af from two-port analysis. (The return ratio $\mathcal{R}$ is referred to as *loop gain* in some textbooks. That name is not associated with $\mathcal{R}$ here to avoid confusion with $T = af$, which is called loop gain in this chapter.) For negative feedback circuits, $\mathcal{R} > 0$. In an ideal feedback circuit, $\mathcal{R} \to \infty$ and the closed-loop gain is A_∞, which typically depends only on passive components. The actual gain of a feedback circuit is close to A_∞ if $\mathcal{R} \gg 1$. Blackman's impedance formula (8.238) gives the closed-loop impedance in terms of two return ratios.

Return-ratio analysis is often simpler than two-port analysis of feedback circuits. For example, return-ratio analysis uses equations that are independent of the type of feedback, and simple manipulations of the circuit allow computation of the various terms in the equations. In contrast, two-port analysis uses different two-port representations for each of the four feedback configurations (series-series, series-shunt, shunt-series, and shunt-shunt). Therefore, the type of feedback must be correctly identified before undertaking two-port analysis. The resulting two-ports for the amplifier and feedback networks must be manipulated to find the open-loop forward gain a and the open-loop input and output impedances. With two-port analysis, the open-loop impedance is either multiplied or divided by $(1 + T)$ to give the closed-loop impedance, depending upon the type of feedback. In contrast, Blackman's formula gives an equation for finding the closed-loop impedance, and this one equation applies to any port in any feedback circuit.

8.9 Modeling Input and Output Ports in Feedback Circuits

Throughout this chapter, the source and load impedances have been included when analyzing a feedback circuit. For instance, the inverting voltage-gain circuit in Fig. 8.47a, with source resistance R_S and load resistance R_L, can be analyzed using the two-port or return-ratio methods described in this chapter. The resulting model is shown in Fig. 8.47b. The source and load resistances do not appear explicitly in the model, but the gain A, input resistance R_i, and output resistance R_o are functions of the source and load resistances. Therefore, use of this model requires that both the source and load resistances are known. However, both the source and load are not always known or fixed in value. For example, a feedback amplifier might have to drive a range of load resistances. In that case, it would be desirable to have a simple model of the amplifier with the following properties: the elements in the model do not depend on the load, and the effect of a load on the gain can be easily calculated.

If only one of the resistances R_S and R_L is known, a useful model can be generated. First, consider Fig. 8.47a when the source resistance is unknown but the load is known. Then, the model in Fig. 8.47c can be used. Here, the key differences from the model in Fig. 8.47b are that R_i' and A' are used rather than R_i and A, R_S is shown explicitly, and the controlling voltage for A' is the voltage v_i across R_i' rather than the source voltage v_s. (The single-prime mark here denotes that the quantity is computed with R_S unknown.) The input resistance R_i' and gain A' are computed with the load connected and with an ideal input driving network. Here, the Thévenin driving network is ideal if the source resistance R_S is zero. (If the input is a Norton equivalent consisting of a current source and parallel source resistance, R_i' and A' are found with $R_S \to \infty$.) The resulting A' and R_i' are not functions of R_S. The source resistance R_S in Fig. 8.47c forms a voltage divider with R_i'; therefore, the overall voltage gain can be found by

$$\frac{v_o}{v_s} = \frac{v_o}{v_i} \cdot \frac{v_i}{v_s} = A' \frac{R_i'}{R_i' + R_S} \qquad (8.257)$$

Next, consider Fig. 8.47a when the load is unknown but the source resistance is known. Fig. 8.47d shows the appropriate model. Here, the key differences from the model in Fig. 8.47b are that R_o'' and A'' are used rather than R_o and A, and R_L is shown explicitly. (The double-prime mark denotes that the quantity is computed with R_L unknown.) The output resistance R_o'' and gain A'' for this Thévenin model are computed assuming an ideal load, which is an open circuit here ($R_L \to \infty$). The load resistance R_L in Fig. 8.47d forms a voltage divider with R_o'', so the loaded voltage gain is

$$\frac{v_o}{v_s} = A'' \frac{R_L}{R_o'' + R_L} \qquad (8.258)$$

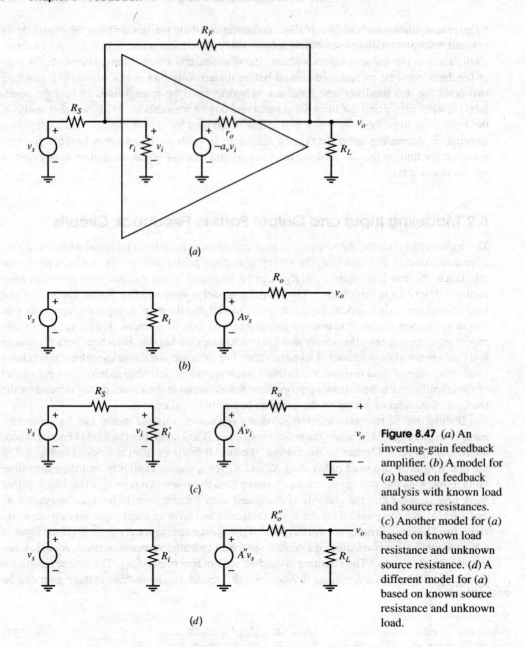

Figure 8.47 (a) An inverting-gain feedback amplifier. (b) A model for (a) based on feedback analysis with known load and source resistances. (c) Another model for (a) based on known load resistance and unknown source resistance. (d) A different model for (a) based on known source resistance and unknown load.

■ **EXAMPLE**

Model Fig. 8.47a with the circuit in Fig. 8.47d, assuming the load resistance is unknown.

The model in Fig. 8.47d can be used when R_L is unknown. Using return-ratio analysis, the calculations of R_o'' and A'' are as follows. With $R_L \to \infty$, the circuit in Fig. 8.47a is identical to Fig. 8.39a, and therefore the return ratio is given by (8.196):

$$\mathcal{R}'' = \mathcal{R}(R_L \to \infty) = \frac{R_S \| r_i}{R_S \| r_i + R_F + r_o} a_v \qquad (8.259)$$

The output resistance with $R_L \to \infty$, R_o'', can be found using Blackman's formula. First, the output resistance with the controlled source set to zero is

$$R_o''(a_v = 0) = r_o || (R_F + r_i) || R_S) \tag{8.260}$$

The return ratio with the output port open is given in (8.259). The return ratio with the output port shorted is zero because shorting the output eliminates the feedback. Therefore, for the closed-loop output resistance, (8.238) gives

$$R_o'' = r_o || (R_F + r_i) || R_S) \frac{1 + 0}{1 + \dfrac{R_S || r_i}{R_S || r_i + R_F + r_o} a_v} \tag{8.261}$$

Calculation of A'' requires that A_∞'' and d'' be found with $R_L \to \infty$.

$$A_\infty'' = \left. \frac{v_o}{v_s} \right|_{a_v = \infty \,\&\, R_L = \infty} = -\frac{R_F}{R_S} \tag{8.262}$$

and

$$d'' = \left. \frac{v_o}{v_s} \right|_{a_v = 0 \,\&\, R_L = \infty} = \frac{(R_F + r_o) || r_i}{R_S + (R_F + r_o) || r_i} \cdot \frac{r_o}{R_F + r_o} \tag{8.263}$$

Using the results for A_∞'', $\mathscr{R}''$, and d'', we can compute

$$A'' = A_\infty'' \frac{\mathscr{R}''}{1 + \mathscr{R}''} + \frac{d''}{1 + \mathscr{R}''} \tag{8.264}$$

[Typically, the $d''/(1 + \mathscr{R}'')$ term is small and can be ignored.] The voltage gain when a resistive load is connected can be found using (8.258). The only element of the model that was not computed is the input resistance R_i. It is a function of R_L, so it can only be computed once R_L is known.

■

The output-port model in Fig. 8.47d could be drawn as a Thévenin or Norton equivalent. With the Thévenin equivalent shown (a controlled voltage source in series with an output resistance), R_o'' and A'' are computed with $R_L \to \infty$. When the output port is modeled by a Norton equivalent with a controlled current source and parallel output resistance, their values are found with $R_L = 0$.

PROBLEMS

Note: In these problems, *loop transmission* is used generically to refer to loop gain $T = af$ or return ratio $\mathscr{R}$.

8.1(a) In a feedback amplifier, forward gain $a = 10,000$ and feedback factor $f = 10^{-3}$. Calculate overall gain A and the percentage change in A if a changes by 10 percent.

(b) Repeat (a) if $f = 0.1$.

8.2 For the characteristic of Fig. 8.2 the following data apply:

$$S_{o2} = 15\text{ V} \quad S_{o1} = 7\text{ V} \quad a_1 = 50{,}000$$

$$a_2 = 20{,}000$$

(a) Calculate and sketch the overall transfer characteristic of Fig 8.3 for the above amplifier when placed in a feedback loop with $f = 10^{-4}$.

(b) Repeat (a) with $f = 0.1$.

8.3(a) For the conditions in Problem 8.2(b), sketch the output voltage waveform S_o and the error voltage waveform S_ϵ if a sinusoidal input voltage S_i with amplitude 1.5 V is applied.

(b) Repeat (a) with an input amplitude of 2 V.

8.4 Verify (8.40), (8.41), and (8.42) for a shunt-series feedback amplifier.

8.5 Verify (8.43), (8.44), and (8.45) for a series-series feedback amplifier.

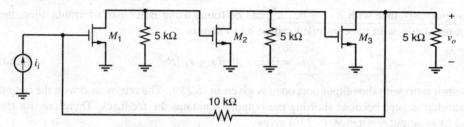

Figure 8.48 An ac schematic of a shunt-shunt feedback amplifier.

8.6 For the shunt-shunt feedback amplifier of Fig. 8.15a, take $R_F = 100$ kΩ and $R_L = 15$ kΩ. For the op amp, assume that $R_i = 500$ kΩ, $R_o = 200$ Ω, and $a_v = 75{,}000$. Calculate input resistance, output resistance, loop transmission, and closed-loop gain:

(a) Using the formulas from two-port analysis (Section 8.5).

(b) Using the formulas from return-ratio analysis (Section 8.8).

8.7 The ac schematic of a shunt-shunt feedback amplifier is shown in Fig. 8.48. All transistors have $I_D = 1$ mA, $W/L = 100$, $k' = 60$ μA/V^2, and $\lambda = 1/(50$ V$)$.

(a) Calculate the overall gain v_o/i_i, the loop transmission, the input impedance, and the output impedance at low frequencies. Use the formulas from two-port analysis (Section 8.5).

(b) If the circuit is fed from a source resistance of 2 kΩ in parallel with i_i, what is the new output resistance of the circuit?

8.8(a) Repeat Problem 8.7(a) with all NMOS transistors in Fig. 8.48 replaced by bipolar *npn* transistors. All collector currents are 1 mA and $\beta = 200$, $V_A = 50$ V, and $r_b = 0$.

(b) If the circuit is fed from a source resistance of 2 kΩ in parallel with i_i, what is the new output resistance of the circuit?

8.9 Repeat Problem 8.7 using the formulas from return-ratio analysis (Section 8.8).

8.10 Repeat Problem 8.8 using the formulas from return-ratio analysis (Section 8.8).

8.11 The half-circuit of a balanced monolithic series-series triple is shown in Fig. 8.18a. Calculate the input impedance, output impedance, loop gain, and overall gain of the half-circuit at low frequencies using the following data:

$$R_{E1} = R_{E2} = 290 \ \Omega \quad R_F = 1.9 \ k\Omega$$

$$R_{L1} = 10.6 \ k\Omega \quad R_{L2} = 6 \ k\Omega$$

For the transistors, $I_{C1} = 0.5$ mA, $I_{C2} = 0.77$ mA, $I_{C3} = 0.73$ mA, $\beta = 120$, $r_b = 0$, and $V_A = 40$ V.

8.12 Repeat Problem 8.11 if the output signal is taken as the voltage at the emitter of Q_3.

8.13 A feedback amplifier is shown in Fig. 8.49. Device data are as follows: $\beta_{npn} = 200$, $\beta_{pnp} = 100$, $|V_{BE(on)}| = 0.7$ V, $r_b = 0$, and $|V_A| = \infty$. If the dc input voltage is zero, calculate the overall gain v_o/v_i, the loop gain, and the input and output impedance at low frequencies. Compare your answers with a SPICE simulation. Also use SPICE to plot the complete large-signal transfer characteristic and find the second and third harmonic distortion in v_o for a sinusoidal input voltage with peak-peak amplitude of 0.5 V at v_i.

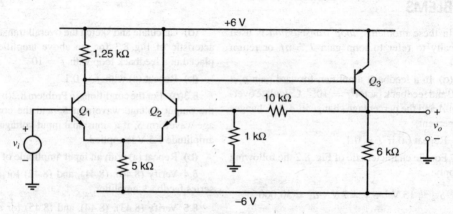

Figure 8.49 Feedback amplifier circuit.

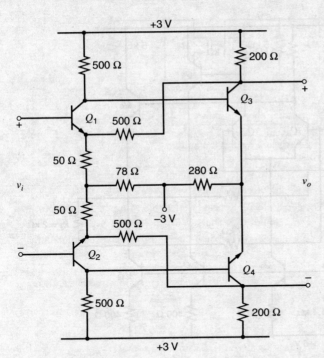

Figure 8.50 Balanced series-shunt feedback amplifier.

8.14 Replace *npn* transistors Q_1-Q_2 in Fig. 8.49 with NMOS transistors M_1-M_2, and replace the *pnp* transistor Q_3 with PMOS transistor M_3. Also, replace the 1.25 kΩ resistor in the drain of M_1 with a 4.35 kΩ resistor. Repeat the calculations and simulations in Problem 8.13. For all transistors, use $W/L = 100$, $\gamma = 0$, and $|\lambda| = 0$. Also, $V_{tn} = -V_{tp} = 1$ V, $k'_n = 60$ μA/V^2, and $k'_p = 20$ μA/V^2.

8.15 A balanced monolithic series-shunt feedback amplifier is shown in Fig. 8.50.

(a) If the common-mode input voltage is zero, calculate the bias current in each device. Assume that β_F is large.

(b) Calculate the voltage gain, input impedance, output impedance, and loop gain of the circuit at low frequencies using the following data:

$$\beta = 100 \quad r_b = 50 \, \Omega \quad V_A = \infty \quad V_{BE(on)} = 0.7 \text{ V}$$

(c) Compare your answers with a SPICE simulation (omit the loop gain) and also use SPICE to plot the complete large-signal transfer characteristic. If the resistors have a temperature coefficient of +1000 ppm/°C, use SPICE to determine the temperature coefficient of the circuit gain over the range -55°C to $+125$°C.

8.16 How does the loop gain $T = af$ of the circuit of Fig. 8.50 change as the following circuit elements change? Discuss qualitatively.

(a) 50 Ω emitter resistor of the input stage
(b) 500 Ω feedback resistor
(c) 200 Ω load resistor on the output

8.17 The ac schematic of a shunt-series feedback amplifier is shown in Fig. 8.31. Element values are $R_F = 1$ kΩ, $R_E = 100$ Ω, $R_{L1} = 4$ kΩ, $R_S = 1/y_S = 1$ kΩ, and $z_L = 0$. Device data: $\beta = 200$, $r_b = 0$, $I_{C1} = I_{C2} = 1$ mA, $V_A = 100$ V.

(a) Calculate the overall gain i_o/i_i, the loop transmission, and the input and output impedances at low frequencies.

(b) If the value of R_{L1} changes by $+10$ percent, what is the approximate change in overall transmission and input impedance?

8.18(a) Repeat Problem 8.17(a) with $R_F = 5$ kΩ, $R_E = 200$ Ω, $R_{L1} = 10$ kΩ, and $y_s = 0$.

(b) If the collector current of Q_1 increases by 20 percent, what will be the approximate change in overall gain and output resistance?

8.19 Calculate the transconductance, input impedance, output impedance, and loop transmission at low frequencies of the local series-feedback stage of Fig. 8.34 with parameters $R_E = 100$ Ω, $\beta = 150$, $I_C = 1$ mA, $r_b = 100$ Ω, and $V_A = 80$ V.

8.20 If the 723 voltage regulator is used to realize an output voltage $V_o = 10$ V with a 1-kΩ load, calculate the output resistance and the loop gain of the regulator. If a 500-Ω load is connected to the

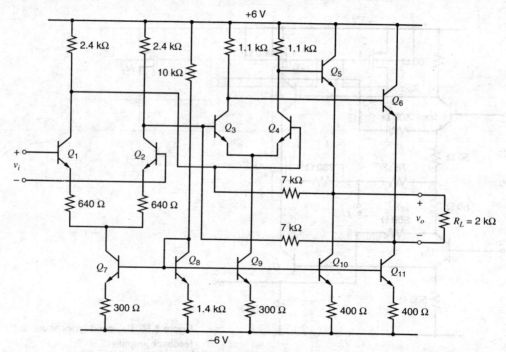

Figure 8.51 Circuit diagram of the 733 wideband monolithic amplifier.

regulator in place of the 1-kΩ load, calculate the new value of V_o. Use SPICE to determine the line regulation and load regulation of the circuit. Use $I_1 = 1$ mA, $\beta = 100$, $V_A = 100$ V, $I_s = 10^{-15}$ A, and $r_b = 0$.

8.21 A commercial wideband monolithic feedback amplifier (the 733) is shown in Fig. 8.51. This consists of a local series-feedback stage feeding a two-stage shunt-shunt feedback amplifier. The current output of the input stage acts as a current drive to the shunt-shunt output stage.

(a) Assuming all device areas are equal, calculate the collector bias current in each device.

(b) Calculate input impedance, output impedance, and overall gain v_o/v_i for this circuit at low frequencies with $R_L = 2$ kΩ. Also calculate the loop gain of the output stage.

Data: $\beta = 100$, $r_b = 0$, $r_o = \infty$.

(c) Compare your answers with a SPICE simulation of the bias currents, input and output impedances, and the voltage gain.

8.22 A variable-gain CMOS amplifier is shown in Fig. 8.52. Note that M_4 represents shunt feedback around M_6. Assuming that the bias value of V_i is adjusted so that $V_{GD6} = 0$ V dc, calculate bias currents in all devices and the small-signal voltage gain and output resistance for V_c equal to 3 V and then 4 V. Compare your answer with a SPICE simulation and use SPICE to plot out the complete large-signal transfer characteristic of the circuit. Use $\mu_n C_{ox} = 60$ μA/V^2, $\mu_p C_{ox} = 30$ μA/V^2, $V_{tn} = 0.8$ V, $V_{tp} = -0.8$ V, $\lambda_n = \lambda_p = 0$, and $\gamma_n = 0.5$ V$^{1/2}$.

8.23 Assume the BiCMOS amplifier of Fig. 3.78 is fed from a current source. Calculate the low-frequency small-signal transresistance v_o/i_i, the loop gain, and the input and output impedances of the circuit. Use data as in Problem 3.17. Compare your answers with a SPICE simulation and also use SPICE to plot the complete large-signal transfer characteristic of the circuit.

8.24 An active-cascode gain stage is shown in Fig. 8.53. Assume the amplifier A_1 has a voltage gain $a = 1 \times 10^3$ and infinite input impedance. For the transistors, $k'_n = 140$ μA/V^2, $V_{ov} = 0.3$ V, $\gamma = 0$, and $\lambda_n = 0.03$ V^{-1}. Assume all transistors are active. Calculate the output resistance using Blackman's impedance formula. Then calculate the voltage gain v_o/v_i.

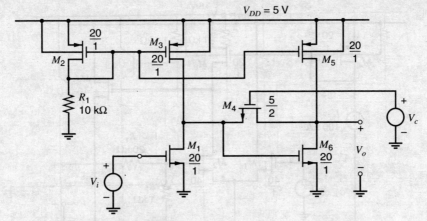

Figure 8.52 Variable-gain CMOS amplifier for Problem 8.22.

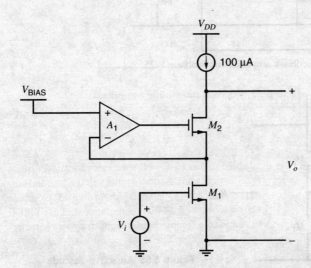

Figure 8.53 An active-cascode gain stage.

8.25 A CMOS feedback amplifier is shown in Fig. 8.54. If the dc input voltage is zero, calculate the overall gain v_o/v_i and the output resistance. Compare your answer with a SPICE simulation. Use $\mu_n C_{ox} = 60 \times 10^{-6}$ A/V^2, $\mu_p C_{ox} = 30 \times 10^{-6}$ A/V^2, $V_{tn} = 0.8$ V, $V_{tp} = -0.8$ V, $\lambda_n = |\lambda_p| = 0.03$ V^{-1}, and $\gamma_n = \gamma_p = 0$.

8.26 Use Blackman's impedance formula to find the output resistance of the active-cascode current source in Fig. 8.55. Express the result in terms of g_{m1}, g_{m2}, r_{o1}, r_{o2}, and a, which is the voltage gain of the op amp. Assume all transistors are active with $(W/L)_1 = (W/L)_2 = (W/L)_3$ and $\gamma = 0$. (The drain of M_2 connects to other circuitry that is not shown.) Also, assume that the op amp has infinite input impedance and zero output impedance.

(a) Carry out the calculations with respect to controlled source g_{m2}.

(b) Repeat the calculations with respect to the voltage-controlled voltage source a in the amplifier.

(c) Compare the results of (a) and (b).

8.27 Use return-ratio analysis and Blackman's impedance formula to find the closed-loop gain, return ratio, input resistance, and output resistance for the inverting gain amplifier in Fig. 8.56. For the op amp, assume that $R_i = 1$ MΩ, $R_o = 10$ kΩ, and $a_v = 200$.

8.28 An ac schematic of a local shunt-shunt feedback circuit is shown in Fig. 8.57. Take $R_F = 100$ kΩ and $R_L = 15$ kΩ. For the MOS transistor, $I_D = 0.5$ mA, $W/L = 100$, $k' = 180$ μA/V^2, and $r_o = \infty$.

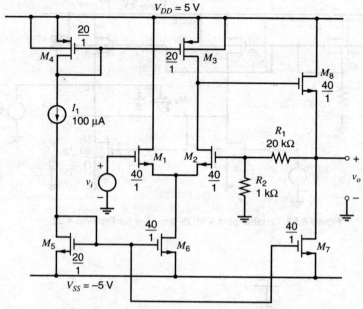

Figure 8.54 CMOS feedback amplifier for Problem 8.25.

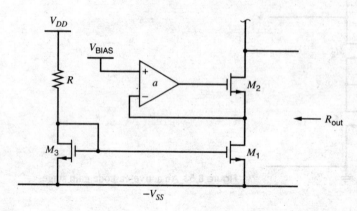

Figure 8.55 An active-cascode current source.

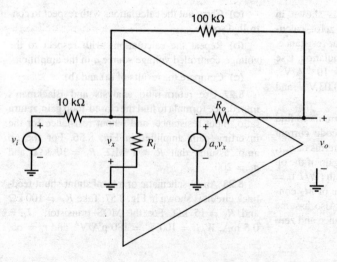

Figure 8.56 An inverting feedback amplifier.

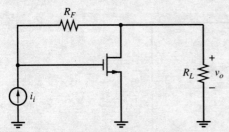

Figure 8.57 A local shunt-shunt feedback amplifier.

Calculate input resistance, output resistance, loop transmission, and closed-loop gain:

(a) Using the formulas from two-port analysis (Section 8.5).

(b) Using the formulas from return-ratio analysis (Section 8.8).

8.29 Replace the MOS transistor in Fig. 8.57 by a *npn* transistor. Take $R_F = 2$ kΩ, $R_L = 2$ kΩ, $\beta = 200$, $I_C = 1$ mA, $r_b = 0$, and $V_A = 100$ V.

(a) Repeat Problem 8.28(a).

(b) Repeat Problem 8.28(b).

8.30 A voltage-follower feedback circuit is shown in Fig. 8.58. For the MOS transistor, $I_D = 0.5$ mA, $k' = 180$ μA/V^2, $r_o = \infty$, $W/L = 100$, $|\phi_f| = 0.3$ V, and $\gamma = 0.3$ V$^{1/2}$. For the op amp, assume that $R_i = 1$ MΩ, $R_o = 10$ kΩ, and $a_v = 1{,}000$. Calculate input resistance, output resistance, loop transmission, and closed-loop gain:

(a) Using the formulas from two-port analysis (Section 8.5).

(b) Using the formulas from return-ratio analysis (Section 8.8).

8.31 Replace the MOS transistor in Fig. 8.58 with a *npn* transistor. For the transistor, $I_C = 0.5$ mA and $r_o = \infty$.

(a) Repeat the calculations in Problem 8.30(a).

(b) Repeat the calculations in Problem 8.30(b).

8.32 For the noninverting amplifier shown in Fig. 8.59, $R_1 = 1$ kΩ, and $R_2 = 5$ kΩ. For the op amp, take $R_i = 1$ MΩ, $R_o = 100$ Ω, and $a_v = 1 \times 10^4$. Calculate input resistance, output resistance, loop transmission, and closed-loop gain:

(a) Using the formulas from two-port analysis (Section 8.5).

(b) Using the formulas from return-ratio analysis (Section 8.8).

8.33 Calculation of return ratio begins by breaking a feedback loop at a controlled source. However, breaking a feedback loop at a controlled source is often impossible in a SPICE simulation because the controlled source (e.g., the g_m source in a transistor's small-signal model) is embedded in a small-signal model. Therefore, it cannot be accessed or disconnected in simulation. A technique that can be used to simulate the return ratio with SPICE is illustrated in Fig. 8.60 for the circuit in Fig. 8.59. First, the independent source V_i is set to zero. Next, ac test signals v_t and

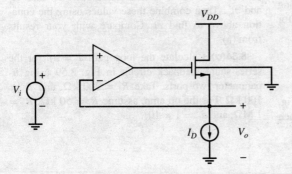

Figure 8.58 A voltage follower.

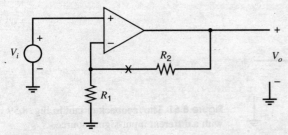

Figure 8.59 A noninverting feedback amplifier.

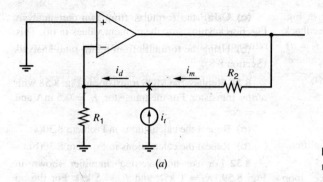

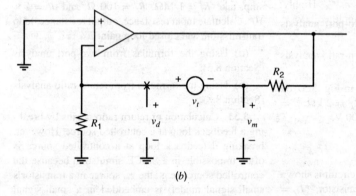

Figure 8.60 The feedback circuit in Fig. 8.59 modified to calculate (a) $\mathcal{R}'_i$ and (b) $\mathcal{R}'_v$.

i_t are injected into the loop at a convenient point (e.g., at the "X" in Fig. 8.59), creating two modified versions of the circuit as shown in Fig. 8.60a and 8.60b. Using Fig. 8.60a, calculate $\mathcal{R}'_i = -i_m/i_d$. Using Fig. 8.60b, calculate $\mathcal{R}'_v = -v_m/v_d$. The amplitudes of test signals i_t and v_t do not affect $\mathcal{R}'_i$ or $\mathcal{R}'_v$. Also, these ac test signals do not affect the dc operating point of the feedback circuit. The return ratio $\mathcal{R}$ for the controlled source is related to $\mathcal{R}'_i$ and $\mathcal{R}'_v$ by[10]

$$\frac{1}{1+\mathcal{R}} = \frac{1}{1+\mathcal{R}'_i} + \frac{1}{1+\mathcal{R}'_v}$$

(a) Compute $\mathcal{R}'_i$ and $\mathcal{R}'_v$ for the circuit in Fig. 8.60. Use element values from Problem 8.32. Then combine these values using the equation above to find $\mathcal{R}$.

(b) Compute $\mathcal{R}$ directly by breaking the loop at the a_v controlled source. Compare the results in (a) and (b).

(c) Carry out a SPICE simulation to find $\mathcal{R}'_i$ and $\mathcal{R}'_v$. Then combine these values using the equation above to find $\mathcal{R}$. Compare with your results from (a).

8.34(a) Calculate the loop gain $T = af$ for the series-shunt feedback circuit in Fig. 8.59 using h-parameter two-ports. Take $R_1 = 200$ kΩ, and $R_2 = 100$ kΩ. For the op amp, assume $R_i = 50$ kΩ, $R_o = 1$ MΩ, and $a_v = 1 \times 10^3$.

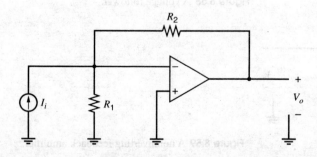

Figure 8.61 The feedback circuit in Fig. 8.59 with a different input-signal source.

(b) If the input source location and type in Fig. 8.59 are changed as shown in Fig. 8.61, the feedback is now shunt-shunt. Calculate the loop gain $T = af$ for this shunt-shunt feedback circuit using y-parameter two-ports. Use the element values in (a).

(c) Calculate the return ratio $\mathcal{R}$ for the circuit in Fig. 8.61, again using the element values in (a). This return ratio is the same as the return ratio for the circuit in Fig. 8.59. Why?

(d) Compare the results in (a), (b), and (c).

REFERENCES

1. C. A. Desoer and E. S. Kuh, *Basic Circuit Theory,* McGraw-Hill, New York, 1969.

2. J. E. Solomon and G. R. Wilson, "A Highly Desensitized, Wideband Monolithic Amplifier," *IEEE J. Solid State Circuits,* Vol. SC-1, pp. 19–28, September 1966.

3. A. B. Grebene, *Analog Integrated Circuit Design,* Van Nostrand Reinhold, New York, 1972, Chapter 6.

4. R. S. Muller and T. I. Kamins, *Device Electronics for Integrated Circuits,* Wiley, New York, 1977.

5. H. W. Bode, *Network Analysis and Feedback Amplifier Design,* Van Nostrand, New York, 1945.

6. E. S. Kuh and R. A. Rohrer, *Theory of Linear Active Networks,* Holden-Day, San Francisco, 1967.

7. S. Rosenstark, *Feedback Amplifier Principles,* MacMillan, New York, 1986.

8. B. Nikolic and S. Marjanovic, "A General Method of Feedback Amplifier Analysis," *IEEE Int'l Symp. on Circuits and Systems,* pp. 415–418, Monterey, CA, 1998.

9. R. B. Blackman, "Effect of Feedback on Impedance," *Bell Sys. Tech. J.,* Vol. 23, pp. 269–277, October 1943.

10. R. D. Middlebrook, "Measurement of Loop Gain in Feedback Systems," *Int. J. Electronics,* Vol. 38, no. 4, pp. 485–512, 1975.

CHAPTER 9

Frequency Response and Stability of Feedback Amplifiers

9.1 Introduction

In Chapter 8, we considered the effects of negative feedback on circuit parameters such as gain and terminal impedance. We saw that application of negative feedback resulted in a number of performance improvements, such as reduced sensitivity of gain to active-device parameter changes and reduction of distortion due to circuit nonlinearities.

In this chapter, we see the effect of negative feedback on the frequency response of a circuit. The possibility of *oscillation* in feedback circuits is illustrated, and methods of overcoming these problems by *compensation* of the circuit are described. Finally, the effect of compensation on the large-signal high-frequency performance of feedback amplifiers is investigated.

Much of the analysis in this chapter is based on the ideal block diagram in Fig. 9.1. This block diagram includes the forward gain a and feedback factor f, which are the parameters used in two-port analysis of feedback circuits in Chapter 8. The equations and results in this chapter could be expressed in terms of the parameters used in the return-ratio analysis in Chapter 8 by an appropriate change of variables, as shown in Appendix A9.1.

The equations and relationships in this chapter are general and can be applied to any feedback circuit. However, for simplicity we will often assume the feedback factor f is a positive, unitless constant. One circuit that has such an f is the series-shunt feedback circuit shown in Fig 8.24. In this circuit, the feedback network is a resistive voltage divider, so f is a constant with $0 \leq f \leq 1$. The forward gain a is a voltage gain that is positive at low frequencies. This circuit gives a noninverting closed-loop voltage gain.

9.2 Relation Between Gain and Bandwidth in Feedback Amplifiers

Chapter 8 showed that the performance improvements produced by negative feedback were obtained at the expense of a reduction in gain by a factor $(1 + T)$, where T is the loop gain. The performance specifications that were improved were also changed by the factor $(1 + T)$.

In addition to the foregoing effects, negative feedback also tends to *broadband* the amplifier. Consider first a feedback circuit as shown in Fig. 9.1 with a simple basic amplifier whose gain function contains a single pole

$$a(s) = \frac{a_0}{1 - \dfrac{s}{p_1}} \qquad (9.1)$$

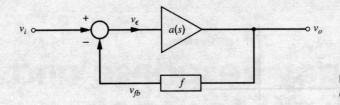

Figure 9.1 Feedback circuit configuration.

where a_0 is the low-frequency gain of the basic amplifier and p_1 is the basic-amplifier pole in radians per second. Assume that the feedback path is purely resistive and thus the feedback function f is a positive constant. Since Fig. 9.1 is an ideal feedback arrangement, the *overall* gain is

$$A(s) = \frac{v_o}{v_i} = \frac{a(s)}{1 + a(s)f} \qquad (9.2)$$

where the loop gain is $T(s) = a(s)f$. Substitution of (9.1) in (9.2) gives

$$A(s) = \frac{\dfrac{a_0}{1 - \dfrac{s}{p_1}}}{1 + \dfrac{a_0 f}{1 - \dfrac{s}{p_1}}} = \frac{a_0}{1 - \dfrac{s}{p_1} + a_0 f} = \frac{a_0}{1 + a_0 f} \cdot \frac{1}{1 - \dfrac{s}{p_1}\dfrac{1}{1 + a_0 f}} \qquad (9.3)$$

From (9.3) the low-frequency gain A_0 is

$$A_0 = \frac{a_0}{1 + T_0} \qquad (9.4)$$

where

$$T_0 = a_0 f = \text{low-frequency loop gain} \qquad (9.5)$$

The -3-dB bandwidth of the feedback circuit (i.e., the new pole magnitude) is $(1 + a_0 f) \cdot |p_1|$ from (9.3). Thus the feedback has reduced the low-frequency gain by a factor $(1 + T_0)$, which is consistent with the results of Chapter 8, but it is now apparent that the -3-dB frequency of the circuit has been *increased* by the same quantity $(1 + T_0)$. Note that the gain-bandwidth product is constant. These results are illustrated in the Bode plots of Fig. 9.2, where the magnitudes of

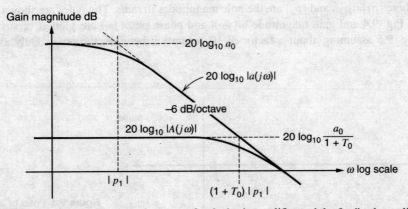

Figure 9.2 Gain magnitude versus frequency for the basic amplifier and the feedback amplifier.

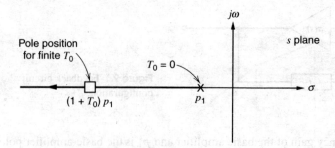

Figure 9.3 Locus of the pole of the circuit of Fig. 9.1 as loop gain T_0 varies.

$a(j\omega)$ and $A(j\omega)$ are plotted versus frequency on log scales. It is apparent that the gain curves for any value of T_0 are contained in an envelope bounded by the curve of $|a(j\omega)|$.

Because the use of negative feedback allows the designer to trade gain for bandwidth, negative feedback is widely used as a method for designing broadband amplifiers. The gain reduction that occurs is made up by using additional gain stages, which in general are also feedback amplifiers.

Let us now examine the effect of the feedback on the pole of the overall transfer function $A(s)$. It is apparent from (9.3) that as the low-frequency loop gain T_0 is increased, the magnitude of the pole of $A(s)$ increases. This is illustrated in Fig. 9.3, which shows the *locus* of the pole of $A(s)$ in the s plane as T_0 varies. The pole starts at p_1 for $T_0 = 0$ and moves out along the negative real axis as T_0 is made positive. Figure 9.3 is a simple *root-locus* diagram and will be discussed further in Section 9.5.

9.3 Instability and the Nyquist Criterion[1]

In the above simple example the basic amplifier was assumed to have a single-pole transfer function, and this situation is closely approximated in practice by internally compensated general-purpose op amps. However, many amplifiers have multipole transfer functions that cause deviations from the above results. The process of compensation overcomes these problems, as will be seen later.

Consider an amplifier with a three-pole transfer function

$$a(s) = \frac{a_0}{\left(1 - \dfrac{s}{p_1}\right)\left(1 - \dfrac{s}{p_2}\right)\left(1 - \dfrac{s}{p_3}\right)} \tag{9.6}$$

where $|p_1|$, $|p_2|$, and $|p_3|$ are the pole magnitudes in rad/s. The poles are shown in the s plane in Fig. 9.4 and gain magnitude $|a(j\omega)|$ and phase ph $a(j\omega)$ are plotted versus frequency in Fig. 9.5 assuming about a factor of 10 separation between the poles. Only asymptotes are

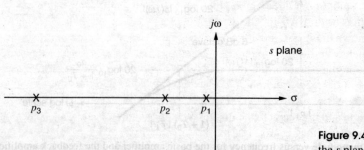

Figure 9.4 Poles of an amplifier in the s plane.

9.3 Instability and the Nyquist Criterion

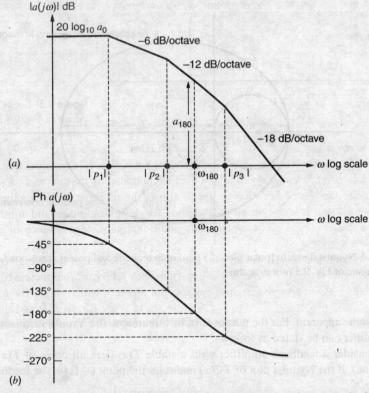

Figure 9.5 Gain and phase versus frequency for a circuit with a three-pole transfer function.

shown for the magnitude plot. At frequencies above the first pole magnitude $|p_1|$, the plot of $|a(j\omega)|$ falls at 6 dB/octave and ph $a(j\omega)$ approaches $-90°$. Above $|p_2|$ these become 12 dB/octave and $-180°$, and above $|p_3|$ they become 18 dB/octave and $-270°$. The frequency where ph $a(j\omega) = -180°$ has special significance and is marked ω_{180}, and the value of $|a(j\omega)|$ at this frequency is a_{180}. If the three poles are fairly widely separated (by a factor of 10 or more), the phase shifts at frequencies $|p_1|, |p_2|,$ and $|p_3|$ are approximately $-45°, -135°,$ and $-225°$, respectively. This will now be assumed for simplicity. In addition, the gain magnitude will be assumed to follow the asymptotic curve and the effect of these assumptions in practical cases will be considered later.

Now consider this amplifier connected in a feedback loop as in Fig. 9.1 with f a positive constant. Since f is constant, the loop gain $T(j\omega) = a(j\omega)f$ will have the same variation with frequency as $a(j\omega)$. A plot of $af(j\omega) = T(j\omega)$ in magnitude and phase on a polar plot (with ω as a parameter) can thus be drawn using the data of Fig. 9.5 and the magnitude of f. Such a plot for this example is shown in Fig. 9.6 (not to scale) and is called a *Nyquist diagram*. The variable on the curve is frequency and varies from $\omega = -\infty$ to $\omega = \infty$. For $\omega = 0$, $|T(j\omega)| = T_0$ and ph $T(j\omega) = 0$, and the curve meets the real axis with an intercept T_0. As ω increases, as Fig. 9.5 shows, $|a(j\omega)|$ decreases and ph $a(j\omega)$ becomes negative and thus the plot is in the fourth quadrant. As $\omega \to \infty$, ph $a(j\omega) \to -270°$ and $|a(j\omega)| \to 0$. Consequently, the plot is asymptotic to the origin and is tangent to the imaginary axis. At the frequency ω_{180} the phase is $-180°$ and the curve crosses the negative real axis. If $|a(j\omega_{180})f| > 1$ at this point, the Nyquist diagram will encircle the point $(-1, 0)$ as shown, and this has particular significance, as will

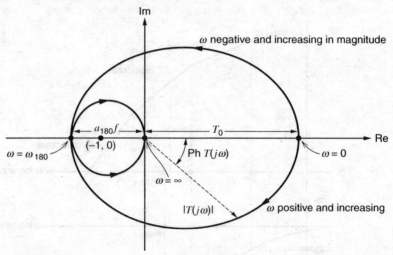

Figure 9.6 Nyquist diagram [polar plot of $T(j\omega)$ in magnitude and phase] corresponding to the characteristic of Fig. 9.5 (not to scale).

now become apparent. For the purposes of this treatment, the *Nyquist criterion* for stability of the amplifier can be stated as follows:

"Consider a feedback amplifier with a stable $T(s)$ (i.e., all poles of $T(s)$ are in the left half-plane). If the Nyquist plot of $T(j\omega)$ encircles the point $(-1, 0)$, the feedback amplifier is unstable."

This criterion simply amounts to a mathematical test for poles of transfer function $A(s)$ in the right half-plane. If the Nyquist plot encircles the point $(-1, 0)$, the amplifier has poles in the right half-plane and the circuit will *oscillate*. In fact the number of encirclements of the point $(-1, 0)$ gives the number of right half-plane poles and in this example there are two. The significance of poles in the right half-plane can be seen by assuming that a circuit has a pair of complex poles at $(\sigma_1 \pm j\omega_1)$ where σ_1 is positive. The transient response of the circuit then contains a term $K_1 \exp \sigma_1 t \sin \omega_1 t$, which represents a *growing* sinusoid if σ_1 is positive. (K_1 is a constant representing initial conditions.) This term is then present even if no further input is applied, and a circuit behaving in this way is said to be *unstable* or *oscillatory*.

The significance of the point $(-1, 0)$ can be appreciated if the Nyquist diagram is assumed to pass through this point. Then at the frequency ω_{180}, $T(j\omega) = a(j\omega)f = -1$ and $A(j\omega) = \infty$ using (9.2) in the frequency domain. The feedback amplifier is thus calculated to have a forward gain of infinity, and this indicates the onset of instability and oscillation. This situation corresponds to poles of $A(s)$ on the $j\omega$ axis in the s plane. If T_0 is then increased by increasing a_0 or f, the Nyquist diagram expands *linearly* and then encircles $(-1, 0)$. This corresponds to poles of $A(s)$ in the right half-plane, as shown in Fig. 9.7.

From the above criterion for stability, a *simpler* test can be derived that is useful in most common cases.

"If $|T(j\omega)| > 1$ at the frequency where ph $T(j\omega) = -180°$, then the amplifier is unstable." The validity of this criterion for the example considered here is apparent from inspection of Fig. 9.6 and application of the Nyquist criterion.

In order to examine the effect of feedback on the stability of an amplifier, consider the three-pole amplifier with gain function given by (9.6) to be placed in a negative-feedback loop with f constant. The gain (in decibels) and phase of the amplifier are shown again in Fig. 9.8, and also plotted is the quantity $20 \log_{10} 1/f$. The value of $20 \log_{10} 1/f$ is approximately equal

9.3 Instability and the Nyquist Criterion

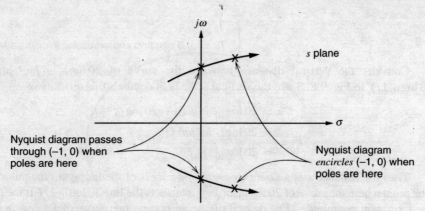

Figure 9.7 Pole positions corresponding to different Nyquist diagrams.

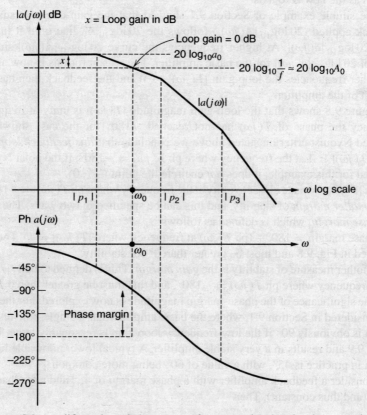

Figure 9.8 Amplifier gain and phase versus frequency showing the phase margin.

to the low-frequency gain in decibels with feedback applied since

$$A_0 = \frac{a_0}{1 + a_0 f} \tag{9.7}$$

and thus

$$\frac{1}{f} \approx A_0 \tag{9.8}$$

if
$$T_0 = a_0 f \gg 1$$

Consider the vertical distance between the curve of $20 \log_{10} |a(j\omega)|$ and the line $20 \log_{10} 1/f$ in Fig. 9.8. Since the vertical scale is in decibels this quantity is

$$x = 20 \log_{10} |a(j\omega)| - 20 \log_{10} 1/f \qquad (9.9)$$
$$= 20 \log_{10} |a(j\omega)f|$$
$$= 20 \log_{10} |T(j\omega)| \qquad (9.10)$$

Thus the distance x is a *direct measure in decibels* of the loop-gain magnitude, $|T(j\omega)|$. The point where the curve of $20 \log_{10} |a(j\omega)|$ intersects the line $20 \log_{10} 1/f$ is the point where the loop-gain magnitude $|T(j\omega)|$ is 0 dB or *unity*, and the curve of $|a(j\omega)|$ in decibels in Fig. 9.8 can thus be considered a curve of $|T(j\omega)|$ in decibels *if the dotted line at* $20 \log_{10} 1/f$ is taken as the new zero axis.

The simple example of Section 9.1 showed that the gain curve versus frequency with feedback applied $(20 \log_{10} |A(j\omega)|)$ follows the $20 \log_{10} A_0$ line until it intersects the gain curve $20 \log_{10} |a(j\omega)|$. At higher frequencies the curve $20 \log_{10} |A(j\omega)|$ simply follows the curve of $20 \log_{10} |a(j\omega)|$ for the basic amplifier. The reason for this is now apparent in that at the higher frequencies the loop gain $|T(j\omega)| \to 0$ and the feedback then has *no influence* on the gain of the amplifier.

Figure 9.8 shows that the loop-gain magnitude $|T(j\omega)|$ is unity at frequency ω_0. At this frequency the phase of $T(j\omega)$ has not reached $-180°$ for the case shown, and using the modified Nyquist criterion stated above we conclude that *this feedback loop is stable*. Obviously $|T(j\omega)| < 1$ at the frequency where ph $T(j\omega) = -180°$. If the polar Nyquist diagram is sketched for this example, it does *not* encircle the point $(-1, 0)$.

As $|T(j\omega)|$ is made closer to unity at the frequency where ph $T(j\omega) = -180°$, the amplifier has a *smaller margin* of stability, and this can be specified in two ways. The most common is the *phase margin*, which is defined as follows:

Phase margin $= 180° +$ (ph $T(j\omega)$ at frequency where $|T(j\omega)| = 1$). The phase margin is indicated in Fig. 9.8 and must be greater than $0°$ for stability.

Another measure of stability is the *gain margin*. This is defined to be $1/|T(j\omega)|$ in decibels at the frequency where ph $T(j\omega) = -180°$, and this must be greater than 0 dB for stability.

The significance of the phase-margin magnitude is now explored. For the feedback amplifier considered in Section 9.1, where the basic amplifier has a single-pole response, the phase margin is obviously $90°$ if the low-frequency loop gain is reasonably large. This is illustrated in Fig. 9.9 and results in a very stable amplifier. A typical lower allowable limit for the phase margin in practice is $45°$, with a value of $60°$ being more common.

Consider a feedback amplifier with a phase margin of $45°$ and a feedback function f that is real (and thus constant). Then

$$\text{ph } T(j\omega_0) = -135° \qquad (9.11)$$

where ω_0 is the frequency defined by

$$|T(j\omega_0)| = 1 \qquad (9.12)$$

Now $|T(j\omega_0)| = |a(j\omega_0)f| = 1$ implies that

$$|a(j\omega_0)| = \frac{1}{f} \qquad (9.13)$$

assuming that f is positive real.

9.3 Instability and the Nyquist Criterion

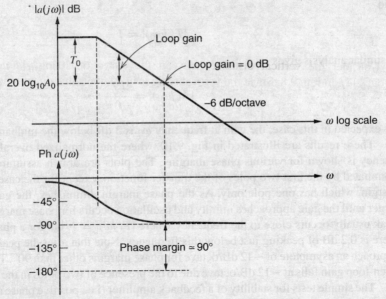

Figure 9.9 Gain and phase versus frequency for a single-pole basic amplifier showing the phase margin for a low-frequency loop gain T_0.

The overall gain is

$$A(j\omega) = \frac{a(j\omega)}{1 + T(j\omega)} \qquad (9.14)$$

Substitution of (9.11) and (9.12) in (9.14) gives

$$A(j\omega_0) = \frac{a(j\omega_0)}{1 + e^{-j135°}} = \frac{a(j\omega_0)}{1 - 0.7 - 0.7j} = \frac{a(j\omega_0)}{0.3 - 0.7j}$$

and thus

$$|A(j\omega_0)| = \frac{|a(j\omega_0)|}{0.76} = \frac{1.3}{f} \qquad (9.15)$$

using (9.13).

The frequency ω_0, where $|T(j\omega_0)| = 1$, is the nominal -3-dB point for a single-pole basic amplifier, but in this case there is 2.4 dB (1.3 ×) of *peaking* above the low-frequency gain of $1/f$.

Consider a phase margin of 60°. At the frequency ω_0 in this case

$$\text{ph } T(j\omega_0) = -120° \qquad (9.16)$$

and

$$|T(j\omega_0)| = 1 \qquad (9.17)$$

Following a similar analysis we obtain

$$|A(j\omega_0)| = \frac{1}{f}$$

In this case there is no peaking at $\omega = \omega_0$, but there has also been no gain reduction at this frequency.

Finally, the case where the phase margin is 90° can be similarly calculated. In this case

$$\text{ph } T(j\omega_0) = -90° \qquad (9.18)$$

and

$$|T(j\omega_0)| = 1 \qquad (9.19)$$

A similar analysis gives

$$|A(j\omega_0)| = \frac{0.7}{f} \qquad (9.20)$$

As expected in this case, the gain at frequency ω_0 is 3 dB below the midband value.

These results are illustrated in Fig. 9.10, where the normalized overall gain versus frequency is shown for various phase margins. The plots are drawn assuming the response is dominated by the first two poles of the transfer function, except for the case of the 90° phase margin, which has one pole only. As the phase margin diminishes, the gain peak becomes larger until the gain approaches infinity and oscillation occurs for phase margin = 0°. The gain peak usually occurs close to the frequency where $|T(j\omega)| = 1$, but for a phase margin of 60° there is 0.2 dB of peaking just below this frequency. Note that after the peak, the gain curves approach an asymptote of -12 dB/octave for phase margins other than 90°. This is because the open-loop gain falls at -12 dB/octave due to the presence of two poles in the transfer function.

The simple tests for stability of a feedback amplifier (i.e., positive phase and gain margins) can only be applied when the phase and gain margins are uniquely defined. The phase margin is uniquely defined if there is only one frequency at which the magnitude of the loop gain equals one. Similarly, the gain margin is uniquely defined if there is only one frequency at which the phase of the loop gain equals $-180°$. In most feedback circuits, these margins are uniquely defined. However, if either of these margins is not uniquely defined, then stability should be checked using a Nyquist diagram and the Nyquist criterion.

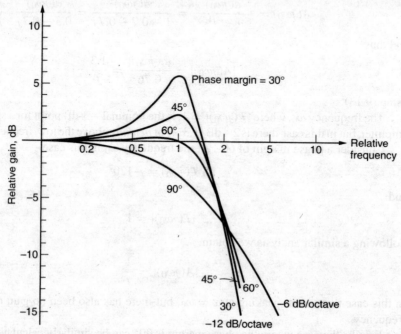

Figure 9.10 Normalized overall gain for feedback amplifiers versus normalized frequency for various phase margins. Frequency is normalized to the frequency where the loop gain is unity.

The loop gain $T = af$ can be examined to determine the stability of a feedback circuit, as explained in this section. Alternatively these measures of stability can be applied to the return ratio $\mathcal{R}$, as explained in Appendix A9.1. Techniques for simulating $\mathcal{R}$[2–5] and $T = af$[4] using SPICE have been developed, based on methods for measuring loop transmission.[6,7] These techniques measure the loop transmission at the closed-loop dc operating point. An advantage of SPICE simulation of the loop transmission is that parasitics that might have an important effect are included. For example, parasitic capacitance at the op-amp input introduces frequency dependence in the feedback network in Fig. 8.24, which may degrade the phase margin.

9.4 Compensation

9.4.1 Theory of Compensation

Consider again the amplifier whose gain and phase is shown in Fig. 9.8. For the feedback circuit in which this was assumed to be connected, the forward gain was A_0, as shown in Fig. 9.8, and the phase margin was positive. Thus the circuit was stable. It is apparent, however, that if the amount of feedback is increased by making f larger (and thus A_0 smaller), oscillation will eventually occur. This is shown in Fig. 9.11, where f_1 is chosen to give a zero phase margin and the corresponding overall gain is $A_1 \simeq 1/f_1$. If the feedback is increased to f_2 (and $A_2 \simeq 1/f_2$ is the overall gain), the phase margin is negative and the circuit will oscillate. Thus if this amplifier is to be used in a feedback loop with loop gain larger than $a_0 f_1$, efforts

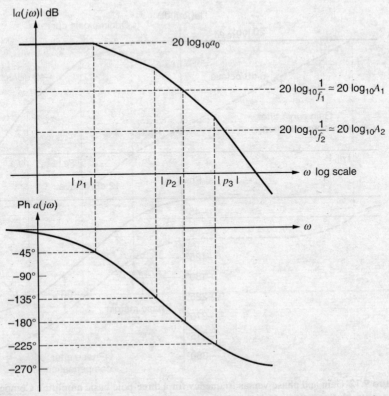

Figure 9.11 Gain and phase versus frequency for a three-pole basic amplifier. Feedback factor f_1 gives a zero phase margin and factor f_2 gives a negative phase margin.

must be made to increase the phase margin. This process is known as *compensation*. Note that without compensation, the forward gain of the feedback amplifier cannot be made less than $A_1 \simeq 1/f_1$ because of the oscillation problem.

The simplest and most common method of compensation is to reduce the bandwidth of the amplifier (often called *narrowbanding*). That is, a dominant pole is deliberately introduced into the amplifier to force the phase shift to be less than $-180°$ when the loop gain is unity. This involves a direct sacrifice of the frequency capability of the amplifier.

If f is constant, the most difficult case to compensate is $f = 1$, which is a unity-gain feedback configuration. In this case the loop-gain curve is identical to the gain curve of the basic amplifier. Consider this situation and assume that the basic amplifier has the same characteristic as in Fig. 9.11. To compensate the amplifier, we introduce a new dominant pole with magnitude $|p_D|$, as shown in Fig. 9.12, and assume that this does not affect the original amplifier poles with magnitudes $|p_1|, |p_2|,$ and $|p_3|$. This is often not the case but is assumed here for purposes of illustration.

The introduction of the dominant pole with magnitude $|p_D|$ into the amplifier gain function causes the gain magnitude to decrease at 6 dB/octave until frequency $|p_1|$ is reached, and over this region the amplifier phase shift asymptotes to $-90°$. If frequency $|p_D|$ is chosen so that the gain $|a(j\omega)|$ is unity at frequency $|p_1|$ as shown, then the loop gain is also unity at frequency $|p_1|$ for the assumed case of unity feedback with $f = 1$. The phase margin in this case is then $45°$, which means that the amplifier is stable. The original amplifier would have been *unstable* in such a feedback connection.

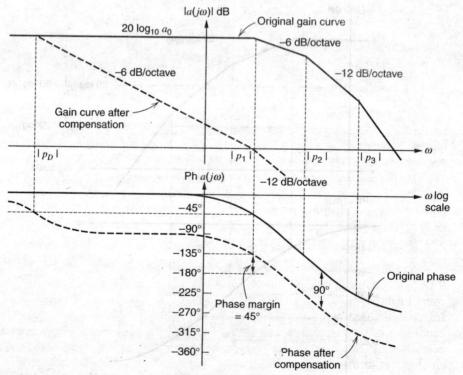

Figure 9.12 Gain and phase versus frequency for a three-pole basic amplifier. Compensation for unity-gain feedback operation ($f = 1$) is achieved by introduction of a negative real pole with magnitude $|p_D|$.

The price that has been paid for achieving stability in this case is that with the feedback removed, the basic amplifier has a unity-gain bandwidth of only $|p_1|$, which is much less than before. Also, with feedback applied, the loop gain now begins to decrease at a frequency $|p_D|$, and all the benefits of feedback diminish as the loop gain decreases. For example, in Chapter 8 it was shown that shunt feedback at the input or output of an amplifier *reduces* the basic terminal impedance by $[1 + T(j\omega)]$. Since $T(j\omega)$ is frequency dependent, the terminal impedance of a shunt-feedback amplifier will begin to *rise* when $|T(j\omega)|$ begins to decrease. Thus the high-frequency terminal impedance will appear *inductive*, as in the case of z_0 for an emitter follower, which was calculated in Chapter 7. (See Problem 9.8.)

■ **EXAMPLE**

Calculate the dominant-pole magnitude required to give unity-gain compensation of the 702 op amp with a phase margin of 45°. The low-frequency gain is $a_0 = 3600$ and the circuit has poles at $-(p_1/2\pi) = 1$ MHz, $-(p_2/2\pi) = 4$ MHz, and $-(p_3/2\pi) = 40$ MHz.

In this example, the second pole p_2 is sufficiently close to p_1 to produce significant phase shift at the amplifier -3-dB frequency. The approach to this problem will be to use the approximate results developed above to obtain an initial estimate of the required dominant-pole magnitude and then to empirically adjust this estimate to obtain the required results.

The results of Fig. 9.12 indicate that a dominant pole with magnitude $|p_D|$ should be introduced so that gain $a_0 = 3600$ is reduced to unity at $|p_1/2\pi| = 1$ MHz with a 6-dB/octave decrease as a function of frequency. The product $|a|\omega$ is constant where the slope of the gain-magnitude plot is -6 dB/octave; therefore

$$\left|\frac{p_D}{2\pi}\right| = \frac{1}{a_0}\left|\frac{p_1}{2\pi}\right| = \frac{10^6}{3600}\text{Hz} = 278 \text{ Hz}$$

This would give a transfer function

$$a(j\omega) = \frac{3600}{\left(1+\dfrac{j\omega}{|p_D|}\right)\left(1+\dfrac{j\omega}{|p_1|}\right)\left(1+\dfrac{j\omega}{|p_2|}\right)\left(1+\dfrac{j\omega}{|p_3|}\right)} \quad (9.21)$$

where the pole magnitudes are in radians per second. Equation 9.21 gives a unity-gain frequency [where $|a(j\omega)| = 1$] of 780 kHz. This is slightly below the design value of 1 MHz because the actual gain curve is 3 dB below the asymptote at the break frequency $|p_1|$. At 780 kHz the phase shift obtained from (9.21) is $-139°$ instead of the desired $-135°$ and this includes a contribution of $-11°$ from pole p_2. Although this result is close enough for most purposes, a phase margin of precisely 45° can be achieved by empirically reducing $|p_D|$ until (9.21) gives a phase shift of $-135°$ at the unity gain frequency. This occurs for $|p_D/2\pi| =$
■ 260 Hz, which gives a unity-gain frequency of 730 kHz.

Consider now the performance of the amplifier whose characteristic is shown in Fig. 9.12 (with dominant pole magnitude $|p_D|$) when used in a feedback loop with $f < 1$ (i.e., overall gain $A_0 > 1$). This case is shown in Fig. 9.13. The loop gain now falls to unity at frequency ω_x and the phase margin of the circuit is now approximately 90°. The -3-dB bandwidth of the feedback circuit is ω_x. The circuit now has more compensation than is needed, and, in fact, bandwidth is being wasted. Thus, although it is convenient to compensate an amplifier for unity gain and then use it unchanged for other applications (as is done in many op amps), this procedure is quite wasteful of bandwidth. Fixed-gain amplifiers that are designed for applications where maximum bandwidth is required are usually compensated for a specified phase margin (typically 45° to 60°) at the required gain value. However, op amps are

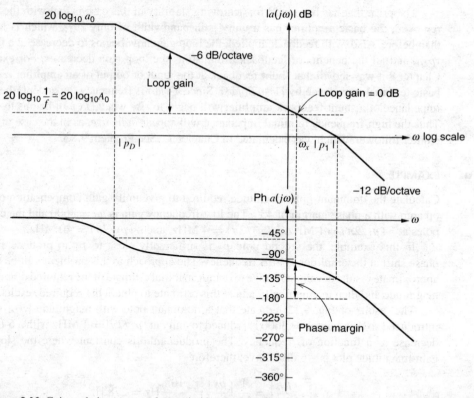

Figure 9.13 Gain and phase versus frequency for an amplifier compensated for use in a feedback loop with $f = 1$ and a phase margin of $45°$. The phase margin is shown for operation in a feedback loop with $f < 1$.

general-purpose circuits that are used with differing feedback networks with f values ranging from 0 to 1. Optimum bandwidth is achieved in such circuits if the compensation is tailored to the gain value required, and this approach gives much higher bandwidths for high gain values, as seen in Fig. 9.14. This figure shows compensation of the amplifier characteristic of Fig. 9.11 for operation in a feedback circuit with forward gain A_0. A dominant pole is added with magnitude $|p'_D|$ to give a phase margin of $45°$. Frequency $|p'_D|$ is obviously $\gg |p_D|$, and the -3-dB bandwidth of the feedback amplifier is nominally $|p_1|$, at which frequency the loop gain is 0 dB (disregarding peaking). The -3-dB frequency from Fig. 9.13 would be only $\omega_x = |p_1|/A_0$ if unity-gain compensation had been used. Obviously, since A_0 can be large, the improvement in bandwidth is significant.

In the compensation schemes discussed above, an additional dominant pole was assumed to be added to the amplifier, and the original amplifier poles were assumed to be unaffected by this procedure. In terms of circuit bandwidth, a much more efficient way to compensate the amplifier is to add capacitance to the circuit in such a way that the original amplifier dominant pole magnitude $|p_1|$ is reduced so that it performs the compensation function. This technique requires access to the internal nodes of the amplifier, and knowledge of the nodes in the circuit where added capacitance will reduce frequency $|p_1|$.

Consider the effect of compensating for unity-gain operation the amplifier characteristic of Fig. 9.11 in this way. Again assume that higher frequency poles p_2 and p_3 are unaffected by this procedure. In fact, depending on the method of compensation, these poles are usually moved up or down in magnitude by the compensation. This point will be taken up later.

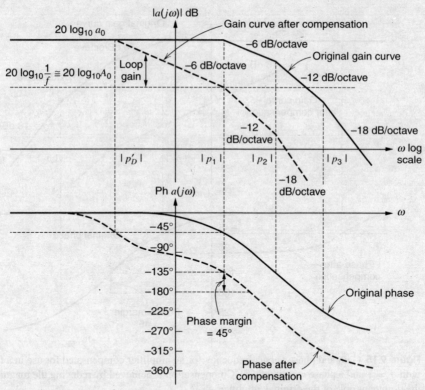

Figure 9.14 Gain and phase versus frequency for an amplifier compensated for use in a feedback loop with $f < 1$ and a phase margin of $45°$. Compensation is achieved by adding a new pole p'_D to the amplifier.

Compensation of the amplifier by reducing $|p_1|$ is shown in Fig. 9.15. For a $45°$-phase margin in a unity-gain feedback configuration, dominant pole magnitude $|p'_1|$ must cause the gain to fall to unity at frequency $|p_2|$ (the second pole magnitude). Thus the nominal bandwidth in a unity-gain configuration is $|p_2|$, and the loop gain is unity at this frequency. This result can be contrasted with a bandwidth of $|p_1|$, as shown in Fig. 9.12 for compensation achieved by adding another pole with magnitude $|p_D|$ to the amplifier. In practical amplifiers, frequency $|p_2|$ is often 5 or 10 times frequency $|p_1|$ and substantial improvements in bandwidth are thus achieved.

The results of this section illustrate why the basic amplifier of a feedback circuit is usually designed with as few stages as possible. Each stage of gain inevitably adds more poles to the transfer function, complicating the compensation problem, particularly if a wide bandwidth is required.

9.4.2 Methods of Compensation

In order to compensate a circuit by the common method of narrowbanding described above, it is necessary to add capacitance to create a dominant pole with the desired magnitude. One method of achieving this is shown in Fig. 9.16, which is a schematic of the first two stages of a simple amplifier. A large capacitor C is connected between the collectors of the input stage. The output stage, which is assumed relatively broadband, is not shown. A differential half-circuit of Fig. 9.16 is shown in Fig. 9.17, and it should be noted that the compensation

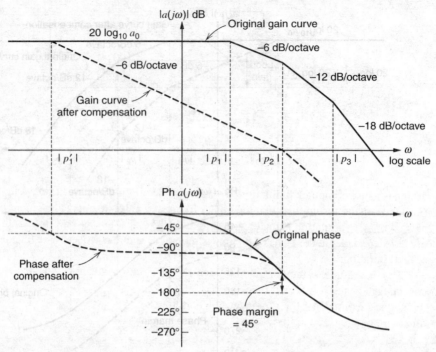

Figure 9.15 Gain and phase versus frequency for an amplifier compensated for use in a feedback loop with $f = 1$ and a phase margin of $45°$. Compensation is achieved by reducing the magnitude $|p_1|$ of the dominant pole of the original amplifier.

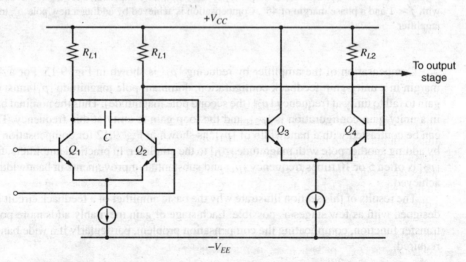

Figure 9.16 Compensation of an amplifier by introduction of a large capacitor C.

capacitor is doubled in the half-circuit. The major contributions to the dominant pole of a circuit of this type (if R_S is not large) come from the input capacitance of Q_4 and Miller capacitance associated with Q_4. Thus the compensation as shown will reduce the magnitude of the dominant pole of the original amplifier so that it performs the required compensation function. Almost certainly, however, the higher frequency poles of the amplifier will also be changed by the addition of C. In practice, the best method of approaching the compensation

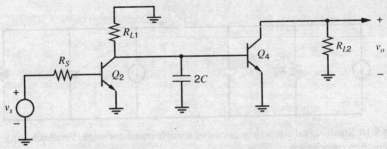

Figure 9.17 Differential half-circuit of Fig. 9.16.

design is to use computer simulation to determine the original pole positions. A first estimate of C is made on the assumption that the higher frequency poles do not change in magnitude and a new computer simulation is made with C included to check this assumption. Another estimate of C is then made on the basis of the new simulation, and this process usually converges after several iterations.

The magnitude of the dominant pole of Fig. 9.17 can be estimated using zero-value time constant analysis. However, if the value of C required is very large, this capacitor will dominate and a good estimate of the dominant pole can be made by considering C only and ignoring other circuit capacitance. In that case the dominant-pole magnitude is

$$|p_D| = \frac{1}{2CR} \qquad (9.22)$$

where

$$R = R_{L1} \| R_{i4} \qquad (9.23)$$

and

$$R_{i4} = r_{b4} + r_{\pi 4} \qquad (9.24)$$

One disadvantage of the above method of compensation is that the value of C required is quite large (typically > 1000 pF) and cannot be realized on a monolithic chip.

Many general-purpose op amps have unity-gain compensation included on the monolithic chip and require no further compensation from the user. (The sacrifice in bandwidth caused by this technique when using gain other than unity was described earlier.) In order to realize an internally compensated monolithic op amp, compensation must be achieved using capacitance less than about 50 pF. This can be achieved using *Miller multiplication* of the capacitance as in the 741 op amp, which uses a 30 pF compensation capacitor and was analyzed in previous editions of this book.

As well as allowing use of a small capacitor that can be integrated on the monolithic chip, this type of compensation has another significant advantage. This is due to the phenomenon of *pole splitting*,[8] in which the dominant pole moves to a lower frequency while the next pole moves to a higher frequency. The splitting of the two low-frequency poles in practical op amps is often a rather complex process involving other higher frequency poles and zeros as well. However, the process involved can be illustrated with the two-stage op-amp model in Fig. 9.18. The input is from from a current i_s, which stems from the transconductance of the first stage times the op-amp differential input voltage. Resistors R_1 and R_2 represent the total shunt resistances at the output of the first and second stages, including transistor input and output resistances. Similarly, C_1 and C_2 represent the total shunt capacitances at the same places. Capacitor C represents transistor collector-base capacitance of the amplifying transistor in the second stage plus the compensation capacitance.

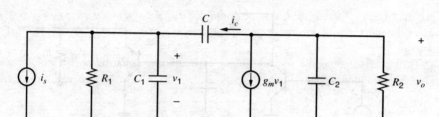

Figure 9.18 Small-signal equivalent circuit of a single transistor stage. Feedback capacitor C includes compensation capacitance.

For the circuit of Fig. 9.18,

$$-i_s = \frac{v_1}{R_1} + v_1 C_1 s + (v_1 - v_o)Cs \tag{9.25}$$

$$g_m v_1 + \frac{v_o}{R_2} + v_o C_2 s + (v_o - v_1)Cs = 0 \tag{9.26}$$

From (9.25) and (9.26)

$$\frac{v_o}{i_s} = \frac{(g_m - Cs)R_2 R_1}{1 + s[(C_2 + C)R_2 + (C_1 + C)R_1 + g_m R_2 R_1 C] + s^2 R_2 R_1 (C_2 C_1 + CC_2 + CC_1)} \tag{9.27}$$

The circuit transfer function has a positive real zero at

$$z = \frac{g_m}{C} \tag{9.27a}$$

which usually has such a large magnitude in bipolar circuits that it can be neglected. This is often not the case in MOS circuits because of their lower g_m. This point is taken up later.

The circuit has a two-pole transfer function. If p_1 and p_2 are the poles of the circuit, then the denominator of (9.27) can be written

$$D(s) = \left(1 - \frac{s}{p_1}\right)\left(1 - \frac{s}{p_2}\right) \tag{9.28}$$

$$= 1 - s\left(\frac{1}{p_1} + \frac{1}{p_2}\right) + \frac{s^2}{p_1 p_2} \tag{9.29}$$

and thus

$$D(s) \simeq 1 - \frac{s}{p_1} + \frac{s^2}{p_1 p_2} \tag{9.30}$$

if the poles are real and widely separated, which is usually true. Note that p_1 is assumed to be the dominant pole.

If the coefficients in (9.27) and (9.30) are equated then

$$p_1 = -\frac{1}{(C_2 + C)R_2 + (C_1 + C)R_1 + g_m R_2 R_1 C} \tag{9.31}$$

and this can be approximated by

$$p_1 \simeq -\frac{1}{g_m R_2 R_1 C} \tag{9.32}$$

since the Miller effect due to C will be dominant if C is large and $g_m R_1, g_m R_2 \gg 1$. Equation 9.31 is the same result for the dominant pole as is obtained using zero-value time constant analysis.

The nondominant pole p_2 can now be estimated by equating coefficients of s^2 in (9.27) and (9.30) and using (9.32).

$$p_2 \simeq -\frac{g_m C}{C_2 C_1 + C(C_2 + C_1)} \tag{9.33}$$

Equation 9.32 indicates that the dominant-pole magnitude $|p_1|$ *decreases* as C *increases*, whereas (9.33) shows that $|p_2|$ *increases* as C *increases*. Thus, increasing C causes the poles to *split apart*. The dominant pole moves to a lower frequency because increasing C increases the time constant associated with the output node of the first stage in Fig. 9.18. The reason the nondominant pole moves to a higher frequency is explained below.

Equation 9.33 can be interpreted physically by associating p_2 with the output node in Fig. 9.18. Then

$$p_2 = -\frac{1}{R_o C_T} \tag{9.33a}$$

where R_o is the output resistance including negative feedback around the second stage through C, and C_T is the total capacitance from the output node to ground. The output resistance is

$$R_o = \frac{R_2}{1+T} \tag{9.33b}$$

where R_2 is the open-loop output resistance, and T is the loop gain around the second stage through capacitor C, which is the open-loop gain, $g_m R_2$, times the feedback factor, f. Therefore,

$$R_o = \frac{R_2}{1 + g_m R_2 f} \simeq \frac{1}{g_m f} \tag{9.33c}$$

assuming that $T = g_m R_2 f \gg 1$. Since p_2 is a high frequency, we will find f at high frequency ω, where $1/\omega C_1 \ll R_1$. Then the feedback around the second stage is controlled by a capacitive voltage divider and

$$f \simeq \frac{C}{C + C_1} \tag{9.33d}$$

Thus,

$$R_o \simeq \frac{C + C_1}{g_m C} \tag{9.33e}$$

The total capacitance from the output node to ground is C_2 in parallel with the series combination of C and C_1:

$$C_T = C_2 + \frac{CC_1}{C + C_1} = \frac{CC_2 + C_1 C_2 + CC_1}{C + C_1} \tag{9.33f}$$

Substituting (9.33e) and (9.33f) into (9.33a) gives (9.33).

Equations 9.33d and 9.33f show that increasing C increases the feedback factor but has little effect on the total capacitance in shunt with the output node because C is in series with C_1. As a result, increasing C reduces the output resistance and increases the frequency of the nondominant pole. In the limit as $C \to \infty$, the feedback factor approaches unity, and $p_2 \to -g_m/(C_2 + C_1)$. In practice, however, (9.33d) shows that the feedback factor is less than unity, which limits the increase in the magnitude of the nondominant pole frequency.

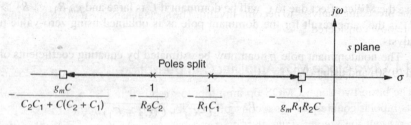

Figure 9.19 Locus of the poles of the circuit of Fig. 9.18 as C is increased from zero, for the case $-1/(R_1 C_1) > -1/(R_2 C_2)$.

On the other hand, with $C = 0$, the poles of the circuit of Fig. 9.18 are

$$p_1 = -\frac{1}{R_1 C_1} \tag{9.34a}$$

$$p_2 = -\frac{1}{R_2 C_2} \tag{9.34b}$$

Thus as C increases from zero, the locus of the poles of the circuit of Fig. 9.18 is as shown in Fig. 9.19.

Another explanation of pole splitting is as follows. The circuit in Fig. 9.18 has two poles. The compensation capacitor across the second stage provides feedback and causes the second stage to act like an integrator. The two poles split apart as C increases. One pole moves to a low frequency (toward dc), and the other moves to a high frequency (toward $-\infty$) to approximate an ideal integrator, which has only one pole at dc.

The previous calculations have shown how compensation of an amplifier by addition of a large Miller capacitance to a single transistor stage causes the nondominant pole to move to a much higher frequency. For the sake of comparison, consider compensating the circuit in Fig. 9.18 without adding capacitance to C by making C_1 large enough to produce a dominant pole. Then the pole can be calculated from (9.31) as $p_1 \simeq -1/R_1 C_1$. The nondominant pole can be estimated by equating coefficients of s^2 in (9.27) and (9.30) and using this value of p_1. This gives $p_2 \simeq -1/R_2(C_2 + C)$. This value of p_2 is approximately the same as that given by (9.34b), which is for $C = 0$ and is *before* pole splitting occurs. Thus, creation of a dominant pole in the circuit of Fig. 9.18 by making C_1 large will result in a second pole magnitude $|p_2|$ that is much smaller than that obtained if the dominant pole is created by increasing C. As a consequence, the realizable bandwidth of the circuit when compensated in this way is much smaller than that obtained with Miller-effect compensation. Also, without using the Miller effect, the required compensation capacitor often would be too large to be included on a monolithic chip. The same general conclusions are true in the more complex situation that exists in many practical op amps.

The results derived in this section are useful in further illuminating the considerations of Section 7.3.3. In that section, it was stated that in a common-source cascade, the existence of drain-gate capacitance tends to cause pole splitting and to produce a dominant-pole situation. If the equivalent circuit of Fig. 9.18 is taken as a representative section of a cascade of common-source stages (C_2 is the input capacitance of the following stage) and capacitor C is taken as C_{gd}, the calculations of this section show that the presence of C_{gd} does, in fact, tend to produce a dominant-pole situation because of the pole splitting that occurs. Thus, the zero-value time constant approach gives a good estimate of ω_{-3dB} in such circuits.

The theory of compensation that was developed in this chapter was illustrated with some bipolar-transistor circuit examples. The theory applies in general to any active circuit, but the unique device parameters of MOSFETs cause some of the approximations that were made in

9.4.3 Two-Stage MOS Amplifier Compensation

The basic two-stage CMOS op amp topology shown in Fig. 6.16 is essentially identical to its bipolar counterpart. As a consequence, the equivalent circuit of Fig. 9.18 can be used to represent the second stage with its compensation capacitance. The poles of the circuit are again given by (9.32) and (9.33) and the zero by (9.27a). In the case of the MOS transistor, however, the value of g_m is typically an order of magnitude lower than for a bipolar transistor, and the break frequency caused by the right half-plane zero in (9.27) may actually fall below the nominal unity-gain frequency of the amplifier. The effect of this is shown in Fig. 9.20. At the frequency $|z|$ the gain characteristic of the amplifier flattens out because of the contribution to the gain of +6 dB/octave from the zero. In the same region the phase is made 90° more *negative* by the positive real zero. As a consequence, the amplifier will have negative phase margin and be unstable when the influence of the next most dominant pole is felt. In effect, the zero halts the gain roll-off intended to stabilize the amplifier and simultaneously pushes the phase in the negative direction. Note also from (9.33) that the low g_m of the MOSFET will tend to reduce the value of $|p_2|$ relative to a bipolar amplifier.

Another way to view this problem is to note from Fig. 9.18 that at high frequencies, feedforward through C tends to overwhelm the normal gain path via g_m of the second stage

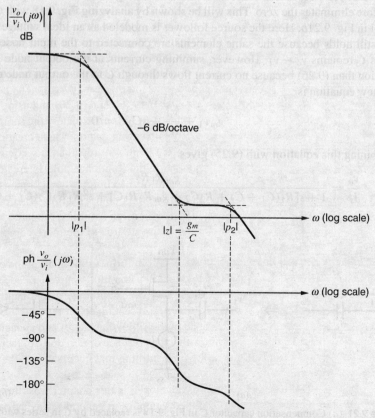

Figure 9.20 Typical gain and phase of the CMOS op amp of Fig. 6.16.

if g_m is small. The feedforward path does not have the 180° phase shift of the normal gain stage, and thus the gain path loses an inverting stage. Any feedback applied around the overall amplifier will then be positive instead of negative feedback, resulting in oscillation. At very high frequencies, C acts like a short circuit, diode-connecting the second stage, which then simply presents a resistive load of $1/g_m$ to the first stage, again showing the loss of 180° of phase shift.

The right half-plane (RHP) zero is caused by the interaction of current from the g_m generator and the frequency-dependent current that flows forward from the input node to the output node through C. The current through C in Fig. 9.18 is

$$i_c = sC(v_o - v_1) \tag{9.35}$$

This current can be broken into two parts: a feedback current $i_{fb} = sCv_o$ that flows from the output back toward the input and a feedforward current $i_{ff} = sCv_1$ that flows forward from the input toward the output. This feedforward current is related to v_1. The current $g_m v_1$ from the controlled source flows out of the output node and is also related to v_1. Subtracting these two currents gives the total current at the output node that is related to v_1:

$$i_{v_1} = (g_m - sC)v_1 \tag{9.36}$$

A zero exists in the transfer function where this current equals zero, at $z = g_m/C$.

Three techniques have been used to eliminate the effect of the RHP zero. One approach is to put a source follower in series with the compensation capacitor,[9] as shown in Fig. 9.21a. The source follower blocks feedforward current through C from reaching the output node and therefore eliminates the zero. This will be shown by analyzing Fig. 9.18 with C replaced by the model in Fig. 9.21b. Here the source follower is modeled as an ideal voltage buffer. Equation 9.25 still holds because the same elements are connected to the input node and the voltage across C remains $v_o - v_1$. However, summing currents at the output node gives a different equation than (9.26) because no current flows through C to the output node due to the buffer. The new equation is

$$g_m v_1 + \frac{v_o}{R_2} + sC_2 v_o = 0 \tag{9.37}$$

Combining this equation with (9.25) gives

$$\frac{v_o}{i_s} = \frac{g_m R_1 R_2}{1 + s[R_1(C_1 + C) + R_2 C_2 + g_m R_2 R_1 C] + s^2 R_1 R_2 C_2 (C_1 + C)} \tag{9.38}$$

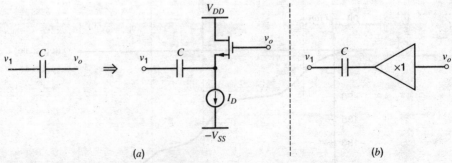

Figure 9.21 (a) Compensation capacitor C in Fig. 9.18 is replaced by C in series with a source follower. (b) A simple model for the capacitor and source follower.

The zero has been eliminated. Assuming $g_m R_1, g_m R_2 \gg 1$ and C is large, the same steps that led from (9.27) to (9.32) and (9.33) give

$$p_1 \approx -\frac{1}{g_m R_2 R_1 C} \tag{9.39a}$$

$$p_2 \approx -\frac{g_m C}{(C_1 + C)C_2} \approx -\frac{g_m}{C_2} \tag{9.39b}$$

The dominant pole p_1 is unchanged, and p_2 is about the same as before if $C_2 \gg C_1$. This approach eliminates the zero, but the follower requires extra devices and bias current. Also, the source follower has a nonzero dc voltage between its input and output. This voltage will affect the output voltage swing since the source-follower transistor must remain in the active region to maintain the desired feedback through C.

A second approach to eliminate the RHP zero is to block the feedforward current through C using a common-gate transistor,[10] as illustrated in Fig. 9.22a. This figure shows a two-stage op amp, with the addition of two current sources of value I_2 and transistor M_{11}. The compensation capacitor is connected from the op-amp output to the source of M_{11}. Here, common-gate M_{11} allows capacitor current to flow from the output back toward the input of the second stage. However, the impedance looking into the drain of M_{11} is very large. Therefore, feedforward current through C is very small. If the feedforward current is zero, the RHP zero is eliminated. A simplified small-signal model for the common-gate stage and compensation capacitor is shown in Fig. 9.22b. Here common-gate M_{11} is modeled as an ideal current buffer. Replacing C in Fig. 9.18 with the model in Fig. 9.22b yields

$$-i_s = \frac{v_1}{R_1} + v_1 C_1 s - v_o C s \tag{9.40a}$$

$$g_m v_1 + \frac{v_o}{R_2} + v_o C s + v_o C_2 s = 0 \tag{9.40b}$$

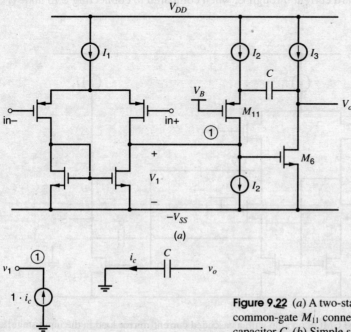

Figure 9.22 (a) A two-stage CMOS op amp with common-gate M_{11} connected to compensation capacitor C. (b) Simple small-signal model for M_{11} and C.

Combining these equations gives

$$\frac{v_o}{i_s} = \frac{g_m R_1 R_2}{1 + s[R_1 C_1 + R_2(C + C_2) + g_m R_1 R_2 C] + s^2 R_1 R_2 C_1 (C_2 + C)} \quad (9.41)$$

The zero has been eliminated. Again assuming $g_m R_1, g_m R_2 \gg 1$ and C is large, the poles are

$$p_1 \approx -\frac{1}{g_m R_2 R_1 C} \quad (9.42a)$$

$$p_2 \approx -\frac{g_m}{C + C_2} \cdot \frac{C}{C_1} \quad (9.42b)$$

The dominant pole is the same as before. However, the nondominant pole p_2 is different. This p_2 is at a higher frequency than in the two previous approaches because $C \gg C_1$ when C and C_2 are comparable. (In this section, we assume that the two-stage MOS op amp in Fig. 9.18 drives a load capacitor C_2 that is much larger than parasitic capacitance C_1; therefore $C_2 \gg C_1$.) Therefore, a smaller compensation capacitor C can be used here for a given load capacitance C_2, when compared to the previous approaches. The increase in $|p_2|$ arises because the input node is not connected to, and therefore is not loaded by, the compensation capacitor. An advantage of this scheme is that it provides better high-frequency negative-power-supply rejection than Miller compensation. (Power-supply rejection was introduced in Section 6.3.6.) With Miller compensation, C is connected from the gate to drain of M_6, and it shorts the gate and drain at high frequencies. Assuming V_{gs6} is approximately constant, high-frequency variations on the negative supply are coupled directly to the op-amp output. Connecting C to common-gate M_{11} eliminates this coupling path. Drawbacks of this approach are that extra devices and dc current are needed to implement the scheme in Fig. 9.22a. Also, if there is a mismatch between the I_2 current sources, the difference current must flow in the input stage, which disrupts the balance in the input stage and affects the input-offset voltage of the op amp.

When the first stage of the op amp uses a cascode transistor, the compensation capacitor can be connected to the source of the cascode device as shown in Fig. 9.23.[11] This connection reduces the feedforward current through C, when compared to connecting C to node Ⓨ, if the

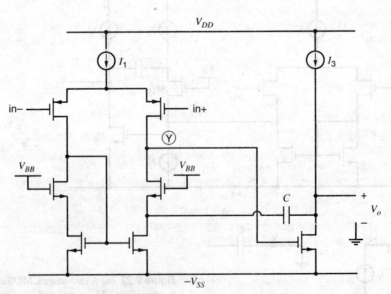

Figure 9.23 A two-stage CMOS op amp with a cascoded current-mirror load in the input stage, and with the compensation capacitor C connected to the cascode node.

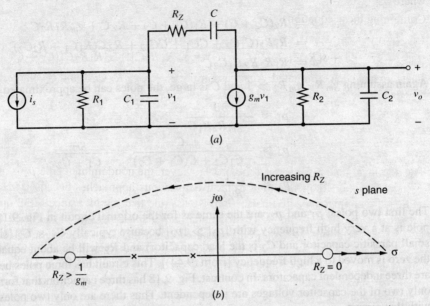

Figure 9.24 (a) Small-signal equivalent circuit of a compensation stage with nulling resistor. (b) Pole-zero diagram showing movement of the transmission zero for various values of R_Z.

voltage swing at the source of the cascode device is smaller than the swing at its drain. This approach eliminates the feedforward path, and therefore the zero, if the voltage swing at the source of the cascode device is zero. An advantage of this approach is that it avoids the extra devices, bias current, and mismatch problems in Fig. 9.22a.

A third way to deal with the RHP zero is to insert a resistor in series with the compensation capacitor, as shown in Fig. 9.24a.[12,13] Rather than eliminate the feedforward current, the resistor modifies this current and allows the zero to be moved to infinity. If the zero moves to infinity, the total forward current at the output node that is related to v_1 must go to zero when $\omega \to \infty$. When $\omega \to \infty$, capacitor C is a short circuit and therefore the feedforward current is only due to R_Z:

$$i_{ff}(\omega \to \infty) = -\frac{v_1}{R_Z} \quad (9.43)$$

When this current is added to the current from the g_m source, the total current at the output node that is related to v_1 is

$$i_{v_1} = \left(g_m - \frac{1}{R_Z}\right) v_1 \quad (9.44)$$

when $\omega \to \infty$. If $R_Z = 1/g_m$, this term vanishes, and the zero is at infinity.

The complete transfer function can be found by carrying out an analysis similar to that performed for Fig. 9.18, which gives

$$\frac{v_o}{i_s} = \frac{g_m R_1 R_2 \left[1 - sC\left(\frac{1}{g_m} - R_Z\right)\right]}{1 + bs + cs^2 + ds^3} \quad (9.45)$$

where

$$b = R_2(C_2 + C) + R_1(C_1 + C) + R_Z C + g_m R_1 R_2 C \quad (9.46a)$$
$$c = R_1 R_2 (C_1 C_2 + CC_1 + CC_2) + R_Z C(R_1 C_1 + R_2 C_2) \quad (9.46b)$$
$$d = R_1 R_2 R_Z C_1 C_2 C \quad (9.46c)$$

Again assuming $g_m R_1, g_m R_2 \gg 1$ and C is large, the poles can be approximated by

$$p_1 \approx -\frac{1}{g_m R_2 R_1 C} \quad (9.47a)$$

$$p_2 \approx -\frac{g_m C}{C_1 C_2 + C(C_1 + C_2)} \approx -\frac{g_m}{C_1 + C_2} \quad (9.47b)$$

$$p_3 \approx -\frac{1}{R_Z C_1} \quad (9.47c)$$

The first two poles, p_1 and p_2, are the same as for the original circuit in Fig. 9.18. The third pole is at a very high frequency with $|p_3| \gg |p_2|$ because typically $C_1 \ll C_2$ (since C_1 is a small parasitic capacitor and C_2 is the load capacitor) and R_Z will be about equal to $1/g_m$ if the zero is moved to a high frequency [from (9.44)]. This circuit has three poles because there are three independent capacitors. In contrast, Fig. 9.18 has three capacitors that form a loop, so only two of the capacitor voltages are independent. Thus there are only two poles associated with that circuit.

The zero of (9.45) is

$$z = \frac{1}{\left(\dfrac{1}{g_m} - R_Z\right) C} \quad (9.48)$$

This zero moves to infinity when R_Z equals $1/g_m$. Making the resistor greater than $1/g_m$ moves the zero into the left half-plane, which can be used to provide positive phase shift at high frequencies and improve the phase margin of a feedback circuit that uses this op amp.[13] The movement of the zero for increasing R_Z is shown in Fig. 9.24b.

Figure 9.25 shows a Miller-compensated op amp using a resistor R_Z in series with the compensation capacitor. In practice, resistor R_Z is usually implemented using a MOS transistor

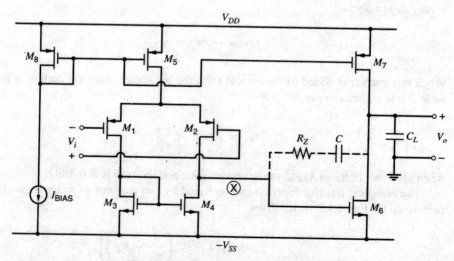

Figure 9.25 A two-stage CMOS op amp.

biased in the triode region. From (1.152), a MOS transistor operating in the triode region behaves like a linear resistor if $V_{ds} \ll 2(V_{GS} - V_t)$. The on-resistance R_Z of the triode device can be made to track $1/g_m$ of common-source transistor M_6 if the two transistors are identical and have the same $V_{GS} - V_t$. When this MOS transistor is placed to the left of the compensation capacitor as shown in Fig. 9.25, its source voltage is set by V_{gs6}, which is approximately constant. Therefore, V_{GS} of the triode transistor can be set by connecting its gate to a dc bias voltage, which can be generated using replica biasing.[13] (See Problem 9.23.)

Another way to shift the zero location that can be used in multistage op amps will be presented in Section 9.4.5.

In all the compensation approaches described so far, the dominant pole is set by compensation capacitor C and is independent of the load capacitor C_2. However, the second pole is a function of C_2. If the op amp will be used in different applications with a range of load capacitors, the compensation capacitor should be selected to give an acceptable phase margin for the largest C_2. Then the phase margin will increase as the load capacitor decreases because $|p_2|$ is inversely proportional to C_2.

■ **EXAMPLE**

Compensate the two-stage CMOS op amp from the example in Section 6.3.5 (Fig. 6.16) to achieve a phase margin of 45° or larger when driving a load capacitance of 5 pF, assuming the op amp is connected in unity-gain feedback.

With the op amp in unity-gain feedback, $f = 1$ and the loop gain $T = af = a$ (or, equivalently, $A_\infty = 1$ and the return ratio $\mathcal{R} = a$). Therefore, the phase and gain margins can be determined from Bode plots of $|a|$ and ph(a).

The two-stage op amp and a simplified model for this op amp are shown in Fig. 9.25. In the model, all capacitances that connect to node $\otimes$ are lumped into C_1, and all capacitances that connect to the output node are lumped into C_2. If we apply an input voltage v_i in Fig. 9.26, a current $i_1 = g_{m1}v_i$ is generated. This i_1 drives a circuit that is the same as the circuit that i_s drives in Fig. 9.18. Therefore, the equations for the two poles and one zero for the circuit in Fig. 9.18 apply here with $i_s = g_{m1}v_i$, $g_m = g_{m6}$, $R_1 = r_{o2}||r_{o4}$, and $R_2 = r_{o6}||r_{o7}$.

We will use Miller compensation with a series resistance to eliminate the zero. To achieve a 45° phase margin, the compensation capacitor C should be chosen so that $|p_2|$ equals the unity-gain frequency (assuming the zero has been eliminated and $|p_3| \gg |p_2|$). Since the gain roll-off from $|p_1|$ to $|p_2|$ is -6 dB/octave, $|a(j\omega)| \cdot \omega$ is constant from $|p_1|$ to $|p_2|$. Therefore,

$$a_o \cdot |p_1| = 1 \cdot |p_2| \tag{9.49}$$

where

$$a_o = g_{m1}(r_{o2}||r_{o4})g_{m6}(r_{o6}||r_{o7}) = g_{m1}R_1 g_{m6}R_2 \tag{9.50}$$

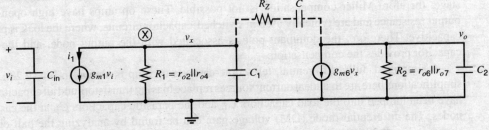

Figure 9.26 A small-signal model for the op amp in Fig. 9.25.

is the dc gain of the op amp. Substitution of (9.47) and (9.50) into (9.49) gives

$$g_{m1} R_1 g_{m6} R_2 \cdot \frac{1}{g_{m6} R_2 R_1 C} = 1 \cdot \frac{g_{m6}}{C_1 + C_2}$$

or

$$\frac{g_{m1}}{C} = \frac{g_{m6}}{C_1 + C_2} \qquad (9.51)$$

The capacitance C_2 at the output is dominated by the 5-pF load capacitance, and the internal parasitic capacitance C_1 is much smaller than 5 pF (SPICE simulation gives $C_1 \approx 120$ fF). Therefore $C_1 + C_2 \approx 5$ pF. From the example in Section 6.3.5, we find

$$g_{m1} = k'_p (W/L)_1 |V_{ov1}| = (64.7 \ \mu A/V^2)(77)(0.2 \ V) = 1 \ mA/V$$

and

$$g_{m6} = k'_n (W/L)_6 (V_{ov6}) = (194 \ \mu A/V^2)(16)(0.5 \ V) = 1.55 \ mA/V$$

Substituting these values into (9.51) and rearranging gives

$$C = \frac{g_{m1}}{g_{m6}}(C_1 + C_2) \approx \frac{1 \ mA/V}{1.55 \ mA/V}(5 \ pF) = 3.2 \ pF$$

To eliminate the zero due to feedforward through C, a resistor R_Z of value $1/g_{m6} = 645 \ \Omega$ can be connected in series with the compensation capacitor C. (In practice, this resistance should be implemented with an NMOS transistor that is a copy of M_6 biased in the triode region, so that $R_Z = 1/g_{m6}$. See Problem 9.23.)

SPICE simulations (using models based on Table 2.4) of the op amp before and after compensation give the magnitude and phase plots shown in Fig. 9.27. Before compensation, the amplifier is unstable and has a phase margin of $-6°$. After compensation with $R_Z = 645 \ \Omega$ and $C = 3.2$ pF the phase margin improves to $41°$ with a unity-gain frequency of 35 MHz, and the gain margin is 15 dB. This phase margin is less than the desired $45°$. The simulated value of g_{m6} is 1.32 mA/V and differs somewhat from the calculated g_{m6}, because the formulas used to calculate g_m are based on square-law equations that are only approximately correct. Changing R_Z to $1/g_{m6}$(SPICE) $= 758 \ \Omega$ gives a phase margin of $46°$ with a unity-gain frequency of 35 MHz, and the gain margin is 22 dB. Without R_Z, the phase margin is $14°$, so eliminating the right-half-plane zero significantly improves the phase margin.

Two earlier assumptions can be checked from SPICE simulations. First, $C_1 \approx 120$ fF from SPICE and $C_2 \approx 5$ pF; therefore, the assumption that $C_1 \ll C_2$ is valid. Also, $|p_3| \gg |p_2|$ follows from $|p_3| \approx 1/(R_Z C_1) = g_{m6}/C_1$, $|p_2| \approx g_{m6}/C_2$, and $C_1 \ll C_2$.

■

9.4.4 Compensation of Single-Stage CMOS Op Amps

Single-stage op amps, such as the telescopic cascode or folded cascode, have only one gain stage; therefore Miller compensation is not possible. These op amps have high open-loop output resistance and are typically used in switched-capacitor circuits, where the load is purely capacitive. Therefore, the dominant pole is associated with the output node, and the load capacitor provides the compensation.

A simplified, fully differential, telescopic-cascode op amp is shown in Fig. 9.28a. The simplifications here are that ideal current sources replace biasing transistors and all capacitances have been lumped into the load capacitors C_L and the parasitic capacitors C_p at the cascode nodes. The differential-mode (DM) voltage gain can be found by analyzing the half-circuit shown in Fig. 9.28b. Since there are two independent capacitors, the DM gain has two poles.

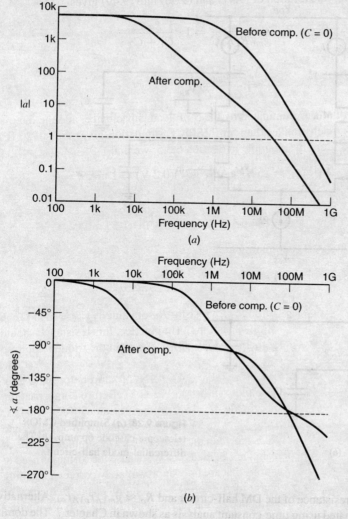

Figure 9.27 Plots of the simulated (a) magnitude and (b) phase of the op-amp gain before and after compensation ($C = 3.2$ pF, $R_Z = 645\ \Omega$) for the op amp in Fig. 9.25.

An exact analysis, ignoring body effect, gives a DM gain of

$$\frac{v_{od}}{v_{id}} = -\frac{g_{m1}r_{o1}(g_{m1A}r_{o1A} + 1)}{1 + s(r_{o1A}C_L + r_{o1}C_p + r_{o1}C_L + g_{m1A}r_{o1A}r_{o1}C_L) + s^2 r_{o1}r_{o1A}C_pC_L} \quad (9.52)$$

If $g_m r_o \gg 1$, (9.52) simplifies to

$$\frac{v_{od}}{v_{id}} = -\frac{g_{m1}r_{o1}g_{m1A}r_{o1A}}{1 + sg_{m1A}r_{o1A}r_{o1}C_L + s^2 r_{o1}r_{o1A}C_pC_L} \quad (9.53)$$

The gain has two poles and no zeros. Assuming widely spaced real poles, the poles can be approximated using (9.29) and (9.30):

$$p_1 \approx -\frac{1}{g_{m1A}r_{o1A}r_{o1}C_L} \approx -\frac{1}{R_oC_L} \quad (9.54a)$$

$$p_2 \approx -\frac{g_{m1A}}{C_p} \quad (9.54b)$$

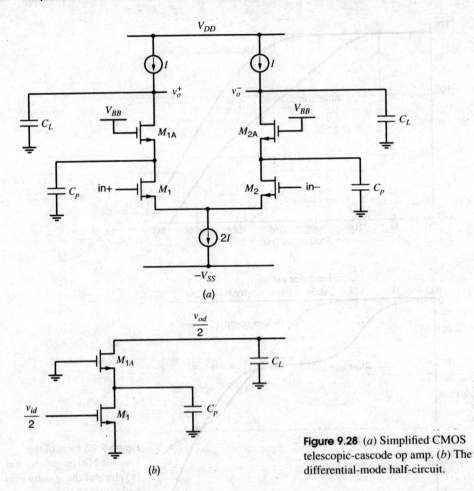

Figure 9.28 (a) Simplified CMOS telescopic-cascode op amp. (b) The differential-mode half-circuit.

where R_o is the output resistance of the DM half-circuit and $R_o \approx g_{m1A} r_{o1A} r_{o1}$. Alternatively, these poles can be estimated using time-constant analysis as shown in Chapter 7. The dominant pole is set by the zero-value time constant for C_L, which is computed with C_p open and equals $R_o C_L$. The nondominant pole can be approximated using the short-circuit time constant for C_p, which is computed with C_L shorted. When C_L is shorted, the resistance seen by C_p is the resistance looking into the source of M_{1A}, which is $1/g_{m1A}$ (ignoring body effect). Typically, $|p_1| \ll |p_2|$ because $R_o \gg 1/g_{m1A}$ and $C_L \gg C_p$. If the phase margin is not large enough for a given feedback application, additional capacitance can be added at the output node to increase C_L, which decreases $|p_1|$ without affecting p_2 and therefore increases the phase margin.

Capacitance C_p consists of C_{gs1A} plus smaller capacitances such as C_{db1} and C_{sb1A}. Assuming $C_p \approx C_{gs1A}$, then $|p_2| \approx g_{m1A}/C_p \approx g_{m1A}/C_{gs1A} \approx \omega_T$ of M_{1A}. Thus, the frequency at which the magnitude of the op-amp gain equals one, which is called the unity-gain bandwidth, can be very high with this op amp.

A simplified, fully differential, folded-cascode op amp is shown in Fig. 9.29a. As above, the simplifications are that ideal current sources replace biasing transistors and all capacitances have been lumped into the load capacitors C_L and the parasitic capacitors C'_p at the cascode nodes. With these simplifications, the DM voltage gain can be found by analyzing the half-circuit shown in Fig. 9.29b. This circuit is identical to Fig. 9.28b except that the cascode device is p-channel rather than n-channel and C'_p replaces C_p. Therefore, the gain is identical to (9.52)

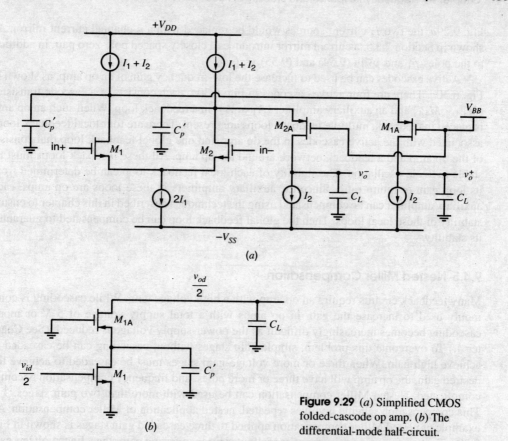

Figure 9.29 (a) Simplified CMOS folded-cascode op amp. (b) The differential-mode half-circuit.

with C_p replaced by C'_p. Hence the dominant pole has the same form as (9.54a)

$$p_1 \approx -\frac{1}{g_{m1A}r_{o1A}r_{o1}C_L} \approx -\frac{1}{R_o C_L} \tag{9.55a}$$

The second pole is associated with C'_p and is approximately given by

$$p_2 \approx -\frac{g_{m1A}}{C'_p} \tag{9.55b}$$

Equations 9.55b and 9.54b look similar, but $|p_2|$ for the folded-cascode op amp will usually be smaller than $|p_2|$ for the telescopic-cascode op amp. The reason is that, while the transconductances of the cascode devices in the two circuits are often comparable, C'_p will be significantly larger than C_p. One cause of the higher capacitance is that more devices are connected to the node associated with C'_p in the folded-cascode op amp than are connected to the node associated with C_p in the telescopic cascode. (Recall that the output of each ideal current source in Fig. 9.29a is the drain of a transistor.) Also, W/L of the p-channel cascode transistor M_{1A} in Fig. 9.29b must be larger than W/L of the n-channel cascode device in Fig. 9.28b to make their transconductances comparable. The larger W/L will cause C'_p to be larger than C_p. The smaller $|p_2|$ for the folded cascode leads to a smaller unity-gain bandwidth, if the two op amps are compensated to give the same phase margin in a given feedback application.

The circuits in Figs. 9.28 and 9.29 are fully differential. These op amps can be converted to single-ended op amps by replacing a pair of matched current sources with a current mirror. In Fig. 9.28a, the two I current sources would be replaced with a p-channel current mirror. In

Fig. 9.29a, the two I_2 current sources would be replaced with a n-channel current mirror. As shown in Section 7.3.5, a current mirror introduces a closely spaced pole-zero pair, in addition to the poles p_1 and p_2 in (9.54) and (9.55).

Active cascodes can be used to increase the low-frequency gain of an op amp, as shown in Fig. 6.30a. There are four active cascodes in Fig. 6.30a; each consists of a cascode transistor ($M_{1A} - M_{4A}$) and an auxiliary amplifier (A_1 or A_2) in a feedback loop. When such an op amp is placed in feedback, multiple feedback loops are present. There are four local feedback loops associated with the active cascodes in the op amp and one global feedback loop that consists of the op amp and a feedback network around the op amp. All these feedback loops must be stable to avoid oscillation. The stability of each local feedback loop can be determined from its loop gain or return ratio. Since the auxiliary amplifiers in these loops are op amps, each auxiliary amplifier can be compensated using the techniques described in this chapter to ensure stability of these local loops. Then the global feedback loop can be compensated to guarantee its stability.

9.4.5 Nested Miller Compensation

Many feedback circuits require an op amp with a high voltage gain. While cascoding is commonly used to increase the gain in op amps with a total supply voltage of 5 V or more, cascoding becomes increasingly difficult as the power-supply voltage is reduced. (See Chapter 4.) To overcome this problem, simple gain stages without cascoding can be cascaded to achieve high gain. When three or more voltage-gain stages must be cascaded to achieve the desired gain, the op amp will have three or more poles, and frequency compensation becomes complicated. Nested Miller compensation can be used with more than two gain stages.[14,15] This compensation scheme involves repeated, nested application of Miller compensation. An example of nested Miller compensation applied to three cascaded gain stages is shown in Fig. 9.30a. Two noninverting gain stages are followed by an inverting gain stage. Each voltage-gain stage is assumed to have a high-output resistance and therefore is labeled as a g_m block. The sign of the dc voltage gain of each stage is given by the sign of the transconductance. Two Miller compensation capacitors are used: C_{m1}, which is placed around the last gain stage, and C_{m2}, which is connected across the last two gain stages. Because the dc gain of the second stage is positive and the dc gain of the third stage is negative, both capacitors are in negative feedback loops.

A simplified circuit schematic is shown in Fig. 9.30b. Each noninverting gain stage is composed of a differential pair with a current-source load. The inverting gain stage consists of a common-source amplifier with a current-source load. A simplified small-signal model is shown in Fig. 9.30c. The main simplification here is that all capacitances associated with the gain stages are modeled by C_0, C_1, and C_2.

Without the compensation capacitors, this amplifier has three real poles that are not widely spaced if the $R_i C_i$ time constants are comparable. When C_{m1} is added, the two poles associated with the output nodes of the second and third stages split apart along the real axis due to the Miller compensation, but the pole associated with output of the first stage does not change. From a design standpoint, the goal of this pole splitting is to cause one pole to dominate the frequency response of the second and third stages together. Assume at first that this goal is met. Then adding C_{m2} across the second and third stages is similar to adding C_{m1} across the third stage. Pole splitting occurs again, and the pole associated with the output node of the first stage becomes dominant because the Miller-multiplied C_{m2} loads this node. Meanwhile, the pole associated with the output of the second stage moves to higher frequency because of negative feedback through C_{m2}. The polarity of this feedback does not become positive at any frequency where the gain around the loop is at least unity because the frequency response of the second and third stages is dominated by one pole.

9.4 Compensation

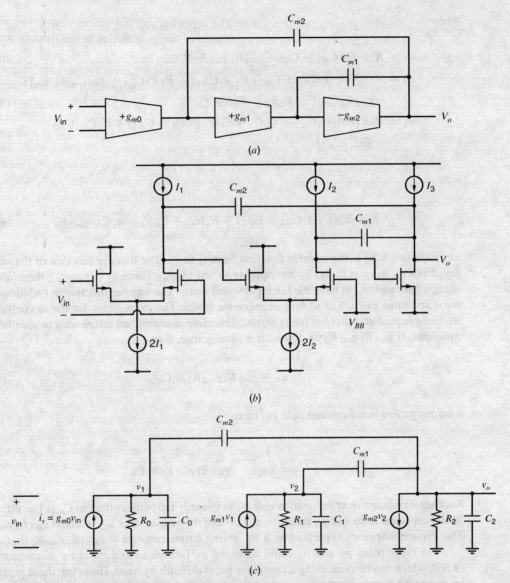

Figure 9.30 (a) Block diagram for a three-stage op amp with nested Miller compensation. (b) A simplified schematic for such an op amp in CMOS. (c) A small-signal model.

In practice, the exact movement of the poles is complicated by the nondominant pole in the feedback loop though C_{m2}. Also, zeros are introduced by feedforward through C_{m1} and C_{m2}. The pole and zero locations can be found from an exact analysis of the small-signal circuit. The analysis can be carried out by summing currents at the outputs of the g_m generators, then manipulating the resulting three equations. These steps are not conceptually difficult but are not shown here. The exact transfer function from the output of the current generator in the input stage, $i_s = g_{m0} v_{in}$, to the output voltage v_o is

$$\frac{v_o}{i_s} = -\frac{N(s)}{D(s)} \qquad (9.56)$$

$$= -\frac{R_0 g_{m1} R_1 g_{m2} R_2 - (g_{m1} R_1 C_{m1} + C_{m2}) R_0 R_2 s - R_0 R_1 R_2 C_{m2}(C_1 + C_{m1}) s^2}{1 + a_1 s + a_2 s^2 + a_3 s^3}$$

where

$$a_1 = K + R_0(C_{m2} + C_0) + g_{m1}R_1 g_{m2}R_2 R_0 C_{m2} \tag{9.57a}$$

$$a_2 = R_1 R_2(C_2 + C_{m1} + C_{m2})(C_1 + C_{m1}) - R_1 R_2 C_{m1}^2 + R_0(C_{m2} + C_0)K$$
$$- g_{m1}R_1 C_{m1} C_{m2} R_0 R_2 - R_0 R_2 C_{m2}^2 \tag{9.57b}$$

$$a_3 = R_0 R_1 R_2[(C_2 C_{m2} + C_0 C_2 + C_0 C_{m2})(C_1 + C_{m1}) + C_1 C_{m1} C_{m2}$$
$$+ C_0 C_1 C_{m1}] \tag{9.57c}$$

with

$$K = R_2(C_2 + C_{m1} + C_{m2}) + R_1(C_1 + C_{m1}) + R_1 C_{m1} g_{m2} R_2 \tag{9.57d}$$

Equation 9.56 is the transfer function from i_s to v_o. The transfer function of the voltage gain from v_{in} to v_o is found by multiplying (9.56) by g_{m0} (since $i_s = g_{m0} v_{\text{in}}$); therefore, the voltage gain and (9.56) have the same poles and zeros. The transfer function in (9.56) has two zeros and three poles. Let us first examine the poles. The expressions for the a_i coefficients are complicated and involve many terms. Therefore assumptions are needed to simplify the equations. If $g_{m1}R_1 g_{m2}R_2 \gg 1$, which is usually true, then

$$a_1 \approx g_{m1}R_1 g_{m2}R_2 R_0 C_{m2} \tag{9.58}$$

Assuming there is a dominant pole p_1, then

$$p_1 \approx -\frac{1}{a_1} = -\frac{1}{g_{m1}R_1 g_{m2}R_2 R_0 C_{m2}} \tag{9.59}$$

Another way to arrive at this estimate of p_1 is to apply the Miller effect to C_{m2}. The effective Miller capacitor is about C_{m2} times the negative of the gain across C_{m2}, which is $g_{m1}R_1 g_{m2}R_2$. This capacitor appears in parallel with R_0, giving a time constant of $(g_{m1}R_1 g_{m2}R_2)R_0 C_{m2}$.

The other poles p_2 and p_3 could be found by factoring the third-order denominator in (9.56), which can be done using a computer but is difficult by hand. However, these poles can be estimated from a quadratic equation under certain conditions. If there is a dominant pole p_1, then $|p_2|, |p_3| \gg |p_1|$. At high frequencies, where $|s| \gg |p_1| \approx 1/a_1$, we have $|a_1 s| \gg 1$, so the denominator in (9.56) can be approximated by dropping the constant "1" to give

$$D(s) \approx a_1 s + a_2 s^2 + a_3 s^3 = a_1 s \left(1 + \frac{a_2}{a_1} s + \frac{a_3}{a_1} s^2\right) \tag{9.60}$$

This equation gives three poles. One pole is at dc, which models the effect of the dominant pole p_1 for frequencies well above $|p_1|$. Poles p_2 and p_3 are the other roots of (9.60). They can be found by concentrating on the quadratic term in parenthesis in (9.60), which is

$$D'(s) = \frac{D(s)}{a_1 s} \approx 1 + \frac{a_2}{a_1} s + \frac{a_3}{a_1} s^2 \approx \left(1 - \frac{s}{p_2}\right)\left(1 - \frac{s}{p_3}\right) \tag{9.61}$$

Assuming that $R_0, R_1, R_2 \gg |1/(g_{m2} - g_{m1})|$ and C_o is small compared to the other capacitors, (9.57b) and (9.57c) simplify to

$$a_2 \approx R_0 R_1 R_2 (g_{m2} - g_{m1}) C_{m1} C_{m2} \tag{9.62}$$

$$a_3 \approx R_0 R_1 R_2 (C_1 C_2 C_{m2} + C_2 C_{m1} C_{m2} + C_1 C_{m1} C_{m2}) \tag{9.63}$$

Using (9.58), (9.62), and (9.63), the coefficients in $D'(s)$ are

$$\frac{a_2}{a_1} \approx \frac{g_{m2} - g_{m1}}{g_{m1} g_{m2}} C_{m1} \tag{9.64}$$

$$\frac{a_3}{a_1} \approx \frac{C_1 C_2 + C_{m1} C_1 + C_2 C_{m1}}{g_{m1} g_{m2}} \tag{9.65}$$

To ensure that the high-frequency poles are in the left half-plane (LHP), a_2/a_1 must be positive (see Appendix A9.2). Therefore, g_{m2} must be larger than g_{m1}. Poles p_2 and p_3 can be real or complex, and in general the quadratic formula must be used to solve for these poles. However, if these poles are real and widely spaced and if $C_{m1} \gg C_1, C_2$, then approximate expressions can be found. If $|p_2| \ll |p_3|$, then $-1/p_2$ is approximately equal to the coefficient of s in $D'(s)$, so

$$p_2 \approx -\frac{a_1}{a_2} = -\frac{g_{m1} g_{m2}}{(g_{m2} - g_{m1}) C_{m1}} \tag{9.66a}$$

Also $1/(p_2 p_3)$ is equal to the coefficient of s^2 in $D'(s)$, so

$$p_3 \approx \frac{a_1}{a_3} \frac{1}{p_2} = -\frac{g_{m1} g_{m2}}{C_1 C_2 + C_{m1} C_1 + C_2 C_{m1}} \cdot \frac{(g_{m2} - g_{m1}) C_{m1}}{g_{m1} g_{m2}} \tag{9.66b}$$

$$= -\frac{(g_{m2} - g_{m1}) C_{m1}}{C_1 C_2 + C_{m1}(C_1 + C_2)} \approx -\frac{g_{m2} - g_{m1}}{C_1 + C_2}$$

The final approximation here follows if C_{m1} is large. Equations 9.66a and 9.66b are accurate if $|p_2| \ll |p_3|$. Substituting (9.66a) and (9.66b) into this inequality produces an equivalent condition

$$|p_2| \approx \frac{g_{m1} g_{m2}}{(g_{m2} - g_{m1}) C_{m1}} \ll \frac{(g_{m2} - g_{m1}) C_{m1}}{C_1 C_2 + C_{m1}(C_1 + C_2)} \approx |p_3| \tag{9.67}$$

If this condition is not satisfied, p_2 and p_3 are either complex conjugates or real but closely spaced. C_{m1} can always be chosen large enough to satisfy the inequality in (9.67). While it is possible to make the high-frequency poles real and widely separated, higher unity-gain bandwidth may be achievable when p_2 and p_3 are not real and widely separated.[16]

In the simplified equations 9.66a and 9.66b, poles p_2 and p_3 are dependent on C_{m1} but not on C_{m2}. In contrast, dominant pole p_1 is inversely proportional to C_{m2} and is independent of C_{m1}. The poles can be positioned to approximate a two-pole op amp by making $|p_1| \ll |p_2| \ll |p_3|$ and positioning $|p_3|$ well beyond the unity-gain frequency of the op amp.

The zero locations can be found by factoring the second-order numerator $N(s)$ in (9.56). The coefficients of s and s^2 in the numerator are negative and the constant term is positive. As a result, the zeros are real. One is positive and the other is negative, as is shown in Appendix A9.2.

The zeros will be found using some simplifying assumptions. First, the numerator of (9.56) can be rewritten as

$$N(s) = R_0 g_{m1} R_1 g_{m2} R_2 \left[1 - s \left(\frac{C_{m1}}{g_{m2}} + \frac{C_{m2}}{g_{m1} R_1 g_{m2}} \right) - s^2 \frac{C_{m2}(C_1 + C_{m1})}{g_{m1} g_{m2}} \right] \tag{9.68}$$

Assuming that $C_{m1} \gg C_1$ and $C_{m1} \gg C_{m2}/(g_{m1}R_1)$, then

$$N(s) \approx R_0 g_{m1} R_1 g_{m2} R_2 \left[1 - s\frac{C_{m1}}{g_{m2}} - s^2\frac{C_{m2}C_{m1}}{g_{m1}g_{m2}}\right] \quad (9.69)$$

The zeros are the roots of $N(s) = 0$. Using the quadratic formula and (9.69), the zeros are

$$z_{1,2} = -\frac{g_{m1}}{2C_{m2}} \pm \sqrt{\left(\frac{g_{m1}}{2C_{m2}}\right)^2 + \frac{g_{m1}g_{m2}}{C_{m1}C_{m2}}} = -\frac{g_{m1}}{2C_{m2}}\left(1 \pm \sqrt{1 + \frac{4g_{m2}C_{m2}}{g_{m1}C_{m1}}}\right) \quad (9.70)$$

Taking the positive square root in the right-most formula in (9.70) yields a value that is larger than one. Adding this value to 1 gives a positive value for the term in parentheses; subtracting this value from 1 gives a negative quantity with a smaller magnitude than the sum. Therefore, one zero is in the LHP and has a magnitude greater than $g_{m1}/(2C_{m2})$. The other zero is in the RHP and has a smaller magnitude than the LHP zero. As a result, the effect of the RHP zero is felt at a lower frequency than the LHP zero.

The magnitude of one or both zeros can be comparable to $|p_2|$. Because the RHP zero is at a lower frequency than the LHP zero, the RHP zero can cause significant negative phase shift for frequencies at or below $|p_2|$, which would degrade the phase margin of a feedback loop. This undesired negative phase shift would not occur if the transfer function did not have zeros. Unfortunately, the three techniques considered in Section 9.4.3 to eliminate a RHP zero have important limitations in a low-supply application. First, the zeros could be eliminated by adding a source-follower buffer between the op-amp output and the right-hand side of capacitors C_{m1} and C_{m2} (as in Fig. 9.21), thereby eliminating the feedforward paths through the capacitors. However, the source follower has a nonzero dc voltage between its input and output. This voltage may limit the op-amp output swing to an unacceptably low value in a low-power-supply application. Second, cascode stages could be used to eliminate the zeros, as shown in Fig. 9.23. However, the requirement that all transistors in the cascode stage operate in the active region may limit the minimum supply voltage. Finally, a series zero-canceling resistance (as in Fig. 9.24a) implemented with a transistor may require a large gate voltage that exceeds the power supply.

The NE5234 op amp uses nested Miller-effect compensation. Figure 9.31 repeats the simplified ac schematic of the high-frequency gain path of the NE5234 shown in Fig. 7.36. Here, the common-mode input voltage is assumed to be low enough that Q_1 and Q_2 in Fig. 6.36 are off. Also, the dc load current is assumed to be $I_L = 1$ mA as in the calculations in Chapter 6. Therefore, Q_{75} in Fig. 6.39 conducts a nearly constant current and is omitted in Fig. 9.31 along with the circuits that control it for simplicity. In practice, these transistors are important under other bias conditions. Also, note that the transconductance of the output stage depends on the bias point assumed. The key point here is that this op amp uses three nested compensation loops: through C_{22}, C_{25}, and C_{65}. The loop through C_{25} includes series resistor $R_{25} = 1.3$ kΩ to reduce the effects of the zero introduced through C_{25} and increase the phase margin.[17] This structure has one more level of nesting than shown in Fig. 9.30. The extra level is introduced through C_{65} in the third stage, and its purpose is explained next.

Chapter 6 pointed out that the output transistors Q_{74} and Q_{75} in Fig. 6.39 are driven by emitter followers to increase the current gain of the output stage and reduce its load on the second stage. Because the integrated-circuit process is optimized to build much higher quality npn transistors than pnp transistors, $\beta_{pnp} < \beta_{npn}$ in practice. To provide adequate current gain when Q_{74} controls the output as shown in Fig. 9.31, two emitter followers Q_{64}-Q_{65} drive Q_{74}. In contrast, Fig. 6.39 shows that only one emitter follower Q_{68} is used to drive Q_{75}. Furthermore, Q_{64} and Q_{65} use opposite polarity transistors to avoid introducing a large dc level shift that would increase the minimum required power-supply voltage.

9.4 Compensation

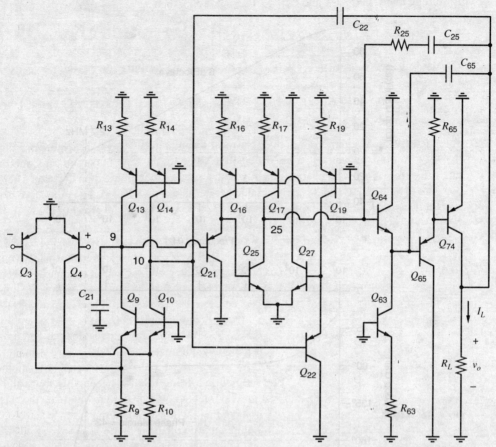

Figure 9.31 An ac schematic of the high-frequency gain path of the NE5234 op amp assuming that the common-mode input voltage is low enough that Q_1 and Q_2 in Fig. 6.36 are off and assuming that Q_{75} in Fig. 6.39 conducts a constant current and can be ignored along with the elements that drive it.

Ideally, these emitter followers give unity gain and do not limit the frequency response of the third stage. In practice, however, they introduce extra poles that contribute unwanted phase shift at high frequency that reduces the phase margin when the op amp is connected in a feedback loop. This problem is especially severe in driving Q_{74} because two emitter followers are used instead of one and because the output transistor and one of the emitter followers are *pnp* transistors, which have much lower f_T than *npn* transistors operating at the same bias currents. If Miller compensation were not applied through C_{65}, the presence of the extra poles in the output stage due to the emitter followers would introduce extra undesired phase shift near the unity-gain frequency of the op amp and significantly reduce the phase margin. To overcome this problem, the extra level of Miller-effect compensation through C_{65} is introduced. It forces one pole to be dominant in the output stage when feedback is applied through C_{25}. The minimum required value of C_{65} must be able to cope with all possible bias currents in the output stage. From a stability standpoint, the worst case is when the bias current and transconductance of Q_{74} are maximum because the op-amp bandwidth is increased in this case, which increases the importance of poles introduced by the emitter followers. In practice, C_{65} is chosen from simulations to be 10 pF.[17] The corresponding capacitor on the *npn* side of the output stage is C_{68} in Fig. 6.39, and this capacitor is only 1 pF. In practice, $C_{68} \ll C_{65}$ because the *npn* side uses only one emitter follower and because the transistors on this side are both *npn* transistors.

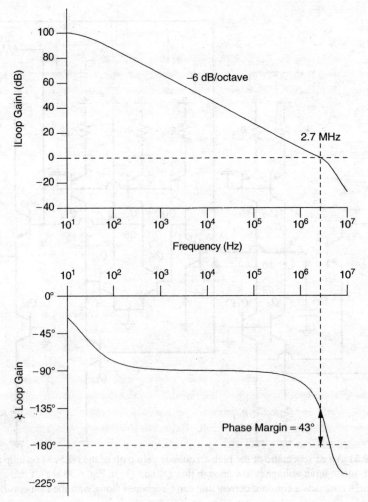

Figure 9.32 Gain and phase versus frequency for the NE5234 op amp from SPICE.

The presence of an extra level of Miller compensation reduces the bandwidth of the output stage to make one pole dominant. Although it allows a simple compensation scheme for the op amp, it also limits the high-frequency performance of the op amp with compensation.

As shown in Chapter 7, the frequency response of the NE5234 is dominated by the Miller multiplied C_{22}. With unity-gain feedback as in Fig. 6.3c, the resulting gain and phase plots for the NE5234 are shown in Fig. 9.32. These plots were generated using SPICE with transistor parameters as shown in Fig. 2.32 except $\beta_F = 40$ and $V_A = 30$ V for *npn* transistors and $\beta_F = 10$ and $V_A = 20$ V for *pnp* transistors. The bias conditions are the same as assumed in Chapter 6. The resulting unity-gain frequency is 2.7 MHz, and the phase margin is 43 degrees. Return ratio simulations give the same results.

Fig. 9.33a shows another technique for eliminating a RHP zero that can be used with cascaded stages in a low-supply application.[16,18] Two gain stages and one Miller compensation capacitor are shown. A transconductance stage, g_{mf}, is included. It provides a feedforward path that can be used to move the zero to infinity. The small-signal circuit is shown in Fig. 9.33b. To allow a simple explanation of this circuit, initially assume that $C_1 = C_2 = 0$. The circuit has one pole due to C_m and one zero due to the feedforward current through C_m. If the zero moves to infinity, the total forward current must go to zero when $\omega \to \infty$. Also, if the zero moves to

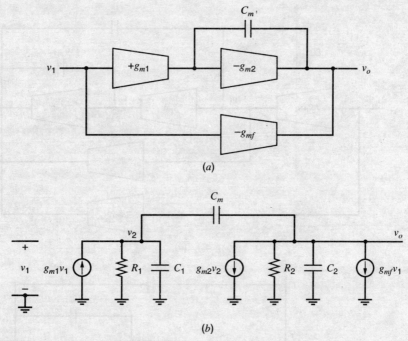

Figure 9.33 (a) Block diagram of a two-stage op amp with Miller compensation and a feedforward transconductor. (b) A small-signal model.

infinity, the output voltage will go to zero as $\omega \to \infty$ due to the pole in the transfer function. When $\omega \to \infty$, capacitor C_m becomes a short circuit, so $v_2 = 0$ when $\omega \to \infty$. Therefore at infinite frequency, the current $g_{m1}v_1$ from the g_{m1} source flows through C_m. Adding this feedforward current to the current $g_{mf}v_1$ from the g_{mf} generator gives the total current at the output node that is related to v_1

$$i_{ff}(\omega \to \infty) = (-g_{m1} + g_{mf})v_1 \qquad (9.71)$$

If $g_{mf} = g_{m1}$, this current equals zero, which means the zero is at infinity.

An exact analysis of the circuit in Fig. 9.33 gives a transfer function

$$\frac{v_o}{v_1} = \qquad (9.72)$$

$$\frac{-g_{m1}R_1 g_{m2} R_2 - g_{mf} R_2 - sR_1 R_2 [g_{mf}(C_1 + C_m) - g_{m1}C_m]}{1 + s[g_{m2}R_1 R_2 C_m + R_2(C_2 + C_m) + R_1(C_1 + C_m)] + s^2 R_1 R_2 (C_1 C_2 + C_1 C_m + C_2 C_m)}$$

The zero can be moved to infinity by choosing g_{mf} so that the coefficient of s in the numerator is zero, which occurs when

$$g_{mf} = g_{m1}\frac{C_m}{C_1 + C_m} = \frac{g_{m1}}{1 + \frac{C_1}{C_m}} \qquad (9.73)$$

This value of g_{mf} depends on the ratio of an internal parasitic capacitance C_1, which is not well controlled, and compensation capacitor C_m. Using $g_{mf} = g_{m1}$ moves the zero into the LHP to about $-g_{m2}/C_1$; the magnitude of this zero is usually above the unity-gain frequency of the op amp. If the g_{m1} stage has a differential input, the $-g_{mf}$ stage can be realized using a replica of the g_{m1} stage with the inputs reversed to change the sign of the transconductance.

This zero-cancellation scheme can be used repeatedly in a three-stage op amp to eliminate the zeros, as shown in Fig. 9.34a. A small-signal model is shown in Fig. 9.34b. Analysis of

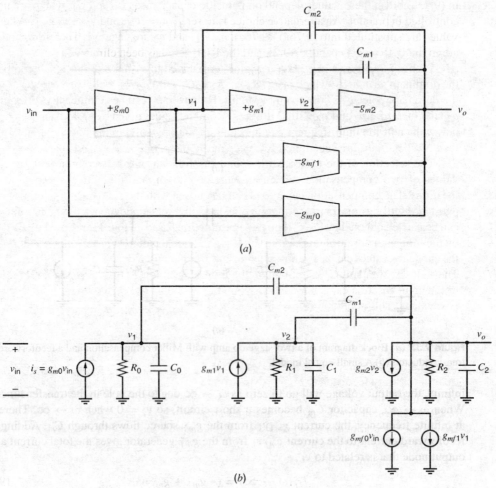

Figure 9.34 (a) Block diagram of a three-stage op amp with nested Miller compensation and two feedforward transconductors. (b) Small-signal model.

this circuit gives a voltage gain of

$$\frac{v_o}{v_{in}} = \frac{R_2(n_0 + n_1 s + n_2 s^2)}{1 + b_1 s + b_2 s^2 + b_3 s^3} \quad (9.74)$$

where b_1-b_3 are related to a_1-a_3 in (9.57) by

$$b_1 = a_1 + g_{mf1} R_0 R_2 C_{m2} \quad (9.75a)$$

$$b_2 = a_2 + g_{mf1} R_0 R_1 R_2 (C_1 + C_{m1}) C_{m2} \quad (9.75b)$$

$$b_3 = a_3 \quad (9.75c)$$

and the coefficients in the numerator are

$$n_0 = -g_{m0} g_{m1} g_{m2} R_0 R_1 - g_{mf0} - g_{m0} g_{mf1} R_0 \quad (9.76a)$$

$$n_1 = g_{m0}(g_{m1} - g_{mf1}) R_0 R_1 C_{m1} + (g_{m0} - g_{mf0}) R_0 C_{m2}$$
$$- g_{mf0} R_1 (C_1 + C_{m1}) - g_{mf0} R_0 C_0 - g_{m0} g_{mf1} R_0 R_1 C_1 \quad (9.76b)$$

$$n_2 = (g_{m0} - g_{mf0}) R_0 R_1 (C_1 + C_{m1}) C_{m2} - g_{mf0} R_0 R_1 (C_1 + C_{m1}) C_0 \quad (9.76c)$$

The coefficients of s and s^2 in the numerator include both positive and negative terms. Therefore, they can be set to zero, which eliminates the zeros, by properly choosing g_{mf0} and g_{mf1}. As in (9.73) above, these values depend on parasitic capacitances C_0 and C_1, which are not well controlled in practice. An alternative choice is to set $g_{mf0} = g_{m0}$ and $g_{mf1} = g_{m1}$. When these values are substituted into (9.76a)–(9.76c), n_0, n_1, and n_2 are negative. Therefore, both zeros are in the LHP (see Appendix A9.2), and the RHP zero has been eliminated.

With $g_{mf1} = g_{m1}$ and $g_{mf0} = g_{m0}$, the term added to a_1 in (9.75a) is small compared to the dominant term in a_1, which is $g_{m1}R_1g_{m2}R_2R_0C_{m2}$, if $g_{m2}R_1 \gg 1$. Therefore $b_1 \approx a_1$, and the dominant pole p_1 is still given by (9.59). However, b_2 can be significantly different from a_2, and therefore p_2 and p_3 will be different from the values given by (9.66a) and (9.66b). The new values of the high frequency poles can be found by solving the quadratic equation that results when b_1-b_3 are substituted for a_1-a_3 in (9.60).

The selection of the nested Miller compensation capacitors is complicated because the values of two compensation capacitors must be chosen, and they affect the pole and zero locations. The compensation capacitors can be chosen with the aid of a computer to achieve a particular settling-time or phase-margin goal in a feedback application. Computer optimization can be carried out on the closed-loop transfer function based on the op-amp transfer function or on the loop gain or return ratio, if the small-signal model parameters are known. Alternatively, the capacitor values can be estimated using approximations and the equations presented above. Then SPICE simulations can be run on the transistor circuit starting with the initial estimates of the compensation capacitors and varying the capacitors by small amounts to determine the best values. This approach is used in the following example.

■ **EXAMPLE**

Design the 3-stage op amp in Fig. 9.34 to give a low-frequency gain of 86 dB and 45° phase margin for unity feedback ($f = 1$) when driving a 5 pF load. Compensate the op amp so that all the poles are real and widely spaced. To simplify this example, assume that the output resistance of each stage is 5 kΩ and the internal node capacitances C_0 and C_1 are each 0.05 pF. Determine the compensation capacitors and the transconductances for the op amp.

The feedforward transconductances g_{mf0} and g_{mf1} will be used to move the zeros to well beyond the unity-gain frequency. To simplify the design equations, let $g_{mf0} = g_{m0}$ and $g_{mf1} = g_{m1}$, based on (9.73)–(9.76) and the assumption that C_0 and C_1 are small compared to C_{m1} and C_{m2}.

When $g_{mf0} = g_{mf1} = 0$, the coefficients a_i of the denominator of the transfer function are given by (9.57). With nonzero g_{mf0} and g_{mf1}, however, the coefficients of s and s^2 in the denominator of the transfer function change and are given by (9.75). From (9.75c), $b_3 = a_3$. Also, as noted in the text following (9.76), the term added to a_1 in (9.75a) is small compared to a_1, so $b_1 \approx a_1$ and p_1 is given by (9.59). Hence, poles p_2 and p_3 are changed due to the added term that includes g_{mf1} in b_2 in (9.75b). Assuming $C_1 \ll C_{m1}$, (9.75b) reduces to

$$b_2 \approx a_2 + g_{mf1}R_0R_1R_2C_{m1}C_{m2}$$

Substituting the approximate expression for a_2 in (9.62) and using $g_{mf1} = g_{m1}$, this equation becomes

$$b_2 \approx g_{m2}R_0R_1R_2C_{m1}C_{m2}$$

Following the analysis from (9.60) to (9.67), we find

$$p_2 \approx -\frac{b_1}{b_2} \approx -\frac{g_{m1}}{C_{m1}} \qquad (9.77a)$$

$$p_3 \approx -\frac{b_1}{b_3}\frac{1}{p_2} \approx -\frac{g_{m2}}{C_2} \qquad (9.77b)$$

To satisfy $|p_2| \ll |p_3|$, let $|p_3| = 10|p_2|$. Substituting (9.77) in this equality and rearranging yields

$$C_{m1} = 10 \frac{g_{m1}}{g_{m2}} C_2 \qquad (9.78)$$

To ensure that C_{m1} is not much larger than $C_2 = 5$ pF we need $g_{m1}/g_{m2} \ll 1$ in (9.78). Here, we chose $g_{m1}/g_{m2} = 0.2$. Substituting this value into (9.78) gives

$$C_{m1} = 10(0.2)(5 \text{ pF}) = 10 \text{ pF}$$

With widely spaced poles, placing $|p_2|$ at the unity-gain frequency gives a 45° phase margin. Since $|gain| \times frequency$ is constant for frequencies between $|p_1|$ and $|p_2|$, we can write

$$|a_0| \cdot |p_1| = 1 \cdot |p_2| \qquad (9.79)$$

where

$$|a_0| = g_{m0} R_0 g_{m1} R_1 g_{m2} R_2 \qquad (9.80)$$

is the low-frequency gain. Substitution of (9.59), (9.77a), and (9.80) into (9.79) gives

$$\frac{g_{m0}}{C_{m2}} = \frac{g_{m1}}{C_{m1}}$$

If the first two gain stages are made identical to reduce the circuit-design effort, $g_{m0} = g_{m1}$, and the last equation reduces to

$$C_{m2} = C_{m1} = 10 \text{ pF}$$

Now the transconductances can be found from the low-frequency gain requirement and (9.80),

$$|a_0| = g_{m0} R_0 g_{m1} R_1 g_{m2} R_2 = \frac{g_{m1}^3}{0.2}(5 \text{ k}\Omega)^3 = 20{,}000 = 86 \text{ dB}$$

since $g_{m0} = g_{m1} = 0.2 g_{m2}$ has been selected. Solving gives $g_{m1} = g_{m0} = g_{mf1} = g_{mf0} = 3.2$ mA/V and $g_{m2} = g_{m1}/0.2 = 16$ mA/V.

SPICE simulation of this op amp gives a dc gain of 86.3 dB and a phase margin of 52 degrees with a unity-gain frequency of 40 MHz. These values are close enough to the specifications to illustrate the usefulness of the calculations. The pole locations are $|p_1|/2\pi = 2.3$ kHz, $|p_2|/2\pi = 59$ MHz, and $|p_3|/2\pi = 464$ MHz. The zero locations are complex with a magnitude much larger than the unity-gain frequency, at $z_{1,2}/2\pi = -345$ MHz $\pm$ $j1.58$ GHz. Running simulations with slight changes to the compensation capacitors, we find that using $C_{m1} = 10.4$ pF and $C_{m2} = 8.3$ pF gives a phase margin of 47 degrees with a unity-gain frequency of 45 MHz.

■

9.5 Root-Locus Techniques[1,19]

To this point the considerations of this chapter have been mainly concerned with calculations of feedback amplifier stability and compensation using frequency-domain techniques. Such techniques are widely used because they allow the design of feedback amplifier compensation without requiring excessive design effort. The *root-locus* technique involves calculation of the actual poles and zeros of the amplifier and of their movement in the s plane as the low-frequency, loop-gain magnitude T_0 is changed. This method thus gives more information about the amplifier performance than is given by frequency-domain techniques, but also requires more computational effort. In practice, some problems can be solved equally well using either method, whereas others yield more easily to one or the other. The circuit designer needs skill in applying both methods. The root-locus technique will be first illustrated with a simple example.

9.5.1 Root Locus for a Three-Pole Transfer Function

Consider an amplifier whose transfer function has three identical poles. The transfer function can be written as

$$a(s) = \frac{a_0}{\left(1 - \frac{s}{p_1}\right)^3} \qquad (9.81)$$

where a_0 is the low-frequency gain and $|p_1|$ is the pole magnitude. Consider this amplifier placed in a negative-feedback loop as in Fig. 9.1, where the feedback network has a transfer function f, which is a constant. If we assume that the effects of feedback loading are small, the overall gain with feedback is

$$A(s) = \frac{a(s)}{1 + a(s)f} \qquad (9.82)$$

Using (9.81) in (9.82) gives

$$A(s) = \frac{\dfrac{a_0}{\left(1 - \dfrac{s}{p_1}\right)^3}}{1 + \dfrac{a_0 f}{\left(1 - \dfrac{s}{p_1}\right)^3}} = \frac{a_0}{\left(1 - \dfrac{s}{p_1}\right)^3 + T_0} \qquad (9.83)$$

where $T_0 = a_0 f$ is the low-frequency loop gain.

The *poles* of $A(s)$ are the *roots* of the equation

$$\left(1 - \frac{s}{p_1}\right)^3 + T_0 = 0 \qquad (9.84)$$

That is

$$\left(1 - \frac{s}{p_1}\right)^3 = -T_0$$

and thus

$$1 - \frac{s}{p_1} = \sqrt[3]{-T_0} = -\sqrt[3]{T_0} \quad \text{or} \quad \sqrt[3]{T_0}e^{j60°} \quad \text{or} \quad \sqrt[3]{T_0}e^{-j60°}$$

Thus the three roots of (9.84) are

$$\begin{aligned} s_1 &= p_1\left(1 + \sqrt[3]{T_0}\right) \\ s_2 &= p_1\left(1 - \sqrt[3]{T_0}e^{j60°}\right) \\ s_3 &= p_1\left(1 - \sqrt[3]{T_0}e^{-j60°}\right) \end{aligned} \qquad (9.85)$$

These three roots are the poles of $A(s)$ and (9.83) can be written as

$$A(s) = \frac{a_0}{1 + T_0} \frac{1}{\left(1 - \dfrac{s}{s_1}\right)\left(1 - \dfrac{s}{s_2}\right)\left(1 - \dfrac{s}{s_3}\right)} \qquad (9.86)$$

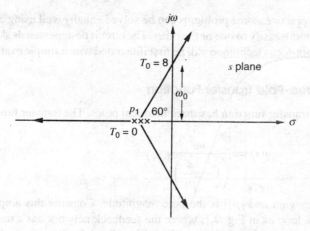

Figure 9.35 Root locus for a feedback amplifier with three identical poles in $T(s)$.

The equations in (9.85) allow calculation of the poles of $A(s)$ for any value of low-frequency loop gain T_0. For $T_0 = 0$, all three poles are at p_1 as expected. As T_0 increases, one pole moves out along the negative real axis while the other two leave the axis at an angle of 60° and move toward the right half-plane. The *locus of the roots* (or the *root locus*) is shown in Fig. 9.35, and each point of this root locus can be identified with the corresponding value of T_0. One point of significance on the root locus is the value of T_0 at which the two complex poles cross into the right half-plane, as this is the value of loop gain causing *oscillation*. From the equation for s_2 in (9.85), this is where $\text{Re}(s_2) = 0$, from which we obtain

$$1 - \text{Re}(\sqrt[3]{T_0} e^{j60°}) = 0$$

That is,

$$\sqrt[3]{T_0} \cos 60° = 1$$

and

$$T_0 = 8$$

Thus, *any* amplifier with three identical poles becomes unstable for low-frequency loop gain T_0 greater than 8. This is quite a restrictive condition and emphasizes the need for compensation if larger values of T_0 are required. Note that not only does the root-locus technique give the value of T_0 causing instability, it also allows calculation of the amplifier poles for values of $T_0 < 8$, and thus allows calculation of both sinusoidal *and* transient response of the amplifier.

The frequency of oscillation can be found from Fig. 9.35 by calculating the distance

$$\omega_0 = |p_1| \tan 60° = 1.732|p_1| \qquad (9.87)$$

Thus, when the poles just enter the right half-plane, their imaginary part has a magnitude $1.732|p_1|$ and this will be the frequency of the increasing sinusoidal response. That is, if the complex poles are at $(\sigma \pm j\omega_0)$ where σ is small and positive, the transient response of the circuit contains a term $Ke^{\sigma t} \sin \omega_0 t$, which represents a *growing* sinusoid. (K is set by an initial condition.)

It is useful to calculate the value of T_0 causing instability in this case by using the frequency-domain approach and the Nyquist criterion. From (9.81) the loop gain is

$$T(j\omega) = \frac{a_0 f}{\left(1 + \dfrac{j\omega}{|p_1|}\right)^3} = \frac{T_0}{\left(1 + j\dfrac{\omega}{|p_1|}\right)^3} \qquad (9.88)$$

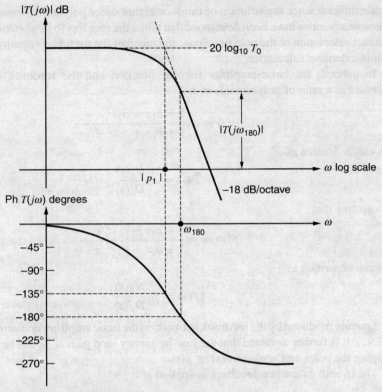

Figure 9.36 Magnitude and phase of $T(j\omega)$ for a feedback amplifier with three identical poles in $T(s)$.

The magnitude and phase of $T(j\omega)$ as a function of ω are sketched in Fig. 9.36. The frequency ω_{180} where the phase shift of $T(j\omega)$ is $-180°$ can be calculated from (9.88) as

$$180° = 3 \arctan \frac{\omega_{180}}{|p_1|}$$

and this gives

$$\omega_{180} = 1.732|p_1| \qquad (9.89)$$

Comparing (9.87) with (9.89) shows that

$$\omega_{180} = \omega_0 \qquad (9.90)$$

The loop-gain magnitude at ω_{180} can be calculated from (9.88) as

$$|T(j\omega_{180})| = \frac{T_0}{\left|1 + j\dfrac{\omega_{180}}{|p_1|}\right|^3} = \frac{T_0}{8} \qquad (9.91)$$

using (9.89). The Nyquist criterion for stability indicates it is necessary that $|T(j\omega_{180})| < 1$. This requires that $T_0 < 8$, the same result as obtained using root-locus techniques.

9.5.2 Rules for Root-Locus Construction

In the above simple example, it was possible to calculate exact expressions for the amplifier poles as a function of T_0, and thus to plot the root loci exactly. In most practical cases this is quite difficult since the solution of third- or higher order polynomial equations is required. Consequently, rules have been developed that allow the root loci to be sketched without requiring exact calculation of the pole positions, and much of the useful information is thus obtained without extensive calculation.

In general, the basic-amplifier transfer function and the feedback function may be expressed as a ratio of polynomials in s.

$$a(s) = a_0 \frac{1 + a_1 s + a_2 s^2 + \cdots}{1 + b_1 s + b_2 s^2 + \cdots} \tag{9.92}$$

This can be written as

$$a(s) = a_0 \frac{N_a(s)}{D_a(s)} \tag{9.93}$$

Also assume that

$$f(s) = f_0 \frac{1 + c_1 s + c_2 s^2 + \cdots}{1 + d_1 s + d_2 s^2 + \cdots} \tag{9.94}$$

This can be written as

$$f(s) = f_0 \frac{N_f(s)}{D_f(s)} \tag{9.95}$$

Loading produced by the feedback network on the basic amplifier is assumed to be included in (9.92). It is further assumed that the low-frequency loop gain $a_0 f_0$ can be changed without changing the poles and zeros of $a(s)$ or $f(s)$.

The overall gain when feedback is applied is

$$A(s) = \frac{a(s)}{1 + a(s)f(s)} \tag{9.96}$$

Using (9.93) and (9.95) in (9.96) gives

$$A(s) = \frac{a_0 N_a(s) D_f(s)}{D_f(s) D_a(s) + T_0 N_a(s) N_f(s)} \tag{9.97}$$

where

$$T_0 = a_0 f_0 \tag{9.98}$$

is the low-frequency loop gain.

Equation 9.97 shows that the *zeros* of $A(s)$ are the *zeros* of $a(s)$ *and* the *poles* of $f(s)$. From (9.97) it is apparent that the *poles* of $A(s)$ are the roots of

$$D_f(s) D_a(s) + T_0 N_a(s) N_f(s) = 0 \tag{9.99}$$

Consider the two extreme cases.

(a) Assume that there is no feedback and that $T_0 = 0$. Then, from (9.99), the *poles* of $a(s)$ are the *poles* of $a(s)$ and $f(s)$. However, the *poles* of $f(s)$ are also *zeros* of $A(s)$ and these cancel, leaving the *poles* of $A(s)$ composed of the *poles* of $a(s)$ as expected. The *zeros* of $A(s)$ are the *zeros* of $a(s)$ in this case.

(b) Let $T_0 \to \infty$. Then (9.99) becomes

$$N_a(s) N_f(s) = 0 \tag{9.100}$$

This equation shows that the *poles* of $A(s)$ are now the zeros of $a(s)$ and the *zeros* of $f(s)$. However, the *zeros* of $a(s)$ are also *zeros* of $A(s)$ and these cancel, leaving the *poles* of $A(s)$ composed of the *zeros* of $f(s)$. The *zeros* of $A(s)$ are the *poles* of $f(s)$ in this case.

Rule 1. The branches of the root locus start at the poles of $T(s) = a(s)f(s)$ where $T_0 = 0$, and terminate on the zeros of $T(s)$ where $T_0 = \infty$. If $T(s)$ has more poles than zeros, some of the branches of the root locus will terminate at infinity.

Examples of loci terminating at infinity are shown in Figs. 9.3 and 9.35. More rules for the construction of root loci can be derived by returning to (9.99) and dividing it by $D_f(s)D_a(s)$. Poles of $A(s)$ are roots of

$$1 + T_0 \frac{N_a(s)}{D_a(s)} \frac{N_f(s)}{D_f(s)} = 0$$

That is

$$T_0 \frac{N_a(s)}{D_a(s)} \frac{N_f(s)}{D_f(s)} = -1$$

The complete expression including poles and zeros is

$$T_0 \frac{\left(1 - \dfrac{s}{z_{a1}}\right)\left(1 - \dfrac{s}{z_{a2}}\right)\cdots\left(1 - \dfrac{s}{z_{f1}}\right)\left(1 - \dfrac{s}{z_{f2}}\right)\cdots}{\left(1 - \dfrac{s}{p_{a1}}\right)\left(1 - \dfrac{s}{p_{a2}}\right)\cdots\left(1 - \dfrac{s}{p_{f1}}\right)\left(1 - \dfrac{s}{p_{f2}}\right)\cdots} = -1 \qquad (9.101)$$

where

$$z_{a1}, z_{a2} \cdots \text{ are zeros of } a(s)$$
$$z_{f1}, z_{f2} \cdots \text{ are zeros of } f(s)$$
$$p_{a1}, p_{a2} \cdots \text{ are poles of } a(s)$$
$$p_{f1}, p_{f2} \cdots \text{ are poles of } f(s)$$

Equation 9.101 can be written as

$$T_0 \frac{(-p_{a1})(-p_{a2})\cdots(-p_{f1})(-p_{f2})\cdots}{(-z_{a1})(-z_{a2})\cdots(-z_{f1})(-z_{f2})\cdots}$$

$$\times \frac{(s - z_{a1})(s - z_{a2})\cdots(s - z_{f1})(s - z_{f2})\cdots}{(s - p_{a1})(s - p_{a2})\cdots(s - p_{f1})(s - p_{f2})\cdots} = -1 \qquad (9.102)$$

If the poles and zeros of $a(s)$ and $f(s)$ are restricted to the left half-plane [this does *not* restrict the poles of $A(s)$], then $-p_{a1}, -p_{a2}$, and so on are *positive* numbers and (9.102) can be written

$$T_0 \frac{|p_{a1}| \cdot |p_{a2}| \cdots |p_{f1}| \cdot |p_{f2}| \cdots}{|z_{a1}| \cdot |z_{a2}| \cdots |z_{f1}| \cdot |z_{f2}| \cdots} \times \frac{(s - z_{a1})(s - z_{a2})\cdots(s - z_{f1})(s - z_{f2})\cdots}{(s - p_{a1})(s - p_{a2})\cdots(s - p_{f1})(s - p_{f2})\cdots} = -1$$

$$(9.103)$$

Values of complex variable s satisfying (9.103) are *poles* of closed-loop function $A(s)$. Equation 9.103 requires the fulfillment of two conditions simultaneously, and these conditions are used to determine points on the root locus.

The *phase condition* for values of s satisfying (9.103) is

$$\underline{/s - z_{a1}} + \underline{/s - z_{a2}} \cdots + \underline{/s - z_{f1}} + \underline{/s - z_{f2}} + \cdots$$

$$- (\underline{/s - p_{a1}} + \underline{/s - p_{a2}} \cdots + \underline{/s - p_{f1}} + \underline{/s - p_{f2}} \cdots) = (2n - 1)\pi \qquad (9.104)$$

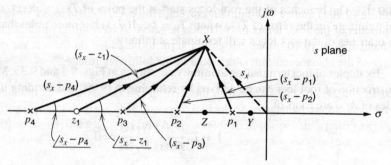

Figure 9.37 Poles and zeros of loop gain $T(s)$ of a feedback amplifier. Vectors are drawn to the point X to determine if this point is on the root locus.

The magnitude condition for values of s satisfying (9.103) is

$$T_0 \frac{|p_{a1}| \cdot |p_{a2}| \cdots |p_{f1}| \cdot |p_{f2}| \cdots}{|z_{a1}| \cdot |z_{a2}| \cdots |z_{f1}| \cdot |z_{f2}| \cdots} \frac{|s - z_{a1}| \cdot |s - z_{a2}| \cdots |s - z_{f1}| \cdot |s - z_{f2}| \cdots}{|s - p_{a1}| \cdot |s - p_{a2}| \cdots |s - p_{f1}| \cdot |s - p_{f2}| \cdots} = 1 \quad (9.105)$$

Consider an amplifier with poles and zeros of $T(s)$ as shown in Fig. 9.37. In order to determine if some arbitrary point X is on the root locus, the phase condition of (9.104) is used. Note that the vectors of (9.104) are formed by drawing lines *from* the various poles and zeros of $T(s)$ to the point X and the angles of these vectors are then substituted in (9.104) to check the phase condition. This is readily done for points Y and Z on the axis.

At Y

$$\angle s_Y - z_1 = 0°$$
$$\angle s_Y - p_1 = 0°$$

and so on. All angles are zero for point Y and thus the phase condition is not satisfied. This is the case for all points to the right of p_1.

At Z

$$\angle s_Z - z_1 = 0°$$
$$\angle s_Z - p_1 = 180°$$
$$\angle s_Z - p_2 = 0°$$
$$\angle s_Z - p_3 = 0°$$
$$\angle s_Z - p_4 = 0°$$

In this case, the phase condition of (9.104) *is* satisfied, and points on the axis between p_1 and p_2 *are* on the locus. By similar application of the phase condition, the locus can be shown to exist on the real axis between p_3 and z_1 and to the left of p_4.

In general, if $T(s)$ has all its zeros and poles in the LHP, the locus is situated along the real axis where there is an odd number of poles and zeros of $T(s)$ to the right. In some cases, however, all the zeros of $T(s)$ are not in the LHP. For example, an op amp that uses Miller compensation can have a RHP zero in $a(s)$ and therefore in $T(s)$. If $a(s)$ has at least one RHP zero, at least one of the $-z_{ai}$ terms in (9.102) is negative, rather than positive as assumed in (9.103). If the number of RHP zeros is even, an even number of $-z_{ai}$ terms that are negative appear in the denominator of (9.102). The product of these negative terms is positive, and therefore (9.103) and (9.104) remain correct. However, if the number of RHP zeros is odd, the product of the $-z_{ai}$ terms in (9.102) is negative. As a result, a minus sign appears on the left-hand side of (9.103) that causes a π term to be added on the left side of (9.104). This change to the phase condition is reflected in the following rule.

Rule 2. If $T(s)$ has all its zeros in the LHP or if $T(s)$ has an even number of RHP zeros, the locus is situated along the real axis wherever there is an odd number of poles and zeros of $T(s)$ to the right. However, if $T(s)$ has an odd number of RHP zeros, the locus is situated along the real axis wherever there is an even number of poles and zeros of $T(s)$ to the right.

Consider again the situation in Fig. 9.37. Rule 1 indicates that branches of the locus must start at p_1, p_2, p_3, and p_4. Rule 2 indicates that the locus exists between p_3 and z_1, and thus the branch beginning at p_3 ends at z_1. Rule 2 also indicates that the locus exists to the left of p_4, and thus the branch beginning at p_4 moves out to negative infinity. The branches beginning at p_1 and p_2 must also terminate at infinity, which is possible only if these branches *break away* from the real axis as shown in Fig. 9.38. This can be stated as follows.

Rule 3. All segments of loci that lie on the real axis between pairs of poles (or pairs of zeros) of $T(s)$ must, at some internal break point, branch out from the real axis.

The following rules can be derived.[20]

Rule 4. The locus is symmetrical with respect to the real axis (because complex roots occur only in conjugate pairs).

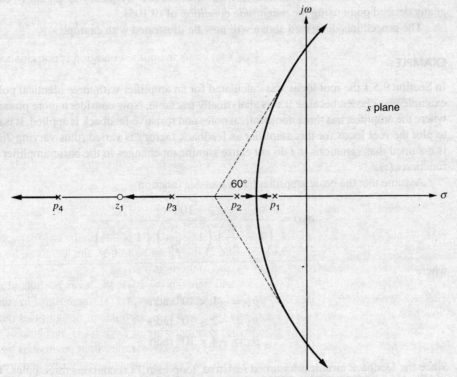

Figure 9.38 Root-locus construction for the poles and zeros of Fig. 9.37.

Rule 5. Branches of the locus that leave the real axis do so at *right angles*, as illustrated in Fig. 9.38.

Rule 6. If branches of the locus break away from the real axis, they do so at a point where the vector sum of reciprocals of distances to the poles of $T(s)$ equals the vector sum of reciprocals of distances to the zeros of $T(s)$.

Rule 7. If $T(s)$ has no RHP zeros or an even number of RHP zeros, branches of the locus that terminate at infinity do so asymptotically to straight lines with angles to the real axis of $[(2n - 1)\pi]/(N_p - N_z)$ for $n = 0, 1, \ldots, N_p - N_z - 1$, where N_p is the number of poles and N_z is the number of zeros. However, if $T(s)$ has an odd number of RHP zeros, the asymptotes intersect the real axis at angles given by $(2n\pi)/(N_p - N_z)$.

Rule 8. The asymptotes of branches that terminate at infinity all intersect on the real axis at a point given by

$$\sigma_a = \frac{\sum[\text{poles of } T(s)] - \sum[\text{zeros of } T(s)]}{N_p - N_z} \tag{9.106}$$

A number of other rules have been developed for sketching root loci, but those described above are adequate for most requirements in amplifier design. The rules are used to obtain a rapid idea of the shape of the root locus in any situation, and to calculate amplifier performance in simple cases. More detailed calculation on circuits exhibiting complicated pole-zero patterns generally require computer calculation of the root locus.

Note that the above rules are all based on the *phase condition* of (9.104). Once the locus has been sketched, it can then be *calibrated* with values of low-frequency loop gain T_0 calculated at any desired point using the *magnitude condition* of (9.105).

The procedures described above will now be illustrated with examples.

■ **EXAMPLE**

In Section 9.5.1 the root locus was calculated for an amplifier with three identical poles. This example was chosen because it was analytically tractable. Now consider a more practical case where the amplifier has three nonidentical poles and resistive feedback is applied. It is required to plot the root locus for this amplifier as feedback factor f is varied (thus varying T_0), and it is assumed that variations in f do not cause significant changes in the basic-amplifier transfer function $a(s)$.

Assume that the basic amplifier has a transfer function

$$a(s) = \frac{100}{\left(1 - \dfrac{s}{p_1}\right)\left(1 - \dfrac{s}{p_2}\right)\left(1 - \dfrac{s}{p_3}\right)} \tag{9.107}$$

where

$$p_1 = -1 \times 10^6 \text{ rad/s}$$
$$p_2 = -2 \times 10^6 \text{ rad/s}$$
$$p_3 = -4 \times 10^6 \text{ rad/s}$$

Since the feedback circuit is assumed resistive, loop gain $T(s)$ contains three poles. The root locus is shown in Fig. 9.39, and, for convenience, the numbers are normalized to 10^6 rad/s.

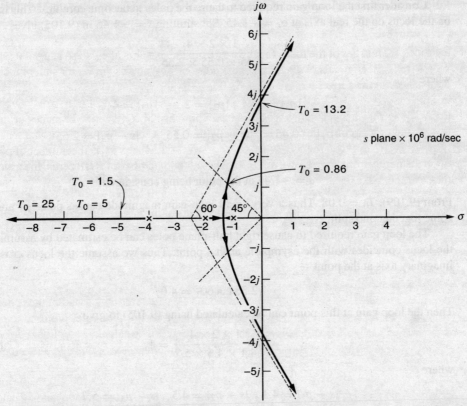

Figure 9.39 Root-locus example for poles of $T(s)$ at -1×10^6, -2×10^6, and -4×10^6 rad/s.

Rules 1 and 2 indicate that branches of the locus starting at poles p_1 and p_2 move toward each other and then split out and asymptote to infinity. The branch starting at pole p_3 moves out along the negative real axis to infinity.

The breakaway point for the locus between p_1 and p_2 can be calculated using rule 6. If σ_i is the coordinate of the breakaway point, then

$$\frac{1}{\sigma_i + 1} + \frac{1}{\sigma_i + 2} + \frac{1}{\sigma_i + 4} = 0 \tag{9.108}$$

Solving this quadratic equation for σ_i gives $\sigma_i = -3.22$ or -1.45. The value -1.45 is the only possible solution because the breakaway point lies between -1 and -2 on the real axis.

The angles of the asymptotes to the real axis can be found using rule 7 and are $\pm 60°$ and $180°$. The asymptotes meet the real axis at a point whose coordinate is σ_a given by (9.106), and using (9.106) gives

$$\sigma_a = \frac{(-1 - 2 - 4) - 0}{3} = -2.33$$

When these asymptotes are drawn, the locus can be sketched as in Fig. 9.39 noting, from rule 5, that the locus leaves the real axis at right angles. The locus can now be calibrated for loop gain by using the magnitude condition of (9.105). Aspects of interest about the locus may be the loop gain required to cause the poles to become complex, the loop gain required for poles with an angle of 45° to the negative real axis, and the loop gain required for oscillation (right half-plane poles).

Consider first the loop gain required to cause the poles to become complex. This is a point on the locus on the real axis at $\sigma_i = -1.45$. Substituting $s = -1.45$ in (9.105) gives

$$T_0 \frac{1 \times 2 \times 4}{0.45 \times 0.55 \times 2.55} = 1 \qquad (9.109)$$

where

$$|p_1| = 1 \qquad |p_2| = 2 \qquad |p_3| = 4$$

$$|s - p_1| = 0.45 \qquad |s - p_2| = 0.55 \qquad |s - p_3| = 2.55$$

and

$$s = -1.45 \text{ at the point being considered}$$

From (9.109), $T_0 = 0.08$. Thus a very small loop-gain magnitude causes poles p_1 and p_2 to come together and split.

The loop gain required to cause right half-plane poles can be estimated by assuming that the locus coincides with the asymptote at that point. Thus we assume the locus crosses the imaginary axis at the point

$$j2.33 \tan 60° = 4.0j$$

Then the loop gain at this point can be calculated using (9.105) to give

$$T_0 \frac{1 \times 2 \times 4}{4.1 \times 4.5 \times 5.7} = 1 \qquad (9.110)$$

where

$$|s - p_1| = 4.1 \qquad |s - p_2| = 4.5 \qquad |s - p_3| = 5.7$$

and

$$s = 4j \text{ at this point on the locus}$$

From (9.110), $T_0 = 13.2$. Since $a_0 = 100$ for this amplifier [from (9.107)], the overall gain of the feedback amplifier to $T_0 = 13.2$ is

$$A_0 = \frac{a_0}{1 + T_0} = 7.04$$

and

$$f = \frac{T_0}{a_0} = 0.132$$

The loop gain when the complex poles make an angle of 45° with the negative real axis can be calculated by making the assumption that this point has the same real-axis coordinate as the breakaway point. Then, using (9.105) with $s = (-1.45 + 1.45j)$, we obtain

$$T_0 \frac{1 \times 2 \times 4}{1.52 \times 1.55 \times 2.93} = 1$$

and thus

$$T_0 = 0.86$$

Finally, the loop gain required to move the locus out from pole p_3 is of interest. When the real-axis pole is at -5, the loop gain can be calculated using (9.105) with $s = -5$ to give

$$T_0 \frac{1 \times 2 \times 4}{1 \times 3 \times 4} = 1$$

That is,

$$T_0 = 1.5$$

When this pole is at -6, the loop gain is

$$T_0 \frac{1 \times 2 \times 4}{2 \times 4 \times 5} = 1$$

and thus

$$T_0 = 5$$

These values are marked on the root locus of Fig. 9.39.

In this example, it is useful to compare the prediction of instability at $T_0 = 13.2$ with the results using the Nyquist criterion. The loop gain in the frequency domain is

$$T(j\omega) = \frac{T_0}{\left(1 + \dfrac{j\omega}{10^6}\right)\left(1 + \dfrac{j\omega}{2 \times 10^6}\right)\left(1 + \dfrac{j\omega}{4 \times 10^6}\right)} \tag{9.111}$$

A series of trial substitutions shows that $\angle T(j\omega) = -180°$ for $\omega = 3.8 \times 10^6$ rad/s. Note that this is close to the value of 4×10^6 rad/s where the root locus was assumed to cross the $j\omega$ axis. Substitution of $\omega = 3.8 \times 10^6$ in (9.111) gives, for the loop gain at that frequency,

$$|T(j\omega)| = \frac{T_0}{11.6} \tag{9.112}$$

Thus, for stability, the Nyquist criterion requires that $T_0 < 11.6$ and this is close to the answer obtained from the root locus. If the point on the $j\omega$ axis where the root locus crossed had been determined more accurately, it would have been found to be at 3.8×10^6 rad/s, and both methods would predict instability for $T_0 > 11.6$.

It should be pointed out that the root locus for Fig. 9.39 shows the movement of the *poles* of the feedback amplifier as T_0 changes. The theory developed in Section 9.5.2 showed that the *zeros* of the feedback amplifier are the *zeros* of the basic amplifier and the *poles* of the feedback network. In this case there are no zeros in the feedback amplifier, but this is not always the case. It should be kept in mind that if the basic amplifier has zeros in its transfer function, these may be an important part of the overall transfer function.

■

The rules for drawing a root locus were presented for varying T_0, assuming that the poles and zeros of $a(s)$ and $f(s)$ do not change when T_0 changes. This assumption is often not valid in practice, since changing the circuit to change $T_0 = a_0 f$ usually affects at least some of the poles and zeros. Alternatively, these rules can be used to draw a root locus of the poles of a transfer function as the value x of an element in the circuit changes if the closed-loop gain $A(s)$ can be written in the form

$$A(s) = \frac{M(s)}{G(s) + xH(s)} \tag{9.113}$$

where $M(s)$, $G(s)$, and $H(s)$ are polynomials in s, and $G(s)$ and $H(s)$ are not functions of x. The poles of $A(s)$ are the roots of $G(s) = 0$ when $x = 0$ and the roots of $H(s) = 0$ when $x \to \infty$. The roots of $G(s) = 0$ are the starting points of the root locus, and the roots of $H(s) = 0$ are the ending points of the root locus. The complete locus for all values of x can be drawn by following the rules given in this section. For example, this approach could be used to plot a locus of the poles of the transfer function in (9.27) as the compensation capacitor C varies. (In this case, $x = C$.)

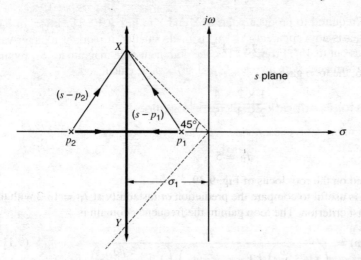

Figure 9.40 Root locus for an op amp with two poles in its transfer function. The feedback is assumed resistive.

9.5.3 Root Locus for Dominant-Pole Compensation

Consider an op amp that has been compensated by creation of a dominant pole at p_1. If we assume the second most dominant pole is at p_2 and neglect the effect of higher order poles, the root locus when resistive feedback is applied is as shown in Fig. 9.40. Using rules 1 and 2 indicates that the root locus exists on the axis between p_1 and p_2, and the breakaway point is readily shown to be

$$\sigma_i = \frac{p_1 + p_2}{2} \tag{9.114}$$

using rule 6. Using rules 7 and 8 shows that the asymptotes are at 90° to the real axis and meet the axis at σ_i.

As T_0 is increased, the branches of the locus come together and then split out to become complex. As T_0 becomes large, the imaginary part of the poles becomes large, and the circuit will then have a high-frequency *peak* in its overall gain function $A(j\omega)$. This is consistent with the previous viewpoint of gain peaking that occurred with diminishing phase margin.

Assume that maximum bandwidth in this amplifier is required, but that little or no peaking is allowed. This means that with maximum loop gain applied, the poles should not go beyond the points marked X and Y on the locus where an angle of 45° is made between the negative real axis and a line drawn from X or Y to the origin. At X, the loop gain can be calculated using (9.105)

$$T_0 \frac{|p_1| \cdot |p_2|}{|s - p_1| \cdot |s - p_2|} = 1 \tag{9.115}$$

If p_1 is a dominant pole, we can assume that $|p_1| \ll |p_2|$ and $\sigma_i = p_2/2$. For poles at 45°, $|s - p_1| = |s - p_2| \simeq \sqrt{2}|p_2|/2$. Thus (9.115) becomes

$$T_0 = \frac{1}{|p_1| \cdot |p_2|} \left(\sqrt{2} \frac{|p_2|}{2} \right)^2$$

This gives

$$T_0 = \frac{1}{2} \frac{|p_2|}{|p_1|} \tag{9.116}$$

for the value of T_0 required to produce poles at X and Y in Fig. 9.40. The effect of narrow-banding the amplifier is now apparent. As $|p_1|$ is made smaller, it requires a larger value of T_0 to move the poles out to 45°. From (9.116), the dominant-pole magnitude $|p_1|$ required to ensure adequate performance with a given T_0 and $|p_2|$ can be calculated.

9.5.4 Root Locus for Feedback-Zero Compensation

The techniques of compensation described earlier in this chapter involved modification of the basic amplifier only. This is the universal method used with op amps that must be compensated for use with a wide variety of feedback networks chosen by the user. However, this method is quite wasteful of bandwidth, as was apparent in the calculations.

In this section, a different method of compensation will be described that involves modification of the feedback path and is generally limited to fixed-gain amplifiers. This method finds application in the compensation of wideband feedback amplifiers where bandwidth is of prime importance. An example is the shunt-series feedback amplifier of Fig. 8.31, which is known as a *current feedback pair*. The method is generally useful in amplifiers of this type, where the feedback is over two stages, and in circuits such as the series-series triple of Fig. 8.18a.

A shunt-series feedback amplifier including a feedback capacitor C_F is shown in Fig. 9.41. The basic amplifier including feedback loading for this circuit is shown in Fig. 9.42. Capacitors C_F at input and output have only a minor effect on the circuit transfer function. The feedback circuit for this case is shown in Fig. 9.43 and feedback function f is given by

$$f = \frac{i_i}{i_2} = -\frac{R_E}{R_F + R_E} \frac{1 + R_F C_F s}{1 + \dfrac{R_E R_F}{R_E + R_F} C_F s} \tag{9.117}$$

Feedback function f thus contains a zero with a magnitude

$$\omega_z = \frac{1}{R_F C_F} \tag{9.118}$$

and a pole with a magnitude

$$\omega_p = \frac{R_E + R_F}{R_E} \frac{1}{R_F C_F} \tag{9.119}$$

Quantity $(R_E + R_F)/R_E$ is approximately the low-frequency gain of the overall circuit with feedback applied, and, since it is usually true that $(R_E + R_F)/R_E \gg 1$, the pole magnitude

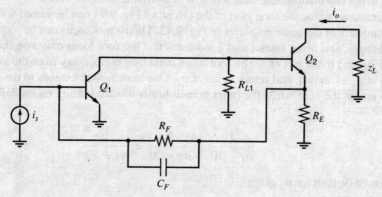

Figure 9.41 Shunt-series feedback amplifier including a feedback capacitor C_F.

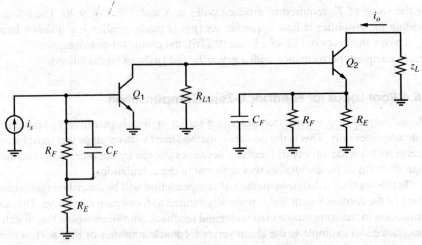

Figure 9.42 Basic amplifier including feedback loading for the circuit of Fig. 9.41.

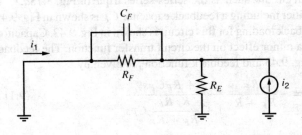

Figure 9.43 Circuit for the calculation of feedback function f for the amplifier of Fig. 9.41.

given by (9.119) is usually much larger than the zero magnitude. This will be assumed and the effects of the pole will be neglected, but if $(R_E + R_F)/R_E$ becomes comparable to unity, the pole will be important and must be included.

The basic amplifier of Fig. 9.42 has two significant poles contributed by Q_1 and Q_2. Although higher magnitude poles exist, these do not have a dominant influence and will be neglected. The effects of this assumption will be investigated later. The loop gain of the circuit of Fig. 9.41 thus contains two forward-path poles and a feedback zero, giving rise to the root locus of Fig. 9.44. For purposes of illustration, the two poles are assumed to be $p_1 = -10 \times 10^6$ rad/s and $p_2 = -20 \times 10^6$ rad/s and the zero is $z = -50 \times 10^6$ rad/s. For convenience in the calculations, the numbers will be normalized to 10^6 rad/s.

Assume now that the loop gain of the circuit of Fig. 9.41 can be varied without changing the parameters of the basic amplifier of Fig. 9.42. Then a root locus can be plotted as the loop gain changes, and using rules 1 and 2 indicates that the root locus exists on the axis between p_1 and p_2, and to the left of z. The root locus must thus break away from the axis between p_1 and p_2 at σ_1 as shown, and return again at σ_2. One branch then extends to the right along the axis to end at the zero while the other branch heads toward infinity on the left. Using rule 6 gives

$$\frac{1}{\sigma_1 + 10} + \frac{1}{\sigma_1 + 20} = \frac{1}{\sigma_1 + 50} \qquad (9.120)$$

Solution of (9.120) for σ_1 gives

$$\sigma_1 = -84.6 \quad \text{or} \quad -15.4$$

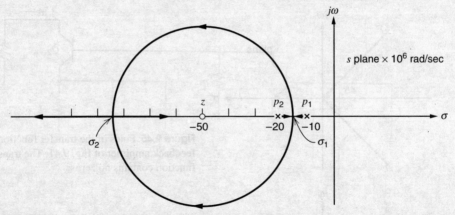

Figure 9.44 Root locus for the circuit of Fig. 9.41 assuming the basic amplifier contributes two poles to $T(s)$ and the feedback circuit contributes one zero.

Obviously $\sigma_1 = -15.4$ and the other value is $\sigma_2 = -84.6$. Note that these points are equidistant from the zero, and, in fact, it can be shown that in this example the portion of the locus that is off the real axis is a *circle* centered on the zero. An aspect of the root-locus diagrams that is a useful aid in sketching the loci is apparent from Fig. 9.39 and Fig. 9.44. The locus tends to *bend toward* zeros as if attracted and tends to *bend away* from poles as if repelled.

The effectiveness of the feedback zero in compensating the amplifier is apparent from Fig. 9.44. If we assume that the amplifier has poles p_1 and p_2 and there is no feedback zero, then when feedback is applied the amplifier poles will split out and move parallel to the $j\omega$ axis. For practical values of loop gain T_0, this would result in "high Q" poles near the $j\omega$ axis, which would give rise to an excessively *peaked* response. In practice, oscillation can occur because higher magnitude poles do exist and these would tend to give a locus of the kind of Fig. 9.39, where the remote poles cause the locus to bend and enter the right half-plane. (Note that this behavior is consistent with the alternative approach of considering a diminished phase margin to be causing a peaked response and eventual instability.) The inclusion of the feedback zero, however, *bends* the locus away from the $j\omega$ axis and allows the designer to position the poles in any desired region.

An important point that should be stressed is that the root locus of Fig. 9.44 gives the *poles* of the feedback amplifier. The zero in that figure is a zero of loop gain $T(s)$ and thus must be included in the root locus. However, the zero is contributed by the *feedback network* and is *not* a zero of the overall feedback amplifier. As pointed out in Section 9.5.2, the *zeros* of the overall feedback amplifier are the *zeros* of basic amplifier $a(s)$ and the *poles* of feedback network $f(s)$. Thus the transfer function of the overall feedback amplifier in this case has two poles and no zeros, as shown in Fig. 9.45, and the poles are assumed placed at 45° to the axis by appropriate choice of z. Since the feedback zero affects the root locus but does *not* appear as a zero of the overall amplifier, it has been called a *phantom zero*.

On the other hand, if the zero z were contributed by the basic amplifier, the situation would be different. For the same zero, the root locus would be identical but the transfer function of the overall feedback amplifier would then *include* the zero as shown in Fig. 9.46. This zero would then have a significant effect on the amplifier characteristics. This point is made simply to illustrate the difference between forward path and feedback-path zeros. There is no practical way to introduce a useful forward-path zero in this situation.

Before leaving this subject, we mention the effect of higher magnitude poles on the root locus of Fig. 9.44, and this is illustrated in Fig. 9.47. A remote pole p_3 will cause the locus to

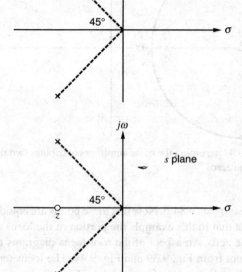

Figure 9.45 Poles of the transfer function of the feedback amplifier of Fig. 9.41. The transfer function contains no zeros.

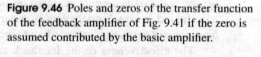

Figure 9.46 Poles and zeros of the transfer function of the feedback amplifier of Fig. 9.41 if the zero is assumed contributed by the basic amplifier.

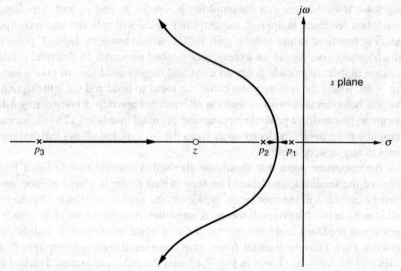

Figure 9.47 Root locus of the circuit of Fig. 9.41 when an additional pole of the basic amplifier is included. (Not to scale.)

deviate from the original as shown and produce poles with a larger imaginary part than expected. The third pole, which is on the real axis, may also be significant in the final amplifier. Acceptable performance can usually be obtained by modifying the value of z from that calculated above.

Finally, the results derived in this chapter explain the function of capacitors C_P and C_F in the circuit of the MC 1553 series-series triple of Fig. 8.21a, which was described in Chapter 8. Capacitor C_P causes pole splitting to occur in stage Q_2 and produces a dominant pole in the basic amplifier, which aids in the compensation. However, as described above, a large value of C_P will cause significant loss of bandwidth in the amplifier, and so a feedback zero is introduced via C_F, which further aids in the compensation by moving the root locus away from the $j\omega$ axis. The final design is a combination of two methods of compensation in an attempt to find an optimum solution.

9.6 Slew Rate[8]

The previous sections of this chapter have been concerned with the small-signal behavior of feedback amplifiers at high frequencies. However, the behavior of feedback circuits with large input signals (either step inputs or sinusoidal signals) is also of interest, and the effect of frequency compensation on the large-signal, high-frequency performance of feedback amplifiers is now considered.

9.6.1 Origin of Slew-Rate Limitations

A common test of the high-frequency, large-signal performance of an amplifier is to apply a step input voltage as shown in Fig. 9.48. This figure shows an op amp in a unity-gain feedback configuration and will be used for purposes of illustration in this development. Assuming the op amp is powered from a single supply between 3 V and ground, the input here is chosen to step from 0.5 V to 2.5 V so that the circuit operates linearly well before and well after the step. Suppose initially that the circuit has a single-pole transfer function given by

$$\frac{V_o}{V_i}(s) = \frac{A}{1+s\tau} \quad (9.121)$$

where

$$\tau = \frac{1}{2\pi f_o} \quad (9.122)$$

and f_o is the -3-dB frequency. Since the circuit is connected as a voltage follower, the low-frequency gain A will be close to unity. If we assume that this is so, the response of the circuit to this step input $[V_i(s) = 2/s]$ is given by

$$V_o(s) = \frac{1}{1+s\tau}\frac{2}{s} \quad (9.123)$$

using (9.121). Equation 9.123 can be factored to the form

$$V_o(s) = \frac{2}{s} - \frac{2}{s+\dfrac{1}{\tau}} \quad (9.124)$$

From (9.124)

$$V_o(t) = 2(1 - e^{-t/\tau}) \quad (9.125)$$

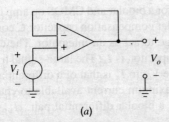

(a)

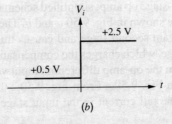

(b)

Figure 9.48 (a) Circuit and (b) input for testing slew-rate performance.

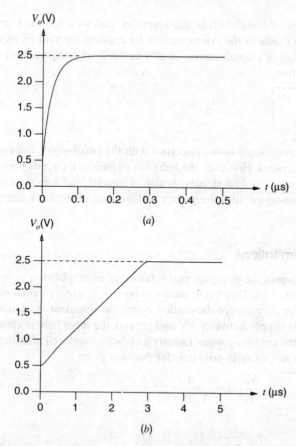

Figure 9.49 Response of the circuit of Fig. 9.48 when a 2-V step input is applied. (a) Response predicted by (9.125) for the NE5234 op amp. (b) Simulated response for the NE5234.

The predicted response from (9.125) is shown in Fig. 9.49a using data for the NE5234 op amp with $f_o \simeq 2.7$ MHz. This shows an exponential rise of $V_o(t)$ by 2 V and the output reaches 90 percent of its final value in about 0.14 μs.

A typical output for the NE5234 op amp in such a test is shown in Fig. 9.49b and exhibits a completely different response. The output voltage is a slow ramp of almost constant slope and takes about 2.6 μs to reach 90 percent of its final value. Obviously the small-signal linear analysis is inadequate for predicting the circuit behavior under these conditions. The response shown in Fig. 9.49b is typical of op-amp performance with a large input step voltage applied. The rate of change of output voltage dV_o/dt in the region of constant slope is called the *slew rate* and is usually specified in V/μs.

The reason for the discrepancy between predicted and observed behavior noted above can be appreciated by examining the circuit of Fig. 9.48a and considering the responses in Fig. 9.49. At $t = 0$, the input voltage steps up by $+2$ V, but the output voltage cannot respond instantaneously and is initially unchanged. Thus the op-amp differential input is $V_{id} = 2$ V, which drives the input stage completely out of its linear range of operation. This can be seen by considering a two-stage op amp; simplified schematics for a bipolar and CMOS op amp for use in this analysis are shown in Fig. 9.50 *a* and *b*. The Miller compensation capacitor C connects around the high-gain second stage and causes this stage to act as an integrator. The current from the input stage, which charges the compensation capacitor, is I_x. The large-signal transfer characteristic from the op-amp differential input voltage V_{id} to I_x is that of a differential pair, which is shown in Fig. 9.50c. From Fig. 9.50c, the maximum current available to charge C is $2I_1$, which is the tail current in the input stage. For a bipolar differential pair, $|I_x| \approx 2I_1$

9.6 Slew Rate

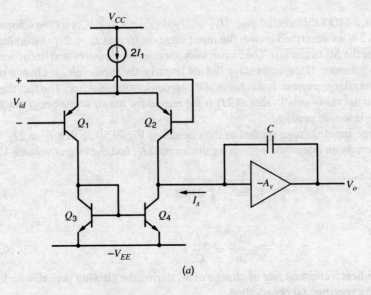

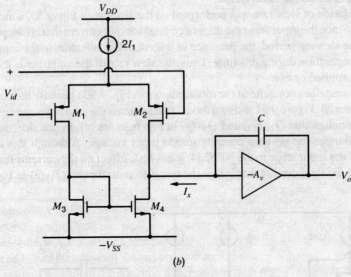

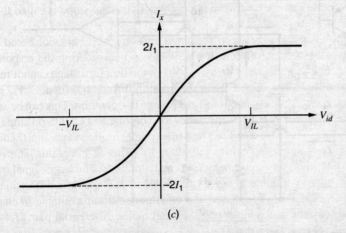

Figure 9.50 Simplified schematics of a two-stage (a) bipolar and (b) MOS op amp for slew rate calculations and (c) approximate large signal transfer characteristic for the input stages in (a) and (b). For the bipolar differential pair, $V_{IL} \approx 3V_T$. For the MOS differential pair, $V_{IL} \approx \sqrt{2}|V_{ov1}|$.

if $|V_{id}| > 3V_T$. For a MOS differential pair, $|I_x| \approx 2I_1$ if $|V_{id}| > \sqrt{2}|V_{ov1}|$. (See Chapter 3.) Thus, when $V_{id} = 2$ V as described above, the input stage *limits* and $I_x \approx 2I_1$ (assuming that $\sqrt{2}|V_{ov1}| < 2$ V for the MOS circuit). The circuit thus operates *nonlinearly*, and linear analysis fails to predict the behavior. If the input stage did act linearly, the input voltage change of 2 V would produce a very large current I_x to charge the compensation capacitor. The fact that this current is limited to the fairly small value of $2I_1$ is the reason for the slew rate being much less than a linear analysis would predict.

Consider a large input voltage applied to the circuits of Fig. 9.50 so that $I_x = 2I_1$. Then the second stage acts as an integrator with an input current $2I_1$, and the output voltage V_o can be written as

$$V_o = \frac{1}{C} \int 2I_1 \, dt \tag{9.126}$$

and thus

$$\frac{dV_o}{dt} = \frac{2I_1}{C} \tag{9.127}$$

Equation 9.127 predicts a constant rate of change of V_o during the slewing period, which is in agreement with the experimental observation.

The above calculation of slew rate was performed on the circuits of Fig. 9.50, which have no overall feedback. Since the input stage produces a constant output current that is independent of its input during the slewing period, the presence of a feedback connection to the input does not affect the circuit operation during this time. Thus, the slew rate of the amplifier is the same whether feedback is applied or not.

The NE5234 op amp does not quite fit the model shown in Fig. 9.50a because the output of its first stage is differential. Figure 9.51 shows a model that assumes the op-amp common-mode input voltage is low enough that Q_1, Q_2, and Q_5-Q_7 in Fig. 6.36 are off. In practice, the input step in Fig. 9.48a changes the op-amp common-mode input voltage. Although this change affects the biasing of the input stage in the NE5234, it has little effect on the currents that limit the slew rate because the total current that biases the two differential pairs Q_1-Q_4 in Fig. 6.36

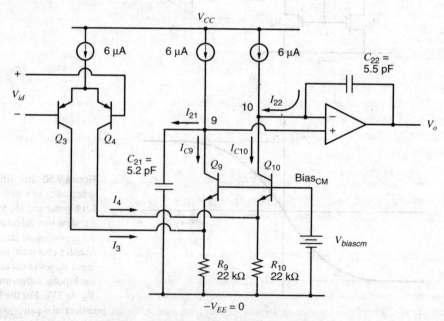

Figure 9.51 Simplified schematic of the NE5234 op amp.

is constant. Therefore, the change in the op-amp common-mode input voltage is ignored here. The three current sources at the top of Fig. 9.51 model the dc currents set by transistors Q_{11}, Q_{13}, and Q_{14} in Fig. 6.36 and are assumed constant here. When $V_{id} = 0$, $I_{C3} = I_{C4} = -3$ μA, ignoring base currents (as is done throughout this analysis for simplicity). The negative signs here stem from the convention that defines transistor collector current as positive when it flows into the collector. To simplify the following description, let $I_3 = -I_{C3}$ and $I_4 = -I_{C4}$. Then with $V_{id} = 0$, $I_3 = I_4 = 3$ μA, $I_{C9} = I_{C10} = 6$ μA, and capacitor currents $I_{21} = I_{22} = 0$, as shown in Fig. 9.51 and calculated in Chapter 6. Immediately after the step input in Fig. 9.48, Q_4 turns off, $I_3 = 6$ μA and $I_4 = 0$. Note that the changes in I_3 and I_4 are differential in the sense that one increases and the other decreases while their average value is constant. So we will ignore the common-mode feedback circuit that controls the average voltage to ground at the first-stage outputs and assume the voltage from node Bias$_{CM}$ to ground (V_{biascm}) is constant.

First, we calculate I_{C9} immediately after the input step using the assumption that V_{biascm} is constant by setting $V_{R9} + V_{be9}$ before and after the step equal to each other.

$$(3 \text{ μA} + 6 \text{ μA})22 \text{ k}\Omega + V_T \ln\left(\frac{6 \text{ μA}}{I_{S9}}\right) = (6 \text{ μA} + I_{C9})22 \text{ k}\Omega + V_T \ln\left(\frac{I_{C9}}{I_{S9}}\right) \quad (9.128a)$$

Simplifying this equation gives

$$0 = (I_{C9} - 3 \text{ μA})22 \text{ k}\Omega + V_T \ln\left(\frac{I_{C9}}{6 \text{ μA}}\right) \quad (9.128b)$$

Solving this equation by trial and error gives $I_{C9} = 3.6$ μA. Then from KCL at node 9, $I_{21} = 6 \text{ μA} - I_{C9} = 2.4$ μA. As a result, $dV_9/dt = 2.4 \text{ μA}/5.2 \text{ pF} = 0.46$ V/μs, where V_9 is the voltage from node 9 to ground.

Next, we calculate I_{C10} immediately after the input step in a similar manner.

$$(3 \text{ μA} + 6 \text{ μA})22 \text{ k}\Omega + V_T \ln\left(\frac{6 \text{ μA}}{I_{S10}}\right) = (I_{C10})22 \text{ k}\Omega + V_T \ln\left(\frac{I_{C10}}{I_{S10}}\right) \quad (9.128c)$$

Simplifying this equation gives

$$0 = (I_{C10} - 9 \text{ μA})22 \text{ k}\Omega + V_T \ln\left(\frac{I_{C10}}{6 \text{ μA}}\right) \quad (9.128d)$$

Solving this equation by trial and error gives $I_{C9} = 8.6$ μA. Then from KCL at node 10, $I_{22} = I_{C10} - 6 \text{ μA} = 2.6$ μA. Therefore, the voltage across C_{22} increases at a rate of $d(V_o - V_{10})/dt = 2.6 \text{ μA}/5.5 \text{ pF} = 0.47$ V/μs, where V_{10} is the voltage from node 10 to ground. In Fig. 9.51, the amplifier that represents the second and third stages in the NE5234 has negative feedback connected around it through capacitor C_{22}. Assuming that the gain of this amplifier is large and that it operates linearly, V_o is driven so that $V_{10} \simeq V_9$. Therefore, the slew rate of the NE5234 is

$$dV_o/dt = dV_9/dt + d(V_o - V_{10})/dt = (0.46 + 0.47) \text{ V/μs} = 0.93 \text{ V/μs} \quad (9.128e)$$

In contrast, the plot in Fig. 9.49b shows that the simulated slew rate is about 0.68 V/μs, and the difference stems partly from ignoring base currents in the calculations above.

9.6.2 Methods of Improving Slew-Rate in Two-Stage Op Amps

In order to examine methods of slew-rate improvement, a more general analysis is required. This can be performed using the circuit of Fig. 9.52, which is a general representation of an op amp circuit. The input stage has a small-signal transconductance g_{mI} and, with a large input voltage, can deliver a maximum current I_{xm} to the next stage. The compensation is shown

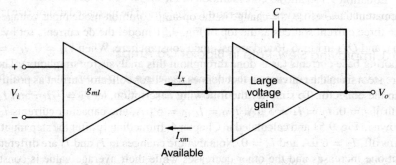

Figure 9.52 Generalized representation of an op amp for slew-rate calculations.

as the Miller effect using the capacitor C, since this representation describes most two-stage integrated-circuit op amps.

From Fig. 9.52 and using (9.127), we can calculate the slew rate for a large input voltage as

$$\frac{dV_o}{dt} = \frac{I_{xm}}{C} \tag{9.129}$$

Consider now *small-signal* operation. For the input stage, the small-signal transconductance is

$$\frac{\Delta I_x}{\Delta V_i} = g_{m1} \tag{9.130}$$

For the second stage (which acts as an integrator) the transfer function at high frequencies is

$$\frac{\Delta V_o}{\Delta I_x} = \frac{1}{sC} \tag{9.131a}$$

and in the frequency domain

$$\frac{\Delta V_o}{\Delta I_x}(j\omega) = \frac{1}{j\omega C} \tag{9.131b}$$

Combining (9.130) and (9.131b) gives

$$\frac{\Delta V_o}{\Delta V_i}(j\omega) = \frac{g_{m1}}{j\omega C} \tag{9.131c}$$

In our previous consideration of compensation, it was shown that the small-signal, open-loop voltage gain $(\Delta V_o/\Delta V_i)(j\omega)$ must fall to unity at or before a frequency equal to the magnitude of the second most dominant pole (ω_2). If we assume, for ease of calculation, that the circuit is compensated for unity-gain operation with 45° phase margin as shown in Fig. 9.15, the gain $(\Delta V_o/\Delta V_i)(j\omega)$ as given by (9.131c) must fall to unity at frequency ω_2. (Compensation capacitor C must be chosen to ensure that this occurs.) Thus from (9.131c)

$$1 = \frac{g_{m1}}{\omega_2 C}$$

and thus

$$\frac{1}{C} = \frac{\omega_2}{g_{m1}} \tag{9.132}$$

Note that (9.132) was derived on the basis of a *small-signal* argument. This result can now be substituted in the *large-signal* equation (9.129) to give

$$\text{Slew rate} = \frac{dV_o}{dt} = \frac{I_{xm}}{g_{m1}}\omega_2 \tag{9.133}$$

Equation 9.133 allows consideration of the effect of circuit parameters on slew rate, and it is apparent that, for a given ω_2, the ratio I_{xm}/g_{mI} must be increased if slew rate is to be increased.

9.6.3 Improving Slew-Rate in Bipolar Op Amps

The analysis of the previous section can be applied to a bipolar op amp that uses Miller compensation. In the case of the op amp in Fig. 9.50a, we have $I_{xm} = 2I_1$, $g_{mI} = qI_1/kT$, and substitution in (9.133) gives

$$\text{Slew rate} = 2\frac{kT}{q}\omega_2 \tag{9.134}$$

Since both I_{xm} and g_{mI} are proportional to bias current I_1, the influence of I_1 cancels in the equation and slew rate is *independent* of I_1 for a given ω_2. However, increasing ω_2 will increase the slew rate, and this course is followed in most high-slew-rate circuits. The limit here is set by the frequency characteristics of the transistors in the IC process, and further improvements depend on circuit modifications as described below.

The above calculation has shown that varying the input-stage bias current of a two-stage bipolar op amp does not change the circuit slew rate. However, (9.133) indicates that for a given I_{xm}, slew rate can be increased by *reducing* the input-stage transconductance. One way this can be achieved is by including emitter-degeneration resistors to reduce g_{mI} as shown in Fig. 9.53. The small-signal transconductance of this input stage can be shown to be

$$g_{mI} = \frac{\Delta I_x}{\Delta V_{id}} = g_{m1}\frac{1}{1 + g_{m1}R_E} \tag{9.135}$$

where

$$g_{m1} = \frac{qI_1}{kT} \tag{9.136}$$

The value of I_{xm} is still $2I_1$. Substituting (9.135) in (9.133) gives

$$\text{Slew rate} = \frac{2kT}{q}\omega_2(1 + g_{m1}R_E) \tag{9.137}$$

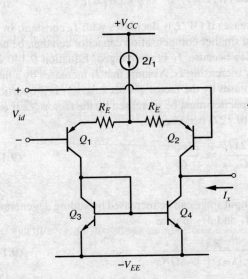

Figure 9.53 Inclusion of emitter resistors in the input stage in Fig. 9.50a to improve slew rate.

Thus the slew rate is increased by the factor $[1 + (g_{m1}R_E)]$ over the value given by (9.134). The fundamental reason for this is that, for a given bias current I_1, reducing g_{m1} reduces the compensation capacitor C required, as shown by (9.132).

The practical limit to this technique is due to the fact that the emitter resistors of Fig. 9.53 have a dc voltage across them, and mismatches in the resistor values give rise to an input dc offset voltage. The use of large-area resistors can give resistors whose values match to within 0.2 percent (1 part in 500). If the maximum contribution to input offset voltage allowed from the resistors is 1 mV, then these numbers indicate that the maximum voltage drop allowed is

$$I_1 R_E|_{\max} = 500 \text{ mV} \tag{9.138}$$

Thus

$$g_{m1} R_E|_{\max} = \frac{q}{kT} I_1 R_E|_{\max} = \frac{500}{26} = 20 \tag{9.139}$$

Using (9.139) in (9.137) shows that given these data, the maximum possible improvement in slew rate by use of emitter resistors is a factor of 21 times.

Finally, in this description of methods of slew-rate improvement, we mention the *Class AB* input stage described by Hearn.[21] In this technique, the small-signal transconductance of the input stage is left essentially unchanged, but the limit I_{xm} on the maximum current available for charging the compensation capacitor is greatly increased. This is done by providing alternative paths in the input stage that become operative for large inputs and deliver large charging currents to the compensation point. This has resulted in slew rates of the order of 30 V/μs in bipolar op amps, and, as in the previous cases, the limitation is an increase in input offset voltage.

9.6.4 Improving Slew-Rate in MOS Op Amps

A two-stage Miller-compensated MOS op amp is shown in Fig. 9.50b, and its slew rate is given by (9.127). From the analysis in Section 9.6.2, (9.133) shows that the slew rate can be increased by increasing ω_2. On the other hand, if ω_2 is fixed, increasing the ratio I_{xm}/g_{mI} improves the slew rate. Using (1.180), (9.133) can be rewritten as

$$\text{Slew rate} = \frac{I_{xm}}{g_{mI}} \omega_2 = \frac{2I_1}{\sqrt{2k'(W/L)_1 I_1}} \omega_2 = \sqrt{\frac{2I_1}{k'(W/L)_1}} \omega_2 \tag{9.140}$$

This equation shows that the slew rate increases if $(W/L)_1$ decreases with I_1 constant. In this case, $g_{mI} = g_{m1}$ decreases. From (9.132), a smaller compensation capacitor can then be used; therefore, the slew rate in (9.127) increases because I_1 is unchanged. Equation 9.140 also shows that the slew rate can be increased by increasing I_1. Assume that I_1 increases by a factor x where $x > 1$. Then the ratio I_{xm}/g_{mI} increases by the factor $\sqrt{x}$ because g_{mI} is proportional to $\sqrt{I_1}$. From (9.132), the compensation capacitor must be increased by the factor $\sqrt{x}$ if ω_2 is fixed. With these changes, the slew rate in (9.127) becomes

$$\frac{dV_o}{dt} = \frac{2xI_1}{C\sqrt{x}} = \frac{2I_1\sqrt{x}}{C} \tag{9.141}$$

Since $x > 1$, the slew rate is increased.

Alternatively, the ratio I_{xm}/g_{mI} of the input stage can be increased by adding degeneration resistors R_S in series with the sources of M_1 and M_2 to give

$$g_{mI} = \frac{g_{m1}}{1 + (g_{m1} + g_{mb1})R_S} \tag{9.142}$$

For fixed I_1, increasing R_S decreases g_{mI} and increases I_{xm}/g_{mI}, which increases the slew rate.

These approaches increase the slew rate but have some drawbacks. First, decreasing g_{mI} of the input stage while keeping its bias current constant will usually lower the dc gain of the first stage and hence reduce the dc gain of the entire op amp. Also, increasing I_1 or reducing $(W/L)_1$ tends to increase the input-offset voltage of the op amp, as can be seen from (3.248). Finally, if source-degeneration resistors are added, mismatch between these resistors degrades the input-offset voltage.

For single-stage MOS op amps, such as the telescopic-cascode and folded-cascode op amps, the slew rate is set by the maximum output current divided by the capacitance that loads the output. The maximum output current is equal to the tail current in these op amps.

■ **EXAMPLE**

Find the output slew rate for the cascode op amp shown in Fig. 9.54.

Assuming the op amp has a large positive differential input voltage applied, M_2 is cutoff and I_{TAIL} flows through M_1. Therefore the drain current in M_{2A} is zero, and the drain current in M_3 is $I_{d3} = -I_{TAIL}$. The current mirror M_3-M_4 forces $I_{d3} = I_{d4}$. It follows that $I_{d4A} = I_{d4} = -I_{TAIL}$. The current flowing into the load capacitor C_L is

$$I_o = -I_{d2A} - I_{d4A} = -0 - (-I_{TAIL}) = I_{TAIL}$$

Therefore the positive output slew rate is

$$\frac{dV_o}{dt} = \frac{I_o}{C_L} = \frac{I_{TAIL}}{C_L} \quad (9.143)$$

Application of a large negative input forces M_1 into cutoff so I_{TAIL} must flow through M_2. Therefore, $I_{d4A} = I_{d4} = I_{d3} = 0$ and $I_{d2A} = I_{d2} = I_{TAIL}$. The current I_o flowing through C_L is

$$I_o = -I_{d2A} - I_{d4A} = -I_{TAIL} - 0 = -I_{TAIL}$$

■ Hence, the negative slew rate is the opposite of the value in (9.143), $-I_{TAIL}/C_L$.

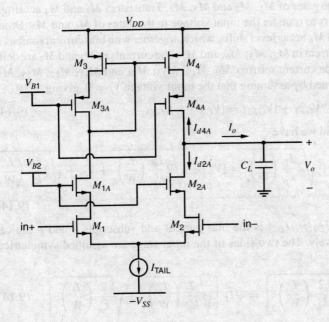

Figure 9.54 A CMOS telescopic-cascode op amp.

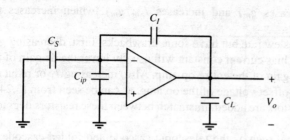

Figure 9.55 An op amp with capacitive load and feedback. This is the switched-capacitor integrator of Fig. 6.10a during ϕ_2, assuming ideal MOS switches.

CMOS op amps are often used without an output stage when the output loading is purely capacitive, as is the case in switched-capacitor circuits. Avoiding an output stage saves power and is possible because low-output resistance is not needed to drive a capacitive load. An example of such a circuit is the switched-capacitor integrator shown in Fig. 6.10a. This circuit is redrawn in Fig. 9.55 when clock phase ϕ_2 is high and ϕ_1 is low, assuming that MOS transistors M_1-M_4 behave like ideal switches. The additional capacitor C_{ip} here models the total parasitic capacitance at the op-amp input and includes the input capacitance of the op amp. A question that arises is: "For the feedback circuit in Fig. 9.55, what value of output load capacitance should be used to compute the slew rate for a single-stage op amp?" When the op amp is slewing, its behavior is nonlinear. Therefore the feedback is not effective and the virtual ground at the negative op-amp input is lost. With the feedback loop broken, the total capacitance seen from the output to ground is

$$C_L + C_I \| (C_S + C_{ip}) \tag{9.144}$$

This is the capacitance seen looking from the op-amp output node to ground, with the connection to the op-amp inverting input replaced with an open circuit. The effective output load capacitance in (9.144) is the same as the output load found when the feedback loop is broken to find the return ratio.

For the CMOS op amps considered so far in this section, the slew rate is proportional to a bias current in the op amp. A CMOS op amp with a Class AB input stage can give a slew rate that is not limited by a dc bias current in the op amp. An example[22,23] is shown in Fig. 9.56. The input voltage is applied between the gates of M_1, M_2 and M_3, M_4. Transistors M_1 and M_4 act simply as unity-gain source followers to transfer the input voltage to the gates of M_6 and M_7. Diode-connected transistors M_5 and M_8 act as level shifts, which, together with bias current sources I_1, set the quiescent Class AB current in M_2, M_3, M_6, and M_7. The currents in M_3 and M_7 are delivered to the output via cascode current mirrors M_9, M_{10}, M_{13}, M_{14} and M_{11}, M_{12}, M_{15}, M_{16}. Bias currents can be determined by assuming that the input voltage $V_i = 0$, giving

$$V_{GS1} + |V_{GS5}| = |V_{GS6}| + V_{GS3} \tag{9.145}$$

Assuming that (1.157) is valid we have

$$V_{tn} + \sqrt{2\frac{I_1}{k'_n}\left(\frac{L}{W}\right)_1} + |V_{tp}| + \sqrt{2\frac{I_1}{k'_p}\left(\frac{L}{W}\right)_5} = |V_{tp}| + \sqrt{2\frac{I_B}{k'_P}\left(\frac{L}{W}\right)_6} + V_{tn} + \sqrt{2\frac{I_B}{k'_n}\left(\frac{L}{W}\right)_3} \tag{9.146}$$

where $I_B = |I_{D6}| = I_{D3} = I_{D2} = |I_{D7}|$ is the bias current and subscripts n and p indicate NMOS and PMOS, respectively. The two sides of the input stage are assumed symmetrical. From (9.146) we have

$$\sqrt{I_B}\left[\sqrt{\frac{2}{k'_p}\left(\frac{L}{W}\right)_6} + \sqrt{\frac{2}{k'_n}\left(\frac{L}{W}\right)_3}\right] = \sqrt{I_1}\left[\sqrt{\frac{2}{k'_p}\left(\frac{L}{W}\right)_5} + \sqrt{\frac{2}{k'_n}\left(\frac{L}{W}\right)_1}\right] \tag{9.147}$$

Equation 9.147 is the design equation for the input-stage bias current I_B.

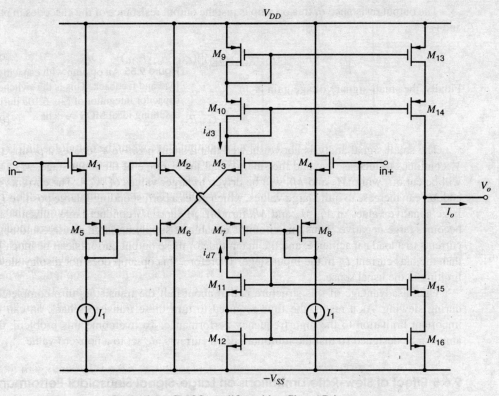

Figure 9.56 CMOS amplifier with a Class AB input stage.

Assuming that the cascode current mirrors in Fig. 9.56 have unity current gain, the bias currents in M_9-M_{16} all equal I_B. To analyze this circuit, we will connect a voltage V_i to the noninverting op-amp input and ground the inverting op-amp input. If a positive V_i is applied, the magnitude of the currents in M_3 and M_6 increase, while the magnitude of the currents in M_2 and M_7 decrease. When mirrored to the output, these changes drive I_o and V_o positive. To calculate the small-signal gain, we neglect body effect. We can consider M_6 to act as source degeneration for M_3. The resistance looking into the source of M_6 is $1/g_{m6}$, thus

$$i_{d3} = \frac{g_{m3}}{1 + \frac{g_{m3}}{g_{m6}}} v_i \tag{9.148}$$

Similarly, M_2 acts as source degeneration for M_7, so

$$i_{d7} = \frac{g_{m7}}{1 + \frac{g_{m7}}{g_{m2}}} v_i = \frac{g_{m2}}{1 + \frac{g_{m2}}{g_{m7}}} v_i \tag{9.149}$$

where the right-most expression is found by rearranging. Thus, the transconductance of the amplifier is

$$G_m = \left.\frac{i_o}{v_i}\right|_{v_o=0} = \frac{i_{d3} + i_{d7}}{v_i} = \frac{2g_{m3}}{1 + \frac{g_{m3}}{g_{m6}}} \tag{9.150}$$

using $g_{m2} = g_{m3}$ and $g_{m6} = g_{m7}$. If $g_{m3} = g_{m6}$, then $G_m = g_{m3}$.

The output resistance of this op amp is just the output resistance of the cascodes in parallel and is

$$R_o \approx (r_{o14}g_{m14}r_{o13}) \| (r_{o15}g_{m15}r_{o16}) \quad (9.151)$$

Finally, the small-signal voltage gain is

$$A_v = G_m R_o \quad (9.152)$$

The small-signal analysis above showed that a small positive V_i causes a positive I_o. If V_i continues to increase beyond the small-signal linear range of the input stage, M_2 and M_7 will be cut off, while M_3 and M_6 will be driven to larger values of $|V_{gs}|$. The currents in M_3 and M_6 can increase to quite large values, which gives a correspondingly large positive I_o. For large negative values of V_i, M_3 and M_6 turn off, M_2 and M_7 conduct large currents, and I_o becomes large negative. Thus this circuit is capable of supplying large positive and negative currents to a load capacitance, and the magnitude of these output currents can be much larger than the bias current I_B in the input stage. Therefore, this op amp does not display slew-rate limiting in the usual sense.

One disadvantage of this structure is that about half the transistors turn completely off during slewing. As a result, the time required to turn these transistors back on can be an important limitation to the high-frequency performance. To overcome this problem, the op amp can be designed so that the minimum drain currents are set to a nonzero value.[24]

9.6.5 Effect of Slew-Rate Limitations on Large-Signal Sinusoidal Performance

The slew-rate limitations described above can also affect the performance of the circuit when handling large sinusoidal signals at higher frequencies. Consider the circuit of Fig. 9.48 with a large sinusoidal signal applied as shown in Fig. 9.57a. Since the circuit is connected as a voltage follower, the output voltage V_o will be forced to follow the V_i waveform. The maximum value of dV_i/dt occurs as the waveform crosses the axis, and if V_i is given by

$$V_i = \hat{V}_i \sin \omega t \quad (9.153)$$

then

$$\frac{dV_i}{dt} = \omega \hat{V}_i \cos \omega t$$

and

$$\left. \frac{dV_i}{dt} \right|_{\max} = \omega \hat{V}_i \quad (9.154)$$

As long as the value of $dV_i/dt|_{\max}$ given by (9.154) is less than the slew-rate limit, the output voltage will closely follow the input. However, if the product $\omega \hat{V}_i$ is greater than the slew-rate limit, the output voltage will be unable to follow the input, and waveform distortion of the kind shown in Fig. 9.57b will result. If a sine wave with $\hat{V}_i$ equal to the supply voltage is applied to the amplifier, slew limiting will eventually occur as the sine-wave frequency is increased. The frequency at which this occurs is called the *full-power* bandwidth of the circuit. (In practice, a value of $\hat{V}_i$ slightly less than the supply voltage is used to avoid clipping distortion of the type described in Chapter 5.)

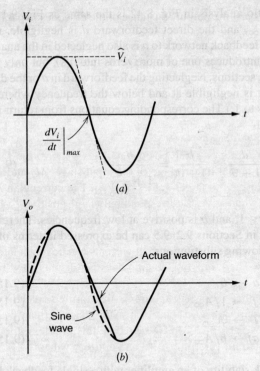

Figure 9.57 (a) Large sinusoidal input voltage applied to the circuit of Fig. 9.48. (b) Output voltage resulting from input (a) showing slew limiting.

■ **EXAMPLE**

Calculate the *full-power* bandwidth of the NE5234. Use $\hat{V}_i = 1$ V. From (9.154) put

$$\omega \hat{V}_i = \text{slew rate}$$

Using the slew rate of 0.68 V/μs found in simulation gives

$$\omega = \frac{0.68 \text{ V/μs}}{1 \text{ V}} = 680 \times 10^3 \text{ rad/s}$$

Thus

$$f = 110 \text{ kHz}$$

This means that a NE5234 op amp with a sinusoidal output of 1 V amplitude will begin to show slew-limiting distortion if the frequency exceeds 110 kHz. ■

APPENDIX

A.9.1 ANALYSIS IN TERMS OF RETURN-RATIO PARAMETERS

Much of the analysis in this chapter is based on the ideal block diagram in Fig. 9.1. This block diagram includes the forward gain a and feedback f, which are the parameters used in two-port analysis of feedback circuits in Chapter 8. The resulting closed-loop gain expression is

$$A = \frac{a}{1+T} = \frac{a}{1+af} \qquad (9.155)$$

The block diagram from return-ratio analysis in Fig. 8.42 is the same as Fig. 9.1 if a is replaced by b, f is replaced by $1/A_\infty$, and the direct feedforward d is negligible. (The contribution of feedforward through the feedback network to a is also neglected in the analysis in Sections 9.2–9.5, since feedforward introduces one or more zeros into $a(s)$, but only one- and two-pole $a(s)$ are considered in these sections. Neglecting the feedforward in a or the direct feedthrough d is reasonable if its effect is negligible at and below the frequency where the magnitude of the loop transmission falls to 1.) The corresponding equations from return-ratio analysis are

$$A = \frac{b}{1+\mathcal{R}} + \frac{d}{1+\mathcal{R}} \approx \frac{b}{1+\mathcal{R}} = \frac{b}{1+\dfrac{b}{A_\infty}} \qquad (9.156)$$

For the circuit in Fig. 8.24, $0 \leq 1/A_\infty \leq 1$, and b is positive at low frequencies. Therefore, the equations, graphs, and relationships in Sections 9.2–9.5 can be expressed in terms of the return-ratio variables by making the following substitutions:

$$a \to b \qquad (9.157a)$$
$$f \to 1/A_\infty \qquad (9.157b)$$
$$T \to \mathcal{R} \qquad (9.157c)$$
$$af \to b/A_\infty \qquad (9.157d)$$

The return ratio can be used to check stability of an amplifier with a single feedback loop because A_∞ and d are stable transfer functions associated with passive networks, and $\mathcal{R}(s)$ is stable because it is the signal transfer around a loop that consists of one gain stage or a cascade of stable gain stages. Therefore the zeros of $1 + \mathcal{R}(s)$, which are poles of the closed-loop gain A, determine the stability of the feedback circuit.[25] From the Nyquist stability criterion, these zeros are in the left half-plane if a polar plot of $\mathcal{R}(j\omega)$ does not encircle the point $(-1,0)$. In most cases, this stability condition is equivalent to having a positive phase margin. The phase margin is measured at the frequency where $|\mathcal{R}(j\omega)| = 1$.

Since the equations for two-port and return-ratio analyses are not identical, $T(s)$ and $\mathcal{R}(s)$ may be different for a given circuit.[26] In general, the phase margins using T and $\mathcal{R}$ may differ, but both will have the same sign and therefore will agree on the stability of the feedback circuit.

A.9.2 ROOTS OF A QUADRATIC EQUATION

A second-order polynomial often appears in the denominator or numerator of a transfer function, and the zeros of this polynomial are the poles or zeros of the transfer function. In this appendix, the relationships between the zeros of a quadratic and its coefficients are explored for a few specific cases of interest. Also, the conditions under which a dominant root exists are derived.

Consider the roots of the quadratic equation

$$as^2 + bs + c = 0 \qquad (9.158)$$

The two roots of this equation, r_1 and r_2, are given by the quadratic formula:

$$r_{1,2} = \frac{-b \pm \sqrt{b^2 - 4ac}}{2a} \qquad (9.159)$$

where it is understood that the square root of a positive quantity is positive. Factoring b out of the square root and rearranging gives

$$r_{1,2} = -\frac{b}{2a}\left(1 \pm \sqrt{1 - \frac{4ac}{b^2}}\right) \quad (9.160a)$$

$$= -\frac{b}{2a}\left(1 \pm \sqrt{D}\right) \quad (9.160b)$$

The quantity under the square root in (9.160a) has been replaced by D in (9.160b), where

$$D = 1 - \frac{4ac}{b^2} \quad (9.161)$$

Now, consider the locations of the roots if coefficients a, b, and c all have the same sign. In this case, both roots are in the left half-plane (LHP), as will be shown next. First, note that if all the coefficients have the same sign, then

$$\frac{b}{2a} > 0 \quad (9.162)$$

and

$$\frac{4ac}{b^2} > 0. \quad (9.163)$$

Let us divide (9.163) into two different regions. First, if

$$0 < \frac{4ac}{b^2} \le 1 \quad (9.164)$$

then D will be positive and less than one. Therefore, $\sqrt{D} < 1$, so $1 + \sqrt{D}$ and $1 - \sqrt{D}$ are both positive. As a result, the roots are both negative and real, because $-b/2a < 0$.

Now, consider the other region for (9.163), which is

$$\frac{4ac}{b^2} > 1 \quad (9.165)$$

In this case, $D < 0$; therefore $\sqrt{D}$ is imaginary. The roots are complex conjugates with a real part of $-b/2a$, which is negative. So the roots are again in the LHP. Therefore, when coefficients a, b, and c all have the same sign, both roots are in the LHP.

Next, consider the locations of the roots if coefficients a and b have the same sign and c has a different sign. In this case, one real root is in the right half-plane (RHP) and the other is in the LHP. To prove this, first note from (9.161) that $D > 1$ here because $4ac/b^2 < 0$. Therefore both roots are real and $\sqrt{D} > 1$, so

$$1 + \sqrt{D} > 0 \quad (9.166a)$$

and

$$1 - \sqrt{D} < 0 \quad (9.166b)$$

Substituting into (9.160), one root will be positive and the other negative (the sign of $-b/2a$ is negative here).

Finally, let us consider the conditions under which LHP roots are real and widely spaced. From (9.160), real LHP roots are widely spaced if

$$-\frac{b}{2a}(1 + \sqrt{D}) \gg -\frac{b}{2a}(1 - \sqrt{D}) \quad (9.167)$$

or

$$1 + \sqrt{D} \gg 1 - \sqrt{D} \quad (9.168)$$

Substituting the expression for D in (9.161) into (9.168) and simplifying leads to an equivalent condition for widely spaced roots, which is

$$\frac{4ac}{b^2} \ll 1 \tag{9.169}$$

Under this condition, one root is

$$r_2 = -\frac{b}{2a}(1 + \sqrt{D}) \approx -\frac{b}{2a}(1 + 1) = -\frac{b}{a} \tag{9.170a}$$

The other root is

$$\begin{aligned}
r_1 &= -\frac{b}{2a}(1 - \sqrt{D}) \\
&= -\frac{b}{2a}\left(1 - \sqrt{1 - \frac{4ac}{b^2}}\right) \\
&\approx -\frac{b}{2a}\left(1 - \left(1 - \frac{4ac}{2b^2}\right)\right) \\
&= -\frac{c}{b}
\end{aligned} \tag{9.170b}$$

where the approximation

$$\sqrt{1-x} \approx 1 - \frac{x}{2} \quad \text{for} \quad |x| \ll 1 \tag{9.171}$$

has been used. Here, $|r_1| \ll |r_2|$ because $|r_1| \approx c/b \ll b/a \approx |r_2|$ (which follows from $4ac/b^2 \ll 1$). If these roots are poles, r_1 corresponds to the dominant pole, and r_2 gives the nondominant pole. Equations 9.170a and 9.170b are in agreement with (9.30) through (9.33).

Table 9.1 summarizes the location of the roots of (9.158) for the cases considered in this appendix. When both roots are in the LHP, the roots are both real if (9.164) is satisfied. These roots are widely spaced if (9.169) is satisfied.

Table 9.1

Sign of Coefficient Values in (9.158)			Roots
a	b	c	
+	+	+	Both in LHP
−	−	−	Both in LHP
+	+	−	One in LHP, one in RHP
−	−	+	One in LHP, one in RHP

PROBLEMS

9.1 An amplifier has a low-frequency forward gain of 200 and its transfer function has three negative real poles with magnitudes 1 MHz, 2 MHz, and 4 MHz. Calculate and sketch the Nyquist diagram for this amplifier if it is placed in a negative feedback loop with $f = 0.05$. Is the amplifier stable? Explain.

9.2 For the amplifier in Problem 9.1, calculate and sketch plots of gain (in decibels) and phase versus frequency (log scale) with no feedback applied. Determine the value of f that just causes instability and the value of f giving a 60° phase margin.

9.3 If an amplifier has a phase margin of 30°, how much does the closed-loop gain peak (above the low-frequency value) at the frequency where the loop-gain magnitude is unity?

9.4 An amplifier has a low-frequency forward gain of 40,000 and its transfer function has three negative real poles with magnitudes 2 kHz, 200 kHz, and 4 MHz.

(a) If this amplifier is connected in a feedback loop with f constant and with low-frequency gain $A_0 = 400$, estimate the phase margin.

(b) Repeat (a) if A_0 is 200 and then 100.

9.5 An amplifier has a low-frequency forward gain of 5000 and its transfer function has three negative real poles with magnitudes 300 kHz, 2 MHz, and 25 MHz.

(a) Calculate the dominant-pole magnitude required to give unity-gain compensation of this amplifier with a 45° phase margin if the original amplifier poles remain fixed. What is the resulting bandwidth of the circuit with the feedback applied?

(b) Repeat (a) for compensation in a feedback loop with a closed-loop gain of 20 dB and 45° phase margin.

9.6 The amplifier of Problem 9.5 is to be compensated by reducing the magnitude of the most dominant pole.

(a) Calculate the dominant-pole magnitude required for unity-gain compensation with 45° phase margin, and the corresponding bandwidth of the circuit with the feedback applied. Assume that the remaining poles do not move.

(b) Repeat (a) for compensation in a feedback loop with a closed-loop gain of 40 dB and 45° phase margin.

9.7 Repeat Problem 9.6 for the amplifier of Problem 9.4.

9.8 An op amp has a low-frequency open-loop voltage gain of 100,000 and a frequency response with a single negative-real pole with magnitude 5 Hz. This amplifier is to be connected in a series-shunt feedback loop with $f = 0.01$ giving a low-frequency closed-loop voltage gain $A_0 \approx 100$. If the output impedance without feedback is resistive with a value of 100 Ω, show that the output impedance of the feedback circuit can be represented as shown in Fig. 9.58, and calculate the values of these elements. Sketch the magnitude of the output impedance of the feedback circuit on log scales from 1 Hz to 100 kHz.

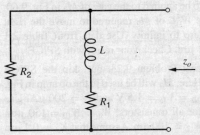

Figure 9.58 Circuit representation of the output impedance of a series-shunt feedback circuit.

9.9 An op amp with low-frequency gain of 108 dB has three negative real poles with magnitudes 30 kHz, 500 kHz, and 10 MHz before compensation. The circuit is compensated by placing a capacitance across the second stage, causing the second most dominant pole to become negligible because of pole splitting. Assume the small-signal transconductance of the second stage is 6.39 mA/V and the small-signal resistances to ground from the input and output are 1.95 MΩ and 86.3 kΩ, respectively. Calculate the value of capacitance required to achieve a 60° phase margin in a unity-gain feedback connection and calculate the frequency where the resulting open-loop gain is 0 dB. Assume that the pole with magnitude 10 MHz is unaffected by the compensation.

9.10 Repeat Problem 9.9 if the circuit is compensated by using shunt capacitance to ground at the input of the second stage. Assume that this affects only the most dominant pole.

9.11 Calculate and sketch the root locus for the amplifier of Problem 9.4 as f varies from 0 to 1. Estimate the value of f causing instability and check using the Nyquist criterion.

9.12 An amplifier has gain $a_0 = 200$ and its transfer function has three negative real poles with magnitudes 1 MHz, 3 MHz, and 4 MHz. Calculate and sketch the root locus when feedback is applied as f varies from 0 to 1. Estimate the value of f causing instability.

9.13 For the circuit of Fig. 9.41, parameter values are $R_F = 5$ kΩ, $R_E = 50$ Ω, and $C_F = 1.5$ pF. The basic amplifier of the circuit is shown in Fig. 9.42 and has two negative real poles with magnitudes 3 MHz and 6 MHz. The low-frequency *current gain* of the basic amplifier is 4000. Assuming that the loop gain of the circuit of Fig. 9.41 can be varied without changing the parameters of the basic amplifier, sketch root loci for this circuit as f varies from 0 to 0.01 both with and without C_F. Estimate the pole positions of the current-gain transfer function of the feedback amplifier of Fig. 9.41 with the values of R_F and R_E specified both with and without C_F. Sketch graphs in each case of gain magnitude versus frequency on log scales from $f = 10$ kHz to $f = 100$ MHz.

9.14 An op amp has two negative real open-loop poles with magnitudes 100 Hz and 120 kHz and a negative real zero with magnitude 100 kHz. The low-frequency open-loop voltage gain of the op amp is 100 dB. If this amplifier is placed in a negative feedback loop, sketch the root locus as f varies from 0 to 1. Calculate the poles and zeros of the feedback amplifier for $f = 10^{-3}$ and $f = 1$.

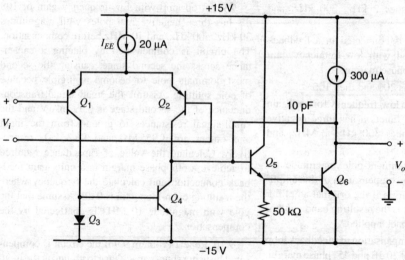

Figure 9.59 Input stages of an op amp.

9.15 Repeat Problem 9.14 if the circuit has negative real poles with magnitudes 100 Hz and 100 kHz and a negative real zero with magnitude 120 kHz.

9.16 The input stages of an op amp are shown in the schematic of Fig. 9.59.

(a) Assuming that the frequency response is dominated by a single pole, calculate the frequency where the magnitude of the small-signal voltage gain $|v_o(j\omega)/v_i(j\omega)|$ is unity and also the output slew rate of the amplifier.

(b) Sketch the response $V_o(t)$ from 0 to 20 μs for a step input at V_i from -5 V to $+5$ V. Assume that the circuit is connected in a noninverting unity-gain feedback loop.

(c) Compare your results with a SPICE simulation using parameters $\beta = 100$, $V_A = 130$ V, and $I_S = 10^{-15}$ A for all devices.

9.17 Repeat Problem 9.16 if the circuit of Fig. 9.59 is compensated by a capacitor of 0.05 μF connected from the base of Q_5 to ground. Assume that the voltage gain from the base of Q_5 to V_o is -500.

9.18 The slew rate of the circuit of Fig. 9.59 is to be increased by using 10 kΩ resistors in the emitters Q_1 and Q_2. If the same unity-gain frequency is to be achieved, calculate the new value of compensation capacitor required and the improvement in slew rate. Check your result with SPICE simulations.

9.19 Repeat Problem 9.18 if PMOS transistors replace Q_1 and Q_2 (with no degeneration resistors). Assume that the PMOS transistors are biased to 300 μA each ($I_{EE} = 600$ μA), at which bias value the MOS transistors have $g_m = 400$ μA/V.

9.20(a) Calculate the full-power bandwidth of the circuit of Fig. 9.59.

(b) If this circuit is connected in a noninverting unity-gain feedback loop, sketch the output waveform V_o if V_i is a sinusoid of 10 V amplitude and frequency 45 kHz.

9.21 For the CMOS operational amplifier shown in Fig. 9.60, calculate the open-loop voltage gain, unity-gain bandwidth, and slew rate. Assume the parameters of Table 2.1 with $X_d = 1$ μm. Assume that the gate of M_9 is connected to the positive power supply and that the W/L of M_9 has been chosen to cancel the right half-plane zero. Compare your results with a SPICE simulation.

9.22 Repeat Problem 9.21 except use the aspect ratios, supply voltages, and bias current given in Fig. 6.58 instead of the values in Fig. 9.60. Also, assume that $X_d = 0.1$ μm for all transistors operating in the active region, and use Table 2.4 for other parameters.

9.23 If the circuit of Fig. 9.61 is used to generate the voltage to be applied to the gate of M_9 in Fig. 9.60, calculate the W/L of M_9 required to move the right half-plane zero to infinity. Use data from Table 2.1 with $X_d = 1$ μm. Check your result with SPICE.

9.24 Repeat Problem 9.23, but skip the SPICE simulation. Here, M_9 will be used in the op amp in Fig. 6.58. Let $V_{DD} = V_{SS} = 1.5$ V and $I_S = 200$ μA. Use $L = 1$ μm for all transistors, $W_8 = W_{10} = 150$ μm, and $W_{11} = W_{12} = 100$ μm. Assume that $X_d = 0.1$ μm for all transistors operating in the active region, and use Table 2.4 for other parameters.

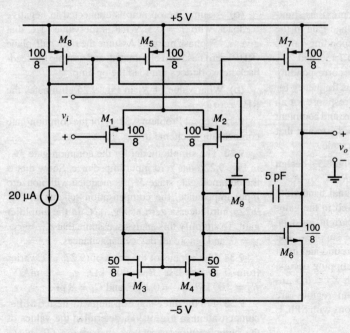

Figure 9.60 Circuit for Problem 9.21.

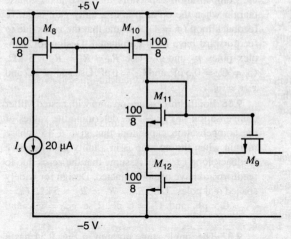

Figure 9.61 Circuit for Problem 9.23.

9.25 Assuming that the zero has been moved to infinity, determine the maximum load capacitance that can be attached directly to the output of the circuit of Fig. 9.60 and still maintain a phase margin of 45°. Neglect all higher order poles except any due to the load capacitance. Use the value of W/L obtained in Problem 9.23 for M_9 with the bias circuit of Fig. 9.61.

9.26 Repeat Problem 9.25 except, for the op amp, use the aspect ratios, supply voltages, and bias current given in Fig. 6.58 instead of the values in Fig. 9.60. Also, for the bias circuit, use the aspect ratios, supply voltages, and bias current given in Problem 9.24. Ignore junction capacitance for all transistors. Also,

assume that $X_d = 0.1$ μm for all transistors operating in the active region, and use Table 2.4 for other parameters.

9.27 For the CMOS op amp of Fig. 9.60, assume that M_9 and the compensation capacitor are removed and the output is loaded with a 1 MΩ resistor. Using the data of Table 2.1, use SPICE to determine the gain and phase versus frequency of the small-signal circuit voltage gain.

The amplifier is to be connected in a negative feedback loop with the 1-MΩ resistor connected from the output to the gate of M_1, and a resistor R_x from the M_1 gate to ground. An input voltage is applied from the gate of M_2 to ground. From your previous simulated data, determine the forward voltage gain of the feedback configuration and the corresponding values of R_x giving phase margins of 80°, 60°, 45°, and 20°. For each case use SPICE to plot out the corresponding overall small-signal voltage gain versus frequency for the feedback circuit and also the step response for an output voltage step of 100 mV. Compare and comment on the results obtained. Assume $X_d = 1$ μm and that the drain and source regions are 2 μm wide.

9.28 Using the basic topology of Fig. 8.53, design a CMOS feedback amplifier with $R_i = \infty$, $R_o < 30$ Ω, $A_v = v_o/v_i = 10$, and small-signal bandwidth $f_{-3dB} > 2$ MHz. No peaking is allowed in the gain-versus-frequency response. Supply current must be less than 2 mA from each of ± 5 V supplies. The circuit operates with $R_L = 1$ kΩ to ground and must be able to swing $V_o = ± 1$ V before clipping occurs. Use the process data of Table 2.1 with $X_d = 0.5$ μm and $\gamma_n = 0.5$ V$^{1/2}$. Source and drain regions are 9 μm wide. Verify your hand calculations with SPICE simulations.

9.29 The CMOS circuit of Fig. 9.56 is to be used as a high-slew-rate op amp. A load capacitance of $C_L = 10$ pF is connected from V_o to ground. Supply voltages are ± 5 V and $I_1 = 20$ μA. Devices M_1–M_4 have $W = 20$ μm and $L = 1$ μm and devices M_5–M_8 have $W = 60$ μm and $L = 1$ μm. All other NMOS devices have $W = 60$ μm and $L = 1$ μm, and all other PMOS devices have $W = 300$ μm and $L = 1$ μm. Device data are $\mu_n C_{ox} = 60$ μA/V^2, $V_{tn} = 0.7$ V, $V_{tp} = -0.7$ V, $\gamma = 0$, and $|\lambda| = 0.05$ V^{-1}.

(a) Calculate the small-signal open-loop gain and unity-gain bandwidth of the circuit. Derive an expression for the large-signal transfer function I_o/V_i when all four devices M_2, M_3, M_6, and M_7 are on, and also for larger V_i when two of them cut off. At what value of V_i does the transition occur?

(b) Connect the circuit in a unity-gain negative feedback loop (V_o to the gate of M_1) and drive the circuit with a voltage step from -1.5 V to $+1.5$ V at the gate of M_4. Calculate and sketch the corresponding output waveform V_o assuming linear operation, and compare all your results with a SPICE simulation. What is the peak current delivered to C_L during the transient?

9.30 Determine the compensation capacitor for the two-stage op amp in the example in Section 9.4.3 that gives a 60° phase margin.

9.31 The Miller-compensated two-stage op amp in Fig. 9.25 can be modeled as shown in Fig. 9.26. In the model, let $g_{m1} = 0.5$ mA/V, $R_1 = 200$ kΩ, $g_{m6} = 2$ mA/V, $R_2 = 100$ kΩ, $C_1 = 0.1$ pF, and $C_2 = 8$ pF.

(a) Assume the op amp is connected in negative feedback with $f = 0.5$. What is the value of C that gives a 45° phase margin? Assume the right half-plane (RHP) zero has been eliminated, and assume the feedback network does not load the op amp.

(b) What value of R_z in Fig. 9.26 eliminates the RHP zero?

9.32 Repeat Problem 9.31(a) for the common-gate compensation scheme in Fig. 9.22a.

9.33 The simple model for the common-gate M_{11} in Fig. 9.22b has zero input impedance. Show that if the common-gate stage M_{11} is modeled with nonzero input impedance, the compensation scheme in Fig. 9.22a introduces a zero at $-g_{m11}/C$ in the amplifier gain. To simplify this analysis, assume that $r_{o11} = \infty$, $\gamma = 0$, and ignore all device capacitances.

9.34 Plot a locus of the poles of (9.27) as C varies from 0 to ∞. Use $R_1 = 200$ kΩ, $g_m = 2$ mA/V, $R_2 = 100$ kΩ, $C_1 = 0.1$ pF, and $C_2 = 8$ pF.

9.35 For the three-stage op amp with nested Miller compensation in Fig. 9.30c, determine the values of the compensation capacitors that give a 60° phase margin when the op amp is in a unity-gain negative feedback loop ($f = 1$). Assume that the zeros due to feedforward have been eliminated. Design for complex poles p_2 and p_3. Use $R_0 = R_1 = R_2 = 5$ kΩ, $C_0 = C_1 = 0.5$ pF, and $C_2 = 6$ pF. Use $g_{m0} = g_{m1}$ and $g_{m2} = 6g_{m1}$.

9.36 For the three-stage op amp with nested Miller compensation in Fig. 9.30c, determine the values of the compensation capacitors that give a 45° phase margin when the op amp is in a unity-gain negative feedback loop ($f = 1$). Assume that the zeros due to feedforward have been eliminated. Design for widely spaced real poles. Take $R_0 = R_1 = R_2 = 5$ kΩ, $C_0 = C_1 = 0.5$ pF, and $C_2 = 6$ pF. Use $g_{m0} = g_{m1}$ and $g_{m2} = 6g_{m1}$.

9.37 The single-stage op amp in Fig. 9.54 has a 45° phase margin when the op amp is in a unity-gain negative feedback loop ($f = 1$) with an output load capacitance $C_L = 1$ pF. What value of C_L will give a 60° phase margin? (Assume that the capacitance at the op-amp output is dominated by C_L and the op-amp gain $a_v(s)$ can be modeled as having two poles.)

9.38 The single-stage op amp in Fig. 9.54 has a nondominant pole p_2 with $|p_2| = 200$ Mrad/s. The op amp is in a unity-gain negative feedback loop ($f = 1$).

(a) If $g_{m1} = 0.5$ mA/V, what value of C_L gives a 45° phase margin? (Assume that the capacitance at the op amp output is dominated by C_L and the op-amp gain $a(s)$ can be modeled as having two poles.)

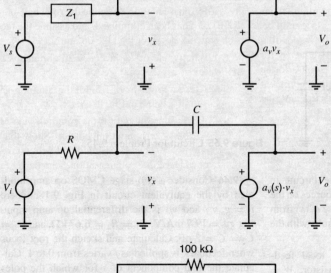

Figure 9.62 Feedback circuit for Problem 9.40.

Figure 9.63 Circuit for Problem 9.41.

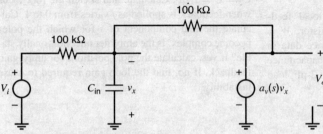

Figure 9.64 Circuit for Problem 9.42.

(b) If $I_{TAIL} = 0.5$ mA, what is the output slew rate with this C_L?

9.39 The feedback circuit in Fig. 9.55 is a switched-capacitor circuit during one clock phase. Assume the op amp is the telescopic-cascode op amp in Fig. 9.54. Take $C_L = 1.5$ pF, $C_I = 4$ pF, $C_S = 0.4$ pF, and $C_{ip} = 0.1$ pF.

(a) If $I_{TAIL} = 0.2$ mA, what is the output slew rate?

(b) Assume that $g_{m1} = 0.1$ mA/V, the loop transmission [loop gain $T(s)$ or return ratio $\mathcal{R}(s)$] can be modeled as having two poles, and the magnitude of the nondominant pole p_2 is $|p_2| = 200$ Mrad/s. What is the phase margin of this feedback circuit?

9.40 Calculate the return ratio for the feedback circuit in Fig. 9.62. Assume that the amplifier voltage gain is constant with $a_v > 0$. Show that this feedback circuit is always stable if each impedance is either a resistor or a capacitor.

9.41 Calculate the return ratio for the integrator in Fig. 9.63. Show that this feedback circuit is stable for all values of R and C if $a_v(s)$ has two left half-plane poles and $a_v(s=0) > 0$.

9.42 Calculate the return ratio for the inverting amplifier in Fig. 9.64. Here, the controlled source and C_{in} form a simple op-amp model. Assume $a_v(s) = 1000/[(1 + s/100)(1 + s/10^6)]$.

(a) Assume the op-amp input capacitance $C_{in} = 0$. What is the frequency at which $|\mathcal{R}(j\omega)| = 1$? How does this frequency compare to the frequency at which $|a_v(j\omega)| = 1$?

(b) Find the phase margin for the cases $C_{in} = 0$, $C_{in} = 4$ pF, and $C_{in} = 20$ pF.

9.43 A technique that allows the return ratio to be simulated using SPICE without disrupting the dc operating point is shown in Fig. 8.60 and explained in Problem 8.33.

(a) Use that technique to simulate the return ratio for the op amp from Problem 9.21 connected in a noninverting unity-gain configuration for $f = 1$ kHz, 100 kHz, 10 MHz, and 1 GHz.

(b) Use that technique to plot the magnitude and phase of the return ratio. Determine the unity-gain frequency for the return ratio and the phase and gain margins. [Note: This calculation requires combining the complex values of $\mathcal{R}'_i(j\omega)$ and $\mathcal{R}'_v(j\omega)$ to find the complex quantity $\mathcal{R}(j\omega)$.]

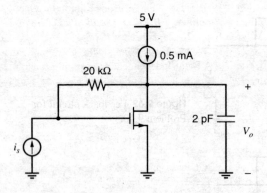

Figure 9.65 Circuit for Problem 9.45.

9.44 Repeat Problem 9.43 for the circuit in Fig. 9.64 with $C_{in} = 4$ pF. Inject the test sources on the left-hand side of the feedback resistor. Use $a_v(s)$ from Problem 9.42. Compare the simulation results with the calculated values from Problem 9.42.

9.45 Repeat Problem 9.43 for the local feedback circuit in Fig. 9.65. For the transistor, $W = 50$ μm and $L_{eff} = 0.6$ μm. Use the device data in Table 2.4. Ignore the drain-body junction capacitance (assuming it is small compared to the 2-pF load capacitor).

9.46 Consider a two-stage CMOS op amp modeled by the equivalent circuit in Fig. 9.18, where $i_s = g_m v_{id}$ and v_{id} is the differential op-amp input. Let $g_m = 19.7$ mA/V, $R_1 = R_2 = 6.67$ kΩ, and $C_1 = C_2 = C = 2$ pF. Calculate and sketch the root locus when feedback is applied as f varies from 0 to 1. Calculate the real component of s for which the poles become complex. Is the amplifier unconditionally stable? If yes, calculate the pole positions for unity-gain feedback. If no, find the loop gain required to cause instability.

REFERENCES

1. K. Ogata. *Modern Control Engineering*, 2nd Edition. Prentice-Hall, Englewood Cliffs, NJ, 1990.

2. P. W. Tuinenga. *SPICE: A Guide to Circuit Simulation and Analysis using PSPICE*, 3rd Edition. Prentice-Hall, Englewood Cliffs, NJ, 1995.

3. G. W. Roberts and A. S. Sedra. *SPICE*, 2nd Edition. Oxford Press, New York, 1997.

4. P. J. Hurst. "Exact Simulation of Feedback Circuit Parameters," *IEEE Trans. on Circuits and Systems*, Vol. CAS-38, No. 11, pp. 1382–1389, November 1991.

5. P. J. Hurst and S.H. Lewis. "Determination of Stability Using Return Ratios in Balanced Fully Differential Feedback Circuits," *IEEE Trans. on Circuits and Systems II*, pp. 805–817, December 1995.

6. S. Rosenstark. *Feedback Amplifier Principles*, MacMillan, New York, 1986.

7. R. D. Middlebrook. "Measurement of Loop Gain in Feedback Systems," *Int. J. Electronics*, Vol. 38, No. 4, pp. 485–512, 1975.

8. J. E. Solomon. "The Monolithic Op Amp: A Tutorial Study," *IEEE J. Solid-State Circuits*, Vol. SC-9, pp. 314–332, December 1974.

9. Y. P. Tsividis and P.R. Gray. "An Integrated NMOS Operational Amplifier with Internal Compensation," *IEEE J. Solid-State Circuits*, Vol. SC-11, pp. 748–753, December 1976.

10. B. K. Ahuja. "An Improved Frequency Compensation Technique for CMOS Operational Amplifiers," *IEEE J. Solid-State Circuits*, Vol. SC-18, pp. 629–633, December 1983.

11. D. B. Ribner and M. A. Copeland. "Design Techniques for Cascoded CMOS Op Amps with Improved PSRR and Common-Mode Input Range," *IEEE J. Solid-State Circuits*, pp. 919–925, December 1984.

12. D. Senderowicz, D. A. Hodges, and P. R. Gray. "A High-Performance NMOS Operational Amplifier," *IEEE J. Solid-State Circuits*, Vol. SC-13, pp. 760–768, December 1978.

13. W. C. Black, D. J. Allstot, and R. A. Reed. "A High Performance Low Power CMOS Channel Filter," *IEEE J. Solid-State Circuits*, Vol. SC-15, pp. 929–938, December 1980.

14. E. M. Cherry. "A New Result in Negative Feedback Theory and Its Application to Audio Power Amplifiers," *Int. J. Circuit Theory*, Vol. 6, pp. 265–288, July 1978.

15. J. H. Huijsing and D. Linebarger. "Low-Voltage Operational Amplifier with Rail-to-Rail Input and Output Ranges," *IEEE J. Solid-State Circuits*, Vol. 20, pp. 1144–1150. December 1985.

16. R. G. H. Eschauzier and J. H. Huijsing. *Frequency Compensation Techniques for Low-Power Operational Amplifiers*. Kluwer, Dordrecht, The Netherlands, 1995.

17. M. J. Fonderie and J. H. Huijsing. *Design of Low-Voltage Bipolar Operational Amplifiers*. Kluwer Academic Publishers, Boston, 1993.

18. F. You, H. K. Embabi, and E. Sanchez-Sinencio. "A Multistage Amplifier Topology with Nested Gm-C Compensation," *IEEE J. Solid-State Circuits*, Vol. 32, pp. 2000–2011, December 1997.

19. P. E. Gray and C. L. Searle. *Electronic Principles: Physics, Models, and Circuits*. Wiley, New York, 1969.

20. J. D'Azzo and C. Houpis. *Linear Control System Analysis and Design: Conventional and Modern*. McGraw-Hill, New York, 1975.

21. W. E. Hearn. "Fast Slewing Monolithic Operational Amplifier," *IEEE J. Solid-State Circuits*, Vol. SC-6, pp. 20–24, February 1971.

22. P. W. Li, M. J. Chin, P. R. Gray, and R. Castello. "A Ratio-Independent Algorithmic Analog-to-Digital Conversion Technique," *IEEE J. Solid-State Circuits*, Vol. SC-19, pp. 828–836, December 1984.

23. E. Seevinck and R. Wassenaar. "A Versatile CMOS Linear Transconductor/Square-Law Function Circuit," *IEEE J. Solid-State Circuits*, Vol. SC-22, pp. 366–377, June 1987.

24. F. N. L. O. Eynde, P. F. M. Ampe, L. Verdeyen, and W. M. C. Sansen. "A CMOS Large-Swing Low-Distortion Three-Stage Class AB Power Amplifier," *IEEE J. Solid-State Circuits*, Vol. SC-25, pp. 265–273, February 1990.

25. H. W. Bode. *Network Analysis and Feedback Amplifier Design*. Van Nostrand, New York, 1945.

26. P. J. Hurst. "A Comparison of Two Approaches to Feedback Circuit Analysis," *IEEE Trans. on Education*, Vol. 35, No. 3, pp. 253–261, August 1992.

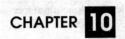

Nonlinear Analog Circuits

10.1 Introduction

Chapters 1 through 9 dealt almost entirely with analog circuits whose primary function is linear amplification of signals. Although some of the circuits discussed (such as Class AB output stages) were actually nonlinear in their operation, the operations performed on the signal passing through the amplifier were well approximated by linear relations.

Nonlinear operations on continuous-valued analog signals are often required in instrumentation, communication, and control-system design. These operations include modulation, demodulation, frequency translation, multiplication, and division. In this chapter, we analyze the most commonly used techniques for performing these operations within a monolithic integrated circuit. We first discuss the use of the bipolar transistor to synthesize nonlinear analog circuits and analyze the Gilbert multiplier cell, which is the basis for a wide variety of such circuits. Next we consider the application of this building block as a small-signal analog multiplier, as a modulator, as a phase comparator, and as a large-signal, four-quadrant multiplier.

Following the multiplier discussion, we introduce a highly useful circuit technique for performing demodulation of FM and AM signals and, at the same time, performing bandpass filtering. This circuit, the phase-locked-loop (PLL), is particularly well-suited to monolithic construction. After exploring the basic concepts involved, we analyze the behavior of the PLL in the locked condition, and then consider the capture transient. Finally, some methods of realizing arbitrary nonlinear transfer functions using bipolar transistors are considered.

10.2 Analog Multipliers Employing the Bipolar Transistor

In analog-signal processing the need often arises for a circuit that takes two analog inputs and produces an output proportional to their product. Such circuits are termed *analog multipliers*. In the following sections we examine several analog multipliers that depend on the exponential transfer function of bipolar transistors.

10.2.1 The Emitter-Coupled Pair as a Simple Multiplier

The emitter-coupled pair, shown in Fig. 10.1, was shown in Chapter 3 to produce output currents that are related to the differential input voltage by

$$I_{c1} = \frac{I_{EE}}{1 + \exp\left(-\dfrac{V_{id}}{V_T}\right)} \qquad (10.1)$$

$$I_{c2} = \frac{I_{EE}}{1 + \exp\left(\dfrac{V_{id}}{V_T}\right)} \qquad (10.2)$$

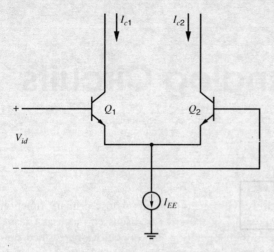

Figure 10.1 Emitter-coupled pair.

where base current has been neglected. Equations 10.1 and 10.2 can be combined to give the *difference* between the two output currents:

$$\Delta I_c = I_{c1} - I_{c2} = I_{EE} \tanh\left(\frac{V_{id}}{2V_T}\right) \quad (10.3)$$

This relationship is plotted in Fig. 10.2 and shows that the emitter-coupled pair by itself can be used as a primitive multiplier. We first assume that the differential input voltage V_{id} is much less than V_T. If this is true, we can utilize the approximation

$$\tanh\frac{V_{id}}{2V_T} \approx \frac{V_{id}}{2V_T} \qquad \frac{V_{id}}{2V_T} \ll 1 \quad (10.4)$$

And (10.3) becomes

$$\Delta I_c \approx I_{EE}\left(\frac{V_{id}}{2V_T}\right) \quad (10.5)$$

The current I_{EE} is actually the bias current for the emitter-coupled pair. With the addition of more circuitry, we can make I_{EE} proportional to a second input signal V_{i2}, as shown in

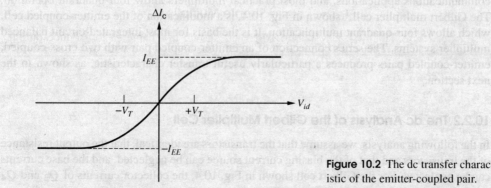

Figure 10.2 The dc transfer characteristic of the emitter-coupled pair.

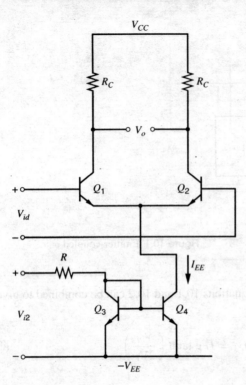

Figure 10.3 Two-quadrant analog multiplier.

Fig. 10.3. Thus we have

$$I_{EE} \approx K_o(V_{i2} - V_{BE(on)}) \tag{10.6}$$

The differential output current of the emitter-coupled pair can be calculated by substituting (10.6) in (10.5) to give

$$\Delta I_c = \frac{K_o V_{id}(V_{i2} - V_{BE(on)})}{2V_T} \tag{10.7}$$

Thus we have produced a circuit that functions as a multiplier under the assumption that V_{id} is small, and that V_{i2} is greater than $V_{BE(on)}$. The latter restriction means that the multiplier functions in only two quadrants of the V_{id}–V_{i2} plane, and this type of circuit is termed a two-quadrant multiplier. The restriction to two quadrants of operation is a severe one for many communications applications, and most practical multipliers allow four-quadrant operation. The Gilbert multiplier cell,[1] shown in Fig. 10.4, is a modification of the emitter-coupled cell, which allows four-quadrant multiplication. It is the basis for most integrated-circuit balanced multiplier systems. The series connection of an emitter-coupled pair with two cross-coupled, emitter-coupled pairs produces a particularly useful transfer characteristic, as shown in the next section.

10.2.2 The dc Analysis of the Gilbert Multiplier Cell

In the following analysis, we assume that the transistors are identical, that the output resistance of the transistors and that of the biasing current source can be neglected, and the base currents can be neglected. For the Gilbert cell shown in Fig. 10.4, the collector currents of Q_3 and Q_4

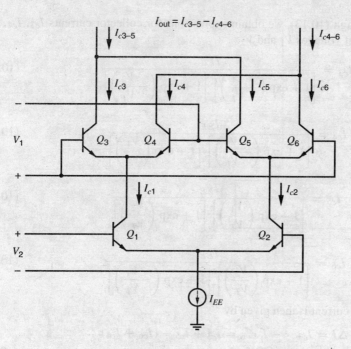

Figure 10.4 Gilbert multiplier circuit.

are, using (10.1) and (10.2),

$$I_{c3} = \frac{I_{c1}}{1 + \exp\left(-\dfrac{V_1}{V_T}\right)} \tag{10.8}$$

$$I_{c4} = \frac{I_{c1}}{1 + \exp\left(\dfrac{V_1}{V_T}\right)} \tag{10.9}$$

Similarly, the collector currents of Q_5 and Q_6 are given by

$$I_{c5} = \frac{I_{c2}}{1 + \exp\left(\dfrac{V_1}{V_T}\right)} \tag{10.10}$$

$$I_{c6} = \frac{I_{c2}}{1 + \exp\left(-\dfrac{V_1}{V_T}\right)} \tag{10.11}$$

The two currents I_{c1} and I_{c2} can be related to V_2 by again using (10.1) and (10.2)

$$I_{c1} = \frac{I_{EE}}{1 + \exp\left(-\dfrac{V_2}{V_T}\right)} \tag{10.12}$$

$$I_{c2} = \frac{I_{EE}}{1 + \exp\left(\dfrac{V_2}{V_T}\right)} \tag{10.13}$$

Combining (10.8) through (10.13), we obtain expressions for collector currents I_{c3}, I_{c4}, I_{c5}, and I_{c6} in terms of input voltages V_1 and V_2.

$$I_{c3} = \frac{I_{EE}}{\left[1 + \exp\left(-\frac{V_1}{V_T}\right)\right]\left[1 + \exp\left(-\frac{V_2}{V_T}\right)\right]} \tag{10.14}$$

$$I_{c4} = \frac{I_{EE}}{\left[1 + \exp\left(-\frac{V_2}{V_T}\right)\right]\left[1 + \exp\left(\frac{V_1}{V_T}\right)\right]} \tag{10.15}$$

$$I_{c5} = \frac{I_{EE}}{\left[1 + \exp\left(\frac{V_1}{V_T}\right)\right]\left[1 + \exp\left(\frac{V_2}{V_T}\right)\right]} \tag{10.16}$$

$$I_{c6} = \frac{I_{EE}}{\left[1 + \exp\left(\frac{V_2}{V_T}\right)\right]\left[1 + \exp\left(-\frac{V_1}{V_T}\right)\right]} \tag{10.17}$$

The differential output current is then given by

$$\Delta I = I_{c3-5} - I_{c4-6} = I_{c3} + I_{c5} - (I_{c6} + I_{c4})$$
$$= (I_{c3} - I_{c6}) - (I_{c4} - I_{c5}) \tag{10.18}$$

$$= I_{EE} \left[\tanh\left(\frac{V_1}{2V_T}\right)\right]\left[\tanh\left(\frac{V_2}{2V_T}\right)\right] \tag{10.19}$$

The dc transfer characteristic, then, is the product of the hyperbolic tangent of the two input voltages.

Practical applications of the multiplier cell can be divided into three categories according to the magnitude relative to V_T of applied signals V_1 and V_2. If the magnitude of V_1 and V_2 are kept small with respect to V_T, the hyperbolic tangent function can be approximated as linear and the circuit behaves as a multiplier, developing the product of V_1 and V_2. However, by including nonlinearity to compensate for the hyperbolic tangent function in series with each input, the range of input voltages over which linearity is maintained can be greatly extended. This technique is used in so-called four-quadrant analog multipliers.

The second class of applications is distinguished by the application to one of the inputs of a signal that is large compared to V_T, causing the transistors to which that signal is applied to behave like switches rather than near-linear devices. This effectively multiplies the applied small signal by a square wave, and in this mode of operation the circuit acts as a modulator.

In the third class of applications, the signals applied to both inputs are large compared to V_T, and all six transistors in the circuit behave as nonsaturating switches. This mode of operation is useful for the detection of phase differences between two amplitude-limited signals, as is required in phase-locked loops, and is sometimes called the phase-detector mode.

We first consider the application of the circuit as an analog multiplier of two continuous signals.

10.2.3 The Gilbert Cell as an Analog Multiplier

As mentioned earlier, the hyperbolic tangent function may be represented by the infinite series:

$$\tanh x = x - \frac{x^3}{3} \cdots \tag{10.20}$$

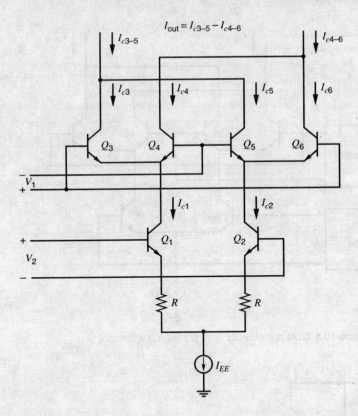

Figure 10.5 Gilbert multiplier with emitter degeneration applied to improve input voltage range on V_2 input.

Assuming that x is much less than one, the hyperbolic tangent can then be approximated by

$$\tanh x \approx x \tag{10.21}$$

Applying this relation to (10.19), we have

$$\Delta I \approx I_{EE} \left(\frac{V_1}{2V_T}\right)\left(\frac{V_2}{2V_T}\right) \qquad V_1, V_2 \ll V_T \tag{10.22}$$

Thus for small-amplitude signals, the circuit performs an analog multiplication. Unfortunately, the amplitudes of the input signals are often much larger than V_T, but larger signals can be accommodated in this mode in a number of ways. In the event that only one of the signals is large compared to V_T, emitter degeneration can be utilized in the lower emitter-coupled pair, increasing the linear input range for V_2 as shown in Fig. 10.5. Unfortunately, this cannot be done with the cross-coupled pairs Q_3-Q_6 because the degeneration resistors destroy the required nonlinear relation between I_c and V_{be} in those devices.

An alternate approach is to introduce a nonlinearity that predistorts the input signals to compensate for the hyperbolic tangent transfer characteristic of the basic cell. The required nonlinearity is an inverse hyperbolic tangent characteristic, and a hypothetical example of such a system is shown in Fig. 10.6. Fortunately, this particular nonlinearity is straightforward to generate.

Referring to Fig. 10.7, we assume for the time being that the circuitry within the box develops a differential output current that is linearly related to the input voltage V_1. Thus

$$I_1 = I_{o1} + K_1 V_1 \tag{10.23}$$

$$I_2 = I_{o1} - K_1 V_1 \tag{10.24}$$

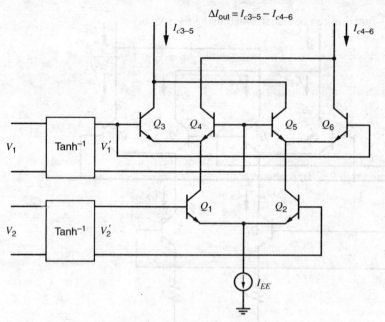

Figure 10.6 Gilbert multiplier with predistortion circuits.

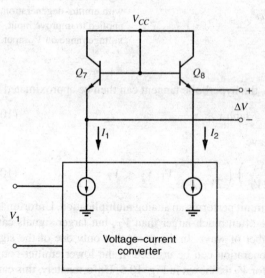

Figure 10.7 Inverse hyperbolic tangent circuit.

Here I_{o1} is the dc current that flows in each output lead if V_1 is equal to zero, and K_1 is the transconductance of the voltage-to-current converter. The differential voltage developed across the two diode-connected transistors is

$$\Delta V = V_T \ln\left(\frac{I_{o1} + K_1 V_1}{I_S}\right) - V_T \ln\left(\frac{I_{o1} - K_1 V_1}{I_S}\right)$$

$$= V_T \ln\left(\frac{I_{o1} + K_1 V_1}{I_{o1} - K_1 V_1}\right) \tag{10.25}$$

This function can be transformed using the identity

$$\tanh^{-1} x = \frac{1}{2} \ln\left(\frac{1+x}{1-x}\right) \quad (10.26)$$

into the desired relationship.

$$\Delta V = 2V_T \tanh^{-1}\left(\frac{K_1 V_1}{I_{o1}}\right) \quad (10.27)$$

Thus if this functional block is used as the compensating nonlinearity in series with each input as shown in Fig. 10.6, the overall transfer characteristic becomes, using (10.19),

$$\Delta I = I_{EE}\left(\frac{K_1 V_1}{I_{o1}}\right)\left(\frac{K_2 V_2}{I_{o2}}\right) \quad (10.28)$$

where I_{o2} and K_2 are the parameters of the functional block following V_2.

Equation 10.28 shows that the differential output current is directly proportional to the product $V_1 V_2$, and, in principle, this relationship holds for all values of V_1 and V_2 for which the two output currents of the differential voltage-to-current converters are positive. For this to be true, I_1 and I_2 must always be positive, and from (10.23) and (10.24), we have

$$-\frac{I_{o1}}{K_1} < V_1 < \frac{I_{o1}}{K_1} \quad (10.29)$$

$$-\frac{I_{o2}}{K_2} < V_2 < \frac{I_{o2}}{K_2} \quad (10.30)$$

Note that the inclusion of a compensating nonlinearity on the V_2 input simply makes the collector currents of Q_1 and Q_2 directly proportional to input voltage V_2 rather than to its hyperbolic tangent. Thus the combination of the pair Q_1-Q_2 and the compensating nonlinearity on the V_2 input is redundant, and the output currents of the voltage-to-current converter on the V_2 input can be fed directly into the emitters of the Q_3-Q_4 and Q_5-Q_6 pairs with exactly the same results. The multiplier then takes on the form shown in Fig. 10.8.

10.2.4 A Complete Analog Multiplier[2]

In order to be useful in a wide variety of applications, the multiplier circuit must develop an output voltage that is referenced to ground and can take on both positive and negative values. The transistors Q_3, Q_4, Q_5, Q_6, Q_7, and Q_8, shown in Fig. 10.8, are referred to as the multiplier core and produce a differential current output that then must be amplified, converted to a single-ended signal, and referenced to ground. An output amplifier is thus required, and the complete multiplier consists of two voltage-current converters, the core transistors, and an output current-to-voltage amplifier. While the core configuration of Fig. 10.8 is common to most four-quadrant transconductance multipliers, the rest of the circuitry can be realized in a variety of ways.

The most common configurations used for the voltage-current converters are emitter-coupled pairs with emitter degeneration as shown in Fig. 10.5. The differential-to-single-ended converter of Fig. 10.8 is often realized with an op amp circuit of the type shown in Fig. 6.4. If this circuit has a transresistance given by

$$\frac{V_{\text{out}}}{\Delta I} = K_3 \quad (10.31)$$

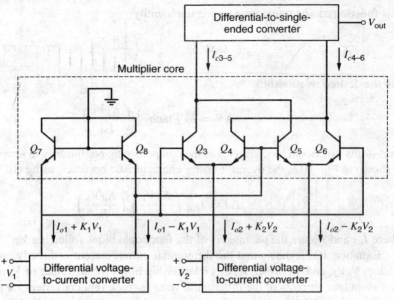

Figure 10.8 Complete four-quadrant multiplier.

then substitution in (10.28) gives for the overall multiplier characteristic

$$V_{\text{out}} = I_{EE} K_3 \frac{K_1}{I_{o1}} \frac{K_2}{I_{o2}} V_1 V_2 \tag{10.32}$$

The output voltage is thus proportional to the product $V_1 V_2$ over a wide range. The constants in (10.32) are usually chosen so that

$$V_{\text{out}} = 0.1 V_1 V_2 \tag{10.33}$$

and all voltages have a ± 10-V range.

10.2.5 The Gilbert Multiplier Cell as a Balanced Modulator and Phase Detector

The four-quadrant multiplier just described is an example of an application of the multiplier cell in which all the devices remain in the active region during normal operation. Used in this way the circuit is capable of performing precise multiplication of one continuously varying analog signal by another. In communications systems, however, the need frequently arises for the multiplication of a continuously varying signal by a square wave. This is easily accomplished with the multiplier circuit by applying a sufficiently large signal (i.e., large compared to $2V_T$) directly to the cross-coupled pair so that two of the four transistors alternately turn completely off and the other two conduct all the current. Since the transistors in the circuit do not enter saturation, this process can be accomplished at high speed. A set of typical waveforms that might result when a sinusoid is applied to the small-signal input and a square wave to the large-signal input is shown in Fig. 10.9. Note that since the devices in the multiplier are being switched on and off by the incoming square wave, the amplitude of the output waveform is independent of the amplitude of the square wave as long as it is large enough to cause the devices in the multiplier circuit to be fully on or fully off. Thus the circuit in this mode does not perform a linear multiplication of two waveforms, but actually causes the output voltage of the circuit produced by the small-signal input to be alternately multiplied by $+1$ and -1.

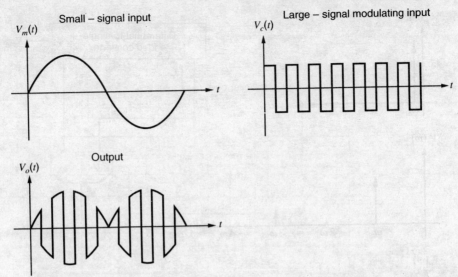

Figure 10.9 Input and output waveforms for a phase detector with large input signals.

The spectrum of the output may be developed directly from the Fourier series of the two inputs. For the low-frequency modulating sinusoidal input,

$$V_m(t) = V_m \cos \omega_m t \qquad (10.34)$$

and for the high-frequency square wave input, which we assume has an amplitude of ± 1 as discussed above,

$$V_c(t) = \sum_{n=1}^{\infty} A_n \cos n\omega_c t, \qquad A_n = \frac{\sin \frac{n\pi}{2}}{\frac{n\pi}{4}} \qquad (10.35)$$

Thus the output signal is

$$V_o(t) = K[V_c(t) V_m(t)] = K \sum_{n=1}^{\infty} A_n V_m \cos \omega_n t \cos n\omega_c t \qquad (10.36)$$

$$= K \sum_{n=1}^{\infty} \frac{A_n V_m}{2} [\cos(n\omega_c + \omega_m)t + \cos(n\omega_c - \omega_m)t] \qquad (10.37)$$

where K is the magnitude of the gain of the multiplier from the small-signal input to the output.

The spectrum has components located at frequencies ω_m above and below each of the harmonics of ω_c, but no component at the carrier frequency ω_c or its harmonics. The spectrum of the input signals and the resulting output signal is shown in Fig. 10.10. The lack of an output component at the carrier frequency is a very useful property of balanced modulators. The signal is usually filtered following the modulation process so that only the components near ω_c are retained.

If a dc component is added to the modulating input, the result is a signal component in the output at the carrier frequency and its harmonics. If the modulating signal is given by

$$V_m(t) = V_m(1 + M \cos \omega_m t) \qquad (10.38)$$

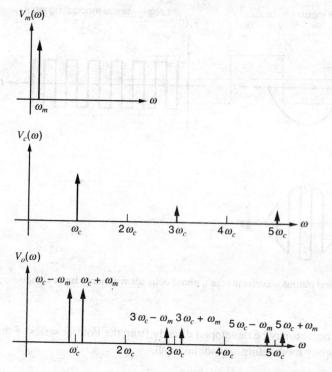

Figure 10.10 Input and output spectra for a balanced modulator.

where the parameter M is called the modulation index, then the output is given by

$$V_o(t) = K \sum_{n=1}^{\infty} A_n V_m \left[\cos(n\omega_c t) + \frac{M}{2} \cos(n\omega_c + \omega_m)t + \frac{M}{2} \cos(n\omega_c - \omega_m)t \right] \quad (10.39)$$

This dc component can be introduced intentionally to provide conventional amplitude modulation or it can be the result of offset voltages in the devices within the modulator, which results in undesired carrier feedthrough in suppressed-carrier modulators.

Note that the balanced modulator actually performs a frequency translation. Information contained in the modulating signal $V_m(t)$ was originally concentrated at the modulating frequency ω_m. The modulator has translated this information so that it is now contained in spectral components located near the harmonics of the high-frequency signal $V_c(t)$, usually called the carrier. Balanced modulators are also useful for performing demodulation, which is the extraction of information from the frequency band near the carrier and retranslation of the information back down to low frequencies.

In frequency translation, signals at two different frequencies are applied to the two inputs, and the sum or the difference frequency component is taken from the output. If unmodulated signals of identical frequency ω_o are applied to the two inputs, the circuit behaves as a *phase detector* and produces an output whose dc component is proportional to the phase difference between the two inputs. For example, consider the two input waveforms in Fig. 10.11, which are applied to the Gilbert multiplier shown in the same figure. We assume first for simplicity that both inputs are large in magnitude so that all the transistors in the circuit are behaving as switches. The output waveform that results is shown in Fig. 10.11c and consists of a dc component and a component at twice the incoming frequency. The dc component of this

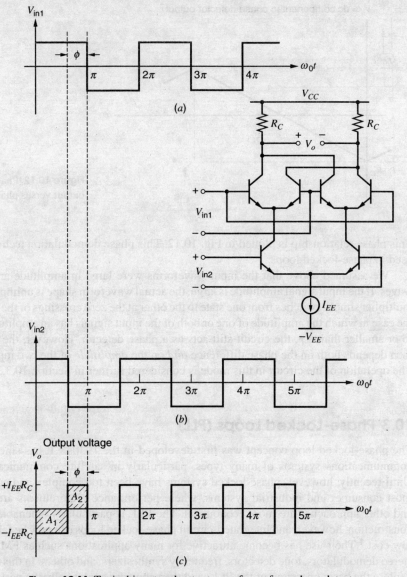

Figure 10.11 Typical input and output waveforms for a phase detector.

waveform is given by

$$V_{\text{average}} = \frac{1}{2\pi} \int_0^{2\pi} V_o(t) d(\omega_0 t) \qquad (10.40)$$

$$= \frac{-1}{\pi}(A_1 - A_2) \qquad (10.41)$$

where areas A_1 and A_2 are as indicated in Fig. 10.11c. Thus,

$$V_{\text{average}} = -\left[I_{EE} R_C \frac{(\pi - \phi)}{\pi} - \frac{I_{EE} R_C \phi}{\pi}\right] \qquad (10.42)$$

$$= I_{EE} R_C \left(\frac{2\phi}{\pi} - 1\right) \qquad (10.43)$$

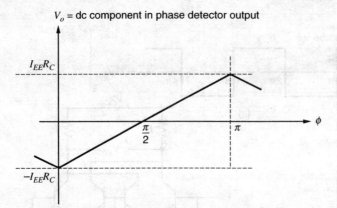

Figure 10.12 Phase detector output versus phase difference.

This phase relationship is plotted in Fig. 10.12. This phase demodulation technique is widely used in phase-locked loops.

We assumed above that the input waveforms were large in amplitude and were square waves. If the input signal amplitude is large, the actual waveform shape is unimportant since the multiplier simply switches from one state to the other at the zero crossings of the waveform. For the case in which the amplitude of one or both of the input signals has an amplitude comparable to or smaller than V_T, the circuit still acts as a phase detector. However, the output voltage then depends both on the phase difference *and* on the *amplitude* of the two input waveforms. The operation of the circuit in this mode is considered further in Section 10.3.3.

10.3 Phase-Locked Loops (PLL)

The phase-locked loop concept was first developed in the 1930s.[3] It has since been used in communications systems of many types, particularly in satellite communications systems. Until recently, however, phase-locked systems have been too complex and costly for use in most consumer and industrial systems, where performance requirements are more modest and other approaches are more economical. The PLL is particularly amenable to monolithic construction, however, and integrated-circuit phase-locked loops can now be fabricated at very low cost.[4] Their use has become attractive for many applications such as FM demodulators, stereo demodulators, tone detectors, frequency synthesizers, and others. In this section we first explore the basic operation of the PLL, and then consider analytically the performance of the loop in the locked condition. We then discuss the design of monolithic PLLs.

10.3.1 Phase-Locked Loop Concepts

A block diagram of the basic phase-locked loop system is shown in Fig. 10.13. The elements of the system are a phase comparator, a loop filter, an amplifier, and a voltage-controlled oscillator. The voltage-controlled oscillator, or VCO, is simply an oscillator whose frequency is proportional to an externally applied voltage. When the loop is locked on an incoming periodic signal, the VCO frequency is exactly equal to that of the incoming signal. The phase detector produces a dc or low-frequency signal proportional to the phase difference between the incoming signal and the VCO output signal. This phase-sensitive signal is then passed through the loop filter and amplifier and is applied to the control input of the VCO. If, for example, the frequency of the incoming signal shifts slightly, the phase difference between

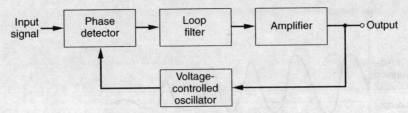

Figure 10.13 Phase-locked-loop system.

the VCO signal and the incoming signal will begin to increase with time. This will change the control voltage on the VCO in such a way as to bring the VCO frequency back to the same value as the incoming signal. Thus the loop can maintain lock when the input signal frequency changes, and the VCO input voltage is proportional to the frequency of the incoming signal. This behavior makes PLLs particularly useful for the demodulation of FM signals, where the frequency of the incoming signal varies in time and contains the desired information. The range of input signal frequencies over which the loop can maintain lock is called the *lock range*.

An important aspect of PLL performance is the capture process, by which the loop goes from the unlocked, free-running condition to that of being locked on a signal. In the unlocked condition, the VCO runs at the frequency corresponding to zero applied dc voltage at its control input. This frequency is called the center frequency, or free-running frequency. When a periodic signal is applied that has a frequency near the free-running frequency, the loop may or may not lock on it depending on a number of factors. The capture process is inherently nonlinear in nature, and we will describe the transient in only a qualitative way.

First assume that the loop is opened between the loop filter and the VCO control input, and that a signal whose frequency is near, but not equal to, the free-running frequency is applied to the input of the PLL. The phase detector is usually of the type discussed in the last section, but for this qualitative discussion we assume that the phase detector is simply an analog multiplier that multiplies the two sinusoids together. Thus the output of the multiplier-phase detector contains the sum and difference frequency components, and we assume that the sum frequency component is sufficiently high in frequency that it is filtered out by the low-pass filter. The output of the low-pass filter, then, is a sinusoid with a frequency equal to the difference between the VCO free-running frequency and the incoming signal frequency.

Now assume that the loop is suddenly closed, and the difference frequency sinusoid is now applied to the VCO input. This will cause the VCO frequency itself to become a sinusoidal function of time. Let us assume that the incoming frequency was lower than the free-running frequency. Since the VCO frequency is varying as a function of time, it will alternately move *closer* to the incoming signal frequency and *farther away* from the incoming signal frequency. The output of the phase detector is a near-sinusoid whose frequency is the *difference* between the VCO frequency and the input frequency. When the VCO frequency moves away from the incoming frequency, this sinusoid moves to a higher frequency. When the VCO frequency moves closer to the incoming frequency, the sinusoid moves to a lower frequency. If we examine the effect of this on the phase detector output, we see that the *frequency* of this sinusoidal difference-frequency waveform is reduced when its incremental amplitude is negative, and increased when its amplitude is positive. This causes the phase detector output to have an asymmetrical waveform during capture, as shown in Fig. 10.14. This asymmetry in the waveform introduces a dc component in the phase detector output that shifts the average VCO frequency toward the incoming signal frequency, so that the difference frequency gradually decreases. Once the system becomes locked, of course, the difference frequency becomes zero and only a dc voltage remains at the loop-filter output.

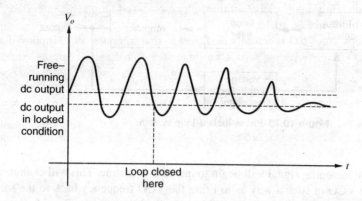

Figure 10.14 Typical phase detector output during capture transient.

The capture range of the loop is that range of input frequencies around the center frequency onto which the loop will become locked from an unlocked condition. The pull-in time is the time required for the loop to capture the signal. Both these parameters depend on the amount of gain in the loop itself, and the bandwidth of the loop filter. The objective of the loop filter is to filter out difference components resulting from interfering signals far away from the center frequency. It also provides a memory for the loop in case lock is momentarily lost due to a large interfering transient. Reducing the loop filter bandwidth thus improves the rejection of out-of-band signals, but, at the same time, the capture range is decreased, the pull-in time becomes longer, and the loop phase margin becomes poorer.

10.3.2 The Phase-Locked Loop in the Locked Condition

Under locked conditions, a linear relationship exists between the output voltage of the phase detector and the phase difference between the VCO and the incoming signal. This fact allows the loop to be analyzed using standard linear feedback concepts when in the locked condition. A block diagram representation of the system in this mode is shown in Fig. 10.15. The gain of the phase comparator is K_D V/rad of phase difference, the loop-filter transfer function is $F(s)$, and any gain in the forward loop is represented by A. The VCO gain is K_o rad/s per volt.

If a constant input voltage is applied to the VCO control input, the output frequency of the VCO remains constant. However, the phase comparator is sensitive to the difference between the *phase* of the VCO output and the *phase* of the incoming signal. The phase of the VCO output is actually equal to the time integral of the VCO output frequency, since

$$\omega_{osc}(t) = \frac{d\phi_{osc}(t)}{dt} \tag{10.44}$$

and thus

$$\phi_{osc}(t) = \phi_{osc}|_{t=0} + \int_0^t \omega_{osc}(t)dt \tag{10.45}$$

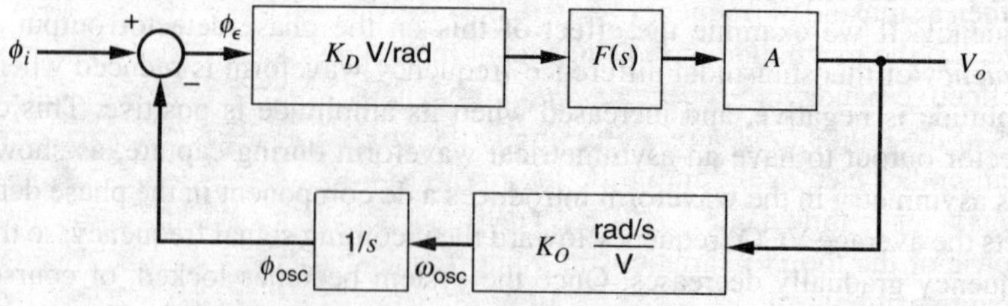

Figure 10.15 Block diagram of the PLL system.

Thus an integration inherently takes place within the phase-locked loop. This integration is represented by the $1/s$ block in Fig. 10.15.

For practical reasons, the VCO is actually designed so that when the VCO input voltage (i.e., V_o) is zero, the VCO frequency is not zero. The relation between the VCO output frequency ω_{osc} and V_o is actually

$$\omega_{\text{osc}} = \omega_o + K_O V_o$$

where ω_o is the free-running frequency that results when $V_o = 0$.

The system can be seen from Fig. 10.15 to be a classical linear feedback control system.[5] The closed-loop transfer function is given by

$$\frac{V_o}{\phi_i} = \frac{K_D F(s) A}{1 + K_D F(s) A \dfrac{K_o}{s}} \tag{10.46}$$

$$= \frac{s K_D F(s) A}{s + K_D K_O A F(s)} \tag{10.47}$$

Usually we are interested in the response of this loop to *frequency* variations at the input, so that the input variable is frequency rather than phase. Since

$$\omega_i = \frac{d\phi_i}{dt} \tag{10.48}$$

then

$$\omega_i(s) = s\phi_i(s) \tag{10.49}$$

and

$$\frac{V_o}{\omega_i} = \frac{1}{s} \frac{V_o}{\phi_i} = \frac{K_D F(s) A}{s + K_D K_O A F(s)} \tag{10.50}$$

We first consider the case in which the loop filter is removed entirely, and $F(s)$ is unity. This is called a first-order loop and we have

$$\frac{V_o}{\omega_i} = \left(\frac{K_v}{s + K_v}\right)\left(\frac{1}{K_O}\right) \tag{10.51}$$

where

$$K_v = K_O K_D A \tag{10.52}$$

Thus the loop inherently produces a first-order, low-pass transfer characteristic. Remember that we regard the input variable as the frequency ω_i of the incoming signal. The response calculated above, then, is really the response from the frequency modulation on the incoming carrier to the loop voltage output.

The constant above (K_v) is termed the loop bandwidth. If the loop is locked on a carrier signal, and the frequency of that carrier is made to vary sinusoidally in time with a frequency ω_m, then a sinusoid of frequency ω_m will be observed at the loop output. When ω_m is increased above K_v, the magnitude of the sinusoid at the output falls. The loop bandwidth K_v, then, is the effective bandwidth for the *modulating* signal that is being demodulated by the PLL. In terms of the loop parameters, K_v is simply the product of the phase detector gain, VCO gain, and any other electrical gain in the loop. The root locus of this single pole as a function of loop gain K_v is shown in Fig. 10.16a. The frequency response is also shown in this figure. The response of the loop to variations in input frequency is illustrated in Fig. 10.16b and by the following example.

720 Chapter 10 ▪ Nonlinear Analog Circuits

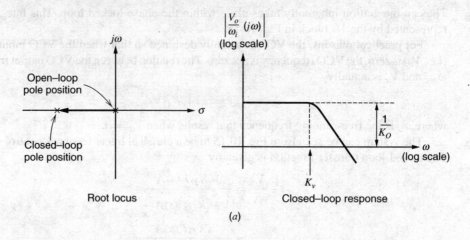

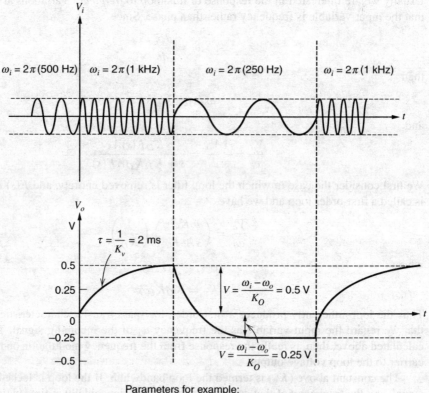

Figure 10.16 (a) Root locus and frequency response of a first-order, phase-locked loop. (b) Response of the loop output voltage to step changes in input frequency, example first-order loop.

EXAMPLE

A PLL has a K_O of 2π (1 kHz/V), a K_v of 500 s^{-1}, and a free-running frequency of 500 Hz.

(a) For a constant input signal frequency of 250 Hz and 1 kHz, find V_o.

$$V_o = \frac{\omega_i - \omega_o}{K_O}$$

where

ω_o = oscillator free-running frequency

At 250 Hz

$$V_o = \frac{2\pi \, (250 \text{ Hz}) - 2\pi \, (500 \text{ Hz})}{2\pi \, (1 \text{ kHz/V})} = -0.25 \text{V}$$

At 1 kHz

$$V_o = \frac{2\pi \, (1 \text{ kHz}) - 2\pi \, (500 \text{ Hz})}{2\pi \, (1 \text{ kHz/V})} = +0.5 \text{V}$$

(b) Now the input signal is frequency modulated, so that

$$\omega_i(t) = (2\pi) \, 500 \text{ Hz} \, [1 + 0.1 \sin(2\pi \times 10^2) \, t]$$

Find the output signal $V_o(t)$. From (10.51) we have

$$\frac{V_o(j\omega)}{\omega_i(j\omega)} = \frac{1}{K_O}\left(\frac{K_v}{K_v + j\omega}\right) = \frac{1}{K_O}\left[\frac{K_v}{K_v + j(2\pi \times 10^2)}\right]$$

$$= \frac{1}{2\pi \, (1 \text{ kHz/V})}\left(\frac{500}{500 + j628}\right)$$

$$= \frac{1}{2\pi \, (1 \text{ kHz/V})}(0.39 - j0.48)$$

The *magnitude* of $\omega_i(j\omega)$ is

$$|\omega_i(j\omega)| = (0.1)(500 \text{ Hz})(2\pi) = (50)(2\pi)$$

Therefore

$$V_o(j\omega) = \frac{50 \text{ Hz}}{1 \text{ kHz}}(0.39 - j0.48) = \frac{50}{1000}(0.62\angle{-51°})$$

and

$$V_o(t) = 0.031 \sin[(2\pi \times 10^2 t) - 51°]$$

Operating the loop with no loop filter has several practical drawbacks. Since the phase detector is really a multiplier, it produces a sum frequency component at its output as well as the difference frequency component. This component at twice the carrier frequency will be fed directly to the output if there is no loop filter. Also, all the out-of-band interfering signals present at the input will appear, shifted in frequency, at the output. Thus, a loop filter is very desirable in applications where interfering signals are present.

The most common configuration for integrated circuit PLLs is the second-order loop. Here, loop filter $F(s)$ is simply a single-pole, low-pass filter, usually realized with a single

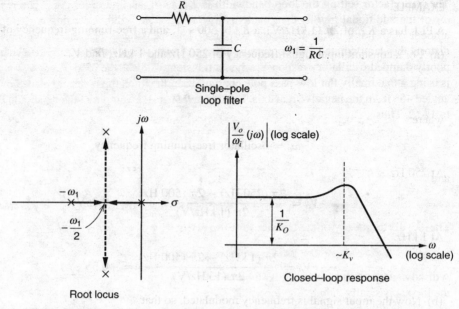

Figure 10.17 Root locus and frequency response of second-order, phase-locked loop.

resistor and capacitor. Thus

$$F(s) = \left(\frac{1}{1 + \frac{s}{\omega_1}} \right) \tag{10.53}$$

By substituting into (10.50), the transfer function becomes

$$\frac{V_o}{\omega_1}(s) = \frac{1}{K_O} \left(\frac{1}{1 + \frac{s}{K_v} + \frac{s^2}{\omega_1 K_v}} \right) \tag{10.54}$$

The root locus for this feedback system as K_v varies is shown in Fig. 10.17, along with the corresponding frequency response. The roots of the transfer function are

$$s = -\frac{\omega_1}{2}\left(1 \pm \sqrt{1 - \frac{4K_v}{\omega_1}}\right) \tag{10.55}$$

Equation 10.54 can be expressed as

$$\frac{V_o}{\omega_i} = \frac{1}{K_O}\left(\frac{1}{\frac{s^2}{\omega_n^2} + \frac{2\zeta}{\omega_n}s + 1}\right) \tag{10.56}$$

where

$$\omega_n = \sqrt{K_v \omega_1} \tag{10.57}$$

$$\zeta = \frac{1}{2}\sqrt{\frac{\omega_1}{K_v}} \tag{10.58}$$

The basic factor setting the loop bandwidth is K_v as in the first-order case. The magnitude ω_1 of the additional pole is then made as low as possible without causing an unacceptable amount of peaking in the frequency response. This peaking is of concern both because it distorts the demodulated FM output and because it causes the loop to ring, or experience a poorly damped oscillatory response, when a transient disturbs the loop. A good compromise is using a maximally flat low-pass pole configuration in which the poles are placed on radials angled 45° from the negative real axis. For this response, the damping factor ζ should be equal to $1/\sqrt{2}$. Thus

$$\frac{1}{\sqrt{2}} = \frac{1}{2}\sqrt{\frac{\omega_1}{K_v}} \tag{10.59}$$

and

$$\omega_1 = 2K_v \tag{10.60}$$

The -3-dB frequency of the transfer function $(V_o/\omega_i)(j\omega)$ is then

$$\omega_{-3dB} = \omega_n = \sqrt{K_v \omega_1} = \sqrt{2} K_v \tag{10.61}$$

A disadvantage of the second-order loop as discussed thus far is that the -3-dB bandwidth of the loop is basically dictated by loop gain K_v as shown by (10.61). As we will show, the loop gain also sets the lock range, so that with the simple filter used above these two parameters are constrained to be comparable. Situations do arise in phase-locked communications in which a wide lock range is desired for tracking large signal-frequency variations, yet a narrow loop bandwidth is desired for rejecting out-of-band signals. Using a very small ω_1 would accomplish this were it not for the fact that this produces underdamped loop response. By adding a zero to the loop filter, the loop filter pole can be made small while still maintaining good loop dampening.

The effect of the addition of a zero on the loop response is best seen by examining the open-loop response of the circuit. Shown in Fig. 10.18a is the open-loop response of the circuit with no loop filter. Because of the integration inherent in the loop, the response has a -20-dB/decade slope throughout the frequency range and crosses unity gain at K_v. In Fig. 10.18b, a loop filter in which ω_1 is much less than K_v has been added, and, as a result, the loop

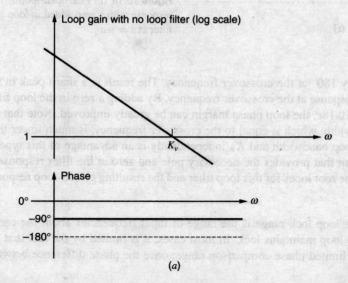

Figure 10.18 (a) PLL open-loop response with no loop filter.

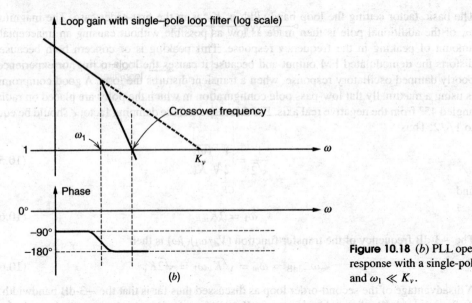

Figure 10.18 (b) PLL open-loop response with a single-pole filter and $\omega_1 \ll K_v$.

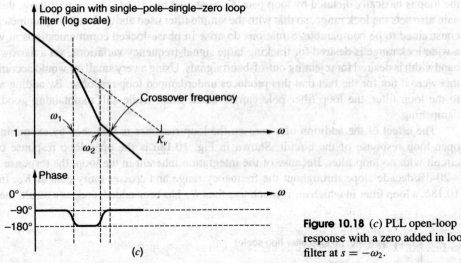

Figure 10.18 (c) PLL open-loop response with a zero added in loop filter at $s = -\omega_2$.

phase shift is very nearly 180° at the crossover frequency. The result is a sharp peak in the closed-loop frequency response at the crossover frequency. By adding a zero in the loop filter at ω_2, as shown in Fig. 10.18c, the loop phase margin can be greatly improved. Note that for this case the loop bandwidth, which is equal to the crossover frequency, is much lower than K_v. This ability to set loop bandwidth and K_v independently is an advantage of this type of loop filter. An R-C circuit that provides the necessary pole and zero in the filter response is shown in Fig. 10.18d. The root locus for this loop filter and the resulting closed-loop response are also shown.

Loop Lock Range. The loop lock range is the range of input frequencies about the center frequency for which the loop maintains lock. In most cases, it is limited by the fact that the phase comparator has a limited phase comparison range; once the phase difference between

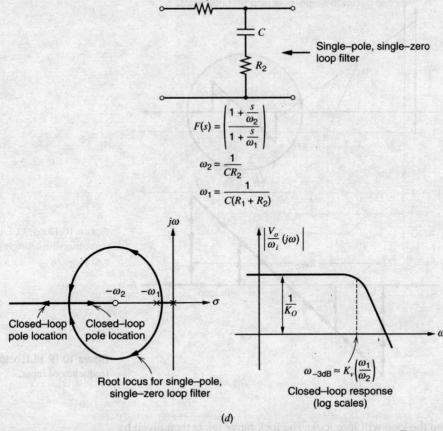

Figure 10.18 (d) Root locus and frequency response of a second-order PLL with a zero. Frequency response shown is for large loop gain such that poles are located as shown in the root locus.

the input signal and the VCO output reaches some critical value, the phase comparator ceases to behave linearly. The transfer characteristic of a typical analog phase comparator is shown in Fig. 10.12. It is clear from this figure that in order to maintain lock, the phase difference between the VCO output and the incoming signal must be kept between zero and π. If the phase difference is equal to either zero or π, then the magnitude of the dc voltage at the output of the phase comparator is

$$V_{o(\max)} = \pm K_D \left(\frac{\pi}{2}\right) \quad (10.62)$$

This dc voltage is amplified by the electrical gain A, and the result is applied to the VCO input, producing a frequency shift away from the free-running center frequency of

$$\Delta \omega_{\text{osc}} = K_D A K_O \left(\frac{\pi}{2}\right) = \left(\frac{K_v \pi}{2}\right) \quad (10.63)$$

If the input frequency is now shifted away from the free-running frequency, more voltage will have to be applied to the VCO in order for the VCO frequency to shift accordingly. However, the phase detector can produce no more dc output voltage to shift the VCO frequency further,

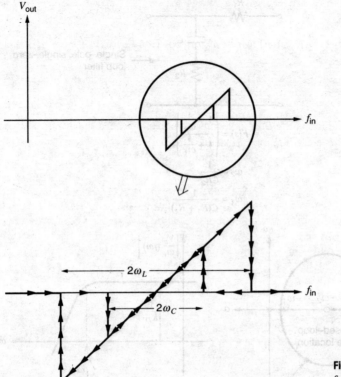

Figure 10.19 PLL output versus frequency of input.

so the loop will lose lock. The lock range ω_L is then given by

$$\omega_L = K_v \frac{\pi}{2} \tag{10.64}$$

This is the frequency range on either side of the free-running frequency for which the loop will track input frequency variations. It is a parameter that depends only on the dc gain in the loop and is independent of the properties of the loop filter. Other types[6] of phase detectors can give larger linear ranges of phase-comparator operation.

The capture range is the range of input frequencies for which the initially unlocked loop will lock on an input signal when initially in an unlocked condition and is always less than the lock range. When the input frequency is swept through a range around the center frequency, the output voltage as a function of input frequency displays a hysteresis effect, as shown in Fig. 10.19. As discussed earlier, the capture range is difficult to predict analytically. As a very rough rule of thumb, the approximate capture range can be estimated using the following procedure: Refer to Fig. 10.13 and assume that the loop is opened at the loop-amplifier output and that a signal with a frequency not equal to the free-running VCO frequency is applied at the input of the PLL. The sinusoidal *difference* frequency component that appears at the output of the phase detector has the value

$$V_p(t) = \frac{\pi}{2} K_D \cos(\omega_i - \omega_{osc})t \tag{10.65}$$

where ω_i is the input signal frequency and ω_{osc} is the VCO free-running frequency. This component is passed through the loop filter, and the output from the loop amplifier resulting from this component is

$$V_o(t) = \frac{\pi}{2} K_D A |F[j(\omega_i - \omega_{osc})]| \cos[(\omega_i - \omega_{osc})t + \phi] \quad (10.66)$$

where

$$\phi = \angle F[j(\omega_i - \omega_{osc})]$$

The output from the loop amplifier thus consists of a sinusoid at the difference frequency whose *amplitude* is reduced by the loop filter. In order for capture to occur, the magnitude of the voltage that must be applied to the VCO input is

$$|V_{osc}| = \frac{\omega_i - \omega_{osc}}{K_O} \quad (10.67)$$

The capture process itself is rather complex, but the capture range can be estimated by setting the magnitudes of (10.66) and (10.67) equal. The result is that capture is likely to occur if the following inequality is satisfied:

$$|(\omega_i - \omega_{osc})| < \frac{\pi}{2} K_D K_O A |F[j(\omega_i - \omega_{osc})]| \quad (10.68)$$

This equation implicitly gives an estimation of the capture range. For the first-order loop, where $F(s)$ is unity, it predicts that the lock range and capture range are approximately equal, and that for the second-order loop the capture range is significantly less than the lock range because $|F[j(\omega_i - \omega_{osc})]|$ is then less than unity.

10.3.3 Integrated-Circuit Phase-Locked Loops

The principal reason that PLLs have come to be widely used as system components is that the elements of the phase-locked loop are particularly suited to monolithic construction, and complete PLL systems can be fabricated on a single chip. We now discuss the design of the individual PLL components.

Phase Detector. Phase detectors for monolithic PLL applications are generally of the Gilbert multiplier configuration shown in Fig. 10.4. As illustrated in Fig. 10.11, if two signals large enough in amplitude to cause limiting in the emitter-coupled pairs making up the circuit are applied to the two inputs, the output will contain a dc component given by

$$V_{average} = -I_{EE} R_C \left(1 - \frac{2\phi}{\pi}\right) \quad (10.69)$$

where ϕ is the phase difference between the input signals. An important aspect of the performance of this phase detector is that if the amplitude of the applied signal at V_{in2} is small compared to the thermal voltage V_T, the circuit behaves as a balanced modulator, and the dc compartment of the output depends on the amplitude of the low-level input. The output waveform is then a sinusoid multiplied by a synchronous square wave, as shown in Fig. 10.20. In the limiting case when the small input is small compared to V_T, the dc component in the output becomes, referring to Fig. 10.20,

$$V_{average} = \frac{1}{\pi} g_m R_C V_i \left[\int_0^\phi (\sin \omega t) \, d(\omega t) - \int_\phi^\pi (\sin \omega t) \, d(\omega t)\right] \quad (10.70)$$

$$= -\frac{2 g_m R_C V_i \cos \phi}{\pi} \quad (10.71)$$

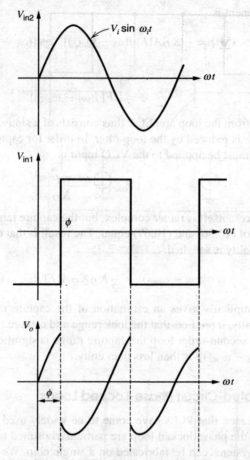

Figure 10.20 Sinusoid multiplied by a synchronous square wave.

where R_C is the collector resistor in the Gilbert multiplier and g_m is the transconductance of the transistors. The phase detector output voltage then becomes proportional to the amplitude V_i of the incoming signal, and if the signal amplitude varies, then the loop gain of the phase-locked loop changes. Thus when the signal amplitude varies, it is often necessary to precede the phase detector with an amplifier/limiter to avoid this problem. In FM demodulators, for example, any amplitude modulation appearing on the incoming frequency-modulated signal will be demodulated, producing an erroneous output.

In PLL applications the frequency response of the phase detector is usually not the limiting factor in the usable operating frequency range of the loop itself. At high operating frequencies, the parasitic capacitances of the devices result in a feedthrough of the carrier frequency, giving an erroneous component in the output at the center frequency. This component is removed by the loop filter, however, and does not greatly affect loop performance. The VCO is usually the limiting factor in the operating frequency range.

Voltage-Controlled Oscillator. The operating frequency range, FM distortion, center-frequency drift, and center-frequency supply-voltage sensitivity are all determined by the performance of the VCO. Integrated-circuit VCOs often are simply R-C multivibrators in which the charging current in the capacitor is varied in response to the control input. We first consider the emitter-coupled multivibrator as shown in Fig. 10.21a, which is typical of those

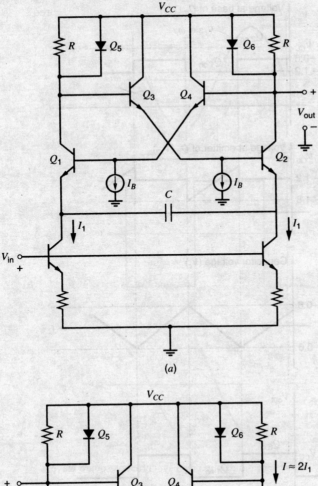

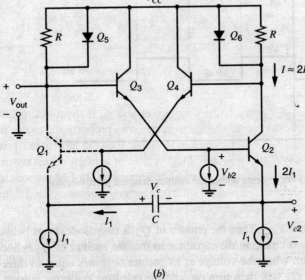

Figure 10.21 (a) Voltage-controlled, emitter-coupled multivibrator.

Figure 10.21 (b) Equivalent circuit during one half-cycle.

used in this application. We calculate the period by first assuming that Q_1 is turned off and Q_2 is turned on. The circuit then appears as shown in Fig. 10.21b. We assume that current I is large so that the voltage drop IR is large enough to turn on diode Q_6. Thus the base of Q_4 is one diode drop below V_{CC}, the emitter is two diode drops below V_{CC}, and the base of Q_1 is two diode drops below V_{CC}. If we can neglect the base current of Q_3, its base is at V_{CC} and its

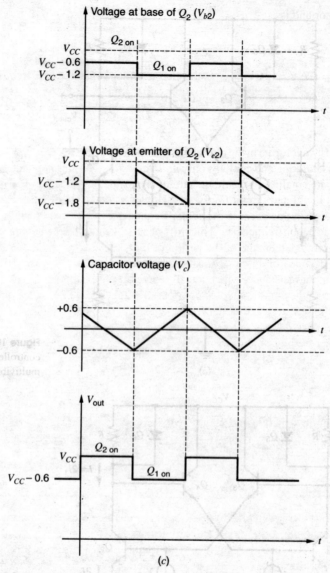

Figure 10.21 (c) Waveforms within the emitter-coupled multivibrator.

emitter is one diode drop below V_{CC}. Thus the emitter of Q_2 is two diode drops below V_{CC}. Since Q_1 is off, the current I_1 is charging the capacitor so that the emitter of Q_1 is becoming more negative. Q_1 will turn on when the voltage at its emitter becomes equal to three diode drops below V_{CC}. Transistor Q_1 will then turn on, and the resulting collector current in Q_1 turns on Q_5. As a result, the base of Q_3 moves in the negative direction by one diode drop, causing the base of Q_2 to move in the negative direction by one diode drop. Q_2 will turn off, causing the base of Q_1 to move positive by one diode drop because Q_6 also turns off. As a result, the emitter-base junction of Q_2 is reverse biased by one diode drop because the voltage on C cannot change instantaneously. Current I_1 must now charge the capacitor voltage in the negative direction by an amount equal to two diode drops before the circuit will switch back again. Since the circuit is symmetrical, the half period is given by the time required to charge the capacitor and is

$$\frac{T}{2} = \frac{Q}{I_1} \tag{10.72}$$

where $Q = C\Delta V = 2CV_{BE(on)}$ is the charge on the capacitor. The frequency of the oscillator is thus

$$f = \frac{1}{T} = \frac{I_1}{4CV_{BE(on)}} \tag{10.73}$$

The various waveforms in the circuit are shown in Fig. 10.21c. This emitter-coupled configuration is nonsaturating and contains only *npn* transistors. Furthermore, the voltage swings within the circuit are small. As a result, the circuit is capable of operating up to approximately 1 GHz for typical integrated-circuit transistors. However, the usable frequency range is limited to a value lower than this because the center frequency drift with temperature variations becomes large at the higher frequencies. This drift occurs because the switching transients themselves become a large percentage of the period of the oscillation, and the duration of the switching transients depends on circuit parasitics, circuit resistances, transistor transconductance, and transistor input resistance, which are all temperature sensitive.

Although the emitter-coupled configuration is capable of high operating speed, it displays considerable sensitivity of center frequency to temperature even at low frequencies, since the period is dependent on $V_{BE(on)}$. Utilizing (10.73) we can calculate the temperature coefficient of the center frequency as

$$\frac{1}{\omega_{osc}} \frac{d\omega_{osc}}{dT} = -\frac{1}{V_{BE(on)}} \frac{dV_{BE(on)}}{dT} = \frac{+2 \text{ mV/}°\text{C}}{600 \text{ mV}} = +3300 \text{ ppm/}°\text{C} \tag{10.74}$$

This temperature sensitivity of center frequency can be compensated by causing current I_1 to be temperature sensitive in such a way that its effect is equal and opposite to the effect of the variation of $V_{BE(on)}$.

10.4 Nonlinear Function Synthesis

The need often arises in electronic systems for circuits with arbitrary nonlinear transfer functions. For example, a common need is for square-law and square-root transfer characteristics in order to generate true rms quantities. The unique, precision exponential transfer characteristic of the bipolar transistor can be used[7,8] to generate these and many other nonlinear functions.

Consider the circuit shown in Fig. 10.22a. We have

$$V_{BE1} + V_{BE2} - V_{BE3} - V_{BE4} = 0 \tag{10.75}$$

Neglecting base currents and assuming all devices are forward active, we find

$$V_T \ln \frac{I_B}{I_{S1}} + V_T \ln \frac{I_i}{I_{S2}} - V_T \ln \frac{I_o}{I_{S3}} - V_T \ln \frac{I_o}{I_{S4}} = 0$$

and thus

$$I_o = \sqrt{I_i}\sqrt{I_B}\sqrt{\frac{I_{S3}I_{S4}}{I_{S1}I_{S2}}} \tag{10.76}$$

This circuit thus realizes a square-root transfer function with a scale factor set only by the bias current I_B (which could be another input signal) and device area ratios. There is no (first-order) dependence on supply voltage or temperature. Not that the input current source

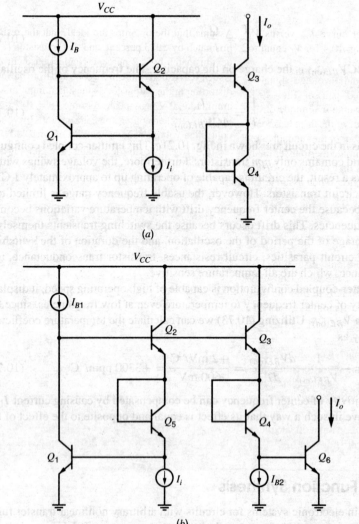

Figure 10.22 Bipolar nonlinear function circuits. (a) Square root. (b) Square law.

I_i must be capable of working into a dc bias voltage of one V_{BE} set by Q_1. The small-signal input impedance of the circuit is very low due to the feedback provided by Q_1 and Q_2. Since all nodes in the circuit are low impedance, parasitic capacitance has little influence, and the bandwidth of the circuit can be on the order of the device f_T.

For the circuit in Fig. 10.22b we have

$$V_{BE1} + V_{BE5} + V_{BE2} = V_{BE3} + V_{BE4} + V_{BE6} \quad (10.77)$$

from which

$$I_o = I_i^2 \frac{I_{B1}}{I_{B2}^2} \frac{I_{S3}I_{S4}I_{S6}}{I_{S1}I_{S5}I_{S2}} \quad (10.78)$$

Thus this circuit realizes the square-law transfer function with very wide bandwidth and insensitivity to temperature and supply voltage. Simpler versions of this circuit can be derived if the current source I_{B1} is used as a signal input. However, this requires that I_{B1} be realized by an active *pnp* current source (or a PMOS current source in BiCMOS technology), which will usually restrict the circuit bandwidth.

PROBLEMS

10.1 Sketch the dc transfer curve I_{out} versus V_2 for the Gilbert multiplier of Fig. 10.4 for V_1 equal to $0.1 V_T$, $0.5 V_T$, and V_T.

10.2 Assume that a sinusoidal voltage signal is applied to the emitter-coupled pair of Fig. 10.1. Determine the maximum allowable magnitude of the sinusoid such that the magnitude of the third harmonic in the output is less than 1 percent of the fundamental. To work this problem, approximate the transfer characteristic of the pair with the first two terms of the Taylor series for the tanh function. Then assume that all the other harmonics in the output are negligible and that the output is approximately

$$I_{out}(t) = I_o(\sin \omega_o t + \delta \sin 3\omega_o t)$$

where δ = fractional third-harmonic distortion. Use SPICE to check your result. For the same sinusoidal input voltage amplitude, use SPICE to find the third harmonic distortion in the output if emitter resistors R_E are added to each device such that $I_{E1}R_E = I_{E2}R_E = 100$ mV.

10.3 For the emitter-coupled pair of Fig. 10.1, determine the magnitude of the dc differential input voltage required to cause the *slope* of the transfer curve to be different by 1 percent from the slope through the origin.

10.4 Determine the worst-case input offset voltage of the voltage-current converter shown in Fig. 10.23. Assume that the op amps are ideal, that the resistors mismatch by ± 0.3 percent, and that transistor I_S values mismatch by ± 2 percent. Neglect base currents. Use SPICE to determine the second and third harmonic distortion in the output for a sinusoidal input drive of amplitude 20 V peak-peak. Assume $I_S = 10^{-16}$ A for the transistor and approximate the op amps by voltage-controlled voltage sources with a gain of 10,000.

10.5 Determine the dc transfer characteristic of the circuit of Fig. 10.24. Assume that $Z = 0.1 XY$ for the multiplier.

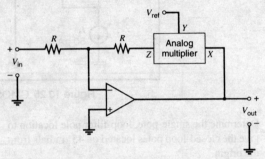

Figure 10.24 Circuit for Problem 10.5. The op amp is ideal.

10.6 A phase-locked loop has a center frequency of 10^5 rad/s, a K_O of 10^3 rad/V-s, and a K_D of 1 V/rad. There is no other gain in the loop. Determine the loop bandwidth in the first-order loop configuration.

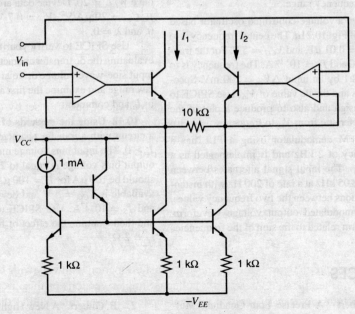

Figure 10.23 Circuit for Problem 10.4. All op amps are ideal.

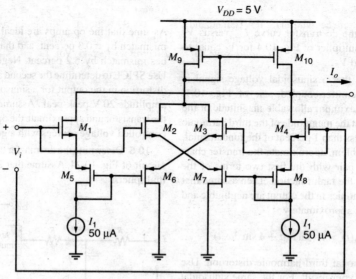

Figure 10.25 CMOS square-law circuit.

Determine the single-pole, loop-filter pole location to give the closed-loop poles located on 45° radials from the origin.

10.7 For the same PLL of Problem 10.6, design a loop filter with a zero that gives a crossover frequency for the loop gain of 100 rad/s. The loop phase shift at the loop crossover frequency should be $-135°$.

10.8 Estimate the capture range of the PLL of Problem 10.7, assuming that it is not artificially limited by the VCO frequency range.

10.9 Design a voltage-controlled oscillator based on the circuit of Fig. 10.21a. The center frequency is to be 10 kHz, $C = 0.01$ μF, and $V_{CC} = 5$ V. For the transistors $\beta = 100$ and $I_S = 10^{-16}$ A. The frequency is to be varied by 2:1 by an input $\Delta V_{in} = 200$ mV. Specify all resistors and the dc value of V_{in}. Use SPICE to check your design and also to produce a plot of the transfer characteristic from V_{in} to frequency.

10.10 An FM demodulator using a PLL has a center frequency of 2 kHz and is implemented as a first-order loop. The input signal alternates between 1.95 kHz and 2.05 kHz at a rate of 200 Hz with instantaneous transitions between the two frequency values. Sketch the demodulated output voltage waveform, ignoring the term related to the sum of the frequencies

at the PLL input and the VCO output. For the PLL, use $A = 1$, $K_D = 2.55$ V/rad and $K_O = 0.93\omega_{osc}$, where ω_{osc} is the free-running frequency in rad/sec.

10.11 Show that the CMOS circuit of Fig. 10.25 realizes a square-law transfer characteristic from V_i to I_o assuming that the MOSFETs have square-law characteristics. Specify the range of V_i over which this holds. (The bias analysis of Section 9.6.4 applies.) All PMOS devices have $W/L = 60$ and all NMOS have $W/L = 20$. Device data are $\mu_n C_{ox} = 60$ μA/V², $\mu_p C_{ox} = 20$ μA/V², $V_{tn} = 0.7$ V, $V_{tp} = -0.7$ V, $\gamma = 0$, and $\lambda = 0$.

Use SPICE to verify your result by plotting and evaluating the dc transfer characteristic. Then apply an input sine-wave voltage drive at V_i within the square-law range and examine the first and second harmonics in I_o and comment.

10.12 Using the methods of Section 10.4, design a circuit with a transfer characteristic $I_o = K I_i^{3/2}$ for $I_i \geq 0$. The input bias voltage must be $\geq V_{BE}$, and the output bias voltage is equal to $2V_{BE}$. The value of I_o should be 100 μA for $I_i = 100$ μA. The supply voltage available is $V_{CC} = 5$ V and device data are $\beta = 100$ and $I_S = 10^{-17}$ A. Use SPICE to verify your design and then examine the effect of finite $r_b = 200$ Ω and $r_e = 2$ Ω.

REFERENCES

1. B. Gilbert. "A Precise Four-Quadrant Multiplier with Subnanosecond Response," *IEEE J. Solid-State Circuits*, Vol. SC-3, pp. 365–373, December 1968.

2. B. Gilbert. "A New High-Performance Monolithic Multiplier Using Active Feedback," *IEEE J. Solid-State Circuits*, Vol. SC-9, pp. 364–373, December 1974.

3. F. M. Gardner. *Phase-Lock Techniques.* Wiley, New York, 1966.

4. A. B. Grebene and H. R. Camenzind. "Frequency Selective Integrated Circuits Using Phase-Locked Techniques," *IEEE J. Solid-State Circuits,* Vol. SC-4, pp. 216–225, August 1969.

5. *Applications of Phase-Locked Loops,* Signetics Corporation, 1974.

6. M. Soyuer and R. G. Meyer. "Frequency Limitations of a Conventional Phase-Frequency Detector," *IEEE J. Solid-State Circuits,* Vol. 25, pp. 1019–1022, August 1990.

7. B. Gilbert. "General Technique for N-Dimensional Vector Summation of Bipolar Signals," *Electronics Letters,* Vol. 12, pp. 504–505, September 16, 1976.

8. E. Seevinck. *Analysis and Synthesis of Translinear Integrated Circuits.* Elsevier, Amsterdam, 1988.

CHAPTER 11

Noise in Integrated Circuits

11.1 Introduction

This chapter deals with the effects of *electrical noise* in integrated circuits. The noise phenomena considered here are caused by the small current and voltage fluctuations that are generated within the devices themselves, and we specifically exclude extraneous pickup of human-made signals that can also be a problem in high-gain circuits. The existence of noise is basically due to the fact that electrical charge is not continuous but is carried in discrete amounts equal to the electron charge, and thus noise is associated with fundamental processes in the integrated-circuit devices.

The study of noise is important because it represents a lower limit to the size of electrical signal that can be amplified by a circuit without significant deterioration in signal quality. Noise also results in an upper limit to the useful gain of an amplifier, because if the gain is increased without limit, the output stage of the circuit will eventually begin to limit (that is, a transistor will leave the active region) on the amplified noise from the input stages.

In this chapter the various sources of electronic noise are considered, and the equivalent circuits of common devices including noise generators are described. Methods of circuit analysis with noise generators as inputs are illustrated, and the noise analysis of complex circuits such as op amps is performed. Methods of computer analysis of noise are examined, and, finally, some common methods of specifying circuit noise performance are described.

11.2 Sources of Noise

11.2.1 Shot Noise[1,2,3,4]

Shot noise is *always* associated with a direct-current flow and is present in diodes, MOS transistors, and bipolar transistors. The origin of shot noise can be seen by considering the diode of Fig. 11.1a and the carrier concentrations in the device in the forward-bias region as shown in Fig. 11.1b. As explained in Chapter 1, an electric field $\mathscr{E}$ exists in the depletion region and a voltage $(\psi_0 - V)$ exists between the p-type and the n-type regions, where ψ_0 is the built-in potential and V_i is the forward bias on the diode. The forward current of the diode I is composed of holes from the p region and electrons from the n region, which have sufficient energy to overcome the potential barrier at the junction. Once the carriers have crossed the junction, they diffuse away as minority carriers.

The passage of each carrier across the junction, which can be modeled as a random event, is dependent on the carrier having sufficient energy and a velocity directed toward the junction. Thus external current I, which appears to be a steady current, is, in fact, composed of

11.2 Sources of Noise

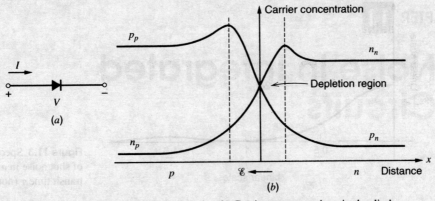

Figure 11.1 (a) Forward-biased *pn* junction diode. (b) Carrier concentrations in the diode (not to scale).

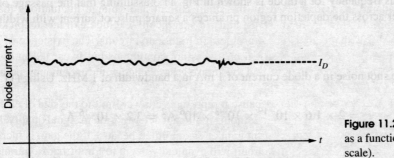

Figure 11.2 Diode current I as a function of time (not to scale).

a large number of random independent current pulses. If the current is examined on a sensitive oscilloscope, the trace appears as in Fig. 11.2, where I_D is the average current.

The fluctuation in I is termed *shot noise* and is generally specified in terms of its mean-square variation about the average value. This is written as $\overline{i^2}$, where

$$\overline{i^2} = \overline{(I - I_D)^2}$$
$$= \lim_{T \to \infty} \frac{1}{T} \int_0^T (I - I_D)^2 \, dt \qquad (11.1)$$

It can be shown that if a current I is composed of a series of random independent pulses with average value I_D, then the resulting noise current has a mean-square value

$$\overline{i^2} = 2qI_D \Delta f \qquad (11.2)$$

where q is the electronic charge (1.6×10^{-19} C) and Δf is the bandwidth in hertz. This equation shows that the noise current has a mean-square value that is *directly proportional* to the bandwidth Δf (in hertz) of the measurement. Thus a noise-current *spectral density* $\overline{i^2}/\Delta f$ (with units square amperes per hertz) can be defined that is *constant* as a function of frequency. Noise with such a spectrum is often called *white noise*. Since noise is a purely random signal, the *instantaneous value* of the waveform cannot be predicted at any time. The only information available for use in circuit calculations concerns the *mean square* value of the signal given by (11.2). Bandwidth Δf in (11.2) is determined by the circuit in which the noise source is acting.

Equation 11.2 is valid until the frequency becomes comparable to $1/\tau$, where τ is the carrier transit time through the depletion region. For most practical electronic devices, τ is extremely

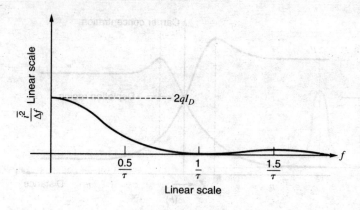

Figure 11.3 Spectral density of shot noise in a diode with transit time τ (not to scale).

small and (11.2) is accurate well into the gigahertz region. A sketch of noise-current spectral density versus frequency for a diode is shown in Fig. 11.3 assuming that the passage of each charge carrier across the depletion region produces a square pulse of current with width τ.

■ **EXAMPLE**

Calculate the shot noise in a diode current of 1 mA in a bandwidth of 1 MHz. Using (11.2) we have

$$\overline{i^2} = 2 \times 1.6 \times 10^{-19} \times 10^{-3} \times 10^6 \text{ A}^2 = 3.2 \times 10^{-16} \text{ A}^2$$

and thus

$$i = 1.8 \times 10^{-8} \text{A rms}$$

■ where i represents the root-mean-square (rms) value of the noise current.

The effect of shot noise can be represented in the low-frequency, small-signal equivalent circuit of the diode by inclusion of a current generator shunting the diode, as shown in Fig. 11.4. Since this noise signal has random phase and is defined solely in terms of its mean-square value, it also has no polarity. Thus the arrow in the current source in Fig. 11.2 has no significance and is included only to identify the generator as a current source. This practice is followed in this chapter where we deal only with noise generators having random phase.

The noise-current signal produced by the shot noise mechanism has an amplitude that varies randomly with time and that can only be specified by a *probability-density* function. It can be shown that the amplitude distribution of shot noise is Gaussian and the probability-density function $p(I)$ of the diode current is plotted versus current in Fig. 11.5 (not to scale). The probability that the diode current lies between values I and $(I + dI)$ at any time is given by $p(I)dI$. If σ is the standard deviation of the Gaussian distribution, then the diode current

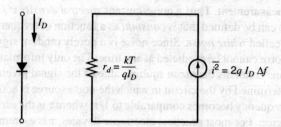

Figure 11.4 Junction diode small-signal equivalent circuit with noise.

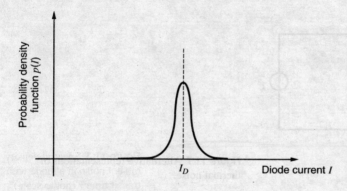

Figure 11.5 Probability density function for the diode current I (not to scale).

amplitude lies between limits $\overline{I}_D \pm \sigma$ for 68 percent of the time. By definition, variance σ^2 is the mean-square value of $(I - I_D)$ and thus, from (11.1),

$$\sigma^2 = \overline{i^2}$$

and

$$\sigma = \sqrt{2qI_D \, \Delta f} \qquad (11.3)$$

using (11.2). Note that, theoretically, the noise amplitude can have positive or negative values approaching infinity. However, the probability falls off very quickly as amplitude increases and an effective limit to the noise amplitude is $\pm 3\sigma$. The noise signal is within these limits for 99.7 percent of the time. A brief description of the Gaussian distribution is given in Appendix A.3.1 in Chapter 3.

It is important to note that the distribution of noise in frequency as shown in Fig. 11.3 is due to the random nature of the hole and electron transitions across the *pn* junction. Consider the situation if all the carriers made transitions with uniform time separation. Since each carrier has a charge of 1.6×10^{-19} C, a 1-mA current would then consist of current pulses every 1.6×10^{-16} s. The Fourier analysis of such a waveform would give the spectrum of Fig. 11.6, which shows an average or dc value I_D and harmonics at multiples of $1/\Delta t$, where Δt is the period of the waveform and equals 1.6×10^{-16} s. Thus the first harmonic is at 6×10^6 GHz, which is far beyond the useful frequency of the device. There would be *no noise produced* in the normal frequency range of operation.

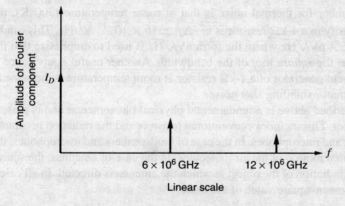

Figure 11.6 Shot-noise spectrum assuming uniform emission of carriers.

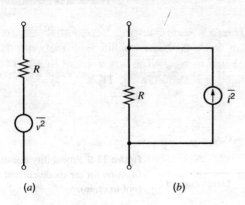

Figure 11.7 Alternative representations of thermal noise.

(a)   (b)

11.2.2 Thermal Noise[1,3,5]

Thermal noise is generated by a completely different mechanism from shot noise. In conventional resistors it is due to the random thermal motion of the electrons and is unaffected by the presence or absence of direct current, since typical electron drift velocities in a conductor are much less than electron thermal velocities. Since this source of noise is due to the thermal motion of electrons, we expect that it is related to absolute temperature T. In fact thermal noise is *directly proportional* to T (unlike shot noise, which is *independent of T*) and, as T approaches zero, thermal noise also approaches zero.

In a resistor R, thermal noise can be shown to be represented by a series voltage generator $\overline{v^2}$ as shown in Fig. 11.7a, or by a shunt current generator $\overline{i^2}$ as in Fig. 11.7b. These representations are equivalent and

$$\overline{v^2} = 4kTR\,\Delta f \tag{11.4}$$

$$\overline{i^2} = 4kT\frac{1}{R}\Delta f \tag{11.5}$$

where k is Boltzmann's constant. At room temperature $4kT = 1.66 \times 10^{-20}$ V-C. Equations 11.4 and 11.5 show that the noise spectral density is again *independent* of frequency and, for thermal noise, this is true up to 10^{13} Hz. Thus thermal noise is another source of white noise. Note that the Norton equivalent of (11.5) can be derived from (11.4) as

$$\overline{i^2} = \frac{\overline{v^2}}{R^2} \tag{11.6}$$

A useful number to remember for thermal noise is that at room temperature (300°K), the thermal noise spectral density in a 1-kΩ resistor is $\overline{v^2}/\Delta f \simeq 16 \times 10^{-18}$ V²/Hz. This can be written in rms form as $v \simeq 4$ nV/$\sqrt{\text{Hz}}$ where the form nV/$\sqrt{\text{Hz}}$ is used to emphasize that the *rms noise voltage* varies as the *square root* of the bandwidth. Another useful equivalence is that the thermal noise-current generator of a 1-kΩ resistor at room temperature is the same as that of 50 μA of direct current exhibiting shot noise.

Thermal noise as described above is a fundamental physical phenomenon and is present in *any* linear passive resistor. This includes conventional resistors and the radiation resistance of antennas, loudspeakers, and microphones. In the case of loudspeakers and microphones, the source of noise is the thermal motion of the air molecules. In the case of antennas, the source of noise is the black-body radiation of the object at which the antenna is directed. In all cases, (11.4) and (11.5) give the mean-square value of the noise.

The amplitude distribution of thermal noise is again Gaussian. Since both shot and thermal noise each have a flat frequency spectrum and a Gaussian amplitude distribution, they are indistinguishable once they are introduced into a circuit. The waveform of shot and thermal noise combined with a sinewave of equal power is shown in Fig. 11.21.

11.2.3 Flicker Noise[6,7,8] (1/f Noise)

This is a type of noise found in all active devices, as well as in some discrete passive elements such as carbon resistors. The origins of flicker noise are varied, but it is caused mainly by traps associated with contamination and crystal defects. These traps capture and release carriers in a random fashion and the time constants associated with the process give rise to a noise signal with energy concentrated at low frequencies.

Flicker noise, which is always associated with a flow of direct current, displays a spectral density of the form

$$\overline{i^2} = K_1 \frac{I^a}{f^b} \Delta f \tag{11.7}$$

where

Δf = small bandwidth at frequency f
I = direct current
K_1 = constant for a particular device
a = constant in the range 0.5 to 2
b = constant of about unity

If $b = 1$ in (11.7), the noise spectral density has a $1/f$ frequency dependence (hence the alternative name $1/f$ *noise*), as shown in Fig. 11.8. It is apparent that flicker noise is most significant at low frequencies, although in devices exhibiting high flicker noise levels, this noise source may dominate the device noise at frequencies well into the megahertz range.

It was noted above that flicker noise only exists in association with a direct current. Thus, in the case of carbon resistors, no flicker noise is present until a direct current is passed through the resistor (however, *thermal* noise *always* exists in the resistor and is *unaffected* by any direct current as long as the temperature remains constant). Consequently, carbon resistors can be used if required as external elements in low-noise, low-frequency integrated circuits as long as they carry no direct current. If the external resistors for such circuits must carry direct current, however, metal film resistors that have no flicker noise should be used.

In earlier sections of this chapter, we saw that shot and thermal noise signals have well-defined mean-square values that can be expressed in terms of current flow, resistance, and a

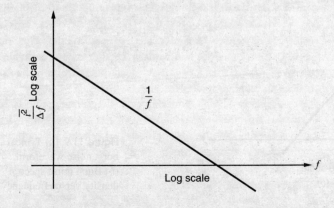

Figure 11.8 Flicker noise spectral density versus frequency.

number of well-known physical constants. By contrast, the mean-square value of a flicker noise signal as given by (11.7) contains an unknown constant K_1. This constant not only varies by orders of magnitude from one device type to the next, but it can also vary widely for different transistors or integrated circuits from the same process wafer. This is due to the dependence of flicker noise on contamination and crystal imperfections, which are factors that can vary randomly even on the same silicon wafer. However, experiments have shown that if a typical value of K_1 is determined from measurements on a number of devices from a given process, then this value can be used to predict average or typical flicker noise performance for integrated circuits from that process.[9]

The final characteristic of flicker noise that is of interest is its amplitude distribution, which is often non-Gaussian, as measurements have shown.

11.2.4 Burst Noise[7] (*Popcorn Noise*)

This is another type of low-frequency noise found in some integrated circuits and discrete transistors. The source of this noise is not fully understood, although it has been shown to be related to the presence of heavy-metal ion contamination. Gold-doped devices show very high levels of burst noise.

Burst noise is so named because an oscilloscope trace of this type of noise shows bursts of noise on a number (two or more) of discrete levels, as illustrated in Fig. 11.9a. The repetition rate of the noise pulses is usually in the audio frequency range (a few kilohertz or less) and produces a *popping* sound when played through a loudspeaker. This has led to the name *popcorn noise* for this phenomenon.

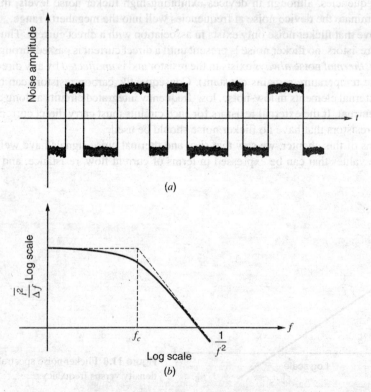

Figure 11.9 (*a*) Typical burst noise waveform. (*b*) Burst noise spectral density versus frequency.

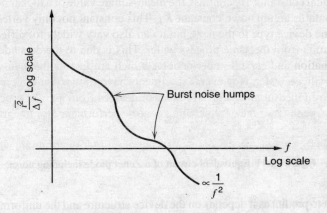

Figure 11.10 Spectral density of combined multiple burst noise sources and flicker noise.

The spectral density of burst noise can be shown to be of the form

$$\overline{i^2} = K_2 \frac{I^c}{1 + \left(\dfrac{f}{f_c}\right)^2} \Delta f \tag{11.8}$$

where

$K_2 =$ constant for a particular device
$I =$ direct current
$c =$ constant in the range 0.5 to 2
$f_c =$ particular frequency for a given noise process

This spectrum is plotted in Fig. 11.9b and illustrates the typical hump that is characteristic of burst noise. At higher frequencies the noise spectrum falls as $1/f^2$.

Burst noise processes often occur with multiple time constants, and this gives rise to multiple humps in the spectrum. Also, flicker noise is invariably present as well so that the composite low-frequency noise spectrum often appears as in Fig. 11.10. As with flicker noise, factor K_2 for burst noise varies considerably and must be determined experimentally. The amplitude distribution of the noise is also non-Gaussian.

11.2.5 Avalanche Noise[10]

This is a form of noise produced by Zener or avalanche breakdown in a *pn* junction. In avalanche breakdown, holes and electrons in the depletion region of a reverse-biased *pn* junction acquire sufficient energy to create hole-electron pairs by colliding with silicon atoms. This process is cumulative, resulting in the production of a random series of large noise spikes. The noise is always associated with a direct-current flow, and the noise produced is much greater than shot noise in the same current, as given by (11.2). This is because a single carrier can start an avalanching process that results in the production of a current burst containing many carriers moving together. The total noise is the sum of a number of random bursts of this type.

The most common situation where avalanche noise is a problem occurs when Zener diodes are used in the circuit. These devices display avalanche noise and are generally avoided in low-noise circuits. If Zener diodes are present, the noise representation of Fig. 11.11 can be used, where the noise is represented by a series voltage generator v^2. The dc voltage V_z is the breakdown voltage of the diode, and the series resistance R is typically 10 to 100 Ω. The

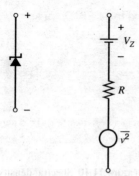

Figure 11.11 Equivalent circuit of a Zener diode including noise.

magnitude of $\overline{v^2}$ is difficult to predict as it depends on the device structure and the uniformity of the silicon crystal, but a typical measured value is $\overline{v^2}/\Delta f \simeq 10^{-14}$ V²/Hz at a dc Zener current of 0.5 mA. Note that this is equivalent to the thermal noise voltage in a 600-kΩ resistor and completely overwhelms thermal noise in R. The spectral density of the noise is approximately flat, but the amplitude distribution is generally non-Gaussian.

11.3 Noise Models of Integrated-Circuit Components

In the above sections, the various physical sources of noise in electronic circuits were described. In this section, these sources of noise are brought together to form the small-signal equivalent circuits including noise for diodes and for bipolar and MOS transistors.

11.3.1 Junction Diode

The equivalent circuit for a forward-biased junction diode was considered briefly in the consideration of shot noise. The basic equivalent circuit of Fig. 11.4 can be made complete by adding series resistance r_s as shown in Fig. 11.12. Since r_s is a physical resistor due to the resistivity of the silicon, it exhibits thermal noise. Experimentally it has been found that any flicker noise present can be represented by a current generator in shunt with r_d, and can be

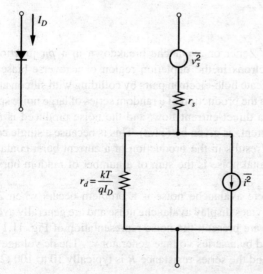

Figure 11.12 Complete diode small-signal equivalent circuit with noise sources.

conveniently combined with the shot-noise generator as indicated by (11.10) to give

$$\overline{v_s^2} = 4kTr_s \, \Delta f \tag{11.9}$$

$$\overline{i^2} = 2qI_D \, \Delta f + K\frac{I_D^a}{f} \Delta f \tag{11.10}$$

11.3.2 Bipolar Transistor[11]

In a bipolar transistor in the forward-active region, minority carriers diffuse and drift across the base region to be collected at the collector-base junction. Minority carriers entering the collector-base depletion region are accelerated by the field existing there and swept across this region to the collector. The time of arrival at the collector-base junction of the diffusing (or drifting) carriers can be modeled as a random process, and thus the transistor collector current consists of a series of random current pulses. Consequently, collector current I_c shows *full shot noise* as given by (11.2), and this is represented by a shot noise current generator $\overline{i_c^2}$ from collector to emitter as shown in the equivalent circuit of Fig. 11.13.

Base current I_B in a transistor is due to recombination in the base and base-emitter depletion regions and also to carrier injection from the base into the emitter. All of these are independent random processes, and thus I_B also shows *full shot noise*. This is represented by shot noise current generator $\overline{i_b^2}$ in Fig. 11.13.

Transistor base resistor r_b is a physical resistor and thus has thermal noise. Collector series resistor r_c also shows thermal noise, but since this is in series with the high-impedance collector node, this noise is negligible and is usually not included in the model. Note that resistors r_π and r_o in the model are *fictitious* resistors that are used for modeling purposes only, and they do *not* exhibit thermal noise.

Flicker noise and burst noise in a bipolar transistor have been found experimentally to be represented by current generators across the internal base-emitter junction. These are conveniently combined with the shot noise generator in $\overline{i_b^2}$. Avalanche noise in bipolar transistors is found to be negligible if V_{CE} is kept at least 5 V below the breakdown voltage BV_{CEO}, and this source of noise will be neglected in subsequent calculations.

The full small-signal equivalent circuit including noise for the bipolar transistor is shown in Fig. 11.13. Since they arise from separate, independent physical mechanisms, all the noise sources are *independent* of each other and have mean-square values:

$$\overline{v_b^2} = 4kTr_b \, \Delta f \tag{11.11}$$

$$\overline{i_c^2} = 2qI_C \, \Delta f \tag{11.12}$$

$$\overline{i_b^2} = \underbrace{2qI_B \, \Delta f}_{\text{Shot noise}} + \underbrace{K_1 \frac{I_B^a}{f} \Delta f}_{\text{Flicker noise}} + \underbrace{K_2 \frac{I_B^c}{1 + \left(\frac{f}{f_c}\right)^2} \Delta f}_{\text{Burst noise}} \tag{11.13}$$

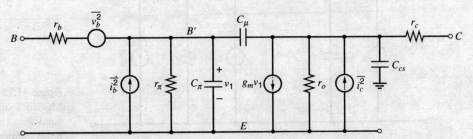

Figure 11.13 Complete bipolar transistor small-signal equivalent circuit with noise sources.

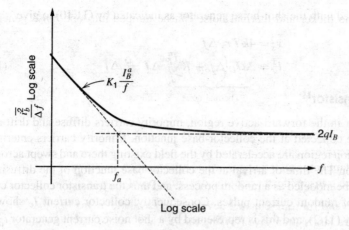

Figure 11.14 Spectral density of the base-current noise generator in a bipolar transistor.

This equivalent circuit is valid for both *npn* and *pnp* transistors. For *pnp* devices, the magnitudes of I_B and I_C are used in the above equations.

The base-current noise spectrum can be plotted using (11.13), and this has been done in Fig. 11.14, where burst noise has been neglected for simplicity. The shot noise and flicker noise asymptotes meet at a frequency f_a, which is called the flicker noise *corner* frequency. In some transistors using careful processing, f_a can be as low as 100 Hz. In other transistors, f_a can be as high as 10 MHz.

11.3.3 MOS Transistor[12]

The structure of MOS transistors was described in Chapter 1. We showed there that the resistive channel under the gate is modulated by the gate-source voltage so that the drain current is controlled by the gate-source voltage. Since the channel material is *resistive*, it exhibits *thermal noise*, which is a major source of noise in MOS transistors. This noise source can be represented by a noise-current generator $\overline{i_d^2}$ from drain to source in the small-signal equivalent circuit of Fig. 11.15.

Another source of noise in MOS transistors is flicker noise. Because MOS transistors conduct current near the surface of the silicon where surface states act as traps that capture and release current carriers, their flicker noise component can be large. Flicker noise in the MOS transistor is also found experimentally to be represented by a drain-source current

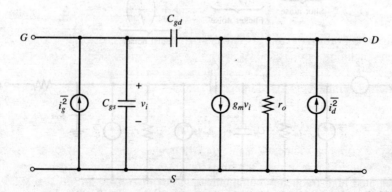

Figure 11.15 MOSFET small-signal equivalent circuit with noise sources.

generator, and the flicker and thermal noise can be lumped into one noise generator $\overline{i_d^2}$ in Fig. 11.15 with

$$\overline{i_d^2} = \underbrace{4kT\left(\frac{2}{3}g_m\right)\Delta f}_{\text{Thermal noise}} + \underbrace{K\frac{I_D^a}{f}\Delta f}_{\text{Flicker noise}} \quad (11.14)$$

where

I_D = drain bias current
K = constant for a given device
a = constant between 0.5 and 2
g_m = device transconductance at the operating point

This equation is valid for long-channel devices. For channel lengths less than 1 μm, thermal noise 2 to 5 times larger than given by the first term in (11.14) has been measured.[13] This increase in thermal noise may be attributed to hot electrons in short-channel devices.

Another source of noise in the MOS transistor is shot noise generated by the gate leakage current. This noise can be represented by $\overline{i_g^2}$ in Fig. 11.15, with

$$\overline{i_g^2} = 2qI_G\Delta f \quad (11.15)$$

This noise current is usually very small since the dc gate current I_G is typically less than 10^{-15} A. The noise terms in (11.14) and (11.15) are *all independent* of each other.

There is one other component of noise that is usually insignificant at low frequencies but important in very high-frequency MOS circuits, such as radio-frequency amplifiers, for example. At an arbitrary point in the channel, the gate-to-channel voltage has a random component due to fluctuations along the channel caused by thermal noise. These voltage variations generate a noisy ac gate current i_g due to the capacitance between the gate and channel. The mean-squared value of this gate current for a long-channel device in the active region is

$$\overline{i_g^2} = \frac{16}{15}kT\omega^2 C_{gs}^2 \Delta f \quad (11.16)$$

where $C_{gs} = (2/3)C_{ox}WL$. The gate-current noise in (11.16) is correlated with the thermal-noise term in (11.14) because both noise currents stem from thermal fluctuations in the channel. The magnitude of the correlation between these currents is 0.39.[12] For channel lengths less than 1 μm, this component of gate-current noise may be larger than given by (11.16) if thermal noise associated with the channel increases due to hot electrons as noted above.[14] The total gate-current noise is the sum of the terms in (11.15) and (11.16).

11.3.4 Resistors

Monolithic and thin-film resistors display thermal noise as given by (11.4) and (11.5), and the circuit representation of this is shown in Fig. 11.7. As mentioned in Section 11.2.3, discrete carbon resistors also display flicker noise, and this should be considered if such resistors are used as external components to the integrated circuit.

11.3.5 Capacitors and Inductors

Capacitors are common elements in integrated circuits, either as unwanted parasitics or as elements introduced for a specific purpose. Inductors are sometimes realized on the silicon die in integrated high-frequency communication circuits. There are *no sources of noise* in *ideal* capacitors or inductors. In practice, real components have parasitic resistance that *does* display noise as given by the thermal noise formulas of (11.4) and (11.5). In the case of integrated-circuit capacitors, the parasitic resistance usually consists of a small value in series with the capacitor. Parasitic resistance in inductors can be modeled by either series or shunt elements.

11.4 Circuit Noise Calculations[15,16]

The device equivalent circuits including noise that were derived in Section 11.3 can be used for the calculation of circuit noise performance. First, however, methods of circuit calculation with noise generators as sources must be established, and attention is now given to this problem.

Consider a noise current source with mean-square value

$$\overline{i^2} = S(f)\Delta f \tag{11.17}$$

where $S(f)$ is the *noise spectral density*. The value of $S(f)$ is plotted versus frequency in Fig. 11.16a for an arbitrary noise generator. In a small bandwidth Δf, the mean-square value of the noise current is given by (11.17), and the rms values can be written as

$$i = \sqrt{S(f)\Delta f} \tag{11.18}$$

The noise current in bandwidth Δf can be represented approximately[15] by a sinusoidal current generator with rms value i as shown in Fig. 11.16b. If the noise current in bandwidth Δf is now applied as an input signal to a circuit, its effect can be calculated by substituting the sinusoidal generator and performing circuit analysis in the usual fashion. When the circuit response to the sinusoid is calculated, the mean-square value of the output sinusoid gives the mean-square value of the output noise in bandwidth Δf. *Thus network noise calculations reduce to familiar sinusoidal circuit-analysis calculations.* The only difference occurs when multiple noise sources are applied, as is usually the case in practical circuits. Each noise source is then represented by a separate sinusoidal generator, and the output contribution of each one is separately calculated. The total output noise in bandwidth Δf is calculated as a *mean-square* value by *adding* the individual *mean-square* contributions from each output sinusoid. This

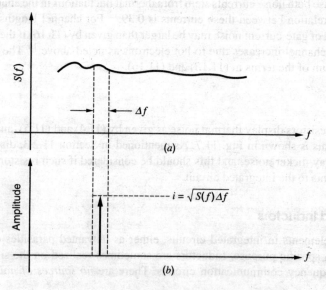

Figure 11.16 Representation of noise in a bandwidth Δf by an equivalent sinusoid with the same rms value.

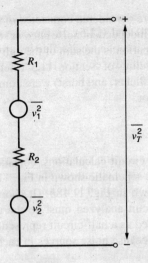

Figure 11.17 Circuit for the calculation of the total noise $\overline{v_T^2}$ produced by two resistors in series.

depends, however, on the original noise sources being *independent*, as will be shown below. This requirement is usually satisfied if the equivalent noise circuits derived in previous sections are used, as all the noise sources except the induced gate noise in (11.16) arise from separate mechanisms and are thus independent.

For example, consider two resistors R_1 and R_2 connected in series as shown in Fig. 11.17. Resistors R_1 and R_2 have respective noise generators

$$\overline{v_1^2} = 4kTR_1 \Delta f \tag{11.19a}$$

$$\overline{v_2^2} = 4kTR_2 \Delta f \tag{11.19b}$$

In order to calculate the mean-square noise voltage $\overline{v_T^2}$ produced by the two resistors in series, let $v_T(t)$ be the instantaneous value of the total noise voltage and $v_1(t)$ and $v_2(t)$ the instantaneous values of the individual generators. Then

$$v_T(t) = v_1(t) + v_2(t) \tag{11.20}$$

and thus

$$\overline{v_T(t)^2} = \overline{[v_1(t) + v_2(t)]^2}$$
$$= \overline{v_1(t)^2} + \overline{v_2(t)^2} + \overline{2v_1(t)v_2(t)} \tag{11.21}$$

Now, since noise generators $v_1(t)$ and $v_2(t)$ arise from separate resistors, they must be *independent*. Thus the *average* value of their product $\overline{v_1(t)v_2(t)}$ will be zero and (11.21) becomes

$$\overline{v_T^2} = \overline{v_1^2} + \overline{v_2^2} \tag{11.22}$$

Thus the mean-square value of the sum of a number of independent noise generators is the sum of the individual mean-square values. Substituting (11.19a) and (11.19b) in (11.22) gives

$$\overline{v_T^2} = 4kT(R_1 + R_2) \Delta f \tag{11.23}$$

Equation 11.23 is just the value that would be predicted for thermal noise in a resistor $(R_1 + R_2)$ using (11.4), and thus the results are consistent. These results are also consistent with the representation of the noise generators by *independent* sinusoids as described earlier. It is easily shown that when two or more such generators are connected in series, the mean-square value of the total voltage is equal to the sum of the individual mean-square values.

In the above calculation, two noise voltage sources were considered connected in series. It can be similarly shown that an analogous result is true for independent noise *current* sources connected in *parallel*. The mean-square value of the combination is the sum of the individual mean-square values. This result was assumed in the modeling of Section 11.3 where, for example, three independent noise-current generators (shot, flicker, and burst) were combined into a single base-emitter noise source for a bipolar transistor.

11.4.1 Bipolar Transistor Noise Performance

As an example of the manipulation of noise generators in circuit calculations, consider the noise performance of the simple transistor stage with the ac schematic shown in Fig. 11.18a. The small-signal equivalent circuit including noise is shown in Fig. 11.18b. (It should be pointed out that, for noise calculations, the equivalent circuit analyzed must be the actual circuit configuration used. That is, Fig. 11.18a cannot be used as a half-circuit representation of a differential pair for the purposes of noise calculation because noise sources in each half of a differential pair affect the total output noise.)

In the equivalent circuit of Fig. 11.18b, the external input signal v_i has been ignored so that output signal v_o is due to noise generators only. C_μ is assumed small and is neglected. Output resistance r_o is also neglected. The transistor noise generators are as described previously and in addition

$$\overline{v_s^2} = 4kTR_S \Delta f \tag{11.24}$$

$$\overline{i_l^2} = 4kT\frac{1}{R_L} \Delta f \tag{11.25}$$

The total output noise can be calculated by considering each noise source in turn and performing the calculation *as if* each noise source were a sinusoid with rms value equal to that of the noise source being considered. Consider first the noise generator v_s due to R_S. Then

$$v_1 = \frac{Z}{Z + r_b + R_S} v_s \tag{11.26}$$

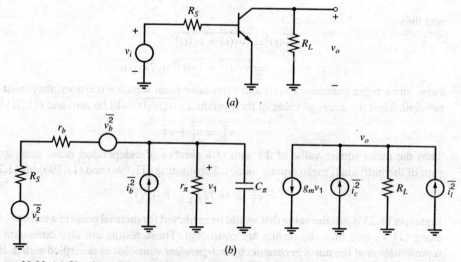

Figure 11.18 (a) Simple transistor amplifier ac schematic. (b) Small-signal equivalent circuit with noise sources.

where

$$Z = r_\pi \left\| \frac{1}{j\omega C_\pi} \right. \quad (11.27)$$

The output noise voltage due to v_s is

$$v_{o1} = -g_m R_L v_1 \quad (11.28)$$

Use of (11.26) in (11.28) gives

$$v_{o1} = -g_m R_L \frac{Z}{Z + r_b + R_S} v_s \quad (11.29)$$

The phase information contained in (11.29) is irrelevant because the noise signal has random phase and the only quantity of interest is the mean-square value of the output voltage produced by v_s. From (11.29) this is

$$\overline{v_{o1}^2} = g_m^2 R_L^2 \frac{|Z|^2}{|Z + r_b + R_S|^2} \overline{v_s^2} \quad (11.30)$$

By similar calculations it is readily shown that the noise voltage produced at the output by $\overline{v_b^2}$ and $\overline{i_b^2}$ is

$$\overline{v_{o2}^2} = g_m^2 R_L^2 \frac{|Z|^2}{|Z + r_b + R_S|^2} \overline{v_b^2} \quad (11.31)$$

$$\overline{v_{o3}^2} = g_m^2 R_L^2 \frac{(R_S + r_b)^2 |Z|^2}{|Z + r_b + R_S|^2} \overline{i_b^2} \quad (11.32)$$

Noise at the output due to $\overline{i_l^2}$ and $\overline{i_c^2}$ is

$$\overline{v_{o4}^2} = \overline{i_l^2} R_L^2 \quad (11.33)$$

$$\overline{v_{o5}^2} = \overline{i_c^2} R_L^2 \quad (11.34)$$

Since all five noise generators are *independent*, the total output noise is

$$\overline{v_o^2} = \sum_{n=1}^{5} \overline{v_{on}^2} \quad (11.35)$$

$$= g_m^2 R_L^2 \frac{|Z|^2}{|Z + r_b + R_S|^2} \left[\overline{v_s^2} + \overline{v_b^2} + (R_S + r_b)^2 \overline{i_b^2} \right] \quad (11.36)$$
$$+ R_L^2 (\overline{i_l^2} + \overline{i_c^2})$$

Substituting expressions for the noise generators we obtain

$$\frac{\overline{v_o^2}}{\Delta f} = g_m^2 R_L^2 \frac{|Z|^2}{|Z + r_b + R_S|^2} [4kT(R_S + r_b) + (R_S + r_b)^2 2qI_B]$$
$$+ R_L^2 \left(4kT \frac{1}{R_L} + 2qI_C \right) \quad (11.37)$$

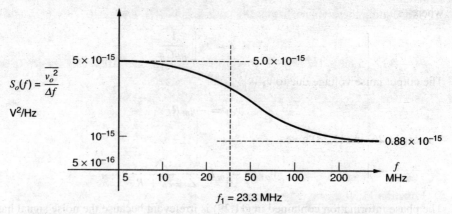

Figure 11.19 Noise voltage spectrum at the output of the circuit of Fig. 11.18.

where flicker noise has been assumed small and neglected. Substituting for Z from (11.27) in (11.37) we find

$$\frac{\overline{v_o^2}}{\Delta f} = g_m^2 R_L^2 \frac{r_\pi^2}{(r_\pi + R_S + r_b)^2} \frac{1}{1 + \left(\dfrac{f}{f_1}\right)^2} [4kT(R_S + r_b) + (R_S + r_b)^2 2qI_B]$$

$$+ R_L^2 \left(4kT\frac{1}{R_L} + 2qI_C\right) \tag{11.38}$$

where

$$f_1 = \frac{1}{2\pi[r_\pi \| (R_S + r_b)]C_\pi} \tag{11.39}$$

The output noise-voltage spectral density represented by (11.38) has a frequency-dependent part and a constant part. The frequency dependence arises because the gain of the stage begins to fall above frequency f_1, and noise due to generators $\overline{v_s^2}$, $\overline{v_b^2}$, and $\overline{i_b^2}$, which appears amplified in the output, also begins to fall. The constant term in (11.38) is due to noise generators $\overline{i_l^2}$ and $\overline{i_c^2}$. Note that this noise contribution would also be frequency dependent if the effect of C_μ had not been neglected. The noise-voltage spectral density represented by (11.38) has the form shown in Fig. 11.19.

■ **EXAMPLE**

In order to give an appreciation of the numbers involved, specific values will now be assigned to the parameters of (11.38), and the various terms in the equation will be evaluated. Assume that

$$I_C = 100 \ \mu\text{A} \quad \beta = 100 \quad r_b = 200 \ \Omega$$
$$R_S = 500 \ \Omega \quad C_\pi = 10 \ \text{pF}$$
$$R_L = 5 \ \text{k}\Omega$$

Substituting these values in (11.38) and using $4kT = 1.66 \times 10^{-20}$ V-C gives

$$\frac{\overline{v_o^2}}{\Delta f} = \left[5.82 \times 10^{-18} \frac{1}{1+\left(\frac{f}{f_1}\right)^2}(700 + 9.4) + 1.66 \times 10^{-20}(5000 + 48{,}080) \right] \text{V}^2/\text{Hz}$$

$$= \left[\frac{4.13 \times 10^{-15}}{1+\left(\frac{f}{f_1}\right)^2} + 0.88 \times 10^{-15} \right] \text{V}^2/\text{Hz} \qquad (11.40)$$

Equation 11.39 gives

$$f_1 = 23.3 \text{ MHz} \qquad (11.41)$$

Equation 11.40 shows the output noise-voltage spectral density is 5.0×10^{-15} V^2/Hz at low frequencies, and it approaches 0.88×10^{-15} V^2/Hz at high frequencies. The major contributor to the output noise in this case is the source resistance R_S, followed by the base resistance of the transistor. The noise spectrum given by (11.40) is plotted in Fig. 11.19.

■ **EXAMPLE**

Suppose the amplifier in the above example is followed by later stages that limit the bandwidth to a sharp cutoff at 1 MHz. Since the noise spectrum as shown in Fig. 11.19 does not begin to fall significantly until $f_1 = 23.3$ MHz, the noise spectrum may be assumed constant at 5.0×10^{-15} V^2/Hz over the bandwidth 0 to 1 MHz. Thus the *total* noise voltage at the output of the circuit of Fig. 11.18a in a 1-MHz bandwidth is

$$\overline{v_{oT}^2} = 5.0 \times 10^{-15} \times 10^6 \text{ V}^2 = 5.0 \times 10^{-9} \text{ V}^2$$

and thus

$$v_{oT} = 71 \text{ μV rms} \qquad (11.42)$$

Now suppose that the amplifier of Fig. 11.18a is *not* followed by later stages that limit the bandwidth but is fed directly to a wideband detector (this could be an oscilloscope or a voltmeter). In order to find the total output noise voltage in this case, the contribution from each frequency increment Δf must be summed at the output. This reduces to *integration* across the bandwidth of the detector of the noise-voltage spectral-density curve of Fig. 11.19. For example, if the detector had a 0 to 50-MHz bandwidth with a sharp cutoff, then the total output noise would be

$$\overline{v_{oT}^2} = \sum_{f=0}^{50 \times 10^6} S_o(f) \Delta f$$

$$= \int_0^{50 \times 10^6} S_o(f) \, df \qquad (11.43)$$

where

$$S_o(f) = \frac{\overline{v_o^2}}{\Delta f} \qquad (11.44)$$

is the noise spectral density defined by (11.40). In practice, the exact evaluation of such integrals is often difficult and approximate methods are often used. Note that if the integration of (11.43) is done graphically, the noise spectral density versus frequency must be plotted on
- linear scales.

11.4.2 Equivalent Input Noise and the Minimum Detectable Signal

In the previous section, the output noise produced by the circuit of Fig. 11.18 was calculated. The significance of the noise performance of a circuit is, however, the limitation it places on the smallest input signals the circuit can handle before the noise degrades the quality of the output signal. For this reason, the noise performance is usually expressed in terms of an *equivalent input noise signal*, which gives the same output noise as the circuit under consideration. In this way, the equivalent input noise can be compared directly with incoming signals and the effect of the noise on those signals is easily determined. For this purpose, the circuit of Fig. 11.18 can be represented as shown in Fig. 11.20, where $\overline{v_{iN}^2}$ is an input noise-voltage generator that produces the same output noise as all of the original noise generators. All other sources of noise in Fig. 11.20 are considered removed. Using the same equivalent circuit as in Fig. 11.18b, we obtain, for the output noise from Fig. 11.20,

$$\overline{v_o^2} = g_m^2 R_L^2 \frac{|Z|^2}{|Z + r_b + R_S|^2} \overline{v_{iN}^2} \qquad (11.45)$$

If this noise expression is equated to $\overline{v_o^2}$ from (11.37), the equivalent input noise voltage for the circuit can be calculated as

$$\frac{\overline{v_{iN}^2}}{\Delta f} = 4kT(R_S + r_b) + (R_S + r_b)^2 2qI_B$$

$$+ \frac{1}{g_m^2 R_L^2} \frac{|Z + r_b + R_S|^2}{|Z|^2} R_L^2 \left(4kT \frac{1}{R_L} + 2qI_C\right) \qquad (11.46)$$

Note that the noise-voltage spectral density given by (11.46) *rises* at high frequencies because of the variation of $|Z|$ with frequency. This is due to the fact that as the *gain* of the device falls with frequency, output noise generators $\overline{i_c^2}$ and $\overline{i_l^2}$ have a larger effect when referred back to the input.

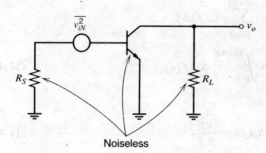

Figure 11.20 Representation of circuit noise performance by an equivalent input noise voltage.

EXAMPLE

Calculate the *total* input noise voltage $\overline{v_{iNT}^2}$ for the circuit of Fig. 11.18 in a bandwidth of 0 to 1 MHz.

This could be calculated using (11.46), derived above. Alternatively, since the total output noise voltage $\overline{v_{oT}^2}$ has already been calculated, this can be used to calculate $\overline{v_{iNT}^2}$ (in a 1-MHz bandwidth) by dividing by the circuit voltage gain squared. If A_v is the low-frequency, small-signal voltage gain of Fig. 11.18, then

$$A_v = \frac{r_\pi}{r_b + r_\pi + R_S} g_m R_L$$

Use of the previously specified data for this circuit gives

$$A_v = \frac{26{,}000}{200 + 26{,}000 + 500} \frac{5000}{260} = 18.7$$

Since the noise spectrum is flat up to 1 MHz, the low-frequency gain can be used to calculate $\overline{v_{iNT}^2}$ as

$$\overline{v_{iNT}^2} = \frac{\overline{v_{oT}^2}}{A_v^2} = \frac{5 \times 10^{-9}}{18.7^2} \text{V}^2 = 14.3 \times 10^{-12} \text{ V}^2$$

Thus we have

$$v_{iNT} = 3.78 \text{ μV rms}$$

The above example shows that in a bandwidth of 0 to 1 MHz, the noise in the circuit *appears to come* from a 3.78-μV rms noise-voltage source in series with the input. This noise voltage can be used to estimate the smallest signal that the circuit can effectively amplify, sometimes called the *minimum detectable signal* (MDS). This depends strongly on the nature of the signal and the application. If no special filtering or coding techniques are used, the MDS

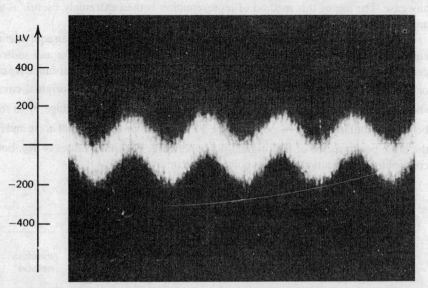

Figure 11.21 Output voltage waveform of the circuit of Fig. 11.18 with a 3.78-μV rms sinewave applied at the input. The circuit bandwidth is limited to 1 MHz, which gives an equivalent input noise voltage of 3.78 μV rms.

can be taken as equal to the equivalent input noise voltage in the passband of the amplifier. Thus, in this case

$$\text{MDS} = 3.78 \; \mu\text{V rms}$$

If a sinewave of magnitude 3.78 μV rms were applied to this circuit, and the output in a 1-MHz bandwidth examined on an oscilloscope, the sine wave would be barely detectable in the noise, as shown in Fig. 11.21. The noise waveform in this figure is typical of that produced by shot and thermal noise.

11.5 Equivalent Input Noise Generators[17]

In the previous section, the equivalent input noise voltage for a particular configuration was calculated. This gave rise to an expression for an equivalent input noise-voltage generator that was dependent on the source resistance R_S, as well as the transistor parameters. This method is now extended to a more general and more useful representation in which the noise performance of *any* two-port network is represented by *two* equivalent input noise generators. The situation is shown in Fig. 11.22, where a two-port network containing noise generators is represented by the *same* network with internal noise sources removed (the *noiseless* network) and with a noise voltage $\overline{v_i^2}$ and current generator $\overline{i_i^2}$ connected at the input. It can be shown that this representation is valid for *any* source impedance, provided that *correlation* between the two noise generators is considered. That is, the two noise generators are not independent in general because they are both dependent on the same set of original noise sources.

The inclusion of correlation in the noise representation results in a considerable increase in the complexity of the calculations, and if correlation is important, it is often easier to return to the original network with internal noise sources to perform the calculations. However, in a larger number of practical circuits, the correlation is small and may be neglected. In addition, if either equivalent input generator $\overline{v_i^2}$ or $\overline{i_i^2}$ dominates, the correlation may be neglected in any case. The use of this method of representation is then extremely useful, as will become apparent.

The need for both an equivalent input noise voltage generator and an equivalent input noise current generator to represent the noise performance of the circuit for any source resistance can be appreciated as follows. Consider the extreme case of source resistance R_S equal to zero or infinity. If $R_S = 0$, $\overline{i_i^2}$ in Fig. 11.22 is shorted out, and since the original circuit will still show output noise in general, we need an equivalent input noise voltage $\overline{v_i^2}$ to represent this behavior. Similarly, if $R_S \to \infty$, $\overline{v_i^2}$ in Fig. 11.22 cannot produce output noise and $\overline{i_i^2}$ represents the noise performance of the original noisy network. For finite values of R_S, both $\overline{v_i^2}$ and $\overline{i_i^2}$ contribute to the equivalent input noise of the circuit.

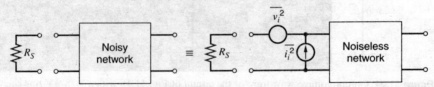

Figure 11.22 Representation of noise in a two-port network by equivalent input voltage and current generators.

11.5 Equivalent Input Noise Generators

The values of the equivalent input generators of Fig. 11.22 are readily determined. This is done by first short-circuiting the input of both circuits and equating the output noise in each case to calculate $\overline{v_i^2}$. The value of $\overline{i_i^2}$ is found by open-circuiting the input of each circuit and equating the output noise in each case. This will now be done for the bipolar transistor and the MOS transistor.

11.5.1 Bipolar Transistor Noise Generators

The equivalent input noise generators for a bipolar transistor can be calculated from the equivalent circuit of Fig. 11.23a. The output noise is calculated with a short-circuited load, and C_μ is neglected. This will be justified later. The circuit of Fig. 11.23a is to be equivalent to that of Fig. 11.23b in that each circuit should give the *same* output noise for *any* source impedance.

The value of $\overline{v_i^2}$ can be calculated by short-circuiting the input of each circuit and equating the output noise i_o. We use rms noise quantities in the calculations, but make no attempt to preserve the signs of the noise quantities as the noise generators are all independent and have random phase. The polarity of the noise generators does not affect the answer. Short-circuiting the inputs of both circuits in Fig. 11.23, assuming that r_b is small ($\ll r_\pi$) and equating i_o, we obtain

$$g_m v_b + i_c = g_m v_i \tag{11.47}$$

which gives

$$v_i = v_b + \frac{i_c}{g_m} \tag{11.48}$$

Since r_b is small, the effect of $\overline{i_b^2}$ is neglected in this calculation.

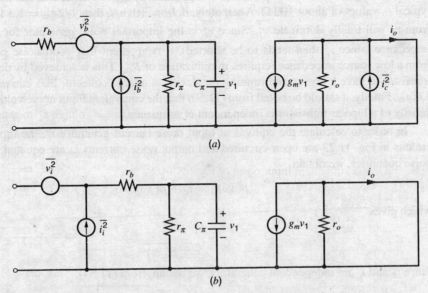

Figure 11.23 (*a*) Bipolar transistor small-signal equivalent circuit with noise generators. (*b*) Representation of the noise performance of (*a*) by equivalent input generators.

Using the fact that v_b and i_c are *independent*, we obtain, from (11.48),

$$\overline{v_i^2} = \overline{v_b^2} + \frac{\overline{i_c^2}}{g_m^2} \tag{11.49}$$

Substituting in (11.49) for $\overline{v_b^2}$ and $\overline{i_c^2}$ from (11.11) and (11.12) gives

$$\overline{v_i^2} = 4kTr_b\Delta f + \frac{2qI_C \Delta f}{g_m^2}$$

and thus

$$\frac{\overline{v_i^2}}{\Delta f} = 4kT\left(r_b + \frac{1}{2g_m}\right) \tag{11.50}$$

The equivalent input noise-voltage spectral density of a bipolar transistor thus appears to come from a resistor R_{eq} such that

$$\frac{\overline{v_i^2}}{\Delta f} = 4kTR_{eq} \tag{11.51}$$

where

$$R_{eq} = r_b + \frac{1}{2g_m} \tag{11.52}$$

and this is called the *equivalent input noise resistance*. Of this fictitious resistance, portion r_b is, in fact, a physical resistor in series with the input, whereas portion $1/2g_m$ represents the effect of collector current shot noise referred back to the input. Equations 11.50 and 11.52 are extremely useful approximations, although the assumption that $r_b \ll r_\pi$ may not be valid at high collector bias currents, and the calculation should be repeated without restrictions in those circumstances.

Equation 11.50 allows easy comparison of the relative importance of noise from r_b and I_C in contributing to $\overline{v_i^2}$. For example, if $I_C = 1$ μA, then $1/2g_m = 13$ kΩ and this will dominate typical r_b values of about 100 Ω. Alternately, if $I_C = 10$ mA, then $1/2g_m = 1.3$ Ω and noise from r_b will totally dominate $\overline{v_i^2}$. Since $\overline{v_i^2}$ is the important noise generator for low source impedance (since $\overline{i_i^2}$ then tends to be shorted), it is apparent that good noise performance from a low source impedance requires minimization of R_{eq}. This is achieved by designing the transistor to have a low r_b and running the device at a large collector bias current to reduce $1/2g_m$. Finally, it should be noted from (11.50) that the equivalent input noise-voltage spectral density of a bipolar transistor is independent of frequency.

In order to calculate the equivalent input noise current generator $\overline{i_i^2}$, the inputs of both circuits in Fig. 11.23 are open-circuited and output noise currents i_o are equated. Using rms noise quantities, we obtain

$$\beta(j\omega)i_i = i_c + \beta(j\omega)i_b \tag{11.53}$$

which gives

$$i_i = i_b + \frac{i_c}{\beta(j\omega)} \tag{11.54}$$

Since i_b and i_c are independent generators, we obtain, from (11.54),

$$\overline{i_i^2} = \overline{i_b^2} + \frac{\overline{i_c^2}}{|\beta(j\omega)|^2} \tag{11.55}$$

where

$$\beta(j\omega) = \frac{\beta_0}{1 + j\dfrac{\omega}{\omega_\beta}} \qquad (11.56)$$

and β_0 is the low-frequency, small-signal current gain. [See (1.122) and (1.126)].

Substituting in (11.55) for $\overline{i_b^2}$ and $\overline{i_c^2}$ from (11.13) and (11.12) gives

$$\frac{\overline{i_i^2}}{\Delta f} = 2q\left[I_B + K_1'\frac{I_B^a}{f} + \frac{I_C}{|\beta(j\omega)|^2}\right] \qquad (11.57)$$

where

$$K_1' = \frac{K_1}{2q} \qquad (11.57a)$$

and the burst noise term has been omitted for simplicity. The last term in parentheses in (11.57) is due to collector current noise referred to the input. At low frequencies this becomes I_C/β_0^2 and is negligible compared with I_B for typical β_0 values. When this is true, $\overline{i_i^2}$ and $\overline{v_i^2}$ do not contain common noise sources and are *totally independent*. At high frequencies, however, the last term in (11.57) increases and can become dominant, and correlation between $\overline{v_i^2}$ and $\overline{i_i^2}$ may then be important since both contain a contribution from $\overline{i_c^2}$.

The equivalent input noise current spectral density given by (11.57) appears to come from a current I_{eq} showing full shot noise, such that

$$\frac{\overline{i_i^2}}{\Delta f} = 2qI_{eq} \qquad (11.58)$$

where

$$I_{eq} = I_B + K_1'\frac{I_B^a}{f} + \frac{I_C}{|\beta(j\omega)|^2} \qquad (11.59)$$

and this is called the *equivalent input shot noise current*. This is a fictitious current composed of the base current of the device plus a term representing flicker noise and one representing collector-current noise transformed to the input. It is apparent from (11.59) that I_{eq} is minimized by utilizing low bias currents in the transistor, and also using high-β transistors. Since $\overline{i_i^2}$ is the dominant equivalent input noise generator in circuits where the transistor is fed from a high source impedance, low bias currents and high β are obviously required for good noise performance under these conditions. Note that the requirement for low bias currents to minimize $\overline{i_i^2}$ conflicts with the requirement for *high* bias current to minimize $\overline{v_i^2}$.

Spectral density $\overline{i_i^2}/\Delta f$ of the equivalent input noise current generator can be plotted as a function of frequency using (11.57). This is shown in Fig. 11.24 for typical transistor parameters. In this case, the spectral density is frequency dependent at both low and high frequencies, the low-frequency rise being due to flicker noise and the high-frequency rise being due to collector-current noise referred to the input. This input-referred noise rises at high frequencies because the transistor current gain begins to fall, and this is the reason for degradation in transistor noise performance observed at high frequencies.

Frequency f_b in Fig. 11.24 is the point where the high-frequency noise asymptote intersects the midband asymptote. This can be calculated from (11.57) as follows:

$$\beta(jf) = \frac{\beta_0}{1 + j\dfrac{f}{f_T}\beta_0} \qquad (11.60)$$

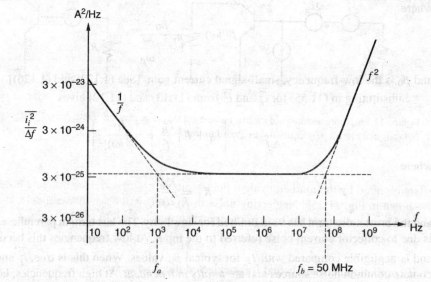

Figure 11.24 Equivalent input noise current spectral density of a bipolar transistor with $I_C = 100\ \mu\text{A}$, $\beta_0 = \beta_F = 100$, $f_T = 500$ MHz. Typical flicker noise is included.

where β_0 is the low-frequency, small-signal current gain. Thus the collector current noise term in (11.57) is

$$2q\frac{I_C}{|\beta(jf)|^2} = 2q\frac{I_C}{\beta_0^2}\left(1 + \frac{f^2}{f_T^2}\beta_0^2\right) \simeq 2qI_C\frac{f^2}{f_T^2} \qquad (11.61)$$

at high frequencies. Equation 11.61 shows that the equivalent input noise current spectrum rises as f^2 at high frequencies. Frequency f_b can be calculated by equating (11.61) to the midband noise, which is $2q[I_B + (I_C/\beta_0^2)]$. For typical values of β_0, this is approximately $2qI_B$, and equating this quantity to (11.61) we obtain

$$2qI_B = 2qI_C\frac{f_b^2}{f_T^2}$$

and thus

$$f_b = f_T\sqrt{\frac{I_B}{I_C}} \qquad (11.62)$$

The large-signal (or dc) current gain is defined as

$$\beta_F = \frac{I_C}{I_B} \qquad (11.63)$$

and thus (11.62) becomes

$$f_b = \frac{f_T}{\sqrt{\beta_F}} \qquad (11.64)$$

Using the data given in Fig. 11.24, we obtain $f_b = 50$ MHz for that example.

Once the above input noise generators have been calculated, the transistor noise performance with any source impedance is readily calculated. For example, consider the simple circuit of Fig. 11.25a with a source resistance R_S. The noise performance of this circuit can be

11.5 Equivalent Input Noise Generators

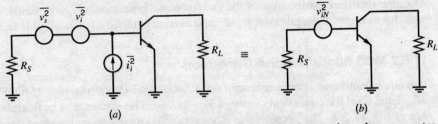

Figure 11.25 Representation of circuit noise by a single equivalent input noise voltage generator. (a) Original circuit. (b) Equivalent representation.

represented by the *total* equivalent noise voltage $\overline{v_{iN}^2}$ in series with the input of the circuit as shown in Fig. 11.25b. Neglecting noise in R_L (this will be discussed later), and equating the total noise voltage at the base of the transistor in Figs. 11.25a and 11.25b, we obtain

$$v_{iN} = v_s + v_i + i_i R_S$$

If correlation between v_i and i_i is neglected this equation gives

$$\overline{v_{iN}^2} = \overline{v_s^2} + \overline{v_i^2} + \overline{i_i^2} R_S^2 \tag{11.65}$$

Using (11.50) and (11.57) in (11.65) and neglecting flicker noise, we find

$$\frac{\overline{v_{iN}^2}}{\Delta f} = 4kTR_S + 4kT\left(r_b + \frac{1}{2g_m}\right) + R_S^2 2q\left[I_B + \frac{I_C}{|\beta(jf)|^2}\right] \tag{11.66}$$

Equation 11.66 is similar to (11.46) if r_b is small, as has been assumed.

■ EXAMPLE

Using data from the example in Section 11.4.1, calculate the total input noise voltage for the circuit of Fig. 11.25a in a bandwidth 0 to 1 MHz neglecting flicker noise and using (11.66). At low frequencies, (11.66) becomes

$$\frac{\overline{v_{iN}^2}}{\Delta f} = 4kT\left(R_S + r_b + \frac{1}{2g_m}\right) + R_S^2 2qI_B$$

$$= [1.66 \times 10^{-20}(500 + 200 + 130) + 500^2 \times 3.2 \times 10^{-19} \times 10^{-6}] \text{ V}^2/\text{Hz}$$

$$= (13.8 + 0.08) \times 10^{-18} \text{ V}^2/\text{Hz}$$

$$= 13.9 \times 10^{-18} \text{ V}^2/\text{Hz}$$

The total input noise in a 1-MHz bandwidth is

$$\overline{v_{iNT}^2} = 13.9 \times 10^{-18} \times 10^6 \text{ V}^2$$

$$= 13.9 \times 10^{-12} \text{ V}^2$$

and thus

$$v_{iNT} = 3.73 \text{ μV rms}$$

This is almost identical to the answer obtained in Section 11.4.1. However, the method described above has the advantage that once the equivalent input generators are known for any particular device, the answer can be written down almost by inspection and requires much less labor.

Also, the relative contributions of the various noise generators are more easily seen. In this case, for example, the equivalent input noise current is obviously a negligible factor.

11.5.2 MOS Transistor Noise Generators

The equivalent input noise generators for a MOS field-effect transistor (MOSFET) can be calculated from the equivalent circuit of Fig. 11.26a. This circuit is to be made equivalent to that of Fig. 11.26b. The output noise in each case is calculated with a short-circuit load and C_{gd} is neglected.

If the input of each circuit in Fig. 11.26 is short-circuited and resulting output noise currents i_o are equated, we obtain

$$i_d = g_m v_i$$

and thus

$$\overline{v_i^2} = \frac{\overline{i_d^2}}{g_m^2} \tag{11.67}$$

Substituting $\overline{i_d^2}$ from (11.14) in (11.67) gives

$$\frac{\overline{v_i^2}}{\Delta f} = 4kT\frac{2}{3}\frac{1}{g_m} + K\frac{I_D^a}{g_m^2 f} \tag{11.68a}$$

The equivalent input noise resistance R_{eq} of the MOS transistor is defined as

$$\frac{\overline{v_i^2}}{\Delta f} = 4kTR_{eq}$$

where

$$R_{eq} = \frac{2}{3}\frac{1}{g_m} + K'\frac{I_D^a}{g_m^2 f} \tag{11.68b}$$

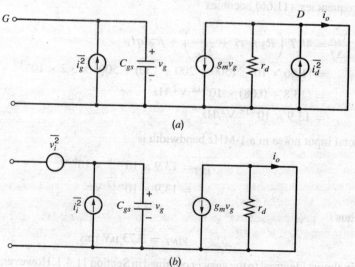

Figure 11.26 (a) MOSFET small-signal equivalent circuit with noise generators. (b) Representation of (a) by two input noise generators.

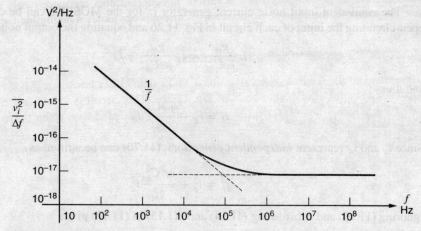

Figure 11.27 Typical equivalent input noise voltage spectral density for a MOSFET.

and

$$K' = \frac{K}{4kT}$$

At frequencies above the flicker noise region, $R_{eq} = (2/3)(1/g_m)$. For $g_m = 1$ mA/V, this gives $R_{eq} = 667\ \Omega$, which is significantly *higher* than for a bipolar transistor at a comparable bias current (about 1 mA). The equivalent input noise-voltage spectral density for a typical MOS transistor is plotted versus frequency in Fig. 11.27. Unlike the bipolar transistor, the equivalent input noise-voltage generator for a MOS transistor contains flicker noise, and it is not uncommon for the flicker noise to extend well into the megahertz region.

In MOS field-effect transistors, the presence of electron energy states at the Si-SiO$_2$ interface tends to result in an input-referred flicker noise component that is larger than the thermal noise component for frequencies below 1 to 10 kHz for most bias conditions and device geometries. Thus, an accurate representation of the input-referred flicker noise component in MOS transistors is important for the optimization of the noise performance of MOS analog circuits.

The physical mechanisms giving rise to $1/f$ noise in MOS transistors have received extensive study.[18] The exact dependence of the magnitude of the input-referred flicker noise on transistor bias conditions and device geometry is dependent on the details of the process that is used to fabricate the device. In most cases, the magnitude of the input-referred flicker noise component is approximately independent of bias current and voltage and is inversely proportional to the active gate area of the transistor. The latter occurs because as the transistor is made larger, a larger number of surface states are present under the gate, so that an averaging effect occurs that reduces the overall noise. It is also observed that the input-referred flicker noise is an inverse function of the gate-oxide capacitance per unit area. This is physically reasonable since the surface states can be thought of as a time-varying component of the surface-state charge Q_{ss}. From (1.139) this produces a time-varying component in the threshold voltage that is inversely proportional to C_{ox}. For a MOS transistor, then, the equivalent input-referred voltage noise can often be written as

$$\frac{\overline{v_i^2}}{\Delta f} = 4kT\frac{2}{3}\frac{1}{g_m} + \frac{K_f}{WLC_{ox}f} \qquad (11.69)$$

A typical value for K_f is 3×10^{-24} V^2-F, or 3×10^{-12} V^2-pF.

The equivalent input noise-current generator i_i^2 for the MOSFET can be calculated by open-circuiting the input of each circuit in Fig. 11.26 and equating the output noise. This gives

$$i_i \frac{g_m}{j\omega C_{gs}} = i_g \frac{g_m}{j\omega C_{gs}} + i_d$$

and thus

$$i_i = i_g + \frac{j\omega C_{gs}}{g_m} i_d \qquad (11.70)$$

Since i_g and i_d represent *independent generators*, (11.70) can be written as

$$\overline{i_i^2} = \overline{i_g^2} + \frac{\omega^2 C_{gs}^2}{g_m^2} \overline{i_d^2} \qquad (11.71)$$

Ignoring (11.16) and substituting (11.14) and (11.15) in (11.17) gives

$$\frac{\overline{i_i^2}}{\Delta f} = 2qI_G + \frac{\omega^2 C_{gs}^2}{g_m^2}\left(4kT\frac{2}{3}g_m + K\frac{I_D^a}{f}\right) \qquad (11.72)$$

In (11.72) the ac current gain of the MOS transistor can be identified as

$$A_I = \frac{g_m}{\omega C_{gs}} \qquad (11.73)$$

and thus the noise generators at the output are divided by A_I^2 when referred back to the input. At low frequencies the input noise-current generator is determined by the gate leakage current I_G, which is very small (10^{-15} A or less). For this reason, MOS transistors have noise performance that is *much superior* to bipolar transistors when the driving source impedance is large. Under these circumstances the input noise-current generator is dominant and is much smaller for a MOS transistor than for a bipolar transistor. It should be emphasized, however, that the input noise-*voltage* generator of a bipolar transistor in (11.50) is typically *smaller* than that of a MOS transistor in (11.68a) because the bipolar transistor has a larger g_m for a given bias current. Thus for *low* source impedances, a bipolar transistor often has noise performance *superior* to that of a MOS transistor.

11.6 Effect of Feedback on Noise Performance

The representation of circuit noise performance with two equivalent input noise generators is extremely useful in the consideration of the effect of feedback on noise performance. This will be illustrated by considering first the effect of ideal feedback on the noise performance of an amplifier. Practical aspects of feedback and noise performance will then be considered.

11.6.1 Effect of Ideal Feedback on Noise Performance

In Fig. 11.28*a* a series-shunt feedback amplifier is shown where the feedback network is ideal in that the signal feedback to the input is a pure voltage source and the feedback network is unilateral. Noise in the basic amplifier is represented by equivalent input generators $\overline{v_{ia}^2}$ and $\overline{i_{ia}^2}$. The noise performance of the overall circuit is represented by equivalent input generators $\overline{v_i^2}$ and $\overline{i_i^2}$ as shown in Fig. 11.28*b*. The value of $\overline{v_i^2}$ can be found by short-circuiting the input of each circuit and equating the output signal. However, since the output of the feedback network has a zero impedance, the current generators in each circuit are then short-circuited and the

11.6 Effect of Feedback on Noise Performance

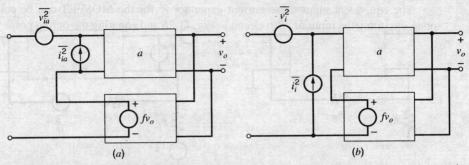

Figure 11.28 (*a*) Series-shunt feedback amplifier with noise generators. (*b*) Equivalent representation of (*a*) with two input noise generators.

two circuits are identical only if

$$\overline{v_i^2} = \overline{v_{ia}^2} \tag{11.74}$$

If the input terminals of each circuit are open-circuited, both voltage generators have a floating terminal and thus no effect on the circuit, and for equal output it is necessary that

$$\overline{i_i^2} = \overline{i_{ia}^2} \tag{11.75}$$

Thus, for the case of *ideal feedback*, the equivalent input noise generators can be moved *unchanged* outside the feedback loop and the feedback has *no effect* on the circuit noise performance. Since the feedback reduces the circuit gain, the *output noise* is reduced by the feedback, but desired signals are reduced by the same amount and the signal-to-noise ratio will be unchanged. The above result is easily shown for all four possible feedback configurations described in Chapter 8.

11.6.2 Effect of Practical Feedback on Noise Performance

The idealized series-shunt feedback circuit considered in the previous section is usually realized in practice as shown in Fig. 11.29a. The feedback circuit is a resistive divider consisting of R_E and R_F. If the noise of the basic amplifier is represented by equivalent input noise generators $\overline{i_{ia}^2}$ and $\overline{v_{ia}^2}$ and the thermal noise generators in R_F and R_E are included, the circuit is as shown in Fig. 11.29b. The noise performance of the circuit is to be represented by two equivalent input generators $\overline{v_i^2}$ and $\overline{i_i^2}$, as shown in Fig. 11.29c.

In order to calculate $\overline{v_i^2}$, consider the inputs of the circuits of Fig. 11.29b and 11.29c short-circuited, and equate the output noise. It is readily shown that

$$v_i = v_{ia} + i_{ia}R + \frac{R_F}{R_F + R_E}v_e + \frac{R_E}{R_F + R_E}v_f \tag{11.76}$$

where

$$R = R_F \| R_E \tag{11.77}$$

Assuming that all noise sources in (11.76) are independent, we have

$$\overline{v_i^2} = \overline{v_{ia}^2} + \overline{i_{ia}^2}R^2 + 4kTR\Delta f \tag{11.78}$$

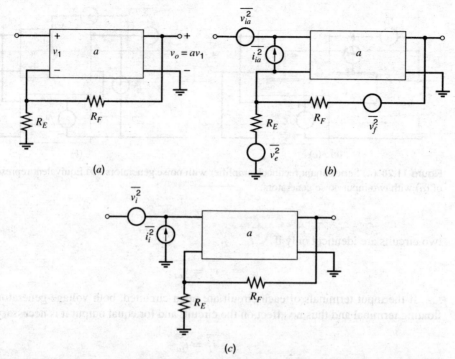

Figure 11.29 (a) Series-shunt feedback circuit. (b) Series-shunt feedback circuit including noise generators. (c) Equivalent representation of (b) with two input noise generators.

where the following substitutions have been made:

$$\overline{v_e^2} = 4kTR_E\,\Delta f \tag{11.79}$$

$$\overline{v_f^2} = 4kTR_F\,\Delta f \tag{11.80}$$

Equation 11.78 shows that in this practical case, the equivalent input noise voltage of the overall amplifier contains the input noise voltage of the basic amplifier plus two other terms. The second term in (11.78) is usually negligible, but the third term represents thermal noise in $R = R_E \| R_F$ and is often significant.

The equivalent input noise current $\overline{i_i^2}$ is calculated by open-circuiting both inputs and equating output noise. It is apparent that

$$\overline{i_i^2} \simeq \overline{i_{ia}^2} \tag{11.81}$$

since noise in the feedback resistors is no longer amplified, but appears only in shunt with the output. Thus the equivalent input noise current is *unaffected* by the application of feedback. The above results are true in general for series feedback at the input. For single-stage series feedback, the above equations are valid with $R_F \rightarrow \infty$ and $R = R_E$.

If the basic amplifier in Fig. 8.29 is an op amp, the calculation is slightly modified. This is due to the fact (shown in Section 11.8 and Fig. 11.39) that an op amp must be considered a three-port device for noise representation. However, if the circuit of Fig. 11.39 is used as the basic amplifier in the above calculation, expressions very similar to (11.78) and (11.81) are obtained.

Consider now the case of shunt feedback at the input, and as an example consider the shunt-shunt feedback circuit of Fig. 11.30a. This is shown in Fig. 11.30b with noise sources

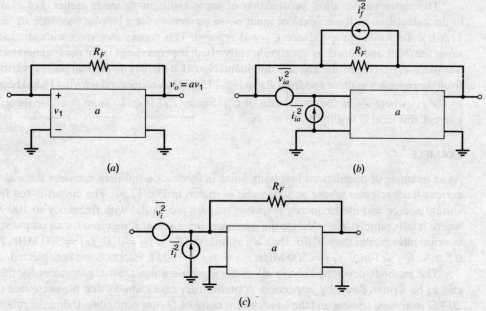

Figure 11.30 (a) Shunt-shunt feedback circuit. (b) Shunt-shunt feedback circuit including noise generators. (c) Equivalent representation of (b) with two input noise generators.

v_{ia}^2 and i_{ia}^2 of the basic amplifier, and noise source i_f^2 due to R_F. These noise sources are referred back to the input to give v_i^2 and i_i^2 as shown in Fig. 11.30c.

Open-circuiting the inputs of Fig. 11.30b and 11.30c, and equating output noise, we calculate

$$i_i = i_{ia} + \frac{v_{ia}}{R_F} + i_f \qquad (11.82)$$

Assuming that all noise sources in (11.82) are independent, we find

$$\overline{i_i^2} = \overline{i_{ia}^2} + \frac{\overline{v_{ia}^2}}{R_F^2} + 4kT\frac{1}{R_F}\Delta f \qquad (11.83)$$

Thus the equivalent input noise current with shunt feedback applied consists of the input noise current of the basic amplifier together with a term representing thermal noise in the feedback resistor. The second term in (11.83) is usually negligible. These results are true in general for shunt feedback at the input. *A general rule* for calculating the equivalent input noise contribution due to thermal noise in the feedback resistors is to follow the two-port methods described in Chapter 8 for calculating feedback-circuit loading on the basic amplifier. Once the shunt or series resistors representing feedback loading at the input have been determined, these same resistors may be used to calculate the thermal noise contribution at the input due to the feedback resistors.

If the inputs of the circuits of Fig. 11.30b and 11.30c are short-circuited, and the output noise is equated, it follows that

$$\overline{v_i^2} \simeq \overline{v_{ia}^2} \qquad (11.84)$$

Equations 11.83 and 11.84 are true in general for shunt feedback at the input. They apply directly when the basic amplifier of Fig. 11.30 is an op amp, since one input terminal of the basic amplifier is grounded and the op amp becomes a two-port device.

The above results allow justification of some assumptions made earlier. For example, in the calculation of the equivalent input noise generators for a bipolar transistor in Section 11.5.1, collector-base capacitance C_μ was ignored. This capacitance represents single-stage *shunt feedback* and thus does *not* significantly affect the equivalent input noise generators of a transistor, *even if* the Miller effect is dominant. Note that there is no thermal noise contribution from the capacitor as there was from R_F in Fig. 11.30. Also, the second term in (11.83) becomes $\overline{v_{ia}^2}/|Z_F|^2$ where Z_F is the impedance of C_μ. Since $|Z_F|$ is quite large at all frequencies of interest, this term is negligible.

■ **EXAMPLE**

As an example of calculations involving noise in feedback amplifiers, consider the wideband current-feedback pair whose ac schematic is shown in Fig. 11.31. The circuit is fed from a current source and the frequency response $|(i_o/i_i)(j\omega)|$ is flat with frequency to 100 MHz, where it falls rapidly. We calculate the minimum input signal i_s required for an output signal-to-noise ratio greater than 20 dB. Data are as follows: $\beta_1 = \beta_2 = 100$, $f_{T1} = 300$ MHz, $I_{C1} = 0.5$ mA, $I_{C2} = 1$ mA, $f_{T2} = 500$ MHz, $r_{b1} = r_{b2} = 100$ Ω. Flicker noise is neglected.

The methods developed above allow the equivalent input noise generators for this circuit to be written down by inspection. A preliminary check shows that the noise due to the 20-kΩ interstage resistor and the base current noise of Q_2 are negligible. Using the rule stated in Section 11.2.2, we find that the 20-kΩ resistor contributes an equivalent noise current of 2.5 μA. The base current of Q_2 is 10 μA. Both of these can be neglected when compared to the 500 μA collector current of Q_1. Thus the input noise generators of the whole circuit are those of Q_1 moved outside the feedback loop, together with the noise contributed by the feedback resistors.

Using the methods of Chapter 8, we can derive the basic amplifier including feedback loading and noise sources for the circuit of Fig. 11.31 as shown in Fig. 11.32. The equivalent input noise-current generator for the overall circuit can be calculated from Fig. 11.32 or by using (11.83) with $R_F = 5.5$ kΩ. Since the circuit is assumed to be driven from a current

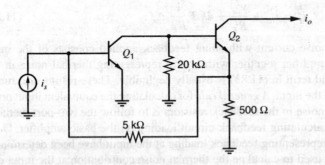

Figure 11.31 An ac schematic of a current feedback pair.

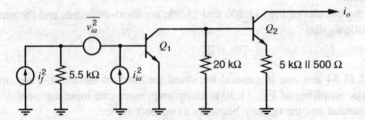

Figure 11.32 Basic amplifier for the circuit of Fig. 11.31 including feedback loading and noise sources.

source, the equivalent input noise voltage is not important. From (11.83)

$$\overline{i_i^2} = \overline{i_{ia}^2} + \frac{\overline{v_{ia}^2}}{(5500)^2} + 4kT\frac{1}{5500}\Delta f \qquad (11.85)$$

Using (11.57) and neglecting flicker noise, we have for $\overline{i_{ia}^2}$

$$\overline{i_{ia}^2} = 2q\left(I_B + \frac{I_C}{|\beta(jf)|^2}\right)\Delta f$$

and thus

$$\frac{\overline{i_{ia}^2}}{\Delta f} = 2q\left(5 + \frac{500}{|\beta|^2}\right) \times 10^{-6} \text{ A}^2/\text{Hz} \qquad (11.86)$$

Substitution of (11.86) in (11.85) gives

$$\frac{\overline{i_i^2}}{\Delta f} = 2q\left(5 + \frac{500}{|\beta|^2}\right) \times 10^{-6} + \frac{\overline{v_{ia}^2}}{(5500)^2 \Delta f} + 2q(9.1) \times 10^{-6} \qquad (11.87)$$

where the noise in the 5.5-kΩ resistor has been expressed in terms of the equivalent noise current of 9.1 μA.

Use of (11.50) gives

$$\frac{\overline{v_{ia}^2}}{\Delta f} = 4kT\left(r_{b1} + \frac{1}{2g_m}\right) = 4kT(126)$$

Division of this equation by $(5500)^2$ gives

$$\frac{\overline{v_{ia}^2}}{(5500)^2 \Delta f} = 4kT\frac{1}{240,000} \qquad (11.88)$$

$$= 2q(0.2) \times 10^{-6} \qquad (11.89)$$

Thus the term involving $\overline{v_{ia}^2}$ in (11.87) is seen to be equivalent to thermal noise in a 240-kΩ resistor using (11.88), and this can be expressed as noise in 0.2 μA of equivalent noise current, as shown in (11.89). This term is negligible in this example, as is usually the case.

Combining all these terms we can express (11.87) as

$$\frac{\overline{i_i^2}}{\Delta f} = 2q\left(5 + \frac{500}{|\beta|^2} + 0.2 + 9.1\right) \times 10^{-6} \text{ A}^2/\text{Hz}$$

$$= 2q\left(14.3 + \frac{500}{|\beta|^2}\right) \times 10^{-6} \text{ A}^2/\text{Hz} \qquad (11.90)$$

Equation 11.90 shows that the equivalent input noise-current spectral density rises at high frequencies (as $|\beta|$ falls) as expected for a transistor. In a single transistor without feedback, the equivalent input noise current also rises with frequency, but because the transistor gain falls with frequency, the output noise spectrum of a transistor without feedback always *falls* as frequency rises (see Section 11.4.1). However, in this case, the negative feedback holds the gain constant with frequency, and thus the *output* noise spectrum of this circuit will *rise* as frequency increases, until the amplifier band edge is reached. This is illustrated in Fig. 11.33, where the input noise-current spectrum, the amplifier frequency response squared, and the output noise-current spectrum (product of the first two) are shown. The current gain of the circuit is $A_I \simeq 11$.

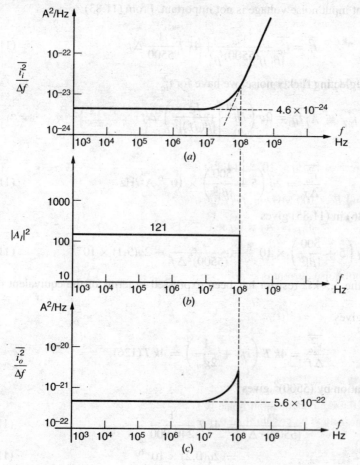

Figure 11.33 Noise performance of the circuit of Fig. 11.32. (a) Equivalent input noise spectrum. (b) Frequency response squared. (c) Output noise spectrum.

The total output noise from the circuit $\overline{i_{oT}^2}$ is obtained by integrating the output noise spectral density, which is

$$\frac{\overline{i_o^2}}{\Delta f} = A_I^2 \frac{\overline{i_i^2}}{\Delta f} \tag{11.91}$$

Thus

$$\overline{i_{oT}^2} = \int_0^B A_I^2 \frac{\overline{i_i^2}}{\Delta f} df$$

$$= A_I^2 \int_0^B 2q \left(14.3 + \frac{500}{|\beta(jf)|^2}\right) \times 10^{-6} df \tag{11.92}$$

where (11.90) has been used and A_I^2 is assumed constant up to $B = 10^8$ Hz as specified earlier. The current gain is

$$\beta(jf) = \frac{\beta_0}{1 + j\dfrac{\beta_0 f}{f_{T1}}} \tag{11.93}$$

and

$$\frac{1}{|\beta(jf)|^2} = \frac{1}{\beta_0^2}\left(1 + \frac{\beta_0^2 f^2}{f_{T1}^2}\right) \tag{11.94}$$

Substitution of (11.94) in (11.92) gives

$$\overline{i_{oT}^2} = A_I^2 2q \times 10^{-6} \int_0^B \left[14.3 + \frac{500}{\beta_0^2}\left(1 + \frac{\beta_0^2 f^2}{f_{T1}^2}\right)\right] df \tag{11.95}$$

$$= A_I^2 2q \times 10^{-6} \left[14.3 f + \frac{500}{\beta_0^2} f + \frac{500}{f_{T1}^2} \frac{f^3}{3}\right]_0^B \tag{11.96}$$

Using $\beta_0 = 100$ and $B = 100$ MHz $= f_{T1}/3$ gives

$$\overline{i_{oT}^2} = A_I^2 \times 2q \times 10^{-6}(14.3B + 18.6B) \tag{11.97}$$

$$\overline{i_{oT}^2} = A_I^2 \times 1.05 \times 10^{-15} \text{ A}^2 \tag{11.98}$$

The equivalent input noise current is

$$\overline{i_{iT}^2} = \frac{\overline{i_{oT}^2}}{A_I^2} = 1.05 \times 10^{-15} \text{ A}^2$$

and from this

$$i_{iT} = 32.4 \text{ nA rms} \tag{11.99}$$

Thus the equivalent input noise current is 32.4 nA rms and (11.97) shows that the frequency-dependent part of the equivalent input noise is dominant. For a 20-dB signal-to-noise ratio, input signal current i_s must be greater than 0.32 µA rms.

■

11.7 Noise Performance of Other Transistor Configurations

Transistor configurations other than the common-emitter and common-source stages considered so far are often used in integrated-circuit design. The noise performance of other important configurations will now be considered. To avoid repetition, the discussion will be directed toward bipolar devices only. However, all the results carry over directly to FET circuits.

11.7.1 Common-Base Stage Noise Performance

The common-base stage is sometimes used as a low-input-impedance current amplifier. A common-base stage is shown in Fig. 11.34a and the small-signal equivalent circuit is shown in Fig. 11.34b together with the equivalent input noise generators derived for a common-emitter stage. Since these noise generators represent the noise performance of the transistor in any connection, Fig. 11.34b is a valid representation of common-base noise performance. In Fig. 11.34c the noise performance of the common-base stage is represented in the standard fashion with equivalent input noise generators $\overline{v_{iB}^2}$ and $\overline{i_{iB}^2}$. These can be related to the common-emitter input generators by alternately short circuiting and open circuiting the circuits of Fig. 11.34b and 11.34c and equating output noise. It then follows that

$$\overline{i_{iB}^2} = \overline{i_i^2} \tag{11.100}$$

$$\overline{v_{iB}^2} = \overline{v_i^2} \tag{11.101}$$

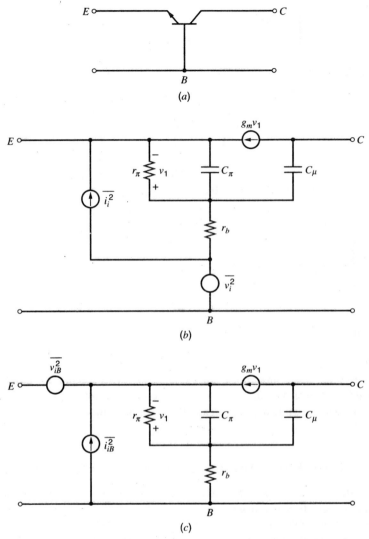

Figure 11.34 (a) Common-base transistor configuration. (b) Common-base equivalent circuit with noise generators. (c) Common-base equivalent circuit with input noise generators.

Thus the equivalent input noise generators of common-emitter and common-base connections are the same and the noise performance of the two configurations is identical, even though their input impedances differ greatly.

Although the noise performances of common-emitter and common-base stages are nominally identical (for the same device parameters), there is one characteristic of the common-base stage that makes it generally *unsuitable* for use as a low-noise input stage. This is due to the fact that its current gain $\alpha \simeq 1$, and thus any noise current at the *output* of the common-base stage is referred directly back to the input *without* reduction. Thus a 10-kΩ load resistor that has an equivalent noise current of 5 μA produces this amount of equivalent noise current at the input. In many circuits this would be the dominant source of input current noise. The equivalent input noise currents of following stages are also referred back unchanged to the input of the common-base stage. This problem can be overcome in discrete common-base circuits by use

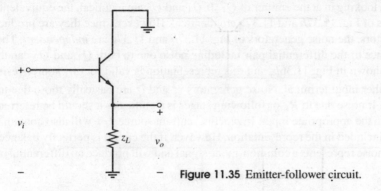

Figure 11.35 Emitter-follower circuit.

of a transformer that gives current gain at the output of the common-base stage. This option is not available in integrated-circuit design unless resort is made to external components.

11.7.2 Emitter-Follower Noise Performance

Consider the emitter follower shown in Fig. 11.35. The noise performance of this circuit can be calculated using the results of previous sections. The circuit can be viewed as a series-feedback stage and the equivalent input noise generators of the transistor can be moved unchanged back to the input of the complete circuit. Thus, if noise in z_L is neglected, the emitter follower has the same equivalent input noise generators as the common-emitter and common-base stages. However, since the emitter follower has unity voltage gain, the equivalent input noise voltage of the following stage is transformed unchanged to the input, thus degrading the noise performance of the circuit. Noise due to z_L must also be included, but since the follower output is taken at the emitter, which is a low impedance point, the noise due to z_L is greatly attenuated compared with its effect on a series-feedback stage.

11.7.3 Differential-Pair Noise Performance

The differential pair is the basic building block of linear integrated circuits and, as such, its noise performance is of considerable importance. A bipolar differential pair is shown in Fig. 11.36a, and the base of each device is generally independently accessible, as shown. Thus this circuit cannot in general be represented as a two-port, and its noise performance cannot be represented in the usual fashion by two input noise generators. However, the techniques developed previously can be used to derive an equivalent noise representation of the circuit that employs two noise generators at *each* input. This is illustrated in Fig. 11.36b, and a simpler version of this circuit, which employs only three noise generators, is shown in Fig. 11.36c.

The noise representation of Fig. 11.36b can be derived by considering noise due to each device separately. Consider first noise in Q_1, which can be represented by input noise generators $\overline{v_i^2}$ and $\overline{i_i^2}$ as shown in Fig. 11.37a. These noise generators are those for a single transistor as given by (11.50) and (11.57). Transistor Q_2 is initially assumed *noiseless* and the impedance seen looking in its emitter is z_{E2}. Note that z_{E2} will be a function of the impedance connected from the base of Q_2 to ground. As described in previous sections, the noise generators of Fig. 11.37a can be moved unchanged to the input of the circuit (independent of z_{E2}) as shown in Fig. 11.37b. This representation can then be used to calculate the output noise produced by Q_1 in the differential pair for *any* impedances connected from the base of Q_1 and Q_2 to ground.

Now consider noise due to Q_2. In similar fashion this can be represented by noise generators $\overline{v_i^2}$ and $\overline{i_i^2}$, as shown in Fig. 11.37c. In this case Q_1 is assumed noiseless and z_{E1} is the impedance seen looking in at the emitter of Q_1. If Q_1 and Q_2 are identical, the equivalent input noise generators of Fig. 11.37b and 11.37c are identical. However, since they are produced by different transistors, the noise generators of Fig. 11.37b and 11.37c are *independent*. The total noise performance of the differential pair including noise due to both Q_1 and Q_2 can thus be represented as shown in Fig. 11.36b, and this representation is valid for *any* source resistance connected to either input terminal. Noise generators $\overline{v_i^2}$ and $\overline{i_i^2}$ are basically those due to each transistor alone. If noise due to R_L or following stages is significant, it should be referred back symmetrically to the appropriate input. In practice, current source I_{EE} will also contain noise, and this can be included in the representation. However, if the circuit is perfectly balanced, the current-source noise represents a common-mode signal and will produce no differential output.

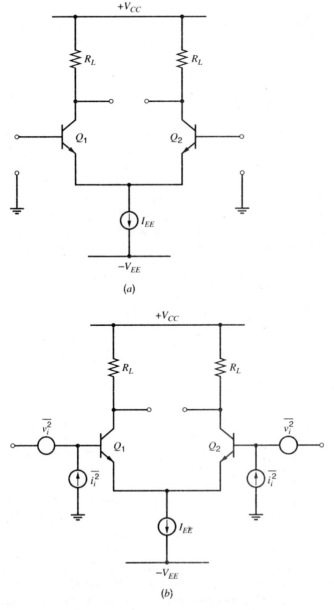

Figure 11.36
(a) Differential-pair circuit.
(b) Complete differential pair noise representation.

11.7 Noise Performance of Other Transistor Configurations

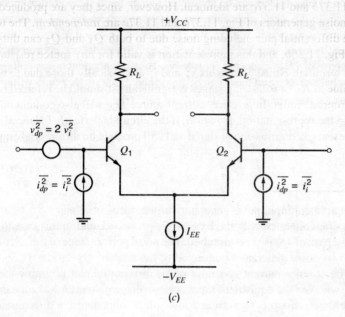

Figure 11.36 (c) Simplified noise representation.

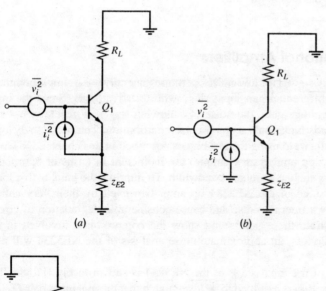

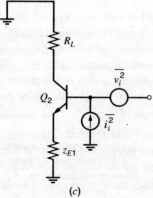

Figure 11.37 (a) ac schematic of a differential pair including noise due to Q_1 only. (b) ac schematic of a differential pair with noise due to Q_1 referred to the input. (c) ac schematic of a differential pair including noise due to Q_2 only.

The noise representation of Fig. 11.36b can be simplified somewhat if the common-mode rejection of the circuit is high. In this case, one of the noise-voltage generators can be moved to the other side of the circuit, as shown in Fig. 11.36c. This can be justified if equal noise-voltage generators are added in series with each input and these generators are chosen such that the noise voltage at the base of Q_2 is canceled. This leaves two independent noise-voltage generators in series with the base of Q_1, and these can be represented as a single noise-voltage generator of value $\overline{2v_i^2}$. Thus for the circuit of Fig. 11.36c we can write

$$\overline{v_{dp}^2} = \overline{2v_i^2} \tag{11.102}$$

$$\overline{i_{dp}^2} = \overline{i_i^2} \tag{11.103}$$

where $\overline{v_{dp}^2}$ and $\overline{i_{dp}^2}$ are the equivalent input noise generators of the differential pair.

The differential pair is often operated with the base of Q_2 grounded, and in this case the noise-current generator at the base of Q_2 is short circuited. The noise performance of the circuit is then represented by the two noise generators connected to the base of Q_1 in Fig. 11.36c. In this case the equivalent input noise-current generator of the differential pair is simply that due to one transistor alone, whereas the equivalent input noise-voltage generator has a mean-square value *twice* that of either transistor. Thus from a low source impedance, a differential pair has an equivalent input noise voltage 3 dB higher than a common-emitter stage with the same collector current as the devices in the pair.

11.8 Noise in Operational Amplifiers

Integrated-circuit amplifiers designed for low-noise operation generally use a simple common-emitter or common-source differential-pair input stage with resistive loads. Since the input stage has both current and voltage gain, the noise of following stages is generally not significant, and the resistive loads make only a small noise contribution. The noise analysis of such circuits is quite straightforward using the techniques described in this chapter. However, circuits of this type (the 725 op amp is an example) are inefficient in terms of optimizing important op amp parameters such as gain and bandwidth. To improve the gain, active loads can be used as in the first stage of the NE5234 op amp. However, by their very nature, active loads amplify their own internal noise and cause considerable degradation of circuit noise performance. To illustrate these points and show the compromises involved in the design of general-purpose circuits, an approximate noise analysis of the NE5234 will now be made.[19]

A simplified schematic of the input stage of the NE5234 is shown in Fig. 11.38a. The common-mode input voltage, V_{IC}, is assumed to be low enough that the *npn* input pair Q_1-Q_2 in Fig. 6.36 is off and can be ignored here. The differential output current is i_o. Components Q_{13} and R_{13} of the active load generate noise and contribute to the output noise at i_o. Since transistor Q_9 presents a high impedance to the active load, the noise due to the active load can be calculated from the isolated portion of the circuit shown in Fig. 11.38b. Noise due to Q_{13} and R_{13} is represented by equivalent input noise generators $\overline{v_{i13}^2}$ and $\overline{i_{i13}^2}$. The impedance looking back into the Bias$_1$ output of the bias circuit is small and ignored. Therefore, $\overline{i_{i13}^2}$ may be neglected. Also, noise generated in the bias circuit may be ignored because it couples into Q_{13} and Q_{14} equally and does not affect the differential output current i_o with perfect matching. Start with (11.78). Using (11.50) for $\overline{v_{ia}^2}$ and neglecting the $\overline{i_{ia}^2}$ term as well as flicker

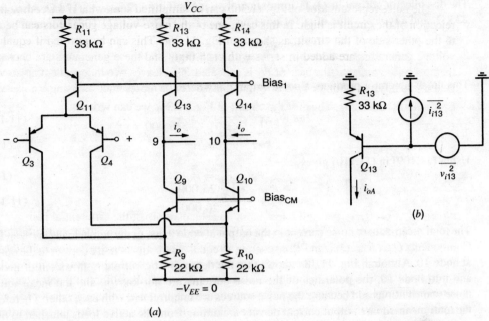

Figure 11.38 (a) Simplified schematic of the NE5234 op amp input stage with $V_{IC} \ll 0.8$ V. (b) ac schematic (including noise) of Q_{13} and R_{13}.

noise gives

$$\frac{\overline{v_{i13}^2}}{\Delta f} = 4kT\left(r_{b13} + \frac{1}{2g_{m13}} + R_{13}\right) \tag{11.104}$$

This noise generator can be evaluated as follows. The dc collector current in Q_{13} is approximately 6 μA, giving $1/2g_{m13} = 2.17$ kΩ. Also, $R_{13} = 33$ kΩ, and a typical value of r_b for *pnp* transistors is 200 Ω. Thus (11.104) becomes

$$\frac{\overline{v_{i13}^2}}{\Delta f} = 4kT(35,000) \tag{11.105}$$

This noise generator produces a noise output i_{oA} where

$$i_{oA} \simeq \frac{1}{\frac{1}{g_{m13}} + R_{13}} v_{i13} \simeq \frac{v_{i13}}{37,000} \tag{11.106}$$

Using (11.105) in (11.106) gives

$$\frac{\overline{i_{oA}^2}}{\Delta f} = 4kT\frac{35,000}{37,000^2} \tag{11.107}$$

A similar analysis done to find $\overline{v_{i9}^2}$, the noise due to Q_9 and R_9 in series with the base of Q_9, gives

$$\frac{\overline{v_{i9}^2}}{\Delta f} = 4kT\left(r_{b9} + \frac{1}{2g_{m9}} + R_9\right) \tag{11.108}$$

The dc collector current in Q_9 is approximately 6 μA, giving $1/2g_{m9} = 2.17$ kΩ. Also, $R_9 = 22$ kΩ, and a typical value of r_b for npn transistors is 400 Ω. Thus (11.108) becomes

$$\frac{\overline{v_{i9}^2}}{\Delta f} = 4kT(25,000) \qquad (11.109)$$

This noise generator produces a noise output i_{oB} where

$$i_{oB} \simeq \frac{1}{\frac{1}{g_{m9}} + R_9} v_{i9} \simeq \frac{v_{i9}}{26,000} \qquad (11.110)$$

Using (11.109) in (11.110) gives

$$\frac{\overline{i_{oB}^2}}{\Delta f} = 4kT \frac{25,000}{26,000^2} \qquad (11.111)$$

The total mean-square noise current in the output at node 9 due to the active loads is $\overline{i_{oA}^2} + \overline{i_{oB}^2}$. Components Q_{14}, R_{14}, Q_{10}, and R_{10} generate an equal mean-square noise current in the output at node 10. Although Fig. 11.38a shows that the deterministic current i_o flows out of node 9 and into node 10, the polarities of the noise currents are not known, and the mean-square noise contributions add because the noise sources are uncorrelated with each other. Therefore, the total mean-square output current density stemming from the active loads attached to both nodes 9 and 10 is

$$\frac{\overline{i_{oAB}^2}}{\Delta f} = 2\left(\frac{\overline{i_{oA}^2}}{\Delta f} + \frac{\overline{i_{oB}^2}}{\Delta f}\right) \qquad (11.112)$$

This result can be referred back to the input of the complete circuit of Fig. 11.38a in the standard manner. Consider the contribution to the equivalent input noise voltage. A voltage applied at the input of the full circuit of Fig. 11.38a gives

$$\frac{i_o}{v_{id}} = G_{m1}$$

where $G_{m1} = 97$ μA/V from (6.186). Therefore,

$$\overline{i_o^2} = G_{m1}^2 \overline{v_{id}^2} \qquad (11.113)$$

Equating output noise current in (11.112) and (11.113), we obtain an equivalent input noise voltage due to the active loads as

$$\frac{\overline{v_{iAB}^2}}{\Delta f} = 4kT\left[2\left(\frac{35,000}{37,000^2} + \frac{25,000}{26,000^2}\right)\right]\left(\frac{1}{97 \times 10^{-6}}\right)^2 = 4kT(13,000) \qquad (11.114)$$

Thus the active load causes a contribution of 13 kΩ to the equivalent input noise resistance of the NE5234 op amp. Finally, the results of Section 11.7.3 show that the equivalent input noise voltage squared of the differential pair Q_3-Q_4 is just the sum of the values from each half of the circuit. This gives an input noise-voltage contribution due to Q_3-Q_4 of

$$\frac{\overline{v_{iC}^2}}{\Delta f} = 4kT\left(\frac{1}{2g_{m3}} + \frac{1}{2g_{m4}} + r_{b3} + r_{b4}\right) \qquad (11.115)$$

Assuming a collector bias current of 3 μA for each device, and $r_b = 200$ Ω in (11.115), we calculate

$$\frac{\overline{v_{iC}^2}}{\Delta f} = 4kT(9,000) \qquad (11.116)$$

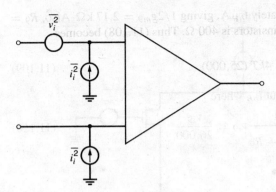

Figure 11.39 Complete op amp noise representation.

For the total input noise voltage of the NE5234, combining (11.114) and (11.116) gives

$$\frac{\overline{v_i^2}}{\Delta f} = \frac{\overline{v_{iAB}^2}}{\Delta f} + \frac{\overline{v_{iC}^2}}{\Delta f} = 4kT(22{,}000) \qquad (11.117)$$

Thus the input noise voltage of the NE5234 is represented by an equivalent input noise resistance of 22 kΩ. This is a large value and is close to the measured and computer-calculated result. Note that the active load is the main contributor to the noise. The magnitude of the noise voltage can be appreciated by noting that if this op amp is fed from a 1-kΩ source resistance for example, the circuit adds twenty-two times as much noise power as is contributed by the source resistance itself.

The calculations above concerned the equivalent input noise voltage of the NE5234. A similar calculation performed to determine the equivalent input noise current shows that it is dominated by the base current of the input device. The current gain of the input differential pair is sufficient to ensure that following stage noise, including noise in the active loads, gives a negligible contribution to input current noise. Since the circuit is differential, the complete noise representation consists of the equivalent input noise voltage calculated above plus two equivalent input noise-current generators as shown in Fig. 11.39. This follows from the discussion of differential-pair noise performance in Section 11.7.3. The equivalent input noise-current generators are (neglecting flicker noise)

$$\frac{\overline{i_i^2}}{\Delta f} \simeq 2qI_B$$

where I_B is the base current of Q_1 or Q_2 in Fig. 11.38a. With $\beta_{F(pnp)} = 10$,

$$\frac{\overline{i_i^2}}{\Delta f} \simeq 2q(0.3 \times 10^{-6}) \qquad (11.118)$$

Increasing β_F reduces the input mean-square current noise. When $V_{IC} \gg 0.8$ V, the input stage of the NE5234 turns off Q_3-Q_4 and turns on Q_1-Q_2 with the same collector bias currents as assumed above. Therefore, the input mean-square current density depends on V_{IC} unless $\beta_{F(npn)} = \beta_{F(pnp)}$.

Consider the case of the CMOS input stage shown in Fig. 11.40. If the noise in each MOS transistor is represented as shown by its equivalent input noise voltage generator, the equivalent input noise voltage of the circuit $\overline{v_{eqT}^2}$ can be calculated by equating the output current noise

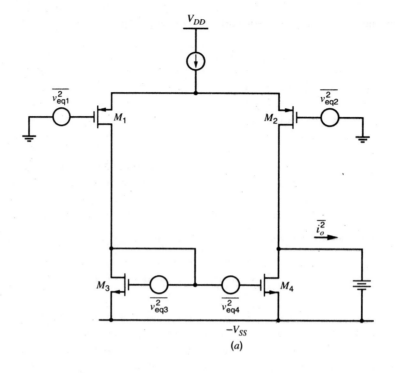

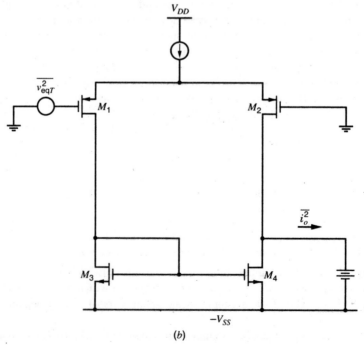

Figure 11.40 (a) CMOS input stage device noise contributions. (b) Equivalent input noise representation.

$\overline{i_o^2}$ for the circuits in Fig. 11.40a and Fig. 11.40b giving

$$\overline{v_{eqT}^2} = \overline{v_{eq1}^2} + \overline{v_{eq2}^2} + \left(\frac{g_{m3}}{g_{m1}}\right)^2 (\overline{v_{eq3}^2} + \overline{v_{eq4}^2}) \tag{11.119}$$

where it has been assumed that $g_{m1} = g_{m2}$ and that $g_{m3} = g_{m4}$. Thus, the input transistors contribute to the input noise directly while the contribution of the loads is reduced by the square of the ratio of their transconductance to that of the input transistors. The significance of this in the design can be further appreciated by considering the input-referred $1/f$ noise and the input-referred thermal noise separately.

The dependence of the $1/f$ portion of the device equivalent input noise-voltage spectrum on device geometry and bias conditions was considered earlier. Considerable discrepancy exists in the published data on $1/f$ noise, indicating that it arises from a mechanism that is strongly affected by details of device fabrication. The most widely accepted model for $1/f$ noise is that, for a given device, the gate-referred equivalent mean-square voltage noise is approximately independent of bias conditions in saturation and is inversely proportional to the gate capacitance of the device. The following analytical results are based on this model, but it should be emphasized that the actual dependence must be verified for each process technology and device type. Thus we let

$$\overline{v_{eq}^2} = \frac{K_f \Delta f}{C_{ox} W L f} \tag{11.120}$$

where the parameter K_f is the flicker-noise coefficient. Utilizing this assumption, and using (11.119), we obtain for the equivalent input $1/f$ noise generator for the circuit of Fig. 11.40a

$$\overline{v_{1/f}^2} = \frac{2K_p}{f W_1 L_1 C_{ox}} \left(1 + \frac{K_n \mu_n L_1^2}{K_p \mu_p L_3^2}\right) \Delta f \tag{11.121}$$

where K_n and K_p are the flicker noise coefficients for the n-channel and p-channel devices, respectively. Depending on processing details, these may be comparable or different by a factor of two or more. Note that the first term in (11.121) is the equivalent input noise of the input transistors alone, and the term in parentheses is the increase in noise over and above this value due to the loads. The second term shows that the load contribution can be made small by simply making the channel lengths of the loads longer than those of the input transistors by a factor of about two or more. The input transistors can then be made wide enough to achieve the desired performance. It is interesting to note that changing the width of the channel in the loads does not effect the $1/f$ noise performance.

The thermal noise performance of the circuit of Fig. 11.40a can be calculated as follows. As discussed in Section 11.5.2, the input-referred thermal noise of a MOS transistor is given by

$$\overline{v_{eq}^2} = 4kT \left(\frac{2}{3g_m}\right) \Delta f \tag{11.122}$$

Utilizing the same approach as for the flicker noise, we obtain for the equivalent input thermal noise of the circuit of Fig. 11.40a

$$\overline{v_{Th}^2} = 4kT \frac{4}{3\sqrt{2\mu_p C_{ox}(W/L)_1 I_D}} \left(1 + \sqrt{\frac{\mu_n (W/L)_3}{\mu_p (W/L)_1}}\right) \Delta f \tag{11.123}$$

where I_D is the bias current in each device. Again, the first term represents the thermal noise from the input transistors and the term in parentheses represents the fractional increase in noise due to the loads. The term in parentheses will be small if the W/L ratios are chosen so that the transconductance of the input devices is much larger than that of the loads. If this condition is satisfied, then the input noise is simply determined by the transconductance of the input transistors.

11.9 Noise Bandwidth

In the noise analysis performed thus far, the circuits considered were generally assumed to have simple gain-frequency characteristics with abrupt band edges as shown in Fig. 11.33b. The calculation of total circuit noise then reduced to an integration of the noise spectral density across this band. In practice, many circuits do not have such ideal gain-frequency characteristics, and the calculation of total circuit noise can be much more complex in those cases. However, if the equivalent input noise spectral density of a circuit is *constant* and independent of frequency (i.e., if the noise is white), we can simplify the calculations using the concept of *noise bandwidth* described below.

Consider an amplifier as shown in Fig. 11.41, and assume it is fed from a low source impedance so that the equivalent input noise voltage $\overline{v_i^2}$ determines the noise performance. Assume initially that the spectral density $\overline{v_i^2}/\Delta f = S_i(f) = S_{i0}$ of the input noise voltage is flat as shown in Fig. 11.42a. Further assume that the magnitude squared of the voltage gain $|A_v(jf)|^2$ of the circuit is as shown in Fig. 11.42b. The output noise-voltage spectral density $S_o(f) = \overline{v_o^2}/\Delta f$ is the product of the input noise-voltage spectral density and the square of the voltage gain and is shown in Fig. 11.42c. The *total* output noise voltage is obtained by summing the contribution from $S_o(f)$ in each frequency increment Δf between zero and infinity to give

$$\overline{v_{oT}^2} = \sum_{f=0}^{\infty} S_o(f)\Delta f = \int_0^\infty S_o(f)df = \int_0^\infty |A_v(jf)|^2 S_{i0}df$$

$$= S_{i0}\int_0^\infty |A_v(jf)|^2 df \qquad (11.124)$$

The evaluation of the integral of (11.124) is often difficult except for very simple transfer functions. However, if the problem is transformed into a *normalized* form, the integrals of common circuit functions can be evaluated and tabulated for use in noise calculations. For this purpose, consider a transfer function as shown in Fig. 11.43 with the same low-frequency value A_{v0} as the original circuit but with an abrupt band edge at a frequency f_N. Frequency f_N is chosen to give the *same* total output noise voltage as the original circuit when the same input noise voltage is applied. Thus

$$\overline{v_{oT}^2} = S_{i0}A_{v0}^2 f_N \qquad (11.125)$$

If (11.124) and (11.125) are equated, we obtain

$$f_N = \frac{1}{A_{v0}^2}\int_0^\infty |A_v(jf)|^2 df \qquad (11.126)$$

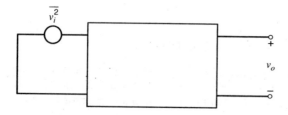

Figure 11.41 Circuit with equivalent input noise voltage generator.

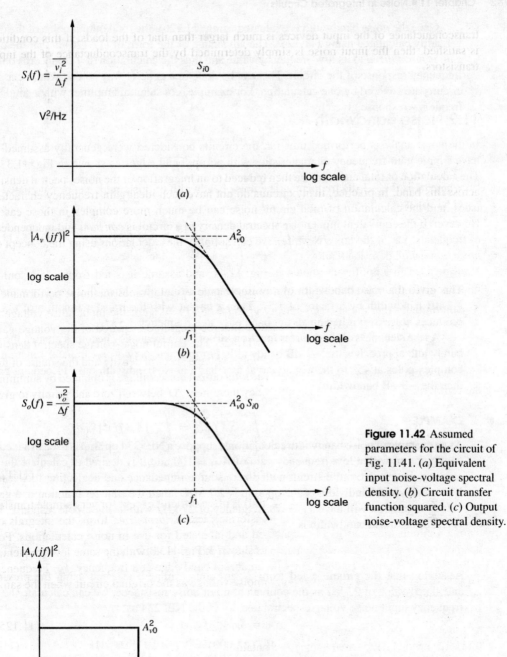

Figure 11.42 Assumed parameters for the circuit of Fig. 11.41. (*a*) Equivalent input noise-voltage spectral density. (*b*) Circuit transfer function squared. (*c*) Output noise-voltage spectral density.

Figure 11.43 Transfer function of a circuit giving the same output noise as a circuit with a transfer function as specified in Fig. 11.42*b*.

where f_N is the *equivalent noise bandwidth* of the circuit. Although derived for the case of a voltage transfer function, this result can be used for any type of transfer function. Note that the integration of (11.126) can be performed numerically if measured data for the circuit transfer function is available.

Once the noise bandwidth is evaluated using (11.126), the total output noise of the circuit is readily calculated using (11.125). The advantage of the form of (11.126) is that the circuit gain is *normalized* to its low-frequency value and thus the calculation of f_N concerns only the frequency response of the circuit. This can be done in a general way so that whole classes of circuits are covered by one calculation. For example, consider an amplifier with a single-pole frequency response given by

$$A_v(jf) = \frac{A_{v0}}{1 + j\frac{f}{f_1}} \quad (11.127)$$

where f_1 is the -3-dB frequency. The noise bandwidth of this circuit can be calculated from (11.126) as

$$f_N = \int_0^\infty \frac{df}{1 + \left(\frac{f}{f_1}\right)^2} = \frac{\pi}{2} f_1 = 1.57 f_1 \quad (11.128)$$

This gives the noise bandwidth of *any* single-pole circuit and shows that it is larger than the -3-dB bandwidth by a factor of 1.57. Thus a circuit with the transfer function of (11.127) produces noise *as if* it had an abrupt band edge at a frequency $1.57 f_1$.

As the steepness of the transfer function fall-off with frequency becomes greater, the noise bandwidth approaches the -3-dB bandwidth. For example, a two-pole transfer function with complex poles at 45° to the negative real axis has a noise bandwidth only 11 percent greater than the -3-dB bandwidth.

■ **EXAMPLE**

As an example of noise bandwidth calculations, suppose a NE5234 op amp is used in a feedback configuration with a low-frequency gain of $A_{v0} = 100$ and it is desired to calculate the total output noise v_{oT} from the circuit with a zero source impedance and neglecting flicker noise. If the unity-gain bandwidth of the op amp is 2.7 MHz, then the transfer function in a gain of 100 configuration will have a -3-dB frequency of 27 kHz with a single-pole response. From (11.128) the noise bandwidth is

$$f_N = 1.57 \times 27 \text{ kHz} = 42 \text{ kHz} \quad (11.129)$$

Assuming that the circuit is fed from a zero source impedance, and using the previously calculated value of 22 kΩ as the equivalent input noise resistance, we can calculate the low-frequency input noise voltage spectral density of the NE5234 as

$$S_{i0} = \frac{\overline{v_i^2}}{\Delta f} = 4kT(22{,}000) = 3.7 \times 10^{-16} \text{ V}^2/\text{Hz} \quad (11.130)$$

Using $A_{v0} = 100$ together with substitution of (11.129) and (11.130) in (11.125) gives, for the total output noise voltage,

$$\overline{v_{oT}^2} = 3.7 \times 10^{-16} \times (100)^2 \times 42 \times 10^3 \text{ V}^2 = 1.6 \times 10^{-7} \text{ V}^2$$

■ and thus $v_{oT} = 390$ μV rms.

The calculations of noise bandwidth considered above were based on the assumption of a flat input noise spectrum. This is often true in practice and the concept of noise bandwidth is useful in those cases, but there are also many examples where the input noise spectrum varies with frequency. In these cases, the total output noise voltage is given by

$$\overline{v_{oT}^2} = \int_0^\infty S_o(f)df \tag{11.131}$$

$$= \int_0^\infty |A_v(jf)|^2 S_i(f)df \tag{11.132}$$

where $A_v(jf)$ is the voltage gain of the circuit and $S_i(f)$ is the input noise-voltage spectral density. If the circuit has a voltage gain A_{vS} at the frequency of the applied input signal, then the total equivalent input noise voltage becomes

$$\overline{v_{iT}^2} = \frac{1}{A_{vS}^2} \int_0^\infty |A_v(jf)|^2 S_i(f)df \tag{11.133}$$

$$= \int_0^\infty \left|\frac{A_v(jf)}{A_{vS}}\right|^2 S_i(f)df \tag{11.134}$$

Equation 11.134 shows that, in general, the total equivalent input noise voltage of a circuit is obtained by integrating the product of the input noise-voltage spectrum and the normalized voltage gain function.

One last topic that should be mentioned in this section is the problem that occurs in calculating the flicker noise in direct-coupled amplifiers. Consider an amplifier with an input noise spectral density as shown in Fig. 11.44a and a voltage gain that extends down to dc and up to $f_1 = 10$ kHz with an abrupt cutoff, as shown in Fig. 11.44b. Then, using (11.134), we

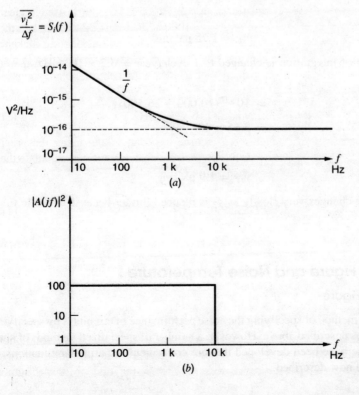

Figure 11.44 (a) Input noise-voltage spectral density for a circuit. (b) Circuit transfer function squared.

can calculate the total equivalent input noise voltage as

$$\overline{v_{iT}^2} = \int_0^{f_1} S_i(f)df$$

$$= \int_0^{f_1} \left(1 + \frac{1000}{f}\right) \times 10^{-16} df$$

$$= 10^{-16}[f + 1000 \ln f]_0^{f_1} \qquad (11.135)$$

Evaluating (11.135) produces a problem since $\overline{v_{iT}^2}$ is infinite when a lower limit of zero is used on the integration. This suggests infinite power in the $1/f$ noise signal. In practice, measurements of $1/f$ noise spectra show a continued $1/f$ dependence to as low a frequency as is measured (cycles per day or less). This problem can be resolved only by noting that $1/f$ noise eventually becomes indistinguishable from thermal drift and that the lower limit of the integration must be specified by the period of observation. For example, taking a lower limit to the integration of $f_2 = 1$ cycle/day, we have $f_2 = 1.16 \times 10^{-5}$ Hz. Changing the limit in (11.135) we find

$$\overline{v_{iT}^2} = 10^{-16}[f + 1000 \ln f]_{f_2}^{f_1}$$

$$= 10^{-16}\left[(f_1 - f_2) + 1000 \ln \frac{f_1}{f_2}\right] \qquad (11.136)$$

Using $f_1 = 10$ kHz and $f_2 = 1.16 \times 10^{-5}$ Hz in (11.136) gives

$$\overline{v_{iT}^2} = 10^{-16}(10{,}000 + 20{,}600)$$

$$= 3.06 \times 10^{-12} \text{ V}^2$$

and thus

$$v_{iT} = 1.75 \text{ μV rms}$$

If the lower limit of integration is changed to 1 cycle/year $= 3.2 \times 10^{-8}$ Hz, then (11.136) becomes

$$\overline{v_{iT}^2} = 10^{-16}(10{,}000 + 26{,}500)$$

$$= 3.65 \times 10^{-12} \text{ V}^2$$

and thus

$$v_{iT} = 1.9 \text{ μV rms}$$

The noise voltage changes very slowly as f_2 is reduced further because of the ln function in (11.136).

11.10 Noise Figure and Noise Temperature

11.10.1 Noise Figure

The most general method of specifying the noise performance of circuits is by specifying input noise generators as described above. However, a number of specialized methods of specifying noise performance have been developed that are convenient in particular situations. Two of these methods are now described.

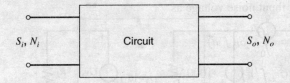

Figure 11.45 Signal and noise power at the input and output of a circuit.

The *noise figure* (F) is a commonly used method of specifying the noise performance of a circuit or a device. Its disadvantage is that it is limited to situations where the source impedance is resistive, and this precludes its use in many applications where noise performance is important. However, it is widely used as a measure of noise performance in communication systems where the source impedance is often resistive.

The definition of the noise figure of a circuit is

$$F = \frac{\text{input } S/N \text{ ratio}}{\text{output } S/N \text{ ratio}} \qquad (11.137)$$

and F is usually expressed in decibels. The utility of the noise-figure concept is apparent from the definition, as it gives a direct measure of the signal-to-noise (S/N) ratio degradation that is caused by the circuit. For example, if the S/N ratio at the input to a circuit is 50 dB, and the circuit noise figure is 5 dB, then the S/N ratio at the output of the circuit is 45 dB.

Consider a circuit as shown in Fig. 11.45, where S represents signal power and N represents noise power. The input noise power N_i is always taken as the noise in the *source resistance*. The output noise power N_o is the total output noise including the circuit contribution and noise transmitted from the source resistance. From (11.137) the noise figure is

$$F = \frac{S_i}{N_i} \frac{N_o}{S_o} \qquad (11.138)$$

For an *ideal noiseless amplifier*, all output noise comes from the source resistance at the input, and thus if G is the circuit power gain, then the output signal S_o and the output noise N_o are given by

$$S_o = GS_i \qquad (11.139)$$

$$N_o = GN_i \qquad (11.140)$$

Substituting (11.139) and (11.140) in (11.138) gives $F = 1$ or 0 dB in this case.

A useful alternative definition of F may be derived from (11.138) as follows:

$$F = \frac{S_i}{N_i} \frac{N_o}{S_o} = \frac{N_o}{GN_i} \qquad (11.141)$$

Equation 11.141 can be written as

$$F = \frac{\text{total output noise}}{\text{that part of the output noise due to the source resistance}} \qquad (11.142)$$

Note that since F is specified by a *power* ratio, the value in decibels is given by $10 \log_{10}$ (numerical ratio).

The calculations of the previous sections have shown that the noise parameters of most circuits vary with frequency, and thus the bandwidth must be specified when the noise figure of a circuit is calculated. The noise figure is often specified for a *small bandwidth* Δf at a frequency f where $\Delta f \ll f$. This is called the *spot noise figure* and applies to tuned amplifiers and also to broadband amplifiers that may be followed by frequency-selective circuits. For broadband amplifiers whose output is utilized over a wide bandwidth, an *average* noise figure is often

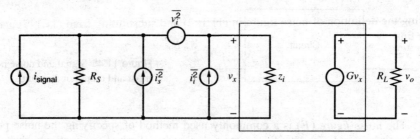

Figure 11.46 Equivalent input noise representation for the calculation of noise figure.

specified. This requires calculation of the total output noise over the frequency band of interest using the methods described in previous sections.

In many cases, the most convenient way to calculate noise figure is to return to the original equivalent circuit of the device with its basic noise generators to perform the calculation. However, some insight into the effect of circuit parameters on the noise figure can be obtained by using the equivalent input noise generator representation of Fig. 11.46. In this figure, a circuit with input impedance z_i and voltage gain $G = v_o/v_x$ is fed from a source resistance R_S and drives a load R_L. The source resistance shows thermal noise $\overline{i_s^2}$, and the noise of the circuit itself is represented by $\overline{i_i^2}$ and $\overline{v_i^2}$, assumed uncorrelated. The noise at the input terminals due to $\overline{v_i^2}$ and $\overline{i_i^2}$ is

$$v_{xA} = v_i \frac{z_i}{z_i + R_S} + i_i \frac{R_S z_i}{R_S + z_i}$$

and thus

$$\overline{v_{xA}^2} = \overline{v_i^2} \frac{|z_i|^2}{|z_i + R_S|^2} + \overline{i_i^2} \frac{|R_S z_i|^2}{|R_S + z_i|^2} \qquad (11.143)$$

The noise power in R_L produced by $\overline{v_i^2}$ and $\overline{i_i^2}$ is

$$N_{oA} = \frac{|G|^2}{R_L} \overline{v_{xA}^2} = \frac{|G|^2}{R_L} \left(\overline{v_i^2} \frac{|z_i|^2}{|z_i + R_S|^2} + \overline{i_i^2} \frac{|R_S z_i|^2}{|R_S + z_i|^2} \right) \qquad (11.144)$$

The noise power in R_L produced by source resistance noise generator $\overline{i_s^2}$ is

$$N_{oB} = \frac{|G|^2}{R_L} \frac{|R_S z_i|^2}{|R_S + z_i|^2} \overline{i_s^2} \qquad (11.145)$$

The noise in the source resistance in a narrow bandwidth Δf is

$$\overline{i_s^2} = 4kT \frac{1}{R_S} \Delta f \qquad (11.146)$$

Substituting (11.146) in (11.145) gives

$$N_{oB} = \frac{|G|^2}{R_L} \frac{|R_S z_i|^2}{|R_S + z_i|^2} 4kT \frac{1}{R_S} \Delta f \qquad (11.147)$$

Using the definition of noise figure in (11.142) and substituting from (11.147) and (11.144), we find

$$F = \frac{N_{oA} + N_{oB}}{N_{oB}}$$

$$= 1 + \frac{N_{oA}}{N_{oB}} \qquad (11.148)$$

$$= 1 + \frac{\overline{v_i^2}}{4kTR_S\,\Delta f} + \frac{\overline{i_i^2}}{4kT\dfrac{1}{R_S}\Delta f} \qquad (11.149)$$

Equation 11.149 gives the circuit *spot* noise figure assuming negligible correlation between $\overline{v_i^2}$ and $\overline{i_i^2}$. Note that F is *independent* of all circuit parameters (G, z_i, R_L) except the source resistance R_S and the equivalent input noise generators.

It is apparent from (11.149) that F has a minimum as R_S varies. For very low values of R_S, the $\overline{v_i^2}$ generator is dominant, whereas for large R_S the $\overline{i_i^2}$ generator is most important. By differentiating (11.149) with respect to R_S, we can calculate the value of R_S giving minimum F:

$$R_{S(opt)}^2 = \frac{\overline{v_i^2}}{\overline{i_i^2}} \qquad (11.150)$$

This result is true in general, even if correlation is significant. A graph of F in decibels versus R_S is shown in Fig. 11.47.

The existence of a minimum in F as R_S is varied is one reason for the widespread use of transformers at the input of low-noise tuned amplifiers. This technique allows the source impedance to be transformed to the value that simultaneously gives the lowest noise figure and causes minimal loss in the circuit.

For example, consider the noise figure of a bipolar transistor at low-to-moderate frequencies, where both flicker noise and high-frequency effects are neglected. From (11.50) and (11.57),

$$\overline{v_i^2} = 4kT\left(r_b + \frac{1}{2g_m}\right)\Delta f$$

$$\overline{i_i^2} = 2qI_B\,\Delta f = 2q\frac{I_C}{\beta_F}\Delta f$$

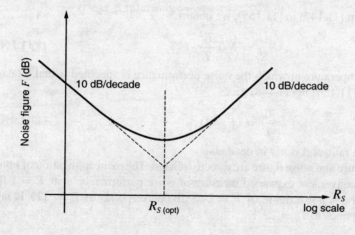

Figure 11.47 Variation in noise figure F with source resistance R_S.

Substitution of these values in (11.150) gives

$$R_{S(\text{opt})} = \frac{\sqrt{\beta_F}}{g_m}\sqrt{1 + 2g_m r_b} \qquad (11.151)$$

At this value of R_S, the noise figure is given by (11.149) as

$$F_{\text{opt}} \simeq 1 + \frac{1}{\sqrt{\beta_F}}\sqrt{1 + 2g_m r_b} \qquad (11.152)$$

At a collector current of $I_C = 1$ mA, and with $\beta_F = 100$ and $r_b = 50\ \Omega$, (11.151) gives $R_{S(\text{opt})} = 572\ \Omega$ and $F_{\text{opt}} = 1.22$. In decibels the value is $10\log_{10} 1.22 = 0.9$ dB. Note that F_{opt} decreases as β_F increases and as r_b and g_m decrease. However, increasing β_F and decreasing g_m result in an *increasing* value of $R_{S(\text{opt})}$, and this may prove difficult to realize in practice.

As another example, consider the MOSFET at low frequencies. Neglecting flicker noise, we can calculate the equivalent input generators from Section 11.5.2 as

$$\overline{v_i^2} \simeq 4kT \frac{2}{3}\frac{1}{g_m}\Delta f \qquad (11.153)$$

$$\overline{i_i^2} \simeq 0 \qquad (11.154)$$

Using these values in (11.149) and (11.150), we find that $R_{S(\text{opt})} \to \infty$ and $F_{\text{opt}} \to 0$ dB. Thus the MOS transistor has excellent noise performance from a high source resistance. However, if the source resistance is low (kilohms or less) and transformers cannot be used, the noise figure for the MOS transistor may be worse than for a bipolar transistor. For source resistances of the order of megohms or higher, the MOS transistor usually has significantly *lower* noise figure than a bipolar transistor.

11.10.2 Noise Temperature

Noise temperature is an alternative noise representation and is closely related to noise figure. The noise temperature T_n of a circuit is defined as the temperature at which the source resistance R_S must be held so that the noise output from the circuit due to R_S equals the noise output due to the circuit itself. If these conditions are applied to the circuit of Fig. 11.46, the output noise N_{oA} due to the circuit itself is unchanged but the output noise due to the source resistance becomes

$$N'_{oB} = \frac{|G|^2}{R_L}\frac{|R_S z_i|^2}{|R_S + z_i|^2}4kT_n \frac{1}{R_S}\Delta f \qquad (11.155)$$

Substituting for N_{oB} from (11.147) in (11.155), we obtain

$$N'_{oB} = N_{oB}\frac{T_n}{T} \qquad (11.156)$$

where T is the circuit temperature at which the noise performance is specified (usually taken as $290°$K). Substituting (11.156) in (11.148) gives

$$\frac{T_n}{T} = (F - 1) \qquad (11.157)$$

where F is specified as a ratio and is *not* in decibels.

Thus noise temperature and noise figure are directly related. The main application of noise temperature provides a convenient expanded measure of noise performance near $F = 1$ for very-low-noise amplifiers. A noise figure of $F = 2$ (3 dB) corresponds to $T_n = 290°$K and $F = 1.1$ (0.4 dB) corresponds to $T_n = 29°$K.

PROBLEMS

11.1 Calculate the noise-voltage spectral density in V^2/Hz at v_o for the circuit in Fig. 11.48, and thus calculate the total noise in a 10-kHz bandwidth. Neglect capacitive effects, flicker noise, and series resistance in the diode.

11.2 If the diode in Fig. 11.48 shows flicker noise, calculate and plot the output noise voltage spectral density at v_o in V^2/Hz on log scales from $f = 1$ Hz to $f = 10$ MHz. Flicker noise data: in (11.7) use $a = b = 1$, $K_1 = 3 \times 10^{-16}$ A.

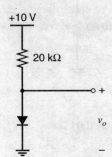

Figure 11.48 Diode circuit for Problems 11.1 and 11.2.

11.3 Repeat Problem 11.2 if a 1000-pF capacitor is connected across the diode. Compare your result with a SPICE simulation.

11.4 The ac schematic of an amplifier is shown in Fig. 11.49. The circuit is fed from a current source i_S and data are as follows:

$$R_S = 1 \text{ k}\Omega \quad R_L = 10 \text{ k}\Omega \quad I_C = 1 \text{ mA}$$

$$\beta = 50 \quad r_b = 0 \quad r_o = \infty$$

Neglecting capacitive effects and flicker noise, calculate the total noise voltage spectral density at v_o in V^2/Hz. Thus calculate the MDS at i_S if the circuit bandwidth is limited to a sharp cutoff at 2 MHz. Compare your result with a SPICE simulation. (This will require setting up a bias circuit.)

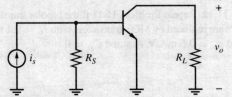

Figure 11.49 Amplifier ac schematic for Problems 11.4 and 11.6.

11.5 Calculate the total input- and output-referred noise voltages at 10 Hz, 100 kHz, and 1 GHz for the common-source amplifier shown in Fig. 7.2b. Assume that $W = 100$ μm, $L = 1$ μm, $I_D = 100$ μA, $V_t = 0.6$ V, $k' = 194$ μA/V^2, $t_{ox} = 80$ Å, $L_d = 0$, $X_d = 0$, and $K_f = 3 \times 10^{-24}$ F-V^2. Ignore gate-current noise in (11.15) and (11.16). Let $R_S = 100$ kΩ and $R_L = 10$ kΩ. Verify your result using SPICE. (Add dc sources for SPICE.)

11.6 Calculate equivalent input noise voltage and current generators for the circuit of Fig. 11.49 (omitting R_S). Using these results, calculate the total equivalent input noise current in a 2-MHz bandwidth for the circuit of Fig. 11.49 with $R_S = 1$ kΩ and compare with the result of Problem 11.4. Neglect correlation between the noise generators.

11.7 Four methods of achieving an input impedance greater than 100 kΩ are shown in the ac schematics of Fig. 11.50.

(a) Neglecting flicker noise and capacitive effects, derive expressions for the equivalent input noise voltage and current generators of these circuits. For circuit (i) this will be on the *source* side of the 100-kΩ resistor.

(b) Assuming that following stages limit the bandwidth to dc-10 kHz with a sharp cutoff, calculate the magnitude of the total equivalent input noise voltage in each case. Then compare these circuits for use as low-noise amplifiers from low source impedances.

11.8 Neglecting capacitive effects, calculate equivalent input noise voltage and current generators for circuit (iv) of Fig. 11.50, assuming that the spectral density of the flicker noise in the MOS transistor drain current equals that of the thermal noise at 100 kHz. Assuming that following stages limit the bandwidth to 0.001 kHz to 10 kHz with a sharp cutoff, calculate the magnitude of the total equivalent input noise voltage. Ignore gate-current noise in (11.15) and (11.16). (Assume $C_{gs} = 0$.)

11.9 A BiCMOS Darlington is shown in Fig. 11.51. Neglecting frequency effects, calculate the equivalent input noise voltage and current generators for this circuit, assuming that the dc value of V_i is adjusted for $I_{C1} = 1$ mA. Device data is $\mu_n C_{ox} = 60$ μA/V^2, $V_t = 0.7$ V, $\lambda = 0$, $\gamma = 0$, $W = 100$ μm, $L = 1$ μm for the MOSFET and $I_S = 10^{-16}$ A, $V_A = \infty$, $\beta = 100$, $r_b = 100$ Ω for the bipolar transistor. Use SPICE to check your calculation. Then add $C_{gs} = 150$ fF, $C_{sb} = 50$ fF for the MOSFET and $C_\mu = 50$ fF, $f_T = 10$ GHz for the bipolar transistor and use SPICE to determine the frequency where the spectral density of the equivalent input noise voltage generator $\overline{v_i^2}$ has doubled. Also use SPICE to determine the equivalent input noise current spectral density at that frequency.

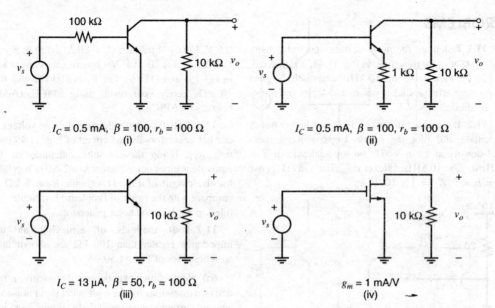

Figure 11.50 Four ac schematics of circuits realizing an input resistance $R_i > 100$ kΩ.

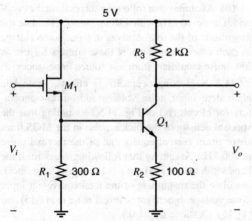

Figure 11.51 BiCMOS Darlington circuit for Problem 11.9.

11.10 The ac schematic of a low-input-impedance common-base amplifier is shown in Fig. 11.52.

(a) Calculate the equivalent noise voltage and current generators of this circuit at the emitter of Q_1

using $I_{C1} = I_{C2} = 1$ mA, $r_{b1} = r_{b2} = 0$, $\beta_1 = \beta_2 = 100$, $f_{T1} = f_{T2} = 400$ MHz. Neglect flicker noise but include capacitive effects in the transistors. Use SPICE to check your result.

(b) If $R_S = 5$ kΩ, and later stages limit the bandwidth to a sharp cutoff at 150 MHz, calculate the value of i_S giving an output signal-to-noise ratio of 10 dB.

11.11 A super-β input stage is shown in Fig. 11.53a.

(a) Neglecting flicker noise and capacitive effects, calculate the equivalent input noise voltage and current generators $\overline{v^2}$ and $\overline{i^2}$ for this stage. *Data:* $I_{EE} = 1$ μA; $\beta_1 = \beta_2 = 5000$; $r_{b1} = r_{b2} = 500$ Ω.

(b) If the circuit is fed from source resistances $R_S = 10$ MΩ as shown in Fig. 11.53b, calculate the total equivalent input noise voltage at v_s in a bandwidth of 1 kHz.

11.12 Repeat Problem 11.11 if the bipolar transistors are replaced by MOS transistors with $I_G = 0.1$ fA and $g_m = 0.5$ mA/V. Assume $C_{gs} = 0$.

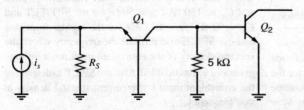

Figure 11.52 An ac schematic of a common-base amplifier for Problem 11.10.

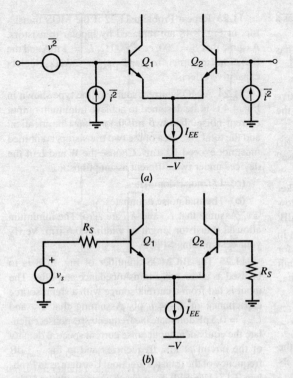

Figure 11.53 Super-β input stage for Problem 11.11.

11.13 If a 10-pF capacitor is connected across the diode in Fig. 11.48, calculate the noise bandwidth of the circuit and thus calculate the *total* output noise at v_o. Neglect flicker noise and series resistance in the diode.

11.14 A differential input stage is shown in Fig. 11.54.

(a) Neglecting flicker noise, calculate expressions for the equivalent input noise voltage and current generators at the base of Q_1. Use SPICE to check your result.

(b) Assuming the circuit has a dominant pole in its frequency response at 30 MHz and $R_S = 50 \, \Omega$, calculate the total equivalent input and output noise voltages. *Data:* $\beta = 100$; $r_b = 200 \, \Omega$.

11.15 Calculate the source resistance giving minimum noise figure and the corresponding noise figure in decibels for a bipolar transistor with parameters

(a) $I_C = 2 \text{ mA} \quad \beta_F = 50 \quad r_b = 100 \, \Omega$

(b) $I_C = 10 \, \mu\text{A} \quad \beta_F = 100 \quad r_b = 300 \, \Omega$

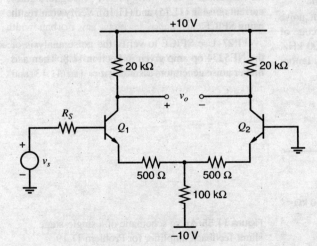

Figure 11.54 Differential-pair input stage for Problem 11.14.

11.16 Repeat Problem 11.15 if the transistor has a 1-kΩ emitter resistor.

11.17 Repeat Problem 11.15 if the transistor has a flicker noise corner frequency of 1 kHz. Calculate spot noise figure at 500 Hz.

11.18 (a) Neglecting flicker noise and capacitive effects, calculate the noise figure in decibels of the circuit of Fig. 11.54 with $R_S = 50\ \Omega$.

(b) If R_S were made equal to (i) 100 Ω or (ii) 200 kΩ, would the noise figure increase or decrease? Explain.

(c) If $R_S = 200$ kΩ, and each device has a flicker noise corner frequency of 10 kHz, calculate the low frequency where the circuit spot noise figure is 20 dB.

11.19 (a) A shunt-feedback amplifier is shown in Fig. 11.55. Using equivalent input noise generators for the device, calculate the spot noise figure of this circuit in decibels for $R_S = 10$ kΩ using the following data.

$$I_C = 0.5\ \text{mA} \quad \beta = 50 \quad r_b = 100\ \Omega$$

Neglect flicker noise and capacitive effects.

(b) If the device has $f_T = 500$ MHz, calculate the frequency where the noise figure is 3 dB above its low-frequency value.

11.20 Neglecting flicker noise, calculate the total equivalent input noise voltage for the MC1553 shown in Figure 8.21a. Use $\beta = 100, r_b = 100\ \Omega$ and assume a sharp cutoff in the frequency response at 50 MHz. Then calculate the average noise figure of the circuit with a source resistance of 50 Ω.

11.21 (a) Neglecting capacitive effects, calculate the noise figure in decibels of the circuit of Fig. 11.52 with $R_S = 5$ kΩ. Use data as in Problem 11.10.

(b) If the flicker noise corner frequency for each device is 1 kHz, calculate the low frequency where the spot noise figure is 3 dB above the value in (a).

11.22 Calculate the total equivalent input noise current for the shunt-shunt feedback circuit of Fig. 8.48 in a bandwidth from 0.01 Hz to 100 kHz. Use the MOS transistor data in Problem 11.5. Ignore gate-current noise in (11.15) and (11.16).

11.23 Repeat Problem 11.22 if the MOS transistors in Fig. 8.48 are replaced by bipolar transistors. Assume that $\beta = 200, r_b = 300\ \Omega, I_C = 1$ mA and the flicker noise corner frequency is $f_a = 5$ kHz. Neglect capacitive effects.

11.24 A MOS current source of the type shown in Fig. 4.4 is to be designed to achieve minimum output current noise. The two transistors must be identical and the total gate area of the two transistors combined must not exceed 10 μm^2. Choose the W and L of the devices under two different assumptions:

(a) $1/f$ noise dominates.

(b) Thermal noise dominates.

Assume that L_d and X_d are zero. The minimum allowed transistor length or width is 0.6 μm. Verify your design using SPICE.

11.25 The BiCMOS amplifier of Fig. 3.78 is to be used as a low-noise transimpedance amplifier. The input is fed from a current source with a shunt source capacitance of $C_S = 1$ pF. Assuming that C_S and $C_{gs1} = 0.5$ pF dominate the frequency response, calculate the equivalent input noise current spectral density of the circuit at low frequencies and at the -3-dB frequency of the transfer function. Use data as in Problem 3.17. Use SPICE to check your result, and also to investigate the effect of adding $r_b = 200\ \Omega$ to the bipolar model.

11.26 Calculate the equivalent input noise voltage at 100 Hz, 1 kHz, and 10 kHz for the CMOS op amp shown in Fig. 6.58. Use the MOS parameters in Table 2.4, and assume that $X_d = 0.1$ μm at the operating point for all transistors operating in the active region. For this problem, assume that the gate of M_9 is attached to the positive power supply, and that the W/L of M_9 has been optimally chosen to cancel the right-half-plane zero. The flicker-noise coefficient for all transistors is $K_f = 3 \times 10^{-24}$ F-V^2. Ignore gate-current noise in (11.15) and (11.16). Verify your result using SPICE.

11.27 Use SPICE to verify the noise analysis of the NE5234 op amp given in Section 11.8. Then add flicker noise generators assuming $a = 1$ in (11.13) and

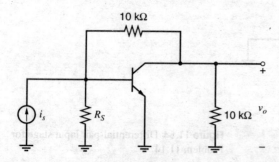

Figure 11.55 An ac schematic of a single-stage shunt-feedback amplifier for Problem 11.19.

the transistor flicker noise corner frequency is 1 kHz. Which device dominates the noise performance at 1 Hz?

11.28 (a) At what frequency are the noise spectral densities, $\overline{i_g^2}/\Delta f$, for the gate-current noise in (11.15) and (11.16) equal? Assume that $W = 50\ \mu\text{m}$, $L = 0.5\ \mu\text{m}$, $I_G = 0.05\ \text{fA}$, $I_D = 100\ \mu\text{A}$, $V_t = 0.6\ \text{V}$, $k' = 194\ \mu\text{A/V}^2$, $t_{ox} = 80\ \text{Å}$, $L_d = 0$, $X_d = 0$, and $K_f = 3 \times 10^{-24}\ \text{F-V}^2$.

(b) For this transistor, considering only the gate-current noise, in what bandwidth (from 0 to f_{BW}) are the total input noise currents due to shot noise in (11.15) and coupled thermal noise in (11.16) equal?

REFERENCES

1. M. Schwartz. *Information Transmission, Modulation, and Noise.* McGraw-Hill, New York, 1959, Chapter 5.

2. W. B. Davenport, Jr., and W. L. Root. *An Introduction to the Theory of Random Signals and Noise.* McGraw-Hill, New York, 1958, Chapter 7.

3. J. L. Lawson and G. E. Uhlenbeck. *Threshold Signals.* McGraw-Hill, New York, 1950, Chapter 4.

4. A. van der Ziel. *Noise.* Prentice-Hall, New York, 1954, Chapter 5.

5. W. B. Davenport, Jr., and W. L. Root, op. cit., Chapter 9.

6. J. L. Plumb and E. R. Chenette. "Flicker Noise Transistors," *IEEE Transactions on Electron Devices,* Vol. ED-10, pp. 304–308, September 1963.

7. R. C. Jaeger and A. J. Broderson. "Low-Frequency Noise Sources in Bipolar Junction Transistors," *IEEE Transactions on Electron Devices,* Vol. ED-17, pp. 128–134, February 1970.

8. M. Nishida. "Effects of Diffusion-Induced Dislocations on the Excess Low-Frequency Noise," *IEEE Transactions on Electron Devices,* Vol. ED-20, pp. 221–226, March 1973.

9. R. G. Meyer, L. Nagel, and S. K. Lui. "Computer Simulation of $1/f$ Noise Performance of Electronic Circuits," *IEEE Journal of Solid-State Circuits,* Vol. SC-8, pp. 237–240, June 1973.

10. R. H. Haitz. "Controlled Noise Generation with Avalanche Diodes," *IEEE Transactions on Electron Devices,* Vol. ED-12, pp. 198–207, April 1965.

11. D. G. Peterson. "Noise Performance of Transistors," *IRE Transactions on Electron Devices,* Vol. Ed-9, pp. 296–303, May 1962.

12. A. van der Ziel. *Noise in Solid State Devices and Circuits.* Wiley, New York, 1986.

13. A. A. Abidi. "High-Frequency Noise Measurements on FETs with Small Dimensions," *IEEE Transactions on Electron Devices,* Vol. 33, no. 11, pp. 1801–1805, November 1986.

14. D. K. Shaeffer and T. H. Lee. "A 1.5-V, 1.5-GHz CMOS Low Noise Amplifier," *IEEE Journal of Solid-State Circuits,* Vol. 32, no. 5, pp. 745–759, May 1997.

15. W. R. Bennett. "Methods of Solving Noise Problems," *Proceedings, IRE,* Vol. 44, pp. 609–638, May 1956.

16. E. M. Cherry and D. E. Hooper. *Amplifying Devices and Low-Pass Amplifier Design.* Wiley, New York, 1968, Chapter 8.

17. H. A. Haus, et al. "Representation of Noise in Linear Twoports," *Proceedings, IRE,* Vol. 48, pp. 69–74, January 1960.

18. M. B. Das and J. M. Moore. "Measurements and Interpretation of Low-Frequency Noise in FETS," *IEEE Transactions on Electron Devices,* Vol. ED-21, pp. 247–257, April 1974.

19. M. J. Fonderie and J. H. Huijsing. *Design of Low-Voltage Bipolar Operational Amplifiers.* Kluwer Academic Publishers, Boston, 1993, pp. 20–22.

CHAPTER 12

Fully Differential Operational Amplifiers

12.1 Introduction

The analysis of integrated-circuit operational amplifiers (op amps) in Chapter 6 focused on op amps with single-ended outputs. The topic of this chapter is fully differential op amps, which have a differential input and produce a differential output. Fully differential op amps are widely used in modern integrated circuits because they have some advantages over their single-ended counterparts. They provide a larger output voltage swing and are less susceptible to common-mode noise. Also, even-order nonlinearities are not present in the differential output of a balanced circuit. (A *balanced* circuit is symmetric with perfectly matched elements on either side of an axis of symmetry.) A disadvantage of fully differential op amps is that they require two matched feedback networks and a common-mode feedback circuit to control the common-mode output voltage.

In this chapter, the properties of fully differential amplifiers are presented first, followed by some common-mode feedback approaches. A number of fully differential CMOS op amps are covered. Some of the terminology used in this chapter was introduced for a simple fully differential amplifier (a differential pair with resistive loads) in Section 3.5. In most of the chapter, the circuits are assumed to be perfectly balanced. The effects of imbalance are considered in Section 12.7. The circuits in this chapter are CMOS; however, most of the techniques and topologies described can be readily extended to bipolar technologies.

12.2 Properties of Fully Differential Amplifiers[1,2]

A fully differential feedback amplifier is shown in Fig. 12.1a. It differs from the single-ended feedback amplifier in Fig. 12.1b in the following two ways. The op amp has two outputs, and two identical resistive networks provide feedback. While many fully differential op-amp topologies exist, the simple fully differential amplifier in Fig. 12.2 will be used for illustration purposes. It consists of a differential pair M_1-M_2, active loads M_3 and M_4, and tail current source M_5.

Fully differential op amps provide a larger output voltage swing than their single-ended counterparts, which is important when the power-supply voltage is small. The larger output voltage swing provided by a fully differential op amp can be explained using the two feedback circuits in Fig. 12.1. Assume that each op-amp output, V_{o1}, V_{o2}, or V_o, can swing up to V_{max} and down to V_{min}. For the single-ended-output circuit in Fig. 12.1b, the peak-to-peak output voltage can be as large as $V_{max} - V_{min}$. In the fully differential circuit in Fig. 12.1a, if V_{o1} swings up to V_{max} and V_{o2} swings down to V_{min}, the peak differential output is $V_{max} - V_{min}$. Therefore, the peak-to-peak differential output is $2(V_{max} - V_{min})$. Thus, the output swing of

12.2 Properties of Fully Differential Amplifiers

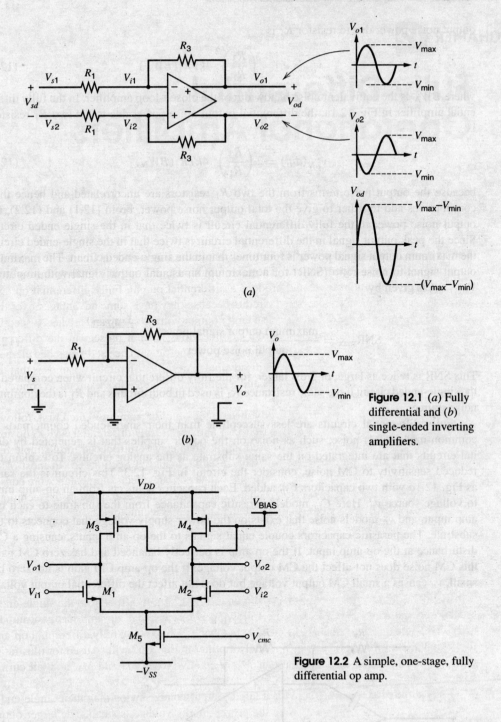

Figure 12.1 (a) Fully differential and (b) single-ended inverting amplifiers.

Figure 12.2 A simple, one-stage, fully differential op amp.

a fully differential op amp is twice as large as that of a similar op amp with a single-ended output.

This larger output swing can result in a higher signal-to-noise ratio. Ignoring the noise from the op amp and from the feedback resistor R_3, we see that thermal noise associated with the R_1 input resistors is the only source of noise. In the single-ended circuit in Fig. 12.1b, the output noise power due to resistor R_1 is

$$\overline{v_{oN}^2}(s.e.) = \left(\frac{R_3}{R_1}\right)^2 4kTR_1(BW_N) \tag{12.1}$$

where BW_N is the equivalent noise bandwidth of the closed-loop amplifier. In the fully differential amplifier in Fig. 12.1a, the differential output noise power due to the two R_1 resistors is

$$\overline{v_{oN}^2}(diff) = 2\left(\frac{R_3}{R_1}\right)^2 4kTR_1(BW_N) \tag{12.2}$$

because the output noise terms from the two R_1 resistors are uncorrelated and hence their contributions add together to give the total output noise power. From (12.1) and (12.2), the output noise power in the fully differential circuit is twice that in the single-ended circuit. Since the peak output signal in the differential circuit is twice that in the single-ended circuit, the maximum output signal power is four times that in the single-ended circuit. The maximum output signal-to-noise ratio (SNR) for a maximum sinusoidal output signal with amplitude $V_{sig(peak)}$ is given by

$$\text{SNR}_{max} = \frac{\text{maximum output signal power}}{\text{output noise power}} = \frac{\frac{V_{sig(peak)}^2}{2}}{\overline{v_{oN}^2}} \tag{12.3}$$

This SNR is twice as large, or 3 dB larger, for the fully differential circuit when compared to the single-ended circuit if the same resistance R_1 is used in both circuits and R_1 is the dominant noise source.

Fully differential circuits are less susceptible than their single-ended counterparts to common-mode (CM) noise, such as noise on the power supplies that is generated by digital circuits that are integrated on the same substrate as the analog circuits. To explain the reduced sensitivity to CM noise, consider the circuit in Fig. 12.3. This circuit is the same as Fig. 12.1a with two capacitors C_{ip} added. Each capacitor connects from an op-amp input to voltage source v_n. Here C_{ip} models parasitic capacitance from the substrate to each op-amp input, and v_n models noise that exists on the power-supply voltage that connects to the substrate. The parasitic capacitors couple equal signals to the op-amp inputs, causing a CM disturbance at the op-amp input. If the op amp is perfectly balanced and has zero CM gain, this CM noise does not affect the CM output voltage. If the op-amp CM gain is nonzero but small, v_n causes a small CM output voltage but does not affect the differential output voltage

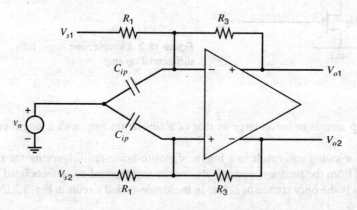

Figure 12.3 The inverting amplifier of Fig. 12.1a including parasitic capacitors C_{ip} and noise source v_n.

if the circuit is perfectly balanced. This capacitive coupling to the op-amp inputs can cause a nonzero differential-mode (DM) output voltage if the circuit is not perfectly balanced. For example, mismatch between the C_{ip} capacitors causes the coupled noise at the two op-amp inputs to be unequal and introduces a differential noise signal across the op-amp inputs.

Even without the capacitive coupling to the substrate in Fig. 12.3, noise on the positive or negative power supply can couple to the op-amp outputs through the transistors in the op amp (see Section 6.2.6). If the circuit is balanced, such coupling is the same at each op-amp output. Therefore, power-supply noise in a balanced circuit alters the CM output but does not affect the DM output.

Even-order nonlinearities are not present in the differential output of a balanced circuit. The cancellation of even-order nonlinearities can be explained using Fig. 12.1a. Assume that the circuit is perfectly balanced but is not perfectly linear. First, consider the case when the inputs are $V_{s1} = V_a$ and $V_{s2} = V_b$, and let the resulting output voltages be $V_{o1} = V_x$ and $V_{o2} = V_y$. In this case, the differential input and output are

$$V_{sd} = V_a - V_b \quad \text{and} \quad V_{od} = V_x - V_y \tag{12.4}$$

Next, consider the circuit with the inputs swapped; that is, $V_{s1} = V_b$ and $V_{s2} = V_a$. Then the output voltages will also be swapped ($V_{o1} = V_y$ and $V_{o2} = V_x$) because the circuit is symmetric. In this second case,

$$V_{sd} = V_b - V_a = -(V_a - V_b) \quad \text{and} \quad V_{od} = V_y - V_x = -(V_x - V_y) \tag{12.5}$$

Equations 12.4 and 12.5 show that changing the polarity of the differential input of a balanced circuit causes only a polarity change in the differential output voltage. Therefore, the differential input-output characteristic $f()$ is an odd function; that is, if $V_{od} = f(V_{id})$, then $-V_{od} = f(-V_{id})$. Hence, the differential transfer characteristic of a balanced amplifier exhibits only odd-order nonlinearities, so only odd-order distortion can appear in the differential output when a differential input is applied. Even-order distortion may exist in each individual output V_{o1} and V_{o2}, but such distortion in these output voltages is identical and cancels when they are subtracted to form V_{od}.

In a balanced, fully differential amplifier, the small-signal differential output voltage is proportional to the small-signal differential input voltage but is independent of the CM input voltage, as was shown in Section 3.5.4. Similarly, the small-signal CM output voltage is proportional to the small-signal CM input voltage but is independent of the differential input voltage.

12.3 Small-Signal Models for Balanced Differential Amplifiers

A balanced signal source driving a balanced, fully differential amplifier is shown in Fig. 12.4. A T-network small-signal model for the balanced signal source is shown in Fig. 12.5. Equations that describe this model are found as follows. Applying KVL from v_{s1} to ground,

$$v_{s1} = \frac{v'_{sd}}{2} + v'_{sc} + \frac{R_{sd}}{2} i_{s1} + \left(\frac{R_{sc}}{2} - \frac{R_{sd}}{4}\right)(i_{s1} + i_{s2}) \tag{12.6}$$

Rearranging gives

$$v_{s1} = \frac{v'_{sd}}{2} + v'_{sc} + \frac{R_{sd}}{2} \frac{i_{s1} - i_{s2}}{2} + R_{sc} \frac{i_{s1} + i_{s2}}{2} \tag{12.7}$$

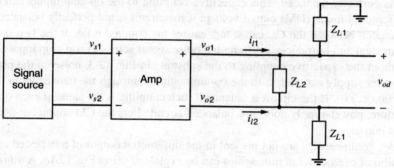

Figure 12.4 A block diagram of a fully differential signal source and amplifier driving a complex load.

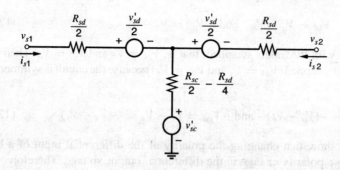

Figure 12.5 A model for a differential signal source.

Define

$$i_{sd} = \frac{i_{s1} - i_{s2}}{2} \quad (12.8)$$

$$i_{sc} = \frac{i_{s1} + i_{s2}}{2} \quad (12.9)$$

Then

$$i_{s1} = i_{sd} + i_{sc} \quad (12.10)$$

$$i_{s2} = -i_{sd} + i_{sc} \quad (12.11)$$

Substituting (12.8) and (12.9) in (12.7) gives

$$v_{s1} = \frac{v'_{sd}}{2} + v'_{sc} + \frac{R_{sd}}{2} i_{sd} + R_{sc} i_{sc} \quad (12.12)$$

A similar analysis applied from v_{s2} to ground yields

$$v_{s2} = -\frac{v'_{sd}}{2} + v'_{sc} - \frac{R_{sd}}{2} i_{sd} + R_{sc} i_{sc} \quad (12.13)$$

The usual definitions apply for the DM and CM source voltages: $v_{sd} = v_{s1} - v_{s2}$ and $v_{sc} = (v_{s1} + v_{s2})/2$. Voltages v'_{sd} and v'_{sc} are the DM and CM open-circuit source voltages, respectively. That is, if $i_{sd} = i_{sc} = 0$, then $v_{sd} = v'_{sd}$ and $v_{sc} = v'_{sc}$. Resistances R_{sd} and R_{sc} are the DM and CM resistances, respectively, associated with the signal source. Subtracting (12.13) from (12.12) and manipulating the result, we can write

$$R_{sd} = \left. \frac{v_{sd}}{i_{sd}} \right|_{v'_{sd} = 0} \quad (12.14)$$

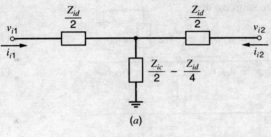

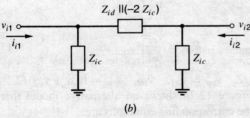

Figure 12.6 Models for the input impedance of a fully differential amplifier: (a) a T-network model and (b) a π-network model.

Adding (12.12) and (12.13), we find

$$R_{sc} = \left. \frac{v_{sc}}{i_{sc}} \right|_{v'_{sc}=0} \tag{12.15}$$

These simple expressions result from the definitions of the CM and DM currents in (12.8) and (12.9), and the standard definitions for the CM and DM voltages.

Two equivalent models for the inputs of the amplifier in Fig. 12.4 are shown in Fig. 12.6. These models are extensions of Fig. 3.61 and (3.197) and (3.198), which were derived to model the inputs for a differential pair with resistive loads. Two equivalent models for the output ports of the amplifier are shown in Fig. 12.7. The equations that describe the model in Fig. 12.7a,

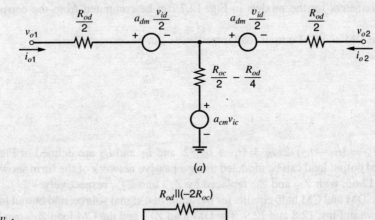

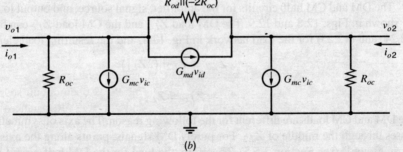

Figure 12.7 Models for the output ports of a fully differential amplifier: (a) a Thévenin-network model and (b) a Norton-network model.

which uses voltage-controlled voltage sources, are

$$v_{o1} = a_{dm}\frac{v_{id}}{2} + a_{cm}v_{ic} + \frac{R_{od}}{2}i_{od} + R_{oc}i_{oc} \tag{12.16}$$

$$v_{o2} = -a_{dm}\frac{v_{id}}{2} + a_{cm}v_{ic} - \frac{R_{od}}{2}i_{od} + R_{oc}i_{oc} \tag{12.17}$$

where

$$a_{dm} = \left.\frac{v_{od}}{v_{id}}\right|_{i_{od}=0} \qquad a_{cm} = \left.\frac{v_{oc}}{v_{ic}}\right|_{i_{oc}=0} \tag{12.18}$$

$$R_{od} = \left.\frac{v_{od}}{i_{od}}\right|_{v_{id}=0} \qquad R_{oc} = \left.\frac{v_{oc}}{i_{oc}}\right|_{v_{ic}=0} \tag{12.19}$$

and $v_{id} = v_{i1} - v_{i2}$, $v_{ic} = (v_{i1} + v_{i2})/2$, $v_{od} = v_{o1} - v_{o2}$, $v_{oc} = (v_{o1} + v_{o2})/2$, $i_{od} = (i_{o1} - i_{o2})/2$, and $i_{oc} = (i_{o1} + i_{o2})/2$. Figure 12.7b shows an alternative model that uses voltage-controlled current sources, and the corresponding equations are

$$i_{o1} = i_{od} + i_{oc} = G_{md}v_{id} + G_{mc}v_{ic} + \frac{v_{od}}{R_{od}} + \frac{v_{oc}}{R_{oc}} \tag{12.20}$$

$$i_{o2} = -i_{od} + i_{oc} = -G_{md}v_{id} + G_{mc}v_{ic} - \frac{v_{od}}{R_{od}} + \frac{v_{oc}}{R_{oc}} \tag{12.21}$$

where

$$G_{md} = \left.\frac{i_{od}}{v_{id}}\right|_{v_{od}=0} \tag{12.22}$$

$$G_{mc} = \left.\frac{i_{oc}}{v_{ic}}\right|_{v_{oc}=0} \tag{12.23}$$

The parameters for the models in Fig. 12.7 can be computed from the corresponding half-circuits.

The DM and CM output-load impedances can be computed using

$$Z_{Ld} = \frac{v_{od}}{i_{ld}} \tag{12.24}$$

and

$$Z_{Lc} = \frac{v_{oc}}{i_{lc}} \tag{12.25}$$

where $i_{ld} = (i_{l1} - i_{l2})/2$, $i_{lc} = (i_{l1} + i_{l2})/2$, and i_{l1} and i_{l2} are defined in Fig. 12.4. A fully balanced output load can be modeled using a passive network of the form shown in Fig. 12.6a or Fig. 12.6b, with Z_{id} and Z_{ic} replaced by Z_{Ld} and Z_{Lc}, respectively.

The DM and CM half-circuits for the amplifier, signal source, and output load in Fig. 12.4 are shown in Figs. 12.8 and 12.9. The DM load Z_{Ld} and the CM load Z_{Lc} can be found using (12.24) and (12.25) for the load network in Fig. 12.4, and the resulting load elements are

$$Z_{Ld} = Z_{L2} \| (2Z_{L1}) \tag{12.26}$$

$$Z_{Lc} = Z_{L1} \tag{12.27}$$

The DM and CM loads are different for the following reason. The axis of symmetry in Fig. 12.4 passes through the middle of Z_{L2}. For purely DM signals, points along the axis of symmetry are ac grounds (see Section 3.5.5). Therefore, the load for the DM half-circuit is half of Z_{L2}

12.3 Small-Signal Models for Balanced Differential Amplifiers

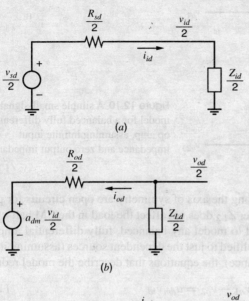

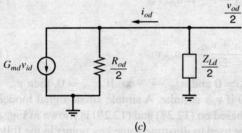

Figure 12.8 DM half-circuits for a fully differential signal source and amplifier: (*a*) the input port, (*b*) the output port using a Thévenin equivalent, (*c*) the output port using a Norton equivalent.

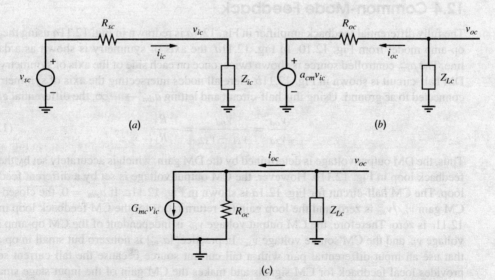

Figure 12.9 CM half-circuits for a fully differential signal source and amplifier: (*a*) the input port, (*b*) the output port using a Thévenin equivalent, (*c*) the output port using a Norton equivalent.

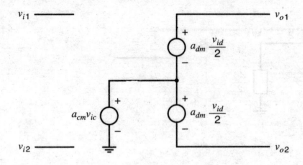

Figure 12.10 A simple small-signal model for a balanced fully differential op amp, assuming infinite input impedance and zero output impedance.

in parallel with Z_{L1}. However, points along the axis of symmetry are open circuits for purely CM signals (see Section 3.5.5). Therefore, Z_{L2} does not affect the load in the CM half-circuit.

These amplifier models can be used to model any balanced, fully differential amplifier, including an op amp. If the model is simplified to just the dependent sources (assuming infinite input impedance and zero output impedance), the equations that describe the model reduce to

$$v_{od} = v_{o1} - v_{o2} = a_{dm} v_{id} \tag{12.28}$$

and

$$v_{oc} = \frac{v_{o1} + v_{o2}}{2} = a_{cm} v_{ic} \tag{12.29}$$

In an ideal, fully differential op amp, $a_{cm} = 0$ and $a_{dm} \to -\infty$. If $a_{cm} = 0$, then $v_{oc} = 0$ if v_{ic} is finite. If $a_{dm} \to -\infty$, then $v_{id} \to 0$ if v_{od} is finite. A simple small-signal model for a balanced, fully differential op amp that is based on (12.28) and (12.29) is shown in Fig. 12.10. Because of its simplicity, this model will be used to illustrate some key points in the following sections.

12.4 Common-Mode Feedback

The fully differential feedback amplifier in Fig. 12.1a is redrawn in Fig. 12.11a using the ideal op-amp model from Fig. 12.10. In Fig. 12.11a, the axis of symmetry is shown as a dashed line. The a_{cm} controlled source is shown twice, once on each side of the axis of symmetry. The DM half-circuit is shown in Fig. 12.11b. Here all nodes intersecting the axis of symmetry are connected to ac ground. Using this half-circuit and letting $a_{dm} \to -\infty$, the differential gain is

$$\frac{v_{od}}{v_{sd}} = \frac{v_{od}}{v_{s1} - v_{s2}} = -\frac{R_3}{R_1} \tag{12.30}$$

Thus, the DM output voltage is determined by the DM gain, which is accurately set by the DM feedback loop in Fig. 12.11b. However, the CM output voltage is set by a different feedback loop. The CM half-circuit for Fig. 12.1a is shown in Fig. 12.11c. If $a_{cm} = 0$, the closed-loop CM gain v_{oc}/v_{sc} is zero, and the loop gain (or return ratio) for the CM feedback loop in Fig. 12.11c is zero. Therefore, the CM output voltage v_{oc} is independent of the CM op-amp input voltage v_{ic} and the CM source voltage v_{sc}. In practice, $|a_{cm}|$ is nonzero but small in op amps that use an input differential pair with a tail current source because the tail current source provides local feedback for CM signals and makes the CM gain of the input stage small. If $|a_{cm}|$ is small, the magnitude of the loop gain (or return ratio) for the CM feedback loop in Fig. 12.11c is small, and this feedback loop exerts little control on the CM output voltage. As a

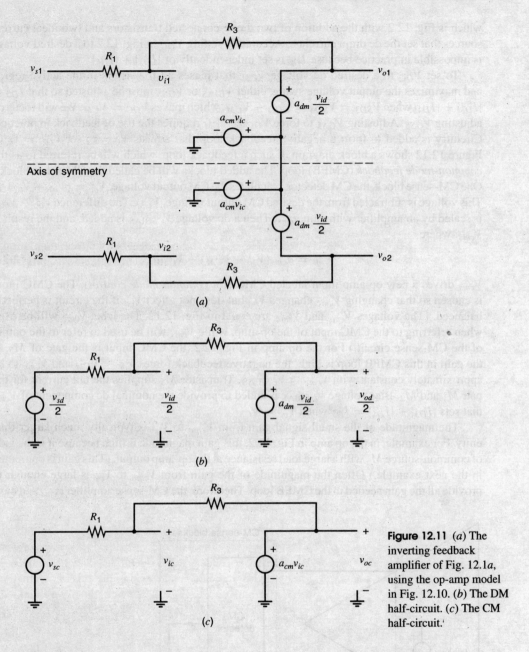

Figure 12.11 (a) The inverting feedback amplifier of Fig. 12.1a, using the op-amp model in Fig. 12.10. (b) The DM half-circuit. (c) The CM half-circuit.

result, a different feedback loop with high loop gain is used to control the CM output voltage, as described next.

12.4.1 Common-Mode Feedback at Low Frequencies

For the op amp in Fig. 12.2, the ideal operating point biases M_1–M_5 in the active region and sets the dc CM output voltage V_{OC} to the value that maximizes the swing at the op-amp outputs for which all transistors operate in the active region. However, as described in Section 4.3.5.1, V_{OC} is very sensitive to mismatches and component variations, and the circuit is not practical from a bias stability standpoint. (Section 4.3.5.1 describes the circuit in Fig. 4.24a, which is Fig. 12.2 with the addition of two diode-connected transistors and two ideal current sources that set the dc drain currents.) Accurately setting V_{OC} in Fig. 12.2 to a desired voltage is impossible in practice because I_{D5} is set independently of $|I_{D3}| + |I_{D4}|$.

To set V_{OC} to a desired dc voltage V_{CM} that biases all transistors in the active region and maximizes the output voltage swing, either V_{BIAS} or V_{GS5} must be adjusted so that $I_{D5} = |I_{D3}| + |I_{D4}|$ when $V_{SD3} = V_{SD4} = V_{DD} - V_{CM}$, which makes $V_{OC} = V_{CM}$. We will focus on adjusting V_{GS5}. Adjusting V_{GS5} to force $V_{OC} = V_{CM}$ requires the use of feedback in practice. Circuitry is added to form a negative feedback loop that adjusts V_{GS5} to set $V_{OC} = V_{CM}$. Figure 12.12 shows a block diagram of such a feedback loop, which will be referred to as the *common-mode feedback* (CMFB) loop. The added blocks will be called the CM-sense blocks. One CM-sense block, the CM detector, calculates the CM output voltage, $V_{oc} = (V_{o1} + V_{o2})/2$. This voltage is subtracted from the desired CM output voltage, V_{CM}. The difference $V_{oc} - V_{CM}$ is scaled by an amplifier with gain a_{cms}. Then a dc voltage V_{CSBIAS} is added, and the result is V_{cms}, where

$$V_{cms} = a_{cms}(V_{oc} - V_{CM}) + V_{CSBIAS} \quad (12.31)$$

V_{cms} drives a new op-amp input labeled CMC (for *common-mode control*). The CMC input is chosen so that changing V_{cmc} changes V_{oc} but does not affect V_{od} if the circuit is perfectly balanced. (The voltages V_{cms} and V_{cmc} are equal in Fig. 12.12. The label V_{cmc} will be used when referring to the CMC input of the op amp, while V_{cms} will be used to refer to the output of the CM-sense circuit.) For the op amp in Fig. 12.2, the CMC input is the gate of M_5. If the gain in this CMFB loop is high, the negative feedback forces $V_{oc} \approx V_{CM}$ and V_{cmc} to be approximately constant with $V_{cmc} \approx V_{CSBIAS}$. Transistor M_5 supplies the tail current for the pair M_1 and M_2. Bias voltage V_{CSBIAS} is added to provide the nominal dc component of V_{cmc} that sets $|I_{D3}| + |I_{D4}| = I_{D5}$ when $V_{oc} = V_{CM}$.

The magnitude of the small-signal gain from V_{cmc} to V_{oc} is typically much larger than unity. For example, in the op amp in Fig. 12.2, this gain magnitude is high because it is the gain of common-source M_5 with a large load resistance at the op-amp output. (This gain is computed in the next example.) Often the magnitude of the gain from V_{cmc} to V_{oc} is large enough to provide all the gain needed in the CMFB loop. Therefore, the CM-sense amplifier a_{cms} can have

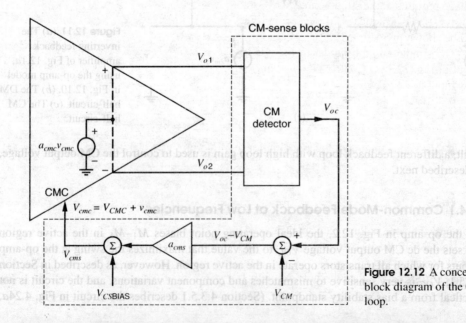

Figure 12.12 A conceptual block diagram of the CMFB loop.

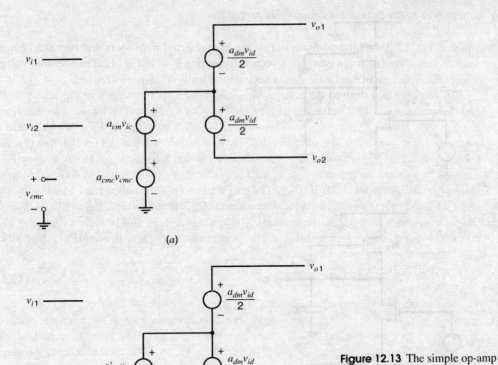

Figure 12.13 The simple op-amp model of Fig. 12.10 (a) including the CMC gain a_{cmc} and (b) replacing the a_{cm} and a_{cmc} controlled sources with an equivalent controlled source a'_{cm}.

low gain, and, as a result, a wide bandwidth. Because the CM-sense amplifier is in the CMFB loop, wide bandwidth in this amplifier simplifies the frequency compensation of the CMFB loop. If $V_{oc} = V_{CM}$ in Fig. 12.12, $V_{cmc} = V_{CSBIAS}$. In practice, the bias voltage V_{CSBIAS} is usually generated in the CM-sense-amplifier circuit. Hence, the CM-sense amplifier is designed so that when its differential input voltage is zero, its output voltage equals the nominal bias voltage required at the single-ended CMC input. For the op amp in Fig. 12.2, this bias voltage is $V_{GS5} - V_{SS}$. Practical CM-sense amplifiers that can generate such an output bias voltage will be shown in Section 12.5.

The simple op-amp model in Fig. 12.10 is modified to include the CM control (CMC) input in Fig. 12.13a. Controlled source a_{cmc} models the small-signal voltage gain from this new input to v_{oc}. That is,

$$a_{cmc} = \left. \frac{v_{oc}}{v_{cmc}} \right|_{v_{ic}=0} \quad (12.32)$$

When this gain is included, the equation for the CM output voltage in (12.29) becomes

$$v_{oc} = a_{cm}v_{ic} + a_{cmc}v_{cmc} \quad (12.33)$$

■ **EXAMPLE**

Compute the three voltage gains in the model in Fig. 12.13a for the op amp in Fig. 12.2. Use the 0.8-μm CMOS model data in Table 2.3 with $|V_{ov}| = |V_{GS} - V_t| = 0.2$ V and $L_{\text{eff}} = 0.8$ μm for all devices. Take $I_{D5} = 200$ μA, $|V_{Ap}| = 20$ V, and $V_{An} = 10$ V. (These V_A values follow from (1.163) and the data in Table 2.3 with $L_{\text{eff}} = 0.8$ μm.) Ignore body effect.

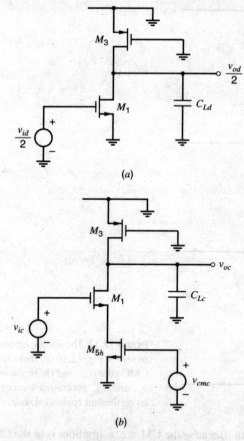

Figure 12.14 (a) The DM half-circuit for Fig. 12.2, explicitly showing the output load capacitance. (b) The CM half-circuit for Fig. 12.2, explicitly showing the output load capacitance.

The DM half-circuit is shown in Fig. 12.14a. It is a common-source amplifier with an active load. The low-frequency DM gain is

$$a_{dm} = \frac{v_{od}}{v_{id}} = -g_{m1}(r_{o1}||r_{o3}) \qquad (12.34)$$

Using (1.181) and $I_{D1} = |I_{D3}|$ in (12.34),

$$a_{dm} = -\frac{2I_{D1}}{V_{ov1}}\left(\frac{V_{A1}}{I_{D1}}||\frac{|V_{A3}|}{|I_{D3}|}\right) = -\frac{2}{V_{ov1}}\frac{V_{A1}|V_{A3}|}{V_{A1}+|V_{A3}|} = -\frac{2}{0.2}\frac{10 \times 20}{10+20} = -66.7$$

The CM half-circuit is shown in Fig. 12.14b. To form this half-circuit, the original circuit was transformed into a symmetric circuit by splitting M_5 into two identical halves in parallel, each called M_{5h} with $(W/L)_{5h} = (W/L)_5/2$ ($W_{5h} = W_5/2$, $L_{5h} = L_5$) and $I_{D5h} = I_{D5}/2$. This half-circuit has two inputs, v_{ic} and v_{cmc}. First, we find the gain from v_{cmc} to v_{oc} with $v_{ic} = 0$. This circuit consists of common-source M_{5h}, common-gate M_1, and active load M_3. The low-frequency gain is

$$a_{cmc} = \frac{v_{oc}}{v_{cmc}} = -g_{m5h}(R_{o(\text{down})}||r_{o3}) \qquad (12.35)$$

where $R_{o(\text{down})}$ is the resistance looking into the drain of M_1, which is given by

$$R_{o(\text{down})} = r_{o1}(1 + g_{m1}r_{o5h}) \approx r_{o1}(g_{m1}r_{o5h}) \qquad (12.36)$$

12.4 Common-Mode Feedback

Using the approximation in (12.36), (1.181), and $I_{D1} = |I_{D3}| = I_{D5h}$ in (12.35), we find

$$a_{cmc} = -\frac{2I_{D5h}}{V_{ov5h}}\left\{\left[\frac{V_{A1}}{I_{D1}}\left(\frac{2I_{D1}}{V_{ov1}}\frac{V_{A5h}}{I_{D5h}}\right)\right]\bigg\|\frac{|V_{A3}|}{|I_{D3}|}\right\} = -\frac{2}{V_{ov5h}}\frac{V_{A1}\left(\frac{2V_{A5h}}{V_{ov1}}\right)|V_{A3}|}{V_{A1}\left(\frac{2V_{A5h}}{V_{ov1}}\right) + |V_{A3}|}$$

$$= -\frac{2}{0.2}\frac{10\left(\frac{2(10)}{0.2}\right)20}{10\left(\frac{2(10)}{0.2}\right) + 20} = -196$$

Finally, we calculate the gain from v_{ic} to v_{oc} with $v_{cmc} = 0$. In this circuit, M_1 is a common-source amplifier with a degeneration resistance that is the output resistance of M_{5h}. The gain is

$$a_{cm} = \frac{v_{oc}}{v_{ic}} = -\frac{g_{m1}}{1 + g_{m1}r_{o5h}}(R_{o(\text{down})}\|r_{o3}) \approx -\frac{1}{r_{o5h}}(R_{o(\text{down})}\|r_{o3}) \quad (12.37)$$

The approximation is accurate if $g_{m1}r_{o5h} \gg 1$. Using the approximation in (12.36), (1.181), and $I_{D1} = |I_{D3}| = I_{D5h}$ in (12.37),

$$a_{cm} = -\frac{I_{D5h}}{V_{A5h}}\left\{\left[\frac{V_{A1}}{I_{D1}}\left(\frac{2I_{D1}}{V_{ov1}}\frac{V_{A5h}}{I_{D5h}}\right)\right]\bigg\|\frac{|V_{A3}|}{|I_{D3}|}\right\} = -\frac{1}{V_{A5h}}\frac{V_{A1}\left(\frac{2V_{A5h}}{V_{ov1}}\right)|V_{A3}|}{V_{A1}\left(\frac{2V_{A5h}}{V_{ov1}}\right) + |V_{A3}|}$$

$$= -\frac{1}{10}\frac{10\left(\frac{2(10)}{0.2}\right)20}{10\left(\frac{2(10)}{0.2}\right) + 20} = -1.96 \quad (12.38)$$

■ In this example, $|a_{cmc}|$ is much larger than $|a_{cm}|$ because the transconductance in (12.35) is much larger than the degenerated transconductance in (12.37).

The CMFB loop uses negative feedback to make $V_{OC} \approx V_{CM}$. If V_{CM} changes by a small amount from its design value due to parameter variations in the circuit that generates V_{CM}, V_{OC} should change by an equal amount so that V_{OC} tracks V_{CM}. The ratio $\Delta V_{OC}/\Delta V_{CM}$ is the closed-loop small-signal gain of the CMFB loop, which from Fig. 12.12 is

$$A_{CMFB} = \frac{\Delta V_{OC}}{\Delta V_{CM}} = \frac{v_{oc}}{v_{cm}} = \frac{a_{cms}(-a_{cmc})}{1 + a_{cms}(-a_{cmc})} \quad (12.39)$$

If $a_{cms}(-a_{cmc}) \gg 1$, $A_{CMFB} \approx 1$ and $\Delta V_{OC} \approx \Delta V_{CM}$.

The CM gain from v_{ic} to v_{oc} is affected by the CMFB loop. This gain can be calculated using the CMFB block diagram in Fig. 12.12 and the op-amp model in Fig. 12.13a. The op-amp CM gain when the CMFB is present, which we will call a'_{cm}, is found with the DM input signal set to zero. Using (12.31), the small-signal CM-sense voltage is related to the small-signal CM output voltage by

$$v_{cms} = a_{cms}v_{oc} \quad (12.40)$$

Using this equation, (12.33), and $v_{cms} = v_{cmc}$, we find

$$a'_{cm} = \frac{v_{oc}}{v_{ic}}\bigg|_{\text{with CMFB}} = \frac{a_{cm}}{1 + a_{cms}(-a_{cmc})} \quad (12.41)$$

Therefore, $|a'_{cm}| \ll |a_{cm}|$ if $|a_{cms}(-a_{cmc})| \gg 1$.

The CM gain for a balanced differential amplifier with CMFB can be found using either the model in Fig. 12.13a or 12.13b with a'_{cm} given by (12.41). These models are equivalent because the effect of the controlled source a_{cmc} that is part of the CMFB loop is included in a'_{cm}.

■ **EXAMPLE**

Compute the CM gain from v_{ic} to v_{oc} when the CMFB loop is active, a'_{cm}, for the op amp in the last example. Assume that $a_{cms} = 1$.

Substituting values from the last example in (12.41) with $a_{cms} = 1$ gives

$$a'_{cm} = \left.\frac{v_{oc}}{v_{ic}}\right|_{\text{with CMFB}} = \frac{-1.96}{1+(1)(196)} = -0.01$$

Comparing this result with (12.38), we see that the CMFB has reduced the CM gain by more than two orders of magnitude. ■

12.4.2 Stability and Compensation Considerations in a CMFB Loop

Since the CMFB loop is a negative feedback loop, stability is a key issue. For illustration purposes, consider the op amp in Fig. 12.2 driving a load capacitance and using the CMFB scheme shown in Fig. 12.12. The gain in the CMFB loop is $(-a_{cmc})a_{cms}$. The dominant pole p_{1c} in the CMFB loop is set by the load capacitance and the output resistance in the CM half-circuit in Fig. 12.14b. Ignoring all nondominant poles and using (12.35), we find

$$a_{cmc}(s) = -\frac{g_{m5h}(R_{o(\text{down})}||r_{o3})}{1+s(R_{o(\text{down})}||r_{o3})C_{Lc}} \tag{12.42}$$

At high frequencies [$\omega \gg |p_{1c}| = 1/(R_{o(\text{down})}||r_{o3})C_{Lc}$], (12.42) reduces to

$$a_{cmc}(j\omega)\Big|_{\omega \gg |p_{1c}|} \approx -\frac{g_{m5h}}{j\omega C_{Lc}} \tag{12.43}$$

This equation follows from the observation that the drain current from M_{5h} flows into the load capacitor at high frequencies. From (12.43), $|a_{cmc}| = 1$ at the frequency

$$\omega_{u,cm} = \frac{g_{m5h}}{C_{Lc}} \tag{12.44}$$

Nondominant poles exist in the CMFB loop, due to capacitance at the source of M_1 in Fig. 12.14b and due to poles in the gain $a_{cms}(s)$ of the CM-sense circuit. If the gain roll-off in (12.43) due to the dominant pole does not provide adequate phase margin for the CMFB loop, the unity-gain frequency for the CMFB loop gain can be decreased to increase the phase margin. From (12.44), increasing the CM load capacitance C_{Lc} decreases $\omega_{u,cm}$. However, adding capacitance to the op-amp outputs increases both the CM and DM load capacitances, as can be seen from (12.26) and (12.27). The CMFB loop gain may need a smaller unity-gain frequency than the DM loop gain because the CMFB loop may have more high frequency poles than the DM loop. For example, poles of $a_{cms}(s)$ of the CM-sense amplifier and the pole associated with the capacitance at the source of M_1 in Fig. 12.14b are poles in the CMFB loop. However, they are not poles in the DM feedback loop since the source of M_1 is an ac ground in the DM half-circuit. As a result, the load capacitance required in (12.43) to provide adequate phase margin for the CMFB loop may result in a larger DM load capacitance than desired, thereby overcompensating the DM feedback loop. While such overcompensation increases the phase margin of the DM feedback loop, it also decreases the unity-gain bandwidth of the DM

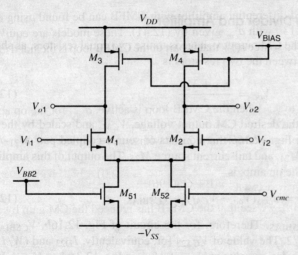

Figure 12.15 The op amp of Fig. 12.2 modified by replacing M_5 with M_{51} with V_{g51} constant, and M_{52} with $V_{g52} = V_{cmc}$.

loop gain and the 3-dB bandwidth of the DM closed-loop gain, which is undesirable when high bandwidth is desired in the DM feedback circuit to amplify a wide-band DM signal.

To overcome this problem, (12.44) shows that decreasing $g_{m5h} = g_{m5}/2$ decreases the unity-gain frequency of the gain a_{cmc} in the CMFB loop and therefore increases the phase margin of the CMFB loop. Assuming that the tail bias current I_{D5} cannot be changed, this decrease could be achieved by decreasing $(W/L)_5$. However, decreasing $(W/L)_5$ increases V_{ov5}, which affects the CM input range of the op amp. Alternatively, a reduction in g_{m5h} can be realized by splitting M_5 into two parallel transistors, which are labeled M_{51} and M_{52} in Fig. 12.15. Transistor M_{51} has its gate connected to a bias voltage and carries a constant drain current. The gate of M_{52} acts as the CMC input. To keep the bias currents in the op amp the same as in Fig. 12.2, we want

$$I_{D51} + I_{D52} = I_{D5} \tag{12.45}$$

To keep the CM input range of the op amp unchanged, we need

$$V_{ov51} = V_{ov52} = V_{ov5} \tag{12.46}$$

From (1.181), (12.45), and (12.46),

$$g_{m52} = \frac{2I_{D52}}{V_{ov52}} < g_{m5} = \frac{2I_{D5}}{V_{ov5}} \tag{12.47}$$

as desired. For the circuit in Fig. 12.15, $g_{m52h} = g_{m52}/2$ replaces $g_{m5h} = g_{m5}/2$ in (12.42), (12.43), and (12.44). A disadvantage of this approach is that reducing the transconductance in (12.42) reduces the magnitude of the CMC gain, $|a_{cmc}|$, at dc.

12.5 CMFB Circuits

In this section, circuits that detect the CM output voltage and generate a signal (a current or a voltage) that is a function of $V_{oc} - V_{CM}$ are described. These circuits are part of the CMFB loop and will be referred to as CM-sense circuits. For simplicity, these circuits are described using the simple, fully differential amplifier shown in Fig. 12.2.

12.5.1 CMFB Using Resistive Divider and Amplifier

A straightforward way to detect the CM output voltage is to use two equal resistors, as shown in Fig. 12.16a.[3,4] The voltage between the two resistors is

$$V_{oc} = \frac{V_{o1} + V_{o2}}{2} \qquad (12.48)$$

This voltage is subtracted from the desired CM output voltage, V_{CM}, and scaled by the differencing CM-sense amplifier in Fig. 12.16b that consists of source-coupled pair M_{21}-M_{22}, diode-connected loads M_{23} and M_{24}, and tail-current source M_{25}. The output of this amplifier, which drives the CMC input of the op amp, is

$$V_{cms} = a_{cms}(V_{oc} - V_{CM}) + V_{CSBIAS} \qquad (12.49)$$

If $V_{oc} = V_{CM}$, then $V_{cms} = V_{CSBIAS}$. Therefore, for the circuit of Fig. 12.16b, $V_{CSBIAS} = V_{GS23} - V_{SS}$ when $I_{D23} = I_{D25}/2$. The value of V_{GS23} [or, equivalently, I_{D23} and $(W/L)_{23}$] is chosen so that I_{D5} is equal to the design value of $|I_{D3}| + |I_{D4}|$ in Fig. 12.2 when $V_{oc} = V_{CM}$.

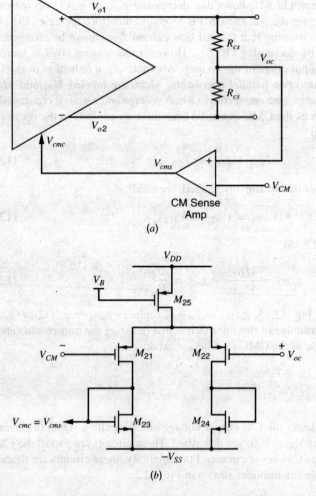

Figure 12.16 (a) CMFB using a resistive divider to detect V_{oc} and a CM-sense amplifier. (b) A schematic for the CM-sense amplifier that can be used with the op amp in Fig. 12.2.

In (12.49), a_{cms} is the small-signal voltage gain of the CM-sense amplifier

$$a_{cms} = \left.\frac{v_{cms}}{v_{oc}}\right|_{\text{CMFB loop open}} = \frac{1}{2}\frac{g_{m21}}{g_{m23}} \qquad (12.50)$$

Here, we have assumed that the CM gain of the CM-sense amplifier is much smaller in magnitude than its DM gain. The factor of $1/2$ multiplies g_{m21}/g_{m23} in (12.50) because the output is taken from only one side of the differential amplifier.

■ EXAMPLE

Determine the value of V_{CM} that maximizes the output swing for the differential op amp in Fig. 12.2. Use the CMFB scheme in Fig. 12.16a, and design the CM-sense amplifier in Fig. 12.16b. Assume that the CM-sense resistors R_{cs} are very large and can be neglected when computing small-signal voltage gains for the op amp. Use the data and assumptions in the next-to-last example with $V_{DD} = V_{SS} = 2.5$ V. Assume $V_{ic} = 0$. Ignore the body effect.

For the op amp in Fig. 12.2, if the magnitude of its DM gain is large and if the op amp operates in a DM negative feedback loop (for example, as shown in Fig. 12.1a), $V_{id} \approx 0$. Therefore, both op-amp inputs will be close to ground since $V_{ic} = 0$; that is, $V_{i1} = V_{ic} + V_{id}/2 \approx 0$ and $V_{i2} = V_{ic} - V_{id}/2 \approx 0$. The output V_{o1} reaches its lower limit when M_1 enters the triode region, which occurs when $V_{gd1} = V_{t1}$; therefore,

$$V_{o1(\min)} = -V_{t1} + V_{i1} \approx -V_{t1} = -0.7 \text{ V}$$

(Body effect would increase V_{t1} and decrease $V_{o1(\min)}$.) The upper output swing limit occurs when M_3 enters the triode region, and

$$V_{o1(\max)} = V_{DD} - |V_{ov3}| = 2.5 - 0.2 = 2.3 \text{ V}$$

To maximize the output swing, the dc CM output voltage V_{OC} should be halfway between the swing limits:

$$V_{OC} = \frac{V_{o1(\max)} + V_{o1(\min)}}{2} = \frac{2.3 + (-0.7)}{2} = 0.8 \text{ V}$$

Therefore, we choose $V_{CM} = 0.8$ V. The resulting peak differential output voltage is

$$V_{od(\text{peak})} = V_{o1(\max)} - V_{o2(\min)} = V_{o1(\max)} - V_{o1(\min)} = 2.3 - (-0.7) = 3.0 \text{ V}$$

To design the CM-sense amplifier in Fig. 12.16b, we must choose a value for its low-frequency gain. In the CMFB loop, the loop gain is $(-a_{cmc})a_{cms}$. From the previous example, $a_{cmc} = -196$. If we design for $a_{cms} = 1$, the CMFB loop gain is 196, and (12.39) gives $A_{CMFB} = 0.995$. Therefore, V_{oc} closely tracks changes in V_{CM}. With this choice of gain in the CM-sense amplifier in Fig. 12.16b, (12.50) gives

$$a_{cms} = \frac{1}{2}\frac{g_{m21}}{g_{m23}} = \frac{1}{2}\frac{\sqrt{2k'_p(W/L)_{21}|I_{D21}|}}{\sqrt{2k'_n(W/L)_{23}I_{D23}}} = \frac{1}{2}\frac{\sqrt{k'_p(W/L)_{21}}}{\sqrt{k'_n(W/L)_{23}}} = 1 \qquad (12.51)$$

The dc output voltage of the CM-sense amplifier when $V_{oc} = V_{CM}$ should equal the dc voltage needed at the CMC op-amp input, which is

$$-V_{SS} + V_{GS5} = -V_{SS} + V_{t5} + V_{ov5}$$

This dc voltage is produced by the CM-sense amplifier if M_5 and M_{23} have equal overdrive voltages. Assuming that $r_o \to \infty$, matching V_{ov5} and V_{ov23} requires that M_5 and M_{23} have equal drain-current-to-W/L ratios:

$$\frac{I_{D5}}{(W/L)_5} = \frac{I_{D23}}{(W/L)_{23}} \tag{12.52}$$

In (12.51) and (12.52), there are three unknowns: $I_{D23}, (W/L)_{23}$, and $(W/L)_{21}$. Therefore, many possible solutions exist. [Note that $(W/L)_5$ can be determined from $V_{ov5} = 0.2$ V (by assumption) and $I_{D5} = 200$ μA.]

One simple solution is $I_{D23} = I_{D5}$ and $(W/L)_{23} = (W/L)_5$. Then $(W/L)_{21}$ can be determined from (12.51). While this solution is simple, it requires as much dc current in the CM-sense amplifier as in the op amp. Equations 12.51 and 12.52 can be solved with $I_{D23} < I_{D5}$, which reduces the power dissipation in the CM-sense amplifier. However, the magnitude of the pole associated with the M_5-M_{23} current mirror decreases as I_{D23} decreases. To illustrate this point, we will ignore all capacitors except the gate-source capacitors for M_5 and M_{23} and assume L_{23} is fixed. Then the magnitude of the nondominant pole associated with the current mirror is

$$|p_{nd}| = \frac{g_{m23}}{C_{gs5} + C_{gs23}} = \frac{\frac{2I_{D23}}{V_{ov23}}}{C_{gs5} + (2/3)C_{ox}W_{23}L_{23}} \tag{12.53}$$

since the small-signal resistance of diode-connected M_{23} is $1/g_{m23}$ (assuming that $r_{o23} \gg 1/g_{m23}$). If I_{D23} is scaled by a factor x ($x < 1$) and if M_5 is unchanged, W_{23} must also scale by the factor x to satisfy (12.52). With this scaling, the pole magnitude in (12.53) decreases because the numerator scales by the factor x, but the denominator scales by a factor greater than x due to the constant C_{gs5} term in the denominator. This pole appears in the CMFB loop gain. Therefore, the phase margin of the CMFB loop decreases as this pole magnitude decreases due to a decrease in I_{D23}.

Finally, we must verify that the CM input range of the CM-sense amplifier includes its CM input voltage, which is $V_{CM} = 0.8$ V. The upper limit of the CM input voltage occurs when M_{25} enters the triode region, when $|V_{DS25}| = |V_{ov25}|$; therefore, we want

$$V_{IC} < V_{DD} - |V_{ov25}| - |V_{GS21}| = V_{DD} - |V_{ov25}| - |V_{tp}| - |V_{ov21}|$$
$$= 2.5 - 0.2 - 0.7 - 0.2 = 1.4 \text{ V}$$

The lower limit of the CM input voltage occurs when M_{21} (or M_{22}) enters the triode region (when $V_{GD21} = V_{t21}$); hence,

$$V_{IC} > -V_{SS} + V_{GS23} - |V_{t21}| = -V_{SS} + (V_{tn} + V_{ov23}) - |V_{tp}|$$
$$= -2.5 + (0.7 + 0.2) - 0.7 = -2.3 \text{ V}$$

The applied CM input voltage of 0.8 V falls between these limits; therefore, all transistors operate in the active region as assumed. ■

In this CMFB approach, the inputs to the CM-sense amplifier are ideally constant, which simplifies its design. One disadvantage of this CM-sense circuit is that the R_{cs} resistors and the input capacitance of the CM-sense amplifier introduce a pole in the transfer function of the CM-sense circuit and therefore in the CMFB loop. A capacitor C_{cs} can be connected in parallel with each sense resistor to introduce a left-half-plane zero in the CM-sense circuit to reduce the effect of the pole at high frequencies. (See Problem 12.18.)

Another disadvantage of this CM-sense circuit is that the sense resistor R_{cs} loads the op-amp output in the DM half-circuit, since the node between the resistors is a DM ac ground. This loading reduces the open-loop differential voltage gain unless R_{cs} is much larger than the output resistance of the DM half-circuit.

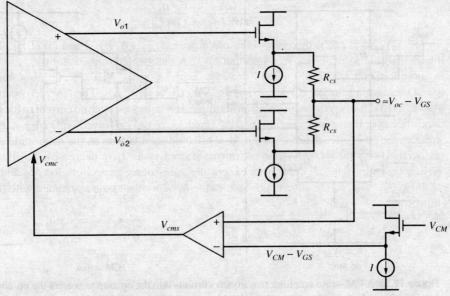

Figure 12.17 The CMFB scheme of Fig. 12.16 with source followers added as buffers between the op-amp outputs and the R_{cs} resistors.

To avoid this resistive output loading, voltage buffers can be added between the op-amp outputs and the R_{cs} resistors. Source followers are used as buffers in Fig. 12.17. One potential problem is that each source follower introduces a dc offset of V_{GS} between its input and output. To avoid a shift in the CM operating point caused by these offsets, voltage V_{CM} can be buffered by an identical source follower so that the op-amp output voltages and V_{CM} experience equal offsets. However, these offsets limit the op-amp output swing since each source-follower transistor that connects to an op-amp output must remain in the active region over the entire output voltage swing.

The CMFB scheme in Fig. 12.16 can be modified to eliminate the M_{23}-M_5 current mirror from the CMFB loop, as shown in Fig. 12.18. This modified CM-sense amplifier directly injects currents to control the op-amp CM output.[5] Here, M_{21} in Fig. 12.16b is split into two matched transistors, M_{21A} and M_{21B}, and the drain of each transistor connects to an op-amp output. The current injected by M_{21A} and M_{21B} into either output is

$$I_{cms} = \frac{I_{26}}{4} - \frac{g_{m21A}}{2}(V_{oc} - V_{CM})$$

Transistors M_3-M_5 act as current sources. The CMFB loop will adjust I_{cms} so that

$$|I_{D3}| + |I_{D4}| + 2I_{cms} = I_{D5}$$

If $V_{oc} = V_{CM}$, M_{21A}, M_{21B}, and M_{22} give $2I_{cms} = I_{26}/2$. Therefore, I_{26} should be chosen so that

$$|I_{D3}| + |I_{D4}| + \frac{I_{26}}{2} = I_{D5}$$

when all devices are active.

An advantage of this approach is that it avoids the pole associated with the M_5-M_{23} current mirror in Figs. 12.2 and 12.16. However, M_{21A} and M_{21B} add resistive and capacitive loading

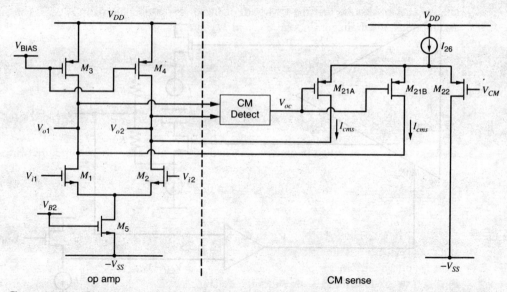

Figure 12.18 A CM-sense amplifier that injects currents into the op amp to control the op-amp CM output voltage. In the CM-sense circuit, $(W/L)_{21A} = (W/L)_{21B} = 0.5(W/L)_{22}$.

at the op-amp outputs. If an op amp uses cascoded devices, M_{21A} and M_{21B} can connect to low-impedance cascode nodes to reduce the impact of this loading.

12.5.2 CMFB Using Two Differential Pairs

A CMFB scheme that uses only transistors is shown in simplified form in Fig. 12.19. Here M_{21}-M_{24} are matched. The source-coupled pairs M_{21}-M_{22} and M_{23}-M_{24} together sense the CM output voltage and generate an output that is proportional to the difference between V_{oc} and V_{CM}.[5,6,7] To show this, assume at first that the differential inputs to the two source-coupled pairs, which are $V_{o1} - V_{CM}$ and $V_{o2} - V_{CM}$, are small enough to allow the use of small-signal

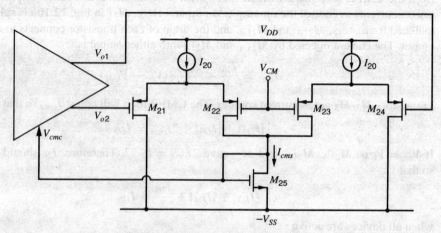

Figure 12.19 A CMFB approach that uses two differential pairs. This circuit can be used with the op amp in Fig. 12.2.

analysis. Also, assume that the CM gain of these source-coupled pairs is zero. Under these assumptions, the drain currents in M_{22} and M_{23} are

$$I_{d22} = -\frac{I_{20}}{2} - g_{m22}\frac{(V_{o2} - V_{CM})}{2} \tag{12.54}$$

$$I_{d23} = -\frac{I_{20}}{2} - g_{m23}\frac{(V_{o1} - V_{CM})}{2} \tag{12.55}$$

These currents are summed in diode-connected M_{25} to give the CM sensor output current

$$I_{cms} = I_{d25} = -I_{d22} - I_{d23} = I_{20} + g_{m22}\left(\frac{V_{o1} + V_{o2}}{2} - V_{CM}\right)$$

$$= I_{20} + g_{m22}(V_{oc} - V_{CM}) \tag{12.56}$$

since $g_{m22} = g_{m23}$. This last expression shows that the current through M_{25} includes a dc term I_{20} plus a term that is proportional to $V_{oc} - V_{CM}$. The current I_{d25} is mirrored by M_5 in Fig. 12.2 to produce the tail current in the op amp, which controls the CM output voltage.

The dc output of the CM-sense circuit should provide the dc voltage needed at the CMC input to give $V_{oc} = V_{CM}$. If $V_{oc} = V_{CM}$, the drain current in M_{25} is

$$I_{D25} = |I_{D22}| + |I_{D23}| = \frac{I_{20}}{2} + \frac{I_{20}}{2} = I_{20} \tag{12.57}$$

Choosing $I_{20} = |I_{D3}| + |I_{D4}|$ and $(W/L)_{25} = (W/L)_5$ is one design option. Again, as for the CM-sense amplifier in Fig. 12.16b, a smaller value of I_{20} can be used, but such current reduction causes the magnitude of the pole associated with the M_5-M_{25} current mirror to decrease. [See the text associated with (12.53).] This scheme does not resistively load the op-amp outputs, but the source-coupled pairs M_{21}-M_{24} capacitively load the op-amp outputs.

The above analysis of this CM-sense circuit assumed that M_{21}-M_{24} always operate in the active region and that voltages $V_{o1} - V_{CM}$ and $V_{o2} - V_{CM}$ could be treated as small-signal inputs. Even if these voltages become large, the CMFB loop continues to operate as long as M_{21}-M_{24} remain on. However, the small-signal analysis is not valid if the transistors leave the active region. If the op-amp outputs become large enough to turn off any of M_{21}-M_{24} during a portion of the output swing, the CMFB loop will not operate properly during that part of the output swing. The requirement that M_{21}-M_{24} remain on during the entire output swing imposes a limit on the output swing of the op amp. The input voltage range for which both transistors in a differential pair remain on is related to the gate overdrive voltage of those transistors. [See (3.161).] Therefore, to keep M_{21}-M_{24} on for a large V_{o1} and V_{o2}, M_{21}-M_{24} require large overdrives. In contrast, the scheme that uses resistors to detect the CM output voltage does not impose such an output-swing limit since the CM-sense amplifier is driven by V_{oc}, which is ideally constant, rather than V_{o1} and V_{o2}, which include CM and DM components.

Equation 12.56 implies that this CM-sense circuit is nearly perfect since it produces an output current that includes a constant term plus a term that is proportional to $V_{oc} - V_{CM}$. This result is based on a linear small-signal analysis. However, the inputs to the differential pairs in the CM-sense circuit can include large signals because the op-amp DM output voltage can be large. Next, a large signal analysis of this circuit is carried out.

The drain current in M_{25} is

$$I_{d25} = -I_{d22} - I_{d23} \tag{12.58}$$

The differential input of the M_{21}-M_{22} pair is $V_{o2} - V_{CM}$. Using (3.159) and (1.166) gives

$$-I_{d22} = \frac{I_{20}}{2} + \frac{k'_p}{4}\left(\frac{W}{L}\right)_{22}(V_{o2} - V_{CM})\sqrt{4V_{ov22}^2 - (V_{o2} - V_{CM})^2}$$

$$\approx \frac{I_{20}}{2} + \frac{k'_p}{4}\left(\frac{W}{L}\right)_{22}\left(-\frac{V_{od}}{2} + V_{oc} - V_{CM}\right)\sqrt{4V_{ov22}^2 - (V_{od}/2)^2 + (V_{oc} - V_{CM})V_{od}}$$

$$= \frac{I_{20}}{2} + \frac{k'_p}{4}\left(\frac{W}{L}\right)_{22}\left(-\frac{V_{od}}{2} + V_{oc} - V_{CM}\right)\sqrt{4V_{ov22}^2 - (V_{od}/2)^2}\sqrt{1 + \frac{(V_{oc} - V_{CM})V_{od}}{4V_{ov22}^2 - (V_{od}/2)^2}}$$

$$\approx \frac{I_{20}}{2} + \frac{k'_p}{4}\left(\frac{W}{L}\right)_{22}\left(-\frac{V_{od}}{2} + V_{oc} - V_{CM}\right)\sqrt{4V_{ov22}^2 - (V_{od}/2)^2}$$

$$\times \left[1 + \frac{1}{2}\left(\frac{(V_{oc} - V_{CM})V_{od}}{4V_{ov22}^2 - (V_{od}/2)^2}\right) - \frac{1}{8}\left(\frac{(V_{oc} - V_{CM})V_{od}}{4V_{ov22}^2 - (V_{od}/2)^2}\right)^2 + \cdots\right] \quad (12.59)$$

where $|V_{oc} - V_{CM}| \ll |V_{od}|$ was assumed in the first approximation above and $\sqrt{1+x} \approx 1 + x/2 - x^2/8 + \cdots$, where $x = [(V_{oc} - V_{CM})V_{od}]/[4V_{ov22}^2 - (V_{od}/2)^2]$, was used in the last line.

The differential input of the M_{23}-M_{24} pair is $V_{o1} - V_{CM}$. A similar analysis to that above for this differential pair yields

$$-I_{d23} \approx \frac{I_{20}}{2} + \frac{k'_p}{4}\left(\frac{W}{L}\right)_{23}\left(\frac{V_{od}}{2} + V_{oc} - V_{CM}\right)\sqrt{4V_{ov23}^2 - (V_{od}/2)^2}$$

$$\times \left[1 - \frac{1}{2}\left(\frac{(V_{oc} - V_{CM})V_{od}}{4V_{ov23}^2 - (V_{od}/2)^2}\right) - \frac{1}{8}\left(\frac{(V_{oc} - V_{CM})V_{od}}{4V_{ov23}^2 - (V_{od}/2)^2}\right)^2 + \cdots\right] \quad (12.60)$$

Substituting (12.59) and (12.60) in (12.58) with $V_{ov22} = V_{ov23}$ and $(W/L)_{22} = (W/L)_{23}$ gives

$$I_{cms} = I_{d25} \approx I_{20} + \frac{k'_p}{2}\left(\frac{W}{L}\right)_{23}(V_{oc} - V_{CM})\sqrt{4V_{ov23}^2 - (V_{od}/2)^2}$$

$$\times \left[1 - \frac{1}{4}\left(\frac{V_{od}^2}{4V_{ov23}^2 - (V_{od}/2)^2}\right) - \frac{1}{8}\left(\frac{(V_{oc} - V_{CM})V_{od}}{4V_{ov23}^2 - (V_{od}/2)^2}\right)^2 + \cdots\right] \quad (12.61)$$

If $|V_{od}/2| \ll |2V_{ov23}|$, this equation reduces to (12.56). To interpret (12.61), first consider the case when $V_{oc} = V_{CM}$. Then (12.61) shows that the CM-sense output current is constant with $I_{cms} = I_{20}$. Whereas I_{cms} is constant, (12.59) and (12.60) show that I_{d22} and I_{d23} are not constant if V_{od} is nonzero and time-varying, and I_{d22} and I_{d23} are nonlinear functions of V_{od} (see the plot in Fig. 3.51). However, the variation in I_{d22} due to nonzero V_{od} is equal and opposite to the variation in I_{d23} due to V_{od}; therefore, these variations cancel when these currents are summed to form I_{cms}. Next, consider the case when $V_{oc} \neq V_{CM}$. V_{oc} may not equal V_{CM} due to device mismatch, finite gain in the CMFB loop, or the presence of an ac component in V_{oc}. Equation 12.61 shows that I_{cms} has terms that include V_{od}^2 that affect I_{cms} when $V_{oc} \neq V_{CM}$. Therefore, even if the transistors are perfectly matched, this CM sensor does not behave like an ideal CM sensor as described by (12.49). The terms that include V_{od}^2 stem from the (square-law) nonlinearity associated with the M_{21}-M_{22} and M_{23}-M_{24} differential pairs that convert $V_{o1} - V_{CM}$ and $V_{o2} - V_{CM}$ into currents. The dependence of I_{cms} on V_{od}^2 can cause a shift in the dc CM output voltage. Moreover, if V_{od} is not constant, this dependence can produce an ac component in V_{oc}.

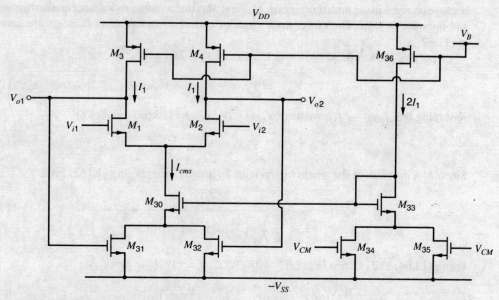

Figure 12.20 A CMFB approach that uses transistors M_{31}, M_{32}, M_{34}, and M_{35} biased in the triode region.

12.5.3 CMFB Using Transistors in the Triode Region

Another CMFB scheme is shown in Fig. 12.20.[8] The simple op amp of Fig. 12.2 is redrawn here, with M_5 replaced by M_{30}-M_{32}. Transistors M_{30}-M_{35} are part of the CMFB loop. Here, M_{31}, M_{32}, M_{34}, and M_{35} operate in the triode region, while M_{30}, M_{33}, and M_{36} operate in the active region. The desired CM output voltage V_{CM} is connected to the gates of M_{34} and M_{35}. To simplify the description of this circuit, assume that M_{30}-M_{35} are matched and ignore body effect. Before this CMFB approach is mathematically analyzed, its operation will be explained intuitively. At first, assume that the op-amp outputs in Fig. 12.20 have only a CM component; that is, $V_{o1} = V_{o2} = V_{oc}$. Then the gate voltages of M_{31} and M_{32} equal V_{oc}. Since $|I_{D3}| = |I_{D4}| = I_1$, the drain current in M_{30} must equal $2I_1$ to satisfy KCL. Transistors M_{30}-M_{35} form a degenerated current mirror and give $I_{D30} = I_{D33} = 2I_1$ when $V_{oc} = V_{CM}$ because M_{30}-M_{35} are matched. Therefore, $V_{oc} = V_{CM}$ is a possible operating point for this circuit. Negative feedback forces the circuit to this operating point, as described below. So far, we have assumed that only CM signals are present. Differential signals do not affect the operation of this feedback loop, as shown by the following analysis.

Since M_{31} and M_{32} are matched and operate in the triode region, the sum of their drain currents I_{cms} is [using (1.152)]

$$I_{cms} = I_{d31} + I_{d32} = k'_n \left(\frac{W}{L}\right)_{31} \left((V_{o1} + V_{SS} - V_{t31})V_{ds31} - \frac{V_{ds31}^2}{2}\right)$$

$$+ k'_n \left(\frac{W}{L}\right)_{32} \left((V_{o2} + V_{SS} - V_{t32})V_{ds32} - \frac{V_{ds32}^2}{2}\right)$$

$$= 2k'_n \left(\frac{W}{L}\right)_{31} \left(V_{oc} + V_{SS} - V_{tn} - \frac{V_{ds31}}{2}\right) V_{ds31} \quad (12.62)$$

since $V_{ds31} = V_{ds32}$, $V_{t31} = V_{t32} = V_{tn}$, and $(W/L)_{31} = (W/L)_{32}$. This equation shows that I_{cms} is dependent on the CM output voltage and independent of the differential output voltage because changes in the drain currents in M_{31} and M_{32} due to nonzero V_{od} are equal in magnitude and opposite in sign. Therefore, these changes cancel when the drain currents are summed in (12.62).

Applying KVL around the lower transistors M_{30}-M_{35} gives

$$V_{ds31} = V_{ds35} + V_{gs33} - V_{gs30} \tag{12.63}$$

Assuming that $I_{d30} \approx I_{d33}$, we have $V_{gs30} \approx V_{gs33}$, and (12.63) reduces to

$$V_{ds31} \approx V_{ds35} \tag{12.64}$$

Since M_{35} operates in the triode region with $I_{D35} = I_1$, rearranging (1.152) gives

$$V_{ds35} = \frac{I_1}{k'_n \left(\frac{W}{L}\right)_{35} \left(V_{CM} + V_{SS} - V_{tn} - \frac{V_{ds35}}{2}\right)} \tag{12.65}$$

Using (12.64) and (12.65) in (12.62) with $(W/L)_{31} = (W/L)_{35}$ gives

$$I_{cms} \approx I_1 \frac{2k'_n \left(\frac{W}{L}\right)_{35}\left(V_{oc} + V_{SS} - V_{tn} - \frac{V_{ds35}}{2}\right)}{k'_n \left(\frac{W}{L}\right)_{35}\left(V_{CM} + V_{SS} - V_{tn} - \frac{V_{ds35}}{2}\right)} = 2I_1 \frac{V_{oc} + V_{SS} - V_{tn} - \frac{V_{ds35}}{2}}{V_{CM} + V_{SS} - V_{tn} - \frac{V_{ds35}}{2}}$$

$$= 2I_1 \frac{V_{CM} + V_{SS} - V_{tn} - \frac{V_{ds35}}{2}}{V_{CM} + V_{SS} - V_{tn} - \frac{V_{ds35}}{2}} + 2I_1 \frac{V_{oc} - V_{CM}}{V_{CM} + V_{SS} - V_{tn} - \frac{V_{ds35}}{2}}$$

$$= 2I_1 + 2I_1 \frac{V_{oc} - V_{CM}}{V_{CM} + V_{SS} - V_{tn} - \frac{V_{ds35}}{2}} \tag{12.66}$$

This last expression shows that the op-amp tail current I_{cms} consists of a constant term $2I_1$ plus a term that depends on $V_{oc} - V_{CM}$. If $|I_{d3}| = |I_{d4}| = I_1$, then KCL requires that $I_{cms} = 2I_1$. Using this value in (12.66) gives $V_{oc} \approx V_{CM}$, as desired. In practice, mismatches can cause V_{oc} to deviate from V_{CM}. For example, if the drain currents in M_3 and M_4 are larger than I_1, then (12.66) shows that V_{oc} must be larger than V_{CM} to force I_{cms} to be larger than $2I_1$.

To see that the CMFB loop here is a negative feedback loop, assume that V_{oc} increases. Then the gate-source voltages on M_{31} and M_{32} increase, which in turn increases I_{cms}. Increasing I_{cms} causes $V_{sd3} = V_{sd4}$ to increase since M_3 and M_4 have fixed gate-source voltages. This increase in $V_{sd3} = V_{sd4}$ causes V_{oc} to fall and counteract the assumed increase in V_{oc}. In steady state, this CMFB loop forces $V_{oc} \approx V_{CM}$.

One limitation of this scheme is that the CMFB loop will not function properly whenever the output voltage swing is large enough to turn off either M_{31} or M_{32}. Therefore, neither op-amp output is allowed to swing within a threshold voltage of $-V_{SS}$. Thus, the op-amp output swing is limited by this CMFB scheme. Another limitation is that the magnitude of the small-signal gain in the CMFB loop is smaller here than in the previous approaches because the transconductance of M_{31} or M_{32} in the triode region is smaller than it is in the active region. (See Problem 12.19.) Reducing the CMFB loop gain reduces the control that the CMFB loop exerts on the CM output voltage. Also, the bandwidth of the CMFB loop is lower here than in other cases due to the low transconductance of M_{31} and M_{32}. Bandwidth requirements for the CMFB loop are considered in Section 12.8.

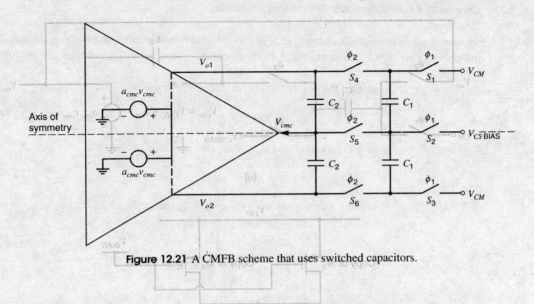

Figure 12.21 A CMFB scheme that uses switched capacitors.

12.5.4 Switched-Capacitor CMFB

To overcome the op-amp output swing limitations imposed by the last two CMFB approaches and to avoid resistive output loading of the op amp, capacitors can be used to detect the CM output voltage. If the CM-sense resistors R_{cs} in Fig. 12.16 are replaced with capacitors, the resistive output loading is eliminated, but these capacitors are open circuits at dc. To avoid a dc bias problem, switched capacitors can be used as the CM detector.[9] A switched-capacitor (SC) CMFB scheme that is often used in switched-capacitor amplifiers and filters (see Section 6.1.7) is shown in Fig. 12.21. Here the network that consists of switches S_1-S_6 and capacitors C_1 and C_2 sense the CM output voltage and subtract it from the desired CM output voltage V_{CM}. Voltage V_{CSBIAS} is a dc bias voltage. As in Fig. 6.8, assume that each switch is on when its control signal is high and is off when its control signal is low. The switches are controlled by two nonoverlapping clocks, ϕ_1 and ϕ_2 (that is, ϕ_1 and ϕ_2 are never high at the same time). In this section, we will assume that these switches are ideal. In practice, switches S_1-S_6 are implemented with MOS transistors. As in the previous section, we use the simple op amp in Fig. 12.2 as the op amp in Fig. 12.21.

The SC CMFB circuit is a linear, balanced, discrete-time circuit. Therefore, all points on the axis of symmetry (shown as a dashed line in Fig. 12.21) operate at ac ground for differential signals. The op-amp CMC input is along the axis of symmetry, so V_{cmc} has a CM component but zero DM component. Therefore, the switched-capacitor circuit is a good CM sensor. To show that voltage V_{cmc} depends on the difference between the actual and desired CM output voltages, consider the CM half-circuit shown in Fig. 12.22a. Capacitor C_2 is not switched and connects from V_{cmc} to V_{oc}. Since V_{cmc} is the gate voltage of M_5 in Fig. 12.2, there is voltage gain from V_{cmc} to V_{oc}, which is modeled by controlled source a_{cmc}. Comparing this half-circuit to Fig. 6.10a, we see that C_2 connected across the gain stage and switched-capacitor C_1 form a switched-capacitor integrator. This integrator is in a negative feedback loop since its output V_{oc} is connected back to a switch that connects to C_1.

When ϕ_1 is high, C_1 charges to $V_{CM} - V_{CSBIAS}$. When ϕ_2 is high, C_1 connects between V_{oc} and V_{cmc}. In steady state, V_{oc} is constant because the applied voltages V_{CM} and V_{CSBIAS} are both dc voltages and because the switched-capacitor integrator operates in a negative feedback loop. After V_{oc} becomes constant, C_1 does not transfer charge onto C_2 when ϕ_2 is high. This

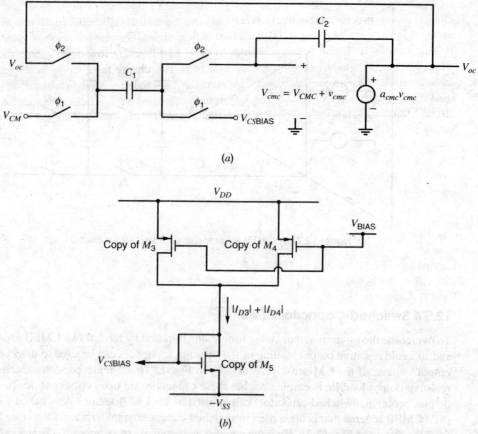

Figure 12.22 (a) A CM half-circuit for Fig. 12.21. (b) Replica bias circuit for generating V_{CSBIAS} for the differential op amp in Fig. 12.2.

condition is satisfied if the charge on C_1 when ϕ_1 is high is the same when ϕ_2 is high, or

$$Q(\phi_1) = C_1(V_{CM} - V_{CSBIAS}) = Q(\phi_2) = C_1(V_{oc} - V_{cmc}) \quad (12.67)$$

This equation reduces to

$$V_{CM} - V_{oc} = V_{CSBIAS} - V_{cmc} \quad (12.68)$$

If V_{CSBIAS} equals the nominal bias voltage required at the CMC input and if $|a_{cmc}| \gg 1$, V_{cmc} is about constant with $V_{cmc} \approx V_{CSBIAS}$. Then (12.68) reduces to

$$V_{oc} \approx V_{CM} \quad (12.69)$$

as desired. For the op amp in Fig. 12.2, bias voltage V_{CSBIAS} could be generated by passing a current equal to $|I_{D3}| + |I_{D4}|$ through a diode-connected copy of M_5 connected to $-V_{SS}$, as shown in Fig. 12.22b. The copies of M_3 and M_4 have the same source and gate connections as in the op amp and duplicate the currents $|I_{D3}|$ and $|I_{D4}|$ that flow in Fig. 12.2. The voltage V_{CSBIAS} is the gate voltage of the copy of M_5. Since this bias circuit uses copies or *replicas* of the transistors in the op amp to generate V_{CSBIAS}, this technique is referred to as *replica biasing*.

An advantage of this CMFB approach is that the op-amp output voltage swing is not limited by this CM-sense circuit because it consists only of passive elements (capacitors) and switches. (If a switch is constructed of n-channel and p-channel transistors in parallel driven by clock ϕ and its inverse, respectively, it can pass any signal that falls between the power-supply voltages if $V_{DD} + V_{SS} > V_{tn} + |V_{tp}|$.[10] In practice, an MOS transistor is not an ideal switch. It must have a W/L that is large enough to give a sufficiently low drain-source resistance when it is on. However, when each transistor turns off, charge from its channel and charge associated with its gate overlap capacitance transfer onto its drain and source nodes. Therefore, the MOS transistors acting as switches will transfer charge onto C_1. Let ΔQ represent the net charge transferred onto C_1 each clock period. Including the effect of this charge, (12.67) becomes

$$C_1(V_{CM} - V_{CSBIAS}) = C_1(V_{oc} - V_{cmc}) + \Delta Q \qquad (12.70)$$

or

$$V_{CM} - V_{oc} = V_{CSBIAS} - V_{cmc} + \frac{\Delta Q}{C_1} \qquad (12.71)$$

Comparing (12.71) with (12.68) shows that $\Delta Q/C_1$ introduces an offset in V_{oc}, making V_{oc} differ from V_{CM} when $V_{CSBIAS} = V_{cmc}$. If V_{CM} was chosen to maximize the op-amp output swing, a shift in V_{oc} will reduce the op-amp output swing. The magnitude of the charge transferred by each switch transistor increases with its width W [since the gate-channel and overlap capacitances are proportional to W as shown in (1.187) and (2.45)], so a trade-off exists between low switch on-resistance and small charge transfer. From (12.71), increasing C_1 reduces the effect of the transferred charge on V_{oc}, but increasing C_1 increases the capacitive loading at the op-amp outputs when ϕ_2 is high.

12.6 Fully Differential Op Amps

Some fully differential op amps are presented in this section. The singled-ended counterpart of each op amp was covered in previous chapters (low-frequency operation in Chapter 6 and compensation in Chapter 9). The two-stage op amp will be covered first, followed by single-stage op amps.

12.6.1 A Fully Differential Two-Stage Op Amp

A fully differential two-stage op amp is shown in Fig. 12.23. Compared to its single-ended counterpart in Fig. 6.16, two differences are the addition of M_9-M_{10}, which is a copy of the common-source stage M_6-M_7, to generate the second output, and the removal of the gate-to-drain connection on M_3 to give a symmetric input stage. The input stage is a complementary version of the differential stage in Fig. 12.2. The common-mode control (CMC) input is the gate of tail current source M_5. If the voltage at the gate of M_5 changes, the magnitudes of the drain currents in M_1-M_4 change by equal amounts. Therefore, V_{ds3} and V_{ds4} change by equal amounts. These voltage changes are amplified by the common-source stages M_6-M_7 and M_9-M_{10} to cause equal changes in output voltages V_{o1} and V_{o2}, which changes V_{oc}. Therefore, the CM output voltage can be controlled by a CMFB loop that connects to the gate of M_5.

In Fig. 12.23, two Miller compensation capacitors C are connected across the symmetric second stages. These capacitors compensate both the DM and CM half-circuits, which are shown in Fig. 12.24. Although not shown, any of the approaches described in Chapter 9 for eliminating the right-half-plane zero associated with feedforward through the compensation capacitor could be used here.

824   Chapter 12 ▪ Fully Differential Operational Amplifiers

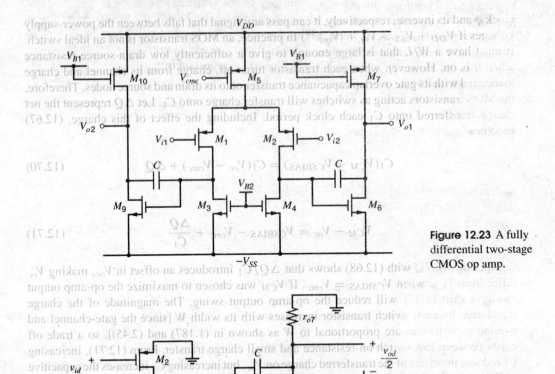

Figure 12.23 A fully differential two-stage CMOS op amp.

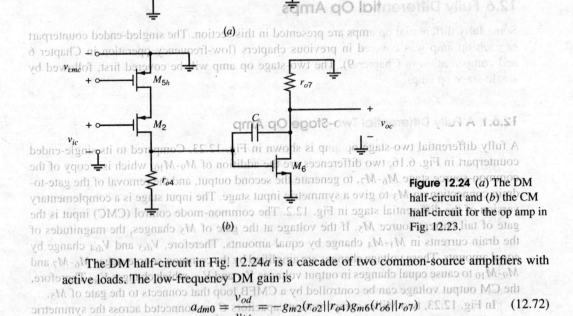

Figure 12.24 (a) The DM half-circuit and (b) the CM half-circuit for the op amp in Fig. 12.23.

The DM half-circuit in Fig. 12.24a is a cascade of two common-source amplifiers with active loads. The low-frequency DM gain is

$$a_{dm0} = \frac{v_{od}}{v_{id}} = -g_{m2}(r_{o2}\|r_{o4})g_{m6}(r_{o6}\|r_{o7}) \quad (12.72)$$

The Miller-compensated second stage can be modeled by the circuit in Fig. 9.21 with $R_1 = r_{o2}\|r_{o4}$, $g_m = g_{m6}$, $R_2 = r_{o6}\|r_{o7}$, $C_1 = C_{1d}$, and $C_2 = C_{2d}$. (The input capacitance C_{1d} of the second stage and load capacitance C_{2d} of the second stage are not shown explicitly in

Fig. 12.24a.) Therefore, the poles p_{1d} and p_{2d} of the DM half-circuit are given by (9.32) and (9.33). Assume that the op amp is operating in a feedback loop, the feedback factor f_{dm} for the DM feedback loop is frequency-independent, and the right-half-plane zero has been eliminated. Then to achieve 45° phase margin, the magnitude of the DM loop gain should be unity at the frequency $|p_{2d}|$. Since $|gain| \times frequency$ is constant from $|p_{1d}|$ to $|p_{2d}|$ due to the one-pole roll-off there, we can write

$$|a_{dm0} f_{dm} p_{1d}| = 1 \cdot |p_{2d}| \qquad (12.73)$$

Substituting (12.72), (9.32), and (9.33) in (12.73) gives

$$\frac{g_{m2}}{C} f_{dm} \approx \frac{g_{m6}}{C_{2d}} \qquad (12.74)$$

assuming that the DM load capacitance C_{2d} and the compensation capacitor C are much larger than the internal node capacitance C_{1d}. If the other values are known, the compensation capacitor is determined by (12.74).

The CM half-circuit is shown in Fig. 12.24b. The first stages of the CM and DM half-circuits are different, but the second stages are identical. To focus on the CMFB loop, we will assume $v_{ic} = 0$. (Nonzero v_{ic} will be considered later.) In the CM half-circuit, the first stage consists of common-source M_{5h} with common-gate M_2 and active load M_4. As in Fig. 12.14b, M_{5h} is one half of M_5, with $(W/L)_{5h} = (W/L)_5/2$ and $I_{D5h} = I_{D5}/2$. The first stage is followed by the common-source second stage, M_6-M_7. The low-frequency CMC gain is

$$a_{cmc0} = \frac{v_{oc}}{v_{cmc}} \approx g_{m5h}[(r_{o2} g_{m2} r_{o5h})||r_{o4}] g_{m6}(r_{o6}||r_{o7}) \qquad (12.75)$$

Capacitance associated with the source of cascode M_1 introduces a pole p_x in the CMC gain. If $|p_x|$ is much larger than the magnitude of the nondominant pole $|p_{2c}|$ in (9.33) from the Miller-compensated second stage, pole p_x can be ignored, and the gain a_{cmc} can be approximated as having two poles that are given by (9.32) and (9.33). These poles can be different than the poles in the DM gain for two reasons. First the output load capacitances in the DM and CM half-circuits can be different, and second the output resistances of the first stages in the half-circuits can be different. The zero due to feedforward is the same as for DM gain and can be eliminated as described in Chapter 9. To simplify the following analysis, we will assume that all poles and zeros in the CMFB loop other than the two poles associated with the Miller compensation can be ignored. To achieve 45° phase margin, the magnitude of the CMFB loop gain should fall to unity at $|p_{2c}|$. Therefore,

$$|a_{cmc0} a_{cms0} p_{1c}| = 1 \cdot |p_{2c}| \qquad (12.76)$$

where a_{cms0} is the low-frequency gain through the CM-sense circuit

$$a_{cms0} = \frac{v_{cmc}}{v_{oc}}\bigg|_{\omega=0,\ \text{CMFB loop open}} = \frac{v_{cms}}{v_{oc}}\bigg|_{\omega=0,\ \text{CMFB loop open}} \qquad (12.77)$$

Substituting (12.75), (9.32), and (9.33) in (12.76) and using $R_1 \approx r_{o2} g_{m2} r_{o5h}$ gives

$$\frac{g_{m5h}}{C}|a_{cms0}| \approx \frac{g_{m6}}{C_{2c}} \qquad (12.78)$$

assuming that the CM load capacitance C_{2c} and the compensation capacitor C are much larger than the internal node capacitance C_{1c}. The compensation capacitor required for the CMFB loop can be found from (12.78).

Ideally, the compensation capacitor values calculated in (12.74) and (12.78) would be equal, and the CMFB and DM loops would each have a phase margin of 45°. In practice, these values are rarely equal. If the value of C is chosen to be the larger of the values given by (12.74) and (12.78), one feedback loop will have a phase margin of 45°, and the other loop will have a phase margin larger than 45° and will be overcompensated. A drawback of overcompensation is that the unity-gain frequency of the loop gain and the closed-loop bandwidth are smaller than they would be if the loop were optimally compensated. If the larger C is required to compensate the DM feedback loop, using that compensation capacitor will overcompensate the CMFB loop. Since the CMFB ideally operates on only dc signals, reducing its bandwidth may be acceptable. (See Section 12.8 for more on this topic.) If the larger C is required to compensate the CMFB feedback loop, using that compensation capacitor will overcompensate the DM loop. However, reducing the bandwidth of the DM feedback loop by overcompensation is usually undesirable because this loop operates on the DM input signal, which may have a wide bandwidth.

An alternative to using the larger C value that optimally compensates the CMFB loop but overcompensates the DM loop is the following. The value of C that gives a 45° phase margin in the DM loop from (12.74) can be used if g_{m5h} is scaled down to satisfy (12.78). This approach gives a 45° phase margin for both feedback loops without sacrificing bandwidth in the DM loop. Scaling of $g_{m5h} = g_{m5}/2$ could be achieved by reducing $(W/L)_5$, but such scaling would reduce the CM input range of the op amp because decreasing $(W/L)_5$ increases $|V_{ov5}|$. Another solution is to split M_5 into two parallel transistors, one that has its gate connected to a bias voltage and the other with its gate connected to CMC, as described in Section 12.4.2 and Fig. 12.15. A drawback of this approach is that reducing g_{m5h} reduces $|a_{cmc0}|$, as can be seen in (12.75).

Ignoring limitations imposed by the CM-sense circuit, we see that each op-amp output in Fig. 12.23 can swing until a transistor in the second stage enters the triode region. The maximum value of V_{o1} is $V_{DD} - |V_{ov7}|$, and its minimum value is $-V_{SS} + V_{ov6}$. Therefore, the peak differential output voltage is

$$V_{od(peak)} = V_{o1(max)} - V_{o2(min)} = V_{o1(max)} - V_{o1(min)} = V_{DD} - |V_{ov7}| - (-V_{SS} + V_{ov6})$$
$$= V_{DD} + V_{SS} - V_{ov6} - |V_{ov7}| \quad (12.79)$$

The CM input range of the op amp is limited in the positive direction by the tail current source, which transitions from active to triode when $|V_{ds5}| = |V_{ov5}|$; therefore, we want

$$V_{IC} < V_{DD} - |V_{GS1}| - |V_{ov5}| \quad (12.80)$$

The lower limit of the CM input range occurs when input transistor M_1 (or M_2) enters the triode region, so

$$V_{IC} > -V_{SS} + V_{GS6} + V_{t1} \quad (12.81)$$

■ **EXAMPLE**

Modify the single-ended two-stage op amp from the examples in Section 6.3.5 and Section 9.4.3 into a fully differential op amp. It will be used in the feedback circuit shown in Fig. 12.25, which represents the connections in a switched-capacitor circuit when one clock is high (assuming the switches are ideal). The capacitor values are $C_S = 2$ pF, $C_F = 5$ pF, and $C_L = 2$ pF. Design for 1-V peak output swing and phase margins of 45° or greater in the DM and CMFB loops. Use $V_{DD} = V_{SS} = 1.65$ V, and design for a CM output voltage of 0 V.

First, we will design the devices to satisfy the bias and low-frequency requirements. Then we will compensate the amplifier. Using the device sizes and bias currents from the example in Section 6.3.5, we have

$$(W/L)_1 = (W/L)_2 = 77 \quad (W/L)_3 = (W/L)_4 = 4 \quad (W/L)_5 = 25$$
$$(W/L)_6 = (W/L)_9 = 16 \quad (W/L)_7 = (W/L)_{10} = 50$$

12.6 Fully Differential Op Amps

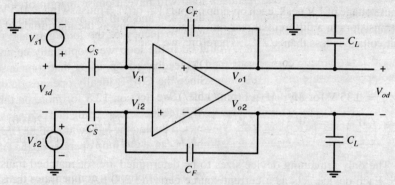

Figure 12.25 A fully differential op amp with capacitive load and feedback.

with $L_{drwn} = 1$ μm, $|I_{D1}| = |I_{D2}| = 100$ μA, and $I_{D6} = 400$ μA. In that example, these values gave a calculated dc gain of $a_{dm0} = -7500$ and a simulated gain of $a_{dm0} = -6200$.

For CMFB, we will use two differential pairs as shown in Fig. 12.26. Since the input stage in Fig. 12.23 is the complement of the op amp in Fig. 12.2, the CMFB circuit in Fig. 12.26 is the complement of the circuit in Fig. 12.19 to allow control of the op-amp tail current I_{D5} through a current mirror formed by M_5 and M_{25}. Also, the CM-sense output V_{cms} is taken from the drains of M_{21} and M_{24}, which makes the gain a_{cms} negative. This inversion is needed here to give negative feedback in the CMFB loop because the CMC gain a_{cmc} is positive at low frequencies in this two-stage op amp. We choose M_{25} to be matched to M_5, so they form a unity-gain current mirror. Since the desired tail current is $|I_{D5}| = 200$ μA, we want

$$|I_{D25}| = I_{D26} = I_{D27} = 200 \text{ μA}$$

Therefore, each transistor M_{21}-M_{24} nominally carries 100 μA of drain current. These transistors must remain active over the entire range of the op-amp output swing. For a differential

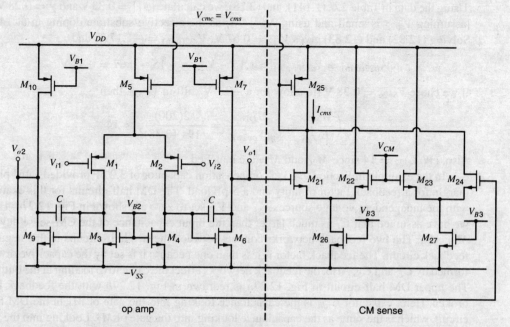

Figure 12.26 A fully differential two-stage CMOS op amp using the CMFB scheme from Fig. 12.19.

output voltage of 1 V peak, each op-amp output (V_{o1} or V_{o2}) must swing $\pm$ 0.5 V. From (3.161), the transistors in a differential pair remain active as long as the magnitude of the differential input voltage is less than $\sqrt{2}V_{ov}$. Therefore, we want

$$\sqrt{2}V_{ov} = 0.5\text{V}$$

or $V_{ov} = 0.35$ V for M_{21}-M_{24}. From (1.157), we get

$$\left(\frac{W}{L}\right)_{21} = \left(\frac{W}{L}\right)_{22} = \left(\frac{W}{L}\right)_{23} = \left(\frac{W}{L}\right)_{24} = \frac{2I_{D21}}{k'_n(V_{ov21})^2} = \frac{2(100)}{(194)(0.35)^2} = 8.4$$

The only remaining device sizes to be determined are for matched transistors M_{26} and M_{27}. Each device acts as a current source carrying 200 µA. For those transistors to act as current sources, they should always be active. Focusing on M_{26}, we want $V_{ov26} < V_{ds26(\text{min})}$. To determine $V_{ds26(\text{min})}$, consider an extreme case when M_{21} just turns off as V_{o1} swings down to its lowest value. In this case, M_{22} carries 200 µA, and

$$V_{gs22(\text{max})} = V_{t22} + \sqrt{\frac{2I_{D22(\text{max})}}{k'_n(W/L)_{22}}} = V_{t22} + \sqrt{\frac{2(200)}{194(8.4)}} = V_{t22} + 0.5 \text{ V}$$

The gate voltage of M_{22} is $V_{CM} = 0$. Therefore, the minimum source-body voltage for M_{22}, which is the minimum drain-source voltage of M_{26}, is

$$V_{sb22(\text{min})} = V_{ds26(\text{min})} = V_{s22(\text{min})} - (-V_{SS}) = V_{CM} - V_{gs22(\text{max})} + V_{SS}$$

$$= 0 - (0.5 + V_{t22}) + 1.65 = 1.15 - V_{t22} \quad (12.82)$$

Since V_{sb22} is not zero, the threshold voltage of M_{22} is given by (1.140) as

$$V_{t22} = V_{tn0} + \gamma\left[\sqrt{V_{sb22} + 2\phi_f} - \sqrt{2\phi_f}\right] \quad (12.83)$$

Using the data in Table 2.4, (1.141), and (2.28), we calculate $|\phi_f| = 0.33$ V and $\gamma = 0.28$ V$^{1/2}$ [assuming V_{sb22} is small and using $N_A + N_{si}$ as the effective substrate doping in (1.141)]. Solving (12.82) and (12.83) gives $V_{t22} = 0.67$ V, $V_{s22(\text{min})} = -1.17$ V, and

$$V_{sb22(\text{min})} = V_{ds26(\text{min})} = 1.15 - V_{t22} = 1.15 - 0.67 = 0.48 \text{ V}$$

If we chose $V_{ov26} = 0.38$ V (to allow for a -0.1 V shift in V_{CM}), then

$$\left(\frac{W}{L}\right)_{26} = \frac{2I_{D26}}{k'_n(V_{ov26})^2} = \frac{2(200)}{(194)(0.38)^2} \approx 14$$

Also, $(W/L)_{27} = 14$ since M_{26} and M_{27} are matched.

In the example in Section 9.4.3, a compensation capacitor of 3.2 pF provided a 45° phase margin for a feedback factor of unity and a 5-pF load. The DM half-circuits for this example with the independent voltage sources V_{s1} and V_{s2} set to zero are shown in Fig. 12.27a. Here, we have assumed that C_L is much larger than the input capacitance of the CM-sense devices M_{21}-M_{24}. The two feedback networks connect between the two half-circuits in this negative feedback circuit. The feedback factor is less than one because it is set by the capacitive divider formed by C_F and C_S. Also, the feedback networks affect the capacitive loading at the outputs. The upper DM half-circuit in Fig. 12.27a is redrawn in Fig. 12.27b with the feedback loop broken. Here, capacitor C_{idh} is the capacitance looking into the gate of M_1 in the DM half-circuit, which is the same as the capacitance looking into the gate of M_2. Looking into the gate of M_2, we see C_{gs2} and overlap capacitance C_{gd2} increased by the Miller effect as in (7.5);

12.6 Fully Differential Op Amps

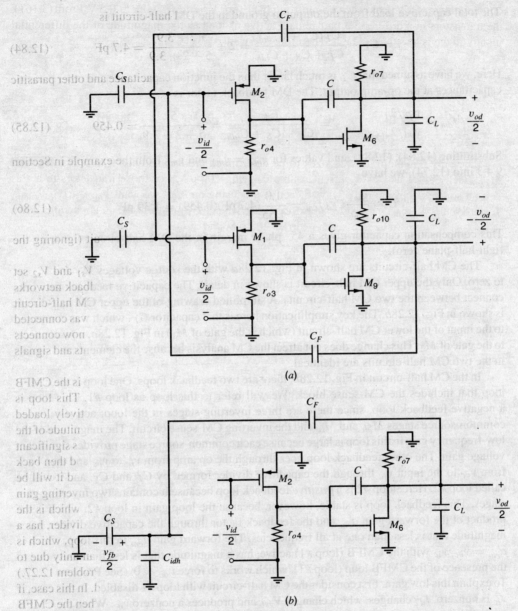

Figure 12.27 (a) The DM half-circuits for Figs. 12.25 and 12.26. (b) The upper DM half-circuit in (a) with the feedback loop broken.

therefore,

$$C_{idh} = C_{gs2} + C_{gd2}(1 - a_{dm1}) \approx \frac{2}{3} C_{ox} W_2 L_2 + C_{ol} W_2 (1 - a_{dm1})$$

$$= \frac{2}{3}\left(4.43 \frac{\text{fF}}{\mu\text{m}^2}\right)(77\ \mu\text{m})(0.82\ \mu\text{m}) + \left(0.35 \frac{\text{fF}}{\mu\text{m}}\right)(77\ \mu\text{m})(1 + 137) = 3.9\ \text{pF}$$

Here, we used $L_2 = 1\mu\text{m} - 2L_d = 0.82\ \mu\text{m}$ and $a_{dm1} = -g_{m2}(r_{o2} \| r_{o4}) = -137$ for the low-frequency gain of the first stage. These values follow from the example in Section 6.3.5.

The total capacitive load from the output to ground in the DM half-circuit is

$$C_{2d} = C_L + \frac{C_F(C_S + C_{idh})}{C_F + C_S + C_{idh}} = 2 + \frac{5(2+3.9)}{5+2+3.9} = 4.7 \text{ pF} \qquad (12.84)$$

Here, we have assumed that C_L is much larger than the junction capacitance and other parasitic capacitances at the op-amp output. The DM feedback factor is

$$f_{dm} = \left.\frac{v_{fb}/2}{v_{od}/2}\right|_{\text{loop broken}} = \frac{C_F}{C_F + C_S + C_{idh}} = \frac{5}{5+2+3.9} = 0.459 \qquad (12.85)$$

Substituting (12.84), (12.85), and values for $g_{m2} = g_{m1}$ and g_{m6} from the example in Section 9.4.3 into (12.74), we have

$$C \approx \frac{g_{m2}}{g_{m6}} C_{2d} f_{dm} = \frac{1.0}{1.55}(4.7 \text{ pF})(0.459) = 1.39 \text{ pF} \qquad (12.86)$$

This compensation capacitor gives a 45° phase margin in the DM half-circuit (ignoring the right-half-plane zero).

The CM half-circuits are shown in Fig. 12.28a with the source voltages V_{s1} and V_{s2} set to zero. Only the upper CM half-circuit is shown in detail. The capacitive feedback networks connect between the two CM half-circuits. A simplified drawing of the upper CM half-circuit is shown in Fig. 12.28b. The key simplification here is that capacitor C_F, which was connected to the input of the lower CM half-circuit (which is the gate of M_1) in Fig. 12.28a, now connects to the gate of M_2. This change does not affect the CM analysis because the elements and signals in the two CM half-circuits are identical.

In the CM half-circuit in Fig. 12.28b, there are two feedback loops. One loop is the CMFB loop that includes the CM-sense block. We will refer to this loop as loop #1. This loop is a negative feedback loop, since there are three inverting stages in the loop: actively loaded common-source stages M_{5h} and M_6, and the inverting CM-sense circuit. The magnitude of the low-frequency gain in this loop is large because each common-source stage provides significant voltage gain. The other feedback loop goes through the op amp from v_{ic} to v_{oc} and then back from v_{oc} to the input v_{ic} through the capacitive divider formed by C_F and C_S, and it will be called loop #2. Here, loop #2 is a *positive* feedback loop because it contains two inverting gain stages. This feedback loop is stable, however, because the loop gain in loop #2, which is the product of the forward gain a'_{cm} and the feedback factor through the capacitive divider, has a magnitude that is less than one at all frequencies. The forward gain a'_{cm} in this loop, which is $a'_{cm} = v_{oc}/v_{ic}$ with the CMFB (loop #1) active, has a magnitude that is less than unity due to the presence of the CMFB loop (loop #1), which works to force $v_{oc} \approx 0$. (See Problem 12.27.) To explain this low gain, first consider the CM half-circuit with loop #1 disabled. In this case, if v_{ic} is nonzero, I_{d2} changes, which changes V_{gs6} and produces a nonzero v_{oc}. When the CMFB loop #1 is enabled, this loop senses any nonzero v_{oc} and adjusts V_{cmc} to produce a change in I_{d5} that counteracts the change in I_{d2} to give $v_{oc} \approx 0$.

The magnitude responses of the loop gains for these two loops are plotted in Fig. 12.29, ignoring any zeros and poles other than the dominant and nondominant poles, p_{1c} and p_{2c}, associated with the Miller-compensated gain stage in Fig. 12.28b. Here, loop #1 is assumed to be compensated so that its unity-gain frequency is equal to $|p_{2c}|$, which gives a 45° phase margin. Since loop #2 is stable, we need only focus on the stability and compensation of the high-gain CMFB loop (loop #1).

This CMFB loop is shown in Fig. 12.28c. Here, the lumped load capacitance C_{2c} includes the output loading due to C_L and the capacitive feedback network, including the capacitance C_{ich} looking into the gate of M_2 in Fig. 12.28b. This capacitance is smaller than C_{idh} because M_2 has a large source degeneration resistance (r_{o5h}) that provides local CM feedback. This

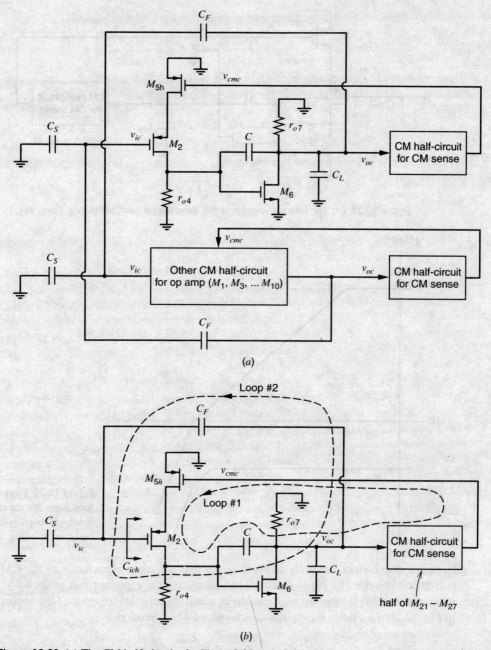

Figure 12.28 (a) The CM half-circuits for Figs. 12.25 and 12.26. (b) The upper CM half-circuit in (a) simplified.

feedback increases the impedance (decreases the capacitance) looking into the gate of M_2. Also, this feedback reduces the magnitude of the voltage gain from the gate to the drain of M_2, which decreases the Miller input capacitance due to C_{gd2}. Therefore, assuming $C_{ich} \ll C_S$, we find

$$C_{2c} = C_L + \frac{C_F(C_S + C_{ich})}{C_F + C_S + C_{ich}} \approx C_L + \frac{C_F C_S}{C_F + C_S} = 2 + \frac{5(2)}{5+2} = 3.43 \text{ pF}$$

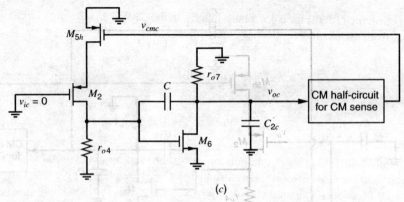

Figure 12.28 (c) The CM half-circuit in (b), focusing on the CMFB loop (loop #1).

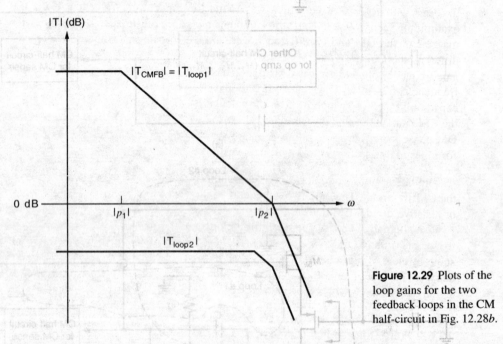

Figure 12.29 Plots of the loop gains for the two feedback loops in the CM half-circuit in Fig. 12.28b.

Here, we also assumed that C_L is much larger than the input capacitance of the CM-sense half-circuit. To use (12.78) to calculate the compensation capacitor that gives a 45° phase margin in the CMFB loop, we must find the dc small-signal gain through the CM-sense circuit. In Fig. 12.26, I_{cms} flows through diode-connected M_{25}; therefore,

$$v_{cmc} = -\frac{i_{cms}}{g_{m25}} \qquad (12.87)$$

if $r_{o25} \gg 1/g_{m25}$. A small-signal version of (12.56) is $i_{cms} = g_{m22}v_{oc} = g_{m21}v_{oc}$. Using this expression and (12.87), the gain a_{cms} at low frequency is

$$|a_{cms0}| = \left.\frac{|v_{cmc}|}{|v_{oc}|}\right|_{\text{CMFB loop open}} = \left.\frac{|v_{cmc}|}{|i_{cms}|}\frac{|i_{cms}|}{|v_{oc}|}\right|_{\text{CMFB loop open}}$$

$$= \frac{g_{m21}}{g_{m25}} = \frac{\dfrac{2I_{D21}}{V_{ov21}}}{\dfrac{2|I_{D25}|}{|V_{ov25}|}} = \frac{\dfrac{2(100)}{0.35}}{\dfrac{2(200)}{0.5}} = 0.71 \qquad (12.88)$$

Solving (12.78) for a 45° phase margin in the CMFB loop, using (12.88) and the value of g_{m6} from the example in Section 9.4.3, gives

$$C = \frac{g_{m5h}}{g_{m6}}|a_{cms0}|C_{2c} = \frac{(g_{m5}/2)}{g_{m6}}|a_{cms0}|C_{2c}$$

$$= \frac{\frac{2(200\ \mu A)}{0.5\ V}\frac{1}{2}}{1.55\ mA/V}(0.71)(3.43\ pF) = 0.63\ pF \tag{12.89}$$

From (12.86) and (12.89), a larger compensation capacitor is required to compensate the DM loop than the CMFB loop. Therefore, using $C = 1.39$ pF will give phase margins of 45° for the DM feedback loop and greater than 45° for the CMFB loop, which is acceptable.

To verify this design, SPICE simulations of this op amp were carried out with $C = 1.39$ pF in series with $R_Z = 758\ \Omega$, which was found to eliminate the right-half-plane zero in the example in Section 9.4.3. The SPICE models are based on the data in Table 2.4. The phase margins of the DM and CMFB loops were simulated using techniques developed for fully differential circuits.[11] The DM feedback loop has a simulated phase margin of 43° and unity-gain frequency of 53 MHz. The CMFB loop has a simulated phase margin of 62° and unity-gain frequency of 24 MHz. Thus, the CMFB loop is overcompensated. Changing the compensation capacitor to 0.63 pF, which is the value calculated from (12.89) to give a 45° phase margin in the CMFB loop, we find the simulated phase margin of the CMFB loop changes to 41°, but the DM phase margin drops to an unacceptably low 29°. These simulation results verify that the formulas in this section give a reasonable estimate for the compensation capacitor.

In the previous example, the op-amp output swing is limited by the linear input range of the CMFB circuit. The output swing could be increased by using either a switched-capacitor or resistive-divider CM detector.

An alternative CMFB approach for the op amp in Fig. 12.23 is to connect the gate of M_5 to a dc bias voltage and to use the gates of M_3-M_4 as the CMC input. In this case, the first gain stage of the CM half-circuit in Fig. 12.28c consists of common-source M_4 with a cascoded active load. Also, the pole associated with the capacitance at the source of M_2 is not in the signal path of the CMFB loop.

12.6.2 Fully Differential Telescopic Cascode Op Amp

A fully differential cascode op amp is shown in Fig. 12.30. Compared to its single-ended complementary counterpart, which is the first stage in Fig. 6.25, the main difference is that diode connections are removed from transistors M_3 and M_{3A}. Also, the gates of cascode transistors M_{1A}-M_{4A} are connected to bias voltages here. The op-amp outputs are taken from the drains of M_{1A} and M_{2A}. The resulting circuit is symmetric with each output loaded by a cascoded current source. An advantage of the topology in Fig. 12.30 is that the DM signal path consists only of n-channel transistors. That is, only the n-channel transistors conduct time-varying currents. The p-channel transistors conduct constant currents. Such a configuration maximizes the op-amp speed because n-channel transistors have higher mobility and f_T than their p-channel counterparts (if the channel lengths and overdrive voltages are the same). One CMFB approach is to set the currents through M_3-M_4 by connecting their gates to a dc bias voltage (set by a diode-connected transistor that is the input of a current mirror) and use the gate of M_5 as the CMC input. In this case, the magnitude of the low-frequency CMC gain $|a_{cmc0}|$ can be large since M_1-M_2 and M_{1A}-M_{2A} provide two levels of NMOS cascoding, and M_{3A}-M_{4A} provide one level of PMOS cascoding. This cascoding increases the output resistance, but the two levels of NMOS cascoding introduce high-frequency poles in $a_{cmc}(s)$.

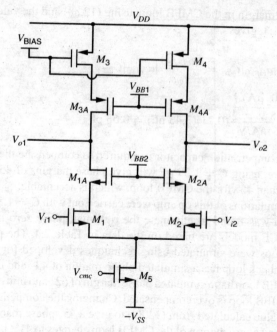

Figure 12.30 A fully differential CMOS telescopic-cascode op amp.

An alternative CMFB approach is to set the current through M_5 by connecting its gate to a dc bias voltage and use the gates of M_3 and M_4 as the CMC input. This approach has one level of cascoding in the CMC gain path, which introduces one high-frequency pole in $a_{cmc}(s)$. Here, however, the amplifying devices in the CMFB loop are p-channel M_3 and M_4, which have lower mobility and f_T than n-channel M_5 if they have the same channel lengths and overdrive voltages.

The DM and CMFB feedback loops are compensated by the load capacitances at the op-amp outputs. The phase margin of the CMFB loop can be changed without changing the load capacitance, by splitting the transistor(s) that connect to the CMC input into parallel transistors, as described in Section 12.4.2 and shown in Fig. 12.15.

12.6.3 Fully Differential Folded-Cascode Op Amp

A fully differential folded-cascode op amp is shown in Fig. 12.31. Compared to its single-ended counterpart shown in Fig. 6.28, the main difference is that the diode connections on M_3 and M_{3A} have been eliminated. The resulting circuit is symmetric, and the outputs are taken from the drains of M_{1A} and M_{2A}.

To satisfy KCL, the sum of the currents flowing through M_3, M_4, and M_5 must equal the sum of the drain currents flowing through M_{11} and M_{12}. To satisfy KCL with all transistors active and to accurately set V_{oc}, CMFB is used. The CMC input could be taken as the gate of M_5, the gates of M_3-M_4, or the gates of M_{11}-M_{12}, which is shown in Fig. 12.31. With this last option, the CMFB loop contains less nodes than the other two cases, and the gain a_{cmc} is provided by common-source n-channel transistor M_{11} (or M_{12}), which has a larger g_m than p-channel M_3 (or M_4) if $(W/L)_{11} \approx (W/L)_3$ because $k'_n > k'_p$ and $I_{D11} > |I_{D3}|$.

The DM and CMFB feedback loops are compensated by the capacitances at the op-amp outputs.

The folded-cascode op amp with active cascodes that is shown in Fig. 6.30 can also be converted to a fully differential op amp.[12] As for the folded-cascode op amp, there are three choices for the CMC input.

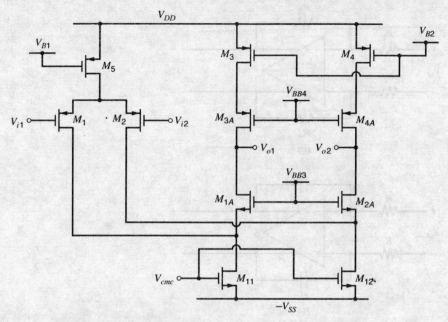

Figure 12.31 A fully differential CMOS folded-cascode op amp.

12.6.4 A Differential Op Amp with Two Differential Input Stages

The fully differential op amps presented above have two input terminals that accept one differential input. Those op amps can be used in the amplifier, integrator, and differentiator shown in Fig. 12.32. To implement a fully differential non-inverting gain stage with a very large input impedance, an op amp with four input terminals is needed. Two inputs connect to the feedback networks and the other two inputs connect to the differential signal source, as shown in Fig. 12.33. Assuming the magnitudes of the op amp gains from v_{id1} and v_{id2} to v_{od} are large, negative feedback forces $v_{id1} \approx 0$ and $v_{id2} \approx 0$. The two pairs of inputs are produced by two source-coupled pairs, as shown for a two-stage op amp in Fig. 12.34. The two source-coupled pairs, M_1-M_2 and M_{1X}-M_{2X}, share a pair of current-source loads. Assuming the input pairs are matched, the differential small-signal voltage gain is the same from either input; therefore,

$$v_{od} = a_{dm}(v_{id1} + v_{id2}) \qquad (12.90)$$

The CM input range for the op amp must be large enough to include the full range of the input signals, V_{s1} and V_{s2}, because they connect directly to op-amp inputs. In this op amp, the CMFB loop could adjust either I_1 or I_2 by controlling the gate voltages of the transistors that generate those currents.

12.6.5 Neutralization

In a fully differential op amp, a technique referred to as *capacitive neutralization* can be used to reduce the component of the op-amp input capacitance due to the Miller effect (see Chapter 7). Reducing the input capacitance increases the input impedance, which is desirable. Neutralization is illustrated in Fig. 12.35*a*. The gate-to-drain overlap capacitances are shown explicitly for M_1 and M_2. First, ignore the neutralization capacitors C_n. Then the DM capacitance looking into either op-amp input (with respect to ground) is

$$C_{idh} = C_{gs1} + C_{gd1}(1 - a_{dm1}) \qquad (12.91)$$

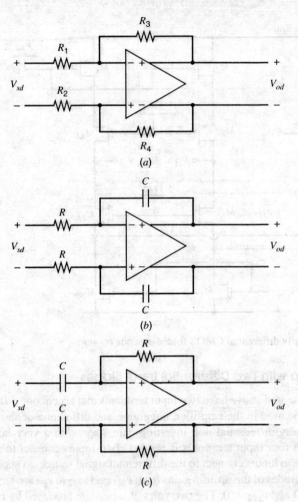

Figure 12.32 Fully differential (a) inverting gain stage, (b) integrator, and (c) differentiator.

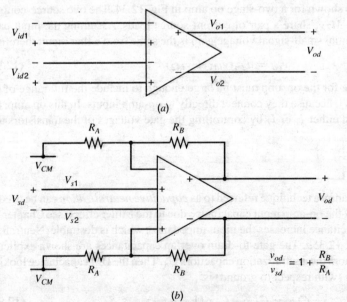

$$\frac{v_{od}}{v_{sd}} = 1 + \frac{R_B}{R_A}$$

Figure 12.33 (a) An op amp with two pairs of inputs. (b) A noninverting fully differential feedback amplifier.

12.6 Fully Differential Op Amps 837

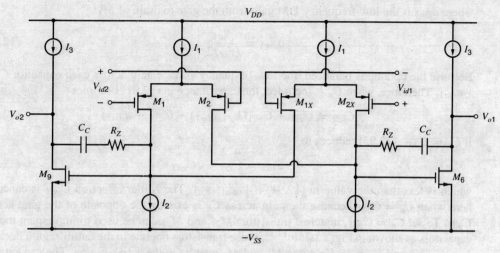

Figure 12.34 A simplified schematic of a two-stage op amp with two pairs of inputs.

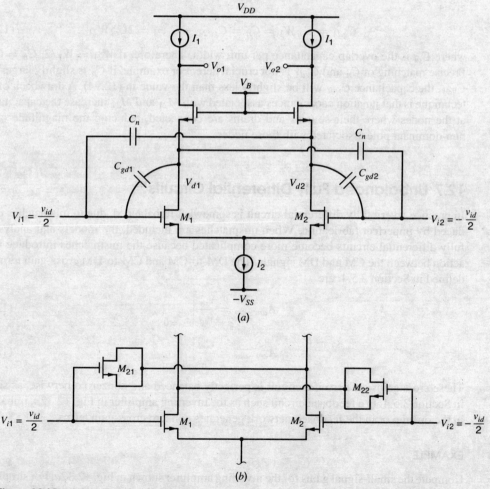

Figure 12.35 (a) An example of capacitive neutralization. (b) Using transistors M_{21} and M_{22} in cutoff to implement the neutralization capacitors.

where a_{dm1} is the low-frequency DM gain from the gate to drain of M_1:

$$a_{dm1} = \frac{v_{d1}}{(v_{id}/2)}\bigg|_{\omega=0} \tag{12.92}$$

Because the op amp is balanced, the low-frequency voltage gain across each capacitor C_n is $-a_{dm1}$. Therefore, when C_n is included, the capacitance in (12.91) becomes

$$C_{idh} = C_{gs1} + C_{gd1}(1 - a_{dm1}) + C_n(1 + a_{dm1}) \tag{12.93}$$

If $C_n = C_{gd1}$, (12.93) reduces to

$$C_{idh} = C_{gs1} + 2C_{gd1} \tag{12.94}$$

which is less than the value in (12.91) if $|a_{dm1}| > 1$. The Miller effect on C_{gd1} is canceled here when $C_n = C_{gd1}$ because the gain across C_n is exactly the opposite of the gain across C_{gd1}. To set $C_n = C_{gd1}$, matched transistors M_{21} and M_{22} can be used to implement the C_n capacitors as shown in Fig. 12.35b.[13,14] These transistors operate in the cutoff region because $V_{GD1} < V_{t1}$ and $V_{GD2} < V_{t2}$ since M_1 and M_2 operate in the active region. The capacitance C_n is the sum of the gate-to-drain and gate-to-source overlap capacitances. Setting $C_n = C_{gd1}$ gives

$$C_{gd1} = C_{ol}W_1 = C_n = C_{gd21} + C_{gs21} = 2C_{ol}W_{21} \tag{12.95}$$

where C_{ol} is the overlap capacitance per unit width. Therefore, if $W_{21} = W_1/2$, $C_n = C_{gd1}$. Precise matching of C_n and C_{gd1} is not crucial here. For example, if C_n is slightly larger than C_{gd1}, the capacitance C_{idh} will be slightly less than the value in (12.94). A drawback of this technique is that junction capacitances associated with M_{21} and M_{22} increase the capacitances at the nodes where their sources and drains are connected, reducing the magnitude of the non-dominant pole associated with those nodes.

12.7 Unbalanced Fully Differential Circuits[1,2]

In practice, every fully differential circuit is somewhat imbalanced, due to mismatches introduced by imperfect fabrication. When mismatches are included, the models and analysis of fully differential circuits become more complicated because the mismatches introduce interaction between the CM and DM signals. The DM-to-CM and CM-to-DM cross-gain terms, as defined in Section 3.5.4, are

$$A_{dm-cm} = \frac{v_{oc}}{v_{id}}\bigg|_{v_{ic}=0} \tag{12.96}$$

$$A_{cm-dm} = \frac{v_{od}}{v_{ic}}\bigg|_{v_{id}=0} \tag{12.97}$$

These cross gains are zero if a circuit is perfectly balanced and nonzero otherwise, as shown in Section 3.5.4. In a feedback circuit such as the inverting amplifier in Fig. 12.32a, imbalance in the op amp or in the feedback network generates nonzero cross-gain terms.

■ **EXAMPLE**

Compute the small-signal gains for the inverting amplifier shown in Fig. 12.32a. For simplicity, assume the op amp is balanced, has infinite input impedance, zero output impedance, $a_{dm} \to -\infty$ and $a'_{cm} = -0.1$. (This a'_{cm} is the CM gain, including the effect of the CMFB loop.) The

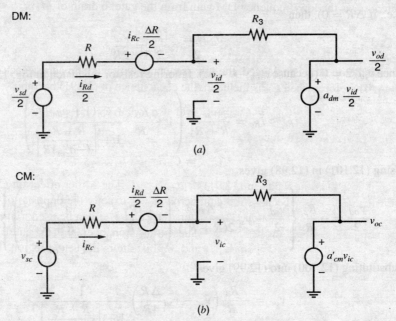

Figure 12.36 Coupled (*a*) DM and (*b*) CM half-circuits for the gain stage in Fig. 12.32*a* with a balanced op amp and mismatch in the feedback network.

resistors across the op amp are matched with $R_3 = R_4 = 5$ kΩ, but the resistors that connect to the signal source are mismatched with $R_1 = 1.01$ kΩ and $R_2 = 0.99$ kΩ. Therefore, the only imbalance in this circuit stems from the mismatch between R_1 and R_2.

We will analyze this circuit using coupled half-circuits, which were introduced in Section 3.5.6.9. The two coupled half-circuits are shown in Fig. 12.36. The circuits are coupled by the nonzero resistor mismatch

$$\Delta R = R_1 - R_2 = 0.02 \text{ k}\Omega$$

The resistance R in the figure is the average value of the mismatched resistors

$$R = \frac{R_1 + R_2}{2} = 1 \text{ k}\Omega$$

Letting $a_{dm} \to -\infty$ in the DM half-circuit gives

$$\frac{v_{od}}{2} = -\frac{R_3}{R}\left(\frac{v_{sd}}{2} - i_{Rc}\frac{\Delta R}{2}\right) \quad (12.98)$$

Analysis of the CM half-circuit gives

$$v_{oc} = -\frac{R_3}{R}\left(v_{sc} - \frac{i_{Rd}}{2}\frac{\Delta R}{2}\right)\frac{1}{1 + \left[\dfrac{R + R_3}{(-a'_{cm})R}\right]} \quad (12.99)$$

From these equations and the coupled half-circuits, exact input-output relationships could be found. However, for small mismatches, the approximate method described in Section 3.5.6.9 simplifies the analysis and provides sufficient accuracy for hand calculations. The key simplification in the approximate analysis is that the currents i_{Rd} and i_{Rc} are estimated from their respective half-circuits, ignoring mismatch effects. If the mismatch is ignored in Fig. 12.36*a* (i.e., if $\Delta R = 0$), then

$$\frac{i_{Rd}}{2} \approx \frac{\hat{i}_{Rd}}{2} = \frac{v_{sd}}{2R} \qquad (12.100)$$

since $v_{id}/2 = 0$ (because $a_{dm} \to -\infty$). Ignoring resistor mismatch in Fig. 12.36b, we find

$$i_{Rc} \approx \hat{i}_{Rc} = \frac{v_{sc}}{R + R_3}\left(1 + \frac{R_3}{R} \cdot \frac{1}{1 + \left[\frac{R + R_3}{(-a'_{cm})R}\right]}\right) \qquad (12.101)$$

Using (12.101) in (12.98) gives

$$\frac{v_{od}}{2} = -\frac{R_3}{R}\left\{\frac{v_{sd}}{2} - v_{sc}\frac{\Delta R}{2(R + R_3)}\left(1 + \frac{R_3}{R} \cdot \frac{1}{1 + \left[\frac{R + R_3}{(-a'_{cm})R}\right]}\right)\right\} \qquad (12.102)$$

Substituting (12.100) into (12.99) gives

$$v_{oc} = -\frac{R_3}{R}\left(v_{sc} - v_{sd}\frac{\Delta R}{4R}\right)\frac{1}{1 + \left[\frac{R + R_3}{(-a'_{cm})R}\right]} \qquad (12.103)$$

From (12.102),

$$A_{dm} = \left.\frac{v_{od}}{v_{sd}}\right|_{v_{sc}=0} = -\frac{R_3}{R} = -\frac{5}{1} = -5$$

$$A_{cm-dm} = \left.\frac{v_{od}}{v_{sc}}\right|_{v_{sd}=0} = \frac{R_3}{R}\frac{\Delta R}{(R + R_3)}\left(1 + \frac{R_3}{R} \cdot \frac{1}{1 + \left[\frac{R + R_3}{(-a'_{cm})R}\right]}\right)$$

$$= \frac{5}{1} \cdot \frac{0.02}{(1 + 5)}\left(1 + \frac{5}{1} \cdot \frac{1}{1 + \left[\frac{1 + 5}{0.1(1)}\right]}\right) = 0.018$$

From (12.103),

$$A_{cm} = \left.\frac{v_{oc}}{v_{sc}}\right|_{v_{sd}=0} = -\frac{R_3}{R}\frac{1}{1 + \left[\frac{R + R_3}{(-a'_{cm})R}\right]} = -\frac{5}{1} \cdot \frac{1}{1 + \left[\frac{1 + 5}{0.1(1)}\right]} = -0.082$$

$$A_{dm-cm} = \left.\frac{v_{oc}}{v_{sd}}\right|_{v_{sc}=0} = \frac{R_3}{R}\frac{\Delta R}{4R}\frac{1}{1 + \left[\frac{R + R_3}{(-a'_{cm})R}\right]} = \frac{5}{1}\frac{0.02}{4(1)}\frac{1}{1 + \left[\frac{1 + 5}{0.1(1)}\right]} = 0.00041$$

The resistor mismatch causes nonzero cross-gain terms in the closed-loop amplifier. Exact analysis of this circuit gives essentially the same gain values as above.

A model for an op amp with mismatch (but assuming infinite input impedance and zero output impedance, for simplicity) is shown in Fig. 12.37. The equations corresponding to this

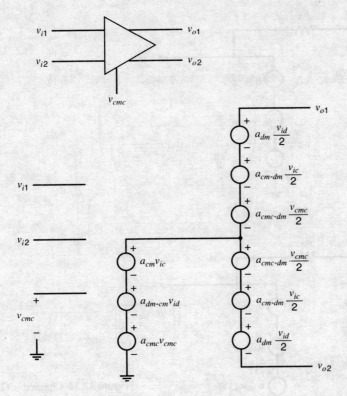

Figure 12.37 A simple small-signal model of a fully differential amplifier including cross-gain terms, assuming infinite input impedance and zero output impedance.

model are

$$v_{od} = a_{dm}v_{id} + a_{cm-dm}v_{ic} + a_{cmc-dm}v_{cmc} \qquad (12.104)$$

$$v_{oc} = a_{cm}v_{ic} + a_{dm-cm}v_{id} + a_{cmc}v_{cmc} \qquad (12.105)$$

The cross-gain terms a_{cm-dm}, a_{dm-cm}, and a_{cmc-dm} are zero when the op amp is perfectly balanced. If the CM-sense circuit is not perfectly balanced, its output v_{cms}, which ideally is proportional to the CM output, contains a component that depends on the DM output:

$$v_{cms} = a_{cms}v_{oc} + a_{dm-cms}v_{od} \qquad (12.106)$$

Also, $v_{cmc} = v_{cms}$ when the CMFB loop is closed.

To illustrate the effect of feedback on the open-loop op-amp gains, consider the inverting amplifier in Fig. 12.32a with a balanced feedback network ($R_1 = R_2$ and $R_3 = R_4$) but with imbalances in the op amp. The circuit could be analyzed exactly to find the closed-loop gains, but the analysis is difficult. Therefore, we will use the approximate, coupled half-circuit analysis that was used in the last example. The coupled DM and CM half-circuits are shown in Fig. 12.38. The imbalances in the op amp are modeled by the a_{cm-dm}, a_{dm-cm}, and a_{cmc-dm} controlled sources in Fig. 12.38, based on (12.104) and (12.105). To simplify the analysis, we will assume that the CM-sense circuit is balanced (i.e., $a_{dm-cms} = 0$). Under this assumption, (12.106) reduces to

$$v_{cmc} = v_{cms} = a_{cms}v_{oc} \qquad (12.107)$$

DM:

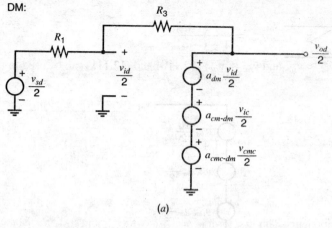

CM:

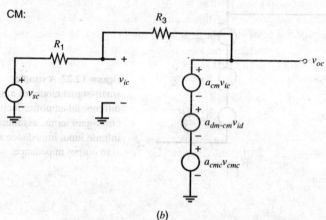

Figure 12.38 Coupled (a) DM and (b) CM half-circuits for the gain stage in Fig. 12.32a with an unbalanced op amp and a balanced feedback network.

Substituting (12.107) in (12.104) and (12.105) gives

$$v_{od} = a_{dm} v_{id} + a_{cm-dm} v_{ic} + a_{cmc-dm} a_{cms} v_{oc} \quad (12.108)$$

$$v_{oc} = a_{cm} v_{ic} + a_{dm-cm} v_{id} + a_{cmc} a_{cms} v_{oc} \quad (12.109)$$

To carry out the approximate analysis, each half-circuit is first analyzed with the coupling between the half-circuits eliminated. Then the results of these analyses are used to find the closed-loop cross gains. The coupling in the DM half-circuit is eliminated by setting $a_{cm-dm} = 0$ and $a_{cmc-dm} = 0$. With these changes, analysis of Fig. 12.38a gives

$$\hat{v}_{id} = -\frac{R_3}{R_1} \cdot \frac{1}{1 + \dfrac{R_1 + R_3}{(-a_{dm})R_1}} \frac{v_{sd}}{a_{dm}} \quad (12.110)$$

Similarly, the coupling in the CM half-circuit is eliminated by setting $a_{dm-cm} = 0$. With this coupling eliminated, analysis of Fig. 12.38b gives

$$\hat{v}_{ic} = \frac{\hat{v}_{oc}}{a'_{cm}} = -\frac{R_3}{R_1} \cdot \frac{1}{1 + \dfrac{R_1 + R_3}{(-a'_{cm})R_1}} \frac{v_{sc}}{a'_{cm}} \quad (12.111)$$

where

$$a'_{cm} = \frac{a_{cm}}{1+(-a_{cmc}a_{cms})} \qquad (12.112)$$

Assuming $\hat{v}_{id} \approx v_{id}$, $\hat{v}_{ic} \approx v_{ic}$, and $\hat{v}_{oc} \approx v_{oc}$, (12.110) and (12.111) can be used in (12.108) to give

$$v_{od} = -\frac{R_3}{R_1} \cdot \frac{1}{1+\frac{R_1+R_3}{(-a_{dm})R_1}} v_{sd} + \frac{a'_{cm-dm}}{1+\frac{(-a_{dm})R_1}{R_1+R_3}} \frac{\frac{R_3}{R_3+R_1}}{1+\frac{(-a'_{cm})R_1}{R_1+R_3}} v_{sc} \qquad (12.113)$$

where

$$a'_{cm-dm} = a_{cm-dm} + a'_{cm}a_{cms}a_{cmc-dm} \qquad (12.114)$$

This gain has two components. The first term is a_{cm-dm}, which is the CM-to-DM gain of the op amp. The second term is the product of three gains: 1) a'_{cm}, which is the gain from v_{ic} to v_{oc} including the effect of the CMFB loop, 2) a_{cms}, which is the CM-sense gain from v_{oc} to $v_{cms} = v_{cmc}$, and 3) a_{cmc-dm}, which is the gain from v_{cmc} to v_{od} (due to mismatch in the op amp). Therefore, the second term in (12.114) is the gain through an indirect path from v_{ic} to v_{od}.

Again assuming $\hat{v}_{id} \approx v_{id}$, $\hat{v}_{ic} \approx v_{ic}$, and $\hat{v}_{oc} \approx v_{oc}$, (12.110) and (12.111) can be used in (12.109) to give

$$v_{oc} = \frac{\frac{a'_{cm}R_3}{R_1+R_3}}{1+\frac{(-a'_{cm})R_1}{R_1+R_3}} v_{sc} + \frac{a'_{dm-cm}}{1+\frac{(-a'_{cm})R_1}{R_1+R_3}} \frac{\frac{R_3}{R_1+R_3}}{1+\frac{(-a_{dm})R_1}{R_1+R_3}} v_{sd} \qquad (12.115)$$

where

$$a'_{dm-cm} = \frac{a_{dm-cm}}{1+(-a_{cmc}a_{cms})} \qquad (12.116)$$

Equations 12.113 and 12.115 relate the DM and CM source and output voltages for the feedback amplifier. The open-loop-gain and cross-gain terms in (12.108) and (12.109) have been modified by the feedback. To allow simplification of these gain terms, define

$$T_{dm} = \frac{(-a_{dm})R_1}{R_1+R_3} \qquad (12.117)$$

$$T_{cm} = \frac{(-a'_{cm})R_1}{R_1+R_3} \qquad (12.118)$$

$$T_{cmfb} = -a_{cmc}a_{cms} \qquad (12.119)$$

which are the loop gains around the DM, CM, and CMFB loops, respectively. Using (12.117)–(12.119), the closed-loop gain terms in (12.113) can be written as

$$A_{dm} = \left.\frac{v_{od}}{v_{sd}}\right|_{v_{sc}=0} = -\frac{R_3}{R_1} \cdot \frac{1}{1+\frac{1}{T_{dm}}} \approx -\frac{R_3}{R_1} \qquad (12.120)$$

where $|T_{dm}| \gg 1$ has been used, and

$$A_{cm-dm} = \left.\frac{v_{od}}{v_{sc}}\right|_{v_{sd}=0} = \frac{a'_{cm-dm}\left(\frac{R_3}{R_3+R_1}\right)}{(1+T_{dm})(1+T_{cm})} \qquad (12.121)$$

As $|T_{dm}|$ increases, $|A_{cm-dm}|$ decreases. Therefore, $|T_{dm}| \gg 1$ is desired to reduce the magnitude of the closed-loop CM-to-DM cross gain. The $(1 + T_{cm})$ term has little effect here since (12.112) usually gives $|a'_{cm}| \ll 1$; therefore, $(1 + T_{cm}) \approx 1$.

Using (12.116)–(12.119), the closed-loop gain terms in (12.115) can be written as

$$A_{cm} = \left.\frac{v_{oc}}{v_{sc}}\right|_{v_{sd}=0} = \frac{\dfrac{a'_{cm} R_3}{R_1 + R_3}}{1 + T_{cm}} \approx \frac{a'_{cm} R_3}{R_1 + R_3} \qquad (12.122)$$

where $(1 + T_{cm}) \approx 1$ has been used, and

$$A_{dm-cm} = \left.\frac{v_{oc}}{v_{sd}}\right|_{v_{sc}=0} = \frac{a'_{dm-cm}}{1 + T_{cm}} \frac{\dfrac{R_3}{R_1 + R_3}}{1 + T_{dm}}$$

$$= \frac{a_{dm-cm} \dfrac{R_3}{R_1 + R_3}}{(1 + T_{cmfb})(1 + T_{dm})(1 + T_{cm})} \qquad (12.123)$$

As $|T_{dm}|$ or $|T_{cmfb}|$ increases, $|A_{dm-cm}|$ decreases. Therefore, the DM and CMFB loops work to reduce the closed-loop DM-to-CM cross gain. Again, the $(1 + T_{cm})$ term has little effect here since $|1 + T_{cm}| \approx 1$.

The analysis in this section shows how the closed-loop cross-gain terms for the feedback amplifier in Fig. 12.32a are affected by imbalances in the op amp and feedback network. In practice, the imbalances are usually caused by random mismatches between components, and the effect of such mismatches on circuit performance is often evaluated through SPICE simulations.

12.8 Bandwidth of the CMFB Loop

Ideally, a fully differential circuit processes a DM input signal and produces a purely DM output signal. The closed-loop bandwidth required for the DM gain is set by the bandwidth of the DM signal. For example, consider the differential gain stage in Fig. 12.32a. To avoid filtering the signal, the closed-loop bandwidth of the DM gain must be larger than the highest frequency in the applied DM input signal. Since the closed-loop bandwidth is approximately equal to the unity-gain frequency of the DM loop gain (or return ratio), this unity-gain frequency should be about equal to the required closed-loop bandwidth. However, the unity-gain frequency required for the CMFB loop is not so easily determined. Consider a fully differential feedback circuit that is linear and perfectly balanced. If no ac CM signals are present in the circuit, the ac CM output voltage will be zero and the CM output voltage is constant. In such a case, the bandwidth of the CMFB loop is unimportant since it only operates on dc signals.

In practice, there are many sources of ac CM signals. For example, an ac CM signal can be present in the signal source, or CM noise can be introduced by coupling from a noisy power supply. Furthermore, even when the signal source is purely differential, an ac CM signal can be created by circuit imbalance that causes DM-to-CM conversion. Regardless of the source of an ac CM signal, the CMFB works to suppress the ac CM output signal and to give a CM output voltage that is about constant. Suppression of the ac CM output component is important for two reasons. First, if the CM output varies, the DM output swing must be reduced to allow for the CM output swing. Second, if two feedback amplifiers are cascaded, any ac CM output voltage from the first amplifier is a CM input voltage for the second amplifier, and any imbalance in the second amplifier will convert some of its CM input voltage into a DM output voltage.

To see how the CMFB suppresses the ac CM output signal, consider the small-signal block diagram in Fig. 12.39. Here, v_n is a CM disturbance (for example, an ac signal on the power

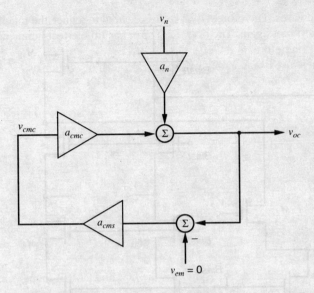

Figure 12.39 A model for the CMFB loop with injected noise v_n.

supply). The signal v_{cm}, which is the small-signal component of the applied dc voltage V_{CM}, is zero. Also, a_n is the voltage gain from the CM disturbance to the CM output voltage when the CMFB loop is disabled. That is,

$$a_n = \left.\frac{v_{oc}}{v_n}\right|_{\text{no CMFB}} = \left.\frac{v_{oc}}{v_n}\right|_{v_{cmc}=0} \qquad (12.124)$$

When the CMFB is active, the small-signal gain from v_n to v_{oc} becomes

$$A_n = \left.\frac{v_{oc}}{v_n}\right|_{\text{with CMFB}} = \frac{a_n}{1 + a_{cms}(-a_{cmc})} \qquad (12.125)$$

The CMFB gives $|A_n| \ll |a_n|$ at frequencies where $|a_{cms}(-a_{cmc})| \gg 1$. Therefore, $|a_{cms}(-a_{cmc})| \gg 1$ is desired at frequencies where a significant ac CM output voltage would be generated without CMFB. One possible objective is to satisfy this condition over the bandwidth of DM input signal, or equivalently to make the unity-gain frequencies of the DM and CMFB loops about equal.[15] While desirable, this goal can be difficult to achieve in practice because the CMFB loop often includes more transistors and has more nondominant poles than the DM loop. In any case, suppression of spurious CM signals is an important consideration in determining the required bandwidth of the CMFB loop in practical fully differential amplifiers.

12.9 Analysis of a CMOS Fully Differential Folded-Cascode Op Amp

In this section, we analyze an example fully differential folded-cascode op amp. The particular op amp considered here is similar to an op amp used in an analog-to-digital converter that uses switched-capacitor gain stages.[16]

The core of the op amp is shown in Fig. 12.40. The op amp is a complementary version of the folded-cascode op amp shown in Fig. 12.31. Transistors M_1-M_2 form the NMOS input differential pair. NMOS transistors are used here because they have larger k' than PMOS devices and therefore yield larger g_m for the same device dimensions and currents. Transistors M_3-M_5 and M_{11}-M_{12} act as current sources. Transistors M_{1A}, M_{2A}, M_{3A} and M_{4A} are cascode

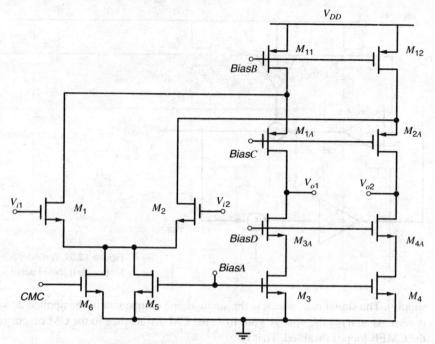

Figure 12.40 Folded-cascode op-amp circuit.

devices that boost the output resistance and also the dc voltage gain of the op amp. Finally, M_6 is a transistor that does not appear in Fig. 12.31. The gate of M_6 is the common-mode control (CMC) input. The common-mode sense circuit, which will be described later, connects to the *CMC* node.

The op amp makes repeated use of unit transistors to assure good matching, accurate ratios of W/L values, and process insensitivity, as described in Section 6.3.3 [see the paragraph after (6.68)]. Multiple unit transistors are connected either in parallel to increase the W/L or in series to decrease the W/L. The unit transistor has $W = 40$ μm and $L = 0.4$ μm.

The bias circuits for the op amp in Fig. 12.40 are shown in Fig. 12.41. The nodes *BiasA* through *BiasD* of the bias circuit connect to the nodes with the same labels in the op amp in Fig. 12.40.

The dimensions of the transistors in Figures 12.40 and 12.41 are given in Tables 12.1 and 12.2, respectively. In these tables, m is a factor that multiplies the W/L of a unit transistor to give the aspect ratio of a transistor. For example, $m = 10$ for M_1 (i.e., $m_1 = 10$), which means that M_1 consists of ten unit transistors connected in parallel. Therefore, $(W/L)_1 = 10 \cdot [W/L$ of a unit transistor]. For M_{23}, $m_{23} = 1/3$, which means that M_{23} consists of three unit transistors connected in series. Therefore, $(W/L)_{23} = (1/3) \cdot [W/L$ of a unit transistor].

The bias circuit consists of three branches: M_{21}-M_{24}, M_{25}-M_{30}, and M_{31}-M_{33}. The bias circuit makes extensive use of the Sooch cascode current mirror, which was described in Section 4.2.5.2 and shown in Fig. 4.12. For simplicity, first consider transistors M_{21}-M_{26}. The topology here is the PMOS counterpart of the Sooch current mirror in Fig. 4.12b. Input bias current I_{BIAS} flows through the left-hand branch that contains M_{21}-M_{24}. Since M_{21}-M_{22} and M_{25}-M_{26} are all identical, these transistors form a 1:1 current mirror, so I_{BIAS} flows through M_{25}-M_{26} and then through M_{27}-M_{30}. Since M_{21}-M_{22} and M_{31}-M_{32} are identical, I_{BIAS} also flows through M_{31}-M_{33}. The value of I_{BIAS} is 0.2 mA.

12.9 Analysis of a CMOS Fully Differential Folded-Cascode Op Amp

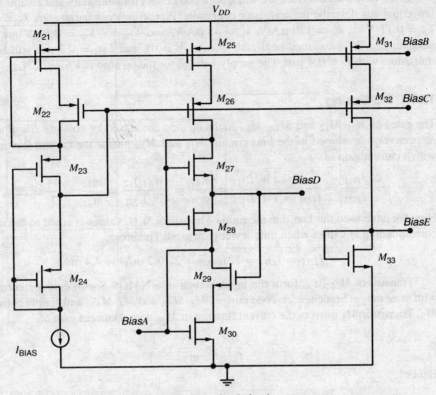

Figure 12.41 Bias circuit for the op amp.

Table 12.1 Op-amp transistor data. A unit transistor has $W = 40$ μm, $L = 0.4$ μm and $L_{\text{eff}} = 0.3$ μm.

	M_1, M_2	M_{1A}, M_{2A}	M_3, M_{3A}, M_4, M_{4A}	M_5, M_6	M_{11}, M_{12}
m (W/L scale factor)	10	24	12	10	44
calculated $|I_D|$ (mA)	2.0	2.4	2.4	2.0	4.4
simulated $|I_D|$ (mA)	2.01	2.40	2.40	2.05, 1.97	4.41

Table 12.2 Bias-circuit transistor data. A unit transistor has $W = 40$ μm, $L = 0.4$ μm and $L_{\text{eff}} = 0.3$ μm.

	$M_{21}, M_{22}, M_{25}, M_{26}, M_{31}, M_{32}$	M_{23}	M_{24}, M_{27}	M_{28}	M_{29}, M_{30}	M_{33}
m (W/L scale factor)	2	1/3	3	1/9	1	1/14
calculated $|I_D|$ (mA)	0.20	0.20	0.20	0.20	0.20	0.20
simulated $|I_D|$ (mA)	0.20	0.20	0.20	0.20	0.20	0.20

In the following analysis, we will ignore body effect for simplicity and assume the square-law equations describe the transistor operation. Typical process parameters $\lambda_n = 0.15$ V^{-1}, $\lambda_p = 0.17$ V^{-1}, $k'_n = 170$ μA/V, $k'_p = 45$ μA/V, and $V_{tn} = -V_{tp} = 0.50$ V are used. For simplicity, ΔW is assumed negligible (i.e., $\Delta W = 0$), and $L_{\text{eff}} = 0.3$ μm will be used for transistors with $L = 0.4$ μm. The supply voltage for the op amp is $V_{DD} = 2$ V.

12.9.1 DC Biasing

The gates of M_{11}-M_{12} and M_{1A}-M_{2A} in the op amp are biased by voltages $BiasB$ and $BiasC$, respectively, developed in the bias circuit. M_{11} and M_{12} mirror the current flowing in M_{21}, with a current gain of

$$\frac{I_{D11}}{I_{D21}} = \frac{I_{D12}}{I_{D21}} = \frac{(W/L)_{11}}{(W/L)_{21}} = \frac{m_{11}}{m_{21}} = \frac{(W/L)_{12}}{(W/L)_{21}} = \frac{m_{12}}{m_{21}} = \frac{44}{2} = 22 \quad (12.126)$$

Here we have used the fact that the ratio of transistor W/L values is equal to the ratio of the corresponding m values when unit devices are used. Therefore,

$$I_{D11} = I_{D12} = 22 I_{BIAS} = 22(0.2 \text{ mA}) = 4.4 \text{ mA} \quad (12.127)$$

Transistors M_{27}-M_{30} form the input branch of a NMOS Sooch cascode current mirror with three output branches: cascode mirrors M_3-M_{3A} and M_4-M_{4A}, and simple current source M_5. Transistor M_5 mirrors the current flowing in M_{30} with a current gain of

$$\frac{I_{D5}}{I_{D30}} = \frac{m_5}{m_{30}} = \frac{10}{1} \quad (12.128)$$

Hence

$$I_{D5} = 10 I_{D30} = 10(0.2 \text{ mA}) = 2.0 \text{ mA} \quad (12.129)$$

Transistors M_3 and M_4 mirror the current flowing in M_{30} with a current gain of

$$\frac{I_{D3}}{I_{D30}} = \frac{I_{D4}}{I_{D30}} = \frac{m_3}{m_{30}} = \frac{m_4}{m_{30}} = \frac{12}{1} \quad (12.130)$$

so

$$I_{D3} = I_{D4} = 12 I_{D30} = 12(0.2 \text{ mA}) = 2.4 \text{ mA} \quad (12.131)$$

The M_{31}-M_{33} branch is included to generate a dc voltage at node $BiasE$. This voltage is used to bias nodes in the switched-capacitor feedback application, which is described later in this section.

In the bias circuit in Fig. 12.41, the drain current of each transistor is 0.2 mA. In the op amp, the drain currents range from 2.0 mA to 4.4 mA. Lower current is used in the bias circuit to reduce power dissipation there. While low-power dissipation in the bias circuit is desirable, the impedances associated with the bias nodes $BiasA$-$BiasE$ increase as the current I_{BIAS} decreases. The device dimensions are scaled based on the drain currents. Two transistors will have the same V_{ov} if they have the same ratios of drain current to W/L, or the same current densities (as defined in Section 6.3.3). The W/L values in the bias circuit were chosen to make the V_{DS} of each transistor in the op amp greater than its V_{ov}, so it operates in the active (or saturation) region with high output resistance as explained in Section 4.2.5.2.

Calculated and simulated drain currents are given in Tables 12.1 and 12.2.

To estimate the output voltage swing of the op amp, the voltages V_{BiasC} and V_{BiasD} at nodes $BiasC$ and $BiasD$, respectively, are needed. To find $BiasD$, we first find the voltage at node $BiasA$, V_{BiasA}:

$$V_{BiasA} = V_{GS30} = V_{t30} + V_{ov30} \quad (12.132)$$

where

$$V_{ov30} = \sqrt{\frac{2(I_{D30})}{k'_n(W/L)_{30}}} = \sqrt{\frac{2(0.2 \text{ mA})}{(170 \text{ μA/V}^2)(40/0.3)}} = 0.13 \text{ V} \quad (12.133)$$

Therefore,

$$V_{BiasA} = V_{GS30} = V_{t30} + V_{ov30} = (0.50 + 0.13) \text{ V} = 0.63 \text{ V} \quad (12.134)$$

Now, we will find V_{BiasD}. Using KVL,

$$V_{BiasD} = V_{GS30} + V_{DS28} \quad (12.135)$$

The drain currents in M_{27} and M_{28} are equal:

$$I_{D27} = I_{D28} \quad (12.136)$$

Since M_{27} operates in the active or saturation region and M_{28} operates in the triode region, (12.136) can be rewritten as

$$\frac{k'_n}{2}\left(\frac{W}{L}\right)_{27}(V_{GS27} - V_{t27})^2 = \frac{k'_n}{2}\left(\frac{W}{L}\right)_{28}\left[2(V_{GS28} - V_{t28})V_{DS28} - V_{DS28}^2\right] \quad (12.137)$$

Using KVL,

$$V_{GS28} = V_{DS28} + V_{GS27} \quad (12.138)$$

Substituting (12.138) into (12.137) and using $V_{t27} = V_{t28}$ yields

$$\frac{k'_n}{2}\left(\frac{W}{L}\right)_{27}(V_{GS27} - V_{t27})^2 = \frac{k'_n}{2}\left(\frac{W}{L}\right)_{28}\left[2(V_{DS28} + V_{GS27} - V_{t27})V_{DS28} - V_{DS28}^2\right] \quad (12.139)$$

Dividing both sides by $k'_n/2$ and replacing $V_{GS27} - V_{t27}$ with V_{ov27} gives

$$\left(\frac{W}{L}\right)_{27}(V_{ov27})^2 = \left(\frac{W}{L}\right)_{28}\left[2(V_{DS28} + V_{ov27})V_{DS28} - V_{DS28}^2\right] \quad (12.140)$$

The unknowns in (12.140) are V_{DS28} and V_{ov27}. The overdrive voltage V_{ov27} is

$$V_{ov27} = \sqrt{\frac{2(I_{D27})}{k'_n(W/L)_{27}}} = \sqrt{\frac{2(0.2 \text{ mA})}{(170 \text{ μA/V}^2)(40/0.3)(3)}} = 0.077 \text{ V} \quad (12.141)$$

Substituting this overdrive voltage and W/L values from Table 12.2 into (12.140) gives

$$V_{DS28} = 0.33 \text{ V} \quad (12.142)$$

Finally, substituting (12.134) and (12.142) into (12.135) gives

$$V_{BiasD} = 0.63 \text{ V} + 0.33 \text{ V} = 0.96 \text{ V} \quad (12.143)$$

A similar analysis could be carried out to find the voltage at node *BiasC*. However, an alternative approach is used next to find this voltage. First, we find the voltage at node *BiasB*, V_{BiasB}, which is V_{SG21} below V_{DD}:

$$V_{BiasB} = V_{DD} - V_{SG21} \quad (12.144)$$

A component of V_{SG21} is

$$|V_{ov21}| = \sqrt{\frac{2|I_{D21}|}{k'_p(W/L)_{21}}} = \sqrt{\frac{2(0.2 \text{ mA})}{(45 \text{ μA/V}^2)(40/0.3)(2)}} = 0.18 \text{ V} \quad (12.145)$$

Hence

$$V_{SG21} = |V_{t21}| + |V_{ov21}| = (0.50 + 0.18) \text{ V} = 0.68 \text{ V} \quad (12.146)$$

Using KVL,

$$V_{BiasC} = V_{BiasB} - V_{SG23\oplus 24} + V_{SG24} \quad (12.147)$$

Here, the symbol $\oplus$ means "in series with." Therefore, $V_{SG23\oplus 24}$ is the gate-to-source voltage of a transistor $M_{23\oplus 24}$ that is equivalent to M_{23} in series with M_{24}. The aspect ratio of this equivalent transistor $M_{23\oplus 24}$ can be found by adding the lengths of the series transistors, after scaling to match their W values:

$$\left(\frac{W}{L}\right)_{23\oplus 24} = \frac{1}{3}\left(\frac{W}{L}\right)_u \oplus \frac{3}{1}\left(\frac{W}{L}\right)_u = \frac{3}{9}\left(\frac{W}{L}\right)_u \oplus \frac{3}{1}\left(\frac{W}{L}\right)_u = \frac{3}{10}\left(\frac{W}{L}\right)_u$$

$$(12.148)$$

Hence

$$|V_{ov23\oplus 24}| = \sqrt{\frac{2|I_{D23\oplus 24}|}{k'_p(W/L)_{23\oplus 24}}} = \sqrt{\frac{2(0.2 \text{ mA})}{(45 \text{ μA/V}^2)(40/0.3)(3/10)}} = 0.47 \text{ V}$$

$$(12.149)$$

The overdrive voltage for M_{24}, which is active, is

$$|V_{ov24}| = \sqrt{\frac{2|I_{D24}|}{k'_n(W/L)_{24}}} = \sqrt{\frac{2(0.2 \text{ mA})}{(45 \text{ μA/V}^2)(40/0.3)(3)}} = 0.15 \text{ V} \quad (12.150)$$

Using (12.144), (12.145), (12.147), (12.149), and (12.150), the voltage at node $BiasC$ is

$$\begin{aligned} V_{BiasC} &= V_{BiasB} - V_{SG23\oplus 24} + V_{SG24} \\ &= V_{DD} - V_{SG21} - (|V_{t23\oplus 24}| + |V_{ov23\oplus 24}|) + (|V_{t24}| + |V_{ov24}|) \\ &= [2.0 - 0.68 - (0.50 + 0.47) + (0.50 + 0.15)] \text{ V} = 1.0 \text{ V} \end{aligned} \quad (12.151)$$

12.9.2 Low-Frequency Analysis

The low-frequency DM gain of the op amp can be found using the DM half-circuit in Fig. 12.42 and is given by

$$a_{dm0} = -g_{m1} \times R_{odh} \quad (12.152)$$

where R_{odh} is the output resistance of the DM half-circuit. This resistance is the parallel combination of the resistance looking into the drain of M_{1A} and the resistance looking into the drain of M_{3A}. It is given by

$$R_{odh} \approx [r_{o1A}g_{m1A}(r_{o11}\|r_{o1})] \| (r_{o3A}g_{m3A}r_{o3}) \quad (12.153)$$

We now compute transconductance values that will be used in this section:

$$g_{m1} = \sqrt{2k'_n(W/L)_1 I_{D1}}$$

$$= \sqrt{2(170 \times 10^{-6})[10(40/0.3)]0.002} \; \frac{\text{A}}{\text{V}} = 30 \text{ mA/V} \quad (12.154)$$

12.9 Analysis of a CMOS Fully Differential Folded-Cascode Op Amp

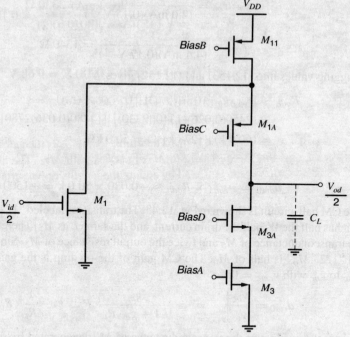

Figure 12.42 DM half-circuit for the op amp.

$$g_{m1A} = \sqrt{2k'_p(W/L)_{1A}|I_{D1A}|}$$

$$= \sqrt{2(45 \times 10^{-6})[24(40/0.3)]0.0024} \; \frac{\text{A}}{\text{V}} = 26 \text{ mA/V} \quad (12.155)$$

$$g_{m3} = g_{m3A} = \sqrt{2k'_n(W/L)_{3A}I_{D3}}$$

$$= \sqrt{2(170 \times 10^{-6})[12(40/0.3)]0.0024} \; \frac{\text{A}}{\text{V}} = 36 \text{ mA/V} \quad (12.156)$$

$$g_{m6} = \sqrt{2k'_n(W/L)_6 I_{D6}}$$

$$= \sqrt{2(170 \times 10^{-6})[10(40/0.3)]0.002} \; \frac{\text{A}}{\text{V}} = 30 \text{ mA/V} \quad (12.157)$$

$$g_{m11} = \sqrt{2k'_p(W/L)_{11}|I_{D11}|}$$

$$= \sqrt{2(45 \times 10^{-6})[44(40/0.3)]0.0044} \; \frac{\text{A}}{\text{V}} = 48 \text{ mA/V} \quad (12.158)$$

The output resistances, using $r_o = 1/(|I_D|\lambda)$, are

$$r_{o1} = \frac{1}{(2.0 \text{ mA})(0.15 \text{ V}^{-1})} = 3330 \; \Omega \quad (12.159)$$

$$r_{o1A} = \frac{1}{(2.4 \text{ mA})(0.17 \text{ V}^{-1})} = 2450 \; \Omega \quad (12.160)$$

$$r_{o3} = r_{o3A} = \frac{1}{(2.4 \text{ mA})(0.15 \text{ V}^{-1})} = 2780 \; \Omega \quad (12.161)$$

$$r_{o5} = r_{o6} = \frac{1}{(2.0 \text{ mA})(0.15 \text{ V}^{-1})} = 3330 \text{ }\Omega \tag{12.162}$$

$$r_{o11} = \frac{1}{(4.4 \text{ mA})(0.17 \text{ V}^{-1})} = 1340 \text{ }\Omega \tag{12.163}$$

Plugging values into (12.153) and (12.152) gives:

$$\begin{aligned} R_{odh} &\approx [r_{o1A}g_{m1A}(r_{o11}||r_{o1})] \,||\, (r_{o3A}g_{m3A}r_{o3}) \\ &= [2450(0.026)(1340||3330)] \,||\, [2780(0.036)2780] \text{ }\Omega \\ &= (60.9 \text{ k}) \,||\, (278 \text{ k}) \text{ }\Omega = 50.0 \text{ k}\Omega \end{aligned} \tag{12.164}$$

and

$$a_{dm0} = -g_{m1} \times R_{odh} \approx -0.030 \times 50.0 \text{ k} = -1500 \tag{12.165}$$

The CM half-circuit is shown in Fig. 12.43. The transistor labeled "$1/2 \cdot M_5$" is half of M_5, that is, it has half the W, half the drain current, and the same L as M_5. Therefore, "$1/2 \cdot M_5$" has half the transconductance of M_5 and twice the output resistance of M_5. Similarly, the transistor labeled "$1/2 \cdot M_6$" is half of M_6. The CM gain of the op amp is the gain of the half-circuit from v_{ic} to v_{oc} with $v_{cmc} = 0$:

$$a_{cm0} = \left.\frac{v_{oc}}{v_{ic}}\right|_{v_{cmc}=0} = -\frac{g_{m1}}{1 + g_{m1}R_t} \cdot R_{och} \approx -\frac{1}{R_t} \cdot R_{och} \tag{12.166}$$

This gain is the product of the transconductance of M_1 degenerated by resistance R_t and the output resistance of the CM half-circuit, R_{och}. Here R_t is the resistance looking down into the drains of "$1/2 \cdot M_5$" and "$1/2 \cdot M_6$"; therefore, R_t is the parallel connection of the output resistances of "$1/2 \cdot M_5$" ($= 2r_{o5}$) and "$1/2 \cdot M_6$" ($= 2r_{o6}$). The approximation in (12.166) is

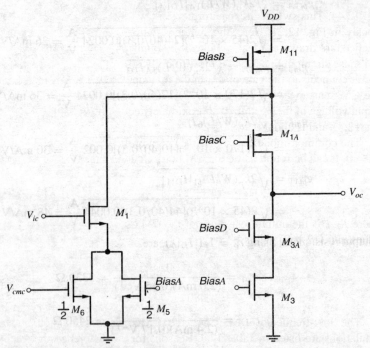

Figure 12.43 CM half-circuit for the op amp, for finding the CM and CMC gains.

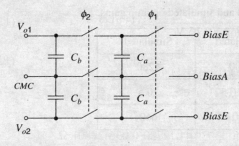

Figure 12.44 The switched-capacitor CM sense network. $C_a = C_b = 0.1$ pF. The nodes labeled V_{CM} are connected to *BiasE* in Fig. 12.41.

based on the assumption that $g_{m1}R_t \gg 1$. Resistance R_{och} is the parallel combination of the resistance looking into the drain of M_{1A}, and the resistance looking into the drain of M_{3A} in the CM half-circuit. Hence

$$R_{och} \approx [r_{o1A}g_{m1A}(r_{o11}||R_o(M_1))] || (r_{o3A}g_{m3A}r_{o3})$$
$$\approx [r_{o1A}g_{m1A}(r_{o11})] || (r_{o3A}g_{m3A}r_{o3})$$
$$= [2450(0.026)1340] || [2780(0.036)2780] \ \Omega$$
$$= (85.0 \text{ k})||(278 \text{ k}) \ \Omega = 65.3 \text{ k}\Omega \quad (12.167)$$

where $R_o(M_1) \approx r_{o1}g_{m1}[(2r_{o5})||(2r_{o6})]$ is the resistance looking into the drain of M_1. The assumption $R_o(M_1) \gg r_{o11}$ was used in (12.167), which yields $r_{o11}||R_o(M_1) \approx r_{o11}$. Substituting values into (12.166) gives

$$a_{cm0} \approx -\frac{1}{R_t} \cdot R_{och} = -\frac{1}{(2r_{o5})||(2r_{o6})} \cdot R_{och} \quad (12.168)$$
$$= -\frac{1}{(2 \cdot 3330)||(2 \cdot 3330)} \cdot (65.3 \text{ k}) = -\frac{1}{3330} \cdot (65.3 \text{ k}) = -19.6$$

The CM sense circuit consists of four capacitors and six switches and is shown in Fig. 12.44. This switched-capacitor network was described in Section 12.5.4 and shown previously in Fig. 12.21. Here, the switches, which are implemented with MOS transistors in practice, are drawn as ideal switches for simplicity. For a purely CM output, the gain of the CM sense circuit in Fig. 12.44 is close to unity, so $v_{cmc} \approx v_{oc}$. The CMC node of the CM sense circuit connects to the gate of M_6 in the op amp, which is also labeled *CMC* in Figure 12.40. The CM sense network and the op amp form a negative feedback loop that forces the CM output voltage of the op amp to approximately equal the voltage V_{BiasE}, which is generated in Fig. 12.41 and is about 0.9 V.

The common-mode control (CMC) gain of the op amp is the gain from v_{cmc} to v_{oc} with $v_{ic} = 0$. It can be found using the CM half-circuit in Fig. 12.43:

$$a_{cmc0} = \left.\frac{v_{oc}}{v_{cmc}}\right|_{v_{ic}=0} = -\frac{g_{m6}}{2} \times R_{och} \quad (12.169)$$

where R_{och} is the output resistance of the CM half-circuit as given by (12.167). Substituting values into (12.169) yields

$$a_{cmc0} = -\frac{g_{m6}}{2} \times R_{och} \approx -\left(\frac{0.030}{2}\right)(65.3 \text{ k}) \quad (12.170)$$
$$= -980$$

The low-frequency DM, CM, and CMC gains calculated above and from SPICE simulations are listed in Table 12.3. Reasons for the differences between calculated and simulated values include short-channel effects (the transistor behavior deviates from the simple

Table 12.3 Calculated and simulated op-amp gains.

	Calculated	Simulated
$|a_{dm0}|$	1500	1280
$|a_{cm0}|$	19.6	20.5
$|a_{cmc0}|$	980	660

square-law equations on which our small-signal formulas are based), errors in r_o (λ_n and λ_p are not constant in practice and depend on V_{DS}), body effect (which was ignored in the calculations), and approximations used to simplify the calculations.

The output voltage swing of the op amp is now estimated. Consider the output swing of V_{o1} in Fig. 12.40. The upper swing limit occurs when M_{1A} enters the triode region. This occurs when the gate-to-drain voltage of M_{1A} equals a threshold voltage. Thus, using (12.151), the upper swing limit of V_{o1} is

$$V_{o1}(\max) = V_{BiasC} + |V_{t1A}| = 1.0 \text{ V} + 0.50 \text{ V} = 1.50 \text{ V} \quad (12.171)$$

The lower swing limit of V_{o1} occurs when M_{3A} enters the triode region. This occurs when the gate-to-drain voltage of M_{3A} equals a threshold voltage. Using (12.143), the lower swing limit of V_{o1} is

$$V_{o1}(\min) = V_{BiasD} - V_{t3A} = 0.96 \text{ V} - 0.50 \text{ V} = 0.46 \text{ V} \quad (12.172)$$

Therefore the peak-to-peak output swing of V_{o1} is

$$V_{o1,pp} = V_{o1}(\max) - V_{o1}(\min) = 1.50 \text{ V} - 0.46 \text{ V} = 1.04 \text{ V} \quad (12.173)$$

if the CMFB circuit biases V_{o1} midway between the swing limits. In that case, the peak-to-peak differential output swing, which is twice the swing of either V_{o1} or V_{o2}, is

$$V_{od,pp} = 2V_{o1,pp} = 2(1.04) \text{ V} = 2.08 \text{ V} \quad (12.174)$$

Simulation gives a differential output swing of 2.4 V.

The common-mode input range (CMIR) of the op amp is the range of CM input voltage over which the transistors associated with the input differential pair remain in the active or saturation region. The lower limit of the CMIR occurs when the CM input voltage forces either M_5 or M_6 into the triode region. Since M_5 and M_6 have equal drain currents and aspect ratios, they both enter the triode region when $V_{DS5} = V_{DS6} = V_{ov5}$. The corresponding CM input voltage is

$$V_{IC}(\min) = V_{GS1} + V_{ov5} = V_{t1} + V_{ov1} + V_{ov5}$$

$$= V_{t1} + \sqrt{\frac{2(I_{D1})}{k'_n(W/L)_1}} + \sqrt{\frac{2(I_{D5})}{k'_n(W/L)_5}}$$

$$= 0.50 \text{ V} + \sqrt{\frac{2(2.0 \text{ mA})}{(170 \text{ }\mu\text{A/V}^2)(40/0.3)(10)}} + \sqrt{\frac{2(2.0 \text{ mA})}{(170 \text{ }\mu\text{A/V}^2)(40/0.3)(10)}}$$

$$= 0.50 \text{ V} + 0.13 \text{ V} + 0.13 \text{ V} = 0.76 \text{ V} \quad (12.175)$$

The upper limit of the CMIR occurs when the CM input voltage forces either M_1 or M_2 into the triode region. Since M_1 and M_2 are matched with equal drain currents, they both enter the triode region when the CM input voltage is a threshold voltage above the voltage at the drain of M_1 or M_2. Considering M_1, its drain voltage is also the voltage at the source of M_{1A}, V_{S1A}. Therefore, using (12.151):

$$\begin{aligned}
V_{IC}(\max) &= V_{S1A} + V_{t1} = V_{BiasC} + V_{SG1A} + V_{t1} \\
&= V_{BiasC} + |V_{t1A}| + |V_{ov1A}| + V_{t1} \\
&= V_{BiasC} + |V_{t1A}| + \sqrt{\frac{2|I_{D1A}|}{k'_p(W/L)_{1A}}} + V_{t1} \quad (12.176) \\
&= 1.0 \text{ V} + 0.50 \text{ V} + \sqrt{\frac{2(2.4 \text{ mA})}{(45 \text{ μA/V}^2)(40/0.3)(24)}} + 0.50 \text{ V} = 2.2 \text{ V}
\end{aligned}$$

From (12.175) and (12.176), the CMIR is from 0.76 V to 2.2 V. The bias circuit generates $V_{BiasE} = 0.9$ V. As described in the next subsection, this voltage sets the CM input voltage of the op amp to 0.9 V, which falls within the CMIR of the op amp. The CMIR for this folded-cascode op amp is fairly large. The CMIR for a folded-cascode op amp is usually larger than the CMIR for a telescopic-cascode op amp designed in the same CMOS technology, due to the "folded" topology.

The input-referred noise of the op amp can be found by considering a DM half-circuit, summing the contributions of the device noise-current generators to the output current when the output node is shorted to small-signal ground, and then referring the output noise to the op-amp input. These calculations are carried out next. Flicker noise is ignored for simplicity here.

Fig. 12.45 shows the DM small-signal half-circuit including a noise generator for each transistor. Here, the output is shorted to ground to allow calculation of the output noise current. For simplicity, let $r_o \to \infty$ for all transistors and ignore all capacitors. The drain currents of M_1, M_3, and M_{11} all flow into the short at the output. Therefore, noise in these currents directly

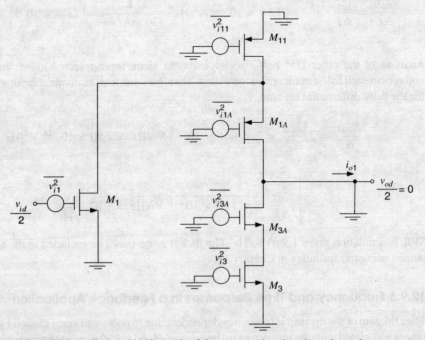

Figure 12.45 DM small-signal half-circuit of the op amp showing the noise-voltage generators.

contributes to the noise in the output current i_{o1}. However, M_{1A} and M_{3A} do not contribute to the output noise current. To understand why, first consider M_{3A}. Using superposition, consider only the noise-voltage source connected to its gate. The drain current of M_{3A} is the product of this noise voltage and the transconductance of M_{3A} degenerated by the resistance r_{o3}. With $r_{o3} \to \infty$, the degenerated transconductance is zero, and therefore $\overline{v_{i3A}^2}$ does not contribute to the noise in the output current i_{o1}. Similarly, M_{1A} does not contribute to the noise in the output current. Therefore, the total noise current at the output is the sum of the noise-current contributions from M_1, M_3, and M_{11} in Fig. 12.45:

$$\frac{\overline{i_{o1}^2}}{\Delta f} = g_{m1}^2 \frac{\overline{v_{i1}^2}}{\Delta f} + g_{m11}^2 \frac{\overline{v_{i11}^2}}{\Delta f} + g_{m3}^2 \frac{\overline{v_{i3}^2}}{\Delta f} \tag{12.177}$$

Using the thermal-noise voltage expression from (11.68a):

$$\frac{\overline{v_i^2}}{\Delta f} = 4kT \frac{2}{3} \frac{1}{g_m} \tag{12.178}$$

for each noise-voltage generator, the total output noise current in (12.177) can be written:

$$\frac{\overline{i_{o1}^2}}{\Delta f} = g_{m1}^2 4kT \frac{2}{3} \frac{1}{g_{m1}} + g_{m11}^2 4kT \frac{2}{3} \frac{1}{g_{m11}} + g_{m3}^2 4kT \frac{2}{3} \frac{1}{g_{m3}} \tag{12.179}$$

At $T = 300°K$, $4kT = 1.66 \times 10^{-16}$ V-C. Substituting $4kT$ and g_m values into (12.179) and evaluating gives

$$\frac{\overline{i_{o1}^2}}{\Delta f} = (3.32 + 5.32 + 3.98) \times 10^{-22} \text{ A}^2/\text{Hz} = 12.6 \times 10^{-22} \text{ A}^2/\text{Hz} \tag{12.180}$$

Referring this output noise current to an equivalent noise voltage v_{eqH} at the input of the DM half-circuit yields

$$\frac{\overline{v_{eqH}^2}}{\Delta f} = \frac{\overline{i_{o1}^2}}{\Delta f} \cdot \frac{1}{g_{m1}^2} = 12.6 \times 10^{-22} \text{ A}^2/\text{Hz} \cdot \frac{1}{(30 \text{ mA/V})^2} = 1.40 \times 10^{-18} \text{ V}^2/\text{Hz} \tag{12.181}$$

Analysis of the other DM half-circuit yields the same input-referred noise, and the noise voltages in each half-circuit are uncorrelated. Therefore, the total equivalent input-noise voltage for the fully differential op amp, v_{eqT}, is

$$\frac{\overline{v_{eqT}^2}}{\Delta f} = 2 \frac{\overline{v_{eqH}^2}}{\Delta f} = 2 \times 1.40 \times 10^{-18} \text{ V}^2/\text{Hz} = 2.80 \times 10^{-18} \text{ V}^2/\text{Hz} \tag{12.182}$$

or

$$\sqrt{\frac{\overline{v_{eqT}^2}}{\Delta f}} = \sqrt{2.80 \times 10^{-18} \text{ V}^2/\text{Hz}} = 1.67 \frac{\text{nV}}{\sqrt{\text{Hz}}} \tag{12.183}$$

SPICE simulation gives 1.9 nV/$\sqrt{\text{Hz}}$. The flicker noise could be included in the above calculations using the formulas in Chapter 11.

12.9.3 Frequency and Time Responses in a Feedback Application

The DM gain of the op amp is frequency dependent due to poles and zeros caused by the capacitances associated with the transistors in the op amp. Hand calculation of the frequency response

12.9 Analysis of a CMOS Fully Differential Folded-Cascode Op Amp

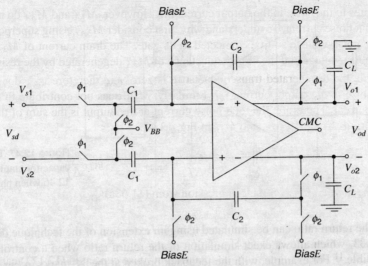

Figure 12.46 Switched-capacitor amplifier that uses the op amp in Figure 12.40. The CM sense circuitry is not shown.

is possible but the result is often inaccurate since the poles and zeros depend on device and parasitic capacitances. Therefore, the frequency response of the op-amp gain is usually simulated. Also, the op amp is always used in feedback, and the stability of the feedback loops, under the loading imposed in part by the passive feedback elements, is a key issue. Therefore, the magnitude and phase responses of the return ratio of each feedback loop is of interest. The feedback loops will be considered later in this section for the feedback circuit that is described next.

Fig. 12.46 shows the op amp used in a fully differential switched-capacitor application. The operation of this circuit is similar to that of the single-ended switched-capacitor gain stage in Fig. 6.8a, and it uses the clocks shown in Fig. 6.8b. Ideally, this circuit provides a voltage gain during ϕ_1 of $-C_1/C_2$. In practice, the MOS transistors act as switches, and the switches are designed so that their resistance when on (and in the triode region) is small enough to be ignored. The ideal value of the CM input and output voltages is V_{BiasE}, which is the voltage at node $BiasE$, and V_{BB} is the CM value of the input sources V_{s1} and V_{s2}. When ϕ_1 is high, the circuit is redrawn in Fig. 12.47. (The CM input and output voltages of the op amp are the same in Fig. 12.46, but they could be different. The CM input voltage must fall within the CM input range, and the CM output voltage is usually chosen to maximize output swing. Sometimes two different bias voltages are used to independently optimize these voltages.) Here, switches driven by the ϕ_1 clock are short circuits, and switches driven by ϕ_2 are open circuits. An op amp can appear in the feedback configuration in Fig. 12.47 in many applications, such as a switched-capacitor gain stage and a switched-capacitor integrator, which is shown in Fig. 6.10a.

Because node $BiasE$ connects to switches in Fig. 12.46, its voltage can experience significant transients after switching. If other bias voltages were produced by the same branch in Fig. 12.41 that generates V_{BiasE}, significant transients could be introduced in these other bias voltages also. To avoid this problem, the branch in Fig. 12.41 that generates V_{BiasE} generates no other bias voltages.

For the feedback circuit in Fig. 12.47, stability of the DM and CM feedback loops is considered next, using a combination of simulations and calculations. Return ratio is used to investigate stability. In the following, $C_1 = C_2 = 1$ pF and the value of the load capacitance C_L varies.

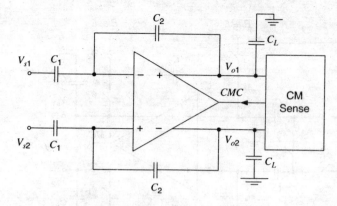

Figure 12.47 The switched-capacitor amplifier of Figure 12.46 when phase ϕ_1 is high.

The return ratio can be simulated using an extension of the technique described in Problem 8.33, which allows exact simulation of the return ratio when a controlled source is not accessible.[11] For example, with the feedback broken at the × marks in Fig. 12.48, the quantities $\mathcal{R}'_{v,dm}$ and $\mathcal{R}'_{i,dm}$ are simulated, and then these quantities are combined to calculate the return ratio $\mathcal{R}_{dm}$ using the equation in Problem 8.33.

For the return-ratio simulation, a dc path to each op-amp input is provided by each resistor R_{BIG} in Fig. 12.48. A very large resistance, $R_{BIG} = 1 \times 10^{12} \Omega$, is used that has little effect on the simulation results at frequencies where $\omega \gg 1/(R_{BIG}C_1)$. The dc voltage at each op-amp input is V_{BiasE}, which is developed at node *BiasE* in the bias circuit in Fig. 12.41. Without these resistors, the dc op-amp input voltages are undefined, and therefore, SPICE would not be able to simulate the circuit. In normal operation, one plate of each switched capacitor is alternately connected to *BiasE* or an op-amp input as shown in Fig. 12.46. This switching makes the dc voltage at the op-amp inputs equal the voltage V_{BiasE}.

The simulated magnitude and phase responses of the return ratio for the DM feedback loop are plotted in Fig. 12.49. Here, the load capacitance in the DM half-circuit is $C_L = 2$ pF. The phase margin is 72°, and the gain margin is 38 dB. The frequency f_r where the magnitude of the return ratio is unity is 390 MHz.

For comparison to simulation, the DM return ratio will now be calculated. The feedback circuit in Fig. 12.47 is shown in Fig. 12.50a with two capacitors C_{ia} added; these capacitors

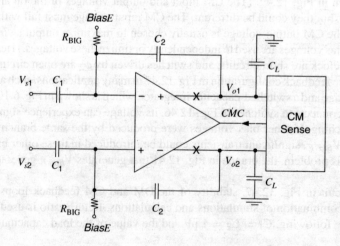

Figure 12.48 The feedback circuit in Figure 12.47, with resistors R_{BIG} added to bias the op-amp inputs.

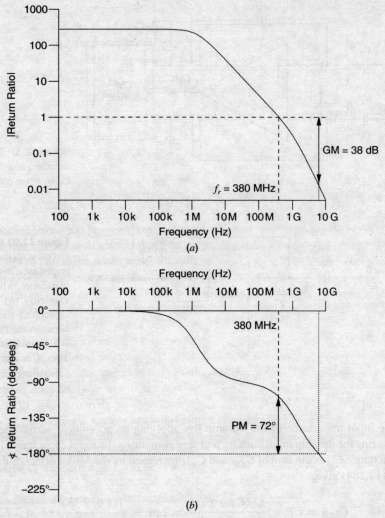

Figure 12.49 Simulated DM return ratio for $C_L = 2$ pF. (a) Magnitude and (b) phase responses.

model the op-amp input impedance. The corresponding DM small-signal half-circuit is shown in Fig. 12.50b, where the op-amp output port is modeled simply by a transconductance g_{m1} and resistance R_{odh} [based on (12.165)]. This model ignores all poles and zeros except the pole associated with the RC time constant at the op-amp output node.

If the feedback is broken at the × in Fig. 12.50b and the input voltage $V_{sd}/2$ is set to zero for calculation of the DM return ratio, the total load capacitance from the op-amp output to ground is

$$C_{Leff} = C_L + \frac{C_2(C_1 + C_{ia})}{C_2 + C_1 + C_{ia}} \quad (12.184)$$

Here the capacitors in the CM sense block are ignored for simplicity; in practice, they are small compared to the capacitors C_1, C_2 and C_L. The right-most term in (12.184) is the capacitance associated with C_2 in series with the parallel connection of C_1 and C_{ia}. Based on a simulation

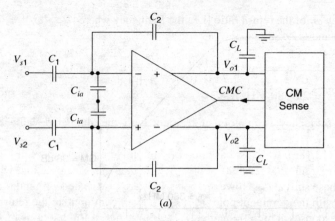

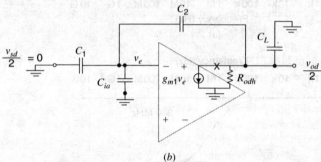

Figure 12.50 (a) The feedback circuit with the op-amp input impedance modeled by the C_{ia} capacitors, which are shown explicitly. (b) The DM small-signal half-circuit. The node between the two C_{ia} capacitors is a ground for DM signals. The × marks a break point for calculating the DM return ratio.

of the input impedance of the op amp (by applying an ac voltage across the op-amp input, measuring the op-amp input current, and then computing the input capacitance), $C_{ia} \approx 2.58$ pF. Capacitance C_{ia} is the sum of C_{gs1} and C_{gd1} increased by the Miller effect. Substituting values into (12.184) gives

$$C_{Leff} = C_L + \frac{C_2(C_1 + C_{ia})}{C_2 + C_1 + C_{ia}} = 2 \text{ pF} + \frac{1(1 + 2.58)}{1 + 1 + 2.58} \text{ pF} = 2.72 \text{ pF} \quad (12.185)$$

For comparison, the DM return ratio can be calculated with the feedback loop broken in the simplified circuit in Fig. 12.50b. Following Section 8.8, the calculated return ratio is:

$$\mathcal{R}_{dm}(\omega) = g_{m1} R_{odh} \cdot \frac{1}{1 + j\omega R_{odh} C_{Leff}} \cdot \frac{C_2}{C_2 + C_1 + C_{ia}} \quad (12.186)$$

The pole here is $-1/(R_{odh} C_{Leff})$, due to the RC time constant at the op-amp output in Fig. 12.50b. Here C_{Leff} is the effective output load capacitance with the loop broken, as given by (12.184). At dc, (12.186) reduces to

$$\mathcal{R}_{dm0} = \mathcal{R}_{dm}(\omega = 0) = g_{m1} R_{odh} \cdot 1 \cdot \frac{C_2}{C_2 + C_1 + C_{ia}} \quad (12.187)$$

At high frequencies $[\omega \gg 1/(R_{odh} C_{Leff})]$, (12.186) reduces to

$$\mathcal{R}_{dm}(\omega) \approx g_{m1} \cdot \frac{1}{j\omega C_{Leff}} \cdot \frac{C_2}{C_2 + C_1 + C_{ia}} \quad (12.188)$$

The unity-gain frequency ω_r of the return ratio [i.e., the frequency where $|\mathcal{R}_{dm}(\omega_r)| = 1$] can be found using (12.185) and (12.188):

$$\omega_r \approx \frac{g_{m1}}{C_{\text{Leff}}} \cdot \frac{C_2}{C_2 + C_1 + C_{ia}} = \frac{0.030}{2.72 \times 10^{-12}} \cdot \frac{1}{1 + 1 + 2.58} \text{ rad/s}$$
$$= 2.4 \text{ Grad/s} = 2\pi(380 \text{ MHz}) \qquad (12.189)$$

This calculated unity-gain frequency is in good agreement with the simulated value of 390 MHz.

The DM return ratio in (12.186), which is based on the simplified op-amp model in Fig. 12.50b, would give a phase margin of at least 90°, since $\mathcal{R}_{dm}(\omega)$ has only one pole that will introduce at most $-90°$ of phase shift at ω_r. However, poles and zeros associated with internal nodes in the op amp, along with the dominant pole $-1/(R_{odh}C_{\text{Leff}})$, contribute to the return-ratio magnitude and phase responses at high frequencies. Therefore, the return ratio may have unacceptable gain or phase margin due to these poles and zeros. In practice, C_L, which is a significant component of C_{Leff} in (12.185), is chosen to assure good gain and phase margins for the CM and DM feedback loops.

For the DM feedback loop in Fig. 12.47 with $C_L = 1$ pF, the phase margin is 65°, and the gain margin is 35 dB. The frequency where the magnitude of the return ratio is unity is $f_r = 550$ MHz.

For the DM feedback loop in Fig. 12.47 with $C_L = 0$ pF, the capacitance C_{Leff} is dominated by capacitors C_1, C_2, and C_{ia}. The phase margin is 55° and the gain margin is 31 dB. The frequency where the magnitude of the return ratio is unity is $f_r = 810$ MHz.

Note that as C_L increases, the phase margin increases and the unity-gain frequency of the DM return ratio ($\omega_r = 2\pi f_r$) decreases. This is as expected, since increasing C_L increases C_{Leff}, which reduces the magnitude of the dominant pole in the return ratio without affecting the other poles and zeros.

The low-frequency closed-loop DM gain is given by

$$A_{cl0,dm} = \frac{v_{od}}{v_{sd}} = A_\infty \cdot \frac{\mathcal{R}_{dm0}}{1 + \mathcal{R}_{dm0}} \qquad (12.190)$$

from (8.209). The direct feedthrough $d = 0$ for this circuit, due to the capacitive feedback elements that are open circuits at dc, and hence block signal feedthrough at dc. The ideal closed-loop gain is $A_\infty = -C_1/C_2 = -1$. Using the simulated value of $\mathcal{R}_{dm0}$ in (12.190) yields

$$A_{cl0,dm} = A_\infty \cdot \frac{\mathcal{R}_{dm0}}{1 + \mathcal{R}_{dm0}} = -1 \cdot \frac{280}{1 + 280} = -0.996 \qquad (12.191)$$

SPICE simulation agrees exactly with this closed-loop gain. The simulated -3-dB bandwidth of the closed-loop gain is 630 MHz with $C_L = 2$ pF.

The CM return ratio can be simulated[11] by breaking the feedback at the $\times$ marks in Fig. 12.48, simulating the quantities $\mathcal{R}'_{v,cm}$ and $\mathcal{R}'_{i,cm}$, and then combining these quantities to calculate the CM return ratio $\mathcal{R}_{cm}$ using the equation in Problem 8.33. There are two CM feedback loops in Fig. 12.48. One includes the CM sense circuit and the gain from the op-amp CMC input to the op-amp CM output voltage. The other includes the feedback elements C_1 and C_2 and the CM gain from the op-amp CM input voltage to the op-amp CM output voltage. When the feedback is broken at the $\times$ marks in Fig. 12.48, both of these CM loops are broken.[11] The simulated CM return ratio is shown in Fig. 12.51 for $C_L = 2$ pF. The phase margin is 76°, and the gain margin is 41 dB. So the CM loop is stable with good margins for $C_L = 2$ pF. The unity-gain frequency for this return ratio is 420 MHz.

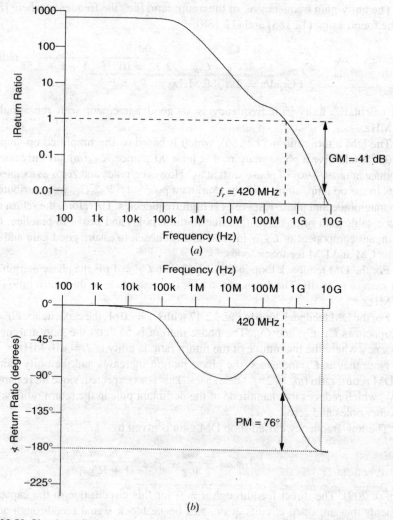

Figure 12.51 Simulated CM return ratio for $C_L = 2$ pF. (*a*) Magnitude and (*b*) phase responses.

The CM return ratio differs from the DM return ratio because the small-signal half-circuits differ; points of symmetry are open circuits for CM signals but are small-signal grounds for DM signals. For example, the node between the two C_{ia} capacitors in Fig. 12.50*a* is a ground for DM signals but is an open circuit for CM signals. Differences between the return ratios can be seen by comparing Figs. 12.49 and 12.51. For example, a left-half-plane zero exists in the CM return ratio at about 60 MHz. It causes the magnitude response to flatten and the phase response to increase. It was determined through simulations that this zero is related to the nonzero output impedances in the bias circuit at the *BiasC* and *BiasD* nodes. When the impedance at nodes *BiasC* and *BiasD* was decreased by connecting a large capacitor from those bias nodes to ground, the effect of the zero was not visible in the magnitude or phase response, and the general shape of the CM return-ratio response resembled the plots in Fig. 12.49. With these bypass capacitors, the CM return-ratio phase margin is 54°, and the gain margin is 16 dB.

The step responses of the DM op-amp output voltage V_{od} for three values of C_L are plotted in Fig. 12.52. The ideal output V_{od} is also shown; ideally V_{od} is the differential input inverted since the closed-loop DM gain is $-C_1/C_2 = -1$. The overshoot/ringing in the step response

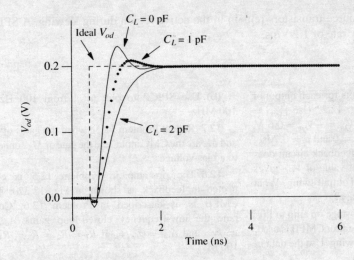

Figure 12.52 Step responses (V_{od} versus t) for $C_L = 0$, 1 and 2 pF. The ideal output is plotted with a dashed line. The input is a -0.2 V step applied at $t = 0.4$ ns.

increases as C_L decreases (and the phase margin decreases). As the phase margin decreases, the poles of the closed-loop transfer function move toward the $j\omega$ axis. (In a two-pole system, the closed-loop poles will be on the $j\omega$ axis when the phase margin is zero. In that case, the step response contains a sinusoid in steady state.) In switched-capacitor applications, the output of the op amp must settle to its final value within one-half clock period. For example, the output voltage of the op amp in Fig. 12.46 must settle during the time when clock ϕ_1 is high. The settling time is usually found by applying a step input. Then the settling time is measured from the time the input changes until the output voltage remains within a specified band around its steady-state value. For the step responses in Fig. 12.52, the 1 percent settling times (the time until the output remains within 1 percent of the final output voltage) are 1.67 ns for $C_L = 2$ pF, 1.35 ns for $C_L = 1$ pF, and 1.37 ns for $C_L = 0$ pF.

Equation 12.189 shows that ω_r can be increased by increasing g_{m1}, by decreasing C_{Leff} (by decreasing some or all of the capacitances that contribute to it), or by increasing the term $C_2/(C_2 + C_1 + C_{ia})$, which is sometimes called "the feedback factor." Assuming a constant phase margin, if ω_r increases, the closed-loop -3-dB bandwidth increases (see Fig. 9.10) and the settling time of the step response decreases. Also, if the phase margin is larger than required, ω_r could be increased (and phase margin decreased) by increasing the magnitude of the dominant pole. This can be achieved by reducing C_L.

During slewing, a large voltage appears between op-amp inputs and causes one of the input transistors (M_1 or M_2) in Fig. 12.40 to turn off while the other remains on. In that case, the sum of the drain currents from M_5 and M_6 flows through the input transistor that is on. This causes a current change in both M_1 and M_2 of magnitude $I_{slew} = (I_{D5} + I_{D6})/2$; this current flows from the op-amp outputs and through the differential load. The op-amp inputs are not virtually shorted during slewing because one of the op-amp input transistors is off, and thus the op-amp operation is highly nonlinear. Therefore, the total load capacitance from each op-amp output to ground is equal to C_{Leff} as given in (12.184). The differential load capacitance between the two op-amp outputs will be half of C_{Leff}, since two C_{Leff} capacitors are in series between the two op-amp outputs. Therefore, the DM output slew rate is

$$\frac{dV_{od}}{dt} = \frac{I_o(\max)}{\frac{C_{Leff}}{2}} = \frac{I_{slew}}{\frac{C_{Leff}}{2}} = \frac{\frac{I_{D5} + I_{D6}}{2}}{\frac{C_{Leff}}{2}} = \frac{\frac{4.0}{2} \text{ mA}}{\frac{2.72 \text{ pF}}{2}} = 1.47 \frac{\text{V}}{\text{ns}} \quad (12.192)$$

assuming all current-source transistors remain in the active region during slewing. A SPICE simulation gives a slew rate of 1.2 V/ns.

PROBLEMS

12.1 What are the swing limits for each output of the differential amplifier in Fig. 12.2? Use $|V_{ov}| = 0.25$ V for all transistors and $V_{tn} = -V_{tp} = 0.6$ V. Assume $V_{DD} = V_{SS} = 2.5$ V, $V_{ic} = 0$ and $\gamma = 0$. Also, assume that the common-mode feedback circuit does not limit the output swing. What value of V_{OC} gives the largest symmetric differential output swing? What is the peak value of V_{od} in this case?

12.2 Repeat 12.1 for the two-stage op amp in Fig. 12.23. Assume that switched-capacitor CMFB is used, which does not limit the output swing. Use the data in Problem 12.1.

12.3 A balanced fully differential circuit displays only odd-order nonlinearity. Use SPICE to verify this fact for the op amp in the example in Section 12.6.1.

(a) Either use SPICE to find the distortion in the output waveform for a low-frequency differential sinusoidal input, or use SPICE to plot the dc transfer characteristic V_{od} versus V_{id} and verify that the characteristic is odd.

(b) Repeat (a) with a 1 percent mismatch between W_1 and W_2. Observe that even-order nonlinearity now exists.

12.4 For the op amp in Fig. 12.2, use the device data and operating point from the first example in Section 12.4.1. Assume all transistors operate in the active region.

(a) Find the element values in the two models for the output ports in Fig. 12.7.

(b) Find the element values in the two models for the input ports in Fig. 12.6. (The elements are capacitors.) Use $C_{gs} = 180$ fF and $C_{gd} = 20$ fF.

12.5 For the op amp in Fig. 12.2, use the data from Problem 12.4 except use $I_{DS} = 100$ µA and $|V_{ov}| = 0.1$ V for all transistors. Assume all transistors operate in the active region with $C_{gs} = 180$ fF and $C_{gd} = 20$ fF.

(a) Find the element values in the two models for the output ports in Fig. 12.7.

(b) Find the element values in the two models for the input ports in Fig. 12.6. (The elements are capacitors.)

(c) Calculate the common-mode control gain a_{cmc}.

12.6 (a) For the op amp in Problem 12.5, calculate a'_{cm}. Assume that the CMFB scheme in Fig. 12.17 is used and that $a_{cms} = 1$. Recall that $a'_{cm} = v_{oc}/v_{ic}$ when the CMFB loop is active.

(b) Use SPICE to plot $|a'_{cm}|$ from 100 Hz to 100 MHz.

12.7 Repeat Problem 12.5c when the gates of M_3 and M_4 are the CMC input, and the gate of M_5 connects to a bias voltage.

12.8 The op amp in Problem 12.5 is connected in feedback as shown in Fig. 12.32a. The CMFB is as described in Problem 12.6. Compute the low-frequency closed-loop gains $A_{dm} = v_{od}/v_{sd}$ and $A_{cm} = v_{oc}/v_{sc}$ if $R_1 = R_2 = R_3 = R_4 = 100$ MΩ.

12.9 Calculate the DM output slew rate dV_{od}/dt for the op amp in Fig. 12.2. Assume $I_{D5} = 100$ µA and a 5-pF capacitor is connected from each op-amp output to ground.

12.10 Calculate the CM output slew rate dV_{oc}/dt for the op amp in Fig. 12.2. Assume $I_{D5} = 100$ µA and a 5-pF capacitor is connected from each op-amp output to ground.

12.11 Compute the output slew rate dV_{od}/dt for the op amp in the example in Section 12.6.1. Use the bias currents from the example and $C = 1.39$ pF.

12.12 For this problem, use the op amp in Fig. 12.23 and the CMFB scheme in Fig. 12.17. Use the complement of the amplifier in Fig. 12.16b as the CM-sense amplifier, modified to give a negative dc gain. Assume the source followers in Fig. 12.17 have a low-frequency gain of 0.95 and $R_{cs} = 15$ kΩ. Use the transistor and op-amp operating-point data given in the example in Section 12.6.1.

(a) Design the CM-sense amplifier so that the total low-frequency gain $a_{cms0} = v_{oc}/v_{cms} = -0.71$, which is the same value as in the example in Section 12.6.1.

(b) With this CMFB circuit, what are the swing limits for each op-amp output voltage (V_{o1} and V_{o2})? Assume that the biasing current source in each source follower in Fig. 12.17 is implemented with a NMOS transistor, and the current-source and source-follower transistors operate with $V_{GS} = 0.8$ V and $V_{ov} = 0.2$ V. For simplicity, assume V_{GS} is constant and take $\gamma = 0$.

(c) What value of V_{CM} gives the largest symmetric output swing?

(d) Verify that this CMFB circuit works correctly by running a SPICE simulation. Use the value of V_{CM} from part (c).

12.13 Compute the op-amp CM and DM load capacitances for the output loading in Fig. 12.53a and

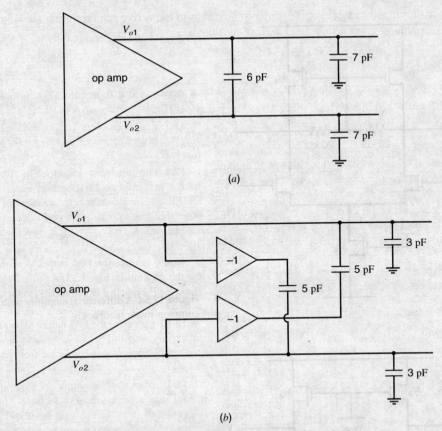

Figure 12.53 Output load networks for Problem 12.13.

12.53b. Assume the inverting voltage buffers in Fig. 12.53b are ideal.

12.14 (a) For the amplifier in Fig. 12.16b, estimate the pole associated with the RC time constant at the V_{cms} output node. Assume $|I_{D25}| = 0.4$ mA, $V_{ov23} = 0.2$ V, and $V_{OC} = V_{CM}$. Ignore all capacitances except C_{gs23} in parallel with a fixed capacitance of 90 fF. Take $L_{eff} = 0.8$ µm. Use the data in Table 2.3.

(b) Repeat (a) except use $|I_{D25}| = 0.1$ mA.

(c) Compare the results in (a) and (b). Explain the difference.

12.15 A differential amplifier with local CMFB is shown in Fig. 12.54. Use $|V_{ov}| = 0.2$ V for all transistors, $V_{tn} = -V_{tp} = 0.6$ V, $I_{D5} = 200$ µA, $V_{An} = 10$ V, $|V_{Ap}| = 20$ V, and $\gamma = 0$. Assume $V_{DD} = V_{SS} = 2.5$ V and $V_{ic} = 0$.

(a) What is the dc common-mode output voltage of this amplifier?

(b) Compute the low-frequency gains a_{dm} and a_{cm}. Compare these gains with the gains calculated in the first example in Section 12.4.1.

12.16 A differential amplifier that does not use a tail current source is shown in Fig. 12.55.

(a) Compute the low-frequency gains a_{dm} and a_{cm}. For all transistors, drain currents are 100 µA and $|V_{ov}| = 0.2$ V. Also, $V_{An} = 10$ V, and $|V_{Ap}| = 20$ V.

(b) Compare these gains with the gains calculated in the first example in Section 12.4.1.

12.17 (a) For the op amp in Fig. 12.2, assume the CM and DM load capacitances are $C_{Lc} = C_{Ld} = 2$ pF. Calculate the frequencies at which $|a_{dm}(j\omega)| = 1$ and $|a_{cmc}(j\omega)| = 1$, ignoring other capacitors. Use $|V_{ov}| = 0.25$ V for all transistors, $V_{tn} = -V_{tp} = 0.6$ V, $I_{D5} = 200$ µA, $V_{An} = 20$ V, $|V_{Ap}| = 25$ V, and $\gamma = 0$.

(b) Calculate the unity-gain frequencies in part (a) if $C_{Lc} = 2$ pF and $C_{Ld} = 4$ pF.

(c) The unity-gain frequency for the CMC gain can be made equal to the DM unity-gain frequency in part (b) by splitting M_5 as shown in Fig. 12.15. How should the 200 µA current be split between M_{51} and M_{52}? Assume $V_{ov51} = V_{ov52} = 0.25$ V.

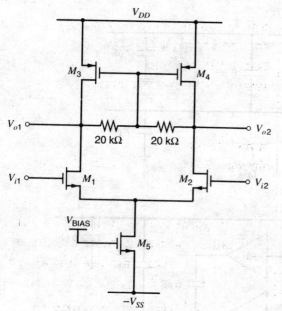

Figure 12.54 A differential amplifier with local common-mode feedback.

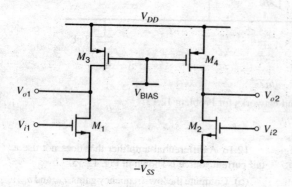

Figure 12.55 A differential amplifier without a tail current source.

12.18 For the CM-detector in Fig. 12.56, find $a_{cms}(s) = v_{cms}(s)/v_{oc}(s)$, assuming the CM-sense amplifier is ideal with unity gain. Then find $a_{cms}(s)$ when a capacitor C_{cs} is connected in parallel with each resistor R_{cs}. What is the effect of the C_{cs} capacitors?

12.19 A NMOS transistor is operating in the triode region. Find a formula for its transconductance $g_m = \partial I_d/\partial V_{gs}$. Compare it with g_m in the active region at the same dc drain current. Which is larger?

12.20 For the fully differential circuit in Fig. 12.32a, assume the op amp is ideal with $R_i = \infty$, $R_o = 0$, $a_{dm} = -\infty$, and $a_{cm} = 0$. Find the closed-loop gains $A_{dm} = v_{od}/v_{sd}$, $A_{cm} = v_{oc}/v_{sc}$, $A_{dm-cm} = v_{oc}/v_{sd}$, and $A_{cm-dm} = v_{od}/v_{sc}$, under the following conditions.

(a) $R_1 = R_2 = 1 \text{ k}\Omega$ and $R_3 = R_4 = 5 \text{ k}\Omega$.
(b) $R_1 = 1.01 \text{ k}\Omega$, $R_2 = 0.99 \text{ k}\Omega$, $R_3 = R_4 = 5 \text{ k}\Omega$.

12.21 For the circuit in Problem 12.20a, the applied source voltage is a single-ended signal with $V_{s1} = 0.2 \text{ V} \sin(100t)$ and $V_{s2} = 0$. Assume a CMFB loop forces $V_{OC} = 0$. What are $V_{o1}(t)$, $V_{o2}(t)$, $V_{od}(t)$, and $V_{oc}(t)$? What are $V_{i1}(t)$, $V_{i2}(t)$, $V_{id}(t)$, and $V_{ic}(t)$?

12.22 The op amp in Problem 12.4 is used with the CMFB scheme shown in Fig. 12.17. The circuit is perfectly balanced except that the CM-sense resistors are mismatched with the upper resistor $R_{cs1} = 10.1 \text{ k}\Omega$ and the lower resistor $R_{cs2} = 9.9 \text{ k}\Omega$. Assume the source followers and the CM-sense amplifier are ideal with gains of unity.

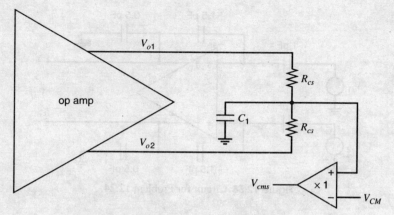

Figure 12.56 Circuit for Problem 12.18.

(a) Compute the gains a_{cms} and a_{dm-cms} in (12.106).

(b) Compute the low-frequency op-amp gains v_{od}/v_{id}, v_{oc}/v_{ic}, v_{od}/v_{ic}, and v_{oc}/v_{id} with the CMFB active.

(c) Use SPICE to simulate these gains.

12.23 A fully differential op amp with CMFB is shown in Fig. 12.57. For M_1, M_{1C}, M_2, and M_{2C}, use $W/L = (64~\mu\text{m})/(0.8~\mu\text{m})$. For M_3-M_4, M_{26}-M_{27} and M_{11}, $W/L = (96~\mu\text{m})/(1.4~\mu\text{m})$. For M_{21}-M_{24}, $W/L = (6~\mu\text{m})/(0.8~\mu\text{m})$. For M_{14}, M_{25}, and M_{52}, $W/L = (16~\mu\text{m})/(0.8~\mu\text{m})$. For M_{13}, $W/L = (1.4~\mu\text{m})/(0.8~\mu\text{m})$. Take $V_{CM} = -0.65$ V.

(a) Choose W values for M_{12} and M_{51} so that $|I_{D13}| = 20~\mu\text{A}$. Use $L = 0.8~\mu\text{m}$ for M_{51} and $L = 1.4~\mu\text{m}$ for M_{12}.

(b) Use SPICE to find the low-frequency op-amp gains v_{od}/v_{id}, v_{oc}/v_{ic}, v_{od}/v_{ic}, and v_{oc}/v_{id} with the CMFB active.

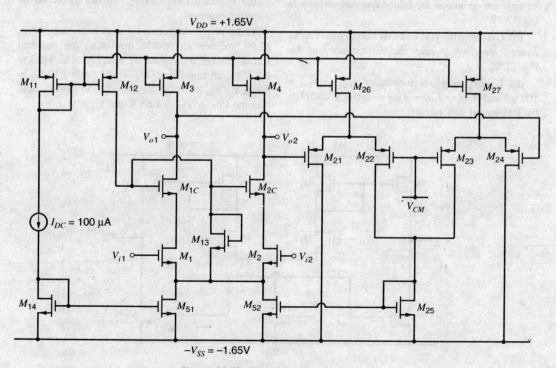

Figure 12.57 Circuit for Problem 12.23.

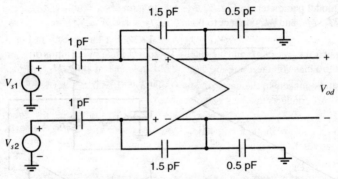

Figure 12.58 Circuit for Problem 12.24.

(c) Calculate the output slew rate dV_{od}/dt if a 4-pF capacitor is connected from each op-amp output to ground.

(d) What is the differential output voltage swing of this op amp? Assume $V_{ic} = V_{CM}$, and ignore body effect for this calculation.

(e) Repeat (b) when the input transistors are mismatched with $W_1 = 63$ μm and $W_2 = 65$ μm. (Note: With mismatch, the op-amp offset voltage is not zero.)

(f) Repeat (b) when the load transistors are mismatched with $W_3 = 95$ μm and $W_4 = 97$ μm.

12.24 The feedback circuit in Fig. 12.58 is a switched-capacitor circuit during one clock phase. Assume the op amp is the folded-cascode op amp in Fig. 12.31.

(a) Calculate the DM and CM output load capacitances, considering only the capacitances in the Fig. 12.58.

(b) If the op-amp bias currents are $|I_{D3}| = |I_{D4}| = 100$ μA and $|I_{D5}| = I_{D11} = I_{D12} = 200$ μA, calculate the DM output slew rate dV_{od}/dt.

(c) If all transistors have $|V_{ov}| = 0.2$ V and $V_{DD} = V_{SS} = 2$ V, what is the maximum peak differential output swing? Assume V_{BB3} and V_{BB4} are chosen to give maximum swing.

12.25 In the switched-capacitor CMFB scheme in Fig. 12.21, $C_1 = 0.1$ pF and $C_2 = 0.2$ pF.

(a) With $V_{CSBIAS} = -1$ V, $V_{OC} = V_{CM} = 0.5$ V. If V_{CSBIAS} changes to -1.1 V, what is the new value of V_{OC}? Assume $|a_{cmc}| \gg 1$.

(b) Ignoring all capacitors except C_1 and C_2, what are the CM and DM output load capacitors when the ϕ_1 switches are on and the ϕ_2 switches are off?

(c) Repeat (b) when the ϕ_2 switches are on and the ϕ_1 switches are off.

12.26 A current-mirror op amp is shown in Fig. 12.59. Assume all NMOS transistors are matched and all PMOS transistors are matched. Use $|V_{ov}| = 0.2$ V for all transistors, $V_{tn} = -V_{tp} = 0.6$ V, $I_{D5} = 200$ μA, $V_{An} = 10$ V, $|V_{Ap}| = 20$ V, and $\gamma = 0$. Assume $V_{DD} = V_{SS} = 1.65$ V and $V_{ic} = 0$.

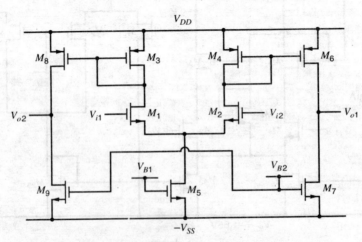

Figure 12.59 A fully differential current-mirror op amp.

(a) Calculate the model parameters in Fig. 12.7a. Assume the gates of M_5, M_7, and M_9 connect to bias voltages.

(b) If the CMC input is the gate of M_5 and the gates of M_7 and M_9 connect to a bias voltage, compute a_{cmc}.

(c) If the CMC input connects to the gates of M_7 and M_9 and the gate of M_5 connects to a bias voltage, compute a_{cmc}.

(d) What are the output swing limits for each output? What value of V_{OC} gives the maximum symmetric output swing?

12.27 Find the low-frequency value of a'_{cm} for the two-stage op amp in the example in Section 12.6.1. Use the data in that example. Recall that $a'_{cm} = v_{oc}/v_{ic}$ when the CMFB loop is active.

12.28 Assume that the CMFB circuit in the example in Section 12.6.1 is changed so that the CM-sense amplifier has a low-frequency gain $|a_{cms0}| = 2.5$. Determine the compensation capacitor C needed in the op amp to assure that the CMC and DM feedback loops in the example have a phase margin of 45° or larger.

12.29 Neutralization capacitors C_n are to be added to cancel the Miller effect on C_{gd1} and C_{gd2} in the two-stage op amp in the example in Section 12.6.1.

(a) Show how the C_n capacitors should be connected in the op amp. What value of C_n should be used?

(b) If these capacitors are constructed from MOS transistors operating in the cutoff region, what type of transistor and what device dimensions should be used?

12.30 Modify the CMFB schematic in Fig. 12.26 to inject currents at the drains of M_1 and M_2, in a manner similar to that shown in Fig. 12.18. Give a set of bias current values on the schematic. Assume $I_{D3} = I_{D4} = 150$ µA and $I_{D6} = I_{D9} = 400$ µA.

12.31 A fully differential op amp with mismatch is connected in the feedback circuit in Fig. 12.32a with $R_1 = R_2 = 10$ kΩ and $R_3 = R_4 = 40$ kΩ. The op amp model is shown in Fig. 12.37 where $a_{dm} = -181$, $a_{cm} = -2.89$, $a_{cmc} = -226$, $a_{dm-cm} = 8.95$, $a_{cm-dm} = 0.15$, and $a_{cmc-dm} = 11.6$. If the gain of the CMFB circuit is $a_{cms} = 0.76$, find the closed-loop gains $A_{dm} = v_{od}/v_{sd}$, $A_{cm} = v_{oc}/v_{sc}$, $A_{dm-cm} = v_{oc}/v_{sd}$, and $A_{cm-dm} = v_{od}/v_{sc}$.

12.32 For the op amp in Fig. 12.40, what are the output voltage swing limits of V_{o1} and V_{o2} if the threshold voltages of only M_{1A}, M_{2A}, M_{3A}, and M_{4A} are changed to $V_{t1A} = V_{t2A} = -0.3$ V and $V_{t3A} = V_{t4A} = 0.3$ V?

12.33 For the op amp in Fig. 12.40, what is the common-mode input range (CMIR) if only M_1 and M_2 are changed to low-threshold devices with threshold voltages $V_{t1} = V_{t2} = 0.3$ V?

12.34 For the feedback circuit in Fig. 12.46, the capacitor values are $C_1 = C_2 = 4$ pF and $C_L = 6$ pF. The op amp is the folded-cascode amplifier in Fig. 12.40 with low-frequency gain $a_{dm0} = 1280$ and $g_{m1} = 30$ mA/V.

(a) What is the low-frequency value of the differential-mode return ratio?

(b) What is the approximate unity-gain frequency of the differential-mode return ratio?

12.35 For the feedback circuit in Fig. 12.46, $C_1 = C_2 = 4$ pF. The op amp is the folded-cascode amplifier in Fig. 12.40 with low-frequency gain $a_{dm0} = 1280$ and $g_{m1} = 30$ mA/V. What value of C_L will give $\omega_r = 1.0$ Grad/s for the unity-gain frequency of the differential-mode return ratio?

12.36 For the feedback circuit in Fig. 12.46 using the op amp in Fig. 12.40, the capacitor values are $C_1 = C_2 = 4$ pF and $C_L = 6$ pF. What is the differential-mode output slew rate?

12.37 If W/L is doubled for $M_1 = M_2$ (i.e., $m_1 = m_2 = 20$), what is the new low-frequency DM op-amp gain a_{dm0} for the op amp in Fig. 12.40? Assume the bias conditions and operating regions do not change.

12.38 Calculate the spectrum of the input-referred thermal noise voltage for the op amp in Fig. 12.40 if W/L is doubled for $M_1 = M_2$ (i.e., $m_1 = m_2 = 20$). Assume the bias conditions and operating regions are not changed.

12.39 Calculate the spectrum of the input-referred $1/f$ voltage noise for the op amp in Fig. 12.40. Use (11.69) with $K_f = 4.8 \times 10^{-25}$ V²-F for NMOS devices and $K_f = 8.3 \times 10^{-26}$ V²-F for PMOS devices. Use $C_{ox} = 6.9 \times 10^{-15}$ F/(µm)².

REFERENCES

1. R. D. Middlebrook. *Differential Amplifiers*. Wiley, New York, 1963.

2. L. J. Giacoletto. *Differential Amplifiers*. Wiley, New York, 1970.

3. W. G. Garrett and T. G. Maxfield. "A Monolithic Differential-Output Operational Amplifier," *Digest of Technical Papers, International Solid-State Circuits Conf.*, pp. 174–175, Philadelphia, PA, February 1972.

4. M. Banu, J. M. Khoury, and Y. Tsividis. "Fully Differential Operational Amplifiers with Accurate Output Balancing," *IEEE Journal of Solid-State Circuits*, Vol. 23, no. 6, pp. 1410–1414, December 1988.

5. P. W. Li, M. J. Chin, P. R. Gray, and R. Castello. "A Ratio Independent Algorithmic Analog-to-Digital Conversion Technique," *IEEE Journal of Solid-State Circuits*, Vol. 19, no. 6, pp. 828–836, December 1984.

6. R. A. Whatley. "Fully Differential Operational Amplifier with DC Common-Mode Feedback," U.S. Patent 4,573,020, February 1986.

7. C. -C. Shih and P. R. Gray. "Reference Refreshing Cyclic Analog-to-Digital and Digital-to-Analog Converters," *IEEE Journal of Solid-State Circuits*, Vol. 21, no. 4, pp. 544–554, August 1986.

8. T. C. Choi, R. T. Kaneshiro, R. W. Broderson, P. R. Gray, W. B. Jett, and M. Wilcox. "High-Frequency CMOS Switched-Capacitor Filters for Communications Application," *IEEE Journal of Solid-State Circuits*, Vol. 18, no. 6, pp. 652–664, December 1983.

9. D. Senderowicz, S. F. Dreyer, J. H. Huggins, C. F. Rahim, and C. A. Laber. "A Family of Differential NMOS Analog Circuits for a PCM Codec Filter Chip," *IEEE Journal of Solid-State Circuits*, Vol. 17, no. 6, pp. 1014–1023, December 1982.

10. D. J. Allstot and W. C. Black, Jr. "Technological Design Considerations for Monolithic MOS Switched-Capacitor Filtering Systems," *Proceedings of the IEEE*, Vol. 71, no. 8, pp. 967–986, August 1983.

11. P. J. Hurst and S. H. Lewis. "Determination of Stability Using Return Ratios in Balanced Fully Differential Feedback Circuits," *IEEE Trans. on Circuits and Systems II*, pp. 805–817, December 1995.

12. K. Bult and G. J. G. M. Geelen. "A Fast-Setting CMOS Op Amp for SC Circuits with 90-dB DC Gain," *IEEE Journal of Solid-State Circuits*, Vol. 25, no. 6, pp. 1379–1384, December 1990.

13. B. -S. Song, M. F. Thompsett, and K. R. Lakshmikumar. "A 12-bit 1-Msample/s Capacitor Error-Averaging Pipelined A/D Converter," *IEEE Journal of Solid-State Circuits*, Vol. 23, no. 6, pp. 1324–1333, December 1988.

14. J. A. Mataya, G. W. Haines, and S. B. Marshall. "IF Amplifier Using C_c Compensated Transistors," *IEEE Journal of Solid-State Circuits*, Vol. 3, pp. 401–407, December 1968.

15. J. F. Duque-Carrillo. "Control of the Common-Mode Component in CMOS Continuous-Time Fully Differential Signal Processing," *Analog Integrated Circuits and Signal Processing, An International Journal*, pp. 131–140, Kluwer Academic Publishers, September 1993.

16. X. Wang, P. Hurst, and S. Lewis. "A 12-bit 20-Msample/s Pipelined Analog-to-Digital Converter with Nested Digital Background Calibration," *IEEE Journal of Solid-State Circuits*, pp. 1799–1808, November 2004.

Index

Abrupt junction, 4
Active cascode, 210
Active-cascode op amp, 446
Active-device parameter summary, 73
Active level shift, 439, 458, 469, 485
Active load, 276, 421, 448, 457, 546
    common-emitter amplifier, 277, 280, 284
    common-mode rejection ratio, 291
    common-source amplifier, 277, 280, 282
    complementary, 277
    current-mirror, 285
    depletion, 280
    diode-connected, 282
    enhancement, 284
    noise, 776
    offset voltage, 330
Advanced bipolar integrated-circuit fabrication, 92, 106
All-*npn* output stage, 375
Amplifier:
    current, 557
    fully differential, 799, 811
    operational, *see* Operational amplifier
    transconductance, 562
    transresistance, 560
    voltage, 557
    wideband, 514, 527, 575, 617
Amplifier stages:
    Class A, 349, 576
    Class AB, 361, 461, 688
    Class B, 362
    multiple-transistor, 201
    single-transistor, 173
Analog multiplier, 708, 711
Anneal, 87
Avalanche breakdown, 7, 21, 49, 71
Avalanche noise, 743
Average power, 345

Balanced circuit, 796
Balanced differential amplifier, 799
Balanced modulator, 712
Band-gap reference, 315, 321
    curvature compensated, 318
Bandwidth of feedback amplifiers, 624
Base-charging capacitance, 27

Base-diffused resistors, 118
Base diffusion, 86
Base-emitter voltage temperature coefficient, 11, 315
Base resistance, 30, 98
Base transport factor, 13
Base width, 12
Beta, *see* Current gain
Bias reference circuit:
    band-gap, 315, 321
    bootstrap, 307
    current routing, 329
    low-current, 297
    proportional to absolute temperature, 321
    self-biased, 307
    start-up circuit, 308, 312
    supply-insensitive, 303
    temperature-insensitive, 315
    threshold-referenced, 307
    V_{BE}-referenced, 308, 311
    voltage routing, 329
    V_T-referenced, 314, 315
BiCMOS, 152
    amplifier, 249, 618, 794
    cascode, 209
    Darlington, 204, 791
    output stage, 377, 398
    technology, 152
Bilateral amplifier, 172
Bipolar transistor:
    advanced technology, 95, 106
    base-charging capacitance, 27
    base resistance, 30, 98
    base transit time, 28
    breakdown voltage, 20, 86
    in CMOS technology, 144, 151
    collector-base resistance, 30
    collector series resistance, 89, 100
    current-mirror matching, 325
    cutoff region, 146
    diffusion profile, 91
    fabrication, 88, 92
    forward-active region, 45
    frequency response, 34
    heterojunction, 153

Bipolar transistor (*Continued*)
    input resistance, 28
    inverse-active region, 16
    large-signal models, 8
    lateral, 140
    noise, 745, 750
    output characteristics, 15
    output resistance, 29
    parameter summary (functional), 73
    parameter summary (numerical), 106, 108, 112, 114
    parasitic capacitance, 31, 106, 111, 116
    reverse-active region, *see* Inverse-active region
    saturation current, 18, 97, 316
    saturation region, 17
    self-aligned structure, 94
    small-signal model, 25
    SPICE model parameters, 162
    transconductance, 26
Bird's beak, 141, 257, 426
Blackman's impedance formula, 608
Body effect in MOS transistors, 53, 139
Boltzmann approximation, 10
Bootstrap bias technique, 307
Breakdown, 6
    in bipolar transistors, 38
    in MOSFETs, 49
    in superbeta transistors, 124
Breakdown voltage:
    base-emitter, 22
    collector-base, 20, 86
    collector-emitter, 21
    drain-substrate, 49
    junction-diode, 7
    MOS transistor, 45
    superbeta transistors, 124
    Zener diode, 8
Built-in potential, 2
Buried layer, 89
Burst noise, 745

Capacitance:
    base-charging, 27
    base-collector, 31, 97
    base-emitter, 31, 96
    channel-substrate, 138
    collector-substrate, 31, 103
    depletion region, 5, 31, 54, 103, 140, 530
    drain-body, 54
    gate-body, 55
    gate-drain, 51
    gate-source, 51
    gate-substrate, *see* Capacitance: gate-body
    overlap, 55, 142
    sidewall, 143
    source-body, 54
Capacitive neutralization, 835
Capacitors:
    in bipolar integrated circuits, 115
    lateral, 150
    in MOS integrated circuits, 148
    MOS transistors, 148
    poly-poly, 146
    vertical, 146
Capture range, 718, 726
Cascode configuration, 205, 207, 439, 442, 446, 527
    active, 210, 446
Cascode current mirror, 261, 690
    Bipolar, 261
    MOS, 264
Cascode frequency response, 527
Channel-length modulation, 43
Class A output stage, 349, 576
Class AB input stage, 688
Class AB output stage, 364
Class B output stage, 361
Clipping, 346, 361
Closed-loop gain, 401, 554, 601
Clubhead, 116
CMFB, *see* Common-mode feedback
CMOS operational amplifier:
    fully differential, 834
    with single-ended output, 421, 439, 442, 446
CMOS output stage, 379
CMOS technology, 46, 128
Collector-base capacitance, 102
Collector-base resistance, 30
Collector-series resistance, 32, 100
Common-base configuration, 182, 184, 185, 514, 529
Common-base frequency response, 514
Common-base stage noise performance, 771
Common centroid geometry, 437
Common-collector-common-collector configuration, 201
Common-collector-common-emitter configuration, 201

Common-collector configuration, 191, 505
Common-drain configuration, 194, 511, 591
Common-emitter-common-base configuration, *see* cascode configuration
Common-emitter configuration, 174, 490, 495
Common-emitter frequency response, 490, 498
Common-emitter output stage, 576
Common-gate configuration, 185, 187, 514
Common-gate frequency response, 514
Common-mode feedback, 227, 286, 458, 804
    circuits, 796
    using a resistive divider, 812
    using switched capacitors, 821
    using transistors in the triode region, 812
    using two differential pairs, 816
Common-mode feedback loop, 804
    bandwidth, 844
    stability and compensation, 810
Common-mode gain, 222, 804
Common-mode half-circuit, 226, 239, 804
Common-mode input range, 854
Common-mode input resistance, 228
Common-mode rejection ratio, 223, 228, 234, 417, 427
    of active-load stage, 293
    frequency response, 501
Common-mode-to-differential-mode gain, 222
Common-source amplifier, 181, 490, 499
Common-source-common-gate configuration, *see* cascode configuration
Common-source frequency response, 490, 498, 500
Compensation:
    of amplifiers, 633
    capacitor, 597, 637, 682
    by feedback zero, 677
    methods, 637
    of MOS amplifiers, 643, 650
    of NE5234 op amp, 658
    nested Miller method, 654
    theory, 633
Complementary load, 277
Complementary output stage, 360
Composite *pnp*, 377

Conductivity, 80
Copper, 156
Correlation, 747, 756, 789
Critical field, 7, 59
Crossover distortion, 362, 365, 368, 375, 384
Crossunder, 117
Crowding, 32, 99
Current amplifier, 562
Current crowding, 32, 99
Current density, 425
Current-feedback pair, 586, 677, 768
Current gain:
    dependence on operating conditions, 23, 110
    forced, 18
    forward, 12
    inverse, 19
    npn transistor, 105
    pnp transistor, 107
    short-circuit, 177
    small-signal, 29
    small-signal high-frequency, 34
    temperature coefficient, 23
Current mirror, 251
    with beta helper, 258
    cascode, 261, 264
    current routing, 329
    with degeneration, 260, 261
    gain error, 252
    general properties, 251
    high-swing cascode, 268
    input voltage, 252
    load, 285
    matching, 325
    minimum output voltage, 255
    output resistance, 252, 254, 257, 258, 260, 261, 264, 266, 272, 275
    simple, 253, 255
    Sooch cascode, 269, 270
    voltage routing, 329
    Wilson, 272, 275, 595, 610
Current routing, 329
Current source, 297, 300, 301, 302, 303, 304, 307
    Peaking, 301, 302
    Widlar, 297, 300, 304
Cutoff, 46

Darlington configuration, 201, 203, 397, 595, 791

dc analysis of the NE5234 op amp, 452
Deadband, 361, 368, 375
Depletion-mode load, 281
Depletion region:
   capacitance, 5, 31, 54, 103, 530
   collector-base, 14, 89
   of *pn* junction, 1
Deviation from ideality in real op amps, 415
Die, 91
Dielectric isolation, 123
Differential amplifier, 220, 404
   perfectly balanced, 222
   unbalanced, 237
Differential-mode gain, 222, 804
Differential-mode half-circuit, 225, 239, 804
Differential-mode input resistance, 228
Differential-mode-to-common-mode gain, 222, 838
Differential pair:
   bipolar transistor, 214
   MOS, 217
   noise performance, 773
   with current-mirror load, 285, 534
Differential signal source, 800
Differential-to-single-ended conversion, 291
Differentiator, 406
Diffused-layer sheet resistance, 82
Diffused resistors, 115
Diffusion constant, 10
Diffusion current, 10
Diffusion of impurities, 80
Diffusion profile of a bipolar transistor, 91, 92, 96, 98
Diode:
   Junction, 121
   Zener, *see* Zener diode
Diode-connected load, 282
Diode-connected MOS transistor, 255
Distortion, 358, 399
Distortion in the source follower, 355
Distortion reduction by negative feedback, 553
Dominant pole, 495, 497, 499, 518, 522, 524
Doping, 79
Double-diffused *npn* transistors, 95
Double-diffused *pnp* transistors, 126
Drift, emitter-coupled pair, 234
   source-coupled pair, 238

Driver stage, 343, 351, 366, 367, 369, 370, 371, 389, 395

Early effect, 16, 21, 262
Early voltage, 15, 29, 43, 52, 255, 258
Ebers-Moll equations, 19
Economics of IC fabrication, 156
Effective channel length, 43, 140
Effective channel width, 141
Efficiency:
   Class A, 347, 349, 350
   Class B, 360, 362, 366, 375
Electromigration, 156
Emitter-coupled multivibrator, 728
Emitter-coupled pair, 214, 704, 773
   emitter degeneration, effect of, 216
   frequency response, 495
   input offset current, 230, 234
   input offset voltage, 230, 231
   offset voltage drift, 233
   small-signal analysis, 220, 237
   transfer characteristics, 214
Emitter degeneration, 196, 216, 260, 587
Emitter-diffused resistors, 117
Emitter-follower, 191
   drive requirements, 351
   frequency response, 505
   noise performance, 773
   output stage, 341, 344, 576
   small-signal properties, 192, 352
   terminal impedances, 507
   transfer characteristics, 341
Emitter injection efficiency, 13
Emitter resistance, 32
Enhancement load, 284
Epitaxial growth, 86
Epitaxial resistor, 119
Equivalent circuit of an op amp, 420
Equivalent input noise generators, 756
Equivalent input noise resistance, 758
Equivalent input shot noise current, 759
Etching, 85

Fabrication of integrated circuits, 78
Feedback, 403, 553
   bandwidth, 624
   common mode, 227, 286, 458, 804
   configurations, 557
   effect on distortion, 555

effect on gain sensitivity, 555
effect on noise, 764
effect on terminal impedances, 558, 561, 562, 563, 607
ideal analysis, 553
latchup, 151
loading, effect of, 563
local, 227, 263
loop gain, 384, 401, 554
practical configurations, 563
return ratio, 599, 612
series-series, 562, 569
series-shunt, 557, 579
shunt-series, 561, 583
shunt-shunt, 560, 563
single-stage, 587
table of relationships, 588
Feedback-zero compensation, 677
Feedforward, 389, 644–650, 660
Field-effect transistor, see MOS transistor
Field region, 128
First-order phase-locked loop, 720
Flicker noise, 741
Flicker noise corner frequency, 746
Folded-cascode op amp, 446, 652
Forward-active region, 8, 26, 45, 112, 175
Free-running frequency, 717, 726
Frequency response:
  bipolar transistor, 33, 56
  cascode, 527
  common-base, 514
  common-drain, 511
  common-emitter, 495
  common-emitter cascode, 527
  common-gate, 514
  common-mode, 501
  common-source, 499
  common-source cascode, 527
  of CMRR, 503
  of current mirror, 251, 256
  differential-mode, 495, 496, 500
  of differential pair with current-mirror load, 290, 534
  emitter follower, 505
  MOS amplifier, 643, 650
  MOS transistor, 55
  multistage amplifier, 518
  NE5234 op amp, 539
  single stage, 490
  source follower, 511

Full-power bandwidth, 692
Fully differential amplifiers, 796
  cross gains, 838
Fully differential op amps, 796, 823, 845
  folded cascode, 834
  telescopic cascode, 855
  two-stage, 823

Gain-bandwidth product, 625
Gain margin, 630
Gaussian distribution, 232, 244, 738
Generation, 79
Gilbert multiplier, 704
g-parameters, 583
Graded junction, 5
Guard ring, 152

Half circuit:
  of balanced amplifier, 225, 226, 799
  coupled, 237
  of unbalanced amplifier, 239, 240
  independent, 238
Harmonic distortion, 357, 399
Heterojunction, 153
High-frequency equivalent circuit of the NE5234 op amp, 540, 659
High-level injection, 36, 110
High-voltage integrated-circuit fabrication, 88
Homojunction, 53
Hot carriers, 49
h-parameter, 579
Hybrid-π model, 30, 52

Ideal feedback analysis, 553
Impact ionization, 71, 209
Impurity concentration, 79
Instability in amplifiers, 626, 666
Instantaneous power, 345–351, 365
Integrated circuit:
  advanced bipolar fabrication, 92, 107
  biasing, 251
  capacitors, 120, 146
  cost considerations, 156, 159
  current mirrors, 251
  device models, 1
  economics, 156
  fabrication, 78
  fabrication yield, 157
  frequency response, 490

Integrated circuit (*Continued*)
  mixed-signal, 221
  noise, 736
  op amps, 402, 796, 823
  output stages, 341
  packaging, 159
  passive components, 115, 146
  phase-locked loop, 727
  resistors, 115, 119, 146
Integrator:
  continuous-time, 406
  switched-capacitor, 412
Interconnect delay, 156
Internal amplifier, 407
Intrinsic carrier concentration, 2, 79
Inverse-active region of bipolar transistors, 19, 45
Inversion, 40
Inverting amplifier, 402
Ion implantation, 85, 87
Isolation diffusion, 89, 90

Junction breakdown, 6, 49
Junction diodes, 122

KCL, 184
Kirk effect, 25, 36
KVL, 184

Large-signal model:
  bipolar transistor, 8
  MOS transistor, 45
Latchup in CMOS, 151, 341
Lateral bipolar transistor, 146
Lateral *pnp* transistor, 108, 368, 376
Layout, 435
Leakage current, 7, 18, 19
  collector base, 19
Level shifting, 194, 196, 439, 485
Lifetime of minority carriers, 11
Line regulation, 595
Load line, 345–351, 364, 365
Load regulation, 595
Local feedback, 587
Local oxidation, 87
Lock range, 717, 724
Logarithmic amplifier, 405
Loop gain, 384, 401, 552
Low-current biasing, 297

Low-level injection, 10, 110
Low-permittivity dielectric, 156

Matching in transistor current mirrors, 325
Miller capacitance, 491, 498, 500, 638, 641, 831
Miller effect, 114, 490, 492, 497, 499, 828
Minimum detectable signal, 754
Minority-carrier lifetime, 11
Mismatch effects in differential amplifiers, 229
Moat, 93
Moat etch, 123
Mobility, 80
Mobility degradation, 65
Model selection for IC analysis, 170
Modulator, 712
MOS transistor:
  active region, 45
  bird's beak, 141, 257, 426
  body effect, 53, 139
  breakdown voltage, 149
  channel-length modulation, 43
  complementary MOS (CMOS), 39, 46, 127
  critical field, 59
  current density, 425
  current-mirror matching, 328
  cutoff region, 46
  depletion mode, 47, 144, 281
  diode connected, 255
  effective channel length, 43, 140
  effective channel width, 141
  enhancement-mode, 38, 127
  fabrication, 127
  field-effect transistor, 41
  field region, 128
  frequency response, 55
  hot carriers, 49
  impact ionization, 71
  input resistance, 52
  inversion, 40
  junction breakdown, 49
  large-signal model, 45
  mobility, 42
  mobility degradation, 65, 135
  moderate inversion, 69
  n-channel, 38, 131, 144
  noise, 750, 762
  ohmic region, 44

operational amplifier, 421
output resistance, 52
output stages, 379
overdrive, 47, 181, 434
oxide breakdown, 49
parameter summary (functional), 73
parameter summary (numerical), 132, 133, 134, 135
parasitic elements, 54
p-channel, 39, 144
punchthrough, 49
saturation region, 44, 51
short-channel effects, 59
silicon gate, 129
small-signal model, 49
source-coupled pairs, 217
source follower, 194
SPICE model parameters, 162
strong inversion, 65
substrate current, 71
subthreshold conduction, see Weak inversion
threshold temperature dependence, 47
threshold voltage, 40, 129, 131
transconductance, 50, 63
transfer characteristic, 38
triode region, 44
weak inversion, 65, 180
velocity saturation, 59
Multiplication factor, 7
Multiplier circuits, 704
Multistage amplifier frequency response, 518

Negative feedback, see Feedback
Neutralization, 835
NMOS, 38
Noise:
  in active loads, 776
  amplitude distribution, 738
  avalanche, 743
  bandwidth, 776
  burst, 742
  in capacitors and inductors, 747
  circuit calculations, 750
  common-base stage, 771
  differential pair, 773
  in diodes, 736
  in direct-coupled amplifiers, 785
  emitter follower, 773
  equivalent input noise, 754
  $1/f$, 741
  feedback, effect of, 764
  figure, 786
  flicker noise, 741
  generator, 738
  models, 744
  MOS amplifier, 746, 762, 779
  in NE5234 op amp, 776
  operational amplifier, 776, 855
  popcorn noise, 742
  in resistors, 747
  shot, 736
  spectral density, 737, 748
  temperature, 786
  thermal, 740
  white, 737
  in Zener diodes, 743
Nondominant pole, 495, 497, 500, 536
  of NE5234 op amp, 542
Noninverting amplifier, 404
Nonlinear analog operations, 405
Nonlinearity, 343, 352, 353, 357, 368, 375, 393, 395, 708
Nonoverlapping clock signals, 409, 412
  charge-transfer phase, 409
  input sample phase, 409
Normal distribution, see Gaussian distribution
Nyquist criterion, 626, 666
Nyquist diagram, 627

Offset current, 416
  emitter-coupled pair, 231, 233
  NE5234 op amp, 477, 481
Offset voltage, 424
  differential pair with active load, 330
  emitter-coupled pair, 231, 233
  emitter-coupled pair with active load, 330
  NE5234 op amp, 477
  source-coupled pair, 235
  source-coupled pair with active load, 337
Open-circuit time constants, 519, 536
  also see Zero-value time constants
Open-loop gain, 401
Operational amplifier:
  active-cascode, 446
  applications, 401
  bipolar, 448
  with cascodes, 438

Operational amplifier (*Continued*)
  CMOS, 421
  common-mode input range, 416, 427
  common-mode rejection ratio, 417, 427
  compensation, 420, 633, 676
  deviations from ideality, 415
  differential amplifier, 404
  effect of overdrive voltages, 434
  equivalent circuit, 420
  folded-cascode, 442, 652, 845
  frequency response, 420
  fully differential op amps, 796, 823
  input bias current, 415
  input resistance, 420, 422
  internal amplifier, 407
  inverting amplifier, 402
  layout considerations, 435
  MOS, 421, 438, 442, 446
  NE5234 op amp analysis, 452
  noise, 776
  offset current, 416
  offset drift, 416
  offset voltage, 416, 424
  open-circuit voltage gain, 422
  output buffer, 407
  output resistance, 420, 422
  output swing, 423
  power-supply rejection ratio, 418, 430, 435
  random offset voltage, 426
  settling time, 411
  with single-ended outputs, 400
  slew rate, 420, 681
  small-signal characteristics, 422
  supply capacitance, 432
  systematic offset voltage, 424
  telescopic-cascode, 439, 650
Oscillation, 624, 666, 679
Output buffer, 407
Output resistance of current mirror, 252, 254, 257, 258, 260, 261, 264, 266, 272, 275
Output stage, 341
  all-*npn*, 375
  BiCMOS, 377, 397
  Class A, 349, 576
  Class AB, 361, 379
  Class B, 360
  CMOS, 379
  combined common-drain common-source, 388, 390
  common-drain, 353, 380
  common emitter, 576
  common-source, with error amplifiers, 381
  complementary, 360
  Darlington, 397
  emitter follower, 341
  NE5234 op amp, 458, 474
  overload protection, 377
  parallel common-source, 390
  push-pull, 359
  quasi-complementary, 376, 381, 384, 388
  709 op amp, 366
  741 op amp, 368
  short-circuit protection, 377
  source follower, 353, 380
Overdrive, 47
Overload protection, 377
Oxidation, 84
Oxide breakdown, 49

Packaging considerations, 159
Parasitic elements:
  bipolar transistor, 31, 98
  MOS transistor, 54, 141
Passive components:
  in bipolar integrated circuits, 115
  in MOS technology, 144
Permittivity, 3
Phase detector, 712, 716, 725, 727
Phase-locked loop:
  basic concepts, 718
  capture range, 718, 726
  first-order loop, 720
  integrated circuit, 721, 727
  lock range, 717, 724
  loop bandwidth, 719, 724
  root locus, 719, 722
  second-order loop, 723
Phase margin, 629–638, 643, 648, 652, 658, 663, 676, 686, 694
Photolithography, 84
Photoresist, 84
Pinch-off in MOSFETs, 43
Pinch resistors, 118, 119
PMOS, 39
pn junction depletion region, 1
Pole splitting, 639, 642, 654, 680

Polysilicon, 87, 93, 129
Popcorn noise, 742
Power conversion efficiency, 347
Power hyperbola, 348
Power output:
   Class A, 349
   Class B, 362
Practical Class B output stages, 368
Predeposition, 81, 84
Probability distribution, 244
Protection of output stages, 377
Punchthrough, 49, 125
Push-pull output stage, 359

Quasi-complementary output stages, 376, 381, 384, 388

Reciprocity condition, 19
Recombination, 10, 11, 79
References, 297
Regulated cascode, *see* Active cascode
Regulator, 593
Replica biasing, 822
Resistivity, 79
Resistor:
   base-diffused, 116
   base-pinch, 118
   diffused, 146
   emitter-diffused, 117
   epitaxial, 119
   MOS device, 146
   in MOS technology, 146
   polysilicon, 146
   well, 144
Return ratio, 599
   simulation of, 621, 857
Reverse-active region, *see* Inverse-active region of bipolar transistors
Reverse injection, 13
Rise time, 543
Root locus, 626, 642, 664
   for dominant-pole compensation, 676
   for feedback-zero compensation, 677
   for three-pole transfer function, 665
Root locus rules, 667

Saturation current, of bipolar transistor, 11, 97, 316

Saturation region:
   of bipolar transistor, 16, 175
   of MOSFET, 44, 45
Scattering-limited velocity, 59, 135
Second-harmonic distortion, 358, 396
Second-order phase-locked loop, 722
Self-aligned structure, 94
Self-biasing, 307
Sensitivity to power-supply voltage, 304
Series-series feedback, 562, 569
Series-series triple, 575, 578, 583
Series-shunt feedback, 557, 579
Sheet resistance, 82
Short-channel effects in MOS Transistors, 59
Short-circuit current gain, 177
Short-circuit protection, 377, 468
Short-circuit time-constants, 536
Shot noise, 736
Shunt-series feedback, 561, 583
Shunt-shunt feedback, 560, 563
Signal-to-noise ratio, 798
Silicide, 95, 147
Silicon dioxide, 84, 87
Silicon nitride, 87
Single-stage feedback, 587
Single-stage frequency response, 490
Slew rate, 681
   effect on sinusoidal signals, 692
   in fully differential folded-cascode op amp, 863
   improvement, 685, 687
Small-signal analysis of NE5234 op amp, 469
Small-signal model:
   bipolar transistor, 25
   MOS transistor, 49
Source-coupled pair, 217
   frequency response, 499
   input offset voltage, 236
   offset voltage drift, 236
   small-signal analysis, 220, 237
   transfer characteristics, 217
Source degeneration, 199, 261
Source follower, 194, 511, 592
   distortion, 355
   frequency response, 511
   output stage, 353
   small-signal properties, 194
   super, 212, 609
   transfer characteristics, 353

Space-charge region, see Depletion region
Spectral density of noise, 737, 748
SPICE parameter summary, 162
Square-law characteristic of MOSFETs, 43
Square-law circuit, 734
Start-up circuit, 308, 310
Step coverage, 153
Step response, 543, 862
Stress migration, 156
Substitutional impurities, 80
Substrate contact, 152
Substrate current in MOSFETs, 71
Substrate *pnp* transistors, 111
Subthreshold conduction, 135
   also see MOS transistor: weak inversion
Summing node, 412, 553
Summing-point constraints, 403, 405
Super source follower, 212, 609
Superbeta transistors, 124
Supply-insensitive biasing, 303
Sustaining voltage, 21
Switched-capacitor amplifier, 407
   parasitic insensitive, 411, 857
Switched-capacitor common-mode feedback, 821, 853
Switched-capacitor filter, 412
Switched-capacitor integrator, 412
Symmetry:
   mirror, 436
   translational, 436

T model, 182, 185
Tail current source, 214, 217
Telescopic-cascode op amp, 439, 650
Temperature coefficient:
   of band-gap reference, 317, 319, 324
   of base-emitter voltage, 13, 311, 315
   of bias reference circuits, 311, 319
   of bipolar transistor current gain, 23
   effective, 319
   of integrated-circuit resistors, 120
   of threshold voltage, 48
   of Zener diodes, 121, 122
Temperature-insensitive biasing, 315
Thermal noise, 747
Thin-film resistors, 127
Third-harmonic distortion, 358, 396
Threshold-referenced bias circuit, 307

Threshold voltage of MOS transistors, 40, 129, 131
Threshold voltage temperature dependence, 47
Time constant, 36
   open circuit, 519, 536
   short circuit, 536
   zero value, 519
Time response, 542
Transconductance amplifier, 563
Transfer characteristic:
   Class B stage, 360
   common-emitter amplifier, 176
   emitter-coupled pair, 214
   emitter-coupled pair with emitter degeneration, 216
   emitter follower, 341
   source-coupled pair, 217
   source follower, 353
Transit time, 28
Transition frequency f_T, 34, 55, 64
Transresistance amplifier, 561
Tunneling, 8, 121
Two-port representation, 171, 563, 584

Unbalanced fully differential circuits, 838
Uniform-base transistor, 10
Unilateral amplifier, 173
Unit devices, 254, 257, 426
Unity-gain feedback configuration, 404, 634, 649, 660, 681

V_{BE}-referenced bias circuit, 307, 311
VCO, 716, 725, 734
Velocity saturation, 59
Virtual ground, 404
Voltage amplifier, 559
Voltage-controlled oscillator, 716, 728, 734
Voltage gain, 176, 422
   open-circuit, 177, 180
Voltage regulator, 593
Voltage regulator analysis, 593
Voltage routing, 329
V_T-referenced bias circuit, 314

Wafer, 79
Well, 46, 128
Well contact, 152
White noise, 737

Wideband amplifier, 514, 527, 575, 618, 677, 753
Widlar current source, 297, 300, 304
Wilson current mirror, 272, 275, 595, 611

Yield considérations, 157
y-parameters, 563

Zener diode, 8, 22, 121, 593, 595, 743
Zener diode noise, 743
Zener diode temperature coefficient, 120, 122
Zener-referenced bias circuit, 595
Zero-value time constant analysis, 519
z-parameters, 569